ELEMENTS OF

REAL ANALYSIS

For B.A. (Mathematics), B.Sc. (Mathematics & Physics), M.A.; M.Sc. (Mathematics) Students of Various Universities/Institutions. Also Useful for the Aspirants of GATE, CSIR-UGC (NET) and Other Competitive Examinations]

SHANTI NARAYAN

Formerly
Dean of Colleges
University of Delhi, and
Principal
Hansraj College, Delhi

Revised by
Dr. M.D. RAISINGHANIA
M.Sc., Ph.D.
Formerly Head
Department of Mathematics
S.D. (Postgraduate) College, Muzaffarnagar (U.P.)

S. CHAND
PUBLISHING

S Chand And Company Limited

(ISO 9001 Certified Company)

S Chand And Company Limited
(ISO 9001 Certified Company)
Head Office: Block B-1, House No. D-1, Ground Floor, Mohan Co-operative Industrial Estate, New Delhi – 110 044 | Phone: 011-66672000

S. CHAND
PUBLISHING

Registered Office: A-27, 2nd Floor, Mohan Co-operative Industrial Estate, New Delhi – 110 044
Phone: 011-49731800
www.**schandpublishing.com**; e-mail: **info@schandpublishing.com**

Branches

Chennai	:	Ph: 23632120; chennai@schandpublishing.com
Guwahati	:	Ph: 2738811, 2735640; guwahati@schandpublishing.com
Hyderabad	:	Ph: 40186018; hyderabad@schandpublishing.com
Jalandhar	:	Ph: 4645630; jalandhar@schandpublishing.com
Kolkata	:	Ph: 23357458, 23353914; kolkata@schandpublishing.com
Lucknow	:	Ph: 4003633; lucknow@schandpublishing.com
Mumbai	:	Ph: 25000297; mumbai@schandpublishing.com
Patna	:	Ph: 2260011; patna@schandpublishing.com

First Edition 1965
Subsequent Editions and Reprints 1966, 69, 74, 76, 79, 80, 83, 85, 87, 89, 92, 95, 98, 2001, 2003, 2007 (Twice), 2009, 2010, 2011, 2012, 2013, 2014
Seventeenth Revised Edition 2015, Reprints 2015, 2016
Reprint with updated content and questions 2017, Reprint 2018 (Twice)
Reprint with updates 2019
Reprint with updates 2020 (Twice)
Reprint with updates 2021

ISBN: 978-81-219-0306-6 **Product Code:** H6RAN61MATH10ENAQ15O

PRINTED IN INDIA

By Vikas Publishing House Private Limited, Plot 20/4, Site-IV, Industrial Area Sahibabad, Ghaziabad – 201 010 and Published by S Chand And Company Limited, A-27, 2nd Floor, Mohan Co-operative Industrial Estate, New Delhi – 110 044.

PREFACE TO THE SEVENTEENTH EDITION

Latest papers of CSIR-UGC (NET) - 2011-2016 and GATE – (2000 – 2016) have been solved and presented in "Miscellaneous results and problems based on the entire book". Chapter 22 has been re-written and new useful results have been included in this chapter.

Suggestion for further improvement of the book will be suggestion gratefully received.

M.D. Raisinghania

PREFACE TO THE EIGHTH EDITION

The book originally written, about 40 years ago, has during the intevening period, been revised and reprinted several times. Due to the new U.G.C. Model syllabus and the demand for more matter from the students and teachers, a thorough revision of the book was overdue. I very humbly took the challenge of revising this perfect well-written book of late Shri Shanti Narayan with whom I had personal contact from 1962 onwards. He was my teacher and guiding star in art of writting a book.

I have tried to meet the rapidly changing demands of students interested in self-study and appearing in various examinations. Accordingly, a large variety of illustrative solved examples have been included in every chapter. References to the latest papers of various universities and I.A.S. examination have been inserted at proper places.

The following new chapters have been added in this present edition

- Countability of sets • The Riemann-Stieltjes Integrals • Uniform convergence of sequences and series of functions • Improper Integrals • Metric spaces

The book, in the present form, is a humble effort to make it more useful to the students and teachers.

I am extremely thankfull to the Managing Director, Shri Ravindra Kumar Gupta, Shri Navin Joshi, Vice President (Publishing) and Advisor, Shri R.S. Saxena for personal intersest throughout the preparation of the book. My sincere thanks are also due to Mr. Shishir Bhatnagar of S. Chand & Company for bringing this book in an excellent form.

All valuable suggestions for further improvement of the book will be highly appreciated

M.D. Raisinghania

PREFACE TO THE FIRST EDITION

This book is an attempt to present Elements of Real Analysis to under-graduate students,* on the basis of the University Grants Commission Review Committee report recommendations and several Universities having provided a course along the lines of these recomendations. This book must not, however, be thought of an abridged edition of the Author's "A Course of Mathematical Analysis" for M.A. students.

Chapter I provides description of Set of Real Number as a complete ordered field and no attempt has been made to construct the Set starting from some Axioms. Chapter II deals with bounds and limit points of sets of real numbers. Chapter III concerns itself with Real sequences defined as functions on the set of Natural numbers into the set of Real numbers. This is followed by Chapter IV of Infinite Series dealing with mostly convergence tests for positive term series as also a test on alternating series. Chapter V deals with the nature of the range of a real valued continuous function with a closed finite interval as its domain. Chapter VI on Derivability deals with the rigorous proof of Role's theorem as also of Lagrange's theorem. Chapter VIII deals with Riemann Integrability.

The book contains some examples and exercises meant only to help a proper undertaking of the text. The book also seeks to adopt comparatively more modern notation for the subject.

It is hoped that this "Elements of Real Analysis" will provide a stimulus for the commencement of the study of Analysis at undergraduate stage which has already been so much delayed.

March, 1965 **SHANTI NARAYAN**

CONTENTS

1.	**SETS AND FUNCTIONS**	...	**1.1–1.9**
1.1.	Introduction	...	1.1
1.2.	Statements	...	1.1
1.3.	Connectives	...	1.2
1.4.	Sets	...	1.4
1.5.	Functions (or mappings)	...	1.6
1.6.	Composite of functions (or product of functions)	...	1.7
1.7.	Inverse function	...	1.8
1.8.	Binary operation	...	1.8
	Objective questions	...	1.9
2.	**THE REAL NUMBERS**	...	**2.1–2.32**
2.1.	Introduction	...	2.1
2.2.	The set **N** of natural numbers	...	2.2
2.3.	The set **I** or **Z** of integers	...	2.3
2.4.	The set **Q** of rational numbers	...	2.4
2.5.	The set of real numbers **R** as a complete ordered field	...	2.6
2.6.	Closed, open, semi-closed and semi-open intervals	...	2.7
2.7.	Set bounded above, set bounded below, l.u.b. (supremum) and g.l.b. (infimum) of a set. The greatest and smallest members of a set. Bounded and unbounded sets	...	2.7
2.8.	Order-completeness of the set of real numbers	...	2.16
2.9.	Equivalent descriptions of the order-completeness. Property of the set of real numbers	...	2.18
2.10.	Explicit statement of the properties of the set of real numbers as a complete ordered field	...	2.20
2.11.	Some important properties of the system of real numbers	...	2.20
2.12.	The denseness property of the set of real numbers **R**	...	2.21
2.13.	The modulus (or absolute value) of a real number	...	2.26
2.14.	Principle of Mathematical Induction	...	2.29
2.15.	Bernoulli's inequality	...	2.29
3.	**NEIGHBOURHOODS AND LIMIT POINTS OF A SET. OPEN AND CLOSED SETS**	...	**3.1–3.36**
3.1.	Introduction	...	3.1
3.2.	Neighbourhood of a point	...	3.1
3.3.	Properties of neighbourhoods	...	3.3
3.4.	Limit (or accumulation or condensation) point of a set	...	3.4
3.5.	Existence of limit points	...	3.8
3.6.	Open and closed sets	...	3.13
3.7.	Basic theorems concerning families of open and closed sets	...	3.15
3.8.	Illustrations of open sets	...	3.18
3.9.	Illustrations of closed sets	...	3.19
3.10.	Interior point and interior of a set	...	3.22
3.11.	Exterior point and exterior of a set	...	3.22
3.12.	Boundary (or frontier) point and boundary (or frontier) of a set	...	3.22
3.13.	Theorems on interior of a set	...	3.23
3.14.	Adherent point (or a contact point) and closure of a set	...	3.25
3.15.	Theorems on closure of a set	...	3.26

3.16.	Isolated points of a set and discrete set	...	3.29
3.17.	Dense (or everywhere dense), dense in itself, nowhere dense (or non-dense) and perfect sets.	...	3.29
3.18.	Cantor nested interval theorem	...	3.30
3.19.	Cover (or covering) of a set	...	3.31
3.20.	Compact set	...	3.31
3.21.	Properties of a compact set	...	3.32
	Objective questions		3.35
4.	**COUNTABILITY OF SETS**	...	**4.1–4.11**
4.1.	Equivalent sets	...	4.1
4.2.	Finite and infinite sets	...	4.1
4.3.	Denumerable (or enumerable or countably infinite), countable and uncountable sets	...	4.1
4.4.	Decimal, ternary and binary representation	...	4.7
4.5.	Cantor set or Cantor ternary set	...	4.8
	Objective questions	...	4.11
5.	**SEQUENCES**	...	**5.1–5.60**
5.1.	Sequence	...	5.1
5.2.	Bounded and unbounded sequences	...	5.1
5.3.	Limit point (or cluster point or point of condensation) of a sequence	...	5.2
5.4.	Convergent sequences. The limit of a sequence	...	5.5
5.5.	Algebra of convergent sequences	...	5.11
5.6.	Bounded non-convergent sequences	...	5.15
5.7.	Some theorems on divergent sequences	...	5.16
5.8.	Some important theorems on limits	...	5.19
5.9.	Cauchy (or fundamental) sequences	...	5.29
5.10.	Convergence of a sequence	...	5.37
5.11.	Monotonic sequences and their convergence	...	5.37
5.12.	Subsequences	...	5.48
5.13.	Limit superior and limit inferior of a sequence	...	5.52
5.14.	Some theorems on limit superior and limit inferior of a sequence	...	5.53
	Objective questions		5.57
6.	**INFINITE SERIES WITH POSITIVE TERMS**		**6.1–6.52**
6.1.	Infinite series, its convergence and sum	...	6.1
6.2.	A necessary condition for the convergence of an infinite series	...	6.2
6.3.	Cauchy's general principle of convergence for series	...	6.3
6.4.	General test for the convergence of positive term series	...	6.4
6.5.	Two important standard series Σr^n *and* $\Sigma(1/n^\lambda)$	...	6.6
6.6.	Comparison tests for the convergence of positive term series	...	6.8
6.7.	Comparison tests of the first type	...	6.9
6.8.	Practical comparison tests of the first type	...	6.9
6.9.	Comparison tests of the second type	...	6.13
6.10.	Practical comparison tests of the second type	...	6.14
	D'Alembert's ratio test	...	6.14
6.11.	Cauchy's nth root test	...	6.19
6.12.	Cauchy's nth root test is superior than D'Alembert's ratio test	...	6.22
6.13.	Raabe's test	...	6.23
6.14.	Logarithmic test	...	6.25
6.15.	De Morgan's and Bertrand's test	...	6.25

6.16.	Second logarithmic ratio test ...	6.27
6.17.	Kummer's test ...	6.36
6.18.	Gauss's test ...	6.37
6.19.	Cauchy's integral test ..	6.41
6.20.	Cauchy's condensation test ...	6.47
6.21	An important auxiliary series ...	6.48
	Objective questions ...	6.49
7.	**INFINITE SERIES WITH POSITIVE AND NEGATIVE TERMS ...**	**7.1–7.30**
7.1.	Introduction ...	7.1
7.2.	Absolute convergence and conditional convergence ...	7.1
7.3.	Alternating series ...	7.2
7.4.	Cauchy principle of convergence for a series ...	7.10
7.5.	Some important theorems on absolutely convergent series ...	7.11
7.6.	Dirichlet's theorem ...	7.12
7.7.	Abel's theorem ...	7.13
7.8.	Re-arrangements of series ...	7.14
7.9.	Re-arrangements of a conditionally convergent series ...	7.16
	Objective questions ...	7.18
7.10	Modified forms of some important theorems ...	7.20
7.11	Additional solved examples ...	7.21
7.12	Cauchy product of two infinite series ...	7.25
7.12A	Solved examples based on Art. 7.12 ...	7.29
8.	**REAL FUNCTIONS. LIMIT AND CONTINUITY ...**	**8.1–8.55**
8.1.	Introduction ...	8.1
8.2.	Algebraic operations on functions ...	8.2
8.3.	Bounded and unbounded functions ...	8.2
8.4.	Limit of a function ...	8.3
8.5.	Algebra of limits ...	8.4
8.6.	One-sided limits — right-hand and left-hand limits ...	8.8
8.7.	Limits at infinity and infinite limits. ...	8.10
8.8.	Characterization of the limit of a function at a point in terms of sequences. Sequential criterion for limits ...	8.14
8.9.	Cauchy's criterion for finite limits ...	8.15
8.10.	The four functional limits at a point ...	8.16
8.11.	Continuous functions ...	8.16
	Discontinutiy of a function ...	8.17
8.12.	Algebra of continuous functions ...	8.26
8.13.	Function of a function. Composite of functions ...	8.28
8.14.	Continuity of the composite function ...	8.29
8.15.	Criteria for continuity. Equivalent definition of continuity ...	8.29
8.16.	Some properties of the continuity of a function at a point ...	8.35
8.17.	Properties of functions continuous in closed finite intervals ...	8.37
8.18.	Existence of the nth root of a given positive real number ...	8.42
8.19.	Uniform continuity ...	8.43
8.20.	Evaluation of $\lim_{x \to o} (\sin x) / x$, x being measured in radians ...	8.48
8.21.	Continuity of the inverses of continuous functions ...	8.49
8.21.	Root function ...	8.50
	Objective questions ...	8.51
9.	**REAL FUNCTIONS. THE DERIVATIVE ....**	**9.1–9.21**
9.1.	Derivability of a function ...	9.1
9.2.	A necessary condition for the existence of a finite derivative ...	9.2
9.3.	Algebra of derivatives ...	9.3
9.4.	Geometrical meaning of the derivative ...	9.6

9.5.	Meaning of the sign of derivative at a point	...	9.15
9.6.	Darboux's theorem (or Intermediate value theorem for derivatives)	...	9.16
10.	**MEAN VALUE THEOREMS**	...	**10.1–10.45**
10.1.	Rolle's theorem	...	10.1
10.2.	Failure of Rolle's theorem	...	10.2
10.3.	Geometrical interpretation of Rolle's theorem	...	10.2
10.4.	Lagrange's mean value theorem or first mean value theorem	...	10.9
10.5.	Increasing and decreasing functions. Monotonic functions		10.10
10.6.	Some useful deductions from the mean value theorem	...	10.11
10.7.	Increasing and decreasing functions and their application in establishing some inequalities		10.19
10.8.	Cauchy's mean value theorem	...	10.24
10.9.	Generalised mean value theorem	...	10.25
10.10.	Higher derivatives	...	10.27
10.11.	Taylor's theorem with Schlomilch and Roche form of remainder	...	10.27
10.11A.	Taylor's theorem with Cauchy's form of remainder	...	10.28
10.11B.	Taylor's theorem with Lagrange's form of remainder	...	10.28
11.12.	Power series representation of functions	...	10.29
10.13.	Maclaurin's infinite series	...	10.30
10.14.	Some standard results	...	10.30
10.15.	Powers series expansions of some standard functions	...	10.31
10.16.	Vortex function	...	10.42
11.	**MAXIMA AND MINIMA**	...	**11.1–11.14**
11.1.	Introduction	...	11.1
11.2.	A necessary condition for the existence of extreme values	...	11.1
11.3.	Sufficient criteria for the existence of extreme values	...	11.2
11.4.	Applications to problems	...	11.7
	Objective questions	...	11.12
12.	**INDETERMINATE FORMS**		**12.1–12.20**
12.1.	Introduction	...	12.1
12.2.	The indeterminate form (0/0)	...	12.1
	L'Hospital's theorem	...	12.2
12.3.	The indeterminate form (∞/∞)	...	12.3
12.4.	Some useful results	...	12.4
12.5.	Working rule to evaluate limit in indeterminate form (0/0)	...	12.5
12.6.	Application of L' Hospital's rule for ∞/∞ form	...	12.9
12.7.	The indeterminate form $0 \times \infty$	...	12.10
12.8.	The indeterminate form $\infty - \infty$	...	12.11
12.9.	The indeterminate forms 0^0, ∞^0 and 1^∞	...	12.12
	Objective questions	...	12.18
13.	**RIEMANN INTEGRABILITY**	...	**13.1–13.62**
13.1	Introduction	...	13.1
13.2	Partitions and Riemann (or Darboux) sums	...	13.1
13.3	Some properties of Darboux sums	...	13.2
13.4.	Upper and lower Riemann integrals. Riemann integral	...	13.6
13.5	Another equivalent definition of integrability and integral	...	13.16
13.6	A second definition of Riemann integrability	...	13.17
	Summation of series Theorem	...	13.19
13.7	Necessary and sufficient condition for integrability	...	13.22

13.8. Particular classes of bounded integrable functions ... 13.25
13.9 Properties of integrable functions ... 13.29
13.10 Integrability of the sum, difference, product and quotient of integrable functions ... 13.30
13.11 Integrability of the modulus of a bounded integrable function ... 13.33
13.12 Definition of $\int_a^b f(x)\,dx$, if $b \leq a$ ... 13.35
13.13 Inequalities for an integral ... 13.36
13.14 Functions defined by definite integrals ... 13.39
13.15 Fundamental theorem of integral calculus ... 13.40
13.16 Generalized mean value theorem or second mean value theorem ... 13.41
13.17 Abel's lemma. ... 13.42
13.18 Second mean value theorem (Weierstrass form) ... 13.42
13.19 Change of variable in an integral ... 13.44
13.20 Integration by parts ... 13.45
Objective questions ... 13.58

14. THE RIEMANN STIELTJES INTEGRAL ... 14.1–14.32

14.1 Introduction ... 14.1
14.2 Partition of a set. Lower and upper Rieman–Stieltjes sums ... 14.1
14.3 The lower and upper Riemann–Stieltjes integrals ... 14.4
14.4. The Riemann–Stieltjes integrals ... 14.5
14.5. The Riemann–Stieltjes integral as a limit of sum ... 14.7
14.6. Some useful inequalities related to R-S integrals ... 14.16
14.7. Algebra of R-S integrable functions ... 14.18
14.8. Reduction of Riemann–Stielijes integral into Riemann integral ... 14.26
14.9 Some useful theorems ... 14.28
Objective questions ... 14.31

15. UNIFORM CONVERGENCE OF SEQUENCES AND SERIES OF FUNCTIONS ... 15.1–15.34

15.1 Introduction ... 15.1
15.2 Cauchy's general principle of uniform convergence ... 15.2
15.3 A test for uniform convergence of sequence of functions ... 15.5
15.4 Countinuity of the uniform limit of a uniformly convergent sequence of continuous functions ... 15.7
15.5 Integrability of uniform limit of a uniformly convergent sequence of integrable functions ... 15.9
15.6 Derivability of the point-wise limit of a sequence of derivable functions if the derivatives are continuous and the sequence of derivatives is uniformly convergent ... 15.13
15.7 Infinite series of functions ... 15.14
15.8 Test for the uniform convergence of a series ... 15.14
15.8.1. Cauchy's general principle of uniform convergence for series... 15.14
15.8.2. Weierstrass' M-Test for uniform convergence ... 15.14
15.9 Abel's test and Dirichlet's test ... 15.18
15.10 Proberties of uniformly convergent series of functions ... 15.21
15.11 The Weierstrass approximation theorem ... 15.29
Objective questions ... 15.31

16. IMPROPER INTEGRALS ... 16.1–16.36

16.1 Proper and improper integrals ... 16.1
16.2 Convergence of improper integrals of the first kind ... 16.1
16.3 Test for convergence at 'a'. Positive integrand ... 16.4

16.4	The necessary and sufficient condition for the convergence of the improper integral ...	16.5
16.5	Comparison of two integrals ...	16.5
16.5A	A practical comparison test ...	16.6
16.6	Useful comparison integrals ...	16.7
16.7	Two useful tests ...	16.7
16.8	$f(x)$ not necessarily positive. General test for convergence ...	16.14
16.9	Absolute and conditionally convergence ...	16.15
16.10	Convergence of improper integrals of the second kind ...	16.16
16.11	Convergence at ∞. The integrand being positive ...	16.18
16.12	Comparison of two integrals (Comparison test) ...	16.19
16.13	A useful comparison integral ...	16.19
16.14	General test. Convergence at ∞. The integrand being not necessarily positive. Cauchy's test for convergence ...	16.25
16.15	Absolute and conditionally convergence of improper integrals of second kind ...	16.25
16.16	Test for the absolute convergence of the integral of a product ...	16.26
16.17	Abel's test ...	16.26
16.18	Dirichlet's test ...	16.27
	Objective questions ...	16.34
17.	**POWER SERIES**	**17.1–17. 20**
17.1	Introduction ...	17.1
17.2	Power seriers ...	17.1
17.3	Some important facts about the power series ...	17.1
17.4	Radius of convergence and interval of convergence ...	17.2
17.5	Formulas for determining the radius of convergence ...	17.2
17.5A	Solved examples based on Art. 17.5 ...	17.3
17.6	Some basis theorems ...	17.5
17.7	Properties of functions expressible as power series ...	17.6
17.8	Abel's theorem ...	17.9
17.9	Some theorems on power series ...	17.12
17.10	Solved examples ...	17.14
17.11	Elementary functions ...	17.17
18.	**DOUBLE SEQUENCES AND SERIES**	**18.1–18.12**
18.1	Double sequence ...	18.1
18.2	Convergence of a double sequence ...	18.1
18.3	Cauchy general principle of convergence of double sequence ...	18.1
18.4	Repeated double limits ...	18.1
18.5	Double series ...	18.2
18.6	Convergence of a double series ...	18.4
18.7	Double series of positive terms ...	18.5
18.8	Some tests for convergence of a double series of positive terms ...	18.6
18.9	Mixed series ...	18.7
18.10	Solved examples ...	18.8
18.11	Taylor's theorem for power series ...	18.11
19.	**METRIC SPACES**	**19.1–19.76**
19.1	Introduction ...	19.1
19.2	Metric space ...	19.1
19.3	Pseudo-metric space ...	19.1
19.4	Some important results for direct applications ...	19.1
19.5	Examples of metric spaces ...	19.2
19.6	Distance between two sets. Diameter of a set ...	19.4
19.7	Bounded and unbounded metric spaces ...	19.4
19.8	Open and closed spheres (or balls) ...	19.5
19.9	Neighbourhood of a point ...	19.5
19.10	Open set ...	19.5

19.11 Properties of open sets ... 19.6
19.12 Equivalent metrics ... 19.7
19.13 Limit Point (or accumulation point or a cluster point or a condensation point) 19.10
19.14 Closed set ... 19.10
19.15 Properties of closed sets ... 19.10
19.16 Subspaces ... 19.11
19.17 Adherent point (or contact point). Closure of a set ... 19.12
19.18 Properties of closure of a set ... 19.12
19.19 Interior point of a set. Interior of a set ... 19.13
19.20 Properties of interior of a set ... 19.14
19.21 Dense (or everywhere dense), dense in itself, nowhere dense (or non-dense) and perfect sets ... 19.15
19.22 Separable space ... 19.15
19.23 Exterior, frontier and boundary points ... 19.16
19.24 Product of metric spaces ... 19.16
19.25 Continuous functions on metric spaces ... 19.17
19.26 Properties of continuous functions ... 19.17
19.27 Uniform continuity ... 19.18
19.28 Homeomorphism ... 19.19
19.29 Isometry ... 19.19
19.30 Sequence in a metric space ... 19.20
19.31 Cauchy sequence ... 19.23
19.32 Complete metric space ... 19.24
19.33 Properties of complete metric spaces ... 19.26
19.34 Cantor's intersection theorem ... 19.27
19.35 Contraction mapping principle ... 19.30
Banach's fixed point theorem ... 19.30
19.36 A subset of first category and second category ... 19.31
19.37 Compact metric space ... 19.32
Finite intersection property ... 19.34
Balzano-Weierstrass property (BWP) ... 19.34
Sequentially compact metric space ... 19.34
Countably compact metric space ... 19.34
19.37A $\in$-net and total boundedness ... 19.35
19.37B Lebesgue number for a cover ... 19.36
Lebesgue covering lemma ... 19.36
19.37C Locally compact metric space ... 19.38
19.38 Connected metric spaces ... 19.40
Separated sets ... 19.40
Connected and disconnected sets ... 19.40
19.39 Components of a metric spaces ... 19.46
19.40 Connectedness of product of connected metric spaces ... 19.47
Objective questions ... 19.47
Additional problem on Chapter 19 ... 19.53

20. BETA AND GAMMA FUNCTIONS **20.1–20.26**
20.1 Introduction ... 20.1
20.2 Euler's integrals. Beta and Gamma functions ... 20.1
20.3 Properties of Gamma function ... 20.1
20.4 Extension of definiton of Gamma function ... 20.2
20.5 Theorem. To show that $\Gamma(1/2) = \sqrt{\pi}$ ... 20.3
20.6 Transformation of Gamma function ... 20.3
20.7 Solved examples based on Gamma function ... 20.4
20.8 Symmetrical properties of Beta function ... 20.8
20.9 Evaluation of $B\ (m, n)$ in an explicit form when m or n is a positive integer 20.8
20.10 Transformation of Beta function ... 20.9
20.11 Relation between Beta and Gamma functions ... 20.12
20.12 Solved examples ... 20.15

20.13	Legendre-duplication formula	...	20.21
20.14	Solved examples	...	20.22
	Miscellaneous problems based on this Chapter	...	20.23
21.	**DIFFERENTIATION UNDER THE INTEGRAL SIGN**	...	**21.1–21.24**
21.1	Introduction	...	21.1
21.2	Leibnitz's rule for differentiation under the integral sign	...	21.1
21.3	General form of Leibnitz's rule of differentiation under the integral sign when the limits of integration are functions of the prarameters	...	21.2
21.4A	Evaluation of integral $\int_a^b f(x,\alpha)dx$, where a and b are independent of α. Working rule	...	21.4
21.4B	Solved examples of type I based on Art 21.4A	...	21.4
21.5A	Evaluation of integral $\int_a^b g(x,\alpha,\beta)\,dx$, where a and b are independent of parameters α and β. Working rule	...	21.14
21.5B	Solved examples of type 2 based on Art. 21.5 A	...	21.14
21.6A	Evaluation of integral $\int_{g(\alpha)}^{h(\alpha)} f(x,\alpha)\,dx$, where $g(\alpha)$ and $h(\alpha)$ are functions of parameter α. Working rule	...	21.19
21.6B	Solved examples of types 3 based on Art. 21.6 A	...	21.19
21.7A	Determination of the value of an integral when certain standard known integral is given with its value. Working rule	...	21.21
21.7B	Solved examples of type 4 based on Art 21.7A	...	21.21
22.	**FOURIER SERIES**	...	**22.1–22.48**
22.1	Introduction	...	22.1
22.2	Euler's formulae for Fourier series expansion of a function $f(x)$	...	22.1
22.3	Dirichlet's conditions for Fourier series	...	22.3
22.4	Convergence of Fourier series	...	22.4
22.5	Some particular cases of theorem of Art. 22.2 for direct applications	...	22.4
22.6	Solved examples based on results of Art. 22.5	...	22.6
22.7	Functions having arbitrary periods. Change of intervals	...	22.10
22.8	Solved examples based on results of Art. 22.7	...	22.12
22.9	Half range series	...	22.13
22.10	Solved examples based on results of Art. 22.9	...	22 14
22.11	Parseval's theorem. Parseval's identity	...	22.15
22.12	Some particular Fourier series	...	22.18
22.13	Some important theorems	...	22.22
22.14	Some typical solved examples	...	22.28
22.15	Additional solved examples	...	22.33
22.16	Some important properties of Fourier coefficients	...	22.37
22.17	An important lemma	...	22.43
22.18	Absolute and uniform convergence of Fourier series	...	22.44
22.19	Differentiation of Fouier Series	...	22.45
22.20	Integration of Fourier Series	...	22.46
	Appendix-A Additional results and problems based on chapter 15	...	**A.1-A.14**
A.1	Some useful theorems	...	A.1
A.2	Uniform convergence and continuity	...	A.3
A.3	Dini's theorems on uniform convergence	...	A.4
A.4	Uniform convergence and integration	...	A.5
A.5	Uniform convergence and integration (particular cases of theorems 1 and 2 discussed in Art. A.4)	...	A.6
A.6	Uniform convergence and differentiation	...	A.8
A.7	Miscellaneous problems on chapter 15	...	A.12
	Miscellaneous results and problems based on the entire book	...	**M.1-M.94**
	INDEX	...	I.1-I.5

SYMBOLS

$\boldsymbol{N}$	the set of natural numbers
$\boldsymbol{Z}$ or $\boldsymbol{I}$	the set of integers
$\boldsymbol{Z}^+$ or $\boldsymbol{I}^+$	the set of positive integers
$\boldsymbol{Q}$	the set of rational numbers
$\boldsymbol{Q}^+$	the set of positive rational numbers
$\boldsymbol{R}$	the set of real numbers
$\boldsymbol{R}^+$	the set of positive real numbers
$\Rightarrow$	implies
$\Leftrightarrow$	is equivalent to
$\{\ \}$	set
$\in$	is an element of
$:$	such that
$\subset$	is contained in (is a subset of)
$\supset$	contains (is a superset of)
A' or $\sim A$ or A^c or $U \sim A$ or $U \backslash A$	complement of A with respect to universal set U
A' or $\sim A$ or A^c or $\mathbf{R} \sim A$ or $\mathbf{R} \backslash A$	complement of A with respect to $\boldsymbol{R}$
$\cup$	union
$\cap$	intersection
ϕ	the empty set
$\exists$	there exists
$\forall$	for all

THE GREEK ALPHABET

Name	Lower	Upper	Name	Lower	Upper
alpha	α	A	xi	ξ	Ξ
beta	β	B	omicron	o	O
gamma	γ	Γ	pi	π	Π
delta	δ	Δ	rho	ρ	P
epsilon	ε	E	sigma	σ	Σ
zeta	ζ	Z	tau	τ	T
eta	η	H	upsilon	υ	Υ
theta	θ	Θ	phi	ϕ	Φ
iota	ι	I	chi	χ	X
kappa	κ	K	psi	ψ	Ψ
lambda	λ	Λ	omega	ω	Ω
mu	μ	M			
nu	ν	N			

Chapter 1

Sets and Functions

1.1. INTRODUCTION

Set-Theoretic Notation and Terminology. What is called *Real Analysis* is a development of the set of real numbers which is reached through a series of successive extensions and generalisations starting from the set of natural numbers. As a matter of fact, starting from the set of natural numbers, we first pass on to the set of integers, then to the set of rational numbers and finally to the set of real numbers. Of course, from the set of real numbers, we can also pass on to the set of complex numbers but since Real Analysis is not concerned with complex numbers, we shall have nothing to do with complex numbers in this course. What is known as *Complex Analysis* is a development of the set of complex numbers.

It is no part of the plan of this book to *construct* the various systems of numbers and to develop their properties axiomatically from a given system of postulates. All that is intended here is to *describe* the various properties of these systems and to bring out the essential differences between them; it being understood that most of these properties are already familiar to the reader. What is important from the point of view of this course is the form in which these properties are stated and the type of emphasis which is thus brought out. While this programme will be undertaken in the following Chapter 2, we propose in this chapter to introduce some new notations and terms pertaining to '*sets*' and '*functions*' which will be found useful for the exposition of the subject proposed to be studied in this book.

1.2. STATEMENTS

In our everyday language, we are concerned with statements which are often distinguished as Interrogative, Imperative, Exclamatory or Declarative. In Mathematics, however, our chief interest is only in those statements which are Declarative and which may be either true or false. Consider, for example, the following statements, some of which are true and some false :

(*i*) The sum of the three angles of a triangle is equal to two right angles.

(*ii*) Every rectangle is a triangle.

(*iii*) The sum of an opposite pair of angles of a cyclic quadrilateral is equal to two right angles.

(*iv*) If two straight lines are perpendicular to the same straight line, then they are parallel to each other.

(*v*) The straight line joining the mid-points of two sides of a triangle is parallel to the third.

(*vi*) If x is 2 and y is 5, then $x + y$ is 7.

(*vii*) If $xy = 0$ and x, y are real numbers, then $x = 0$ and $y = 0$.

(*viii*) If $xy = 1$ and x, y are natural numbers, then $x = 1$ and $y = 1$.

(*ix*) If $xy = 1$ and x, y are rational numbers, then $x = 1$ and $y = 1$.

(*x*) If $x = 3$, then $x^2 = 9$.

(*xi*) If $x^2 = 9$, then $x = 3$.

(*xii*) Every pair of natural numbers admits of a highest common factor.

(*xiii*) Every natural number admits of an infinite number of factors.

1.3. CONNECTIVES : $\Rightarrow$, $\Leftrightarrow$, $\wedge$, $\vee$, ~

In the study of Mathematics, we are concerned with logical inter-connections between statements. Also on the basis of given statements, we build up new statements. There are a few symbols which are found useful for describing logical inter-connections between given statements and for building up new statements from the old ones and we now proceed to introduce these symbols.

The Connective $\Rightarrow$: If P and Q be two statements such that the truth of the statement P implies the truth of the statement Q, we exhibit this relationship between the two statements symbolically as $P \Rightarrow Q$ so that the symbol, $\Rightarrow$, stands for '*Implies*'. Thus, we have

(*i*) $x = 3 \Rightarrow x^2 = 9$. (*ii*) x *is a real number* $\Rightarrow x^2 \geq 0$.

(*iii*) *ABCD is a parallelogram* $\Rightarrow AB = CD$. (*iv*) *ABC is a triangle* $\Rightarrow AB + BC > AC$

(*v*) $AB \parallel CD$ *and* $CD \parallel EF \Rightarrow AB \parallel EF$.

The statement $P \Rightarrow Q$ formulates what is often expressed in any one of the following ways :

(*i*) If P then Q.

(*ii*) A necessary condition for the truth of P is the truth of Q.

(*iii*) A sufficient condition for the truth of Q is the truth of P.

The symbol $\Rightarrow$ is a *connective* in as much as connecting the two statements P, Q, it generates a third statement, *viz.*, the statement $P \Rightarrow Q$.

The Connective $\Leftrightarrow$: If P, Q are two statements such that we have $P \Rightarrow Q$ as well as $Q \Rightarrow P$, we write $P \Leftrightarrow Q$ and say that P implies and is implied by Q or that P is *equivalent* to Q.

Thus, *ABCD is a parallelogram* $\Leftrightarrow$ $AB = CD$ *and* $BC = AD$.

$x^2 = 9$, x, y are real numbers $\Leftrightarrow x \in \{3, -3\}$.

The statement $P \Leftrightarrow Q$ expresses what is also described in one of the following ways :

(*i*) P if and only if Q. (*ii*) Q if and only if P.

(*iii*) A necessary and sufficient condition for the truth of P is the truth of Q.

(*iv*) A necessary and sufficient condition for the truth of Q is the truth of P.

(*v*) P and Q are equivalent statements

The Connectives $\wedge$, $\vee$: The symbols $\wedge$, $\vee$, stand for *and*, *or* respectively.

If P and Q be two statements, then with the help of these connectives, we form two new statements $P \wedge Q$, $P \vee Q$.

The statement $P \wedge Q$ is true if and only if P is true *as well as* Q is true. Thus, if the statements P, Q are *both* true, then the statement $P \wedge Q$ is true and *vice-versa*.

The statement $P \vee Q$ is true if and only if P is true or Q is true, *i.e.*, if and only if at least one of P and Q is true.

For example, *ABCD* is a parallelogram $\Leftrightarrow AB = CD \wedge BC = AD$.

$x^2 - 5x + 6 = 0 \Rightarrow x = 2 \vee x = 3$.

$xy = 0$, x, y are real numbers $\Leftrightarrow x = 0 \vee y = 0$.

$x = 2 \wedge y = 3 \Rightarrow x + y = 5$.

Negation ~ : If P denotes a statement, then ~ P, denotes the *negation* or the *denial* of P. Let P denote the statement $x = 4$. Then, ~ P, denotes the statement $x \neq 4$.

For example, *ABC* is a triangle $\Rightarrow$ ~ $[AB + BC \leq AC]$.

We may note that the statement $\sim P$ is true or false according as the statement P is false or true.

It may be noticed that given any statement P, true or false, the statement $P \vee \sim P$ is always true and the statement $P \wedge \sim P$ is always false.

Existential Quantifiers $\forall, \exists$: We shall use the symbols $\forall, \exists$ called *existential quantifiers* to stand for *for all*, *there exists* respectively.

EXERCISES

1. Which of the following statements are false :

(*a*) (*i*) A, B, C are three collinear points $\Rightarrow AB + BC = AC$.

(*ii*) $AB + BC = AC \Rightarrow A, B, C$ are three collinear points.

(*iii*) A, B, C are three collinear points $\Leftrightarrow AB + BC = AC$.

(*b*) (*i*) A, B, C are three collinear points

$$\Rightarrow (AB + BC - CA)(BC + CA - AB)(CA + AB - BC) = 0.$$

(*ii*) $(AB + BC - CA)(BC + CA - AB)(CA + AB - BC) = 0$

$$\Rightarrow A, B, C \text{ are three collinear points.}$$

(*iii*) A, B, C are three collinear points

$$\Leftrightarrow (AB + BC - CA)(BC + CA - AB)(CA + AB - BC) = 0.$$

(*c*) (*i*) ABC is a right-angled triangle $\Rightarrow AB^2 + BC^2 = AC^2$.

(*ii*) $AB^2 + BC^2 = AC^2 \Rightarrow ABC$ is a right-angled triangle.

(*iii*) ABC is a right-angled triangle $\Leftrightarrow AB^2 + BC^2 = AC^2$.

(*iv*) $AB^2 + BC^2 = AC^2 \Leftrightarrow ABC$ is a right-angled triangle.

(*d*) (*i*) ABC is a right-angled triangle

$$\Rightarrow (AB^2 + BC^2 - AC^2)(BC^2 + CA^2 - AB^2)(CA^2 + AB^2 - BC^2) = 0.$$

(*ii*) $(AB^2 + BC^2 - AC^2)(BC^2 + CA^2 - AB^2)(CA^2 + AB^2 - BC^2) = 0$

$$\Rightarrow ABC \text{ is a right-angled triangle.}$$

(*iii*) ABC is a right-angled triangle

$$\Leftrightarrow (AB^2 + BC^2 - AC^2)(BC^2 + CA^2 - AB^2)(CA^2 + AB^2 - BC^2) = 0.$$

2. Which of the following statements are true; x denoting a real number :

(*a*) (*i*) $\sim (x > 1) \Rightarrow x < 1$. (*ii*) $\sim (x < 1) \Rightarrow x > 1$.

(*iii*) $\sim (x > 1) \Leftrightarrow x < 1$. (*iv*) $x < 1 \Leftrightarrow \sim (x > 1)$.

(*b*) (*i*) $\sim (x = 0) \Rightarrow \sim (x^2 = 0)$. (*ii*) $\sim (x^2 = 0) \Rightarrow \sim (x = 0)$.

(*iii*) $\sim (x = 0) \Leftrightarrow \sim (x^2 = 0)$. (*iv*) $\sim (x^2 = 0) \Leftrightarrow \sim (x = 0)$.

(*c*) (*i*) $\sim (x > 0) \Rightarrow \sim (x^2 < 0)$. (*ii*) $\sim (x^2 < 0) \Rightarrow \sim (x > 0)$.

(*iii*) $\sim (x > 0) \Leftrightarrow \sim (x^2 < 0)$. (*iv*) $\sim (x^2 < 0) \Leftrightarrow \sim (x > 0)$.

(*d*) (*i*) $x = 0 \Rightarrow \sin x = 0$. (*ii*) $\sim (x = 0) \Rightarrow \sim (\sin x = 0)$.

(*iii*) $\sim (\sin x = 0) \Rightarrow \sim (x = 0)$.

3. Which of the following are false statements; x denoting a real number :

(*a*) (*i*) $x \geq 3 \Leftrightarrow x > 3 \vee x = 3$. (*ii*) $x \geq 3 \Leftrightarrow x > 3 \wedge x = 3$.

(*b*) (*i*) $x \leq 1 \Leftrightarrow x < 1 \wedge x = 1$. (*ii*) $x \leq 1 \Leftrightarrow x < 1 \vee x = 1$.

(*c*) (*i*) $x \neq 5 \Leftrightarrow x > 5 \wedge x < 5$. (*ii*) $x \neq 5 \Leftrightarrow x > 5 \vee x < 5$.

(*d*) (*i*) $-2 < x < -1 \Leftrightarrow -2 < x \vee x < -1$.

(*ii*) $x > 1 \wedge x < 2 \Leftrightarrow 1.2 < x < 1.8$.

4. Which of the following statements are true; x denoting a real number :

(*a*) (*i*) $x + y = 3 \wedge x - y = 1 \Leftrightarrow x = 2 \wedge y = 1.$

(*ii*) $x + y = 3 \vee x - y = 1 \Leftrightarrow x = 2 \vee y = 1.$

(*iii*) $x + y = 3 \vee x - y = 1 \Leftrightarrow x = 2 \wedge y = 1.$

(*iv*) $x + y = 3 \wedge x - y = 1 \Leftrightarrow x = 2 \vee y = 1.$

(*b*) (*i*) $x > 2 \wedge x < 4 \Rightarrow x = 3.$

(*ii*) $x = 3 \Rightarrow x > 2 \wedge x < 4.$

(*iii*) $x = 3 \Rightarrow x > 2 \vee x < 4.$

(*c*) (*i*) $(x - 1)(x - 2) = 0 \Leftrightarrow (x - 1 = 0) \wedge (x - 2 = 0).$

(*ii*) $(x - 1)(x - 2) = 0 \Leftrightarrow (x - 1 = 0) \vee (x - 2 = 0).$

1.4. SETS

If S be a set or a collection of objects and x a member of S, *i.e.*, if x is an object which belongs to S, we shall write $x \in S$ and say that "*x is an element of S*" or that "*x is a member of S*" or that "*x belongs to S*".

Also we shall write $S = \{x : P(x)\}$, which indicates that S is a set of objects x for which the statement $P(x)$ involving x is true.

If an object x is *not* an element of a set S, we write $x \notin S$.

It is also sometimes possible to describe the set by listing its elements. Thus, $S = \{2, 4, 5, 9\}$ denotes a set consisting of 2, 4, 5, 9 only as its elements.

Thus, while $4 \in \{2, 4, 5, 9\}$, $7 \notin \{2, 4, 5, 9\}$.

We also have to often use the notation

$$T = \{1, 2, 3, 4, 5, 6, ...\}$$

where the dots ..... stand for 'and so on'. Here T is the set consisting of all the natural numbers.

Void set. It is found convenient to utilise the notion of a set such that, given any object whatsoever; it is not a member of the set, *i.e.*, to say, no object belongs to the same. This set called the *void set*, *empty set* or the *null set* is often denoted by the symbol ϕ.

We may, for example, write $\phi = \{x : x \neq x\}$.

Singleton. A set consisting of a single element is called a *singleton*. For example, $\{a\}$ is a singleton set.

Equality of sets. If S and T are two sets, we write $S = T$ to indicate that each element of S is an element of T and that each element of T is an element of S.

Sub-sets. If A and B are two sets such that an element of A is as well as an element of B, we say that A is a *sub-set* of B and denote this by writing $A \subset B$ or equivalently $B \supset A$.

If A is a sub-set of B, we also say that A is contained in B. We also sometimes say that B is a *super-set* of A.

Thus, if A is a sub-set of B, there is no element of A which is not as well as an element of B.

Sometimes we also say that A is a *proper* sub-set of B, if A is a sub-set of B without being the same as B. For example, $\{2, 3, 4, 5\}$ is a proper sub-set of $\{4, 3, 2, 6, 5\}$.

Universal set. In any mathematical discussion, we consider all the sets to be the sub-sets of a given fixed set known as a *universal set*. It is generally denoted by U or X. For example, while studying sets containing natural numbers only, **N** (*i.e.*, set of all natural numbers) is taken as the universal set.

OPERATIONS ON SETS

Union. If A and B are two sets, then the set consisting of objects which belong to A or to B or to both is called the *union* of the sets A and B. The union of sets A and B is denoted by the symbol $A \cup B$. Thus, we have

$$A \cup B = \{x : x \in A \text{ or } x \in B \text{ or } x \in \text{ both } A \text{ and } B\}.$$

Intersection. The set consisting of elements which belong to A as well as B is called the *intersection* of A and B. This intersection of sets A and B is denoted by $A \cap B$. Thus, we have

$$A \cap B = \{x : x \in A \text{ and } x \in B\}.$$

We have $A \cap B = \phi$, if the sets A and B have no element in common.

In this case, we say that the sets A and B are *disjoint*.

Indexed family of sets. Let Λ be a non-empty set. If for each $\lambda \in \Lambda$, we are given a set A_λ, then we say that $\{A_\lambda\}_{\lambda \in \Lambda}$ is a family of sets indexed by the set Λ.

If $\{A_\lambda\}_{\lambda \in \Lambda}$ be an arbitrary family of sets, then the union and intersection of this arbitrary family of sets are denoted and defined by

$$\bigcup_{\lambda \in \Lambda} A_\lambda = \{x : x \in A_\lambda \text{ for at least one } \lambda \in \Lambda\}$$

and
$$\bigcap_{\lambda \in \Lambda} A_\lambda = \{x : x \in A_\lambda \text{ for every } \lambda \in \Lambda\}.$$

Difference of sets. Let A and B be two sets. The *difference of* A and B, denoted by $A - B$ or $A \sim B$, is the set of all those elements of A which do not belong to B. Thus, we have

$$A - B = A \sim B = \{x : x \in A \text{ and } x \notin B\}.$$

For example, if $A = \{1, 2, 3, 4, 5\}$ and $B = \{2, 4, 6, 8\}$, then $A - B = \{1, 3, 5\}$ and $B - A = \{6, 8\}$. Clearly, $A - B \neq B - A$.

Complement of a set. Let U be the universal set and $A \subset U$. Then the *complement of* A with respect to U is denoted by A' or A^c or $U - A$ or $U \setminus A$ or $\sim A$ and defined as the set of all those elements of U which are not in A. Thus, we have

$$A' = A^c = U - A = U \setminus A = \sim A = \{x : x \in U \text{ and } x \notin A\}$$

or
$$A' = A^c = U - A = U \setminus A = \sim A = \{x \in U : x \notin A\}.$$

LAWS OF ALGEBRA OF SETS

(*i*) *Idempotent laws.* For any set A, $A \cup A = A$, $A \cap A = A$.

(*ii*) *Identity laws.* For any set A, $A \cup \phi = A$, $A \cap U = A$.

(*iii*) *Commutative laws.* For any two sets A and B, we have

$$A \cup B = B \cup A, \quad A \cap B = B \cap A.$$

(*iv*) *Associative laws.* If A, B and C are any three sets, then

$$(A \cup B) \cup C = A \cup (B \cup C), \quad (A \cap B) \cap C = A \cap (B \cap C).$$

(*v*) *Distributive laws.* If A, B and C are any three sets, then

$$A \cup (B \cap C) = (A \cup B) \cap (A \cup C), \quad A \cap (B \cup C) = (A \cap B) \cup (A \cap C).$$

(*vi*) *De-Morgan's laws.* For any two sets A and B, we have

$$(A \cup B)' = A' \cap B', \quad (A \cap B)' = A' \cup B'.$$

Ordered pair. An *ordered pair* consists of two elements in a given *order*. For example, if A and B are any two sets, then by an ordered pair of elements, we mean a pair (a, b) in that order, where $a \in A$ and $b \in B$.

Equality of ordered pairs. Two ordered pairs (a_1, b_1) and (a_2, b_2) are equal if and only if $a_1 = a_2$ and $b_1 = b_2$.

Cartesian product of sets. Let A and B be any two non-empty sets. The set of all ordered pairs (a, b) such that $a \in A$ and $b \in B$ is called the *cartesian product* of the sets A and B and is denoted by $A \times B$. Thus, we have

$$A \times B = \{(a, b) : a \in A \text{ and } b \in B\}.$$

For example, if $\mathbf{R}$ be the set of all real numbers, then

$$\mathbf{R} \times \mathbf{R} = \{(x, y) : x \in \mathbf{R}, y \in \mathbf{R}\},$$

i.e., $\mathbf{R} \times \mathbf{R}$ is the set of all points in the cartesian plane. It is also denoted by $\mathbf{R}^2$.

Note 1. If A, B and C be three non-empty sets, then cartesian product of A, B and C is denoted and defined as

$$A \times B \times C = \{(a, b, c) : a \in A, b \in B, c \in C\}.$$

For example, if $\mathbf{R}$ be the set of all real numbers, then

$$\mathbf{R} \times \mathbf{R} \times \mathbf{R} = \{(x, y, z) : x \in \mathbf{R}, y \in \mathbf{R}, z \in \mathbf{R}\},$$

i.e., $\mathbf{R} \times \mathbf{R} \times \mathbf{R}$ is the set of all points in the Euclidean plane. It is also denoted by $\mathbf{R}^3$.

Note 2. Cartesian product of n non-empty sets $A_1, A_2, \ldots, A_n$ is denoted and defined as

$$A_1 \times A_2 \times \ldots \times A_n = \{(a_1, a_2, \ldots, a_n) : a_i \in A_i, 1 \le i \le n\}$$

where $(a_1, a_2, \ldots, a_n)$ is known as an ordered n-tuple.

1.5. FUNCTIONS (OR MAPPINGS)

Let X and Y be two non-empty sets. If there exists a rule 'f' which associates to every element $x \in X$, a unique element $y \in Y$, then such a rule 'f' is known as a *function* (or *mapping*) from the set X to the set Y. If f is a function from X to Y, then we write $f : X \to Y$ or $X \xrightarrow{f} Y$, which is read as f is a function from X to Y or f maps X to Y. If an element $x \in X$ is associated to an element of $y \in Y$, then y is called '*f-image of x*' or '*image of x under f*' or '*the value of the function f at x*'. Also, x is called the *pre-image* of y under the mapping f. We write it as $y = f(x)$.

The set X is called the *domain* of f and Y is called the *co-domain* of f. The set of all f-images is called the *range* of f and is denoted by $f(X)$. Thus, we have

$$f(X) = \text{Range of } f = \{f(x) : x \in X\}. \text{ Clearly, } f(X) \subseteq Y.$$

Equal functions. Two functions $f : X \to Y$ and $g : X \to Y$ are said to be equal if and only if

$$f(x) = g(x) \ \forall \ x \in X.$$

KINDS OF FUNCTIONS

(*i*) **One-one function (or injection).** A function $f : X \to Y$ is said to be a *one-one function* or an *injection* if different elements of X have different images in Y. Thus, we have

$$f : X \to Y \text{ is one-one } \Leftrightarrow x_1 \ne x_2 \Rightarrow f(x_1) \ne f(x_2) \ \forall \ x_1, x_2 \in X$$

or equivalently $\quad f : X \to Y$ is one-one $\Leftrightarrow f(x_1) = f(x_2) \Rightarrow x_1 = x_2 \ \forall \ x_1, x_2 \in X.$

(*ii*) **Many-one function.** A function $f : X \to Y$ is said to be *many-one function* if two or more elements of X have the same image in Y.

Thus, $f : X \to Y$ is a many-one function if $\exists \ x_1, x_2 \in X$ such that $x_1 \ne x_2 \Rightarrow f(x_1) = f(x_2)$.

(*iii*) **Onto function (or surjection).** A function $f : X \to Y$ is said to be an *onto function* or a *surjection* if every element of Y is the f-image of some element of X, *i.e.*, if $f(X) = Y$ or range of f is the co-domain of f.

Thus, $f : X \to Y$ is an onto function if and only if for each $y \in Y$, $\exists \ x \in X$ such that

$$f(x) = y.$$

(*iv*) **Into function.** A function $f : X \to Y$ is said to be an *into function* if there exists at least one element $y \in Y$ which has no pre-image under f. Clearly,

$$f \text{ is into function } \Rightarrow f(X) \subset Y.$$

(*v*) **One-one onto function (or bijection).** A *one-one onto* function is said to be a *bijection*.

(*vi*) **Identity function.** A function $f : X \to X$ is said to be an *identity function* if $f(x) = x$ $\forall\ x \in X$. Identity function on X is denoted as I_X and is clearly a one-one onto function.

(*vii*) **Constant function.** A function $f : X \to Y$ is said to be a *constant function* if every element of X has the same image, *i.e.*, $f(x) = c\ \forall\ x \in X$, where $c \in Y$. Also, then $f(X) = \{c\}$.

Example. *Give an example each of the following : (i) One-one into function (ii) Many-one onto function (iii) One-one onto function (iv) Many-one into function.*

Solution. (*i*) Let **N** denote the set of all natural numbers.

Let $f : \mathbf{N} \to \mathbf{N}$ be defined by $f(n) = 2n\ \forall\ n \in \mathbf{N}$.

Then, $\quad f(n_1) = f(n_2) \Rightarrow 2n_1 = 2n_2 \Rightarrow n_1 = n_2\ \forall\ n_1, n_2 \in \mathbf{N}.$

Hence, f is one-one.

Here $1 \in \mathbf{N}$. If possible, let n be its pre-image under f. Then

$$f(n) = 1 \Rightarrow 2n = 1 \Rightarrow n = 1/2 \notin \mathbf{N},$$

showing that f is not onto.

(*ii*) Let **R** and $\mathbf{R}^+$ denote the set of all real numbers and the set of all positive real numbers respectively.

Let $f : \mathbf{R} \to \mathbf{R}^+$ be defined by $\quad f(x) = x^2\ \forall\ x \in \mathbf{R}$.

Here $\pm 1 \in \mathbf{R}$ and $1 \neq -1$, but $f(1) = 1 = f(-1)$. So, f is many-one.

Again for every $y \in \mathbf{R}^+$, $\exists \sqrt{y} \in \mathbf{R}$ such that $f(\sqrt{y}) = (\sqrt{y})^2 = y$, showing that every element of $\mathbf{R}^+$ has pre-image under f. Hence, f is onto.

(*iii*) Let $f : \mathbf{R} \to \mathbf{R}$ be defined by $f(x) = x + 2\ \forall\ x \in \mathbf{R}$.

Then, $\quad f(x_1) = f(x_2) \Rightarrow x_1 + 2 = x_2 + 2 \Rightarrow x_1 = x_2\ \forall\ x_1, x_2 \in \mathbf{R}.$

Hence, f is one-one.

Next, let y be any element of **R** and let x be its pre-image under f. Then, we have

$$f(x) = y \Rightarrow x + 2 = y \Rightarrow x = y - 2 \in \mathbf{R},$$

showing that for every $y \in \mathbf{R}$, $\exists\ x \in \mathbf{R}$ such that $f(x) = y$. So, f is onto.

(*iv*) Let $f : \mathbf{R} \to \mathbf{R}$ be defined by $f(x) = x^2\ \forall\ x \in \mathbf{R}$.

Here $\pm 1 \in \mathbf{R}$ and $1 \neq -1$, but $f(1) = 1 = f(-1)$. So, f is many-one.

Here $-1 \in \mathbf{R}$. Let x be pre-image of -1 under f. Then

$$f(x) = -1 \Rightarrow x^2 = -1 \Rightarrow x = \pm i \notin \mathbf{R}. \text{ So, } f \text{ is into.}$$

1.6. COMPOSITE OF FUNCTIONS (OR PRODUCT OF FUNCTIONS)

Let $f : X \to Y$ and $g : Y \to Z$ be two functions such that

$$f(x) = y \text{ and } g(y) = z, \text{ where } x \in X, y \in Y, g \in Z.$$

Then the function $h : X \to Z$ such that

$$h(x) = z = g(y) = g(f(x))\ \forall\ x \in X$$

is known as the *composite* of f and g and is denoted by *gof*.

i.e., $\quad gof : X \to Z$ is defined by $(gof)(x) = g(f(x))\ \forall\ x \in X$.

PROPERTIES OF COMPOSITE OF FUNCTIONS

(*i*) The composite of functions is not always commutative, *i.e.*, $fog \neq gof$.

(*ii*) The composite of functions is associative, *i.e.*, $(fog)oh = fo(goh)$.

(*iii*) The product of any function with the identity function is the function itself, *i.e.*, if $f : X \to Y$, then $fof_X = f = f_Yof$.

(*iv*) The composite of two bijections is also a bijection.

1.7. INVERSE FUNCTION

Let $f : X \to Y$ be a one-one onto function. Then the function $g : Y \to X$ which associates to each element $y \in Y$ the unique element $x \in X$ such that $f(x) = y$ is known as the *inverse function* of *f*. The inverse function *g* of *f* is denoted by f^{-1}. Then, we have

$$f^{-1} : Y \to X \text{ such that } f^{-1}(y) = x, \text{ where } f(x) = y.$$

SOME USEFUL RESULTS

(*i*) A function is invertible if and only if it is one-one onto.

(*ii*) The inverse function, if it exists, is unique.

(*iii*) $(f^{-1})^{-1} = f$ (*iv*) $(gof)^{-1} = f^{-1}og^{-1}$.

1.8. BINARY OPERATION

Let *S* be a non-empty set. Then the mapping $f : S \times S \to S$ is known as a *binary operation* on *S*.

Hence, given $(x, y) \in S \times S \Rightarrow f(x, y) \in S$.

This unique element $f(x, y)$, for the sake of convenience, is written as $x * y$ in place of $f(x, y)$, where $*$ denotes a binary operation on *S*.

For example, addition is a binary operation on the sets **N, Z, Q, R** and **C** whereas subtraction is not a binary operation on *S*.

EXERCISES

1. Given $A = \{4, 8, 12, 16, 20\}$, $B = \{5, 10, 15, 20, 25\}$, $C = \{6, 12, 18, 24, 30\}$, describe the sets :

(*i*) $(A \cup B) \cap (B \cup C)$. (*ii*) $(A \cap B) \cup (B \cap C)$. (*iii*) $(A \cup B) \cap C$. (*iv*) $(A \cap B) \cup C$.

2. Given that A = {Quadrilaterals}, B = {Trapeziums}, C = {Parallelograms}, D = {Rhombuses}, E = {Rectangles}, F = {Squares} which of the following statements are true :

$$F \subset E, \quad C \subset D, \quad D \subset C, \quad B \subset D, \quad E \subset C.$$

3. Show that whatever be the sets A, B, C,

(*i*) $A \cup B = B \cup A$. (*ii*) $A \cap B = B \cap A$.

(*iii*) $(A \cup B) \cup C = A \cup (B \cup C)$. (*iv*) $(A \cap B) \cap C = A \cap (B \cap C)$.

(*v*) $(A \cup B) \cap C = (A \cap C) \cup (B \cap C)$. (*vi*) $(A \cap B) \cup C = (A \cup C) \cap (B \cup C)$.

4. Let $f : X \to Y$ be a mapping. Let $A \subset X$ and $B \subset Y$.

Define : $f(A) = \{f(x) : x \in A\}$ and $f^{-1}(B) = \{x \in X : f(x) \in B\}$.

Prove that (*i*) $A \subset f^{-1}(f(A))$ (*ii*) $f(f^{-1}(B)) \subset B$.

5. Let $f : X \to Y$, $A_1 \subset X$ and $A_2 \subset X$. Then, prove that

(*i*) $f(\phi) = \phi$ (*ii*) $A_1 \subset A_2 \Rightarrow f(A_1) \subset f(A_2)$

(*iii*) $f(A_1 \cup A_2) = f(A_1) \cup f(A_2)$ (*iv*) $f(A_1 \cap A_2) \subset f(A_1) \cap f(A_2)$.

6. Let $f: X \to Y$, $B_1 \subset Y$ and $B_2 \subset Y$. Then, prove that

(*i*) $f^{-1}(\phi) = \phi$ and $f^{-1}(Y) = X$ (*ii*) $B_1 \subset B_2 \Rightarrow f^{-1}(B_1) \subset f^{-1}(B_2)$

(*iii*) $f^{-1}(B_1 \cup B_2) = f^{-1}(B_1) \cup f^{-1}(B_2)$ (*iv*) $f^{-1}(B_1 \cap B_2) = f^{-1}(B_1) \cap f^{-1}(B_2)$.

OBJECTIVE QUESTIONS

Multiple Choice Type : *Select (a), (b), (c) or (d), whichever is correct.*

1. For the function $f(x) = 1/x$, the image of the interval [2, 7] is :

(*a*) [0, 1/8] (*b*) [1/7, 1/2] (*c*) [0, 1/7] (*d*) [0, 1/2]. **(Kanpur, 2004)**

2. If $f: S \to T$ be a function and if $X \subset S$, then :

(*a*) $X \subseteq f^{-1}(f(X))$ (*b*) $X \supset f^{-1}(f(X))$ (*c*) $X = f^{-1}(f(X))$ (*d*) None of these.

(Bharathiar Univ., 2004)

3. If $f: X \to Y$ and $B \subset Y$, then :

(*a*) $B \subseteq f(f^{-1}(B))$ (*b*) $B = f(f^{-1}(B))$ (*c*) $f(f^{-1}(B)) \subseteq B$ (*d*) None of these.

4. f and g are one-one and onto, then gof is :

(*a*) only one-one (*b*) only onto (*c*) one-one and onto (*d*) None of these.

5. Let A and B be two finite sets with m and n elements respectively. Then number of elements in $A \times B$ is :

(*a*) $m + n$ (*b*) $m \times n$ (*c*) m/n (*d*) None of these.

6. Which one of the following statements is always correct for arbitrary sets A, B, C

(*a*) If $A \in B$ and $B \subseteq C$, then $A \in C$ (*b*) If $A \in B$ and $B \subseteq C$, then $A \subseteq C$

(*c*) If $A \subseteq B$ and $B \in C$, then $A \in C$

(*d*) If $A \subseteq B$ and $B \in C$, then $A \subseteq C$ **[I.A.S. Prel. 2006]**

7. Let $X = \{n : n \text{ is a positive integer, } n \le 50\}$. If $A = \{n \in X ; n \text{ is even}\}$ and $B = \{n \in X : n$ is a multiple of 7}, then what is the number of elements in the smallest subset of X containing both A and B.

(*a*) 28 (*b*) 29 (*c*) 32 (*d*) 35 **[I.A.S. Prel, 2007]**

8. Let $A = \{t \in \mathbf{N} : 12$ and t are relatively prime} and $B = \{t \in \mathbf{N} : t \le 24\}$. What is the number of elements in $A \cap B$.

(*a*) 10 (*b*) 8 (*c*) 7 (*d*) 4 **[I.A.S. Prel, 2007]**

9. If $f(x+1) + f(x-1) = 2f(x)$ and $f(0) = 0$, then what is $f(n)$ where $n \in \mathbf{N}$?

(*a*) $nf(1)$ (*b*) $[f(1)]^n$ (*c*) 0 (*d*) n **[I.A.S. Prel 2009**

10. If $f(x) = 1/(1-x)$, $g(x) = f[(f(x)]$ and $h(x) = f[g(x)]$ then what is $f(x)\, g(x)\, h(x)$ equai to

(*a*) –1 (*b*) 0 (*c*) 1 (*d*) 2 **[I.A.S. Prel 2009]**

ANSWERS. **1.** (*b*) **2.** (*a*) **3.** (*c*) **4.** (*c*) **5.** (*b*) **6.** (*a*) **7.** (b) **8.** (d) **9.** (*a*) **10.** (*a*)

MISCELLANEOUS PROBLEM ON CHAPTER 1

Ex. 1. *Define power set of a given set A. Show that if A contains n elements, then the power set P(A) of A will contain 2^n elements.*

Sol. Power set. Definition. The set of all subsets of the given set A is denoted by $P(A)$ and is called the power set of A. Thus, $P(A) = \{B : B \subseteq A\}$.

To find the number of elements in power set $P(A)$ of the given finite set A.

Suppose A contains n elements. Then, observe that

I_0 : The null set ϕ is a subset of A, i.e.,

the number of subsets containing no element = ${}^nC_0$

I_1 : The number of subsets containing one element = ${}^nC_1$

I_2 : The number of subsets containing two element = ${}^nC_2$

....

I_n : The number of subsets containing n elements = ${}^nC_n$

Hence, the total number of subsets of $A = {}^nC_0 + {}^nC_1 + {}^nC_2 + \cdots + {}^nC_n = (1 + 1)^n = 2^n$, showing that the number of elements in power set of A is 2^n if A contains n elements.

Chapter

2

The Real Numbers

2.1. INTRODUCTION

In this chapter, we shall introduce the system of real numbers as a *complete ordered field structure*. While we shall give a precise description of what is meant by a complete ordered field, we shall in this manner actually only *define* the system of real numbers. This implies that we are laying down this definition of the system of real numbers on an *axiomatic basis* for discussion and development of Real Analysis. It will be seen that the sets of Natural numbers, Integers and Rational numbers will arise as sub-sets of the sets of Real numbers.

To facilitate an easy understanding of what we have said in the previous paragraph, we firstly describe the *known* properties of the systems of natural numbers, integers and rational numbers. This will help the reader to acquire familiarity with the various concepts which underlie the formulation of the system of real numbers as complete ordered field.

Note. The system of real numbers in an axiomatic manner starting from the system of rational numbers has been constructed by different mathematicians. We mention in this connection the names of Geerg Cantor (1845-1918), Richard Dedekind (1831-1916) and Karl Weierstrass (1815-1897).

(Readers interested in an axiomatic formulation of Real Numbers starting from Natural Numbers may refer to the book *Number Systems* by the author.)

2.2. THE SET N OF NATURAL NUMBERS

The natural numbers are 1, 2, 3, 4, 5, ..., so that we have N = {1, 2, 3, 4, 5, ...}.

Algebraic structure of the set of natural numbers. To each pair a, b of natural numbers, there corresponds

(*i*) a natural number denoted by $a + b$ and called the *sum* of a and b.

(*ii*) a natural number denoted by ab and called the *product* of a and b.

This process of associating to each pair of natural numbers a natural number, called their sum, is called *Addition Composition* in the set of natural numbers and that of associating to each pair of natural numbers a natural number called their product, is called *Multiplication Composition* in the set of natural numbers. The fact of these existing in the set N of natural numbers these compositions is referred to as the set N of natural numbers possessing an *Algebraic Structure.*

BASIC PROPERTIES OF THE TWO COMPOSITIONS IN N

Here a, b, c etc. denote members of N.

I. Commutativity of addition and multiplication

$$a + b = b + a;\ ab = ba\ \forall\ a, b \in \mathbf{N}.$$

II. Associativity of addition and multiplication

$$a + (b + c) = (a + b) + c;\ a\,(bc) = (ab)\,c\ \forall\ a, b, c \in \mathbf{N}.$$

In view of the associativity of the two compositions, we often omit the brackets and write $a + b + c$ in place of $(a + b) + c$ and abc in place of $(ab)\,c$.

III. Cancellation Laws $a + c = b + c \Rightarrow a = b;\ ac = bc \Rightarrow a = b.$

IV. Distributivity of addition with respect to multiplication

$$a\,(b + c) = ab + ac\ \forall\ a, b, c \in \mathbf{N}.$$

V. Multiplication property of 1 $a \cdot 1 = a\ \forall\ a \in \mathbf{N}.$

Because of this property, 1 is called the *Multiplicative Identity.*

ORDER STRUCTURE OF THE SET N OF NATURAL NUMBERS

The relation 'Greater than' : Given any two *different* natural numbers a, b, we have, either $a > b$ or $b > a$, *i.e.*, either 'a is greater than b' or 'b is greater than a'.

Instead of the relation 'greater than' we also consider the relation 'smaller than' such that $a > b \Leftrightarrow b < a$, *i.e.*, a is greater than $b \Leftrightarrow b$ is smaller than a.

The relation 'greater than' between different natural numbers is known as an '*Order relation*' in the set of natural numbers and the presence of this relation in **N** is referred to as **N** having an *order structure.*

Well ordering property of natural numbers. Every non-empty subset of **N** has a least element.

PROPERTIES OF THE ORDER RELATION

VI. Transitivity of the relation $[a > b] \wedge [b > c] \Rightarrow a > c.$

This property is referred to as the *transitivity* of the order relation.

COMPATIBILITY OF ALGEBRAIC STRUCTURE WITH ORDER STRUCTURE

The following two properties relate the order structure with each of the two compositions separately.

VII. Compatibility of the order relation with the addition composition

$$a > b \Rightarrow a + c > b + c.$$

VIII. Compatibility of the order relation with the multiplication composition

$$a > b \Rightarrow ac > bc.$$

As a result of the properties VII and VIII, we say that the order structure in **N** is *compatible* with its algebraic structure or *vice-versa.*

IX. Principle of finite induction

Let $n \in \mathbf{N}$ and let $P(n)$ denote a statement pertaining to n. If

(i) $P(1)$ is true, i.e., the statement is true for $n = 1$ and

(ii) $P(n)$ is true $\Rightarrow$ $P(n + 1)$ is true, then $P(n)$ is true for every natural number n.

We say that the set **N** of natural numbers satisfies the *principle of finite induction.*

INVERSE OPERATIONS AND THE CORRESPONDING LIMITATIONS. SUBTRACTION AND DIVISION IN N

Subtraction. *Given two members a, c of* **N**, *does there exist $x \in$* **N** *such that $a + x = c$?*

It is easy to see that x, if it exists, is unique. This is a consequence of the cancellation principle in as much as $a + x = a + y \Rightarrow x = y$.

Also x exists if and only if $c > a$.

For example, if $a = 5$, $c = 8$ so that $c = 8 > 5 = a$, we have $x = 3$.

If, however, we take $a = 5$, $c = 3$ so that $c = 3 < 5 = a$, there exists no $x \in \mathbf{N}$ such that $5 + x = 3$.

In case $c > a$, so that there exists x such that $a + x = c$, we denote this x by $c - a$.

The symbol $c - a$ denotes the natural number which when added to a gives c. This symboi is meaningful if and only if $c > a$.

Division. *Given two natural numbers a and c, does there exist a natural number x such that* $ax = c$?

The number x, if it exists, is unique. This is a consequence of the cancellation law which states that $ax = ay \Rightarrow x = y$.

Also the number x exists if a is a divisor of c.

For example, if $a = 3$, $c = 15$ so that 3 is a divisor of 15, we have $x = 5$.

If, however, we take $a = 3$, $c = 14$ so that $a = 3$ is *not* a divisor of $c = 14$, there exists *no* natural number x such that $3x = 14$.

In case a is a divisor of c so that there exists x such that $ax = c$, we denote this x by $c \div a$.

Note. From above, we see that if a, c be two given natural numbers, the symbol $c - a$ is meaningful if and only if c is greater than a and the symbol $c \div a$ is meaningful if and only if a is a divisor of c.

2.3. THE SET I OR Z OF INTEGERS

The set **I** or Z of integers consists of the numbers ....., – 3, – 2, – 1, 0, 1, 2, 3, ..., so that we have

$$\mathbf{Z} = \mathbf{I} = \{0, -1, 1, -2, 2, -3, 3, -4, 4, ...\}.$$

In this section a, b, c etc., refer to arbitrary members of **I,** *viz.*, arbitrary integers.

Algebraic structure of the set I of integers, Addition Composition in I. Addition composition in **I** which associates to each pair of members a, b of **I** a number called their sum and denoted by $a + b$ has the following basic properties :

(*i*) $a + b = b + a \ \forall \ a, b \in \mathbf{I}$. *Commutativity.*

(*ii*) $(a + b) + c = a + (b + c) \ \forall \ a, b, c \in \mathbf{I}$. *Associativity.*

(*iii*) The number $0 \in \mathbf{I}$ is such that $a + 0 = a \ \forall \ a \in \mathbf{I}$.

The number '0', because of this relation, is referred to as the **Additive Identity.**

(*iv*) To each $a \in \mathbf{I}$ there corresponds another, *viz.*, $-a \in \mathbf{I}$ such that $a + (-a) = 0$.

The integer, $-a$, is said to be the *negative* of the integer a or the **Additive Inverse** of a.

Inverse of addition. The equation $a + x = b$, $a \in \mathbf{I}$, $b \in \mathbf{I}$ admits of a unique solution x, *viz.*, $b - a \in \mathbf{I}$.

Subtraction in I. It will be seen that *subtraction is always possible in* **I,** *i.e.*, given any two members, a, b of **I,** $a - b$ is again a member of **I** for all a, b so that here we have a property of **I** which does *not* hold for **N.**

Ex. Deduce from the above properties of the addition composition in **I,** that the cancellation law holds for addition in **I,** *viz.*, that $a + c = b + c \Rightarrow a = b$.

Multiplication composition in I. Multiplication composition in **I** which associates to each pair of members a, b of **I** a member of **I** denoted by ab and called their product has the following basic properties :

(*i*) $ab = ba \ \forall \ a, b \in \mathbf{I}$. *Commutativity.*

(*ii*) $(ab)\, c = a\, (bc) \ \forall \ a, b, c \in \mathbf{I}$. *Associativity.*

(*iii*) The number '1' $\in \mathbf{I}$ is such that $a \cdot 1 = a \ \forall \ a \in \mathbf{I}$.

Because of the property (*iii*), the integer '1' is known as the *Multiplicative Identity.*

(*iv*) $[ab = ac \wedge a \neq 0] \Rightarrow b = c$. *Cancellation law for multiplication.*

The following law relates the two compositions :

$$a(b+c) = ab + ac \ \forall \ a, b, c \in \mathbf{I}. \textit{ Distributivity.}$$

We refer to this law by saying that *Multiplication distributes addition in* **I.**

Ex. Deduce from above the following basic properties of Addition and Multiplication in **I.**

(*i*) $ab = 0 \Leftrightarrow a = 0 \vee b = 0.$ (*ii*) $a(-b) = -(ab), (-a)(b) = -(ab), (-a)(-b) = ab.$

Division in I. Factors and multiples. If a, b are two non-zero members of **I,** we say that a is a factor of b if there exists $c \in \mathbf{I}$ such that $b = ac$.

It will be seen that $b \div a$ is meaningful if and only if $a \neq 0$ and a is a factor of b or that a is a divisor of b.

ORDER STRUCTURE OF I

Given any two *different* members $a, b \in \mathbf{I}$, we have either $a > b$ or $b > a$.

The 'Greater than' relation is *transitive* in as much as $a > b \wedge b > c \Rightarrow a > c$.

Also $$a > b \Rightarrow a + c > b + c$$

and $$a > b, c > 0 \Rightarrow ac > bc.$$

Thus, the system **I** of integers has, what has already been referred to, an order structure compatible with its algebraic structure.

Ex. Show that $a > b \wedge c < 0 \Rightarrow ac < bc$.

2.4. THE SET Q OF RATIONAL NUMBERS

The rational numbers are of the form p/q where p, q are arbitrary integers with $q \neq 0$.

Algebraic structure of Q. As in **I,** the set **Q** of rational numbers admits of two compositions, *viz.*, Addition and Multiplication. We give below the basic properties of these two compositions. Here a, b, c etc., denote arbitrary members of the set **Q** of rational numbers.

1. The addition composition is commutative, associative, admits of an additive identity, *viz.*, 0 and each element a admits of an additive inverse, *viz.*, $-a$.

2. The multiplication composition is commutative, associative, admits of a multiplicative identity, *viz.*, 1 and each non-zero element p/q admits of multiplicative inverse, *viz.*, q/p.

3. Multiplication distributes addition.

DEFINITIONS : Subtraction. Division. *Let a, b be two given rational numbers. We write*

$$a - b = a + (-b).$$

Thus, $a - b$ is obtained by adding to a the additive inverse $-b$, of b. Also, if $b \neq 0$, we write $a \div b = a(1/b)$ so that $a \div b$ is obtained on multiplying a with the multiplicative inverse $1/b$ of the non-zero b.

Ex. 1. Show that

$$a(b - c) = ab - ac.$$

$$ab = 0 \Leftrightarrow a = 0 \vee b = 0.$$

$$a(-b) = -(ab), (-a)(b) = -(ab), (-a)(-b) = ab.$$

Ex. 2. Show that if $a \neq 0$, the equation $ax + b = 0$ admits of a unique solution in **Q;** given that $a \in \mathbf{Q}, b \in \mathbf{Q}$.

Show also that if $a = 0, b \neq 0$, the equation has no solution and that if $a = 0, b = 0$, every member of **Q** is a root of the equation.

Ex. 3. Show that $ab = ac \wedge a \neq 0 \Rightarrow b = c$.

ORDER STRUCTURE OF Q

Given any two *different* rational numbers a, b, we have either $a > b$ or $b > a$.

Moreover, the order relation is transitive and compatible with the addition and multiplication compositions, *i.e.*, we have

(*i*) $a > b \wedge b > c \Rightarrow a > c.$

(*ii*) $a > b \Rightarrow a + c > b + c.$

(*iii*) $a > b \wedge c > 0 \Rightarrow ac > bc.$

Ex. 1. (*i*) Show that $x > y \wedge z < 0 \Rightarrow xz < yz.$

(*ii*) Show that $x^2 \geq 0 \; \forall \; x \in \mathbf{Q}.$

Ex. 2. Given two different rational numbers a, b such that $a < b$; show that there exist an infinite number of rational numbers c such that $a < c < b$.

FIELDS OF NUMBERS

A set K of numbers containing at least two members is called a *field*, if it is such that when a, b are arbitrary members of K, then $a + b$, ab, $a - b$ are also members of K and if $b \neq 0$, then $a \div b$ is also a member of K.

It will be seen that while the set **Q** of rational numbers is a field, the sets **I** and **N** are *not* fields.

Ex. Show that no proper sub-set of the field of rational numbers is a field.

ORDERED FIELDS OF NUMBERS

Since the set **Q** of rational numbers, besides having a field structure, also has an order structure compatible with its field structure, we say that the set **Q** of rational numbers is an *ordered field*.

EXAMPLES

Example 1. *Show that there is no rational number whose square is 2.*

(Calicut, 2004; Delhi B.Sc. (Hons.) I, 2016)

Solution. Let us assume that there exists a rational number whose square is 2. Let p, q be two integers *without a common factor* such that the square of the rational number p/q, is 2, *i.e.*,

$$(p/q)^2 = 2.$$

We have $$(p/q)^2 = 2 \Rightarrow p^2 = 2q^2.$$

Now, q is an integer, so is q^2 and $2q^2$. Thus, p^2 is an integer divisible by 2. As such p must itself be divisible by 2, for otherwise, its square would *not* be divisible by 2. Let $p = 2n$ where n is an integer.

We have $$p^2 = 2q^2 \Rightarrow q^2 = 2n^2, \text{ as } p^2 = 4n^2.$$

so that the integer q is also divisible by 2.

Thus, we conclude that p, q have a common factor 2 and this conclusion is contrary to the hypothesis that p, q are integers without a common factor. It follows that there exists no rational number whose square is 2.

Example 2. *Show that there exists no rational number whose square is 8.*

Solution. We assume that there exists a rational number, whose square is 8. Let q be the smallest positive integer such that for some integer p, $(p/q)^2 = 8$.

Now, $$(p/q)^2 = 8 \Rightarrow 2 < p/q < 3.$$

Also $$2 < (p/q) < 3 \Rightarrow 2q < p < 3q$$
$$\Rightarrow 0 < p - 2q < q.$$

Thus, $p - 2q$ is a positive integer smaller than q and as such $(p/q)\,(p - 2q)$ is *not* an integer.

Again $$(p/q)\,(p - 2q) = (p^2/q) - 2p$$
$$= (p^2/q^2)\,q - 2p = 8q - 2p$$

$$\Rightarrow \quad (p/q)\,(p-2q) \text{ is an integer.}$$

Thus, we arrive at a contradiction. Hence, $\sqrt{8}$ is not a rational number.

Note. In the above solved examples 1 and 2, we have considered $\sqrt{n}$, where $n \in \mathbf{N}$ and n is not a perfect square. If n is prime, then use the method explained in Ex. 1 and if n is a composite number, then use the method explained in Ex. 2.

Irrational number. Definition : A real number which is not rational is called irrational. For example, $\sqrt{2}$, $\sqrt{8}$, π and e etc. are irrational numbers.

EXERCISES

1. Show that $\sqrt{3}$ and $\sqrt{15}$ are irrational numbers.

2. Show that if x^2 is irrational, then x is irrational. Hence or otherwise, prove that $\sqrt{2} + \sqrt{3}$ is irrational.

3. If p is any prime number, prove that $\sqrt{p}$ is not a rational number.

4. Prove that if x is rational and y irrational, then $x + y$ is irrational and if $x \neq 0$, xy is irrational.

2.5. THE SET OF REAL NUMBERS AS A COMPLETE ORDERED FIELD

We now describe the basic properties of the set **R** of real numbers. These properties will be described in three stages. The set of properties included in the first stage will describe the *Field structure* of the set of real numbers. We shall then proceed to describe at the second stage the *Order structure* of the set of real numbers such that the properties enumerated in two stages will describe the set of real numbers as an *Ordered field*. It will be seen that the set of rational numbers and the set of real numbers are both ordered fields. At the third stage, we shall describe a property of the order structure of the set of real numbers which is **not** a property possessed by the ordered field of rational numbers. This property will be referred to by saying that the *Field of real numbers is* **order-complete.**

On the basis of the properties of the set of real numbers enumerated in the three stages, we say that the *set of real numbers is a complete ordered field.* The set of rational numbers is an ordered field alright but not a complete ordered field.

Every property of the set of real numbers can be derived as a consequence of the basic character of the set of real numbers as a complete ordered field.

The character of the set of real numbers as a complete ordered field will now be described.

Field Structure. Addition Composition. To each ordered pair of real numbers, there corresponds a real number called their *sum* and denoted by $a + b$. This process of associating to each ordered pair of real numbers a real number called their sum is known as addition composition in the set. In the following a, b, c etc. denote real numbers. This addition composition has the following properties :

A.1. $a + b = b + a \ \forall \ a, b \in \mathbf{R}.$ *Commutativity.*

A.2. $(a + b) + c = a + (b + c) \ \forall \ a, b, c \in \mathbf{R}.$ *Associativity.*

A.3. There exists a real number, *viz.*, '0' such that $\forall \ a \in \mathbf{R}$

$$a + 0 = a \qquad \text{[Existence of additive identity.]}$$

A.4. To each real number a there corresponds a real number, *viz.*, $-a$, such that

$$a + (-a) = 0 \qquad \text{[Existence of additive inverse.]}$$

Multiplication Composition. To each ordered pair, a, b of real numbers, there corresponds a real number called their *product* and denoted by ab. This process of associating to each ordered pair of real numbers a real number called their product is known as multiplication composition in the set. This multiplication composition has the following properties :

M.1. $ab = ba \ \forall \ a, b, \in \mathbf{R}$. [*Commutativity.*]

M.2. $(ab)\ c = a\ (bc) \ \forall \ a, b, c \in \mathbf{R}$. [*Associativity.*]

M.3. There exists a real number, *viz.*, '1' such that $\forall \ a \in \mathbf{R}$

$$a \cdot 1 = 1$$ [*Existence of multiplicative identity.*]

M.4. To each real number $a \neq 0$, there corresponds another, *viz.*, $1/a$ such that

$$a\ (1/a) = 1$$ [*Existence of multiplicative inverse.*]

There is also a law known as Distributive Law which relates the two compositions, *viz.*,

AM $a\ (b + c) = ab + ac \ \forall \ a, b, c \in \mathbf{R}$. [*Distributive law.*]

The fact of the set of real numbers admitting of two compositions satisfying the *nine* properties mentioned above is referred to as the set of real numbers having a field structure.

ORDER STRUCTURE OF R

'Greater than' relation. Given any two *different* real numbers a, b we have either $a > b$ or $b > a$. This 'Greater than' relation has the following property :

$$a > b \wedge b > c \ \Rightarrow \ a > c.$$ [*Transitivity*].

ALGEBRAICO-ORDER STRUCTURE

We have a property pertaining jointly to the addition composition and the 'Greater than' relation referred to by saying that the addition composition is *compatible* with the 'Greater than' relation.

AO $$a > b \ \Rightarrow \ a + c > b + c.$$

Also we have a property pertaining jointly to the multiplication composition and the 'Greater than' relation referred to by saying that the multiplication composition is compatible with the 'Greater than' relation.

MO $$a > b \wedge c > 0 \ \Rightarrow \ ac > bc.$$

The *twelve* properties referred to above complete the description of the set of real numbers as an **Ordered field.**

Note. We shall often use the following notations :

(*i*) $a \geq b \ \Leftrightarrow \ a > b \vee a = b$, (*ii*) $a < b \ \Leftrightarrow \ b > a$, (*iii*) $a \leq b \ \Leftrightarrow \ a < b \vee a = b$.

Ex. Show that $a \geq b \wedge b \geq a \ \Leftrightarrow \ a = b$.

2.6. CLOSED, OPEN, SEMI-CLOSED AND SEMI-OPEN INTERVALS

Closed interval. If a, b be two given real numbers such that $a < b$, the set $\{x : a \leq x \leq b\}$ is called a *closed interval* denoted by the symbol $[a, b]$.

Open interval. We write $]a, b[= \{x : a < x < b\}$ and refer to the set $]a, b[$ as an *open interval*.

Semi-open or semi-closed intervals. We write $[a, b[= \{x : a \leq x < b\}$, $]a, b] = \{x : a < x \leq b\}$ and call these sets as *semi-open* or *semi-closed* intervals.

2.7. SET BOUNDED ABOVE, SET BOUNDED BELOW, l.u.b. (SUPREMUM) AND g.l.b. (INFIMUM) OF A SET. THE GREATEST AND SMALLEST MEMBERS OF A SET [*Delhi BA (Prog.) III 2010, 11; Delhi B.Sc. (Hons.) I, 2016; Delhi B.Sc. I (Prog.) 2007,09; Delhi B.Sc. III (Prog) 2010, 11, 12*]

Set bounded above. Definition. Let S be a sub-set of real numbers. We say that S *is bounded above* if there exists a real number b, not necessarily a member of S, such that

$$x \in S \ \Rightarrow \ x \leq b \ \forall \ x \in S \quad ...(1)$$

The number b is called an *upper bound of S*.

If there exists no real number b satisfying (1), then the set S is said to be *not bounded* or *unbounded above.* Thus, a set is unbounded above if, however, large a real number b may be chosen, there exists at least one $x \in S$ such that $x > b$.

Set bounded below. Definition

Let S be a sub-set of real numbers. We say that S *is bounded below* if there exists a real number a, not necessarily a member of S, such that

$$x \in S \Rightarrow x \geq a \ \forall \ x \in S \quad ...(2)$$

The number a is called a *lower bound* of S.

If there exists no real number a satisfying (2), then the set S is said to be *not bounded* or *unbounded below*. Thus, a set is unbounded below if however, small a real number a may be chosen, there exists at least one $x \in S$ such that $x < a$.

Note 1. The set bounded above has an infinite number of upper bounds in as much as every number greater than an upper bound of a set is itself an upper bound of the set. Thus, every set bounded above determines another non-empty set, *viz.*, the set of its upper bounds. The set of upper bounds of a set S which is bounded above is itself bounded below in as much as every member of the set S is a lower bound thereof.

Note 2. Every set bounded below determines an infinite set, *viz.*, the set of its lower bounds. The set of lower bounds of a set S which is bounded below is itself bounded above in as much as every member of the set S is an upper bound thereof.

Least upper bound or supremum. Definition ***[Kanpur 2011; Delhi B.A. (Prog.) III 2012]***

If the set of all upper bounds of a set S of real numbers has a *smallest member*, k, say, then k is said to be a *least upper bound* or a *supremum* of S and is abbreviated as l.u.b. or Sup S respectively.

Greatest lower bound or infimum. Definition ***[Kanpur 2011; Delhi Maths 2007]***

If the set of all lower bounds of a set S of real numbers has a *greatest member*, l, say, then l is said to be a *greatest lower bound* or an *infimum* of S and is abbreviated as g.l.b. or Inf S respectively.

Theorem I. *Supremum (or l.u.b.) of a non-empty set S of real numbers, whenever it exists, is unique.* ***[Delhi Maths (H), 1994, 95; Kanpur, 1997]***

Proof. Let, if possible, k_1 and k_2 be two suprema of S. Since both s_1 and s_2 are suprema of S, hence by definition both of them are upper bounds of S.

Now, k_1 is a supremum and k_2 is an upper bound of S

$$\Rightarrow k_1 \leq k_2 \quad ...(1)$$

Again, k_2 is a supremum and k_1 is an upper bound of S

$$\Rightarrow k_2 \leq k_1. \quad ...(2)$$

$\therefore$ (1) and (2) $\Rightarrow$ $k_1 = k_2$. Hence, the set S has a unique supremum.

Theorem II. *Infimum (or g.l.b.) of a non-empty set S of real numbers, whenever it exists, is unique.* ***(Delhi B.Sc. (Prog) III 2011, Meerut, 2003)***

Proof. Left as an exercise for the reader.

Note. In view of the uniqueness of supremum and infimum of a set S, henceforth we shall write 'the supremum' (or 'the least upper bound') and 'the infimum' (or 'the greatest upper bound') of a set in place of a supremum (or a least upper bound) and an infimum (or a greatest lower bound).

Characteristic properties of supremum and infimum.

Theorem III. *The necessary and sufficient condition for a real number 's' to be the supremum of a bounded above set S is that 's' must satisfy the following conditions :*

(*i*) $x \le s \;\forall\; x \in S$

(*ii*) *For each positive real number* ε, *there exists a real number* $x \in S$ *such that* $x > s - \varepsilon$.

[Delhi B.Sc. (Hons.) I, 2015, 2017]

Proof. *The condition is necessary.* Let $s = \sup S$. Then, by definition, s is an upper bound of S and so
$$x \le s \;\forall\; x \in S.$$

Again, if $\varepsilon > 0$ be given, then $s - \varepsilon$ is smaller than the supremum 's' of S. Hence, $s - \varepsilon$ cannot be an upper bound of S. Hence, there exists at least one element $x \in S$ such that $x > s - \varepsilon$.

Hence the condition is necessary.

The condition is sufficient. Let s be any real number satisfying conditions (*i*) and (*ii*) of the statement of the theorem.

Now, the condition (*ii*), namely, $x \in s \;\forall\; x \in S \;\Rightarrow\; s$ is an upper bound of S.

In order to show that s is the supremum of S, we must prove that no real number less than s can be an upper bound of S.

Let, if possible, s' be another real number such that $s' < s$. Then, $(s - s') > 0$. Let $\varepsilon = s - s'$ so that $\varepsilon > 0$. Hence, by the condition (*ii*), there exists some $x \in S$ such that
$$x > s - \varepsilon, \text{ i.e., } x > s - (s - s'), \text{ i.e., } x > s',$$
showing that s' is not an upper bound of S.

Hence, we find that s is an upper bound of S and no real number less than s is an upper bound of S. Therefore, by definition, s is the supremum.

Hence the condition is sufficient.

Theorem IV. *The necessary and sufficient condition for a real number 't' to be the infimum of a bounded below set S is that 't' must satisfy the following conditions :*

(*i*) $x \ge t \;\forall\; x \in S$

(*ii*) *For each positive real number* ε, *there exists a real number* $x \in S$ *such that* $x < t + \varepsilon$.

[Delhi B.A. (Prog.) III, 2007; Delhi B.Sc. (Hons.) I, 2018]

Proof. Left as an exercise for the reader.

Note 1. A set may or may not have the supremum or/and infimum. Of course, the supremum and infimum, if they exist are unique.

Note 2. The supremum and infimum of a set may or may not belong to the set.

Greatest member of a set bounded above. Definition

A number ξ is called the *greatest member* of a set S, if

(*i*) $\xi \in S$, *i.e.*, ξ is itself a member of S.

(*ii*) No member of S is greater than ξ, *i.e.*, ξ is an upper bound of S, so that
$$x \in S \;\Rightarrow\; x \le \xi.$$

Thus, the greatest member of a set is a member of the set bounded above and is as well as an upper bound of the set.

Smallest member of a set bounded below. Definition

A number η is called the *smallest member* of a set S, if

(*i*) $\eta \in S$, *i.e.*, η is itself a member of S.

(*ii*) No member of S is smaller than η, *i.e.*, η is a lower bound of S, so that

$$x \in S \Rightarrow x \geq \eta.$$

Thus, the smallest member of a set is a member of the set bounded below and is as well as a lower bound of the set.

Theorem V. *The greatest member of a set, if it exists, is the supremum (l.u.b.) of the set.* (*Delhi B.Sc. III 2009*)

Proof. Let ξ be the greatest member of S. Then, by definition $x \leq \xi \ \forall \ x \in S$, showing that ξ is an upper bound of S.

Let k be any number less than ξ. Then there must exist at least one member ξ' of S which is greater than k. Hence, we find that no number less than ξ can be an upper bound of S. Therefore, ξ is the least of all upper bounds of S and so $\xi = \sup S$.

Theorem VI. *The smallest member of a set, if it exists, is the infimum (g.l.b.) of the set.*

Proof. Left as an exercise for the reader. (*Delhi B.Sc. (Prog) III 2010*)

Existence of the greatest member. A set which is bounded above may or may not have a greatest number. Of course, the greatest member of a set, if it exists, is unique.

We give below various possibilities in regard to the existence of the greatest members of sets.

Every finite set admits of a greatest member.

An infinite set which is not bounded above *cannot* have a greatest member.

The cases of finite sets and of infinite sets which are not bounded above are thus trivial in regard to the existence of greatest members. The case of importance which remains is that of infinite sets which are bounded above. Such a set may or may not have a greatest member and no general statement can as such be made.

Existence of the smallest member. A set which is bounded below may or may not have a smallest member. Of course, the smallest member of a set, if it exists, is unique.

Every finite set admits of a smallest member. An infinite set which is not bounded below cannot have a smallest member. Again, infinite sets which are bounded below may or may not have a smallest member and no general statement can as such be made.

Bounded and unbounded sets (*Delhi 2012; Kanpur 2011, Chennai 2011*)

A set S of real numbers is said to be *bounded*, if it is bounded above as well as below. Thus, when S is bounded, there exist two real numbers k and K such that

$$k \leq x \leq K \ \forall \ x \in S.$$

A set is said to be *unbounded* if it is not bounded.

Illustrations : [*Delhi B.Sc. (Prog) III 2010*]

1. The set **N** of natural numbers is bounded below but not bounded above. It admits of a smallest member, *viz.*, 1. Thus, inf **N** = 1.

2. The sets **I, Q** are neither bounded above nor bounded below.

3. Each of the intervals $]a, b[$, $]a, b]$, $[a, b[$, $[a, b]$ is bounded set.

For $]a, b[$, the Supremum = b, the Infimum = a. Both do not belong to $]a, b[$

For $]a, b]$, the Supremum = b, the Infimum = a. Only $b \in]a, b]$

For $[a, b[$, the Supremum = b, the Infimum = a. Only $a \in [a, b[$

For $[a, b]$, the Supremum = b, the Infimum = a. Both a and b belong to $[a, b]$.

The closed set $[a, b]$ admits of the greatest member, *viz.*, b and the smallest member, *viz.*, a.

The semi-open set $[a, b[$ admits of the smallest member but not the greatest.

EXAMPLES

Example 1. *Give an example each of a bounded set which contains its*

(*i*) *g.l.b. but does not contain its l.u.b.*

(*ii*) *l.u.b. but does not contain its g.l.b.* **(Meerut, 1995)**

Solution. (*i*) Let $S = \{x : x \in \mathbf{R} \text{ and } a \le x < b\}$. Then the g.l.b. of S is a which is a member of S whereas the l.u.b. of S is b which is not a member of S.

(*ii*) Let $S = \{x : x \in \mathbf{R} \text{ and } a < x \le b\}$. Then the g.l.b. of S is a which is not a member of S whereas the l.u.b. of S is b which is a member of S.

Example 2. (*i*) *If S is a bounded non-empty sub-set of* **R**, *then prove that* $\inf S \le \sup S$.

[Delhi B.Sc. (Prog.) II, 2015, Delhi Maths (H), 1999, 2008]

(*ii*) *Let S be a non-empty bounded sub-set of real numbers such that* $\sup S = \inf S$. *What can be said about the set S ?* **[Delhi Maths (H), 2002]**

Solution. (*i*) Since S is non-empty, we can choose some $x \in S$. Then, by definition of $\inf S$ and $\sup S$, we must have

$$\inf S \le x \text{ and } x \le \sup S \text{ so that } \inf S \le \sup S.$$

(*ii*) Let $\sup S = \inf S = k$, say.

Then, by definition, k is an upper bound as well as a lower bound for the set S. Hence, we must have

$$x \le k \ \forall \ x \in S \quad \text{and} \quad x \ge k \ \forall \ x \in S,$$

which implies that $x = k \ \forall \ x \in S$, *i.e.*, k is the only element of S. Hence, S is the singleton $\{k\}$.

Thus, if $\sup S = \inf S$, then S must be a singleton set.

Example 3. *Let S and T be two sub-sets of real numbers such that* $S \le T$ *and* $S \ne \phi$. *Then, prove that* (*i*) *If T is bounded above, then* $\sup S \le \sup T$.

(*ii*) *If T is bounded below, then* $\inf T \le \inf S$.

Solution. (*i*) Let $\sup T = k$. Then, by definition, $k \ge y \ \forall \ y \in T$.

Since $S \le T, x \in S \Rightarrow x \in T$. So, $k \ge x \ \forall \ x \in S$,

showing that k is an upper bound of S and hence $\sup S \le k$.

Thus, $\sup S \le \sup T$, as $k = \sup T$.

(*ii*) Let $\inf T = l$. Then, by definition, $l \le y \ \forall \ y \in T$.

Since $S \le T, x \in S \Rightarrow x \in T$.

So, $l \le x \ \forall \ x \in S$, showing that l is a lower bound of S and hence $l \le \inf S$.

Thus, $\inf T \le \inf S$, as $l = \inf T$.

Example 4. *If S and T are non-empty bounded sub-sets of real numbers, then show that*

$$S \le T \Rightarrow \inf T \le \inf S \le \sup S \le \sup T.$$

Solution. From part (*ii*) of Ex. 3, $\inf T \le \inf S$ ...(1)

From part (*i*) of Ex. 2, $\inf S \le \sup S$ ...(2)

From part (*i*) of Ex. 3, $\sup S \le \sup T$ ...(3)

Then, from (1), (2) and (3), $\inf T \le \inf S \le \sup S \le \sup T$.

Example 5. *Which of the following sets are bounded above, bounded below or otherwise ? Also find the supremum (l.u.b.) and infimum (g.l.b.), if they exist. Which of these belong to the set*

(*i*) $\{1/n : n \in \mathbf{N}\}$ **(Kanpur, 1993; Meerut, 2001, 02)** *ii*) $\left\{(-1)^n \frac{1}{n} : n \in \mathbf{N}\right\}$ **[*Meerut, 2003*; *Delhi B.A. (Prog) III, 2010, 11*]**

iii) $\{(-1)^n\, n : n \in \mathbf{N}\}$ ***(Kanpur 2014***

(*iv*) $\left\{1+(-1)^n \frac{1}{n} : n \in \mathbf{N}\right\}$ ***(Utkal, 2003)***

(*v*) $\{(4n+3)/n : n \in \mathbf{N}\}$

(*vi*) $\{n/(n+1) : n \in \mathbf{N}\}$ ***(Avadh, 1995; Gorakhpur, 1993; Purvanchal, 1995)***

(*vii*) $\left\{-2, -\frac{3}{2}, -\frac{4}{3}, -\frac{5}{4}, \ldots\ldots, -\frac{n+1}{n}, \ldots\ldots\right\}$

(*viii*) $\left\{1, 1+\frac{1}{2}, 1+\frac{1}{2}+\frac{1}{2^2}, \ldots\ldots, 1+\frac{1}{2}+\frac{1}{2^2}+\ldots\ldots+\frac{1}{2^{n-1}}, \ldots\ldots\right\}$

(*ix*) *Set of all real numbers.* ***(Meerut, 1995)***

Solution. (*i*) Let $S = \left\{\frac{1}{n} : n \in \mathbf{N}\right\} = \left\{1, \frac{1}{2}, \frac{1}{3}, \frac{1}{4}, \ldots\ldots\right\}$.

Clearly, S is bounded because $0 \le x \le 1\ \forall\ x \in S$. Here $\inf S = 0 \notin S$ and $\sup S = 1 \in S$. Here the greatest member of S is 1 whereas the smallest member of S does not exist.

(*ii*) Let $S = \left\{(-1)^n \frac{1}{n} : n \in \mathbf{N}\right\} = \left\{-1, \frac{1}{2}, -\frac{1}{3}, \frac{1}{4}, -\frac{1}{5}, \ldots\ldots\right\}$.

Here S is bounded because $-1 \le x \le 1/2\ \forall\ x \in S$. $\inf S = -1 \in S$ and $\sup S = 1/2 \in S$. The greatest member is 1/2 and the smallest member is -1.

(*iii*) Let $S = \{(-1)^n n : n \in \mathbf{N}\} = \{-1, 2, -3, 4, -5, \ldots\ldots\} = \{\ldots\ldots, -5, -3, -1, 2, 4, 6, \ldots\ldots]$.

Clearly, S is neither bounded above nor bounded below. So S is an unbounded set. Sup S and inf S do not exist.

(*iv*) Let $S = \left\{1+(-1)^n \frac{1}{n}\right\} = \left\{0, \frac{3}{2}, \frac{2}{3}, \frac{5}{4}, \frac{4}{5}, \frac{7}{6}, \frac{6}{7}, \ldots\ldots\right\}$.

Rewriting, $S = \left\{\frac{0}{1}, \frac{2}{3}, \frac{4}{5}, \ldots\ldots, \frac{2n-2}{2n-1}, \ldots\ldots\right\} \cup \left\{\frac{3}{2}, \frac{5}{4}, \frac{7}{6}, \ldots\ldots, \frac{2n+1}{2n}, \ldots\ldots\right\}$.

Note that the proper fractions 0/1, 2/3, 4/5, ..., are increasing and tending to 1 whereas the improper fractions 3/2, 5/4, 7/6, ..., are decreasing and tending to 1. Thus, there exist two numbers 1 and 3/2 such that $1 \le x \le 3/2\ \forall\ x \in S$. Hence, S is a bounded set. Also, $\inf S = 1 \notin S$ and $\sup S = 3/2 \in S$.

(*v*) Let $S = \left\{4+\frac{3}{n} : n \in \mathbf{N}\right\} = \left\{7, 5\frac{1}{2}, 5, 4\frac{3}{4}, 4\frac{3}{5}, 4\frac{3}{6}, \ldots\ldots\right\}$.

Here $4 \le x \le 7\ \forall\ x \in S$ and hence S is bounded. $\inf S = 4 \notin S$ whereas $\sup S = 7 \in S$.

(*vi*) Let $S = \left\{\frac{n}{n+1} : n \in \mathbf{N}\right\} = \left\{\frac{1}{2}, \frac{2}{3}, \frac{3}{4}, \frac{4}{5}, \ldots\ldots\right\}$.

Here $1/2 \le x \le 1\ \forall\ x \in S$ and hence S is bounded. $\inf S = 1/2 \in S$ and $\sup S = 1 \notin S$.

(*vii*) Let $S = \left\{-2, -\frac{3}{2}, -\frac{4}{3}, -\frac{5}{4}, \ldots\ldots\right\}$.

Here $-2 \le x \le -1\ \forall\ x \in G$. Hence, S is bounded. $\inf S = -2 \in S$ and $\sup S = -1 \notin S$.

(*viii*) Let $S = \left\{1, 1+\frac{1}{2}, 1+\frac{1}{2}+\frac{1}{2^2}, \ldots\ldots, 1+\frac{1}{2}+\frac{1}{2^2}+\ldots\ldots+\frac{1}{2^n}+\ldots\ldots\right\}$.

Here $1+\frac{1}{2}+\frac{1}{2^2}+\ldots\ldots+\frac{1}{2^{n-1}} = \frac{1\times(1-1/2^n)}{1-(1/2)} = 2\left(1-\frac{1}{2^n}\right) = 2-\frac{1}{2^{n-1}}$.

Thus, $S = \left\{2-\frac{1}{2^{n-1}} : n \in \mathbf{N}\right\} = \left\{2, 2-\frac{1}{2}, 2-\frac{1}{2^2}, \ldots\ldots, 2-\frac{1}{2^{n-1}}, \ldots\ldots\right\}$.

Clearly, $1 \le x \le 2 \ \forall\ x \in S$. So S is bounded. Here $\inf S = 1 \in S$ and $\sup S = 2 \notin S$.

(*ix*) Unbounded.

Example 6. *If A and B are bounded sub-sets of real numbers, then prove that $A \cap B$ and $A \cup B$ are also bounded.*

Solution. Since A is bounded, there exist two real numbers l_1 and k_1 such that

$$l_1 \le a \le k_1,\ \forall\ a \in A.$$

Again, since B is bounded, there exist two real numbers l_2 and k_2 such that

$$l_2 \le b \le k_2,\ \forall\ b \in B.$$

Let $l = \min\{l_1, l_2\}$ and $k = \max\{k_1, k_2\}$.

$\therefore \quad l \le x \le k \ \forall\ x \in A \cap B$ and $l \le x \le k \ \forall\ x \in A \cup B$.

Hence $A \cap B$ and $A \cup B$ are bounded.

Example 7. *If S and T are arbitrary non-empty bounded sub-sets of real numbers, then show that*

(*a*) *Sup $(S \cup T)$ = max (sup S, sup T)* **[Delhi B.Sc. (Hons) I 2011]**

(*b*) *Inf $(S \cup T)$ = min (inf S, inf T).*

Solution. (*a*) Let $\sup S = s$ and $\sup T = t$.

Let x be an arbitrary element of $S \cup T$.

$\therefore\ x \in S$ or $x \in T$

$\Rightarrow\ x \le \sup S$ or $x \le \sup T$

$\Rightarrow\ x \le s$ or $x \le t$.

$\Rightarrow\ x \le \max(s, t)\ \forall\ x \in S \cup T$.

$\Rightarrow$ max (s, t) is an upper bound of $S \cup T$.

Now, max (s, t), *i.e.*, max (sup S, sup T) will be the supremum of S if we show that no real number less than max (s, t) can be an upper bound of $S \cup T$. Let k be any real number less than max (s, t).

Now, $\quad k < \max(s, t) \Rightarrow k < s$ or $k < t$.

Accordingly, two situations arise as discussed below.

Case (*i*). Let $k < s$, *i.e.*, $k < \sup S$. Then, by definition,

k cannot be an upper bound of S

$\Rightarrow\ \exists$ some $x \in S$ such that $x > k$

$\Rightarrow\ \exists$ some $x \in S \cup T$ such that $x > k$, as $x \in S \Rightarrow x \in S \cup T$

$\Rightarrow\ k$ is not an upper bound of $S \cup T$.

Thus, we see that max (s, t) is an upper bound of $S \cup T$ and no real number less than max (s, t) is an upper bound of $S \cup T$. Hence, by definition

$$\sup(S \cup T) = \max(s, t), \quad i.e., \quad \sup(S \cup T) = \max(\sup S, \sup T) \qquad \ldots(1)$$

Case (*ii*). Let $k < t$, *i.e.*, $k < \sup T$. Then, proceed as in case (*i*) and prove the same result (1).

(*b*) Left as an exercise for the reader.

Example 8. *Let S and T be non-empty sub-sets of* **R** *with the property* :

$$s \le t \ \forall \ s \in S \text{ and } t \in T.$$

(*i*) *Show that S is bounded above and T is bounded below.*

(*ii*) *Show that* $\sup S \le \inf T$. **[*Delhi Maths (H), 1997, 99*]**

Solution. Given that $\quad s \le t \ \forall \ s \in S$ and $\forall \ t \in T$. ...(1)

Let t be an arbitrary element of T. Then, from (1),

$$s \le t \ \forall \ s \in S \text{ so that } t \text{ is an upper bound of } S.$$

Hence, S is bounded above.

Similarly, let s be an arbitrary element of S. Then, from (1),

$$s \le t \ \forall \ t \in T \text{ so that } s \text{ is a lower bound of } S.$$

Hence, T is bounded below.

Again, $s \le t \ \forall \ s \in S$ and T is bounded below

$\Rightarrow \quad s \le \inf T \ \forall \ s \in S$

$\Rightarrow \quad \inf T$ is an upper bound of S

$\Rightarrow \quad \sup S \le \inf T$, by definition of sup S.

EXERCISES

1. Show that every sub-set of a set bounded above (bounded below, bounded) is bounded above (bounded below, bounded).
2. Give two examples of sets which are (*i*) bounded above but not bounded below.
   (*ii*) bounded below but not bounded above.
   (*iii*) neither bounded above nor bounded below
   (*iv*) bounded. (*v*) not bounded. (*Delhi B.A. (Prog.) III 2012*]
3. Show that the greatest member of a set, if it exists, is the smallest member of the set of upper bounds of the set and the smallest member of the set, if it exists, is the greatest member of lower bounds of the set.
4. Examine the following sets for the existence of greatest and smallest members :

   (*i*) $A = \left\{\dfrac{1}{n},\ n \in \mathbf{N}\right\}$ (*ii*) $B = \left\{1 + \dfrac{(-1)^n}{n},\ n \in \mathbf{N}\right\}$ (*iii*) $C = \left\{\dfrac{1}{n},\ n \in \mathbf{I}^*\right\}$.

   Here **I*** denotes the set of all non-zero integers.
5. Give an example of a non-empty sub-set S of **R** whose supremum and infimum both belong to $\mathbf{R} \sim S$. **[Delhi Maths (H), 2002]**
6. Give an example of a set of irrational numbers that has a rational number supremum. **(Utkal, 2003)**
7. Find the g.l.b. and l.u.b. of the set $S = \{x \in \mathbf{Z} : x^2 \le 25\}$, where **Z** is the set of all integers.
8. Which of the following sets are bounded above, bounded below, bounded or unbounded ? Also find the supremum (l.u.b.) and infimum (g.l.b.), it they exist. Which of these belong to the set ?

   (*i*) $S = \{2, 3, 5, 10\}$ (*ii*) $S = \{2, 2^2, 2^3, \ldots.., 2^n, \ldots..\} = \{2^n : n \in \mathbf{N}\}$.

   (*iii*) $S = \{\pi + 1/2, \pi + 1/4, \pi + 1/8, \ldots..\}$. **(Meerut, 2003)** (*iv*) $S = \{-1, -1/2, -1/3, ...\} = \{-(1/n) : n \in \mathbf{N}\}$.

(*v*) $S = (1, 2] \cup [3, 8)$. **(Meerut, 2003)**

(*vi*) $S = \{1 + (1/n) : n \in \mathbf{N}\}$. (*vii*) $S = \{1 - (1/n) : n \in \mathbf{N}\}$. **(Delhi B.Sc. (Hons.) I; 2018)**

(*viii*) $S = \left\{(-1)^n \left(\frac{1}{4} - \frac{4}{n}\right) : n \in \mathbf{N}\right\}$. (*ix*) $S = \left\{\left(1 - \frac{1}{n}\right) \sin \frac{n\pi}{2} : n \in \mathbf{N}\right\}$.

(Meerut 2018)

(*x*) $S = \left\{\sin \frac{\pi}{6}, \sin \frac{2\pi}{6}, \sin \frac{3\pi}{6}, ..., \sin \frac{n\pi}{6}, ...\right\}$. (*xi*) $S = \left\{m + \frac{1}{n} : m, n \in \mathbf{N}\right\}$.

(*xii*) $S = \{x \in \mathbf{I} : x^2 \le 25\}$. **(Meerut, 2001, 02)**

9. Show that the set $\mathbf{R}^+$ of positive real numbers is not bounded above. **(Meerut, 2003)**

10. If $a > 1$, then prove that the set $\{a^n : n \in \mathbf{N}\}$ is unbounded above. **(Utkal, 2003)**

11. Let A and B be two non-empty subsets of $\boldsymbol{R}$ such that $a \le b \ \forall a \in A, b \in B$. Show that $\text{Sup}\, A \le \text{Sup}\, B$. Illustrate for $A = \{(n-1)/n : n \in N\}$ *and* $B = \{(n+1)/n : n \in N\}$

[Delhi B.Sc. I (Prog) 2010]

12. Find lub and glb (if they exist) of the set $S = \{\sin x - 2\cos x;\ x \in R\}$

(G.N.D.U. Amritsar 2010)

ANSWERS

2. (*i*) The set of all negative integers = $\{....., -3, -2, -1\}$ and $\{\log(1/n) : n \in \mathbf{N}\}$.

(*ii*) The set of all natural numbers = $\{1, 2, 3, ...\}$ and $\{1, 2, 2^2, ...\}$

(*iii*) The sets I and Q are neither bounded above nor bounded below.

(*iv*) Any finite set, for example, $\{1, 3, 8, 10\}$ and any closed interval $[a, b]$.

(*v*) The set of all integers = $\{....., -3, -2, -1, 1, 2, 3, ...\}$ and $\{(-1)^n n : n \in \mathbf{N}\}$.

4. (*i*) A admits of the greatest member, *viz.*, 1 but not the smallest member.

(*ii*) B admits of the greatest member, *viz.*, 3/2 but not the smallest member.

(*iii*) C admits of the greatest member, *viz.*, 1 and the smallest member, *viz.*, – 1.

5. For the set $S = \{x : x \in \mathbf{R} \text{ and } 1 < x < 2\}$, we have

$$\sup S = 2 \in \mathbf{R} \sim S \text{ and } \inf S = 1 \in \mathbf{R} \sim S.$$

6. $S = \{\sqrt{n+1} - \sqrt{n} : n \in \mathbf{N}\}$ is a set of irrational numbers. Its sup $S = 1$, which is a rational number.

7. g.l.b. = – 5 and l.u.b. = 5.

8. (*i*) Bounded. Sup $S = 10 \in S$ and Inf $S = 2 \in S$.

(*ii*) Bounded below. Inf $S = 2 \in S$. Not bounded above. Unbounded set.

(*iii*) Bounded. Sup $S = \pi + 1/2 \in S$, Inf $S = \pi \notin S$.

(*iv*) Bounded. Sup $S = 0 \notin S$, Inf $S = -1 \in S$.

(*v*) Bounded. Sup $S = 8 \notin S$, Inf $S = 1 \notin S$.

(*vi*) Bounded. Sup $S = 2 \in S$, Inf $S = 1 \notin S$.

(*vii*) Bounded. Sup $S = 1 \notin S$, Inf $S = 0 \in S$.

(*viii*) Bounded. Sup $S = 15/4$, Inf $S = -7/4$.

(*ix*) Bounded. Sup $S = 1 \notin S$, Inf $S = -1 \notin S$.

(*x*) Bounded. Sup $S = 1 \in S$, Inf $S = -1 \in S$.

(*xi*) Bounded below. Inf $S = 1 \notin 1$. Not bounded above. Unbounded set.

(*xii*) Bounded. Sup $S = 5 \in S$, Inf $S = -5 \in S$.

12. lub $= -\sqrt{5}$, glb $= \sqrt{5}$

2.8. ORDER-COMPLETENESS (OR COMPLETENESS PROPERTY) OF THE SET OF REAL NUMBERS

[Delhi B.Sc. (Hons) I 2011; Delhi B.A (Prog) III 2008, 17; Delhi B.Sc. (Prog.) II, 2016, 17]

We have seen that the set **R** of real numbers is an *ordered field* and so is the set **Q** of rational numbers. We will now state a property which establishes a *distinction* between these two ordered fields.

Order-completeness of R. *The set of upper bounds of a non-empty set of real numbers which is bounded above has a smallest number.*

Equivalently, we may say that *every non-empty set of real numbers which is bounded above admits of a least upper bound* (i.e., supremum).

The property of the set **R** of real numbers, *viz., that every non-empty sub-set of* **R** *which is bounded above admits of least upper bound* is referred to as its *order-completeness* or *completeness property* of real numbers. This property is also called the supremum property of **R**.

This property states that if S be a set of real numbers which is bounded above, there exists the smallest of the upper bounds of S. The fact of a number b being the smallest of the upper bounds of S may also be described by the following two properties :

I. *The number b is an upper bound of S, i.e.,* no member of S is greater than b, *i.e.,*

$$x \in S \Rightarrow x \leq b \Leftrightarrow x \leq b \ \forall \ x \in S.$$

II. *No number less than b is an upper bound of S, i.e.,* if b' be a number less than b so that b' is not an upper bound of S, there exists at least one member $x \in S$ such that $x > b'$. Thus,

$$\text{if } b' < b \text{ then } \exists \ x \in S \text{ such that } x > b'.$$

Thus, we have completed the description of the set of real numbers as a **complete ordered field.**

To bring about a distinction between **Q** and **R,** we prove the following theorem.

Theorem. *The ordered field of rational numbers is not order-complete.*

[Delhi B.Sc. (Prog.) II, 2016, 17; Delhi B.A. (Prog.) III, 2013, 17; Kanpur, 2003, 09]

It would now be shown that the property of order-completeness of **R** does *not* hold good for the ordered field **Q** of rational numbers, so that while *the ordered field* **R** *is order-complete, the ordered field* **Q** *is not order-complete.* To establish this, we shall give an example of a set of rational numbers which is bounded above but for which there exists no rational number which is the smallest of its upper bounds, *i.e.,* for which supremum does not exist.

Let us consider a sub-set of the set **Q** of the rational numbers defined as follows :

$$S = \{x : x \in \mathbf{Q}, x \geq 0 \wedge x^2 < 2\}, \qquad \text{...(1)}$$

so that S consists of all those non-negative rational numbers whose squares are less than 2.

Since $1 > 0$ and $1^2 < 2$, it follows that $1 \in S$ and hence S is a non-empty set. Again, $3 > 0$, but $3^2 = 9 > 2$ and so $3 \notin S$. Thus, $x < 3 \ \forall \ x \in S$ and so 3 is an upper bound of S. Hence, S is bounded above.

We now show that the set of upper bounds of S has no least member. Of course, here we are taking account of the set of rational numbers only.

Let, if possible, a rational number k be the least upper bound of S. Then k is positive. Now, by the law of trichotomy which holds good for the set of rational numbers, one and only one of the following three possibilities hold :

(*i*) $k^2 = 2$ (*ii*) $k^2 > 2$ (*iii*) $k^2 < 2$.

We now discuss these situations one by one.

Case (*i*). Let $k^2 = 2$, *i.e.,* square of a rational number k is equal to 2. This is not possible because it can be shown that there is no rational number whose square is 2.

(For complete proof, give solution of Example 1, given on page 2.5.)

Case (*ii*). Let $k^2 > 2$. Consider the rational number 'y' given by

$$y = (4 + 3k)/(3 + 2k) \quad \text{...(2)}$$

Then, $$2 - y^2 = 2 - \left(\frac{y+3k}{3+2k}\right)^2 = \frac{2-k^2}{(3+2k)^2} < 0, \text{ as } k^2 > 2 \quad \text{...(3)}$$

Thus, $$2 - y^2 < 0 \text{ and hence } y^2 > 2,$$

showing that y is an upper bound of S ...(4)

Again, $$k - y = k - \frac{4+3k}{3+2k} = \frac{2(k^2-2)}{3+2k} > 0, \text{ as } k^2 > 2 \quad \text{...(5)}$$

Thus, $$k - y > 0 \text{ and hence } y < k. \quad \text{...(6)}$$

From (4) and (6), it follows that there exists an upper bound of S smaller than the assumed l.u.b. of S and as such we have a contradiction.

Case (*iii*). Let $k^2 < 2$. As before, consider a rational number y given by (2). Then, proceeding as in case (*ii*), we have

$$2 - y^2 = 2 - \left(\frac{y+3k}{3+2k}\right)^2 = \frac{2-k^2}{(3+2k)^2} > 0, \text{ as now } k^2 < 2$$

$$\Rightarrow 2 - y^2 > 0 \Rightarrow y^2 < 2 \Rightarrow y \in S, \text{ by (1)} \quad \text{...(7)}$$

and $$k - y = k - \frac{4+3k}{3+2k} = \frac{2(k^2-2)}{3+2k} < 0, \text{ as now } k^2 < 2.$$

$$\Rightarrow k - y < 0 \text{ so that } y > k. \quad \text{...(8)}$$

From (7) and (8), it follows that the member y of S is greater than the assumed l.u.b. of S, so that k cannot be an upper bound of S and as such we have a contradiction.

Since, by the law of trichotomy, the three above mentioned possibilities are mutually exclusive and exhaustive, therefore, it follows that if k is any rational number, then k cannot be the l.u.b. of S.

We thus see that for the bounded set S of rational numbers as defined above, there does *not* exist any rational number which is the least of all its upper bounds. This illustrates the fact that the set of rational numbers is not order-complete.

Note 1. It will be seen in the following that order-completeness of the system of real numbers has very far-reaching consequences. As a first consequence, *we show the existence of a real number with square 2.*

Let, $$S = \{x : x \in \mathbf{R} \wedge x \geq 0 \wedge x^2 < 2\}.$$

Now, S is not empty inasmuch as $0 \in S$. Also, S is bounded above in asmuch as every positive real number with square greater than 2 is an upper bound of S. In particular, 2 is an upper bound of S.

By the order-completeness of **R**, the set S admits of a least upper bound, say, k. It will be shown that $k^2 = 2$.

Now, $$k^2 < 2 \Rightarrow [(4 + 3k)^2/(3 + 2k)^2 < 2] \wedge [(4 + 3k)/(3 + 2k) > k],$$

showing that there exists a member $(4 + 3k)/(3 + 2k)$ of S greater than its upper bound k.

Again, $$k^2 > 2 \Rightarrow [(4 + 3k)^2/(3 + 2k)^2 > 2] \wedge [(4 + 3k)/(3 + 2k) < k],$$

showing that there exists an upper bound $(4 + 3k)/(3 + 2k)$ of S smaller than its assumed least upper bound k.

Thus, it follows that $k^2 = 2$.

As the square of no rational number is 2, it follows that the real number k whose square is 2 is not rational.

The real number k whose square is 2 is known as irrational number. Similarly, we can show the existence of real numbers whose squares are 3, 10, 15, ..., etc. This establishes the existence of irrational numbers.

Irrational number. Definition. A real number which is not rational is called irrational. For example, $\sqrt{2}$, $\sqrt{3}$, $\sqrt{10}$, $\sqrt{15}$ etc. are all irrational numbers.

Note 2. A more powerful result than the one stated and proved above which can be shown to be true in **R** as a result of its order-completeness is as follows :

Given a positive real number a and a natural number n, there exists one and only one positive real number b such that $b^n = a$.

While this result can be proved as a direct consequence of the order-completeness property of **R,** we shall give a different and apparently similar proof in a later chapter.

2.9. EQUIVALENT DESCRIPTIONS OF THE ORDER-COMPLETENESS PROPERTY OF THE SET OF REAL NUMBERS

We now give a property of the set of real numbers, which is equivalent to the order-completeness property described above. The description of this property is straightforward in as much as it is obtained on replacement of the '*less than*' relation by '*greater than*' and *vice-versa*.

I. Theorem. *A non-empty sub-set of real numbers which is bounded below has the greatest lower bound (infimum) in* **R** *(the set of all real numbers).*

[*Delhi B.Sc. (Prog.) II, 2015; Delhi Maths (H), 2002, 07, 08 ; Kanpur, 1994*]

Proof. Let S be a non-empty sub-set of real numbers which is bounded below.

We define a set T as follows : $T = \{-x : x \in S\}$. ...(1)

Then, clearly T is also a non-empty sub-set of real numbers.

First of all, we shall prove that T is bounded above.

Since S is bounded below, there exists a lower bound a for S such that

$$x \geq a \ \forall \ x \in S$$

$$\Rightarrow \quad -x \leq -a \ \forall \ x \in S$$

$$\Rightarrow \quad (-x) \leq -a \ \forall \ (-x) \in T, \text{ by (1)}$$

$$\Rightarrow \quad T \text{ is bounded above with } (-a) \text{ as its upper bound.}$$

Hence, by the order-completeness property of real numbers, the sub-set T of real numbers has the supremum (l.u.b.), say, t.

We shall now show that $\inf S$ = g.l.b. of $S = -t$.

Since t is the l.u.b. of T, it follows that $-t$ is a lower bound of S. Let b be any lower bound of S. Then $-b$ is an upper bound of T.

Now, $t = \sup T$ and $-b$ is an upper bound of $T \Rightarrow t \leq -b$, *i.e.*, $-t \geq b$, showing that $-t$ must be the g.l.b. of S, *i.e.*, $\inf S = -t$.

Thus, if S is bounded below, then it must possess g.l.b.

The following two properties of **R** are thus equivalent in that each implies the order.

(*i*) The set of upper bounds of a non-empty set of real numbers which is bounded above has a *least member*.

(*ii*) The set of lower bounds of a non-empty set of real numbers which is bounded below has a *greatest member*.

Note. In the following we shall state what is called *Dedekind property* and establish its equivalence to that of the *order-completeness* property.

II. Dedekind Property. *Let L and U be two non-empty sub-sets of the ordered field* **R** *such that*

(*i*) $L \cup U = \mathbf{R}$.

(*ii*) *each member of L is less than each member of U, i.e.,* $x \in L \wedge y \in U \Rightarrow x < y$.

Then either the sub-set L has a greatest member or the sub-set U has a smallest member.

The property of the ordered field **R** referred to above is often described as the *Dedekind property*.

We shall now prove the equivalence of the Dedekind property with the order-completeness property.

Suppose that the ordered field **R** satisfies the Dedekind property. We shall deduce that it also satisfies the order-completeness property.

Let S be a set of real numbers bounded above.

We divide the set **R** into two sub-sets L and U defined as follows :

$U = \{x : x$ is an upper bound of $S\}$, $L = \{x : x$ is not an upper bound of $S\}$.

It may be seen that

$$L \neq \phi,\ U \neq \phi$$
$$L \cup U = \mathbf{R}$$
$$x \in L \wedge y \in U \Rightarrow x < y.$$

Then either L has a greatest member or U has a least member.

We shall show that L cannot have greatest member. Let, if possible, L have a greatest member say, ξ

$\xi \in L \Rightarrow \xi$ is not an upper bound of $S \Rightarrow \exists\ a \in S$ such that $a > \xi$.

Let b be a real number such that $\xi < b < a$.

Now, $\xi < b \Rightarrow b \in U \Rightarrow b$ is an upper bound of S.

It follows that there exists a member a of S greater than its upper bound b. Thus, we arrive at contradictory conclusions and as such it follows that L has no greatest member. Then by the Dedekind property, the set U which consists of all the upper bounds of S has smallest member.

We see that Dedekind property $\Rightarrow$ Order-completeness property.

We have now to prove the converse. Let **R** have the order-completeness property. We shall show that it also has the Dedekind property. Let L, U be any two sub-sets of **R** such that

(*i*) $L \neq \phi$, $U \neq \phi$,

(*ii*) $L \cup U = \mathbf{R}$,

(*iii*) each member of L is less than each member of U.

The set L is bounded above. If L has a greatest member, we have finished. If L does not admit of a greatest member, then the set of all the upper bounds of L coincides with U. Thus, the

Order-completeness property $\Rightarrow$ Dedekind property.

We have thus proved the equivalence of the Dedekind and the Order-completeness properties of **R**.

2.10. EXPLICIT STATEMENT OF THE PROPERTIES OF THE SET OF REAL NUMBERS AS A COMPLETE ORDERED FIELD

[Delhi B.A. (Prog) III 2011, 12; Delhi B.Sc. III (Prog.) 2009; Delhi Maths (H), 2002, 08]

We give below at one place the description of the set of real numbers as a complete ordered field. In the following a, b, c denote members of **R**, *viz.*, the real numbers.

A. To each pair of members, a, b of **R**, there corresponds a member $a + b$ of **R**, such that the following conditions are satisfied :

A.1. $a + b = b + a \ \forall \ a, b \in \mathbf{R}$.

A.2 $a + (b + c) = (a + b) + c \ \forall \ a, b, c \in \mathbf{R}$.

A.3. $a + 0 = a \ \forall \ a \in \mathbf{R}$.

A.4. To each $a \in \mathbf{R}$, there corresponds $-a \in \mathbf{R}$ such that $a + (-a) = 0$.

M. To each pair of members a, b of **R**, there corresponds a member ab of **R** such that the following conditions are satisfied :

M.1. $ab = ba \ \forall \ a, b \in \mathbf{R}$.

M.2. $a\,(bc) = (ab)\,c \ \forall \ a, b, c \in \mathbf{R}$.

M.3. $a \cdot 1 = a \ \forall \ a \in \mathbf{R}$.

M.4. To each $a \neq 0 \in \mathbf{R}$, there corresponds $a^{-1} \in \mathbf{R}$ such that $aa^{-1} = 1$.

AM $a\,(b + c) = ab + ac \ \forall \ a, b, c \in \mathbf{R}$.

O. For each distinct pair of members a, b of **R**, we have either $a < b$ or $b < a$.

O.1. $a < b \wedge b < c \ \Rightarrow \ a < c$,

OA $a < b \ \Rightarrow \ a + c < b + c$,

OM $a < b \wedge 0 < c \ \Rightarrow \ ac < bc$.

OC. The set of upper bounds of every non-empty set of real numbers which is bounded above has a smallest member.

Definition. *Subtraction and Division in* **R**.

By definition, $a - b = a + (-b)$,

and $a \div b = a\,(b^{-1}),\ b \neq 0$. ($a \div b$ is also written as a/b.)

Note. The multiplicative inverse of $b \neq 0$ may be denoted as b^{-1} or $1/b$.

2.11. SOME IMPORTANT PROPERTIES OF THE SYSTEM OF REAL NUMBERS

Archimedean property of R. ***[Delhi B.Sc. (Prog.) II, 2016, Delhi B.Sc. (Hons) 2009]***

We shall now state and prove the Archimedean property of **R**.

It will be seen that the Archimedean property of **R** is a consequence of the order-completeness property of **R**.

Theorem I. *If x and y are two given real numbers with $x > 0$, then there exists a natural number n such that $nx > y$.*

[Purvanchal, 1995; Delhi B.Sc. (Hons.) I, 2016; Delhi Maths (G), 2004]

Proof. We *assume* the *negation* of the conclusion to be established and show that we arrive at a contradiction.

Thus, we suppose that there exists *no* natural number n such that $nx > y$ so that

$$nx \leq y \ \forall \ n \in \mathbf{N}.$$

Consider the set $A = \{nx : n \in \mathbf{N}\}$.

This set A is, in accordance with our assumption, bounded above and y is an upper bound of the same. By the order-completeness of **R**, the set A admits of the least upper bound.

Let α be the least upper bound of the set A. We have

$$nx \leq \alpha \ \forall \ n \in \mathbf{N}$$
$$\Rightarrow \ (m+1)\,x \leq \alpha \ \forall \ m \in \mathbf{N}$$
$$\Rightarrow \ mx \leq \alpha - x \ \forall \ m \in \mathbf{N}$$
$$\Rightarrow \ \alpha - x \text{ is an upper bound of } A.$$

Since $\alpha - x < \alpha$, we contradict the statement that α is the least upper bound of A.

Thus, the set A is *not* bounded above and as such there exists a natural number n such that

$$nx > y.$$

Corollary 1. *For any real number a there exists a positive integer n such that $n > a$.*

[Delhi B.Sc. (Hons) I 2011; Delhi B.A. (Prog.) III, 2006]

Proof. Taking the two real numbers 1 and a, the result follows from Theorem I.

Corollary 2. *For any $\varepsilon > 0$ there exists a positive integer n, such that $1/n < \varepsilon$.*

Proof. Take $a = 1/\varepsilon$ in Corollary 1. Then, $n > 1/\varepsilon$ or $1/n < \varepsilon$.

Corollary 3. *For any real number x, there exist two integers m and n such that $m < x < n$.*

Proof. Left as an exercise for the reader.

Corollary 4. *For any real number x, there exists a unique integer n such that $n \leq x < n + 1$.*

Proof. Left as an exercise for the reader.

Corollary 5. *For any real number x, there exists a unique integer n such that $x - 1 < n < x$.*

Proof. Left as an exercise for the reader.

ARCHIMEDEAN ORDERED FIELD

An ordered field is said to be Archimedean if it has the Archimedean property.

For example, the field **R** of set of all real numbers is an Archimedean ordered field.

Theorem II. *The set* **N** *of natural numbers is not bounded above.*

(Delhi B.Sc. (Prog.) III 2011)

Proof. Method I : Let, if possible, **N** be bounded above.

Since $1 \in \mathbf{N}$, so $\mathbf{N} \neq \phi$. Since **N** is a non-empty and bounded above sub-set of **R**, so by order-completeness property of real numbers, **N** must have the supremum, b, say.

Then $n \leq b \ \forall \ n \in \mathbf{N}$

$\Rightarrow$ $(n+1) \leq b \ \forall \ n \in \mathbf{N}$, as $n \in \mathbf{N} \Rightarrow (n+1) \in \mathbf{N}$

$\Rightarrow$ $n \leq b - 1 \ \forall \ n \in \mathbf{N}$

$\Rightarrow$ $(b-1)$ is an upper bounded of **N**.

Now, $(b-1) < b$ and sup $\mathbf{N} = b$, by our assumption. Thus, we get an upper bound of **N** which is less than b. This contradicts the fact that b is the supremum of **N**.

Hence, **N** is not bounded above.

Second Method. By Archimedean property in real numbers, for each positive real number x, there exists $n \in \mathbf{N}$, such that $n > x$.

Hence, there exists no positive real number x such that $n \leq x \ \forall \ n \in \mathbf{N}$

$\Rightarrow$ No positive real number is an upper bound of **N**

$\Rightarrow$ **N** is not bounded above.

2.12. THE DENSENESS PROPERTY OF THE SET OF REAL NUMBERS R

Theorem I The density theorem. *Between any two distinct real numbers there always lies a rational number.* **(Delhi B.Sc. (Hons.) I, 2016, 2017, 18)**

(Kanpur 2009, 2014; Bhopal, 2004)

Proof. We shall show that if x, y are real numbers and $x < y$, then there exists a rational number r such that $x < r < y$.

We firstly suppose that $x > 0$.

Also $$y > x \Rightarrow y - x > 0.$$

By the Archimedean property of **R**, there exists a natural number q such that
$$q(y - x) > 1 \Rightarrow (1/q) < y - x.$$

Again, by the Archimedean property of **R**, there exists a natural number n such that $n > qx$.

Let p be the smallest such natural number n so that p is the smallest natural number such that $qx < p$.

If $p = 1$ so that $p - 1 = 0$, we have $p - 1 = 0 < qx < 1 = p \Rightarrow p - 1 < qx < p$.

If $p \geq 2$ so that $p - 1$ is a natural number, we have by the definition of p, $p - 1 < qx$.

Now, $qx < p = (p - 1) + 1 < qx + q(y - x) = qy \Rightarrow qx < p < qy \Rightarrow x < p/q < y$.

We now consider the case when $x \leq 0$.

By Archimedean property of **R**, there exists a natural number k such that
$$k = k \cdot 1 > -x \Rightarrow k + x > 0.$$

There exists, therefore, a rational number, say s, such that
$$k + x < s < k + y \Rightarrow x < s - k < y.$$

Here $s - k$ is a rational number.

Thus, we have proved the theorem.

Theorem II. *If x, y are two real numbers with $x < y$, then there exists an irrational number i such that $x < i < y$.*

Proof. We have to show that, there exists an irrational number between any two given different real numbers.

Case I. *Let x be an irrational number.*

By the preceding result, there exists a rational number r such that $x < r < y$.

Consider now the number $(x + r)/2$, which, as may be easily shown, is irrational. We may also show that $$x < (x + r)/2 < y.$$

Thus, $(x + r)/2$ is an irrational number between x and y.

Case II. *Let x be a rational number.*

We shall make use of the known fact that $\sqrt{2}$ is an irrational number.

Let r be a rational number between x and y so that $x < r < y$.

As $r - x > 0$, there exists a natural number n such that
$$n(r - x) > \sqrt{2} \Rightarrow \sqrt{2}/n < r - x.$$

Let $$i = x + \sqrt{2}n \text{ so that } i < r.$$

Now, i is an irrational such that $x < i < r < y \Rightarrow x < i < y$.

Hence, the result.

Corollary. Between two different real numbers there lie an infinite number of irrational numbers.

Proof. Left as an exercise for the reader.

Theorem III. *Between any two distinct real numbers, there lie an infinite number of real numbers.*

Proof. The proof follows from above Theorems I and II.

EXAMPLES

Example 1. *Show that sup $\{r \in \mathbf{Q} : r < a\} = a$, for each $a \in \mathbf{R}$.* **[*Delhi B.Sc. (Prog.) III, 2012*]**

Solution. Let a denote any real number and let

$$S = \{r : r \text{ is a rational number and } r < a\} \quad ...(1)$$

From (1), $r < a \ \forall \ r \in S$. Hence, by definition, a is an upper bound of S. Hence, S is bounded above and so by the order-completeness of **R**, S has the l.u.b. (*i.e.*, the supremum). Let sup $S = s$.

Let, if possible, $s < a$. Since s and a are real numbers, there exists a rational number r such that $s < r < a$ (Refer Theorem I, Art. 2.12).

In view of (1), $\qquad r < x \ \Rightarrow \ r \in S,$

which contradicts the assumption that s is an upper bound of S.

Hence, $\qquad a = s$, *i.e.*, sup $S = a$.

Example 2. *Let A and B be two non-empty sub-sets of* **R** *and let* $C = \{x + y : x \in A$ *and* $y \in B\}$. *Then,*

(*i*) *if each of A and B has a supremum, show that C has a supremum and* sup C = sup A + sup B. **[*Delhi B.Sc. (Hons.)I, 2016*]**

(*ii*) *if each of A and B has an infimum, show that C has an infimum and* inf C = inf A + inf B.

Proof. (*i*) Given $\qquad C = \{x + y : x \in A \text{ and } y \in B\}$ ...(1)

Let $\qquad$ sup $A = a$ and sup $B = b$ ...(2)

Then (2) $\Rightarrow \ x \le a \ \forall \ x \in A$ ...(3)

and $\qquad y \le b \ \forall \ y \in B$ ...(4)

Now, (3) and (4) $\Rightarrow \ x + y \le a + b \ \forall \ x \in A$ and $\forall \ y \in B$ ...(5)

Let $z \in C$. Then $z = x + y$ for some $x \in A$ and $y \in B$ ...(6)

Then (5) and (6) $\Rightarrow \ z \le a + b \ \forall \ z \in C$,

showing that $a + b$ is an upper bound of C. ...(7)

Let d be any real number such that $\qquad d < a + b$ ...(8)

From (8), $d - b < a$. Since sup $A = a$ and $(d - b) < a$, so $(d - b)$ cannot be an upper bound of A. Hence, there exists some $x \in A$ such that

$$d - b < x \text{ so that } d - x < b.$$

Now, sup $B = b$ and $(d - x) < b$, so $(d - x)$ cannot be an upper bound of B and hence there exists some $y \in B$ such that

$$d - x < y \quad \text{or} \quad d < x + y, \text{ where } x + y \in C.$$

$\Rightarrow d$ cannot be an upper bound of C. ...(9)

From (7), (8) and (9), we see that $a + b$ is an upper bound of C and no real number less than $a + b$ is an upper bound for S. Hence, $a + b$ is the l.u.b. of C.

Thus, $\qquad$ sup $C = a + b$ $\quad$ or $\quad$ sup C = sup A + sup B.

(*ii*) Left as an exercise for the reader.

Example 3. *For a real number* λ, *let* λA *denote the set* $\lambda A = \{\lambda x : x \in A\}$. *Prove that if A is a bounded sub-set of* **R**, *then* λA *is also bounded and*

(*i*) *Sup* $\lambda A = \lambda$ *sup A, if* $\lambda > 0$

(*ii*) *Sup* $\lambda A = \lambda$ *inf A, if* $\lambda < 0$

(*iii*) *Inf* $\lambda A = \lambda$ *inf A, if* $\lambda > 0$ **(*Bhopal, 2004*)**

(*iv*) *Inf* $\lambda A = \lambda$ *sup A, if* $\lambda < 0$. **(*Bhopal, 2004*)**

Proof. Given $\lambda A = \{\lambda x : x \in A\}$. ...(1)

Since A is bounded, there exists a real number k such that

$$|x| \leq k \ \forall \ x \in A. \quad ...(2)$$

Now, $|\lambda x| = |\lambda|\,|x| \leq k \ \forall \ \lambda x \in \lambda A$, using (2).

$\Rightarrow \lambda A$ is bounded and so it has supremum and infimum.

(*i*) Let $\sup A = a$ so that $x \leq a \ \forall \ x \in A$

$\Rightarrow \ \lambda x \leq \lambda a \ \forall \ \lambda x \in \lambda A$, if $\lambda > 0$

$\Rightarrow \ \lambda a$ is an upper bound of λA

$\Rightarrow \ \sup \lambda A \leq \lambda a$ ...(3)

Now, for any $\varepsilon > 0$ and $\lambda > 0$, we have $\varepsilon/\lambda > 0$. So $a - (\varepsilon/\lambda) < a$.

$\therefore \ a - \dfrac{\varepsilon}{\lambda}$ is not an upper bound of A

$\Rightarrow$ there exists some $x \in A$ such that $x > a - \dfrac{\varepsilon}{\lambda}$

$\Rightarrow \ \lambda x > \lambda a - \varepsilon$, where $\lambda x \in \lambda A$

$\Rightarrow$ any number $< \lambda a$ cannot be an upper bound of λA

$\Rightarrow \ \sup \lambda A \geq \lambda a$ ...(4)

Now, (3) and (4) $\Rightarrow$ $\sup \lambda A = \lambda a$, *i.e.*, $\sup \lambda A = \lambda \sup A$, $\lambda > 0$.

(*ii*) Let $\inf A = b$ so that $x \geq b \ \forall \ x \in A$.

$\Rightarrow \ \lambda x \leq \lambda b \ \forall \ \lambda x \in \lambda A$, if $\lambda < 0$

$\Rightarrow \ \lambda b$ is an upper bound of λA

$\Rightarrow \ \sup \lambda A \leq \lambda b$ ...(5)

Now, for any $\varepsilon < 0$ and $\lambda < 0$, we have $\varepsilon/\lambda < 0$. So $b - (\varepsilon/\lambda) > b$.

$\therefore \ b - \dfrac{\varepsilon}{\lambda}$ is not a lower bound of A

$\Rightarrow$ there exists some $x \in A$ such that $x < b - \dfrac{\varepsilon}{\lambda}$

$\Rightarrow \ \lambda x > \lambda b - \varepsilon$, where $\lambda x \in \lambda A$ and $\lambda < 0$

$\Rightarrow$ any number $< \lambda b$ cannot be an upper bound of λA

$\Rightarrow \ \sup A \geq \lambda b$ ...(6)

Now, (5) and (6) $\Rightarrow$ $\sup \lambda A = \lambda b$, *i.e.*, $\sup \lambda A = \lambda \inf A$, if $\lambda < 0$.

Parts (*iii*) and (*iv*). Left as exercises for the reader.

Example 4. *Define a lower bound and infimum of a set of real numbers. State the order-completeness property of real numbers and apply it to prove that any non-empty set of real numbers which is bounded below has an infimum.* **[*Delhi Maths* (*G*), *2002*]**

[**Hint.** Refer Art. 2.7, Art. 2.8 and Art. 2.9]

Example 5. *If $a \geq 0$ and $a < \varepsilon$ for every $\varepsilon > 0$, then prove that $a = 0$.* **(*Utkal, 2003*)**

Solution. Let $a > 0$. Take $\varepsilon = a/2 > 0$.

Now, $a > a/2 \ \Rightarrow \ a > \varepsilon$, which is a contradiction

($\because$ by hypothesis, $a < \varepsilon$ for every $\varepsilon > 0$)

Hence, $a > 0$ is not possible. Thus, we must have $a = 0$.

EXERCISES

1. Show that a real number admits only one additive inverse. Also show that every non-zero real number admits of only one multiplicative inverse.

2. Show that given a non-zero number a and two numbers b, c the equation $ax + b = c$; $a, b, c \in \mathbf{R}$ possesses one and only one solution.

3. Show that, $-(a+b) = -a - b \ \forall\ a, b \in \mathbf{R}$; $(ab)^{-1} = a^{-1}b^{-1} \ \forall\ a, b \in \mathbf{R} \sim \{0\}$.

4. Show that, $a - (b - c) = (a - b) + c$; $(a - b) - c = a - (b + c)$.

5. Show that, $\forall\ a \in \mathbf{R}$, $\forall\ b \in \mathbf{R} \sim \{0\}$, $(a \div b) \div c = a \div (bc)$; $a \div (b \div c) = (ac) \div b$.

6. If a, b, c, d are four given real numbers and $b \neq 0, d \neq 0$, then

$$\frac{a}{b} + \frac{c}{d} = \frac{ad + bc}{bd}, \quad \frac{a}{b} \cdot \frac{c}{d} = \frac{ac}{bd}.$$

7. Show that $\dfrac{a}{b} \div \dfrac{c}{d} = \dfrac{ad}{bc}$ if $b \neq 0, d \neq 0, c \neq 0$.

8. Show that $a(-b) = -(ab)$, $(-a)(-b) = ab \ \forall\ a, b \in \mathbf{R}$, $ab = 0 \Leftrightarrow a = 0 \vee b = 0$.

9. Show that

(*i*) $a > 0 \wedge b > 0 \Rightarrow ab > 0, a < 0 \wedge b < 0 \Rightarrow ab > 0$.

(*ii*) $a < 0 \wedge b > 0 \Rightarrow ab < 0$. (*iii*) $a < 0 \Leftrightarrow -a > 0$; $a > 0 \Leftrightarrow -a < 0$.

(*iv*) $a^2 \geq 0 \ \forall\ a \in \mathbf{R}$. (*v*) $[a > 0] \wedge [b > 0] \wedge [a > b] \Rightarrow a^2 > b^2$.

(*vi*) $a > 0 \Rightarrow a > -a$; $a < 0 \Rightarrow a < -a$.

(*vii*) $a^2 + b^2 \geq 2ab \ \forall\ a, b \in \mathbf{R}$. (*viii*) $a^2 + b^2 = 0 \Leftrightarrow a = 0 \wedge b = 0$.

(*ix*) $x = y \Rightarrow x^2 = y^2$. Is the converse true ? (*x*) $x = y \Leftrightarrow x^3 = y^3$.

10. Show that

$$a > 0 \wedge b \geq 0 \Rightarrow a + b > 0,$$
$$a < b \wedge c < d \Rightarrow a + c < b + d,$$
$$a > b \Leftrightarrow a - b > 0; \ a < b \Leftrightarrow a - b < 0.$$

11. Given that $a > 0 \wedge b > 0$, show that $a > b \Leftrightarrow 1/a < 1/b$.

12. Which of the following statements are true ?

(*i*) $xy = 0 \Leftrightarrow x^2 + y^2 = 0$. (*ii*) $x^2 + y^2 = 0 \Leftrightarrow x^4 + y^4 = 0$.

(*iii*) There exists one and only one real number x such that $x^3 = 8$.

13. Show that $x^2 + xy + y^2 \geq 0 \ \forall\ x, y \in \mathbf{R}$.

14. Show that $(x - 3)(x - 4) \geq 0 \Leftrightarrow x \leq 3 \vee x \geq 4$.

15. Show, how starting from the set of real numbers as a complete ordered field, you can construct the systems of natural numbers, integers, rational numbers as sub-sets of the set of real numbers.

16. Show that $\inf\{1/n : n \in N\} = 0$ and hence show that if $t > 0$, then $\exists\ n \in N$ such that $0 < 1/n < t$. **[Delhi B.Sc. I (Hons.) 2010]**

17. Neither the set of natural numbers nor the set of integers is an ordered field. Explain and justify this statement.

18. (*i*) Give an example of a non-empty sub-set of rational numbers which is bounded above but for which the supremum does not exist.

(*ii*) Give an example of a non-empty sub-set of rational numbers which is bounded but does not have greatest lower bound.

19. Let $-A$ denote the set $-A = \{-x : x \in A\}$. Prove that if A is bounded sub-set of **R**, then $-A$ is also bounded and (*i*) Sup $(-A) = -$ inf A (*ii*) Inf $(-A) = -$ sup A.

20. Let A and B be two non-empty bounded sets of positive real numbers and let

$$C = \{xy : x \in A \text{ and } y \in B\}.$$

Show that C is bounded and that

(*i*) $\sup C = \sup A \cdot \sup B$ (*ii*) $\inf C = \inf A \cdot \inf B$.

21. (*i*) Prove that the set $\{x : x \in \mathbf{Q}^+ \text{ and } 0 < x^2 < 3\}$ does not have any least upper bound in **Q**, where $\mathbf{Q}^+$ denotes the set of all positive rational numbers.

(*ii*) Prove that the set $\{x : x \in \mathbf{Q}^+ \text{ and } x^2 < 5\}$ does not have infimum in **Q**.

[**Hint.** (*i*) Proceed as in theorem of Art. 2.8 by taking $y = 3\,(k + 1)/(k + 3)$

(*ii*) Proceed as in theorem of Art. 2.8 by taking $y = 5\,(k + 1)/(k + 5)$]

22. If x and y are real numbers and $x > 1$, prove that there exists a positive integer n such that $x^n > y$.

23. For a given positive real number a and a natural number n, show that there exists one and only one positive real number b such that $b^n = a$.

24. State and prove Archimedean property of real numbers. Deduce that for every real number x there exists a positive integer n such that $-n < x$. **[Delhi Maths (H), 2000]**

25. Defining order-completeness property of the set of real numbers, prove Archimedean property of real numbers as a consequence of completeness. **[Delhi Maths (H), 2004]**

2.13. THE MODULUS (OR ABSOLUTE VALUE) OF A REAL NUMBER

Let x be any real number. We have the following possibilities : $x > 0$, $x < 0$, $x = 0$.

We write
$$|x| = \begin{cases} x \text{ if } x \ge 0 \\ -x \text{ if } x < 0 \end{cases}$$

and call $|x|$ the *modulus* or the *absolute value* of the real number x.

It would be seen that $\forall\, x \in \mathbf{R}$, we have $|x| \ge 0$, $|x| \ge x$ and $|x| \ge -x$.

In fact
$$|x| = \max\{x, -x\},$$
where $\max\{x, -x\}$ denotes the greater of the two numbers x and $-x$.

It can be easily verified that $|x| = \sqrt{x^2}$.

SOME RESULTS INVOLVING MODULI

Theorem 1 : *If a, b are any two real numbers, then*

(*i*) $|a + b| \le |a| + |b|$ **[Delhi Maths (H), 2001, 02; Delhi B.A. (Prog.) III, 2014]**

(*ii*) $|a - b| \ge \big||a| - |b|\big|$ **[Delhi B.Sc. (Prog.) II, 2015; Delhi B.A. (Prog.) III 2014]**

(*iii*) $|ab| = |a|\,|b|$.

Proof. (*i*) $a \le |a|,\ b \le |b| \Rightarrow a + b \le |a| + |b|$.

Similarly, we may show that $-(a + b) \le |a| + |b|$.

It follows that $|a + b| \le |a| + |b|$.

Thus, we have proved the result (*i*).

(*ii*) We have $a = a - b + b \Rightarrow |a| = |(a - b) + b|$

$\le |a - b| + |b|$, using result (*i*)

Thus, $$|a - b| \ge |a| - |b| \qquad \text{...(1)}$$

Interchanging a and b in (1), we have

$$| b - a | \geq | b | - | a |$$

$$\Rightarrow | a - b | \geq | b | - | a | \qquad ...(2)$$

From (2) and (3), we have $| a - b | \geq || a | - | b ||$.

(*iii*) $a \geq 0 \wedge b \geq 0 \Rightarrow ab \geq 0 \Rightarrow | ab | = ab = | a | | b |$.

Again $a \geq 0 \wedge b \leq 0 \Rightarrow ab \leq 0 \Rightarrow | ab | = -(ab) = a(-b) = | a | | b |$.

The two remaining cases may be similarly disposed of.

Theorem 2. $| a - b | < k \Leftrightarrow b - k < a < b + k$. ***[Delhi Maths (H), 1997, 2000, 08, 09]***

Proof. We have $| a - b | = \max \{a - b, b - a\} < k$,

$$\Leftrightarrow a - b < k \wedge b - a < k$$

$$\Leftrightarrow a < b + k \wedge b - k < a$$

$$\Leftrightarrow b - k < a < b + k.$$

Hence, the result.

EXAMPLES

Example 1. *Prove that for all real numbers x and y*

$$| x + y |^2 + | x - y |^2 = 2 | x |^2 + 2 | y |^2.$$

[Delhi Maths (G), 1995; Delhi Maths (H), 2001, 04]

Solution. We know that for all $x \in \mathbf{R}$, $| x |^2 = x^2$

$$\therefore \quad | x + y |^2 = (x + y)^2 = x^2 + y^2 + 2xy = | x |^2 + | y |^2 + 2xy \qquad ...(1)$$

and $$| x - y |^2 = (x - y)^2 = x^2 + y^2 - 2xy = | x |^2 + | y |^2 - 2xy \qquad ...(2)$$

Adding (1) and (2), $| x + y |^2 + | x - y |^2 = 2 | x |^2 + 2 | y |^2$.

Example 2. *Show that $| a + b | = | a | + | b |$ if and only if $ab \geq 0$.*

[Delhi B.Sc. (Hons) I 2011; Delhi Maths (H), 1994, 2004]

Solution. Let $ab \geq 0$. We have $| a + b |^2 = (a + b)^2 = a^2 + b^2 + 2ab \qquad ...(1)$

But $a^2 = | a |^2$, $b^2 = | b |^2$. Also $ab \geq 0 \Rightarrow | ab | = ab$.

$\therefore$ (1) reduces to $$| a + b |^2 = | a |^2 + | b |^2 + 2 | ab |$$

or $$| a + b |^2 = | a |^2 + | b |^2 + 2 | a | | b |$$

or $$| a + b |^2 = (| a | + | b |)^2. \qquad ...(2)$$

But since $| a + b |$ and $(| a | + | b |)$ are both non-negative real numbers, therefore, taking square root of both sides of (2), we get

$$| a + b | = | a | + | b |.$$

Conversely, let $$| a + b | = | a | + | b |$$

$\therefore \quad | a + b |^2 = (| a | + | b |)^2$

$\Rightarrow \quad | a |^2 + | b |^2 + 2ab = | a |^2 + | b |^2 + 2 | a | | b |$, as before

$\Rightarrow \quad ab = | ab | \Rightarrow ab \geq 0$.

Example 3. *Show that for all real numbers a and b,* $\dfrac{| a + b |}{1 + | a + b |} \leq \dfrac{| a |}{1 + | a |} + \dfrac{| b |}{1 + | b |}$.

[Delhi Maths (H) 2009; Delhi Maths (H), 1996, 2003]

Solution. We know that $| a + b | \leq | a | + | b |$

$\Rightarrow \quad 1 + | a + b | \leq 1 + | a | + | b |$

$$\Rightarrow \frac{1}{1+|a+b|} \geq \frac{1}{1+|a|+|b|} \quad \Rightarrow \quad -\frac{1}{1+|a+b|} \leq -\frac{1}{1+|a|+|b|}$$

$$\Rightarrow 1-\frac{1}{1+|a+b|} \leq 1-\frac{1}{1+|a|+|b|} \quad \Rightarrow \quad \frac{|a+b|}{1+|a+b|} \leq \frac{|a|+|b|}{1+|a|+|b|}$$

$$\Rightarrow \frac{|a+b|}{1+|a+b|} \leq \frac{|a|}{1+|a|+|b|}+\frac{|b|}{1+|a|+|b|} \qquad \text{...(1)}$$

Now, $1+|a|+|b| \geq 1+|a| \Rightarrow \dfrac{|a|}{1+|a|+|b|} \leq \dfrac{|a|}{1+|a|}$ ...(2)

and $1+|a|+|b| \geq 1+|b| \Rightarrow \dfrac{|b|}{1+|a|+|b|} \leq \dfrac{|b|}{1+|b|}$ ...(3)

Adding (2) and (3), we get

$$\frac{|a|}{1+|a|+|b|}+\frac{|b|}{1+|a|+|b|} \leq \frac{|a|}{1+|a|}+\frac{|b|}{1+|b|} \qquad \text{...(4)}$$

Now, from (1) and (4), $\dfrac{|a+b|}{1+|a+b|} \leq \dfrac{|a|}{1+|a|}+\dfrac{|b|}{1+|b|}$.

Example 4. *Show that for any two real numbers a and b,*

(*i*) $\max\{a, b\} = \frac{1}{2}(a+b+|a-b|)$ (*ii*) $\min\{a, b\} = \frac{1}{2}(a+b-|a-b|)$.

Solution. (*i*) When $a > b$, then $a - b > 0$ so that $|a - b| = a - b$.

$\therefore \quad \frac{1}{2}(a+b+|a-b|) = \frac{1}{2}(a+b+a-b) = a = \max\{a, b\}$

When $a < b$, then $a - b < 0$ so that $|a - b| = -(a - b) = b - a$.

$\therefore \quad \frac{1}{2}(a+b+|a-b|) = \frac{1}{2}(a+b+b-a) = b = \max\{a, b\}$.

(*ii*) Left as an exercise for the reader.

EXERCISES

1. Prove that (*i*) $\sqrt{a^2+b^2} \leq |a|+|b|$ (*ii*) $\sqrt{|a+b|} \leq \sqrt{|a|}+\sqrt{|b|}$

[Delhi Maths (H), 1998]

(*iii*) $|a - b| < c$ iff $b - c < a < b + c \ \forall \ a, b, c \in \mathbf{R}$. **[Delhi Maths (H), 1997]**

2. Determine and sketch the set of pairs (x, y) in $(\boldsymbol{R}\times\boldsymbol{R})$ that satisty $|x| + |y| = 1$.

[Delhi B.Sc. I (Hons) 2010]

3. If $a_1, a_2, \ldots, a_n$ are real numbers, then show that

(*i*) $|a_1 + a_2 + \ldots + a_n| \leq |a_1| + |a_2| + \ldots + |a_n|$

(*ii*) $|a_1 \cdot a_2 \cdot a_3 \cdot \ldots \cdot a_n| = |a_1|\,|a_2|\,|a_3| \ldots |a_n|$.

4. Show that for all real numbers x and y, $\dfrac{|x-y|}{1+|x-y|} = \dfrac{|x|}{1+|x|}+\dfrac{|y|}{1+|y|}$.

[Delhi Maths (H), 1999]

5. Show that for all real numbers x and y,

(*i*) $|x-y| \leq |x|+|y|$ (*ii*) $|x+y| \geq \big||x|-|y|\big|$ (*iii*) $|x| < \varepsilon \Leftrightarrow -\varepsilon < x < \varepsilon$.

6. Prove that for real numbers x and y,

(*i*) $| x + y | = | x | + | y |$ if and only if $xy \geq 0$.

(*ii*) $| x + y | < | x | + | y |$ if and only if $xy < 0$.

7. Which of the following statements are true ?

(*i*) $\{x : | 3 - x | < 4\} = \{x : -1 < x < 7\}$ (*ii*) $\{x : | 4 - x | < 1\} = \{x : 3 < x < 5\}$

(*iii*) $\{x : | 1 - x | < 2\} = \{x : 1 < x < 3\}$ **[Ans.** (*i*) True (*ii*) True (*iii*) False.]

8. For a real number x and $\varepsilon > 0$, show that

(*i*) $|x| < \varepsilon \Leftrightarrow \ \varepsilon < x < \varepsilon$ (*ii*) $|x \ \ a| < \varepsilon \Leftrightarrow a \ \ \varepsilon < x < a + \varepsilon$

[Delhi B.Sc. (Prog.) III 2010]

9. Prove that for all real numbers x and y,

$$\frac{|x+y|}{2+|x|+|y|} \leq \frac{|x|}{1+|x|} + \frac{|y|}{1+|y|}$$

[Delhi B.Sc. (Hons.) I 2008]

2.14. PRINCIPLE OF MATHEMATICAL INDUCTION

For each $n \in \mathbf{N}$, let $P(n)$ be a statement about n. Suppose that

(*i*). $P(1)$ is true

(*ii*). For every $m \in \mathbf{N}$, if $P(m)$ is true, then $P(m + 1)$ is true

Then $P(n)$ is true for all $n \in \mathbf{N}$

Remark. It may happen that $P(n)$ are false for certain natural numbers, but then are true for all $n \geq p$ for some particular p. In order to deal with such situations, the principle of Mathematical Induction is modified in the following form:

PRINCIPLE OF MATHEMATICAL INDUCTION (SECOND VERSION)

Let $p \in \mathbf{N}$ and let $P(n)$ be a statement for each natural number $n \geq p$. Suppose that

(*i*). The statement $P(p)$ is true.

(*ii*). For every $m \geq p$, the truth of $P(m)$ implies the truth of $P(m + 1)$.

Then, $P(n)$ is true for all $n \geq p$.

2.15. BERNOULLI'S INEQUALITY

If $x > -1$, then $(1 + x)^n \geq 1 + nx$ for all $n \in \mathbf{N}$

Proof. We are to prove $(1 + x)^n \geq 1 + nx, \ \forall n \in \mathbf{N}$, if $x > -1$ ... (1)

In order to prove the result (1), we shall use Mathematical Induction.

For $n = 1$, (1) reduces to $1 + x \geq 1 + x$. Since $1 + x = 1 + x$ is true, it follows that the statement (1) is true for $n = 1$.

Next, we assume that (1) true for $m \in \mathbf{N}$ so that

$$(1 + x)^m \geq 1 + mx \qquad \text{... (2)}$$

Since $x > -1$, $(1 + x) > 0$ and so multiplying both sides of (2) by $(1 + x)$, we have

$$(1 + x)^{m+1} \geq (1 + x)(1 + mx)$$

or

$$(1 + x)^{m+1} \geq 1 + (m + 1)x + mx^2 \qquad \text{... (3)}$$

Since $x^2 \geq 0$ and $m \in \mathbf{N}$, we have $mx^2 \geq 0$

Hence,

$$1 + (m + 1)x + mx^2 \geq 1 + (m + 1)x \qquad \text{... (4)}$$

From (3) and (4), $(1 + x)^{m+1} \geq 1 + (m + 1)x$,

showing that (1) is true for $n = m + 1$ if it is true for $n = 1$. So (1) is true $\forall n \in \mathbf{N}$.

OBJECTIVE QUESTIONS

Multiple Choice Type : *Select (a), (b), (c) or (d), whichever is correct.*

1. Greatest lower bound of set of all positive even integers is :
(a) 0 (b) 2 (c) 1 (d) None of these. **(Kanpur, 2004)**

2. If x and y are real numbers, which one of the following is always true ?
(a) $|x-y| \le |x| - |y|$ (b) $|x-y| \ge |x| + |y|$
(c) $|x-y| \ge ||x| - |y||$ (d) $|x-y| = |x| - |y|$. **(I.A.S. Prel., 1993)**

3. If **N** stands for the set of natural numbers, then of the following the unbounded set is :
(a) $\{x : x = 1/n, n \in \mathbf{N}\}$ (b) $\{x : x = 1/2^n, n \in \mathbf{N}\}$
(c) $\{x : x = 2^n, n \in \mathbf{N}\}$ (d) None of these. **(I.A.S. Prel., 1993)**

4. The least upper bound of the set $\{1/n, n \in \mathbf{N}\}$ is :
(a) 1 (b) 0 (c) -1 (d) None of these. **(Kanpur, 2004)**

5. Between any two distinct real numbers there exist :
(a) only one rational number (b) finite number of rational numbers
(c) infinitely many rational numbers (d) None of these.

6. If u is an upper bound of a set A of real numbers and $u \in A$, then u is :
(a) infimum of A (b) both infimum and supremum of A
(c) supremum of A (d) neither infimum nor supremum of A.

7. Any non-empty sub-set of real numbers which is bounded below has :
(a) infimum (b) both infimum and supremum
(c) supremum (d) neither infimum nor supremum.

8. A sub-set A of real numbers is said to be bounded if it is :
(a) bounded above (b) bounded above as well as bounded below
(c) bounded below (d) None of these.

ANSWERS

1. (b) **2.** (c) **3.** (c) **4.** (a) **5.** (c) **6.** (c) **7.** (a) **8.** (b)

II. Fill in the blanks :

1. If x is a positive real number and y is any real number, then there exists a positive integer n such that ..... .

2. The supremum of a set A of real numbers, if it exists, is ..... .

3. The least upper bound of the set $\left\{\dfrac{3n+2}{2n+1} : n \in \mathbf{N}\right\}$ is ...... **(Meerut, 2003)**

4. The set **N** of natural numbers is ..... bounded above.

5. The g.l.b. of the set $\left\{x \in \mathbf{Q} : x = \dfrac{(-1)^n}{n}, n \in \mathbf{N}\right\}$ is ......

ANSWERS

1. $nx > y$ **2.** unique **3.** 5/3 **4.** not **5.** -1.

MISCELLANEOUS PROBLEMS ON CHAPTER 2

1. Find the supremum and infimum of the following sets :

(*i*) All rational numbers between $\sqrt{2}$ and $\sqrt{3}$

(*ii*) $\{-1, 1/2, -1/3, 1/4, \ldots\ldots, (-1)^n/n, \ldots\ldots\ldots\}$ **[Delhi Maths (G) 2006]**

Answers. (*i*) Does not exist (*ii*) supremum = 1/2, infimum = –1

2. Show that e is irrational **[M.S. Univ. T.N. 2006]**

3. Define infimum of a subset of real numbers. Prove that a non-empty subset of real numbers which is bounded below has an infimum in R. **[Delhi Maths (H) 2007]**

4. Define a bounded set, its supremum and infimum. Find the supremum and infimum of following sets :

(*i*) All rational numbers between $\sqrt{3}$ and $\sqrt{5}$

(*ii*) $\{(n+1)/n : n = 1, 2, 3, \ldots.\}$ **[Delhi Maths 2007]**

5. State Archimedian property of real numbers. Use it to prove that if $a, b \in \boldsymbol{R}$ such that $a \leq b + 1/n$ for all $n \in N$, then $a \leq b$ **[Delhi B.Sc. I (Prog) 2009]**

[**Sol.** If possible, let $a \leq b + 1/n \;\forall\; n \in N$ does not imply $a \leq b$. Then, by Trichotomy law, $a > b$ so that $a = b + x$, $x > 0$. Using Archimedian property to numbers 1 and x, we see that there must exist $n \in N$ such that $nx > 1$ so that $x > 1/n$. Hence, we have $a = b + x > b + 1/n$ for some $n \in N$. But this contradicts the given hypothesis, namely, $a \leq b + 1/n \;\forall\; n \in N$. Hence we cannot assume that $a > b$. So $a \leq b$]

6. Prove that for all real numbers x and y,

$$|x+y|/(2+|x|+|y|) \leq |x|/(1+|x|)+|y|/(1+|y|)$$ **[Delhi 2008]**

7. State Archimedian property of real numbers. Use it to prove that if for any $\varepsilon > 0, |b-a| < \varepsilon$, then $b = a$. **[Delhi B.Sc. III (Prog) 2010]**

8. Prove that the set of rational numbers is an ordered field but not a complete ordered field.

Sol. Refer Art. 2.5 and Art. 2.8. **[Delhi B.A. (Prog) III, 2013]**

9. If $S = \{1/n - 1/m : n, m \in \mathbf{N}\}$, find inf S and sup S. **[Delhi B.S. (Hons.) I, 2016]**

Ans. inf $S = -1$, sup $S = 1$.

10. If $a = -3$ and $S = \{x : |x| < 2\}$. Is sup $(aS) = a$ sup (S) ? Justify.

[Delhi B.Sc. (Prog.) II, 2015]

Sol. Here $S = \{x : |x| < 2\} = \{x : -2 < x < 2\}$. Hence, by the definition of supremum of S, we have sup $(S) = 2$ and hence

$$a \sup (S) = (-3) \times 2 = -6 \quad \ldots (1)$$

Now, from part (*ii*) of Ex. 3, page 2.23, we have

$$\sup (aS) = a \inf (S), \text{ if } a < 0. \quad \ldots (2)$$

Here $a = -3 < 0$ and inf (S) = –2, by definition. Hence, (2) reduces to

$$\sup (aS) = (-3) \times (-2) = 6 \quad \ldots (3)$$

From (1) and (3), we have $\sup (aS) \neq a \sup (S)$.

MISCELLANEOUS PROBLEMS ON CHAPTER 2

1. Find the supremum and infimum of the following sets:

(i) All rational numbers between $\sqrt{2}$ and $\sqrt{3}$.

(ii) $\{-1, 1/2, -1/3, 1/4, \ldots, (-1)^n/n, \ldots\}$ [Delhi Maths (G) 2006]

Answers. (i) Does not exist (ii) supremum $= 1/2$, infimum $= -1$

2. Show that $\mathbf{R}$ is [illegible]. [M.S. Delhi [illegible] 2006]

3. Define infimum of a subset of real numbers. Prove that a non-empty subset of real numbers which is bounded below has an infimum in $\mathbf{R}$. [Delhi Maths (G) 2007]

4. Define a bounded set, its supremum and infimum. Find the supremum and infimum of following sets:

(a) All rational numbers between $\sqrt{2}$ and $\sqrt{5}$.

(b) [illegible] [Delhi Maths 2007]

5. State Archimedean property of real numbers. Use it to prove that if $a, b \in \mathbf{R}$ such that $a \le b + 1/n$ for all $n \in \mathbf{N}$, then $a \le b$. [Delhi B.Sc. I (Hons) 2009]

[**Sol.** If possible, let $a > b$. Then, by trichotomy law, $a - b > 0$ so that $a - b = x > 0$. Using Archimedean property to numbers 1 and x, we see that there must exist $m \in \mathbf{N}$ such that $mx > 1$ so that $x > 1/m$. Hence, we have $a - b > 1/m$ for some $m \in \mathbf{N}$. But this contradicts the given hypothesis, namely $a \le b + 1/n$ for $n \in \mathbf{N}$. Hence we cannot assume that $a > b$. So $a \le b$.]

6. Prove that for all real numbers x and y,

$$\max\{x, y\} = \frac{1}{2}(x + y + |x - y|).$$ [Delhi 2008]

7. State Archimedean property of real numbers. Use it to prove that if for any $\varepsilon > 0$, $a < b + \varepsilon$, then $a \le b$. [Delhi B.Sc. I (Prog.) 2010]

8. Prove that the set of rational numbers is an ordered field but not a complete ordered field.

Sol. Refer Art. 2.3 and Art. 2.6. [Delhi B.Sc. (Prog) III 2013]

9. If $S = \{2/n : n \in \mathbf{N}\}$, find $\inf S$ and $\sup S$. [Delhi B.Sc. (Hons) I 2014]

Ans. $\inf S = 0$ and $\sup S = 2$

10. If $a = -3$ and $S = \{x \in \mathbf{R} : |x| < 2\}$, check whether $\sup (aS) = a \sup (S)$ holds.

[Delhi B.Sc. (Prog) II 2015]

Sol. Here $S = \{x : |x| < 2\} = \{x : -2 < x < 2\}$. Hence by the definition of supremum of S, we have $\sup (S) = 2$ and hence

$$a \sup (S) = (-3) \times 2 = -6 \qquad \ldots(1)$$

Now, from part (ii) of Ex. [illegible], page 2.21, we have

$$\sup (aS) = a \inf (S), \text{ if } a < 0 \qquad \ldots(2)$$

Here $a = -3 < 0$ and $\inf (S) = -2$, by definition. Hence (2) reduces to

$$\sup (aS) = (-3) \times (-2) = 6 \qquad \ldots(3)$$

From (1) and (3), we have $\sup (aS) \ne a \sup (S)$.

Chapter

3

Neighbourhoods and Limit Points of a Set. Open and Closed Sets

3.1. INTRODUCTION

In this chapter, we shall be connected with the important notion of limit points of a set of real numbers and shall prove the *Bolzano-Weierstrass theorem* which lays down a sufficient condition for the existence of the limit points of a set.

We shall also deal briefly with the definition and properties of open and closed sets of real numbers.

A study of the notion of *limits points* of a set necessitates the consideration of the *neighbourhoods* of a point which we shall now introduce.

In what follows, we shall deal only with real numbers and sets of real numbers unless otherwise stated.

3.2. NEIGHBOURHOOD OF A POINT

[*Kanpur 2008 Delhi B.A. (Prog.) III, 2014, 2015; Delhi B.Sc. (Hons.) I 2007, 08*]

Definition. A set N is called a neighbourhood of a point p, if there exists an open interval I containing p and contained in the set, *i.e.*, $p \in I \in N$.

It follows that an open interval is a neighbourhood of each of its points. While open intervals containing a point are not the only neighbourhoods of the point, *such neighbourhoods as are open intervals prove adequate in relation to our use of neighbourhoods*.

It may be remarked that the open interval $]p - \varepsilon, p + \varepsilon[$ is a neighbourhood of p for every positive number ε. This nbd of p is called the ε-nbd of p and is denoted by $V_\varepsilon(p)$. Thus in symbols, we have $V_\varepsilon(p) = \{x \in \mathbf{R}: |x - p| < \varepsilon\} = (p - \varepsilon, p + \varepsilon)$

Deleted neighbourhood of a point. Definition. **[*Kanpur 2010*]**

If from the neighbourhood of a point, the point itself is excluded or deleted, we obtain the so called *deleted neighbourhood* of that point.

Thus, if N is a neighbourhood of a point p, then the set $N - \{p\}$ is a deleted neighbourhood of p.

Deleted symmetrical neighbourhood of p is given by

$$]p - \varepsilon, p + \varepsilon[- \{p\} =]p - \varepsilon, p[\cup]p, p + \varepsilon[= \{x : 0 < |x - p| < \varepsilon\}.$$

In what follows, we shall use the abbreviated form 'nbd' for the word neighbourhood.

ILLUSTRATIONS

(*i*) *Any open interval* $]a, b[$ *is a nbd of each of its points.* **[*Delhi B.A. (Prog.) III 2012, 15*]**

Let x be an arbitrary point of the given open interval $]a, b[$. Since every set is a sub-set of itself, we have

$$x \in]a, b[\subset]a, b[,$$

showing that $]a, b[$ is a nbd of x. But x is an arbitrary point of $]a, b[$. Hence $]a, b[$ is a nbd of each of its points.

(*ii*) *A closed set* $[a, b]$ *is a nbd of each of its points except the two end points a and b.* [*Delhi B.A. (Prog.) III 2012*]

Let x be an arbitrary point of the open interval $]a, b[$. Then

$$x \in]a, b[\subset [a, b],$$

showing that $[a, b]$ is a nbd of each point of $]a, b[$.

Now $[a, b]$ cannot be a nbd of a because to be a nbd of a, it must contain an open interval $[a - \varepsilon, a + \varepsilon[$, *i.e.*, $[a, b]$ has to contain points less than a, which it does not. Similarly, $[a, b]$ cannot be a nbd of b because to be a nbd of b, it must contain an open interval $]b - \varepsilon, b + \varepsilon[$ *i.e.*, $[a, b]$ has to contain points greater than b, which it does not.

Thus, $[a, b]$ is a nbd of each of its points except the two end points a and b.

(*iii*) $[a, b[$ *is a nbd of each of its points except a and* $]a, b]$ *is a nbd of each of its points except b.*

(*iv*) *A non-empty finite set cannot be a nbd of any of its points.* **(*Meerut*, 2003)**

Let S be any non-empty finite set and let x be an arbitrary point of S. Then S will be a nbd of x if there exists $\varepsilon > 0$, such that $x \in]x - \varepsilon, x + \varepsilon[\subset S$. But the open interval $]x - \varepsilon, x + \varepsilon[$ is an infinite set and so it cannot be a sub-set of S. Hence condition $]x - \varepsilon, x + \varepsilon[\subset S$ cannot be satisfied for any $\varepsilon > 0$.

Thus S is not a nbd of x. Since x is an arbitrary point of S, so S is not a nbd of any of its points.

Note. It is worth noting that each nbd of a point is an infinite set but every infinite set need not be a nbd.

(*v*) *The set* **N** *of natural numbers is not a nbd of any of its points.* **(*Delhi Maths* 2007; *Meerut*, 2001, 02)**

Let n be an arbitrary natural number. Then for any $\varepsilon > 0$, $n - \varepsilon$ and $n + \varepsilon$ are two distinct real numbers and we know that between any two distinct real numbers there lie infinite real numbers which are not members of **N.** Thus, we cannot find $\varepsilon > 0$, such that

$$n \in]n - \varepsilon, n + \varepsilon[\subset \mathbf{N},$$

showing that **N** is not a nbd of n. Since n is an arbitrary point of **N,** hence **N** is not a nbd of any of its points.

(*vi*) *The set* **I** (*or* **Z**) *of all integers is not a nbd of any of its points.* **[Delhi 2006, 2015]**

(*vii*) *The set* **Q** *of rational numbers is not a nbd of any of its points.*

Let q be any arbitrary rational number. Then for any $\varepsilon > 0$, $q - \varepsilon$ and $q + \varepsilon$ are two distinct real numbers and we know that between two distinct real numbers there lie infinite irrational numbers which are not members of **Q.** Thus, we cannot find $\varepsilon > 0$, such that

$$q \in]q - \varepsilon, q + \varepsilon[\subset \mathbf{Q},$$

showing that **Q** is not a nbd of q. Since q is an arbitrary point of **Q,** hence **Q** is not a nbd of any of its points.

(*viii*) *The set* **Q'** *of all irrational numbers is not a nbd of any of its points.*

(*ix*) *The set* **R** *of real numbers is a nbd of each of its points.* **[*Kanpur* 2008]**

Let x be any real number. Then $\forall\ \varepsilon > 0$, we have

$$x \in]x - \varepsilon, x + \varepsilon[\subset \mathbf{R},$$

showing that **R** is a nbd of **R.** Since x is any point of **R,** so **R** is a nbd of each of its points.

(*x*) *Any set N cannot be a nbd of any point of* **R** – *N.* **[*Delhi Maths* (*H*), 2003]**

Let x be any arbitrary point of **R** – N. Then, $x \notin N$. If possible, let N be a nbd of x. Then, by definition there exists some $\varepsilon > 0$ such that

$$x \in \,]x - \varepsilon, x + \varepsilon[\, \subset N,$$

showing that $x \in N$. This contradicts the fact that $x \notin N$. Hence N cannot be a nbd of $\mathbf{R} - N$. Since x is an arbitrary point of $\mathbf{R} - N$, it follows that N cannot be a nbd of any point of $\mathbf{R} - N$.

3.3. PROPERTIES OF NEIGHBOURHOODS

Theorem I. *If M is a neighbourbood of a point x and $N \supset M$, then N is also a neighbourhood of the point x.* **[Delhi Maths (G) 2006; Delhi Math (H), 1997, 2008]**

Proof. Since M is a neighbourhood of point x, therefore, there exists an open interval $]x - \varepsilon, x + \varepsilon[$, $\varepsilon > 0$ such that

$$x \in \,]x - \varepsilon, x + \varepsilon[\, \subset M$$
$$\Rightarrow \; x \in \,]x - \varepsilon, x + \varepsilon[\, \subset M \subseteq N$$
$$\Rightarrow \; x \in \,]x - \varepsilon, x + \varepsilon[\, \subset N$$

Hence, N is a neighbourhood of x.

Theorem II. (*i*) *If M and N are neighbourhoods of a point x, then $M \cap N$ is also a neighbourhood of x.* **[Delhi B.Sc. III 2008; Delhi B.A. (Prog.) III 2014]**

(*ii*) *Intersection of a finite number of neighbourhoods of a point is also a nbd of the point.* **[Kanpur 2014; Delhi Maths (P), 1997; Delhi Maths (G), 2001, 02]**

Proof. (*i*) Since M and N are neighbourhoods of x, therefore, there exist $\varepsilon_1 > 0$ and $\varepsilon_2 > 0$ such that

$$x \in \,]x - \varepsilon_1, x + \varepsilon_1[\, \subset M,$$

and
$$x \in \,]x - \varepsilon_2, x + \varepsilon_2[\, \subset N. \quad ...(1)$$

Let $\varepsilon = \min \{\varepsilon_1, \varepsilon_2\}$ so that $\varepsilon \le \varepsilon_1$ and $\varepsilon \le \varepsilon_2$. Thus

$$x \in \,]x - \varepsilon, x + \varepsilon[\, \subset \,]x - \varepsilon_1, x + \varepsilon_1[\, \subset M$$
$$x \in \,]x - \varepsilon, x + \varepsilon[\, \subset \,]x - \varepsilon_2, x + \varepsilon_2[\, \subset N$$
$$\therefore\; x \in \,]x - \varepsilon, x + \varepsilon[\, \subset M \cap N$$

Hence, $M \cap N$ is a neighbourhood of x.

Proof. (*ii*) Left as an exercise.

Note. The conclusion of the above theorem II may not be true when arbitrary collection of neigbourhoods of a point is taken in place of a finite collection of neighbourhood as shown below.

Let $\quad I_n = \,]-1/n, 1/n[, \; \forall \; n \in N.$

Then $\forall \; n \in N, I_n$ is a nbd of 0. Also, we have

$$\bigcap_{n=1}^{\infty} I_n = \{0\}$$

But $\{0\}$ is not a nbd of 0. Hence, we have proved that the intersection of an arbitrary collection of neighbourhoods of a point may not be a nbd of that point. **[Delhi B.A. (Prog.) III, 2014]**

Theorem III. *A set N is a nbd of the point p if and only if there exists a positive integer n such that*

$$\left(p - \frac{1}{n}, \; p + \frac{1}{n}\right) \subset N$$ **[Avadh, 1996; Delhi Maths (H), 2003; Meerut, 1996]**

Proof. Let the set N be a nbd of a point p. Then there exists $\varepsilon > 0$ such that

$$p \in (p - \varepsilon, p + \varepsilon) \subset N \quad ...(1)$$

Now, $\varepsilon > 0 \Rightarrow 1/\varepsilon > 0$. Also $1/\varepsilon$ is a real number. Since it is always possible to find a positive integer greater than any real number (Archimedean property), we can choose a positive integer n such that

$$n > \frac{1}{\varepsilon} \text{ so that } \frac{1}{n} < \varepsilon \text{ and } -\varepsilon < -\frac{1}{n}$$

$$\Rightarrow \quad p + \frac{1}{n} < p + \varepsilon \text{ and } p - \varepsilon < p - \frac{1}{n}$$

$$\Rightarrow \quad \left(p - \frac{1}{n}, p + \frac{1}{n}\right) \subset (p - \varepsilon, p + \varepsilon) \subset N, \text{ by (1)}$$

So if N is a nbd of p then there exists a +ve integer n such that

$$\left(p - \frac{1}{n}, p + \frac{1}{n}\right) \subset N$$

Conversely, let
$$\left(p - \frac{1}{n}, p + \frac{1}{n}\right) \subset N, \quad \ldots(2)$$

where n is a positive integer.

Since $\left(p - \frac{1}{n}, p + \frac{1}{n}\right)$ is an open interval containing p, it follows from (2) and definition of nbd that N is a nbd of p.

EXERCISES

1. Give example of each of the following :
   (*i*) a set which is not a nbd of any of its points.
   (*ii*) a set which is a nbd of each of its points with the exception of one point.
   (*iii*) a set which is a nbd of each of its points with the exception of two points.
   (*iv*) a set which is a nbd of each of its points.
   (*v*) a set which is not an interval but is a nbd of each of its points.
2. Show that the intersection of the family of all neighbourhoods of a point x is $\{x\}$.
3. Let $I_n = \left(-\frac{1}{n}, 1 + \frac{1}{n}\right)$ be an open interval such that for each $n \in \mathbf{N}$, the set of natural numbers. Show that $\bigcap_{n=1}^{\infty} I_n$ is a nbd of each of its points with the exception of two points.
4. Show that every point has an infinite number of neighbourhoods.
5. If x and y are any two distinct numbers, then show that there exist neighbourhoods of x and y which are disjoint. (This is known as *Haousdroff property*.)
6. Let x denote any point of intersection M of the neighbourhoods N_1 and N_2 of x_1 and x_2. Then show that there exists a nbd of x which is entirely contained in M.
7. Let x be any point of the nbd N of y. Then show that there exists a nbd of x which is entirely contained in N.

ANSWERS

1. (*i*) Any non-empty finite set (*ii*) $[a, b[$ is a nbd of each of its points except b (*iii*) $[a, b]$ is a nbd of each of its points except a and b (*iv*) $]a, b[$ (*v*) $]1, 2[\cup]3, 4[$

3.4. LIMIT (OR ACCUMALATION OR CONDENSATION OR CLUSTER POINT) POINT OF A SET

[*Delhi B.A (Prog) III 2011, 12, 13, 14, 15, 17; Delhi B.Sc. (Hons.) I, 2016; Purvanchal 2006; Kanpur 2014; Delhi B.Sc. (Hons) I 2011; Delhi B.Sc. (Prog) II 2009, 10, 17*]

Definition. A number p is a *limit point* of set S of real numbers if every neighbourhood of p contains a point of S different from p.

A limit point of a set is also sometimes described as an *accumulation point* or *condensation point* or *cluster point of the set.*

Theorem. *A point p is a limit point of a set S if and only if every neighbourhood of p contains infinitely many points of S.*

[*Delhi B.Sc. (Prog.) III 2012; Delhi Maths (H), 2002; Avadh, 1994; Meerut, 1994*]

Proof. The condition is necessary. Let p be a limit point of S. Then we wish to prove that every nbd of p contains infinitely many points of S. Let, if possible, every nbd of p contains a finite number of points $p_1, p_2, \ldots, p_n$ different from p.

Let $\varepsilon = \min \{|p - p_1|, |p - p_2|, \ldots, |p - p_n|\} > 0$.

Then $]p - \varepsilon, p + \varepsilon[$ is a nbd of p which contains no point of S other than p and so p is not a nbd of S by definition of limit point. Thus, we arrive at a contradiction. Hence our supposition that every nbd of p contains only finite number of points of S is wrong. Hence every nbd of p must contain infinitely many points of S.

The condition is sufficient. Suppose every nbd of p contains infinitely many points of S. Then obviously every nbd of p contains a point of S different from p and hence p is a limit point of S.

Another definition of limit point. In view of the above theorem, an alternative definition of limit point can be given as follows :

A point p is a limit point of a set S if every neighbourhood of p contains infinitely many points of S.

Note 1. p is a limit point of a set S if and only if given any nbd N of p, the set $N \cap S$ is an infinite set.

Note 2. Symbolically, a point p is a limit point of a set S if for each nbd N of p,

$$(N \cap S) - \{p\} \neq \phi$$

Note 3. In light of the above mentioned two definitions of limit point of a set, the following three rules can be used to show that p is not a limit of a set S.

Rule I. p is not a limit point of a set S if there exists at least one nbd of p containing at the most of finite number of members of S.

Rule II. p is not a limit point of a set S if there exists at least one nbd of p containing no point of S other than p.

Rule III. A point p is not a limit point of a set S, if there exists at least one nbd N of p such that

$$N \cap S = \{p\} \quad \text{or} \quad N \cap S = \phi.$$

Note 4. A limit point of a set may or may not be a member of the set. Again a set may have no limit point, a unique limit point or a finite or infinite number of limit points.

Derived set of a set. Definition. ***(Ponchi 2010)***

The set of all the limit points of a set A is known as its *derived set* and is denoted by A' or $D(A)$.

Note. The derived set of A' is known as the second derived set of A and is denoted by A'' or $D^2(A)$. In general, the nth derived set of A is denoted by $A^{(n)}$ or $D^{(n)}(A)$.

A set is said to be of *first species* if it has only a finite number of derived sets. It is said to be of *second species* if the number of its derived sets is infinite.

ILLUSTRATIONS

1. *Every point of the set* **R** *of real numbers is a limit point and hence the derived set of* **R** *is* **R**, *i.e.*, **R′** = **R**. **[Delhi B.A. (Prog.) III 2012]**

Let p be any real number. Then for each $\varepsilon > 0$, the nbd $]p - \varepsilon, p + \varepsilon[$ contains infinitely many real numbers. Hence p is a limit point of **R**. Since p is an arbitrary real number, so every point of **R** is a limit point and the derived set of **R** = **R′** = **R**.

2. *The limit points of the set* **Q** *of all rational numbers and the derived set of* **Q** *is* **R**, *i.e.*, **Q′** = **R**. **[Delhi B.A. (Prog.) III 2012, 2015]**

Let p be any real number and let $\varepsilon > 0$ be given. Then $p - \varepsilon$ and $p + \varepsilon$ are two different real numbers and hence there exist infinitely many rational numbers between them. So for each $\varepsilon > 0$, nbd $]p - \varepsilon, p + \varepsilon[$ contains infinitely many rational numbers. Hence p is a limit point of **Q.** Since p is an arbitrary real number, so every point of **R** is a limit point of **Q** and the derived set of **Q** = **Q′** = **R.**

3. *The limit points of the set* **R** – **Q** *of all irrational numbers and the derived set of* **R** – **Q** *is* **R**. **(Delhi B.A. (Prog.) III, 2010; Meerut, 1996)**

Proceed as in illustration 2 to show that every real number is a limit point of the set of irrational numbers and so its derived set is **R.**

4. *The set* **Z** *of integers has no limit point so that the derived set of* **Z** = **Z′** = ϕ *(empty set).* **[Delhi B.Sc. (Hons) I 2011; Delhi B.A. (Prog.), III, 2014, 15; Delhi Maths (G), 2003, 07]**

The following two cases arise :

(*i*) Let p be any integer. Then the nbd $]p - 1/2, p + 1/2[$ contains no point of **Z** other than p. Hence p cannot be a limit point of **Z.** Since p is any integer, so no integer is a limit point of **Z.**

(*ii*) Let p be any real number and $p \notin$ **Z.** Then, there exist $m \in$ **Z** such that $m < p < m + 1$. Hence there exists a nbd $]m, m + 1[$ containing no point of **Z.** So p cannot be a limit point of **Z.**

Hence, in view of cases (*i*) and (*ii*), we see that **Z** has no limit point and so derived set of **Z** is the empty set.

5. *The set* **N** *of natural numbers has no limit point so that the derived set of* **N** *is the empty set.* **[Delhi B.A. (Prog.) III 2012]**

6. *A finite set has no limit points.* **[Agra 2010; Kanpur 2009; Delhi 2009]**

Let S be a finite set. Let p be any real number. Then for $\varepsilon > 0$, there exists a nbd $]p - \varepsilon, p + \varepsilon[$ containing only finite number of members of the given finite set S. Hence p is not a limit point of S. Since p is any real number, it follows that the finite set S has no limit points and its derived set = $S' = \phi$.

7. *The set of limit points of* (*i*) $]a, b[$ (*ii*) $[a, b]$ (*iii*) $]a, b]$ (*iv*) $[a, b[$, *where* $a, b \in$ **R.** *Also find derived set for these intervals.*

(*i*) Let p be any real number.

If $p < a$, then for $0 < \varepsilon < a - p$, there exists a nbd $]p - \varepsilon, p + \varepsilon[$ containing no point of $]a, b[$. Hence any real number p ($< a$) cannot be a limit point of $]a, b[$.

If $p > b$, then for $0 < \varepsilon < p - b$, there exists a nbd $]p - \varepsilon, p + \varepsilon[$ containing no point of $]a, b[$. Hence any real number p ($> b$) cannot be a limit point of $]a, b[$.

Finally, if $p \in [a, b]$, then for any $\varepsilon > 0$, $]p - \varepsilon, p + \varepsilon[$ is a nbd of p such that

$$]p - \varepsilon, p + \varepsilon[\cap]a, b[=]c, d[,$$

where $c = \max \{p - \varepsilon, a\}$ and $d = \min \{p + \varepsilon, b\}$.

Then, since $]c, d[$ contains infinitely many points of $]a, b[$, it follows that p is a limit point of $]a, b[$. Since p is any point of $[a, b]$ so each point of $[a, b]$ is a limit point of $]a, b[$.

Thus, the set of all limit points of $]a, b[$ is $[a, b]$, *i.e.*, the derived set of $]a, b[$ is $[a, b]$.

(*ii*) Left as an exercise. **Ans.** $[a, b]$

(*iii*) Left as an exercise. **Ans.** $[a, b]$

(*iv*) Left as an exercise. **Ans.** $[a, b]$

8. *The set $S = \{1/n : n \in N\}$ has only one limit point, namely 0. Thus, derived set of $S = S' = \{0\}$.* **[Agra, 2001; Delhi B.Sc. (Prog.) II, 2016; Meerut, 2000, 01]**

The following six cases arise :

(*i*) For $\varepsilon > 0$, $]-\varepsilon, \varepsilon[$ is a nbd of 0.

By Archimedean property of real numbers, for each $\varepsilon > 0$, there exist $m \in \mathbf{N}$ such that $m > 1/\varepsilon$.

Now, $$m > \frac{1}{\varepsilon} \Rightarrow \frac{1}{m} < \varepsilon \Rightarrow -\varepsilon < 0 < \frac{1}{m} < \varepsilon \Rightarrow \frac{1}{m} \in]-\varepsilon, \varepsilon[,$$
showing that each nbd $]-\varepsilon, \varepsilon[$ of 0 contains a point $1/m$ of S other than 0. Hence 0 is a limit point of S.

(*ii*) Let $p < 0$. Then there exists a nbd $]-\infty\ 0[$ of p such that $]-\infty, 0[\cap S = \phi$. Hence $p\ (< 0)$ is not a limit point of S.

(*iii*) Let $p > 1$. Then there exists a nbd $]1, \infty[$ of p such that $]1, \infty[\cap S = \phi$. Hence $p\ (> 1)$ is not a limit point of S.

(*iv*) Let $0 < p < 1$ such that $p \notin S$. Then $1/p > 0$. Hence there exists a unique $m \in \mathbf{N}$ such that
$$m < \frac{1}{p} < m+1 \Rightarrow \frac{1}{m+1} < p < \frac{1}{m},$$
showing that there exists a nbd $\left]\frac{1}{m+1}, \frac{1}{m}\right[$ of p which contains no point of S. Hence p is not a limit point of S.

(*v*) Let $p = 1$. Then there exists a nbd $]1/2, \infty[$ of 1 which contains no point of S other than 1. Hence 1 is not a limit point of S.

(*vi*) Let $p \neq 1$ and $p \in S$. Let $p = 1/m$, where $m \in \mathbf{N}$ and $m \neq 1$. Then there exists a nbd $\left]\frac{1}{m+1}, \frac{1}{m-1}\right[$ of p which contains no point of S other than p. Hence p is not a limit point of S.

Therefore, in view of the cases (*i*) to (*vi*), it follows that no real number other than 0, is a limit point of S. Thus, the derived set of $S = S' = \{0\}$.

9. *Examples of derived sets of first and second species.*

Consider $S = \{1/n : n \in N\}$. In illustration 9 we have shown that $S' = \{0\}$. Again S'' = derived set of $S' = \phi$.

Hence the set S is of the first species.

Consider $T = [a, b]$. In illustration 7 (*ii*), we have shown that $T' = [a, b] = T$. So $T'' = T' = T = [a, b]$ and so on. Thus $T^{(n)} = [a, b]$ for every positive integer n. Hence $[a, b]$ is a set of second species.

10. *If ϕ be the empty set, then its derived set is also ϕ.*

Let p be any point of $\mathbf{R}$. Then $\mathbf{R}$ is a nbd of p containing no point of ϕ because ϕ is an empty set. Hence p is not a limit point of ϕ. Thus no point of $\mathbf{R}$ is a limit point of ϕ and so derived set of $\phi = \phi$.

3.5. EXISTENCE OF LIMIT POINTS

A finite set does not have a limit point and an infinite set may or may not have a limit point. For example, the infinite set **I** of integers has no limit point but the infinite set **Q** has. We shall now have a following theorem which gives sufficient conditions for an infinite set to possess a limit point.

Theorem I. **Bolzano-Weierstrass theorem.* ***[Delhi B.Sc. III, 2008, 09, 10, 12; I.A.S. 2009]***

Every bounded and infinite set has a limit point. ***(Rajasthan 2010)***

[Delhi B.A. (Prog) III, 2011, 13, 15; Purvanchal 2006; M.S. Univ. T.N. 2006; Agra, 2002; Meerut, 1998; Kanpur 2011, 13; Delhi Maths (H), 2001, 03, 04, 06, 08, 09]

Proof. We shall need to use the property of *Order-completeness* of the set of real numbers for the proof of the theorem.

Let A be an infinite and bounded set. There exists an interval $[k, K]$ such that

$$A \subset [k, K].$$

We define a set S as follows :

$$x \in S$$

if and only if it exceeds at the most a finite number of members of the set A.

Thus, while $k \in S$ and $K \notin S$.

Now $S = \{x : x$ exceeds at the most of finite number of members of $A\}$.

The set S is *not* void, for $k \in S$.

Also S is bounded above in as much as K is an upper bound of the same.

Let ξ be the least upper bound of S. Surely it exists by the order-completeness of **R**. We shall show that ξ is a limit point of A.

Consider a neighbourhood, say B of ξ. There exists an open interval $]a, b[$ such that

$$\xi \in\ ]a, b[\ \subset B.$$

Now the number a which is less than the least upper bound ξ of the set *is not* an upper bound of S. Thus, there exists a number, say η, of S such that

$$a < \eta \leq \xi, \eta \in S.$$

Also η, being a member of S, exceeds at the most a finite number of members of A. It follows that the number a also exceeds at the most a finite number of members of A.

Again the number b which is greater than ξ is an upper bound of S without being a member of S. Thus, *b must exceed an infinite number of members of A*.

It follows that

(*i*) a exceeds at the most a finite number of members of S.

(*ii*) b exceeds an infinite number of members of S.

Thus, $]a, b[$ contains an infinite number of members of A so that ξ is a limit point of the set S.

Hence the theorem.

Note 1. The condition of 'boundedness' of the above theorem I is only sufficient condition for the existence of a limit point of an infinite set A. This condition is not necessary for a set A may have a limit point. Even an infinite unbounded set may have a limit point. For examples, the set $A = \{1/2, 2, 1/3, 3, 1/4, 4, 1/5, 5, ...\}$ is infinite and unbounded and has the limit point 0. Similarly, the set **Q** of all rational numbers is an infinite and unbounded set and still every real number is a limit point of **Q**.

* Bolzano (1778-1848) and Weierstrass (1815-94) were German mathematicians.

Note 2. *Examples to show that the conditions of the above theorem cannot be relaxed.* **[*Delhi B.A. (Prog III) 2013, 15; Delhi B.Sc. (Prog.) III 2012; Delhi Maths (H), 2003, 08, 09*]**

Condition for the set to be infinite cannot be dropped because a finite set has no limit point.

Condition for the set to be bounded cannot be dropped because the set **N** of natural numbers is an infinite unbounded set and it has no limit point. Similarly, the set **Z** of integers is also infinite and unbounded set and it has no limit point.

Example. *State Bolzano-Weierstrass theorem. Prove that the set* $S = \{3^n + 1/3^n : n \in \mathbf{N}\}$ *has no limit point. Does it contradict Bolzano-Weierstrass theorem ?* **[*Delhi Maths (P), 1999*]**

Solution. For statement of theorem, refer Art. 3.5. Clearly the given set S is infinite and unbounded. It can be easily verified that S has no limit point. The absence of a limit point does not contradict Bolzano-Weierstrass theorem because here S is infinite as well as unbounded whereas the result of the theorem holds if it is bounded and infinite.

Theorem II. *If A and B are sets of real numbers, then*

(*i*) $A \subset B \Rightarrow A' \subset B'$ **[*Kanpur 2006*]**

(*ii*) $(A \cup B)' = A' \cup B'$ **(*Agra, 2002; Avadh, 1993*)**

(*iii*) $(A \cap B)' \subset A' \cap B'$ **(*Kanpur 2007*)**

Give an example to show that $(A \cap B)'$ *and* $A' \cap B'$ *may not be equal.* **[*Delhi Maths (G), 1997*]**

Solution. (*i*) Let $A \subset B$. We shall prove that $A' \subset B'$.

Let $x \in A'$ so that x is a limit point of A. Thus, for each $\varepsilon > 0$, $]x - \varepsilon, x + \varepsilon[$ contains a point $x_1 \in A$ such that $x_1 \neq x$. Since $A \subset B$, $x_1 \in A \Rightarrow x_1 \in B$.

Consequently, $]x - \varepsilon, x + \varepsilon[$ contains a point $x_1 \in B$, $x_1 \neq x$, and so x is a limit point of B, *i.e.*, $x \in B'$.

$\therefore \quad x \in A' \Rightarrow x \in B'$. Hence, $A' \subset B'$.

(*ii*) We know that $A \subset A \cup B$

$\Rightarrow A' \subset (A \cup B)'$, by part (*i*)

Similarly, $B \subset A \cup B \Rightarrow B' \subset (A \cup B)'$

$$\therefore \quad A' \cup B' \subset (A \cup B)' \quad \text{...(1)}$$

Now, we shall prove that $(A \cup B)' \subset A' \cup B'$ ...(2)

Let $x \in (A \cup B)'$

$\Rightarrow$ x is a limit point of $A \cup B$

$\Rightarrow$ every neighbourhood of x contains a point $y \in A \cup B, y \neq x$

$\Rightarrow$ every neighbourhood of x contains a point $(y \in A$ or $y \in B), y \neq x$

$\Rightarrow$ every neighbourhood of x contains a point $y \in A, y \neq x$,

or every neighbourhood of x contains a point $y \in B, y \neq x$.

$\Rightarrow$ x is a limit point of A or x is a limit point of B.

$\Rightarrow x \in A'$ or $x \in B'$

$\Rightarrow x \in A' \cup B'$.

From (1) and (2), we have

$$(A \cup B)' = A' \cup B'.$$

(*iii*) We know that $A \cap B \subset A$

$\Rightarrow (A \cap B)' \subset A'$, by part (*i*) ..(3)

and $A \cap B \subset B \Rightarrow (A \cap B)' \subset B'$ ...(4)

$\therefore$ From (3) and (4), $(A \cap B)' \subset A' \cap B'$

To prove that the equality does not hold, we take

$$A =]1, 2[\text{ and } B =]2, 3[\text{ so that } A \cap B = \phi$$

$$\Rightarrow A' = [1, 2], B' = [2, 3] \text{ and } (A \cap B)' = \phi' = \phi$$

Now $\quad A' \cap B' = \{2\}$ and $(A \cap B)' = \phi$

Thus, $\quad (A \cap B)' \neq A' \cap B'$

Theorem III. (*a*) *If a non-empty sub-set S of* **R** *which is bounded above has no maximum member, then show that its supremum is a limit point of the set S.*

(*b*) *If a non-empty sub-set S of* **R** *which is bounded below has no minimum member, then show that its infimum is a limit point of the set S.*

Solution. (*a*) Since S is bounded above, so by the order-completeness property of real numbers, S has a supremum (*i.e.*, least upper bound) in **R**.

Let $\quad \sup S = b$.

Since S has no maximum member, so $b \notin S$.

Let $\varepsilon > 0$. Since $\sup S = b$, so $b - \varepsilon$ cannot be an upper bound for S and hence there exists $x \in S$ such that $x > b - \varepsilon$.

Again $\sup S = b$ and $b + \varepsilon > b$, so $b + \varepsilon$ is also an upper bound for S. Thus, $x \in S \Rightarrow x < b + \varepsilon$.

Hence for each $\varepsilon > 0$, there exists $x \in S$ such that $b - \varepsilon < x < b + \varepsilon$.

Also, $b \notin S \Rightarrow x \neq b$.

Hence for each $\varepsilon > 0$, nbd $]b - \varepsilon, b + \varepsilon[$ of b contains a point x of S which is different from b. Hence b is a limit point of S, *i.e.*, $\sup S$ = limit point of S.

(*b*) Left as an exercise.

Theorem IV. *Show that the derived set of any bounded set is also a bounded set.*

Solution. Let S be a bounded set. So there exist $h, k \in$ **R** such $S \subset [h, k]$.

In what follows, we shall prove that no element of S' (*i.e.*, no limit point of S) is less than h or greater than k.

If $p > k$, then for $\varepsilon = x - k > 0$, $]p - \varepsilon, p + \varepsilon[$ is a nbd of p containing no element of $[h, k]$ and hence containing no element of S {as $S \subset [h, k]$}. So $p \notin S'$. ...(1)

Similarly, if $p < h$, then we can show that $p \notin S'$.

Thus, we have shown that

$$x \notin [h, k] \Rightarrow x \notin S'$$

$\Rightarrow$ all the limit points of S lie in $[h, k]$

$\Rightarrow S' \subset [h, k]$

$\Rightarrow$ Derived set S' of S is bounded.

Theorem V. *Every infinite bounded set has the greatest and the smallest limit points, i.e., the derived set of an infinite bounded set attains its bounds.*

Or

Let S be a bounded infinite non-empty sub-set of **R**. *Then the derived set S has the smallest and the greatest members.* **[Delhi Maths (G), 1999]**

Solution. Let S be an infinite bounded set. Since S is bounded, so there exists $h, k \in$ **R** such that $S \subset [h, k]$.

Now, $S \subset [h, k] \Rightarrow$ derived set of $S \subset$ derived set of $[h, k]$

But, derived set of $[h, k] = [h, k]$.

Hence, $S' \subset [h, k]$ and so S' is bounded.

Since S is an infinite bounded set, so by Bolzano-Weierstrass theorem, S has a limit so that $S' \neq \phi$.

Hence S' is non-empty and bounded and so by order-completeness property of **R**, S' has both infimum and supremum.

Let $\inf S' = a$ and $\sup S' = b$.

In what follows, we shall prove that both a and b belong to S', *i.e.*, both a and b are limit points of S.

Now, $a = \inf S'$

$\Rightarrow$ for any $\varepsilon > 0$, there exists some $x \in S'$ such that $a \leq x < a + \varepsilon$

$\Rightarrow a - \varepsilon < a \leq x < a + \varepsilon$

$\Rightarrow]a - \varepsilon, a + \varepsilon[$ is a nbd of $x \in S'$

$\Rightarrow]a - \varepsilon, a + \varepsilon[$ is a nbd of a limit point of S

$\Rightarrow]a - \varepsilon, a + \varepsilon[$ contains infinitely many points of S

$\Rightarrow$ every nbd of a contains infinitely many points of S

$\Rightarrow a$ is a limit point of S

Similarly, we can prove that b is a limit point of S.

Thus both a and b, *i.e.*, $\inf S'$ and $\sup S'$ belong to S'. Hence a is the smallest and b the greatest member of S'. In other words, the set S has smallest and greatest limit points.

Note. From the above theorem V, it follows that the smallest and the greatest members a and b of the derived set S' of an infinite and bounded set S always exist. They are usually denoted by $\underline{\lim}\, S$ and $\overline{\lim}\, S$ respectively and are known as the *inferior* (*or lower*) *limit* of S and the *superior* (*or upper*) *limit* of S.

EXERCISES

1. Show that a necessary condition for a set to have a limit point is that it is an infinite set. Give an example to show that this condition is not sufficient.

2. Show that ξ is a limit point of a set A if and only if every neighbourhood of ξ contains at least one point of A other than ξ.

3. Show that every limit point of a sub-set B of A is as well as limit point of A.

4. Obtain the derived set of the set $\{1/m + 1/n : m, n \in \mathbf{N}\}$. **(Bundelkhand, 1996)**

5. Give an example of a set whose derived set is

(*i*) void, (*ii*) sub-set of the given set,

(*iii*) super-set of the given set, (*iv*) neither a sub-set nor a super-set of the given set,

(*v*) same as the given set.

6. Prove that the set of a real numbers is the derived set of the set of rational numbers.

7. Prove that the set of a real numbers is the derived set of the set of irrational numbers.

8. If $A, B \subseteq \mathbf{R}$, show that $x \in \mathbf{R}$, is a limit point of $A \cup B$ if and only if x is a limit point of A or x is a limit point of B. **[Delhi Maths (P), 1999]**

9. Show that if $x \in \mathbf{R}$ is a limit point of $A \cap B$, then x is a limit point of A and x is a limit point of B. Is the converse true ? Justify your answer. **[Delhi Maths (P), 1999]**

10. If x is a limit point of a set $S \subset \mathbf{R}$, is x a limit point of $S - \{x\}$? Justify your answer. **[Delhi Maths (H), 1999]**

11. Find the limit points of the following sets :

(*i*) $S = \left\{\dfrac{n}{n+1} : n \in \mathbf{N}\right\}$. **(Meerut, 2002)**

(*ii*) $]0, 1[$. **[Delhi Maths (G), 2000; Meerut, 2000]**

12. Prove that $x_0 \in \mathbf{R}$ is a limit point of the set $S \subset \mathbf{R}$ if and only if for every nbd N of x_0,

$$(S \cap N) - \{x_0\} \neq \phi.$$ **(Avadh, 1995)**

13. Find the limit points of the set

$$S = \left\{1, -1, 1\frac{1}{2}, -1\frac{1}{2}, 1\frac{1}{3}, -1\frac{1}{3}, \ldots\right\}.$$ **(Meerut, 1996)**

14. (*a*) Give examples of each of the following (Justifying your answers) :

(*i*) a set having no limit point **[Delhi Maths (H) 2007, 09; Delhi Maths (P), 2000]** **[Delhi B.Sc. (Prog) III, 2009, 10]**

(*ii*) every point of the set is its limit point **(Meerut, 1994)**

(*iii*) an infinite number of limit points **[Delhi Maths (H) 2007]**

(*iv*) exactly one limit point **[Delhi B.Sc. (Prog) III 2009, 10]**

(*v*) exactly two limit points **[Delhi B.Sc. (Prog) III, 2009, 10]**

(*vi*) Finite number of limit points **[Delhi Maths (H) 2007]**

(*b*) *Find the derived sets of the following sets* :

(*i*) $\{x : 0 \leq x < 1\}$ **[Delhi Maths (P), 2004]**

(*ii*) $\{x : 0 < x < 1, x \text{ is a rational number}\}$ **[Delhi Maths (P), 2004]**

(*iii*) $]a, \infty[$

(*iv*) $\{1/m + 1/n;\ m \in \mathbf{N}, n \in \mathbf{N}\}$

(*v*) $\{1 + 1/2^n, n \in \mathbf{N}\}$

(*vi*) $\{1/n : n \in \mathbf{Z}, n \neq 0\}$

15. Show that the only limit point of $S = \{a + 1/n : n \in \mathbf{N}\}$ is a.

16. Discuss the existence of a limit point of the following sets :

(*i*) $S = \left\{\dfrac{1}{n} : n \in \mathbf{N}\right\}$ (*ii*) $S = \left\{(-1)^n \dfrac{n}{n+1} : n \in \mathbf{N}\right\}$

Hint : Show that each of these sets is an infinite and bounded set. Hence by Bolzano-Weierstrass theorem, they possess a limit point.

17. Prove that a point $p \in \mathbf{R}$ is a limit point of a set S iff for each positive rational number r, $(]p - r, p + r[\cap S) \sim \{p\} \neq \phi$.

18. Prove that a point $p \in \mathbf{R}$ is a limit point of a set S iff for each positive integer n, the open interval $]p - 1/n, p + 1/n[$ contains a point of S other than p.

19. If A be any sub-set of $\mathbf{R}$, then prove that $x \in D(A) \Rightarrow x \in D(A - \{x\})$.

20. Find a set with only $\sqrt{2}$ as its limit point.

21. Find the derived sets of the following sets :

(*i*) $\left\{1 + \dfrac{(-1)^n}{n} : n \in \mathbf{N}\right\}$ (*ii*) $\left\{\dfrac{1 + (-1)^n}{n} : n \in \mathbf{N}\right\}$

(*iii*) $\left\{(-1)^n + \dfrac{1}{n} : n \in \mathbf{N}\right\}$ (*iv*) $\left\{2^n + \dfrac{1}{2^n} : n \in \mathbf{N}\right\}$

22. Give example of each of the following : (Justifying your answer)

(*i*) a bounded set having no limit point

(*ii*) a bounded set having limit points

(*iii*) an unbounded set having no limit point

(*iv*) an unbounded set having limit points

(*v*) an infinite set having a finite number of limit points.

23. Find the limit points of the set $S = \{1/2m + 1/3n : m, n = 1, 2, 3, ...\}$.

24. Find the limit points of the set $\{3^{-n} + 5^{-n} : n \in N\}$ **[Delhi B.A. (Prog.) III, 2014]**

ANSWERS/HINTS

4. $\{1/n : n \in \mathbf{N}\} \cup \{0\}$ or $\{1/m : m \in \mathbf{N}) \cup \{0\}$.

5. (*i*) The set **N** of all natural numbers has no limit point, *i.e.*, $\mathbf{N}' = \phi$.

(*ii*) The set $S = \left\{1, -1, 1\frac{1}{2}, -1\frac{1}{2}, 1\frac{1}{3}, -1\frac{1}{3}, ...\right\}$ has only two limit points 1 and – 1. So $S' = \{1, -1\}$ and $S' \subset S$.

(*iii*) The set $S =]a, b[$ has the set $[a, b]$ for its limit points. Thus, here $S' = [a, b]$ and so S' is super-set of S, *i.e.*, $S' \supset S$.

(*iv*) The only limit point of the set $S = \{1/n : n \in \mathbf{N}\}$ is 0. Also $0 \notin S$. Thus, $S' = \{0\}$ which is neither a sub-set nor a super-set of S.

(*v*) If $S = [a, b]$. Then its derived set $= S' = [a, b]$. So here the derived set is same as the set S itself.

11. (*i*) 1 is the only limit point.

(*ii*) Closed interval [0, 1] gives the set of all limit points.

13. 1 and – 1 are only two limit points.

14. **(*a*)** (*i*) **N** (*ii*) $[a, b]$ (*iii*) $]a, b[$

(*iv*) $\left\{\frac{1}{n} : n \in \mathbf{N}\right\}$ (*v*) $\left\{\frac{1}{2}, -\frac{1}{2}, \frac{2}{3}, -\frac{2}{3}, ..., \frac{n}{n+1}, -\frac{n}{n+1}, ...\right\}$ (*vi*) $\{1/n : n \in \mathbf{N}\}$

(*b*) (*i*) [0, 1] (*ii*) [0, 1] (*iii*) $[a, \infty[$

(*iv*) $\{1/m : m \in N\} \cup \{0\}$ (*v*) $\{1\}$ (*vi*) $\{1, -1\}$

20. $\{\sqrt{2} + 1/n : n \in \mathbf{N}\}$ has $\sqrt{2}$ as its only limit point.

21. (*i*) $\{1\}$ (*ii*) $\{0\}$ (*iii*) $\{-1, 1\}$ (*iv*) ϕ

22. (*i*) Any finite set (*ii*) $[a, b]$ (*iii*) **N**, set of natural numbers

(*iv*) **Q**, the set of rational numbers (*v*) $\{1/n : n \in \mathbf{N}\}$.

23. $\{1/3m : m = 1, 2, 3, ...\} \cup \{1/2n : n = 1, 2, 3, ...\} \cup \{0\}$. **24.** 0

3.6. OPEN AND CLOSED SETS

[Delhi B.Sc. I (Hons) 2010, Delhi B.A. (Prog) III, 2010, 11; Ranchi 2010, 07, 09; Delhi B.Sc. (Prog) III 2011]

Open Sets. Def. *A set is said to be open if it is a neighbourhood of each of its points.*

Thus, if A be an open set and x is any member of A, then by the definition of an open set, there exists an open interval $]a, b[$ such that

$$x \in]a, b[\subset A.$$

Equivalently, A is open if for each $x \in A$, there exists $\varepsilon > 0$ such that $]x - \varepsilon, x + \varepsilon[\subset A$.

Note. In order to show that A is not open we should prove that there exist at least one point of A of which A is not a nbd, *i.e.*, there exists some $x \in A$ such that for each $\varepsilon > 0$, however small, $]x - \varepsilon, x + \varepsilon[$ is not a sub-set of A.

Theorem I. *Every open interval is an open set.* **[Delhi Maths (H), 2004]**

Proof. In fact, let y be a point of an open interval $]a, b[$ so that we have

$$a < y < b.$$

Let c, d be two real numbers such that

$$a < c < y, \qquad y < d < b$$

$$\text{so that } a < c < y < d < b$$

$$\Rightarrow y \in]c, d[\subset]a, b[$$

$$\Rightarrow]a, b[\text{ is a neighbourhood of } y.$$

Hence, the result.

Closed Sets. Def. ***[Kanpur 2011; Ranchi 2010; Delhi B.A. (Prog.) III 2014]***

A set is said to be closed if each of its limit points is a member of the set.

Thus, a set A is closed if and only if

$$A' \subset A;$$

A' denoting the derived set of A.

The reader would do well to notice that the concepts of open and closed sets are not mutually exclusive so that there may be a set which is both open and closed. For example, the set **R** is both open and closed. Similarly, ϕ is also both open and closed.

The concepts of open and closed sets are also not exhaustive so that there may be a set which is neither open nor closed. For example, the set **Q** of rational numbers as a sub-set of the set **R** of real numbers is neither open nor closed.

The relationship between the open and closed sets is brought out in the following theorem on the basis of the notion of *complement* of a set.

Theorem II : *A set is closed if and only if its complement is open.*

(Kanpur 2010, 13; M.S. Univ. T.N. 2006; G.N.D.U., 1998; Nagpur, 2003)

Let A be a closed set. We shall show that its complement

$$\mathbf{R} \sim A = B$$

is open.

Let x be a point of the complement B of A.

As $x \in B$, we see that x is not a member of A. Again since A is closed and x is not a member of A, we see that x is not a limit point of A.

Thus, there exists a neighbourhood say, $]a, b[$ of x not containing any point of A so that we have

$$x \in]a, b[\subset B,$$

and as such there exists a neighbourhood of $x \in B$ contained in B. Also x is an arbitrary point of B. It follows that B is an open set.

Now suppose that B is an open set. We shall show that its complement.

$$\mathbf{R} \sim B = A$$

is a closed set.

Let x be a limit point of the set A. We have to show that x belongs to A or equivalently that x does not belong to the complement B of A.

Surely x cannot be a member of the open set B; for, if it were so, there would exist a neighbourhood of x contained in B and thus containing no point of A. As a result it would follow that x is not a limit point of A. Thus, x is not a member of B and as such x belongs to A. It follows that A is a closed set.

Note. Let A be a sub-set of **R.** Then, in what follows, unless stated otherwise, we shall denote the complement $\mathbf{R} \sim A$ of **A** in **R** by the notation A^c.

3.7. BASIC THEOREMS CONCERNING FAMILIES OF OPEN AND CLOSED SETS

Theorem I. *The intersection of two open sets is open.* **[Delhi B.Sc. I (Hons) 2010]**
[Delhi B.Sc. (Hons.) II 2006; Delhi B.A. (Prog.) III, 2004, 06; Kanpur, 2015]

Proof. Let G_1 and G_2 be two open sets.

If $G_1 \cap G_2 = \phi$, then $G_1 \cap G_2$ is open because ϕ is open.

If $G_1 \cap G_2 \neq \phi$, let a be any point of $G_1 \cap G_2$.

Then $a \in G_1 \cap G_2$

$\Rightarrow$ $a \in G_1$ and $a \in G_2$

$\Rightarrow$ G_1 and G_2 are neighbourhoods of a. ($\because$ G_1 and G_2 are open sets)

$\Rightarrow$ $G_1 \cap G_2$ is a neighbourhood of a.

$\Rightarrow$ $G_1 \cap G_2$ is a neighbourhood of a for each $a \in G_1 \cap G_2$.

Thus, $G_1 \cap G_2$ is an open set.

Theorem II. *The intersection of a finite number of open sets is an open set.*
[Delhi B.A (Prog) III 2013, 2015; Meerut 2018; Kanpur, 2009, 2019]

Proof. Let $G_1, G_2, \ldots\ldots, G_n$ be n open sets.

If $\bigcap_{i=1}^{n} G_i = \phi$, then $\bigcap_{i=1}^{n} G_i$ is open because ϕ is open.

If $\bigcap_{i=1}^{n} G_i \neq \phi$, let a be any point of $\bigcap_{i=1}^{n} G_i$.

Then, $a \in \bigcap_{i=1}^{n} G_i$

$\Rightarrow$ $a \in G_i$ for each $i = 1, 2, 3, \ldots\ldots, n$

$\Rightarrow$ G_i is a nbd of a for each $i = 1, 2, \ldots\ldots, n$ ($\because$ G_i is open for each $i = 1, 2, \ldots\ldots, n$)

$\Rightarrow$ $\bigcap_{i=1}^{n} G_i$ is a nbd of a

$\Rightarrow$ $\bigcap_{i=1}^{n} G_i$ is a nbd of a for each $a \in \bigcap_{i=1}^{n} G_i$

Thus, $\bigcap_{i=1}^{n} G_i$, *i.e.*, the intersection of a finite number n of open sets is an open set.

Note. *The intersection of an arbitrary family of open sets may not be open as is clear from the following example.* **[Delhi B.Sc. (Prog) III 2011, 13, 15; Kanpur 2007; Delhi B.A. (Prog) III 2008, 11; Delhi B.Sc. (Hons.) II 2016]**

Let $\mathbf{I}_n =]-1/n, 1/n[$, $n \in \mathbf{N}$. Then $\{\mathbf{I}_n\}_{n \in \mathbf{N}}$ is an infinite family of open sets.

Now $\bigcap_{n=1}^{\infty} \mathbf{I}_n = \{0\}$, which being a non-empty finite set is not an open set.

Theorem III. *The union of two open sets is an open set.* **(Kanpur 2002; 2014**

Proof. Let G_1 and G_2 be two open sets. Let a be any element of $G_1 \cup G_2$.

Now, $a \in G_1 \cup G_2$

$\Rightarrow$ $a \in G_1$ or $a \in G_2$

$\Rightarrow$ G_1 is a nbd of a or G_2 is a nbd of a [$\because$ G_1 and G_2 are open sets]

But $G_1 \subset G_1 \cup G_2$ and $G_2 \subset G_1 \cup G_2$ and any super-set of the nbd of a point is also a nbd of that point

Hence $G_1 \cup G_2$ is a nbd of a

Since a is any point of $G_1 \cup G_2$, hence $G_1 \cup G_2$ is a nbd of each of its points.

Therefore $G_1 \cup G_2$ is an open set.

Theorem IV. *The union of an arbitrary family of open sets is open.*

[Delhi Maths (P), 2005; Delhi Maths (G), 2000, 03, 04; Delhi Maths (H), 2001; Kanpur, 1992]

Proof. Let $\{G_\lambda : \lambda \in \Lambda\}$ be an arbitrary family of open set, where Λ is the indexing set.

Let $G = \bigcup_{\lambda \in \Lambda} G_\lambda$. We have to show that G is an open set.

Let a be any element of $G = \bigcup_{\lambda \in \Lambda} G_\lambda$.

Then $a \in G_\lambda$ for some $\lambda \in \Lambda$

Since G_λ is given to be an open set and $a \in G_\lambda$, so G_λ is a neighbourhood of a.

Thus, there exist some $\varepsilon > 0$ such that

$$a \in]a - \varepsilon, a + \varepsilon[\subset G_\lambda$$

$$\Rightarrow \quad a \in]a - \varepsilon, a + \varepsilon[\subset G_\lambda \subset \bigcup_{\lambda \in \Lambda} G_\lambda = G$$

$$\Rightarrow \quad a \in]a - \varepsilon, a + \varepsilon[\subset G \ \forall\ a \in G$$

$\Rightarrow$ G is a neighbourhood of each of its points and consequently $G = \bigcup_{\lambda \in \Lambda} G_\lambda$ is an open set.

Theorem V. *The union of two closed sets is a closed set.* **[Kanpur 2005]**

Proof. Let G_1 and G_2 be two closed sets.

$\Rightarrow$ G_1^c and G_2^c are open sets [$\because$ Complement of a closed set is an open set]

$\Rightarrow$ $G_1^c \cap G_2^c$ is an open set, by theorem I

$\Rightarrow$ $(G_1 \cup G_2)^c$ is an open set [$\because$ By De Morgan's law, $G_1^c \cap G_2^c = (G_1 \cup G_2)^c$]

$\Rightarrow$ $G_1 \cup G_2$ is a closed set [$\because$ Complement of an open set is a closed set]

Theorem VI. *The union of a finite number of closed sets is a closed set.*

Purvanchel 2006; Delhi B.Sc. (Hons) I 2011, 1996; Delhi B.A. III 2010; Kanpur, 2008

Proof. Let $G_1, G_2, \ldots, G_n$ be n closed sets.

$\Rightarrow$ $G_1^c, G_2^c, \ldots, G_n^c$ are open sets. [$\because$ Complement of a closed set is an open set]

$\Rightarrow$ $\bigcap_{i=1}^{n} G_i^c$ is an open set

[$\because$ the intersection of a finite number of open sets is an open set]

$\Rightarrow \left(\bigcup_{i=1}^{n} G_i\right)^c$ is an open set $\left[\because \text{ By De Morgan's law, } \bigcap_{i=1}^{n} G_i^c = \left(\bigcup_{i=1}^{n} G_i\right)^c\right]$

$\Rightarrow \bigcup_{i=1}^{n} G_i$ is a closed set [∵ Complement of an open set is a closed set]

Note. *The union of an arbitrary family of closed sets, may not be closed as is clear from the following example.* **[*Delhi B.Sc. (Hons.) I 2011;, Kanpur 2008; Delhi B.A. III 2010*]**

Let $\mathbf{I}_n = [1/n, 1], n \in \mathbf{N}$. Then $\{\mathbf{I}_n\}_{n \in \mathbf{N}}$ is an infinite family of closed sets.

Now, $\bigcup_{n=1}^{\infty} \mathbf{I}_n =]\,0, 1\,]$,

which is not a closed set.

Theorem VII. *The intersection of two closed sets is a closed set.* **[*Kanpur 2005*]**

Proof. Let G_1 and G_2 be two closed sets.

$\Rightarrow G_1^c$ and G_2^c are two open sets. [∵ Complement of a closed set is an open set]

$\Rightarrow G_1^c \cup G_2^c$ is an open set [∵ Union of two open sets is an open set]

$\Rightarrow (G_1 \cap G_2)^c$ is an open set [∵ By De Morgan's law, $G_1^c \cup G_2^c = (G_1 \cap G_2)^c$]

$\Rightarrow G_1 \cap G_2$ is a closed set [∵ Complement of an open set is a closed set]

Theorem VIII. *The intersection of an arbitrary family of closed sets is closed.*

[*Delhi Maths (Prog) 2009; Delhi Maths (H), 2004, 07, 09*]

Proof. Let $\{G_\lambda : \lambda \in \Lambda\}$ be an arbitrary family of closed sets where λ is the indexing set.

$\Rightarrow \{G_\lambda^c : \lambda \in \Lambda\}$ is an arbitrary family of open sets

[∵ Complement of a closed set is an open set]

$\Rightarrow \bigcup_{\lambda \in \Lambda} G_\lambda^c$ is an open set

[∵ The union of an arbitrary family of open sets is open]

$\Rightarrow \left(\bigcap_{\lambda \in \Lambda} G_\lambda\right)^c$ is an open set $\left[\because \text{ By De Morgan's law, } \bigcup_{\lambda \in \Lambda} G_\lambda^c = \left(\bigcap_{\lambda \in \Lambda} G_\lambda\right)^c\right]$

$\Rightarrow \bigcap_{\lambda \in \Lambda} G_\lambda$ is a closed set [∵ Complement of an open set is a closed set]

Theorem IX. *Every open set is a union of open intervals.* **[*Ranchi 2010*]**

Proof. Let A be a non-empty set and a_λ be its any point. Considering A as a union of singletons like $\{a_\lambda\}$, we have

$$A = \bigcup_{\lambda \in \Lambda} \{a_\lambda\}, \lambda \text{ being the indexing set} \qquad \ldots(1)$$

Since A is an open set, so it is a nbd of each of its points. Hence for each $a_\lambda \in A$, there exists an open interval $\mathbf{I}_{a_\lambda}$ such that

$$a_\lambda \in \mathbf{I}_{a_\lambda} \subset A.$$

Now, $\mathbf{I}_{a_\lambda} \subset A$ for each $\lambda \in \Lambda$

$$\Rightarrow \bigcup_{\lambda \in \Lambda} \mathbf{I}_{a_\lambda} \subset S \qquad \ldots(2)$$

Again, $a_\lambda \in \mathbf{I}_{a_\lambda} \Rightarrow \{a_\lambda\} \subset \mathbf{I}_{a_\lambda}$

$\therefore$ $\bigcup_{\lambda \in \Lambda} \{a_\lambda\} \subset \bigcup_{\lambda \in \Lambda} \mathbf{I}_{a_\lambda}$

i.e., $S \subset \bigcup_{\lambda \in \Lambda} \mathbf{I}_{a_\lambda}$, using (1) ...(3)

From (2) and (3), $S = \bigcup_{\lambda \in \Lambda} \mathbf{I}_{a_\lambda}$,

showing that the open set S can be expressed as union of open intervals $\mathbf{I}_{a_\lambda}$

3.8. ILLUSTRATIONS OF OPEN SETS

1. Every open interval $]a, b[$ is an open set. (See theorem I, Art. 3.6.)

2. The interval $[a, b[$ is not an open set because it is not a nbd of a. Similarly $]a, b]$ is not an open set because it is not a nbd of b. Also, $[a, b]$ is not an open set because it is not a nbd of a and b.

3. The set $\mathbf{R}$ of all real numbers is an open set. For, if x is an arbitrary point of $\mathbf{R}$, then there exists $\varepsilon > 0$ such that $]x - \varepsilon, x + \varepsilon[\subset \mathbf{R}$ and so $\mathbf{R}$ is nbd of x. Since x is an arbitrary point of $\mathbf{R}$, it follows that $\mathbf{R}$ is nbd of each of its points. Hence $\mathbf{R}$ must be an open set.

4. The empty set ϕ is an open set because there is no point in ϕ of which it is not a nbd.

5. Sets $]a, \infty[$ and $]-\infty, a[$ are open sets.

6. Sets $[a, \infty[$ and $]-\infty\ a]$ are not open sets.

7. The set $\mathbf{Q}$ of all rational numbers is not an open set as shown below.

[Delhi B.A. (Prog) III 2011, Purvanchal 2006; Delhi Maths (H), 2004]

Let x be any point of $\mathbf{Q}$. Now, $\mathbf{Q}$ will be open if there exists $\varepsilon > 0$ such that

$$]x - \varepsilon, x + \varepsilon[\subset \mathbf{Q}.$$

But $]x - \varepsilon, x + \varepsilon[$ contains infinitely many (irrational) numbers which do not belong to $\mathbf{Q}$

Hence $\mathbf{Q}$ cannot be a nbd of x. But x is any point of $\mathbf{Q}$, so $\mathbf{Q}$ is not a nbd of any of its points. Therefore $\mathbf{Q}$ is not open.

8. The set $\mathbf{Q}^c$ of all irrational numbers is not an open set. To prove it give similar arguments as in illustration 7.

9. The set $\mathbf{N}$ of all natural numbers is not an open set because it is not nbd of any of its points. (Prove it !)

10. The set $\mathbf{Z}$ of all integers is not an open set because it is not nbd of any of its points. (Prove it !)

11. A non-empty finite set is not open. **[Delhi Maths (G), 2002; Delhi Maths (H), 2004]**

Let S be a non-empty finite set. Let x be any point of S. Now, S will be open if there exists $\varepsilon > 0$ such that $]x - \varepsilon, x + \varepsilon[\subset S$.

Now, S being finite, $]x - \varepsilon, x + \varepsilon[$ will contain infinitely many numbers which do not belong to S.

Hence S cannot be a nbd of x. But x is any point of S, so S is not a nbd of any of its points. Therefore, S is not open.

12. The set $G =]-1, 0[\cup]1, 2[$ is open. **[Delhi Maths (P) 2005]**

Since an open interval is open set so G is union of two open sets. Hence G is open by theorem III of Art. 3.7.

3.9. ILLUSTRATIONS OF CLOSED SETS (Kanpur 2010)

1. Every closed interval $[a, b]$ is a closed set. For, if $S = [a, b]$, then

$$\mathbf{R} \sim S =]-\infty, a[\cup]b, \infty[$$

being the union of two open sets is itself open. Since $\mathbf{R} \sim S$ is open, therefore S is a closed set.

2. Every open interval $]a, b[$ is not a closed set. For, if $S =]a, b[$, then

$$\mathbf{R} \sim S =]-\infty, a] \cup [b, \infty[$$

which is not a nbd of a or b. Hence $\mathbf{R} \sim S$ is not open.

3. $]a, b]$ and $[a, b[$ are both closed sets. (Prove it !)

4. $\mathbf{R}^c = \phi \;\Rightarrow\;$ $\mathbf{R}$ is closed, as ϕ is closed.

5. $\phi^c = \mathbf{R} \;\Rightarrow\;$ ϕ is closed, as $\mathbf{R}$ is closed.

6. $]-\infty, a]$ is a closed set because its complement, namely, $]-\infty, a[$ is an open set. Similarly, $[a, \infty[$ is a closed set.

7. $]-\infty, a[$ and $]a, \infty[$ are not closed sets. (Prove it !)

8. The set $\mathbf{Q}^c$ of all irrational numbers is not a closed set because its complement $\mathbf{Q}$ is not an open set. (Use illustration 7 of Art. 3.8.)

9. The set $\mathbf{Q}$ of all rational numbers is not a closed set because its complement $\mathbf{Q}^c$ of all irrational numbers is not an open set.

10. The set $\mathbf{N}$ of natural numbers is a closed set. **[Delhi B.A. (Prog.) III, 2014]**

We have $\mathbf{N}^c = \mathbf{R} - \mathbf{N} =]-\infty, 1[\cup]1, 2[\cup]2, 3[\cup .....,$

Now, the set on the R.H.S., being a union of infinite number of open intervals (and hence open sets), is itself an open set. Hence $\mathbf{N}^c$ is an open set and so its complement $\mathbf{N}$ is a closed set.

Similarly, the reader can verify that the set $\mathbf{Z}$ of all integers is also a closed set.

11. A non-empty finite set is a closed set. **[Kanpur 2006]**

Let $S = \{a_1, a_2, ..., a_n\}$ be a non-empty, finite set.

Then, $S^c = \mathbf{R} - S =]-\infty, a_1[\cup]a_1, a_2[\cup ..... \cup]a_n, \infty[$

Now, the set on the R.H.S., being a union of finite number of open intervals (and hence open sets), is itself an open set. Hence S^c is an open set and so its complement S is a closed set.

In particular a singleton set $\{a_1\}$ is a closed set.

EXAMPLES

Example 1. *Give an example of*

(*i*) *an open set which is not an interval* **[Delhi B.Sc. (Prog) III 2011]**

(*ii*) *an interval which is an open set*

(*iii*) *an interval which is not an open set*

(*iv*) *a set which is neither an interval nor an open set.*

Solution. (*i*) $]a, b[$ and $]c, d[$ are open sets and so their union $]a, b[\cup]c, d[$ is also an open set but it is not an interval.

(*ii*) $]a, b[$ is an interval and it is an open set.

(*iii*) $[a, b]$ is an interval but it is not an open set.

(*iv*) The set $\mathbf{N}$ of all natural numbers is neither an interval nor an open set.

Example 2. *Give an example of*

(*i*) *an interval which is a closed set*

(*ii*) *an interval which is not a closed set*

(iii) a closed set which is not an interval

(iv) a set which is open as well as closed

(v) a set which is neither open nor closed **[Delhi B.A. Pass III 2009]**

(vi) a set which is neither an interval nor a closed set.

Solution. (*i*) $[a, b]$ (*ii*) $]a, b[$ (*iii*) Any non-empty finite set (*iv*) **R**, the set of all real numbers (*v*) $]a, b]$ (*vi*) $]a, b[\cup]c, d[$

Example 3. *Show that a sub-set of real numbers is closed iff it contains all its limit points.*

[Delhi Maths (G), 2002, 03; Delhi Maths (H), 2008]

Solution. Let S be any sub-set of R. Then

S is a closed set

$\Leftrightarrow$ $\mathbf{R} \sim S$ is an open set

$\Leftrightarrow$ $\mathbf{R} \sim S$ is a nbd of each of its points

$\Leftrightarrow$ each point of $\mathbf{R} \sim S$ has a nbd $\mathbf{R} \sim S$ such that

$$(\mathbf{R} \sim S) \cap S = \phi$$

$\Leftrightarrow$ no point of $\mathbf{R} \sim S$ can be a limit point of S

$\Leftrightarrow$ all limit points of S, if any, belong to S.

Example 4. *Prove that the derived set of any set is closed.*

[Kanpur 2013; Ranchi 2010; Agra, 2001; Delhi Maths (H), 2003]

Solution. Let S be any sub-set of **R** and let S' be its derived set. In order to prove that S' is a closed set, we shall prove that its complement $(S')^c$ is an open set.

Let x be an arbitrary element of $(S')^c$. Then

$x \in (S')^c$

$\Rightarrow$ $x \notin S'$, *i.e.*, x is not a limit point of S

$\Rightarrow$ there exists a nbd $]x - \varepsilon, x + \varepsilon[$ of x such that

$$(]x - \varepsilon, x + \varepsilon[\cap S) - \{x\} = \phi, \text{ where } \varepsilon > 0$$

Let $y \in]x - \varepsilon, x + \varepsilon[$, which being an open interval is itself an open set.

$\therefore$ $y \in]x - \varepsilon, x + \varepsilon[$

$\Rightarrow$ $]x - \varepsilon, x + \varepsilon[$ is a nbd of y such that $(]x - \varepsilon, x + \varepsilon[\cap S) - \{x\} = \phi$

$\Rightarrow$ y is not a limit point of S so that $y \notin S'$

$\Rightarrow$ $y \in (S')^c$

Thus, $y \in]x - \varepsilon, x + \varepsilon[\Rightarrow y \in (S')^c$

$\Rightarrow$ $]x - \varepsilon, x + \varepsilon[\subset (S')^c$

$\Rightarrow$ $(S')^c$ is a nbd of x

Since x is an arbitrary point of $(S')^c$, so $(S')^c$ is a nbd of each of its points. Hence $(S')^c$ is an open set and so its complement S' is a closed set.

EXERCISES

1. *Answer in Yes or No, justifying your answer in each case either by a proof or by counter example :*

   (*i*) Is a sub-set of an open set open ?

   (*ii*) Is a super-set of an open set open ?

   (*iii*) Can a finite set be open ? **[Delhi Maths (G), 2002]** (*iv*) Can a non-empty finite set be open ?

(*v*) Is every infinite set open ? **[Delhi Maths (H), 1997, 98]**

(*vi*) Is the union of an arbitrary collection of open sets open ? **[Delhi 2002]**

(*vii*) Is the intersection of an arbitrary collection of open sets open ? **[Delhi 2002]**

2. *Answer in Yes or No, justifying your answer in each case either by a proof or by counter example* :

(*i*) Is a sub-set of a closed set closed ?

(*ii*) Is a super-set of a closed set closed ?

(*iii*) Can a finite set be closed ?

(*iv*) Can a non-empty finite set be closed ?

(*v*) Is every infinite set closed ?

(*vi*) Is the union of an arbitrary collection of closed sets closed ?

(*vii*) Is the intersection of an arbitrary collection of closed sets closed ?

3. Let $A = [1, 2] \cup (3, 4) \cup \{5\}$. Is A open ? Justify your answer and find all limit points of A . **[Delhi B.Sc. I (Hons) 2010]**

Ans. Noto open; set of limit point is $[1,2] \cup [3, 4]$

4. Which of the following sets are closed, open, neither closed nor open :

(*i*) $[a, b] \cup [c, d]$ (*ii*) $]a, b[\cup]c, d[$ (*iii*) $[a, b] \cap]c, d[$ (*iv*) $[a, b] \cup [c, d[$

5. Prove that a sub-set S of $\mathbf{R}$ is an open set iff :

(*i*) for each $x \in S$, there exists a positive integer n such that $]x - 1/n, x + 1/n[\subset S$

(*ii*) for each $x \in S$, there exists a positive rational number ·· such that

$$]x - r, x + r[\subset S$$

(*iii*) for each $x \in S$, there exists a positive integer n such that $]x - 2^{-n}, x + 2^{-n}[\subset S$.

6. (*a*) If S is a closed bounded set, then show that every infinite sub-set of S has a limit point in S.

(*b*) Prove that a sub set A of real number set $\mathbf{R}$ is an open set if and only if its complement $\sim A$ is closed. **[Rajasthan 2010]**

(*c*) Is the closed interval $[a, b]$ open ? Justify your answer. **[Ranchi 2010]**

7. Given a family $\{[a_n, b_n] : n \in \mathbf{N}\}$ of closed intervals such that $[\ n \in \mathbf{N}$

$$]a_{n+1}, b_{n+1}[\subset [a_n, b_n],$$

show that the intersection of the family is not void.

8. Let $\mathbf{I}_n = \left] x - \frac{1}{n}, x + \frac{1}{n} \right[$, $n = 1, 2, \ldots\ldots$

Prove that $\mathbf{I}_n$ is an open set for each positive integer n and $\bigcap_{n=1}^{\infty} \mathbf{I}_n = \{x\}$.

9. (*a*) Show that the set $S = \left\{1, \frac{1}{2}, \frac{1}{3}, \frac{1}{4}, \ldots\ldots\right\}$ is neither open nor closed.

(*b*) Which of the following sets are open? Give arguments in support of your answer. (*i*) The set Q of rational numbers (*ii*) The interval [0, 2].**[Delhi B.A. (Prog) III 2010]**

10. Show that the set $S = \left\{1, -1, \frac{1}{2}, -\frac{1}{2}, \frac{1}{3}, -\frac{1}{3}, \ldots\ldots\right\}$ is neither open nor closed.

11. Show that the set $S = \left\{1, -1, 1\frac{1}{2}, -1\frac{1}{2}, 1\frac{1}{3}, -1\frac{1}{3}, ...\right\}$ is closed but not open.

12. Prove that a non-empty bounded closed set contains its supremum as well as its infimum. **[Delhi Maths (G), 1998; Delhi Maths (H), 1999]**

13. Give examples of each of the following with justification :

(*i*) an interval which is not an open set. **[Delhi Maths (P), 2000]**

(*ii*) an open set which is not an interval. **[Delhi Maths (P), 2000]**

(*iii*) a set which is neither open nor closed. **[Delhi Maths (P), 2002]**

14. Define the derived set S' of a set S of real numbers. For an open set G, show that
$$G \cap S = \phi \Rightarrow G \cap S' = \phi.$$
[Delhi Maths (H), 2005]

ANSWERS/HINTS

1. (*i*) No, for example $[2, 3] \subset]1, 4[$ (*ii*) No, for example $[2, 3] \supset]2, 3[$
(*iii*) Yes, ϕ is finite and ϕ is open (*iv*) No
(*v*) No, for example the set **Q** (*vi*) Yes. Refer theorem IV of Art. 3.7
(*vii*) Not necessary. Refer note of theorem II of Art. 3.7.

2. (*i*) No (*ii*) No (*iii*) Yes (*iv*) Yes (*v*) No
(*vi*) Not necessary. Refer note of theorem VI of Art. 3.7
(*vii*) Yes. Refer theorem VIII of Art. 3.7.

4. (*i*) Close (*ii*) Open (*iii*) Neither close nor open (*iv*) Neither close nor open.

9. Yes **13.** (*i*) $[a, b]$ (*ii*) $]a, b[\cup]c, d[$ (*iii*) $]a, b]$.

3.10. INTERIOR POINT AND INTERIOR OF A SET

[Kanpur 2008, 2012; Delhi Maths (Prog) 2008; Purvanchal 2006]

A point p is called an *interior point* of a set S if S is a nbd of p.

Equivalently, a point p is called an *interior point* of a set S if there exists an $\varepsilon > 0$ such that $]p - \varepsilon, p + \varepsilon[\subset S$

The set of all interior points of S is called the *interior* of S and is denoted by int S or S^o.

3.11. EXTERIOR POINT AND EXTERIOR OF A SET ***[Kanpur 2009]***

A point p is called an *exterior point* of a set S if there exists a nbd N of p such that
$$N \cap S = \phi.$$

The set of all exterior points of S is called the *exterior* of S and is denoted by ext (S) or $e(S)$.

3.12. BOUNDARY (OR FRONTIER) POINT AND BOUNDARY (OR FRONTIER) OF A SET ***[Kanpur 2009]***

A point p is called a *boundary point* of a set S if it is neither an interior point nor an exterior point of S.

The set of all boundary points of S is called the *boundary* of S and is denoted by $b(S)$ or Fr (S).

ILLUSTRATIONS

(*i*) Every point of $]a, b[$ is an interior point of $]a, b[$ and int $]a, b[=]a, b[$.

(*ii*) a and b are not interior points of $[a, b]$, because $[a, b]$ is not nbd of a or b. Also
$$\text{int } [a, b] =]a, b[.$$

(*iii*) Every point of **R** is an interior point of **R**. So int **R** = **R**.

(*iv*) No point of **N** is an interior point of **N** because **N** is not nbd of any of its points. So $\mathbf{N}^o = \phi$. Similarly, $\mathbf{Z}^o = \phi$.

(*v*) If A is a non-empty finite set then A cannot be a nbd of any of its points. Hence $A^o = \phi$.

(*vi*) Int $\mathbf{Q} = \phi$, ext $(\mathbf{Q}) = \phi$ and Fr $(\mathbf{Q}) = \mathbf{R}$, where **Q** is the set of all rational numbers and **R** is the set of real numbers.

(*vii*) If $A = [a, b]$, then $b\ (A) = \{a, b\}$.

(*viii*) If S denotes the set of all rational points in $[a, b]$ then $b\ (A) = \{a, b\}$.

EXERCISES

1. For any sub-set S of **R**, prove that

(*i*) $b\ (A) = b\ (\mathbf{R} - A)$ (*ii*) $b\ (A) = \overline{A} \cap (\mathbf{R} - \overline{A})$ (*iii*) $\overline{A} = A \cup b\ (A)$

2. For any two sub-sets A and B of **R**, prove that

(*i*) ext $(A \cup B) =$ ext $(A) \cap$ ext (B) (*ii*) ext $(A \cap B) \supset$ ext $(A) \cup$ ext (B)

3.13. THEOREMS ON INTERIOR OF A SET [Purvanchal 2006]

Theorem I. *For any set S, S^o is open.*

Proof. Let p be any element of S. Then, we have

$$p \in S \Rightarrow S \text{ is a nbd of } p$$
$$\Rightarrow \text{there exists } \varepsilon > 0 \text{ such that }]p - \varepsilon, p + \varepsilon[\subset S$$

Let q be any point of $]p - \varepsilon, p + \varepsilon[$.

Since an open set is a nbd of each of its points, it follows that $]p - \varepsilon, p + \varepsilon[$ is a nbd of q.

Now, $]p - \varepsilon, p + \varepsilon[$ is a nbd of q and $S \supset\ ]p - \varepsilon, p + \varepsilon[\quad \Rightarrow S$ is a nbd of q

But q is any point of $]p - \varepsilon, p + \varepsilon[$. Hence S is nbd of each point of $]p - \varepsilon, p + \varepsilon[$ and so every point of $]p - \varepsilon, p + \varepsilon[$ is an interior point of S.

$\therefore \quad]p - \varepsilon, p + \varepsilon[\subset S^o$, *i.e.*, $p \in\ ]p - \varepsilon, p + \varepsilon[\subset S^o \quad \Rightarrow S^o$ is a nbd of p.

Since p is any point of S^o, so S^o is a nbd of each of its points, Therefore, by definition, S^o is an open set.

Theorem II. *The interior of a set S is a sub-set of S, i.e., $S^o \subset S$.*

Proof. Let p be any point of S^o. Then, we have

$$p \in S^o \Rightarrow S \text{ is a nbd of } p \Rightarrow p \in S$$

Thus, $p \in S^o \Rightarrow p \in S$. Therefore, $S^o \subset S$.

Theorem III. *S^o is the largest open set contained in S.*

By theorem I, S^o is an open set. Hence in order to prove the theorem, we have to show that if T is any set such that $T \subset S$, then $T \subset S^o$.

Let p be any point of T. Since T is an open set, so T is a nbd of p. But $T \subset S$ and hence S is also a nbd of p and so $p \in S^o$.

Thus, $p \in T \Rightarrow p \in S^o$ and so $T \subset S^o$,

showing that every open set contained in S is a sub-set of S^o.

Hence, S^o is the largest open set contained in S.

Corollary. *Interior of a set S is the union of all open sub-sets of S.*

Proof. Left as an exercise.

Theorem IV. *A set S is open iff $S^o = S$.*

Proof. Let S be an open set.

By theorem II, $S^o \subset S.$ ...(1)

By theorem III, S^o is the largest open set contained in S. Also S being open, S is an open set contained in S. Hence

$$S \subset S^o \quad ...(2)$$

From (1) and (2), $S^o = S.$

Conversely, let us suppose that $S^o = S.$ ...(3)

By theorem I, S^o is an open set. So (3) $\Rightarrow$ S is also an open set.

Theorem V. *For any set S,* $(S^o)^o = S^o$.

Proof. Left as an exercise.

Theorem VI. If $S \subset T$, then $S^o \subset T^o$.

Proof. Let p be any point of S^o. Then, we have

$$\begin{aligned} p \in S^o &\Rightarrow S \text{ is a nbd of } p \\ &\Rightarrow T \text{ is a nbd of } p \\ &\Rightarrow p \subset T^o \end{aligned}$$

Thus, $p \in S^o \Rightarrow p \in T^o$ and hence $S^o \subset T^o$.

Theorem VII. *If S and T are any two sets, then* $(S \cap T)^o = S^o \cap T^o$.

Now, $S \cap T \subset S$ and $S \cap T \subset T$

$$\Rightarrow (S \cap T)^o \subset S^o \text{ and } (S \cap T)^o \subset T^o$$

$$\Rightarrow (S \cap T)^o \subset S^o \cap T^o \quad ...(1)$$

Let p be any point of $S^o \cap T^o$. Then, we have

$$\begin{aligned} & p \in S^o \cap T^o \\ \Rightarrow\ & p \in S^o \text{ and } p \in T^o \\ \Rightarrow\ & p \text{ is a nbd of } S \text{ and } p \text{ is a nbd of } T \\ \Rightarrow\ & S \cap T \text{ is a nbd of } p \\ \Rightarrow\ & p \in (S \cap T)^o \end{aligned}$$

Hence, $S^o \cap T^o \subset (S \cap T)^o.$ ...(2)

From (1) and (2), $(S \cap T)^o = S^o \cap T^o.$

Theorem VIII. *Let S and T be any sets. Then* $S^o \cup T^o \subset (S \cup T)^o$. *Give an example to show that* $S^o \cup T^o \neq (S \cup T)^o$.

Proof. Now, $S \subset S \cup T$ and $T \subset S \cup T$

$$\Rightarrow S^o \subset (S \cup T)^o \text{ and } T^o \subset (S \cup T)^o$$

$$\Rightarrow S^o \cup T^o \subset (S \cup T)^o$$

Example to show that, in general, $S^o \cup T^o \neq (S \cup T)^o$

Let $S = [1, 2]$ and $T = [2, 3]$.

Then $S^o =]1, 2[$ and $T^o =]2, 3[$

Here, $S \cup T = [1, 3]$ and so $(S \cup T)^o =]1, 3[$ ...(1)

But $S^o \cup T^o =]1, 2[\cup]2, 3[=]1, 3[- \{2\}$ ...(2)

From (1) and (2), $S^o \cup T^o \neq (S \cup T)^o.$

Example. *Find the interior of the following sets*

(*i*) $[1, 2] \cup [3, 4]$ (*ii*) $\{1/n : n \in \mathbf{N}\}$ (*iii*) $\mathbf{R} - \mathbf{Q}$

Solution. (*i*) Since $[1, 2] \cup [3, 4]$ is a nbd of each of its points, so

$$\text{int}\{[1, 2] \cup [3, 4]\} =]1, 2[\cup]3, 4[.$$

(*ii*) Given set = $S = \{1, 1/2, 1/3, ...\}$ is not a nbd of any of its points. So $S^o = \phi$.

(*iii*) **R** – **Q** is the set of all irrational numbers. Since **R** – **Q** is not a nbd of any of its points, so $(\mathbf{R} - \mathbf{Q})^o = \phi$.

EXERCISES

1. Show by means of an example that two distinct sets A and B may have the same interior.

2. Give an example of two sets A and B such that $A \subset B$ and $A^o = B^o$.

3. Prove that x is an interior point of a set S iff there exists a positive integer n such that

$$]x - 1/n, x + 1/n[\subset S.$$

4. Prove that x is an interior point of a set S iff there exists a rational number r such that

$$]x - r, x + r[\subset S.$$

ANSWERS/HINTS

1. Let $A = [a, b[$ and $B =]a, b]$. Then $A \neq B$, yet $A^o = B^o =]a, b[$.

2. Let $A =]a, b[$ and $B = [a, b]$. Then $A \subset B$, yet $A^o = B^o =]a, b[$.

3.14. ADHERENT POINT (OR A CONTACT POINT) AND CLOSURE OF A SET

A point p is called an *adherent point* of a set S if every nbd of p contains a point of S *i.e.*, a point p is an *adherent point* of S iff for each nbd N of p, $N \cap S \neq \phi$.

Equivalently, a point p is called an *adherent point* of a set S if for each $\varepsilon > 0$, $]p - \varepsilon, p + \varepsilon[$ contains a point of S.

Note. The distinction between an adherent point and a limit point of a set must be carefully noted. If $p \in S$, then p is an adherent point of S, since each nbd of p contains p which belongs to S. If S' be the derived set of S and if $p \in S'$, then p is a limit point of S and so each nbd of p contains a point of S other than p. Hence p is an adherent point of S.

Thus, p is an adherent point of $S \Leftrightarrow p \in S$ or $p \in S'$.

Thus each point of S and each limit of S will be an adherent point. Of course each adherent point of S need not be always a limit of S.

Closure of a set. The set of all adherent points of S, called the closure of S, is denoted by cl S or $\overline{S}$.

Theorem. *If S be any set, then* $\overline{S} = S \cup S'$.

Proof. By definition, every point of S is an adherent point of S and so

$$S \subset \overline{S} \qquad ...(1)$$

Again each limit point of S is also an adherent point of S and so

$$S' \subset \overline{S} \qquad ...(2)$$

From (1) and (2),
$$S \cup S' \subset \overline{S} \qquad ...(3)$$

We now proceed to show that

$$\overline{S} \subset S \cup S' \qquad ...(4)$$

Let p be any element of $\overline{S}$. If $p \in S$, then (4) is true. Now, consider the situation when $p \in \overline{S}$ and $p \notin S$. Then we shall prove that $p \in S'$.

Let N be any nbd of p. Since $p \in \overline{S}$, $N \cap S \neq \phi$. Now, $p \notin S \Rightarrow p \notin N \cap S$. Hence there exists a nbd of p containing a point other than p. Since N is any nbd of p, it follows that each nbd of p contains a point other than p. Hence p is a limit point of S, *i.e.*, $p \in S'$ and so (4) is true.

From (3) and (4),
$$\overline{S} = S \cup S'.$$

ILLUSTRATIONS

1. $\overline{\mathbf{N}}$ = closure of $\mathbf{N} = \mathbf{N} \cup \mathbf{N}' = \mathbf{N} \cup \phi = \mathbf{N}$, as $\mathbf{N}' = \phi$

   Similarly, $\overline{\mathbf{Z}} = \mathbf{Z}$.

2. $\overline{\mathbf{R}}$ = closure of $\mathbf{R} = \mathbf{R} \cup \mathbf{R}' = \mathbf{R} \cup \mathbf{R} = \mathbf{R}$, as $\mathbf{R}' = \mathbf{R}$.
3. $\overline{\mathbf{Q}} = \mathbf{Q} \cup \mathbf{Q}' = \mathbf{Q} \cup \mathbf{R} = \mathbf{R}$, as $\mathbf{Q}' = \mathbf{R}$.
4. $\overline{\phi} = \phi \cup \phi' = \phi \cup \phi = \phi$, as $\phi' = \phi$.
5. Let $S = \{1/n : n \in \mathbf{N}\}$. Then $S' = \{0\}$ and so

   $\overline{S} = S \cup S' = \{1/n : n \in \mathbf{N}\} \cup \{0\}$.

3.15. THEOREMS ON CLOSURE OF A SET

Theorem I. *For any two sets S and T,* $S \subset T \Rightarrow \overline{S} \subset \overline{T}$.

Proof. Let $S \subset T$ and p be any element of $\overline{S}$. Then, we have

$$p \in \overline{S} \Rightarrow p \in S \cup S' \Rightarrow p \in S \text{ or } p \in S'$$

$$\Rightarrow p \in T \text{ or } p \in T', \text{ as } S \subset T \Rightarrow S' \subset T' \text{ and so}$$

$$p \in S \Rightarrow p \in T \text{ and } p \in S' \Rightarrow p \in T'$$

$$\Rightarrow p \in T \cup T' \Rightarrow p \in \overline{T}$$

Thus, $$S \subset T \Rightarrow \overline{S} \subset \overline{T}.$$

Theorem II. *For any two sets S and T,* $\overline{S \cup T} = \overline{S} \cup \overline{T}$.

Proof. $$S \subset S \cup T \text{ and } T \subset S \cup T$$

$$\Rightarrow \overline{S} \subset \overline{S \cup T} \text{ and } \overline{T} \subset \overline{S \cup T}, \text{ by theorem I}$$

$$\Rightarrow \overline{S} \cup \overline{T} \subset \overline{S \cup T} \quad \text{...(1)}$$

We now propose to prove that

$$\overline{S \cup T} \subset \overline{S} \cup \overline{T}. \quad \text{...(2)}$$

In order to prove (2), we shall prove that if a point does not belong to $\overline{S} \cup \overline{T}$, then it cannot belong to $\overline{S \cup T}$. To this end, let p be any point such that $p \notin \overline{S} \cup \overline{T}$ so that $p \notin \overline{S}$ and $p \notin \overline{T}$.

Now, $p \notin \overline{S} \Rightarrow$ there exists a nbd N_1 of p such that

$$N_1 \cap S = \phi \quad \text{...(3)}$$

Similarly, $p \notin \overline{T} \Rightarrow$ there exists a nbd N_2 of p such that

$$N_2 \cap T = \phi \quad \text{...(4)}$$

Again, N_1 and N_2 are nbds of $p \Rightarrow N_1 \cap N_2$ is a nbd of p.

Now, (3) and (4) $\Rightarrow (N_1 \cap N_2) \cap S = \phi$ and $(N_1 \cap N_2) \cap T = \phi$

$$\Rightarrow (N_1 \cap N_2) \cap (S \cap T) = \phi.$$

$$\Rightarrow p \notin \overline{S \cup T}$$

Thus, $p \notin \overline{S} \cup \overline{T} \Rightarrow p \notin \overline{S \cup T}$ and so (2) is true.

Now, (1) and (2) $\Rightarrow \overline{S \cup T} = \overline{S} \cup \overline{T}$.

Theorem III. *For any two sets S and T,* $\overline{S \cap T} \subset \overline{S} \cap \overline{T}$. *Give an example to show that* $\overline{S \cap T} \neq \overline{S} \cap \overline{T}$. **[Delhi Maths (H), 2005]**

Proof. $S \cap T \subset S$ and $S \cap T \subset T$

$\Rightarrow \overline{S \cap T} \subset \overline{S}$ and $\overline{S \cap T} \subset \overline{T}$, by theorem I

$\Rightarrow \overline{S \cap T} \subset \overline{S} \cap \overline{T}$

Example to show that sets S and T exist such that

$$\overline{S \cap T} \neq \overline{S} \cap \overline{T}.$$

Let $S =]a, b[$ and $T =]b, c[$.

Then $\overline{S} = [a, b]$ and $\overline{T} = [b, c]$

$\overline{S} \cap \overline{T} = \{b\}$ and $S \cap T = \phi \Rightarrow \overline{S \cap T} = \overline{\phi} = \phi$.

Thus, $\overline{S} \cap \overline{T} \neq \overline{S \cap T}$.

Theorem IV. *The closure of a set S is a closed super-set of S.*

Proof. $\overline{S} = S \cup S' \Rightarrow \overline{S}$ is a super-set of S. ...(1)

We shall now prove that $\overline{S}$ is a closed set. Now, we have

$\overline{S}$ is closed $\Leftrightarrow$ $\mathbf{R} - \overline{S}$ is open,

where $\mathbf{R} - \overline{S}$ is the complement of $\overline{S}$ in $\mathbf{R}$.

Let p be any point of $\mathbf{R} - \overline{S}$. Then $p \notin \overline{S}$ and so there exists $\varepsilon > 0$ such that $]p - \varepsilon, p + \varepsilon[$ contains no point of S.

Let q be any point of $]p - \varepsilon, p + \varepsilon[$. Since $]p - \varepsilon, p + \varepsilon[$ is a nbd of q containing no points of S, it follows that $q \notin \overline{S}$. Since q is an arbitrary point of $]p - \varepsilon, p + \varepsilon[$, we see that no point of $]p - \varepsilon, p + \varepsilon[$ can belong to $\overline{S}$.

$\therefore$ $]p - \varepsilon, p + \varepsilon[\subset \mathrm{R} - \overline{S}$

$\Rightarrow \mathrm{R} - \overline{S}$ is a nbd of p

But p is any point of $\mathbf{R} - \overline{S}$ and so $\mathbf{R} - \overline{S}$ is nbd of each of its limits.

Then $\mathbf{R} - \overline{S}$ is an open set and so $\overline{S}$ is a closed set.

In view of (1), $\overline{S}$ is a closed super-set of S.

Theorem V. *The closure of a set S is the smallest closed super-set of S.*

Proof. By theorem IV, $\overline{S}$ is a closed super-set of S. Hence in order to prove the theorem we shall show that $\overline{S}$ is the smallest out of all closed super-sets of S.

To this end, let T be any closed super-set of S. Then, we show that $\overline{S} \subset T$.

Now, T is a closed super-set of S

$\Rightarrow \mathbf{R} - T$ is an open set such that $(\mathbf{R} - T) \cap S = \phi$

$\Rightarrow$ no point of $\mathbf{R} - T$ is an adherent point of S

$\Rightarrow \overline{S} \subset T$.

Theorem VI. *A set S is closed* $\Leftrightarrow$ $\overline{S} = S$.

Proof. Let S be a closed set. Then the smallest closed super-set of S will be S itself and so by theorem V, $\overline{S} = S$.

Conversely, let $\overline{S} = S$. By theorem IV, $\overline{S}$ is a closed set and hence S must be a closed set.

Theorem VII. *The closure of a set S is the intersection of all closed super-sets of S.*

[Delhi Maths (H), 2005]

Proof. Let F be the family of closed super-set of S, *i.e.*,

$F = \{A_\lambda : A_\lambda$ is closed set such that $A_\lambda \supset S$ and $\lambda \in \Lambda\}$, where Λ is the indexing set.

Let
$$T = \bigcap_{\lambda \in \Lambda} A_\lambda \quad \text{...(1)}$$

We wish to prove that
$$\overline{S} = T.$$

Now $\overline{S}$ is a closed super-set of S and so $\overline{S} \in F$. Also, T is a sub-set of every member of F and hence

$$T \subset \overline{S} \quad \text{...(2)}$$

Now, by (1), T being the intersection of a family of closed sets, is itself a closed set. Since each member of the family F contains S, hence T must contain S, *i.e.*, $T \supset S$. Hence T is a closed super-set of S and therefore

$$T \supset \overline{S} \quad \text{or} \quad \overline{S} \subset T \quad \text{...(3)}$$

From (2) and (3),
$$\overline{S} = T.$$

Theorem VIII. *If S is a bounded set, then $\overline{S}$ is also bounded.*

Proof. S is a bounded set.

$\Rightarrow$ there exist real numbers h, k such that $A \subset [h, k]$

$\Rightarrow$ (derived set of A) $\subset$ (derived set of $[h, k]$) $\quad [\because S \subset T \Rightarrow S' \subset T']$

$\Rightarrow A' \subset [h, k]$

Now, $\quad A \subset [h, k]$ and $A' \subset [h, k]$

$\Rightarrow A \cup A' \subset [h, k] \Rightarrow \overline{A} \subset [h, k]$

$\Rightarrow \overline{A}$ is a bounded set.

EXERCISES

1. Find the closure of the following sets :

(*i*) $[a, b[$ (*ii*) $]a, b]$ (*iii*) $]a, \infty[$ (*iv*) $]-\infty, a[$

2. Find the set of all adherent points of the following sets :

(*i*) $\{1 - 2/n : n \in \mathbf{N}\}$ (*ii*) $\mathbf{R} - \{1/n : n \in \mathbf{N}\}$

(*iii*) $\{4 + (-1/10)^n : n \in \mathbf{N}\}$ (*iv*) $\{1/m + 1/n : m, n \in \mathbf{N}\}$

3. Give an example of two sets A and B such that A is a proper sub-set of B, and yet $\overline{A} = \overline{B}$.

4. Show that a point p is an adherent point of S iff for each positive integer n,
$$p \in]p - 1/n, p + 1/n[.$$

5. Show that a point p is an adherent point of S iff for each positive rational number r,
$$]p - r, p + r[\neq \phi.$$

6. Show that if S is any set and T is an open set, then $S \cap T = \phi \Rightarrow \overline{S} \cap T = \phi$.

7. Show that a point p is an adherent point of a set S iff every open set containing p, contains a point of S.

ANSWERS/HINTS

1. (*i*) $[a, b]$ (*ii*) $[a, b]$ (*iii*) $[a, \infty[$ (*iv*) $]-\infty, a]$

2. (*i*) $\{1 - 2/n : n \in \mathbf{N}\} \cup \{1\}$ (*ii*) $\mathbf{R}$

(*iii*) $\{4 + (-1/10)^n : n \in \mathbf{N}\} \cup \{4\}$

(*iv*) $\{1/n : n \in \mathbf{N}\} \cup \{1/m + 1/n : m, n \in \mathbf{N}\} \cup \{0\}$

3.16. ISOLATED POINTS OF A SET AND DISCRETE SET

If a point $p \in S$ is not a limit point of S, then it is known as an *isolated point* of S. In other words, $p \in S$ is an isolated point of S if there exists a nbd of p which contains no point of S other than p.

A set S is known as a *discrete set* if all its points are isolated points. For example, the set $S = \{1, 1/2, 1/3, 1/4, ...\}$ is a discrete set because all the points of this set are isolated points.

Note. From the above definition, it follows that each point of any sub-set S of $\mathbf{R}$ is either a limit point of S or an isolated point of S.

3.17. DENSE (OR EVERYWHERE DENSE), DENSE IN ITSELF, NOWHERE DENSE (OR NON-DENSE) AND PERFECT SETS. [Kanpur 2012]

A set S is said to be *dense* in $\mathbf{R}$ if $\overline{S} = \mathbf{R}$.

A set S is said to be *dense in itself* if every point of S is a limit point of S, *i.e.*, if $S \subseteq S'$.

A set S is said to be *nowhere dense* relative to $\mathbf{R}$ if no nbd in $\mathbf{R}$ is contained in $\overline{S}$ or equivalently if the complement of $\overline{S}$ is dense in $\mathbf{R}$, *i.e.*, $\overline{(\overline{S})^c} = \mathbf{R}$.

Note that if S is an interval or contains an interval, then it cannot be nowhere dense set, since there exists an interval, say $I \subset \mathbf{R}$, such that $I \cap S \neq \phi$. Also, there exist sets containing no interval and yet they are not nowhere dense sets; for example the set of all rational numbers and the set of all irrational numbers are both nowhere dense sets.

A set S is said to be *perfect* if $S = S'$ or equivalently a set S is perfect when S is closed and dense in itself.

ILLUSTRATIONS

1. The set $\mathbf{Q}$ of rational numbers is dense in $\mathbf{R}$ as $\overline{\mathbf{Q}} = \mathbf{R}$. Again, the set $\mathbf{R} - \mathbf{Q}$ of irrational numbers is dense in $\mathbf{R}$.

2. The set $\mathbf{Q}$ of rational numbers is dense in itself but not closed. Similarly, the set $\mathbf{R} - \mathbf{Q}$ of irrational numbers is dense in itself but not closed.

3. A finite set is closed but not dense in itself.

4. $]a, b[$ is dense in itself.

5. $[a, b]$ is closed and dense in itself and so it is a perfect set. **[Kanpur 2012]**

6. If $S = \{1, 1/2, 1/3, ..., 1/n, ...\}$, $n \in \mathbf{N}$, then $S' = \{0\}$. S is neither closed nor dense in itself.

7. If $S = \{0, 1/2, 1/3, ..., 1/n, ...\}$, $n \in \mathbf{N}$, then $S' = \{0\}$. S is nowhere dense in $\mathbf{R}$ because 0 is the only limit point of S and no nbd of 0 in $\mathbf{R}$ is contained in $\overline{S}$.

8. The sets $\mathbf{R}$ and ϕ are perfect sets.

9. The sets $\mathbf{N}$ and $\mathbf{Z}$ are not dense in itself.

10. The closed unbounded intervals $[a, \infty[$ and $]-\infty, a]$ are perfect sets.

11. The sets $\mathbf{N}$ and $\mathbf{Z}$ are not dense in $\mathbf{R}$. **(Utkal, 2003)**

3.18. CANTOR NESTED INTERVAL THEOREM

For each $n \in \mathbf{N}$, let $I_n = [a_n, b_n]$ be a non-empty closed bounded interval of real numbers such that

$$I_1 \supset I_2 \supset I_3 \supset \ldots\ldots \supset I_n \supset I_{n+1} \supset \ldots\ldots$$

and $$\lim_{n\to\infty} (b_n - a_n) = \lim_{n\to\infty} l(I_n) = 0,$$

where $l(I_n)$ denotes the length of the interval I_n.

Then $\bigcap_{n=1}^{\infty} I_n$ contains precisely one point.

Proof. Given that $$I_n \supset I_{n+1} \;\forall\; n \in \mathbf{N} \qquad \ldots(1)$$

and $$\lim_{n\to\infty} (b_n - a_n) = \lim_{n\to\infty} l(I_n) = 0, \qquad \ldots(2)$$

where $$I_n = [a_n, b_n] \text{ and } l(I_n) = b_n - a_n. \qquad \ldots(3)$$

In view of (1) and (3), we have

$$a_1 < a_2 < a_3 < \ldots\ldots < a_n < \ldots\ldots \quad \text{and} \quad b_1 > b_2 > b_3 > \ldots\ldots > b_n > \ldots\ldots \qquad \ldots(4)$$

Then, $m > n \Rightarrow a_m < b_m \le b_n$ and $m \le n \Rightarrow a_m \le a_n < b_n$

Hence, we have $$a_m < b_n \;\forall\; m, n \in \mathbf{N}. \qquad \ldots(5)$$

Thus, each b_n is an upper bound for the set $S = \{a_1, a_2, a_3, \ldots\ldots\}$. Let $x = \sup S$.

Since each b_n is an upper bound of S and $x = \sup S$, it follows that

$$x \le b_n \;\forall\; n \in \mathbf{N}. \qquad \ldots(6)$$

Again, since x is an upper bound of S, so

$$a_n \le x. \qquad \ldots(7)$$

From (6) and (7), $a_n \le x \le b_n$ for each $n \in \mathbf{N}$ so that $x \in I_n$ for each $n \in \mathbf{N}$.

Hence, $$x \in \bigcap_{n=1}^{\infty} I_n. \qquad \ldots(8)$$

We now prove that there cannot be more than one point in $\bigcap_{n=1}^{\infty} I_n$. If possible, let y be another point such that

$$y \in \bigcap_{n=1}^{\infty} I_n. \qquad \ldots(9)$$

Then, from (9), we have $$a_n \le y \le b_n \;\forall\; n \in \mathbf{N} \qquad \ldots(10)$$

Let $|x - y| = \varepsilon$. Since $l(I_n) \to 0$ as $n \to \infty$, there exists an interval $[a_m, b_m]$ of the nest such that $b_m - a_m < \varepsilon$, which is impossible because $a_m \le x \le b_m$ and $a_m \le y \le b_m$.

Note 1. If the intervals I_n are not closed, the above theorem may not hold. For example, consider the nest of open intervals $I_n =]0, 1/n[$, $n \in \mathbf{N}$. Then $\bigcap_{n=1}^{\infty} I_n = \phi$.

Note 2. If $l(I_n)$ does not tend to zero as $n \to \infty$, then the intersection $\bigcap_{n=1}^{\infty} I_n$ may contain more than one point. For example, if $I_n =]-(n+1)/n, (n+1)/n[$. Then $l(I_n) = 2(1 + 1/n)$ which does not tend to zero as $n \to \infty$ and hence $\bigcap_{n=1}^{\infty} I_n = \{-1, 1\}$, showing that the intersection does not contain precisely one point.

3.19. COVER (OR COVERING) OF A SET

Let S be a non-empty sub-set of **R**. A family

$$F = \{G_\lambda : \lambda \in \Lambda\}$$

of sub-sets of **R** is said to be a *cover* of S if

$$S \subseteq \cup \{G_\lambda : \lambda \in \Lambda\},$$

Λ being an index set.

Equivalently, F is said to be cover of S if every point of S belongs to some member of the family F.

We say that F is an *open cover* if every member of F is an open set.

If there exist $\Lambda' \subset \Lambda$ such that Λ' is finite and the family of sub-sets $F' = \{G_\lambda : \lambda \in \Lambda'\}$ is also a cover of S, then the cover F' is said to be a *finite sub-cover* of the cover F of the set S.

ILLUSTRATIONS

1. Let $G_n =]-n, n[\ \forall\ n \in \mathbf{N}$. Then the family $F = \{G_n : n \in \mathbf{N}\}$ is an open cover of **R**. Let $H_n =]-2n, 2n[\ \forall\ n \in \mathbf{N}$. Then the family $F' = \{H_n : n \in \mathbf{N}\}$ is also an open cover of **R**. Since F' is a sub-family of F, it follows that F' is a sub-cover of F.

Clearly there exists no finite sub-cover of F.

2. Let $S = [-10, 10]$. Then the family $F = \{]-n, n[: n \in \mathbf{N}\}$ is an open cover of **R**. Again, the family $F' = \{]-n, n[: 1 \leq n \leq 11\}$ is a finite sub-family of F that covers S. Thus, there exists a finite sub-cover F' of F.

3.20. COMPACT SET

Let S be a non-empty sub-set of **R**. Then S is said to be *compact* if every open cover of S admits of a finite sub-cover.

ILLUSTRATIONS

1. *Every finite sub-set of* **R** *is compact.*

Let $S = \{a_i : 1 \leq i \leq n\}$ be a finite sub-set of **R**.

Let F be any open cover of S. Then, for each $a_i \in S$, there must exist an open set $G_i \in F$ such that $a_i \in G_i\ \forall\ i = 1, 2, \ldots, n$. Then the family $F' = \{G_i : 1 \leq i \leq n\}$ is clearly a finite open cover of S consisting of members of F. Hence the open cover F admits of a finite sub-cover F' and so F is a compact set.

2. *The set of real numbers* **R** *is not a compact set.*

Consider the family $F = \{]-n, n[: n \in \mathbf{N}\}$ of open sets.

Then, clearly $\cup \{]-n, n[: n \in \mathbf{N}\} = \mathbf{R}$,

showing that F is an open cover of **R**.

Let $F' = \{]-n_1, n_1[,]-n_2, n_2[, \ldots,]-n_m, n_m[\}$ be any sub-family of F.

Let $\max\{n_1, n_2, \ldots, n_m\} = k$, then

$$]-n_1, n_1[\cup]-n_2, n_2[\cup \ldots \cup]-n_m, n_m[=]-k, k[,$$

showing that F' is not a cover of **R**. Thus, we see that no sub-family of F can be a cover of **R**. Thus, there exists an open cover F of **R** which does not admit of a finite sub-cover and hence **R** is not a compact set.

3.21. PROPERTIES OF A COMPACT SET

Theorem I. *Every compact sub-set of* **R** *is bounded.*

Proof. Let S be a non-empty compact sub-set of **R.** Since the family of open sets $F = \{]-n, n[: n \in \mathbf{N}\}$ is an open cover of **R,** it follows that F is also an open cover of S.

Now, S is compact and F is its open cover

$\Rightarrow$ there exists a finite sub-cover of S, F' (say) given by

$$F' = \{]-n_1, n_1[,]-n_2, n_2[, \ldots,]-n_m, n_m[\}$$

$$\therefore \quad S \subset]-n_1, n_1[\cup]-n_2, n_2[\cup \ldots \cup]-n_m, n_m[\qquad \ldots(1)$$

Let $k = \max\{n_1, n_2, \ldots, n_m\}$. Then, (1) gives

$S \subset]-k, k[$ and hence S is bounded.

Theorem II. *Every compact sub-set of* **R** *is closed.*

Proof. Let S be a non-empty compact sub-set of **R.**

Let p be any point of $\mathbf{R} - S$. For each point $q \in S$, let $|p - q|/2 = \varepsilon_q$.

Let $G_q =]p - \varepsilon_q, p + \varepsilon_q[$ and $H_q =]q - \varepsilon_q, q + \varepsilon_q[$. Then $G_q \cap H_q = \phi$ and $p \in G_q$, $q \in H_q$.

Clearly, the family $F = \{H_q : q \in S\}$ is an open cover of S, for $S \subset \cup \{H_q : q \in S\}$.

S is compact and F is its open cover.

$\Rightarrow$ there exists a finite sub-cover of S, F' (say), given by

$F' = \left\{H_{q_1}, H_{q_2}, \ldots, H_{q_n}\right\}$, such that

$$S \subset H_{q_1} \cup H_{q_2} \cup \ldots \cup H_{q_n} \qquad \ldots(1)$$

Let $$G = G_{q_1} \cap G_{q_2} \cap \ldots \cap G_{q_n} \qquad \ldots(2)$$

and $$H = H_{q_1} \cup H_{q_2} \cup \ldots \cup H_{q_n} \qquad \ldots(3)$$

From (1) and (3), $$S \subset H. \qquad \ldots(4)$$

Since each of the sets, $G_{q_1}, G_{q_2}, \ldots, G_{q_n}$ contains p, so from (2) we have

$$p \in G. \qquad \ldots(5)$$

From the construction of open sets G_q and H_q, we have

$$G_{q_i} \cap H_{q_i} = \phi, \text{ where } i = 1, 2, \ldots, n \qquad \ldots(6)$$

Now, $G \cap H_{q_i} = (G_{q_1} \cap G_{q_2} \cap \ldots G_{q_n}) \cap H_{q_i}$

or $G \cap H_{q_i} = (G_{q_1} \cap H_{q_i}) \cap (G_{q_2} \cap H_{q_i}) \cap \ldots \cap (G_{q_n} \cap H_{q_i})$

or $G \cap H_{q_i} = \phi$, where $i = 1, 2, \ldots, n$ [using (6)] ...(7)

From (3) and (7), $$G \cap H = \phi \qquad \ldots(8)$$

Now, (4) and (8) $\Rightarrow G \cap S = \phi \Rightarrow G \subset \mathbf{R} \sim S$ ..(9)

Then, from (5) and (9), $$p \in G \subset \mathbf{R} - S,$$

showing that $\mathbf{R} - S$ is a nbd of p. Since p is any point of $\mathbf{R} - S$, it follows that $\mathbf{R} - S$ is a nbd of each of its points and hence $\mathbf{R} - S$ is an open set. Therefore, its complement S is a closed set.

Note. *Combining the above two theorems I and II, we see that a compact sub-set of* **R** *is closed and bounded.*

In what follows, we propose to show that if a sub-set of **R** is closed and bounded, then it must be a compact set. We begin with the well-known *Heine-Borel theorem.*

Theorem III. Heine-Borel theorem. *The closed and bounded interval* $[a, b]$ *is compact.*

[*M.S. Univ. T.N. 2006; Meerut, 2004; Delhi Maths (H), 2004*]

Proof. Let us write $a = a_1$, $b = b_1$ and $I_1 = [a, b] = [a_1, b_1]$.

Let $F = \{]c_\lambda, d_\lambda[: \lambda \in \Lambda\}$ be an open cover of I_1, Λ being an index set.

In order to prove that I_1 is compact, we shall show that there exists a finite sub-cover of F.

If possible, let I_1 be not compact. Then there exists no finite sub-cover of F.

Let us divide I_1 into two equal closed intervals

$$\left[a_1, \frac{a_1+b_1}{2}\right] \text{ and } \left[\frac{a_1+b_1}{2}, b_1\right]$$

Then, by our hypothesis, at least one of these two intervals cannot be covered by any finite sub-family of the open cover F. Let that particular interval be denoted by I_2 as follows :

$$I_2 = [a_2, b_2], \text{ where}$$

$$[a_2, b_2] = \left[a_1, \frac{a_1+b_1}{2}\right] \text{ or } \left[\frac{a_1+b_1}{2}, b_1\right].$$

As before, divide I_2 into two equal closed intervals

$$\left[a_2, \frac{a_2+b_2}{2}\right] \text{ and } \left[\frac{a_2+b_2}{2}, b_2\right].$$

In view of our hypothesis regarding non-occurrence of a finite sub-cover of F_1 at least one of these two intervals will not be covered by any finite sub-family of the open cover F. Let that particular interval be denoted by I_3 as follows :

$$I_3 = [a_3, b_3], \text{ where}$$

$$[a_3, b_3] = \left[a_2, \frac{a_2+b_2}{2}\right] \text{ or } \left[\frac{a_2+b_2}{2}, b_2\right].$$

On repeating the above process an infinite number of times, we arrive at a sequence of intervals $I_1, I_2, I_3, \ldots, I_n, \ldots$ and satisfying the following conditions :

(*i*) $I_1 \supset I_2 \supset I_3 \supset \ldots \supset I_n \supset I_{n+1} \supset \ldots$ *i.e.* $I_n \supset I_{n+1}\ \forall\ n \in \mathbf{N}$

(*ii*) $I_n = [a_n, b_n]$ is a closed interval $\forall\ n \in \mathbf{N}$

(*iii*) $\lim_{n\to\infty} l(I_n) = 0$, where $l(I_n)$ denotes the length of the interval I_n

(*iv*) I_n is not covered by any finite sub-family of F

In view of the conditions (*i*), (*ii*) and (*iii*), we find that the sequence of closed intervals $< I_n >$ satisfies all the conditions of Cantor nested interval theorem (refer Art. 3.18) and hence

$$\bigcap_{n=1}^{\infty} I_n \text{ contains precisely one point, say } p_0$$

$\therefore$ $$p_0 \in I_n\ \forall\ n \in \mathbf{N}. \qquad \ldots(1)$$

In view of the condition (*iii*), for a given $\varepsilon > 0$, there exists, $n_0 \in \mathbf{N}$ such that $l(I_{n_0}) < \varepsilon$.

Also, using (1), $$p_0 \in I_{n_0}$$

Let min. $\{l[p_0 - a_1, p_0 + b_1], l[p_0 - a_2, p_0 + b_2], \ldots, l[p_0 - a_n, p_0 + b_n], \ldots\} = \varepsilon$, where $l[p_0 - a_n, p_0 + b_n]$ denotes the length of the interval $[p_0 - a_n, p_0 + b_n]$ for each $n \in \mathbf{N}$.

Hence $$l(I_{n_0}) < \varepsilon \Rightarrow I_{n_0} \subset]p_0 - \varepsilon, p_0 + \varepsilon[,$$

which is contrary to the condition (*iv*) of the intervals I_n.

Thus we arrive at a contradiction. Hence our hypothesis is wrong. Therefore every open cover of $[a, b]$ will admit of a finite sub-cover and so $[a, b]$ is compact.

Theorem IV. *Every closed and bounded sub-set of R is compact.* ***(Agra, 2001)***

Proof. Let S be a non-empty sub-set of **R.** Let S be closed and bounded and let $F = \{G_\lambda : \lambda \in \Lambda\}$ be an open cover of S, Λ being an index set.

S is bounded $\Rightarrow \exists\, k > 0$ such that $|x| \le k\ \forall\ x \in S$

$\therefore$ $$S \subset [-k, k].$$

Also, S is closed $\Rightarrow$ $\mathbf{R} \sim S$ is an open set.

Let G denote the family of open sets consisting of $\mathbf{R} \sim S$ and all members of the family F. Then, G is an open cover of **R** and hence G is also an open cover of $[-k, k]$. By Heine-Borel theorem, the open cover G must admit of a finite sub-cover G' (say) of $[-k, k]$.

Since G' is a cover of $[-k, k]$ and $S \subset [-k, k]$, it follows that G' also covers S.

Two cases arise !

Case (*i*). Let $\mathbf{R} \sim S$ be not a member of the cover G' of S. Then, $G' \subset F$ and hence G' is a finite sub-cover of S by members of F.

Case (*ii*). Let $\mathbf{R} \sim S$ be a member of G'. Then, in view of $(\mathbf{R} \sim S) \cap S = \phi$, the collection G'' obtained by deleting the set $\mathbf{R} \sim S$ from G' is also an open cover of S. This new finite open cover of S consists of members all of which belong to the open cover F of S.

Thus, in both cases, the open cover F admits of a finite sub-cover and hence S is compact.

Theorem V. *If B is an infinite sub-set of a compact set A, then B has a limit point in A.* ***(G.N.D.U., 1998; Nagpur, 2003)***

Proof. Since A is a compact set, so A is closed and bounded.

Now, $B \subset A$ and A is bounded $\Rightarrow$ B is bounded.

Now B is an infinite bounded set. Hence, by Bolzano-Weierstrass theorem, there exists a limit point of B. Let p be a limit point of B, *i.e.*, $p \in B'$.

Now, $$B \subset A \Rightarrow B' \subset A'$$

$\therefore$ $$p \in B' \Rightarrow p \in A'.$$

Since A is closed, so $$A' \subset A$$

$\therefore$ $$p \in A' \Rightarrow p \in A,$$

showing that B has a limit point p belonging to A.

EXAMPLES

Example 1. *Which of the following sets are compact* :

(*i*) **N** (*ii*) **Q** (*iii*) $[a, \infty[$

(*iv*) $[a, b]$ (*v*) $]a, b]$ (*vi*) $[a, b] \cup [c, d]$

Solution. (*i*) **N** is not bounded, so it is not compact.

(*ii*) **Q** is neither closed nor bounded, so it is not compact.

(*iii*) $[a, \infty[$ is not bounded, so it is not compact.

(*iv*) $[a, b]$ is closed and bounded, so it is compact.

(*v*) $]a, b]$ is not closed, so it is not compact.

(*vi*) The set $[a, b] \cup [c, d]$ being the union of two closed and bounded sets is closed and bounded. Hence $[a, b] \cup [c, d]$ is a compact set.

Example 2. (*i*) *Give an example of a closed non-compact sub-set of* **R.**

(*ii*) *Show by means of an example that a bounded sub-set of* **R** *may fail to be compact.*

Hint. (*i*) **N,** the set of all natural numbers. (*ii*) [*a*, *b*[

Example 3. *If F is closed and K is compact, then prove that* $F \cap K$ *is compact.* **(*Nagpur, 2003*)**

Solution. Left as an exercise.

Example 4. *Prove that the union of a finite family of compact sets is compact.*

Solution. Left as an exercise.

OBJECTIVE QUESTIONS

I. Multiple Choice Type Questions : *Select* (*a*), (*b*), (*c*) *or* (*d*), *whichever is correct.*

1. Which is not nbd of 3. (*a*)]2, 4[(*b*)]1, 4[(*c*)]1, 3[(*d*)]0, 3[**(Kanpur, 2002)**

2. Nbd of 1/2 is the set

(*a*) [– 1/2, 1/2] (*b*) (0, 1/2] (*c*) (– ∞, ∞) (*d*) None of these. **(Kanpur, 2002)**

3. The set of irrational numbers **Q′** is nbd of :

(*a*) 1/3 (*b*) 1/2 (*c*) $\sqrt{2}$ (*d*) None of these. **(Kanpur, 2004)**

4. Which of the following is nbd of each of its points ?

(*a*) Set **Q** of rational numbers (*b*) Set **Q′** of irrational numbers

(*c*) Set **N** of natural numbers (*d*) Open interval (*a*, *b*). **(Kanpur, 2004)**

5. Which one of the following statements are true ?

(*a*) [2, 3] is a nbd of 1 (*b*) [4, 5] is a nbd of 5

(*c*) [2, 3[is a nbd of 3 (*d*) None of these.

6. The set of limit points of {1, 3, 7, 11} is :

(*a*) {1} (*b*) {11} (*c*) {1, 3, 7, 11} (*d*) None of these. **(Meerut, 2003)**

7. Let $f : \mathbf{R} \to \mathbf{R}$ be a constant function. Then image of **R** under *f* is :

(*a*) open (*b*) not open (*c*) empty (*d*) infinite. **(Bharathiar, 2004)**

8.]0, 1] is : (*a*) an open set (*b*) a closed set (*c*) neither open nor closed (*d*) None of these. **(Utkal, 2003)**

9. For any set *A*, A° is : (*a*) open (*b*) neither open nor closed (*c*) closed (*d*) None of these.

10. For any set *A*, $\overline{A}$ is : (*a*) open (*b*) neither open nor closed (*c*) closed (*d*) None of these.

11. The set $\{1/n\}_{n \in N}$ is (*i*) closed (*ii*) open (*iii*) neither open nor closed (*iv*) open but not closed. **(Ranchi 2010)**

II. Fill in the blanks :

1. If every limit point of a set belongs to the set, it is said to be ..... . **(Agra, 2002)**

2. The derived set of any set is ..... .

ANSWERS

Multiple Choice Type Questions :

1. (*a*) **2.** (*c*) **3.** (*c*) **4.** (*d*) **5.** (*c*) **6.** (*d*) **7.** (*b*) **8.** (*c*) **9.** (*a*) **10.** (*b*) **11.** (*iii*)

Fill in the blanks :

1. Closed set **2.** Closed set

MISCELLANEOUS PROBLEMS ON CHAPTER 3

1. State Bolzano-Weierstrass theorem for sets. Is it applicable to the set of all rational numbers ? Justify. **[Delhi Maths (H) 2006]**

2. Define a closed set. Show that a closed set contains all its limit points. Give an example to show that an arbitrary union of closed sets may not be a closed set **(Delhi Maths (G) 2006).**

3. Show that a set M is a nbd of a point p iff there exists a positive rational number r s.t. $]p - r, p + r[\subset$ M. **[Delhi Maths (G) 2006]**

4. Prove that every finite sub-set of $\boldsymbol{R}$ is a closed set. **[Kanpur 2006]**

5. Let $A, B, \subseteq \boldsymbol{R}$. Prove that $\overline{A \times B} = \overline{A} \times \overline{B}$. **[M.S. Univ. T.N. 2006]**

6. Show that the derived set of a set is a closed set **[Purvanchal 2006]**

7. Define nbd of a point $x \in \boldsymbol{R}$. When is a real number ξ a limit point of a set $S \subseteq \boldsymbol{R}$?

[Delhi Maths (H) 2007]

8. Let $S_n = [a + 1/n,\ a + 2],\ n \in \boldsymbol{N},\ a \in \boldsymbol{R}$ be a subset of $\boldsymbol{R}$. Is $\bigcup_{n=1}^{\infty} S_n$ a closed set ?

Explain. **[Delhi Maths (H) 2007]**

[Sol. Being a closed interval , S_n is a closed set $\forall n \in \boldsymbol{N}$ Now $\bigcup_{n=1}^{\infty} S_n =]a, a + 2]$. Clearly, $a \notin \bigcup_{n=1}^{\infty} S_n$, because $\forall n \in \boldsymbol{N}$, $a + 1/n > a$. Now, $]a, a + 2]$ being semi-open and semi-closed intervals is not a closed set.]

9. Let $T = \{1/n, n \in \mathbf{N}\} \cup \{1 + 3/2n, n \in \mathbf{N}\} \cup \{6 - 1/3n, n \in \mathbf{N}]$. Find derived set T' of T. Also find supremum of T and greatest number of T. **[I.A.S. 2008]**

10. Find all interior points of the following sets ? (*i*) S = $[a, b)$ (*ii*) $S = \mathbf{Q}$.

[Delhi Maths (Prog) 2008)

11. Which of the following sets are closed. Justify answer. (*i*) $S = \mathbf{N}$ (*ii*) $S = [a, b]$ (*iii*) $S = \mathbf{Q}$) **[Delhi Maths Prog. 2008]**

12. Define neighbourhood of a point. Show that the intersection of two neighbourhoods of a point of **R** is again a neighbourhood of the same point. What happens to the conclusion of the statement if arbitrary collection of neighbourhoods is taken. Justify. **[Delhi 2008]**

13. Show that neighbourhood of every point is an infinite set. **[Delhi B.A. III 2009]**

14. Define nested sequence of intervals and show this with a figure on real line. Also give an example of a nested sequence <I_n) with $\bigcap_{n=1}^{\infty} I_n = 0$. **[Delhi B.Sc. I (Hon) 2010]**

Ex. 15. *Find the limit points of {1,2,3,4}* **[Delhi B.A. (Prog.) II, 2014]**

Sol. Proceed as in illustration 6, page 3.6 to show that the given set has no limit points.

Ex. 16. *Is {1, 2, 3} closed ?* **[Delhi B.A. (Prog.) III, 2014]**

Sol. Proceed as in illustration 11, page 3.19 to show that the given set is closed.

Chapter

4

Countability of Sets

4.1. EQUIVALENT SETS

Two sets A and B are said to be equivalent if there exists a one-one onto mapping from A to B. If A and B are equivalent, we denote this relation by the symbol ~.

Thus $$A \sim B \Leftrightarrow A \text{ and } B \text{ are equivalent.}$$

For example, the sets $\mathbf{N} = \{1, 2, 3, ...\}$ of natural numbers and $E = \{2, 4, 6, ...\}$ of all even natural numbers are equivalent because there exists the mapping $f : \mathbf{N} \to E$ defined by $f(n) = 2n \ \forall \ n \in \mathbf{N}$ which is one-one from $\mathbf{N}$ onto E.

Theorem. *The relation '~' is an equivalence relation.*

Proof. (*i*) **'~' is reflexive.** Since for any set A, the identity mapping $I_A : A \to A$ is one-one and onto, it follows that $A \sim A$ for any set.

(*ii*) **'~' is symmetric.** Let $A \sim B$. Then there exists a mapping $f : A \to B$ which is one-one and onto. Its inverse mapping $f^{-1} : B \to A$ is also one-one onto so $B \sim A$. Thus $A \sim B \Rightarrow B \sim A$. Hence '~' is symmetric.

(*iii*) **'~' is transitive.** Let $A \sim B$ and $B \sim C$. Hence there exists one-one and onto mappings $f : A \to B$ and $g : B \to C$. Then we know that the composite mapping $gof : A \to C$ is also one-one and onto so that $A \sim C$. Thus, $A \sim B$ and $B \sim C \Rightarrow A \sim C$ and hence '~' is transitive.

Since '~' is reflexive, symmetric and transitive, it follows that '~' is an equivalence relation.

4.2. FINITE AND INFINITE SETS [Delhi B.Sc. (Prog.) II 2008, 2017]

A set is called *finite* and is said to contain n elements if $A \sim \{1, 2, 3, ..., n\}$. If a set is not finite, then it is said to be *infinite*.

If $A \sim \{1, 2, 3, ..., n\}$, then n is called the ***cardinal number*** of A.

ILLUSTRATIONS

1. The set ϕ is a finite set.
2. The set of all primes less than 100 is a finite set.
3. The set of all natural numbers is an infinite set.
4. The set of all positive even integers is an infinite set.

4.3. DENUMERABLE (OR ENUMERABLE OR COUNTABLY INFINITE), COUNTABLE AND UNCOUNTABLE SETS

[Kanpur 2000; Meerut 2003, 04; Delhi B.Sc. (Prog) II 2011; 2012; 2017]

A set A is said to be *denumerable* if there exists a one-one mapping from the set $\mathbf{N}$ of all natural numbers onto the set A, *i.e.*, if $A \sim \mathbf{N}$.

A set is said to be *countable* if either A is finite or A is denumerable, *i.e.*, if either A is finite or $A \sim \mathbf{N}$. If a set is not countable, then it is said to be *uncountable*.

ILLUSTRATIONS

(*i*) The empty set is countable.

(*ii*) The set of all natural numbers is denumerable, because identity mapping $I_N : \mathbf{N} \to \mathbf{N}$ is one-one onto from **N** to **N**.

Note: The following statements are equivalent:

(*i*) A is a countable set (*ii*) There exists a surjection of **N** on *A* (*iii*) There exists injection of all *A* into **N**.

Theorem I. (*i*) *Every sub-set of a finite set is finite.*

(*ii*) *Every super-set of an infinite set is infinite.*

(*iii*) *If A and B are finite sets, then* $A \cap B$ *is also a finite set.*

(*iv*) *If A and B are finite sets, then* $A \cup B$ *is also a finite set.*

Theorem II. *Every sub-set of a countable set is countable.* **[Kanpur 2015, 16; Meerut 2003; Delhi B.Sc. (Prog.) II 2016; Delhi B.Sc. (Prog.) III 2012]**

Proof. Let *A* be a countable set and let *B* be a sub-set of *A*. Since *A* is a countable, so *A* is either finite or denumerable.

Case I. Let *A* be a finite set. Since every sub-set of a finite set is finite, it follows that *B* is finite and hence countable.

Case II. Let *A* be an infinite denumerable set. Let $A = \{a_1, a_2, a_3, ...\}$.

Two cases arise :

Case IIA. If *B* is finite, then *B* is countable.

Case IIB. If *B* is infinite, then each element of *B* will be a_r for some positive integer *r*. Let n_1 be least positive integer such that $a_{n_1} \in B$. Since *B* is infinite, so $B \neq \{a_{n_1}\}$. Let n_2 be the least positive integer such that $a_{n_2} \in B$ and $a_{n_2} \in A - \{a_{n_1}\}$. Since B is infinite, so $B \neq \{a_{n_1}, a_{n_2}\}$. Let n_3 be the least positive integer such that $a_{n_3} \in B$ and $a_{n_3} \in A - \{a_{n_1}, a_{n_2}\}$.

Continuing likewise, we find $B = \{a_{n_1}, a_{n_2}, a_{n_3}, ...\}$ where $n_1 < n_2 < n_3 < .....$.

Define a mapping $\quad f : \mathbf{N} \to B$ by $f(k) = a_{n_k} \;\; \forall\ k \in \mathbf{N}$.

Since *f* is one-one and onto, so *B* is denumerable and hence *B* is countable.

Corollary I. *Every infinite sub-set of a denumerable set is denumerable.* **(Chennai 2011)**

Corollary II. *If A and B are countable sets, then* $A \cap B$ *is also a countable set.*

Proof. Left as exercises.

Theorem III. *Every super-set of an uncountable set is uncountable.*

Proof. Let *A* be an uncountable set and let *B* be any super-set of *A*. Let, if possible, *B* be a countable set. Then *A* being a sub-set of a countable set must be countable (by theorem II), which is a contradiction of the given fact that *A* is an uncountable set. Hence *B* must be uncountable.

Theorem IV. *Every infinite set has a denumerable sub-set.*

Proof. Let *A* be an infinite set. Let $a_1 \in A$.

Since *A* is infinite, $A \neq \{a_1\}$. Hence there exists $a_2 \neq a_1$ such that $a_2 \in A$. Since *A* is infinite, $A \neq \{a_1, a_2\}$. So there exists $a_3 \neq a_2 \neq a_1$ such that $a_3 \in A$.

Proceeding likewise, we obtain a sub-set $B = \{a_1, a_2, a_3, ...\}$ of *A*.

Define a mapping $\quad f : \mathbf{N} \to B$ by $f(k) = a_k \;\; \forall\ k \in \mathbf{N}$.

Since *f* is one-one and onto, so *B* is denumerable.

Note. From the above theorem, it follows that countably infinite sets are the smallest infinite sets.

Theorem V. *Every infinite set is equivalent to one of its proper sub-sets.*

Proof. Let A be an infinite set. Then, by theorem IV, A has a countable sub-set, C (say), given by

$$C = \{a_1, a_2, a_3, \ldots\ldots, a_n, \ldots\ldots\}$$

Let $D = A - \{a_1\}$ and let $B = (A - C) \cup D$.

Since $a_1 \in A$ and $a_1 \notin B$, it follows that B is a proper sub-set of A.

Define a mapping $f : A \to B$ by

$$f(x) = x \;[\; x \in A - C \text{ and } f(a_r) = a_{r+1} \text{ for } a_r \in C, r = 1, 2, 3, \ldots\ldots.$$

Since f is one-one mapping from A to B, so $A \sim B$. Hence the result.

Theorem VI. *The countable union of countable sets is countable* **[Rajasthan 1998, Garhwal 2003; Kanpur 2000, 02; 10; Meerut 1995; 97: Kurukshetra 1997]**

or The union of a countable family of pairwise disjoint sets is countable.**(Kanpur 2003, 11)**

or *If $A_1, A_2, A_3, \ldots\ldots$ are countable sets, then $\bigcup_{n=1}^{\infty} A_n$ is countable.* ***(Nagpur 2003)***

Proof. Let $\{A_i, i \in \mathbf{N}\}$ be a countable collection of countable sets. Since $A_1, A_2, A_3, \ldots\ldots$ are countable, they can be represented as follows :

$$A_1 = \{a_{11}, a_{12}, a_{13}, \ldots\ldots, a_{1n}, \ldots\ldots\}$$
$$A_2 = \{a_{21}, a_{22}, a_{23}, \ldots\ldots, a_{2n}, \ldots\ldots\}$$
$$A_3 = \{a_{31}, a_{32}, a_{33}, \ldots\ldots, a_{3n}, \ldots\ldots\}$$
$$\cdots\cdots\cdots\cdots\cdots\cdots\cdots\cdots\cdots\cdots$$
$$A_n = \{a_{n1}, a_{n2}, a_{n3}, \ldots\ldots, a_{nn}, \ldots\ldots\}$$
$$\cdots\cdots\cdots\cdots\cdots\cdots\cdots\cdots\cdots\cdots$$

Then a_{ij} is the jth element of A_i. We observe that there is only one element a_{11} the sum of whose suffices is 2. There are only two elements a_{12}, a_{21} the sum of whose suffices is 3. There are three elements a_{13}, a_{22}, a_{31}, the sum of whose suffices is 4. In this manner we find that for any positive integer $m \geq 2$, there are $(m - 1)$ elements, sum of whose suffices is m.

Therefore we can arrange the elements of $A_1, A_2, A_3, \ldots\ldots$ according to their sum of the suffices as

$$a_{11}, a_{12}, a_{21}, a_{31}, a_{22}, a_{13}, \ldots\ldots,$$

where we have removed the elements a_{ij} which have already occurred. Thus all the elements can be counted out and hence $\bigcup_{n=1}^{\infty} A_n$ is countable.

Pictorially, the above counting process can be represented as follows :

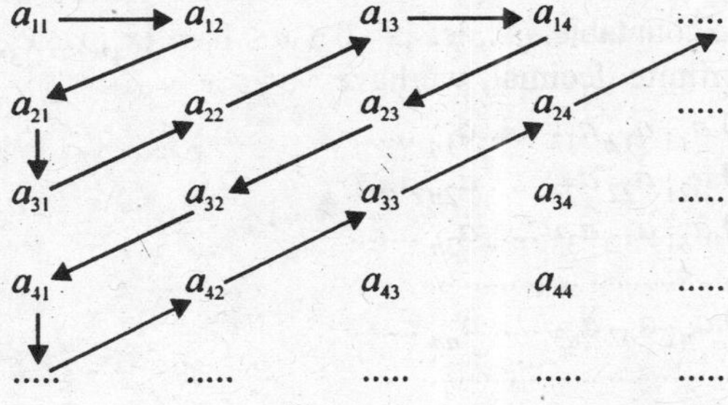

Theorem VII. *The set of all rational numbers is countable.*

[*Kanpur 2019; Meerut 2003; M.S. Univ. T.N. 2006; Purvanchal 2003*]

Proof. Let A_n be the set of all rational numbers which can be written with denominator n.

Thus,
$$A_1 = \left\{\frac{0}{1}, \frac{-1}{1}, \frac{1}{1}, \frac{-2}{1}, \frac{2}{1}, \ldots\right\}$$
$$A_2 = \left\{\frac{0}{2}, \frac{-1}{2}, \frac{1}{2}, \frac{-2}{2}, \frac{2}{2}, \ldots\right\}$$
$$\ldots\ldots\ldots\ldots\ldots\ldots\ldots\ldots$$
$$A_n = \left\{\frac{0}{n}, \frac{-1}{n}, \frac{1}{n}, \frac{-2}{n}, \frac{2}{n}, \ldots\right\}$$
$$\ldots\ldots\ldots\ldots\ldots\ldots\ldots\ldots$$

Then the set of all rational numbers $= \mathbf{Q} = \cup \{A_i : i \in \mathbf{N}\}$

Let us consider the mapping $f : \mathbf{N} \to A_n$ defined by
$$f(m) = (m-1)/2n, \text{ when } m \text{ is odd}$$
$$= (-m)/2n, \quad \text{when } m \text{ is even}$$

Then clearly, f is one-one and onto. Hence $A_n \sim \mathbf{N}$ and so A_n is countable for each $n \in \mathbf{N}$.

Since $\mathbf{Q} = \cup \{A_i : i \in \mathbf{N}\}$ is the countable union of countable sets, hence $\mathbf{Q}$ is a countable set.

Corollary 1. *The set of all positive rational numbers is countable.*

[*Delhi B.Sc. (Prog.) II, 2016*]

Proof. Let $\mathbf{Q}$ and $\mathbf{Q}^+$ denote respectively the set of all rational numbers and the set of positive rational numbers. Then $\mathbf{Q}^+ \subset \mathbf{Q}$. Since every sub-set of a countable set is countable and $\mathbf{Q}$ is countable, so $\mathbf{Q}^+$ is also countable.

Corollary 2. *The set of all negative rational numbers is countable.*

Proof. Proceed as in Corollary 1 above.

Corollary 3. *The set of all rational numbers in [0, 1] is countable.*

Proof. Let A be the set of all rational numbers in [0, 1]. Then $A \subset \mathbf{Q}$. Since every sub-set of a countable set is countable and $\mathbf{Q}$ is countable, so A is also countable. Hence proved.

Theorem VIII. *The unit interval [0, 1] is uncountable*, i.e., *the set of all real numbers in the closed interval [0, 1] is not enumerable*

or *The set of real numbers x such that $0 \le x \le 1$ is not countable.*

[*Garhwal 2003; Himachal 2002; Meerut 2004; Kanpur 2016, 18, 19; Utkal 2003*]

Proof. We know that every real number can be expressed uniquely as a non-terminating decimal of the form $0.a_1 a_2 a_3 \ldots a_n \ldots$ where $0 \le a_i \le 9$ for each i. Each decimal expansion is supposed to contain an infinite number of non-zero digits; finite expansions may be written with repeated nine's at the end such as 1 can be written as 0.9999 .... and 0.5 can be written as 0.49999 .... .

Let, if possible, [0, 1] be countable, *i.e.*, let $\{x : 0 \le x \le 1\} = \{x_1, x_2, x_3, \ldots, x_n, \ldots\}$. Expanding each x_i in the form of an infinite decimal, we have
$$x_1 = 0.a_{11}\, a_{12}\, a_{13} \ldots a_{1n} \ldots$$
$$x_2 = 0.a_{21}\, a_{22}\, a_{23} \ldots a_{2n} \ldots$$
$$x_3 = 0.a_{31}\, a_{32}\, a_{33} \ldots a_{3n} \ldots$$
$$\ldots\ldots\ldots\ldots\ldots\ldots\ldots\ldots$$
$$x_n = 0.a_{n1}\, a_{n2}\, a_{n3} \ldots a_{nn} \ldots$$
$$\ldots\ldots\ldots\ldots\ldots\ldots\ldots\ldots$$

where each $a_{ij} \in \{0, 1, 2, 3, \ldots, 9\}$ and each decimal contains an infinite number of non-zero elements.

We now proceed to construct a real number

$$y = 0.b_1b_2b_3 \,.....\, b_n \,.....$$

which will belong to [0, 1] and $y \neq x_n$, $\forall\ n \in \mathbf{N}$.

Let us choose b_n for each $n \in \mathbf{N}$, as follows

$$b_n = 1, \text{ when } a_{nn} \neq 1 \text{ and } b_n = 2, \text{ when } a_{nn} = 1$$

Thus we choose $b_1 = 1$ when $a_{11} \neq 1$ and $b_1 = 2$ when $a_{11} = 1$. Then $b_n \neq a_{nn}$ $\forall\ n \in \mathbf{N}$. Now y differs from x_1 in the first decimal place because $b_1 \neq a_{11}$, it differs from x_2 in the second decimal place because $b_2 \neq a_{22}$, ..., it differs from x_n in the nth decimal place because $b_n \neq a_{nn}$. Moreover the decimal expansion of y is unique because $b_n \notin \{0, 9\}$. Hence $y \neq x_n$ $\forall\ n \in \mathbf{N}$. Thus we find that $b \in [0, 1]$ and it escapes assumed enumeration $\{x_1, x_2, x_3, ..., x_n, ...\}$ of [0, 1]. So the assumption that [0, 1] is countable leads to a contradiction. Hence [0, 1] is not countable.

Corollary 1. *The set of real numbers is not countable.* ***[Kanpur 2000, 02, 09, 14]***

Proof. Let, if possible, the set **R** of real numbers be countable. Then, [0, 1] which is a sub-set of **R** must be countable because every sub-set of a countable set is countable. But [0, 1] is not countable. Thus the assumption that **R** is countable leads to a contradiction. Hence **R** cannot be countable.

Corollary 2. *The set of irrational numbers is uncountable.* ***[Garhwal 2003; Chennai 2011]***

Proof. Let A denote the set of irrational numbers. Let, if possible, A be countable. We know that the set **Q** of rational numbers is countable.

Since A and **Q** are countable, so their union $A \cup \mathbf{Q}$, *i.e.*, **R** must be countable. But **R** is not countable. Thus the assumption that A is countable leads to a contradiction. Hence A, *i.e.*, the set of irrational numbers is uncountable.

Theorem IX. *The set* $\mathbf{N} \times \mathbf{N}$ *is countable, where* **N** *is the set of natural numbers.*

(Kanpur 2011, 13)

Proof. Let us consider the following countable collection of sets.

$$A_1 = \{(1, 1), (1, 2), (1, 3), ..., (1, n), ...\}$$
$$A_2 = \{(2, 1), (2, 2), (2, 3), ..., (2, n), ...\}$$
$$...$$
$$A_n = \{(n, 1), (n, 2), (n, 3), ..., (n, n), ...\}$$
$$...$$

Consider the mapping $f : A_n \to \mathbf{N}$ defined by $f(n, m) = m$.

Clearly f is one-one and onto. Hence $A_n \sim \mathbf{N}$ and so A_n is countable.

Also, $$\mathbf{N} \times \mathbf{N} = \cup \{A_n : n \in \mathbf{N}\},$$

showing that $\mathbf{N} \times \mathbf{N}$ is a union of countable collection of countable sets and hence $\mathbf{N} \times \mathbf{N}$ is itself countable.

Corollary. *If A and B are countable sets, prove that* $A \times B$ *is countable.*

(Purvanchal 2004)

Or

The cartesian product of two countable sets is countable.

Proof. Let A and B be countable sets and $A \times B = \{(a, b) : a \in A \text{ and } b \in B\}$ be their cartesian product. Three cases arise :

Case I. Let one of the two sets A and B be empty. Then $A \times B = \phi$ and the result follows.

Case II. Let one of the two sets A and B be finite. Then, if A is finite with m elements, then the cartesian product of $A = \{a_1, a_2, ..., a_m\}$ and $B = \{b_1, b_2, ..., b_n, ...\}$ is given by

$$A \times B = \{(a_1, b_1), (a_1, b_2), \ldots\ldots, (a_1, b_n), \ldots\ldots$$
$$(a_2, b_1), (a_2, b_2), \ldots\ldots, (a_2, b_n), \ldots\ldots$$
$$\ldots\ldots\ldots\ldots\ldots\ldots\ldots\ldots\ldots\ldots\ldots\ldots$$
$$(a_m, b_1), (a_m, b_2), \ldots\ldots, (a_m, b_n), \ldots\ldots\}$$

Let us list the elements of $A \times B$ as follows :

$(a_1, b_1), (a_2, b_1), \ldots, (a_m, b_1); (a_1, b_2), (a_2, b_2), \ldots, (a_m, b_2); \ldots, (a_1, b_n), (a_2, b_n), \ldots, (a_m, b_n); \ldots$

Thus, $A \times B \sim \mathbf{N}$ and hence $A \times B$ is countable.

Case III. Let $A = \{a_1, a_2, \ldots\ldots, a_n, \ldots\ldots\}$ and $B = \{b_1, b_2, \ldots\ldots, b_n, \ldots\ldots\}$ be two countably infinite sets. Consider a mapping $f : \mathbf{N} \times \mathbf{N} \to A \times B$ defined by

$$f(i, j) = (a_i, b_j)\ \forall\ (i, j) \in \mathbf{N} \times \mathbf{N} \qquad \ldots(1)$$

Let $x = (m, n)$ and $y = (p, q)$ be any two arbitrary elements of $\mathbf{N} \times \mathbf{N}$. Then, we have

$$f(x) = f(y) \Rightarrow f(m, n) = f(p, q) \Rightarrow (a_m, b_n) = (a_p, b_q)$$
$$\Rightarrow a_m = a_p \text{ and } b_n = b_q \Rightarrow m = p \text{ and } n = q$$
$$\Rightarrow (m, n) = (p, q) \Rightarrow x = y$$

Thus, $\qquad f(x) = f(y) \Rightarrow x = y,$

showing that f is a one-one mapping.

We now show that f is an onto mapping. Let (a_i, b_j) be an arbitrary element of $A \times B$. Then, by definition (1), we have $(i, j) \in \mathbf{N} \times \mathbf{N}$ such that $f(i, j) = (a_i, b_j)$. Hence f is onto. Thus, f is one-one and onto and so $A \times B \sim \mathbf{N} \times \mathbf{N}$. Since $\mathbf{N} \times \mathbf{N}$ is countable, it follows that $A \times B$ is also countable.

Theorem X. *Any open interval* $]a, b[$ *is equivalent to any other open interval* $]c, d[$.

Proof. Consider a mapping $f :]a, b[\to]c, d[$ defined by

$$f(x) = c + \frac{d-c}{b-a}(x-a),\ \forall\ x \in]a, b[\qquad \ldots(1)$$

We easily see that if $x \in]a, b[$, then $f(x) \in]c, d[$ as follows. Clearly, $x = (a + b)/2 \in]a, b[$. Then, from (1), we have

$$f(x) = c + \frac{d-c}{b-a}\left(\frac{a+b}{2} - a\right) = c + \frac{d-c}{2} = \frac{c+d}{2} \in]c, d[$$

Also, we can easily verify that f is one-one and onto mapping. Hence $]a, b[\sim]c, d[$.

Theorem XI. (*i*) The intervals $]0, 1[$ and $[0, 1]$ are equivalent **[I.A.S. 2008]**

(*ii*) The intervals $[0, 1]$ and $[0, 1[$ are equivalent

(*iii*) The intervals $[0, 1]$ and $]0, 1]$ are equivalent

(*iv*) Any open interval is equivalent to $[0, 1]$.

Proof. (*i*) Let $A = [0, 1] - \{0, 1, 1/2, 1/3, \ldots\ldots\} =]0, 1[- \{1/2, 1/3, 1/4, \ldots\ldots\}$

Then, we have

$$[0, 1] = A \cup \{0, 1, 1/2, 1/3, \ldots\ldots\}$$

and $\qquad]0, 1[= A \cup \{1/2, 1/3, 1/4, \ldots\ldots\}$

Consider the mapping $f : [0, 1] \to]0, 1[$ defined as follows :

$$f(x) = x, \text{ when } x \in A$$
$$= 1/2, \text{ when } x = 0$$
$$= 1/(n+2), \text{ if } x = 1/n \text{ where } n \in \mathbf{N}$$

Then the mapping f is one-one and onto. Hence $[0, 1] \sim]0, 1[$.

(*ii*) Using the notation of part (*i*) above, let us define a mapping $f: [0, 1] \to [0, 1[$ as follows :

$$f(x) = \begin{cases} 1/(n+1), \text{ when } x = 1/n, \text{ where } n \in \mathbf{N} \\ x, \text{ when } x \neq 1/n, \text{ where } n \in \mathbf{N} \end{cases}$$

Then the function f is one-one and onto. Hence $[0, 1] \sim [0, 1[$.

(*iii*) Let us define a mapping $f: [0, 1[\to]0, 1]$ as follows :

$$f(x) = 1 - x$$

Then f is one-one and onto. Hence $[0, 1[\sim]0, 1]$.

Again from part (*ii*), $[0, 1] \sim [0, 1[$. Thus, we have

$$[0, 1] \sim [0, 1[\quad \text{and} \quad [0, 1[\sim]0, 1] \Rightarrow [0, 1] \sim]0, 1],$$

because an equivalence relation is transitive.

(*iv*) From theorem X, any two open intervals are equivalent. So $]a, b[\sim]0, 1[$.

But $\qquad]0, 1[\sim [0, 1]$ by part (*i*).

So $]a, b[\sim]0, 1[$ and $]0, 1[\sim [0, 1] \Rightarrow]a, b[\sim [0, 1]$ as an equivalence relation is transitive.

Theorem XII. *Any interval is equivalent to the set* **R** *of real numbers. In particular* **R** *is equivalent to* $[0, 1]$.

Proof. Consider an interval $I =]-\pi/2, \pi/2[$. Define a mapping $f: I \to \mathbf{R}$ as follows :

$$f(x) = \tan x, \ \forall\, x \in I.$$

Then f is one-one and onto mapping and so $I \sim \mathbf{R}$. Let I' be any interval. Then, we know that $I' \sim I$.

Thus, $\qquad I' \sim I \quad \text{and} \quad I \sim \mathbf{R} \Rightarrow I' \sim \mathbf{R}$.

Taking $I' = [0, 1]$, the result of the second part of the theorem follows.

4.4. DECIMAL, TERNARY AND BINARY REPRESENTATION

(*i*) **Decimal representation.** The digits 0, 1, 2, 3, ..., 9 are said to be the *decimal digits*. When the series

$$\frac{d_1}{10} + \frac{d_2}{10^2} + \frac{d_3}{10^3} + \ldots\ldots + \frac{d_n}{10^n} + \ldots\ldots$$

converges to x with each d_n as a decimal digit, then a *decimal representation* of x is given by

$$x = 0.d_1 d_2 d_3 \ldots\ldots$$

For example, $\dfrac{2}{10} + \dfrac{5}{10^2} + \dfrac{7}{10^3} + \dfrac{1}{10^4} + \ldots\ldots$ represents $0.2571 \ldots\ldots$

(*ii*) **Ternary representation.** The digits 0, 1, 2 are said to be *ternary digits*. When the series

$$\frac{t_1}{3} + \frac{t_2}{3^2} + \frac{t_3}{3^3} + \ldots\ldots + \frac{t_n}{3^n} + \ldots\ldots$$

converges to x with each t_n as a ternary digit, then a *ternary representation* of x is given by

$$x = 0.t_1 t_2 t_3 \ldots\ldots$$

For example, $\qquad \dfrac{0}{3} + \dfrac{2}{3^2} + \dfrac{0}{3^4} + \dfrac{2}{3^4} + \ldots\ldots = \dfrac{2/9}{1-(1/9)} = \dfrac{1}{4}$

Thus, $\qquad 1/4 = 0.020202 \ldots\ldots$

Similarly, $\qquad 3/4 = 0.202020 \ldots\ldots$

Observe that the ternary expansions of 1/4 and 3/4 are non-terminating and hence in such cases the ternary representation is unique.

Now, $\frac{1}{3} = 0.1 = 0.10000$ .....

Also, $\frac{2}{3^2} + \frac{2}{3^3} + \frac{2}{3^4} + ... = \frac{2/9}{1-1/3} = \frac{1}{3}$ and so $\frac{1}{3} = 0.0222$ .....

Thus we see that some numbers (like 1/3 etc.) can be represented in two ways. Reader can verify that 2/3 has also two ternary representations, namely, 2/3 = 0.1222 ..... = 0.2000 ..... .

(iii) **Binary representation.** The digits 0 and 1 are said to be *binary digits*. When the series

$$\frac{b_1}{2} + \frac{b_2}{2^2} + \frac{b_3}{2^3} + ... + \frac{b_n}{2^n} + ...$$

converges to x with each b_n as a binary digit, then a *binary representation* of x is given by

$$x = 0.b_1 b_2 b_3 .....$$

For example, $\frac{0}{2} + \frac{1}{2^2} + \frac{0}{2^3} + \frac{1}{2^4} + \frac{0}{2^5} + \frac{1}{2^6} + ... = \frac{1/4}{1-1/4} = \frac{1}{3}$

Thus, $1/3 = 0.01010101$ .....

Similarly, $1/4 = 0.010000$ ..... .

Theorem I. *Let A be a set of all sequences whose elements are the digits 0 and 1. Then the set A is countable.* ***(G.N.D.U. Amritsar 1998; Nagpur 2003)***

Proof. By hypothesis, the members of the given set A are sequences of type <1, 0, 0, 1, 0, 0, ···>. Let B be any countable sub-set of A so that $B = \{s_1, s_2, s_3, ...\}$, where each s_i is a sequence with 0 and 1 as its elements.

We shall now construct a sequence s such that $s \in A$ but $s \notin B$. If the nth digit in s_n is 1, then we shall take the nth digit of s to be 0 and *vice-versa*. In this manner, the sequence s will be different from each member of B in at least one place. Thus we see that there exists a sequence s such that $s \in A$ but $s \notin B$ and hence B is a proper sub-set of A. Since B is any sub-set of A, so every countable sub-set of A is a proper sub-set of A. Consequently, A is uncountable. For otherwise A would be a proper sub-set of A and hence A would become countable which is not possible.

4.5. CANTOR SET OR CANTOR TERNARY SET *(Meerut 2011)*

We shall construct a set, known as Cantor set. Divide the closed interval [0, 1] into three equal parts and remove the middle third open interval]1/3, 2/3[. This leaves two disjoint closed intervals [0, 1/3] and [2/3, 1] each having length 1/3. We now divide each of the two closed intervals [0, 1/3] and [2/3, 1] into three equal parts and remove their middle third open intervals]1/9, 2/9[and]7/9, 8/9[. This leaves four disjoint closed intervals [0, 1/9], [2/9, 1/3], [2/3, 7/9] and [8/9, 1].

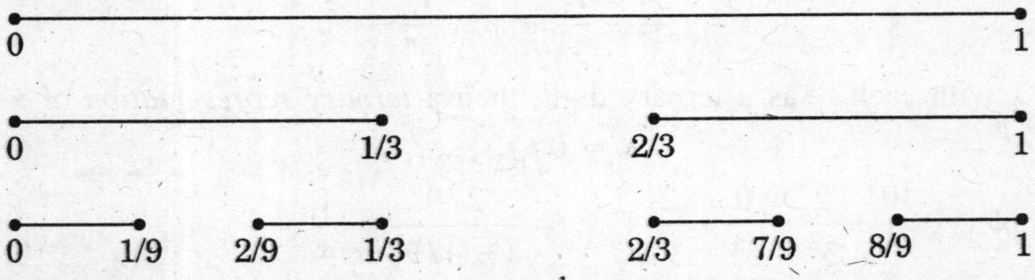

In general, at the nth stage we remove 2^{n-1} open intervals and are left with 2^n closed intervals, each having length $1/3^n$. The above process may be carried on indefinitely.

For describing the Cantor set, we shall use the ternary representation of real numbers $x = 0.t_1 t_2 t_3$ ..... . But the Cantor's set consists of only those ternary representations in which t's have the value 0 and 2 and never the value 1.

Recall that the open interval]1/3, 2/3[is removed in the first stage. Since 1/3 = 0.1 and 2/3 = 0.2, all the points of the open interval]1/3, 2/3[are of the form $0.1t_2t_3$ ..... .

Again, the open intervals]1/9, 2/9[and]7/9, 8/9[are removed in the second stage. Now, it is easy to verify that 1/9 = 0.0100 ..... and 2/9 = 0.2000 ..... . Hence all the numbers in the open interval]1/9, 2/9[will be of the form $0.01t_3t_4$ ..... always having 1 at the second place after the decimal. Similarly, since 7/9 = 0.21000 ..... and 8/9 = 0.22000 ..... all numbers lying in the open interval (7/9, 8/9) will be of the form $0.21t_3t_4$ ..... always having 1 at the second place after the decimal. Proceeding likewise, we see that all the numbers removed in the *n*th stage will have 1 in the *n*th position in the ternary scale. Hence all the left out numbers will not have 1 in their expansions. Thus the Cantor set is the set of all numbers $x \in [0, 1]$ which have a ternary representation without digit 1, where we suppose that if the end point like 1/3 has two representations, then we shall consider that representation which is free from digit 1.

The set of points K given by the end points 1/3, 2/3, 1/9, 2/9, 7/9, 8/9, ... of the intervals that have been removed together with limiting points of these end points is known as *Cantor ternary set* or simply *Cantor set.*

Note 1. When the above process of construction of Cantor set is repeated indefinitely, the number of portions removed and their lengths are given in the following table.

Number of portions removed	: 1	2	2^2	2^3	.....	2^{n-1}	.....
Length of each portions	: 1/3	$1/3^2$	$1/3^3$	$1/3^4$	.....	$1/3^n$	.....

Now, the sum of the length of intervals removed in [0, 1]

$$= \frac{1}{3} + 2 \times \frac{1}{3^2} + 2^2 \times \frac{1}{3^3} + 2^3 \times \frac{1}{3^4} + \ldots\ldots + 2^{n-1} \times \frac{1}{3^n} + \ldots\ldots$$

$$= \frac{1}{3} + \frac{1}{3}\cdot\frac{2}{3} + \frac{1}{3}\cdot\left(\frac{2}{3}\right)^2 + \frac{1}{3}\cdot\left(\frac{2}{3}\right)^3 + \ldots\ldots \text{ad inf} = \frac{1/3}{1-(2/3)} = 1$$

Note 2. It can be easily proved that the Cantor set is (*i*) Closed and bounded (*ii*) Compact (*iii*) non-dense.

Theorem I. *The Cantor set is equivalent to [0, 1].*

Proof. We know that the Cantor set consists of points whose ternary representation is possible without the use of 1. Thus, if x be a member of the Cantor set, then $x = 0.t_1t_2t_3$ ..... where t's have the values 0 and 2. Hence a one-to-one correspondence can be established between the Cantor set and the set of all numbers in [0, 1] in the binary scale. For example, $(0.02)_3 \to (0.01)_2$, $(0.0202)_3 \to (0.0101)_2$ etc.

Since the set of all real numbers in [0, 1] in the binary scale is equivalent to [0, 1] and the set of all numbers in Cantor set are equivalent to the set of all real numbers in [0, 1] in the binary scale, it follows that the Cantor set is equivalent to [0, 1] (remembering that an equivalent relation is transitive).

Corollary. *The Cantor set is not countable.* **(*Kanpur 2003*)**

Proof. Since Cantor set is equivalent to [0, 1] and [0, 1] is not countable, it follows that the Cantor set is not countable.

EXAMPLES

Example 1. *Show that the set of all integers* **Z** *is countable.*

Solution. Consider the mapping $f : \mathbf{N} \to \mathbf{Z}$ defined by

$$f(n) = (n-1)/2, \quad \text{if } n = 1, 3, 5, \ldots\ldots$$
$$= -(n/2), \quad \text{if } n = 2, 4, 6, \ldots\ldots$$

Then f is one-one and onto and so $\mathbf{N} \sim \mathbf{Z}$ and hence **Z** is countable.

Example 2. *Prove that the set* $\{\pm 1, \pm 4, \pm 9, \pm 16, ...\}$ *is countable.* **(Kanpur 2004)**

Solution. Let $A = \{1^2, 2^2, 3^2, 4^2, ...\}$ and $B = \{-1^2, -2^2, -3^2, -4^2, ...\}$

Consider the mapping $f : \mathbf{N} \to A$ defined by $f(n) = n^2$. Then f is one-one onto and so $\mathbf{N} \sim A$ and hence A is countable. Similarly, B is also countable. Since union of two countable sets is countable, it follows that $A \cup B$, *i.e.*, $\{\pm 1, \pm 4, \pm 9, \pm 16, ...\}$ is also countable.

Example 3. *Show that if B is countable sub-set of an uncountable set A, then A – B is uncountable.*

Solution. Let, if possible, $A - B$ be countable. Then $(A - B) \cup B$ would be countable, being the union of two countable sets. But $B \subset A \Rightarrow (A - B) \cup B = A$. Hence A would be countable, which contradicts the given hypothesis. Therefore, $A - B$ must be uncountable.

Example 4. (*a*) *If A is countable set and f is function from A onto B, prove that B is countable.*

(*b*) *If* $f : A \to B$ *and the range of f is uncountable, prove that the domain of f is uncountable.*

Solution. (*a*) Let $b \in B$. Since f is onto, there exists at least one pre-image $a \in A$ for $b \in B$ such that $f(a) = b$. Consider the mapping $g : B \to A$ so that $g(b) = a$. Then g is one-to-one and hence B is equivalent to a sub-set of a countable set. Therefore B must be countable.

(*b*) Here $f(A)$ is uncountable. Let, if possible, A be countable. Since A is countable and f is function from A onto $f(A)$, so by part (*a*), $f(A)$ must be countable. This contradicts the given hypothesis that $f(A)$ is uncountable. Hence A cannot be countable, *i.e.*, the domain A of f is uncountable.

EXERCISES

1. Indicate countable and uncountable sets from the following :

(*i*) $\{3, 3^2, 3^3, 3^4, ..., 3^n, ...\}$ (*ii*) The set of positive rational numbers

(*iii*) $]0, 1], [0, 1[,]0, 1[$ (*iv*) The set of prime numbers.

2. Show that if B is a countable sub-set of an uncountable set, then $A - B$ is uncountable.

3. Show that the set of all intervals with rational end points is a countable set.

4. Prove that if $f : A \to B$ is a function defined on a countable set, then the range of f is countable.

5. Prove that any set containing an uncountable set is uncountable.

6. Prove that any sub-set of $\mathbf{R}$ which contains $]0, 1[$ is uncountable.

7. Prove that the set of all finite sub-sets of a countable set is countable.

8. Let P_n be the set of polynomial functions f of degree n defined by relations of the form

$$f(x) = C_0 x^n + C_1 x^{n-1} + ... + C_n,$$

where n is a fixed non-negative integer, the coefficients $C_0, C_1, ..., C_n, ...$ are all integers and $C_0 \neq 0$. Show that the set P_n is countable.

9. Show the set P of polynomial functions with integer coefficients is countable.

10. Let $\mathbf{Q}_n$ be the set of polynomial functions of the form $C_0 x^n + C_1 x^{n-1} + ... + C_n$, where $C_0, C_1, ..., C_n$ are rational numbers and $C_0 \neq 0$. Prove that $\mathbf{Q}_n$ is countable. Also, show that $\bigcup_{n=1}^{\infty} \mathbf{Q}_n$ is countable.

11. (*a*) Show that the set of algebraic numbers is countable. **[Kanpur 2001]**

(*b*) Show that the set of transcendental numbers is uncountable. **[Garhwal 2001]**

ANSWERS

1. (*i*) countable (*ii*) countable (*iii*) all are uncountable (*iv*) countable

OBJECTIVE QUESTIONS

Multiple Choice Type Questions : *Select (a), (b), (c) or (d), whichever is correct.*

1. Cantor set is : (*a*) Countable (*b*) Dense in [0, 1] (*c*) nbd of 1/2 (*d*) None of these.

(Kanpur, 2003)

2. Which of the following statements is true ? Set of rational numbers is :

(*a*) Countable (*b*) nbd of 1/2 (*c*) Not dense in **R** (*d*) Not countable.

(Kanpur, 2004)

3. Which of the following sets is countable ?

(*a*) [0, 1] (*b*) [0, 1[

(*c*)]0, 1[(*d*) $\{0, \pm 1/n, n = 1, 2, 3, ...\}$

4. Which of the following sets is uncountable ?

(*a*) {1, 4, 9, 16, 25, ...} (*b*) $\{2n : n \in \mathbf{N}\}$

(*c*) all rational numbers (*d*) all irrational numbers.

5. Point out the wrong statement out of the following :

(*a*) The countable union of countable sets is countable

(*b*) If A and B are countable, then $A \times B$ is countable

(*c*) The cartesian product $\mathbf{N} \times \mathbf{N}$ is uncountable

(*d*) Every infinite set is equivalent to one of its proper sub-sets.

ANSWERS

1. (*d*) **2.** (*a*) **3.** (*d*) **4.** (*d*) **5.** (*c*)

MISCELLANEOUS PROBLEMS ON CHAPTER 4

1. Prove that union of two countable sets is countable. **[Kanpur 2005]**

2. Prove that the union of countable collection of countable sets is also countable

[Gerhwal 2001, 02; Kanpur 2001, 02; Meerut 2001]

3. If A_i is countable infinite set, then prove that $\bigcup_{i=1}^{\infty} A_i$ is also countable infinite.

[Garhwal 1998; Gorakhpur 2003, 05]

4. Prove that the set of all real numbers in the open interval]0, 1[is not enumerable

[Himanchal 2000; Kanpur 2001,03, 04; Meerut 1996, Rajasthan 1995]

5. Define a perfect set and show that every non-empty perfect set is uncountable.

[M.S. Univ. T.N. 2006]

6. Show that every set of countable sets is countable. **[Delhi B.Sc. (Prog.) II, 2016]**

7. (*i*) Show that a set A is finite if and only if it is not equivalent to any proper subset of A.

(*ii*) Prove that enumerable union of enumerable sets is enumerable.

8. Explain with the help of examples, a finite set, denumerable set, countable and uncountable sets. Prove that every sub-set of a countable set is countable. **[Delhi B.Sc. (Prog) III 2008]**

9. Is every superset of a countable set countable ? **[Delhi B.Sc. (Prog.) II, 2016]**

Sol. The set **R** of all real numbers is a superset of the set **Q** of all rational numbers. But **Q** is countable (see theorem VII, page 4.4) whereas **R** is not countable (see corollary I, page 4.5). Thus, every superset of a countable set is not necessarily a countable set.

10. Using the fact that countable union of countable sets is countable, prove that the set $\{(1/n, 1/m) : n, m \in \mathbf{N}\}$ is countable. **[Delhi B.Sc. (Prog.) II, 2015]**

Chapter

5

Sequences

5.1. SEQUENCE

A function whose domain is the set of natural numbers **N** and range a sub-set of real numbers **R** is called a *real sequence*.

Since in this chapter we shall be concerned with real sequences only, we shall refer to them as just sequences.

Since the domain of all sequences is **N**, it follows that a sequence is completely determined if $f(n)\ \forall\ n \in \mathbf{N}$ is known. For the sake of simplicity, $f(n)$ is, generally, denoted by f_n. The sequence f is denoted by

$$<f_n> \text{ or } \{f_n\} \text{ or } \{f_n\}_{n=1}^{\infty} \text{ or } (f_n) \text{ or } \{f_n : n \in \mathbf{N}\} \text{ or } <f_1, f_2, f_3, \ldots\ldots, f_n, \ldots\ldots>,$$

where $f_1, f_2, f_3, \ldots\ldots, f_n, \ldots\ldots$ are known as the first term, second term, third term, ..., nth term, respectively of the sequence $<f_n>$. The mth and nth terms f_m and f_n for $m \neq n$ are treated as distinct even if $f_m = f_n$. Thus the terms of a sequence $<f_n>$ are arranged in a definite order as first, second, third, ..., nth term and the terms occurring at different positions are treated as distinct even though they have the same value.

Range of a sequence. The set of all distinct terms of a sequence is called its range.

Note. In a sequence $<f_n>$, since $n \in \mathbf{N}$ and **N** is an infinite set, the number of terms of a sequence is always infinite. However, the range of a sequence may be finite.

ILLUSTRATIONS

(*i*) $<1/n>$ is the sequence $<1, 1/2, 1/3, 1/4, \ldots\ldots, 1/n, \ldots\ldots>$ with infinite range $\{1, 1/2, 1/3, \ldots\ldots, 1/n, \ldots\ldots\}$.

(*ii*) $<(-1)^n/n>$ is the sequence $<-1, 1/2, -1/3, 1/4, \ldots\ldots>$ with infinite range $\{-1, 1/2, -1/3, 1/4, \ldots\ldots\}$.

(*iii*) $<(-1)^n>$ is the sequence $<-1, 1, -1, 1, \ldots\ldots>$ with finite range $= \{1, -1\}$.

(*iv*) $<1 + (-1)^n>$ is the sequence $<0, 2, 0, 2, \ldots\ldots>$ with finite range $= \{0, 2\}$.

(*v*) A sequence may be defined with help of a recursion formula.

Let $f_1 = 1, f_2 = 2, \quad f_{n+2} = f_{n+1} + 3f_n\ \forall\ n \geq 1$.

Then, putting $n = 1, 2, 3, \ldots\ldots$ in $f_{n+2} = f_{n+1} + 3f_n$ and using $f_1 = 1$ and $f_2 = 2$, we get $f_3 = f_2 + 3f_1 = 2 + 3 \times 1 = 5, f_4 = f_3 + 3f_2 = 5 + 3 \times 2 = 11, f_5 = f_4 + 3f_3 = 26$ etc.

$\therefore\quad <f_n>$ is the sequence $<1, 2, 5, 11, 26, 59, \ldots>$ with infinite range $\{1, 2, 5, 11, 26, 59, \ldots\}$.

5.2. BOUNDED AND UNBOUNDED SEQUENCES

(Delhi B.A. (Prog.) III, 2014; Kanpur 2008; Delhi B.Sc. III (Prog) 2009)

We have already discussed the boundedness of a set in Chapter 2. A sequence is said to be bounded if and only if its range (which is a set) is bounded.

Bounded above sequence. A sequence $<f_n>$ is said to be bounded above if there exists a real number K such that $f_n \le K \ \forall \ n \in \mathbf{N}$
i.e., if the range of the sequence is bounded above.

Bounded below sequence. A sequence $<f_n>$ is said to be bounded below if there exists a real number k such that $f_n \ge k \ \forall \ n \in \mathbf{N}$
i.e., if the range of the sequence is bounded below.

Bounded sequence. A sequence is said to be bounded if it is bounded above as well as below. Hence a sequence $<f_n>$ is bounded if there exist two real numbers k and K, such that $k \le f_n \le K$ $\forall \ n \in \mathbf{N}$. Alternately, $<f_n>$ is bounded if there exists a real number l such that $| f_n | \le l$ for $\forall \ n \in \mathbf{N}$, *i.e.*, if the range of the sequence is bounded.

Unbounded sequence. A sequence is said to be unbounded if it is not bounded.

Unbounded above sequence. A sequence $<f_n>$ is said to be unbounded above if it is not bounded above, *i.e.*, if for every real number K, there exists $m \in \mathbf{N}$ such that $f_m > K$.

Unbounded below sequence. A sequence $<f_n>$ is said to be unbounded below if it is not bounded below, *i.e.*, if for every real number k, there exists $m \in \mathbf{N}$ such that $f_m < k$.

Note. The least upper bound (*i.e.*, the supremum) and the greatest lower bound (*i.e.*, the infimum) of the range of a bounded sequence may be referred as its greatest lower bound and the least upper bound respectively.

ILLUSTRATIONS

1. $<2^n>$ is bounded below but not bounded above as $2^n \ge 2 \ \forall \ n \in \mathbf{N}$.

2. $<(-1)^n/n>$ is bounded as $| (-1)^n/n | \le 1 \in \mathbf{N}$. **(Agra, 2001)**

3. $<(-1)^n n>$, *i.e.*, $<-1, 2, -3, 4, -5, 6, ...>$ is neither bounded above nor bounded below. **(Agra, 2001)**

4. $\left\langle \dfrac{n}{n+1} \right\rangle$ is bounded since $\dfrac{1}{2} \le \dfrac{n}{n+1} < 1 \ \forall \ n \in \mathbf{N}$.

5. $<-n^2>$ is bounded above by -1 but not bounded below.

6. Define unbounded sequence giving an example. ***[Delhi 2009]***

5.3. LIMIT POINT (OR CLUSTER POINT OR POINT OF CONDENSATION) OF A SEQUENCE

We have already introduced the notion of a limit point of an arbitrary set of real numbers and shall now introduce the corresponding notion of a limit point of a sequence; the two notions having much in common.

Def. *A number ξ is said to be a limit point of a sequence $<f_n>$ if, given any neighbourhood of ξ, f_n belongs to the same for an infinite number of values of n.*

This means that ξ is a limit point of sequence $<f_n>$ if and only if given any positive number ε,

$$f_n \in \,]\xi - \varepsilon, \xi + \varepsilon[$$

for an infinite number of values of n; the open interval $]\xi - \varepsilon, \xi + \varepsilon[$ being a neighbourhood of ξ.

It will be seen that if ξ is *not* a limit point of a sequence $<f_n>$, there exists $\varepsilon > 0$ such that

$$f_n \in \,]\xi - \varepsilon, \xi + \varepsilon[$$

for *at the most finite number* of values of n.

ILLUSTRATIONS

1. Consider $<f_n> = <(-1)^n> = <-1, 1, -1, 1, ...>$ so that $f_n = -1$ when n is odd and $f_n = 1$ when n is even. Now, any nbd of -1 will contain all the odd terms (since each odd term $= -1$) of the sequence. So -1 is a limit point of $<(-1)^n>$.

Again, every nbd of 1 will contain all the even terms (since each even term = 1) of the sequence. So 1 is a limit point.

2. Consider $<n>$, *i.e.*, $<1, 2, 3, ..., n, ...>$. Let p be any real number. Then the nbd $]p - 1/4, p + 1/4[$ of p will contain at most one term of the sequence $<n>$. Hence p is not a limit point of $<n>$. Since p is any real number, it follows that $<n>$ has no limit point.

3. Consider $<1/n>$, *i.e.*, $<1, 1/2, 1/3, ..., 1/n, ...>$.

For $\varepsilon > 0$, there exists $m \in \mathbf{N}$ such that $1/m < \varepsilon$ ...(1)

$\therefore$ For $n \geq m$, $0 < 1/n \leq 1/m < \varepsilon$, by (1)

$\Rightarrow -\varepsilon < 0 < 1/n < \varepsilon \ \forall \ n \geq m$

$\Rightarrow 1/n \in]-\varepsilon, \varepsilon[\ \forall \ n \geq m$

$\Rightarrow$ every nbd of 0 contains infinitely many terms of $<1/n>$

$\Rightarrow$ 0 is a limit point of $<1/n>$.

4. If $<f_n>$ is a sequence such that $f_n = k$, for infinitely many values of n then k is a limit point of $<f_n>$.

Theorem I. (Bolzano-Weierstrass theorem for sequences)

Every bounded sequence has a limit point. **[Delhi B.Sc. (Prog.) II, 2016]**

[Agra, 2000, 01, 02; Delhi Maths (H), 1999; Kanpur, 2005, 09]

Proof. Let $<f_n>$ be given bounded sequence with A as its range. Then we distinguish between the following two possibilities.

(*i*) A is a finite set (*ii*) A is an infinite set.

We first consider the case (*i*). Of the different members of the finite set A, there must exist at least one, say, ξ, such that $f_n = \xi$ for an infinite number of values of n.

This means that if $]\xi - \varepsilon, \xi + \varepsilon[$ be a nbd of ξ then

$$f_n = \xi \in]\xi - \varepsilon, \xi + \varepsilon[$$

for an infinite number of values of n. Thus, ξ is a limit point of the sequence $<f_n>$.

We now consider the case (*ii*). The sequence being bounded, the range is bounded. The range A, being infinite and bounded, by Bolzano-Weierstrass theorem for a set (see chapter 3), there exist at least one limit point, say η. Then an infinite number of members of the set A belong to $]\eta - \varepsilon, \eta + \varepsilon[$, $\in$ being any given positive number. Hence, we have

$$f_n \in]\eta - \varepsilon, \eta + \varepsilon[$$

for an infinite number of values of n and as such η is a limit point of the sequence $<f_n>$.

Theorem II. *The set of all limit points of a bounded sequence is bounded.*

[Delhi Maths (H), 1999]

Proof. Let $<f_n>$ be a bounded sequence. Then there exist real numbers k and K $(k \leq K)$ such that

$$k \leq f_n \leq K \ \forall \ n \in \mathbf{N} \quad ...(1)$$

$$(1) \Rightarrow f_n \notin]-\infty, k[\text{ and } f_n \notin]K, \infty[\quad ...(2)$$

Let p be any real number.

If $p \in]-\infty, k[$, then by (2), $]-\infty, k[$ contains no term of $<f_n>$ and so p is not a limit point.

Similarly, if $p \in]K, \infty[$, then p is not a limit point of $<f_n>$.

Thus no point outside $[k, K]$ is a limit point of $<f_n>$.

$\Rightarrow$ all the limit points of $<f_n>$ lie in $[k, K]$

$\Rightarrow$ the set of all limit points of $<f_n>$ is bounded.

Theorem III. *Every bounded sequence has the greatest and the least limit points.*

Proof. Let $<f_n>$ be a bounded sequence. Then by theorem II, the set S of limit points of $<f_n>$ is also bounded.

Again, by Bolzano-Weierstrass theorem, the bounded sequence $<f_n>$ has at least one limit point and so S is non-empty.

By the order completeness axiom of real numbers, the set S has infimum and supremum. Let inf $S = a$ and sub $S = b$. Then we shall show that $a, b \in S$.

For $\varepsilon > 0$, let $]b - \varepsilon, b + \varepsilon[$ be a nbd of b.

Now, $b = \sup S \Rightarrow$ there exists $p \in S$ such that $b - \varepsilon < p \leq b < b + \varepsilon$

$\Rightarrow p \in]b - \varepsilon, b + \varepsilon[$

$\Rightarrow]b - \varepsilon, b + \varepsilon[$ is a nbd of p

Since $p \in S$, so p is a limit point of $<f_n>$.

$\therefore$ Every nbd of p contains infinitely many terms of $<f_n>$ and so $]b - \varepsilon, b + \varepsilon[$ contains infinitely many terms of $<f_n>$.

Since it holds for every $\varepsilon > 0$, hence every nbd of b contains infinitely many terms of $<f_n>$.

Therefore b is a limit point of $<f_n>$ and so $b \in S$.

Similarly, we can show that $a \in S$.

Thus the bounded sequence has the greatest and smallest limit points.

EXERCISES

Give the limit points of the sequences (in exercises 1, 2 and 3) defined as follows :

1. (*a*) $< f_n > = \sin \frac{1}{2} n\pi$, (*b*) $< g_n > = \sin \frac{1}{3} n\pi$, (*c*) $< h_n > = \sin \frac{1}{4} n\pi$.

2. (*a*) $< f_n > = \sin \frac{1}{2} n\pi + \cos \frac{1}{3} n\pi$, (*b*) $< g_n > = \sin \frac{1}{2} n\pi + \cos \frac{1}{4} n\pi$.

3. (*i*) $n \to (-1)^n/n$ (*ii*) $n \to 1 + (-1)^n$ (*iii*) $n \to (-1)^n (1 + 1/n)$

(*iv*) $n \to (4n + 1)/4^n$ or $(1 - 4^n)/4^n$, according as n is even or odd.

(*v*) $< f_n > = \begin{cases} (n+1)/n & \text{when } n = 3m, \\ (n+2)/2n & \text{when } n = 3m+1, \\ 1/(n+1) & \text{when } n = 3m+2. \end{cases}$

4. Show that every point of the range of a sequence is as well as a limit point of the sequence but not necessarily conversely.

5. Show that ξ is a limit point of a sequence $<f_n>$ if and only if given any positive number ε and a positive integer m, there exists an integer $m' \geq m$ such that

$$f_{m'} \in]\xi - \varepsilon, \xi + \varepsilon[.$$

6. Give an example of an unbounded sequence with exactly two limit points.

[Delhi Maths (H), 1999]

Hint. $<f_n> = <0, 1, 2, 0, 1, 2^2, 0, 1, 2^3, 0, 1, 2^4, ...>$ is an unbounded sequence with exactly two limit points 0 and 1.

7. Give an example of a bounded sequence having exactly two limit points.

Hint. $<(-1)^n>$ has 1 and -1 as two limit points.

8. Show that the set of limit points of a sequence is a closed set.

9. Show that the set of limit points of a bounded sequence is a compact set.

10. Given that $<f_n>$ is a bounded sequence with a *single* limit point l and $]l-\varepsilon, l+\varepsilon[$ is a given neighbourhood of l, which, of any, of the following statements are true :

(*i*) $f_n \in\,]l-\varepsilon, l+\varepsilon[$ for an infinite number of values of n.

(*ii*) $f_n \notin\,]l-\varepsilon, l+\varepsilon[$ for an infinite number of values of n.

11. Give a sequence such that $-1, 1$ are its two limit points and no member of the sequence belongs to the interval $[-1, 1]$.

5.4. CONVERGENT SEQUENCES. THE LIMIT OF A SEQUENCE

[Delhi B.A. (Prog.) III, 2014; Delhi B.Sc. (Hons.) I, 2016]

Def. *A sequence* $<f_n>$ *is said to be convergent, if there exists a number* l *for which the following property holds* :

To each given $\varepsilon > 0$, *there corresponds a natural number* m *such that*

$$|f_n - l| < \varepsilon \;\; \forall \; n \geq m.$$

We also say that l is the limit of $<f_n>$.

Also if a sequence $<f_n>$ is convergent and l is its limit, we say that the sequence $<f_n>$ converges to l and, in symbols, write

$$\lim_{n \to 0} f_n = l \quad \text{or} \quad \lim f_n = l \quad \text{or} \quad f_n \to l \text{ as } n \to \infty.$$

Moreover, if a sequence $<f_n>$ is convergent, we say that $\lim f_n$ exists.

Divergent sequence. Definition.

If a sequence has no limit, we say that the sequence is divergent.

Note 1. In order to understand the difference between a limit point of $<f_n>$ and the limit of $<f_n>$, the following observations should be taken care of :

(*i*) If l is a limit point of $<f_n>$, then every nbd of l containing an infinite number of its members does not exclude the possibility of an infinite number of members of $<f_n>$ lying outside that nbd.

(*ii*) If l is the limit of $<f_n>$, then every nbd of l contains all but a finite number of its members.

(*iii*) The limit of $<f_n>$ is a limit point of $<f_n>$ but a limit point of $<f_n>$ need not be always the limit of $<f_n>$.

Note 2. Null sequence. A sequence $<f_n>$ is said to be a null sequence if it converges to zero,

i.e., if $\lim_{n \to 0} f_n = 0.$

Theorem I. *Limit of a sequence, if it exists, is unique.* **[Kanpur 2010; Agra 2007]**

Or **[G.N.D.U. Amritsar 2010]**

A sequence cannot converge to more than one limit. **[Delhi B.A. (Prog.) III, 2013]**

[*Agra, 2001*; *Delhi B.Sc. (Hons) I 2011*; *Kanpur, 2003, 05, 07*; *Patna, 2003*]

Proof. Let, if possible, the sequence $<f_n>$ converges to two distinct numbers say l_1 and l_2. Since by assumption, $l_1 \neq l_2$, so $| l_1 - l_2 | > 0$.

Let $$\varepsilon = \frac{1}{2} | l_1 - l_2 | > 0 \quad ...(1)$$

Now, $<f_n>$ converges to $l_1 \Rightarrow$ there exists $m_1 \in \mathbf{N}$ such that
$$| f_n - l_1 | < \varepsilon, \forall \ n \geq m_1 \quad ...(2)$$
and $<f_n>$ converges to $l_2 \Rightarrow$ there exists $m_2 \in \mathbf{N}$ such that
$$| f_n - l_2 | < \varepsilon, \forall \ n \geq m_2 \quad ...(3)$$
Let $$m = \max \{m_1, m_2\}. \text{ Then} \quad ...(4)$$
$$(2) \text{ and } (4) \Rightarrow | f_n - l_1 | < \varepsilon, \forall \ n \geq m \quad ...(5)$$
and $$(3) \text{ and } (4) \Rightarrow | f_n - l_2 | < \varepsilon, \forall \ n \geq m \quad ...(6)$$
Then, $$| l_1 - l_2 | = | l_1 - f_n + f_n - l_2 |$$
$$\leq | l_1 - f_n | + | f_n - l_2 |$$
$$< \varepsilon + \varepsilon = 2\varepsilon, \forall \ n \geq m, \text{ using (5) and (6).}$$
or $$| l_1 - l_2 | < \frac{1}{2} | l_1 - l_2 |, \text{ using (1)}$$
which is absurd. Hence our assumption that $l_1 \neq l_2$ is wrong and hence a sequence cannot converge to more than one limit, *i.e.*, the limit of a sequence is unique.

Note. In view of uniqueness of limit of a sequence we shall use the phrase "the limit" in place of "a limit" of a sequence.

Theorem II. *Every convergent sequence is bounded but the converse is not true.*
[*Delhi B.Sc. (Prog) III 2012; Delhi B.Sc.* (*Hons*) *I 2016*; *Delhi B.A.* (*Prog*) III, *2013, 15*; *Pune 2010; Kakatiya, 2001*; *Kanpur, 2008, 14*; *Meerut, 2003, 05*; *Purvanchal, 1994*]

Proof. Let $<f_n>$ be a sequence which converges to l. Let $\varepsilon = 1$. Then there exist $m \in \mathbf{N}$ such that
$$| f_n - l | < 1, \forall \ n \geq m$$
$$\Rightarrow l - 1 < f_n < l + 1, \forall \ n \geq m. \quad ...(1)$$
Let $$k = \min \{a_1, a_2, \ldots, a_{m-1}, l - 1\} \quad ...(2)$$
and $$K = \max \{a_1, a_2, \ldots, a_{m-1}, l + 1\} \quad ...(3)$$
From (1), (2) and (3), $k \leq f_n \leq K \ \forall \ n \in \mathbf{N}$

Hence $<f_n>$ is bounded.

To show that the converse of the above theorem does not hold.

Consider the sequence $<f_n> = <(-1)^n>$, *i.e.*, $<-1, 1, -1, 1, \ldots>$
which is bounded since here $f_n = (-1)^n$ and $-1 \leq f_n \leq 1 \ \forall \ n \in \mathbf{N}$.

We now show that $<f_n>$ is not a convergent sequence.

Let, if possible, $<f_n>$ be convergent and let $\lim\limits_{n \to 0} f_n = l$.

Then for $\varepsilon = 1/2$, there exists a $m \in \mathbf{N}$ such that
$$| f_n - l | < 1/2 \ \forall \ n \geq m$$
i.e., $$| (-1)^n - l | < 1/2 \ \forall \ n \geq m \quad ...(1)$$
From (1) we obtain $$| 1 - l | < 1/2, \forall \ n \geq m \text{ and } n \text{ is even} \quad ...(2)$$
and $$| -1 - l | < 1/2, \forall \ n \geq m \text{ and } n \text{ is odd}$$
i.e., $$| 1 + l | < 1/2, \forall \ n \geq m \text{ and } n \text{ is odd} \quad ...(3)$$

Now, $2 = |(1 + l) + (1 - l)| \leq |1 + l| + |1 - l| < 1/2 + 1/2$, by (2) and (3)

Thus, $2 < 1$, which is absurd. Hence our initial assumption $f_n \to l$ as $n \to \infty$ is wrong. Hence $< f_n >$, *i.e.*, $<(-1)^n>$ is not convergent.

Note. *From the above theorem II, it follows that an unbounded sequence must be divergent.*

Theorem III. *If* $\lim_{n\to\infty} f_n = l$, *then* l *is the unique limit point of* $<f_n>$.

Proof. First of all we shall prove that l is a limit point of $<f_n>$.

Now, $\lim_{n\to\infty} f_n = l$.

$\Rightarrow$ for $\varepsilon > 0$, there exists $m \in \mathbf{N}$ such that $|f_n - l| < \varepsilon \ \forall \ n \geq m$

$\Rightarrow$ $l - \varepsilon < f_n < l + \varepsilon$ for infinitely many values of n

$\Rightarrow$ $f_n \in \,]l - \varepsilon, l + \varepsilon[$ for infinitely many values of n

$\Rightarrow$ l is a limit point of $<f_n>$, by definition.

Now we shall prove that if l' is any other limit point of $<f_n>$ then $l' = l$.

Let $\varepsilon > 0$ be arbitrary real number. Since $\lim_{n\to\infty} f_n = l$, so there exists $m_1 \in \mathbf{N}$ such that

$$|f_n - l| < \varepsilon/2 \ \forall \ n \geq m_1 \quad \text{...(1)}$$

Again, since l' is a limit point of $<f_n>$, so there exists $m_2 \in \mathbf{N}$ such that $m_2 > m_1$ and

$$|f_{m_2} - l'| < \varepsilon/2 \quad \text{...(2)}$$

Since $m_2 > m_1$, setting $n = m_2$ in (1) gives $\left|f_{m_2} - l\right| < \varepsilon/2$ ... (3)

Now, $|l - l'| = |(f_{m_2} - l') - (f_{m_2} - l)| \leq |f_{m_2} - l'| + |f_{m_2} - l|$

or $|l - l'| < \varepsilon/2 + \varepsilon/2$ or $|l - l'| < \varepsilon$, using (2) and (3)

Now, ε is arbitrarily small and $|l - l'| < \varepsilon$

$$\Rightarrow \quad |l - l'| = 0 \ \Rightarrow \ l = l',$$

showing that l is the unique limit point of $<f_n>$.

Hence the theorem.

Note. *The converse of the above theorem may not be always true, i.e., a sequence having unique (only one) limit point may not converge.*

For example, let $<f_n>$ be a sequence, where

$$f_n = \begin{cases} 1/n, & \text{if } n \text{ is even} \\ n, & \text{if } n \text{ is odd} \end{cases}$$

It can be easily verified that $<f_n>$ has only one limit point, namely, 0, and yet $<f_n>$ is not convergent.

Theorem IV. *Prove that a bounded sequence with a unique limit point is convergent.*

[*Bangalore, 2004*]

Proof. *Let* $<f_n>$ *be a bounded sequence and let* l *be its only limit point.* We shall show that $<f_n>$ converges to l.

Let $\varepsilon > 0$ be any given positive number. Consider the neighbourhood $]l + \varepsilon, l + \varepsilon[$ of l.

Now, l being the only limit point of the bounded sequence $<f_n>$, there can exist only a finite number of values of n such that the corresponding f_n does not belong to $]l - \varepsilon, l + \varepsilon[$. Let $m - 1$ be the greatest of such exceptional values of n. We then have

$$f_n \in \,]l - \varepsilon, l + \varepsilon[\ \forall \ n \geq m$$

$$\Leftrightarrow |f_n - l| < \varepsilon \ \forall \ n \geq m.$$

Thus, the sequence is convergent with l as its limit.

Combining theorems II, III and IV, we obtain the following necessary and sufficient condition for the convergence of a sequence.

Theorem VI. *A sequence is convergent if and only if it is bounded and has only one limit point.*

Keeping the above theorem VI in mind, an alternative definition of a convergent sequence is given below :

Definition. *A sequence is said to be convergent if it is bounded and has a unique limit point.*

In view of the above definition, the following theorem follows :

Theorem VII. *A necessary and sufficient condition for a sequence $<f_n>$ to converge to l is that for each $\varepsilon > 0$ there corresponds a $m \in \mathbf{N}$ such that*

$$|f_n - l| < \varepsilon, \ \forall \ n \geq m.$$

Proof. Left as an exercise.

EXAMPLES

Example 1. *Show that the constant sequence $<f_n>$, where $f_n = c \ \forall \ n \in \mathbf{N}$, converges to c.*

Solution. Since $f_n = c \ \forall \ n \in \mathbf{N}$ so $|f_n - c| = 0 \ \forall \ n \geq 1$

Let ε be given. Then, taking $m = 1$, we have

$$|f_n - c| < \varepsilon \ \forall \ n \geq m$$

Hence, by definition of a convergent sequence, the given sequence converges to c.

Example 2. *Show that $<1/3^n>$ converges to zero.*

Solution. Let $f_n = 1/3^n$ and let $\varepsilon > 0$ be given.

Now, $|f_n - 0| = |1/3^n - 0| = 1/3^n < \varepsilon$ if $3^n > 1/\varepsilon$

i.e., if $n \log 3 > \log(1/\varepsilon)$ or $n > \dfrac{\log(1/\varepsilon)}{\log 3}$

Let m be any positive integer greater than $\dfrac{\log(1/\varepsilon)}{\log 3}$.

Then for $\varepsilon > 0$, there exists a positive integer m such that

$$|f_n - 0| < \varepsilon, \ \forall \ n \geq m$$

$$\Rightarrow <f_n>, \text{ i.e., } <1/3^n> \text{ converges to } 0.$$

Example 3. *Show that the sequence $<f_n>$ defined by $f_n = \{\sqrt{n+1} - \sqrt{n}\} \ \forall \ x \in \mathbf{N}$ is convergent.* **[Delhi B.Sc. (Hons.) I, 2018; Meerut 2018]**

Solution. Here $f_n = \sqrt{n+1} - \sqrt{n}$

$$= \frac{(\sqrt{n+1} - \sqrt{n})(\sqrt{n+1} + \sqrt{n})}{\sqrt{n+1} + \sqrt{n}} = \frac{1}{\sqrt{n+1} + \sqrt{n}} < \frac{1}{\sqrt{n} + \sqrt{n}} = \frac{1}{2\sqrt{n}}$$

Let $\varepsilon > 0$ be given. Then

$$|f_n - 0| = \left| \frac{1}{2\sqrt{n}} - 0 \right| = \frac{1}{2\sqrt{n}} < \varepsilon, \text{ if } \sqrt{n} > \frac{1}{2\varepsilon} \text{ or } n > \frac{1}{4\varepsilon^2}.$$

Let m be any positive integer greater than $1/4\varepsilon^2$.

Then for $\varepsilon > 0$, there exists a positive integer m such that

$$| f_n - 0 | < \varepsilon, \ \forall \ n \geq m$$

$$\Rightarrow < f_n > \text{ is convergent and } \lim_{n \to \infty} f_n = 0.$$

Example 4. *Show that* $\lim_{n \to \infty} \dfrac{n^2 + 3n + 5}{2n^2 + 5n + 7} = \dfrac{1}{2}$.

Solution. Let $f_n = (n^2 + 3n + 5)/(2n^2 + 5n + 7)$

Let $\varepsilon > 0$ be given. Then

$$\left| f_n - \frac{1}{2} \right| = \left| \frac{n^2 + 3n + 5}{2n^2 + 5n + 7} - \frac{1}{2} \right| = \frac{n + 3}{2\,(2n^2 + 5n + 7)}$$

$$\leq \frac{n + 3n}{2\,(2n^2 + 5n + 7)} = \frac{2n}{2n^2 + 5n + 7} < \frac{2n}{2n^2} = \frac{1}{n} < \varepsilon, \text{ if } n > \frac{1}{\varepsilon}.$$

Let m be any positive integer greater than $1/\varepsilon$.

Then for $\varepsilon > 0$, there exists a positive integer m such that

$$| f_n - 1/2 | < \varepsilon, \ \forall \ n \geq m$$

$$\Rightarrow \quad < f_n > \text{ is convergent and } \lim_{n \to \infty} f_n = \frac{1}{2}, \text{ i.e., } \lim_{n \to \infty} \frac{n^2 + 3n + 5}{2n^2 + 5n + 7} = \frac{1}{2}$$

Example 5. *Show that the sequences* $< r^n >$ *converges to zero if* $| r | < 1$.

[Delhi B.Sc. III 2009; Delhi Physics (H), 1993; Delhi Maths (G), 2000]

or

Show that $\lim_{n \to \infty} r^n = 0$, *if* $| r | < 1$.

Solution. Let $f_n = r^n$. Since $| r | < 1$, there exists some $h > 0$ such that

$$| r | < 1/(1 + h) \qquad \text{...(1)}$$

Using the binomial theorem, we have

$$(1 + h)^n = 1 + nh + \frac{n\,(n - 1)}{2!} h^2 + \ldots\ldots + h^n, \ \forall \, n \in \mathbf{N}$$

or $$(1 + h)^n > 1 + nh, \text{ since } h > 0$$

or $$\frac{1}{(1 + h)^n} < \frac{1}{1 + nh}, \ \forall \, n \in \mathbf{N} \qquad \text{...(2)}$$

Now, $$| f_n - 0 | = | r^n | = | r |^n = 1/(1 + h)^n, \text{ using (1)} \qquad \text{...(3)}$$

From (2) and (3), $$| f_n - 0 | < \frac{1}{1 + nh}, \ \forall \ n \in \mathbf{N}$$

Let $\varepsilon > 0$ be given. Then

$$| f_n - 0 | < \varepsilon, \text{ if } \frac{1}{1 + nh} < \varepsilon \quad \text{or} \quad n > \frac{(1/\varepsilon - 1)}{h}.$$

Let m be any positive integer greater than $(1/\varepsilon - 1)/h$.

Then for $\varepsilon > 0$, there exists a positive integer m such that

$$| f_n - 0 | < \varepsilon, \ \forall \ n \geq m$$

$$\Rightarrow \ < f_n > \text{ i.e., } < r^n > \text{ converges to 0.}$$

Example 6. *Show that* $\lim\limits_{n \to \infty} n^{1/n} = 1.$ ***[Agra 2009; Delhi B.A. (Prog) III 2010]***

[G.N.D.U. Amritsar 2010; Delhi Maths (H), 1997; Delhi Maths (P), 2004

Solution. Let $f_n = n^{1/n} = 1 + h_n$, where $h_n \geq 0$ ...(1)

From (1), $n = (1+h_n)^n = 1 + nh_n + \dfrac{n(n-1)}{2!} h_n^2 + \ldots\ldots\ldots + h_n^n$, by binomial theorem

$$\Rightarrow n > \frac{n(n-1)}{2} h_n^2 \ \forall\, n \in \mathrm{N}, \text{ since } h_n \geq 0$$

$$\Rightarrow h_n^2 < \frac{2}{n-1} \text{ for } n \geq 2 \Rightarrow |\, h_n \,| < \sqrt{\frac{2}{n-1}}, \text{ for } n \geq 2 \qquad \text{...(2)}$$

Let $\varepsilon > 0$ be given. Then using (1) and (2), we get

$$|\, f_n - 1 \,| = |\, h_n \,| < \sqrt{\frac{2}{n-1}} < \varepsilon, \text{ if } \frac{2}{n-1} < \varepsilon^2 \quad \text{or} \quad n > 1 + \frac{2}{\varepsilon^2}.$$

Let m be a positive integer greater than $1 + 2/\varepsilon^2$. Then

$$|\, f_n - 1 \,| < \varepsilon \ \forall\ n \geq m \quad \text{or} \quad |\, n^{1/n} - 1 \,| < \varepsilon \ \forall\ n \geq m,$$

showing that $\lim\limits_{n \to \infty} n^{1/n} = 1.$

Example 7. *Prove that* $\lim\limits_{n \to \infty} \dfrac{1}{\sqrt{n!}} = 0.$ ***[Delhi Maths (H), 2001]***

Solution. Let $f_n = 1/\sqrt{n!}$ and let $\varepsilon > 0$ be given.

Now, $|\, f_n - 0 \,| = \left| \dfrac{1}{\sqrt{n!}} - 0 \right| = \dfrac{1}{\sqrt{n!}} < \dfrac{2}{n}$ as $\sqrt{n!} > \dfrac{n}{2} \ \forall\ n \geq 1$

$\Rightarrow$ $|\, f_n - 0 \,| < \varepsilon$, if $2/n < \varepsilon$ or $n > 2/\varepsilon$.

Let m be any positive integer greater than $2/\varepsilon$.

Then for $\varepsilon > 0$, there exists a positive integer m such that

$$|\, f_n - 0 \,| < \varepsilon \ \forall\ n \geq m \quad \text{or} \quad \left| 1/\sqrt{n!} - 0 \right| < \varepsilon \ \forall\ n \geq m,$$

showing that $\lim\limits_{n \to \infty} \dfrac{1}{\sqrt{n!}} = 0.$

Example 8. *Show that the sequence* $<n>$ *does not have a limit.*

Solution. Let $f_n = n$. Let if possible, let there exists a real number l such that $f_n \to l$ as $n \to \infty$. Hence for any $\varepsilon > 0$, there exists a positive integer m such that

$$|\, f_n - l \,| < \varepsilon \ \forall\ n \geq m \qquad \text{...(1)}$$

Choose $\varepsilon = 1$ and let m_0 be the corresponding value of m for which (1) is true. Then, we have

$$|\, f_n - l \,| < 1 \ \forall\ n \geq m_0 \Rightarrow f_n \in \,]l-1, l+1[\ \forall\ n \geq m_0$$

i.e., $n \in \,]l-1, l+1[\ \forall\ n \geq m_0,$

showing that the values of n greater than m_0 always lie between $l - 1$ and $l + 1$ for larger values of n which is absurd. Hence our initial assumption is wrong. Thus $<f_n>$ *i.e.*, $<n>$ does not have a limit.

EXERCISES

1. From definition, show that :

(*i*) $\lim \frac{1}{n} = 0$ (*ii*) $\lim \frac{2n+3}{3n+4} = \frac{2}{3}$ (*iii*) $\lim \frac{1}{n^2} = 0$ (*iv*) $\lim \frac{1+n^2}{2+3n^2} = \frac{1}{3}$

2. From definition show that :

(*i*) $\lim \frac{3+2\sqrt{n}}{\sqrt{n}} = 2$ (*ii*) $\lim_{n \to \infty} \frac{5 \times 3^n}{3^n - 2} = 5$ **[Delhi Physics (H), 2000]**

(*iii*) $< \sqrt{n^2+1} - n >$ is a null sequence **(Bangalore, 2002)**

(*iv*) $\lim_{n \to \infty} \{1 + (-1)^n / n\} = 1$ (*v*) $\lim_{n \to \infty} \frac{1}{n^p} = 0$, if $p > 0$

(*vi*) $\lim_{n \to \infty} \frac{1+2+3+.....+n}{n^2} = \frac{1}{2}$ (*vii*) $\lim_{n \to \infty} \frac{1+3+5+.....+(2n-1)}{n^2} = 1$

(*viii*) $<1 + (-1)^n>$ does not have a limit.

(*ix*) Show that the sequence $\{5 - 1/2^{n-1}\}$ converges to 5. **(Nagpur 2010)**

3. Examine the truth of the following statements :

(*i*) $\lim f_n = l$ if, given any neighbourhood of l, f_n belongs to the same for an infinite number of values of n.

(*ii*) $\lim f_n = l$ if, given any neighbourhood of l, f_n belongs to the same for all except at the most a finite number of values of n.

(*iii*) Every convergent sequence is bounded.

(*iv*) Every bounded sequence is convergent.

4. Given that $\lim f_n = l > 0$, show that there exists a positive integer m such that $f_n \geq 0$ $\forall\ n \geq m$.

5. If f_n is a convergent sequence such that $f_n \geq 0\ \forall\ n$, show that $\lim f_n \geq 0$.

6. Show that (*i*) if $a_n \to a$, then $a_n^{\ 2} \to a^2$. Is the converse true ?

(*ii*) If $a_n^2 \to 0$, then $a_n \to 0$ as $n \to \infty$. **[Delhi Maths (H), 1999]**

7. Show that if $f_n^{\ 2} \to l^2$, then $| f_n | \to | l |$ as $n \to \infty$.

8. If $<f_n>$ is bounded and $\lim g_n = 0$, then prove that $\lim <f_n\, g_n> = 0$.

9. (*a*) Show that if $p > 0$, then $\lim_{n \to \infty} p^{1/n} = 1$. **[Delhi B.Sc. (Hons.) I, 2016]**

(*b*) If a is a number, greater than 1, prove that $\lim_{n \to \infty} a^{1/n} = 1$. Is the result true if $0 < a \leq 1$? **[Delhi B.Sc. (H) Phy, 1997]**

5.5. ALGEBRA OF CONVERGENT SEQUENCES

Definitions. If $<f_n>$ and $<\phi_n>$ be two given sequences, then the sequence whose nth terms are $f_n + \phi_n$, $f_n - \phi_n$ and $f_n\, \phi_n$ are respectively known as the *sum*, *difference* and *product* of the sequences $<f_n>$ and $<\phi_n>$. We denote these sequences by $<f_n + \phi_n>$, $<f_n - \phi_n>$ and $<f_n\, \phi_n>$ respectively.

Let $\phi_n \neq 0 \ \forall \ n \in \mathbf{N}$. Then the sequence whose nth term is $1/\phi_n$ is the *reciprocal* of the sequence $<\phi_n>$ and the sequence whose nth term is f_n/ϕ_n is known as the *quotient* of the sequence $<f_n>$ by the sequence $<\phi_n>$. We denote these sequences by $<1/\phi_n>$ and $<f_n/\phi_n>$ respectively.

Let k be any number. Then the sequence whose nth term is $k f_n$ is known as the *scalar multiple* of $<f_n>$ by k and will be denoted as $<k f_n>$.

We now state and prove that relations of two convergent sequences $<f_n>$ and $<\phi_n>$ with respect to their sum, product etc.

Theorem I. *If* $\lim f_n = k$ *and* $\lim \phi_n = l$, *then*

$$\lim (f_n + \phi_n) = k + l = \lim f_n + \lim \phi_n,$$

i.e., the limit of the sum of two convergent sequences is the sum of their limits.

[*Meerut 2017; Delhi B.A. (Prog.) II, 2003; Delhi B.Sc. (Hons) I, 2010*]

Proof. We have, $\forall \ n \in \mathbf{N}$

$$\begin{aligned} |(f_n + \phi_n) - (k + l)| &= |(f_n - k) + (f_n - l)| \\ &\leq |f_n - k| + |\phi_n - l| \end{aligned} \qquad ...(1)$$

Let $\varepsilon \geq 0$ be given. Now

$\lim f_n = k \ \Rightarrow$ there exists a positive integer m_1 such that

$$|f_n - k| \leq \varepsilon/2 \ \forall \ n \geq m_1$$

$\lim \phi_n = l \ \Rightarrow$ there exists a positive integer m_2 such that

$$|\phi_n - l| < \varepsilon/2 \ \forall \ n \geq m_2$$

Let $m = \max \{m_1, m_2\}$ so that $\forall \ n \geq m$, we have

$$|f_n - k| < \varepsilon/2 \text{ and } |\phi_n - l| < \varepsilon/2 \qquad ...(2)$$

From (1) and (2), we have

$$|(f_n + \phi_n) - (k + l)| < \varepsilon \ \forall \ n \geq m$$

It follows that to each $\varepsilon > 0$, there corresponds an integer m such that

$$|(f_n + \phi_n) - (k + l)| < \varepsilon \ \forall \ n \geq m$$

Thus, $\lim (f_n + \phi_n) = k + l = \lim f_n + \lim \phi_n$.

Theorem II. *If* $\lim f_n = k$ *and* $\lim \phi_n = l$, *then*

$$\lim (f_n - \phi_n) = k - l = \lim \phi_n - \lim \phi_n$$

i.e., the limit of the difference of two convergent sequences is the difference of their limits.

(*Gorakhpur, 1997*)

Proof. We have, $\forall \ n \in \mathbf{N}$

$$\begin{aligned} |(f_n - \phi_n) - (k - l)| &= |(f_n - k) + (l - \phi_n)| \\ &\leq |f_n - k| + |\phi_n - l| \end{aligned}$$

The result now follows as in theorem I above.

Theorem III. *If* $\lim f_n = k$ *and* $\lim \phi_n = l$, *then*

$$\lim (f_n \phi_n) = kl = (\lim f_n)(\lim \phi_n).$$

[*Delhi B.Sc. (Hons.) I, 2016; Delhi B.Sc. (Prog.) II, 2015; Nagpur 2010; Meerut 2006, 2017; Chennai 2011*]

Proof. We have $\forall \ n \in \mathbf{N}$

$$\begin{aligned} |f_n \phi_n - kl| &= |\phi_n (f_n - k) + k(\phi_n - l)| \\ &\leq |\phi_n||f_n - k| + |k||\phi_n - l| \end{aligned} \qquad ...(1)$$

The sequence $<\phi_n>$, being convergent, is bounded so that there exists $\varepsilon > 0$ such that

$$|\phi_n| \leq c \qquad ...(2)$$

From (1) and (2), we have

$$| f_n \phi_n - kl | \le c | f_n - k | + | k | | \phi_n - l | \ \forall \ n \in \mathbf{N} \quad ...(3)$$

Let $\varepsilon > 0$ be given. Now

$\lim f_n = k \Rightarrow$ there exists a positive integer m_1 such that

$$| f_n - k | < \varepsilon/2c \ \forall \ n \ge m_1 \quad ...(4)$$

$\lim \phi_n = l \Rightarrow$ there exists a positive integer m_2 such that

$$| \phi_n - l | < \frac{\varepsilon}{2(|k|+1)} \ \forall \ n \ge m_2 \quad ...(5)$$

Let $m = \max \{m_1, m_2\}$. Then, $\forall \ n \ge m$, from (3), (4) and (5), we have

$$| f_n \phi_n - kl | < c \times \frac{\varepsilon}{2c} + | k | \times \frac{\varepsilon}{2(|k|+1)} < \varepsilon$$

Thus $\qquad \lim (f_n \phi_n) = kl = (\lim f_n)(\lim \phi_n).$

Theorem IV. *If* $\lim f_n = k$ *and* $\lim \phi_n = l$, *where* $l \ne 0$ *then*

$$\lim (f_n/\phi_n) = k/l = (\lim f_n)/(\lim \phi_n).$$ **[*Delhi Maths (G), 1990*]**

Proof. We have, $\forall \ n \in \mathbf{N}$

$$\left| \frac{f_n}{\phi_n} - \frac{k}{l} \right| = \left| \frac{l(f_n - k) - k(\phi_n - l)}{l \phi_n} \right|$$

$$\le \frac{| l | | f_n - k | + | k | | \phi_n - l |}{| l | | \phi_n |} \quad ...(1)$$

Now $\lim \phi_n = l \ne 0 \Rightarrow$ there exists a positive integer m_1 such that

$$| \phi_n - l | < | l |/2 \ \forall \ n \ge m_1$$

Thus $\forall \ n \ge m_1$, we have

$$| | l | - | \phi_n | | < | \phi_n - l | < | l |/2$$

$$\Rightarrow | \phi_n | \ge | l |/2 \quad ...(2)$$

From (1) and (2), we see that $\forall \ n \ge m_1$

$$\left| \frac{f_n}{\phi_n} - \frac{k}{l} \right| \le \frac{2}{| l |} | f_n - k | + \frac{2 | k |}{l^2} | \phi_n - l | \quad ...(3)$$

Let $\varepsilon > 0$ be given. Then there exist positive integers m_2 and m_3 such that

$$| f_n - k | < \frac{1}{4} | l | \varepsilon \ \forall \ n \ge m_2$$

and

$$| \phi_n - l | < \frac{1}{4} \frac{| l |^2}{| k | + 1} \varepsilon \ \forall \ n \ge m_3 \quad ...(5)$$

Let $m = \max \{m_1, m_2, m_3\}$. Then $\forall \ n \ge m$, from (3), (4) and (5), we have

$$| f_n/\phi_n - k/l | < \varepsilon$$

Thus we have proved that

$$\lim (f_n/\phi_n) = k/l = (\lim f_n)/(\lim \phi_n).$$

Note 1. The theorems proved above are also sometimes referred to by saying that the operation of limit is compatible with the addition, multiplication, subtraction and division compositions in the set of all real sequences.

Note 2. As a particular case of the theorems proved above, we see that if sequences f_n, ϕ_n be convergent, then the sequences

$$f_n + \phi_n,\ f_n - \phi_n,\ f_n \phi_n$$

are also convergent. Further, if $\lim \phi_n \neq 0$, then the sequence f_n/ϕ_n is also convergent.

The converse, however, may not be true as the following examples show :

(*i*) Let f_n and ϕ_n be defined by

$$f_n = (-1)^n,\ \phi_n = (-1)^{n+1}$$

so that $$f_n + \phi_n = 0 \ \forall\ n \in \mathbf{N}$$

Here neither f_n nor ϕ_n is convergent but $f_n + \phi_n$ is convergent.

It may be also seen that each of $f_n - \phi_n$, $f_n \phi_n$, f_n/ϕ_n is convergent.

Theorem V. *If* $\lim f_n = k$ *and* $k \neq 0$, *then there exists a positive number c and a positive integer m such that* $| f_n | > c \ \forall\ n \geq m$. ***[Delhi Maths (G), 1999; Delhi Maths (H), 1995]***

Proof. Since $k \neq 0$, we take $\varepsilon = | k |/2$ as a given positive number. Now, $\lim f_n = k$, so there exists $m \in \mathbf{N}$ such that

$$| f_n - k | < \varepsilon \ \forall\ n \geq m \qquad ...(1)$$

Now $$| k | = | (k - f_n) + f_n | \leq | k - f_n | + | f_n | \qquad ...(2)$$

From (1) and (2), $$| k | < \varepsilon + | f_n | \ \forall\ n \geq m \qquad ...(3)$$

From (3), $$| f_n | > | k | - \varepsilon \ \forall\ n \geq m$$

$$\Rightarrow | f_n | > | k |/2 \text{ (as } \varepsilon = | k |/2),\ \forall\ n \geq m$$

showing that a positive number $c\ (= | k |/2)$ and a positive integer m can be found such that

$$| f_n | > c \ \forall\ n \geq m.$$

Theorem VI. *If* $\lim f_n = k$, $f_n \neq 0\ \forall\ n \in \mathbf{N}$ *and* $k \neq 0$, *then*

$$\lim (1/f_n) = 1/k = 1/(\lim f_n).$$ ***(Chennai 2011)***

Proof. We have, $\forall\ n$

$$\left| \frac{1}{f_n} - \frac{1}{k} \right| = \frac{| k - f_n |}{| f_n || k |} \leq \frac{| f_n - k |}{| f_n || k |} \qquad ...(1)$$

Since $k \neq 0$, there exists a positive integer m_1 and a positive constant c, such that

$$| f_n | > c \text{ for } n \geq m_1 \qquad ...(2)$$

Also, $\lim f_n = k \Rightarrow$ there exists a positive integer m_2 such that

$$| f_n - k | < c\, | k |\, \varepsilon \ \forall\ n \geq m_2 \qquad ...(3)$$

Let $m = \max \{m_1, m_2\}$. Then, from (1), (2) and (3), we have

$$\left| \frac{1}{f_n} - \frac{1}{k} \right| < \frac{1}{| k |} \cdot \frac{1}{c} \cdot c\, | k |\, \varepsilon \ \forall\ n \geq m$$

i.e., $$| 1/f_n - 1/k | < \varepsilon \ \forall\ n \geq m$$

Thus $$\lim (1/f_n) = 1/k = 1/(\lim f_n).$$

Theorem VII. *Let* $\lim f_n = k$ *and c be any fixed real numbers. Then*

$$\lim (c f_n) = ck = c (\lim f_n).$$

Proof. Let $\phi_n = c \ \forall\ n \in \mathbf{N}$ so that $<\phi_n>$ is a constant sequence. Then

$$\lim \phi_n = c \qquad ...(1)$$

Now, $$\lim (c f_n) = \lim (\phi_n f_n)$$
$$= (\lim \phi_n)(\lim f_n), \text{ by theorem III}$$
$$= ck = c \lim f_n.$$

5.6. BOUNDED NON-CONVERGENT SEQUENCES

A bounded sequence which *does not converge* is said to *oscillate finitely*.

A finitely oscillating sequence must necessarily have at least two limit points.

Unbounded sequences. We shall now consider the case of unbounded sequences. In the case of sequences which are not bounded, we distinguish the following *three* types of behaviours :

(*i*) **Divergence to ∞.** A sequence $<f_n>$ is said to be divergent to ∞, if to each given positive number Δ, there corresponds an integer *m* such that

$$f_n > \Delta \quad \forall\ n \geq m,$$

and, in symbols, we have

$$\lim f_n = \infty \quad \text{or} \quad \lim_{n \to \infty} f_n = \infty.$$

(*ii*) **Divergence to – ∞.** A sequence f_n is said to be divergent to – ∞, if to each given positive number Δ, there corresponds an integer *m* such that

$$f_n < -\Delta \quad \forall\ n \geq m$$

and in symbols, we write

$$\lim f_n = -\infty \quad \text{or} \quad \lim_{n \to \infty} f_n = -\infty$$

(*iii*) **Infinite oscillation.** A sequence f_n is said to *oscillate infinitely*, if it is unbounded and is divergent neither to ∞ nor to – ∞.

Note. From the above it follows that for any given sequence, we distinguish between five types of different behaviours.

1. A bounded sequence may (*i*) converge or (*ii*) oscillate. II. An unbounded sequence may (*iii*) diverge to + ∞ (*iv*) diverge – ∞ (*v*) or oscillate infinitely.

ILLUSTRATIONS

1. $<n^2> = <1, 4, 9, 16, n^2, ...>$ diverges to + ∞ ***(Kanpur 2019)***

2. $<2^n> = <2, 2^2, 2^3, ..., 2^n, ...>$ diverges to + ∞

3. $<-2n> = <-2, -4, -6, ..., -2n, ...>$ diverges to – ∞

4. $<-x^n> = <-x, -x^2, -x^3, ..., -x^n, ...>$ diverges to – ∞, if $x > 1$

5. $<(-1)^n\, n> = <-1, 2, -3, 4, -5, 6, ...>$ neither diverges to ∞ nor to – ∞

6. $<(-1)^n> = <-1, 1, -1, 1, -1, 1, ..., (-1)^n, ...>$ oscillates finitely ***(Meerut 2018)***

7. $<1 + (-1)^n> = <0, 2, 0, 2, ..., 1 + (-1)^n, ...>$ oscillates finitely

8. $<(-1)^n\, n> = <-1, 2, -3, 4, ..., (-1)^n\, n, ...>$ oscillates infinitely

EXERCISES

1. When is a sequence of real number said to diverge to (*i*) + ∞ (*ii*) – ∞.

[Delhi Maths (H), 2004]

2. Show that the sequence :

(*i*) $<3^n>$ diverges to + ∞ (*ii*) $<-e^n>$ diverges to – ∞ ***(Agra, 2001)***

Solution. (*i*) $\Delta > 0$ be given and let $f_n = 3^n$.

$$f_n > \Delta \;\Rightarrow\; 3^n > \Delta \;\Rightarrow\; n \log 3 > \log \Delta \;\Rightarrow\; n > \frac{\log \Delta}{\log 3}$$

∴ If *m* be a positive integer > (log Δ)/log 3, then

$$f_n > \Delta\ \forall\ n \geq m \;\Rightarrow\; \lim_{n \to \infty} f_n = \infty \;\Rightarrow\; \lim_{n \to \infty} 3^n = \infty.$$

(*ii*) Let $\Delta > 0$ be given. Let $f_n = -e^n$.

$\therefore\ f_n < -\Delta \Rightarrow -e^n < -\Delta \Rightarrow e^n > \Delta \Rightarrow n > \log \Delta$

$\therefore$ If m be a positive integer $> \log \Delta$, then

$$f_n < -\Delta\ \forall\ n \geq m \Rightarrow \lim f_n = -\infty.$$

$\therefore$ $<f_n>$, *i.e.*, $<-e^n>$ diverges to $-\infty$.

3. Show that the sequence $<\log(1/n)>$ diverges to $-\infty$.

5.7. SOME THEOREMS ON DIVERGENT SEQUENCES

Theorem I. *If the sequences* $<f_n>$ *and* $<\phi_n>$ *diverge to infinity then* $<f_n + \phi_n>$ *and* $<f_n\, \phi_n>$ *both diverge to infinity.*

Proof. We have, $\forall\ n \in \mathbf{N}$

$\lim f_n = \infty \Rightarrow$ for a given positive number Δ_1, there exists $m_1 \in \mathbf{N}$ such that

$$f_n > \Delta_1 \quad \forall\ n \geq m_1 \qquad ...(1)$$

and $\lim \phi_n = \infty \Rightarrow$ for a given positive number Δ_2, there exists $m_2 \in \mathbf{N}$ such that

$$\phi_n > \Delta_2 \quad \forall\ n \geq m_2 \qquad ...(2)$$

Let $m = \max\{m_1, m_2\}$. Then, from (1) and (2), we have

$$f_n + \phi_n > \Delta_1 + \Delta_2 > \Delta_1 \quad \forall\ n \geq m$$

and

$$f_n\, \phi_n > \Delta_1\, \Delta_2 > \Delta_1 \quad \forall\ n \geq m$$

$$\Rightarrow <f_n + \phi_n> \text{ and } <f_n\, \phi_n> \text{ both diverge to infinity.}$$

Theorem II. *If* $<f_n>$ *diverges to infinity and* $<\phi_n>$ *is bounded then* $<f_n + \phi_n>$ *diverges to infinity.*

Proof. The sequence $<\phi_n>$ is bounded $\Rightarrow$ there exists $\Delta_1 > 0$ such that

$$|\phi_n| < \Delta_1 \quad \forall\ n \in \mathbf{N} \qquad ...(1)$$

Now, $<f_n>$ diverges to infinity $\Rightarrow$ for $\Delta_2 > 0$ there exists $m \in \mathbf{N}$ such that

$$f_n > \Delta_1 + \Delta_2 \quad \forall\ n \geq m \qquad ...(2)$$

Hence, for $\forall\ n \geq m$, we have

$$f_n + \phi_n \geq f_n - |\phi_n|, \text{ as } \phi_n \geq -|\phi_n|$$

$$> (\Delta_1 + \Delta_2) - \Delta_1, \text{ by (1) and (2)}$$

Thus, for $\Delta_2 > 0$, there exists $m \in \mathbf{N}$ such that

$$f_n + \phi_n > \Delta_2 \quad \forall\ n \geq m,$$

showing that $<f_n + \phi_n>$ diverges to infinity.

Theorem III. *If* $<f_n>$ *diverges to infinity and* $<\phi_n>$ *converges, then* $<f_n + \phi_n>$ *diverges to infinity.*

Proof. Since $<\phi_n>$ is a convergent sequence, so it must be a bounded. Now proceed as theorem II above.

Theorem IV. (*i*) $<\phi_n>$ *diverges to* ∞ *and* $f_n \geq \phi_n\ \forall\ n \in \mathbf{N}$, *then* $<f_n>$ *diverges to* ∞.

(*ii*) *If* $<\phi_n>$ *diverges to* $-\infty$ *and* $f_n \leq \phi_n\ \forall\ n \in \mathbf{N}$, *then* $<\phi_n>$ *diverges to* $-\infty$.

Proof. (*i*) $<\phi_n>$ diverges to $\infty \Rightarrow$ for $\Delta > 0$, there exists $m \in \mathbf{N}$ such that

$$\phi_n > \Delta \quad \forall\ n \geq m \qquad ...(1)$$

$$\therefore \quad f_n > \phi_n \Rightarrow f_n > \Delta\ \forall\ n \geq m, \text{ using (1)}$$

Hence $<f_n>$ diverges to ∞.

(*ii*) Left as an exercise.

Theorem V. *If $<f_n>$ diverges to ∞ and c be non-zero constant, then*

(i) $<c f_n>$ *diverges to* $+\infty$ *if* $c > 0$; *(ii)* $<c f_n>$ *diverges to* $-\infty$ *if* $c < 0$.

Proof. Left as an exercise.

Theorem VI. *If* $f_n > 0\ \forall\ n \in \mathbf{N}$, *then*

$$f_n \to \infty \text{ as } n \to \infty \iff 1/f_n \to 0 \text{ as } n \to \infty$$

Proof. Let $f_n \to \infty$ as $n \to \infty$

Let $\varepsilon > 0$ be given. Then

$\lim_{n \to \infty} f_n = \infty \Rightarrow$ for given $1/\varepsilon > 0$ there exists $m \in \mathbf{N}$ such that

$$f_n > 1/\varepsilon \quad \forall\ n \geq m$$

$$\Rightarrow \qquad \frac{1}{f_n} < \varepsilon \quad \forall\ n \geq m \Rightarrow \left| \frac{1}{f_n} - 0 \right| < \varepsilon \quad \forall\ n \geq m,$$

showing that $1/f_n \to 0$ as $n \to \infty$.

Conversely, let $1/f_n \to 0$ as $n \to \infty$.

Let $k > 0$ be a given constant. Then

$\lim_{n \to \infty} 1/f_n = 0 \Rightarrow 1/k > 0$ there exists $m \in \mathbf{N}$ such that

$$|1/f_n - 0| < 1/k, \ \forall\ n \geq m$$

$$\Rightarrow \ 1/f_n < 1/k \quad \forall\ n \geq m \Rightarrow f_n > k \quad \forall\ n \geq m,$$

showing that $f_n \to \infty$ as $n \to \infty$.

EXAMPLES

Example 1. (*a*) *Let* $<f_n>$ *and* $<g_n>$ *be two sequences such that* $<f_n + g_n>$ *and* $<f_n g_n>$ *converge. Show, by an example, that the sequences* $<f_n>$ *and* $<g_n>$ *may fail to converge.*

[Delhi B.A. (Prog.) III, 2014]

(*b*) *Let* $<f_n>$ *and* $<g_n>$ *be two sequences such that* $<f_n - g_n)$ *and* $<f_n/g_n>$ *converge. Show, by an example, that the sequences* $<f_n>$ *and* $<g_n>$ *was fail to converge.*

Solution. (*a*) Consider two sequences $<f_n>$ and $<g_n>$ given by

$<f_n> = <(-1)^n> = <-1, 1, -1, 1, \ldots..>$ and $<g_n> = <(-1)^{n+1}> = <1, -1, 1, -1, \ldots.>$

Then, $\quad <f_n + g_n> = <0, 0, 0, \ldots..>$, which converges to 0

and $\quad <f_n g_n> = <-1, -1, -1, \ldots..>$, which converges to -1.

But $<f_n>$ and $<g_n>$ both are not convergent sequences.

(*b*) **Hint.** Take $<f_n> = <g_n> = <(-1)^n>$.

Example 2. *Prove that the sequence* $\left\langle \dfrac{n^2 + 3n + 5}{2n^2 + 5n + 7} \right\rangle$ *converges to* $\dfrac{1}{2}$. **(Meerut, 1996)**

Solution. Let $f_n = (n^2 + 3n + 5)/(2n^2 + 5n + 7)$. Then

$$\lim_{n \to \infty} f_n = \lim_{n \to \infty} \frac{n^2 + 3n + 5}{2n^2 + 5n + 7} = \lim_{n \to \infty} \frac{1 + 3 \times (1/n) + 5 \times (1/n^2)}{2 + 5 \times (1/n) + 7 \times (1/n^2)}$$

$$= \frac{1 + 3 \times 0 + 5 \times 0}{2 + 5 \times 0 + 7 \times 0}, \text{ as } \lim_{n \to \infty} \frac{1}{n} = 0 \text{ and } \lim_{n \to \infty} \frac{1}{n^2} = 0$$

$$= 1/2$$

$\Rightarrow$ the given sequence converges to 1/2.

Example 3. *Show that* $< f_n >$, *where* $f_n = \sqrt{n}\,(\sqrt{n+1} - \sqrt{n})$ *converges.*

Solution. Left as an exercise.

Example 4. *Prove that as* $f_n \to 0 \Leftrightarrow |f_n| = 0$.

Hence or otherwise prove that $\lim\limits_{n \to \infty} \dfrac{\sin n\pi}{n} = 0$.

Solution. Let $\varepsilon > 0$ be given. Then, since

$$|f_n - 0| = |f_n| = ||f_n| - 0|,$$

so
$$|f_n - 0| < \varepsilon \Leftrightarrow ||f_n| - 0| < \varepsilon \quad \forall\ n \geq m$$

$\therefore$
$$f_n \to 0 \Leftrightarrow |f_n| \to 0.$$

Here
$$\left|\frac{\sin n\pi}{n}\right| \leq \frac{1}{n} \to 0 \text{ as } n \to \infty \qquad [\because\ |\sin n\pi| \leq 1]$$

Thus,
$$\left|\frac{\sin n\pi}{n}\right| \to 0 \text{ and so } \frac{\sin n\pi}{n} \to 0.$$

Example 5. *Show that the sequence* $<f_n>$, *where*

$$f_n = 1 + \frac{1}{3} + \frac{1}{3^2} + \ldots.. + \frac{1}{3^{n-1}} \textit{ converges. Find } \lim_{n \to \infty} f_n.$$

Solution. Here $f_n = \dfrac{1 - (1/3)^n}{1 - 1/3} = \dfrac{3}{2}\left(1 - \dfrac{1}{3^n}\right) = \dfrac{3}{2} - \dfrac{3}{2} \times \left(\dfrac{1}{3}\right)^n.$

$$\left[\because\ a + ar + ar^2 + \ldots.. + ar^{n-1} = \frac{a(1 - r^n)}{1 - r}\right]$$

$\therefore$
$$\lim_{n \to \infty} f_n = \lim\left[\frac{3}{2} - \frac{3}{2} \times \left(\frac{1}{3}\right)^n\right] = \frac{3}{2} - \frac{3}{2}\lim\left(\frac{1}{3}\right)^n$$

$$= 3/2 - 0 \text{ as } 1/3 < 1 \text{ so } \lim_{n \to \infty} (1/3)^n = 0$$

$$= 3/2 \qquad \left[\because\ \lim_{n \to \infty} r^n = 0, \text{ if } |r| < 1\right]$$

So $< f_n >$ is convergent and $\lim\limits_{n \to \infty} f_n = 3/2$.

EXERCISES

1. Show that the sequence $<f_n>$ where

   (i) $f_n = (-1)^n$, (ii) $f_n = 1 + (-1)^n$

   oscillates finitely.
2. Show that a divergent sequence can have no limit point.
3. If $\lim f_n = 0$ and $<\phi_n>$ oscillates finitely, show that $\lim (f_n \phi_n) = 0$.
4. If $<f_n>$ is convergent and $<\phi_n>$ is divergent, show that $\lim (f_n \div \phi_n) = 0$.
5. If $<f_n>$ is convergent and $<\phi_n>$ is divergent, show that $<f_n + \phi_n>$ is divergent.
6. Given that $<f_n>$ and $<g_n>$ are two convergent sequences and two sequences $<\phi_n>$ and $<\psi_n>$ are defined as follows :

   $$\phi_n = \max\{f_n, g_n\}, \qquad \psi_n = \min\{f_n, g_n\}.$$

   Show that the sequences $<\phi_n>$ and $<\psi_n>$ are also convergent and that

   $$\lim \phi_n = \max\{\lim f_n, \lim g_n\}, \qquad \lim \psi_n = \min\{\lim f_n, \lim g_n\}.$$

7. If $<f_n>$ is a convergent sequence and k is a number such that $f_n \le k \ \forall \ n$, then show that $\lim f_n \le k$. Give an example to show the possibility of equality.

8. Given that f_n is a convergent sequence, such that $f_n \ge 0 \ \forall \ n$, show that

$$\lim \sqrt{f_n} = \sqrt{\lim f_n}.$$

9. If $f_n \to +\infty$ and $g_n \to -\infty$, then show by examples that $<f_n + g_n>$ may

(*i*) converge, (*ii*) diverge to $+\infty$, (*iii*) diverge to $-\infty$ and (*iv*) oscillate.

10. Give examples of sequences $<f_n>$ and $<g_n>$ such that $<f_n>$ diverges to $+\infty$ and $<g_n>$ converges but $<f_n g_n>$

(*i*) diverges to $+\infty$ (*ii*) converges (*iii*) oscillates.

11. Give an example of pair of sequences $\{s_n\}_{n=1}^{\infty}$ and $\{t_n\}_{n=1}^{\infty}$ for which as $n \to \infty$, $s_n \to \infty$, $t_n \to -\infty$, and $s_n + t_n \to \infty$ ***(Pune 2010)***

ANSWERS/HINTS

9. (*i*) Use sequences $<f_n> = <n^2>$ and $<g_n> = <-n^2>$

(*ii*) Use sequences $<f_n> = <2n>$ and $<g_n> = <-n>$

(*iii*) Use sequences $<f_n> = <n>$ and $<g_n> = <-2n>$

(*iv*) Use sequences $<g_n> = <-n>$ and $<f_n>$ defined by

$$f_n = \begin{cases} n^2, & \text{if } n \text{ is odd} \\ n+1, & \text{if } n \text{ is even} \end{cases}$$

Then $<f_n> = <1, 3, 3^2, 5, 5^2, ..., n^2, n+1, ...>$

Clearly, $f_n \to +\infty$ and $g_n \to -\infty$ but

$<f_n + g_n> = <0, 1, 6, 1, 20, 1, ...>$, which oscillates infinitely.

10. (*i*) Use sequences $<f_n> = <n^2>$ and $<g_n> = <1/n>$

(*ii*) Use sequences $<f_n> = <n>$ and $<g_n> = <1/n^2>$

(*iii*) Use sequences $<f_n> = <n>$ and $<g_n> = <(-1)^n/n>$

Then $<f_n>$ diverges to $+\infty$ and $<g_n>$ converges to 0 but $<f_n g_n> = <(-1)^n>$ oscillates between -1 and 1.

5.8. SOME IMPORTANT THEOREMS ON LIMITS

Theorem I. *$\lim f_n = l \Rightarrow \lim |f_n| = |l|$ but the converse is not true.*

Proof. Since $\lim f_n = l$, so for a given $\varepsilon > 0$, there exists a positive integer m such that

$$|f_n - l| < \varepsilon \quad \forall \ n \ge m \qquad ...(1)$$

Now $$||f_n| - |l|| < |f_n - l| \qquad ...(2)$$

From (1) and (2), $$||f_n| - |l|| < \varepsilon \quad \forall \ n \ge m$$

Hence $$\lim |f_n| = |l|$$

We now show that the converse need not be true, *i.e.*, $\lim |f_n| = |l|$ may not imply that $\lim f_n = l$.

Let $$<f_n> = <(-1)^n> = <-1, 1, -1, 1, ...>$$

Then $<f_n>$ does not converge to any limit whereas $<|f_n|>$ *i.e.*, $<|(-1)^n|>$, *i.e.*, $<1, 1, 1, ...>$ converges to 1.

Thus, $\lim |f_n| = \lim |(-1)^n| = |1|$ does not imply that $\lim f_n = \lim (-1)^n = 1$.

Theorem II. *If* $f_n \geq 0\ \forall\ n \in \mathbf{N}$ *and* $\lim f_n = l$, *then* $l \geq 0$.

[Delhi B.Sc. (Prog.) II 2016; B.Sc. (Hons.) I, 2016, 17]

Proof. If possible, let $l < 0$. Now

$\lim f_n = l \Rightarrow$ for a given $\varepsilon > 0$, there exists a positive integer m such that

$$|f_n - l| < \varepsilon \quad \forall\ n \geq m, \quad i.e., \quad l - \varepsilon < f_n < l + \varepsilon \quad \forall\ n \geq m \qquad ...(1)$$

Since $l < 0$, we may take $\varepsilon = -(l/2) > 0$ in (1).

Hence, $l + \frac{l}{2} < f_n < l - \frac{l}{2}\quad \forall\ n \geq m \qquad i.e., \qquad f_n < l/2 < 0 \quad \forall\ n \geq m,$

which contradicts the fact that $f_n \geq 0\ \forall\ n \in \mathbf{N}$. Hence our assumption $l < 0$ is wrong and so we must have $l \geq 0$.

Theorem III. *If* $<f_n>$ *and* $<g_n>$ *be two convergent sequences, then*

$f_n \leq g_n\ \forall\ n \in \mathbf{N} \Rightarrow \lim f_n \leq \lim g_n$. **[Delhi B.Sc. (Hons.) I, 2016]**

Proof. Let $\lim f_n = l$ and $\lim g_n = l'$.

We suppose that $l > l'$ and show that we arrive at a contradiction. Let

$$l - l' = 3\varepsilon,$$

and consider the neighbourhoods $]l' - \varepsilon, l' + \varepsilon[,\ ]l - \varepsilon, l + \varepsilon[$

of l' and l respectively. It may be seen that these neighbourhoods are disjoint. In fact, as may be verified $l - l' = 3\varepsilon \Rightarrow l' + \varepsilon < l - \varepsilon$

Now there exists $m \in \mathbf{N}$ such that $\forall\ n \geq m, f_n \in\ ]l - \varepsilon, l + \varepsilon[,\ g_n \in\ ]l' - \varepsilon, l' + \varepsilon[$

so that $\forall\ n \geq m$, we have $f_n < g_n$

and as such we arrive at a contradiction.

Hence, we have shown that $f_n \geq g_n \quad \forall\ n \in \mathbf{N} \Rightarrow \lim f_n \leq \lim g_n$.

Corollary to theorem III. *If* $< f_n >$ *is a convergent sequence and if* $a \leq f_n \leq b$ *for all* $n \in \mathbf{N}$, *then* $a \leq \lim f_n \leq b$.

Proof. Let $< g_n >$ be the constant sequence $< b, b, b, ... >$. Then, we have $\lim g_n = b$. Since $f_n \leq b$, theorem III implies that $\lim f_n \leq \lim g_n$, *i.e.*, $\lim f_n \leq b$. Similarly, we can show that $a \leq \lim f_n$. Thus, $a \leq \lim f_n \leq b$.

Theorem IV. Sandwich Theorem or Squeeze theorem (principle)

If $<f_n>$, $<g_n>$ *and* $<h_n>$ *are three sequences such that* (i) $\lim f_n = \lim h_n = l$

(ii) *For some positive integer* p, $f_n \leq g_n \leq h_n \quad \forall\ n \geq p$, *then* $\lim g_n = l$.

[Kanpur 2012; Delhi B.A. (Prog) III, 2008, 2014; Delhi B.Sc. (Hons.) I, 2018]

Proof. Let $\varepsilon > 0$ be given. Now, we have

$\lim f_n = l$ and $\lim h_n = l$

$\Rightarrow$ there exist positive integers m_1 and m_2 such that

$|f_n - l| < \varepsilon\ \forall\ n \geq m_1$ and $|h_n - l| < \varepsilon \quad \forall\ n \geq m_2$

$\Rightarrow l - \varepsilon < f_n < l + \varepsilon \quad \forall\ n \geq m_1$ and $l - \varepsilon < h_n < l + \varepsilon \quad \forall\ n \geq m_2$ ...(1)

Also, given that $f_n \leq g_n \leq h_n \quad \forall\ n \geq p$ ...(2)

Let $m = \max\{m_1, m_2, p\}$. Then from (1) and (2), we have

$$l - \varepsilon < f_n < l + \varepsilon \quad \forall\ n \geq m \qquad ...(3)$$

$$l - \varepsilon > h_n < l + \varepsilon \quad \forall\ n \geq m \qquad ...(4)$$

and

$$f_n \leq g_n \leq h_n \quad \forall\ n \geq m \qquad ...(5)$$

From (3), (4) and (5), we have

$$l - \varepsilon < f_n \le g_n \le h_n < l + \varepsilon \quad \forall\, n \ge m$$
$$\Rightarrow l - \varepsilon < g_n < l + \varepsilon \quad \forall\, n \ge m,$$
$$\Rightarrow | g_n - l | < \varepsilon \quad \forall\, n \ge m,$$

showing that $\lim g_n = l$.

Corollary. *If $<f_n>$ and $<g_n>$ are two sequences such that $| f_n | \le | g_n |\ \forall\, n \ge m$ where $m \in \mathbf{N}$ and $\lim g_n = 0$, then $\lim f_n = 0$.*

Proof. Given that $\lim g_n = 0$. So $\lim | g_n | = 0$ and $\lim (-| g_n |) = 0$.

Now, $$| f_n | \le | g_n | \quad \forall\, n \ge m$$
$$\Rightarrow -| g_n | \le f_n \le | g_n | \quad \forall\, n \ge m$$

Hence by the above theorem, $\lim f_n = 0$.

Example. Let $f_n = (\cos n\pi)/n$. Then

$$| f_n | = \left| \frac{\cos n\pi}{n} \right| = \frac{| \cos n\pi |}{n} \le \frac{1}{n}$$

Thus, $$| f_n | \le \left| \frac{1}{n} \right| \text{ and } \lim \frac{1}{n} = 0.$$

So by above corollary, $\lim f_n = \lim \dfrac{\cos n\pi}{n} = 0.$

Theorem V. (Cauchy's first theorem on limits)

If $\lim\limits_{n \to \infty} f_n = l$, *then* $\lim\limits_{n \to \infty} \dfrac{f_1 + f_2 + \ldots\ldots + f_n}{n} = l.$ **[*G.N.D.U. Amritsar 2010*]**

[*Delhi B.Sc. (Prog.) II 2017; Agra 2011; Delhi B.A. (Prog) III, 2013; Meerut, 2005; Kanpur 2011, 2013*]

Proof. Define a sequence $<g_n>$ such that

$$g_n = f_n - l \quad \text{or} \quad f_n = l + g_n \quad \forall\, n \in \mathbf{N} \qquad \ldots(1)$$

Then $\lim\limits_{n \to \infty} g_n = \lim\limits_{n \to \infty} f_n - l = l - l = 0$, as $\lim\limits_{n \to \infty} f_n = l$.

Thus, $$\lim_{n \to \infty} g_n = 0 \qquad \ldots(2)$$

Using $f_n = l + g_n \quad \forall\, n \in \mathbf{N}$, we have

$$\frac{f_1 + f_2 + \ldots\ldots + f_n}{n} = \frac{(l + g_1) + (l + g_2) + \ldots\ldots + (l + g_n)}{n}$$

or $$\frac{f_1 + f_2 + \ldots\ldots + f_n}{n} = l + \frac{g_1 + g_2 + \ldots\ldots + g_n}{n} \qquad \ldots(3)$$

Hence in order to prove the theorem, we must prove that

$$\lim_{n \to \infty} \frac{g_1 + g_2 + \ldots\ldots + g_n}{n} = 0. \qquad \ldots(4)$$

Let $\varepsilon > 0$ be given. Since by (2), $\lim g_n = 0$, so there exists a positive integer m such that

$$| g_n - 0 | < \varepsilon/2 \quad \forall\, n \ge m, \quad i.e., \quad | g_n | < \varepsilon/2 \quad \forall\, n \ge m \qquad \ldots(5)$$

Again, $\lim g_n = 0 \Rightarrow <g_n>$ is a convergent sequence

$\Rightarrow <g_n>$ is a bounded sequence

$\Rightarrow$ there exists a number $K > 0$ such that $| g_n | \le K\ \forall\, n \in \mathbf{N}$ ...(6)

Now, $\left|\dfrac{g_1+g_2+.....+g_m}{n}\right| \le \dfrac{1}{n}(|g_1|+|g_2|+.....|g_m|) \le \dfrac{mK}{n}$, using (6)

Thus, $\left|\dfrac{g_1+g_2+.....+g_m}{n}\right| < \dfrac{\varepsilon}{2}$, if $\dfrac{mK}{n} < \dfrac{\varepsilon}{2}$, *i.e.*, if $n > \dfrac{2mK}{\varepsilon}$.

Let m' be a positive integer $> (2mK)/\varepsilon$. Then

$$\left|\frac{g_1+g_2+.....+g_m}{n}\right| < \frac{\varepsilon}{2} \quad \forall\ n \ge m' \qquad ...(7)$$

Again,
$$\left|\frac{g_{m+1}+g_{m+2}+.....+g_n}{n}\right| \le \frac{1}{n}(|g_{m+1}|+|g_{m+2}|+.....+|g_n|)$$
$$\le \frac{1}{n}\left(\frac{\varepsilon}{2}+\frac{\varepsilon}{2}+.....+\frac{\varepsilon}{2}\right), \text{ by (5)}$$
$$= \frac{n-m}{n}\times\frac{\varepsilon}{2} < \frac{\varepsilon}{2} \quad \forall\ n \ge m$$

Thus,
$$\left|\frac{g_{m+1}+g_{m+2}+.....+g_n}{n}\right| < \frac{\varepsilon}{2} \quad \forall\ n \ge m \qquad ...(8)$$

Let $M = \max\{m, m'\}$. Then, from (7) and (8), we have

$$\left|\frac{g_1+g_2+.....+g_m}{n}\right| < \frac{\varepsilon}{2} \text{ and } \left|\frac{g_{m+1}+g_{m+2}+.....+g_n}{n}\right| < \frac{\varepsilon}{2} \quad \forall\ n \ge M \qquad ...(9)$$

Now
$$\left|\frac{g_1+g_2+.....+g_n}{n}-0\right| = \left|\frac{g_1+g_2+.....+g_m}{n}+\frac{g_{m+1}+g_{m+2}+.....+g_n}{n}\right|$$
$$\le \left|\frac{g_1+g_2+.....+g_m}{n}\right|+\left|\frac{g_{m+1}+g_{m+2}+.....g_n}{n}\right|$$
$$< \varepsilon/2+\varepsilon/2,\ \forall\ n \ge M, \text{ using (9)}$$

Thus,
$$\left|\frac{g_1+g_2+.....+g_n}{n}-0\right| < \varepsilon \quad \forall\ n \ge M$$

Hence $\lim\limits_{n\to\infty}\dfrac{g_1+g_2+.....+g_n}{n} = 0$ and hence from (3), we have

$$\lim_{n\to\infty}\frac{f_1+f_2+.....+f_n}{n} = l.$$

Note. The converse of the above theorem need not be true. Consider the sequence $<f_n>$, where $f_n = (-1)^n$. For this sequence, we have

$$\frac{f_1+f_2+.....+f_n}{n} = \begin{cases} 0, & \text{if } n \text{ is even} \\ -1/n, & \text{if } n \text{ is odd} \end{cases}$$

Hence
$$\lim\frac{f_1+f_2+.....+f_n}{n} = 0$$

whereas $<f_n>$, *i.e.*, $<(-1)^n>$ is not a convergent sequence.

ILLUSTRATIONS

Example 1. *Show that* $\lim_{n\to\infty} \frac{1}{n}(1 + 2^{1/2} + 3^{1/3} + ... + n^{1/n}) = 1.$

[Delhi B.Sc. (Prog.) II, 2015; Delhi B.A. (Prog.) III, 2013]
[Delhi B.Sc. III 2009; Delhi Maths (H), 2003; Meerut, 2002, 03]

Solution. Let $f_n = n^{1/n}$. Then, by example 6, page 5.10, we have

$$\lim_{n\to\infty} f_n = \lim_{n\to\infty} n^{1/n} = 1.$$

Hence by Cauchy's first theorem on limits, we have

$$\lim_{n\to\infty} \frac{f_1 + f_2 + ... + f_n}{n} = 1$$

$$\Rightarrow \lim_{n\to\infty} \frac{1}{n}(1 + 2^{1/2} + 3^{1/3} + ... n^{1/n}) = 1.$$

Example 2. *Show that* (*i*) $\lim_{n\to\infty}\left[\frac{1}{\sqrt{n^2+1}} + \frac{1}{\sqrt{n^2+2}} + ... + \frac{1}{\sqrt{n^2+n}}\right] = 1.$

(Kanpur 2012; Bundelkhand, 1996; Meerut, 2001)

(*ii*) $\lim_{n\to\infty}\left[\frac{1}{\sqrt{2n^2+1}} + \frac{1}{\sqrt{2n^2+2}} + ... \frac{1}{\sqrt{2n^2+n}}\right] = \frac{1}{\sqrt{2}}.$ **(Meerut, 2002)**

Solution. (*i*) Let $f_n = \frac{n}{\sqrt{n^2+n}}$. Then $\lim_{n\to\infty} f_n = \lim_{n\to\infty} \frac{1}{\sqrt{1+1/n}} = 1$

Hence by Cauchy's first theorem on limits, we have

$$\lim_{n\to\infty} \frac{f_1 + f_2 + ... + f_n}{n} = 1$$

$$\Rightarrow \lim_{n\to\infty} \frac{1}{n}\left[\frac{n}{\sqrt{n^2+1}} + \frac{n}{\sqrt{n^2+2}} + ... + \frac{n}{\sqrt{n^2+n}}\right] = 1$$

$$\Rightarrow \lim_{n\to\infty}\left[\frac{1}{\sqrt{n^2+1}} + \frac{1}{\sqrt{n^2+2}} + ... + \frac{1}{\sqrt{n^2+n}}\right] = 1.$$

(*ii*) Left as an exercise for the reader.

Theorem VI. (Cauchy's second theorem on limits) **[Ranchi 2010]**

If $\lim_{n\to\infty} f_n = l$, *where* $f_n > 0 \quad \forall\ n \in \mathbf{N}$, *then*

$$\lim_{n\to\infty} (f_1 f_2 ..... f_n)^{1/n} = l.$$

[Delhi Maths (H), 2004; Delhi Maths (G), 2002, 05; Meerut, 2001, 02]

Proof. Let $\langle g_n \rangle$ be another sequence defined by

$$g_n = \log f_n \quad \forall\ n \in \mathbf{N}.$$

Now, $\lim_{n\to\infty} f_n = l \Rightarrow \lim_{n\to\infty} \log f_n = \log l$, when $f_n > 0 \ \forall\ n \in \mathbf{N}$ and $l > 0$

i.e., $$\lim_{n\to\infty} g_n = \log l.$$

Hence by Cauchy's first theorem on limits,

$$\lim_{n\to\infty} \frac{g_1 + g_2 + ... + g_n}{n} = \log l$$

$$\Rightarrow \lim_{n\to\infty} \frac{\log f_1 + \log f_2 + ... + \log f_n}{n} = \log l$$

$$\Rightarrow \lim_{n\to\infty} \frac{1}{n} \log (f_1\, f_2\, ...\, f_n) = \log l$$

$$\Rightarrow \lim_{n\to\infty} \log (f_1\, f_2\, ...\, f_n)^{1/n} = \log l$$

and hence $$\lim_{n\to\infty} (f_1\, f_2\, ...\, f_n)^{1/n} = l.$$

Corollary. *If $<f_n>$ is a sequence such that $f_n > 0\ \forall\ n \in \mathbf{N}$ and*

$$lim \frac{f_{n+1}}{f_n} = l,\ l > 0. \text{ Then } \lim_{n\to\infty} (f_n)^{1/n} = l.$$

Proof. Let us define another sequences $<g_n>$ as follows :

$$g_1 = f_1,\ g_2 = f_2/f_1,\ g_3 = f_3/f_2,\ .....,\ g_n = f_n/f_{n-1}.$$

Then, $$g_1 g_2\ .....\ g_n = f_n \qquad ...(1)$$

Now, $$\lim_{n\to 0} \frac{f_{n+1}}{f_n} = l \Rightarrow \lim_{n\to\infty} \frac{f_n}{f_{n-1}} = l \Rightarrow \lim_{n\to\infty} g_n = l, \text{ using (1)}$$

Also, $$f_n > 0\ \forall\ n \in \mathbf{N} \Rightarrow g_n > 0\ \forall\ n \in \mathbf{N}$$

Thus there exists a sequence $<g_n>$ such that $g_n > 0\ \forall\ n \in \mathbf{N}$ and $\lim_{n\to\infty} g_n = l$. Hence by Cauchy's second theorem on limits, we have

$$\lim_{n\to\infty} (g_1\, g_2\, ...\, g_n)^{1/n} = l, \text{ i.e., } \lim_{n\to\infty} (f_n)^{1/n} = l, \text{ by (1).}$$

ILLUSTRATIONS

Example 1. *Prove that if a sequence $<a_n>$ of positive numbers converge to a limit l, then $\lim_{n\to\infty} (a_1\, a_2\, ...\, a_n)^{1/n} = l$. Deduce that $\lim_{n\to\infty} n^{1/n} = 1$.* **[*Delhi Maths (H), 1998, 2004*]**

Solution. For first part, refer theorem VI.

Deduction. Let use define a sequence $<a_n>$ as given below :

$$a_1 = 1,\ a_2 = 2/1,\ a_3 = 3/2,\ .....,\ a_n = n/(n-1).$$

Then, $$a_1\, a_2\, ...\, a_n = n \text{ and } \lim_{n\to\infty} a_n = \lim_{n\to\infty} \frac{n}{n-1} = \lim_{n\to\infty} \frac{1}{1-(1/n)} = 1$$

Since $$\lim_{n\to\infty} (a_1\, a_2\, ...\, a_n)^{1/n} = l, \text{ hence } \lim_{n\to\infty} n^{1/n} = 1.$$

Example 2. *If $\lim_{n\to\infty} \left(1+\frac{1}{n}\right)^n = e$, then show that*

$$\lim_{n\to\infty} \left[\frac{2}{1}\left(\frac{3}{2}\right)^2 \left(\frac{4}{3}\right)^3 .....\left(\frac{n+1}{n}\right)^n\right]^{1/n} = e.$$

[*Delhi Maths (H), 1993; Meerut, 1996*]

Solution. Let $$f_n = \left(1+\frac{1}{n}\right)^n = \left(\frac{n+1}{n}\right)^n \qquad ...(1)$$

Then $f_n > 0 \ \forall \ n \in \mathbf{N}$ and $\lim_{n \to \infty} f_n = \lim_{n \to \infty} \left(1+\frac{1}{n}\right)^n = e > 0$

Hence, by Cauchy's second theorem on limits

$$\lim_{n \to \infty} (f_1 f_2 \ldots\ldots f_n)^{1/n} = e$$

i.e., $$\lim_{n \to \infty} \left[\frac{2}{1}\left(\frac{3}{2}\right)^2 \left(\frac{4}{3}\right)^3 \ldots\ldots \left(\frac{n+1}{n}\right)^n\right]^{1/n} = e, \text{ using (1).}$$

Example 3. *Prove that* $\lim_{n \to \infty} \left(\frac{n^n}{n!}\right)^{1/n} = e$, *i.e.,* $\lim_{n \to \infty} \frac{n}{(n!)^{1/n}} = e.$

[*Agra, 2000, 10; Delhi Maths (P), 1999, 2001; Meerut, 2001, 09*]

Solution. Let $f_n = \frac{n^n}{n!}$ so that $f_{n+1} = \frac{(n+1)^{n+1}}{(n+1)!}$

Then, $$\frac{f_{n+1}}{f_n} = \frac{(n+1)^{n+1}}{(n+1)!} \times \frac{n!}{n^n} = \frac{(n+1)^{n+1} \times n!}{(n+1)\, n!\, n^n} = \left(\frac{n+1}{n}\right)^n = \left(1+\frac{1}{n}\right)^n$$

$$\therefore \quad \lim_{n \to \infty} \frac{f_{n+1}}{f_n} = \lim_{n \to \infty} \left(1+\frac{1}{n}\right)^n = e > 0$$

$\therefore$ By the corollary of theorem VI, $\lim_{n \to \infty} (f_n)^{1/n} = e$, *i.e.,* $\lim_{n \to \infty} \left(\frac{n^n}{n!}\right)^{1/n} = e.$

Example 4. (*a*) *Prove that if* $f_n = \frac{1}{n}\{(n+1)(n+2)\ldots\ldots(n+n)\}^{1/n}$, *then the sequence* $<f_n>$ *converges to* $4/e$. **(*Delhi Maths 1997; Meerut 2011*)**

(*b*) Show that the sequence $<b_n^{1/n}\}$ is convergent and find its limit, where $b_n = n^n / \{(n+1)(n+2)\ldots\ldots(n+n)\}$ **[*Delhi B.Sc. (Prog) III 2009*]**

Solution. (*a*) Let us define a sequence $<g_n>$ as follows :

$$g_n = \frac{(n+1)(n+2)\ldots\ldots(n+n)}{n^n} \quad \text{so that} \quad f_n = (g_n)^{1/n} \qquad ...(1)$$

From (1), $$g_{n+1} = \frac{(n+2)(n+3)\ldots\ldots 2n\,(2n+1)(2n+2)}{(n+1)^{n+1}}$$

$$\frac{g_{n+1}}{g} = \frac{n^n}{(n+1)^n} \times \frac{(n+2)(n+3)\ldots\ldots 2n\,(2n+1)(2n+2)}{(n+1)^2\,(n+2)\ldots\ldots 2n}$$

$$\Rightarrow \lim_{n \to \infty} \frac{g_{n+1}}{g_n} = \lim_{n \to \infty} \left(\frac{n}{n+1}\right)^n \times \frac{(2n+1)(2n+2)}{(n+1)^2}$$

$$= \lim_{n \to \infty} \frac{1}{(1+1/n)^n} \times \frac{(2+1/n)(2+2/n)}{(1+1/n)^2} = \frac{4}{e} \quad \text{as} \quad \lim_{n \to \infty} (1+1/n)^n = e$$

Hence by corollary to theorem VI, we have

$$\lim_{n \to \infty} (g_n)^{1/n} = 4/e \quad \text{...(2)}$$

From (1) and (2), $\lim_{n \to \infty} f_n = \lim_{n \to \infty} (g_n)^{1/n} = 4/e.$

(*b*) Proceed as in part (a) and verify that $\lim_{n\to\infty} b_n^{1/n} = e/4$

Theorem VII. Cesaro's Theorem

If $\lim f_n = l$ *and* $\lim g_n = l'$, *then* $\quad \lim \dfrac{f_1\, g_n + f_2\, g_{n-1} + \ldots.. + f_n\, g_1}{n} = ll'.$

Proof. Let $\quad f_n = l + h_n$ and $|h_n| = H_n$. ...(1)

Then, $\quad \lim h_n = 0$ and so $\lim H_n = 0$.

∴ By Cauchy's first theorem on limits, we get

$$\lim_{n \to \infty} \frac{1}{n}(H_1 + H_2 + \ldots.. + H_n) = 0 \quad \text{...(2)}$$

Now, $\dfrac{1}{n}(f_1\, g_n + f_2\, g_{n-1} + \ldots.. + f_n\, g_1)$

$$= \frac{1}{n}[(l + h_1)\, g_n + (l + h_2)\, g_{n-1} + \ldots.. + (l + h_n)\, g_1], \text{ using (1)}$$

$$= \frac{l}{n}(g_1 + g_2 + \ldots..\, g_n) + \frac{1}{n}(h_1\, g_n + h_2\, g_{n-1} + \ldots.. + h_n\, g_1) \quad \text{...(3)}$$

Now, $\quad <g_n>$ is convergent

⇒ $<g_n>$ is bounded

⇒ there exists a positive number k such that

$$|g_n| < k \;[\; n \in \mathbf{N} \quad \text{...(4)}$$

Then $\quad 0 \le \left| \dfrac{1}{n}(h_1\, g_n + h_2\, g_{n-1} + \ldots.. + h_n\, g_1) \right|$

$$\le \frac{1}{n}\{|h_1| \times |g_n| + |h_2| \times |g_{n-1}| + \ldots.. + |h_n| \times |g_1|\}$$

$$< \frac{k}{n}(|h_1| + |h_2| + \ldots.. + |h_n|, \text{ using (4)}$$

$$= (k/n) \times (H_1 + H_2 + \ldots.. + H_n), \text{ using (1)} \quad \text{...(5)}$$

Using (2), (5) and Sandwich theorem IV, we obtain

$$\lim \frac{1}{n}(h_1\, g_n + h_2\, g_{n-1} + \ldots.. + h_n\, g_1) = 0 \quad \text{...(6)}$$

Since $\lim_{n \to \infty} g_n = l'$, so by Cauchy's first theorem on limits, we have

$$\lim_{n \to \infty} \frac{1}{n}(g_1 + g_2 + \ldots.. + g_n) = l'. \quad \text{...(7)}$$

Using (6) and (7) in (3), we finally have

$$\lim_{n \to \infty} \frac{1}{n}(f_1\, g_n + f_2\, g_{n-1} + \ldots.. + f_n\, g_n) = ll'.$$

Theorem VIII. *If* $<f_n>$ *is a sequence such that*

$$\lim_{n \to \infty} \frac{f_{n+1}}{f_n} = l, \textit{ where } |l| < 1, \textit{ then } \lim_{n \to \infty} f_n = 0.$$

[*Delhi Maths (G), 1998; Delhi Physics (H) 1995, 97, 99*]

Proof. Since $| l | < 1$, we can choose a positive number, ε, so small that

$$| l | + \varepsilon < 1 \text{ or } k < 1, \text{ where } k = | l | + \varepsilon \quad ...(1)$$

Now, $$\lim_{n \to \infty} \frac{f_{n+1}}{f_n} = l$$

$\Rightarrow$ there exists a positive integer m such that $\forall\ n \geq m$

$$\left| \frac{f_{n+1}}{f_n} - l \right| < \varepsilon$$

$$\Rightarrow \quad \left| \frac{f_{n+1}}{f_n} \right| - | l | < \left| \frac{f_{n+1}}{f_n} - l \right| < \varepsilon, \text{ as } | a | - | b | \leq | a - b |, \text{ if } a, b \in \mathbf{R}$$

$$\Rightarrow \quad \left| \frac{f_{n+1}}{f_n} \right| < | l | + \varepsilon \text{ or } \left| \frac{f_{n+1}}{f_n} \right| < k, \text{ by (1)} \quad ...(2)$$

Changing n to $m, m + 1, \ldots, (n - 1)$ in (2) and multiplying, we get

$$\left| \frac{f_{m+1}}{f_m} \right| \times \left| \frac{f_{m+2}}{f_{m+1}} \right| \times \ldots \times \left| \frac{f_n}{f_{n-1}} \right| < \underbrace{k \cdot k \ldots k}_{(n-m) \text{ factors}}$$

$$\Rightarrow \quad \left| \frac{f_n}{f_m} \right| < k^{n-m} \Rightarrow | f_n | < \frac{| f_m |}{k^m} k^n \quad ...(3)$$

Since $0 < k < 1$, so $\lim_{n \to \infty} k^n = 0$ ...(4)

$\therefore$ (3) $\Rightarrow \lim_{n \to \infty} | f_n | = 0$, using (4)

$\Rightarrow \lim_{n \to \infty} f_n = 0$.

Theorem IX. *If $< f_n >$ is a sequence such that*

$$\lim \frac{f_{n+1}}{f_n} = l > 1, \text{ then } \lim_{n \to \infty} f_n = \infty.$$

Proof. Since $l > 1$, we can choose a positive number, ε, so small that

$$l - \varepsilon > 1 \text{ or } k > 1 \text{ where } k = l - \varepsilon \quad ...(1)$$

Now, $$\lim_{n \to \infty} \frac{f_{n+1}}{f_n} = l$$

$\Rightarrow$ there exists a positive integer m such that

$$\left| \frac{f_{n+1}}{f_n} - l \right| < \varepsilon \quad \forall\ n \geq m$$

$$\Rightarrow \quad l - \varepsilon < \frac{f_{n+1}}{f_n} < l + \varepsilon, \quad \forall\ n \geq m$$

$$\Rightarrow \quad k < \frac{f_{n+1}}{f_n} < l + \varepsilon, \quad \forall\ n \geq m \text{ using (1)}$$

Thus, $$\frac{f_{n+1}}{f_n} > k \quad \forall\, n \geq m \qquad ...(2)$$

Changing n to $m, m+1, ..., (n-1)$ is (2) and multiplying, we get

$$\frac{f_{m+1}}{f_m} \times \frac{f_{m+2}}{f_{m+1}} \times ..... \times \frac{f_n}{f_{n-1}} > \underbrace{k \cdot k \, ... \, k}_{(n-m)\text{ factors}}$$

or $$\frac{f_n}{f_m} > k^{n-m} \quad \text{or} \quad f_n > \frac{f_m}{k^m} \cdot k^n \qquad ...(3)$$

Since $k > 1$, so $\lim\limits_{n \to \infty} k^n = \infty$. Then (3) $\Rightarrow \lim\limits_{n \to \infty} f_n = \infty$.

ILLUSTRATIONS

Example 1. *Show that* $\lim\limits_{n \to \infty} 2^{-n} n^2 = 0.$ **[*Delhi B.Sc.* (*H*) *Physics, 1995*]**

Solution. Let $$f_n = \frac{n^2}{2^n} \text{ so that } f_{n+1} = \frac{(n+1)^2}{2^{n+1}}$$

$$\therefore \quad \lim_{n \to \infty} = \frac{f_{n+1}}{f_n} = \lim_{n \to \infty} \frac{(n+1)^2}{2^{n+1}} \times \frac{2^n}{n^2} = \lim_{n \to \infty} \frac{1}{2}\left(1 + \frac{1}{n}\right)^2 = \frac{1}{2} < 1$$

Hence, by theorem VIII, $\lim\limits_{n \to \infty} f_n = \lim\limits_{n \to \infty} 2^{-n} n^2 = 0.$

Example 2. *Show that* $\lim\limits_{n \to \infty} \dfrac{m(m-1)(m-2).....(m-n+1)}{n!} x^n = 0$, *if* $|x| < 1.$

[*Delhi Maths* (*H*), *2000*]

Solution. Let $$f_n = \frac{m(m-1).....(m-n+1)}{n!} x^n$$

so that $$f_{n+1} = \frac{m(m-1).....(m-n+1)(m-n)}{(n+1)!} x^{n+1}$$

$$\therefore \quad \frac{f_{n+1}}{f_n} = \frac{(m-n)\, n!\, x}{(n+1)!} = \frac{(m-n)\, x}{n+1} = \frac{(m/n) - 1}{1 + (1/n)} x$$

$$\Rightarrow \quad \lim_{n \to \infty} \frac{f_{n+1}}{f_n} = -x, \text{ where } |-x| = |x| < 1 \text{ (given)}$$

Hence by theorem VIII, $\lim\limits_{n \to \infty} f_n = 0$. Hence the result.

EXERCISES

1. Show that

(*i*) $\lim\limits_{n \to \infty} \dfrac{1}{n}\left(1 + \dfrac{1}{2} + \dfrac{1}{3} + ... + \dfrac{1}{n}\right) = 0$ (*ii*) $\lim\limits_{n \to \infty} \dfrac{1}{n}\left(1 + \dfrac{1}{3} + \dfrac{1}{5} + ... + \dfrac{1}{2n-1}\right) = 0$

(*iii*) $\lim\limits_{n \to \infty} \left[\dfrac{1}{(n+1)^2} + \dfrac{1}{(n+2)^2} + ... + \dfrac{1}{(2n)^2}\right] = 0$ **(Meerut, 2000)**

(iv) $\lim_{n\to\infty}\left[\frac{1}{n^2}+\frac{1}{(n+1)^2}+.....+\frac{1}{(2n)^2}\right]=0$ **[Purvanchal 2006]**

2. Show that $\lim_{n\to\infty}\left\{\frac{(3n)\,!}{(n\,!)^3}\right\}^{1/n}=27.$

3. Show that for any number x, $\lim_{n\to\infty}\frac{x^n}{n!}=0.$ **[Meerut 2016]**

4. Show that $\lim_{n\to\infty}\frac{(1+y)^n}{n!}=0$, for all y.

5. Prove that if $p>0$, then $\lim_{n\to\infty}\frac{n^k}{(1+p)^n}=0$, k being a fixed real number.

6. If $|x|>$ and $k>0$, show that $\lim_{n\to\infty}\frac{n^k}{x^n}=0.$

7. Prove that $\lim_{n\to\infty}\left[\frac{1}{(n+1)^\lambda}+\frac{1}{(n+2)^\lambda}+.....+\frac{1}{(2n)^\lambda}\right]=0$, if $\lambda>1$.

8. Prove that $\lim_{n\to\infty}\frac{(n!)^{1/n}}{n}=\frac{1}{e}.$

9. If $s_n=(1+1/n)^n$, prove $\lim_{n\to\infty}(s_1\,s_2\cdots s_n)=e$ **[Meerut 2015]**

5.9. CAUCHY (OR FUNDAMENTAL) SEQUENCES [Delhi B.Sc. I (Hons) 2017, 18; Delhi BSc. (Prog) II, 2016, 17; Rajasthan 2010; Kanpur, 2002; Meerut 2011]

Definition 1. *A sequence $<f_n>$ is said to be a Cauchy sequence if given $\varepsilon>0$, there exists a positive integer m such that*

$$|f_n-f_m|<\varepsilon\quad\forall\ n\geq m.$$

Note. The above definition can also rewritten in the following two equivalent forms.

Definition 2. *A sequence $<f_n>$ is said to be a Cauchy sequence if given $\varepsilon>0$, there exists a positive integer m such that*

$$|f_{m+p}-f_m|<\varepsilon\quad\forall\ p>0.$$

Definition 3. *A sequence $<f_n>$ is said to be Cauchy sequence if given $\varepsilon>0$, there exists a positive integer m such that*

$$|f_p-f_q|<\varepsilon\quad\forall\ p,q\geq m$$

From the above definition 3, it follows that, roughly, a sequence $<f_n>$ is Cauchy if f_p and f_q are close together when p and q are large.

ILLUSTRATIONS

1. *The sequence <1/n> is a Cauchy sequence.*

Let $\varepsilon>0$ be given, let $n>m$ and $f_n=1/n$.

Now, $|f_n-f_m|=\left|\frac{1}{n}-\frac{1}{m}\right|=\frac{1}{m}-\frac{1}{n}$, as $n>m\Rightarrow\frac{1}{m}>\frac{1}{n}$

and so $\left|\frac{1}{n}-\frac{1}{m}\right|=\left|-\left(\frac{1}{m}-\frac{1}{n}\right)\right|=\frac{1}{m}-\frac{1}{n}$

Thus, $|f_n - f_m| < 1/m < \varepsilon$, if $m > 1/\varepsilon$.

Let m be a positive integer $> 1/\varepsilon$. Then

$$|f_n - f_m| < \varepsilon \quad \forall\ n \geq m$$

$\Rightarrow$ $<f_n>$, *i.e.*, $<1/n>$ is a Cauchy sequence.

2. *The sequence* $<n^2>$ *is not a Cauchy sequence.*

Let $f_n = n^2$. Let $p = 2m + 1$ and $q = 2m$ so that $p, q > m$

Then $|f_p - f_q| = |p^2 - q^2| = |(p+q)(p-q)|$

$= |4m + 1| = 4m + 1 > 1\ \forall\ m \in \mathbf{N}$

If $\varepsilon = 1/2$, then it is not possible to find any positive integer such that

$$|f_p - f_q| < \varepsilon \quad \forall\ p, q > m$$

$\Rightarrow$ $<f_n>$, *i.e.*, $<n^2>$ is not a Cauchy sequence.

Theorem I. *Every Cauchy sequence is bounded.*

[*Agra, 2009; Meerut 2009; Delhi B.Sc. (Hons.) I, 2018; Delhi B.A. (Prog.) III, 2016; Chennai 2011; Kanpur, 2018*]

Proof. Let $<f_n>$ be a Cauchy sequence. For $\varepsilon = 1$, there exists a positive integer m such that

$$|f_n - f_m| < 1 \quad \forall\ n \geq m$$

$$\Rightarrow f_m - 1 < f_n < f_m + 1 \quad \forall\ n \geq m$$

Let $k = \min\{f_1, f_2, \ldots\ldots, f_{m-1}, f_m - 1\}$ and $K = \max\{f_1, f_2, \ldots\ldots, f_{m-1}, f_m + 1\}$

Then $k \leq f_n \leq K \quad \forall\ n \in \mathbf{N}$

Hence $<f_n>$ is bounded.

Note. *The converse of the above theorem need not be always true, i.e., every bounded sequence need not be always a Cauchy sequence.* **[*Agra 2009; Delhi 2006, Kanpur 2006*]**

For example, the sequence $<(-1)^n> = <-1, 1, -1, 1, \ldots\ldots>$ is bounded as

$$-1 \leq (-1)^n \leq 1 \quad \forall\ n \in \mathbf{N}.$$

Let $f_n = (-1)^n$. We now show that $<f_n>$ is not a Cauchy sequence.

Let $p = 2m + 1$ and $q = 2m$ so that $p, q > m$.

Now, $|f_p - f_q| = |(-1)^{2m+1} - (-1)^{2m}| = |(-1) - (1)| = 2.$

If $\varepsilon = 1/2$, it is not possible to find any $m \in \mathbf{N}$ such that

$$|f_p - f_q| < \varepsilon \quad \forall\ p, q > m$$

Hence $<f_n>$, *i.e.*, $<(-1)^n>$ is not a Cauchy sequence.

Theorem II. Cauchy's criterion for convergence. **[*Kanpur 2013, 14*]**

A sequence converges if and only if it is a Cauchy sequence.

[*Delhi B.A. (Prog.) III 2012, 17; G.N.D.U. Amritsar 2010; Ranchi 2010; Kanpur, 2015, 16; Delhi B.Sc. (Prog.) II, 2015, 17 Meerut, 2012; Rajasthan 2010; Chennai 2011; Delhi B.Sc. (Hons.) II, 2018*]

Proof. First, let $<f_n>$ be a convergent sequence and $\lim f_n = l$. We shall now show that it is a Cauchy sequence.

Let $\varepsilon > 0$ be given then there exists $m \in \mathbf{N}$ such that

$$|f_n - l| < \varepsilon/2 \quad \forall\ n \geq m \qquad \ldots(1)$$

In particular, for $n = m$, (1) reduces to

$$|f_m - l| < \varepsilon/2 \qquad \ldots(2)$$

Now $|f_n - f_m| = |(f_n - l) - (f_m - l)| \leq |f_n - l| + |f_m - l| < \varepsilon/2 + \varepsilon/2 = \varepsilon\ \forall\ n \geq m$

[using (1) and (2)]

Thus $$|f_n - f_m| < \varepsilon \ \forall \ n \geq m$$

Hence $<f_n>$ is a Cauchy sequence.

Conversely, let $<f_n>$ be a Cauchy sequence. Then we shall prove that $<f_n>$ is a convergent sequence.

Since every Cauchy sequence is bounded, so, $<f_n>$ is bounded.

Again, since every bounded sequence has a limit point, so $<f_n>$ has a limit point, say l. We show that $<f_n>$ converges to l.

Let $\varepsilon > 0$ be given. Since $<f_n>$ is a Cauchy sequence, there exists a positive integer m such that

$$|f_n - f_m| < \varepsilon/3 \quad \forall \ n \geq m \qquad ...(3)$$

Since l is a limit point of $<f_n>$, every nbd of l contains infinitely many terms of $<f_n>$.

$\Rightarrow$ $f_n \in \,]l - \varepsilon/3, l + \varepsilon/3[$ for infinitely many values of n.

In particular, we can find a positive integer $k > m$ such that

$$f_k \in \,]l - \varepsilon/3, l + \varepsilon/3[$$

or $$|f_k - l| < \varepsilon/3 \text{ where } k > m \qquad ...(4)$$

Since $k > m$, hence, from (3), we have

$$|f_k - f_m| < \varepsilon/3 \qquad ...(5)$$

Now, $$\begin{aligned} |f_n - l| &= |(f_n - f_m) + (f_m - f_k) + (f_k - l)| \\ &\leq |f_n - f_m| + |f_k - f_m| + |f_k - l| \\ &< \varepsilon/3 + \varepsilon/3 + \varepsilon/3 \quad \forall \ n \geq m \end{aligned}$$ [using (3), (4) and (5)]

Thus $$|f_n - l| < \varepsilon \quad \forall \ n \geq m$$

Hence the sequence $<f_n>$ convergence to l.

Remark. The result of theorem II is very useful to prove the convergence of a sequence.

Theorem III. Cauchy's general principle of convergence.[*Delhi B.Sc. (Prog) III 2009, 10; Kanpur 2017; Delhi B.A. (Prog.) III 2012)*]

A necessary and sufficient condition for a sequence $<f_n>$ to be convergent is that to each $\varepsilon > 0$, there corresponds a positive integer m such that

$$|f_{n+p} - f_n| < \varepsilon \quad \forall \ n \geq m, \forall \ p \geq 1.$$

Proof. *The condition is necessary.* Let the sequence f_n be convergent so that there exists a number l such that $$\lim f_n = l.$$

Let $\varepsilon > 0$ be a given number. There exists $m \in \mathbf{N}$ such that

$$|f_n - l| < \frac{1}{2}\varepsilon \quad \forall \ n \geq m, \qquad ...(1)$$

$$\Rightarrow \quad |f_{n+p} - l| < \frac{1}{2}\varepsilon \quad \forall \ n \geq m, \forall \ p \geq 1 \qquad ...(2)$$

Thus, we have

$$\begin{aligned} |f_{n+p} - f_n| &= |f_{n+p} - l + l - f_n| \\ &\leq |f_{n+p} - l| + |l - f_n| \\ &< \frac{1}{2}\varepsilon + \frac{1}{2}\varepsilon \ \forall \ n \geq m, \ \forall \ p \geq 1. \end{aligned}$$ [using (1) and (2)]

$$\Rightarrow |f_{n+p} - f_n| < \varepsilon \quad \forall \ n \geq m, \forall \ p \geq 1.$$

Thus, the condition is necessary.

The condition is sufficient. First we show that boundedness of the sequence is a consequence of the given condition.

We take 1 for ε. Now there exists $r \in \mathbf{N}$ such that

$$|f_{r+p} - f_n| < 1 \quad \forall\ n \geq r \quad \text{and} \quad \forall\ p \geq 1.$$

Taking r for n, we have

$$|f_{r+p} - f_r| < 1 \quad \forall\ p \geq 1$$

$$\Rightarrow \quad f_r - 1 < f_{r+p} < f_r + 1 \quad \forall\ p \geq 1.$$

Let $k = \min \{f_r - 1, f_1, f_2, \ldots.., f_{r-1}\}$ and $K = \max \{f_r + 1, f_1, f_2, \ldots.., f_{r-1}\}$

Thus, we have $\qquad k \leq f_n \leq K \quad \forall\ n \in \mathbf{N}.$

so that the sequence f is bounded.

The sequence, being bounded, has at least one limit point.

Let l be a limit point. We shall show that l is the limit of $<f_n>$. Let $\varepsilon > 0$ be given

In accordance with the given condition, there exists a positive integer m such that

$$|f_{n+p} \quad f_n| < \varepsilon/2 \quad \forall\ n \geq m,\ \forall\ p \geq 1.$$

As l is a limit point of the sequence $<f_n>$, the neighbourhood $]l \quad \varepsilon/2, l + \varepsilon/2[$ of l contains an infinite number of members of $<f_n>$. There exists, therefore, a natural number $m_1 > m$ such that

$$| f_{m_1} | \in]l \quad \varepsilon/2, l + \varepsilon/2[\quad \Leftrightarrow \quad | f_{m_1} \quad l | < \varepsilon/2.$$

Thus, we have $\qquad | f_{m_1+p} \quad f_{m_1} | < \varepsilon/2 \quad \forall\ p \geq 1$ ...(3)

and $\qquad | f_{m_1} \quad l | < \varepsilon/2$ ...(4)

From (3) and (4), it follows that

$$| f_{m_1+p} \quad l | < \varepsilon \quad \forall\ p \geq 1$$

or equivalently $\qquad |f_n - l| < \varepsilon \quad \forall\ n \geq m.$

It follows that l is the limit of the sequence $<f_n>$ so that the sequence $<f_n>$ converges.

Note 1. The importance of Cauchy's general principle of convergence lies in the fact that it decides the convergence or otherwise of a sequence without any knowledge of the limit of a sequence and is concerned only with the terms of the sequence.

Note 2. Cauchy's general principle of convergence can also be stated as follows :

A sequence $<f_n>$ is convergent if and only if for each $\varepsilon > 0$, there exists a positive integer m such that

$$|f_n - f_m| < \varepsilon \quad \forall\ n > m$$

i.e., $\qquad |f_{m+1} + f_{m+2} + \ldots.. + f_n| < \varepsilon \quad \forall\ n > m$

Theorem IV. *Let $<f_n>$ and $<g_n>$ be Cauchy sequences. Then*

(*i*) *sum of $<f_n>$ and $<g_n>$, i.e., $<f_n + g_n>$ is a Cauchy sequence*

(*ii*) *difference of $<f_n>$ and $<g_n>$, i.e., $<f_n - g_n>$ is a Cauchy sequence*

(*iii*) *product of $<f_n>$ and $<g_n>$, i.e., $<f_n\, g_n>$ is a Cauchy sequence*

(*iv*) *If $f_n \neq 0\ \forall\ n \in \mathbf{N}$, then $<1/f_n>$ is a Cauchy sequence*

Proof. (*i*) Since $<f_n>$ and $<g_n>$ are Cauchy sequences, so, for $\varepsilon > 0$, there exists $m_1, m_2 \in \mathbf{N}$ such that

$$| f_n \quad f_{m_1} | < \varepsilon/2 \quad \forall\ n \geq m_1 \qquad \text{...(1)}$$

and $\qquad | g_n \quad g_{m_2} | < \varepsilon/2 \quad \forall\ n \geq m_2$ ...(2)

Let $m = \max\{m_1, m_2\}$. Then, from (1) and (2), we have

$$|f_n - f_m| < \varepsilon/2 \quad \text{and} \quad |g_n - g_m| < \varepsilon/2 \quad \forall\, n \geq m \qquad ...(3)$$

Now, $$|(f_n + g_n) - (f_m + g_m)| = |(f_n - f_m) + (g_n - g_m)|$$
$$\leq |f_n - f_m| + |g_n - g_m|$$
$$< \varepsilon/2 + \varepsilon/2 \;\forall\, n \in \mathbf{N} \qquad \text{[using (3)]}$$

Thus, $$|(f_n + g_n) - (f_m + g_m)| < \varepsilon \;\forall\, n \geq m$$

Hence $<f_n + g_n>$ is a Cauchy sequence

(*ii*) Left as an exercise.

(*iii*) $<f_n>$ is a Cauchy sequence

$\Rightarrow$ $<f_n>$ is bounded and so there exists a positive number k such that

$$|f_n| \leq k \quad \forall\, n \in \mathbf{N} \qquad ...(1)$$

Since $<f_n>$ and $<g_n>$ and Cauchy sequences, so far a given $\varepsilon > 0$, we can choose $m \in \mathbf{N}$ such that

$$|g_n - g_m| < \varepsilon/2k \quad \forall\, n \geq m \qquad ...(2)$$

and $$|f_n - f_m| < \frac{\varepsilon}{2(|g_m| + 1)} \quad \forall\, n \geq m \qquad ...(3)$$

Now, $$|f_n g_n - f_m g_m| = |f_n (g_n - g_m) + g_m (f_n - f_m)|$$
$$\leq |f_n|\,|g_n - g_m| + |g_m|\,|f_n - f_m|$$
$$\leq k \cdot \frac{\varepsilon}{2k} + |g_m| \times \frac{\varepsilon}{2(|g_m| + 1)} \;\forall\, n \geq m \qquad \text{[using (1), (2) and (3)]}$$

Thus $$|f_n g_n - f_m g_m| < \varepsilon \quad \forall\, n \geq m$$

Hence $<f_n g_n>$ is a Cauchy sequence.

(*iv*) Left as an exercise.

EXAMPLES

Example 1. *Show that* $<f_n>$, *where* $f_n = (-1)^n/n$, *is a convergent sequence.*

[Kanpur 2014; Delhi B.A. (Prog) III, 2008]

Solution. Let $\varepsilon > 0$ be given and let $n > m$.

Now $$|f_n - f_m| = \left|\frac{(-1)^n}{n} - \frac{(-1)^m}{m}\right| \leq \left|\frac{(-1)^n}{n}\right| + \left|\frac{(-1)^m}{m}\right| = \frac{1}{n} + \frac{1}{m}$$

$$< \frac{1}{m} + \frac{1}{m} = \frac{2}{m} < \varepsilon, \text{ if } m > \frac{2}{\varepsilon}$$

Thus, for each $\varepsilon > 0$, there exists $m \in \mathbf{N}$ such that

$$|f_n - f_m| < \varepsilon \quad \forall\, n \geq m$$

Hence $<f_n>$ is a Cauchy sequence and so by Cauchy's criterion of convergence $<f_n>$ is a convergent sequence.

Example 2. *Using Cauchy's general principle of convergence, show that the sequence* $<f_n>$ *where* $f_n = 1 + 1/2 + 1/3 + ... + 1/n$ *is not convergent.* **[Delhi B.A. (Prog.) III 2008, 12; Delhi B.Sc. I (Hons) 2010, 12; Delhi Maths (G) 2006]**

Solution. For $n > m$, we have

$$|f_n - f_m| = |f_{m+1} + f_{m+2} + ... + f_n|$$

$$= \left|\frac{1}{m+1} + \frac{1}{m+2} + ... + \frac{1}{n}\right| = \frac{1}{m+1} + \frac{1}{m+2} + ... + \frac{1}{n}$$

Taking $n = 2m$, we get

$$| f_{2m} - f_m | = \frac{1}{m+1} + \frac{1}{m+2} + ... + \frac{1}{2m}$$

$$> \frac{1}{2m} + \frac{1}{2m} + ... + \frac{1}{2m} = \frac{1}{2m} \times m = \frac{1}{2}$$

Thus, $$| f_{2m} - f_m | > 1/2 \qquad ...(1)$$

If $<f_n>$ is convergent, then by Cauchy's general principle of convergence, for each $\varepsilon > 0$, there exists a positive integer m such that

$$| f_n - f_m | < \varepsilon \quad \forall \ n \geq m \qquad ...(2)$$

In particular, for $\varepsilon = 1/2$ and $n = 2m$ in (2), for $<f_n>$ to be a convergent sequence, we must have

$$| f_{2m} - f_m | < 1/2,$$

which contradicts (1). Hence $<f_n>$ cannot converge.

Example 3. *Show that the sequence* $<f_n>$ *where* $f_n = 1 + 1/3 + 1/5 + ... + 1/(2n-1)$ *is not a Cauchy sequence. Is it convergent ?* **[Delhi B.Sc. (Prog.) II, 2015, 2016, 2018; (Delhi B.Sc. (Hons.) II, 2009, 18; Delhi B.A. (Prog) III 2010, 12)**

Solution. Given $$f_n = 1 + \frac{1}{3} + \frac{1}{5} + ... + \frac{1}{2n-1}.$$

Let $p = 2m$ and $q = m$ so that $p, q \geq m$

$$\therefore \quad | f_p - f_q | = | f_{2m} - f_m |$$

$$= \left| \left(1 + \frac{1}{3} + ... + \frac{1}{2m-1} + \frac{1}{2m+1} + \frac{1}{2m+3} + ... + \frac{1}{2m+2m-1}\right) - \left(1 + \frac{1}{3} + ... + \frac{1}{2m-1}\right) \right|$$

$$= \frac{1}{2m+1} + \frac{1}{2m+3} + ... + \frac{1}{2m+2m-1}$$

(m terms)

$$> \frac{1}{2m+2m} + \frac{1}{2m+2m} + ... + \frac{1}{2m+2m} = \frac{m}{4m} = \frac{1}{4}$$

Thus, $$| f_p - f_q | > 1/4, \ \forall \ p, q \geq m$$

Hence, if $\varepsilon = 1/4$, it is not possible to find any positive integer m such that

$$| f_p - f_q | < \varepsilon \quad \forall \ p, q \geq m.$$

$\Rightarrow$ $<f_n>$ is not a Cauchy sequence and hence by Cauchy's criterion of convergence, $<f_n>$ is not a convergent sequence.

Example 4. *Using Cauchy's criterion of convergence, examine the convergence of the sequence* $<f_n>$, *where* $f_n = 1 + \frac{1}{1!} + \frac{1}{2!} + ... + \frac{1}{n!}$.

[Delhi B.Sc. (Hons.) I, 2017; Delhi B.Sc. (Prog.) II, 2017]

Solution. Here, for $n \geq m$, we have

$$| f_n - f_m |$$

$$= \left| \left(1 + \frac{1}{1!} + \frac{1}{2!} + ... + \frac{1}{m!} + \frac{1}{(m+1)!} + \frac{1}{(m+2)!} + ... + \frac{1}{n!}\right) - \left(1 + \frac{1}{1!} + ... + \frac{1}{m!}\right) \right|$$

$$= \frac{1}{(m+1)!} + \frac{1}{(m+2)!} + \ldots + \frac{1}{n!}$$

$$\leq \frac{1}{2^m} + \frac{1}{2^{m+1}} + \ldots + \frac{1}{2^{n-1}}, \text{ as } n! \geq 2^{n-1} \quad \forall\, n \in \mathbf{N}$$

$$= \frac{1}{2^m}\left[1 + \frac{1}{2} + \frac{1}{2^2} + \ldots + \frac{1}{2^{(n-m)-1}}\right] = \frac{1}{2^m} \times \frac{1-(1/2)^{n-m}}{1-(1/2)}$$

$$= \frac{1}{2^{m-1}}\left[1 - \frac{1}{2^{n-m}}\right] = \frac{1}{2^{m-1}} - \frac{1}{2^{n-1}} \leq \frac{1}{2^{m-1}}$$

$$\left[\, n \geq m \;\Rightarrow\; n-1 \geq m-1 \;\Rightarrow\; 2^{n-1} \geq 2^{m-1} \;\Rightarrow\; \frac{1}{2^{n-1}} \leq \frac{1}{2^{m-1}} \right]$$

Hence for $\varepsilon > 0$, $\quad | f_n - f_m | < \varepsilon$ if $\quad \dfrac{1}{2^{m-1}} < \varepsilon \quad$ or $\quad 2^{m-1} > \dfrac{1}{\varepsilon}$

Now, $\quad 2^{m-1} > 1/\varepsilon \Rightarrow (m-1)\log 2 > \log(1/\varepsilon)$

$$\Rightarrow m > \{1 + \log(1/\varepsilon) \times (\log 2)^{-1}\} \qquad \text{...(i)}$$

$\therefore$ For each $\varepsilon > 0$, there exists $m \in \mathbf{N}$ such that

$$| f_n - f_m | < \varepsilon \quad \forall\, n \geq m, \text{ where } m \text{ is given by (1).}$$

Hence $<f_n>$ is a Cauchy sequence and so by Cauchy's criteria of convergence, $<f_n>$ must be a convergent sequence.

Example 5. *If $<f_n>$ is a sequence of positive numbers such that*

$$f_n = \frac{1}{2}(f_{n-1} + f_{n-2}), \text{ for all } n \geq 3,$$

then show that $<f_n>$ converges to $(f_1 + 2f_2)/3$.

Solution. Given that $\quad f_n = (f_{n-1} + f_{n-2})/2 \quad \forall\, n \geq 3 \qquad$ ...(1)

If $f_2 = f_1$, then from (1), we find $f_n = f_1\ \forall\, n \in \mathbf{N}$. Hence $<f_n>$ converges to f_1. Let us now consider the situation when $f_2 \neq f_1$.

Now, $\quad | f_n - f_{n-1} | = \left| \dfrac{1}{2}(f_{n-1} + f_{n-2}) - f_{n-1} \right| = \dfrac{1}{2} | f_{n-2} - f_{n-1} |$, by (1)

Thus, $\quad | f_n - f_{n-1} | = (1/2) \times | f_{n-1} - f_{n-2} | \qquad$ ...(2)

Replacing n by $n-1$ in (2), we get

$$| f_{n-1} - f_{n-2} | = (1/2) \times | f_{n-2} - f_{n-3} | \qquad \text{...(3)}$$

From (2) and (3), $\quad | f_n - f_{n-1} | = (1/2)^2 \times | f_{n-2} \times f_{n-3} |$

Proceeding likewise, we find

$$| f_n - f_{n-1} | = \frac{1}{2^{n-2}} | f_2 - f_1 |, \forall\, n \geq 3 \qquad \text{...(4)}$$

$\therefore$ For $n \geq m$, we have

$$| f_n - f_m | = | (f_n - f_{n-1}) + (f_{n-1} - f_{n-2}) + \ldots + (f_{m+1} - f_m) |$$

$$\leq | f_n - f_{n-1} | + | f_{n-1} - f_{n-2} | + \ldots + | f_{m+1} - f_m |$$

$$= \frac{1}{2^{n-2}} | f_2 - f_1 | + \frac{1}{2^{n-3}} | f_2 - f_1 | + \ldots + \frac{1}{2^{m-1}} | f_2 - f_1 |, \text{ by (4)}$$

$$= \left(\frac{1}{2^{m-1}} + \frac{1}{2^m} + \ldots + \frac{1}{2^{n-2}} \right) | f_2 - f_1 |$$

[which is a Geometric progression of $(n-2) - (m-2)$, *i.e.*, $n - m$ terms with first term $= 1/2^{m-1}$ and common ratio $= 1/2$]

$$= \frac{(1/2^{m-1}) \times \{1-(1/2)^{n-m}\}}{1-(1/2)} | f_2 - f_1 | = \frac{1}{2^{m-2}}\left(1 - \frac{1}{2^{n-m}}\right) | f_2 - f_1 |$$

$$< \frac{1}{2^{m-2}} | f_2 - f_1 |$$

Let ε > 0 be given. Then we can choose a positive integer m such that

$$\frac{1}{2^{m-2}} | f_2 - f_1 | < \varepsilon, \quad i.e., \quad 2^{m-2} > \frac{| f_2 - f_1 |}{\varepsilon}$$

$$i.e., \ (m-2)\log 2 > \log \frac{| f_2 - f_1 |}{\varepsilon}, \quad i.e., \quad m > 2 + \frac{\log \{| f_2 - f_1 | / \varepsilon\}}{\log 2} \qquad ...(5)$$

Thus, for the value of m given by (5), we have

$$| f_n - f_m | < \varepsilon \quad \forall\ n \geq m$$

showing that $<f_n>$ is a Cauchy sequence and so by Cauchy's convergence criterion, $<f_n>$ is a convergent sequence.

Let $<f_n>$ converges to l, that is, $\lim_{n \to \infty} f_n = l.$ ...(6)

Putting n = 3, 4, ..., n in (1), we have

$$\left.\begin{aligned} f_3 &= (1/2) \times (f_2 + f_1) \\ f_4 &= (1/2) \times (f_3 + f_2) \\ f_5 &= (1/2) \times (f_4 + f_3) \\ &\ldots\ldots\ldots\ldots\ldots\ldots \\ &\ldots\ldots\ldots\ldots\ldots\ldots \\ f_{n-1} &= (1/2) \times (f_{n-2} + f_{n-3}) \\ f_n &= (1/2) \times (f_{n-1} + f_{n-2}) \end{aligned}\right\} \qquad ...(7)$$

Adding the corresponding sides of the relations (7), we get

$$f_n + f_{n-1} = \frac{1}{2} f_1 + f_2 + \frac{1}{2} f_{n-1} \quad \text{or} \quad f_n + \frac{1}{2} f_{n-1} = \frac{1}{2}(f_1 + 2f_2).$$

Taking limit as $n \to \infty$, we obtain

$$\lim_{n \to \infty} f_n + \frac{1}{2} \lim_{n \to \infty} f_{n-1} = \frac{1}{2}(f_1 + 2f_2) \quad \text{or} \quad l + \frac{1}{2} l = \frac{1}{2}(f_1 + 2f_2), \text{ by (6)}$$

Thus $l = (f_1 + 2f_2)/3$ and so $<f_n>$ converges to $(f_1 + 2f_2)/3$.

EXERCISES

1. *Prove, by definition that the sequences whose nth terms are given below are Cauchy sequences :*

(*i*) $1/n^2$ (*ii*) $n/(n+1)$ (*iii*) $(n+1)/n$ (*iv*) $1/2^n$

2. *Prove that the sequences whose nth terms are given below are not Cauchy sequences :*

(*i*) n (*ii*) $(-1)^n n$

(*iii*) $1 + \frac{1}{6} + \frac{1}{11} + \ldots.. + \frac{1}{5n-1}$ (*iv*) $\frac{1}{9} + \frac{1}{13} + \frac{1}{17} + \ldots.. + \frac{1}{4n+5}$

3. Apply Cauchy's principle of convergence to prove that the sequence $<a_n>$ defined by

$$a_n = 1 + \frac{1}{4} + \frac{1}{7} + ... + \frac{1}{3n-2}$$ is not convergent.

[Delhi B.Sc. (Prog) III 2009; Delhi Maths (H), 1996; Delhi Maths (G), 2002]

4. Show that the sequence $<a_n>$ defined by $a_n = \frac{1}{2} + \frac{1}{5} + \frac{1}{8} + ... + \frac{1}{3n-1}$, $n \geq 1$ does not converge. **(Delhi Maths (G), 2002)**

5. Show that the sequence $<a_n>$ defined as $a_n = 1 + 1/5 + 1/9 + ... + 1/(4n-3)$ does not converge, whereas $<b_n>$ defined as $b_n = (1/n) \times a_n$ converges to 0.

6. Show that the sequence $\left\langle 1 + \frac{1}{4} + \frac{1}{7} + ... + \frac{1}{3n-2} \right\rangle$ is not convergent while

$\left\langle \frac{1}{n}\left(1 + \frac{1}{4} + \frac{1}{7} + ... + \frac{1}{3n-2}\right) \right\rangle$ is convergent.

7. Show that the sequence $<f_n>$, where $f_n = 1 - \frac{1}{2} + \frac{1}{3} - \frac{1}{4} + ... + \frac{(-1)^{n-1}}{n}$ is a Cauchy sequence.

8. Let $<a_n>$ be a sequence of real numbers and for each $n \in \mathbf{N}$, let $f_n = a_1 + a_2 + ... + a_n$ and $g_n = |a_1| + |a_2| + ... + |a_n|$. Prove that if $<g_n>$ is a Cauchy sequence, then so is $<f_n>$.

5.10. CONVERGENCE OF A SEQUENCE

It is important to distinguish between the fact of convergence of a sequence from that of its convergence to a given limit. While we have in Art. 5.4 examined the convergence of a sequence to a given limit, we shall in the following consider conditions for *just* convergence of a sequence. In this context, we shall consider two theorems, one of which relates to the convergence of monotonic sequences and the other to that of arbitrary sequences.

5.11. MONOTONIC SEQUENCES AND THEIR CONVERGENCE

[Kanpur 2013; Delhi B.Sc. (Prog) III 2009, 2010, 2015; Delhi BA (Prog) III 2010]

A sequence $<f_n>$ is said to be *monotonically increasing*, if

$$f_{n+1} \geq f_n \ \forall \ n \in \mathbf{N},$$

and *monotonically decreasing* if

$$f_{n+1} \leq f_n \ \forall \ n \in \mathbf{N}.$$

Thus, for example, the sequence $<f_n>$, defined by

$$f_n = n$$

is monotonically increasing and that defined by

$$f_n = 1/n$$

is monotonically decreasing.

Also $<f_n>$ defined by $f_n = (-1)^n$ is neither monotonically increasing nor decreasing.

A sequence which is either monotonically increasing or decreasing is called a *monotonic sequence*.

Theorem I. (*i*) *Every monotonically increasing sequence which is bounded above converges to its least upper bound.* **[Delhi B.Sc. (Hons.) I, 2017; Bhopal, 2004; Delhi B.A. (Prog) III 2010, 11, 15; Kanpur, 2007; Meerut, 2016]**

(*ii*) *Every monotonically decreasing sequence which is bounded below converges to its greatest lower bound.* **(Delhi B.Sc. (Hons.) I, 2016; Bangalore, 2004)**

Proof. (*i*) Suppose that $<f_n>$ is a bounded monotonically increasing sequence.

Let l be the least upper bound of the sequence, *i.e.*, the least upper bound of the range of the sequence. We shall show that $\lim f_n = l = \sup \{f_n : n \in \mathbf{N}\}$

Let $\varepsilon > 0$ be given. Since l is the least upper bound of the sequence, there exists positive integer m such that

$$f_m > l - \varepsilon \qquad \text{...(1)}$$

Also $<f_n>$ is monotonically increasing $\Rightarrow$ $f_n \geq f_m \quad \forall\ n \geq m$ ...(2)

$\therefore$ From (1) and (2), $f_n > l - \varepsilon\ \forall\ n \geq m$ ...(3)

Again, l is the least upper bound of $<f_n>$

$\Rightarrow$ $f_n \leq l < l + \varepsilon \quad \forall\ n \in \mathbf{N} \Rightarrow f_n < l + \varepsilon \quad \forall\ n \in \mathbf{N}$ ...(4)

From (3) and (4), we have

$$l - \varepsilon < f_n < l + \varepsilon \quad \forall\ n \geq m$$

$$\Rightarrow |f_n - l| < \varepsilon\ \forall\ n \geq m$$

Thus we see that to each $\varepsilon > 0$ there corresponds a positive integer m such that $|f_n - l| < \varepsilon \quad \forall\ n \geq m$ so that by Art. 5.4, we have

$\lim f_n = l$, *i.e.*, $<f_n>$ converges to l.

(*ii*) Suppose that $<f_n>$ is a bounded monotonically decreasing sequence.

Let l' be the greatest lower bound of the sequence, *i.e.*, the greatest lower bound of the range of the sequence. We shall show that $\lim f_n = l' = \inf \{f_n : n \in \mathbf{N}\}$

Let $\varepsilon > 0$ be given. Since l' is the greatest lower bound of the sequence, there exists positive integer m such that

$$f_m < l' + \varepsilon. \qquad \text{...(5)}$$

Also $<f_n>$ in monotonically decreasing

$\Rightarrow$ $f_n \leq f_m\ \forall \quad n \geq m$ ...(6)

$\therefore$ From (5) and (6), $f_n < l + \varepsilon \quad \forall\ n \geq m$...(7)

Again, l' is the greatest lower bound of $<f_n>$

$\Rightarrow$ $f_n \geq l' > l - \varepsilon \quad \forall\ n \in \mathbf{N} \Rightarrow f_n > l' - \varepsilon \quad \forall\ n \in \mathbf{N}$ ...(8)

$\therefore$ From (7) and (8), $l' - \varepsilon < f_n < l' + \varepsilon\ \forall\ n \geq m$

$$\Rightarrow |f_n - l'| < \varepsilon\ \forall\ n \geq m$$

Thus we see that to each $\varepsilon > 0$ there corresponds a positive integer m such that $|f_n - l'| < \varepsilon\ \forall\ n \geq m$ so that by Art. 5.4, we have

$\lim f_n = l'$, *i.e.*, $<f_n>$ converges to l'.

Theorem II. Monotone Convergence Theorem. *A necessary and sufficient condition for a monotonic sequence to be convergent is that it is bounded.*

[Delhi B.A. (Prog.) III, 2017; Purvanchal 2006; Agra, 2002; Kanpur, 2012, 17; Patna, 2003]

Proof. *The condition is necessary.* Let $<f_n>$ be a monotonic and convergent sequence. Then, proceed as in theorem II of Art. 5.4 to prove that $<f_n>$ is bounded.

The condition is sufficient. Let $<f_n>$ be monotonic and bounded sequence. Then we shall prove that it is a convergent sequence.

Since $<f_n>$ is bounded, so it is bounded above as well as below. Again $<f_n>$ is monotonic sequence implies that it is either monotonically increasing or monotonically decreasing sequence. Accordingly, the following two cases arise.

Case (*i*). *When $<f_n>$ is monotonically increasing sequence which is bounded above.* Then proceeding as in part (*i*) of theorem I, show that $<f_n>$ is convergent.

Case (*ii*). *When $<f_n>$ is monotonically decreasing sequence which is bounded below.* Then proceeding as in part (*ii*) of theorem I, show that $<f_n>$ is convergent.

Theorem III. (*i*) *Every monotonically increasing sequence which is not bounded above diverges to* ∞. **(Chennai 2011; G.N.D.U. Amritsar 2010; Meerut, 1997)**

(*ii*) *Every monotonically decreasing sequence which is not bounded below diverges to* $-\infty$. **[Delhi Maths (*H*), 2004]**

Proof. (*i*) Let $<f_n>$ be a monotonically increasing sequence which is not bounded above. Let Δ be any positive number, however large. Then there exists some positive integer m such that

$$f_m > \Delta \qquad \text{...(1)}$$

Now, $<f_n>$ is a monotonically increasing sequence $\Rightarrow f_n \geq f_m \quad \forall\ n \geq m$ ...(2)

From (1) and (2), $f_n > \Delta \quad \forall\ n \geq m$

$$\Rightarrow \lim_{n \to \infty} f_n = \infty, \text{ i.e., } <f_n> \text{ diverges to } \infty.$$

(*ii*) Let $<f_n>$ be a monotonically decreasing sequence which is not bounded below. Let Δ be any positive number, however large. Then there exists some positive integer m such that

$$f_m < -\Delta \qquad \text{...(3)}$$

Now, $<f_n>$ is a monotonically decreasing sequence $\Rightarrow f_n \leq f_m \quad \forall\ n \geq m$ ...(4)

From (3) and (4), $f_n < -\Delta \quad \forall\ n \geq m$

$$\Rightarrow \lim_{n \to \infty} f_n = -\infty, \text{ i.e., } <f_n> \text{ diverges to } -\infty.$$

Theorem IV. (*i*) *A monotonically increasing sequence either converges to its supremum* (*least upper bound*) *or diverges to* $+\infty$ **[Delhi Maths (Prog) 2008; Delhi. Maths (*H*), 2001]**

(*ii*) *A monotonically decreasing sequence either converges to its infimum* (*greatest lower bound*) *or diverges to* $-\infty$. **[Delhi B.Sc. (*H*) Phy, 1998]**

Proof. (*i*) Let $<f_n>$ be a monotonically increasing sequence. Two situations arise :

Case I. Let $<f_n>$ be bounded above. Then, proceed as in part (*i*) of theorem I to show that $<f_n>$ converges to its least upper bound (*i.e.*, supremum).

Case II. Let $<f_n>$ be not bounded above. Then, proceed as in part (*ii*) of theorem III to show that $<f_n>$ diverges to $+\infty$.

(*ii*) Let $<f_n>$ be a monotonically decreasing sequence. Two situations arise :

Case I. Let $<f_n>$ be bounded below. Then, proceed as in part (*ii*) of theorem I to show that $<f_n>$ converges to its greatest lower bound (*i.e.*, infimum).

Case II. Let $<f_n>$ be not bounded below. Then, proceed as in part (*ii*) of theorem III to show that $<f_n>$ diverges to $-\infty$.

EXAMPLES

Example 1. *Prove that the sequence* $\left\langle \dfrac{2n-7}{3n+2} \right\rangle$.

(*i*) *is monotonically increasing* (*ii*) *is bounded and* (*iii*) *tends to the limit 2/3.*

Solution. Let $f_n = (2n - 7)/(3n + 2)$ ...(1)

(*i*) From (1), we have

$$f_{n+1} - f_n = \frac{2n-5}{3n+5} - \frac{2n-7}{3n+2} = \frac{25}{(3n+5)(3n+2)} \ \forall \ n \in \mathbf{N}, \text{ on simplification}$$

$\Rightarrow \ f_{n+1} > f_n \quad \forall \ n \in \mathbf{N}$

$\Rightarrow \ <f_n>$ is monotonically increasing.

(*ii*) Given sequence $= <f_n> = \left\langle -1, -\frac{3}{8}, -\frac{1}{11}, \frac{1}{14}, \frac{3}{17}, \ldots\right\rangle$

Thus, we have $f_n \geq -1 \quad \forall \ n \in \mathbf{N}$ ...(2)

Again, $1 - f_n = 1 - \dfrac{2n-7}{3n+2} = \dfrac{n+9}{3n+2} > 0 \quad \forall \ n \in \mathbf{N}$

$\Rightarrow \ f_n < 1 \quad \forall \ n \in \mathbf{N}$ ...(3)

From (2) and (3), $-1 \leq f_n < 1 \ \forall \ n \in \mathbf{N}$

$\Rightarrow \ <f_n>$ is a bounded sequence.

(*iii*) Since $<f_n>$ is monotonically increasing sequence and bounded above, so it converges.

Now, $\lim\limits_{n\to\infty} f_n = \lim\limits_{n\to\infty} \dfrac{2n-7}{3n+2} = \lim\limits_{n\to\infty} \dfrac{2-(7/n)}{3+(2/n)} = \dfrac{2}{3}$

Thus, $<f_n>$ converges to 2/3.

Example 2. *Prove that the sequence $<a_n>$ where $a_n = \frac{1}{1!} + \frac{1}{2!} + \ldots + \frac{1}{n!}$ is convergent and that $2 \leq \lim\limits_{n\to\infty} a_n \leq 3$.* **[*Agra, 2002; Garhwal, 2001; Kanpur, 1996*]**

Solution. We have

$$a_{n+1} - a_n = 1 + \frac{1}{1!} + \frac{1}{2!} + \ldots + \frac{1}{n!} + \frac{1}{(n+1)!} - \left(1 + \frac{1}{1!} + \frac{1}{2!} + \ldots + \frac{1}{n!}\right)$$

$\Rightarrow \ a_{n+1} - a_n = 1/(n+1)! > 0 \ \forall \ n \in \mathbf{N}$

$\Rightarrow \ <a_n>$ is a monotonically increasing sequence.

Again, $a_n = 1 + 1 + \dfrac{1}{1\cdot 2} + \dfrac{1}{1\cdot 2\cdot 3} + \ldots + \dfrac{1}{1\cdot 2\cdot 3 \ldots n}$

$$< 1 + 1 + \frac{1}{2} + \frac{1}{2^2} + \ldots + \frac{1}{2^{n-1}},$$

which is a G.P. with 1 as its first term and 1/2 as common ratio

$$= 1 + \frac{1-(1/2)^n}{1-(1/2)}$$

Thus, $a_n < 3 - (1/2^{n-1})$ or $a_n < 3 \quad \forall \ n \in \mathbf{N}$

Hence $<a_n>$ is monotonically increasing and bounded above and so $<a_n>$ converges to its least upper bound which is ≤ 3.

$\therefore \quad \lim\limits_{n\to\infty} a_n \leq 3.$ ...(1)

Again. $a_n = 1 + 1 + \dfrac{1}{2!} + \dfrac{1}{3!} + \ldots + \dfrac{1}{n!} \ \Rightarrow \ a_n \leq 2 \quad \forall \ n \in \mathbf{N}$

$\therefore$ $$\lim_{n\to\infty} a_n \geq 2. \qquad ...(2)$$

From (1) and (2), $$2 \leq \lim_{n\to\infty} a_n \leq 3.$$

Example 3. *Show that the sequence $<f_n>$ defined by $f_n = (1 + 1/n)^n$ is convergent and that $\lim_{n\to\infty} (1+1/n)^n$ lies between 2 and 3.* **[Delhi B.Sc. (Prog.) II, 2016, 18; Delhi B.A. (Prog) III 2011; Agra, 2012; Delhi B.Sc. (Hons.) II, 2005; Kanpur, 2013, 19; Meerut, 2015]**

Solution. We have, by the binomial theorem, for positive integral index n,

$$f_n = \left(1+\frac{1}{n}\right)^n = 1+n\cdot\frac{1}{n}+\frac{n(n-1)}{2!}\cdot\frac{1}{n^2}+...+\frac{n(n-1)(n-2)...2.1}{n!}\cdot\frac{1}{n^n}$$

or $$f_n = 1+1+\frac{1}{2!}\left(1-\frac{1}{n}\right)+\frac{1}{3!}\left(1-\frac{1}{n}\right)\left(1-\frac{2}{n}\right)+...+\frac{1}{n!}\left(1-\frac{1}{n}\right)\left(1-\frac{2}{n}\right)...\left(1-\frac{n-1}{n}\right) \quad ...(1)$$

Replacing n by $n + 1$ in (1), we get

$$f_{n+1} = 1+1+\frac{1}{2!}\left(1-\frac{1}{n+1}\right)+\frac{1}{3!}\left(1-\frac{1}{n+1}\right)\left(1-\frac{2}{n+1}\right)+.....$$
$$+\frac{1}{(n+1)!}\left(1-\frac{1}{n+1}\right)\left(1-\frac{2}{n+1}\right).....\left(1-\frac{n}{n+1}\right) \quad ...(2)$$

Now, $$n+1 > n \Rightarrow -(n+1) < -n \Rightarrow -\frac{1}{n+1} > -\frac{1}{n}$$

$$\Rightarrow 1-\frac{1}{n+1} > 1-\frac{1}{n},\ 1-\frac{2}{n+1} > 1-\frac{2}{n} \text{ and so on.} \quad ...(3)$$

In view of (3), from (1) and (2), we find that

$$f_{n+1} \geq f_n \quad \forall\ n \in \mathbf{N}$$
$$\Rightarrow <f_n> \text{ is monotonically increasing.}$$

Also, from (1), we have $\forall\ n \in \mathbf{N}$

$$f_n = \left(1+\frac{1}{n}\right)^n \leq 1+1+\frac{1}{2!}+\frac{1}{3!}+.....+\frac{1}{n!}$$
$$= 1+1+\frac{1}{2}+\frac{1}{2^2}+.....+\frac{1}{2^n}, \text{ as } n! \geq 2^{n-1}\ \forall\ n \in \mathbf{N}$$
$$= 1+\frac{1-(1/2)^n}{1-(1/2)} = 3-\frac{1}{2^{n-1}} < 3 \quad ...(4)$$

Thus, $f_n < 3\ \forall\ n \in \mathbf{N}$ and so $<f_n>$ is bounded above.

Now, $<f_n>$ being bounded above and monotonically increasing sequence, is convergent.

Again, from (1), $$f_n \geq 2 \quad \forall\ n \in \mathbf{N}. \quad ...(5)$$

From (4) and (5), $$2 \leq a_n < 3 \quad \forall\ n \in \mathbf{N}.$$

$$\Rightarrow \quad 2 \leq \lim_{n\to\infty} f_n \leq 3 \quad \text{or} \quad 2 \leq \lim_{n\to\infty} (1+1/n)^n \leq 3.$$

Note. The actual value of $\lim_{n\to\infty} (1+1/n)^n$ is defined to be equal to e. Hence e lies between 2 and 3. Taking limit of both sides of (1) as $n \to \infty$, we get

$$e = \lim_{n\to\infty} (1+1/n)^n = 1+\frac{1}{1!}+\frac{1}{2!}+.....+\frac{1}{n!}+..... \text{ ad inf}$$

Example 4. *If* $g_n = \sum_{k=0}^{n} \frac{1}{k!}$ *and* $f_n = \left(1+\frac{1}{n}\right)^n$, $n \geq 0$ *show that* $<g_n>$ *and* $<f_n>$ *converge to the same limit.* **(Bhopal, 2004)**

Solution. Refer example 2 and 3.

Example 5 (*a*). *Show that the sequence* $<f_n>$ *defined recursively by* $f_1 = \sqrt{2}$, $f_{n+1} = \sqrt{2f_n}$ *converges to 2.* **[Pune 2010; G.N.D.U. Amrisar 2010]**

[Bangalore, 2004; Delhi Maths (H), 1999; Delhi B.A. III 2010; Meerut, 2001, 02]

(*b*) *Give an example of a strictly monotonically increasing sequence converging to the limit 2.*

(*c*) *Show that the sequence* $<s_n>$ *defined by the formula* $s_{n+1} = \sqrt{3s_n}$, $s_1 = 1$ *converges to 3.* **[Delhi Maths (H) 2007; I.A.S., 1996; Meerut, 1996]**

(*d*) *Let* $<a_n>$ *be sequence defined by* $a_1 = 1$, $a_{n+1} = \sqrt{7a_n}$, $n \geq 1$. *Show that* $<a_n>$ *converges. What is the limit of the sequence.* **(Delhi Maths (G), 2004)**

Solution (*a*). It may be seen that different members of the sequence $<f_n>$ are

$$\sqrt{2},\ \sqrt{2\sqrt{2}},\ \sqrt{2\sqrt{2\sqrt{2}}},\ .....$$

We have
$$2\sqrt{2} > 2$$

implying
$$\sqrt{2\sqrt{2}} > \sqrt{2} \Leftrightarrow f_2 > f_1.$$

Again, $f_{n+1} > f_n \Rightarrow \sqrt{2f_{n+1}} > \sqrt{2f_n} \Rightarrow f_{n+2} > f_{n+1}$.

By mathematical induction we deduce that the sequence $<f_n>$ is monotonically increasing.

Clearly, we have
$$f_1 < 2.$$

Again $f_n < 2 \Rightarrow \sqrt{2f_n} < \sqrt{2 \cdot 2} = 2 \Rightarrow f_{n+1} < 2$

so that by mathematical induction again, we deduce that $\forall\ n \in \mathbf{N}$
$$f_n < 2.$$

The sequence $<f_n>$ being bounded above and monotonically increasing, is convergent.

Let
$$\lim f_n = l. \qquad ...(1)$$

We have
$$f_{n+1} = \sqrt{2f_n} \Rightarrow (f_{n+1})^2 = 2f_n$$

Also
$$\lim f_n = l \Rightarrow \lim f_{n+1} = l.$$

Thus, we have
$$(1) \Rightarrow l^2 = 2l \Rightarrow l \in \{0, 2\}.$$

The limit, however, cannot be equal to zero. Thus
$$l = 2 \Leftrightarrow \lim f_n = 2.$$

It follows that
$$\lim f_n = 2.$$

(*b*) Refer part (*a*).

(*c*) Left as an exercise. Ans. $\lim s_n = 3$ **(*d*) Left as an exercise. Ans.** $\lim a_n = 7$

Example 6. *A sequence* $<a_n>$ *is defined as* $a_1 = 1$, $a_{n+1} = (4 + 3a_n)/(3 + 2a_n)$, $n \geq 1$. *Show that* $<a_n>$ *is loounded, monotonic and convergent. Also, find its limit.*

[Delhi B.A. (Prog.) III 2012; Rajasthan 2010; Meerut, 2006, 10]

Solution. Here $a_1 = 1$ and $a_2 = (4 + 3a_1)/(3 + 2a_1) = 7/5 > 1 \Rightarrow a_2 > a_1$.

Let us assume that $a_{n+1} > a_n$ ...(1)

Then, $$a_{n+2} - a_{n+1} = \frac{4+3a_{n+1}}{3+2a_{n+1}} - \frac{4+3a_n}{3+2a_n} = \frac{a_{n+1} - a_n}{(3+2a_{n+1})(3+2a_n)} \quad ...(2)$$

Since $a_{n+1} > a_n$ by (1) and $a_n > 0 \ \forall \ n \in \mathbf{N}$, so from (2)

$$a_{n+2} > a_{n+1}$$

Hence by mathematical induction, $a_{n+1} > a_n \ \forall \ n \in \mathbf{N}$ and so $<a_n>$ is monotonically increasing sequence.

Again, $$a_{n+1} = \frac{4+3a_n}{3+2a_n} = \frac{3}{2} - \frac{1}{2(2a_n+3)} \Rightarrow a_{n+1} < \frac{3}{2} \ \forall \ n \in \mathbf{N}.$$

Also $a_1 = 1 < 3/2$. Thus, $a_n < 3/2 \ \forall \ n \in \mathbf{N}$ and hence $<a_n>$ is bounded above.

Since $<a_n>$ is monotonically increasing and bounded above, it is convergent.

Let $\lim a_n = l$. Then $\lim a_{n+1} = l$.

Now, $$a_{n+1} = \frac{4+3a_n}{3+2a_n} \Rightarrow \lim a_{n+1} = \frac{4+3\lim a_n}{3+2\lim a_n}$$

or $$l = (4+3l)/(3+2l) \quad \text{or} \quad l^2 = 2 \quad \text{or} \quad l = \pm\sqrt{2}.$$

Since all the terms of the given sequence are positive, it follows that l cannot be negative. Hence $l = \sqrt{2}$ or $\lim a_n = \sqrt{2}$.

Example 7. *Give an example of a sequence of rational terms converging to irrational limit with reason to your answer.* **(Utkal, 2003)**

Solution. Refer solved example 6.

Example 8. *Give an example of a sequence of irrational terms converging to a rational limit with reason to your answer.* **(Utkal, 2003)**

Solution. Refer solved example 5.

Example 9. *Let $<a_n>$ be a sequence defined as : $a_1 = 3/2$, $a_{n+1} = 2 - (1/a_n)$, $\forall \ n \geq 1$. Show that $<a_n>$ is monotonic and bounded and converges to 1.*

[Delhi Maths (G), 2003; Delhi Maths (H), 2000; Delhi B.Sc. (H) Physics, 1998]

Solution. Given that $a_1 = 3/2$ and $a_{n+1} = 2 - (1/a_n) \ \forall \ n \geq 1$ ...(1)

Using mathematical induction, we shall prove that $a_n > 1 \ \forall \ n \in \mathbf{N}$ ...(2)

From (1), $a_1 > 1$. So (2) is true for $n = 1$.

Let (2) be true for some positive integer m, *i.e.*, let

$a_m > 1$, where $m \in \mathbf{N}$ ...(3)

Now, $$a_m > 1 \Rightarrow \frac{1}{a_m} < 1 \Rightarrow -\frac{1}{a_m} > -1 \Rightarrow 2 - \frac{1}{a_m} > 2 - 1 \Rightarrow a_{m+1} > 1, \text{ by (1)}$$

Thus, we find that if (2) is true for $n = m$, then it is also true for $n = m + 1$. Also (2) is true for $n = 1$. Hence by mathematical induction, (2) is true $\forall \ n \in \mathbf{N}$.

Now, from (i), $a_1 = 3/2$ and $a_2 = 2 - (1/a_1) = 2 - (2/3) = 4/3$

Thus, $a_2 < a_1$ ...(4)

Using mathematical induction, we shall prove that $a_{n+1} < a_n \ \forall \ n \in \mathbf{N}$ ...(5)

From (4) and (5), we see that (5) is true for $n = 1$.

Let (5) be true for some positive integer m, *i.e.*, let

$$a_{m+1} < a_m \quad ...(6)$$

Then, (6) $\Rightarrow$ $\dfrac{1}{a_{m+1}} > \dfrac{1}{a_m} \Rightarrow -\dfrac{1}{a_{m+1}} < -\dfrac{1}{a_m} < 2 - \dfrac{1}{a_{m+1}} < 2 - \dfrac{1}{a_m}$

Thus, $$a_{m+2} < a_{m+1},$$

showing that if (5) is true for $n = m$, then it is also true for $n = m + 1$. Since (5) is true for $n = 1$ so by mathematical induction, (5) is true $\forall\ n \in \mathbf{N}$.

From (2) and (5), we find that $<a_n>$ is monotonically decreasing sequence and bounded below and so it is convergent.

Let $$\lim a_n = l \quad \text{so that} \quad \lim a_{n+1} = l.$$

Then, $$a_{n+1} = 2 - \frac{1}{a_n} \Rightarrow \lim a_{m+1} = 2 - \frac{1}{\lim a_n} \Rightarrow l = 2 - \frac{1}{l}$$

$\therefore$ $l^2 - 2l + 1 = 0$ or $(l-1)^2 = 0$ or $l = 1$. Thus, $\lim a_n = 1$.

Example 10. *If $<s_n>$ is a bounded sequence such that*

$$s_1 = a > 0,\ s_{n+1} = \sqrt{\frac{ab^2 + s_n^{\ 2}}{a+1}},\ b > a,\ \forall\ n \geq 1,$$

then show that the sequence $<s_n>$ is an increasing sequence and $\lim s_n = b$.

[*Delhi Maths (G), 2003; Delhi Maths (H), 2001, Meerut, 1996*]

Solution. Since also $a > 0$ and $b > a$, we have

$$s_2^{\ 2} - s_1^{\ 2} = \frac{ab^2 + a^2}{a+1} - a^2 = \frac{a(b^2 - a^2)}{a+1} > 0$$

$\Rightarrow$ $$s_2^{\ 2} > s_1^{\ 2} \Rightarrow s_2 > s_1, \text{ as } s_1 > 0 \text{ and } s_2 > 0. \quad ...(1)$$

Using mathematical induction, we shall prove that $s_{n+1} > s_n\ \forall\ n \in \mathbf{N}$ ...(2)

From (1) and (2), we see that (2) is true for $n = 1$.

Let (2) be true for positive integer m, *i.e.*, let

$$s_{m+1} > s_m \quad ...(3)$$

Then, (3) $\Rightarrow$ $s_{m+1}^2 > s_m^2 \Rightarrow ab^2 + s_{m+1}^2 > ab^2 + s_m^2$

$$\Rightarrow \sqrt{\frac{ab^2 + s_{m+1}^2}{a+1}} > \sqrt{\frac{ab^2 + s_m^2}{a+1}} \Rightarrow s_{m+2} > s_{m+1},$$

showing that if (2) is true for $n = m$ then it is also true for $n = m + 1$. Since (2) is true for $n = 1$. Hence, by mathematical induction, (2) is true $\forall\ n \in \mathbf{N}$.

From (1), $$s_1 = a < b. \quad ...(4)$$

Using mathematical induction, we shall show that $s_n < b\ \forall\ n \in \mathbf{N}$ ...(5)

From (4) and (5), we see that (5) is true for $n = 1$.

Let (5) be true for some positive integer m, *i.e.*, let

$$s_m < b. \quad ...(6)$$

Then, (6) $\Rightarrow$ $s_m^2 < b^2 \Rightarrow ab^2 + s_m^2 < ab^2 + b^2 \Rightarrow ab^2 + s_m^2 < b^2(a+1)$

$$\Rightarrow \sqrt{\frac{ab^2 + s_m^2}{a+1}} < \sqrt{\frac{b^2(a+1)}{a+1}} \Rightarrow s_{m+1} < b,$$

showing that if (5) is true for $n = m$, then it is true for $n = m + 1$. Also (5) is true for $n = 1$. So by mathematical induction, (5) is true $\forall\ n \in \mathbf{N}$.

From (2) and (5) we find that $<s_n>$ is monotonically increasing sequence bounded above and so it is convergent.

Let $\lim s_n = l$ so that $\lim s_{n+1} = l$.

Then, $s^2_{n+1} = \dfrac{ab + s_n^2}{a+1} \Rightarrow (\lim s_{n+1})^2 = \dfrac{ab + (\lim s_n)^2}{a+1}$

$\therefore$ $l^2 = (ab + l^2)/(a+1)$ or $al^2 = ab^2$ or $l = \pm b$.

Since all the terms of the given sequence are positive, it follows that l cannot be negative.

So $l = b$. *i.e.,* $\lim s_n = b$.

Example 11. *If $a_{n+1} = \sqrt{k + a_n}$, where k, a_1 are positive, prove that the sequence $<a_n>$ is increasing or decreasing according as a_1 is less than or greater than the positive root of the equation $x^2 = x + k$ and has, in either case, this root as its limit.*

Solution. We have $a_{n+1} = \sqrt{k + a_n}$ ...(1)

$\therefore$ $(a_{n+1})^2 - a_n^2 = (k + a_n) - (k + a_{n-1}) = a_n - a_{n-1}$.

Thus, we see that

$$a_n > a_{n-1} \Rightarrow a_{n+1} > a_n \text{ and } a_n < a_{n-1} \Rightarrow a_{n+1} < a_n$$

and thus $<a_n>$ is a monotonic sequence.

Also it is an increasing or a decreasing sequence according as $a_2 > a_1$ or $a_2 < a_1$

The equation $x^2 - x - k = 0, k > 0$ has one root positive and other negative. Let α, $-\beta$ be these roots where α and β are positive.

Now, $x^2 - x - k = (x - \alpha)(x + \beta) \; \forall \; x$ ...(2)

$\Rightarrow a_1^2 - a_1 - k = (a_1 - \alpha)(a_1 + \beta)$. ...(3)

Let $a_1 > \alpha$. Then, from (3), $a_1^2 - a_1 - k > 0 \Rightarrow a_2 = (a_1 + k)^{1/2} < a_1$.

Thus, $a_1 > \alpha \Rightarrow a_2 < a_1 \Rightarrow <a_n>$ is monotonically decreasing :

Now, $a_n^2 = a_{n-1} + k > a_n + k \Rightarrow a_n^2 - a_n - k > 0$.

From (2), it follows that $a_n > \alpha \; \forall \; n \in \mathbf{N}$.

Thus, $<a_n>$ is monotonically decreasing sequence which is bounded below.

Hence, $\lim <a_n>$ exists. Let it be l. We have $l \geq \alpha$.

Taking limits in (1), we get $l^2 = l + k$

implying that l is equal to the positive root, α, of the equation $x^2 = x + k$.

The case when $a_1 < \alpha$ may be similarly treated.

If $a_n = \alpha$ then as may be easily seen that $a_n = \alpha \; \forall \; n \in \mathbf{N}$.

Example 12. *Show that the sequence $<x^n>$ is convergent if and only if $-1 < x \leq 1$.*

Solution. (*i*) Let $x > 1$. We write $x = 1 + h$ so that h is positive.

By mathematical induction, it may easily be shown that

$$x^n = (1 + h)^n > 1 + nh.$$

Let Δ be a positive number. We have

$$1 + nh > \Delta \Leftrightarrow n > (\Delta - 1)/h.$$

Taking m as any positive integer $> (\Delta - 1)/h$, we see that

$$x^n > \Delta \; \forall \; n \geq m \Rightarrow \lim x^n = \infty.$$

(*ii*) Let $x = 1$. Clearly, in this case, $\lim x^n = 1$.

(*iii*) Let $0 < x < 1$. We write $x = 1/(1 + h)$ so that h is positive. We have

$$0 < x^n = 1/(1 + h)^n < 1/(1 + nh).$$

Let ε be a positive number. We have

$$1/(1 + nh) < \varepsilon \Leftrightarrow n > (1/\varepsilon - 1)/h.$$

Taking m as any integer $> (1/\varepsilon - 1)/h$, we see that

$$0 < x^n < \varepsilon \Leftrightarrow |x^n| < \varepsilon \quad \forall\, n \geq m$$

$$\Rightarrow \lim x^n = 0.$$

(*iv*) Let $x = 0$. Clearly, $\lim x^n = 0$.

(*v*) Let $-1 < x < 0$. We write $x = -\alpha$ so that $0 < \alpha < 1$. We have $x^n = \alpha^n$.
It now follows from (*iii*) that $\lim x^n = 0$.

(*vi*) Let $x = -1$. Obviously x^n oscillates finitely.

(*vii*) Let $x < -1$. We write $x = \alpha$ so that $x > 1$.

As $n \to \infty$, x^n takes values, both positive and negative greater in absolute value than any assigned positive number. Hence, x^n oscillates infinitely.

EXERCISES

1. Show that the sequence $<f_n>$ defined by

$$f_n = \frac{1}{1.2} + \frac{1}{2.3} + \frac{1}{3.4} + \ldots.. + \frac{1}{n(n+1)}$$

is monotonically increasing and that $\lim f_n = 1$.

2. Consider the sequence $<g_n>$ defined as follows :

$$g_n = 1 + \frac{1}{2^2} + \frac{1}{3^2} + \ldots.. + \frac{1}{n^2}.$$

Show that g_n is monotonically increasing and bounded and that $\lim g_n \leq 2$.

3. Consider the sequence $<h_n>$ defined by

$$h_n = \frac{1}{n+1} + \frac{1}{n+2} + \ldots.. + \frac{1}{n+n}$$

and show that $<h_n>$ is monotonically increasing and bounded above and that

$$1/2 \leq \lim h_n \leq 1.$$

[Kanpur 2014]

Hint. $h_n - h_{n-1} = \dfrac{1}{(2n-1)\,2n} > 0 \Rightarrow h_n > h_{n-1};\ n \dfrac{1}{2n} < h_n < \dfrac{n}{n+1} \Rightarrow \dfrac{1}{2} < h_n < 1$

4. Given that $\qquad f_1 = 1,\ f_{n+1} = \sqrt{4 + 3f_n};$

show that $<f_n>$ is monotonic and bounded and find $\lim f_n$.

5. If $<f_n>$ is defined by $\qquad f_n = a/[1 + f_{n-1}]$

where a and f_1 are positive, show that the sequence $<f_n>$ is convergent and converges to the positive root of the equation $x^2 + x = a$.

6. Given that $<f_n>$ is a sequence such that $f_2 \leq f_4 \leq f_6 \leq \ldots.. \leq f_5 \leq f_3 \leq f_1$ and a sequence $<\phi_n>$ defined by $\phi_n = f_{2n-1} - f_{2n}$ converges to 0, show that $<f_n>$ is convergent.

7. Show that sequence defined by $f_n = 1 + 1/2 + 1/3 + \ldots.. + 1/n$ is monotonic but not bounded. Is this convergent ?

8. A sequence $<\phi_n>$ is defined as follows : $\phi_n = 1/1^4 + 1/2^4 + 1/3^4 + \ldots.. + 1/n^4$.

Show that $<\phi_n>$ is monotonically increasing and bounded above and hence convergent.

9. Let $x_n = 1/1^2 + 1/2^2 + \ldots + 1/n^2$ for $n \in N$. Prove that the sequence $<x_n>$ is increasing and bounded and hence convergent. **[Delhi B.Sc. I (Hons) 2010]**

10. If $f_n \geq 0$ and $f_{n+1} \leq a f_n$ where $0 < a < 1$, then show that $\lim f_n = 0$.

11. Examine the convergence of the sequence <cos n)/n>?. **[Delhi B.Sc. I (Prog) 2010]**

12. If $\lim \dfrac{f_{n+1}}{f_n} = l$ and $|l| < 1$, show that $\lim f_n = 0$.

13. Find $\lim n^{-n-1}(n+1)^n$.

14. Discuss the behaviour as $n \to \infty$ of $n^r x^n$; r being any positive integer. In the same way discuss $n^{-r} x^n$.

15. Show that each of the sequence $<a_n>$ defined as follows are convergent :

(*i*) $a_n = \dfrac{1}{n+1} + \dfrac{1}{n+2} + \ldots + \dfrac{1}{n+n}$. **(Purvanchal 2006; Kanpur, 1994; Madras, 1998)**

(*ii*) $a_1 = 1,\ a_n = 1 + \dfrac{1}{1!} + \dfrac{1}{2!} + \ldots + \dfrac{1}{(n-1)!},\ n \geq 2.$ **[Delhi B.Sc. (Prog) III 2008]**

[Delhi B.A (Prog.) III, 2013; Delhi Maths (H), 2002, 06]

(*iii*) $a_1 = 1,\ a_{n+1} = (3 + 2a_n)/2 + a_n,\ n \geq 1$. Also show that $\lim a_n = \sqrt{3}$.

16. Prove that the sequence $<s_n>$ defined by recursion formula $s_{n+1} = \sqrt{7 + s_n},\ s_1 = \sqrt{7}$ converges to the positive root of $x^2 - x - 7 = 0$. **(K.U. BCA II 2007)**

(Delhi Maths (Prog) 2008, 09; Delhi Maths (G), 2003)

17. Show that the sequence defined by $a_{n+1} = (1/2) \times (a_n + 1/a_n),\ n \geq 1,\ a_1 = 1$ is convergent and find $\lim a_n$. **[Delhi B.Sc. (H) Physics, 1998]**

18. Show that the sequence $<a_n>$ defined by $a_{n+1} = 1 - \sqrt{1 - a_n},\ n \geq 1$ and $a_1 < 1$ converges to 0. **[Delhi B.Sc. (H) Physics, 1995]**

19. Find the limit of sequence $<a_n>$ where

(*i*) $a_n = \left(1 + \dfrac{1}{n}\right)^{n+1}$ (*ii*) $\dfrac{(n+1)^n}{n^{n+1}}$ **(Bangalore, 2004)**

[Ans. (*i*) e (*ii*) 0**]**

20. Let $s_n = \dfrac{1 \cdot 3 \cdot 5 \ldots (2n-1)}{2 \cdot 4 \cdot 6 \ldots 2n}$, $\forall\ n \in \mathbf{N}$. Prove that $<s_n>$ is convergent and $\lim s_n \leq 1/2$.

(Rohilkhand, 1997)

21. A sequence $<s_n>$ defined by $s_1 = 1,\ s_{n+1} = \{(3 + s_n^2)/2\}^{1/2}$. Show that $<s_n>$ is a bounded monotonically increasing sequence and converges to $\sqrt{3}$. **[Delhi Maths (H), 1996]**

22. Two sequences $<f_n>$ and $<g_n>$ are defined by $f_1 = 1/2$ and $g_1 = 1$ and

$$f_n = \sqrt{f_{n-1}\, g_{n-1}},\ n = 2, 3, 4, \ldots,\quad \frac{1}{g_n} = \frac{1}{2}\left(\frac{1}{f_n} + \frac{1}{g_{n-1}}\right),\ n = 2, 3, 4, \ldots.$$

Prove that $f_{n-1} < f_n < g_n < g_{n-1},\ n = 2, 3, 4, \ldots$ and deduce that both the sequences converge to the same limit l where $1/2 < l < 1$.

23. Prove that (*i*) $\lim\left(1+\frac{1}{n+1}\right)^n = e$ (*ii*) $\lim\limits_{n\to\infty}\left(1+\frac{2}{n}\right)^n = e^2$.

5.12. SUBSEQUENCES (Rajasthan 2010)

Definition. *Let* $<f_n>$ *be a given sequence. If* $<n_k>$ *is a strictly increasing sequence of natural numbers (i.e.,* $n_1 < n_2 < n_3 < \ldots\ldots$*), then* $< f_{n_k} >$ *is called a subsequence of* $<f_n>$.

ILLUSTRATIONS

1. The sequences $<1, 3, 5, \ldots, 2n-1, \ldots>$, $<2, 4, 6, \ldots, 2n, \ldots>$ and $<1, 4, 9, 16, \ldots, n^2, \ldots>$ are all subsequences of $<n>$.

2. The sequence of primes $<2, 3, 5, 7, 11, \ldots.>$ is a subsequence of natural numbers $<1, 2, 3, 4, \ldots..>$.

Note. The terms of a sequence occur in the same order in which they occur in the original sequence. Accordingly, $<3, 1, 5, 7, \ldots..>$ is not a subsequence of $<1, 2, 3, 4, 5, 6, 7, \ldots..>$.

Theorem I. *If a sequence* $<f_n>$ *converges to l, then every subsequence* of $<f_n>$ *converges to l.* **(*Calicut, 2004*)**

Proof. Let $< f_{n_k} >$ be a subsequence of $<f_n>$ and let $<f_n>$ converge to l.

$\therefore$ Given $\varepsilon > 0$, we can find $m \in \mathbf{N}$ such that

$$| f_n - l | < \varepsilon \quad \forall\ n \geq m$$

$$\Rightarrow \quad | f_{n_k} - l | < \varepsilon \quad \forall\ n_k \geq m$$

$$\Rightarrow \quad < f_{n_k} > \text{ converges to } l.$$

Corollary. All subsequences of a convergent sequence converge to the same limit.

Proof. Left as an exercise.

Note 1. *The converse of the above theorem is not true.* For example the subsequence $<1, 1, 1, \ldots..>$ of $<(-1)^n>$, *i.e.*, $<-1, 1, -1, 1, \ldots..>$ is convergent whereas $<(-1)^n>$ is not convergent.

Note 2. In the order to prove that a sequence is not convergent, it is sufficient to show that any two of its subsequence converge to different limits. For example, consider the two subsequences $<1, 1, 1, \ldots.>$ and $<-1, -1, -1, \ldots.>$ of $<(-1)^n>$. Here the subsequences converges to two different limits 1 and -1 and so $<(-1)^n>$ is not convergent.

Theorem II. *If the subsequences* $<f_{2n-1}>$ *and* $<f_{2n}>$ *of a sequence* $<f_n>$ *converge to the same limit l, then* $<f_n>$ *also converges to l.*

Proof. Since $f_{2n-1} \to l$ and $f_{2n} \to l$, so for any $\varepsilon > 0$, there exist positive integers m_1 and m_2 such that

$$| f_{2n-1} - l | < \varepsilon \quad \forall\ n \geq m_1 \text{ and } | f_{2n} - l | < \varepsilon \quad \forall\ n \geq m_2 \qquad \ldots(1)$$

Let $m = \max\{m_1, m_2\}$. Then (1) reduces to

$$| f_{2n-1} - l | < \varepsilon \quad \forall\ n \geq m \text{ and } | f_{2n} - l | < \varepsilon \quad \forall\ n \geq m \qquad \ldots(2)$$

From (2), $\quad | f_n - l | < \varepsilon \quad \forall\ n \geq m$

Hence $<f_n>$ converges to l.

Theorem III. (*i*) *If a sequence* $<f_n>$ *diverges to* ∞*, then every subsequence of* $<f_n>$ *also diverges to* ∞.

(*ii*) *If a sequence* $<f_n>$ *diverges to* $-\infty$*, then every subsequence of* $<f_n>$ *also diverges to* $-\infty$.

Proof. (*i*) Let $< f_{n_k} >$ be a subsequence of $<f_n>$.

Since $<f_n>$ diverges to ∞. So for every positive number Δ, however large, there exists $m \in \mathbf{N}$ such that

$$f_n > \Delta \quad \forall\ n \geq m$$

$$\Rightarrow f_{n_k} > \Delta \quad \forall\ n_k \geq m$$

$$\Rightarrow < f_{n_k} > \text{ diverges to } \infty.$$

(*ii*) Left as an exercise.

Note. The converse of the above theorem is not true. For example, consider a sequence $<f_n>$,

where
$$f_n = (-1)^n n = \begin{cases} -n, & \text{if } n \text{ is odd} \\ n, & \text{if } n \text{ is even} \end{cases}$$

Clearly the subsequence $<f_{2n-1}>$ diverges to $-\infty$ whereas the subsequence $<f_{2n}>$ diverges to ∞. But the sequence $<f_n>$ is an oscillatory sequence.

Theorem IV. *A real number l is a limit point (or a cluster point) of a sequence $<f_n>$ if and only if a subsequence of $<f_n>$ converges to l.* **[*Delhi Maths (H), 1999, 2006*]**

Proof. Let l be a limit point of the sequence of $<f_n>$. Then we shall establish the existence of a subsequence of $<f_n>$ which converges to l.

Since l is a limit point of $<f_n>$, so for a chosen $\varepsilon = 1$ and $m = 1$, there exists a positive integer $n_1 > 1$ such that

$$| f_{n_1} - l | < 1$$

Again, for a chosen $\varepsilon = 1/2$ and $n = n_1$, there exists a positive integer $n_2 > n_1$ such that

$$| f_{n_2} - l | < 1/2$$

Proceeding likewise, we obtain a subsequence $< f_{n_1}, f_{n_2}, f_{n_3}, \ldots\ldots >$ such that

$$| f_{n_k} - l | < 1/k. \qquad \ldots(1)$$

After getting $f_{n_1}, f_{n_2}, f_{n_3}, \ldots\ldots$, if we choose $\varepsilon = 1/(k+1)$ and $m = n_k$, then it is possible to find a positive integer $n_{k+1} > n_k$ such that

$$| f_{n_{k+1}} - l | < 1/(k+1).$$

Thus, by induction, the subsequence $< f_{n_k} >$ of $<f_n>$ has been constructed. We now prove that $f_{n_k} \to l$.

Given $\varepsilon > 0$, we can find a positive integer p such that $1/p < \varepsilon$. Then, using (1), we have

$$| f_{n_k} - l | < 1/k \leq 1/p < \varepsilon, \ \forall\ k \geq p$$

$$\Rightarrow < f_{n_k} > \text{ converges to } l.$$

Conversely, let $< f_{n_k} >$ be a subsequence of $<f_n>$ such that $f_{n_k} \to l$. Then, we shall show that l is a limit point of $<f_n>$.

l is the limit of $< f_{n_k} >$

$\Rightarrow$ every nbd of l contain infinite terms of $< f_{n_k} >$

$\Rightarrow$ every nbd of l contain infinite terms of $<f_n>$

$\Rightarrow$ l is a limit point of $<f_n>$

Theorem V. Bolzano Weierstrass theorem for sequences. *Every bounded sequence has a convergent subsequence.* ***[Pune 2010; Delhi B.Sc. (Hons.) I, 2016, 2017]***

Proof. Let $<f_n>$ be a bounded sequence. Then by Bolzano-Weierstrass theorem, $<f_n>$ has a limit point, say l. Now proceed as in the first part of proof of the above theorem to establish the existence of a subsequence of $<f_n>$ which converges to l.

Theorem VI. *Every sequence has a monotonic subsequence.* ***[Delhi B.Sc. (Hons) I, 2011]***

Proof. Left as an exercise.

EXAMPLES

Example 1. *If $<f_n>$ be a sequence of positive numbers such that $f_n = (f_{n-1} + f_{n-2})/2$, $\forall$ n > 2, then show that $<f_n>$ converges. Also find $\lim f_n$.*

[Delhi B.Sc. (Hons.) I, 2003]

Solution. Given that $$f_n = (f_{n-1} + f_{n-2})/2 \quad \text{...(1)}$$

Case I. Let $f_2 = f_1$. Then, from (1), $f_3 = (f_2 + f_1)/2 = f_1$ and so on.

Hence $f_n = f_1$ $\forall$ $n \in \mathbf{N}$ and so $<f_n>$ converges to f_1.

Case II. Let $f_2 \neq f_1$ and let $f_1 < f_2$.

Replacing n by 3, 4, 5, 6, ... in (1), we have

$$\left.\begin{aligned} f_3 &= (f_2 + f_1)/2 \;\Rightarrow\; f_1 < f_3 < f_2, \text{ as } f_1 < f_2 \\ f_4 &= (f_3 + f_2)/2 \;\Rightarrow\; f_3 < f_4 < f_2, \text{ as } f_3 < f_2 \\ f_5 &= (f_4 + f_3)/2 \;\Rightarrow\; f_3 < f_5 < f_4, \text{ as } f_3 < f_4 \\ f_6 &= (f_5 + f_4)/2 \;\Rightarrow\; f_5 < f_6 < f_4, \text{ as } f_5 < f_4 \\ &\dots\dots\dots\dots\dots\dots\dots\dots \end{aligned}\right\} \quad \text{...(2)}$$

From (2), $$f_1 < f_3 < f_5 < \dots\dots \text{ and } \dots\dots < f_6 < f_4 < f_2 \quad \text{...(3)}$$

From (1), $$f_{n+2} - f_n = (f_{n+1} + f_n)/2 - f_n = (f_{n+1} - f_n)/2 \quad \text{...(4)}$$

and $$f_{n+2} - f_n = \frac{1}{2}\left\{\frac{1}{2}(f_n + f_{n-1}) - \frac{1}{2}(f_{n-1} + f_{n-2})\right\} = \frac{1}{4}(f_n - f_{n-2}) \quad \text{...(5)}$$

From (3) and (5), we find that the subsequence of odd terms $<f_{2n-1}>$ is monotonic increasing whereas the subsequence of even terms $<f_{2n}>$ is monotonic decreasing sequence.

Now, using (4) for $n = 2m$, we have

$$f_{2m+2} - f_{2m} = (f_{2m+1} - f_{2m})/2 \text{ so that } f_{2m+2} < f_{2m} \Rightarrow f_{2m+1} < f_{2m}.$$

Since $f_{2m} < f_{2m-2} < \dots\dots < f_6 < f_4 < f_2$, it follows that every odd term is less than every even term. Thus, we have

$$f_1 < f_3 < f_5 < \dots\dots < f_{2m+1} < f_{2m} < f_{2m-1} < \dots\dots < f_6 < f_4 < f_2,$$

showing that $<f_{2n-1}>$ is monotonically increasing and bounded above by f_2 and $<f_{2n}>$ is monotonically decreasing and bounded below by f_1. Hence both the subsequences $<f_{2n-1}>$ and $<f_{2n}>$ must be convergent.

Let $$\lim f_{2n-1} = l \quad \text{and} \quad \lim f_{2n} = l'. \quad \text{...(6)}$$

From (1), we have $$f_{2n} = (f_{2n-1} + f_{2n-2})/2.$$

On taking limits as $n \to \infty$ and using (6), we get

$$l' = (l + l')/2 \quad \text{so that} \quad l = l',$$

showing that both the subsequences of $<f_n>$ converge to the same limit l and hence $<f_n>$ converges to l.

Now we shall find the value of l.

Replacing n by 3, 4, 5, ..., $n-1$, n is (1) we have

$$f_3 = (f_2+f_1)/2$$
$$f_4 = (f_3+f_2)/2$$
$$f_5 = (f_4+f_3)/2$$
....
$$f_{n-1} = (f_{n-2}+f_{n-3})/2$$
$$f_n = (f_{n-1}+f_{n-2})/2$$

Adding the corresponding sides and simplifying, we get

$$f_n + \frac{1}{2} f_{n-1} = \frac{1}{2}(f_1 + 2f_2)$$

Taking the limit as $n \to \infty$ and noting that $\lim f_n = \lim f_{n-1} = l$, we have

$$l + l/2 = (f_1 + 2f_2)/2 \quad \text{or} \quad l = (f_1 + 2f_2)/3.$$

Thus, $$\lim f_n = (f_1 + 2f_2)/3.$$

Case III. Let $f_2 \neq f_1$ and let $f_1 > f_2$. Proceeding as in case II above, we can easily show that $\lim f_n = (f_1 + 2f_2)/3$.

Example 2. *If x_1, y_1 are two positive unequal numbers and*

$$x_n = (x_{n-1} + y_{n-1})/2 \text{ and } y_n = \sqrt{x_{n-1}\, y_{n-1}},\ \forall\ n \geq 2.$$

Prove that the sequences $<x_n>$ and $<y_n>$ are monotonic and they converge to the same limit.

(Madras, 1998)

Solution. Since x_1 and y_1 are unequal positive numbers, so let $x_1 > y_1$.

Now $x_1 > y_1$ and A.M. > G.M. $\Rightarrow x_1 > y_1$ and $(x_1 + y_1)/2 > \sqrt{x_1 y_1}$

$$\Rightarrow x_2 > y_2$$

Similarly, $x_3 > y_3$ and so on.

Thus, $$x_n > y_n \ \forall\ n \in \mathbf{N}. \qquad \text{...(1)}$$

Given that $$x_n = (x_{n-1} + y_{n-1})/2 \text{ and } y_n = \sqrt{x_{n-1}\, y_{n-1}},\ \forall\ n \geq 2 \qquad \text{...(2)}$$

From (1) and (2), $$x_{n+1} = (x_n + y_n)/2 < (x_n + x_n)/2$$

Thus, $$x_{n+1} < x_n \ \forall\ n \in \mathbf{N}, \qquad \text{...(3)}$$

showing that $<x_n>$ is a monotonically decreasing sequence.

Again, from (1) and (2), $$y_{n+1} = \sqrt{x_n\, y_n} > \sqrt{y_n\, y_n}$$

Thus, $$y_{n+1} > y_n \ \forall\ n \in \mathbf{N}, \qquad \text{...(4)}$$

showing that $<y_n>$ is a monotonically increasing sequence.

From (1) and (2), $$x_n = (x_{n-1} + y_{n-1})/2 > (y_{n-1} + y_{n-1})/2 \quad \text{or} \quad x_n > y_{n-1} \qquad \text{...(5)}$$

From (4) and (5), $$x_n > y_{n-1} > y_{n-2} > \ldots > y_1 \Rightarrow x_n > y_1 \ \forall\ n \in \mathbf{N},$$

showing that $<x_n>$ is bounded below.

Again, from (1) and (2), $$y_n = \sqrt{x_{n-1}\, y_{n-1}} < \sqrt{x_{n-1}\, x_{n-1}} \quad \text{or} \quad y_n < x_{n-1} \qquad \text{...(6)}$$

From (3) and (6), $$y_n < x_{n-1} < x_{n-2} < \ldots < x_1,$$

showing that $<y_n>$ is bounded above.

Thus we have shown that $<x_n>$ is monotonically decreasing and bounded below whereas $<y_n>$ is monotonically increasing and bounded above. Hence both $<x_n>$ and $<y_n>$ are convergent.

Let $\lim x_n = l$ and $\lim y_n = l'$. ...(7)

On taking limits on both sides of $x_n = (x_{n-1} + y_{n-1})/2$ as $n \to \infty$, we have

$$l = (l + l')/2 \quad \text{so that} \quad l' = l,$$

showing that $<x_n>$ and $<y_n>$ both converge to the same limit.

EXERCISES

1. If s_1 and s_2 are positive real numbers and $s_{n+2} = \sqrt{s_{n+1} s_n}$, then prove that $<s_n>$ is convergent and find its limit. **[Delhi Maths (H), 1998]** **[Ans. $(s_1 s_2^2)^{1/3}$]**
2. If $a_1 > 0$, $a_2 > 0$ and $a_n = (2a_{n-1}\, a_{n-2})/(a_{n-1} + a_{n-2})$, $n > 2$ then show that $<f_n>$ converges to $(3a_1\, a_2)/\,(2a_1 + a_2)$.
3. If f_1 and g_1 are positive numbers and if $\forall\ n \geq 1$,
$$f_{n+1} = \frac{f_n + g_n}{2} \quad \text{and} \quad \frac{2}{g_{n+1}} = \frac{1}{f_n} + \frac{1}{g_n}$$
then show that $<f_n>$ and $<g_n>$ are monotonic sequences which converge to the same limit l, where $l^2 = f_1\, g_1$.
4. Let S be a closed and bounded sub-set of the set **R** of real numbers. Prove that every sequence in S has a subsequence converging to some point of S.
5. If **I** is a closed interval, then show that every sequence in **I** has a subsequence converging to a point of **I**.
6. Prove that a sub-set K of **R** is a compact if and only if, every sequence in K has a subsequence that converges to a point in K. Use this to establish that (0, 1) is not compact in **R**. **(Calicut, 2004)**
7. Prove that a Cauchy sequence of real numbers is convergent if and only if it has a convergent subsequence. **[Delhi Maths (H), 2005]**

5.13. LIMIT SUPERIOR AND LIMIT INFERIOR OF A SEQUENCE

[Delhi Maths (G), 2005; Delhi Maths (H), 2005, 06, 07. 0[illegible]

We have already shown that every bounded sequence has the greatest and the smallest limit points (refer theorem III of Art. 5.3).

Limit inferior (or lower limit). Definition

Let $<f_n>$ be a bounded sequence. The smallest limit point of $<f_n>$ is called the *limit inferior* of $<f_n>$ and is denoted by

$$\lim_{n\to\infty} \inf f_n \quad \text{or} \quad \underline{\lim}_{n\to\infty} f_n.$$

Limit superior (or upper limit). Definition

Let $<f_n>$ be a bounded sequence. The greatest limit point of $<f_n>$ is called the *limit superior* of $<f_n>$ and is denoted by

$$\lim_{n\to\infty} \sup f_n \quad \text{or} \quad \overline{\lim}_{n\to\infty} f_n.$$

Note 1. Since the smallest limit point of a sequence $<f_n> \leq$ the greatest limit point of $<f_n>$, we have

$$\lim_{n\to\infty} \inf f_n \leq \lim_{n\to\infty} \sup f_n.$$

[Delhi Maths (H) 2007]

Note 2. If $<f_n>$ is not bounded above, we write $\lim_{n\to\infty} \sup f_n = \infty$

If $<f_n>$ is not bounded below, we write $\lim_{n\to\infty} \inf f_n = -\infty$

ILLUSTRATIONS

1. Consider the sequence $<f_n>$ where $f_n = (-1)^n \ \forall \ n \in \mathbf{N}$. Its set of limit points is $\{1, -1\}$ which is bounded. The greatest and the smallest limit points are 1 and -1 respectively. So by definitions, we have

$$\lim_{n\to\infty} \inf f_n = -1 \quad \text{and} \quad \lim_{n\to\infty} \sup f_n = 1.$$

2. Consider the sequence $<f_n>$ where $f_n = 1/n$, $n \in \mathbf{N}$. It has exactly one limit point, namely 0. The set $\{0\}$ of limit points is bounded.

$$\therefore \quad \lim_{n\to\infty} \inf f_n = 0 = \lim_{n\to\infty} \sup f_n.$$

3. Consider the sequence $<f_n>$ where

$$f_n = \begin{cases} 1, & \text{when } n \text{ is odd} \\ n, & \text{when } n \text{ is even} \end{cases}$$

Here 1 is a limit point of $<f_n>$ which is not bounded above

$$\therefore \quad \lim_{n\to\infty} \inf f_n = 1 \quad \text{and} \quad \lim_{n\to\infty} \sup f_n = \infty.$$

4. Consider the sequence $<f_n> = \left\langle -1, -2, -\frac{1}{3}, -3, -\frac{1}{5}, \ldots \right\rangle$.

Here 0 is a limit point of $<f_n>$ which is not bounded below

$$\therefore \quad \lim_{n\to\infty} \inf f_n = -\infty \quad \text{and} \quad \lim_{n\to\infty} \sup f_n = 0.$$

5.14. SOME THEOREMS ON LIMIT SUPERIOR AND LIMIT INFERIOR OF A SEQUENCE

Theorem I. *A number l is the limit superior of a bounded sequence $<f_n>$ if and only if the following conditions are satisfied*

(*i*) *For each $\varepsilon > 0$, $f_n > l - \varepsilon$ for infinitely many values of n*

(*ii*) *For each $\varepsilon > 0$, $f_n < l + \varepsilon$ for all except finitely many values of n*

[*Delhi B.Sc. (Hons.)*, *2003, 06, 08*]

Proof. The condition is necessary. Let l be the limit superior of a bounded sequence $<f_n>$ and let $\varepsilon > 0$ be given.

Since l is a limit point of $<f_n>$, so the relation

$$l - \varepsilon < f_n < l + \varepsilon$$

is true for infinitely many values of n. So, in particular,

$$f_n > l - \varepsilon \text{ for infinitely many values of } n. \quad \ldots(1)$$

Since l is the greatest limit point of $<f_n>$ and $l + \varepsilon > l$, it follows that $l + \varepsilon$ cannot be a limit point of $<f_n>$ and so we must have

$$f_n \geq l + \varepsilon \text{ for only finitely many values of } n, \quad \ldots(2)$$

because if $f_n \geq l + \varepsilon$ for infinitely many values of n, then $<f_n>$ will possess a limit point, say, $p \geq l + \varepsilon$ which is impossible, since l is the greatest limit point of $<f_n>$. Hence from (2), we have

$$f_n < l + \varepsilon \text{ for all except finitely many values of } n \quad \ldots(3)$$

(2) and (3) together give the required necessary conditions (*i*) and (*ii*).

The condition is sufficient. Let us assume that l satisfies both the conditions (*i*) and (*ii*). Now, conditions (*i*) and (*ii*)

$\Rightarrow$ $l - \varepsilon < f_n < l + \varepsilon$ for infinitely many values of n

$\Rightarrow$ $f_n \in \,]l - \varepsilon, l + \varepsilon[$ for infinitely many values of n

$\Rightarrow$ l is a limit point of $<f_n>$.

Now, we shall show that l is the greatest limit point of $<f_n>$.

Let l' be any number greatest than l. Let p and q be two numbers such that $l < p < l' < q$

Choosing $\varepsilon = p - l > 0$, condition (*ii*) reduces to

$f_n < p$ for all except finitely many values of n

$\Rightarrow$ nbd $]p, q[$ of l' contains f_n for only finitely many values of n

$\Rightarrow$ l' cannot be a limit point of $<f_n>$, where $l' > l$

$\Rightarrow$ l is the greatest limit of $<f_n>$

$\Rightarrow$ l is the limit superior of $<f_n>$

Theorem II. *A number l is the limit inferior of a bounded sequence $<f_n>$ if and only if the following conditions are satisfied*

(*i*) *For each $\varepsilon > 0$, $f_n < l + \varepsilon$ for infinitely many value of n*

(*ii*) *For each $\varepsilon > 0$, $f_n > l - \varepsilon$ for all except finitely many values of n.*

[*Delhi Maths (G), 1995*

Proof. The condition is necessary. Let l be the limit inferior of a bounded sequence $<f_n>$ and let $\varepsilon > 0$ be given.

Since l is a limit point of $<f_n>$, so the relation

$$l - \varepsilon < f_n < l + \varepsilon$$

is true for infinitely many values of n. So, in particular,

$$f_n < l + \varepsilon \text{ for infinitely many values of } n \qquad ...(1)$$

Since l is the smallest limit of $<f_n>$ and $l - \varepsilon < l$, it follows that $l - \varepsilon$ cannot be a limit point of $<f_n>$ and so we must have

$$f_n \le l - \varepsilon \text{ for only finitely many values of } n, \qquad ...(2)$$

because if $f_n \le l - \varepsilon$ for infinitely many values of n, then $<f_n>$ will possess a limit point, say, $p \le l - \varepsilon$ which is impossible since l is the smallest limit point of $<f_n>$. Hence from (2), we have

$$f_n > l - \varepsilon \text{ for all except finitely many values of } n. \qquad ...(3)$$

(2) and (3) together give the required necessary conditions (*i*) and (*ii*).

The condition is sufficient. Let us assume that l satisfies both the conditions (*i*) and (*ii*). Now, conditions (*i*) and (*ii*)

$\Rightarrow$ $l - \varepsilon < f_n < l + \varepsilon$ for infinitely many values of n

$\Rightarrow$ $f_n \in \,]l - \varepsilon, l + \varepsilon[$ for infinitely many values of n

$\Rightarrow$ l is a limit point of $<f_n>$

Now, we shall show that l is the smallest limit point.

Let l' be another number smaller than l. Let p and q be two numbers such that $p < l' < q < l$.

Choosing $\varepsilon = l - q > 0$, condition (*ii*) reduces to

$f_n > q$ for all except finitely many values of n

$\Rightarrow$ nbd $]p, q[$ of l' contains f_n for only finitely many values of n

$\Rightarrow$ l' is not a limit point $<f_n>$, when $l' < l$

$\Rightarrow$ l is the smallest limit of $<f_n>$

$\Rightarrow$ l is the limit inferior of $<f_n>$.

Theorem III. *A bounded sequence* $<f_n>$ *converges to* l *if and only if*

$$\lim_{n\to\infty} \sup f_n = \lim_{n\to\infty} \inf f_n = l.$$

[*Delhi B.Sc. (H) Physics, 1999; Delhi Maths (H), 2001, 02; Delhi Maths (G), 2001, 02, 03; Meerut, 1996*]

Proof. The condition is necessary. Let the bounded sequence $<f_n>$ converges to l.

Then, by theorem III of Art. 5.4 it follows that l is the only limit point of $<f_n>$. Hence l is the greatest as well as the smallest limit point of $<f_n>$.

Therefore, $$\lim_{n\to\infty} \sup f_n = \lim_{n\to\infty} \inf f_n = l.$$

The condition is sufficient. Suppose that

$$\lim_{n\to\infty} \sup f_n = \lim_{n\to\infty} \inf f_n = l.$$

Now, $\lim_{n\to\infty} \sup f_n = l \Rightarrow f_n < l + \varepsilon$ for all except finitely many values of n

$\Rightarrow$ there exists $m_1 \in \mathbf{N}$ such that $f_n < l + \varepsilon \ \forall \ n \geq m_1$ ...(1)

Again, $\lim_{n\to\infty} \inf f_n = l \Rightarrow f_n > l - \varepsilon$ for all except finitely many values of n

$\Rightarrow$ there exists $m_2 \in \mathbf{N}$ such that $f_n > l - \varepsilon \ \forall \ n \geq m_2$ ...(2)

Let $m = \max \{m_1, m_2\}$. Then, from (1) and (2), we have

$$l - \varepsilon < f_n < l + \varepsilon \ \forall \ n \geq m$$

$$\Rightarrow | f_n - l | < \varepsilon \ \forall \ n \geq m \Rightarrow \lim_{n\to\infty} f_n = l.$$

Theorem IV. *If* $<f_n>$ *and* $<g_n>$ *are bounded sequences such that* $f_n \leq g_n \ \forall \ n \in \mathbf{N}$, *then*

(*i*) $\lim_{n\to\infty} \sup f_n \leq \lim_{n\to\infty} \sup g_n$ (*ii*) $\lim_{n\to\infty} \inf f_n \leq \lim_{n\to\infty} \inf g_n$

[*Delhi Maths (H), 1995, 2000*]

Proof. (*i*) Let $$\lim_{n\to\infty} \sup g_n = l.$$

Then, by theorem I, for any $\varepsilon > 0$, there exists $m \in \mathbf{N}$ such that

$g_n < l + \varepsilon \ \forall \ n \geq m$

$\Rightarrow f_n < l + \varepsilon \ \forall \ n \geq m$, as $f_n \leq g_n \ \forall \ n \in \mathbf{N}$

$\Rightarrow$ all the limit points of $<f_n>$ are $\leq l + \varepsilon$

$\therefore$ In particular, the greatest limit point of $<f_n>$ is $\leq l + \varepsilon$

$$\therefore \sup_{n\to\infty} \sup f_n \leq \lim_{n\to\infty} \sup g_n + \varepsilon$$

Since $\varepsilon > 0$ is arbitrary, it follows that

$$\sup_{n\to\infty} \sup f_n \leq \lim_{n\to\infty} \sup g_n.$$

(*ii*) Use theorem III and complete the proof.

Theorem V. *If* $<f_n>$ *and* $<g_n>$ *are bounded sequence, then*

(*i*) $\lim_{n\to\infty} \sup (f_n + g_n) \leq \lim_{n\to\infty} \sup f_n + \lim_{n\to\infty} \sup g_n$

[*Delhi Maths (H), 1997, 2000; Delhi Maths (G), 1998, 99; 2004, 05*]

(*ii*) $\lim_{n\to\infty} \inf (f_n + g_n) \geq \lim_{n\to\infty} \inf f_n + \lim_{n\to\infty} \inf g_n$.

Proof. (*i*) Let $\lim_{n\to\infty} \sup f_n = f$ and $\lim_{n\to\infty} g_n = g.$

$\Rightarrow$ for any $\varepsilon > 0$, there exist $m_1, m_2 \in \mathbf{N}$ such that (see theorem I)

$$f_n < f + \varepsilon/2 \ \forall\ n \geq m_1 \text{ and } g_n < g + \varepsilon/2 \ \forall\ n \geq m_2 \qquad ...(1)$$

Let $m = \max\{m_1, m_2\}$ so that $m \geq m_1$ and $m \geq m_2$.

$$\therefore \quad (1) \Rightarrow f_n < f + \varepsilon/2 \text{ and } g_n + \varepsilon/2 \ \forall\ n \geq m$$

$$\Rightarrow f_n + g_n < f + g + \varepsilon \ \forall\ n \geq m$$

So all limit points of $<f_n + g_n>$ are $\leq f + g + \varepsilon$

In particular, the greatest limit point of $<f_n + g_n>$ is $\leq f + g + \varepsilon$

Thus, $\lim_{n\to\infty} \sup (f_n + g_n) \leq \lim_{n\to\infty} f_n + \lim_{n\to\infty} g_n + \varepsilon.$

Since $\varepsilon > 0$ is arbitrary, it follows that

$$\lim_{n\to\infty} \sup (f_n + g_n) \leq \lim_{n\to\infty} \sup f_n + \lim_{n\to\infty} \sup g_n.$$

(*ii*) Use theorem III and complete the proof.

Note. We now show by means of examples that strict inequality may hold in results (*i*) and (*ii*) of theorem V.

Let $<f_n> = <0, 1, -1, 0, 1, -1, ...>$ and $<g_n> = <-1, 0, 1, -1, 0, 1, ...>$.

Then, $<f_n + g_n> = <-1, 1, 0, -1, 1, 0, ...>$

Now, $\lim_{n\to\infty} \sup f_n = \lim_{n\to\infty} g_n = \lim_{n\to\infty} \sup (f_n + g_n) = 1,$

showing that here $\lim_{n\to\infty} \sup (f_n + g_n) < \lim_{n\to\infty} \sup f_n + \lim_{n\to\infty} \sup g_n$, as $1 < 2$

Next, $\lim_{n\to\infty} \inf f_n = \lim_{n\to\infty} \inf g_n = \lim_{n\to\infty} \inf (f_n + f_n) = -1,$

showing that here $\lim_{n\to\infty} \inf (f_n + g_n) > \lim_{n\to\infty} \inf f_n + \lim_{n\to\infty} \inf g_n$, as $-1 > -2$.

EXERCISES

1. Find the limit superior and limit inferior of each of the following sequences :

(*i*) $<1, 2, 1, 2, ...>$ (*ii*) $<1, 3, 5, 1, 3, 5, ...>$

(*iii*) $\left\langle 1 + \frac{(-1)^n}{n} \right\rangle$ (*iv*) $\left\langle \frac{(-1)^n}{n^2} \right\rangle$

(*v*) $\left\langle (-1)^n \left(1 + \frac{1}{n}\right) \right\rangle$ **[Delhi Maths (H), 1998, 2005; Delhi Maths (G), 2005 Delhi B.A. 2009]**

(*vi*) $<(-1)^n n>$ (*vii*) $<\sin (n\pi/3)>$ (*viii*) $<f_n>$ where $f_n = \left(1 + \frac{1}{n}\right)^{n+1} \ \forall \in \mathbf{N}$

(*ix*) $<f_n>$ where $f_n = (-2)^{-n}\left(1 + \frac{1}{n}\right) \ \forall \in \mathbf{N}$

2. If $<f_n>$ is any bounded sequence, then prove that

(*i*) $\lim_{n\to\infty} \inf (-f_n) = -\lim_{n\to\infty} \sup f_n$ (*ii*) $\lim_{n\to\infty} \sup (-f_n) = -\lim_{n\to\infty} \inf f_n$

[Delhi B.Sc. I (Hons) 2010]

3. If $<f_n>$ and $<g_n>$ are two bounded sequences of non-negative real numbers, then prove that (*i*) $\lim_{n\to\infty} \inf (f_n g_n) \geq (\lim_{n\to\infty} \inf f_n) \times (\lim_{n\to\infty} \inf g_n)$

(*ii*) $\lim_{n\to\infty} \sup (f_n g_n) \leq (\lim_{n\to\infty} \sup f_n) \times (\lim_{n\to\infty} \sup g_n)$

Show by means of examples that strict inequality may hold in (*i*) and (*ii*) of the above results.

4. If $<f_n>$ is a real bounded sequence and $<g_n>$ is a convergent sequence, then

(*i*) $\lim_{n\to\infty} \sup (f_n + g_n) = \lim_{n\to\infty} \sup f_n + \lim_{n\to\infty} g_n$ **[Delhi Maths (G), 1999]**

(*ii*) $\lim_{n\to\infty} \inf (f_n + g_n) = \lim_{n\to\infty} \inf f_n + \lim_{n\to\infty} g_n$

5. If $<f_n>$ be any bounded sequence of positive numbers and $<g_n>$ is a convergent sequence of positive numbers, then prove that

(*i*) $\lim_{n\to\infty} \sup (f_n g_n) = (\lim_{n\to\infty} \sup f_n) \times (\lim_{n\to\infty} g_n)$

(*ii*) $\lim_{n\to\infty} \inf (f_n g_n) = (\lim_{n\to\infty} \inf f_n) \times (\lim_{n\to\infty} g_n)$

6. If $\lim_{n\to\infty} \sup f_n = M$, then prove that

(*i*) there is a subsequence of $<f_n>$ that converges to M

(*ii*) the limit superior of no subsequence of $<f_n>$ can exceed M

(*iii*) no subsequence of $<f_n>$ can converge to a number greater than M

7. If $\lim_{n\to\infty} \inf f_n = m$, then prove that

(*i*) there is a subsequence of $<f_n>$ that converges to m

(*ii*) the limit inferior of no subsequence of $<f_n>$ can be less than m

(*iii*) no subsequence of $<f_n>$ can converge to a number less than m

[Delhi Maths (H), 1999]

8. Define : $\lim_{n\to\infty} \sup a_n$ and $\lim_{n\to\infty} \inf a_n$ for a bounded sequence $<a_n>$ of real numbers and show that these are cluster points of $<a_n>$. **[Delhi Maths (H), 2005]**

ANSWERS

1. (*i*) 2, 1 (*ii*) 5, 1 (*iii*) 1, 1 (*iv*) 0, 0 (*v*) 1, – 1 (*vi*) ∞, – ∞ (*vii*) $\sqrt{3}/2, -\sqrt{3}/2$ (*viii*) *e*, *e* (*ix*) 0, 0.

OBJECTIVE QUESTIONS

Multiple Choice Type Questions : *Select (a), (b), (c) or (d), whichever is correct.*

1. The sequence $<(-1)^{n-1}>$ is

(*a*) Bounded above, not bounded below (*b*) Not bounded above, bounded below
(*c*) Bounded above, bounded below (*d*) None of these. **(Kanpur, 2001)**

2. The sequence $<1 + (-1)^n/n>$ is

(*a*) Convergent (*b*) Divergent (*c*) Oscillatory (*d*) None of these. **(Kanpur, 2001)**

3. Sequence <1, – 1/2, 1/3, – 1/4, 1/5, ...> is

(*a*) Convergent (*b*) Divergent (*c*) Oscillatory (*d*) None of these. **(Kanpur, 2002)**

4. The sequence $<(-1)^n>$ is

(*a*) Convergent (*b*) Divergent (*c*) Oscillatory (*d*) None of these. **(Meerut, 2003)**

5. An example of oscillatory sequence is

(*a*) $<(-1)^n/n>$ (*b*) $<(-1)^n\, n>$ (*c*) $<(-1)^{n^2}>$ (*d*) $<(-1)^n\, n^2>$. **(Kanpur, 2003)**

6. The sequence $<(-1)^n\, n>$ is

(*a*) Bounded below (*b*) Bounded above

(*c*) Bounded below as well as bounded above

(*d*) Neither bounded below nor bounded above. **(Meerut, 2003)**

7. The sequence $<1, -1, 1, -1, 1, -1, \ldots>$ has

(*a*) no limit point (*b*) only one limit point

(*c*)Two limit points (*d*) An infinite number of limit points ***(Agra 2010)***

Hint: Refer illustration on page 5.3.

8. For the given sequence $<(-1)^n\,(1 + 1/n)$ which one of the following statements is correct ?

(*a*) Limit superior = limit inferior

(*b*) Neither limit superior nor limit inferior exists

(*c*) Limit superior = 1 and limit inferior = – 1

(*d*) Limit superior = 1 and limit inferior = 0. **[I.A.S. (Prel.), 2003]**

9. The least upper bound of the sequence $<1 - (1/n)>$ is :

(*a*) 0 (*b*) – 1 (*c*) 1 (*d*) None of these. **(Kanpur, 2002)**

10. Which of the following statements is true ?

(*a*) For any positive number ε, there is a natural number *n* such that $1/n < \varepsilon$

(*b*) Between any two real numbers there is no irrational number

(*c*) Convergent sequence is not bounded

(*d*) None of the above is true. **(Kanpur, 2003)**

11. The decreasing sequence is :

(*a*) $\left\langle 2+\dfrac{1}{n}\right\rangle$ (*b*) $\left\langle 2-\dfrac{1}{n}\right\rangle$ (*c*) $\left\langle 1-\dfrac{1}{n}\right\rangle$ (*d*) $\left\langle -\dfrac{1}{n}\right\rangle$ **(Kanpur, 2004)**

12. The sequence $<s_n>$ where $s_n = (1+2/n)^{n+3}$ converges to

(*a*) e (*b*) e^2 (*c*) $e + 3$ (*d*) $e^2 + 3$ **(Agra 2010)**

True or false type questions. *Write 'T' or 'F' according as the following statements are true or false.*

1. The sequence $<2, -2, 2, -2, \ldots..>$ is bounded. **(Meerut, 2003)**

2. A sequence can converge to more than one limit.

3. Every bounded monotonically decreasing sequence is divergent.

4. $<1, 1/2, 1/3, \ldots.., 1/n, \ldots..>$ is a Cauchy sequence.

Fill in the blanks correctly :

1. Every bounded monotonic sequence is ..... **(Meerut, 2004)**
2. $\lim_{n\to\infty} \frac{1}{n}(1+2^{1/2}+3^{1/3}+.....+n^{1/n}) = ...$ **(Meerut, 2004)**
3. If $<s_n> = l$ and $<f_n> = m$, then $\lim <s_n + f_n> =$ **(Meerut, 2003)**
4. The value of $\lim_{n\to\infty} (1+1/n)^n$ lies between ..... **(Meerut, 2003)**

ANSWERS

Multiple Choice Type Questions

1. (*c*)	**2.** (*a*)	**3.** (*c*)	**4.** (*c*)	**5.** (*b*)	**6.** (*d*)
7. (*b*)	**8.** (*c*)	**9.** (*c*)	**10.** (*a*)	**11.** (*a*)	**12.** (*b*)

True or False Type Questions

1. T **2.** F **3.** F **4.** T

Fill in the blanks

1. Convergent **2.** 1 **3.** $l + m$ **4.** 2 and 3

MISCELLANEOUS PROBLEMS ON CHAPTER 5

1. Let $<a_n>$ be a sequence defined by $a_1 = 1$, $a_{n+1} = (3+2a_n)/(2+a_n)$, $n = 1, 2, 3, ...$ Show that $<a_n>$ converges to $\sqrt{3}$. **(Delhi Maths (G) 2006)**

2. (*a*) Discuss whether the sequence $<3^n/n!)$ is monotonic, bounded and convergent. **[Kanpur 2006]**

(*b*) Show that the sequence $<3^n/n^3>$ is divergent. **[Agra 2010]**

3. If $s_i = \sqrt{2}$ and $s_{n+1} = \sqrt{2+\sqrt{s_n}}$ $(n = 1, 2, 3, ...)$, prove that $<s_n>$ converges and $s_n < 2$ for $n = 1, 2, 3, ...$ **[M.S. Univ. T.N. 2006]**

4. Show that 1/2 is not a limit point of the sequence $\{n/(n+1)\}_{n\in N}$ **[Ranchi 2010]**

5. Prove that a monotonic increasing and bounded above real sequence is convergent. **[Kanpur 2007]**

6. Discuss whether the sequence $<2^n/n!>$ is monotone, bounded and convergent. **[Kanpur 2007]**

7. Prove that the sequence $<a_n>$ of positive numbers such that $a_n = \sqrt{a_{n-1}\,a_{n-2}}$, $n > 2$ converges. Find its limit. **[Ranchi 2010]** **[Ans.** $(a_1 a_2^2)^{1/3}$ **]**

8. *Show that the sequence* $< s_n >$, *where* $s_n = 1 + 1/3 + 1/5 + ... + 1/(2n+1)$ *is a divergent sequence.* **[Delhi Maths (H) 2007]**

9. (*a*) *Find the limit superior and limit inferior of the sequence* $<a_n>$ *defined by* $a_n = 1 + (-1)^n$, $n \in \mathbf{N}$. **[Ans.** 2, 0**]** **[Delhi Maths (H) 2007]**

(*b*) Find the limit superior of the sequence $\{\sin(n\pi/2)\}_{n=1}^{\infty}$. **[Pune 2010]** **[Ans.** 1**]**

10. *Show that the sequence $< a_n >$ defined by the recursion formula $a_{n+1} = \sqrt{3a_n}$, $a_1 = 1$ is bounded monotonic increasing sequence. What is $\lim_{n\to\infty} a_n$?* **[Delhi Maths (H) 2007]**

11. *Prove that a convergent sequence has a unique limit. What about the converse. Justify.* **[Delhi Maths (H) 2007]**

12. *Prove that every bounded monotonic sequence is convergent.* **[Delhi Maths (H) 2007]**

13. *(a) Prove that every convergent sequence of real numbers is a cauchy sequence*

(b) Prove that Canchy sequence in R is convergent. **[Kanpur 2014]**

14. *If in a sequence $<a_n>$, $a_n = n!/n^n$, prove that the sequence converges to 0.* **[Agra 2008]**

15. *Which one of the following sequences is convergent*

(i) $<\sin n\pi/2>$ *(ii)* $<n^2>$ *(iii)* $<1/n>$ *(iv)* $<(-1)^{n-1}>$ **[Agra 2008]**

16. Show that the sequence $<1/3^n>$ converges to 0. [K.U. BCA (II) 2008]

17. If $<a_n>$ is a null sequence and $<b_n>$ is a bounded sequence, show that $<a_n b_n>$ is a null sequence. **[K.U. B.C.A. II 2006]**

18. If $<a_n>$ is a sequence of positive terms, then $\lim_{n\to\infty} (a_n)^{1/n} = \lim_{n\to\infty} \frac{a_{n+1}}{a_n}$ provided the limt on the R.H.S. exists, whether finite or infinite. **[KU. BCA (II) 2008]**

19. Show that the sequence $<1/2^n>$ is a Cauchy sequence.

20. Show that the sequence $<(n+1)/n>$ is monotonically decreasing and bounded below. **[KU. BCA (II) 2008]**

21. Prove that if a sequence $<a_n>$ converges to l, then l is the only cluster point of $<a_n>$ **[Delhi Maths (Hons) 2009]**

22. Show that the sequence $<r^n>$ converges if $-1 < r \leq 1$. What happen if $r = -1$? **[Delhi B.Sc. (Prog) III 2009, 10]**

[Hint : As shown in Ex. 5 page 5.9, $<r^n>$ converges if $-1 < r < 1$. When $r = 1$, $<r^n> = <1, 1, 1, \ldots>$ and $\lim_{n\to\infty} r^n = 1$ and so $<r^n>$ converges for $r = 1$. Next, when $r = -1$, sequence $<(-1)^n>$ is bounded and has two limit points, namely 1 and -1. Hence $<(-1)^n>$ oscillates finitely.]

23. Find lim sup and lim inf of sequence $<a_n>$ where $a_n = \sin(n\pi/4)$ and justify your answer.

Ans. $\limsup a_n = 1$ *and* $\liminf a_n = -1$ **[Delhi B.Sc. I (Hons) 2010]**

24. Let $<x_n>$ be a bounded sequence and $s = \sup \{x_n : n \in N\}$. Show that there exists a subsequence $<x_{n_k}>$ of $<x_n>$ that converges to s. Illustrate it for $x_n = 1 + (-1)^n/n$. **[Delhi B.Sc. I (Hons) 2010]**

35. Show that the sequences $<a_n^{1/n}>$ and $<b_n^{1/n}>$ where $a_n = (3n)!/(n!)^3$ and $b_n = n^n/\{(n+1)(n+2)\ldots\ldots(n+n)\}$ converge and find their limit. **[Delhi B.Sc. III (Prog) 2010]**

Ans. 27 and $e/4$ respectively

Chapter

6

Infinite Series With Positive Terms

6.1. INFINITE SERIES, ITS CONVERGENCE AND SUM

(Delhi B.Sc. (Prog.) II, 2016; Delhi B.Sc. (Prog.) III 2011; Meerut 2009)

Let $<u_n>$ be a given sequence. Then a symbol of the form

$$u_1 + u_2 + ... + u_n + ...$$

is called an *infinite series.* Another way of writing this infinite series is

$$\sum_{n=1}^{\infty} u_n \quad \text{or} \quad \text{simply } \Sigma\, u_n.$$

To the given sequence $<u_n>$ which underlines the infinite series referred to above, we associate another sequence $<S_n>$ defined as follows :

$$S_n = u_1 + u_2 + ... + u_n.$$

Thus S_n denotes the sum of the first n terms of the infinite series $\Sigma\, u_n$. The sequence $<S_n>$ is said to be sequence of partial sums of the series $\Sigma\, u_n$.

Definitions. The infinite series $\Sigma\, u_n$ is said to be *convergent* if the sequence $<S_n>$ of its partial sums is convergent. If $\lim\limits_{n \to \infty} S_n = S$, then S is called the *sum of the series* $\Sigma\, u_n$ and we write

$$S = \Sigma\, u_n.$$

The series $\Sigma\, u_n$ is said to be divergent, if the sequence $<S_n>$ of partial sums of $\Sigma\, u_n$ is divergent.

The series $\Sigma\, u_n$ is said to oscillate, if the sequence $<S_n>$ of partial sums of $\Sigma\, u_n$ oscillates.

In this chapter, we shall obtain a few tests for the convergence of infinite series. As is to be expected, it will be seen that this chapter leans heavily on developments of the preceding chapter on sequences.

The following theorems can be easily proved in view of the above definition.

Theorem I. The replacement, addition or omission of finite number of terms of a series $\Sigma\, u_n$ has no effect on its convergence.

Theorem II. The convergence of a series remains unchanged it each of its term is multiplied by a non-zero constant.

ILLUSTRATIONS

1. Consider the series $\displaystyle\sum_{n=1}^{\infty} \frac{1}{n\,(n+1)}$, *i.e.*, $\displaystyle\frac{1}{1\cdot 2} + \frac{1}{2\cdot 3} + \frac{1}{3\cdot 4} + ...$

(Delhi B.Sc. (Prog) II, 2016; G.N.D.U. Amritsar 2011)

Here u_n = nth term of the series $= \dfrac{1}{n\,(n+1)} = \dfrac{1}{n} \quad \dfrac{1}{n+1}$

$$\therefore \quad u_1 = 1 \quad \frac{1}{2},\; u_2 = \frac{1}{2} \quad \frac{1}{3},\; .....,\; u_{n \quad 1} = \frac{1}{n \quad 1} \quad \frac{1}{n},\; u_n = \frac{1}{n} \quad \frac{1}{n+1}$$

S_n = sum of the first n terms of the series

or $$S_n = u_1 + u_2 + ... + u_{n-1} + u_n = 1 - \frac{1}{n+1}$$

$$\therefore \quad \lim_{n\to\infty} S_n = \lim_{n\to\infty} \left(1 - \frac{1}{n+1}\right) = 1$$

$\Rightarrow$ $<S_n>$ converges to 1 $\Rightarrow$ Σu_n is convergent with sum 1.

2. Consider the series $\sum_{n=1}^{\infty} n^2$.

Here $u_n = n^2$ and $S_n = u_1 + u_2 + ... + u_n = 1 + 2^2 + ... + n^2$

Thus, $$S_n = \frac{n(n+1)(2n+1)}{6} \quad \text{and} \quad \lim_{n\to\infty} S_n = \infty.$$

$\Rightarrow$ $<S_n>$ diverges to ∞ $\Rightarrow$ Σu_n is divergent.

3. Consider the series $\sum_{n=1}^{\infty} (-1)^{n-1}$.

Here $u_n = (-1)^{n-1}$ and so we have

$S_1 = 1$, $S_2 = 1 - 1 = 0$, $S_3 = 1 - 1 + 1 = 1$, and so on.

$\therefore$ $<S_n> = <1, 0, 1, 0, 1, 0, ...>$, which oscillates

$\Rightarrow$ The given series is oscillatory.

6.2. A NECESSARY CONDITION FOR THE CONVERGENCE OF AN INFINITE SERIES

Theorem. *If the series* $\sum_{n=1}^{\infty} u_n$ *converges, then* $\lim_{n\to\infty} u_n = 0$.

Show, by an example, that the converse is not true. **[Delhi B.Sc. (Hons) I 2011; Delhi Maths (Prog) 2008; Delhi Mahts (G), 2004; Kanpur, 2012; Lucknow, 2001; Utkal, 2003, Agra 2011, 12; G.N.D.U. Amritsar 2010; Ranchi 2010; Delhi B.Sc. (Prog) III 2008, 09, 10, 11]**

Proof. Let $S_n = u_1 + u_2 + ... + u_n$ and let $\lim S_n = S$ so that $\lim S_{n-1} = S$. We have

$$u_n = S_n - S_{n-1}$$

$$\Rightarrow \lim u_n = \lim S_n - \lim S_{n-1} = S - S = 0.$$

Hence the result.

Note. The above condition is only necessary and *not* sufficient. Thus, there exist non-convergent series Σu_n such that $\lim u_n = 0$. **[Delhi B.Sc. (Prog) II, 2015]**

The converse may not be true. For example, consider the series

$$1 + \frac{1}{\sqrt{2}} + \frac{1}{\sqrt{3}} + ... + \frac{1}{\sqrt{n}} + ...$$

Here $u_n = 1/\sqrt{n}$ and so $\lim u_n = 0$. But the series does not converge as shown below.

$$S_n = 1 + \frac{1}{\sqrt{2}} + ... + \frac{1}{\sqrt{n}} > \frac{1}{\sqrt{n}} + \frac{1}{\sqrt{n}} + ... + \frac{1}{\sqrt{n}} = \frac{n}{\sqrt{n}} = \sqrt{n}$$

Thus $S_n > \sqrt{n}$, which tends to ∞ as $n \to \infty$. Hence the series is divergent.

Corollary. *If* $\lim u_n \neq 0$, *then series cannot converge.*

EXAMPLES

Example 1. *Show that the series*

$$\sqrt{\frac{1}{4}}+\sqrt{\frac{2}{6}}+\ldots..+\sqrt{\frac{n}{2(n+1)}}+\ldots..$$

does not converge. **[Delhi B.A. (Prog.) III, 2004]**

Solution.
$$u_n=\sqrt{\frac{n}{2(n+1)}}=\frac{1}{\sqrt{2}}\sqrt{\frac{1}{1+(1/n)}}$$

$$\Rightarrow \lim_{n\to\infty} u_n=\frac{1}{\sqrt{2}}\neq 0$$

$\Rightarrow$ The given series does not converge.

Example 2. *Show that the series* $\sum_{n=1}^{\infty} \cos(1/n^2)$ *is not convergent.*

[Delhi B. Sc. (Prog.) II, 2018]

Solution. Here $u_n = \cos(1/n^2)$. So $\lim u_n = \lim \cos(1/n^2) = 1 \neq 0$. Hence the given series is not convergent.

EXERCISES

1. Show that each of the following series is not convergent.

(i) $\frac{1}{2}+\frac{2}{3}+\frac{3}{4}+\ldots..\frac{n}{n+1}+\ldots..$ **[Delhi B.Sc. III 2009]**

(ii) $\sqrt{\frac{1}{2}}+\sqrt{\frac{2}{3}}+\sqrt{\frac{3}{4}}+\ldots..+\sqrt{\frac{n}{n+1}}+\ldots..$

(iii) $\sum_{n=1}^{\infty} \cos(1/n)$, **[Kanpur 2007; Delhi B.A. (Prog.) III, 2013, 14]**

(iv) $\sum_{n=1}^{\infty}\left(\frac{1}{n}\right)^{1/n}$ (v) $\sum_{n=1}^{\infty}\frac{n}{1+2^{-n}}$ **[Delhi B.A. (Prog.) III, 2014, 2017]**

(vi) $\sum_{n=1}^{\infty}\sqrt{\frac{3n^2+5n+4}{4n^2+1}}$ **[Delhi B.A. (Prog.) III, 2003]**

(vii) $\sum_{n=1}^{\infty}\frac{n^2-1}{n^2+1}$ **[Meerut 2003]**

(viii) $\sum_{n=1}^{\infty}\sqrt{\frac{n}{n+1}}$ **[Delhi B.Sc. (Prog) III 2011; Delhi B.Sc. (Prog.) II, 2018]**

(ix) $\sum_{n=1}^{\infty}\sqrt{\frac{n}{2(n+1)}}$ (x) $\sum_{n=1}^{\infty}\left[\frac{2}{(-1)^n-3}\right]^n$ **[Delhi B.Sc. (Prog) III 2010]**

6.3. CAUCHY'S GENERAL PRINCIPLE OF CONVERGENCE FOR SERIES

(Cauchy's Convergence Criterion) **[Delhi B.Sc. III 2008, 09; Agra 2005]**

Theorem. *A necessary and sufficient condition* for a *series* Σu_n *to converge is that to each* $\varepsilon > 0$, *there exists a positive integer m such that*

$$| u_{m+1}+u_{m+2}+\ldots..+u_n | < \varepsilon \quad \forall\ n > m$$ **[Delhi B.A. (Prog) III, 2008, 13]**

[Delhi B.Sc. (Hons.) II, 2008; Delhi B.Sc. I (Hons.), 2010; Kanpur, 2003]

Proof. Let $<S_n>$ be the sequence of partial sums of the series Σu_n. Now, we have

Σu_n converges

$\Leftrightarrow$ $<S_n>$ converges

$\Leftrightarrow$ to each $\varepsilon > 0$, there exists a positive integer m such that

$| S_n - S_m | < \varepsilon \quad \forall n > m$, by Cauchy's principle of convergence for sequences

$\Leftrightarrow | (u_1 + u_2 + ... + u_m + u_{m+1} + ... + u_n) - (u_1 + u_2 + ... + u_m) | < \varepsilon \quad \forall n > m$

$\Leftrightarrow | u_{m+1} + u_{m+2} + ... + u_n | < \varepsilon \quad \forall n > m$ ...(1)

From (1), the following alternative form of criterion follows

$| u_{n+1} + u_{n+2} + ... + u_{n+p} | < \varepsilon \quad \forall n > m$ and $\forall p \geq 1$

Example. *Show that the series* $\sum \frac{1}{n} = 1 + \frac{1}{2} + \frac{1}{3} + ... + \frac{1}{n} + ...$ *does not converge.*

[*Kanpur 2013; Delhi Maths (Prog) 2008; Delhi Maths (H), 2004*]

Proof. If possible, let the series converge. Then for $\varepsilon = 1/4$, by Cauchy's general principle of convergence, we can find a positive integer *m* such that

$$\frac{1}{m+1} + \frac{1}{m+2} + ... + \frac{1}{n} < \frac{1}{4} \quad \forall n \geq m \qquad ...(1)$$

Taking $n = 2m$, we have

$$\frac{1}{m+1} + \frac{1}{m+2} + ... + \frac{1}{n} = \frac{1}{m+1} + \frac{1}{m+2} + ... + \frac{1}{2m}$$

$$> \frac{1}{2m} + \frac{1}{2m} + ... + \frac{1}{2m} = m \times \frac{1}{2m} = \frac{1}{2}.$$

Thus, $$\frac{1}{m+1} + \frac{1}{m+2} + ... + \frac{1}{n} > \frac{1}{2} \text{ for } n > m$$

This contradicts result (1). Hence the given series is not convergent.

6.4. GENERAL TEST FOR THE CONVERGENCE OF POSITIVE TERM SERIES

As a consequence of the fact that it is comparatively easier to deal with monotonic sequences, we shall see that corresponding is the case with the consideration of the series whose terms are all non-negative, *i.e.*, with series

$$\sum_{n=1}^{\infty} u_n \quad \text{such that} \quad u_n \geq 0 \quad \forall n \in \mathbf{N}$$

In case $u_n \geq 0 \ \forall n \in \mathbf{N}$, the sequence of the partial sums of the series is monotonically increasing. In fact if $u_n \geq 0 \ \forall n \in \mathbf{N}$, we have

$$S_{n+1} - S_n = u_{n+1} \geq 0$$

and accordingly $$S_{n+1} \geq S_n \quad \forall n \in \mathbf{N}.$$

Note. In this chapter we shall consider positive term series and in the following, series with both positive and negative terms.

CONVERGENCE OF POSITIVE TERM SERIES

Theorem I. *A necessary and sufficient condition for the convergence of a positive term series* $\sum_{n=1}^{\infty} u_n$ *is that the sequence* $<S_n>$ *of the partial sums of the series defined by*

$$S_n = u_1 + u_2 + ... + u_n \textit{ is bounded above.}$$

[*Delhi B.Sc. (Prog) III 2010; Delhi Maths (H), 2002, 07; G.N.D.U. Amritsar 2010*]

Proof. As $u_n \geq 0 \ \forall n \in \mathbf{N}$, the sequence $<S_n>$ is monotonically increasing.

A necessary and sufficient condition for the series to be convergent is that the sequence of its partial sums is convergent. Again a necessary and sufficient condition for a sequence $<S_n>$ which is monotonically increasing to be convergent is that it is bounded above. Hence the result.

A positive term series is divergent if and only if the sequence of its partial sums is not bounded above.

Corollary. *A series Σu_n of positive terms converges if and only if there exists a number k such that* $u_1 + u_2 + ... + u_n < k \; \forall \; n \in \mathbf{N}$ **[Delhi Math (Hons) 2009]**

Theorem II. *A series of positive terms in divergent if each term after a fixed stage is greater than some fixed positive number*

Theorem III. *If each term of a series Σu_n of positive terms, does not exceed the corresponding term of a convergent series Σv_n of positive terms, then Σu_n is convergent.*

Also, if each term of Σu_n exceeds (or equals) the corresponding term of a divergent series Σv_n of positive terms, then Σu_n is divergent.

Theorem IV. *If Σu_n is a series of positive terms such that $\lim\limits_{n \to \infty} u_n \neq 0$, then Σu_n diverges.*

Theorem V. *If Σu_n and Σv_n are two convergent series, then $\Sigma (u_n + v_n)$ and $\Sigma (u_n - v_n)$ are also convergent*

Theorem VI. ***If Σu_n converges and Σv_n diverges, then $\Sigma (u_n + v_n)$ diverges.***

[Delhi B.Sc. I (Hons) 2010; Delhi Maths (H), 1999]

Proof. Proofs of theorems II to VI are left as exercises for the reader.

Example. *Show that the series* $\sum\limits_{n=1}^{\infty} (1/n)^{1/n}$ *diverges.* **[Delhi Maths (H), 1995]**

Solution. Here $u_n = (1/n)^{1/n} \Rightarrow \log u_n = (1/n) \log (1/n) = -(\log n)/n$

$$\therefore \quad \lim_{n \to \infty} \log u_n = -\lim_{n \to \infty} \frac{\log n}{n} \qquad \left[\frac{\infty}{\infty}\right] \text{ form}$$

$$= -\lim_{n \to \infty} \frac{1/n}{1}, \text{ by L' Hospital's rule}$$

Thus, $\lim\limits_{n \to \infty} \log u_n = 0 \Rightarrow \log (\lim\limits_{n \to \infty} u_n) = 0 \Rightarrow \lim\limits_{n \to \infty} u_n = e^0 = 1 \neq 0$

Hence the given series is divergent.

Note 1. A series $\sum\limits_{n=1}^{\infty} u_n$ whose terms are not necessarily positive may fail to be convergent even if the sequence $<S_n>$ is bounded above. For example, consider

$$u_n = (-1)^n \text{ so that we have}$$

$$S_n = \begin{cases} -1 \text{ if } n \text{ is odd} \\ \;\;\,0 \text{ if } n \text{ is even.} \end{cases}$$

The sequence $<S_n>$, even though bounded above, is obviously not convergent and as such the series is not convergent. The sequence $<S_n>$ has two limit points, *viz.*, – 1 and 0.

It should be seen that, in general, boundedness of the sequence of partial sums of a series is only a necessary but not a sufficient condition for the convergence of the series Σu_n, and it is only for positive term series that this condition of boundedness of the sequence of its partial sums is as well a sufficient condition for the convergence of the corresponding series.

Note 2. A positive term series which does not converge will diverge to plus infinity. In fact the sequence of partial sums of a positive term series being monotonically increasing either tends to a finite limit or to plus infinity. Thus, whereas there are *five* possible behaviours for an arbitrary term series, there are only *two* for a positive term series.

EXERCISES

1. Given that $\Sigma\, u_n$ is a positive term series and k is a given positive number, show that each of the two series $\Sigma\, ku_n$, $\Sigma\, [u_n + k]$

   is convergent or divergent according as the series $\Sigma\, u_n$ is convergent or divergent.
2. Show that a series each of whose term is non-positive is convergent if and only if the sequence of its partial sums is bounded below.
3. Use Cauchy's convergence theorem to prove that the necessary condition for the convergence of an infinite series Σu_n is that $u_n \to 0$ as $n \to \infty$ **[Ranchi 2010]**

6.5. TWO IMPORTANT STANDARD SERIES $\Sigma\, r^n$ AND $\Sigma\, 1/n^\lambda$

We shall now consider the convergence of two positive term series which will play an important part in the following.

Geometric Series. *The positive term infinite geometrical series*

$1 + r + r^2 + r^3 + \ldots + r^n + \ldots;\ 0 \le r$ is convergent if and only if $r < 1$.

[*Delhi B.Sc.(Prog) II, 2015; Agra 2009; Delhi Maths* (*H*), *2002, 07; Nagpur 2010*]

We have
$$S_n = \begin{cases} \dfrac{1-r^n}{1-r} & \text{if } r \ne 1, \\ n & \text{if } r = 1. \end{cases}$$

Case I. Let $0 \le r < 1$. Then,
$$S_n = \frac{1-r^n}{1-r} = \frac{1}{1-r} - \frac{r^n}{1-r} \le \frac{1}{1-r}\ \forall\, n \in \mathbf{N}$$

so that the sequence $<S_n>$ is bounded above and as such the given series is convergent.

Since $0 \le r < 1 \Rightarrow \lim r^n = 0$, we see that in this case the sum of the infinite geometrical series is $1/(1-r)$.

Case II. Let $r = 1$. Then, $S_n = n$

so that the sequence $<S_n>$ is not bounded and as such the series is not convergent.

Case III. Let $r > 1$. We have, then $r^n > 1\ \forall\, n \in \mathbf{N} \Rightarrow S_n > n\ \forall\, n \in \mathbf{N}$

The series is again, therefore, not convergent.

Note. The series $\Sigma\, r^n$ is not a positive term series if r is negative. In fact the terms of the series $\Sigma\, r^n$ are alternatively positive and negative if r is negative. Such series which are known as *alternating series* will be discussed in the following chapter.

The λ-series or generalised Harmonic series. An important comparison test

The positive term series $\Sigma\, 1/n^\lambda$ is convergent if and only if $\lambda > 1$.

[*Kanpur 2005, 07; Delhi Maths* (*H*), *2001, 09*; *Meerut, 2003, 04, 05*; *Patna, 2003*]

Case I. Let $\lambda > 1$.

It will be shown that the series is convergent in this case.

We have $2^n > n\ \forall\, n \in \mathbf{N}$

If S_n denotes the sum of the first n terms of the given series, we have, the terms being all positive
$$S_n < S_{2^n} \qquad \ldots (1)$$

Now, $$S_{2^n} = \frac{1}{1^\lambda}+\frac{1}{2^\lambda}+.....+\frac{1}{(2^n)^\lambda}.$$

We have

$$S_{2^{n+1}-1} = \frac{1}{1^\lambda}+\frac{1}{2^\lambda}+.....+\frac{1}{(2^{n+1}-1)^\lambda}$$

$$= \frac{1}{1^\lambda}+\left(\frac{1}{2^\lambda}+\frac{1}{3^\lambda}\right)+\left(\frac{1}{4^\lambda}+\frac{1}{5^\lambda}+\frac{1}{6^\lambda}+\frac{1}{7^\lambda}\right)+.....$$

$$+\left[\frac{1}{(2^n)^\lambda}+\frac{1}{(2^n+1)^\lambda}.....+.....+\frac{1}{(2^{n+1}-1)^\lambda}\right]$$

Now, $$\frac{1}{2^\lambda}+\frac{1}{3^\lambda} < \frac{2}{2^\lambda} = \frac{1}{2^{\lambda-1}}$$

$$\frac{1}{4^\lambda}+\frac{1}{5^\lambda}+\frac{1}{6^\lambda}+\frac{1}{7^\lambda} < \frac{4}{4^\lambda} = \frac{1}{2^{2(\lambda-1)}}$$

....

....

$$\frac{1}{(2^n)^\lambda}+\frac{1}{(2^n+1)^\lambda}+.....+\frac{1}{(2^{n+1}-1)^\lambda} < \frac{2^{n+1}-2^n}{(2^n)^\lambda} = \frac{2^n}{(2^n)^\lambda} = \frac{1}{2^{n(\lambda-1)}}.$$

Thus, we have $$S_{2^{n+1}-1} < \frac{1}{1^\lambda}+\frac{1}{2^{\lambda-1}}+\frac{1}{2^{2(\lambda-1)}}+.....+\frac{1}{2^{n(\lambda-1)}}.$$

The series on the right being a geometric series with common ratio $1/2^{\lambda-1}$, its sum is

$$\frac{\left[1-\left(\frac{1}{2^{\lambda-1}}\right)^{n+1}\right]}{1-\frac{1}{2^{\lambda-1}}} < \frac{2^{\lambda-1}}{2^{\lambda-1}-1} \ \forall\ n \in \mathbf{N}.$$

Thus, $$S_{2^{n+1}-1} < 2^{\lambda-1}/(2^{\lambda-1}-1),\ \forall\ n \in N \qquad ...(2)$$

Now, $$2^{n+1}-1 > 2^n \Rightarrow S_{2^{n+1}-1} > S_{2^n} \qquad ...(3)$$

Thus, we have $\forall\ n \in \mathbf{N}$ $$S_n < S_{2^n} < S_{2^{n+1}-1} < \frac{2^{\lambda-1}}{2^{\lambda-1}-1},$$

and as such we see that the sequence $<S_n>$ is bounded above.

Thus, *the series* $\Sigma\, 1/n^\lambda$ *is convergent if* $\lambda > 1$.

Case II. *Let* $\lambda = 1$. *It will be shown that in this case the series is divergent.*

We have the series $$1+\frac{1}{2}+\frac{1}{3}+\frac{1}{4}+.....$$

Consider

$$S_{2^n} = 1+\frac{1}{2}+\frac{1}{3}+\frac{1}{4}+.....+\frac{1}{2^n} = \left(1+\frac{1}{2}\right)+\left(\frac{1}{3}+\frac{1}{4}\right)+\left(\frac{1}{5}+\frac{1}{6}+\frac{1}{7}+\frac{1}{8}\right)$$

$$+.....+\left(\frac{1}{2^{n-1}+1}+\frac{1}{2^{n-1}+2}+.....+\frac{1}{2^n}\right).$$

We have
$$1+\frac{1}{2}>\frac{1}{2}$$
$$\frac{1}{3}+\frac{1}{4}>\frac{2}{4}=\frac{1}{2}$$
$$\frac{1}{5}+\frac{1}{6}+\frac{1}{7}+\frac{1}{8}>\frac{4}{8}=\frac{1}{2}$$
$$\dots\dots\dots\dots\dots\dots$$
$$\dots\dots\dots\dots\dots\dots$$
$$\frac{1}{2^{n-1}+1}+\frac{1}{2^{n-1}+2}+\dots\dots+\frac{1}{2^n}>\frac{1}{2^n}(2^n-2^{n-1})=\frac{2^{n-1}}{2^n}=\frac{1}{2}.$$

We thus see that $S_{2^n} > n/2$

Also by taking n sufficiently large, we can make $n/2$ as large as we like. This shows that the sequence $<S_n>$ of partial sums is *not* bounded above. As such the series $\Sigma\, 1/n$ is *not* convergent.

Case III. *Let* $\lambda < 1$. It will be shown that the series in this case is divergent.

We have $\forall\ n \in \mathbf{N}$ $\quad \lambda < 1 \Rightarrow n^\lambda < n \Rightarrow 1/n^\lambda > 1/n.$

Thus, we see that each term of the series
$$\frac{1}{1^\lambda}+\frac{1}{2^\lambda}+\dots\dots+\frac{1}{n^\lambda}+\dots\dots$$
is greater than the corresponding term of the divergent series $1/1 + 1/2 + \dots\dots + 1/n$ and as such, the series $\Sigma\, 1/n^\lambda$ is divergent, if $\lambda < 1$.

Conclusion. *The series* $\Sigma\, 1/n^\lambda$ *is convergent if and only if* $\lambda > 1$.

Some important limits to be remembered for direct applications

1. $\lim\limits_{n\to\infty} n^{1/n} = 1$ **2.** $\lim\limits_{n\to\infty} \dfrac{\log n}{n} = 0$ **3.** $\lim\limits_{n\to\infty}\left(1+\dfrac{x}{n}\right)^n = e^x$

4. $\lim\limits_{n\to\infty}\left(1+\dfrac{x}{n}\right)^{n+p} = e^x$, $\lim\limits_{n\to\infty}\left(1+\dfrac{x}{n}\right)^{p} = 1$, where p is a finite number.

5.
$$\lim_{n\to\infty}\frac{c_0 n^p + c_1 n^{p-1} + c_2 n^{p-2} + \dots\dots + c_{p-1} n + c_p}{d_0 n^q + d_1 n^{q-1} + d_2 n^{q-2} + \dots\dots + d_{q-1} n + d_q}$$
$$= \begin{cases} c_0/d_0, & \text{if } p = q \\ 0, & \text{if } q > p \\ \infty, & \text{if } p > q \text{ and } c_0 > 0,\ d_0 > 0 \end{cases}$$

6.6. COMPARISON TESTS FOR THE CONVERGENCE OF POSITIVE TERM SERIES

The most important technique to discuss the convergence of a positive term series is to compare the given series with another suitably chosen series with known behaviour. The series
$$\Sigma\, 1/n^\lambda \quad \text{and} \quad \Sigma\, r^n$$
already discussed will be found very useful as comparison series.

We have two types of comparison tests : one of these relates to *the comparison of the general term of one with the general term of the other* and the other to *the comparison of the ratio of consecutive terms of one with the ratio of the corresponding consecutive terms* of the other.

These two types of general comparison tests will now be given. Later on, we shall develop a few *special* tests for the convergence of positive term series on the basis of choices of special series with known behaviour as comparison series.

It is necessary to develop several tests for the determination of the convergence of positive term series in that while one test fails, we may have to have recourse to another. In fact, there exist series for which none of the tests we shall develop may prove decisive.

6.7. COMPARISON TESTS OF THE FIRST TYPE [Delhi B.Sc. (Hons) I 2011, 17]

I. Test for convergence. *Let Σu_n and Σv_n be two positive term series such that*

(*i*) *Σv_n is convergent and*

(*ii*) *there exists $m \in \mathbf{N}$ such that $u_n \le v_n \ \forall \ n \ge m$.*

Then Σu_n is convergent.

We write $\quad S_n = u_1 + u_2 + ... + u_n, \quad T_n = v_1 + v_2 + ... + v_n$

and suppose that $n \ge m$.

We write $\quad u_1 + u_2 + ... + u_m = a$ and $v_1 + v_2 + ... + v_m = b.$

We may then see that $\forall \ n \ge m$

$$S_n - a \le T_n - b \Rightarrow S_n \le T_n + a - b. \qquad ...(i)$$

As Σv_n is convergent, the sequence $<T_n>$ of its partial sums is convergent and therefore bounded so that there exists a number k such that

$$T_n \le k \ \forall \ n \in \mathbf{N}. \qquad ...(ii)$$

From (*i*) and (*ii*), we see that

$$S_n \le k + a - b \ \forall \ n \ge m.$$

We thus see that the sequence $<S_n>$ *of partial sums of the infinite series Σu_n is bounded and as such the series Σu_n is convergent.*

II. Test for divergence. *Let Σu_n and Σv_n be two positive term series such that*

(*i*) *Σv_n is divergent.*

(*ii*) *there exists $m \in \mathbf{N}$ such that $u_n \ge v_n \ \forall \ n \ge m$.*

Then Σu_n is divergent.

The proof which is simple is left to the reader.

6.8. PRACTICAL COMPARISON TESTS OF THE FIRST TYPE

Let Σu_n and Σv_n be two positive term series such that $\lim \dfrac{u_n}{v_n} = l \ne 0$. *Then,*

(i) The two series Σu_n and Σv_n have identical behaviours in relation to convergence. **[Delhi B.Sc. (Hons) I 2011, 2016; Delhi B.Sc. (Prog) III 2009, 12]**

(ii) If $l = 0$ and if Σv_n is convergent, then Σv_n is convergent. **[Delhi B.Sc. (Hons.) I 2018]**

Proof. (*i*). Surely l is positive. Let ε be a positive number less than l so that

$$0 < \varepsilon < l.$$

Now, $\lim [u_n/v_n] = l \Rightarrow$ there exists $m \in \mathbf{N}$ such that $\forall \ n \ge m$

$$l - \varepsilon < u_n/v_n < l + \varepsilon \qquad \Rightarrow (l - \varepsilon) v_n < u_n < (l + \varepsilon) v_n$$

Now suppose that the series Σv_n is convergent.

Then the series $\Sigma (l + \varepsilon) v_n$ is as well convergent.

Now since $\forall \ n \ge m \qquad u_n < (l + \varepsilon) v_n$

and the series $\Sigma (l + \varepsilon) v_n$ is convergent, we deduce from the result I of Art. 6.7 that the series Σu_n is convergent.

Now suppose that $\Sigma\, v_n$ is divergent. The series $\Sigma\,(l - \varepsilon)\, v_n$ is as well divergent.

Now since $\forall\ n \geq m \qquad u_n > (l - \varepsilon)\, v_n$

and the series $\Sigma\,(l - \varepsilon)\, v_n$ is divergent, we deduce from the result II of Art. 6.7 that the series $\Sigma\, u_n$ is divergent.

Part (*ii*): If $l = 0$, then there exists $m \in \mathbf{N}$ such that $0 \leq u_n \leq v_n\ \forall\ n \geq m$. From Art. 6.7, it follows that if $\Sigma\, v_n$ is convergent, then $\Sigma\, u_n$ is also convergent.

SOLVED EXAMPLES

Example 1. *Examine for convergence the infinite series* :

(*i*) $\sum \dfrac{1}{n^2 + a^2}$ (*ii*) $\sum \dfrac{bn - a}{bn^2 + a^2}$

(*iii*) $\sum \dfrac{1}{\sqrt{n} + \sqrt{(n+1)}}$ (*iv*) $\sum \sqrt{\left(\dfrac{n}{n^4 + 2}\right)}$

Solution. (*i*) We write $\qquad u_n = \dfrac{1}{n^2 + a^2}$, and take $v_n = \dfrac{1}{n^2}$.

Now $\qquad \dfrac{u_n}{v_n} = \dfrac{n^2}{n^2 + a^2} = \dfrac{1}{1 + a^2/n^2} \Rightarrow \lim \dfrac{u_n}{v_n} = 1.$

Also the series $\Sigma\, v_n$ is convergent.

Thus, we see that the given series $\Sigma\, u_n$ is convergent.

(*ii*) We write $\qquad u_n = \dfrac{bn - a}{bn^2 + a^2}$ and take $v_n = \dfrac{1}{n}$.

Now, as may be easily seen, $\qquad \lim \dfrac{u_n}{v_n} = 1.$

Because of the divergence of the series $\Sigma\, v_n$, we see that the given series $\Sigma\, u_n$ is as well divergent.

(*iii*) We write $\qquad u_n = \dfrac{1}{\sqrt{n} + \sqrt{(n+1)}}$ and take $v_n = \dfrac{1}{\sqrt{n} + \sqrt{n}} = \dfrac{1}{2\sqrt{n}}$

We may see that $\qquad \lim \dfrac{u_n}{v_n} = 1.$

The series $\Sigma\, v_n$ being divergent, we see that the given series $\Sigma\, u_n$ is also divergent.

(*iv*) We write $\qquad u_n = \sqrt{\left(\dfrac{n}{n^4 + 2}\right)}$ and take $v_n = \sqrt{\left(\dfrac{n}{n^4}\right)} = \dfrac{1}{n^{3/2}}$.

We may see that $\qquad \lim \dfrac{u_n}{v_n} = 1.$

The series $\Sigma\, v_n$ being convergent, we see that the series $\Sigma\, u_n$ is also convergent.

Example 2. *Test for convergence the series* $1 + \dfrac{1}{2^2} + \dfrac{1}{3^3} + \dfrac{1}{4^4} + \ldots.. + \dfrac{1}{n^n} + \ldots..$

Solution. Clearly, $n^n > 2^n$, for $n > 2$.

$\therefore$ $\frac{1}{n^n} < \frac{1}{2^n}$, for $n > 2$.

Since $\sum \frac{1}{2^n} = \frac{1}{2} + \frac{1}{2^2} + \frac{1}{2^3} + \ldots$.. is a geometric series with common ratio $\frac{1}{2} < 1$, so $\sum \frac{1}{2^n}$ is convergent. Hence, by First Comparison Test $\sum \frac{1}{n^n}$ is convergent.

Example 3. *Test for convergence the series :*

(*i*) $\sum_{n=1}^{\infty} \frac{1}{\sqrt{n!}}$ **[*Delhi Maths* (*H*), *1999*]**

(*ii*) $\sum_{n=2}^{\infty} \frac{1}{n^2 \log n}$ **[*Delhi Maths (Prog.) 2007; Delhi Maths* (*G*), *1998*]**

Solution. (*i*) We know $\frac{1}{\sqrt{n\,!}} \le \frac{1}{2^{(n-1)/2}}, \ \forall\ n \ge 2.$

The series of the right hand side being a geometric series with common ratio $1/\sqrt{2} < 1$ is convergent.

Hence, $\Sigma\, 1/\sqrt{n}$ is convergent.

(*ii*) We know $\frac{1}{n^2 \log n} < \frac{1}{n^2}\ \forall\ n \ge 3.$

Since $\Sigma\, 1/n^2$ converges, so by First Comparison Test, $\sum_{n=2}^{\infty} \frac{1}{n^2 \log n}$ converges.

Example 4. *Examine the convergence of the infinite series*

(*i*) $\sum \left(\sqrt{n^4+1} - \sqrt{n^4-1}\right)$ **[*Agra, 1994; Delhi, Maths* (*G*), *1998; Delhi Maths,* (*H*) *2001; Kanpur, 2008; Lucknow, 1994; Meerut, 2003, 05*]**

(*ii*) $\Sigma\, \{(n^3+1)^{1/3} - n\}$ **[*Delhi Maths* (*H*), *1996; Delhi B.Sc. Physics* (*H*), *2000; Delhi B.Sc. (Prog) III 2009; Purvanchal, 2000; Rohilkhand, 2003*]**

Solution. (*i*) Here, $u_n = \sqrt{n^4+1} - \sqrt{n^4-1}$. Rewriting it, we have

$$u_n = \frac{\left(\sqrt{n^4-1} - \sqrt{n^4-1}\right)\left(\sqrt{n^4+1} + \sqrt{n^4-1}\right)}{\sqrt{n^4+1} + \sqrt{n^4-1}} = \frac{2}{\sqrt{n^4+1} + \sqrt{n^4-1}}$$

and take $$v_n = \frac{2}{\sqrt{n^4} + \sqrt{n^4}} = \frac{2}{2n^2} = \frac{1}{n^2}$$

Now, $$\lim_{n\to\infty} \frac{u_n}{v_n} = \lim_{n\to\infty} \frac{2n^2}{\sqrt{n^4+1} + \sqrt{n^4-1}} = \lim_{n\to\infty} \frac{2}{\sqrt{1+(1/n)^4} + \sqrt{1-(1/n)^4}}$$

Thus, $\lim_{n\to\infty} (u_n/v_n) = 1$, which is finite and non-zero.

The series $\Sigma\, v_n$ being convergent, we see that the series $\Sigma\, u_n$ is also convergent.

(*ii*) Here, $u_n = (n^3 + 1)^{1/3} - n = n\ \{(1 + 1/n^3)^{1/3} - 1\}$

$$= n\left\{\left(1+\frac{1}{3}\cdot\frac{1}{n^3}-\frac{1}{9n^6}\ .....\right)-1\right\}=\frac{1}{3n^2}-\frac{1}{9n^5}+.....$$

(By binomial theorem for any index)

and take $v_n = 1/n^3$. Then $\lim_{n\to\infty}(u_n / v_n) = 1/3$, which is finite and non-zero. The series $\Sigma\ v_n$ being convergent, we see that $\Sigma\ u_n$ is also convergent.

Example 5. *Examine the convergence of the series.*

(*i*) $\sum sin\frac{1}{n}$ **(*Delhi Maths (H) 2007; Calicut, 2004; Kanpur, 1996*)**

(*ii*) $\sum\frac{1}{n^{\alpha+1/n}}$ **(*Utkal 2003*)** (*iii*) $\Sigma\frac{1}{1+1/n}$ **(*Meerut 2008*)**

(*iv*) $\sum_{n=1}^{\infty}\sin(1/n^2)$. **[*Delhi B.Sc. (Prog.) III 2012*]**

Solution. (*i*) Here $u_n = \sin(1/n)$. We take $v_n = 1/n$.

Then $$\lim_{n\to\infty}\frac{u_n}{v_n}=\lim_{n\to\infty}\frac{\sin(1/n)}{(1/n)}=1,$$

which is finite and non-zero. Since $\Sigma\ v_n$ is divergent, hence $\Sigma\ u_n$ is also divergent.

(*ii*) Here $u_n = \dfrac{1}{n^{\alpha+1/n}}$, we take $v_n = \dfrac{1}{n^{\alpha}}$.

Then $$\lim_{n\to\infty}\frac{u_n}{v_n}=\lim_{n\to\infty}\frac{n^{\alpha}}{n^{\alpha}\times n^{1/n}}=\lim\frac{1}{n^{1/n}}=1,$$

which is finite and non-zero. Again, $\Sigma\ v_n = \Sigma\ 1/n^{\alpha}$ is convergent, if $\alpha > 1$ and divergent if $\alpha \le 1$. Hence the given series $\Sigma\ u_n$ is convergent, if $\alpha > 1$ and divergent if $\alpha \le 1$.

(*iii*) Taking $\alpha = 1$. do as in part (*ii*). (*iv*) convergent **Ans.** divergent

EXERCISES

1. Examine the convergence of the series.

(*i*) $\sum\frac{1}{n}\sin\frac{1}{n}$ **(Kanpur 2005)** (*ii*) $\sum\frac{1}{n^2}\sin\frac{1}{n}$ (*iii*) $\sum\tan\frac{1}{n}$

(*iv*) $\sum\frac{1}{\sqrt{n}}\tan\frac{1}{n}$ (*v*) $\sum\tan\frac{1}{n^2}$ (*vi*) $\sum\sqrt{\left(\frac{n}{2+3n^3}\right)}$

(*vii*) $\sum\frac{1}{(2n-1)^p}$ (*viii*) $\frac{1}{5}+\frac{\sqrt{2}}{7}+\frac{\sqrt{3}}{9}+\frac{\sqrt{4}}{11}+.....$ **[Delhi B.A. (Prog) III; 2013]**

(*ix*) $\frac{1}{4\cdot 6}+\frac{\sqrt{3}}{6\cdot 8}+\frac{\sqrt{5}}{8\cdot 10}+.....$ **(Delhi B.A. (Prog.) III 1999)**

(*x*) $\frac{1\cdot 2}{3^2\cdot 4^2}+\frac{3\cdot 4}{5^2\cdot 6^2}+\frac{5\cdot 6}{7^2\cdot 8^2}+.....$ **[Delhi B.Sc. (H) Physics, 1998]**

(*xi*) $\sum\frac{n+1}{n^p}$ (*xii*) $1+\frac{1}{2^2}+\frac{2^2}{3^3}+\frac{3^3}{4^4}+\frac{4^4}{5^5}+.....$

(*xiii*) $\sum_{n=1}^{\infty}\frac{n}{n^2+3}$ **[Delhi B.Sc. I (Hons) 2010]** (*xiv*) $\frac{1}{1+\sqrt{2}}+\frac{1}{\sqrt{2}+\sqrt{3}}+\frac{1}{\sqrt{3}+\sqrt{4}}+.....$ **(Meerut, 2003)**

(*xv*) $\sum_{n-1}^{\infty}\frac{x^{n-1}}{1+x^n},\ x>0$ **(K.U. B.C.A II 2008; Patna, 2003)**

(*xvi*) $\sum_{n=1}^{\infty}(4n^3-n+3)/(n^3+2n)$ **[Delhi B.Sc. (Prog) III 2008; Delhi B.A. Prog 2008]**

2. Test the convergence of the series whose *n*th term is :

(*i*) $(\sqrt{n+1}-\sqrt{n-1})/n,\ \sqrt{n^3+1}-\sqrt{n^3-1})$

[Delhi B.Sc. (Prog.) II, 2016; Agra 2009]

(*ii*) $\sqrt{n+1}-\sqrt{n}$ ***[Delhi B.A. (Prog) III 2011, 15]*** (*iii*) $\sqrt{n^2+1}-n$ **[Agra 2006]**

(*iv*) $\sqrt{n^4+1}-n^2$ ***(Kanpur 2011)*** (*v*) e^{-n^2} **[Delhi Maths (H), 2001]**

(*vi*) $\sqrt{n}/(n^2+1)$ ***(Kanpur 2006)***; $\sqrt{n}/(n^3+1)$ **[Kanpur 2007]**

(*vii*) $(n^3+1)^{1/2}-n^{3/2}$ **[Meerut 2011; Delhi Maths (Prog) 2008]**

3. If $a_n>0\ \forall\ n\in\mathbf{N}$, show that $\sum_{n=1}^{\infty}a_n$ and $\sum_{n=1}^{\infty}\frac{a_n}{1+a_n}$ converge and diverge together.

4. Test the convergence of $\frac{1}{3\cdot7}+\frac{1}{4\cdot9}+\frac{1}{5\cdot11}+\frac{1}{6\cdot13}+....$ **[Delhi B.A. (Prog.) III, 2015]**

ANSWERS

1. (*i*) Convergent (*ii*) Convergent (*iii*) Convergent (*iv*) Convergent (*v*) Convergent (*vi*) Divergent (*vii*) Convergent if $p>1$, divergent if $p\le 1$ (*viii*) Divergent (*ix*) Convergent (*x*) Convergent (*xi*) Convergent if $p>2$, divergent if $p\le 2$ (*xii*) Divergent (*xiii*) Divergent (*xiv*) Convergent. (*xv*) Convergent if $x<1$, divergent if $x\ge 1$ (*xvi*) Divergent

2. (*i*) Convergent (*ii*) Convergent (*iii*) Divergent (*iv*) Conv. (*v*) Conv. (*vi*) conv. (*vii*) conv.

4. Convergent

6.9. COMPARISON TESTS OF THE SECOND TYPE

I. Test for convergence. *Let* $\Sigma\, u_n$ *and* $\Sigma\, v_n$ *be two positive term series such that*

(*i*) $\Sigma\, v_n$ *is convergent,*

(*ii*) *there exists* $m\in\mathbf{N}$ *such that* $\forall\ n\ge m$

$$\frac{u_{n+1}}{u_n}\le\frac{v_{n+1}}{v_n}.$$

Then $\Sigma\, v_n$ *is convergent.*

We write $S_n=u_1+.....+u_n,\ \ T_n=v_1+.....+v_n,$

and suppose that $n\ge m$.

We also write $u_1+.....+u_m=a$ and $v_1+.....+v_m=b$.

We have $\forall\ n\ge m$

$$S_n=u_1+.....+u_m+u_{m+1}\left[1+\frac{u_{m+2}}{u_{m+1}}+.....+\frac{u_n}{u_{m+1}}\right]$$

Now,
$$\frac{u_{m+2}}{u_{m+1}}\le\frac{v_{m+2}}{v_{m+1}}$$

$$\frac{u_{m+3}}{u_{m+1}}=\frac{u_{m+3}}{u_{m+2}}\frac{u_{m+2}}{u_{m+1}}\le\frac{v_{m+3}}{v_{m+2}}\frac{v_{m+2}}{v_{m+1}}=\frac{v_{m+3}}{v_{m+1}}$$

Proceeding thus, we may see that

$$\frac{u_n}{u_{m+1}} \le \frac{v_n}{v_{m+1}}$$

Thus, we have

$$S_n \le a + u_{m+1}\left[1 + \frac{v_{m+2}}{v_{m+1}} + \ldots.. + \frac{v_n}{v_{m+1}}\right]$$

$$= a + \frac{u_{m+1}}{v_{m+1}}(T_n - b)$$

$$\Rightarrow \quad S_n \le a - b\frac{u_{m+1}}{v_{m+1}} + T_n \frac{u_{m+1}}{v_{m+1}} \;\forall\; n \ge m.$$

Since $\Sigma\, v_n$ is convergent, the sequence $<T_n>$ is bounded. Thus, the sequence $<S_n>$ is also bounded and as such the series $\Sigma\, u_n$ is convergent.

II. Test for divergence. *Let $\Sigma\, u_n$ and $\Sigma\, v_n$ be two positive term series such that*

(*i*) *$\Sigma\, v_n$ is divergent,*

(*ii*) *there exists $m \in \mathbf{N}$ such that $\forall\; n \ge m$,*

$$\frac{u_{n+1}}{u_n} \ge \frac{v_{n+1}}{v_n}$$

Then $\Sigma\, u_n$ is divergent.

The proof which is similar to that of the first test is left to the reader.

6.10. PRACTICAL COMPARISON TESTS OF THE SECOND TYPE

D'Alembert's Ratio Test. *Let $\Sigma\, u_n$ be a positive term series such that*

(*a*) $$\lim_{n\to\infty} \frac{u_{n+1}}{u_n} = l.$$

Then the series is

(*i*) *convergent if $l < 1$,* (*ii*) *divergent if $l > 1$,*

and (*iii*) *no firm decision is possible if $l = 1$, i.e., the series may converge or diverge if $l = 1$.*

[Delhi B.Sc. (Prog.) II, 2016; Purvanchal 2006; M.S. Univ. T.N. 2006; Meerut 2006, 09, 10, 11; Delhi B.Sc. (Prog) III 2008, 12]

(*b*) $\lim_{n\to\infty} \frac{u_{n+1}}{u_n} = \infty$, *then $\Sigma\, u_n$ is divergent.* ***[Agra, 2003; Delhi Maths (G), 2000, 03; Delhi Maths (H), 2001, 03; Delhi Maths, 2002, 04; Garhwal, 2001; Kanpur, 2002, 03, 04, 05]***

Proof. Case I. Let $l < 1$. Let ρ be any number such that $l < \rho < 1$.

As $$\lim \frac{u_{n+1}}{u_n} = l,$$

there exists m such that $\forall\; n \ge m$

$$\frac{u_{n+1}}{u_n} < \rho = \frac{\rho^{n+1}}{\rho^n}$$

Now ρ being less than 1, we see that the geometric series $\Sigma\, \rho^n$ is convergent. Thus, by the comparison test of the second type (refer case I of Art. 6.9), it follows that in this case series $\Sigma\, u_n$ is convergent.

Case II. Let $l > 1$. Let α be a number such that $1 < \alpha < l$.

There exists m such that $\forall\ n \geq m$

$$\frac{u_{n+1}}{u_n} > \alpha = \frac{\alpha^{n+1}}{\alpha^n}.$$

Now α being greater than 1, the geometric series $\Sigma\, \alpha^n$ is divergent. Thus, by the comparison test of the second type (refer case II of Art. 6.9), it follows that in this case the series $\Sigma\, u_n$ is divergent.

Case III. We shall give examples of *two* series — one convergent and the other divergent for both of which

$$\lim \frac{u_{n+1}}{u_n} = 1.$$

(*i*) $\sum \frac{1}{n}$, (*ii*) $\sum \frac{1}{n^2}$.

(*i*) $\sum \frac{1}{n}$ is divergent and $\lim\limits_{n \to \infty} \frac{u_{n+1}}{u_n} = \lim\limits_{n \to \infty} \frac{1/(n+1)}{1/n} = 1.$

(*ii*) $\sum \frac{1}{n^2}$ is divergent and $\lim\limits_{n \to \infty} \frac{u_{n+1}}{u_n} = \lim\limits_{n \to \infty} \frac{1/(n+1)^2}{1/n^2} = 1.$

(*b*) Since $\lim\limits_{n \to \infty} \frac{u_{n+1}}{u_n} = \infty$, so there exists a positive integer m such that

$$\frac{u_{n+1}}{u_n} > 2\ \forall\ n \geq m, \quad i.e., \quad u_{n+1} > 2u_n\ \forall\ n \geq m.$$

Replacing n by $m, m + 1, m + 2, \ldots\ldots, n - 1$ and multiplying the corresponding sides of the resulting inequalities, we have

$$u_n > 2^{n-m}\, u_m\ \forall\ n \geq m$$

$$\Rightarrow u_n > (u_m/2^m) \times 2^n\ \forall\ n \geq m$$

Since the geometric series $\Sigma\, 2^n$ is divergent, so, by the comparison test $\Sigma\, u_n$ is also divergent.

Note. It may be seen that D'Alembert's ratio test which we have discussed above is essentially a development of the comparison test of the second type, when we adopted a geometric series for comparison purposes.

Note. Another equivalent form of D'Alembert's ratio test. Let $\Sigma\, u_n$ be a series of positive terms such that

(*a*) $\lim\limits_{n \to \infty} \frac{u_n}{u_{n+1}} = l$. Then,

(*i*) if $l > 1$, the series converges (*ii*) if $l < 1$, the series diverges

(*iii*) if $l = 1$, the series may converge or diverge.

(*b*) $\lim\limits_{n \to \infty} \frac{u_n}{u_{n+1}} = \infty$. Then $\Sigma\, u_n$ converges. **[*Delhi B.Sc. I (Hons.) 2008*]**

In what follows, we shall use the above form of ratio test.

EXAMPLES

Example 1. *Test the convergence of the following series*

(*i*) $\frac{2!}{3}+\frac{3!}{3^2}+\frac{4!}{3^3}+.....+\frac{(n+1)!}{3^n}+.....$ **[*Delhi Maths* (*G*), *1997*]**

(*ii*) $\sum_{n=1}^{\infty}\frac{2^{n-1}}{3^n+1}$ **[*Delhi Maths* (*G*), *1997*]**

(*iii*) $\frac{1^2\cdot 2^2}{1!}+\frac{2^2\cdot 3^2}{2!}+\frac{3^2\cdot 4^2}{3!}+.....$ **[*Delhi Maths* (*H*), *2000*; *Delhi Maths* (*G*), *2001, 06*]**

(*iv*) $1+\frac{2^p}{2!}+\frac{3^p}{3!}+\frac{4^p}{4!}+.....$ **(*Kanpur*, *2004*)**

Solution. (*i*) Here $\quad u_n=\frac{(n+1)!}{3^n}\quad$ and $\quad u_{n+1}=\frac{(n+2)!}{3^{n+1}}=\frac{(n+2)\,(n+1)!}{3^{n+1}}$

Thus, $\quad u_n/u_{n+1}=3/(n+2)$

and hence $$\lim_{n\to\infty}\frac{u_n}{u_{n+1}}=\lim_{n\to\infty}\frac{3}{n+2}=0,\text{ which is } < 1.$$

Hence, by ratio test, the given series diverges.

(*ii*) Here $\quad u_n=\frac{2^{n-1}}{3^n+1}\quad$ and $\quad u_{n+1}=\frac{2^n}{3^{n+1}+1}$

$$\therefore\quad \lim_{n\to\infty}\frac{u_n}{u_{n+1}}=\lim_{n\to\infty}\frac{1}{2}\times\frac{3^{n+1}+1}{3^{n+1}}=\lim_{n\to\infty}\frac{1}{2}\times\frac{3+(1/3^n)}{1+(1/3^n)}=\frac{3}{2},\text{ which is } > 1.$$

Hence by ratio test, the given series converges.

(*iii*) Here $\quad u_n=\frac{n^2\,(n+1)^2}{n|}\quad$ and $\quad u_{n+1}=\frac{(n+1)^2\,(n+2)^2}{(n+1)!}$

$$\therefore\quad \lim_{n\to\infty}\frac{u_n}{u_{n+1}}=\lim_{n\to\infty}\frac{(n+1)!}{n!}\cdot\left(\frac{n}{n+2}\right)^2=\lim_{n\to\infty}(n+1)\cdot\frac{1}{\{1+(2/n)\}^2}=\infty$$

Hence, by ratio test, the given series converges

(*iv*) Here $\quad u_n=\frac{n^p}{n!}\quad$ and $\quad u_{n+1}=\frac{(n+1)^p}{(n+1)!}$

$$\therefore\quad \lim_{n\to\infty}\frac{u_n}{u_{n+1}}=\lim_{n\to\infty}\frac{(n+1)!}{n!}\cdot\left(\frac{n}{n+1}\right)^p=\lim_{n\to\infty}\frac{n+1}{(1+1/n)^p}=\infty$$

Hence, by ratio test, the given series converges.

Example 2. *Test the convergence of the following series* :

(*i*) $\sum\frac{n!}{n^n}$ **[Delhi B.Sc. (Prog.) II, 2015]**

(*ii*) $\frac{1}{3}+\frac{1\cdot 2}{3\cdot 5}+\frac{1\cdot 2\cdot 3}{3\cdot 5\cdot 7}+\frac{1\cdot 2\cdot 3\cdot 4}{3\cdot 5\cdot 7\cdot 9}+.....$ **[Delhi B.A. (Prog) III 2010, 2015]**

Solution. (*i*) Here $\quad u_n=\frac{n!}{n^n}\quad$ and $\quad u_{n+1}=\frac{(n+1)!}{(n+1)^{n+1}}$

$$\therefore \quad \frac{u_n}{u_{n+1}} = \frac{n!}{(n+1)!} \times \frac{(n+1)^{n+1}}{n^n} = \frac{1}{n+1}\left(\frac{n+1}{n}\right)^n (n+1) = \left(1+\frac{1}{n}\right)^n$$

So $$\lim_{n\to\infty} \frac{u_n}{u_{n+1}} = \lim_{n\to\infty}\left(1+\frac{1}{n}\right)^n = e = 2.718 > 1.$$

Hence by ratio test, the given series converges.

(*ii*) Here 3, 5, 7, 9, ... is an A.P. where *n*th term = $3 + (n - 1) \times 2 = 2n + 1$

$$\therefore \quad u_n = \frac{1\cdot2\cdot3\cdot4\ldots\ldots n}{3\cdot5\cdot7\cdot9\cdot\ldots\ldots(2n+1)}, \quad u_{n+1} = \frac{1\cdot2\cdot3\cdot4\ldots\ldots n(n+1)}{3\cdot5\cdot7\cdot9\ldots\ldots(2n+1)(2n+3)}$$

So $$\lim_{n\to\infty}\frac{u_n}{u_{n+1}} = \lim_{n\to\infty}\frac{2n+3}{n+1} = \lim_{n\to\infty}\frac{2+3/n}{1+1/n} = 2, \text{ which is } > 1.$$

Hence, by ratio test, the given series converges.

Example 3. *If* $\alpha > 0$, $\beta > 0$, *test the convergence of the series*

$$1+\frac{\alpha+1}{\beta+1}+\frac{(\alpha+1)(2\alpha+1)}{(\beta+1)(2\beta+1)}+\frac{(\alpha+1)(2\alpha+1)(3\alpha+1)}{(\beta+1)(2\beta+1)(3\beta+1)}+\ldots..$$

Solution. Here, we have

$$u_n = \frac{(\alpha+1)(2\alpha+1)\ldots..\{(n-1)\alpha+1\}}{(\beta+1)(2\beta+1)\ldots..\{(n-1)\beta+1\}} \quad \ldots(1)$$

and $$u_{n+1} = \frac{(\alpha+1)(2\alpha+1)\ldots..\{(n-1)\alpha+1\}(n\alpha+1)}{(\beta+1)(2\beta+1)\ldots..\{(n-1)\beta+1\}(n\beta+1)}$$

$$\therefore \quad \lim_{n\to\infty}\frac{u_n}{u_{n+1}} = \lim_{n\to\infty}\frac{n\beta+1}{n\alpha+1} = \lim_{n\to\infty}\frac{\beta+1/n}{\alpha+1/n} = \frac{\beta}{\alpha}$$

Hence, by ratio test, the given series converges if $\beta/\alpha > 1$, *i.e.*, if $\beta > \alpha$, diverges if $\beta/\alpha < 1$, *i.e.*, $\beta < \alpha$, and the test fails if $\beta/\alpha = 1$, *i.e.*, if $\beta = \alpha$.

When $\beta = \alpha$, from (*i*), $u_n = 1$ so $\lim_{n\to\infty} u_n = 1 \neq 0$, showing that the series diverges when $\beta = \alpha$.

Thus the given series converges if $\beta > \alpha$ and diverges if $\beta \le \alpha$.

Example 4. *Test the series* $1 + x^2/2 + x^4/4 + x^6/6 + \ldots..$ *for convergence for all positive values of x.* **[Delhi Maths (H), 2003]**

Solution. Omitting the first term of the series as it will not affect the nature of the series, we have

$$u_n = \frac{x^{2n}}{2n} \quad \text{and} \quad u_{n+1} = \frac{x^{2n+2}}{2n+2}$$

$$\therefore \quad \lim_{n\to\infty}\frac{u_n}{u_{n+1}} = \lim_{n\to\infty}\frac{1}{x^2}\times\frac{2n+2}{2n} = \lim_{n\to\infty}\frac{1}{x^2}\left(1+\frac{1}{n}\right) = \frac{1}{x^2}$$

Hence, by ratio test, the given series converges if $1/x^2 > 1$, *i.e.*, if $x^2 < 1$, *i.e.*, if $x < 1$ (as $x > 0$), diverges if $1/x^2 < 1$, *i.e.*, if $x^2 > 1$, *i.e.*, if $x > 1$, and the test fails if $1/x^2 = 1$, *i.e.*, if $x = 1$.

Now, when $x = 1$, $u_n = 1/2n$. We take $v_n = 1/n$.

$$\therefore \quad \lim_{n\to\infty}\frac{u_n}{v_n} = \lim_{n\to\infty}\frac{1}{2} = \frac{1}{2}, \text{ which is finite and non-zero.}$$

Since Σv_n diverges, so by comparison test, the given series also diverges if $x = 1$.
Thus the given series converges if $x < 1$ and diverges if $x \geq 1$.

Example 5. *Test for convergence the series* $\frac{x}{1\cdot 2}+\frac{x^2}{2\cdot 3}+\frac{x^3}{3\cdot 4}+\frac{x^4}{4\cdot 5}+\ldots, x > 0.$

(*Meerut 2009*)

Solution. Here, we have $u_n = \frac{x^n}{x\,(n+1)}$ and $u_{n+1} = \frac{x^{x+1}}{(n+1)\,(n+2)}$

$$\therefore \quad \lim_{n\to\infty} \frac{u_n}{u_{n+1}} = \lim_{n\to\infty} \frac{n+2}{n}\times\frac{1}{x} = \lim_{n\to\infty} \frac{1}{x}\times\left(1+\frac{2}{n}\right) = \frac{1}{x}.$$

The series is therefore convergent if $1/x > 1$, *i.e.*, if $x < 1$ and divergent if $1/x < 1$, *i.e.*, if $x > 1$. For $x = 1$, the ratio test fails. In this case,

$$u_n = \frac{1}{n\,(n+1)}. \text{ We take } v_n = \frac{1}{n^2}.$$

Then $\lim_{n\to\infty} \frac{u_n}{v_n} = \lim_{n\to\infty} \frac{n^2}{n\,(n+1)} = \lim_{n\to\infty} \frac{1}{1+1/n} = 1$, which is finite and non-zero. Since Σu_n is convergent, so by comparison test the given series is also convergent when $x = 1$.

Thus the given series is convergent if $x \leq 1$ and divergent if $x > 1$.

EXERCISES

1. Test the convergence of the following series

(*i*) $\sum_{n=1}^{\infty} \frac{r^n}{n!}, r > 0$ (*ii*) $\sum_{n=1}^{\infty} \frac{r^n}{n^n}, r > 0$ (*iii*) $\sum_{n=1}^{\infty} \frac{5^n}{n^2+5}$

(*iv*) $\frac{1}{2}+\frac{1}{2.2^2}+\frac{1}{3.2^3}+\frac{1}{4.2^4}+\ldots$ **[Delhi 2008, 13]** (*v*) $\sum_{n=1}^{\infty} \frac{n^3+a}{2^n+a}$

(*vi*) $\frac{1}{2}+\frac{2!}{8}+\frac{3!}{32}+\frac{4!}{128}+\ldots$ (*vii*) $\sum_{n=1}^{\infty} \frac{2^n\,n!}{n^n}$

2. Test the following series for convergence, for all positive values of x.

(*i*) $\frac{x}{1\cdot 3}+\frac{x^2}{2\cdot 4}+\frac{x^3}{3\cdot 5}+\frac{x^4}{4\cdot 6}+\ldots$

(*ii*) $\frac{x^2}{2\sqrt{1}}+\frac{x^3}{3\sqrt{2}}+\frac{x^4}{4\sqrt{3}}+\ldots$ **(Purvanchal 2006; Delhi B.A. (Prog.) III, 2014)**

(*iii*) $\sum_{n=1}^{\infty} \frac{n^n x^n}{n!}$ **(Delhi Maths (G), 1999)**

(*iv*) $1+\frac{x}{2}+\frac{x^2}{5}+\ldots+\frac{x^n}{n^2+1}+\ldots$

(*v*) $x+\frac{x^3}{31}+\frac{x^5}{51}+\ldots$ **[Delhi B.A. (Prog.) III, 2013; Delhi B.Sc. (Prog) III 2011]**

(*vi*) $1+\frac{2}{5}x+\frac{6}{9}x^2+\frac{14}{17}x^3+\ldots+\frac{2^n-2}{2^n+1}x^{n-2}+\ldots$ **[Delhi Maths (H), 2005; Delhi Maths (Prog) 2007]** (*vii*) $\sum_{n=1}^{\infty} \frac{n^2-1}{n^2+1} x^n$ **(Delhi B.Sc. (Prog) III 2009** (*viii*) $\sum_{n=1}^{\infty} \frac{x^n}{n(n+1)}$ **(Pune 2010)**

(*ix*) $\sum_{n=1}^{\infty} \frac{1}{2^n+x}, x \geq 0$ **[Delhi 2004]** (*x*) $\sum_{n=1}^{\infty} \frac{x^n}{n^2}, x > 0$ **[Delhi 2008]**

3. Test the convergence of the series with *n*th term :

(*i*) $\sqrt{\frac{n-1}{n^3+1}}\, x^n, x > 0$ **[Delhi Maths (G), 2004]** (*ii*) $\frac{\sqrt{n}\, x^n}{\sqrt{n^2+1}}, x > 0$

(*iii*) $\{\sqrt{n^2+1}-n\}\, x^{2n}$ (*iv*) $\frac{1}{x^n+x^{-n}}$ **(Meerut, 2000)**

ANSWERS

1. (*i*) Convergent (*ii*) Convergent (*iii*) Divergent (*iv*) Convergent (*v*) Convergent (*vi*) Divergent (*vii*) Convergent (*viii*) Convergent.

2. (*i*) Convergent if $x \leq 1$, Divergent if $x > 1$ (*ii*) Convergent if $x \leq 1$, Divergent if $x > 1$ (*iii*) Convergent if $x < 1$, Divergent if $x \geq 1$ (*iv*) Convergent if $x \leq 1$, Divergent if $x > 1$ (*v*) Convergent for all $x > 0$ (*vi*) Convergent if $x < 1$, Divergent if $x \geq 1$ (*vii*) Convergent if $x < 1$, Divergent, if $x \geq 1$. (*viii*) convegent if $x \leq 1$, divergent if $x > 1$

(*ix*) convergent if $x = 0$, divergent if $x \neq 0$ (*x*) convergent if $x \leq 1$, divergent if x > 1

3. (*i*) Convergent if $x < 1$, Divergent if $x \geq 1$ (*ii*) Convergent if $x < 1$, divergent, if $x \geq 1$. (*ii*) Convergent if $x < 1$, divergent if $x \geq 1$ (*iv*) Convergent if $x < 1$ or $x > 1$, divergent if $x = 1$.

6.11. CAUCHY'S *n*th ROOT TEST **[M.S. Univ. T.N. 2006]**

Let $\Sigma\, u_n$ be a positive term series and let $\lim_{n \to \infty} [u_n]^{1/n} = l.$

Then the series is

(*i*) *convergent if $l < 1$;*

(*ii*) *divergent if $l > 1$;*

and (*iii*) *no firm decision is possible if $l = 1$.***[*Delhi Maths* (*H*), *2003, 06*; *Kanpur, 2004, 07*; *Delhi B.A. (Prog) III 2011, 12; Chennai 2011; Meerut, 2003, 05, 07, 09*; *Delhi B.Sc. (Prog.) II, 2016; Kakatiya, 2003*; *Patna, 2003*; *Purvanchal, 2006*]**

Proof. Case I. Suppose that $l < 1$. Let ρ be a number such that $l < \rho < 1$

Then there exists *m* such that $\forall\ n \geq m$

$$[u_n]^{1/n} < \rho \Rightarrow u_n < \rho^n.$$

Now ρ being less than 1, the geometric series $\Sigma\, \rho^n$ is convergent. Thus, by the comparison test of the first type, it follows that the series $\Sigma\, u_n$ is convergent.

Case II. Suppose that $l > 1$. Let α be a number such that $1 < \alpha < l$.

Then there exists *m* such that $\forall\ n \geq m$

$$[u_n]^{1/n} > \alpha \Rightarrow u_n > 1.$$

It follows that the given series is divergent.

Case III. Suppose that $l = 1$. Consider the two series

(*i*) $\sum \frac{1}{n}$, (*ii*) $\sum \frac{1}{n^2}$.

The series (*i*) is divergent and the series (*ii*) is convergent and we have

$$\lim\left(\frac{1}{n}\right)^{1/n} = 1 \quad \text{and} \quad \lim\left(\frac{1}{n^2}\right)^{1/n} = 1$$

Thus, here we have examples of two series for each of which $\lim (u_n)^{1/n} = 1$, but while one series is convergent the other is divergent.

EXAMPLES

Example 1. *Test the convergence of the series*

(*i*) $\sum_{n=1}^{\infty}\left(1+\frac{1}{n}\right)^{-n^2}$ **[Utkal, 2003]**

(*ii*) $\sum_{n=1}^{\infty}\frac{n^{n^2}}{(n+1)^{n^2}}$ **[Delhi B.Sc. (Hons.) I 2018; Delhi B.A. (Prog) III 2014]**

Solution. (*i*) Here $u_n = \left(1+\frac{1}{n}\right)^{-n^2} = \left\{\left(1+\frac{1}{n}\right)^{-n}\right\}^n$

$$\therefore \quad \lim_{n\to\infty} (u_n)^{1/n} = \lim_{n\to\infty}\left(1+\frac{1}{n}\right)^{-n} = \lim_{n\to\infty}\frac{1}{(1+1/n)^n} = \frac{1}{e} < 1 \quad [\because e = 2.718]$$

Hence, by Cauchy's root test, the given series is convergent

(*ii*) Here $u_n = \left(\frac{n}{n+1}\right)^{n^2} = \left\{\left(\frac{n}{n+1}\right)^n\right\}^n$

$$\therefore \quad \lim_{n\to\infty} (u_n)^{1/n} = \lim_{n\to\infty}\left(\frac{n}{n+1}\right)^n = \lim_{n\to\infty}\frac{1}{(1+1/n)^n} = \frac{1}{e} < 1$$

Hence, by Cauchy's root test, the given series is convergent.

Example 2. *Test the convergence of the series*

$$\left(\frac{2^2}{1^2}-\frac{2}{1}\right)^{-1}+\left(\frac{3^3}{2^3}-\frac{3}{2}\right)^{-2}+\left(\frac{4^4}{3^4}-\frac{4}{3}\right)^{-3}+\ldots..$$

[Avadh, 1994; Delhi Maths (G), 2002; Purvanchal, 1993]

Solution. The *n*th term u_n of the given series is given by

$$u_n = \left\{\left(\frac{n+1}{n}\right)^{n+1} - \left(\frac{n+1}{n}\right)\right\}^{-n}$$

$$\therefore \quad (u_n)^{1/n} = \left[\left(\frac{n+1}{n}\right)^{n+1} - \left(\frac{n+1}{n}\right)\right]^{-1} = \left(\frac{n+1}{n}\right)^{-1}\left\{\left(\frac{n+1}{n}\right)^n - 1\right\}^{-1}$$

$$\therefore \quad \lim_{n\to\infty} (u_n)^{1/n} = \lim_{n\to\infty}\left(1+\frac{1}{n}\right)^{-1}\left\{\left(1+\frac{1}{n}\right)^n - 1\right\}^{-1} = 1\times(e-1)^{-1} = \frac{1}{e-1} < 1$$

$[\because \; e = 2.718]$

Hence, by Cauchy's root test, the given series is convergent.

Example 3. *Examine the convergence of the following series*

$$(i)\ \sum \left(\frac{nx}{n+1}\right)^n \qquad (ii)\ \sum \frac{(1+nx)^n}{n^n}$$

Solution. (*i*) Here $u_n = \left(\frac{nx}{n+1}\right)^n$ so that $(u_n)^{1/n} = \frac{nx}{n+1}$

$$\therefore \quad \lim_{n\to\infty} (u_n)^{1/n} = \lim_{n\to\infty} \frac{x}{1+(1/n)} = x.$$

Hence, by Cauchy's root test, the given series is convergent if $x < 1$ and divergent if $x > 1$. When $x = 1$, the Cauchy's root test fails. In this case

$$u_n = \left(\frac{n}{n+1}\right)^n \quad \text{and} \quad \lim_{n\to\infty} u_n = \lim_{n\to\infty} \frac{1}{(1+1/n)^n} = \frac{1}{e} \neq 0$$

Hence the given series is divergent.

Thus, the given series is convergent if $x < 1$ and divergent if $x \geq 1$.

(*ii*) Left as an exercise. **Ans.** Convergent if $x < 1$ and divergent if $x \geq 1$.

EXERCISES

1. *Test the convergence of the series whose nth term is*

(*i*) $(1+1/\sqrt{n})^{-n^{3/2}}$ **[Delhi Maths (H), 2000]**

(*ii*) $2^{-n-(-1)^n}$ **[Delhi B.A. (Prog) III 2011, 2015]**

(*iii*) $5^{-n-(-1)^n}$

(*iv*) $\dfrac{(n-\log n)^n}{2^n n^n}$

(*v*) $\left\{\dfrac{\log n}{\log(n+1)}\right\}^{n^2 \log n}$ **(Lucknow, 1992; Purvanchal, 1994)**

2. Test $\sum_{n=1}^{\infty} (n^{1/n} + x)^n$ for all positive values of x. **[Delhi B.Sc. (G), 2000]**

3. Test the convergence of the series

(*i*) $\sum_{n=1}^{\infty} \dfrac{1}{n^n}$ **[Delhi Maths (G), 2004; Delhi Math., 1999; Meerut, 1996]**

(*ii*) $\sum_{n=2}^{\infty} \dfrac{1}{(\log n)^n}$ **[Delhi B.A. (Prog) III 2015]** (*iii*) $\sum_{n=1}^{\infty} \dfrac{n}{3^n}$ **[Delhi B.Sc. I (Hons.) 2010]**

(*iv*) $\sum_{n=1}^{\infty} \dfrac{x^n}{n^n}$, if $x > 0$ (*v*) $\dfrac{1}{2} + \dfrac{2}{3}x + \left(\dfrac{3}{4}\right)^2 x^2 + \left(\dfrac{4}{5}\right)^3 x^3 + \ldots,$ if $x > 0$

(*vi*) $\dfrac{2}{1^2}x + \dfrac{3^2}{2^3}x^2 + \dfrac{4^3}{3^4}x^3 + \ldots + \dfrac{(n+1)^n x^n}{n^{n+1}} + \ldots,$ if $x > 0$

(*vii*) $\sum \left(\dfrac{n+1}{n+2}\right)^n x^n,\ x > 0$ (*viii*) $\sum_{n=1}^{\infty} e^{-n^2}$ (*ix*) $\sum_{n=1}^{\infty} \dfrac{(1+1/n)^{2n}}{e^n}$ **[Pune 2010]**

ANSWERS

1. (*i*) Convergent (*ii*) Convergent (*iii*) Convergent (*iv*) Convergent (*v*) Convergent
2. Divergent
3. (*i*) Convergent (*ii*) Convergent (*iii*) Convergent (*iv*) Convergent (*v*) Convergent if $x < 1$, divergent if $x \geq 1$ (*vi*) Convergent if $x < 1$, divergent if $x \geq 1$ (*vii*) Convergent if $x < 1$, divergent if $x \geq 1$ (*viii*) Convergent (*ix*) convergent

6.12. CAUCHY'S *n*th ROOT TEST IS SUPERIOR THAN D'ALEMBERT'S RATIO TEST (G.N.D.U. Amritsar 2010; Patna, 2003; Utkal, 2003)

In the present article we propose to show with help of examples that Cauchy's *n*th root test is superior than D'Alembert's ratio test in the sense that when D'Alembert's ratio test fails, the Cauchy's *n*th root test succeeds.

Example. Consider the series $\Sigma\, 3^{-n-(-1)^n}$. Here $u_n = 3^{-n-(-1)^n}$

$$\therefore \quad \lim_{n\to\infty} (u_n)^{1/n} = \lim_{n\to\infty} 3^{-1-\frac{(-1)^n}{n}} = 3^{-1} = \frac{1}{3} < 1$$

Hence, by Cauchy's *n*th root test, the given series is convergent.

Let us now try to examine the convergence of the given series by using D'Alembert's ratio test.

When n is even (so that $n + 1$ is odd), $u_n = 3^{-n-1}$ and $u_{n+1} = 3^{-(n+1)+1} = 3^{-n}$

$$\therefore \quad \lim_{n\to\infty} \frac{u_n}{u_{n+1}} = \lim_{n\to\infty} \frac{3^{-n-1}}{3^{-n}} = \frac{1}{3} < 1$$

Again, when n is odd (so that $n + 1$ is even), $u_n = 3^{-n+1}$, $u_{n+1} = 3^{-(n+1)+1} = 3^{-n-2}$

$$\therefore \quad \lim_{n\to\infty} \frac{u_n}{u_{n+1}} = \lim_{n\to\infty} \frac{3^{-n+1}}{3^{-n-2}} = 3^3 = 27 > 1.$$

Thus $\lim_{n\to\infty} \dfrac{u_n}{u_{n+1}}$ does not tend to a unique limit when n is even or odd. Hence D'Alembert's ratio test fails in this case.

Thus Cauchy's *n*th root test establishes the convergence of the given series while D'Alembert's ratio test fails to do so. Hence Cauchy's root test is superior than ratio test.

EXERCISES

1. Show that the Cauchy's root test establishes the convergence of the series $\Sigma\, u_n$ where

$$u_n = \begin{cases} 2^{-n}, & \text{if } n \text{ is odd} \\ 2^{-n+2}, & \text{if } n \text{ is even} \end{cases}$$

while D'Alembert's ratio test fails to do so.

2. With help of the series $\dfrac{1}{2} + \dfrac{1}{3^2} + \dfrac{1}{2^3} + \dfrac{1}{3^4} + \ldots..$

show that Cauchy's *n*th root test is superior than D'Alembert's ratio test.

ANSWERS

1. Convergent **2.** Convergent

6.13. RAABE'S TEST [Delhi Maths(H) 2009; Delhi B.A. III 2010; Kanpur 2006, 09]

Let Σu_n be a positive term series and let

$$\lim n\left[\frac{u_n}{u_{n+1}}-1\right]=l.$$

Then the series is

(*i*) *convergent if $l > 1$,*

(*ii*) *divergent if $l < 1$,*

(*iii*) *no firm decision is possible if $l = 1$.*

[*Delhi B.Sc.* (*G*), *2001, 02, 05*; *Delhi B.Sc.* (*H*), *1998, 2005*; *Kanpur, 2003; Purvanchal 2006*]

Proof. Case I. $l > 1$. Let α be a number such that $1 < \alpha < l$.

Then there exists, m such that $\forall\ n \geq m$

$$n\left[\frac{u_n}{u_{n+1}}-1\right]>\alpha$$

$$\Rightarrow\ nu_n - nu_{n+1} > \alpha u_{n+1}$$

$$\Rightarrow\ nu_n - (n+1)\,u_{n+1} > (\alpha - 1)\,u_{n+1}$$

Replacing n by $m, m+1, \ldots\ldots, n$ and adding, we get

$$mu_m - (n+1)\,u_{n+1} > (\alpha - 1)\,[u_{m+1} + \ldots\ldots + u_{n+1}]$$

$$\Rightarrow\ (\alpha - 1)\,[u_{m+1} + \ldots\ldots + u_{n+1}] < mu_m$$

$$\Rightarrow\ u_{m+1} + \ldots\ldots + u_{n+1} < mu_m/(\alpha - 1).$$

Thus, $\forall\ n \in \mathbf{N}$, we have

$$S_n \leq [u_1 + \ldots\ldots + u_{m-1}] + mu_m]/(\alpha - 1),$$

so that the sequence $<S_n>$ is bounded and as such the series is convergent.

Case II. Let
$$\lim n\left[\frac{u_n}{u_{n+1}}-1\right]=l<1.$$

Let α be a number such that $l < \alpha < 1$.

Now there exists m such that $\forall\ n \geq m$,

$$n\left[\frac{u_n}{u_{n+1}}-1\right]<\alpha.$$

It follows, that $\forall\ n \geq m$

$$n\left[\frac{u_n}{u_{n+1}}-1\right]<1$$

$$\Rightarrow\ \frac{u_n}{u_{n+1}}<1+\frac{1}{n}=\frac{n+1}{n}$$

$$\Rightarrow\ \frac{u_{n+1}}{u_n}>\frac{n}{n+1}$$

Replacing n by $m, m+1, m+2, \ldots\ldots, n-1$ and multiplying them together, we see that

$$\frac{u_n}{u_m} > \frac{m}{n} \ \forall\ n \geq m$$

$$\Rightarrow\ u_n > k/n \text{ where } k = mu_m.$$

Also the series $\Sigma\ 1/n$ being divergent, the result follows.

It follows by the comparison test of the second type that $\Sigma\ u_n$ is divergent.

Note. It may be seen that Raabe's test is stronger than D'Alembert's test in much as

(*i*) Raabe's test will certainly be conclusively applicable if D'Alembert's is applicable, as shown in the following :

$$\lim \frac{u_n}{u_{n+1}} < 1 \Rightarrow \lim n\left[\frac{u_n}{u_{n+1}} - 1\right] = -\infty < 1$$

$$\lim \frac{u_n}{u_{n+1}} > 1 \Rightarrow \lim n\left[\frac{u_n}{u_{n+1}} - 1\right] = \infty > 1$$

(*ii*) Raabe's test may be conclusive if D'Alembert's test fails as is shown by the following example.

Example. *Examine the following infinite series for convergence* :

$$\sum \frac{1 \cdot 3 \cdot 5 \cdot \ldots\ldots (2n-1)}{2 \cdot 4 \cdot 6 \ldots\ldots 2n} \cdot \frac{x^{2n}}{2n};$$

x being non-negative.

Solution. We have

$$u_n = \frac{1 \cdot 3 \cdot 5 \ldots\ldots (2n-1)}{2 \cdot 4 \cdot 6 \ldots\ldots 2n} \cdot \frac{x^{2n}}{2n} \quad \text{and} \quad u_{n+1} = \frac{1 \cdot 3 \cdot 5 \ldots\ldots (2n+1)}{2 \cdot 4 \cdot 6 \ldots\ldots (2n+2)} \cdot \frac{x^{2n+2}}{2n+2}.$$

$$\therefore \quad \frac{u_n}{u_{n+1}} = \frac{2n+2}{2n+1} \cdot \frac{2n+2}{2n} \cdot \frac{1}{x^2}$$

$$\therefore \quad \lim_{n \to \infty} \frac{u_n}{u_{n+1}} = \frac{(1+1/n)^2}{1+1/2n} \times \frac{1}{x^2} = \frac{1}{x^2}$$

and as such by D'Alembert's ratio test the series converges if

$$1/x^2 > 1 \Leftrightarrow x^2 < 1 \Leftrightarrow x < 1; x \text{ being non-negative}$$

and diverges if

$$1/x^2 < 1 \Leftrightarrow x^2 > 1 \Leftrightarrow x > 1.$$

Now suppose that $x^2 = 1$. In this case D'Alembert's test is *not* helpful. We have, if $x^2 = 1$

$$\frac{u_n}{u_{n+1}} = \frac{(2n+2)^2}{2n\,(2n+1)}$$

$$\Rightarrow \lim n\left[\frac{u_n}{u_{n+1}} - 1\right] = \lim \frac{n\,(6n+4)}{2n\,(2n+1)} = \frac{3}{2} > 1.$$

$\Rightarrow$ the series converges if $x^2 = 1$, *i.e.*, if $x = 1$.

Thus, the given series converges if $x \leq 1$ and diverges if $x > 1$.

6.14. LOGARITHMIC TEST

Let Σu_n be a positive term series and let

$$\lim_{n \to \infty} \left(n \log \frac{u_n}{u_{n+1}} \right) = l.$$

Then the series is

(*i*) *convergent if $l > 1$*

(*ii*) *divergent if $l < 1$*

(*iii*) *no firm decision is possible if $l = 1$.* ***(Agra, 1994; Gorakhpur, 1994; Kanpur, 1993)***

Proof. Let $l > 1$ and let us choose $\varepsilon > 0$ such that $l - \varepsilon > 1$. Let $l - \varepsilon = \lambda$ so that $\lambda > 1$.

Now, $$\lim_{n \to \infty} \left(n \log \frac{u_n}{u_{n+1}} \right) = l$$

$\Rightarrow$ there exists a positive integer m such that

$$l - \varepsilon < n \log \frac{u_n}{u_{n+1}} < l + \varepsilon, \ \forall \ n \geq m$$

$$\Rightarrow \qquad n \log \frac{u_n}{u_{n+1}} > \lambda, \ \forall \ n \geq m$$

$$\Rightarrow \qquad u_n / u_{n+1} > e^{\lambda/n}, \ \forall \ n \geq m \qquad \ldots(1)$$

We know that the sequence $<(1 + 1/n)^n>$ converges to e and hence

$$e \geq (1 + 1/n)^n, \ \forall \ n \in \mathbf{N}$$

$$\Rightarrow \qquad e^{\lambda/n} \geq (1 + 1/n)^{\lambda} \qquad \ldots(2)$$

From (1) and (2), we have

$$\frac{u_n}{u_{n+1}} > \left(1 + \frac{1}{n}\right)^{\lambda} = \frac{(n+1)^{\lambda}}{n^{\lambda}} = \frac{v_n}{v_{n+1}}, \ \forall \ n \geq m \qquad \ldots(3)$$

where $$v_n = 1/n^{\lambda}.$$

Since $\lambda > 1$, so Σv_n converges. Then, using comparison test of second type (refer Art. 6.9), it follows that the given series Σu_n also converges.

(*ii*) Prove as in part (*i*).

Note 1. The above Logarithmic test is alternative to Raabe's test and should be used when D'Alembert's ratio test fails and when either e occurs in u_n/u_{n+1} or n occurs as an exponent in u_n/u_{n+1}.

Note 2. When Raabe's test fails we may use De Morgan's and Bertrand's test (refer Art. 6.15) given below. Again, when logarithmic test fails we may use second logarithmic ratio test (refer Art. 6.16). In what follows, we now state and prove these two tests.

6.15. DE MORGAN'S AND BERTRAND'S TEST

Let Σu_n be a series of positive terms such that

$$\lim \left[\left\{ n \left(\frac{u_n}{u_{n+1}} - 1 \right) - 1 \right\} \log \ n \right] = l$$

Then the series is

(i) convergent if $l > 1$

(ii) divergent if $l < 1$

(iii) no firm decision is possible if $l = 1$ **(*Patna, 2003*)**

Proof. In what follows, we shall compare Σu_n with the auxiliary series

$$\sum_{n=2}^{\infty} v_n = \sum_{n=2}^{\infty} \frac{1}{n(\log n)^p},$$

which is known (refer Art. 6.21) to be convergent if $p > 1$ and divergent if $p \le 1$. Now, we have

$$\frac{v_n}{v_{n+1}} = \frac{(n+1)\{\log(n+1)\}^p}{n(\log n)^p} = \frac{n+1}{n}\left[\frac{\log(n+1)}{\log n}\right]^p$$

$$= \left(1+\frac{1}{n}\right)\left[\frac{\log\{n(1+1/n)\}}{\log n}\right]^p = \left(1+\frac{1}{n}\right)\left\{\frac{\log n + \log(1+1/n)}{\log n}\right\}^p$$

$$= \left(1+\frac{1}{n}\right)\left\{1+\frac{1}{\log n}\times\log\left(1+\frac{1}{n}\right)\right\}^p$$

$$= \left(1+\frac{1}{n}\right)\left\{1+\frac{1}{\log n}\left(\frac{1}{n}-\frac{1}{2n^2}+\frac{1}{3n^3}-.....\right)\right\}^p$$

$$= \left(1+\frac{1}{n}\right)\left\{1+\frac{p}{\log n}\left(\frac{1}{n}-\frac{1}{2n^2}+\frac{1}{3n^3}-.....\right)\right\}, \text{ by binomial theorem}$$

$$= 1+\frac{1}{n}+\frac{p}{\log n}\left(\frac{1}{n}-\frac{1}{2n^2}+\frac{1}{3n^3}-.....\right)+\frac{p}{n\log n}\left(\frac{1}{n}-\frac{1}{2n^2}+\frac{1}{3n^3}-.....\right)$$

Thus,
$$\frac{v_n}{v_{n+1}} = 1+\frac{1}{n}+\frac{p}{n\log n}\left(1+\frac{1}{2n}-\frac{1}{6n^2}+.....\right)+..... \quad ...(1)$$

Case (*i*). Let $l > 1$. Choose a number p such that $l \ge p > 1$.

By comparison test of the second type (refer Art. 6.9), the series Σu_n will be convergent if there exists a positive integer m such that $\forall\ n \ge m$

$$\frac{u_n}{u_{n+1}} \ge \frac{v_n}{v_{n+1}}$$

i.e., if
$$\frac{u_n}{u_{n+1}} \ge 1+\frac{1}{n}+\frac{p}{n\log n}\left(1+\frac{1}{2n}-\frac{1}{6n^2}+.....\right)+....., \text{ by (1)}$$

i.e., if
$$\frac{u_n}{u_{n+1}}-1 \ge \frac{1}{n}+\frac{p}{n\log n}\left(1+\frac{1}{2n}-\frac{1}{6n^2}+.....\right)+.....$$

i.e., if
$$n\left(\frac{u_n}{u_{n+1}}-1\right) \ge 1+\frac{p}{n\log n}\left(1+\frac{1}{2n}-\frac{1}{6n^2}+.....\right)+.....$$

i.e., if
$$n\left(\frac{u_n}{u_{n+1}}-1\right)-1 \ge \frac{p}{\log n}\left(1+\frac{1}{2n}-\frac{1}{6n^2}+.....\right)+.....$$

i.e., if $\quad \left\{n\left(\dfrac{u_n}{u_{n+1}}-1\right)-1\right\}\log n \geq p +$ terms containing n or $\log n$ in the denominator ...(2)

Taking limits on both sides of (2) as $n \to \infty$ we get

$$\lim_{n\to\infty}\left[\left\{n\left(\frac{u_n}{u_{n+1}}-1\right)-1\right\}\log n\right] > p$$

$\Rightarrow \qquad l > p \;\Rightarrow\; l > 1$ as $p > 1$.

Thus, $\Sigma\, u_n$ converges if $l > 1$.

Case (*ii*). Proceed as in case (*i*) yourself.

6.16. SECOND LOGARITHMIC RATIO TESTS

Let $\Sigma\, u_n$ be a series of positive terms such that

$$\lim_{n\to\infty}\left\{\left(n\log\frac{u_n}{u_{n+1}}-1\right)\log n\right\} = l.$$

Then the series is

(*i*) *convergent if $l > 1$*

(*ii*) *divergent if $l < 1$*

(*iii*) *no firm decision is possible if $l = 1$.*

Proof. In what follows, we shall compare $\Sigma\, u_n$ with the auxiliary series

$$\sum_{n=2}^{\infty} v_n = \sum_{n=2}^{\infty}\frac{1}{n(\log n)^p},$$

which is known (refer Art. 6.21) to be convergent if $p > 1$ and divergent if $p \leq 1$. Now, we have

$$\frac{v_n}{v_{n+1}} = \frac{(n+1)\{\log(n+1)\}^p}{n(\log n)^p} = \frac{n+1}{n}\left[\frac{\log(n+1)}{\log n}\right]^p$$

$$= \left(1+\frac{1}{n}\right)\left[\frac{\log\{n(1+1/n)\}}{\log n}\right]^p = \left(1+\frac{1}{n}\right)\left\{\frac{\log n + \log(1+1/n)}{\log n}\right\}^p$$

$$= \left(1+\frac{1}{n}\right)\left\{1+\frac{1}{\log n}\log\left(1+\frac{1}{n}\right)\right\}^p = \left(1+\frac{1}{n}\right)\left\{1+\frac{1}{\log n}\left(\frac{1}{n}-\frac{1}{2n^2}+\frac{1}{3n^3}-\ldots\right)\right\}^p$$

$$= \left(1+\frac{1}{n}\right)\left\{1+\frac{p}{\log n}\left(\frac{1}{n}-\frac{1}{2n^2}+\frac{1}{3n^3}-\ldots\right)\right\}, \text{ by binomial theorem}$$

Thus, $$\frac{v_n}{v_{n+1}} = 1+\frac{1}{n}+\frac{p}{n\log n}+\ldots \qquad ...(1)$$

Case (*i*). Let $l > 1$. Choose a number p such that $l \geq p > 1$.

By comparison test of the second type, the series $\Sigma\, u_n$ will be convergent if there exists a positive integer m such that $\forall\; n \geq m$

$$\frac{u_n}{u_{n+1}} \geq \frac{v_n}{v_{n+1}}$$

i.e., if $$\frac{u_n}{u_{n+1}} \geq 1+\frac{1}{n}+\frac{p}{n\log n}+\ldots\ldots, \text{ using (1)}$$

i.e., if $$\log\frac{u_n}{u_{n+1}} \geq \log\left\{1+\left(\frac{1}{n}+\frac{p}{n\log n}+\ldots\ldots\right)\right\}$$

i.e., if $$\log\frac{u_n}{u_{n+1}} \geq \left(\frac{1}{n}+\frac{p}{n\log n}+\ldots\ldots\right)-\frac{1}{2}\left(\frac{1}{n}+\frac{p}{n\log n}+\ldots\ldots\right)^2+\ldots\ldots$$

$[\because \log(1+x) = x - x^2/2 + x^3/3 + \ldots\ldots\ldots\ldots]$

i.e., if $$n\log\frac{u_n}{u_{n+1}} \geq n\left(\frac{1}{n}+\frac{p}{n\log n}-\frac{1}{2n^2}+\ldots\ldots\right)$$

i.e., if $$n\log\frac{u_n}{u_{n+1}}-1 \geq \frac{p}{n\log n}-\frac{1}{2n}+\ldots\ldots$$

i.e., if $$\left(n\log\frac{u_n}{u_{n+1}}-1\right)\log n \geq p-\frac{1}{2}\left(\frac{\log n}{n}\right)+\ldots\ldots$$

$$\therefore \qquad \lim_{n\to\infty}\left\{\left(n\log\frac{u_n}{u_{n+1}}-1\right)\log n\right\} \geq p, \text{ as } \lim_{n\to\infty}\frac{\log n}{n} = 0$$

or $$l \geq p. \text{ But } p > 1, \text{ so } l > 1.$$

Hence the given series Σu_n converges if $l > 1$.

(*ii*) Proceed as in case (*i*) yourself.

Solved examples based on Raabe's test, logarithmic test, De-Morgan's and Bertrand's and Second logarithmic ratio test (refer results of Articles 6.13, 6.14, 6.15 and 6.16).

Example 1. *Examine the convergence of the following series :*

(*i*) $\displaystyle\sum_{n=1}^{\infty}\frac{2\cdot 4\cdot 6\,\ldots\ldots\,2n}{1\cdot 3\cdot 5\,\ldots\ldots\,(2n+1)}$ **(*Delhi Maths (G), 1999*)**

(*ii*) $\displaystyle\sum_{n=1}^{\infty}\frac{1\cdot 3\cdot 5\,\ldots\ldots\,(4n-5)\,(4n-3)}{2\cdot 4\cdot 6\,\ldots\ldots\,(4n-4)\,(4n-2)}\,\frac{x^{2n}}{4n},\ x>0$

[*Delhi Maths (H), 2000; Delhi Maths, 2002, 04*]

(*iii*) $1+\dfrac{x}{1}+\dfrac{1}{2}\dfrac{x^3}{3}+\dfrac{1\cdot 3}{2\cdot 4}\dfrac{x^5}{5}+\dfrac{1\cdot 3\cdot 5}{2\cdot 4\cdot 6}\dfrac{x^7}{7}+\ldots$ **(*Agra 2008; Meerut, 2001, 03*)**

(*iv*) $x+\dfrac{1}{2}\cdot\dfrac{x^2}{4}+\dfrac{1\cdot 3\cdot 5}{2\cdot 4\cdot 6}\cdot\dfrac{x^4}{8}+\dfrac{1\cdot 3\cdot 5\cdot 7\cdot 9}{2\cdot 4\cdot 6\cdot 8\cdot 10}\cdot\dfrac{x^6}{12}+\ldots\ldots$ **(*Patna, 2003*)**

Solution. (*i*) Here, we have

$$u_n = \frac{2\cdot 4\cdot 6\,\ldots\ldots\,2n}{1\cdot 3\cdot 5\,\ldots\ldots\,(2n+1)},\quad u_{n+1} = \frac{2\cdot 4\cdot 6\,\ldots\ldots\,2n\,(2n+2)}{1\cdot 3\cdot 5\,\ldots\ldots\,(2n+1)\,(2n+3)}.\ \text{So}\ \frac{u_n}{u_{n+1}} = \frac{2n+3}{2n+2}$$

Here, $$\lim_{n\to\infty}\frac{u_n}{u_{n+1}} = \lim_{n\to\infty}\frac{2+(3/n)}{2+(2/n)} = 1.$$

Hence ratio test fails and we now apply Raabe's test.

$$\lim_{n\to\infty} n\left(\frac{u_n}{u_{n+1}}-1\right) = \lim_{n\to\infty} n\left(\frac{2n+3}{2n+2}-1\right) = \lim_{n\to\infty} \frac{n}{2n+2} = \frac{1}{2} < 1$$

Hence, by Raabe's test, the given series diverges.

(*ii*) $u_n = \frac{1\cdot 3\cdot \ldots (4n-3)}{2\cdot 4\cdot \ldots (4n-2)}\cdot\frac{x^{2n}}{4n},\ u_{n+1} = \frac{1\cdot 3\cdot \ldots (4n-3)(4n+1)}{2\cdot 4\cdot \ldots (4n-2)(4n+2)}\cdot\frac{x^{2n+2}}{4n+4}$

$$\therefore \quad \frac{u_n}{u_{n+1}} = \frac{(4n+2)(4n+4)}{4n(4n+1)}\cdot\frac{1}{x^2} \qquad \ldots(1)$$

and
$$\lim_{n\to\infty}\frac{u_n}{u_{n+1}} = \lim_{n\to\infty}\frac{(4+2/n)(4+4/n)}{4(4+1/n)}\frac{1}{x^2} = \frac{1}{x^2}$$

Hence by ratio test, Σu_n converges if $1/x^2 > 1$, *i.e.*, $x^2 < 1$, *i.e.*, $x < 1$ (as $x > 0$) and diverges if $1/x^2 < 1$, *i.e.*, $x^2 > 1$, *i.e.*, $x > 1$. The test fails if $x = 1$. In that case, from (1), we have

$$\frac{u_n}{u_{n+1}} = \frac{(4n+2)(4n+4)}{4n(4n+1)}$$

$$\therefore \quad n\left(\frac{u_n}{u_{n+1}}-1\right) = n\left\{\frac{(4n+2)(4n+4)}{4n(4n+1)}-1\right\} = \frac{20n+8}{4(4n+1)}$$

and hence
$$\lim_{n\to\infty} n\left(\frac{u_n}{u_{n+1}}-1\right) = \lim_{n\to\infty}\frac{20+8/n}{4(4+1/n)} = \frac{20}{16} > 1.$$

Hence, by Raabe's test, Σu_n is convergent.

Thus the given series converges if $x \le 1$ and diverges if $x > 1$.

(*iii*) Omitting the first term of the given series as it will not change the nature of the series, we obtain

$$u_n = \frac{1\cdot 3\cdot 5\ldots (2n-1)}{2\cdot 4\cdot 6\ldots (2n)}\cdot\frac{x^{2n+1}}{2n+1} \quad\text{and so}\quad u_{n+1} = \frac{1\cdot 3\cdot 5\ldots (2n-1)(2n+1)}{2\cdot 4\cdot 6\ldots (2n)(2n+2)}\cdot\frac{x^{2n+3}}{2n+3}$$

$$\therefore \quad \frac{u_n}{u_{n+1}} = \frac{(2n+2)(2n+3)}{(2n+1)^2}\frac{1}{x^2} = \frac{4n^2+10n+6}{4n^2+4n+1}\cdot\frac{1}{x^2} \qquad \ldots(1)$$

Now,
$$\lim_{n\to\infty}\frac{u_n}{u_{n+1}} = \lim_{n\to\infty}\frac{4+(10/n)+(6/n^2)}{4+(4/n)+(1/n^2)}\cdot\frac{1}{x^2} = \frac{1}{x^2}$$

$\therefore$ By ratio test Σu_n converges if $1/x^2 > 1$, *i.e.*, if $x^2 < 1$ and diverges if $1/x^2 < 1$, *i.e.*, if $x^2 > 1$. When $1/x^2 = 1$, *i.e.*, $x^2 = 1$, the ratio test fails. We shall now apply Raabe's test.

From (1), for $x^2 = 1$,
$$\frac{u_n}{u_{n+1}} = \frac{4n^2+10n+6}{4n^2+4n+1}$$

$$\therefore \quad \lim_{n\to\infty} n\left(\frac{u_n}{u_{n+1}}-1\right) = \lim_{n\to\infty} n\left(\frac{4n^2+10n+6}{4n^2+4n+1}-1\right) = \lim_{n\to\infty}\frac{6n^2+5n}{4n^2+4n+1} = \frac{3}{2} > 1$$

Hence, by Raabe's test, Σu_n converges for $x^2 = 1$.

Thus the given series converges if $x^2 \le 1$ and diverges if $x^2 > 1$.

(*iv*) Omitting the first term, the *n*th term of the resulting series is

$$u_n = \frac{1 \cdot 3 \cdot 5 \cdot 7 \cdot 9 \ldots\ldots (4n-3)}{2 \cdot 4 \cdot 6 \cdot 8 \cdot 10 \ldots\ldots (4n-2)} \cdot \frac{x^{2n}}{4n}$$

[Here the *n*th term of the A.P., 1, 5, 9, ... is $1 + (n-1) \times 4$, *i.e.*, $4n-3$ and the *n*th term of the A.P., 2, 6, 10, ... is $2 + (n-1) \times 4$, *i.e.*, $4n-2$]

and so $$u_{n+1} = \frac{1 \cdot 3 \cdot 5 \ldots\ldots (4n-3)(4n-1)(4n+1)}{2 \cdot 4 \cdot 6 \ldots\ldots (4n-2)(4n)(4n+2)} \cdot \frac{x^{2n+2}}{4n+4}$$

$$\therefore \quad \frac{u_n}{u_{n+1}} = \frac{4n(4n+2)}{(4n-1)(4n+1)} \times \frac{4n+4}{4n} \times \frac{1}{x^2} = \frac{16n^2+24n+8}{16n^2-1} \cdot \frac{1}{x^2} \qquad \ldots(1)$$

So $$\lim_{n\to\infty} \frac{u_n}{u_{n+1}} = \lim_{n\to\infty} \frac{16+(24/n)+(8/n^2)}{16-(1/n^2)} \cdot \frac{1}{x^2} = \frac{1}{x^2}$$

$\therefore$ By ratio test Σu_n converges if $1/x^2 > 1$, *i.e.*, if $x^2 < 1$ and diverges if $1/x^2 < 1$, *i.e.*, if $x^2 > 1$. When $1/x^2 = 1$, *i.e.*, $x^2 = 1$, the ratio test fails. We shall now apply Raabe's test.

From (1), for $x^2 = 1$, $$\frac{u_n}{u_{n+1}} = \frac{16n^2+24n+8}{16n^2-1}$$

$$\therefore \quad \lim_{n\to\infty} n\left(\frac{u_n}{u_{n+1}} - 1\right) = \lim_{n\to\infty} n\left(\frac{16n^2+24n+8}{16n^2-1} - 1\right) = \lim_{n\to\infty} \frac{24n^2+9n}{16n^2-1} = \frac{3}{2} > 1$$

Hence, by Raabe's test, Σu_n converges.

Thus the given series converges if $x^2 \le 1$ and diverges if $x^2 > 1$.

Example 2. *Test the convergence of the following series*

(*i*) $1 + \dfrac{x}{1!} + \dfrac{2^2 x^2}{2!} + \dfrac{3^3 x^3}{3!} + \ldots\ldots$, *for* $x > 0$ **(Delhi Maths (H) 2007; Kanpur, 2003, 06)**

(*ii*) $\dfrac{(a+x)}{1!} + \dfrac{(a+2x)^2}{2!} + \dfrac{(a+3x)^3}{3!} + \ldots\ldots$

Solution. (*i*) Omitting the first term, we have

$$u_n = \frac{n^n x^n}{n!} \quad \text{and} \quad u_{n+1} = \frac{(n+1)^{n+1} x^{n+1}}{(n+1)!}$$

Now, $$\frac{u_n}{u_{n+1}} = \frac{(n+1)!}{n!} \times \frac{n^n}{(n+1)^{n+1}} \times \frac{1}{x} = \left(\frac{n}{n+1}\right)^n \times \frac{1}{x} \qquad \ldots(1)$$

$$\therefore \quad \lim_{n\to\infty} \frac{u_n}{u_{n+1}} = \lim_{n\to\infty} \frac{1}{(1+1/n)^n} \cdot \frac{1}{x} = \frac{1}{ex}, \text{ as } \lim_{n\to\infty} \left(1+\frac{1}{n}\right)^n = e$$

By ratio test, Σu_n converges if $1/ex > 1$, *i.e.*, if $x < 1/e$ and diverges if $1/ex < 1$, *i.e.*, if $x > 1/e$. When $1/ex = 1$, *i.e.*, $x = 1/e$, the test fails. Since u_n/u_{n+1} will involve e for $x = 1/e$, we shall apply logarithmic test.

For $x = \dfrac{1}{e}$, from (1), $$\frac{u_n}{u_{n+1}} = \left(\frac{n}{n+1}\right)^n \times e$$

$$\therefore \qquad \lim_{n\to\infty} n\log\frac{u_n}{u_{n+1}} = \lim_{n\to\infty} n\log\left\{\left(\frac{n}{n+1}\right)^n e\right\} = \lim_{n\to\infty} n\log\left\{e\times\left(\frac{n+1}{n}\right)^{-n}\right\}$$

$$= \lim_{n\to\infty} n\{\log e - n\log(1+1/n)\}$$

$$= \lim_{n\to\infty} n\left\{1-n\left(\frac{1}{n}-\frac{1}{2n^2}+\frac{1}{3n^3}+\ldots..\right)\right\} = \frac{1}{2} < 1$$

Hence, by logarithmic test, Σu_n diverges.

Thus the given series converges if $x < 1/e$ and diverges if $x \geq 1/e$.

(*ii*) Here $u_n = \dfrac{(a+nx)^n}{n!}$ and $u_{n+1} = \dfrac{\{a+(n+1)x\}^{n+1}}{(n+1)!}$

Now,
$$\frac{u_n}{u_{n+1}} = \frac{(n+1)!}{n!}\times\frac{(a+nx)^n}{\{a+(n+1)x\}^{n+1}} = \frac{(a+nx)^n(n+1)}{\{a+(n+1)x\}^{n+1}}$$

$$= \frac{n^n x^n (a/nx+1)^n\, n(1+1/n)}{(n+1)^{n+1}\, x^{n+1}\{a/(n+1)x+1\}^{n+1}}$$

$$= \frac{1}{x}\cdot\frac{n^{n+1}(1+a/nx)^n(1+1/n)}{n^{n+1}(1+1/n)^{n+1}\{1+a/(x+1)x\}^{n+1}}$$

$$= \frac{1}{x}\cdot\frac{\left\{1+\frac{(a/x)}{n}\right\}^n}{\left(1+\frac{1}{n}\right)^n\left\{1+\frac{(a/x)}{n+1}\right\}^{n+1}} \qquad \ldots(1)$$

Hence
$$\lim_{n\to\infty}\frac{u_n}{u_{n+1}} = \frac{1}{x}\cdot\frac{e^{a/x}}{e\times e^{a/x}} = \frac{1}{ex}, \text{ as } \lim_{n\to\infty}\left(1+\frac{p}{n}\right)^n = e^p.$$

By ratio test, Σu_n converges if $1/ex > 1$, *i.e.*, if $x < 1/e$ and diverges if $1/ex < 1$, if $x > 1/e$. When $1/ex = 1$, *i.e.*, $x = 1/e$, the test fails. We now apply logarithmic test.

For $x = \dfrac{1}{e}$, from (1),
$$\frac{u_n}{u_{n+1}} = \frac{e(1+ea/n)^n}{(1+1/n)^n[1+ae/(n+1)]^{n+1}}$$

$$\log\frac{u_n}{u_{n+1}} = \log e + n\log\left(1+\frac{ea}{n}\right) - n\log\left(1+\frac{1}{n}\right) - (n+1)\log\left(1+\frac{ae}{n+1}\right)$$

$$= 1 + n\left(\frac{ea}{n}-\frac{e^2a^2}{n^2}+\frac{e^3a^3}{n^3}\ldots..\right) - n\left(\frac{1}{n}-\frac{1}{2n^2}+\frac{1}{3n^3}\ldots..\right)$$

$$-(n+1)\left[\frac{ea}{n+1}-\frac{e^2a^2}{2(n+1)^2}+\frac{e^3a^3}{3(n+1)^3}\ldots..\right]$$

$$= \frac{1}{n}\left(\frac{1}{2}-\frac{e^2a^2}{2}\right)+\frac{e^2a^2}{2(n+1)}+\frac{1}{n^2}\left(\frac{e^3a^3}{3}-\frac{1}{3}\right)+\ldots..$$

$$\Rightarrow \quad n \log \frac{u_n}{u_{n+1}} = \frac{1}{2} - \frac{e^2 a^2}{2} + \frac{n e^2 a^2}{2(n+1)} + \frac{e^3 a^3 - 1}{3n} + \ldots..$$

$$\Rightarrow \quad \lim_{n \to \infty} n \log \frac{u_n}{u_{n+1}} = \frac{1}{2} - \frac{e^2 a^2}{2} + \frac{e^2 a^2}{2} = \frac{1}{2} < 1, \quad \text{as} \quad \lim_{n \to \infty} \frac{n}{1+n} = \lim_{n \to \infty} \frac{1}{1+(1/n)} =$$

Hence, by logarithmic test, Σu_n diverges.

Thus the given series converges if $x < 1/e$ and diverges if $x \geq 1/e$.

Example 3. *Test for converges the following series :*

(i) $\dfrac{1^2}{2^2} + \dfrac{1^2 \cdot 3^2}{2^3 \cdot 4^2} + \dfrac{1^2 \cdot 3^2 \cdot 5^2}{2^2 \cdot 4^2 \cdot 6^2} + \ldots..$ ***(Ranchi 2010; Meerut, 2002)***

(ii) $\dfrac{a}{b} + \dfrac{a(a+1)}{b(b+1)} + \dfrac{a(a+1)(a+2)}{b(b+1)(b+2)} + \ldots..$ **[Kanpur 2011**

Solution (*i*) Here, we have

$$u_n = \frac{1^2 \cdot 3^2 \cdot 5^2 \ldots..(2n-1)^2}{2^2 \cdot 4^2 \cdot 6^2 \ldots..(2n)^2}, \quad u_{n+1} = \frac{1^2 \cdot 3^2 \cdot 5^2 \ldots..(2n-1)^2 (2n+1)^2}{2^2 \cdot 4^2 \cdot 6^2 \ldots..(2n)^2 (2n+2)^2}$$

$$\therefore \quad \frac{u_n}{u_{n+1}} = \frac{(2n+2)^2}{(2n+1)^2} = \frac{4n^2 + 8n + 4}{4n^2 + 4n + 1} \quad \ldots(1)$$

Here $$\lim_{n\to\infty} \frac{u_n}{u_{n+1}} = \lim_{n\to\infty} \frac{4 + (8/n) + (4/n^2)}{4 + (4/n) + (1/n^2)} = 1$$

Hence the ratio test fails and we now apply Raabe's test.

Using (1), $$n\left(\frac{u_n}{u_{n+1}} - 1\right) = n\left(\frac{4n^2 + 8n + 4}{4n^2 + 4n + 1} - 1\right) = \frac{4n^2 + 3n}{4n^2 + 4n + 1} \quad \ldots(2)$$

$$\therefore \quad \lim_{n \to \infty} n\left(\frac{u_n}{u_{n+1}} - 1\right) = \lim_{n \to \infty} \frac{4 + (3/n)}{4 + (4/n) + (1/n^2)} = 1$$

Hence the Raabe's test fails and we now apply De-Morgan's and Bertrand's test.

From (2), $$n\left(\frac{u_n}{u_{n+1}} - 1\right) - 1 = \frac{4n^2 + 3n}{4n^2 + 4n + 1} - 1 = \frac{-n-1}{4n^2 + 4n + 1}$$

$$\therefore \quad \left\{n\left(\frac{u_n}{u_{n+1}} - 1\right) - 1\right\} \log n = -\frac{(n+1)\log n}{4n^2 + 4n + 1} = -\frac{1 + (1/n)}{4 + (4/n) + (1/n^2)} \cdot \frac{\log n}{n}$$

and so $$\lim_{n \to \infty} \left\{n\left(\frac{u_n}{u_{n+1}} - 1\right) - 1\right\} \log n = -\frac{1}{4} \times 0 = 0 < 1, \text{ as } \lim_{n \to \infty} \frac{\log n}{n} = 0$$

Hence, by De-Morgan's and Bertrand's test, Σu_n diverges.

(*ii*) Here $$u_n = \frac{a(a+1)(a+2)\ldots..(a+n-1)}{b(b+1)(b+2)\ldots..(b+n-1)},$$

and $$u_{n+1} = \frac{a(a+1)(a+2)\ldots..(a+n-1)(a+n)}{b(b+1)(b+2)\ldots..(b+n-1)(b+n)}$$

$$\therefore \qquad \frac{u_n}{u_{n+1}} = \frac{b+n}{a+n} \quad \text{and so} \quad \lim_{n\to\infty} \frac{b+n}{a+n} = \lim_{n\to\infty} \frac{1+b/n}{1+a/n} = 1 \qquad ...(1)$$

Hence the ratio test fails and we now apply Raabe's test.

Using (1), $$n\left(\frac{u_n}{u_{n+1}} - 1\right) = n\left(\frac{b+n}{a+n} - 1\right) = \frac{n\,(b-a)}{a+n} \qquad ...(2)$$

$$\therefore \qquad \lim_{n\to\infty} n\left(\frac{u_n}{u_{n+1}} - 1\right) = \lim_{n\to\infty} \frac{b-a}{(a/n)+1} = b-a$$

Hence, by Raabe's test, $\Sigma\, u_n$ converges if $b - a > 1$ and diverges if $b - a < 1$. When $b - a = 1$, Raabe's test fails and we now apply De-Morgan's and Bertrand's test.

For $b - a = 1$, from (2), $$n\left(\frac{u_n}{u_{n+1}} - 1\right) = \frac{n}{n+a}$$

$$\therefore \qquad n\left(\frac{u_n}{u_{n+1}} - 1\right) - 1 = \frac{n}{n+a} - 1 = -\frac{a}{n+1}$$

$$\therefore \qquad \left\{n\left(\frac{u_n}{u_{n+1}} - 1\right) - 1\right\} \log n = \frac{a \log n}{n+1} = \frac{-a}{1+(1/n)} \times \frac{\log n}{n}$$

So $$\lim_{n\to\infty} \left\{n\left(\frac{u_n}{u_{n+1}} - 1\right) - 1\right\} \log n = -a \times 0 = 0 < 1, \text{ as } \lim_{n\to\infty} \frac{\log n}{n} = 0$$

Hence, by De-Morgan's and Bertrand's test, $\Sigma\, u_n$ diverges. Thus the given series converges if $b - a > 1$ and diverges if $b - a \le 1$.

Example 4. *Test for convergence the following series*

(i) $1^p + \left(\frac{1}{2}\right)^p + \left(\frac{1\cdot 3}{2\cdot 4}\right)^p + \left(\frac{1\cdot 3\cdot 5}{2\cdot 4\cdot 6}\right)^p + ...$ ***(Agra 2009)***

(ii) $x + x^{1+1/2} + x^{1+1/2+1/3} + x^{1+1/2+1/3+1/4} + ...$ ***(Agra 2010; Meerut, 2001, 02)***

Solution. (i) Omitting the first term, we have

$$u_n = \left[\frac{1\cdot 3\cdot 5 \,.....\, (2n-1)}{2\cdot 4\cdot 6 \,.....\, (2n)}\right]^p, \qquad u_{n+1} = \left[\frac{1\cdot 3\cdot 5 \,.....\, (2n-1)\,(2n+1)}{2\cdot 4\cdot 6 \,.....\, (2n)\,(2n+2)}\right]^p$$

$$\therefore \qquad \frac{u_n}{u_{n+1}} = \left(\frac{2n+2}{2n+1}\right)^p = \left(\frac{1+1/n}{1+1/2n}\right)^p \qquad ...(1)$$

Here $$\lim_{n\to\infty} \frac{u_n}{u_{n+1}} = 1 \text{ and so ratio test fails.}$$

We shall now apply logarithmic test. Using (1), we get

$$\log \frac{u_n}{u_{n+1}} = p \log\left(\frac{1+1/n}{1+1/2n}\right) = p\, \{\log(1+1/n) - \log(1+1/2n)\}$$

$$= p\left\{\left(\frac{1}{n}-\frac{1}{2n^2}+\frac{1}{3n^3}+.....\right)-\left(\frac{1}{2n}-\frac{1}{2\cdot 2^2 n^2}+\frac{1}{3\cdot 2^3 n^3}+.....\right)\right\}$$

$$= p\,(1/2n - 3/8n^2 + 7/24n^3 + ...)$$

$$\therefore \qquad n \log \frac{u_n}{u_{n+1}} = p\left(\frac{1}{2}-\frac{3}{8n}+\frac{7}{24n^2}+.....\right) \qquad ...(2)$$

So $$\lim_{n\to\infty} n \log \frac{u_n}{u_{n+1}} = \frac{p}{2}.$$

Hence $\Sigma\, u_n$ converges if $p/2 > 1$, *i.e.*, if $p > 2$ and diverges if $p/2 < 1$, *i.e.*, if $p < 2$ and the test fails if $p = 2$.

For $p = 2$, from (2), $$n \log \frac{u_n}{u_{n+1}} = 2\left(\frac{1}{2}-\frac{3}{8n}+\frac{7}{24n^2}+...\right) = 1-\frac{3}{4n}+\frac{7}{12n^2}+...$$

$$\therefore \qquad n \log \frac{u_n}{u_{n+1}} - 1 = -\frac{3}{4n}+\frac{7}{12n^2}+..... = \left(-\frac{3}{4}+\frac{7}{12n}+.....\right)\times\frac{1}{n}$$

$$\therefore \quad \lim_{n\to\infty}\left(n \log \frac{u_n}{u_{n+1}} - 1\right)\log n = \lim_{n\to\infty}\left(-\frac{3}{4}+\frac{7}{12n}+.....\right)\cdot\frac{\log n}{n} = -\frac{3}{4}\times 0 = 0 < 1$$

Hence by second logarithmic ratio test (see Art. 6.16), $\Sigma\, u_n$ diverges for $p = 2$.

Thus the given series converges if $p > 1$ and diverges if $p \le 2$.

(*ii*) Here, $\quad u_n = x^{1 + 1/2 + ... + 1/n}, \qquad u_{n+1} = x^{1 + 1/2 + ... + 1/n + 1/(n+1)}$

$$\therefore \qquad \frac{u_n}{u_{n+1}} = \frac{1}{x^{1/(n+1)}} \quad \text{and so} \quad \lim_{n\to\infty}\frac{u_n}{u_{n+1}} = \frac{1}{x^0} = 1 \qquad ...(1)$$

Hence ratio test fails and we now apply logarithmic test.

From (1), $$n \log \frac{u_n}{u_{n+1}} = n \log\left(\frac{1}{x}\right)^{1/(n+1)} = \frac{n}{n+1}\log\frac{1}{x} = \frac{1}{1+(1/n)}\log\frac{1}{x} \qquad ...(2)$$

$$\therefore \qquad \lim_{n\to\infty} n \log \frac{u_n}{u_{n+1}} = \log\frac{1}{x}$$

Hence $\Sigma\, u_n$ converges if $\log(1/x) > 1$, *i.e.*, if $1/x > e$, *i.e.*, if $x < 1/e$ and $\Sigma\, u_n$ diverges if $\log(1/x) < 1$, *i.e.*, if $x > 1/e$. When $x = 1/e$, the test fails and we shall now apply second logarithmic test (see Art. 6.16).

For $x = \dfrac{1}{e}$, from (2), $$n \log \frac{u_n}{u_{n+1}} = \frac{1}{1+(1/n)}, \text{ as } \log e = 1$$

$$\lim_{n\to\infty}\left(n \log \frac{u_n}{u_{n+1}} - 1\right)\log n = \lim_{n\to\infty}\left(\frac{n}{n+1}-1\right)\log n = \lim_{n\to\infty}\frac{(-1)}{n+1}\log n$$

$$= \lim_{n\to\infty}\left\{\frac{-1}{1+(1/n)}\cdot\frac{\log n}{n}\right\} = -1\times 0 = 0 < 1$$

Hence $\Sigma\, u_n$ is divergent for $x = 1/e$.

Thus, the given series converges if $x < 1/e$ and diverges if $x \ge 1/e$.

EXERCISES

1. *Test for the convergence of the following series :*

(i) $\sum \frac{1\cdot3\cdot5\,.....\,(2n-1)}{2\cdot4\cdot6\,.....\,(2n)}\cdot\frac{1}{n}$ **[Delhi B.A. (Prog II) 2008, 09, 15]**

(ii) $\sum_{n=1}^{\infty} \frac{3\cdot6\cdot9\,.....\,(3n)}{7\cdot10\cdot13\,.....\,(3n+4)}\,x^n,\ x>0$ **[Delhi B.Sc. (Prog) III 2010]**

(iii) $\sum_{n=1}^{\infty} \frac{(2n)!}{(n!)^2}\,x^n,\ x>0$ **[Delhi Maths (G), 1994]**

(iv) $\sum_{n=1}^{\infty} \frac{1\cdot3\cdot5\,.....\,(2n-1)}{2\cdot4\cdot6\,.....\,(2n)}\,\frac{x^{2n}}{2n},\ x>0\,;\ \sum_{n=1}^{\infty} \frac{1\cdot3\cdot5\,.....\,(2n-1)}{2\cdot4\cdot6\,.....\,(2n)}\,\frac{x^{2n+1}}{2n+1},\ x>0$

[Delhi B.Sc. (Prog) III 2010, 12]

(v) $x^2+\frac{2^2}{3\cdot4}x^4+\frac{2^2\cdot4^2}{3\cdot4\cdot5\cdot6}x^6+\frac{2^2\cdot4^2\cdot6^2}{3\cdot4\cdot5\cdot6\cdot7\cdot8}x^8+.....$

(vi) $1+a+\frac{a\,(a+1)}{1\cdot2}+\frac{a\,(a+1)\,(a+2)}{1\cdot2\cdot3}+.....$ **[Agra 2006]**

(vii) $\frac{1}{2}x+\frac{1\cdot3}{2\cdot4}x^2+\frac{1\cdot3\cdot5}{2\cdot4\cdot6}x^3+.....,\ x>0$ **[I.A.S. 2008; Kanpur 2011, 2013]**

(viii) $1^2+2^2x+3^2x^2+4^2x^3+..........$ **(Nagpur 2010)**

2. *Test for the convergence of the following series :*

(i) $1+\frac{2x}{2!}+\frac{3^2\,x^2}{3!}+\frac{4^3\,x^3}{4!}+\frac{5^4\,x^4}{5!}+.....$ **[Agra 2005, 07; Meerut, 2005]**

(ii) $x+\frac{2^2\,x^2}{2!}+\frac{3^3\,x^3}{3!}+\frac{4^4\,x^4}{4!}+\frac{5^5\,x^5}{5!}+.....$

(iii) $1+\frac{1!}{2}x+\frac{2!}{3^2}x^2+\frac{3!}{4^3}x^3+\frac{4!}{5^4}x^4+.....$ **[Kanpur 2012]**

(iv) $1+\frac{2!}{2^2}x+\frac{3!}{3^3}x^2+.....,\ x>0$

(v) $\frac{1}{(\log 2)^p}+\frac{1}{(\log 3)^p}+.....+\frac{1}{(\log n)^p}+.....$

(vi) $x^2\,(\log 2)^p+x^3\,(\log 3)^p+x^4\,(\log 4)^p+.....$

3. (i) $1+\frac{2^2}{3^2}+\frac{2^2\cdot4^2}{3^2\cdot5^2}+\frac{2^2\cdot4^2\cdot6^2}{3^2\cdot5^2\cdot7^2}+.....$

(ii) $1+\frac{\alpha\cdot\beta}{1\cdot\gamma}x+\frac{\alpha\,(\alpha+1)\,\beta\,(\beta+1)}{1\cdot2\cdot\gamma\cdot(\gamma+1)}x^2+\frac{\alpha\,(\alpha+1)\,(\alpha+2)\,\beta\,(\beta+1)\,(\beta+2)}{1\cdot2\cdot3\cdot\gamma\cdot(\gamma+1)\cdot(\gamma+2)}x^3+.....$

4. If $\frac{u_n}{u_{n+1}}=\frac{n^k+An^{k-1}+Bn^{k-2}+Cn^{k-3}+.....}{n^k+an^{k-1}+bn^{k-2}+cn^{k-3}+.....}$, where k is a positive integer, show that the series $\Sigma\,u_n$ is convergent if $A-a-1>0$ and divergent if $A-a-1\le 0$.

ANSWERS

1. (*i*) Convergent
   (*ii*) Converges if $x \le 1$, diverges if $x > 1$
   (*iii*) Converges if $x < 1/4$, diverges if $x \ge 1/4$
   (*iv*) Converges if $x \le 1$, diverges if $x > 1$
   (*v*) Converges if $x^2 \le 1$, diverges if $x^2 > 1$
   (*vi*) Converges if $a \le 0$, diverges if $a > 0$
   (*vii*) Converges if $0 < x < 1$ and diverges if $x \ge 1$
   (*viii*) Converges if $x < 1$ and diverges if $x \ge 1$
2. (*i*) Converges if $x \le 1/e$, diverges if $x > 1/e$
   (*ii*) Converges if $x < 1/e$, diverges if $x \ge 1/e$
   (*iii*) Converges if $x < e$, diverges if $x \ge e$
   (*iv*) Converges if $x < e$, diverges if $x \ge e$
   (*v*) Divergent for all values of p
   (*vi*) Converges if $x < 1$, diverges if $x \ge 1$
3. (*i*) Divergent
   (*ii*) Converges if $x < 1$ and diverges if $x > 1$. When $x = 1$, the series converges if $\gamma > \alpha + \beta$ and diverges if $\gamma \le \alpha + \beta$.

6.17. KUMMER'S TEST

Let Σu_n be a positive term series and let $\Sigma (1/d_n)$ be a divergent series of positive terms suc that

$$\lim_{n \to \infty} \left(d_n \frac{u_n}{u_{n+1}} - d_{n+1} \right) = l.$$

Then (i) Σu_n diverges if $l < 0$
(ii) Σu_n converges if $l > 0$. **(Patna, 2003)**

Proof. Given that
$$\lim_{n \to \infty} \left(d_n \frac{u_n}{u_{n+1}} - d_{n+1} \right) = l.$$

Hence for a given $\varepsilon > 0$, there exist a positive integer m such that

$$l - \varepsilon < d_n \frac{u_n}{u_{n+1}} - d_{n+1} < l + \varepsilon, \; \forall \, n \ge m \qquad \text{...(1)}$$

(*i*) Let $l < 0$. Since ε is always at our choice, let $\varepsilon = -l$.
Substituting this value of ε in the second part of (1), we get

$$d_n \frac{u_n}{u_{n+1}} - d_{n+1} < 0, \quad \forall \, n \ge m$$

or
$$d_n u_n < d_{n+1} u_{n+1}, \quad \forall \, n \ge m \qquad \text{...(2)}$$

Replacing n by m, $m + 1$, $m + 2$, ..., $n - 1$ in succession in (2), we get

$$d_m u_m < d_{m+1} u_{m+1} < d_{m+2} u_{m+2} < ..... < d_n u_n < .....$$
$$\Rightarrow u_n > d_m u_m \times (1/d_n), \text{ whenever } n > m \qquad \text{...(3)}$$

Since $d_m u_m$ is a positive constant and $\Sigma (1/d_n)$ is a known divergent series, so by comparison test (3) shows that Σu_n also diverges.

(*ii*) Let $l > 0$. Choosing $\varepsilon = l/2$, the first part of (1) gives

$$\frac{l}{2} < d_n \frac{u_n}{u_{n+1}} - d_{n+1} \quad \forall\ n \geq m$$

or $$(l/2) \times u_{n+1} < d_n\, u_n - d_{n+1}\, u_{n+1} \quad \forall\ n \geq m \qquad ...(4)$$

Replacing n by $m, m+1, m+2, \ldots, p-1$ in succession in (4) and adding the corresponding sides of the inequalities, we have

$$(l/2) \times (u_{m+1} + u_{m+2} + \ldots + u_p) < d_m\, u_m - d_p\, u_p$$

or $$(l/2) \times (u_{m+1} + u_{m+2} + \ldots + u_p) < d_m\, u_m$$

or $$u_{m+1} + u_{m+2} + \ldots + u_p < (2/l) \times d_m\, u_m$$

or $$u_1 + u_2 + \ldots + u_p < u_1 + u_2 + \ldots + u_m + (2/l) + d_m\, u_m \quad \forall\ n \geq m$$

$\Rightarrow \displaystyle\sum_1^p u_n$ is bounded for all $p \Rightarrow \Sigma\, u_n$ is a convergent series.

Corollary 1. As a particular case, take $d_n = 1$. Then $\Sigma\,(1/d_n) = 1 + 1 + 1 + \ldots$ is a divergent series. Then, Kummer's test reduces to $\displaystyle\lim_{n \to \infty} \left(\frac{u_n}{u_{n+1}} - 1 \right) = l$ and so the conclusions are :

For the convergent series $\displaystyle\lim_{n \to \infty} \left(\frac{u_n}{u_{n+1}} - 1 \right) < 0$, *i.e.*, $\displaystyle\lim_{n \to \infty} \frac{u_n}{u_{n+1}} < 1$

and for the divergent series $\displaystyle\lim_{n \to \infty} \left(\frac{u_n}{u_{n+1}} - 1 \right) > 0$, *i.e.*, $\displaystyle\lim_{n \to \infty} \frac{u_n}{u_{n+1}} > 1.$

This is statement of well-known D'Alembert's ratio test.

Corollary 2. Let $d_n = n$. Then $\Sigma\,(1/d_n) = \Sigma\,(1/n)$ is divergent. Then, Kummer's test reduces to

$$\lim_{n \to \infty} \left\{ n \frac{u_n}{u_{n+1}} - (n+1) \right\} = l \quad \text{or} \quad \lim_{n \to \infty} \left\{ n \left(\frac{u_n}{u_{n+1}} - 1 \right) - 1 \right\} = l.$$

$\therefore$ for convergent series $\displaystyle\lim_{n \to \infty} \left\{ n \left(\frac{u_n}{u_{n+1}} - 1 \right) - 1 \right\} < 0$ or $\displaystyle\lim_{n \to \infty} n \left(\frac{u_n}{u_{n+1}} - 1 \right) < 1$

and for divergent series $\displaystyle\lim_{n \to \infty} \left\{ n \left(\frac{u_n}{u_{n+1}} - 1 \right) - 1 \right\} > 0$ or $\displaystyle\lim_{n \to \infty} n \left(\frac{u_n}{u_{n+1}} - 1 \right) > 1$

This is statement of well-known Raabe's test.

6.18. GAUSS'S TEST

Let $\Sigma\, u_n$ be a positive term series and let there exist two positive numbers ρ, δ and a bounded sequence $<a_n>$ such that $$\frac{u_n}{u_{n+1}} = 1 + \frac{\rho}{n} + \frac{a_n}{n^{1+\delta}}$$

Then the series $\Sigma\, u_n$ converges if $\rho > 1$ and diverges if $\rho \leq 1$. **(*Meerut, 1996*)**

Proof. Given that $$\frac{u_n}{u_{n+1}} = 1 + \frac{\rho}{n} + \frac{a_n}{n^{1+\delta}},\ \rho > 0,\ \delta > 0$$

$$\Rightarrow \quad n\left(\frac{u_n}{u_{n+1}}-1\right) = \rho + \frac{a_n}{n^\delta} \quad \Rightarrow \quad \lim_{n\to\infty} n\left(\frac{u_n}{u_{n+1}}-1\right) = \lim_{n\to\infty}\left(\rho + a_n \cdot \frac{1}{n^\delta}\right) = \rho$$

[$\because$ $\langle a_n \rangle$ is bounded and $1/n^\delta \to 0$ when $n \to \infty$ as $\delta > 0$]

so that by Raabe's test, the series converges if $\rho > 1$ and diverges if $\rho < 1$.

We have now to consider the case $\rho = 1$. We have in this case

$$\frac{u_n}{u_{n+1}} = 1 + \frac{1}{n} + \frac{a_n}{n^{1+\lambda}} \qquad \ldots(1)$$

We compare the given series Σu_n with the divergent series Σv_n, where $v_n = 1/(n \log n)$. For proof of divergence of Σv_n, refer the next article 6.19.

Now, we have

$$\frac{u_n}{u_{n+1}} - \frac{v_n}{v_{n+1}} = 1 + \frac{1}{n} + \frac{a_n}{n^{1+\lambda}} - \frac{(n+1)\log(n+1)}{n \log n}, \text{ using (1) and } v_n = \frac{1}{n \log n}$$

$$= \frac{a_n}{n^{1+\lambda}} - \frac{n+1}{n}\left\{\frac{\log(n+1)}{\log n} - 1\right\}$$

Thus,
$$\frac{u_n}{u_{n+1}} - \frac{v_n}{v_{n+1}} = \frac{1}{n^{1+\delta}}\left\{a_n - (n+1)\log\left(1+\frac{1}{n}\right)\cdot\frac{n^\delta}{\log n}\right\} \qquad \ldots(2)$$

But
$$\lim_{n\to\infty} (n+1)\log\left(1+\frac{1}{n}\right) = \lim_{n\to\infty}\left\{\log\left(1+\frac{1}{n}\right) + \log\left(1+\frac{1}{n}\right)^n\right\} = 1$$

Also,
$$\lim_{n\to\infty} \frac{n^\delta}{\log n} = \infty$$

We thus see that for sufficiently large values of n

$$a_n - (n+1)\log\left(1+\frac{1}{n}\right)\frac{n^\delta}{\log n}$$

remains negative, *i.e.*, there exists positive integer m such that $\forall\ n \geq m$

$$a_n - (n+1)\log\left(1+\frac{1}{n}\right)\frac{n^\delta}{\log n} < 0 \qquad \ldots(3)$$

Now, (2) and (3) $\Rightarrow$ $\dfrac{u_n}{u_{n+1}} - \dfrac{v_n}{v_{n+1}} < 0 \Rightarrow \dfrac{u_n}{u_{n+1}} < \dfrac{v_n}{v_{n+1}}\ \forall\ n \geq m$

Thus, by comparison test of the second type (refer Art. 6.9), the series Σu_n is divergent.

Corollary. *If there exists $\delta > 0$ such that*

$$\lim_{n\to\infty}\left[n^\delta\left\{n\left(\frac{u_n}{u_{n+1}} - 1\right)\right\}\right]$$

exists finitely, then the series Σu_n is divergent.

Proof. Left as an exercise for the reader.

Note 1. Gauss's test can be applied after the failure of ratio test and when it is possible to expand u_n/u_{n+1} in powers of $1/n$ by binomial theorem

$$(1+x)^n = 1 + nx + \frac{n(n-1)}{2!}x^2 + \frac{n(n-1)(n-2)}{3!}x^3 + \ldots\ldots$$

or by any other method.

Note 2. If the terms of the given Σu_n involve 'x' then we apply ratio test and then try Gauss's test. But if the terms of Σu_n are free from 'x', then Gauss's test should be applied directly.

EXAMPLES

Test for convergence the following series :

(i) $1 + \frac{2^2}{3^2} + \frac{2^2 \cdot 4^2}{3^2 \cdot 5^2} + \frac{2^2 \cdot 4^2 \cdot 6^2}{3^2 \cdot 5^2 \cdot 7^2} + \ldots\ldots$

(ii) $\sum \frac{1^2 \cdot 3^2 \cdot 5^2 \ldots\ldots (2n-1)^2}{2^2 \cdot 4^2 \ldots\ldots (2n)^2} x^{n-1},\ x > 0$

(iii) $1 + \left(\frac{1}{2}\right)^p + \left(\frac{1 \cdot 3}{2 \cdot 4}\right)^p + \left(\frac{1 \cdot 3 \cdot 5}{2 \cdot 4 \cdot 6}\right)^p + \ldots\ldots$

(iv) $1 + \frac{\alpha\beta}{1 \cdot \gamma}x + \frac{\alpha(\alpha+1)\beta(\beta+1)}{1 \cdot 2 \cdot \gamma(\gamma+1)}x^2 + \frac{\alpha(\alpha+1)(\alpha+2)\beta(\beta+1)(\beta+2)}{1 \cdot 2 \cdot 3 \cdot \gamma(\gamma+1)(\gamma+2)}x^3 + \ldots\ldots$

Solution. (*i*) Omitting the first term, we have

$$u_n = \frac{2^2 \cdot 4^2 \cdot 6^2 \cdot \ldots\ldots (2n)^2}{3^2 \cdot 5^2 \cdot 7^2 \cdot \ldots\ldots (2n+1)^2}, \quad u_{n+1} = \frac{2^2 \cdot 4^2 \cdot 6^2 \cdot \ldots\ldots (2n)^2 (2n+2)^2}{3^2 \cdot 5^2 \cdot 7^2 \cdot \ldots\ldots (2n+1)^2 (2n+3)^2}$$

$$\therefore \quad \frac{u_n}{u_{n+1}} = \frac{(2n+3)^2}{(2n+2)^2} = \frac{(1+3/2n)^2}{(1+1/n)^2} \quad \ldots(1)$$

Hence $\lim_{n \to \infty} \frac{u_n}{u_{n+1}} = 1$ and so ratio test fails.

From (1), $$\frac{u_n}{u_{n+1}} = \left(1 + \frac{3}{2n}\right)^2 \left(1 + \frac{1}{n}\right)^{-2} = \left(1 + \frac{3}{n} + \frac{9}{4n^2}\right)\left(1 - \frac{2}{n} + \frac{3}{n^2} + \ldots\ldots\right)$$

[on expanding by binomial theorem]

$$= 1 + \frac{1}{n} + \frac{1}{n^2}\left(-\frac{15}{4} + \ldots\ldots\right) = 1 + \frac{1}{n} + O\left(\frac{1}{n^2}\right),$$

where $O(1/n^2)$ stands for terms of order two or more in $1/n$.

Here ρ = coeff. of $1/n = 1$. So, by Gauss's test, the given series diverges.

(*ii*) Here $u_n = \frac{1^2 \cdot 3^2 \cdot \ldots\ldots (2n-1)^2}{2^2 \cdot 4^2 \cdot \ldots\ldots (2n)^2} x^{n-1}$, $\quad u_{n+1} = \frac{1^2 \cdot 3^2 \cdot \ldots\ldots (2n-1)^2 (2n+1)^2}{2^2 \cdot 4^2 \cdot \ldots\ldots (2n)^2 (2n+2)^2} x^n$

$$\therefore \quad \frac{u_n}{u_{n+1}} = \frac{(2n+2)^2}{(2n+1)^2} \cdot \frac{1}{x} = \frac{(1+1/n)^2}{(1+1/2n)^2} \cdot \frac{1}{x} \quad \ldots(1)$$

Hence $\lim_{n \to \infty} \frac{u_n}{u_{n+1}} = \frac{1}{x}$ and so by ratio test, Σu_n converges if $x < 1$ and diverges if $x > 1$.

Test fails when $x = 1$. In that case, (1) becomes

$$\frac{u_n}{u_{n+1}} = \left(1+\frac{1}{n}\right)^2\left(1+\frac{1}{2n}\right)^{-2} = \left(1+\frac{2}{n}+\frac{1}{n^2}\right)\left(1-\frac{1}{n}+\frac{3}{4n^2}+\ldots..\right)$$

$$= 1+\frac{1}{n}+\frac{1}{n^2}\left(-\frac{1}{4}+\ldots..\right) = 1+\frac{1}{n}+O\left(\frac{1}{n^2}\right).$$

Here ρ = coeff. of $1/n = 1$. So by Gauss's test, the given series diverges.

Thus the given series converges if $x < 1$ and diverges if $x \geq 1$.

(*iii*) Omitting the first term of the given series, we have

$$u_n = \left\{\frac{1\cdot 3\cdot\ldots..(2n-1)}{2\cdot 4\cdot\ldots..2n}\right\}^p, \quad u_{n+1} = \left\{\frac{1\cdot 3\cdot\ldots..(2n-1)(2n+1)}{2\cdot 4\cdot\ldots..(2n)(2n+1)}\right\}^p$$

$$\therefore \qquad \frac{u_n}{u_{n+1}} = \frac{(2n+2)^p}{(2n+1)^p} = \frac{(1+1/n)^p}{(1+1/2n)^p} \qquad \ldots(1)$$

Hence $\lim\limits_{n\to\infty} \dfrac{u_n}{u_{n+1}} = 1$ and so ratio test fails. Then, by (1)

$$\frac{u_n}{u_{n+1}} = \left(1+\frac{1}{n}\right)^p\left(1+\frac{1}{2n}\right)^{-p}$$

$$= \left\{1+\frac{p}{n}+\frac{p(p-1)}{2!}\cdot\frac{1}{n^2}+\ldots..\right\}\times\left\{1-\frac{p}{2n}+\frac{p(p+1)}{2!}\cdot\frac{1}{4n^2}+\ldots..\right\}$$

$$= 1+\frac{p}{2n}+O\left(\frac{1}{n^2}\right).$$

Here ρ = coeff. of $1/n = p/2$. So, by Gauss's test, the given series converges if $p/2 > 1$, *i.e.*, $p > 2$ and diverges if $p/2 \leq 1$, *i.e.*, $p \leq 2$.

Thus the given series converges if $p > 2$ and diverges if $p \leq 2$.

(*iv*) Omitting the first term, we have

$$u_n = \frac{\alpha(\alpha+1)\ldots..(\alpha+n-1)\,\beta(\beta+1)\ldots..(\beta+n-1)}{1\cdot 2\cdot 3\cdot\ldots..n\cdot\gamma(\gamma+1)(\gamma+2)\ldots..(\gamma+n-1)}\,x^n$$

and $$u_{n+1} = \frac{\alpha(\alpha+1)\ldots..(\alpha+n-1)(\alpha+n)\,\beta(\beta+1)\ldots..(\beta+n-1)(\beta+n)}{1\cdot 2\cdot 3\cdot\ldots..n(n+1)\cdot\gamma(\gamma+1)\ldots..(\gamma+n-1)(\gamma+n)}\,x^{n+1}$$

$$\therefore \qquad \frac{u_n}{u_{n+1}} = \frac{(n+1)(\gamma+n)}{(\alpha+n)(\beta+n)}\cdot\frac{1}{x} = \frac{(1+1/n)(1+\gamma/n)}{(1+\alpha/n)(1+\beta/n)}\cdot\frac{1}{x} \qquad \ldots(1)$$

Here $$\lim_{n\to\infty}\frac{u_n}{u_{n+1}} = \frac{1}{x}.$$

Hence, by D'Alembert's ratio test, $\Sigma\, u_n$ converges if $1/x > 1$, *i.e.*, $x < 1$ and diverges if $1/x < 1$, *i.e.*, $x > 1$. For $x = 1$, the test fails. In that case, from (1), we have

$$\frac{u_n}{u_{n+1}} = \left(1+\frac{1}{n}\right)\left(1+\frac{\gamma}{n}\right)\left(1+\frac{\alpha}{n}\right)^{-1}\left(1+\frac{\beta}{n}\right)^{-1}$$

$$= \left(1+\frac{1}{n}\right)\left(1+\frac{\gamma}{n}\right)\left(1-\frac{\alpha}{n}+\frac{\alpha^2}{n^2}+\ldots..\right)\left(1-\frac{\beta}{n}+\frac{\beta^2}{n^2}+\ldots..\right)$$

[on expanding by binomial theorem]

$$= \left(1+\frac{1}{n}+\frac{\gamma}{n}+\frac{\gamma}{n^2}\right)\left(1-\frac{\alpha}{n}-\frac{\beta}{n}+\frac{\alpha\beta}{n^2}+\frac{\alpha^2}{n^2}+\frac{\beta^2}{n^2}+.....\right)$$

$$= 1+\frac{1}{n}(1+\gamma-\alpha-\beta)+O\left(\frac{1}{n^2}\right)$$

Here ρ = coeff. of $1/n = 1 + \gamma - \alpha - \beta$. So, by Gauss's test, the given series converges if $1+\gamma-\alpha-\beta > 1$, *i.e.*, $\gamma > \alpha + \beta$ and diverges if $1 + \gamma - \alpha - \beta \le 1$, *i.e.*, if $\gamma \le \alpha + \beta$.

Thus the given series converges if $x < 1$ and diverges if $x > 1$. If $x = 1$, then the series converges if $\gamma > \alpha + \beta$ and diverges if $\gamma \le \alpha + \beta$.

EXERCISES

Examine the following series for convergence :

1. $\dfrac{1\cdot 3\cdot 5\cdot ...(2n-1)}{2\cdot 4\cdot 6\cdot ...2n}\cdot\dfrac{1}{2n+1}$ **2.** $\left(\dfrac{1}{3}\right)^2+\left(\dfrac{1\cdot 2}{3\cdot 5}\right)^2+\left(\dfrac{1\cdot 2\cdot 3}{3\cdot 5\cdot 7}\right)^2+...$

[Delhi B.A. (Prog.) III, 2014]

3. $1+a+\dfrac{a(a+1)}{1\cdot 2}+\dfrac{a(a+1)(a+2)}{1\cdot 2\cdot 3}+..., a>0$ **4.** $\dfrac{a}{b}+\dfrac{a(a+1)}{b(b+1)}+\dfrac{a(a+1)(a+2)}{b(b+1)(b+2)}+...$

5. $\dfrac{1}{(\log 2)^p}+\dfrac{1}{(\log 3)^p}+\dfrac{1}{(\log 4)^p}+...$ **6.** $\dfrac{1}{2}\cdot\dfrac{x^2}{4}+\dfrac{1\cdot 3\cdot 5}{2\cdot 4\cdot 6}\cdot\dfrac{x^4}{8}+\dfrac{1\cdot 3\cdot 5\cdot 7\cdot 9}{2\cdot 4\cdot 6\cdot 8\cdot 10}\cdot\dfrac{x^6}{12}+...$

7. Show that $\left(\dfrac{1}{3}\right)^2+\left(\dfrac{2\cdot 4}{3\cdot 6}\right)^2+...+\left(\dfrac{1\cdot 4\cdot 7......(3n-2)}{3\cdot 6\cdot 9.....3n}\right)^2+......$ converges ***(I.A.S 2009)***

ANSWERS

1. Convergent **2.** Convergent **3.** Divergent
4. Convergent if $b > a + 1$, divergent if $b \le a + 1$
5. Divergent **6.** Convergent if $x^2 \le 1$, divergent if $x^2 > 1$.

6.19. CAUCHY'S INTEGRAL TEST

Before formulating the integral test which is based on the comparison of a positive term series with an infinite integral, we give a few preliminaries which we shall need.

Let f be a real valued function with domain $]1, \infty[$, *i.e.*, the set $\{x : 1 \le x\}$. We suppose that f is such that

$$\int_1^t f(x)\,dx$$

has a meaning $\forall\ t \ge 1$, and say that the integral

$$\int_1^\infty f(x)\,dx$$

is convergent or that it exists if

$$\lim_{t\to\infty}\int_1^t f(x)\,dx$$

exists finitely. Again, assuming the existence of this limit, we write

$$\int_1^\infty f(x)\,dx = \lim_{t\to\infty}\int_1^t f(x)\,dx$$

Of course, we say that the integral

$$\int_1^{\infty} f(x)\, dx$$

does not exist or that it is not convergent if

$$\lim_{t \to \infty} \int_1^t f(x)\, dx$$

does not exist finitely.

The integral

$$\int_1^{\infty} f(x)\, dx$$

is called an **infinite integral.**

Note. It will be easily seen that instead of 1, we may consider any other number.

ILLUSTRATIONS

1. Let f be defined by $f(x) = 1/\sqrt{x}$.

We have $\int_1^t f(x)\, dx = \int_1^t \frac{1}{\sqrt{x}}\, dx = \Big| 2\sqrt{x} \Big|_1^t = 2\sqrt{t} - 2 \to \infty$ as $t \to \infty$.

We thus see that the infinite integral $\int_1^{\infty} \frac{1}{\sqrt{x}}\, dx$ is *not* convergent.

2. Let f be defined by $f(x) = 1/x^2$.

We have $\int_1^t f(x)\, dx = \int_1^t \frac{1}{x^2}\, dx = -\frac{1}{t} + 1 \to 1$ as $t \to \infty$.

We thus see that the infinite integral $\int_1^{\infty} \frac{1}{x^2}\, dx$ is convergent.

Test for convergence of the infinite integral $\int_1^{\infty} f(x)\, dx$, when $f(x) \geq 0 \ \forall\ x \geq 1$.

We have $\int_a^b f(x)\, dx \geq 0$ when $b > a$.

Now if $y > x$, we have

$$\int_a^y f(x)\, dx = \int_a^x f(x)\, dx + \int_x^y f(x)\, dx$$

$$\geq \int_a^x f(x)\, dx, \text{ for } \int_x^y f(x)\, dx \geq 0.$$

Thus, we see that $y > x \Rightarrow \int_a^y f(x)\, dx \geq \int_a^x f(x)\, dx,$

or, in other words, we see that

$$\int_a^t f(x)\, dx$$

is a monotonically increasing function of t.

We *assume* that because of the monotonic character of $\int_a^t f(x)\, dx$ as a function of t,

$$\lim_{t \to \infty} \int_a^t f(x)\, dx$$

will exist if and only if it is bounded above, *i.e.*, if and only if there exists k such that $\forall\ t \geq a$

$$\int_a^t f(x)\,dx \leq k.$$

Thus, we have the following test :

Let $f(x) \geq 0\ \forall\ x \geq 1$. Then the infinite integral

$$\int_1^\infty f(x)\,dx$$

is convergent if and only if there exists k such that $\forall\ t \geq 1$,

$$\int_1^t f(x)\,dx \leq k.$$

Some basic definitions. The function $f(x)$ is said to be monotonically decreasing on interval $[1, \infty[$, if

$$y \geq x \;\Rightarrow\; f(y) \leq f(x) \qquad \forall\ x, y \in [1, \infty[$$

The function $f(x)$ is said to be non-negative in $[1, \infty[$, if

$$f(x) \geq 0 \qquad \forall\ x \geq 1$$

For example, $f(x) = 1/x^2$ is a non-negative and monotonically decreasing for $\forall\ x \geq 1$.

CAUCHY'S INTEGRAL TEST

(Delhi B.Sc. (Hons.) I, 2016)

If $u(x)$ is a non-negative decreasing integrable function such that $u(n) = u_n$ for all positive integral values of n, then $\sum_{n=1}^{\infty} u_n$ is convergent if and only if the infinite integral $\int_1^\infty u(x)\,dx$ is convergent. **[*Delhi Maths (H), 2001, 03*; *Delhi B.Sc. (H) Physics, 2000, 01*]**

Proof. Since $u(x)$ is monotonically decreasing, we have

$$\therefore \quad u(n+1) \leq u(x) \leq u(n) \text{ when } n+1 \geq x \geq n$$

Again, since $u(x)$ is non-negative and integrable, we have

$$\therefore \quad \int_n^{n+1} u(n+1)\,dx \leq \int_n^{n+1} u(x)\,dx \leq \int_n^{n+1} u(n)\,dx$$

or
$$u(n+1)\int_n^{n+1} dx \leq \int_n^{n+1} u(x)\,dx \leq u(n)\int_n^{n+1} dx$$

or
$$u_{n+1} \leq \int_n^{n+1} u(x)\,dx \leq u_n, \text{ as } u(n) = u_n \qquad \text{...(1)}$$

Substituting 1, 2, 3, ..., $n-1$ successively for n in (1) and adding, we obtain

$$u_2 + u_3 + \ldots + u_n \leq \int_1^2 u(x)\,dx + \int_2^3 u(x)dx + \ldots + \int_{n-1}^n u(x)\,dx \leq u_1 + u_2 + \ldots + u_{n-1}$$

or
$$S_n - u_1 \leq \int_1^n u(x)\,dx \leq S_n - u_n, \qquad \text{...(2)}$$

where
$$S_n = u_1 + u_2 + \ldots + u_n$$

Thus, from (2), we conclude that $\forall\ n \in \mathbf{N}$

$$S_n \leq u_1 + \int_1^n u(x)\,dx \qquad \text{...(3)}$$

and
$$\int_1^n u(x)\,dx < S_n, \text{ as } u_n \geq 0 \qquad \text{...(4)}$$

Now suppose that the infinite series $\sum\limits_{n=1}^{\infty} u_n$ is convergent. The terms of the series being all positive, the sequence $<S_n>$ of its partial sums is bounded above, *i.e.*, there exists a number k such that

$$S_n \le k \ \forall\ n \in \mathbf{N}$$

Also we have shown in (4) that

$$\int_1^n u(x)\,dx \le S_n \ \ \forall\ n \in \mathbf{N}$$

Thus, we conclude that $\forall\ n \in \mathbf{N}$,

$$\int_1^n u(x)\,dx \le k$$

$\Rightarrow$ $\int_1^n u(x)\,dx$ is bounded above

$\Rightarrow$ the infinite integral $\int_1^{\infty} u(x)\,dx$ is convergent.

To prove the converse, we suppose that the infinite integral $\int_1^{\infty} u(x)\,dx$ is convergent so that there exists a number K such that

$$\int_1^n u(x)\,dx \le K \ \forall\ n \in \mathbf{N}$$

Also we have seen in (3), that $\forall\ n \in \mathbf{N}$

$$S_n \le u_1 + \int_1^n u(x)\,dx \ \Rightarrow\ S_n \le u_1 + K \ \forall\ n \in \mathbf{N}$$

Thus $<S_n>$ is bounded above and as such the series is convergent. Hence the result.

EXAMPLES

Example 1. *Apply Cauchy's integral test to examine the convergence of the following series*

(*i*) $\sum\limits_{n=1}^{\infty} \dfrac{1}{n^2+1}$ (*ii*) $\sum\limits_{n=1}^{\infty} \dfrac{1}{n(n+1)}$ ***(Delhi Maths (G), 2003)***

Solution. (*i*) Here $u_n = 1/(n^2+1) = u(n)\ \forall\ n \in \mathbf{N}$ so that $u(x) = 1/(1+x^2)$.

For $x \ge 1$, $u(x)$ is a non-negative and monotonically decreasing function. Hence Cauchy's integral test is applicable.

Now, $$\int_1^t u(x)\,dx = \int_1^t \frac{dx}{1+x^2} = \left[\tan^{-1} x\right]_1^t = \tan^{-1} t - \pi/4$$

$$\therefore \quad \int_1^{\infty} u(x)\,dx = \lim_{t\to\infty} \int_1^t \frac{dx}{1+x^2} = \lim_{t\to\infty} (\tan^{-1} t - \pi/4) = \pi/2 - \pi/4 = \pi/4 = \text{finite.}$$

Hence $\int_1^{\infty} u(x)\,dx$ converges and so by Cauchy's integral test, $\sum\limits_{n=1}^{\infty} u_n$, *i.e.*, $\sum\limits_{n=1}^{\infty} \dfrac{1}{n^2+1}$ also converges.

(*ii*) Here $$u_n = \frac{1}{n(n+1)} = u(n)\ \forall\ n \in \mathbf{N} \text{ so that } u(x) = \frac{1}{x(x+1)}$$

For $x \geq 1$, $u(x)$ is a non-negative and monotonically decreasing function. Hence Cauchy's integral test is applicable.

Now, $$\int_1^t u(x)\,dx = \int_1^t \frac{1}{x(x+1)}\,dx = \int_1^t \left(\frac{1}{x} - \frac{1}{x+1}\right) dx = [\log x - \log(x+1)]_1^t$$

$$= \left[\log \frac{x}{x+1}\right]_1^t = \log \frac{t}{t+1} - \log \frac{1}{2}$$

$$\therefore \quad \int_1^\infty u(x)\,dx = \lim_{t\to\infty} \int_1^t \frac{dx}{x(1+x)} = \lim_{t\to\infty} \left[\log \frac{t}{t+1} - \log \frac{1}{2}\right]$$

$$= \lim_{t\to\infty} \left[\log \frac{1}{1+1/t} - \log \frac{1}{2}\right] = \log 1 - \log 2^{-1} = \log 2 = \text{finite}$$

Hence $\int_1^\infty u(x)\,dx$ converges and so by Cauchy's integral test, $\sum_{n=1}^{\infty} u_n$, *i.e.*, $\sum_{n=1}^{\infty} \frac{1}{n(n+1)}$ also converges.

Example 2. *Show that* $\sum_{n=1}^{\infty} \frac{1}{n^p}$, $p > 0$, *is convergent if* $p > 1$ *and divergent if* $p \leq 1$.

[*Delhi B.Sc. (Hons.) I, 2016*; *Meerut, 2003, 04*; *Patna, 2003*]

Solution. Here $u_n = 1/n^p = u(n)\ \forall\, n \in \mathbf{N}$ so that $u(x) = 1/x^p$. For $x \geq 1$, $u(x)$ is a non-negative and monotonically decreasing function. Hence Cauchy's integral test is applicable.

Case I. Let $p = 1$. Then $f(x) = 1/x$. In this case, we have

$$\int_1^t u(x)\,dx = \int_1^t \frac{1}{x}\,dx = [\log x]_1^t = \log t$$

$$\therefore \quad \int_1^\infty u(x)\,dx = \lim_{t\to\infty} \int_1^t \frac{1}{x}\,dx = \lim_{t\to\infty} \log t = \infty$$

$$\Rightarrow \quad \int_1^\infty u(x)\,dx \text{ diverges} \Rightarrow \sum_{n=1}^{\infty} u_n \text{ diverges} \qquad \text{(Using Cauchy's integral test)}$$

Case II. Let $p > 1$ so that $(p-1) > 0$. Then, we have

$$\int_1^t u(x)\,dx = \int_1^t x^{-p}\,dx = \left[\frac{x^{1-p}}{1-p}\right]_1^t = \left[-\frac{x^{1-p}}{p-1}\right]_1^t = \frac{1}{p-1}\left[1 - \frac{1}{t^{p-1}}\right]$$

$$\therefore \quad \int_1^\infty u(x)\,dx = \lim_{t\to\infty} \int_1^t u(x)\,dx = \lim_{t\to\infty} \frac{1}{p-1}\left(1 - \frac{1}{t^{p-1}}\right) = \frac{1}{p-1} = \text{finite}$$

$$\Rightarrow \quad \int_1^\infty u(x)\,dx \text{ converges} \Rightarrow \sum_{n=1}^{\infty} u_n \text{ converges}$$

Case III. Let $0 < p < 1$ so that $(1-p) > 0$. Then, we have

$$\int_1^t u(x)\,dx = \int_1^t x^{-p}\,dx = \left[\frac{x^{1-p}}{1-p}\right]_1^t = \frac{1}{1-p}(t^{1-p} - 1)$$

$$\therefore \quad \int_1^\infty u(x)\,dx = \lim_{t\to\infty} \int_1^t u(x)\,dx = \lim_{t\to\infty} \frac{1}{1-p}(t^{1-p} - 1) = \infty$$

$$\Rightarrow \qquad \int_1^{\infty} u(x)\,dx \text{ diverges } \Rightarrow \sum_{n=1}^{\infty} u_n \text{ diverges.}$$

Thus $\sum_{n=1}^{\infty} u_n$ converges if $p > 1$ and diverges if $0 < p \leq 1$.

Example 3. *Using integral test, show that the series* $\sum_{n=2}^{\infty} \frac{1}{n(\log n)^p}$ *is convergent if* $p > 1$ *and divergent if* $0 < p \leq 1$. **[*Agra 2008, 12; M.S. Univ. T.N. 2006*]**

Solution. Here $u_n = \dfrac{1}{n(\log n)^p} = u(n)\ \forall\ n \in \mathbf{N}$. So $u(x) = \dfrac{1}{x(\log x)^p}$.

For $n \geq 2$, $u(x)$ is a non-negative and monotonically decreasing function. Hence Cauchy's integral test is applicable.

Case I. Let $p = 1$. Then $u(x) = 1/(x \log x)$. In this case, we have

$$\int_2^t u(x)\,dx = \int_2^t \frac{1}{x \log x}\,dx = [\log \log x]_2^t = \log \log t - \log \log 2$$

$$\therefore \qquad \int_2^{\infty} u(x)\,dx = \lim_{t \to \infty} \int_2^t u(x)\,dx = \lim_{t \to \infty} (\log \log t - \log \log 2) = \infty$$

$$\Rightarrow \qquad \int_2^{\infty} u(x)\,dx \text{ diverges} \Rightarrow \sum_{n=2}^{\infty} u_n \text{ diverges} \qquad \text{[Using Cauchy's integral test]}$$

Case II. Let $p > 1$ so that $(p - 1) > 0$. Then, we have

$$\int_2^t u(x)\,dx = \int_2^t \frac{dx}{x(\log x)^p} = \left[\frac{(\log x)^{1-p}}{1-p}\right]_2^t = \left[-\frac{(\log x)^{1-p}}{p-1}\right]_2^t$$

$$= \frac{1}{p-1}\left[\frac{1}{(\log 2)^{p-1}} - \frac{1}{(\log t)^{p-1}}\right]$$

$$\therefore \qquad \int_2^{\infty} u(x)\,dx = \lim_{t \to \infty} \int_2^t u(x)\,dx = \lim_{t \to \infty} \frac{1}{p-1}\left[\frac{1}{(\log 2)^{p-1}} - \frac{1}{(\log t)^{p-1}}\right]$$

$$= \frac{1}{(p-1)(\log 2)^{p-1}} = \text{finite}$$

$$\Rightarrow \qquad \int_2^{\infty} u(x)\,dx \text{ converges} \Rightarrow \sum_2^{\infty} u_n \text{ converges}$$

Case III. Let $0 < p < 1$ so that $(1 - p) > 0$. Then, we have

$$\int_2^t u(x)\,dx = \int_2^t \frac{dx}{x(\log x)^p} = \left[\frac{(\log x)^{1-p}}{1-p}\right]_2^t = \frac{1}{1-p}[(\log t)^{1-p} - (\log 2)^{1-p}]$$

$$\therefore \qquad \int_2^{\infty} u(x)\,dx = \lim_{t \to \infty} \int_2^t u(x)\,dx = \lim_{t \to \infty} \frac{1}{1-p}[(\log t)^{1-p} - (\log 2)^{1-p}] = \infty$$

$$\Rightarrow \qquad \int_2^{\infty} u(x)\,dx \text{ diverges} \Rightarrow \sum_2^{\infty} u_n \text{ diverges}$$

Thus $\sum_{n=2}^{\infty} u_n$ converges if $p > 1$ and diverges if $0 < p \leq 1$.

EXERCISES

Using integral test, discuss the convergence of the following series

1. (*i*) $\sum_{n=1}^{\infty} \frac{1}{\sqrt{n}}$ (*ii*) $\sum_{n=1}^{\infty} \frac{1}{n^{3/2}}$ (*iii*) $\sum_{n=1}^{\infty} ne^{-n^2}$ **[Delhi B.Sc. (Hons) I, 2001]**

(*iv*) $\sum_{n=2}^{\infty} \frac{1}{n(\log n)}$ **[Delhi B.Sc. (Hons) I, 2018, 2019]**

(*v*) $\sum_{n=2}^{\infty} \frac{1}{n(\log n)^3}$ **[Delhi 2009]**

(*vi*) $\sum_{n=2}^{\infty} \frac{1}{n\sqrt{n^2-1}}$ (*vii*) $\sum_{n=2}^{\infty} \frac{1}{n(\log n)^2}$ **[Delhi Maths (H) 2009]**

2. Show that the infinite series

$$\sum_{n=3}^{\infty} \frac{1}{n \log n (\log \log n)^p} \quad \text{and} \quad \sum_{n=27}^{\infty} \frac{1}{n \log n \log \log n (\log \log \log n)^p}$$

are convergent if and only if $p > 1$.

ANSWERS

1. (*i*) Divergent (*ii*) Convergent (*iii*) Convergent
(*iv*) Divergent (*v*) Convergent (*vi*) Convergent (*vii*) Convergent

6.20. CAUCHY'S CONDENSATION TEST

*If $f(n)$ is a positive monotonically decreasing function of n for $\forall\ n \in$ **N**, then the two infinite series $\Sigma f(n)$ and $\Sigma a^n f(a^n)$ converge or diverge together, a being a positive integer greater than unity.* **[Ranchi 2010]**

Proof. Rewriting the series $\Sigma f(n)$, we have

$$\begin{aligned}\Sigma f(n) = \ & f(1) + f(2) + \ldots + f(a)\\ & + f(a+1) + f(a+2) + \ldots + f(a^2)\\ & + f(a^2+1) + f(a^2+2) + \ldots + f(a^3)\\ & + \ldots\ldots\ldots\ldots\ldots\ldots\\ & + f(a^k+1) + f(a^k+2) + \ldots + f(a^{k+1})\\ & + \ldots\ldots\ldots\ldots\ldots\ldots \end{aligned} \quad \ldots(1)$$

The terms in the kth group are $f(a^k+1) + f(a^k+2) + \ldots + f(a^{k+1})$ ...(2)

and the number of terms in this kth group $= a^{k+1} - a^k = a^k(a-1)$

Since $<f(n)>$ is a decreasing sequence, it follows that $f(a^{k+1})$ is the smallest term in the kth group.

$$\therefore \quad f(a^k+1) + f(a^k+2) + \ldots + f(a^{k+1}) > \underbrace{f(a^{k+1}) + f(a^{k+1}) + \ldots + f(a^{k+1})}_{a^k(a-1) \text{ terms}}$$

i.e., $$f(a^k+1) + f(a^k+2) + \ldots + f(a^{k+1}) > a^k(a-1) f(a^{k+1})$$

or $$f(a^k+1) + f(a^k+2) + \ldots + f(a^{k+1}) > \frac{a-1}{a}\{a^{k+1} f(a^{k+1})\} \quad \ldots(3)$$

Putting $k = 0, 1, 2, 3, \ldots$ in succession in (3), we get

$$\left.\begin{aligned} f(2) + f(3) + \ldots + f(a) &> \frac{a-1}{a}\{a f(a)\}\\ f(a+1) + f(a+2) + \ldots + f(a^2) &> \frac{a-1}{a}\{a^2 f(a^2)\}\\ f(a^2+1) + f(a^2+2) + \ldots + f(a^3) &> \frac{a-1}{a}\{a^3 f(a^3)\}\\ \ldots\ldots\ldots\ldots\ldots\ldots\\ \ldots\ldots\ldots\ldots\ldots\ldots \end{aligned}\right\} \quad \ldots(4)$$

Adding the above inequalities, we have

$$\sum_{n=1}^{\infty} f(n) - f(1) > \frac{a-1}{a} \sum_{n=1}^{\infty} a^n f(a^n) \quad \text{...(5)}$$

If $\Sigma a^n f(a^n)$ is divergent, then by comparison test, (5) shows that $\Sigma f(n)$ is also divergent.

Again, consider the terms is the kth group given by (2). Since $<f(n)>$ is a decreasing sequence, it follows that $f(a^k)$ is greater than each term in the kth group (2).

$\therefore\ f(a^k+1) + f(a^k+2) + \ldots\ldots + f(a^{k+1}) < f(a^k) + f(a^k) + \ldots\ldots + f(a^k)$

i.e., $f(a^k+1) + f(a^k+2) + \ldots\ldots + f(a^{k+1}) < (a-1)\{a^k f(a^k)\}$, as before ...(6)

Putting $k = 0, 1, 2, \ldots\ldots$ in succession in (6), we use

$$\left.\begin{array}{l} f(2) + f(3) + \ldots\ldots + f(a) < (a-1) f(1) \\ f(a+1) + f(a+2) + \ldots\ldots + f(a^2) < (a-1)\{a f(a)\} \\ f(a^2+1) + f(a^2+2) + \ldots\ldots + f(a^3) < (a-1)\{a^2 f(a^2)\} \\ \ldots\ldots\ldots\ldots\ldots\ldots\ldots\ldots\ldots\ldots\ldots\ldots \\ \ldots\ldots\ldots\ldots\ldots\ldots\ldots\ldots\ldots\ldots\ldots\ldots \end{array}\right\} \quad \text{...(7)}$$

Adding the above inequalities (7), we have

$$\sum_{n=1}^{\infty} f(n) - f(1) < (a-1) \sum_{n=1}^{\infty} a^n f(a^n) + (a-1) f(1)$$

or

$$\sum_{n=1}^{\infty} f(n) < a f(1) + (a-1) \sum_{n=1}^{\infty} a^n f(a^n) \quad \text{...(8)}$$

If $\Sigma a^n f(a^n)$ is convergent, then by comparison test, (8) shows that $\Sigma f(n)$ is also convergent. Thus $\Sigma f(n)$ and $\Sigma a^n f(a^n)$ both converge or diverge together.

6.21. An important auxiliary series $\sum_{n=2}^{\infty} \frac{1}{n(\log n)^p}$ *converges if $p > 1$ and diverges if $p \le 1$.*

(Kanpur, 2005)

Proof. Let $f(n) = 1/\{n(\log n)^p\}$, for $\forall\ n > 1$

Case I. Let $p > 0$. Then $f(n)$ is a positive monotonically decreasing function of n for all $n \ge 2$. Hence Cauchy's condensation test is applicable.

Here $a^n f(a^n) = a^n \cdot \frac{1}{a^n (\log a^n)^p} = \frac{1}{(n \log a)^p} = \frac{1}{(\log a)^p} \cdot \frac{1}{n^p}$

$$\therefore \sum_{n=2}^{\infty} a^n f(a^n) = \frac{1}{(\log a)^p} \sum_{n=2}^{\infty} \frac{1}{n^p}, \text{ where } \frac{1}{(\log a)^p} \text{ is a constant} \quad \text{...(1)}$$

Since the series $\Sigma(1/n^p)$ converges if $p > 1$ and diverges if $p \le 1$, so, from (1), it follows that $\Sigma a_n f(a^n)$ converges if $p > 1$ and diverges if $0 < p \le 1$. Again, by Cauchy's condensation test, the series $\Sigma f(n)$ and $\Sigma a^n f(a^n)$ converge and diverge together. Hence $\Sigma f(n)$, *i.e.*, $\sum \frac{1}{n(\log n)^p}$ converges if $p > 1$ and diverges if $0 < p < 1$.

Case II. Let $p \le 0$. Then $\frac{1}{n(\log n)^p} \ge \frac{1}{n}\ \forall\ n > 1$ ...(2)

But $\sum_{n=1}^{\infty} \frac{1}{n}$ is divergent. Hence, by comparison test, (2) shows that $\sum \frac{1}{n(\log n)^p}$ diverges if $p \le 0$.

Thus $\sum_{2}^{\infty} \frac{1}{n(\log n)^p}$ converges if $p > 1$ and diverges if $p \le 1$.

Example. *Examine the following series for convergence*

$$\frac{(\log 2)^2}{2^2} + \frac{(\log 3)^2}{3^2} + \frac{(\log 4)^2}{4^2} + \ldots\ldots + \frac{(\log n)^2}{n^2} + \ldots\ldots.$$

(*Purvanchal, 2001*; *Gorakhpur, 1994*; *Kanpur, 1993*)

Solution. Adding an additional term $(\log 1)^2/1^2$ to the given series, the nth term $f(n)$ of the resulting series is given by

$$f(n) = (\log n)^2/n^2$$

$$\therefore \quad a^n f(a^n) = a^n \cdot \frac{(\log a^n)^2}{(a^n)^2} = \frac{(n \log a)^2}{a^n} = \frac{n^2 (\log a)^2}{a^n}, \text{ where } a > 1.$$

Let $\qquad \Sigma u_n = \Sigma a^n f(a^n) = \sum \frac{n^2 (\log a)^2}{a^n}.$

Then $\qquad u_n = \frac{n^2 (\log a)^2}{a^n}$ and $u_{n+1} = \frac{(n+1)^2 (\log a)^2}{a^{n+1}}$

$$\therefore \quad \lim_{n \to \infty} \frac{u_n}{u_{n+1}} = \lim_{n \to \infty} \left\{ a \left(\frac{n}{n+1} \right)^2 \right\} = \lim_{n \to \infty} \frac{a}{\{1 + (1/n)\}^2} = a > 1$$

Hence by D'Alembert's ratio test, Σu_n, *i.e.*, $\Sigma a^n f(a^n)$ is convergent. Then, by Cauchy's condensation test, the given series $\Sigma f(n)$ is convergent.

EXERCISES

Test for convergence the following series using Cauchy condensation test :

1. $\sum_{n=1}^{\infty} (\log n)/n$ **2.** $\sum_{n=2}^{\infty} \{1/(\log n)^p\}$

3. $\sum_{n=2}^{\infty} \{1/(n \log n)\}$ **4.** $\sum_{n=2}^{\infty} \{1/(n \log n)^p\}$ **5.** $\sum_{n=1}^{\infty} (1/n)$ **(Pune 2010)**

ANSWERS

1. Divergent **2.** Divergent
3. Divergent **4.** Convergent if $p > 1$, divergent if $p \le 1$ **5.** Divergent

OBJECTIVE QUESTIONS

Multiple Choice Type Questions : *Select* (*a*), (*b*), (*c*) *or* (*d*), *whichever is correct.*

1. The series $1 + r + r^2 + r^3 + \ldots..$ is oscillatory if :

(*a*) $r < 1$ (*b*) $r > 1$ (*c*) $r = 1$ (*d*) $r = -1$. **(Rohilkhand, 2001)**

2. Infinite series $\Sigma (1/n^p)$ is convergent if :

(*a*) $p < 1$ (*b*) $p > 1$ (*c*) $p = 1$ (*d*) $p \le 1$. **(Agra 2007; Rohilkhand, 2001)**

3. Σu_n is the series of positive terms and $\lim_{n \to \infty} (u_n)^{1/n} > 1$ then the series is :

(*a*) Divergent (*b*) Convergent (*c*) Oscillatory (*d*) None of these. **(Meerut, 2003)**

4. Series Σu_n of positive terms is divergent if $\lim_{n\to\infty} n\left(\frac{u_n}{u_{n+1}}-1\right)$ is :

(*a*) < 1 (*b*) ≥ 1 (*c*) $= 1$ (*d*) ≤ 1.

5. The series $\sum \frac{1}{n(\log n)^p}$ is divergent if :

(*a*) $p > 1$ (*b*) $p \leq 1$ (*c*) $p < 1$ (*d*) $p = 1$. **(Agra 2008; Rohilkhand, 2001)**

6. The series $\Sigma 1/n^p$ is :

(*a*) Convergent (*b*) Divergent (*c*) Oscillatory (*d*) None of these. **(Meerut, 2004)**

7. The series Σu_n, where $u_n = \sqrt{n^2+1}-n$ is :

(*a*) Convergent (*b*) Divergent (*c*) Oscillatory (*d*) None of these.

8. The series $1+\frac{3}{1!}+\frac{5}{3!}+\frac{7}{4!}+\ldots$ is :

(*a*) Convergent (*b*) Divergent (*c*) Oscillatory (*d*) None of these.

9. The series $\frac{1}{3^p}+\frac{1}{5^p}+\frac{1}{7^p}+\frac{1}{9^p}+\ldots$ converges if :

(*a*) $p < 1$ (*b*) $p = 1$ (*c*) $p > 1$ (*d*) None of these.

10. The series $\Sigma 1/n^{3/4}$ is :

(*a*) Convergent (*b*) Divergent (*c*) Oscillatory (*d*) None of these.

ANSWERS

1. (*d*)	**2.** (*b*)	**3.** (*a*)	**4.** (*a*)	**5.** (*b*)
6. (*a*)	**7.** (*b*)	**8.** (*a*)	**9.** (*c*)	**10.** (*b*)

MISCELLANEOUS PROBLEMS ON CHAPTER 6

1. State and prove *p*-test for convergence and divergence of an infinite series.
Hint : Refer Art. 6.5, page 6.6 **[Kanpur 2005]**

2. Test the convergence of $\sum_{2}^{\infty} 1/(\log n)$. **[Agra 2005]**

3. Test the convergence of $\sum_{n=1}^{\infty} \{(n^2-1)^{1/2}-n\}$. **[Agra 2006]**

4. Select (*a*), (*b*), (*c*), (*d*) whichever is correct.

(*i*) If $\sum u_n$ is a positive term series such that $\lim_{n\to\infty} \frac{u_{n+1}}{u_n} = k$, then Σu_n converges if

(*a*) $k < 1$ (*b*) $k = 1$ (c) $k > 1$ (d) none of these. **[Agra 2006]**

(*ii*) Σu_n is a positive term series such that $\lim_{n\to\infty} (u_n)^{1/n} = k$, then Σu_n is convergent if (*a*) $k > 1$ (*b*) $k < 1$ (*c*) $k = 1$ (*d*) None of these **[Agra 2006]**

5. Test the convergence of the series :

$$\frac{\sqrt{2}-1}{3^3-1}+\frac{\sqrt{3}-1}{4^3-1}+\frac{\sqrt{4}-1}{5^3-1}+\ldots..$$ **[Delhi Maths (G) 2006]**

6. Discuss the convergence of the series

$$\frac{1}{2}x+\frac{1\cdot 3}{2\cdot 4}x^3+\frac{1\cdot 3\cdot 5}{2\cdot 4\cdot 6}x^3+\ldots..+\frac{1\cdot 3\cdot 5\ldots.(2n-1)}{2\cdot 4\cdot 6\ldots..(2n)}x^n+\ldots..$$

for all positive values of x, **[Delhi Maths (G) 2006]**

7. *Test for convergence* $1/(\sqrt{2}+\sqrt{3})+1/(\sqrt{3}+\sqrt{4})+1/(\sqrt{4}+\sqrt{5})+\ldots.$

[Delhi Maths (Prog) 2007]

8. *Test for convergence* $\sum_{n=1}^{\infty}\frac{1\cdot 2\cdot 3\cdots\cdots n}{7\cdot 10\cdots\cdots(3n+4)}$. **[Delhi B.A. (Prog.) III 2012]**

9. *Test the convergence of the following series :*

(*i*) $\sum_{n=1}^{\infty}\sqrt{n}/(n^3+1)$ (*ii*) $\sum_{n=1}^{\infty}\left(\frac{n^2-1}{n^2+1}\right)x^n$

(*iii*) $1+\frac{1+\alpha}{1+\beta}+\frac{(1+\alpha)(2+\alpha)}{(1+\beta)(2+\beta)}+\ldots.$ (*iv*) $\sum_{n=1}^{\infty}\frac{2\cdot 4\cdot 6\cdots\cdots(2n+2)}{3\cdot 5\cdot 7\cdots\cdots(2n+3)}x^n,\ x>0$

(*v*) $\sum_{n=1}^{\infty}\{1/(2n-1)(2n+1)\}$ (*vi*) $\sum_{n=1}^{\infty}2^{-n}$

10. *Prove that a necessary condition for a series* Σa_n *to converge is* $\lim_{n\to\infty}a_n=0$. What about the sufficiency ? Justify **[Delhi Maths (Prog) 2008]**

11. *State Cauchy's convergence criterion for an infinite series and using this principle check the convergence of the series* $\Sigma(1/n)$. **[Delhi Maths (Prog) 2008]**

12. *Test the convergence of the series* $\sum_{2}^{\infty}(1/\log n)$ **[Agra 2005]**

13. Test the convergence of $\sum_{n=1}^{\infty}\frac{1}{e^{n^3}}$ **[K.U. BCA (II) 2007]**

14. Test the convergence of $\sum_{n=1}^{\infty}\frac{1}{3^n+x},\ x\geq 0.$ **[Delhi Maths (Hons) 2009]**

15. Test the convergence of the series whose n^{th} term is (*i*) $\tan^{-1}(1/n)$. **[Kanpur 2008]**

(*ii*) $(2n^2-1)^{1/3}/(3n^3+2n+5)^{1/4}$ **[Kanpur 2009]**

16. Test for convergence (*i*) $1+(x/1)+(x^2/5)+\ldots+x^n/(n^2+1)+\cdots,\ x>0$

(*ii*) $\frac{1}{2\sqrt{1}}+\frac{x^2}{3\sqrt{2}}+\frac{x^4}{4\sqrt{3}}+\frac{x^6}{5\sqrt{4}}+\ldots..$ **[Meerut 2010]**

(*iii*) $1+x/2^2+x^2/3^2+x^3/4^2+\ldots.$ **[Kanpur 2010]**

(*iv*) $1/5+(2!)/5^2+(3!)/5^3+\ldots.$ **[Delhi B.A. (Prog.) III, 2014]**

(*iv*) $1+x^2/2!+x^2/4!+\ldots.$ **[Delhi B.A. (Prog.) III, 2013]**

ANSWERS

2. Conv. **3.** Conv. **4.** (*i*) (*a*); (*ii*) (*b*) **5.** Conv.
6. Conv. if $x < 1$, Div. if $x \geq 1$ **7.** Div. **8.** conv. **9.** (*i*) Conv. (*ii*) Conv. if $x < 1$, Div. if $x \geq 1$, (*iii*) Conv. if $\beta > \alpha + 1$; Div. if $\beta \leq \alpha + 1$ (*iv*) Conv. if $x < 1$ Div. if $x \geq 1$ (*v*) Conv. (*vi*) Conv **11.** Div. **12.** Div 13. Conv. **14.** conv. **15.** (*i*) Div (*ii*) Div. **16.** (*i*) Conv. if $x \leq 1$, div if $x > 1$ (*ii*) Conv. if $x^2 \leq 1$, div. if $x^2 > 1$ (*iii*) Conv. if $x \leq 1$, div. if $x > 1$ (*iv*) Div. (*v*) Conv. for each x

Chapter

7

Infinite Series With Positive and Negative Terms

7.1. INTRODUCTION

In this chapter, we shall be concerned with the consideration of the convergence of series whose terms are *not* all positive or negative and as such which contain both positive and negative terms. The test of boundedness of the partial sum sequences which holds good for the convergence of the positive term series fails for series with both positive and negative terms. As an illustration, we consider the series $\Sigma\,(-1)^n$ which when written in full is

$$-1+1-1+1-1+1\ .....$$

This series is *not* convergent even though the sequence of its partial sums is bounded.

We shall in this chapter also introduce the important concepts of absolute convergence and conditional convergence of a series and examine their implications for the re-arrangements of the terms of series and for the multiplication of series.

7.2. ABSOLUTE CONVERGENCE AND CONDITIONAL CONVERGENCE (or non-absolute convergence or semi convergence)

[Meerut 2009; Kanpur 2006, 07; M.S. Univ. T.N. 2006; Delhi B.A. (Prog.) III, 2013; Pune 2010; Delhi B.Sc. (Prog) III 2008, 09, 11, 12; Delhi Maths (H), 2005; Purvanchal 2006]

Def. *A series $\Sigma\, u_n$ is said to be absolutely convergent if the positive term series $\Sigma\,|\,u_n\,|$ formed by the moduli of the terms of the series is convergent.*

For example, the series

$$\Sigma\, u_n = 1-\frac{1}{2}+\frac{1}{2^2}-\frac{1}{2^3}+.....$$

is absolutely convergent, because

$$\Sigma\,|\,u_n\,| = 1+\frac{1}{2}+\frac{1}{2^2}+\frac{1}{2^3}+.....$$

is an infinite geometric series of positive terms with common ratio = $1/2 < 1$ and hence it is convergent.

A series is said to be *conditionally convergent, if it is convergent without being absolutely convergent.*

For example, the series $\Sigma\, u_n = 1-\frac{1}{2}+\frac{1}{3}-\frac{1}{4}+.....$ is conditionally convergent, because

$$\Sigma\,|\,u_n\,| = 1+\frac{1}{2}+\frac{1}{3}+\frac{1}{4}+.....$$

is not a convergent series. However $\Sigma\, u_n$ is convergent by Leibnitz's test (see Art. 7.3 and example based on it).

An important question at once arises about the relationship between the absolute convergence of a series and its convergence. It will be proved that *every absolutely convergent series is as well as convergent.*

Two proofs of this result will be given; one of these is rather elementary in character and the other is based on the Cauchy general principle of convergence of a series to be given in Art. 7.5.

Theorem. *Every absolutely convergent series is convergent.* **(Delhi B.A (Prog) III 2007) (Meerut 2012; Delhi B.A (Prog) III 2010, 13, 14; Delhi B.Sc. (Prog) III 2010)**

Proof. Let Σu_n be any absolutely convergent series. We associate with the series two positive term series Σv_n, Σw_n defined as follows :

$$v_n = \begin{cases} u_n \text{ if } u_n \geq 0; \\ 0 \text{ if } u_n < 0. \end{cases} \quad \text{and} \quad w_n = \begin{cases} -u_n \text{ if } u_n \leq 0; \\ 0 \text{ if } u_n > 0. \end{cases}$$

It may be seen that

$$|u_n| = v_n + w_n \qquad ...(i)$$

and

$$u_n = v_n - w_n \qquad ...(ii)$$

Now $\forall\ n \in \mathbf{N}$, we have

$$v_n \leq |u_n|, \qquad w_n \leq |u_n|$$

The series $\Sigma |u_n|$ being given to be convergent, by comparison test, it follows that Σv_n and Σw_n are both convergent. From (*ii*), it now follows that Σu_n is convergent. Hence the theorem.

Note 1. If Σu_n is convergent without being absolutely convergent, *i.e.*, if Σu_n is conditionally convergent, then each of the positive term series Σv_n, Σw_n diverges to $+\infty$.

This follows from

$$v_n = \frac{1}{2}\{|u_n| + u_n\} \quad \text{and} \quad w_n = \frac{1}{2}\{|u_n| - u_n\}.$$

Note 2. We shall now prove two tests one of which is of a particular type and the other is general. It may be remarked that there are no comparison tests for the convergence of conditionally convergent series.

7.3. ALTERNATING SERIES (Delhi B.A. (Prog) III 2009, 2015)

A series whose terms are alternately positive and negative is referred to as an alternating series.

A series of the form $u_1 - u_2 + u_3 - u_4 + ...$, where $u_n > 0\ \forall\ n \in \mathbf{N}$ is an alternating series and is denoted by $\sum_{n=1}^{\infty} (-1)^{n-1} u_n$. For example, $1 - 1/2 + 1/3 - 1/4 + ...$ is an alternating series.

Leibnitz's theorem. (or Altering series theorem)

Let $\langle u_n \rangle$ be a sequence such that $\forall\ n \in \mathbf{N}$

(a) $u_n \geq 0$ *(b)* $u_{n+1} \leq u_n$ *(c)* $\lim_{n \to \infty} u_n = 0$

Then the alternating series $\Sigma (-1)^{n-1} u_n$ is convergent. ***[Delhi B.Sc. (Hons) I 2012, 16; Purvanchal 2006; Kanpur, 2005; 07; Delhi B.Sc. II, 2015, 16; Ranchi 2010; Delhi B.A. (Prog) III, 2008, 2009, 2010, 2015; Meerut, 2004]***

Proof. We write $S_n = u_1 - u_2 + u_3 - ..... + (-1)^{n-1} u_n$,

so that $\langle S_n \rangle$ is the sequence of partial sums of the given series.

In order to prove that the given series converges, we shall show that $\langle S_n \rangle$ converges. To this end we shall first prove that the subsequences $\langle S_{2n} \rangle$ and $\langle S_{2n+1} \rangle$ both converge to the same limit.

Now, $S_{2n} = u_1 - u_2 + u_3 - u_4 + ... + u_{2n-1} - u_{2n}$

and $S_{2n+2} = u_1 - u_2 + u_3 - u_4 + ... + u_{2n-1} - u_{2n} + u_{2n+1} - u_{2n+2}$

$\therefore$ $S_{2n+2} - S_{2n} = u_{2n+1} - u_{2n+2} \geq 0, \forall\ n \in \mathbf{N}$, since $u_{n+1} \leq u_n\ \forall\ n \in \mathbf{N}$

$\Rightarrow$ $S_{2n+2} \geq S_{2n}$

$\Rightarrow$ $<S_{2n}>$ is a monotonically increasing sequence.

Also, we have

$$S_{2n} = (u_1 - u_2) + (u_3 - u_4) + ... + (u_{2n-1} - u_{2n})$$

Each of the expressions

$$u_1 - u_2,\ u_3 - u_4,\ .....,\ u_{2n-1} - u_{2n}$$

being positive, we see that S_{2n} is positive. Also

$$S_{2n} = u_1 - [(u_2 - u_3) + (u_4 - u_5) + ... + (u_{2n-2} - u_{2n-1}) + u_{2n}]$$

$\therefore$ $S_{2n} \leq u_1$, as $u_{n+1} \leq u_n$ and $u_n > 0\ \forall\ n \in \mathbf{N}$

$\Rightarrow$ $<S_{2n}>$ is bounded above.

Thus we have shown that $<S_{2n}>$ is a bounded monotonically increasing sequence and is as such convergent.

Let $$\lim_{n \to \infty} S_{2n} = l. \qquad ...(1)$$

We shall now show that $\lim_{n \to \infty} S_{2n+1}$ also exists and is the same as $\lim_{n \to \infty} S_{2n}$.

Now, $S_{2n+1} = u_1 - u_2 + u_3 - u_4 + ... + u_{2n-1} - u_{2n} + u_{2n+1}$

$\Rightarrow$ $S_{2n+1} = S_{2n} + u_{2n+1} \Rightarrow \lim_{n \to \infty} S_{2n+1} = \lim_{n \to \infty} S_{2n} + \lim_{n \to \infty} u_{2n+1}$

$\therefore$ $\lim_{n \to \infty} S_{2n+1} = l + 0$, as by the given condition (c), $\lim_{n \to \infty} u_{2n+1} = 0$

Thus, $$\lim_{n \to \infty} S_{2n+1} = l \qquad ...(2)$$

From (1) and (2), we see that the sequences $<S_{2n}>$ and $<S_{2n+1}>$ both converge to the same limit l. We shall now show that

$$\lim_{n \to \infty} S_n = l.$$

Let $\varepsilon > 0$ be given. Now

$\lim_{n \to \infty} S_{2n+1} = l \Rightarrow$ there exists a positive integer m_1 such that

$$| S_{2n+1} - l | < \varepsilon \quad \forall\ n \geq m_1 \qquad ...(3)$$

Again, $\lim_{n \to \infty} S_{2n} = l \Rightarrow$ there exists a positive integer m_2 such that

$$| S_{2n} - l | < \varepsilon \quad \forall\ n \geq m_2 \qquad ...(4)$$

From (3) and (4), $$| S_n - l | < \varepsilon \quad \forall\ n \geq m, \qquad ...(5)$$

where $m = \max \{m_1, m_2\}$.

From (5), it follows that $<S_n>$ converges to l implying that the given series is convergent.

Note 1. It may be noted that $<S_{2n}>$ is monotonically increasing and $<S_{2n+1}>$ is monotonically decreasing. Of course, both converge to the same limit.

Note 2. The alternating series $\Sigma (-1)^{n-1} u_n$ will not converge if either $\lim u_n \neq 0$ or $u_{n+1} \not< u_n\ \forall\ n \in \mathbf{N}$.

Example showing that a convergent series may not be absolutely convergent.

[Delhi B.Sc. (Prog) III 2011; G.N.D.U. Amritsar 2010; Kanpur 2006]
[Delhi; B.A. (Prog.) III, 2016; Delhi B.Sc. (Hons.) I 2011, 17; Meerut 2012]

Consider the series $\sum_{n=1}^{\infty} (-1)^{n-1} \frac{1}{n} = 1 - \frac{1}{2} + \frac{1}{3} - \frac{1}{4} + \ldots + (-1)^{n-1} \frac{1}{n} + \ldots$

Here $u_n = \frac{1}{n}$ so that $\lim_{n \to \infty} u_n = \lim_{n \to \infty} \frac{1}{n} = 0$. Also $\frac{1}{n} > \frac{1}{n+1} \Rightarrow u_n > u_{n+1} \; \forall \; n \in \mathbf{N}$

Hence by Leibnitz's test $\sum_{n=1}^{\infty} (-1)^{n-1} \frac{1}{n}$ is convergent.

However, the series $\sum_{n=1}^{\infty} \left| (-1)^{n-1} \frac{1}{n} \right|$, *i.e.*, $\sum_{n=1}^{\infty} \frac{1}{n}$ is divergent. Thus, there exists a convergent series $\sum_{n=1}^{\infty} \{(-1)^{n-1} \times (1/n)\}$ which is not an absolutely convergent series.

SOLVED EXAMPLES

Example 1. *Test for convergence, absolute convergence and conditional convergence the series :*

$$\sum_{n=1}^{\infty} \frac{(-1)^{n+1}}{n^p} = 1 - \frac{1}{2^p} + \frac{1}{3^p} - \frac{1}{4^p} + \ldots \text{ for } p > 0.$$

[Delhi B.A. (Prog) III 2009, 2015; B.Sc. (Prog.) II, 2015; Delhi B.Sc. (Hons) I 2011]

Solution. Let the given series be denoted by $\sum_{n=1}^{\infty} (-1)^{n-1} u_n$.

Here $p > 0$ and $(n+1)^p > n^p$, Therefore, we have

$$\frac{1}{(n+1)^p} < \frac{1}{n^p} \Rightarrow u_{n+1} < u_n \; \forall \; n \in N$$

Also, $\lim_{n \to \infty} u_n = \lim_{n \to \infty} \frac{1}{n^p} = 0$, since $p > 0$

$\therefore$ By Leibnitz Test, $\sum_{n=1}^{\infty} \frac{(-1)^{n+1}}{n^p}$ is convergent.

Now, $\sum_{n=1}^{\infty} \left| \frac{(-1)^{n+1}}{n^p} \right| = \sum_{n=1}^{\infty} \frac{1}{n^p}$ is convergent if $p > 1$ and divergent if $p \le 1$.

Hence, the given series is absolutely convergent if $p > 1$ and conditionally convergent if $0 \le p \le 1$.

Example 2. *Test for absolute convergence the series* $\sum_{n=1}^{\infty} (-1)^{n-1} \frac{n^2}{(n+1)!}$.

Solution. We have $\lim_{n \to \infty} \frac{|u_n|}{|u_{n+1}|} = \lim_{n \to \infty} \frac{n^2}{(n+1)!} \times \frac{(n+2)!}{(n+1)^2} = \lim_{n \to \infty} \frac{(n+2)}{(1+1/n)^2} = \infty$

By ratio test, $\Sigma \, | u_n |$ is convergent. Hence, the given series is absolutely convergent.

Example 3. *Test for convergence, absolute convergence and conditional convergence of the series* $\sum_{n=1}^{\infty} \frac{(-1)^{n+1}}{\log (n+1)}$, *i.e.,* $\frac{1}{\log 2} - \frac{1}{\log 3} + \frac{1}{\log 4} - \ldots$ ***[Delhi B.Sc. (Prog) III 2010]***

Solution. We have $u_n = \frac{1}{\log (n+1)} \; \forall \; n \in \mathbf{N}$. Clearly, $\lim_{n \to \infty} u_n = \lim_{n \to \infty} \frac{1}{\log (n+1)} = 0$.

Since log n is an increasing function for all $n > 0$, therefore

$$\log(n+2) > \log(n+1) \qquad (\because\ n+2 > n+1)$$

$$\Rightarrow \qquad \frac{1}{\log(n+2)} < \frac{1}{\log(n+1)}\ \forall\ n \in \mathbf{N}$$

$$\Rightarrow \qquad u_{n+1} < u_n\ \forall\ n \in \mathbf{N}$$

Thus, the conditions of Leibnitz's test are satisfied and so the given series is convergent.

Now, we test the absolute convergence of the given series.

We have $\sum\limits_{n=2}^{\infty} |u_n| = \sum\limits_{n=2}^{\infty} \frac{1}{\log n}$ which is divergent.

Hence, Σu_n is not absolutely convergent, *i.e.*, Σu_n is conditionally convergent.

Example 4. *Show that the series*

$$\frac{\log 2}{2^2} - \frac{\log 3}{3^2} + \frac{\log 4}{4^2} - \ldots\ldots \qquad \textit{i.e.,} \qquad \sum_{t=2}^{\infty}(-1)^k \frac{\log k}{k^2} \quad \textit{converges,}$$

[*Delhi Maths (H), 2000, 02; Delhi B.Sc. (Hons) I, 2011; Delhi Maths (G), 2006*]

Solution. We have $u_n = \frac{\log(n+1)}{(n+1)^2}\ \forall\ n \in \mathbf{N}$.

Since $\lim\limits_{n\to\infty} \frac{\log n}{n^2} = 0$, we have $\lim\limits_{n\to\infty} u_n = 0$.

Now we shall prove that $u_{n+1} < u_n\ \forall\ n \in \mathbf{N}$.

Let $u(x) = \frac{\log x}{x^2}$ so that

$$u'(x) = \frac{x^2 \times (1/x) - 2x\log x}{x^4} \quad \Rightarrow \quad u'(x) = \frac{1 - 2\log x}{x^3} < 0\ \forall\ x > e^{1/2}$$

$$\left(\because\ x > e^{1/2} \Leftrightarrow \log x > \frac{1}{2} \Leftrightarrow 1 - 2\log x < 0\right)$$

$\Rightarrow$ $u(x)$ is a decreasing function $\forall\ x > e^{1/2}$

$$\Rightarrow\ u_{n+2} \le u_{n+1}\ \forall\ n \in \mathbf{N} \qquad (\because\ n+2 > n+1 > e^{1/2}\ \forall\ n \in \mathbf{N})$$

$$\Rightarrow \quad \frac{\log(n+2)}{(n+2)^2} \le \frac{\log(n+1)}{(n+1)^2}\ \forall\ n \in \mathbf{N} \quad \text{so that} \quad u_{n+1} < u_n\ \forall\ n \in \mathbf{N}$$

Thus, both the conditions of Leibnitz's test are satisfied and so the given series is convergent.

Example 5. *Show that the series* $x + \frac{x^2}{2!} + \frac{x^3}{3!} + \ldots$ *converges absolutely for all values of* x.

Solution. We have $u_n = \frac{x^n}{n!}$, $u_{n+1} = \frac{x^{n+1}}{(n+1)!}$

Now, $\lim\limits_{n\to\infty} \frac{|u_n|}{|u_{n+1}|} = \lim\limits_{n\to\infty} \frac{(n+1)!}{n!\,|x|} = \lim\limits_{n\to\infty} \frac{n+1}{|x|} \to \infty$, provided $x \neq 0$.

By ratio test, $\Sigma |u_n|$ is convergent for all values of x. Hence, the given series converges absolutely for all values of x.

Example 6. (*a*) *Prove that the series* $x - x^3/3 + x^5/5 - \ldots$ *converges if and only if* $-1 \le x \le 1$.
[*Purvanchal 2006; Delhi Maths (G), 1995; Delhi Maths (H), 1993*]

(*b*) Discuss the convergence of a series $\sum_{n=1}^{\infty} ar^{n-1}$ where $a \neq 0, r \neq 0$ and *a, r* are fixed real numbers. (*G.N.D.U. Amritsar 2010*)

Solution. (*a*) Here $| u_n | = \left| \dfrac{x^{2n-1}}{2n-1} \right|$ and $| u_{n+1} | = \left| \dfrac{x^{2n+1}}{2n+1} \right|$

$$\Rightarrow \qquad \lim_{n \to \infty} \frac{| u_n |}{| u_{n+1} |} = \lim_{n \to \infty} \frac{1}{x^2} \cdot \frac{2n+1}{2n-1} = \frac{1}{x^2}$$

By ratio test, $\Sigma \, | u_n |$ is convergent if

$$1/x^2 > 1 \Rightarrow x^2 < 1 \Rightarrow | x | < 1,$$

and $\Sigma \, | u_n |$ is divergent if $| x | > 1$.

Thus, $\Sigma \, u_n$ is absolutely convergent if $| x | < 1$.

$\Rightarrow \Sigma \, u_n$ is convergent if $| x | < 1$.

If $| x | = 1$, the ratio test fails.

Now $| x | = 1 \Rightarrow x = 1$ or $x = -1$

Putting $x = 1$, the given series becomes

$$1 - \frac{1}{3} + \frac{1}{5} - \frac{1}{7} - \ldots\ldots$$

which converges by Leibnitz's test.

Putting $x = -1$, the given series becomes

$$-1 + \frac{1}{3} - \frac{1}{5} + \frac{1}{7} - \ldots\ldots$$

which is again convergent by Leibnitz's test.

Hence, the given series is convergent if $| x | \le 1$ or $-1 \le x \le 1$.

If $x > 1$, then u_n does not tend to zero as $n \to \infty$, and so $\Sigma \, u_n$ does not converge when $x > 1$.

Since $\Sigma \, | u_n |$ diverges when $x < -1$ (*i.e.*, $| x | > 1$), so $\Sigma \, u_n$ also diverges when $x < -1$.

Hence, $\Sigma \, u_n$ converges iff $-1 \le x \le 1$.

(*b*) Try yourself

Example 7. *Test for absolute convergence and convergence the following series :*

(*i*) $\sum_{n=1}^{\infty} \dfrac{(-)^n \cos n\alpha}{n\sqrt{n}}$, *α being real* (*ii*) $\sum_{n=1}^{\infty} \dfrac{(-1)^n \sin n\alpha}{n^3}$, *α being real*

[*Delhi B.Sc. (Prog.) III 2011*]

(*iii*) $\sum_{n=1}^{\infty} \dfrac{\cos nx}{n^2}$ (*iv*) $\sum_{n=1}^{\infty} \dfrac{\sin nx}{n^2}$ **[*Meerut, 2003; Delhi Maths (H), 1996*]**

Solution. (*i*) Here $u_n = \dfrac{(-1)^n \cos n\alpha}{n^{3/2}}$

$$\therefore \quad | u_n | = \left| \frac{(-1)^n \cos n\alpha}{n^{3/2}} \right| = \frac{| \cos n\alpha |}{n^{3/2}} \le \frac{1}{n^{3/2}}, \text{ as } | \cos n\alpha | \le 1$$

Also, $\Sigma\,(1/n^{3/2})$ is convergent.

$\therefore$ By comparison test, the series $\Sigma\,|\,u_n\,|$ converges

$\Rightarrow$ The given series is absolutely convergent.

Since every absolutely convergent series is convergent, the given series is convergent.

(ii), (iii) (iv). Try yourself as in part (*i*). All are convergent.

Example 8. (*a*) *Show that* $\sum (-1)^n \dfrac{n+2}{2^n+5} \cos nx$ *is convergent for all real values of x.*

[*Delhi Maths* (G), *2001, 02*]

(*b*) *Test for convergence the series* $\sum\limits_{n=1}^{\infty} (-1)^n \dfrac{n+2}{2^n+5}$. **(*Meerut, 2001, 02*)**

Solution. (*a*) Here $$u_n = (-1)^n \frac{n+2}{2^n+5} \cos nx$$

$$\therefore \quad |\,u_n\,| = \left| (-1)^n \frac{n+2}{2^n+5} \cos nx \right| = \frac{n+2}{2^n+5} |\cos nx| \le \frac{n+2}{2^n+5}, \text{ as } |\cos nx| \le 1 \qquad \ldots(1)$$

Let $v_n = (n+2)/(2^n+5)$. Then $v_{n+1} = (n+3)/(2^{n+1}+5)$.

$$\therefore \quad \frac{v_n}{v_{n+1}} = \frac{n+2}{n+3} \cdot \frac{2^{n+1}+5}{2^n+5} = \frac{1+2/n}{1+3/n} \cdot \frac{2+5/2^n}{1+5/2^n} \quad \Rightarrow \quad \lim_{n\to\infty} \frac{v_n}{v_{n+1}} = 2 > 1$$

$\therefore$ By ratio test, $\Sigma\, v_n$ is convergent. Also, we have $|\,u_n\,| \le v_n$, by (1).

Hence, by comparison test, $\Sigma\,|\,u_n\,|$ is convergent.

$\Rightarrow$ the given series $\Sigma\, u_n$ is absolutely convergent.

Since every absolutely convergent series is convergent, the given series is convergent.

(*b*) Here $u_n = (-1)^n \dfrac{n+2}{2^n+5}$ and so $u_{n+1} = (-1)^{n+1} \dfrac{n+1+2}{2^{n+1}+5}$

$$\therefore \quad \lim_{n\to\infty} \frac{|\,u_n\,|}{|\,u_{n+1}\,|} = \text{iim}_{n\to\infty} \left| \frac{n+2}{2^n+5} \times \frac{2^{n+1}+5}{n+3} \right| = \lim_{n\to\infty} \left| \frac{1+2/n}{1+3/n} \cdot \frac{2+5/2^n}{1+5/2^n} \right| = 2 < 1$$

Hence, by ratio test, $\Sigma\,|\,u_n\,|$ converges. Since every absolutely convergent series is convergent, so $\Sigma\, u_n$ is convergent.

Example 9. *Test for convergence the following series* :

(*a*) $\log\left(\dfrac{2}{1}\right) - \log\left(\dfrac{3}{2}\right) + \log\left(\dfrac{4}{3}\right) - \log\left(\dfrac{5}{4}\right) + \ldots$ (*b*) $\log\left(\dfrac{1}{2}\right) - \log\left(\dfrac{2}{3}\right) + \log\left(\dfrac{3}{4}\right) - \log\left(\dfrac{4}{5}\right) + \ldots$

Solution. (*a*) Let the given series be denoted by $\sum\limits_{n=1}^{\infty} (-1)^{n-1}\, u_n$.

Then, $$u_n = \log \frac{n+1}{n} = \log\left(1 + \frac{1}{n}\right) > 0, \ \forall\, n \in \mathbf{N}$$

$$\therefore \quad u_{n+1} - u_n = \log \frac{n+2}{n+1} - \log \frac{n+1}{n} = \log\left(\frac{n+2}{n+1} \times \frac{n}{n+1}\right) = \log \frac{n^2+2n}{n^2+2n+1}$$

Since $0 < (n^2+2n)/(n^2+2n+1) < 1$, so $\log\,\{n^2+2n)/(n^2+2n+1)\} < 0$

Thus, $u_{n+1} - u_n < 0$ and so $u_{n+1} < u_n \ \forall \ n \in \mathbf{N}$

Again, $\lim_{n\to\infty} u_n = \lim_{n\to\infty} \log(1+1/n) = \log 1 = 0$

Hence by Leibnitz's test, the given series converges.

(*b*) Left as an exercise. **Ans.** Convergent.

EXERCISES

1. Show that the following series are convergent. Which of the series is absolutely convergent and which is conditionally convergent ?

(*i*) $\frac{1}{1^2} - \frac{1}{2^2} + \frac{1}{3^2} - \frac{1}{4^2} + \ldots\ldots$ i.e., $\sum_{n=1}^{\infty} \frac{(-1)^{n-1}}{n^2}$ **[Delhi B.Sc. (Hons.) I, 2016]**

(*ii*) $\frac{1}{2\log 2} - \frac{1}{3\log 3} + \frac{1}{4\log 4} - \ldots\ldots$, i.e., $\sum_{k=2}^{\infty} (-1)^k \frac{1}{k\log k}$

[Delhi B.Sc. (Hons) I 2010, 2011]

2. Test for convergence and absolute convergence the infinite series whose general terms are given below :

(*i*) $(-1)^{n+1} \frac{1}{3n-2}$ (*ii*) $(-1)^{n+1} \frac{1}{2n}$ **[Delhi B.A. (Prog) III, 2011, 2015]**

(*iii*) $(-1)^{n+1} \frac{1}{2n(2n-1)}$ (*iv*) $(-1)^{n+1} \frac{1}{(2n+1)^2}$ (*v*) $(-1)^{n+1} \frac{1}{\sqrt{n}}$

(Delhi B.Sc. (Prog) III 2010, Meerut 2010; Ranchi 2010)

(*vi*) $\sum (-1)^{n-1} \frac{1}{n\sqrt{n}}$, *i.e.*, $1 - \frac{1}{2\sqrt{2}} + \frac{1}{3\sqrt{3}} - \frac{1}{4\sqrt{4}} + \ldots\ldots$ **(Meerut, 2003, 04, 11)**

Is the series conditionally convergent. **(Delhi, B.Sc. (Prog.) II, 2016)**

3. Show that the series $\sum \frac{(-1)^n}{n^p}$ is absolutely convergent $\forall \ p > 0$ and conditionally convergent $\forall \ p < 0$.

4. Prove that the series $1 - \frac{1}{3\cdot 2^2} + \frac{1}{5\cdot 3^2} - \frac{1}{7\cdot 4^2} + \frac{1}{9\cdot 5^2} + \ldots\ldots$ is convergent. Is it absolutely convergent ?

5. Show that the series $1 - \frac{1}{3\cdot 4} + \frac{1}{5\cdot 4^2} - \frac{1}{7\cdot 4^3} + \ldots\ldots$ is convergent.

6. Prove that the series $\sum_{n=1}^{\infty} \frac{(-1)^n \log n}{n^2}$ is convergent. Is it absolutely convergent ?

7. Show that the following series are conditionally convergent :

(*i*) $\sum \frac{(-1)^{n+1}}{\sqrt{n}}$ **(Kanpur 2007)** (*ii*) $\sum \frac{(-1)^{n+1}}{n}$ (*iii*) $\sum \frac{(-1)^{n+1}}{\log(n+1)}$

(*iv*) $\sum \frac{(-1)^n}{n+2}$ (*v*) $\sum \frac{(-1)^{n+1}}{\sqrt{n+1}}$ (*vi*) $\sum_{n=2}^{\infty} \frac{(-1)^{n+1}}{n\log n}$

8. Give examples of series which satisfy conditions :

(*i*) (*a*) and (*b*) but not (*c*), (*ii*) (*b*) and (*c*) but not (*a*), (*iii*) (*c*) and (*a*) but not (*b*).

of the Leibnitz's theorem on alternating series.

9. Consider for the convergence of the following series; a being positive

(i) $\sum \frac{(-1)^n}{n+a}$ *(ii)* $\sum \frac{(-1)^n}{\sqrt{n+a}}$ *(iii)* $\sum \frac{(-1)^n}{\sqrt{n}+\sqrt{a}}$ *(iv)* $\sum \frac{(-1)^n}{(\sqrt{n}+\sqrt{a})^2}$

10. Examine for convergence the series :

(i) $\sum \sin\frac{a}{n}$ *(ii)* $\sum \frac{-}{n}\sin\frac{a}{n}$ *(iii)* $\sum (-1)^n \sin\frac{a}{n}$

(iv) $\sum \left(1-\cos\frac{a}{n}\right)$ *(v)* $\sum (-1)^n\, n\left(1-\cos\frac{a}{n}\right)$

11. Show that $\Sigma (-1)^n (n+a)^{-s}$, $a > 0$ is absolutely convergent if $s > 1$ and conditionally convergent if $0 < s \le 1$.

12. *Show that the following series are absolutely convergent :*

(i) $\sum (-1)^{n-1}\left\{\frac{1}{n^2}+\frac{1}{(n+1)^2}\right\}$ **(Purvanchal 2006)** *(ii)* $\sum (-1)^{n-1}\left\{\frac{1}{n^{5/2}}+\frac{1}{(n+1)^{5/2}}\right\}$

13. *Test for convergence the following series :*

(i) $\sum_{n=1}^{\infty} \frac{(-1)^{n-1}}{n}(\sqrt{n+1}-\sqrt{n-1})$ **[Delhi Physics (H), 1995]**

(ii) $\sum_{n=1}^{\infty} (-1)^{n+1}\frac{\sqrt{n+1}-\sqrt{n}}{n}$, *i.e.,* $\frac{\sqrt{2}-\sqrt{1}}{1}-\frac{\sqrt{3}-\sqrt{2}}{2}+\frac{\sqrt{4}-\sqrt{3}}{3}-\ldots$

(iii) $\frac{1}{2-\sqrt{2}}-\frac{1}{3-\sqrt{3}}+\frac{1}{4-\sqrt{4}}-\ldots$

14. Show that the series

$$\sum_{n=2}^{\infty} (-1)^n \frac{1}{n(\log n)^p} \quad i.e., \quad \frac{1}{2(\log 2)^p}-\frac{1}{3(\log 3)^p}+\frac{1}{4(\log 4)^p}-\ldots$$

is absolutely convergent if $p > 1$ and conditionally convergent if $0 \le p \le 1$.

[Delhi Maths (H), 2005]

15. *Test for convergence, absolute convergence and conditional convergence of the following series :*

(i) $1/\sqrt{1}-1/\sqrt{3}+1/\sqrt{5}-1/\sqrt{7}+\ldots$ **[Delhi B.Sc. I (Hons) 2010]**

(ii) $\frac{1}{1\cdot 2}-\frac{1}{3\cdot 4}+\frac{1}{5\cdot 6}-\frac{1}{7\cdot 8}+\ldots$ **[Meerut 2009, 10]**

(iii) $\left(\frac{1}{2}\right)^2-\left(\frac{1\cdot 3}{2\cdot 4}\right)^2+\left(\frac{1\cdot 3\cdot 5}{2\cdot 4\cdot 6}\right)^2-\ldots$ *(iv)* $\sum_{n=1}^{\infty} (-1)^{n-1}\frac{\sin nx+\cos nx}{n^{3/2}}$

16. Test for convergence and absolute convergence the series :

$$\sum_{n=1}^{\infty} \frac{(-1)^{n+1}}{2n+3}$$

[Delhi Maths (G), 2005]

17. Prove that $1-(1/\sqrt{2})+(1/\sqrt{3})-(1/\sqrt{4})+\ldots$ is not absolutely convergent

(Meerut 2010)

ANSWERS

1. (*i*) Convergent, absolutely convergent (*ii*) Convergent, conditionally convergent
2. (*i*) Convergent, not absolutely convergent (*ii*) Convergent, not absolutely convergent
(*iii*) Convergent, absolutely convergent (*iv*) Convergent, absolutely convergent
(*v*) Convergent, not absolutely convergent (*vi*) Convergent, absolutely convergent
13. (*i*) Convergent, absolutely convergent (*ii*) Convergent, absolutely convergent
(*iii*) Convergent, conditionally convergent
15. (*i*) Convergent, conditionally convergent (*ii*) Convergent, absolutely convergent
(*iii*) Convergent, conditionally convergent (*iv*) Convergent, absolutely convergent
16. Convergent but not absolutely convergent

7.4. CAUCHY PRINCIPLE OF CONVERGENCE FOR A SERIES

We shall now deduce the Cauchy's general principle of convergence for a series from the corresponding principle for a sequence. Consider the series Σu_n.

We write
$$S_n = u_1 + u_2 + \ldots.. + u_n \quad \ldots(1)$$
so that $<S_n>$ denotes the sequence of partial sums of the given series.

Now, by definition, the series Σu_n is convergent if and only if the sequence $<S_n>$ of its partial sums is convergent.

Again by the Cauchy's general principle of convergence for a sequence, a necessary and sufficient condition for the convergence of a sequence $<S_n>$ is that to each given positive number ε, there corresponds a positive integer m such that
$$|S_{n+p} - S_n| < \varepsilon \quad \forall\, n \geq m,\ \forall\, p \geq 1.$$
Let us now interpret this condition in terms of the series Σu_n.

Now,
$$S_{n+p} = u_1 + u_2 + \ldots.. + u_n + u_{n+1} + \ldots.. + u_{n+p} \quad \ldots(2)$$
$\therefore$ (1) and (2) $\Rightarrow$ $S_{n+p} - S_n = u_{n+1} + \ldots.. + u_{n+p}$

Thus, we see that
$$|S_{n+p} - S_n| < \varepsilon,\ \forall\, n \geq m,\ \forall\, p \geq 0 \Leftrightarrow |u_{n+1} + \ldots + u_{n+p}| < \varepsilon,\ \forall\, n \geq m,\ \forall\, p \geq 1$$
It follows that a necessary and sufficient condition for the series Σu_n to be convergent is that to each ε > 0, there corresponds a positive integer m such that
$$|u_{n+1} + u_{n+2} + \ldots.. + u_{n+p}| < \varepsilon\ \forall\, n \geq m,\ \forall\, p \geq 1.$$
Exercise. Deduce Leibnitz's theorem for alternating series from the Cauchy's principle of convergence of series.

Illustration 1. *Show with the help of the Cauchy's principle that the series*
$$1 - 1/2 + 1/3 - 1/4 + \ldots$$
is convergent.

While the convergence of this alternating series is an easy and direct consequence of the Leibnitz's test, we shall also establish the convergence alternatively with the help of the general principle of convergence.

Proof. We have
$$u_{n+1} + u_{n+2} + \ldots.. + u_{n+p} = (-1)^n \left\{\frac{1}{n+1} - \frac{1}{n+2} + \ldots.. + (-1)^{p+1} \frac{1}{n+p}\right\}$$
$$= (-1)^n \left\{\frac{1}{n+1} - \left(\frac{1}{n+2} - \frac{1}{n+3}\right) - \left(\frac{1}{n+4} - \frac{1}{n+5}\right) + \ldots..\right\}$$

$$\Rightarrow |u_{n+1} + u_{n+2} + \ldots + u_{n+p}| \le \frac{1}{n+1} \;\forall\; p.$$

Let $\varepsilon > 0$ be given. There exists $m \in \mathbf{N}$ such that $1/(n+1) < \varepsilon \;\forall\; n \ge m$.

Thus we see that to $\varepsilon > 0$, there corresponds $m \in \mathbf{N}$ such that

$$|u_{n+1} + u_{n+2} + \ldots + u_{n+p}| < \varepsilon \;\;\forall\; n \ge m, \forall\; p \ge 1$$

so that by the Cauchy's principle the series is convergent.

7.5. SOME IMPORTANT THEOREMS ON ABSOLUTELY CONVERGENT SERIES

Theorem I. *Every absolutely convergent series is convergent. The converse need not be true.*

[*Bhopal, 2004; Delhi Maths (H), 2000, 02; Delhi Maths (G), 2000, 02; Meerut, 1998; Patna, 2003; G.N.D.U. amritsar 2010; Kanpur 2010; Delhi Maths (Prog) 2007,08*]

While this result has already been proved in an elementary manner in Art. 7.2, we shall now obtain the result as a consequence of the Cauchy's principle. Thus, we have to show that

$$\Sigma\, |u_n| \text{ is convergent } \Rightarrow \Sigma\, u_n \text{ is convergent.}$$

Proof. Let $\Sigma\, u_n$ be an absolutely convergent series, *i.e.*, let $\Sigma\, |u_n|$ be convergent.

Let $\varepsilon > 0$ be given. We have to show that $\exists\; m \in \mathbf{N}$ such that

$$|u_{n+1} + u_{n+2} + \ldots + u_{n+p}| < \varepsilon, \;\forall\; n \ge m, \;\forall\; p \ge 1$$

We are given, however, that $\Sigma\, |u_n|$ is convergent. By Cauchy's principle of convergence, there exists $m \in \mathbf{N}$ such that

$$|u_{n+1}| + |u_{n+2}| + \ldots + |u_{n+p}| < \varepsilon, \;\forall\; n \ge m, \;\forall\; p \ge 1 \qquad \ldots(1)$$

Again, we have $\forall\; n \in \mathbf{N}$ and $\forall\; p \ge 0$

$$|u_{n+1} + u_{n+2} + \ldots + u_{n+p}| \le |u_{n+1}| + |u_{n+2}| + \ldots + |u_{n+p}| \qquad \ldots(2)$$

From (1) and (2), $\quad |u_{n+1} + u_{n+2} + \ldots + u_{n+p}| < \varepsilon, \;\forall\; n \ge m, \forall\, p \ge 1$

Hence by Cauchy's general principle of convergence the given series $\Sigma\, u_n$ is convergent.

The converse of the above theorem is not true.

Consider the series $\quad \displaystyle\sum \frac{(-1)^{n-1}}{n} = 1 - \frac{1}{2} + \frac{1}{3} - \frac{1}{4} + \ldots\ldots$

Here $u_n = 1/n$ and $\lim\limits_{n\to\infty} u_n = 0$. Also $1/(n+1) < 1/n \Rightarrow u_{n+1} < u_n \;\forall\; n \in \mathbf{N}$

$\therefore$ By Leibnitz's test, $\displaystyle\sum \frac{(-1)^{n-1}}{n}$ is convergent. But the series $\displaystyle\sum \left|\frac{(-1)^{n-1}}{n}\right| = \sum \frac{1}{n}$ is divergent. Hence a convergent series need not be absolutely convergent.

Exercise. If $\sum\limits_{n=1}^{\infty} |a_n|$ converges, then $\sum\limits_{n=1}^{\infty} a_n$ converges. Is the converse true ? Justify.

(G.N.D.U. Amritsar 2010)

Theorem II. *If* $\sum\limits_{n=1}^{\infty} u_n$ *is an absolutely convergent series, then the series of its positive terms and the series of its negative terms are both convergent.*

Proof. Let S_n and σ_n be the nth partial sums of the series $\Sigma\, u_n$ and $\Sigma\, |u_n|$ respectively.

$\therefore \quad S_n = u_1 + u_2 + \ldots + u_n \quad$ and $\quad \sigma_n = |u_1| + |u_2| + \ldots + |u_n|$

Let P_n and $-Q_n$ denote the sum of positive and negative terms in S_n. Then, we have

$$S_n = P_n - Q_n \text{ and } \sigma_n = P_n + Q_n \qquad \ldots(1)$$

From (1), $P_n = (\sigma_n + S_n)/2$ and $Q_n = (\sigma_n - S_n)/2$ ...(2)

Now, Σu_n is absolutely convergent (given)

$\Rightarrow \Sigma u_n$ and $\Sigma | u_n |$ are both convergent

$\Rightarrow \langle S_n \rangle$ and $\langle \sigma_n \rangle$ are both convergent sequences

Let $\lim_{n \to \infty} S_n = S$ and $\lim_{n \to \infty} \sigma_n = \sigma$ ...(3)

Now, from (2), $\lim_{n \to \infty} P_n = \lim_{n \to \infty} \frac{1}{2}(\sigma_n + S_n) = \frac{1}{2}(\sigma + S)$, by (3) ...(4)

and $\lim_{n \to \infty} Q_n = \lim_{n \to \infty} \frac{1}{2}(\sigma_n - S_n) = \frac{1}{2}(\sigma - S)$, by (3) ...(5)

From (4) and (5), it follows that $\langle P_n \rangle$ and $\langle Q_n \rangle$ are convergent. Hence the series of positive terms and the series of negative terms are separately convergent.

Corollary. *If $\sum_{n=1}^{\infty} u_n$ is conditionally convergent, then the series of its positive terms and the series of its negative terms are both divergent.*

Proof. Σu_n is conditionally convergent (given)

$\Rightarrow \Sigma u_n$ is convergent and $\Sigma | u_n |$ is divergent

$\Rightarrow \langle S_n \rangle$ is convergent while $\langle \sigma_n \rangle$ is divergent.

$\Rightarrow \lim_{n \to \infty} S_n = S$, (say) and $\lim_{n \to \infty} \sigma_n = \infty$ ...(6)

Now, from (2) and (6), we have

$$\lim_{n \to \infty} P_n = \infty \text{ and } \lim_{n \to \infty} Q_n = \infty,$$

showing that $\langle P_n \rangle$ and $\langle Q_n \rangle$ are divergent and hence the series of positive terms and the series of negative terms are separately divergent.

7.6. DIRICHLET'S THEOREM

Let (a) $\langle u_n \rangle$ be a positive monotonically decreasing sequence with limit 0 and

(b) $\sum_{n=1}^{\infty} a_n$ be an infinite series such that the sequence of its partial sums is bounded.

Then the series $\sum_{n=1}^{\infty} a_n u_n$ is convergent. [Meerut 2011]

Proof. This theorem will be deduced from the Cauchy's general principle of convergence.

We write $\sigma_n = a_1 + a_2 + ... + a_n$ .... (1)

Now, the sequence $\langle \sigma_n \rangle$ being bounded, there exists a positive number k such that

$| \sigma_n | \le k \ \forall \ n \in \mathbf{N}$, using condition (*b*). ... (2)

We write $S_n = a_1 u_1 + a_2 u_2 + ... + a_n u_n$ ... (3)

Now, for $\forall \ p \ge 1$ we have

$$\begin{aligned} S_{n+p} - S_n &= a_{n+1} u_{n+1} + a_{n+2} u_{n+2} + ... + a_{n+p} u_{n+p} \\ &= (\sigma_{n+1} - \sigma_n) u_{n+1} + (\sigma_{n+2} - \sigma_{n+1}) u_{n+2} \\ &\quad + ... + (\sigma_{n+p} - \sigma_{n+p-1}) u_{n+p}, \text{ using (1)} \\ &= (u_{n+1} - u_{n+2}) \sigma_{n+1} + (u_{n+2} - u_{n+3}) \sigma_{n+2} \\ &\quad + ... + (u_{n+p-1} - u_{n+p}) \sigma_{n+p-1} - u_{n+1} \sigma_n + u_{n+p} \sigma_{n+p} \end{aligned}$$

Thus, we have

$$| S_{n+p} - S_n | \le | u_{n+1} - u_{n+2} | | \sigma_{n+1} | + | u_{n+2} - u_{n+3} | | \sigma_{n+2} |$$
$$+ \ldots\ldots\ldots + \ldots\ldots\ldots$$
$$+ | u_{n+p-1} - u_{n+p} | | \sigma_{n+p-1} | + | u_{n+1} | | \sigma_n | + | u_{n+p} | | \sigma_{n+p} |$$
$$\le k \{(u_{n+1} - u_{n+2}) + (u_{n+2} - u_{n+3}) + \ldots + (u_{n+p-1} - u_{n+p})\}$$
$$+ k (u_{n+1} + u_{n+p}), \text{ by (2)}$$
$$\le k (u_{n+1} - u_{n+p}) + k (u_{n+1} + u_{n+p}) \quad \ldots (4)$$

Now the sequence $<u_n>$ by the given condition (*a*), is positive monotonically decreasing with limit 0. Let $\varepsilon > 0$ be given. There exists $m \in \mathbf{N}$ such that

$$| u_n | < \varepsilon / 4k \ \forall \ n \ge m. \quad \ldots (5)$$

Thus, we see that $\forall \ n \ge m$, (4) and (5) imply that

$$| S_{n+p} - S_n | < k \left(\frac{\varepsilon}{4k} + \frac{\varepsilon}{4k} + \frac{\varepsilon}{4k} + \frac{\varepsilon}{4k} \right) = \varepsilon, \forall \ p \ge 1$$

so that by the Cauchy's general principle of convergence, the series $\sum_{n=1}^{\infty} S_n$, *i.e.*, $\sum_{n=1}^{\infty} u_n a_n$ is convergent.

Deduction of Leibnitz's theorem. The theorem on alternating series is a simple consequence of the Dirichlet's theorem. We take $a_n = (-1)^{n+1}$
so that the sequence of the partial sum of the series $\Sigma a_n = 1 - 1 + 1 - 1 + 1 - 1 \ldots$ is bounded.

Also $<u_n>$ is a positive monotonically decreasing sequence with limit 0. Thus, the series $\Sigma a_n u_n$ *i.e.*, the alternating series $u_1 - u_2 + u_3 - u_4 + \ldots$ is convergent.

7.7. ABEL'S THEOREM (Kanpur 2004, 05, 09; Delhi Maths (H) 2005)

Abel's theorem is obtained from Dirichlet's theorem by *weakening* the condition on the sequence $<u_n>$ and strengthening the condition on the series Σa_n.

Abel's theorem. *If $<u_n>$ is a positive monotonically decreasing sequence and Σa_n is a convergent series, then the series $\Sigma a_n u_n$ is convergent.* **[Delhi Maths (H) 2008]**

We will deduce Abel's theorem from the Dirichlet's theorem.

Let $\lim u_n = l$ so that $\lim (u_n - l) = 0$.

We have $a_n (u_n - l) = a_n u_n - l a_n$

By Dirichlet's theorem, $\Sigma a_n (u_n - l)$ is a convergent series. Also Σa_n being a convergent series, it follows that the series $\Sigma a_n u_n$ is convergent.

EXAMPLES

Example 1. *Test the convergence of the series*

$$\sum \frac{(n^3 + l)^{1/3} - n}{\log n}.$$ ***[Delhi Maths (H), 2005]***

Solution. Let $u_n = (n^3 + 1)^{1/3} - n$ and $a_n = 1/\log n$.

Then the given series can be rewritten as $\Sigma a_n u_n$.

Here $u_n = n (1 + 1/n^3)^{1/3} - n = n \{(1 + 1/n^3)^{1/3} - n\}$

$$= n \left\{ \left(1 + \frac{1}{3n^3} - \frac{1}{9n^6} + \ldots \right) - 1 \right\} = \frac{1}{n^2} \left(\frac{1}{3} - \frac{1}{9n^3} + \ldots \right)$$

We take $v_n = 1/n^2$. Then, we have

$$\lim_{n\to\infty} \frac{u_n}{v_n} = \lim_{n\to\infty} \left(\frac{1}{3} - \frac{1}{9n^3} + ...\right) = \frac{1}{3}, \text{ which is finite and non-zero.}$$

Also, $\Sigma\, v_n = \Sigma\,(1/n^2)$ is convergent. Hence by comparison test, $\Sigma\, u_n$ is convergent.

Again $<a_n>$, *i.e.*, $<1/\log n>$ is a positive monotonically decreasing sequence. Hence, by Abel's test the given series is convergent.

Example 2. *Show that* $1 - \frac{1}{4\cdot 3} + \frac{1}{4^2\cdot 5} - \frac{1}{4^3\cdot 7} + ...$ *is convergent.*

Solution. Rewriting, the given series is

$$\sum \frac{(-1)^{n-1}}{4^{n-1}} \cdot \frac{1}{2n-1}$$

Let $\quad u_n = \dfrac{(-1)^{n-1}}{4^{n-1}} \quad$ and $\quad a_n = \dfrac{1}{2n-1}.$

Then, the given series can be rewritten as $\Sigma\, a_n\, u_n$.

Now, $$\Sigma\,|\, u_n \,| = \sum \left| \frac{(-1)^{n-1}}{4^{n-1}} \right| = \sum \left(\frac{1}{4}\right)^{n-1},$$

which is a geometric series with common ratio 1/4 (< 1) and so $\Sigma\,|\, u_n\,|$ is convergent. Hence $\Sigma\, u_n$ is an absolutely convergent series.

Since every absolutely convergent series is convergent, it follows that $\Sigma\, u_n$ is convergent.

Again $<a_n>$, *i.e.*, $<1/(2n - 1)>$ is a positive monotonically decreasing sequence. Hence, by Abel's test, the given series is convergent.

7.8. RE-ARRANGEMENTS OF SERIES [Delhi Maths (H) 2009]

Definition. *A series* $\Sigma\, u_n$ *is said to be a re-arrangement of a series* $\Sigma\, v_n$, *if every term of* $\Sigma\, u_n$ *is a term of* $\Sigma\, v_n$ *and vice-versa.*

Clearly if $\Sigma\, u_n$ is a re-arrangement of $\Sigma\, v_n$, then $\Sigma\, v_n$ is as well a re-arrangement of $\Sigma\, u_n$.

For example, the series

$$1 + \frac{1}{3} - \frac{1}{2} + \frac{1}{5} + \frac{1}{7} - \frac{1}{4} + ...$$

is re-arrangement of the series

$$1 - \frac{1}{2} + \frac{1}{3} - \frac{1}{4} + \frac{1}{5} - \frac{1}{6} + ...$$

The question now arises as to whether any re-arrangement of a series has the same behaviour as the given series. It will be now shown that so far as an absolutely convergent series is concerned, an arrangement does not alter its behaviour. For conditionally convergent series, however, the behaviour can be altered at will by a suitable re-arrangement. Thus, we shall arrive at a fundamental distinction between absolute and conditional convergence. We shall take up this question in the following.

RE-ARRANGEMENT OF A POSITIVE TERM SERIES

Dirichlet's Theorem. *Every re-arrangement of a positive term convergent series is convergent and the sum also does not change.* **[Delhi Maths (H) 2009]**

Let $\Sigma\, u_n$ be a given positive term convergent series with sum σ and let $\Sigma\, v_n$ be a re-arrangement of $\Sigma\, u_n$.

It will be shown that

(*i*) Σv_n is convergent and

(*ii*) the sum of Σv_n is the same as the sum of Σu_n.

Consider the sum of the first m terms of Σv_n. As Σu_n is a re-arrangement of Σv_n, there exists p such that the m terms of Σv_n are included in the first p terms of Σu_n. Thus we have

$$\sum_{n=1}^{m} v_n \leq \sum_{n=1}^{p} u_n.$$

Now σ being the sum of the positive term convergent series Σu_n, we have

$$\sum_{n=1}^{p} u_n \leq \sigma \quad \forall\ p \in \mathbf{N}.$$

It follows that
$$\sum_{n=1}^{m} v_n \leq \sigma \quad \forall\ m \in \mathbf{N}.$$

Thus, Σv_n is convergent and its sum η satisfies the relation

$$\eta \leq \sigma.$$

Thus, Σu_n being also a re-arrangement of Σv_n, by reversing the argument, it follows that

$$\sigma \leq \eta.$$

Thus, it follows that $\eta = \sigma$. Hence the result.

Note 1. This result may be restated as follows :

The sum of a convergent series of positive terms is the same in whatever order the terms are taken.

Note 2. In the following the result will be extended to absolutely convergent series.

RE-ARRANGEMENT OF AN ABSOLUTELY CONVERGENT SERIES

Theorem. *Every re-arrangement of an absolutely convergent series is convergent and the sum also does not change.* **[Delhi Maths (H) 2003, 04, 06, 07; Jiwaji 1998; Sagar 1999]**

Let Σu_n be a given absolutely convergent series and let Σv_n be any re-arrangement thereof. As in Art. 7.2 we associate two positive term series Σa_n and Σb_n with Σu_n and the two positive term series Σc_n and Σd_n with Σv_n.

Thus
$$a_n = \begin{cases} u_n & \text{if } u_n \geq 0, \\ 0 & \text{if } u_n < 0. \end{cases} \quad \text{and} \quad b_n = \begin{cases} -u_n & \text{if } u_n \leq 0, \\ 0 & \text{if } u_n > 0. \end{cases}$$

Similarly,
$$c_n = \begin{cases} v_n & \text{if } v_n \geq 0, \\ 0 & \text{if } v_n < 0. \end{cases} \quad \text{and} \quad d_n = \begin{cases} -v_n & \text{if } v_n \leq 0, \\ 0 & \text{if } v_n > 0. \end{cases}$$

Thus, we have

$$|u_n| = a_n + b_n,\ u_n = a_n - b_n \quad \text{and} \quad |v_n| = c_n + d_n,\ v_n = c_n - d_n.$$

As seen in Art. 7.2 the positive term series Σa_n and Σb_n are both convergent and the positive term series Σc_n and Σd_n are as well convergent.

The positive term convergent series Σc_n, Σd_n being re-arrangements of the positive term convergent series Σa_n, Σb_n, we have

$$\Sigma c_n = \Sigma a_n \quad \text{and} \quad \Sigma d_n = \Sigma b_n.$$

It follows that $\Sigma |v_n|$ is convergent so that Σv_n is absolutely convergent. Thus

$$\Sigma v_n = \Sigma c_n - \Sigma d_n = \Sigma a_n - \Sigma b_n = \Sigma u_n.$$

Hence the result.

7.9. RE-ARRANGEMENTS OF A CONDITIONALLY CONVERGENT SERIES

***Riemann's theorem.** *To a given conditionally convergent series and to any given number, there corresponds a re-arrangement of the given series which is convergent and whose sum is the given number.* **[Bilaspur 1997; Jiwaji 1999; Ravishankar 2000; Rewa 2000]**

Let Σu_n be a given conditionally convergent series and let k be a given number.

We suppose that k is positive. As in Art. 7.2, we associate two positive term series Σa_n and Σb_n to the series Σu_n. Thus

$$a_n = \begin{cases} u_n & \text{if } u_n \geq 0; \\ 0 & \text{if } u_n < 0. \end{cases} \quad \text{and} \quad b_n = \begin{cases} -u_n & \text{if } u_n \leq 0; \\ 0 & \text{if } u_n > 0. \end{cases}$$

We have, $\quad |u_n| = a_n + b_n \quad$ and $\quad u_n = a_n - b_n$ so that

$$a_n = \{|u_n| + u_n\}/2 \quad \text{and} \quad b_n = \{|u_n| - u_n\}/2.$$

The series Σu_n being conditionally convergent, the sum of Σu_n is finite and that of $\Sigma |u_n|$ is infinity. It follows that each of the series Σa_n and Σb_n diverges to infinity.

We now give the mode of construction of the required re-arrangement of Σu_n.

We add the positive terms of the series Σu_n, *i.e.*, the terms of the series Σa_n so that the sum of the finite number of terms thus obtained is *just* greater than k.

This process is possible because of the series Σa_n being divergent to infinity.

We now add the negative terms of the series Σu_n, *i.e.*, the terms of the series $\Sigma (-b_n)$ so that the total sum of the positive and negative terms used is *just* less than k. This process is possible because of the series Σb_n diverging to infinity.

Again we consider the positive terms and continue with the terms already obtained so that the sum of all the terms so far obtained is just greater than k.

We again consider the negative terms of the series and continue with the process.

Surely this new series is a re-arrangement of the given series which will be shown to be convergent with sum k.

Now the series Σu_n being convergent

$$\lim u_n = 0$$

so that if ε is a given positive number, there exists $m \in \mathbf{N}$ such that

$$|u_n| < \varepsilon \qquad \forall\ n \geq m.$$

By the method of formation of the new series, we see that in due course, we would have included in the re-arrangement all those terms of the given series for which $n \geq m$.

Thus, $\forall\ n \geq m$, the sum of the re-arrangement differs from k by $< \varepsilon$ so that the re-arrangement is convergent with sum k.

Illustration. Consider the conditionally convergent series

$$A : 1 - \frac{1}{2} + \frac{1}{3} - \frac{1}{4} + ...$$

and its re-arrangement

$$B : 1 + \frac{1}{3} - \frac{1}{2} + \frac{1}{5} + \frac{1}{7} - \frac{1}{4} + ...$$ **[Ravishanker 2001]**

where we have two positive terms followed by one negative term.

It will be shown that the sum of the series A is log 2 and that of its re-arrangement B is $(3 \log 2)/2$.

* Riemann (1826-66), was a German Mathematician.

To prove these results, we shall make use of the fact that the sequence $<\gamma_n>$ where

$$\gamma_n = 1 + \frac{1}{2} + ... + \frac{1}{n} - \log n,$$

is convergent. We denote $\lim \gamma_n$ by γ.

We rewrite the sum σ_{2n} of the first $2n$ terms of the series A in the form as

$$\left(1 + \frac{1}{2} + \frac{1}{3} + \frac{1}{4} + ... + \frac{1}{2n}\right) - 2\left(\frac{1}{2} + \frac{1}{4} + ... + \frac{1}{2n}\right) = (\gamma_{2n} + \log 2n) - (\gamma_n + \log n)$$
$$= \gamma_{2n} - \gamma_n + \log 2,$$

which tends to $\gamma - \gamma + \log 2 = \log 2$.

Thus, the sum of the series A is $\log 2$.

The sum S_{3n} of the first $3n$ terms of the re-arranged series B is

$$S_{3n} = \left(1 + \frac{1}{3} - \frac{1}{2}\right) + \left(\frac{1}{5} + \frac{1}{7} - \frac{1}{4}\right) + ...$$
$$= \left(1 + \frac{1}{3} + \frac{1}{5} + \frac{1}{7} + ... + \frac{1}{4n-1}\right) - \left(\frac{1}{2} + \frac{1}{4} + \frac{1}{6} + ... + \frac{1}{2n}\right)$$
$$= \left(1 + \frac{1}{2} + \frac{1}{3} + \frac{1}{4} + ... + \frac{1}{4n}\right) - \left(\frac{1}{2} + \frac{1}{4} + \frac{1}{6} + ... + \frac{1}{4n}\right) - \left(\frac{1}{2} + \frac{1}{4} + \frac{1}{6} + ... + \frac{1}{2n}\right)$$
$$= \gamma_{4n} + \log 4n - (1/2) \times (\gamma_{2n} + \log 2) - (1/2) \times (\gamma_n + \log n)$$
$$= \gamma_{4n} - (1/2) \times \gamma_{2n} - (1/2) \times \gamma_n + \log 4 - (1/2) \times \log 2n - (1/2) \times \log n$$
$$= \gamma_{4n} - (1/2) \times \gamma_{2n} - (1/2) \times \gamma_n + \log 4 - (1/2) \times \log 2$$

which leads to $\gamma - \frac{1}{2}\gamma - \frac{1}{2}\gamma + \frac{3}{2}\log 2 = \frac{3}{2}\log 2$ as its limit. Hence the result.

EXERCISES

1. Prove that the sum of the series

$$1 + \frac{1}{3} - \frac{1}{2} + \frac{1}{5} + \frac{1}{7} - \frac{1}{4} + \frac{1}{9} + \frac{1}{11} - \frac{1}{6} + ...$$

is $(3 \log 2)/2$. **(Patna, 2003)**

2. (*i*) If a series Σu_n is convergent, then prove that any series obtained from it by grouping the terms also converges to the same sum.

(*ii*) By an example prove that the converse is not true. **(Calicut, 2004)**

3. If Σu_n is a convergent series, then show that each of the following series is also convergent

(*i*) $\sum \left(\frac{n+1}{n}\right) u_n$ (*ii*) $\Sigma n^{1/n} u_n$ (*iii*) $\sum \frac{u_n}{\log_a n}$

4. Show that the series $\sum_{n=1}^{\infty} \left(1 + \frac{1}{2} + \frac{1}{3} + ... + \frac{1}{n}\right) \frac{\sin n\alpha}{n}$ converges absolutely for $\alpha = m\pi$, m being an integer and conditionally convergent for all other real values of α.

5. If the series Σu_n converges and $<v_n>$ is a positive monotonically decreasing sequence, then show that the series $\Sigma u_n v_n$ converges. **[Delhi Maths (H), 2005]**

6. (*i*) Show that the series $\sum_{n=1}^{\infty}(-1)^{n-1}\frac{1}{n}$ is conditionally convergent

(*ii*) $\sum_{n=1}^{\infty}\left(\frac{1}{n}+\frac{(-1)^n}{\sqrt{n}}\right)$ is divergent **[Delhi Maths (Prog) 2007]**

OBJECTIVE QUESTIONS

Multiple Choice Type Questions : *Select (a), (b), (c) or (d), whichever is correct.*

1. An absolutely convergent series out of the following series is :

(*a*) $1-\frac{x^2}{2!}+\frac{x^4}{4!}-\frac{x^6}{6!}+.....$ (*b*) $1-\frac{1}{2}+\frac{1}{3}+\frac{1}{4}+.....$

(*c*) $\sum\frac{(-1)^{n+1}}{3n-2}$ (*d*) $\sum\frac{(-1)^{n+1}}{\sqrt{n}}$. **(Kanpur, 2003)**

2. Which of the following is a convergent series ?

(*a*) $\sum_{n=1}^{\infty}(-1)^{n-1}\frac{1}{n}$ (*b*) $\sum_{n=1}^{\infty}a_n$, where $a_n=2,\ \forall\ n\in\mathbf{N}$

(*c*) $\sum_{n=1}^{\infty}\frac{1}{n^{3/4}}$ (*d*) $\sum_{n=1}^{\infty}3^n$. **(Kanpur, 2004)**

3. The series $1-\frac{1}{3}+\frac{1}{3^2}-\frac{1}{3^3}+.....$ is :

(*a*) Divergent (*b*) Conditionally convergent

(*c*) Absolutely convergent (*d*) None of these.

4. The series $1-\frac{1}{\sqrt{2}}+\frac{1}{\sqrt{3}}-\frac{1}{\sqrt{4}}+.....$ is :

(*a*) Divergent (*b*) Conditionally convergent

(*c*) Absolutely convergent (*d*) None of these.

5. The series $1-\frac{1}{2^2}+\frac{1}{3^2}-\frac{1}{4^2}+.....$ is :

(*a*) Oscillatory (*b*) Divergent

(*c*) Absolutely convergent (*d*) Conditionally convergent.

6. The geometric series $1+x+x^2+....+x^{n-1}+.....$ is convergent if

(*a*) $|x|>1$ (*b*) $x=1$ (*c*) $x<1$ (*d*) $x=-1$ **(Agra 2009)**

ANSWERS

1. (*a*) **2.** (*a*) **3.** (*c*) **4.** (*b*) **5.** (*c*) **6.** (*c*)

MIISCELLANEOUS PROBLEMS

1. Discuss the convergence of 1 – (1/2 !) + (1/4 !) – (1/6!) + ... **[Agra 2006]**

[**Sol.** Given series $= \sum_{n=1}^{\infty} \frac{(-1)^{n-1}}{(2n-2)!} = \sum_{n=1}^{\infty} (-1)^{n-1} u_n$, say

Therefore, $\quad u_n = 1/(2n-2)! > 0 \; \forall \; n \in N$

Also, $\quad u_{n+1} - u_n = \frac{1}{(2n)!} - \frac{1}{(2n-2)!} = \frac{1-2n(2n-1)}{(2n)!\,(2n-2)!} < 0, \; \forall \; n \in N$

because $\quad 1-2n(2n-1) = -(4n^2-2n-1) = -\{(2n-1)^2 + 2(n-1)\} < 0, \;\; \forall \in N$

Thus, $\quad u_{n+1} < u_n \, \forall \, n \in N$. Also, $\lim_{n\to\infty} u_n = 0$

Hence the given series converges by Leibnitz's test.

2. Show that 1 – (1/2) + (1/3) – (1/4) + ... is convergent **[Meerut 2006]**

3. State and prove Dirichlet test for the convergence of a series of arbitrary terms. **[Meerut 2006]**

4. Show that the series 1 – (1/3) + (1/5) – (1/7) + ... converges **[Delhi B.A. (Prog.) III, 2018]**

5. Show that 1 – (1/5) + (1/9) – (1/13) + ... is conditionally convergent **[Delhi 2008]**

6. Show that 1 – (1/3) – (1/7) + ... is conditionally convergent **[Delhi B.Sc. (Prog) III 2009; Delhi B.A. (Prog) III 2010, 16]**

7. Let $< a_k >$ be a sequence of real numbers and $|a_{k+1}| \geq |a_k| > 0$, for all k, then show that $\sum_k a_k$ is divergent **[Delhi B.Sc. I (Hons) 2010]**

8. Let $a_k, b_k \in \boldsymbol{R}$ such that $|a_k| \leq b_k \; \forall \; k \in N$. If $\sum_k b_k$ is convergent, then show that $\sum_k a_k$ is absolutely convergent. **[Delhi B.Sc. I (Hons) 2010]**

9. Consider the power series $\sum x^n/\sqrt{n}$ and $\sum x^n/n$. Then (*a*) both converge on (–1, 1] (*b*) both converge on [–1, 1) (*c*) exactly one of them converges on (–1, 1] (d) none of them converges on [–1, 1) **[GATE 2010]** **Ans.** (*b*)

10. Show that if $\sum_{n=1}^{\infty} |a_n|$ converges and $b_n = \begin{cases} a_n, \text{ if } a_n \geq 0 \\ 0, \text{ if } a_n < 0 \end{cases}$ then $\sum_{n=1}^{\infty} b_n$ converges. **(G.N.D.U. Amritsar 2010)**

11. For what values of *x*, the series $\sum x^n/n^n$ converges. **(Pune 2010)**

12. Prove that $(1 - 2) - (1 - 2^{1/2}) + (1 - 2^{(1/3)}) - (1 - 2^{1/4})$ ... is a convergent series. **(Channai 2011)**

13. Test the following series for absolute convergence : $1 - x^2/2! + x^4/4! - x^6 / 6! + ...$ **[Delhi B.Sc. (Prog.) II, 2017; Delhi B.A. (Prog.) III, 2017; Lucknow 2010]**

[**Sol.** We have $\quad |u_n| = x^{2n-2}/\,(2n-2)!$ $\quad$ and $\quad |u_{n+1}| = x^{2n}/(2n)|$. So, we have

$$\lim_{n\to\infty} \left|\frac{u_n}{u_{n+1}}\right| = \lim_{n\to\infty} \frac{((2n)!}{(2n-2)!}\frac{1}{x^2} = \lim_{n\to\infty} \frac{(2n)(2n-1)}{x^2} \to \infty \text{ provided } x \neq 0$$

Hence, by ratio test $\sum |u_n|$ is convergent for all values of *x*. Therefore the given series converges absolutely for all *x*.

7.10. Modified forms of some important theorems

Theorem I. *A series of positive terms, if convergent has a sum independent of its terms, but if divergent, it remains divergent, however, if its terms are re-arranged.* **[Delhi Maths (H) 2004]**

Proof. First part. Refer Dirichlet's theorem on page 7.14.

Second part. Let Σu_n be a given positive term divergent series and let Σv_n be a re-arrangement of Σu_n.

If possible, let Σv_n be convergent. Then, from the first part, it follows that Σu_n (regarded as a re-arrangement of Σv_n) must be convergent, which contradicts the given hypothesis. Hence Σv_n must be divergent.

Theorem II. Riemann theorem. **[Sagar 2000; Ravishankar 2006]**

Let Σu_n be a given series of real numbers which converges, but not absolutely. Suppose that $-\infty \leq \alpha \leq \beta \leq \infty$. Then there exists a re-arrangement Σv_n with sequence of partial sums $< s_n >$ such that $\liminf_{n\to\infty} s_n = \alpha$, $\limsup_{n\to\infty} s_n = \beta$.

Proof. Let us associate two positive term series Σa_n and Σb_n to the series Σu_n as follows:

$$a_n = (|u_n| + u_n)/2 \quad \text{and} \quad b_n = (|u_n| - u_n)/2 \qquad \text{... (1)}$$

Since Σu_n converges, but not absolutely, it follows that the sum of Σu_n is finite while that of $\Sigma |u_n|$ is infinity. Hence from (1), we see that each of the series Σa_n and Σb_n diverges to infinity.

Now, let $P_1, P_2, P_3, ...$ and $Q_1, Q_2, Q_3, ...$ be sequences of the positive terms and of the absolute value of the non-negative terms respectively of Σu_n, in order of original occurrence.

Since $a_1 + a_2 + ... + a_n \leq P_1 + P_2 + ... + P_n$ and $b_1 + b_2 + ... + b_n \leq Q_1 + Q_2 + ... Q_n \quad \forall n \in N$, it follows that ΣP_n and ΣQ_n diverges to infinity.

Moreover, $P_n \to 0$ and $Q_n \to 0$ as $n \to \infty$, because

$$u_n \to 0 \quad \Rightarrow \text{ if } \varepsilon > 0 \text{ then there exists } m \in N \text{ such that } |u_n| < \varepsilon \;\; \forall n \geq m$$

$$\Rightarrow \quad |P_n| < \varepsilon \quad \text{and} \quad |Q_n| < \varepsilon \;\; \forall n \geq m$$

Let us choose two sequences $< \alpha_n >$ and $< \beta_n >$ with $\beta_1 > 0$ and $\alpha_n < \beta_n$ such that $\alpha_n \to \alpha$ and $\beta_n \to \beta$

Now, we construct two sequences $< m_n >$ and $< k_n >$ as follows. Let us construct re-arrangement

$$P_1 + ... + P_{m_1} - Q_1 - .... - Q_{k_1} + P_{m_1+1} + ... + P_{m_2} - Q_{k_1+1} - ... - Q_{k_2} + ... \qquad \text{... (1)}$$

of Σu_n such that m_1, k_1 are the least positive integers satisfying

$$P_1 + ... + P_{m_1} > P_1 \quad \text{and} \quad P_1 + ... + P_{m_1} - Q_1 - Q_{k_1} < \alpha_1$$

Again, suppose that m_2, k_2 are the smallest integers such that

$$P_1 + ... + P_{m_1} - Q_1 - ... - Q_{k_1} + P_{m_1+k} + ... + P_{m_2} > \beta_2$$

and $$P_1 + ... + P_{m_1} - Q_1 - ... - Q_{k_1} + P_{m_1+1} + ... + P_{m_2} - Q_{k_1+1} - ... - Q_{k_2} < \alpha_2$$

and continue likewise.

Let $< s_n' >$ and $< s_n'' >$ be sequences of partial sums of (1) whose last terms are P_{m_n} and $-Q_{k_n}$. Also, clearly $< s_n >$ is the sequence of partial sums of (1). Then, we have

$$0 < s_n' - \beta_n < P_{m_n} \quad \text{and} \quad 0 < \alpha_n - s_n'' < Q_{k_n}$$

Since $P_{m_n} \to 0$ and $Q_{k_n} \to 0$, we have $s_n' \to \beta$ and $s_n'' \to \alpha$

Also, $\liminf_{n\to\infty} s_n = \lim_{n\to\infty} s_n'' = \alpha$ and $\limsup_{n\to\infty} s_n = \lim_{n\to\infty} s_n' = \beta$

Hence, (1) is the desired re-arrangement Σv_n of Σu_n

Corollary. *Let Σu_n be a conditionally convergent series. Then by a suitable re-arragement of terms, Σu_n can be made*

(*i*) *to converge to any real number.*
(*ii*) *to diverge to ∞ or $-\infty$*
(*iii*) *to oscillate finitely or infinitely*

[Bilaspur 1997; Jiwaji 1999; Ravishankar 2000; Rewa 2000]

Proof. Taking $\alpha = \beta$ (finite, ∞ or $-\infty$) and using the above theorem II, the required results follow.

7.11. Additional solved example

Ex. 1. *Show that any re-arrangement of a convergent series of positive terms converges to the same sum. Show by an example that result may not be true if the series is not that of positive terms.* **[Delhi Maths (H) 2002]**

Hint. *First part.* Refer Dirichlet theorem on page 7.14

Second part. See illustrative example on page 7.16

Ex. 2. *Examine the convergence of the series*

$$1 + 1/3^2 - 1/2^2 + 1/5^2 - 1/4^2 + 1/3^2 + \ldots\ldots$$

Sol. The given series can be obtained by re-arrangement of the series

$$1 - 1/2^2 + 1/3^2 - 1/4^2 + \ldots\ldots \qquad \ldots(1)$$

Re-writing (1), here $\Sigma u_n = \Sigma(-1)^{n-1} \times (1/n^2)$. Hence $\Sigma |u_n| = \Sigma(1/n^2)$, which is convergent. Therefore, by definition, (1) is absolutely convergent.

Again, every re-arrangement of an absolutely convergent series is convergent. Hence the given series (which can be obtained from (1) by re-arrangement) is also convergent.

Ex. 3. *Show that the series* $1 - 1 + 1/2 - 1/2 + 1/3 - 1/3 + \ldots$ *is convergent and the following re-arrangement of this series* $1 + 1/2 - 1 + 1/3 + 1/4 - 1/2 + 1/5 + 1/6 - 1/3 + \ldots$ *converges to* $\log 2$. **[Delhi Maths (H) 2001]**

Sol. Given series is $\quad 1 - 1 + 1/2 - 1/2 + 1/3 - 1/3 + \ldots\ldots \qquad \ldots(1)$

In series (1), one positive term is followed by one negative term. Let us discuss $2n$ terms of (1). We have,

$$s_{2n} = (1-1) + \left(\frac{1}{2} - \frac{1}{2}\right) + \left(\frac{1}{3} - \frac{1}{3}\right) + \ldots + \left(\frac{1}{n} - \frac{1}{n}\right)$$

or $$s_{2n} = \left(1 + \frac{1}{2} + \frac{1}{3} + \ldots + \frac{1}{n}\right) - \left(1 + \frac{1}{2} + \frac{1}{3} + \ldots + \frac{1}{n}\right) \qquad \ldots(2)$$

In what follows, we shall make use of the fact that the sequence $<\gamma_n>$ where

$$\gamma_n = 1+1/2+1/3+\ldots+1/n-\log n \quad \ldots (3)$$

is convergent. Also, $$\lim_{n\to\infty} \gamma_n = \gamma = \text{Euler constant} \quad \ldots (4)$$

Using (3), (2) can be re-written as

$$\therefore\ s_{2n} = \log n + \gamma_n - (\log n + \gamma_n) = 0 \qquad \text{so that} \qquad \lim_{n\to\infty} s_{2n} = 0$$

Again, $$s_{2n+1} = s_{2n} + 1/(2n+1) \quad \text{so that} \quad \lim_{n\to\infty} s_{2n+1} = \lim_{n\to\infty} s_{2n} + 0 = 0$$

Hence $\lim_{n\to\infty} s_n = 0$ and therefore (1) is convergent

Second part. Consider the second series, namely,

$$1 + 1/2 - 1 + 1/3 + 1/4 - 1/2 + 1/5 + 1/6 - 1/3 + \ldots.. \quad \ldots (5)$$

In the above series two positive terms are followed by one negative term. Let us discuss the first $3n$ terms of (5). We have,

$$s_{3n} = \left(1+\frac{1}{2}-1\right)+\left(\frac{1}{3}+\frac{1}{4}-\frac{1}{2}\right)+\ldots+\left(\frac{1}{2n-1}+\frac{1}{2n}-\frac{1}{n}\right)$$

$$= \left(1+\frac{1}{2}+\frac{1}{3}+\frac{1}{4}+\ldots+\frac{1}{2n-1}+\frac{1}{2n}\right)-\left(1+\frac{1}{2}+\frac{1}{3}+\ldots+\frac{1}{n}\right)$$

$= \log 2n + \gamma_{2n} - (\log n + \gamma_n)$, using (3)

$= \log 2 + \gamma_{2n} - \gamma_n$, *as* $\log 2n - \log n = \log(2n/n) = \log 2$

Hence $$\lim_{n\to\infty} s_{3n} = \log 2 + \lim_{n\to\infty}(\gamma_{2n} - \gamma_n) = \log 2 + (0-0), \text{ by (4)}$$

Thus, $$\lim_{n\to\infty} s_{3n} = \log 2 \quad \ldots (6)$$

Also, $$s_{3n+1} = s_{3n} + \frac{1}{2n+1} \quad \Rightarrow \quad \lim_{n\to\infty} s_{3n+1} = \lim_{n\to\infty}\left(s_{3n} + \frac{1}{2n+1}\right)$$

or $$\lim_{n\to\infty} s_{3n+1} = \lim_{n\to\infty} s_{3n} + 0 = \log 2, \text{ using (6)}$$

Finally, $$s_{3n+2} = s_{3n} + \frac{1}{2n+1} + \frac{1}{2n+2}$$

This $$\Rightarrow \lim_{n\to\infty} s_{3n+2} = \lim_{n\to\infty} s_{3n} + 0 + 0 = \log 2, \text{ using (6)}$$

Thus, $$\lim_{n\to\infty} s_{3n} = \lim_{n\to\infty} s_{3n+1} = \lim_{n\to\infty} s_{3n+2} = \log 2$$

$$\Rightarrow \lim_{n\to\infty} s_n = \log 2 \quad \Rightarrow \quad 1+\frac{1}{2}-1+\frac{1}{3}+\frac{1}{4}-\frac{1}{2}+\frac{1}{5}+\frac{1}{6}-\frac{1}{3}+\ldots = \log 2$$

Ex. 4. *Prove that the series* $1-1+1/2-1/2+1/3-1/3+\ldots..$ *is convergent and its sum is zero while the sum of the re-arranged series* $1 + 1/2 + 1/3 - 1 + 1/4 + 1/5 + 1/6 - 1/2 + \ldots.$ is $\log 3$.

[Sagar 1998; Rewa 1998]

Sol. *First part.* Refer first part of Ex. 3

Second part. In the second given series three positive terms are followed by one negative term. Again, the second series is a rearrangement of the first series and so we shall consider the first $4n$ terms of this series. We, have

$$s_{4n}=\left(1+\frac{1}{2}+\frac{1}{3}-1\right)+\left(\frac{1}{4}+\frac{1}{5}+\frac{1}{6}-\frac{1}{2}\right)+...+\left(\frac{1}{3n-2}+\frac{1}{3n-1}+\frac{1}{3n}-\frac{1}{n}\right)$$

or $$s_{4n}=\left(1+\frac{1}{2}+\frac{1}{3}+\frac{1}{4}+...+\frac{1}{3n}\right)-\left(1+\frac{1}{2}+\frac{1}{3}+...+\frac{1}{n}\right) \qquad ...(1)$$

We know that the sequence $<\gamma_n>$ is convergent, where

$$\gamma_n=1+1/2+1/3+....+1/n-\log n \qquad ...(2)$$

and $$\lim_{n\to\infty}\gamma_n=\gamma=\text{Euler constant} \qquad ...(3)$$

Using (2), (1) may be re-written as

$$s_{4n}=\log 3n+\gamma_{3n}-(\log n+\gamma_n)=\log 3+\gamma_{3n}-\gamma_n$$

$$\Rightarrow \quad \lim_{n\to\infty}s_{4n}=\log 3+\lim_{n\to\infty}(\gamma_{3n}-\gamma_n)=\log 3+\gamma-\gamma, \text{ using (3)}$$

Thus, we have $$\lim_{n\to\infty}s_{4n}=\log 3$$

Again, $$s_{4n+1}=s_{4n}+\frac{1}{3n+1} \Rightarrow \lim_{n\to\infty}s_{4n+1}=\log 3, \text{ using (4)}$$

$$s_{4n+2}=s_{4n}+\frac{1}{3n+1}+\frac{1}{3n+2} \Rightarrow \lim_{n\to\infty}s_{4n+2}=\log 3, \text{ using (4)}$$

and $$s_{4n+3}=s_{4n}+\frac{1}{3n+1}+\frac{1}{3n+2}+\frac{1}{3n+3} \Rightarrow \lim_{n\to\infty}s_{4n+3}=\log 3, \text{ using (4)}$$

Thus, $$\lim_{n\to\infty}s_{4n}=\lim_{n\to\infty}s_{4n+1}=\lim_{n\to\infty}s_{4n+2}=\lim_{n\to\infty}s_{4n+3}=\log 3$$

Hence, $$\lim_{n\to\infty}s_n=\log 3 \quad \Rightarrow \quad 1+\frac{1}{2}+\frac{1}{3}-1+\frac{1}{4}+\frac{1}{5}+\frac{1}{6}-\frac{1}{2}+....=\log 3$$

Ex. 5. *Show that* $1-\frac{1}{2}-\frac{1}{4}+\frac{1}{3}-\frac{1}{6}-\frac{1}{8}+\frac{1}{5}-\frac{1}{10}-\frac{1}{12}+....=\frac{1}{2}\log 2$

Sol. In the given series one positive term is followed by two negative terms. We shall consider the first $3n$ terms of the series. We, have

$$s_{3n}=\left(1-\frac{1}{2}-\frac{1}{4}\right)+\left(\frac{1}{3}-\frac{1}{6}-\frac{1}{8}\right)+...+\left(\frac{1}{2n-1}-\frac{1}{4n-2}-\frac{1}{4n}\right)$$

$$=\left(1+\frac{1}{3}+\frac{1}{5}+...+\frac{1}{2n-1}\right)-\left(\frac{1}{2}+\frac{1}{6}+...+\frac{1}{4n-2}\right)-\left(\frac{1}{4}+\frac{1}{8}+...+\frac{1}{4n}\right)$$

$$=\left(1+\frac{1}{3}+\frac{1}{5}+...+\frac{1}{2n-1}\right)-\frac{1}{2}\left(1+\frac{1}{3}+\frac{1}{5}+...+\frac{1}{2n-1}\right)-\frac{1}{4}\left(1+\frac{1}{2}+\frac{1}{3}+...+\frac{1}{n}\right)$$

$$=\frac{1}{2}\left(1+\frac{1}{3}+\frac{1}{5}+...+\frac{1}{2n-1}\right)-\frac{1}{4}\left(1+\frac{1}{2}+\frac{1}{3}+...+\frac{1}{n}\right)$$

$$=\frac{1}{2}\left\{\left(1+\frac{1}{2}+\frac{1}{3}+...+\frac{1}{2n}\right)-\left(\frac{1}{2}+\frac{1}{4}+\frac{1}{6}+...+\frac{1}{2n}\right)\right\}-\frac{1}{4}\left(1+\frac{1}{2}+\frac{1}{3}+...\frac{1}{n}\right)$$

$$=\frac{1}{2}\left\{\left(1+\frac{1}{2}+\frac{1}{3}+...+\frac{1}{2n}\right)-\frac{1}{2}\left(1+\frac{1}{2}+\frac{1}{3}+...+\frac{1}{n}\right)\right\}-\frac{1}{4}\left(1+\frac{1}{2}+\frac{1}{3}+...+\frac{1}{n}\right)$$

$$\therefore \qquad s_{3n}=\frac{1}{2}\left(1+\frac{1}{2}+\frac{1}{3}+...+\frac{1}{2n}\right)-\frac{1}{2}\left(1+\frac{1}{2}+\frac{1}{3}+...+\frac{1}{n}\right) \qquad ...(1)$$

We know that the sequence $<\gamma_n>$ is convergent, where

$$\gamma_n=1+1/2+1/3+...+1/n-\log n \qquad ...(2)$$

and $$\lim \gamma_n=\gamma=\text{Euler constant} \qquad ...(3)$$

Using (2), (1) may be re-written as

$$s_{3n}=(1/2)\times(\log 2n+\gamma_{2n})-(1/2)\times(\log n+\gamma_n)=(1/2)\times\log 2+(1/2)\times(\gamma_{2n}-\gamma_n)$$

$$\Rightarrow \lim_{n\to\infty} s_{3n}=(1/2)\times\log 2+\lim_{n\to\infty}(1/2)\times(\gamma_{2n}-\gamma_n)=(1/2)\times\log 2+(1/2)\times(\gamma-\gamma), \text{ using (3)}$$

Thus, $$\lim_{n\to\infty} s_{3n}=(1/2)\times\log 2 \qquad ...(4)$$

Also, $$s_{3n+1}=s_{3n}+\frac{1}{2n+1} \Rightarrow \lim_{n\to\infty} s_{3n+1}=\frac{1}{2}\log 2, \text{ using (4)}$$

and $$s_{3n+2}=s_{3n}+\frac{1}{2n+1}-\frac{1}{4n-2} \Rightarrow \lim_{n\to\infty} s_{3n+2}=\frac{1}{2}\log 2, \text{ using (4)}$$

Thus, $$\lim_{n\to\infty} s_{3n}=\lim_{n\to\infty} s_{3n+1}=\lim_{n\to\infty} s_{3n+2}=(1/2)\times\log 2$$

Hence, $$\lim_{n\to\infty} s_n=\frac{1}{2}\log 2 \Rightarrow 1-\frac{1}{2}-\frac{1}{4}+\frac{1}{3}-\frac{1}{6}-\frac{1}{8}+...=\frac{1}{2}\log 2$$

Ex. 6. *Prove that if* $|x|<1$, *then*

$$1+x^2+x+x^4+x^6+x^3+x^8+x^{10}+x^5+....=1/(1-x)$$

Sol. For $|x|<1$, the series $$1+x+x^2+x^3+ \qquad .....(1)$$
converges absolutes to $1/(1-x)$. Hence, any re-arranged will also converge to the same sum $1/(1-x)$. Since the given series in a rearrangement of (1), it must converge to the same sum $1/(1-x)$.

EXERCISES

1. *If* $s=1-1/2+1/3-....$, *prove that*

(*i*) $\frac{1}{2}s=1-\frac{1}{2}-\frac{1}{4}+\frac{1}{3}-\frac{1}{6}-\frac{1}{8}+...$ (*ii*) $\frac{2}{3}s=1-\frac{1}{2}-\frac{1}{4}+\frac{1}{5}+\frac{1}{7}-\frac{1}{8}-\frac{1}{10}+...$

2. *If* $s=1-1/3+1/5-1/7+....$, *then prove that*

$1+1/5-1/7-1/11+1/13-1/17-1/19+....=4s/3$

3. *Find a derangement of* $\Sigma(-1)^{n-1}/n$ *which converges to* $(3/2) \times \log 2$

4. *By observing that* $\log 2 = 1 - 1/2 + 1/3 - 1/4 + 1/5 - 1/6 + \ldots$ *prove that*

(*i*) $1+\frac{1}{3}-\frac{1}{2}-\frac{1}{4}-\frac{1}{6}+\frac{1}{5}+\frac{1}{7}-\frac{1}{8}-\frac{1}{10}-\frac{1}{12}+\ldots=\frac{1}{2}\log\frac{8}{3}$

(*ii*) $1-\frac{1}{2}-\frac{1}{4}-\frac{1}{6}+\frac{1}{3}-\frac{1}{8}-\frac{1}{10}-\frac{1}{12}+\frac{1}{5}-\frac{1}{14}-\frac{1}{16}-\frac{1}{18}+\ldots=\frac{1}{2}\log\frac{4}{3}$

(*iii*) $1+\frac{1}{3}-\frac{1}{2}-\frac{1}{4}+\frac{1}{5}+\frac{1}{7}-\frac{1}{8}-\frac{1}{10}+\frac{1}{11}+\frac{1}{13}+\ldots=\frac{2}{3}\log 2$

(*iv*) $\frac{1}{3}-\frac{1}{5}-\frac{1}{7}-\frac{1}{5}-\frac{1}{9}-\frac{1}{11}+\frac{1}{7}-\frac{1}{13}-\frac{1}{15}+\frac{1}{9}-\frac{1}{17}-\frac{1}{19}+\ldots=\frac{1}{3}-\frac{1}{2}\log 2$

5. *Prove that* $1 - 1 + 1/2 - 1/2 + 1/3 - 1/3 + \ldots$ *is convergent. By rearranging the terms a new series* $1 + 1/2 + 1/3 - 1 + 1/4 + 1/5 + 1/6 - 1/2 + \ldots$ *is obtained. Find its sum.*

[Hint : Refer solved Ex. 3 of Art. 7.11] **[Sagar 1996]**

6. *Show that the sum of the series, obtained be re-arranging the series* $1 - 1/2 + 1/3 - 1/4 + 1/5 - \ldots$ *such that p positive alternate with q negative, is* $\log 2 + (1/2) \times \log (1/2)$.

[Revishankar 2004]

7. (*a*) *If the series* $1 - 1/2 + 1/3 - 1/4 + 1/5 - \ldots$ *is so altered that in first n terms the ratio of the number of positive terms to the number of negative terms is* a^2, *then prove that the sum of the series is* $\log 2a$.

(*b*) *If the terms of the series* $1 - 1/2 + 1/3 - 1/4 + \ldots$ *be rearranged so that the ratio of the number of positive terms to the number of negative terms in succession is ultimately k, then prove that the sum of the rearranged series will be* $(1/2) \times \log 4k$.

7.12 Cauchy product of two infinite series **[Delhi Maths (H) 2003, 07]**

Definition. *Given* $\sum_{n=1}^{\infty} a_n$ and $\sum_{n=1}^{\infty} b_n$, *we put*

$$c_n = a_1 b_n + a_2 b_{n-1} + \ldots + a_n b_1 = \sum_{r=1}^{n} a_r b_{n-r+1} \quad (n = 1, 2, 3, \ldots)$$

and call $\sum_{n=1}^{\infty} c_n$ *the Cauchy product or simply product of two given series.*

Note. In what follows Σa_n will stand for $\sum_{n=1}^{\infty} a_n$ and so on.

Theorem I. *If* Σa_n and Σb_n *are absolutely convergent series, prove that their Cauchy product* Σc_n, (*where* $c_n = \sum_{r=1}^{n} a_r b_{n-r+1}$) *is also absolutely convergent and sum of the Cauchy product series is the product of the sums, i.e.,* $\Sigma c_n = (\Sigma a_n)(\Sigma b_n)$ **[Delhi Maths (H) 2004]**

or *Prove that if* Σa_n *and* Σb_n *are absolutely convergent with* $\Sigma a_n = \alpha$ *and* $\Sigma b_n = \beta$, *their Cauchy product* Σc_n *also converges absolutely with* $\Sigma c_n = \alpha\beta$

[Delhi Maths (H) 1999, 2000, 03]

Proof. Let Σa_n and Σb_n converge absolutely to α and β respectively.

Let $$c_n = \sum_{r=1}^{n} a_r b_{n-r+1} \quad (n = 1, 2, 3, \ldots)$$

Then the Cauchy product of the given two series is Σc_n.

Let $\alpha_n = \sum_{r=1}^{n} a_r$, $\beta_n = \sum_{r=1}^{n} b_r$, $\sum_{n=1}^{\infty} a_n = \alpha$ and $\sum_{n=1}^{\infty} b_n = \beta$

Then $<\alpha_n>$ and $<\beta_n>$ are the sequences of partial sums of Σa_n and Σb_n respectively. Again, α and β are the sums of Σa_n and Σb_n respectively.

In order to establish the absolute convergence of the product series, consider the following series

$$a_1b_1 + a_1b_2 + a_2b_2 + a_2b_1 + a_1b_3 + a_2b_2 + a_3b_2 + a_3b_1 + ... \quad ...(1)$$

Then, $|a_1 b_1| + |a_1 b_2| + |a_2 b_2| + |a_2 b_1| + ...$ to n terms

$$\leq (|a_1| + |a_2| + ... + |a_n|)\ (|b_1| + |b_2| + ... + |b_n|) \leq \alpha\beta, \ \forall n \in N$$

Hence, the series (1) must be absolutely convergent. Therefore, by Dirichlet's theorem, re-arrangement of (1), namely,

$$a_1b_1 + a_1b_2 + a_2b_1 + a_1b_3 + a_2b_2 + a_3b_1 + ...$$

is also absolutely convergent. Hence by grouping, we see that

$$\sum_{n=1}^{\infty}\left(\sum_{r=1}^{n} a_r b_{n-r+1}\right), \quad i.e., \quad \sum_{n=1}^{\infty} c_n \quad ...(2)$$

is also absolutely convergent.

Since the sum of the first n^2 terms of (1) is $\alpha_n\beta_n$ and $\alpha_n\beta_n \to \alpha\beta$ as $n \to \infty$, therefore, the sum of the series (1) is $\alpha\beta$. Thus, the series (2) which is a grouping of a re-arrangement of the absolutely convergent series (1), must converge to the same sum.

Thus, $\Sigma c_n = (\Sigma a_n)(\Sigma b_n)$, as required.

Theorem II. (Merten's theorem) *Suppose*

(i) $\sum_{n=1}^{\infty} a_n$ *converges absolutely* *(ii)* $\sum_{n=1}^{\infty} a_n = \alpha$ *(iii)* $\sum_{n=1}^{\infty} b_n = \beta$

(iv) $c_n = \sum_{r=1}^{n} a_r b_{n-r+1}$ $(n = 1, 2, 3, ...)$

Then, $\sum_{n=1}^{\infty} c_n = \alpha\beta$

i.e., the product of two convergent series converges, and to the right value, if at least one of the two series converges absolutely. **[Delhi Maths (H) 1999, 2000]**

Proof. Let $\alpha_n = \sum_{r=1}^{n} a_r$, $\beta_n = \sum_{r=1}^{n} b_r$, $\gamma_n = \sum_{r=1}^{n} c_r$ and $\delta_n = \beta_n - \beta$ ...(1)

where $c_n = \sum_{r=1}^{n} a_r b_{n-r+1}$ (n = 1, 2, 3, ...) ...(2)

so that $\sum_{n=1}^{\infty} c_n$ is the Cauchy product of $\sum_{n=1}^{\infty} a_n$ and $\sum_{n=1}^{\infty} b_n$. Then, we have

$$c_n = a_1b_1 + (a_1b_2 + a_2b_1) + ... + (a_1b_n + a_2b_{n-1} + ... + a_nb_1)$$
$$= a_1\beta_n + a_2\beta_{n-1} + ... + a_n\beta_1, \text{ using (1)}$$
$$= a_1(\beta + \delta_n) + a_2(\beta + \delta_{n-1}) + ... + a_n(\beta + \delta_1)$$
$$= \beta(a_1 + a_2 + ... + a_n) + (a_1\delta_n + a_2\delta_{n-1} + ... + a_n\delta_1)$$

Thus, $$c_n = \alpha_n\beta + (a_1\delta_n + a_2\delta_{n-1} + ... + a_n\delta_1) \quad ... (3$$

Let $$e_n = a_1\delta_n + a_2\delta_{n-1} + ... + a_n\delta_1 \quad ... (4)$$

We wish to show that $c_n \to \alpha\beta$. Since $\alpha_n\beta \to \alpha\beta$, it suffices to prove that

$$\lim_{n\to\infty} e_n = 0 \quad ... (5)$$

Assume that $$\sum_{n=1}^{\infty} |a_n| = k, \quad ... (6)$$

where we have used the given fact (*i*), that is, $\sum_{n=1}^{\infty} a_n$ is absolutely convergent

Let $\varepsilon > 0$ be given. From the given fact (*iii*) of the theorem, we have $\beta_n \to \beta$ and hence, (1) $\Rightarrow$ $\delta_n \to 0$. Therefore, we can choose $m \in N$ such that

$$|\delta_n| \le \varepsilon \text{ for } n \ge m \quad ... (7)$$

Now, from (4), $$|e_n| \le |\delta_1 a_n + ... + \delta_m a_{n-m+1}| + |\delta_{m+1}a_{n-m+2} + ... + \delta_n a_1|$$

or $$|e_n| \le |\delta_1 a_n + ... + \delta_m a_{n-m+1}| + \varepsilon k, \text{ using (6) and (7)}$$

where we have used the fact that $a_n \to 0$ as $n \to \infty$. Since ε is arbitrary the above inequality leads to (5).

Now, (3) and (4) $\Rightarrow$ $$\sum_{n=1}^{\infty} c_n = \sum_{n=1}^{\infty} (\alpha_n\beta + e_n) = \beta \sum_{n=1}^{\infty} a_n + \sum_{n=1}^{\infty} e_n \quad ... (8)$$

Letting $n \to \infty$ and using (5), (8) $\Rightarrow$ $\sum_{n=1}^{\infty} c_n = \alpha\beta$

Theorem III. Let $\sum_{n=1}^{\infty} a_n$ and $\sum_{n=1}^{\infty} b_n$ *converge to* α *and* β *respectively. Let* $\sum_{n=1}^{\infty} c_n$ *be Cauchy sequence of the given series, where* $c_n = \sum_{r=1}^{n} a_r b_{n-r+1}$. *If* $\sum_{n=1}^{\infty} c_n$ *converge to* γ, *then* $\gamma = \alpha\beta$.

Proof. Let $\alpha_n = \sum_{r=1}^{n} a_r$, $\quad \beta_n = \sum_{r=1}^{n} b_r \quad$ and $\quad \gamma_n = \sum_{r=1}^{n} c_r$.

Then, $$\gamma_n = a_1b_1 + (a_1b_2 + a_2b_1) + ... + (a_1b_n + a_2b_{n-1} + ... + a_nb_1)$$

or $$\gamma_n = a_1\beta_n + a_2\beta_{n-1} + ... + a_n\beta_1 \quad ... (1)$$

Using (1), we have

$$(\gamma_1 + \gamma_2 + ... + \gamma_n)/n = (\alpha_1\beta_n + \alpha_2\beta_{n-1} + ... + \alpha_n\beta_1)/2 \quad ... (2)$$

Now, as $n \to \infty$, $\alpha_n \to \alpha$, $\beta_n \to \beta$ and $\gamma_n \to \gamma$. Hence letting $n \to \infty$ and using Cesaro's theorem (refer theorem VII, page 5.26), (2) reduces to $\gamma = \alpha\beta$, as required.

Remark. The following examples show that the conditions of the above theorems cannot be relaxed.

Example 1. *If* $a_n = b_n = (-1)^{n-1}/\sqrt{n}$ *and* $c_n = \sum_{r=1}^{n} a_r b_{n-r+1}$. *Then show that* $\sum_{n=1}^{\infty} a_n$ *and* $\sum_{n=1}^{\infty} b_n$ *are conditionally convergent and their Cauchy product* Σc_n *is non-convergent.*

Sol. Here $\sum_{n=1}^{\infty} |a_n| = \sum_{n=1}^{\infty} |b_n| = \sum_{n=1}^{\infty} \left|(-1)^{n-1}/\sqrt{n}\right| = \sum_{n=1}^{\infty} 1/\sqrt{n}$, which is divergent. Hence $\sum_{n=1}^{\infty} a_n$ and $\sum_{n=1}^{\infty} b_n$ are conditionally convergent.

We know that $$c_n = \sum_{r=1}^{n} a_r b_{n-r+1} = a_1 b_n + a_2 b_{n-1} + \ldots + a_n b_1$$

$$\therefore \quad c_n = (-1)^n \left(\frac{1}{\sqrt{n \times 1}} + \frac{1}{\sqrt{2 \times (n-1)}} + \ldots + \frac{1}{\sqrt{(n-1) \times 2}} + \frac{1}{\sqrt{1 \times n}} \right)$$

$\Rightarrow |c_n| \geq 1/\sqrt{n^2} + 1/\sqrt{n^2} + \ldots$ to n terms $\Rightarrow |c_n| \geq 1 \Rightarrow \Sigma c_n$ is non-convergent

Example 2. *Give an example to prove that the product of two convergent series may actually diverge.*

Sol. Let $\sum_{n=1}^{\infty} a_n = \sum_{n=1}^{\infty} b_n = \sum_{n=1}^{\infty} (-1)^n/\sqrt{n}$. Then, as usual show that $\sum_{n=1}^{\infty} a_n$ and $\sum_{n=1}^{\infty} b_n$ both converge while their product diverges as shown is example 1.

Example 3. Show that the series $\sum_{n=1}^{\infty} a_n = 1 - (3/2) - (3/2)^2 - (3/2)^3 - \ldots$ and $\sum_{n=1}^{\infty} b_n = 1 + \{2 + (1/2)^2\} + (3/2) \times \{2^2 + (1/2)^3\} + \ldots$ are divergent but their Cauchy product $\sum_{n=1}^{\infty} c_n$ converges absolutely.

Sol. Since $-(3/2)^n < -1 \ \forall n \geq 2$ and $1 \leq b_n \forall n \in N$, it follows that Σa_n diverges to $-\infty$ while Σb_n diverges to ∞. Again, by definition of Cauchy product, we have

$$c_n = (3/2)^{n-2} \times \{2^{n-1} + (1/2)^n\} - (3/2)^{n-2} \times \{2^{n-2} + (1/2)^{n-1}\} - \ldots - (3/2)^{n-2} \times \{2 + (1/2)^2\} - (3/2)^{n-1}$$

$$= (3/2)^{n-2} \times \left[2^{n-1} + (1/2)^n - \{2^{n-2} + 2^{n-3} + \ldots + 2 + (1/2^{n-1} + 1/2^{n-2} + \ldots + 1/2^n) + 3/2\}\right]$$

$$= (3/2)^{n-2} \times \{2^{n-1} + (1/2)^n - (2^{n-1} - 2) - (1/2 - 1/2^{n-1}) - 3/2\}$$

$$= (3/2)^{n-2} \times \{(1/2)^n + (1/2)^{n-1}\} = (3/2)^{n-2} \times (1/2)^{n-1} \times (1/2 + 1) = (3/4)^{n-1}$$

Thus, $\sum_{n=1}^{\infty} c_n = \sum_{n=1}^{\infty} (3/4)^{n-1} = 1 + (3/4) + (3/4)^2 + \ldots$, which converges absolutely.

7.12A Solved example based on Art. 7.12

Ex. 1. *Prove that for* $|x|<1$, $\dfrac{1}{2}\left(x-\dfrac{x^2}{2}+\dfrac{x^3}{3}-....\right)^2=\sum_{n=2}^{\infty}(-1)^n\dfrac{x^n}{n}\left(1+\dfrac{1}{2}+\dfrac{1}{3}+...+\dfrac{1}{n-1}\right)$

[Delhi Maths (Hons), 2002, 05, 07]

Sol. The series $x - x^2/2 + x^3/3$ .... is absolutely convergent for $|x| < 1$. Hence the Cauchy product of the given series by itself will converge absolutely to the square of its sum. Thus,

$$(x-x^2/2+x^3/3-.....)^2=x^2-\left(\frac{1}{2}+\frac{1}{2}\right)x^3+\left(\frac{1}{1\times3}+\frac{1}{2\times2}+\frac{1}{3\times1}\right)x^4+.....$$

$$+(-1)^n\left\{\frac{1}{1\times(n-1)}+\frac{1}{2\times(n-2)}+...+\frac{1}{(n-2)\times2}+\frac{1}{(n-1)\times1}\right\}x^n+....$$

$$=\sum_{n=2}^{\infty}(-1)^n\left\{\frac{1}{1\times(n-1)}+\frac{1}{2\times(n-2)}+...+\frac{1}{(n-2)\times2}+\frac{1}{(n-1)\times1}\right\}x^n$$

$$=\sum_{n=2}^{\infty}(-1)^n c_n x^n,\text{ say} \qquad ...(1)$$

Then, $$c_n=\frac{1}{1\times(n-1)}+\frac{1}{2\times(n-2)}+....+\frac{1}{(n-2)\times2}+\frac{1}{(n-1)\times1}$$

$$=\frac{1}{n}\left\{\frac{n}{1\times(n-1)}+\frac{n}{2\times(n-2)}+....+\frac{n}{(n-2)\times2}+\frac{n}{(n-1)\times1}\right\}$$

$$=\frac{1}{n}\left\{\left(1+\frac{1}{n-1}\right)+\left(\frac{1}{2}+\frac{1}{n-2}\right)+...+\left(\frac{1}{2}+\frac{1}{n-2}\right)+\left(\frac{1}{n-1}+1\right)\right\}$$

Thus, $$c_n=\frac{2}{n}\left(1+\frac{1}{2}+\frac{1}{3}+...+\frac{1}{n-1}\right) \qquad ...(2)$$

Using (2), (1) reduces to

$$\left(x-\frac{x^2}{2}+\frac{x^3}{3}-....\right)^2=2\sum_{n=2}^{\infty}(-1)^n\frac{x^n}{n}\left(1+\frac{1}{2}+\frac{1}{3}+...+\frac{1}{n-1}\right)$$

$$\Rightarrow\quad \frac{1}{2}\left(x-\frac{x^2}{2}+\frac{x^3}{3}-...\right)^2=\sum_{n=2}^{\infty}(-1)^n\frac{x^n}{n}\left(1+\frac{1}{2}+\frac{1}{3}+...+\frac{1}{n-1}\right)$$

Ex. 2. *If* $\sum_{n=1}^{\infty}a_n$ *and* $\sum_{n=1}^{\infty}b_n$ *are positive term convergent series with sums 'a' and 'b' respectively, then show that their Cauchy product* $\sum_{n=1}^{\infty}c_n$ *converges and* $\sum_{n=1}^{\infty}c_n=\left(\sum_{n=1}^{\infty}a_n\right)\left(\sum_{n=1}^{\infty}b_n\right)$.

Examine this inequality for $\sum_{n=1}^{\infty}a_n=\sum_{n=1}^{\infty}b_n=\sum_{n=1}^{\infty}(1/n^2)$ **[Delhi Maths (H) 2006]**

Sol. Let Σa_n and Σb_n be positive term serves converging to α and β respectively.

Let $$c_n=\sum_{r=1}^{n}a_r b_{n-r+1} \qquad (n=1, 2, 3, ...)$$

Then the Cauchy product of the given two series is Σc_n.

Let $\alpha_n = \sum_{r=1}^{n} a_r$, $\beta_n = \sum_{r=1}^{n} b_r$, $\sum_{n=1}^{\infty} a_n = \alpha$ and $\sum_{n=1}^{\infty} b_n = \beta$

Then $<\alpha_n>$ and $<\beta_n>$ are the sequences of partial sums of Σa_n and Σb_n respectively. Again, α and β are the sums of Σa_n and Σb_n respectively.

In order to establish the absolute convergence of the product series, consider the following series
$$a_1b_1 + a_1b_2 + a_2b_2 + a_2b_1 + a_1b_3 + a_2b_2 + a_3b_2 + a_3b_1 + ... \quad ...(1)$$

Then, $|a_1b_1| + |a_1b_2| + |a_2b_2| + |a_2b_1| + ...$ to n terms
$$\le (|a_1| + |a_2| + ... + |a_n|)(|b_1| + |b_2| + ... + |b_n|) \le \alpha\beta, \forall n \in N$$

Hence the series (1) must be convergent. Therefore, by Dirichlet's theorem (See page 7.14), re-arrangement of (1), namely,
$$a_1b_1 + a_1b_2 + a_2b_1 + a_1b_3 + a_2b_2 + a_3b_1 + ...$$
is also convergent. Hence by grouping, we see that
$$\sum_{n=1}^{\infty}\left(\sum_{r=1}^{n} a_r b_{n-r+1}\right), \quad i.e., \quad \sum_{n=1}^{\infty} c_n \quad ...(2)$$
is also convergent.

Since the sum of the first n^2 terms of (1) is $\alpha_n\beta_n$ and $\alpha_n\beta_n \to \alpha\beta$ as $n \to \infty$, therefore, the sum of the series (1) is $\alpha\beta$. Thus, the series (2) which is a grouping of a re-arrangement of the convergent series of positive terms (1), must converge to the same sum.

Thus, $$\sum_{n=1}^{\infty} c_n = \left(\sum_{n=1}^{\infty} a_n\right)\left(\sum_{n=1}^{\infty} b_n\right) \quad ...(3)$$

EXERCISES

1. *Define Cauchy product of two infinite series.* **[Delhi Maths (H) 2002, 03, 07]**

2. *If two series converge to α and β respectively and their Cauchy's product converge to γ, prove that $\gamma = \alpha\beta$.*

3. *Show that Cauchy's product of two conditionally convergent or divergent series may or may not be convergent.*

4. *Prove that for $-1 < x \le 1$,*

(*i*) $$\left(x - \frac{x^2}{2} + \frac{x^3}{3} - ....\right)^2 = \frac{1}{2}x^2 - \frac{1}{3}\left(1 + \frac{1}{2}\right)x^3 + \frac{1}{4}\left(1 + \frac{1}{2} + \frac{1}{3}\right)x^4 + ... = \{\log(1+x)\}^2$$

(*ii*) $$(\tan^{-1} x)^2 = \left(x - \frac{1}{3}x^3 + \frac{1}{5}x^5 - ...\right)^2 = \sum_{n=1}^{\infty} (-1)^{n-1}\frac{x^{2n}}{n}\left(1 + \frac{1}{3} + ... + \frac{1}{2n-1}\right)$$

(*iii*) $$\left(\sum_{n=1}^{\infty} (-1)^{n-1}\frac{x^n}{2n-1}\right)^2 = x^2 - \frac{1}{2}\left(1 + \frac{1}{3}\right)x^3 + \frac{1}{3}\left(1 + \frac{1}{3} + \frac{1}{5}\right)x^4 - ....$$

5. Prove : $$\sum_{n>0}^{\infty}(-1)^n\left[\frac{1}{(n+1)\cdot 1} + \frac{1}{n\cdot 2} + ... + \frac{1}{1\cdot(n+1)}\right] = (\log 2)^2$$ **[Delhi Maths (Hons) 2009]**

Chapter

8

Real Functions : Limit and Continuity

8.1. INTRODUCTION

While in the preceding chapters, we considered functions with the set of natural numbers as their domain, we shall in this chapter be concerned with real valued functions having intervals, open or closed, as their domains.

ILLUSTRATIONS

1. If $f(x) = 0$ when x is rational and $f(x) = 1$ when x is irrational, then f is a real valued function whose domain is the entire set **R** of real numbers. The range of this function is the set $\{0, 1\}$ with two elements.

2. If $f(x) = [x]$ where $[x]$ denotes the greatest integer not greater than x, then f is a function whose domain is **R** and whose range is the set **I** of integers.

3. If $f(x) = a_0x^n + a_1x^{n-1} + \ldots\ldots + a_n$, where $a_0, \ldots\ldots, a_n$ are given real numbers, then f is a function with **R** as its domain. What can be said about the range of the function ?

4. If $f(x) = \log x$, the domain of f is the set of all positive real numbers and the range is **R**.

5. The domain and range of the function f defined by $f(x) = e^x$ are $]-\infty, \infty[$ and $[0, \infty[$ respectively.

6. The domain and range of the function f defined by $f(x) = \sin x$ are $]-\infty, \infty[$ and $[-1, 1]$ respectively.

Ex. 1. Describe the domains and ranges of the functions defined as follows :

(i) $f(x) = \cos x$; *(ii)* $f(x) = \tan x$; *(iii)* $f(x) = \sec x$;
(iv) $f(x) = \sin^{-1} x$; *(v)* $f(x) = \tan^{-1} x$; *(vi)* $f(x) = \sec^{-1} x$;
(vii) $f(x) = (x^2 - 1)/(x - 1)$; *(viii)* $f(x) = x + 1$; *(ix)* $f(x) = 1/(x + 1)$;
(x) $f(x) = \sqrt{(x-1)}$; *(xi)* $f(x) = \sqrt{\left(\dfrac{2x-3}{4-5x}\right)}$; *(xii)* $f(x) = \sqrt{(2x-3)(4-5x)}$.

Ex. 2. If x denotes a real number, which sub-sets of **R** are the domains of the following functions :

(i) $f : x \to 1/(x + 1)$; *(ii)* $f : x \to x/(x - 1)(x - 2)$;
(ii) $f : x \to \sqrt{x}$; *(iv)* $f : x \to \sqrt{(x+1)}/(x-1)$;
(v) $f : x \to \sqrt{(x-1)} + x/(x-2)$; *(vi)* $f : x \to \sqrt{(x^2 - 5x + 6)}$.

Ex. 3. Obtain the ranges of the functions given as follows :

(i) $f(x) = x/(x - 1)(x + 2)$; *(ii)* $f(x) = (x + 1)(x^2 + 1)$;
(iii) $f(x) = (x + 1)(x^2 - 1)$; *(iv)* $f(x) = x^2 + x + 3$.

Ex. 4. State the ranges and the domains of the following functions :

(*i*) $f : x \to [x]$; (*ii*) $f : x \to \sqrt{[x]}$; (*iii*) $f : x \to \sqrt{|x|}$; (*iv*) $f : x \to |x|$.

Ex. 5. Compare the domains of the following pairs of functions :

(*i*) $x \to x/x$, $x \to 1$. (*ii*) $x \to (2 \sin x)/\sin x$, $x \to 2$.

8.2. ALGEBRAIC OPERATIONS ON FUNCTIONS

Let f_1, f_2 be two functions; with domains D_1, D_2 respectively. We suppose that D_1, D_2 are not disjoint, *i.e.*, $D_1 \cap D_2$ is not the void set. We shall now define the functions

$$f_1 + f_2, \quad f_1 - f_2, \quad f_1 f_2$$

It will be seen that $D_1 \cap D_2$ is the domain of each of these functions. We have, by definition, $x \in D_1 \cap D_2$,

$(f_1 + f_2)(x) = f_1(x) + f_2(x)$, $(f_1 - f_2)(x) = f_1(x) - f_2(x)$ and $(f_1 f_2)(x) = f_1(x) f_2(x)$.

Now suppose that $A = \{x : f_2(x) = 0\}$

so that A is the set of roots of the equation $f_2(x) = 0$. Then $f_1 \div f_2$ is defined as follows :

$$(f_1 \div f_2)(x) = f_1(x) \div f_2(x) \quad \forall x \in D_1 \cap (D_2 \sim A).$$

Ex. 1. Given that f_1, f_2 are the functions defined as follows : $x \to \cos x$, $x \to \sec x$, give the domain of $f_1 + f_2$, $f_1 \div f_2$.

How does the function $x \to 1 \ \forall x$ differ from the function $f_1 f_2$?

Ex. 2. Repeat the preceding exercise with f_1, f_2 defined as $x \to \tan x$, $x \to \cot x$

8.3. BOUNDED AND UNBOUNDED FUNCTIONS **[Meerut 2005]**

A function whose range is bounded is said to be a bounded function and otherwise unbounded. Moreover the greatest lower bound (g.l.b.) of the range of a bounded function is said to be the greatest lower bound of the function; similarly about the least upper bound (l.u.b.).

ILLUSTRATIONS

1. The function f defined as

$$f(x) = \begin{cases} 1/x & \text{when } x \neq 0 \\ 0 & \text{when } x = 0 \end{cases}$$

is *not* bounded.

2. Consider the function defined by $f(x) = x/(x+1)$ with $[0, \infty[$ as its domain. This function is bounded with 0 and 1 as its two bounds. While the greatest lower bound 0 is a value of the function, the least upper bound 1 is not.

EXERCISES

Ex. 1. Which of the following functions are bounded for the given domains :

(*i*) $x \to x^2 + 1$; $x \in [-1, 1]$, (*ii*) $x \to 1/x$; $x > 0$,

(*iii*) $x \to 1/(|x| - 1)$; $x \in \mathbf{R} \sim \{-1, 1\}$, (*iv*) $x \to [x]$; $x \in [-5, 7]$,

(*v*) $x \to x/(x+1)$; $x \in \mathbf{R} \sim \{-1\}$, (*vi*) $x \to \sqrt{x}$; $x \in [0, 4]$.

Also obtain in each case the l.u.b. and the g.l.b. whichever may exist.

Ex. 2. Given that f_1, f_2 are two bounded functions, show that the functions $f_1 + f_2$, $f_1 - f_2$, and $f_1 f_2$ are also bounded.

Ex. 3. Give two bounded functions f_1, f_2 such that $f_1 \div f_2$ is not bounded.

Monotonic functions. Let f be a function with domain $[a, b]$.

We say that f is *monotonically increasing* in $[a, b]$, if $x < y \Rightarrow f(x) \leq f(y)$.

In case $x < y \Rightarrow f(x) < f(y)$,
we say that f is *strictly* monotonically increasing.

We may similarly define *monotonically decreasing* and *strictly monotonically decreasing functions*.

EXERCISES

Ex. 1. Show that

(*i*) $x \to x^2$ is strictly monotonically increasing in $[a, b]$, where a, b are positive numbers.

(*ii*) $x \to x^2 + 2x$ is strictly monotonically decreasing in $[a, -1]$; a being a negative number less than -1.

Ex. 2. Give a few examples of monotonically decreasing (increasing) functions.

8.4. LIMIT OF A FUNCTION (Kanpur, 2001)

Let f be a function defined for all points in some neighbourhood of a point a except *possibly* at the point a itself.

We say that *the function f tends to the limit l as x tends to* (*or approaches*) *a, or symbolically*

$$\lim_{x \to a} f(x) = l,$$

if, to each given positive number ε, *there corresponds a positive number* δ *such that*

$$|f(x) - l| < \varepsilon \quad \textit{when} \quad 0 < |x - a| < \delta$$

i.e., $f(x) \in\,]l - \varepsilon, l + \varepsilon[$ for all those values of x (except possibly a) which $\in\,]a - \delta, a + \delta[$.

Note 1. It may be seen that the inequality $0 < |x - a|$ excludes the possibility $0 = |x - a| \Leftrightarrow x = a$.

Thus $$0 < |x - a| \Leftrightarrow x \neq a.$$

Note 2. Existence of $\lim_{x \to a} f(x)$ implies that the domain of the function f includes the intervals

$$[a - \delta, a[,\ ]a, a + \delta]$$

for some positive δ. It may be noticed that a may or may not belong to the domain of f.

Note 3. The union $[a - \delta, a[\cup]a, a + \delta]$ which is the same as $[a - \delta, a + \delta] \sim \{a\}$ is called a *Deleted Neighbourhood of a*.

Example. Using the definition of limit, prove that

(*i*) $\lim_{x \to a} \dfrac{x^2 - a^2}{x - a} = 2a$ (*i*)′ $\lim_{x \to 2} \dfrac{x^2 - 4}{x - 2} = 4$ **(G.N.D.U. Amritsar 2010)** (*ii*) $\lim_{x \to 0} x \sin \dfrac{1}{x} = 0$

Solution. (*i*) Here $f(x) = (x^2 - a^2)/(x - a),\ x \neq a$.

Let $\varepsilon > 0$ be given. Then we want to find a $\delta > 0$ such that

$$|f(x) - 2a| < \varepsilon \quad \text{whenever} \quad 0 < |x - a| < \delta.$$

Now, $$|f(x) - 2a| = \left| \frac{x^2 - a^2}{x - a} - 2a \right| = \left| \frac{(x - a)^2}{x - a} \right| = |x - a|$$

We can choose $\delta = \varepsilon$. Then we have

$$|f(x) - 2a| < \varepsilon \quad \text{whenever} \quad 0 < |x - a| < \varepsilon$$

Hence $$\lim_{x \to 2a} \frac{x^2 - a^2}{x - a} = 2a.$$

(*i*)′ Proceed as in part (*i*).

(*ii*) Here $f(x) = x \sin(1/x)$.

Let $\varepsilon > 0$ be given. Then we want to find a $\delta > 0$ such that

$$|f(x) - 0| < \varepsilon \quad \text{whenever} \quad 0 < |x - 0| < \delta.$$

Now, $$|f(x) - 0| = \left| x \sin\frac{1}{x} \right| = |x| \left| \sin\frac{1}{x} \right| \le |x|, \text{ as } \left| \sin\frac{1}{x} \right| \le 1$$

We can choose $\delta = \varepsilon$. Then we have

$$|f(x) - 0| < \varepsilon \quad \text{whenever} \quad 0 < |x| < \varepsilon$$

Hence, $$\lim_{x \to 0} x \sin\frac{1}{x} = 0.$$

We now prove a basic property of limit of a function.

Theorem. Uniqueness of limit. *The limit of a function at a point, if it exists, is unique, i.e., if* $\lim_{x \to a} f(x) = l$ *and* $\lim_{x \to a} f(x) = l'$, *then* $l = l'$.

Proof. Let $$\lim_{x \to a} f(x) = l \quad \text{and} \quad \lim_{x \to a} f(x) = l' \qquad ...(1)$$

We shall show that $l = l'$. Let $\varepsilon > 0$ be given.

From equation (1), there exists $\delta_1 > 0$ and $\delta_2 > 0$ such that

$$|f(x) - l| < \varepsilon/2, \text{ when } 0 < |x - a| < \delta_1 \qquad ...(2)$$

and $$|f(x) - l'| < \varepsilon/2, \text{ when } 0 < |x - a| < \delta_2 \qquad ...(3)$$

Let $\delta = \min(\delta_1, \delta_2)$. Then, by equations (2) and (3), we get

$$|f(x) - l| < \varepsilon/2, \text{ when } |x - a| < \delta$$

and $$|f(x) - l'| < \varepsilon/2, \text{ when } |x - a| < \delta$$

Now $$|l - l'| = |l - f(x) + f(x) - l'|$$
$$\le |l - f(x)| + |f(x) - l'| < \varepsilon/2 + \varepsilon/2 = \varepsilon$$

$\therefore$ $$|l - l'| < \varepsilon, \text{ when } 0 < |x - a| < \delta.$$

Since ε is arbitrarily small, $|l - l'| = 0$. Hence, $l = l'$.

8.5. ALGEBRA OF LIMITS

We now present some properties of limits which are similar to those of limits of sequences.

Theorem 1. *If* $\lim_{x \to a} f(x) = l$ *and* $\lim_{x \to a} g(x) = m$, *then* $\lim_{x \to a} [f(x) \pm g(x)] = l \pm m$.

(Meerut, 2003)

Proof. Let $\varepsilon > 0$ be given. Then in view of the given limits, there exist $\delta_1 > 0, \delta_2 > 0$

such that $$|f(x) - l| < \varepsilon/2, \text{ when } 0 < |x - a| < \delta_1$$

and $$|g(x) - m| < \varepsilon/2, \text{ when } 0 < |x - a| < \delta_2$$

Let $\delta = \min(\delta_1, \delta_2)$. Then, we have

$$|f(x) - l| < \varepsilon/2, \text{ when } 0 < |x - a| < \delta,$$

and $$|g(x) - m| < \varepsilon/2, \text{ when } 0 < |x - a| < \delta.$$

Now $$|\{f(x) \pm g(x)\} - (l \pm m)| = |(f(x) - l) \pm (g(x) - m)|$$
$$\le |f(x) - l| + |g(x) - m| < \varepsilon/2 + \varepsilon/2 = \varepsilon,$$

when $0 < |x - a| < \delta$

$\therefore$ $$|\{f(x) \pm g(x)\} - (l \pm m)| < \varepsilon, \text{ whenever } 0 < |x - a| < \delta$$

Hence, $$\lim_{x \to a} [f(x) \pm g(x)] = l \pm m.$$

Theorem 2. *If* $\lim_{x \to a} f(x) = l$ *and* $\lim_{x \to a} g(x) = m$, *then* $\lim_{x \to a} [f(x)\, g(x)] = lm.$

[Delhi Maths (H) 2008, Meerut, 1993; Delhi Maths (G), 2006]

Proof. We have $\quad | f(x)\, g(x) - lm | = | g(x)\,(f(x) - l) + l\,(g(x) - m) |$

$$\leq | g(x) |\, | f(x) - l | + | l |\, | g(x) - m | \qquad \text{...(1)}$$

Since $\lim_{x \to a} f(x) = l$ and $\lim_{x \to a} g(x) = m$, for $0 < \varepsilon' < 1$, there exists some $\delta > 0$, such that

$$| f(x) - l | < \varepsilon' \text{ and } | g(x) - m | < \varepsilon', \text{ when } 0 < | x - a | < \delta \qquad \text{...(2)}$$

Now $\quad | g(x) | = | m + g(x) - m | \leq | m | + | g(x) - m |$

or $\quad | g(x) | < | m | + \varepsilon'$, when $0 < | x - a | < \delta.$ ...(3)

From equations (1), (2) and (3), we get,

$$| f(x)\, g(x) - lm | < (| m | + \varepsilon')\, \varepsilon' + | l |\, \varepsilon', \text{ when } 0 < | x - a | < \delta.$$

$$< (| m | + | l | + 1)\, \varepsilon', \text{ when } 0 < | x - a | < \delta. \qquad (\because \ \varepsilon' < 1)$$

Let us choose ε', such that $\quad \varepsilon' < \dfrac{\varepsilon}{(| m | + | l | + 1)}$, $\quad$ where $\varepsilon > 0$.

$\therefore \quad | f(x)\, g(x) - lm | < \varepsilon$, when $0 < | x - a | < \delta$.

Hence, $\quad \lim_{x \to a} [f(x)\, g(x)] = lm = \lim_{x \to a} f(x) \cdot \lim_{x \to a} g(x).$

Theorem 3. *If* $\lim_{x \to a} f(x) = l$ *and* $\lim_{x \to a} g(x) = m$, *and* $m \neq 0$, *then* $\lim_{x \to a} \left[\dfrac{f(x)}{g(x)} \right] = \dfrac{l}{m}.$

[Delhi Maths (G), 1993; Meerut, 1993]

Proof. We have

$$\left| \frac{f(x)}{g(x)} - \frac{l}{m} \right| = \frac{| mf(x) - lg(x) |}{| g(x) | \cdot | m |} = \frac{| mf(x) - lm + lm - lg(x) |}{| g(x) | \cdot | m |}$$

$$= \frac{| m\,(f(x) - l) + l\,(m - g(x) |}{| g(x) | \cdot | m |}$$

$$\leq \frac{| m | \cdot | f(x) - l | + | l |\, | g(x) - m |}{| m | \cdot | g(x) |} \qquad \text{...(1)}$$

Since $\lim_{x \to a} g(x) = m \neq 0$, there exists a $\delta_1 > 0$ such that

$$| g(x) | > | m |/2, \text{ when } 0 < | x - a | < \delta_1$$

$$\Rightarrow \qquad \frac{1}{| g(x) |} < \frac{2}{| m |}, \text{ when } 0 < | x - a | < \delta_1 \qquad \text{...(2)}$$

Since $\lim_{x \to a} g(x) = l$, there exists some $\delta_2 > 0$ such that

$$| f(x) - l | < \varepsilon\, | m |/4, \text{ when } 0 < | x - a | < \delta_2 \qquad \text{... (3)}$$

Since $\lim_{x \to a} g(x) = m$, there exists some $\delta_3 > 0$ such that

$$| g(x) - m | < \varepsilon\, | m |^2 / 4\, | l |, \text{ when } 0 < | x - a | < \delta_3 \qquad \text{...(4)}$$

Let $\delta = \min(\delta_1, \delta_2, \delta_3)$. Then, from equations (1), (2), (3) and (4), we get

$$\left| \frac{f(x)}{g(x)} - \frac{l}{m} \right| < \frac{2}{| m |^2} \left[\frac{1}{4} \varepsilon\, | m |^2 + \frac{1}{4} \varepsilon\, | m |^2 \right] = \varepsilon, \text{ when } 0 < | x - a | < \delta.$$

Hence, $$\lim_{x \to a} \left[\frac{f(x)}{g(x)}\right] = \frac{l}{m},\ m \neq 0.$$

Theorem 4. *Let f be defined on D and let $f(x) \geq 0$ for all $x \in D$. If $\lim_{x \to a} f(x)$ exists, then*

$$\lim_{x \to a} f(x) \geq 0.$$

Proof. If possible, let $\lim_{x \to a} f(x) = l$, where $l < 0$.

Then, for a given $\varepsilon = -(l/2) > 0$, we can find a $\delta > 0$ such that

$|f(x) - l| < -(l/2)$ whenever $0 < |x - a| < \delta$

$\Rightarrow$ $3l/2 < f(x) < l/2 < 0$, whenever $0 < |x - a| < \delta$

This is impossible, because we are given that $f(x) \geq 0 \quad \forall x \in D$. Hence, l cannot be negative.

Therefore, $$\lim_{x \to a} f(x) \geq 0.$$

Corollary. *Let f be defined on D and let $f(x) > 0 \ \forall x \in D$. If $\lim_{x \to a} f(x)$ exists, then*

$$\lim_{x \to a} f(x) \geq 0.$$

Proof. Here $f(x) > 0 \Rightarrow f(x) \geq 0$, and so by theorem 4, the required result follows.

Theorem 5. *Let f and g be defined on D and let $f(x) \geq g(x) \ \forall x \in D$. Then*

$$\lim_{x \to a} f(x) \geq \lim_{x \to a} g(x), \text{ provided these limits exist.}$$

Proof. Let $$\lim_{x \to a} f(x) = l \quad \text{and} \quad \lim_{x \to a} g(x) = m.$$

Let $h(x) = f(x) - g(x) \quad \forall x \in D$. Then, we have

(*i*) $h(x) \geq 0 \quad \forall x \in D$

(*ii*) $\lim_{x \to a} h(x)$ exists and $\lim_{x \to a} h(x) = l - m$, by theorem 1

(*iii*) $\lim_{x \to a} h(x) \geq 0$, by theorem 4

Now, from (*ii*) and (*iii*), we have

$$l - m \geq 0, \quad i.e., \quad l \geq m, \quad i.e., \quad \lim_{x \to a} f(x) \geq \lim_{x \to a} g(x).$$

Corollary. *Let f and g be defined on D and let $f(x) > g(x) \quad \forall x \in D$. Then*

$$\lim_{x \to a} f(x) \geq \lim_{x \to a} g(x), \text{ provided these limits exist.}$$

Proof. Here $f(x) > 0 \Rightarrow f(x) \geq 0$ and so by theorem 5, the required result follows.

Theorem 6. Squeeze theorem or principle. *Let f, g, h be defined on D and let $f(x) \leq g(x) \leq h(x)$ for all x. Let $\lim_{x \to a} f(x) = \lim_{x \to a} h(x)$. Then $\lim_{x \to a} g(x)$ exists, and*

$$\lim_{x \to a} g(x) = \lim_{x \to a} f(x) = \lim_{x \to a} h(x).$$

[Delhi B.Sc. (Hons.) II, 2016, 2018]

Solution. Let $$\lim_{x \to a} f(x) = \lim_{x \to a} h(x) = l \qquad ...(1)$$

Then for a given $\varepsilon > 0$, there exist positive numbers δ_1 and δ_2 such that

$|f(x) - l| < \varepsilon$ for $0 < |x - a| < \delta_1$ and $|h(x) - l| < \varepsilon$ for $0 < |x - a| < \delta_2$

$\Rightarrow$ $l - \varepsilon < f(x) < l + \varepsilon$ for $0 < |x - a| < \delta_1$ and $l - \varepsilon < h(x) < l + \varepsilon$ for $0 < |x - a| < \delta_2$

Let $\delta = \min\{\delta_1, \delta_2\}$. Then, we have

$$l - \varepsilon < f(x) < l + \varepsilon \quad \text{and} \quad l - \varepsilon < h(x) < l + \varepsilon \quad \text{for} \quad 0 < |x - a| < \delta \qquad ...(2)$$

Also, we are given that $\quad f(x) \le g(x) \le h(x) \qquad ...(3)$

From (2) and (3), $\quad l - \varepsilon < f(x) \le g(x) \le h(x) < l + \varepsilon \quad \text{for} \quad 0 < |x - a| < \delta$

$$\Rightarrow l - \varepsilon < g(x) < l + \varepsilon \quad \text{for} \quad 0 < |x - a| < \delta$$

$$\Rightarrow |g(x) - l| < \varepsilon \quad \text{for} \quad 0 < |x - a| < \delta$$

$$\Rightarrow \lim_{x \to a} g(x) \text{ exists and } \lim_{x \to a} g(x) = l. \qquad ...(4)$$

From (1) and (2), we have $\quad \lim_{x \to a} g(x) = \lim_{x \to a} f(x) = \lim_{x \to a} h(x).$

Theorem 7. *If* $\lim_{x \to a} f(x) = l$, *then* $\lim_{x \to a} |f(x)| = |l|$. *But the converse is not true.*

(*Srivenkateshwara, 2003*)

Proof. Since $\lim_{x \to a} f(x) = l$, for a given $\varepsilon > 0$, there exists a positive number δ such that

$$|f(x) - l| < \varepsilon \quad \text{whenever} \quad 0 < |x - a| < \delta \qquad ...(1)$$

Since $\quad |a - b| \ge ||a| - |b||$, we have

$$|f(x) - l| \ge ||f(x)| - |l|| \qquad ...(2)$$

From (1) and (2), we have

$$||f(x)| - |l|| < \varepsilon \quad \text{whenever} \quad 0 < |x - a| < \delta$$

$$\Rightarrow \lim_{x \to a} |f(x)| \text{ exist and } \lim_{x \to a} |f(x)| = |l|.$$

We now show that the converse is not true.

Consider $\quad f(x) = \begin{cases} -1, & \text{if } x < a \\ 1, & \text{if } x \ge a \end{cases}$

Then $\quad \lim_{x \to a+0} f(x) = \lim_{x \to a+0} 1 = 1, \quad \lim_{x \to a-0} (-1) = -1$, using Art. 8.6

Since $\quad \lim_{x \to a+0} f(x) \ne \lim_{x \to a-0} f(x)$, so $\lim_{x \to a} f(x)$ does not exist.

Again $\quad |f(x)| = 1 \;\; \forall\, x$ and so $\lim_{x \to a} |f(x)| = 1$ exists.

Theorem 8. *If* $\lim_{x \to a} f(x) = 0$ *and* $g(x)$ *is bounded in some deleted neighbourhood of a, then*

$$\lim_{x \to a} f(x)\, g(x) = 0.$$

Proof. $g(x)$ is bounded in some deleted neighbourhood of a

$\Rightarrow$ there exists numbers $k > 0$ and $\delta_1 > 0$ such that

$$|g(x)| \le k \text{ whenever } 0 < |x - a| < \delta_1 \qquad ...(1)$$

Since $\lim_{x \to a} f(x) = 0$, for a given $\varepsilon > 0$, there exists $\delta_2 > 0$ such that

$$|f(x) - 0| < \varepsilon / k \quad \text{whenever} \quad 0 < |x - a| < \delta_2 \qquad ...(2)$$

Let $\delta = \min(\delta_1, \delta_2)$. Then from (1) and (2), we have

$$|g(x)| \le k \quad \text{and} \quad |x| < \varepsilon / k \quad \text{for} \quad 0 < |x - a| < \delta \qquad ...(3)$$

Now, $\quad |f(x)\, g(x) - 0| = |f(x)|\,|g(x)| < (\varepsilon / k) \times k = \varepsilon$, by (3)

$$\Rightarrow \lim_{x \to 0} f(x)\, g(x) = 0.$$

Example. Evaluate $\lim_{x \to 0} x \sin(1/x)$.

Here $\lim_{x \to 0} x = 0$ and $|\sin(1/x)| \le 1$ for all $x \ne 0$, *i.e.*, $\sin(1/x)$ is bounded in some deleted neighbourhood of zero. So taking $f(x) = x$ and $g(x) = \sin(1/x)$ in above theorem, we have

$$\lim_{x \to 0} f(x)\, g(x) = 0 \Rightarrow \lim_{x \to 0} x \sin(1/x) = 0$$

8.6. ONE-SIDED LIMITS — RIGHT-HAND AND LEFT-HAND LIMITS

While defining the limit of a function f as x approaches a, we have considered the values of f in the deleted neighbourhood of a. If we consider the behaviour of f for those values of x greater than a, we say that x *approaches a from the right or from above.* We denote this as $x \to a+$ or $x \to a + 0$. Similarly, if we consider the values of f for x less than a, then we say that x *approaches a from the left or from below.* We denote this as $x \to a-$ or $x \to a - 0$.

Right-hand limit. Definition. We say that the function f tends to l as x tends to a through values greater than a if to each $\varepsilon > 0$ there corresponds $\delta > 0$ such that

$$|f(x) - l| < \varepsilon \quad \text{when} \quad a < x < a + \delta$$

so that $$x \in]a, a + \delta[\;\Rightarrow\; f(x) \in]l - \varepsilon, l + \varepsilon[.$$

Also then we write

$$\lim_{x \to a+0} f(x) = l \quad \text{or} \quad f(a+0) = l \quad \text{or} \quad f(a+) = l.$$

This limit is known as a right-hand limit.

Left-hand limit. Definition. We say that the function f tends to l as x tends to a through values less than a, if to each $\varepsilon > 0$ there corresponds $\delta > 0$ such that

$$|f(x) - l| < \varepsilon \quad \text{when} \quad a - \delta < x < a$$

so that $$x \in]a - \delta, a[\;\Rightarrow\; f(x) \in]l - \varepsilon, l + \varepsilon[.$$

Also then we write

$$\lim_{x \to a-0} f(x) = l \quad \text{or} \quad f(a-0) = l \quad \text{or} \quad f(a-) = l.$$

This limit is known as a left-hand limit.

WORKING RULE TO COMPUTE LEFT-HAND AND RIGHT-HAND LIMITS

Method to compute right-hand limit. Put $x = a + h$, $h > 0$ in $f(x)$ and then take the limit as $h \to 0+$

$$\therefore \quad \text{Right-hand limit} = f(a+0) = \lim_{x \to a+0} f(x) = \lim_{h \to 0+} f(a+h)$$

Method to compute left-hand limit. Put $x = a - h$, $h > 0$ in $f(x)$ and then take limit as $h \to 0+$

$$\therefore \quad \text{Left-hand limit} = f(a-0) = \lim_{x \to a-0} f(x) = \lim_{h \to 0+} f(a-h).$$

Theorem. *Prove that* $\lim_{x \to c} f(x)$ *exists and is equal to a number* l *if and only if both left limit* $\lim_{x \to c-0} f(x)$ *and right limit* $\lim_{x \to c+0} f(x)$ *exist and are equal to* l.

[*Delhi Maths (G), 1996*]

Or

Let f *be defined on a deleted neighbourhood of* c. *Show that* $\lim_{x \to c} f(x)$ *exists and equals* l *iff* $f(c+0)$, *and* $f(c-0)$ *both exist and are equal to* l. **[*Delhi Maths (H), 2002*]**

Proof. *Condition is necessary.*

Let $\lim_{x \to c} f(x) = l$. Then for any $\varepsilon > 0$, there exists some $\delta > 0$ such that

$$|f(x) - l| < \varepsilon, \text{ when } 0 < |x - c| < \delta$$

or $$|f(x) - l| < \varepsilon, \text{ when } c - \delta < x < c + \delta,\ x \neq c.$$

It follows that $$|f(x) - l| < \varepsilon, \text{ when } c - \delta < x < c \quad ...(1)$$

and $$|f(x) - l| < \varepsilon, \text{ when } c < x < c + \delta. \quad ...(2)$$

From equations (1) and (2), we get

$\lim_{x \to c-0} f(x)$ and $\lim_{x \to c+0} f(x)$ both exist and are equal to l.

Condition is sufficient. Let $\lim_{x \to c+0} f(x) = l = \lim_{x \to c-0} f(x)$

Then for any $\varepsilon > 0$, there exist some $\delta_1 > 0$ and $\delta_2 > 0$ such that

$$|f(x) - l| < \varepsilon, \text{ when } c < x < c + \delta_1 \quad ...(3)$$

and $$|f(x) - l| < \varepsilon, \text{ when } c - \delta_2 < x < c. \quad ...(4)$$

Let $\delta = \min(\delta_1, \delta_2)$. Then $\delta \leq \delta_1$ and $\delta \leq \delta_2$

$\Rightarrow\ c + \delta \leq c + \delta_1$ and $c - \delta \geq c - \delta_2$ (or $c - \delta_2 \leq c - \delta$) ...(5)

From equations (3) (4) and (5), we get

$|f(x) - l| < \varepsilon$, when $c < x < c + \delta$ and $|f(x) - l| < \varepsilon$, when $c - \delta < x < c$

$\therefore$ $|f(x) - l| < \varepsilon$, when $c - \delta < x < c + \delta$, $x \neq c$

or $|f(x) - l| < \varepsilon$, when $0 < |x - c| < \delta$.

Hence, $$\lim_{x \to c} f(x) = l.$$

Note. The result of the above theorem is used to establish the existence of $\lim_{x \to a} f(x)$. Thus,

(*i*) If $f(a+0)$ and $f(a-0)$ both exist and are equal, then their common value is the value of $\lim_{x \to a} f(x)$.

(*ii*) If $f(a + 0)$ or $f(a - 0)$ or both do not exist, then $\lim_{x \to a} f(x)$ does not exist.

(*iii*) If $f(a + 0)$ and $f(a - 0)$ both exist but are unequal, then $\lim_{x \to a} f(x)$ does not exist.

Example 1. *Show that (a)* $\lim_{x \to 0} \dfrac{x - |x|}{x}$ *does not exist.* ***(Meerut, 2003)***

(*b*) $\lim_{x \to 0} (|x|/x)$ does not exist ***(Chennai 2011)***

Solution. (*a*) Left-hand limit $= \lim_{x \to 0-} \dfrac{x - |x|}{x} = \lim_{x \to 0} \dfrac{x - (-x)}{0} = \lim_{x \to 0} \dfrac{2x}{x} = 2$

and right-hand limit $= \lim_{x \to 0+} \dfrac{x - |x|}{x} = \lim_{x \to 0} \dfrac{x - x}{0} = \lim_{x \to 0} 0 = 0$

Since left-hand limit $\neq$ right-hand limit, so $\lim_{x \to 0} \dfrac{x - |x|}{x}$ does not exist.

(*b*) Try yourself as in part (*a*)

Example 2. *If* $f : \mathbf{R} \to \mathbf{R}$ *is defined by*

$$f(x) = \begin{cases} 3x - 2, & x < 1 \\ 4x^2 - 3x, & x > 1 \end{cases}, \text{ then find } \lim_{x \to 1} f(x).$$ ***(Osmania, 2004)***

Solution. Left-hand limit $= f(1-0) = \lim\limits_{x\to 1-0} f(x) = \lim\limits_{h\to 0+} f(1-h)$, where $h > 0$

$$= \lim_{h\to 0+} \{3(1-h)-2\} = \lim_{h\to 0+} (1-3h) = 1$$

Again, right-hand limit $= f(1+0) = \lim\limits_{x\to 1+0} f(x) = \lim\limits_{h\to 0+} f(1+h)$, where $h > 0$

$$= \lim_{h\to 0+} \{4(1+h)^2 - 3(1+h)\} = \lim_{h\to 0+} (1+h+4h^2) = 1$$

Since $f(1+h)$ and $f(1-h)$ both exist and are equal to 1, so

$\lim\limits_{x\to 0} f(x)$ exists and is equal to 1.

8.7. LIMITS AT INFINITY AND INFINITE LIMITS. DEFINITIONS

(Kanpur, 2001)

(*i*) A function f is said to tend to l as $x \to \infty$ if given $\varepsilon > 0$, there exists a positive number k such that $|f(x) - l| < \varepsilon$ whenever $x > k$

Also then we write $\lim\limits_{x\to\infty} f(x) = l$ or $f(x) \to l$ as $x \to \infty$.

(*ii*) A function f is said to tend to l as $x \to -\infty$ if given $\varepsilon > 0$, there exists a positive number k such that $|f(x) - l| < \varepsilon$ whenever $x < -k$.

Also then we write $\lim\limits_{x\to -\infty} f(x) = l$ or $f(x) \to l$ as $x \to -\infty$.

(*iii*) A function f is said to tend to ∞ as x tends to a, if given $k > 0$, however large, there exists a positive number δ such that $f(x) > k$ whenever $0 < |x - a| < \delta$.

Also then we write $\lim\limits_{x\to a} f(x) = \infty$ or $f(x) \to \infty$ as $x \to a$.

(*iv*) A function f is said to tend to $-\infty$ as x tends to a, if given $k > 0$, however large, there exists a positive number δ such that $f(x) < -k$ whenever $0 < |x - \infty| < \delta$

Also then we write $\lim\limits_{x\to a} f(x) = -\infty$ or $f(x) \to -\infty$ as $x \to a$.

(*v*) A function f is said to tend to ∞ as $x \to \infty$ if given $k > 0$, however large, there exists a positive number K such that $f(x) > k$ whenever $x > K$

Also then we write $\lim\limits_{x\to\infty} f(x) = \infty$ or $f(x) \to \infty$ as $x \to \infty$.

(*vi*) A function f is said to tend to $-\infty$ as $x \to \infty$ if given $k > 0$, however large, there exists a positive number K such that $f(x) < -k$ whenever $x > K$

Also then we write $\lim\limits_{x\to\infty} f(x) = -\infty$ or $f(x) \to -\infty$ as $x \to \infty$.

(*vii*) A function f is said to tend to ∞ as $x \to -\infty$, if given $k > 0$, however large, there exists a positive number K such that $f(x) > k$ whenever $x < -K$

Also then we write $\lim\limits_{x\to -\infty} f(x) = \infty$ or $f(x) \to \infty$ as $x \to -\infty$.

(*viii*) A function f is said to tend to $-\infty$ as $x \to -\infty$, if given $k > 0$, however large, there exists a positive number K such that $f(x) < -k$ whenever $x < -K$.

EXAMPLES

Example 1. *Prove that* $\lim\limits_{x\to\infty} \dfrac{1}{x+1} = 0.$

Solution. Let $f(x) = 1/(x+1)$ and let $\varepsilon > 0$ be given. Then

$$|f(x) - 0| = \frac{1}{x+1} < \frac{1}{x} < \varepsilon \text{ if } x > \frac{1}{\varepsilon}$$

Taking $k = 1/\varepsilon$. Then, we see that

$$\left|\frac{1}{x+1} - 0\right| < \varepsilon \text{ whenever } x > k \Rightarrow \lim_{x\to\infty} \frac{1}{x+1} = 0.$$

Example 2. *Evaluate* (*i*) $\lim_{x\to 0} \frac{1}{|x|}$ (*ii*) $\lim_{x\to 0} \frac{1}{x}$ **[*Delhi B.Sc. (H) Physics, 2000*]**

Solution. (*i*) Let $f(x) = 1/|x|$. Let $k > 0$ be given and let $\delta = 1/k$. Then

$$0 < |x - 0| < \delta \Rightarrow |x| < \delta \Rightarrow |x| < 1/k \Rightarrow 1/|x| > k$$

Thus, $f(x) > k$ whenever $0 < |x - 0| < \delta \Rightarrow \lim_{x\to 0} \frac{1}{|x|} = \infty$

(*ii*) Let $f(x) = 1/x$. Let us first find right-hand limit $\lim_{x\to 0+0} (1/x)$.

Let $k > 0$ be given and let $\delta = 1/k$. Then

$$0 < x < \delta \Rightarrow 1/x > 1/\delta \Rightarrow f(x) > k$$

Thus, $f(x) > k$ whenever $0 < x < \delta \Rightarrow \lim_{x\to 0+0} f(x) = \infty$,

$$i.e., \quad \lim_{x\to 0+0} (1/x) = \infty \qquad ...(1)$$

We now find left-hand limit $\lim_{x\to 0-0} (1/x)$.

Now $-\delta < x < 0 \Rightarrow -\frac{1}{\delta} > \frac{1}{x} \Rightarrow -k > \frac{1}{x} \Rightarrow \frac{1}{x} < -k$

Thus, $f(x) < -k$ whenever $-\delta < x < 0 \Rightarrow \lim_{x\to 0-0} f(x) = -\infty$,

$$i.e., \quad \lim_{x\to 0-0} (1/x) = -\infty. \qquad ...(2)$$

From (1) and (2), we find that left-hand and right-hand limits are not equal. Hence

$\lim_{x\to 0} (1/x)$ does not exist.

Example 3. *Let* $f(x) = 1/(x^2 - 1)$ $\forall\, x \in \mathbf{R} - \{1, -1\}$. *Then show that* $\lim_{x\to 1+0} f(x) = \infty$.

[*Delhi Maths (H), 2003*]

Solution. Let k be any positive number (however large). Then to prove the required result, we must find a $\delta > 0$ such that

$$0 < x - 1 < \delta \Rightarrow f(x) > k.$$

Now,
$$\begin{aligned} 0 < x - 1 < \delta &\Rightarrow 1 < x < 1 + \delta \\ &\Rightarrow 1 < x^2 < (1+\delta)^2 \\ &\Rightarrow 0 < x^2 - 1 < 2\delta + \delta^2 \\ &\Rightarrow \frac{1}{x^2 - 1} > \frac{1}{2\delta + \delta^2} \qquad ...(1) \end{aligned}$$

$$\therefore \quad \frac{1}{x^2 - 1} > k \quad \text{whenever} \quad \frac{1}{2\delta + \delta^2} > k. \qquad ...(2)$$

Now, $\dfrac{1}{2\delta+\delta^2} > k \;\Rightarrow\; \delta^2 + 2\delta < \dfrac{1}{k} \;\Rightarrow\; (\delta+1)^2 < 1 + \dfrac{1}{k}$

$$\Rightarrow \quad -(1+1/k)^{1/2} < \delta + 1 < (1+1/k)^{1/2}$$

$$\Rightarrow \quad -1-(1+1/k)^{1/2} < \delta < -1+(1+1/k)^{1/2}.$$

Thus, if we choose $\delta = (1/2) \times \{-1 + (1 + 1/k)^{1/2}\}$, then from (1) and (2), we have

$$f(x) > k \quad \text{whenever} \quad 1 < x < 1 + \delta$$

$$\Rightarrow \quad \lim_{x \to 1+0} f(x) = \infty \quad i.e. \quad \lim_{x \to 1+0} \frac{1}{x^2-1} = \infty.$$

Example 4. *Find* (*i*) $\displaystyle \lim_{x\to 0} \frac{e^{1/x}}{e^{1/x}+1}$ (*ii*) $\displaystyle \lim_{x\to 0} [x]$,

where $[x]$ *denotes the greatest integer not greater than* x.

Solution. (*i*) Right-hand limit

$$= \lim_{x\to 0+0} \frac{e^{1/x}}{e^{1/x}+1} = \lim_{h\to 0+} \frac{e^{1/h}}{e^{1/h}+1}, \text{ putting } x = 0+h, \text{ where } h > 0$$

$$= \lim_{h\to 0+} \frac{1}{1+e^{-1/h}}, \text{ by dividing the numerator and denominator by } e^{1/h}$$

$$= \frac{1}{1+0} = 1, \text{ as } e^{-1/h} \to 0 \text{ as } h \to 0+$$

Again, left-hand limit

$$= \lim_{x\to 0-0} \frac{e^{1/x}}{e^{1/x}+1} = \lim_{h\to 0+} \frac{e^{-1/h}}{e^{-1/h}+1} = \frac{0}{0+1} = 0, \text{ putting } x = 0-h,\ h > 0.$$

Since right-hand limit ≠ left-hand limit, so $\displaystyle \lim_{x\to 0} \frac{e^{1/x}}{e^{1/x}+1}$ does not exist.

(*ii*) Right-hand limit $= \displaystyle \lim_{x\to 0+0} [x] = \lim_{h\to 0+} [0+h] = 0$

and Left-hand limit $= \displaystyle \lim_{x\to 0-0} [x] = \lim_{h\to 0+} [0-h] = -1$

Since right-hand limit ≠ left-hand limit, so $\displaystyle \lim_{x\to 0} [x]$ does not exist.

Example 5. *Find* $\displaystyle \lim_{x\to a} f(x)$, *where* $f(x) = \begin{cases} (x^2/a) - a, & \text{for } 0 < x < a \\ 0, & \text{for } x = a \\ a - (a^3/x^2), & \text{for } x > a \end{cases}$

Solution. Right-hand limit $= \displaystyle \lim_{h\to 0+} f(a+h) = \lim_{h\to 0+} \left\{a - \frac{a^3}{(a+h)^2}\right\} = a - \frac{a^3}{a^2} = 0$

Left-hand limit $= \displaystyle \lim_{h\to 0+} f(a-h) = \lim_{h\to 0+} \left\{\frac{(a-h)^2}{a} - a\right\} = \frac{a^2}{a} - a = 0.$

Here, right-hand limit = left-hand limit. So $\displaystyle \lim_{x\to a} f(x) = 0$.

EXERCISES

1. Evaluate the following :

(*i*) $\lim\limits_{x \to a} \dfrac{x^2 - a^2}{x - a}$ (*ii*) $\lim\limits_{x \to 0+} \dfrac{|x|}{x}$ (*iii*) $\lim\limits_{x \to 0-} \dfrac{|x|}{x}$

(*iv*) $\lim\limits_{x \to 0} \dfrac{x\, e^{1/x}}{1 + e^{1/x}}$ (*v*) $\lim\limits_{x \to 0} \dfrac{e^{1/x} - 1}{e^{1/x} + 1}$ (*vi*) $\lim\limits_{x \to 0} \dfrac{e^{1/x} - e^{-1/x}}{e^{1/x} + e^{-1/x}}$

2. Using definition, prove that

(*i*) $\lim\limits_{x \to 0} \dfrac{1}{x^2} = \infty$ (*ii*) $\lim\limits_{x \to \infty} \sqrt{x} = \infty$ (*iii*) $\lim\limits_{x \to 0} \dfrac{1}{\sqrt{|x|}} = \infty$

3. If $\lim\limits_{x \to a} f(x) = l$ and $l \neq 0$, then prove that $\lim\limits_{x \to a} \dfrac{1}{f(x)} = \dfrac{1}{l}$.

4. Prove that

(*i*) $\lim\limits_{x \to \infty} f_1(x) = l_1,\ \lim\limits_{x \to \infty} f_2(x) = l_2 \ \Rightarrow\ \lim\limits_{x \to \infty} [f_1(x) + f_2(x)] = l_1 + l_2$

(*ii*) $\lim\limits_{x \to -\infty} f_1(x) = l_1,\ \lim\limits_{x \to -\infty} f_2(x) = l_2 \ \Rightarrow\ \lim\limits_{x \to -\infty} [f_1(x) + f_2(x)] = l_1 + l_2$

(*iii*) $\lim\limits_{x \to \infty} f_1(x) = \infty,\ \lim\limits_{x \to \infty} f_2(x) = \infty \ \Rightarrow\ \lim\limits_{x \to \infty} [f_1(x) + f_2(x)] = \infty$

Give examples to show that if $\lim\limits_{x \to \infty} f_1(x) = \infty$ and $\lim\limits_{x \to \infty} f_2(x) = -\infty$ then $f_1(x) + f_2(x)$ may tend to a finite limit, $+\infty$, or $-\infty$ or oscillate.

5. Given that $\lim\limits_{x \to a} f(x) = l \neq 0$, show that there exists *a* deleted neighbourhood of *a* at no point of which $f(x)$ is zero.

6. Show that

(*i*) $\lim\limits_{x \to 1} (3x + 1) = 4$. Also obtain values of δ corresponding to $\varepsilon = .1, .01, .001$.

(*ii*) $\lim\limits_{x \to 2} \dfrac{x^2 - 4}{x - 2} = 4$. Give values of δ corresponding to $\varepsilon = .01, .001$.

(*iii*) $\lim\limits_{x \to 1} 1/x = 1$. Give δ for $\varepsilon = .1, .001$.

(*iv*) $\lim\limits_{x \to (1-0)} [x] = 0,\ \lim\limits_{x \to (1+0)} [x] = 1$.

(*v*) $\lim\limits_{x \to 1} (x^2 + 2x) = 3$.

(*vi*) $\lim\limits_{x \to (2-0)} (x + [x]) = 3,\ \lim\limits_{x \to (2+0)} (x + [x]) = 4$. Does $\lim\limits_{x \to 2} (x + [x])$ exist ?

7. Show that $\lim\limits_{x \to 1} (x^3 + x^2 - x + 1) = 2,\ \lim\limits_{x \to 0} \dfrac{x + 1}{x^2 + 1} = 1,\ \lim\limits_{x \to 2} \dfrac{x^2 + 1}{x} = \dfrac{5}{2}$

and give in each case a δ corresponding to a given positive ε.

ANSWERS

1. (*i*) $2a$ (*ii*) 1 (*iii*) -1 (*iv*) 0 (*v*) does not exist (*vi*) does not exist.

8.8. CHARACTERIZATION OF THE LIMIT OF A FUNCTION AT A POINT IN TERMS OF SEQUENCES. SEQUENTIAL CRITERION FOR LIMITS

Theorem. *Let f be a function defined for all points in some neighbourhood of a point a except possibly at the point a itself. Then $f(x) \to l$ as $x \to a$ if and only if the limit of the sequence $\langle f(x_n)\rangle$ exists and is equal to l for any sequence $\langle x_n\rangle$, $x_n \neq a$ for any $n \in$* **N**, *converging to a.*

[Delhi B.Sc. (Hons.) II, 2016, 2017]

Proof. The condition is necessary. Let $\lim_{x \to a} f(x) = l.$

Then by definition, for a given $\varepsilon > 0$, there exists $\delta > 0$, such that

$$|f(x) - 1| < \varepsilon \text{ whenever } 0 < |x - a| < \delta \qquad ...(1)$$

Also, suppose that there is a sequence $\langle x_n\rangle$, $x_n \neq a$ for any $n \in$ **N**, converging to a. Then by definition, there exists $m \in$ **N** such that

$$|x_n - a| < \delta, \quad \forall\ n \geq m \qquad ...(2)$$

From (1) and (2), $\quad |f(x_n) - l| < \varepsilon, \quad \forall\ n \geq m,$

showing that the sequence $\langle f(x_n)\rangle$ converges to l. Again, since this property is true for any sequence $\langle x_n\rangle$, $x_n \neq a$, converging to a, the required result follows.

The condition is sufficient. Suppose that the sequence $\langle f(x_n)\rangle$ converges to l for any sequence $\langle x_n\rangle$, $x_n \neq a$ for any $n \in$ **N**, converging to a. If possible, let $\lim_{x \to a} f(x) \neq l$. Then, by definition, there must exist at least one value of ε, say ε_1 for which there is no corresponding δ. Hence for any δ, there exists a value $x = x(\delta)$ such that

$$x(\delta) \in \{x : 0 < |x - a| < \delta\}$$

and $\quad |f(x(\delta)) - l| \geq \varepsilon_1$

Choose $\delta = 1, 1/2, 1/3, ...$ in succession. Then for such value of δ, there is a value x_n such that

$$0 < |x_n - a| < 1/n,\ x_n \neq a \quad \text{and} \quad |f(x_n) - l| \geq \varepsilon_1,\ n = 1, 2, 3, ...$$

$$\Rightarrow x_n \to a \text{ but } f(x_n) \text{ does not tend to } l.$$

This is a contradiction. Hence our assumption $\lim_{x \to a} f(x) \neq l$ is wrong.

Therefore, $\lim_{x \to a} f(x) = l$. Hence the theorem.

Note. The above criterion can be used to show the non-existence of limit of some function.

Example 1. *Show that* $\lim_{x \to 0} \sin \frac{1}{x}$ *does not exist.*

Solution. Let $f(x) = \sin(1/x)$. Let $x_n = \dfrac{1}{2n\pi}$ and $x'_n = \dfrac{1}{2n\pi + \pi/2}\ \forall\ n \in$ **N.**

Then the both sequences $\langle x_n\rangle$ and $\langle x'_n\rangle$ tend to 0 as $n \to \infty$.

Again, $\quad f(x_n) = \sin(1/x_n) = \sin 2n\pi = 0 \quad \forall\ n \in$ **N**

and $\quad f(x'_n) = \sin(1/x'_n) = \sin(2n\pi + \pi/2) = 1 \quad \forall\ n \in$ **N**

Hence $f(x_n) \to 0$ as $n \to \infty$ whereas $f(x'_n) \to 1$ as $n \to \infty$, showing that every sequence does not approach the same limit. Hence $f(x)$ cannot tend to any limit as $x \to 0$.

Example 2. *Let* $\quad f(x) = \begin{cases} 1, & \text{when } x \text{ is rational} \\ 0, & \text{when } x \text{ is irrational} \end{cases}$

Show that $\lim_{x \to a} f(x)$ *does not exist for any real number a.*

Solution. Let $<x_n>$ be a sequence of rational numbers such that $x_n \to a$ and $x_n \neq a$ for any $n \in \mathbf{N}$. Let $<x'_n>$ be another sequence of irrational numbers such that $x'_n \to a$ and $x'_n \neq a$ for any $n \in \mathbf{N}$. Then by definition of $f(x)$, we have

$$f(x_n) = 1 \quad \text{and} \quad f(x'_n) = 0 \quad \forall\, n \in \mathbf{N}$$

$$\Rightarrow \quad f(x_n) \to 1 \text{ as } n \to \infty \qquad \text{whereas} \qquad f(x'_n) \to 0 \text{ as } n \to \infty,$$

showing that every sequence does not approach the same limit. Hence $\lim_{x \to a} f(x)$ does not exist. Since a was arbitrary, it follows that $\lim_{x \to a} f(x)$ does not exist for any real number a.

8.9. CAUCHY'S CRITERION FOR FINITE LIMITS

Theorem I. *A function f tends to a finite limit as x tends to a if and only if for every* $\varepsilon > 0$ *there exists a neighbourhood N of a such that*

$$| f(x') - f(x'') | < \varepsilon \quad \forall\, x', x'' \in N;\ x', x'' \neq a.$$

Proof. The condition is necessary. Let N denote the deleted neighbourhood of a so that

$$N =]a - \delta, a + \delta[- \{a\}.$$

Let $\lim_{x \to a} f(x) = l$, where l is a finite number. Then, for a given $\varepsilon > 0$, there exists a number δ such that

$$| f(x) - l | < \varepsilon/2 \text{ whenever } 0 < | x - a | < \delta \text{ i.e., } | f(x) - l | < \varepsilon/2 \text{ whenever } x \in N \quad ...(1)$$

Let $x', x'' \in N$. Then, from (1), we have

$$| f(x') - l | < \varepsilon/2 \quad \text{and} \quad | f(x'') - l | < \varepsilon/2 \quad ...(2)$$

$$\therefore \quad | f(x') - f(x'') | = | \{f(x') - l\} + \{l - f(x'')\} |$$
$$\leq | f(x') - l | + | l - f(x'') |$$
$$< \varepsilon/2 + \varepsilon/2 = \varepsilon, \text{ using (2)}$$

Thus, $\quad | f(x') - f(x'') | < \varepsilon \quad \forall\, x', x'' \in N$

The condition is sufficient. Suppose that for any $\varepsilon > 0$, there exists a deleted neighbourhood N such that

$$| f(x') - f(x'') | < \varepsilon \quad \forall\, x', x'' \in N \quad ...(3)$$

Let $<x_n>$, $x_n \neq a$ for any positive integer n be any sequence tending to a such that there exists a positive integer p such that $x_n, x_m \in N$ for $n, m \geq p$. Then by (3)

$$| f(x_n) - f(x_m) | < \varepsilon, \quad \forall\, n, m \geq p \quad ...(2)$$

In view of Cauchy's general principle of convergence, (2) shows that $<f(x_n)>$ tends to a limit. We shall now show that this limit is unique.

Let, if possible, $<x_n>$ and $<x'_n>$, $x_n \neq a$, $x'_n \neq a$ be two sequences such that $x_n \to a$ and $x'_n \to a$ but $f(x_n) \to l$ and $f(x'_n) \to l'$. We shall now prove that $l = l'$.

Construct the sequence $<x_1, x'_1, x_2, x'_2, ...>$ which converges to a. Hence, as before, the sequence $<f(x_1), f(x'_1), f(x_2), f(x'_2), ...>$ converges which can happen only if $l = l'$.

Theorem II. *A function f tends to a finite limit as* $x \to \infty$ *if and only if for each* $\varepsilon > 0$ *there exists* $k > 0$ *such that*

$$| f(x') - f(x'') | < \varepsilon \quad \forall\, x', x'' > k.$$

Proof. Left to the reader.

8.10. THE FOUR FUNCTIONAL LIMITS AT A POINT

Let a function f be defined on $]a, b[$. Let $c \in]a, b[$ and $h > 0$. We give to h a sequence of diminishing values $<h_n>$ such that $h_n \to 0$ as $n \to \infty$.

Consider the right hand neighbourhood $(c, c + h)$ of the point c. Let $M(h_n)$ and $m(h_n)$ be the supremum and infimum of f in $(c, c + h_n)$. Then, clearly,

$$M(h_1) \geq M(h_2) \geq M(h_3) \geq \ldots\ldots \quad \text{and} \quad m(h_1) \leq m(h_2) \leq m(h_3) \leq \ldots\ldots,$$

showing that the sequences $<M(h_n)>$ and $<m(h_n)>$ are monotonic non-increasing and non-decreasing respectively. Hence $\lim_{n\to\infty} M(h_n)$ and $\lim_{n\to\infty} m(h_n)$ exist. Then we write

$$\overline{f(c+0)} = \lim_{n\to\infty} M(h_n) \quad \text{and} \quad \underline{f(c+0)} = \lim_{n\to\infty} m(h_n).$$

These limits are respectively known as the *upper* and *lower limits of f at c on the right*. If $\overline{f(c+0)} = \underline{f(c+0)}$, then their common value is the right-hand limit of f at c and is denoted by $f(c + 0)$.

Similarly, consider the left hand neighbourhood of c. Let $M'(h_n)$ and $m'(h_n)$ be the supremum and infimum of f in $(c - h_n, c)$. As before, $\lim_{n\to\infty} M'(h_n)$ and $\lim_{n\to\infty} m'(h_n)$ both exist. Then we write

$$\overline{f(c-0)} = \lim_{n\to\infty} M'(h_n) \quad \text{and} \quad \underline{f(c-0)} = \lim_{n\to\infty} m'(h_n).$$

These limits are respectively known as the *upper* and *lower limits of f at c on the left*. If $\overline{f(c-0)} = \underline{f(c-0)}$, then their common value is the left-hand limit of f at c and is denoted by $f(c - 0)$.

The limits $\overline{f(c+0)}$, $\underline{f(c+0)}$, $\overline{f(c-0)}$, $\underline{f(c-0)}$ are called the four functional limits of f at the point c. If all these four limits are equal, their common value is the limit of f at c.

Example 1. If $f(x) = (x-a)\sin\dfrac{1}{x-a}$, $x \neq a$, then

$$\overline{f(a+0)} = \underline{f(a+0)} = \overline{f(a-0)} = \underline{f(a-0)} = 0. \text{ So } \lim_{x\to a} f(x) = 0.$$

Example 2. If $f(x) = \dfrac{1}{x-a}\sin\dfrac{1}{x-a}$, $x \neq a$, then

$$\overline{f(a+0)} = \infty,\ \underline{f(a+0)} = -\infty,\ \overline{f(a-0)} = \infty,\ \underline{f(a-0)} = -\infty$$

$\therefore$ $\lim_{x\to a} f(x)$ does not exist.

8.11. CONTINUOUS FUNCTIONS

[Avadh, 2000; Kanpur 2013; Delhi B.A. (Prog) III 2008, 2014]

Let f be a function defined in an interval.

Continuity at an interior point. *The function f is said to be continuous at an interior point c, a < c < b, if*

$$\lim_{x\to c} f(x) = f(c).$$

i.e., if $\lim_{x\to c} f(x)$ exists and equal the value $f(c)$ of the function for $x = c$.

We thus see that a function f is continuous at c if to each given $\varepsilon > 0$, there corresponds $\delta > 0$ such that

$$|x - c| < \delta \Rightarrow |f(x) - f(c)| < \varepsilon.$$

Continuity at an end point. The function f is said to be continuous at the left end point a if
$$\lim_{x \to (a+0)} f(x) = f(a)$$
i.e., if to each $\varepsilon > 0$ there corresponds $\delta > 0$ such that
$$|f(x) - f(a)| < \varepsilon \text{ when } a < x < a + \delta.$$
Again the function f is said to be continuous at the right end point b if
$$\lim_{x \to (b-0)} f(x) = f(b)$$
i.e., if to each $\varepsilon > 0$ there corresponds $\delta > 0$ such that
$$|f(x) - f(b)| < \varepsilon \text{ when } b - \delta < x < b.$$
Ex. A function f is continuous at an interior point of its domain; show that it is bounded in some neighbourhood of the point. What can you say if the point of continuity is at an end of the domain.

Continuity in an interval. *A function is said to be continuous in an interval if it is continuous at every point thereof.*

DISCONTINUITY OF A FUNCTION

A function is said to be discontinuous at a point of its domain if it is not continuous at the point. Moreover the point where a function is discontinuous is said to be a *point of discontinuity* of the function.

The discontinuity of a function f at a point c of its domain can arise in either of the following two ways :

(*i*) $\lim_{x \to c} f(x)$ exists but is different from the value $f(c)$.

(*ii*) $\lim_{x \to c} f(x)$ does not exist.

Types of discontinuities with an example.

(Agra, 2000, 02; Delhi Maths (H), 2003; Kumaun, 1998; Patna, 2003)

(*i*) **Removable discontinuity.** If $\lim_{x \to c} f(x)$ exists but is not equal to $f(c)$ (which may or may not exist), *i.e.*, $f(c-0) = f(c+0) \neq f(c)$ then f is said to have a removable discontinuity at $x = c$. Such a discontinuity can be removed by assigning a suitable value to the function at $x = c$.

Example. Consider the function f defined by
$$f(x) = \begin{cases} (\sin x)/x, & \text{if } x \neq 0 \\ 0, & \text{if } x = 0 \end{cases}$$
(Utkal, 2003)

Now, $$\lim_{x \to 0} f(x) = \lim_{x \to 0} \frac{\sin x}{x} = 1 \quad \text{so that} \quad \lim_{x \to 0} f(x) \neq f(0),$$
showing that f has a removable discontinuity at $x = 0$.

If we redefine the given function as follows
$$f(x) = \begin{cases} (\sin x)/x, & \text{if } x \neq 0 \\ 1, & \text{if } x = 0 \end{cases}$$
then this new function is continuous at $x = 0$. **(Agra, 2004; Kanpur 2007)**

(*ii*) **Discontinuity of the first kind (or jump discontinuity).** A function f is said to have a discontinuity of the first kind at $x = c$, if $f(c+0)$ and $f(c-0)$ both exist but are not equal.

A function f is said to have a discontinuity of the first kind from the left at $x = c$ if $f(c - 0)$ exists but is not equal to $f(c)$, *i.e.*, if $f(c - 0) \neq f(c) = f(c + 0)$.

A function f is said to have a discontinuity of the first hand from the right at $x = c$ if $f(c + 0)$ exists but is not equal to $f(c)$, *i.e.*, if $f(c + 0) \neq f(c) = f(c - 0)$.

EXAMPLES

1. Consider the following function $f(x) = \begin{cases} 1, & x < 0 \\ -1, & x > 0 \end{cases}$

Here $f(0 + 0) = -1$ and $f(0 - 0) = 1$. Since $f(0 + 0) \neq f(0 - 0)$, f has a discontinuity of the first kind or jump discontinuity.

2. Let
$$f(x) = \frac{e^{1/x} - e^{-1/x}}{e^{1/x} + e^{-1/x}}, \ x \neq 0; \ f(0) = 1.$$

Here
$$f(0-0) = \lim_{h \to 0} f(0-h) = \lim_{h \to 0} \frac{e^{-1/h} - e^{1/h}}{e^{-1/h} + e^{1/h}}, \text{ where } h > 0$$
$$= \lim_{h \to 0} \frac{e^{-2/h} - 1}{e^{-2/h} + 1} = -1$$

Also,
$$f(0+0) = \lim_{h \to 0} f(0+h) = \lim_{h \to 0} \frac{e^{1/h} - e^{-1/h}}{e^{1/h} + e^{-1/h}} = \lim_{h \to 0} \frac{1 - e^{-2/h}}{1 + e^{-2/h}} = 1.$$

Thus, $f(0 - 0) \neq f(0) = f(0 + 0)$ and hence f has a discontinuity of the first kind from the left at $x = 0$.

3. Let
$$f(x) = \frac{e^{1/x}}{1 + e^{1/x}}, \ x \neq 0; \ f(0) = 0.$$ **(Agra, 2002)**

Here
$$f(0+0) = \lim_{h \to 0} f(0+h) = \lim_{h \to 0} \frac{e^{1/h}}{1 + e^{1/h}}, \ h > 0$$
$$= \lim_{h \to 0} \frac{1}{e^{-1/h} + 1} = 1$$

Again
$$f(0-0) = \lim_{h \to 0} (0-h) = \lim_{h \to 0} \frac{e^{-1/h}}{1 + e^{-1/h}} = 0$$

Thus, $f(0 - 0) = f(0) \neq f(0 + 0)$. Hence f has a discontinuity of the first kind from the right at $x = 0$.

(*iii*) **Discontinuity of the second kind.** A function f is said to have a discontinuity of the second kind at $x = c$ if neither $f(c - 0)$ nor $f(c + 0)$ exists.

A function f is said to have a discontinuity of the second kind from the left at $x = c$ if $f(c - 0)$ does not exist.

A function f is said to have a discontinuity of the second kind from the right at $x = c$ if $f(c + 0)$ does not exist.

Example. Consider the function f defined by
$$f(x) = \begin{cases} \sin(1/x), & x \neq 0 \\ 0, & x = 0 \end{cases}$$ **[Delhi Maths (Hons), 2003; Kanpur, 2001**

Let us first prove that $f(0+0)$ does not exist. Since $|\sin(1/x)| \le 1 \ \forall\ x \in \mathbf{R}$, so if $f(0+0)$ exists then $-1 < f(0+0) < 1$. Let l be any arbitrary, but fixed real number such that $-1 \le l \le 1$. We wish to prove that $f(0+0) \ne l$.

Let, if possible, $f(0+0) = l$. Let $\varepsilon = 1$. Then we shall prove that there exists no corresponding $\delta > 0$. If possible, suppose δ' is the corresponding value of δ, *i.e.*, let

$$0 < x < \delta' \Rightarrow |f(x) - l| < 1 \qquad \text{...(1)}$$

By virtue of the Archimedean property of real numbers, we can select $m \in N$ such that

$$2m\pi + \pi/2 > 1/\delta'.$$

Let $\quad x_1 = \dfrac{1}{2m\pi + \pi/2} \quad$ and $\quad x_2 = \dfrac{1}{2m\pi + 3\pi/2}$

Then, since $0 < x_1 < \delta'$ and $0 < x_2 < \delta'$, (1) gives

$$|\sin(1/x_1) - l| < 1 \quad \text{and} \quad |\sin(1/x_2) - l| < 1 \qquad \text{...(2)}$$

Now, $\quad |\sin(1/x_1) - \sin(1/x_2)| = |\{\sin(1/x_1) - l\} - \{\sin(1/x_2) - l\}|$

$$\le |\sin(1/x_1) - l| + |\sin(1/x_2) - l|$$

$$< 2, \text{ using (2)} \qquad \text{...(3)}$$

Again, $\quad \sin(1/x_1) = \sin(2m\pi + \pi/2) = 1, \quad \sin(1/x_2) = \sin(2m\pi + 3\pi/2) = -1$

$\therefore$
$$|\sin(1/x_1) - \sin(1/x_2)| = 2. \qquad \text{...(4)}$$

From (3) and (4), we obtain $2 < 2$, which is absurd. So $f(0+0)$ cannot exist. Since $f(x)$ oscillates between -1 and 1, and takes the values -1 and 1 in every interval $]0, \delta'[$ to the right of 0, hence it follows that

$$\overline{f(0+0)} = 1 \quad \text{and} \quad \underline{f(0+0)} = -1$$

Similarly, we can show that $f(0-0)$ cannot exist and

$$\overline{f(0-0)} = 1 \quad \text{and} \quad \underline{f(0-0)} = -1$$

Thus, neither $f(0+0)$ nor $f(0-0)$ exist and so $f(x)$ has a discontinuity of the second kind.

Again since $f(0+0)$ does not exist, so f has a discontinuity of the second kind from the right.

Also, since $f(0-0)$ does not exist, so f has a discontinuity of the second kind from the left also.

(*iv*) **Mixed discontinuity.** A function f is said to have the mixed discontinuity at $x = c$, if f has a discontinuity of the second kind on one side of $x = c$ and on the other side, a discontinuity of the first kind or may be continuous.

Example. Consider the function f defined by

$$f(x) = \begin{cases} 0, & \text{if } x \le 0 \\ \sin(1/x), & \text{if } x > 0 \end{cases}$$

Here $f(0-0) = 0 = f(0)$ and so $f(x)$ is continuous from the left at $x = 0$. Proceeding as in the example of discontinuity of the second kind, we have $\overline{f(0+0)} = 1$ and $\underline{f(0+0)} = -1$. Hence $f(0+0)$ does not exist and so $f(x)$ has a discontinuity of the second kind from the right.

Thus $f(x)$ has discontinuity of second kind from the right of $x = 0$ and it is continuous from the left of $x = 0$. Hence $f(x)$ has a mixed discontinuity at $x = 0$.

(*v*) **Infinite discontinuity.** A function f is said to have an infinite discontinuity at $x = c$ if one or more of the functional limits $\overline{f(c+0)}$, $\underline{f(c+0)}$, $\overline{f(c-0)}$ and $\underline{f(c-0)}$ is $+\infty$ or $-\infty$. It is easy to verify that if f is discontinuous at $x = c$ and is unbounded in every neighbourhood of c, then f has an infinite discontinuity at $x = c$.

Example. Consider the function f defined by

$$f(x) = \begin{cases} \dfrac{1}{x-c} \operatorname{cosec} \dfrac{1}{x-c}, & \text{if } x \neq c \\ 1, & \text{if } x = c \end{cases}$$

Here $f(c+0) = \lim_{h \to 0} f(c+h) = \lim_{h \to 0} \dfrac{1}{h \sin(1/h)}$, where $h > 0$

Since $h \sin(1/h)$ assumes positive and negative values in every interval $]0, \delta[$, however small, and since $\lim_{h \to 0} h \sin(1/h) = 0$, so it follows that

$$\frac{1}{h \sin(1/h)} \text{ oscillates between } -\infty \text{ and } \infty$$

and so $\overline{f(c+0)} = +\infty$ and $\underline{f(c+0)} = -\infty$

Similarly, $\overline{f(c-0)} = +\infty$ and $\underline{f(c-0)} = -\infty$

Hence f has an infinite discontinuity of the second kind on both sides of $x = 0$.

EXAMPLES

Example 1. *Show that the following functions are continuous at $x = 0$*

(*i*) $f(x) = \begin{cases} x \sin(1/x), & x \neq 0 \\ 0, & x = 0 \end{cases}$ **(*Agra, 2000, 02, 11, 12; Kanpur, 2001, 09; Delhi B.Sc. (Prog) I 2011*)**

(*ii*) $f(x) = \begin{cases} x \cos(1/x), & x \neq 0 \\ 0, & x = 0 \end{cases}$ **(*Kanpur 2006; Agra, 2001, 11*)**

Solution. Here, $f(0) = 0$. Also, we have

$$\begin{aligned} f(0+0) &= \lim_{h \to 0} f(0+h) = \lim_{h \to 0} h \sin(1/h), \text{ where } h > 0 \\ &= 0 \times k, \text{ where } -1 \leq k \leq 1 \\ &= 0 \\ f(0-0) &= \lim_{h \to 0} f(0-h) = \lim_{h \to 0} (-h) \sin(-1/h), \; h > 0 \\ &= 0 \times k, \text{ where } -1 \leq k \leq 1 \end{aligned}$$

Thus $f(0+0) = f(0-0) = f(0)$ and so f is continuous at $x = 0$

(*ii*) Try as in part (*i*).

Example 2. *Obtain the points of discontinuity of a function f defined on $[0, 1]$ as follows : $f(0) = 0$, $f(x) = (1/2) - x$, if $0 < x < 1/2$, $f(1/2) = 1/2$, $f(x) = (2/3) - x$, if $1/2 < x < 1$ and $f(1) = 1$. Also examine the types of discontinuities.* **(*Meerut 2005; Rohilkhand, 1993*)**

Solution. To test for continuity at $x = 0$. Here $f(0) = 0$.

Also, $f(0+0) = \lim_{h \to 0} f(0+h) = \lim_{h \to 0} \{(1/2) - (0+h)\}$, $h > 0$

Thus, $f(0+0) = 1/2$ and so $f(0+0) \neq f(0)$.

Hence, $x = 0$ is a point of discontinuity and f has a discontinuity of the first kind from the right at $x = 0$.

To test for continuity at $x = 1/2$. Here $f(1/2) = 1/2$

$$f(1/2+0) = \lim_{h \to 0} f(1/2+h) = \lim_{h \to 0} \{(2/3)-(1/2+h)\} = 1/6$$

and
$$f(1/2-0) = \lim_{h \to 0} f(1/2-h) = \lim_{h \to 0} \{(1/2)-(1/2-h)\} = 0.$$

Thus $f(1/2 + 0)$ and $f(1/2 - 0)$ both exist but are unequal and so $x = 1/2$ is a point of discontinuity and f has discontinuity of the first kind.

To test for continuity at $x = 1$. Here $f(1) = 1$.

$$f(1-0) = \lim_{h \to 0} f(1-h) = \lim_{h \to 0} \{(2/3)-(1-h)\} = -1/3 \neq f(1)$$

Since $f(1 - 0) \neq f(1)$, $x = 1$ is a point of discontinuity and f has a discontinuity of the first kind from the left at $x = 1$.

Example 3. *Show that the function f on $[0, 1]$ defined as*

$$f(x) = 1/2^n, \text{ when } 1/2^{n+1} < x \le 1/2^n, \; n = 0, 1, 2, \ldots$$
$$f(0) = 0$$

is discontinuous at $1/2$, $(1/2)^2$, $(1/2)^3$, ... **[Delhi Maths (Prog) 2007; Delhi Maths (G), 1998]**

Solution. Putting $n = 0, 1, 2, \ldots$ in succession, the given function is defined as

$$\begin{aligned} f(x) &= 1, && \text{if } 1/2 < x \le 1 \\ &= 1/2, && \text{if } 1/2^2 < x \le 1/2. \\ &= 1/2^2, && \text{if } 1/2^3 < x \le 1/2^2 \\ &\ldots\ldots\ldots\ldots \\ &\ldots\ldots\ldots\ldots \end{aligned}$$

Test of continuity at $x = 1/2$. We have

$$\text{Left-hand limit} = \lim_{x \to (1/2)-0} f(x) = \lim_{x \to (1/2)-0} \frac{1}{2} = \frac{1}{2}$$

and
$$\text{right-hand limit} = \lim_{x \to (1/2)+0} f(x) = \lim_{x \to (1/2)+0} 1 = 1$$

Since left-hand limit $\neq$ right-hand limit, $\lim_{x \to 1/2} f(x)$ does not exist and hence $f(x)$ is discontinuous at $x = 1/2$.

Proceeding likewise we easily see that

$$\lim_{x \to 1/2^n + 0} f(x) = \frac{1}{2^n} \quad \text{and} \quad \lim_{x \to 1/2^n - 0} f(x) = \frac{1}{2^{n-1}}$$

$\Rightarrow f$ is not continuous at $x = 1/2^n$, $n = 1, 2, 3, \ldots$.

Example 4. *Determine the constants a and b so that the function f defined below is continuous everywhere*

$$f(x) = \begin{cases} 2x+1, & \text{if } x \le 1 \\ ax^2+b, & \text{if } 1 < x < 3 \\ 5x+2a, & \text{if } x \ge 3 \end{cases}$$

Solution. Since f is continuous everywhere, it must be continuous $x = 1$ and $x = 3$, the breaking points of the domain.

Continuity at $x = 1$. We have $f(1) = 2 \times 1 + 1 = 3$.

$$f(1-0) = \lim_{h \to 0+} f(1-h) = \lim_{h \to 0+} \{2(1-h)+1\} = 3$$

and
$$f(1+0) = \lim_{h \to 0+} f(1+h) = \lim_{h \to 0+} \{a(1+h)^2+b\} = a+b.$$

Since f is continuous at $x = 1$, we have

$$f(1-0) = f(1+0) = f(1) \Rightarrow a + b = 3 \quad ...(1)$$

Continuity at $x = 3$. We have $f(3) = 5 \times 3 + 2a = 2a + 15$

$$f(3-0) = \lim_{h \to 0+} f(3-h) = \lim_{h \to 0+} \{a(3-h)^2 + b\} = 9a + b$$

and $$f(3+0) = \lim_{h \to 0+} f(3+h) = \lim_{h \to 0+} \{5(3+h) + 2a\} = 2a + 15.$$

Since f is continuous at $x = 3$, we have

$$f(3-0) = f(3+0) = f(3) \Rightarrow 9a + b = 2a + 15 \Rightarrow 7a + b = 15 \quad ...(2)$$

From (1) and (2), we have $a = 2$, $b = 1$.

Example 5. *Discuss the continuity of the function $f(x) = [x]$ at the points $x = 1/2$ and 1, where $[x]$ denotes the largest integer $\leq x$.*

Solution. To test for continuity at $x = 1/2$.

$$f(1/2+0) = \lim_{h \to 0+} f(1/2+h) = \lim_{h \to 0+} [1/2+h] = \lim_{h \to 0+} 0 = 0$$

and $$f(1/2-0) = \lim_{h \to 0+} f(1/2-h) = \lim_{h \to 0+} [1/2-h] = \lim_{h \to 0+} 0 = 0$$

Also, $f(1/2) = [1/2] = 0$. Thus, $f(1/2+0) = f(1/2-0) = f(1/2)$

Hence f is continuous at $x = 1/2$.

To test for continuity at $x = 1$

$$f(1+0) = \lim_{h \to 0+} f(1+h) = \lim_{h \to 0+} [1+h] = \lim_{h \to 0+} 1 = 1$$

and $$f(1-0) = \lim_{h \to 0+} f(1-h) = \lim_{h \to 0+} [1-h] = \lim_{h \to 0+} 0 = 0$$

Since $f(1+0) \neq f(1-0)$ so f is not continuous at $x = 1$.

Example 6. *Show that $f(x) = |x| + |x-1|$ is continuous at $x = 0$ and $x = 1$.*

Solution. Rewriting the given function, we have

$$f(x) = -x - (x-1) = 1 - 2x, \quad \text{if } x < 0 \quad ...(1)$$

$$f(x) = x - (x-1) = 1, \quad \text{if } 0 \leq x < 1 \quad ...(2)$$

$$f(x) = x + x - 1 = 2x - 1, \quad \text{if } x \geq 1 \quad ...(3)$$

To test for continuity of $x = 0$

$$f(0+0) = \lim_{h \to 0+} f(0+h) = \lim_{h \to 0} 1 = 1, \text{ by (2)}$$

and $$f(0-0) = \lim_{h \to 0+} f(0-h) = \lim_{h \to 0} \{1 - 2(-h)\} = 1, \text{ by (1)}$$

Also $f(0) = 1$. Thus, $f(0+0) = f(0-0) = f(0)$

Hence f is continuous at $x = 0$.

To test for continuity at $x = 1$

$$f(1+0) = \lim_{h \to 0+} f(1+h) = \lim_{h \to 0+} \{2(1+h) - 1\} = 1, \text{ by (3)}$$

$$f(1-0) = \lim_{h \to 0+} f(1-h) = \lim_{h \to 0+} 1 = 1, \text{ by (2)}$$

Also, $f(1) = 1$. So $f(1+0) = f(1-0) = f(1)$

Hence f is continuous at $x = 1$.

Example 7. *Let f be a real valued function defined as follows :*

$$f(x) = x, \text{ if } -1 \le x \le 1$$
$$f(x+2) = f(x), \text{ for all } x \in \mathbf{R}.$$

Show that f is discontinuous at every odd integer. **(*I.A.S., 2003*)**

Solution. Let $x = 2n + 1$, where n is any integer so that $x = \pm 1, \pm 3, \pm 5, ...$, *i.e.*, any odd integer. We have

$$f(2n+1+0) = \lim_{h \to 0+} f(2n+1+h) = \lim_{h \to 0+} f(1+h) = \lim_{h \to 0+} f(-1+h)$$
$$[\because f(x+2) = f(x) \ \forall \ x \in R]$$
$$= \lim_{h \to 0+} (-1+h) = -1$$

$$f(2n+1-0) = \lim_{h \to 0+} f(2n+1-h) = \lim_{h \to 0+} f(1-h) = \lim_{h \to 0+} (1-h) = 1, \text{ as before}$$

Hence $f(2n + 1 + 0) \ne f(2n + 1 - 0)$ and so f is discontinuous at $x = 2n + 1$, *i.e.*, at every odd integer.

EXERCISES

1. Show that the function f defined by $f(x) = \begin{cases} -x, & x < -1 \\ 1, & -1 \le x \le 1 \\ x, & x > 1 \end{cases}$

is continuous at $x = 1$ and $x = -1$ **[Delhi B.A. (Prog) 2009]**

2. (*a*) Show that the function f defined by

$$f(x) = \frac{e^{1/x} - 1}{e^{1/x} + 1}, \text{ if } x \ne 0 \quad \text{and} \quad f(x) = 0, \text{ if } x = 0$$

s discontinuous at $x = 0$. **(Kanpur 2008, 11; Purvanchal 2006; Meerut, 2003)**

2 (*b*) Show that the function f defined by

$$f(x) = \frac{e^{1/x}}{1 + e^{1/x}}, \text{ if } x \ne 0 \quad \text{and } f(x) = 0, \quad \text{if } x = 0$$

is discontinuous at $x = 0$. **(Kanpur 2008; Agra, 2002)**

3 (*a*) Show that the function f defined by

$$f(x) = x\frac{e^{1/x} - e^{-1/x}}{e^{1/x} + e^{-1/x}}, \text{ if } x \ne 0, \text{ and } f(x) = 0, \textit{ if } x = 0$$

is continuous at $x = 0$. **(Purvanchal 2006)**

(*b*) Show that the function f defined by

$$f(x) = \frac{e^{1/x^2}}{1 - e^{1/x^2}}, \text{ if } x \ne 0 \text{ and } f(x) = 0, \text{ if } x = 0$$

is discontinuous at $x = 0$. **[Delhi Maths (Hons), 1998]**

4. Prove that the function $f(x) = x \log \sin^2 x, x \ne 0$ *and* $f(x) = 0$, if $x = 0$ is continuous at $x = 0$ **(Agra 2010)**

5. Show that the function $f(x) = (1/x) \times \sin^{-1} x$ for $x \ne 0$, $f(0) = 1$ is continuous at the origin. **(Agra 2010)**

6. Show that the function f defined by

$$f(x) = \begin{cases} 2x \sin(1/x), & \text{if } x > 0 \\ 0, & \text{if } x = 0 \\ 3x \sin(1/x), & \text{if } x < 0 \end{cases}$$

is continuous at $x = 0$.

7. Let $f(x) = \begin{cases} 1-2x, & x < 0 \\ 0, & x = 0 \\ 1+3x, & x > 0 \end{cases}$. Show that $\lim_{x \to 0} f(x)$ exists and is equal to 1.

[*Delhi B.Sc. (Hons.) I 2010*]

8. Let $x \in R$ and $[x]$ be the greatest integer function. Determine the points of continuity of the function $f(x) = x[x]$ **[*Delhi B.Sc. (Hons.) I 2010; Delhi B.Sc. (Hons.) II, 2018*]**

Ans. Continuous for all x except when $x = \pm 1, \pm 2, \pm 3, ...$

9. Examine the continuity of the function f defined by

$$f(x) = \frac{2[x]}{3x - [x]} \text{ at } x = -\frac{1}{2} \text{ and } x = 1,$$

where $[x]$ denotes the greatest integer not greater than x.

10. Let

$$f(x) = \frac{[x^2] - 1}{x^2 - 1}, \text{ if } x^2 \neq 1$$
$$= 0 \qquad \text{if } x^2 = 1$$

where $[x]$ denotes the greatest integer function. Discuss the continuity of f at $x = 1$.

11. Consider the function $f(x) = [x]$, where $[x]$ denotes the greatest integer not exceeding x. Show that f is discontinuous at the points $x = 0, \pm 1, \pm 2, \pm 3$, ... and is continuous at every other point. **[*Delhi B.Sc. (Hons.) II, 2016*]**

12. Consider the function $f(x) = x - [x]$, where x is a positive variable and $[x]$ denotes the integral part of x. Show that f is discontinuous for integral values of x and continuous for all others. **[*Delhi B.Sc. (Hons.) II, 2017*]**

13. Let f be the function defined on [0, 1] by setting

$$f(x) = 2rx, \text{ if } 1/(r+1) \leq x < 1/r, \; r = 1, 2, 3, ...$$
$$f(0) = 0 \text{ and } f(1) = 1.$$

Examine for continuity the function f at the points 1, 1/2, 1/3, ..., $1/r$, ..., 0.

14. Let f be the function defined on [0, 1] by setting

$$f(x) = (-1)^r, \text{ if } 1/(r+1) \leq x < 1/r, \; r = 1, 2, 3, ...$$
$$f(0) = 0 \text{ and } f(1) = 1.$$

Examine for the continuity the function f at the points 1, 1/2, 1/3, ..., $1/r$, ..., 0.

15. Show that the following functions are continuous at the given points :

(*i*) $x \to x^2$; $x = 3$, (*ii*) $x \to x^3 + x^2 + x + 1$; $x = 1$

(*iii*) $x \to 1/(x^2 + 1)$; $x = -1$, (*iv*) $x \to (x^2 - 1)/(x^2 + 1)$; $x = 0$,

(*v*) $x \to \sqrt{(x+1)}$; $x = 0$.

16. (*a*) Show that the function f such that $f(x) = \begin{cases} x & \text{when } 0 \le x < 1/2 \\ 1 & \text{when } x = 1/2 \\ 1-x & \text{when } 1/2 < x \le 1 \end{cases}$

has a discontinuity of the first kind at $x = 1/2$.

(*b*) Test the continuity of the function $f(x) = \begin{cases} (x-|x|)/x, & x \ne 0 \\ 3, & x = 0 \end{cases}$ **[Nagpur 2010]**

17. show that the function f such that $f(x) = \begin{cases} \sin(1/x) & \text{when } x \ne 0 \\ 0 & \text{when } x = 0, \end{cases}$

has a discontinuity of the second kind at $x = 0$.

18. Show, from definition, that a polynomial function

$$x \to a_0 x^n + a_1 x^{n-1} + a_2 x^{n-2} + \ldots\ldots + a_{n-1} x + a_n$$

is continuous for $x = 1$.

19. Show that the function $f(x) = \tan^{-1}(1/x)$, $x \ne 0$ is discontinuous at $x = 0$. Also state the type of discontinuity. **(Kanpur, 2003)**

20. Examine the nature of discontinuity a function f defined by $f(x) = x/|x|$, $x \ne 0$; $f(0) = 2$ at $x = 0$. **(Bharathiar, 2004)**

21. Show that the function f defined on **R** by setting

$$f(x) = |x|^m \sin(1/x), \text{ if } x \ne 0;\ f(0) = 0,$$

is continuous at $x = 0$ whenever $m > 0$.

22. Let f be the function defined on **R** by setting

$f(x) = x - [x] - (1/2)$, when x is not an integer

$f(x) = 0$, when x is an integer

Show that f is continuous at all points of $\mathbf{R} \sim \mathbf{Z}$, and is discontinuous whenever $x \in \mathbf{Z}$.

23. Let f be defined on **R** by setting $f(t) = \begin{cases} t, & \text{if } 0 \le t < 1/2 \\ 0, & \text{if } t = 1/2 \\ t-1, & \text{if } 1/2 < t \le 1 \end{cases}$

and $f(n + t) = f(t)$, when n is any integer. Determine the points of discontinuity of f.

24. Prove that

(*i*) every polynomial function is continuous for every real number.

(*ii*) every rational function is continuous for every real number other than the zeros of the denominator of the polynomial function.

25. Determine the positions of discontinuity of the function f defined by

$$\begin{aligned} f(x) &= -x^2, && \text{for } x \le 0 \\ &= 5x - 4, && \text{for } 0 < x \le 1 \\ &= 4x^2 - 3x, && \text{for } 1 < x < 2 \\ &= 3x + 4, && \text{for } x \ge 2. \end{aligned}$$

(Avadh, 2001; Purvanchal, 1995; Rohilkhand, 1994)

26. Show that $f(x) = \begin{cases} (x^2/a) - a & \text{for } x \le a \\ a - (a^2/x) & \text{for } x > a \end{cases}$

is continuous at $x = a$. **(Agra 2009, 2011; Avadh, 1999)**

27. Show that $f(x) = \sin^2 x$ is continuous for every value of x. **(Manipur, 2002)**

28. Examine for continuity the function $f(x) = [x] + [-x]$ at $x = 0$ **(G.N.D.U. Amritsar 2010)**

ANSWERS

9. Continuous at $x = 1$; discontinuous at $x = -1/2$

13. Discontinuous at $x = 1, 1/2, 1/3, \ldots, 1/n, \ldots; 0$

14. Discontinuous at $x = 1, 1/2, 1/3, \ldots, 1/n, \ldots; 0$ **16.** (*b*) Discontinuous

20. Discontinuity of first kind

23. $n + (1/2)$, where n is any integer

25. $x = 0$ is the only position of discontinuity of $f(x)$ **28.** Discontinuous

8.12. ALGEBRA OF CONTINUOUS FUNCTIONS

Theorem I. *Let f and g be defined on an interval I. If f and g are continuous at $a \in I$, then $f + g$ is continuous at a.* **(*Garhwal, 2001; Meerut, 2003*)**

Proof. f and g are continuous at a

$$\Rightarrow \lim_{x \to a} f(x) = f(a) \text{ and } \lim_{x \to a} g(x) = g(a) \qquad \ldots(1)$$

Now, $$\lim_{x \to a} (f + g)(x) = \lim_{x \to a} \{f(x) + g(x)\} = \lim_{x \to a} f(x) + \lim_{x \to a} g(x),$$

$$= f(a) + g(a) = (f + g)(a), \text{ by (1)}$$

$\Rightarrow$ $f + g$ is continuous at a.

Theorem II. *Let f and g be defined on an interval I. If f and g are continuous at $a \in I$, then $f - g$ is continuous at a.*

Proof. Proceed as in theorem I.

Theorem III. *Let f and g be defined on an interval I. If f and g are continuous at $a \in I$, then fg is continuous at a.*

Or

Product of two continuous functions is continuous.

(*Delhi Maths (G) 2002, 03; Garhwal, 2001; Meerut, 2003*)

Proof. f and g are continuous at a

$$\Rightarrow \lim_{x \to a} f(x) = f(a) \text{ and } \lim_{x \to a} g(x) = g(a) \qquad \ldots(1)$$

Now, $$\lim_{x \to a} (fg)(x) = \lim_{x \to a} \{f(x)\, g(x)\} = \left\{\lim_{x \to a} f(x)\right\}\left\{\lim_{x \to a} g(x)\right\}$$

$$= f(a)\, g(a) = (fg)(a), \text{ by (1)}$$

$\Rightarrow$ fg is continuous at a.

Theorem IV. *Let f and g be defined on an interval I and let $g(a) \neq 0$. If f and g are continuous at $a \in I$, then f/g is continuous at a.*

Proof. f and g are continuous at a

$$\Rightarrow \lim_{x \to a} f(x) = f(a) \text{ and } \lim_{x \to a} g(x) = g(a), \text{ where } g(a) \neq 0$$

Now, $$\lim_{x \to a} \left(\frac{f}{g}\right)(x) = \lim_{x \to a} \frac{f(x)}{g(x)} = \frac{\lim_{x \to a} f(x)}{\lim_{x \to a} g(x)} = \frac{f(a)}{g(a)} = \left(\frac{f}{g}\right)(a)$$

$\Rightarrow$ f/g is continuous at a.

Theorem V. *If f is continuous at a point 'a' and c be any real number, the cf is continuous at 'a'.*

Proof. Left as an exercise.

Theorem VI. *If f is continuous at a, then | f | is also continuous at a, but not conversely*

[Delhi (Hons), 1998; Kanpur, 2002, 08, 10]

Proof. Since f is continuous at a, hence for a given $\varepsilon > 0$, there exists a $\delta > 0$ such that

$$| f(x) - f(a) | < \varepsilon \text{ whenever } | x - a | < \delta \qquad ...(1)$$

If a and b are any real numbers, then we know that

$$| a - b | \geq | | a | - | b | | \quad \text{or} \quad | | a | - | b | | \leq | a - b | \qquad ...(2)$$

Using (2), $\quad | | f(x) | - | f(a) | | \leq | f(x) - f(a) |$ ...(3)

From (1) and (3), $\quad | | f(x) | - | f(a) | | < \varepsilon$ whenever $| x - a | < \delta$

$\Rightarrow \quad | f |$ is continuous at a.

To show that the converse may not be true.

Consider the function f defined on set of real numbers **R**.

$$f(x) = \begin{cases} 1, & \text{if } x \geq 0 \\ -1, & \text{if } x < 0 \end{cases} \quad \text{so that} \quad | f |(x) = | f(x) | = 1 \ \forall \ x \in \mathbf{R}$$

Then $\quad \lim_{x \to 0} | f |(x) = \lim_{x \to 0} 1 = 1 = | f |(0)$

$\Rightarrow \quad | f |$ is continuous at 0

But $\quad f(0-0) = \lim_{x \to 0-0} f(x) = \lim_{x \to 0-0} -1 = -1$

and $\quad f(0+0) = \lim_{x \to +0} f(x) = \lim_{x \to 0+0} 1 = 1$

Since $\quad f(0-0) \neq f(0+0)$, f is not continuous at 0.

Thus $| f |$ is continuous at 0 while f is not continuous at 0.

Theorem VII. *If f and g are two continuous functions at a, then the function max {f, g} and min {f, g} are both continuous at a.*

Proof. We have

$$\max. \{f, g\} = \frac{1}{2}(f+g) + \frac{1}{2}| f - g |, \ \min. \{f, g\} = \frac{1}{2}(f+g) - \frac{1}{2}| f - g |. \qquad ...(1)$$

Now, $\quad f$ and g are continuous at a

$\Rightarrow \quad f + g$ is continuous at a, by theorem I

$\Rightarrow \quad (f+g)/2$ is continuous at a, by theorem V ...(2)

Again, f and g are continuous at a

$\Rightarrow \quad f - g$ is continuous at a, by theorem II

$= \quad | f - g |$ is continuous at a, by theorem VI

$= \quad | f - g |/2$ is continuous at a, by theorem V ...(3)

Now, $\quad (f+g)/2$ and $| f - g |/2$ are both continuous at a, by (2) and (3)

$\Rightarrow \quad (f+g)/2 + | f - g |/2$ and $(f+g)/2 - | f - g |/2$ are both continuous at a

$\Rightarrow \quad \max \{f, g\}$ and $\min \{f, g\}$ are continuous at a, by (1)

Example. *Using theorems of algebra of continuous functions, show that the function f defined on* **R** *by setting*

$$f(x) = | x | + \frac{x}{1 + e^{1/x}}, \ x \neq 0; \ f(0) = 0$$

is continuous at x = 0. **[Delhi Maths (Hons), 2000]**

Solution. Let $\quad g(x) = |x|$ and $h(x) = \dfrac{x}{1+e^{1/x}}$

Then $f(x) = g(x) + h(x)$ so that $f(0) = g(0) + h(0)$

But $f(0) = 0$ and $g(0) = 0$. Hence $h(0) = 0$.

Then $\quad g(0+0) = \lim\limits_{h \to 0+} g(0+h) = \lim\limits_{h \to 0+} |h| = 0$

and $\quad g(0-0) = \lim\limits_{h \to 0+} g(0-h) = \lim\limits_{h \to 0+} |-h| = 0$

Thus, $\quad g(0+0) = g(0-0) = g(0) \Rightarrow g$ is continuous at $x = 0$.

Now, $\quad h(0+0) = \lim\limits_{h \to 0+} \dfrac{h}{1+e^{1/h}} = \lim\limits_{h + 0+} \dfrac{h e^{-1/h}}{e^{-1/h}+1} = 0$

and $h(0-0) = \lim\limits_{h \to 0+} \dfrac{-h}{1+e^{-1/h}} = 0$

Thus, $\quad h(0+0) = h(0-0) = h(0) \Rightarrow h$ is continuous at $x = 0$.

Now $f(x) = g(x) + h(x)$, where $g(x)$ and $h(x)$ are both continuous at $x = 0$. Hence $f(x)$ is also continuous at $x = 0$.

8.13. FUNCTION OF A FUNCTION. COMPOSITES OF FUNCTIONS

Let f and g be two functions such that

$$\text{Domain } f = [a, b] \text{ and domain of } g = [\alpha, \beta]$$

We suppose that the range of the function g is a sub-set of the domain of the function f, *i.e.*, Range $g \subset$ domain f.

Now $\quad t \in [\alpha, \beta]$

$\Rightarrow g(t) \in$ Range $g \Rightarrow g(t) \in$ Domain $f \Rightarrow f(g(t))$ has a meaning.

We have thus a new real valued function with $[\alpha, \beta]$ as its domain.

This new function is called a *function of function* and is also denoted as $f \circ g$ and called the *composite* of f and g. Thus, we have

$$(fog)(t) = f(g(t)).$$

It may be emphasized that the composite function $f \circ g$ *has a meaning if and only if the range of the function g is a sub-set of the domain of the function f.*

EXERCISE

1. Let f, g be two functions defined as follows :

$$f(x) = \sqrt{x} \quad \forall\, x \geq 0, \qquad g(x) = x^2 + 1 \quad \forall\, x \in \mathbf{R}.$$

Show that $\quad (f \circ g)(x) = \sqrt{(x^2+1)}.$

What is the domain of the function $f \circ g$?

2. If $\quad f(x) = x^3 + x - 2, \qquad g(x) = 1/(x+1)$

give explicit definitions of $f \circ g$ and $g \circ f$ giving also their domains.

3. Let $\quad f(x) = x^2 + 1, \qquad g(x) = x^4.$

Show that $f \circ g \neq g \circ f$.

4. Let $\quad f(x) = \sqrt{x}, \qquad g(x) = 1/(x^2 - 1),$

determine $g \circ f$ and $f \circ g$ with their domains.

5. Given that $f(x) = \dfrac{1}{\sqrt{x}},\ x > 0, \quad g(x) = \dfrac{x^2+1}{x^2-1},\ x \in \mathbf{R} \sim \{-1, 1\};$

give the values of $(f \circ g)(2)$ and $(g \circ f)(2)$.

6. If x denotes a real number, give the domains of $f, g, f \circ g$ and $g \circ f$ where f and g are given as follows :

(*i*) $f(x) = \sqrt{x},\ g(x) = x^2$ (*ii*) $f(x) = 1/(2x+3),\ g(x) = 2x+3.$

7. Given $f(x) = |x|$, show that $f[f(x)] = f(x)$.

8. For each of the following functions F give some function f and g such that the function F may be expressed as the composite $f \circ g$.

(*i*) $F(x) = \sqrt{(x^2+2)}$, (*ii*) $F(x) = (x^2+1)^3 + 2(x^2+1)^2 + x^2 \quad [x \in \mathbf{R},$

(*iii*) $F(x) = \sin^2 x \quad [x \in \mathbf{R}.$

8.14. CONTINUITY OF THE COMPOSITE FUNCTION

If g is continuous at a point t_0 of $[\alpha, \beta]$ and f is continuous at the corresponding point $x_0 = g(t_0)$, then the composite function $f \circ g$ is continuous at t_0. **(Nagpur 2010)**

We write $\quad x = g(t), \quad y = f(x).$

Let $\varepsilon > 0$ be given. Let $\quad y_0 = f(x_0) = f(g(t_0)).$

As f is continuous at x_0, there exists δ_0 such that

$$|y - y_0| = |f(x) - f(x_0)| < \varepsilon \text{ when } |x - x_0| < \delta_0.$$

As g is continuous at t_0, there exists δ such that

$$|x - x_0| = |g(t) - g(t_0)| < \delta_0 \text{ when } |t - t_0| < \delta.$$

Thus, we have $\quad |y - y_0| < \varepsilon \text{ when } |t - t_0| < \delta$

$$\Rightarrow |(f \circ g)(t) - (f \circ g)(t_0)| < \varepsilon \text{ when } |t - t_0| < \delta$$

$$\Rightarrow f \circ g \text{ is continuous at } t_0.$$

Hence the result.

Ex.1. Let $y = f(x)$ be continuous at $x = a$ and let $z = g(y)$ be continuous at $y = b$ where $b = f(a)$. Then, prove that $z = g(f(x))$ is continuous at $x = a$. **(Nagpur 2010)**

Ex. 2. If a certain f is continuous at a *and* $\langle g_n \rangle$ is a sequence such that $\lim g_n = a$, then

$$\lim f(g_n) = f(a).$$

Also if for every sequence $\langle g_n \rangle$

$$\lim g_n = a \Rightarrow \lim f(g_n) = f(a),$$

the function f is continuous at a.

8.15. CRITERIA FOR CONTINUITY. EQUIVALENT DEFINITION OF CONTINUITY

Theorem I. (Heine's definition of continuity). *A function f defined on an interval I is continuous at $a \in I$ if and only if for every sequence $\langle a_n \rangle$ in I which converges to a, we have* $\lim\limits_{n \to \infty} f(a_n) = f(a).$ **[*Delhi Maths (Hons), 2000, 01, 04, 06, 08; Delhi B.Sc. (Hons.) I 2011 Kanpur 2010; Delhi B.Sc. I (Hons) 2010; Delhi Maths (Prog) 2007*]**

Solution. The condition is necessary. Let f be continuous at $a \in I$ and let $\langle a_n \rangle$ be a sequence in I such that $\quad \lim\limits_{n \to \infty} a_n = a.$

We shall prove that $\quad \lim\limits_{n \to \infty} f(a_n) = f(a).$

Since f is continuous at a, so, for a given $\varepsilon > 0$, there exists $\delta > 0$ such that

$$|f(x) - f(a)| < \varepsilon \text{ whenever } |x - a| < \varepsilon \quad \text{...(1)}$$

Also, since $\lim_{n \to \infty} a_n = a$, so, there exists a positive integer m such that

$$|a_n - a| < \delta \quad \forall\ n \geq m \quad \text{...(2)}$$

Putting $x = a_n$ in (1), we have

$$|f(a_n) - f(a)| < \varepsilon \text{ whenever } |x - a| < \delta \quad \text{...(3)}$$

From (2) and (3), we have $\quad |f(a_n) - f(a)| < \varepsilon \quad \forall\ n \geq m$

$$\Rightarrow \lim_{n \to \infty} f(a_n) = f(a), \text{ as required.}$$

The condition is sufficient.

Let $<f(a_n)>$ converge to $f(a)$ whenever the $<a_n>$ converges to a. ...(4)

We shall prove that f is continuous at a.

Suppose f is not continuous at a. Then there is an $\varepsilon > 0$ such that for every $\delta > 0$ there is an x such that $\quad |f(x) - f(a)| \geq \varepsilon$ when $|x - a| < \delta$

Therefore, by taking $\delta = 1/n$, we find that for each positive integer n, there is a $a_n \in I$, such that $\quad |f(a_n) - f(a)| \geq \varepsilon$ when $|a_n - a| < 1/n$

$\Rightarrow$ $<a_n>$ converges to a whereas $<f(a_n)>$ does not converges to $f(a)$

This contradicts the given fact (4). Hence our assumption that f is not continuous at a is wrong. Therefore f is continuous at a.

Theorem II. *A function f defined on* **R** *is continuous on* **R** *if and only if for each open set G in* **R**, *$f^{-1}(G)$ is open in* **R**. ***[Delhi Maths (G), 2003]***

Proof. The condition is necessary. Let f be continuous on **R** and let G be any open set We shall prove that $f^{-1}(G)$ is open.

If $f^{-1}(G)$ is empty, then it is open and we get the required result. If $f^{-1}(G)$ is non-empty, let $a \in f^{-1}(G)$. Then $f(a) \in G$. Since G is an open set containing $f(a)$, hence there exists an $\varepsilon > 0$ such that

$$]f(a) - \varepsilon, f(a) + \varepsilon[\subset G \quad \text{...(1)}$$

Since f is continuous at a, so there exists $\delta > 0$ such that

$$|x - a| < \delta \Rightarrow |f(x) - f(a)| < \varepsilon$$

i.e., $\quad x \in]a - \delta, a + \delta[\Rightarrow f(x) \in]f(a) - \varepsilon, f(a) + \varepsilon[$

or $\quad x \in]a - \delta, a + \delta[\Rightarrow f(x) \in G$, using (1)

or $\quad x \in]a - \delta, a + \delta[\Rightarrow x \in f^{-1}(G)$

$$\Rightarrow]a - \delta, a + \delta[\subset f^{-1}(G)$$

$$\Rightarrow f^{-1}(G) \text{ is a neighbourhood of } a$$

Since a is an arbitrary point of $f^{-1}(G)$, it follows that $f^{-1}(G)$ is open.

The condition is sufficient. Let $f^{-1}(G)$ be open whenever G is open. Then we shall prove that f is continuous on **R**.

Let a be an arbitrary point of **R**. Let $\varepsilon > 0$ be given. Now $]f(a) - \varepsilon, f(a) + \varepsilon[$ is an open set containing $f(a)$ and so by our hypothesis, $f^{-1}(]f(a) - \varepsilon, f(a) + \varepsilon[)$ is also an open set containing a. Hence, there exists a $\delta > 0$ such that

$$]a - \delta, a + \delta[\subset f^{-1}(]f(a) - \varepsilon, f(a) + \varepsilon[)$$

$$\Rightarrow f(]a - \delta, a + \delta[) \subset]f(a) - \varepsilon, f(a) + \varepsilon[$$

Thus, for a given $\varepsilon > 0$, we have $\delta > 0$ such that

$$x \in]a-\delta, a+\delta[\Rightarrow f(x) \in]f(a)-\varepsilon, f(a)+\varepsilon[$$

i.e.,

$$|x-a| < \delta \Rightarrow |f(x)-f(a)| < \varepsilon$$

$\Rightarrow f$ is continuous at a.

Since a is an arbitrary point of **R**, it follows that f is continuous on **R**.

Theorem III. *A function f defined on* **R** *is continuous on* **R** *if and only if for each closed set H in* **R**, $f^{-1}(H)$ *is also closed in* **R**.

Proof. The condition is necessary. Let f be continuous on **R** and let H be any closed set in **R**. Then $\mathbf{R} \sim H$ is an open set in **R**. Hence by theorem II, $f^{-1}(\mathbf{R} \sim H)$ is an open set in **R**.

But $\quad f^{-1}(\mathbf{R} \sim H) = f^{-1}(\mathbf{R}) - f^{-1}(H) = \mathbf{R} \sim f^{-1}(H)$

$\therefore \quad \mathbf{R} \sim f^{-1}(H)$ is open in **R** and so $f^{-1}(H)$ is a closed set in **R**.

The condition is sufficient. Let $f^{-1}(H)$ be closed whenever H is closed. Then we shall prove that f is continuous. Let G be any open set in **R**. Then $\mathbf{R} \sim G$ is closed set in **R** and so by hypothesis $f^{-1}(\mathbf{R} \sim G)$ is closed set in **R**. But $f^{-1}(\mathbf{R} \sim G) = \mathbf{R} - f^{-1}(G)$. Hence $\mathbf{R} - f^{-1}(G)$ is a closed set in **R** and hence $f^{-1}(G)$ is an open set in **R**. Thus $f^{-1}(G)$ is open in **R** whenever G is open in **R**. Therefore, by theorem II, f is continuous.

EXAMPLES

Example 1. *Let f be a function on* **R** *defined by*

$$f(x) = \begin{cases} 1, \text{ when } x \text{ is rational i.e., when } x \in \mathbf{Q} \\ -1, \text{ when } x \text{ is irrational i.e., when } x \in R - \mathbf{Q} \end{cases}$$

Show that f is discontinuous at every point of **R**.

This is known as Dirichlet's function.

[*Delhi B.Sc. (Hons.) I 2011; Delhi B.A. (Prog.) III 2011, 2015; Kanpur 2011, 2013*]

Solution. Case I. Let a be any rational number, so that $f(a) = 1$.

Then the neighbourhood $]\, a - 1/n, a + 1/n\, [$ of a contains an irrational number a_n for each $n \in \mathbf{N}$.

i.e., $\quad a_n \in]\, a - 1/n, a + 1/n\, [\ \forall\, n \in \mathbf{N}$

$\Rightarrow \quad a - 1/n < a_n < a + 1/n\ \forall\, n \in \mathbf{N} \Rightarrow |a_n - a| < 1/n\ \forall\, n \in \mathbf{N}$

$\Rightarrow \quad |a_n - a| \to 0$ as $n \to \infty \Rightarrow a_n - a \to 0 \Rightarrow$ as $n \to \infty$

$\Rightarrow \quad \langle a_n \rangle$ converges to a.

Now $\quad f(a_n) = -1$ for all n, as a_n is irrational and $f(a) = 1$.

$\therefore \quad \lim_{n \to \infty} f(a_n) = -1 \neq f(a)$

i.e., $\quad \langle f(a_n) \rangle$ does not converges to $f(a)$ when $a_n \to a$.

Thus, f is discontinuous at all rational points a.

Case II. Let b be any irrational number so that $f(b) = -1$.

As explained above, we can choose a rational number b_n such that

$$|b_n - b| < 1/n \Rightarrow b_n \to b \text{ as } n \to \infty$$

$\Rightarrow \langle b_n \rangle$ converges to b.

Now $\quad f(b_n) = 1$ for all n, as b_n is rational and $f(b) = -1$.

$\therefore \quad \lim_{n \to \infty} f(b_n) = 1 \neq f(b).$

$\Rightarrow$ $<f(b_n)>$ does not converge to $f(b)$ when $b_n \to b$.

Thus, the function f is discontinuous at all irrational points b.

Hence, the given function f is discontinuous at all points of **R**.

Example 2. (*a*) *Show that the function f defined on* **R** *by*

$$f(x) = \begin{cases} x, & \text{when } x \text{ is irrational} \\ -x, & \text{when } x \text{ is rational} \end{cases}$$

is continuous only at x = 0. **[*Delhi B.A. (Prog.) III, 2009;*]**

(*b*) *Let f be a function defined on* **R** *by*

$$f(x) = \begin{cases} x, & \text{when } x \text{ is rational} \\ -x, & \text{when } x \text{ is irrational} \end{cases}$$

Show that f is discontinuous everywhere except at x = 0.

Solution. (*a*) **Case I.** Let $a \neq 0$ be any rational number so that $f(a) = -a$.

Then any neighbourhood $]a - 1/n, a + 1/n[$ of a contains an irrational number a_n for each $n \in$ **N**, *i.e.*,

$$a_n \in \;]a - 1/n, a + 1/n[$$

$$\Rightarrow \quad | a_n - a | < 1/n$$

$$\Rightarrow \quad | a_n - a | \to 0 \text{ as } n \to \infty \Rightarrow \; < a_n > \to a.$$

Now $\quad f(a_n) = a_n \quad \forall\, n \; (\because a_n \text{ is irrational})$

$$\lim_{n \to \infty} f(a_n) \quad \lim_{n \to \infty} a_n \quad a \quad f(a) \qquad [\because f(a) \quad a]$$

$$\Rightarrow \quad <f(a_n)> \text{ does not converge to } f(a), \text{ when } <a_n> \to a.$$

So, f is discontinuous at all non-zero rational points.

Case II. Let $b \neq 0$ be any irrational number so that $f(b) = b$.

As explained earlier, there exists a rational number $b_n \;\; \forall\, n \in$ **N** such that

$$| b_n - b | < 1/n \quad \Rightarrow \quad | b_n - b | \to 0 \text{ as } n \to \infty$$

$$\Rightarrow \quad < b_n > \to b.$$

Now, $\quad f(b_n) = -b_n \quad \forall\, n \in \mathbf{N} \; (\because b_n \text{ is rational number})$

$$\Rightarrow \quad \lim_{n \to \infty} f(b_n) = -\lim_{n \to \infty} b_n = -b \neq f(b)$$

Thus, $<b_n> \to b$, but $<f(b_n)>$ does not converge to $f(b)$.

So f is discontinuous at all non-zero irrational points.

We shall now prove that f is continuous only at the point $x = 0$.

Let $\varepsilon > 0$ be any number, and $\delta = \varepsilon/2$.

Then $\quad | x | < \delta \Rightarrow | f(x) - f(0) | = | x | < \delta < \varepsilon$, where x is irrational.

and $\quad | x | < \delta \Rightarrow | f(x) - f(0) | = | -x | = | x | < \delta < \varepsilon$, where x is rational.

$\therefore \quad | f(x) - f(0) | < \varepsilon$, whenever $| x - 0 | < \delta$.

Hence, the function f is continuous only at $x = 0$.

(*b*) Proceed as in part (*a*).

Example 3. *Show that the function f defined by*

$$f(x) = \begin{cases} x, & \text{if } x \text{ is rational} \\ 0, & \text{if } x \text{ is irrational} \end{cases}$$

is continuous only at x = 0. **[*Delhi B.Sc. (Hons.) II, 2017*]**

Solution. Case I. Let $a \neq 0$ be any rational number, so that $f(a) = a$.

Then any neighbourhood $]a - 1/n, a + 1/n[$ of a contains an irrational number a_n for each $n \in \mathbf{N}$, *i.e.*,

$$a_n \in \,]\, a - 1/n, a + 1/n\, [\;\Rightarrow\; |\, a_n - a \,| < 1/n$$

$$\Rightarrow |\, a_n - a \,| \to 0 \text{ as } n \to \infty \;\Rightarrow\; \langle a_n \rangle \to a.$$

Now, $\quad f(a_n) = 0 \quad \forall\, n \in \mathbf{N} \qquad (\because a_n \text{ is irrational})$

$$\Rightarrow \lim_{n \to \infty} f(a_n) = 0 \neq f(a) \qquad [\because f(a) = a \neq 0]$$

Thus, f is discontinuous at every non-zero rational point.

Case II. Let $b \neq 0$ be any irrational number, so that $f(b) = 0$.

Then there exists a rational number $b_n\ \forall\, n \in \mathbf{N}$, such that

$$|\, b_n - b \,| < 1/n \;\Rightarrow\; |\, b_n - b \,| \to 0 \text{ as } n \to \infty$$

$$\Rightarrow\; \langle b_n \rangle \to b.$$

Now $\quad f(b_n) = b_n \quad \forall\, n \in \mathbf{N} \qquad (\because b_n \text{ is rational})$

$$\Rightarrow \lim_{n \to \infty} f(b_n) = \lim_{n \to \infty} b_n = b \neq 0$$

$\Rightarrow \quad \langle f(b_n) \rangle$ does not converge to $f(b)$. $\qquad [\because f(b) = 0]$

Thus, f is discontinuous at every non-zero irrational point.

Now we shall prove that f is continuous at $x = 0$.

We have $f(0) = 0$ and

$$|\, f(x) - f(0) \,| = |\, f(x) \,| = |\, x \,|, \text{ if } x \text{ is rational.}$$
$$= 0, \text{ if } x \text{ is irrational.}$$

Let $\varepsilon > 0$. Then $|\, f(x) - f(0) \,| < \varepsilon$, for $|\, x - 0 \,| < \delta$, taking $\delta = \varepsilon$

Hence, f is continuous only at $x = 0$.

Example 4. *Show that the function f defined as*

$$f(x) = \begin{cases} x, & \text{if } x \text{ is rational} \\ 1 - x, & \text{if } x \text{ is irrational} \end{cases}$$

is continuous only at $x = 1/2$.

Kanpur 2007; Delhi Physics (H), 2000; Delhi Maths (Hons), 2002; I.A.S., 2001, 04.

Or

Give example of a function which is continuous at only one point of the domain.

(Kanpur, 2004)

Solution. Case I. Let $a \neq 1/2$ be any rational number. Then $f(a) = a$.

We can find an irrational number a_n for each $n \in \mathbf{N}$ such that

$$a_n \in \,]a - 1/n, a + 1/n[$$

$$\Rightarrow |\, a_n - a \,| < 1/n \;\Rightarrow\; \langle a_n \rangle \to a.$$

Now, $\quad f(a_n) = 1 - a_n \quad \forall\, n \in \mathbf{N} \qquad (\because a_n \text{ is irrational})$

$$\Rightarrow \lim_{n \to \infty} f(a_n) = 1 - \lim_{n \to \infty} a_n = 1 - a \neq a.$$

$\Rightarrow \langle f(a_n) \rangle$ does not converge to $f(a)$. $\qquad [\because f(a) = a]$

Thus, f is discontinuous at all rational points $a \neq 1/2$.

Case II. Let b be any irrational number so that $f(b) = 1 - b$.

We can find a rational number b_n for each $n \in \mathbf{N}$ such that

$$b_n \in]b - 1/n, b + 1/n[$$

$$\Rightarrow |b_n - b| < 1/n \Rightarrow \langle b_n \rangle \to b$$

Now $$f(b_n) = b_n \quad \forall n \qquad (\because b_n \text{ is rational})$$

$$\Rightarrow \lim_{n\to\infty} f(b_n) = \lim_{n\to\infty} b_n = b \neq 1 - b \qquad (\because b_n \text{ is irrational})$$

$$(\because b = 1 - b \Rightarrow b = 1/2 \Rightarrow b \text{ is rational, a contradiction})$$

$\Rightarrow \langle f(b_n) \rangle$ does not converge to $f(b)$, where $\langle b_n \rangle \to b$.

Thus, f is discontinuous at all irrational points b.

Now we shall show that f is continuous at $x = 1/2$ only.

We have $|f(x) - f(1/2)| = |x - 1/2|$, if x is rational. $[\because f(1/2) = 1/2]$

$= |1 - x - 1/2| = |1/2 - x| = |x - 1/2|$, if x is irrational.

Thus, $$|f(x) - f(1/2)| = |x - 1/2|$$

Let $\varepsilon > 0$ be given. Choose $\delta = \varepsilon > 0$.

Then $$|f(x) - f(1/2)| < \varepsilon, \text{ when } |x - 1/2| < \delta$$

Hence, f is continuous only at $x = 1/2$.

Example 5. *If a continuous function of x satisfies the functional equation* $f(x + y) = f(x) + f(y)$, *then show that* $f(x) = \alpha x$, *where* α *is a constant.* ***(I.A.S., 2003)***

Solution. Given $$f(x + y) = f(x) + f(y) \qquad ...(1)$$

Case I. Taking $x = 0 = y$ in (1), we obtain

$$f(0) = f(0) + f(0) \quad \text{so that} \quad f(0) = 0 \qquad ...(2)$$

Case II. If x is any positive integer, we have

$$f(x) = f(1 + 1 + \ldots + 1) = f(1) + f(1) + \ldots + f(1)$$

or $$f(x) = x f(1) \quad \text{or} \quad f(x) = \alpha x, \text{ say} \qquad ...(3)$$

where $\alpha = f(1)$.

Case III. Let x be any negative integer and let $x = -y$. Then $y\ (= -x)$ is a positive integer.

From Case I, $$0 = f(0) \quad \text{or} \quad 0 = f(y - y) = f(y + (-y))$$

or $$0 = f(y) + f(-y) \quad \text{so that} \quad f(-y) = -f(y) \qquad ...(4)$$

Then, $$f(x) = f(-y) = -f(y) = -y f(1), \text{ by Case II}$$

or $$f(x) = x f(1) \quad \text{or} \quad f(x) = \alpha x \qquad [\because y = -x \text{ and } f(1) = \alpha]$$

Case IV. Let $x = p/q$ be a rational number; q being positive. We have

$$f(p) = f\left(\frac{p}{q} q\right) = f\left(\frac{p}{q}\right) + f\left(\frac{p}{q}\right) + \ldots q \text{ times} = q f\left(\frac{p}{q}\right) \qquad ...(5)$$

Since p can be positive or negative integer, so by Cases II or III, we have $f(p) = \alpha p$. Then (5) becomes

$$\alpha p = q f(p/q) \quad \text{or} \quad f(p/q) = \alpha (p/q) \quad \text{or} \quad f(x) = \alpha x.$$

Case V. Let x be any real number. Let $\langle x_n \rangle$ be a sequence of rational numbers such that

$$\lim_{n\to\infty} x_n = x \qquad ...(6)$$

We have, x_n, being rational, by Case IV

$$f(x_n) = \alpha x_n \quad \text{...(7)}$$

Let $n \to \infty$. As f is continuous, we obtain from (7)

$$\lim_{n\to\infty} f(x_n) = \lim_{n\to\infty} (\alpha x_n) = \alpha \lim_{n\to\infty} x_n = \alpha x, \text{ using (6)}$$

or

$$f(x) = \alpha x.$$

Thus, $f(x) = \alpha x \quad \forall\ x \in \mathbf{R}$, by cases I, II, III, IV and V.

EXERCISES

1. Show that the function f defined by

$$f(x) = \begin{cases} 1, & \text{if } x \text{ is rational,} \\ 0, & \text{if } x \text{ is irrational,} \end{cases}$$

is discontinuous at every point.

2. Let f be the function defined on $[-1, 1]$ by

$$f(x) = \begin{cases} x, & \text{if } x \text{ is irrational,} \\ 0, & \text{if } x \text{ is rational.} \end{cases}$$

Prove that f is continuous only at $x = 0$.

3. Show that a function f defined on $[0, 1]$ by

$$f(x) = \begin{cases} x, & \text{when } x \text{ is rational,} \\ 2x, & \text{when } x \text{ is irrational,} \end{cases}$$

is continuous only at $x = 0$. **[Delhi B.A. (Prog.) III, 2014]**

4. Let f be a function defined on $]0, 1[$ by

$$f(x) = \begin{cases} 0, & \text{if } x \text{ is irrational,} \\ 1/q, & \text{if } x = p/q, \end{cases}$$

where p and q are positive integers having no common factor.

Prove that f is continuous at each irrational point and discontinuous at each rational point. **(Calicut, 2004)**

5. Let f be the function defined on $[0, \infty[$ by setting $f(x) = \lim\limits_{n\to\infty} \dfrac{x^n}{1+x^n}\ \forall\ n \geq 0$. Show that f is continuous for all points of $[0, \infty[$ except $x = 1$.

6. Show that $f(x) = \lim\limits_{n\to\infty} \dfrac{x^{2n}}{1+x^{2n}}$ is continuous at all points of $\mathbf{R}$ except $x = \pm 1$.

7. Show that $f(x) = \lim\limits_{n\to\infty} \dfrac{x^{2n}-1}{x^{2n}+1}$ is continuous at all points of $\mathbf{R}$ except $x = \pm 1$.

8.16. SOME PROPERTIES OF THE CONTINUITY OF A FUNCTION AT A POINT

Boundedness as a consequence of the continuity at a point.

Property I. *If f is continuous at the end point a of an interval* $[a, b]$, *there exists* $c \in [a, b]$ *such that f is bounded in* $[a, c]$.

Proof. Let $\varepsilon > 0$ be given. There exists $\delta > 0$ such that

$$| f(x) - f(a) | < \varepsilon \text{ when } a \leq x < a + \delta.$$

Let $c \in [a, a + \delta]$ so that c is a number less than $a + \delta$.

It follows that $\quad f(a) - \varepsilon < f(x) < f(a) + \varepsilon \quad \forall\, x \in [a, c]$.

Thus, f is bounded in $[a, c]$. Hence the result.

Property II. *If f is continuous at the end point b of an interval $[a, b]$, there exists $d \in [a, b]$ such that f is bounded in $[d, b]$.*

The proof is similar to that of property I.

Property III. *If f is continuous at an interior point ξ of $[a, b]$, there exists an interval $[\alpha, \beta]$ containing ξ in which f is bounded.*

Proof. Let $\varepsilon > 0$ be given. There exists $\delta > 0$ such that

$$| f(x) - f(\xi) | < \varepsilon \text{ when } \xi - \delta < x < \xi + \delta.$$

Let $\quad \alpha \in]\xi - \delta, \xi],\ \beta \in [\xi, \xi + \delta[.$

Thus, we see that $\quad | f(x) - f(\xi) | < \varepsilon \quad \forall\, x \in [\alpha, \beta].$

Hence the result.

Preservation of sign as a consequence of continuity at a point.

Property I. *If a function f is continuous at the end point a of $[a, b]$ and $f(a) > 0$, there exists $c \in [a, b]$ such that* $\quad f(x) > 0 \quad \forall\, x \in [a, c].$

Proof. Because of the continuity of f at a there exists $\delta > 0$ such that

$$| f(x) - f(a) | < (1/2) \times f(a) \text{ when } a \le x < a + \delta$$

so that $(1/2) \times f(a) = f(a) - (1/2) \times f(a) < f(x) < f(a) + (1/2) \times f(a) = (3/2) \times f(a)$ when $a \le x < a + \delta$.

Let $c \in [a, a + \delta]$. We thus see that

$$f(x) > 0 \quad \forall\, x \in [a, c].$$

We may similarly show that if a function f is continuous at the end point a of $[a, b]$ and $f(a) < 0$, then there exists $c \in [a, b]$ such that

$$f(x) < 0 \quad \forall\, x \in [a, c].$$

We note that $\quad f(a) < 0 \Rightarrow -f(a) > 0.$

There exists $\delta > 0$ such that

$$| f(x) - f(a) | < (-1/2) \times f(a) \text{ when } a \le x < a + \delta,$$

so that $(3/2) \times f(a) = f(a) + (1/2) \times f(a) < f(x) < f(a) - (1/2) \times f(a) = (1/2) \times f(a)$ when $a \le x < a + \delta$.

The result now follows.

Property II. *We may state and prove similar results if f is continuous at the end point b of $[a, b]$ and $f(b) > 0$ or $f(b) < 0$.*

Property III. *If f is continuous at an interior point ξ of $[a, b]$ and $f(\xi) > 0$, there exists an interval $[\alpha, \beta]$ containing ξ, such that*

$$f(x) > 0 \quad \forall\, x \in [\alpha, \beta].$$

Proof. There exists $\delta > 0$ such that

$$| f(x) - f(\xi) | < (1/2) \times f(\xi) \text{ when } \xi - \delta < x < \xi + \delta.$$

Let $\quad \alpha \in]\xi - \delta, \xi],\ \beta \in [\xi, \xi + \delta[.$

It follows that

$$(1/2) \times f(\xi) < f(x) < (3/2) \times f(\xi)\ \forall\, x \in [\alpha, \beta]$$

so that $f(x)$ is positive $\quad \forall\, x \in [\alpha, \beta].$

Property IV. *If f is continuous at an interior point ξ, of $[a, b]$ and $f(\xi) < 0$, then there exists an interval $[\alpha, \beta]$ containing ξ such that*

$$f(x) < 0 \ \forall \ x \in [\alpha, \beta].$$

This is left to be proved by the reader.

8.17. PROPERTIES OF FUNCTIONS CONTINUOUS IN CLOSED FINITE INTERVALS

Let f be a function continuous in a closed interval $[a, b]$. We shall now obtain some properties of such a function.

This study will relate to enquiry into the *nature of the range* of the function and it will be seen that the range of such a function will itself be a closed finite interval.

Boundedness of a function continuous in a closed interval.

Theorem I. *(Boundedness Theorem) If a function f is continuous in a closed bounded interval $[a, b]$, then it is bounded.* **[Delhi B.Sc. (Hons.) II, 2017; Kanpur 2014; Meerut, 1995; Delhi B.A. (Prog.) III 2011, 2015; Osmania, 2004; Patna, 2003; Rajasthan 2010]**

Proof. We define a set S as follows :

$$S = \{x : f \text{ is bounded in } [a, x]\}.$$

Surely the set S is non-empty in that, because of the continuity of f at the end point a, there exists an interval $[a, c]$ in which f is bounded so that $c \in S$.

Also S is bounded above for every member of S is $\leq b$.

We now apply the *order-completeness of* **R** to the set S.

Let ξ be the least upper bound of S. Now either $\xi = b$ or $\xi \neq b$.

Firstly, suppose that $\xi \neq b$ so that $\xi \in [a, b[$. Because of the continuity of f at ξ, there exists an interval $[\alpha, \beta]$ containing ξ in which f is bounded.

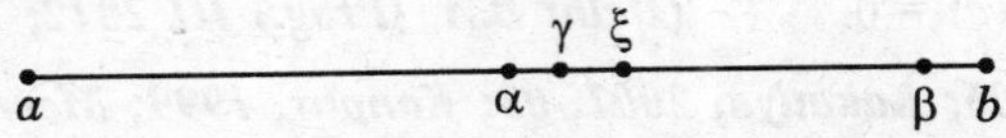

Since ξ is the l.u.b. of S and $\alpha < \xi$, there exists a member γ of S which belongs to $[\alpha, \xi]$.

Now $\gamma \in S \Rightarrow f$ is bounded in $[a, \gamma]$. Also f is bounded in $[\alpha, \beta]$.

It follows that f is bounded in $[a, \beta]$ so that $\beta \in S$.

Thus, there exists a member β of S greater than its l.u.b. ξ and as such we arrive at a contradiction. It follows that $\xi \neq b$ so that $\xi = b$.

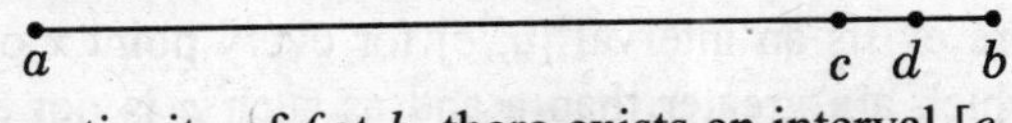

Now because of the continuity of f at b, there exists an interval $[c, b]$ in which f is bounded.

Also because of b being the l.u.b. of S, there exists a member d of S which belongs to $[c, b]$ so that f is bounded in $[a, d]$.

Now f is bounded in $[a, d]$ and $[c, b]$ and as such it follows that f is bounded in $[a, b]$.

Remark. A bounded function on $[a, b]$ need not be continuous on $[a, b]$. For example, $f(x) = 1/x$, $x \neq 0$ and $f(0) = 0$ is bounded on $[0, 1]$ but it is not continuous on $[0, 1]$ since it is discontinuous at $x = 0$.

Theorem II. *If a function f is continuous in a closed interval $[a, b]$, then it possesses greatest and least values.* **[Osmania, 2004; Delhi B.A. (Prog.) III, 2005]**

Proof. We have to show that the range of a continuous function whose domain is a closed finite interval attains its bounds, *i.e.*, the bounds of the range are themselves values of the function.

We have already seen that f is bounded. Let m, M be respectively the greatest lower bound and the least upper bound of f. It will be shown that there exist points α and β of $[a, b]$ such that

$$f(\alpha) = M, \quad f(\beta) = m.$$

Let us consider the case of the least upper bound. Suppose that the function f does not have the value of M for any $x \in [a, b]$ so that

$$M - f(x) \neq 0 \text{ for any } x \in [a, b].$$

By Art. 8.11, page 8.16, we deduce that the function $\quad x \to 1/(M - f(x)) \quad$ ...(*i*)
is continuous and as such bounded in $[a, b]$.

Let k be a given number. Since M is the least upper bound of f, there exists $\xi \in [a, b]$ such that $\quad f(\xi) > M - 1/k.$

This implies that there exists $\xi \in [a, b]$ such that $\quad 1/[M - f(\xi)] > k$

Now k being any given number, we deduce that the function (*i*) is *not* bounded.

We thus arrive at contradictory statements. It is thus proved that the function f attains its least upper bound.

It may similarly be shown that the function f also attains its greatest lower bound.

Note. As a consequence of the two preceding theorems, it follows that if f is continuous in $[a, b]$, there exists an interval $[m, M]$ such that $f(x) \in [m, M]$ for each $x \in [a, b]$ and that m, M are themselves the values of the function. It thus follows that the range of f is a sub-set of $[m, M]$. We shall now show in the following that $[m, M]$ *itself* is the range of, *i.e.*, every point of $[m, M]$ is a value of the function f. Before proceeding to consider this result, we prove a theorem of which this result will be an immediate conclusion.

Theorem III. ***(Location of roots theorem).*** *If a function f is continuous in a closed interval $[a, b]$ and $f(a)$, $f(b)$ have opposite signs, then there exists $c \in [a, b]$ such that $f(c) = 0$.*

or Prove that if a function f is continuous on $[a, b]$ and $f(a) f(b) < 0$, then there exists some point $c \in [a,b]$ much that $f(c) = 0$. ***(Delhi B.A. (Prog.) III 2012; Delhi B.Sc. I (H) 2010 Purvanchal 2006; Kakatiya, 2001, 03; Kanpur, 1999; Meerut, 1998; Patna, 2003]***

Proof. We suppose that $f(a) > 0$ and $f(b) < 0$. Define a set S as follows :

$$S = \{x : f(y) > 0 \; [\; y \in [a, x]\}.$$

The set S is not empty in that $a \in S$. Also S is bounded.

Let ξ be the l.u.b. of S. We shall show that $f(\xi) = 0$.

We firstly see that $\xi \neq a$, $\xi \neq b$.

Now, as $f(a) > 0$, there exists an interval $[a, c]$ for every point x of which $f(x) > 0$. Thus, there are members of S which are greater than a and as such a is not even an upper bound of S. It follows that $\xi \neq a$.

Again as $f(b) < 0$ there exists an interval $[a, b]$ for every point x of which $f(x) < 0$. Thus, no point of S belongs to $[a, b]$ and as such $\xi \neq b$. Thus, $\xi \in \,]a, b[$.

Let $\quad f(\xi) > 0.$

There exists an interval $[\alpha, \beta]$ containing ξ for every point x of which $f(x)$ is positive.

As ξ is the l.u.b. of S, there exists a member γ of S such that $\gamma \in [\alpha, \xi]$.

Now f is positive in $[a, \gamma]$ as also in $[\alpha, \beta]$. It follows that f is positive in $[a, \beta]$ and as such $\beta \in S$. Thus, we conclude that there is a member of S greater than its l.u.b. ξ and as such we arrive at a contradiction.

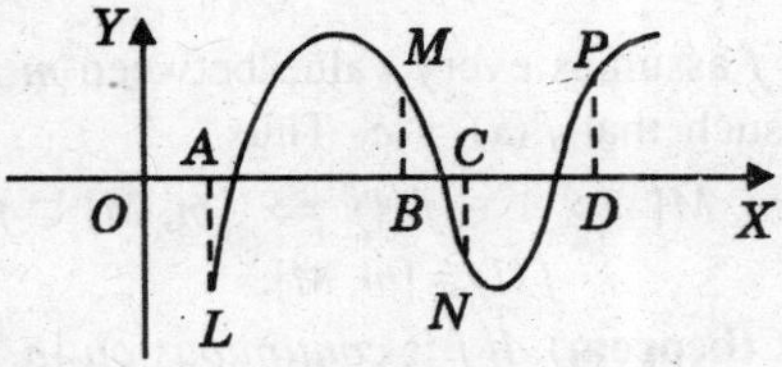

It follows that our assumption, *viz.*, $f(\xi) > 0$ is not true.

We may similarly show that the assumption $f(\xi) < 0$ is also not true. It follows that

$$f(\xi) = 0.$$

Hence the theorem.

Cor. 1. Bolzano's Intermediate Value Theorem. IV. *If a function f is continuous in an interval* $[a, b]$ *and* $f(a) \neq f(b)$, *then f assumes every value between* $f(a)$ *and* $f(b)$.

[*Agra 2010; Bharathiar, 2004; Delhi* B.Sc. I (H) 2010; *Delhi Maths (Prog) 2008*]

Proof. Let κ be a number between $f(a)$ and $f(b)$.

Consider a function ϕ with domain $[a, b]$ such that $\qquad \phi(x) = f(x) - k.$

Now ϕ is continuous in $[a, b]$ and $\phi(a) = f(a) - k$, $\phi(b) = f(b) - k$ are of opposite signs. Thus, there exists $\xi \in [a, b]$ such that $\phi(\xi) = 0$.

Also $\qquad \phi(\xi) = 0 \Rightarrow f(\xi) = k.$

Note. If the function f is not continuous on a closed interval, then the conclusion of the Intermediate value theorem may not hold as shown below.

Consider a function f defined on $[0, 1]$ as follows :

$$f(x) = \begin{cases} x+1, & \text{if } x \in]0, 1] \\ 0, & \text{if } x = 0 \end{cases}$$

$x + 1$ being a polynomial, f is continuous on $]0, 1]$. Now we test $f(x)$ for continuity at $x = 0$.

Here $\qquad f(0+0) = \lim_{h \to 0+} f(0+h) = \lim_{h \to 0+} (h+1) = 1$

and $\qquad f(0-0) = \lim_{h \to 0+} f(0-h) = \lim_{h \to 0+} (-h+1) = 1.$

Thus $f(0+0) = f(0-0) \neq f(0)$ and so f is not continuous at $x = 0$. Hence f is not continuous in the closed interval $[0, 1]$.

Now, $f(0) = 0$ and $f(1) = 2$. Then $f(0) < 1 < f(1)$. But we cannot find any $x \in [0, 1]$ such that $f(x) = 1$.

Cor. 2. *A function which is continuous in a closed interval so that it is bounded and attains its bounds assumes every value between the bounds.* **[*Delhi Maths (G), 1998*]**

Proof. Let f be continuous in $[a, b]$ with m, M as its bounds. Let ξ, η be two points such that $\qquad f(\xi) = m, \quad f(\eta) = M.$

Now f is continuous in $[\xi, \eta]$ or $[\eta, \xi]$ as the case may be and as such, by the cor. 1, it assumes every value between its values at the end points, *i.e.*, between m and M.

It follows that *the range of a continuous function whose domain is a closed interval is as well a closed interval.*

Theorem V. *The image of a closed interval under a continuous function is a closed set.*

Proof. Let f be a continuous function on a closed interval $I = [a, b]$.

Let M = l.u.b. of f on I and m = g.l.b. of f on I.

Then $\qquad m \leq f(x) \leq M \quad \forall\, x \in I$

$$\Rightarrow f(I) \subset [m, M] \qquad ...(1)$$

Since f is continuous on I, f assumes every value between m and M. Hence, if $c \in [m, M]$, then there exists some $x \in I$ such that $f(x) = c$. Thus

$$c \in [m, M] \Rightarrow c \in f(I) \Rightarrow [m, M] \subset f(I). \qquad ...(2)$$

From (1) and (2), we get $f(I) = [m, M]$.

Theorem VI. (Fixed point theorem). *If f is continuous on $[a, b]$ and $f(x) \in [a, b]$ for every $x \in [a, b]$, then f has a fixed point, that is, there exists a point $c \in [a, b]$ such that $f(c) = c$.*

[Delhi B.Sc. (Physics), 1998]

Proof. Let f be continuous on $[a, b]$ and $f(x) \in [a, b]\ \forall\ x \in [a, b]$. If $f(a) = a$ or $f(b) = b$, then the theorem is proved. So let us assume that $f(a) > a$ and $f(b) < b$.

Define $$g(x) = f(x) - x \quad \forall\ x \in [a, b] \qquad ...(1)$$

Since $f(x)$ and x are both continuous on $[a, b]$, so $g(x)$ is also continuous on $[a, b]$. Using our assumption, we have $g(a) > 0$ and $g(b) < 0$. Now 0 is an intermediate value of g on $[a, b]$. Hence by intermediate value theorem, there exists a point $c \in [a, b]$ such that $g(c) = 0$. Then (1) gives $f(c) = c$, as required.

Borel's theorem VII. *If f is a continuous function on the closed interval $[a, b]$, then the interval can always be divided up into a finite number of sub-intervals such that, given $\varepsilon > 0$,*

$$| f(x_1) - f(x_2) | < \varepsilon,$$

where x_1 and x_2 are any two points of the same sub-interval.

(Agra, 2000, 02; Kumaun, 1999)

Alternative statement of Borel's theorem (or Oscillation theorem). *If f is a continuous function on $[a, b]$, then for any $\varepsilon > 0$, the interval can be divided into a finite number of sub-intervals such that in each of which the variation of $f(x)$ is less than ε.*

Proof. We shall prove the theorem by contradiction. Let us assume that the theorem is false.

Divide $[a, b]$ into two equal parts $[a, c]$ and $[c, b]$, where $c = (a + b)/2$. Since the theorem is false on $[a, b]$, it must be false either on at least one of them. Denote this interval by $[a_1, b_1]$. It may be just possible that the theorem is false in both sub-intervals. In that case we take the right hand half interval as $[a_1, b_1]$.

Again divide $[a_1, b_1]$ into two equal parts. The theorem must be false on at least one of these sub-intervals. Let us denote this interval by $[a_2, b_2]$, where

$$\text{length of } [a_2, b_2] = (1/2) \times \text{length of } [a_1, b_1] = (1/2^2) \times (b - a).$$

Continuing this process of repeated bisection indefinitely, we get a sequence of closed intervals $<[a_n, b_n]>$, such that

(*i*) the theorem is false in each of $<[a_n, b_n]>$

(*ii*) $[a, b] \supset [a_1, b_1] \supset [a_2, b_2] \ ..... \supset [a_n, b_n]$

(*iii*) length of interval $[a_n, b_n] = b_n - a_n = (b - a)/2^n \to 0$ as $n \to \infty$

Since the $<[a_n, b_n]>$ satisfies all the conditions of Cantor nested interval theorem, it follows that

$$\bigcap_{n=1}^{\infty} [a_n, b_n] \text{ is a singleton } = \{\alpha\}, \text{ say.}$$

Thus, $$\alpha \in [a_n, b_n] \subset [a_1, b_1], \quad \forall\ n \in \mathbf{N} \qquad ...(1)$$

Let us assume that $\alpha \neq a$ or b. Since f is continuous on $[a, b]$, f is continuous at α also. Hence for a given $\varepsilon > 0$, there exists $\delta > 0$ such that

$$x_1, x_2 \in]\alpha - \delta, \alpha + \delta[\ \Rightarrow\ | f(x_1) - f(x_2) | < \varepsilon \qquad ...(2)$$

Since $(b_n - a_n) \to 0$ as $n \to \infty$, for a given $\delta > 0$, there exists $m \in \mathbf{N}$ such that

$$| (b_n - a_n) - 0 | < \delta \quad \forall\ n \geq m$$

and so in particular, $b_m - a_m < \delta$

Thus, $\alpha \in [a_m, b_m] \Rightarrow [a_m, b_m] \subset]\alpha - \delta, \alpha + \delta[$

$\therefore$ $x_1, x_2 \in [a_m, b_m] \Rightarrow x_1, x_2 \in]\alpha - \delta, \alpha + \delta[$ ...(3)

From (1) and (2), $x_1, x_2, \in [a_m, b_m] \Rightarrow | f(x_1) - f(x_2) | < \varepsilon$

This contradicts the fact that the theorem is false on $[a_m, b_n]$. Hence the theorem must be true.

A slight modification would establish this fact even when $\alpha = a$ or b.

Note 1. Since $\alpha = a$ or b is also possible, hence the interval from a to b must be closed.

Note 2. $| f(x_1) - f(x_2) |$ is called variation of $f(x)$. Hence the alternative form of Borel's theorem is proved.

EXERCISES

1. Show that a continuous function on a closed bounded interval is bounded. Give an example to show that conclusion fails if the interval is not closed.

[Delhi Maths (G), 1996]

2. Let f be continuous on $[a, b]$. Prove that f attains its supremum and infimum on $[a, b]$.

[Delhi Maths (G), 1999, 2000]

3. (*a*) If a function f is continuous in $[a, b]$ and $c \in [a, b]$ such that $f(c) \neq 0$, then there exists some $\delta > 0$ such that $f(x)$ has the same sign as $f(c)$ for all $x \in]c - \delta, c + \delta[$.

[Delhi B.A. (Prog.) III, 2013; Meerut, 1998]

(*b*) Let f be continuous on $[a, b]$ and x_0 be such that $a < x_0 < b$. If $f(x_0) < 0$, then show that $f(x) < 0 \;\forall\; x \in]x_0 - \delta_0, x_0 + \delta_0[$ for some $\delta_0 > 0$.

[Delhi Maths (G), 1993]

4. If a function f is continuous in $[a, b]$ and $f(a)\, f(b) < 0$, then there exists a point $c \in [a, b]$ such that $f(c) = 0$.

Or

If a function f is continuous in $f(a) > 0 > f(b)$, then there exists a point $c \in [a, b]$ such that $f(c) = 0$.

[Delhi Maths (Hons), 1997; Delhi Physics (Hons), 1996; Delhi Maths (G), 1994]

5. Let f be continuous on $[a, b]$ and c be any real number between $f(a)$ and $f(b)$, then show that there exists a real number x in $]a, b[$ such that $f(x) = c$. **[Delhi Maths (H), 2002]**

6. Let f be continuous on $[a, b]$ and $x_1, x_2, ..., x_n$ be points of $[a, b]$. Show that there exists a point $c \in [a, b]$ such that $f(c) = [f(x_1) + f(x_2) + ... + f(x_n)]/n$.

[Delhi Maths (G), 1999]

7. Let $f(x) = \lim\limits_{n \to \infty} \dfrac{\log(2 + x) - x^{2n} \sin x}{1 + x^{2n}}, x \geq 0.$

Show that $f(0)$ and $f(\pi/2)$ differ in sign. Why does $f(x)$ not vanish in $[0, \pi/2]$? Explain.

8. Show that the function $f(x) = \lim\limits_{n \to \infty} \dfrac{x^{2n+2} - \cos x}{x^{2n} + 1}$ does not vanish anywhere in the interval $[0, 2]$, though $f(0)$ and $f(2)$ differ in sign.

9. Show that $f(x) = \dfrac{1}{1 + e^{1/(n!\,\pi x)}}$ can be made discontinuous at any rational point in the interval $(0, 1)$ by the proper selection of n. **(Garhwal, 1994)**

10. Let $f(x) = x\{1 + (1/3) \times \sin \log x^2\}$ for $x \neq 0, f(x) = 0$ for $x = 0$. Show that $f(x)$ is continuous and monotone.

11. If a function $f: \mathbf{R} \to \mathbf{R}$ satisfy the equation $f(x + y) = f(x) + f(y) \quad \forall\ x, y \in \mathbf{R}$. Show that if f is continuous at the point $x = a$, then it is continuous for all $x \in \mathbf{R}$.

12. If $f: \mathbf{R} \to \mathbf{R}$ is a continuous function and satisfies the relation $f(x + y) = f(x)\, f(y)$ [$x, y \in \mathbf{R}$, then show that either $f(x) = 0$ [$x \in \mathbf{R}$ or there exists an $a > 0$ such that $f(x) = a^x$, [$x \in \mathbf{R}$.

13. Show that the function $f: \mathbf{R} \to \mathbf{R}$ defined by

$$f(x) = \lim_{t \to \infty} \frac{(1 + \sin \pi x)^t - 1}{(1 + \sin \pi x)^t + 1}$$

is discontinuous at the points $x = 0, 1, 2, \ldots\ldots, n, \ldots\ldots$. **(Kumaun, 1999)**

14. Consider the function $f(x) = x - [x]$, where x is a positive variable and $[x]$ denotes the greatest integer in x; and show that

(*i*) $f(x)$ is discontinuous for all integral values of x, and continuous for all other values

(*ii*) In every interval which includes an integer, $f(x)$ is bounded between 0 and 1, and

(*iii*) The lower bound 0 is attained but the upper bound is not.

15. Let f and g be continuous functions on an interval I, let $f(x) \neq 0$ for any $x \in I$, and let $[f(x)]^2 = [g(x)]^2$ [$x \in I$. Prove that either $f(x) = g(x)$, [$x \in I$ or $f(x) = -g(x)$ [$x \in I$.

16. Let f be continuous function on $[-1, 1]$ such that $(f(x))^2 + x^2 = 1$ [$x \in [-1, 1]$. Show that either $f(x) = \sqrt{1 - x^2}\ \forall\, x \in [-1, 1]$ or $f(x) = -\sqrt{1 - x^2}\ \forall\, x \in [-1, 1]$.

[Delhi Maths (Hons) 2009]

8.18. EXISTENCE OF THE *n*th ROOT OF A GIVEN POSITIVE REAL NUMBER

Theorem. *Given a positive real number a and any natural number n, there exists one and only one positive real number ξ such that* $\xi^n = a$.

Proof. Let a be a given positive number. We shall show that there exists a unique real number ξ such that $\xi^n = a$.

Consider the function f defined by

$$f(x) = x^n, n \in \mathbf{N}.$$

Suppose that $a < 1$.

The function f is continuous in $[0, 1]$ and

$$f(0) = 0, 1 = f(1).$$

Of course $0 < a < 1$. By the Intermediate Value Theorem, there exists $\xi \in [0, 1]$ such that

$$\xi^n = a.$$

If η is also a positive number with $\eta^n = a$, we have

$$\xi < \eta \Rightarrow \xi^n < \eta^n \Rightarrow a < a \quad \text{and} \quad \xi > \eta \Rightarrow \xi^n > \eta^n \Rightarrow a > a.$$

It follows that $\xi = \eta$.

Now suppose that $a > 1$ so that $1/a < 1$ and as shown above there exists ξ such that

$$\xi^n = 1/a \Rightarrow (1/\xi)^n = a.$$

Note. The unique positive real number satisfying the equation

$$x^n = a;\ a > 0 \text{ and } n \in \mathbf{N}$$

is called the *n*th root of a and is denoted by $a^{1/n}$.

8.19. UNIFORM CONTINUITY

[Delhi B.A. (Prog) III 2014; Meerut 2006; Delhi B.Sc. (Prog) II 2011; G.N.D.U. Amritsar 2010; Nagpur, 2003]

We shall now introduce the notion of uniform continuity of function. It should be seen that while the notion of continuity of a function is essentially *local* in character in that reference to continuity of a function at a point is meaningful, the same is *not* the case in respect of uniform continuity. The notion of uniform continuity of a function is *global* in character inasmuch as we can only talk of uniform continuity of a function only over its domain.

Let f be a function with domain D. Let $c \in D$. We know that f is continuous at c if to each $\varepsilon > 0$, there corresponds $\delta > 0$ such that

$$|f(x) - f(c)| < \varepsilon \quad \forall\ x \in D \text{ such that } |x - c| < \delta.$$

Suppose that f is continuous at each of D and ε is any given positive number. Then there exists a δ corresponding to each point of the domain D. The question now arises, "*Does there exist* δ *which holds uniformly for every point of the domain* ?" It may appear to a beginner that the greatest lower bound of the set consisting of all the values of δ would serve the purpose. The fallacy, however, lies in the fact that greatest lower bound may be zero. In general, therefore, a choice of positive δ which may hold good uniformly for each point of domain D may not be possible. We now define the notion of uniform continuity.

Definition. A function f defined on an interval I is said to be uniformly continuous on I, if to each $\varepsilon > 0$, there corresponds $\delta > 0$ such that $|f(x) - f(y)| < \varepsilon$ where x, y are arbitrary points of I for which $|x - y| < \delta$.

Note 1. In case of continuity of a function at a point a, the choice of $\delta > 0$ depends upon $\varepsilon > 0$ and the point a whereas in the case of uniform continuity of a function in an interval, the choice of $\delta > 0$ depends only on $\varepsilon > 0$ and not on a pair of points of the given interval I.

Note 2. A function f is not uniformly continuous on I, if there exists some $\varepsilon > 0$ for which no $\delta > 0$ serves, *i.e.*, for any $\delta > 0$, there exist $x, y \in I$ such that

$$|f(x) - f(y)| \geq \varepsilon \text{ when } |x - y| < \delta.$$

Note 3. From definition, it follows that if f is uniformly continuous on an interval I, then it is also uniformly continuous on any subinterval of I.

Note 4. Uniformly continuous function on a subset set A of $\mathbf{R}$ is defined by replacing interval I by set A in the above definition.

Theorem I. *Every uniformly continuous function on an interval is continuous on that interval, but the converse is not true.* **(Purvanchal 2006; Delhi B.Sc. (Hons.) II, 2017; Delhi B.Sc. (Hons) I 2011; Delhi B.A. (Prog) III 2010, 2012, 2014; Agra 2007, 2011]**

Proof. Let a function f be uniformly continuous on an interval I. Then given $\varepsilon > 0$, there exists some of $\delta > 0$ such that

$$|f(x) - f(y)| < \varepsilon \text{ whenever } |x - y| < \delta, \text{ where } x, y \in I \quad \text{...(1)}$$

Let c be any point of I. Taking $y = c$ in (1), we have

$|f(x) - f(c)| < \varepsilon$ whenever $|x - c| < \delta$ $\quad \Rightarrow f$ is continuous at the point c.

Since c is any point of I, it follows that f is continuous at every point of I. Hence f is continuous on I.

Converse of the above theorem need not be true, i.e., every continuous function need not be uniformly continuous. **[Delhi B.Sc. (Prog) II 2010, 16; Agra 2007; Kanpur 2012]**

Consider the function f defined on $]0, 1]$ as follows

$$f(x) = 1/x, \quad \forall\ x \subset]0, 1] \quad \text{...(1)}$$

Let c be any point of $]0, 1]$. Then

$$\lim_{x \to c} f(x) = \lim_{x \to c} \frac{1}{x} = \frac{1}{c} = f(c) \quad \Rightarrow f(x) \text{ is continuous at } x = c.$$

Since c is any point of $]0, 1]$, it follows that f is continuous on $]0, 1]$.

Now we shall prove that f is not uniformly continuous in $]0, 1]$. For any $\delta > 0$, we can find a positive integer n such that $1/n < \delta$.

Let $x = 1/n$ and $y = 1/2n$. Then $x, y \in\ ]0, 1]$. Also, we have

$$| x - y | = \left| \frac{1}{n} - \frac{1}{2n} \right| = \left| \frac{1}{2n} \right| = \frac{1}{2n} < \frac{1}{n} < \delta$$

and $$| f(x) - f(y) | = \left| \frac{1}{x} - \frac{1}{y} \right| = | n - 2n | = | -n | = n > \frac{1}{2}$$

[$\because$ n is a positive integer, so $n \geq 1$]

Hence if we take $\varepsilon = 1/2 > 0$, then whatever $\delta > 0$ there exists $x, y \in\ ]0, 1]$ such that

$$| f(x) - f(y) | > \varepsilon \text{ whenever } | x - y | < \delta$$

Hence $f(x) = 1/x$ is not uniformly continuous in $]0, 1]$.

Thus, $f(x) = 1/x$ is continuous in $]0, 1]$ but not uniformly continuous.

[*Calicut, 2004*; *Kanpur, 2001, 2016*]

Theorem II Uniform continuity theorem. *If a function f is continuous on a closed interval $[a, b]$, then it is uniformly continuous on $[a, b]$.* **[*Kanpur 2009; Delhi B.A. (Prog.) III 2007; Delhi B.Sc. (Hons.) II, 2017*; *Srivenkateshwara, 2003; Delhi B.Sc. (Prog) II 2011*]**

Proof. Let, if possible, f be not uniformly continuous on $[a, b]$. Then there exists an $\varepsilon > 0$ such that whatever $\delta > 0$ we take, we can find $x, y \in [a, b]$, for which

$$| f(x) - f(y) | \geq \varepsilon \text{ when } | x - y | < \delta$$

In particular, for each positive integer n, we can find real numbers x_n, y_n in $[a, b]$ such that

$$| f(x_n) - f(y_n) | \geq \varepsilon \text{ when } | x_n - y_n | < 1/n \quad \text{...(1)}$$

Since $a \leq x_n \leq b$ and $a \leq y_n \leq b\ \forall\ n \in \mathbf{N}$, the sequences $<x_n>$ and $<y_n>$ are bounded and so by Bolzano-Weierstrass theorem, each has at least one limit point, say α and β respectively. Hence α and β are limit points of $[a, b]$. Since a closed interval is a closed set and a closed set contains all its limiting points, so $\alpha, \beta \in [a, b]$.

Again since α is a limit point of $<x_n>$, there exists a convergent subsequence $< x_{n_k} >$ of $<x_n>$ such that

$$x_{n_k} \to \alpha \text{ as } k \to \infty \quad \text{...(2)}$$

Similarly since β is a limit point of $<y_n>$, there exists convergent sequence $< y_{n_k} >$ of $<y_n>$ such that

$$y_{n_k} \to \beta \text{ as } k \to \infty \quad \text{...(3)}$$

Now, from (1), for all k, we have

$$| f(x_{n_k}) - f(y_{n_k}) | \geq \varepsilon \text{ when} \left| x_{n_k} - y_{n_k} \right| < 1/n_k \quad \text{...(4)}$$

From the second inequality of the inequalities (4), we have

$$\lim_{k \to \infty} x_{n_k} = \lim_{k \to \infty} y_{n_k} \text{ so that } \alpha = \beta \text{ by (2) and (3)}$$

But from the first of the inequalities (4), we see that in case the sequences $< f(x_{n_k}) >$ and $< f(y_{n_k}) >$ converge, the limits to which they converge and different.

Thus we have shown that there exist two sequences $< x_{n_k} >$ and $< y_{n_k} >$ both of which converge to α but $< f(x_{n_k}) >$ and $< f(y_{n_k}) >$ do not converge to the same limit. So by theorem I,

Art. 8.15, f is not continuous at α, for otherwise the two sequences $< f(x_{n_k}) >$ and $< f(y_{n_k}) >$ would converge to the same point $f(\alpha)$.

Thus we arrive at a contradiction and hence the hypothesis that f is not uniformly continuous on $[a, b]$ is false.

Hence f must be uniformly continuous on $[a, b]$.

Theorem III. *A function defined on a closed interval is continuous if and only if it is uniformly continuous therein.* **[Delhi Maths (Hons), 2000]**

Refer theorems I and II for complete proof.

SOLVED EXAMPLES

Example 1. *Is the function $f(x) = x/(x+1)$ uniformly continuous for $x \in [0, 2]$? Justify your answer.* **[Delhi B.A. (Prog.) III, 2013; Kanpur 2014; Delhi Maths (H), 1995]**

Solution. Let x, y be two arbitrary points in [0, 2]. Then $x \geq 0$, $y \geq 0$.

$$\Rightarrow x + 1 \geq 1 \quad \text{and} \quad y + 1 \geq 1$$

$$\Rightarrow (x + 1)(y + 1) \geq 1 \qquad \text{...(1)}$$

Now, $$| f(x) - f(y) | = \left| \frac{x}{x+1} - \frac{y}{y+1} \right| = \frac{| x - y |}{(x+1)(y+1)} \leq | x - y |, \text{ by (1)}$$

Let $\varepsilon > 0$ be given. Taking $\delta = \varepsilon$, we get

$$| f(x) - f(y) | < \varepsilon, \text{ whenever } | x - y | < \delta \quad \forall \; x, y \in [0, 2]$$

Hence, f is uniformly continuous in [0, 2].

Alternative solution. Here $f(x)$ is a rational function and so it is continuous for every real number other than zeros of the denominator of polynomial function $x + 1$. But $x + 1 = 0 \Rightarrow x = -1$, which does not belong to [0, 2]. Hence $f(x)$ is continuous on the closed interval [0, 2].

Since every function which is continuous in a closed interval is also uniformly continuous is that interval, it follows that $f(x) = x/(x + 1)$ is uniformly continuous in [0, 2].

Example 2. *Let $f(x) = x^2$, $x \in$* **R**. *Show that f is uniformly continuous on every closed and finite interval but is not uniformly continuous on* **R**.

[G.N.D.U., 1997; Delhi Maths (H) 2004, 07, 09; Delhi Maths (G), 1999, 2005; Agra, 2002]

Solution. Let $[a, b]$ be any closed and finite interval.

Let $x_1, x_2 \in [a, b]$. We have

$$| f(x_2) - f(x_1) | = | x_2^2 - x_1^2 | = | x_2 - x_1 | \, | x_2 + x_1 |$$

$$\therefore \quad | f(x_2) - f(x_1) | \leq | x_2 - x_1 | \{| x_1 | + | x_2 |\}$$

Let $\quad k = \max \{| x_1 |, | x_2 |\}$. Then $k > 0$

$$\therefore \quad | f(x_2) - f(x_1) | \leq 2k \, | x_2 - x_1 |$$

Let $\varepsilon > 0$ be given and let $\delta = \varepsilon/2k > 0$.

Then $\quad | f(x_2) - f(x_1) | < \varepsilon$, whenever $\quad | x_2 - x_1 | < \delta \quad \forall \; x_1, x_2 \in [a, b]$.

Hence f is uniformly continuous on $[a, b]$.

Now we shall show that f is not uniformly continuous on **R**.

The given function will be uniformly continuous on **R** if for a given $\varepsilon > 0$, there exist $\delta > 0$ uch that for any $x_1, x_2 \in$ **R** we have

$$| f(x_2) - f(x_1) | < \varepsilon \text{ whenever } | x_2 - x_1 | < \delta \qquad \text{...(1)}$$

We shall show that for the given function $f(x) = x^2$, the above condition (1) is not satisfied.

By the axiom of Archimedes, for any $\delta > 0$, there exists a positive integer n such that

$$n\delta^2 > \varepsilon \qquad ...(2)$$

Let us take $\qquad x_1 = n\delta \quad$ and $\quad x_2 = n\delta + \delta/2$. Then

$$| x_2 - x_1 | = \delta/2 < \delta$$

whereas $\quad | f(x_2) - f(x_1) | = | x_2^{\,2} - x_1^{\,2} | = | x_2 - x_1 | \, | x_2 + x_1 |$

$$= \frac{1}{2}\delta\left(2n\delta + \frac{1}{2}\delta\right) = n\delta^2 + \frac{1}{4}\delta^2 > \varepsilon, \text{ by (2)}$$

Thus there exists $x_1\ x_2 \in \mathbf{R}$ such that

$$| f(x_2) - f(x_1) | > \varepsilon \text{ whenever } | x_2 - x_1 | < \delta,$$

which contradicts (1). Hence f is not uniformly continuous on **R**.

Example 3. *Show that the function $f(x) = 1/x^2$ is uniformly continuous on $[a, \infty[$, where $a > 0$, but not uniformly continuous on $]0, \infty[$.* **[Delhi Maths (H), 2006]**

Solution. *To show that $f(x) = 1/x^2$ is uniformly continuous on $[a, \infty[$ where $a > 0$.*

For $x, y \geq a > 0$, we obtain

$$| f(x) - f(y) | = \left| \frac{1}{x^2} - \frac{1}{y^2} \right| = \left| \frac{1}{x} - \frac{1}{y} \right| \left| \frac{1}{x} + \frac{1}{y} \right|$$

or $\quad | f(x) - f(y) | = \left| \dfrac{y - x}{xy} \right| \left(\dfrac{1}{x} + \dfrac{1}{y} \right) \leq \dfrac{2 | x - y |}{a | xy |} \qquad (\because\ x, y \geq a > 0)$

$$\leq \frac{2}{a^3} | x - y | \qquad \left(\because x \geq a, y \geq a \Rightarrow xy \geq a^2 \Rightarrow \frac{1}{xy} \leq \frac{1}{a^2} \right)$$

Let $\varepsilon > 0$ be given. Let $\delta = \varepsilon a^3/2$. Then

$$| f(x) - f(y) | < \varepsilon, \text{ when } | x - y | < \delta \quad \forall\ x, y \geq a.$$

Hence, f is uniformly continuous on $[a, \infty[$.

Now we show that f is not uniformly continuous in $]0, \infty[$.

Let $\varepsilon = 1/2$ and δ be any positive number. We can always choose a positive integer n such that

$$n > 1/2\delta \quad \text{or} \quad 1/2n < \delta \qquad ...(1)$$

Let $x_1 = 1/\sqrt{n}$ and $x_2 = 1/\sqrt{n+1} \in]0, \infty[$.

Then $\quad | f(x_1) - f(x_2) | = \left| \dfrac{1}{x_1^{\,2}} - \dfrac{1}{x_2^{\,2}} \right| = | n - (n+1) | = 1 > \varepsilon,$

and $\quad | x_1 - x_2 | = \left| \dfrac{1}{\sqrt{n}} - \dfrac{1}{\sqrt{n+1}} \right| = \dfrac{| \sqrt{n+1} - \sqrt{n} |}{\sqrt{n}\,\sqrt{n+1}}$

$$= \frac{1}{\sqrt{n}\,\sqrt{n+1}\,(\sqrt{n+1} + \sqrt{n})}, \text{ on rationalising.}$$

$$< \frac{1}{\sqrt{n} \cdot 2\sqrt{n}} \qquad [\because\ \sqrt{n}\,\sqrt{n+1} > \sqrt{n} \text{ and } \sqrt{n+1} + \sqrt{n} > 2\sqrt{n}]$$

$$= 1/2n < \delta, \text{ by (1)}$$

Thus, $\qquad | f(x_1) - f(x_2) | > \varepsilon, \text{ when } | x_1 - x_2 | < \delta.$

Hence, f is not uniformly continuous on $]0, \infty[$.

Example 4. *Show that sin x is uniformly continuous on* $[0, \infty[$.

[*Agra 2008; Delhi B.A. (Prog.) III, 2017; Kanpur 2019*]

Solution. Let $\varepsilon > 0$ be given and x, y be any two points in $[0, \infty[$.

Let $f(x) = \sin x$. Then

$$|f(x) - f(y)| = |\sin x - \sin y|$$

$$= \left| 2\sin\frac{x-y}{2}\cdot\cos\frac{x+y}{2} \right| = 2\left|\sin\frac{x-y}{2}\right|\left|\cos\frac{x+y}{2}\right|$$

$$\le 2\left|\frac{x-y}{2}\right|\cdot 1 \qquad [\because |\sin\theta| \le |\theta|, \ |\cos\theta| \le 1 \ \forall\ \theta]$$

$\therefore$ $$|f(x) - f(y)| \le |x-y|$$

$\Rightarrow$ $|f(x) - f(y)| < \varepsilon$, when $|x-y| < \delta$, $\forall\ x, y \in [0, \infty[$, taking $\delta = \varepsilon$.

Hence, $f(x) = \sin x$ is uniformly continuous on $[0, \infty[$.

Example 5. *Prove that* $f(x) = \sin x^2$ *is not uniformly continuous on* $[0, \infty[$.

Solution. Let $\varepsilon = 1/2$ and δ be any positive number. We can choose a positive integer n such that

$$n > \pi/\delta^2 \qquad ...(1)$$

Let $$x_1 = \sqrt{\frac{n\pi}{2}},\ x_2 = \sqrt{(n+1)\frac{\pi}{2}} \in [0, \infty[.$$

Then $$|f(x_2) - f(x_1)| = |\sin x_2^2 - \sin x_1^2| = \left|\sin(n+1)\frac{\pi}{2} - \sin\frac{n\pi}{2}\right|$$

$$= \begin{cases} |0 - (\pm 1)| = 1, \text{ if } n \text{ is odd,} \\ |(\pm 1) - 0| = 1, \text{ if } n \text{ is even.} \end{cases}$$

$\therefore$ $$|f(x_2) - f(x_1)| = 1 > \varepsilon,$$

and $$|x_2 - x_1| = \left|\frac{x_2^2 - x_1^2}{x_2 + x_1}\right| = \frac{\pi/2}{\sqrt{(n+1)\frac{\pi}{2}} + \sqrt{\frac{n\pi}{2}}} < \frac{\pi}{2\cdot 2\sqrt{\frac{n\pi}{2}}} < \frac{\pi}{\sqrt{n\pi}}$$

Thus, $$|x_2 - x_1| < \sqrt{\pi/n} < \delta, \text{ by (1)}$$

$\therefore$ $$|f(x_2) - f(x_1)| > \varepsilon, \text{ when } |x_2 - x_1| < \delta.$$

Hence, $f(x) = \sin x^2$ is not uniformly continuous on $[0, \infty[$.

EXERCISES

1. (*a*) Show that $f(x) = x$ is uniformly continuous in $[-2, 2]$ **[*Kanpur 2017, 19*]**

(*b*) Show that $f(x) = x^2$ is uniformly continuous in $[0, 1]$. **[*Lucknow 2010; Kanpur 2018*]**

2. Show that the function f defined by $f(x) = x^3$ is uniformly continuous in $[0, 3]$.

3. Show that $f(x) = 1/x$ is not uniformly continuous on $]0, 1]$, but it is uniformly continuous on $[a, \infty[$, where $a > 0$. **[Delhi B.A. (Prog.) (I) 2012; Delhi B.Sc. (Prog) I 2010]**

4. (*a*) Show that $f(x) = \sqrt{x}$ is uniformly continuous on $[1, 3]$. **[*Delhi B.A. (Prog.) III 2010*]**

(*b*) Show that $f(x) = \sqrt{x}$ is uniformly continuous on $[0, 1]$. **[*Kanpur 2010, 2013*]**

5. Show that $f(x) = x$ is uniformly continuous on **R**. **[*Delhi Maths (G), 2002*]**

6. Give an example of a continuous function which is not uniformly continuous.

7. Show that $\sin(1/x)$ is not uniformly continuous on $]0, \infty[$.

8. Show that $f(x) = \cos x$ is uniformly continuous on **R**. **[Delhi Maths (G), 2002]**

9. Show that the function f defined by

$$f(x) = \begin{cases} \sin(1/x), & x > 0 \\ 0, & x = 0 \end{cases}$$

is continuous but not uniformly continuous on $[0, \infty[$. **(Meerut, 1998)**

10. State a set of conditions under which a continuous function is uniformly continuous. **(Delhi Maths (G), 2003)**

11. (*a*) Show that $f(x) = x^2$ is uniformly continuous on $[0, a]$, $a \in \mathbf{R}^+$ but $g(x) = 1/x$ is not uniformly continuous on $A = \{x \in \mathbf{R} : x > 0\}$. **[Delhi B.Sc. I (Hons) 2010]**

(*b*) Show that the function defined by $f(x) = x^2, -1 \le x \le 1$ is uniformly continuous **[Delhi B.Sc. (Prog) II 2011]**

12. Show that the function of f defined by $f(x) = 2x^2 - 3x + 5$ is uniformly continuous on $[-2, 2]$.

13. Show that the function $f(x) = x/(1 + x^2)$ is uniformly continuous on **R**.

14. Determine whether the function $y = \tan x$ is uniformly continuous in the open interval $]0, \pi/2[$. **(I.A.S., 1999)**

15. If f and g are uniformly continuous on the same interval, prove that $f + g$ and $f - g$ are also uniformly continuous on that interval.

16. Justify with an example that the product of two uniformly continuous functions may not be uniformly continuous.

17. Prove that if f is uniformly continuous on a bounded interval I, then f is bounded on I.

18. Prove that if f and g are each uniformly continuous on the bounded open interval $]a, b[$, then the product fg is uniformly continuous on $]a, b[$.

19. Show that $f(x) = \tan^{-1} x$ is uniformly continuous on **R**.

20. Show that $f(x) = (\sin x)/x$ is uniformly continuous on **R**.

21. If f is uniformly continuous in an interval I and $<x_n>$ is a Cauchy sequence of elements of I, then prove that $<f(x_n)>$ is also a Cauchy sequence.

8.20. EVALUATION OF $\lim_{x \to 0} (\sin x)/x$, x BEING MEASURED IN RADIANS

[Agra 2007; Kanpur 2009]

Proof Let us first prove that $\displaystyle\lim_{x \to 0+} \frac{\sin x}{x} = 1.$

Let $0 < x < \pi/2$. Draw a circle of unit radius and centre O. Consider two points A and P oin the circle such that $\angle AOP = x$ radians. Let AQ be perpendicular to OA and let OP produced meat AQ in Q.

From right angled ΔOAQ, $\quad AQ / OA = \tan x$

Hence, $\quad AQ = OA \tan x = \tan x$, as $OA = 1$

Now, area of $\Delta OAP = (1/2)\times OA \times OP \times \sin x = (1/2) \times \sin x$, as $OP = 1$

area of $\Delta OAQ = (1/2)\times OA \times AQ = (1/2)\times \tan x$, as $AQ = \tan x$

and area of sector $OAP = (1/2)\times OA^2 \times x = (1/2)\times x$

[$\because$ Area of a sector = (1/2) × (radius)^2 × (angle of sector)]

From the above figure, it is easy to note that

Area of ΔOAP < Area of sector OAP < area of ΔOAQ

Thus, $(1/2)\times \sin x < x/2 < (1/2)\times \tan x$ or $\sin x < x < (\sin x)/\cos x$

Dividing by $\sin x$, $1 < x/(\sin x) < 1/\cos x$, as $0 < x < \pi/2 \Rightarrow \sin x > 0$

Taking reciprocals of the above inequalities, we get

$1 > (\sin x)/x > \cos x$ so that $\cos x < (\sin x)/x < 1$ ...(1)

Now, $\lim_{x\to 0+} \cos x = 1 = \lim_{x\to 0+} 1$. Hence, using squeeze principle (refer theorem 6, page 8.6), we have

$$\lim_{x\to 0+} (\sin x)/x = 1. \quad \text{... (2)}$$

Next, $\displaystyle\lim_{x\to 0-} \frac{\sin x}{x} = \lim_{y\to 0+} \frac{\sin(-y)}{(-y)}$, putting $x = -y$, where $y > 0$ so that $y \to 0+$ when $x \to 0-$

$= \lim_{y\to 0+} (\sin y)/y = 1$, using (2)

Since $\lim_{x\to 0+} (\sin x)/x = 1 = \lim_{x\to 0-} (\sin x)/x$, we have $\lim_{x\to 0} (\sin x)/x = 1$

Deductions, I, $\displaystyle\lim_{x\to 0} \frac{x}{\sin x} = \lim_{x\to 0} \frac{1}{(\sin x)/x} = \frac{1}{\lim_{x\to 0}(\sin x)/x} = 1$

II. $\displaystyle\lim_{x\to 0} \frac{\tan x}{x} = \lim_{x\to 0} \frac{\sin x}{x} \times \frac{1}{\cos x} = 1$, **III.** $\displaystyle\lim_{x\to 0} \frac{x}{\tan x} = \lim_{x\to 0} \frac{x}{\sin x} \times \cos x = 1$

8.21. CONTINUITY OF THE INVERSES OF CONTINUOUS FUNCTIONS

Theorem I. *A continuous and strictly monotonic function is invertible and the inverse function is also continuous.*

We suppose that f is continuous and strictly monotonically increasing in $[a, b]$.

Let m, M be the g.l.b. and l.u.b. respectively of the function f.

Since f is continuous and strictly increasing in $[a, b]$, we have $m = f(a)$, $M = f(b)$.

Consider the interval $[m, M]$. Because of the continuity of f in a closed interval every number between $[m, M]$ is a value of f. Also because of the strictly increasing character of f, every value between m and M is attained only once.

We have $y = f(x) \Leftrightarrow x = g(y).$

Now we show that g is continuous in $[m, M]$.

Let $\eta \in [m, M]$ and let $g(\eta) = \xi$.

We have $g(\eta) = \xi \Rightarrow f(\xi) = \eta.$

Let $\varepsilon > 0$ be given. We have to show that there exists $\delta > 0$ such that

$$|g(y) - g(\eta)| < \varepsilon \text{ when } |y - \eta| < \delta \quad \Leftrightarrow \quad |x - \xi| < \varepsilon \text{ when } |y - \eta| < \delta.$$

Now f being a strictly monotonic continuous function

$$\xi - \varepsilon < x < \xi + \varepsilon \Leftrightarrow \varepsilon f(\xi - \varepsilon) < f(x) < f(\xi + \varepsilon).$$

Let $f(\xi - \varepsilon) = \eta - \delta_1, f(\xi + \varepsilon) = \eta + \delta_2$; δ_1, δ_2 being both positive.

Thus, there exist two positive numbers δ_1 and δ_2 such that

$$\xi - \varepsilon < x < \xi + \varepsilon \Leftrightarrow \eta - \delta_1 < y < \eta + \delta_2.$$

Thus $\eta - \delta_1 < y < \eta + \delta_2 \Rightarrow \xi - \varepsilon < x < \xi + \varepsilon.$

Let $\delta = \min(\delta_1, \delta_2)$. It follows that

$$\eta - \delta < y < \eta + \delta \Rightarrow \xi - \varepsilon < x < \xi + \varepsilon,$$

so that we see that there exists $\delta > 0$ such that

$$|y - \eta| < \delta \Rightarrow |x - \xi| < \varepsilon.$$

Hence the result.

Theorem II. (Inverse function theorem) *If f be a continuous one-to-one function on the closed interval $[a, b]$, then f^{-1} is also continuous.* ***(Delhi B.Sc. (Hons.) II 2012)***

Proof. Left as an exercise to the reader.

8.21. ROOT FUNCTION

$$\boldsymbol{x \to x^{1/n};\ n \in \mathbf{N},\ x \geq 0.}$$

We shall introduce the root function $x \to x^{1/n}$ as the *inverse* of the function $x \to x^n$.

First we show that the function $x \to x^n$; $n \in \mathbf{N}$, $x \geq 0$ is invertible.

Now the function $x \to x^n$ is continuous $\forall\ x \geq 0$. Also it is strictly monotonically increasing in that if $x_1 \geq 0$ and $x_2 > 0$, then

$$x_1 > x_2 \Rightarrow x_1^n > x_2^n.$$

Thus, the function $x \to x^n$, $n \in \mathbf{N}$ is continuous and strictly monotonically increasing in every interval $[0, b]$; b being a positive number.

Thus, by the preceding theorem of Art 8.20, the function $x \to x^n$ is invertible and its inverse is continuous. The inverse of this function is denoted by $x \to x^{1/n}$.

Let f, g denote $x \to x^n, x \to x^{1/n}$ respectively so that $f(x) = x^n$, $g(x) = x^{1/n}$.

We have $(g \circ f)(x) = g(f(x)) = g(x^n) = x.$

MISCELLANEOUS EXERCISES

1. Let f be continuous in $[-1, 1]$ and assume only rational values and let $f(0) = 0$. Show that
$$f(x) = 0 \quad \forall\ x \in [-1, 1].$$

2. Discuss the continuity of f where

$$f(x) = \begin{cases} 0 & \text{when } x \text{ is 0 or irrational,} \\ 1/q^3 & \text{when } x \text{ is the rational number } p/q \text{ in its lowest terms.} \end{cases}$$

3. Given $f(x) = (x^2 - 4)/(x - 2)$; $x \neq 2$, define a function g which is continuous for every $x \in \mathbf{R}$ and which coincides with f for every point which belongs to the domain of each f and g.

4. Given that f is continuous in $[a, b]$ and $f(x) = 0 \ \forall \ x \in \mathbf{Q}$.

Show that $$f(x) = 0 \quad \forall \ x \in [a, b].$$

5. Let f be continuous in **R**. Show that the set $A = \{x : f(x) = 0\}$ is closed.

6. Consider $$y = x + x^5;\ x \geq 0.$$

Show that given any number $y \geq 0$, there exists one and only one $x \geq 0$, such that $y = x + x^5$.

7. State with reasons for your conclusion which of the following circumstances are sufficient and which of them are not to determine the value $f(0)$ of f, giving the value, if possible.

(*i*) f is continuous at $x = 0$ and takes in any neighbourhood of $x = 0$ both positive and negative values.

(*ii*) To each $\varepsilon > 0$ there corresponds $\delta > 0$ such that $|f(x)| < \varepsilon$ for $0 < |x| < \delta$.

(*iii*) $[f(h) + f(-h) - 2f(0)]/h \to l$ and $f(h) \to a$ when $h \to 0$.

8. Show that the sum function of the infinite series
$$x^2 + x^2(1 - x^2) + x^2(1 - x^2)^2 + \ldots.. + x^2(1 - x^2)^n + \ldots..$$
is not continuous at $x = 0$ even though each term of the series is continuous thereat.

9. Examine the continuity at $x = 0$ of the sum functions of the following series :

(*i*) $$\frac{x}{x+1} + \frac{x}{(x+1)(2x+1)} + \frac{x}{(2x+1)(3x+1)} + \ldots..$$

(*ii*) $$\frac{x^2}{(1+x^2)(1+2x^2)} + \frac{x^2}{(1+2x^2)(1+3x^2)} + \frac{x^2}{(1+3x^2)(1+4x^2)} + \ldots..$$

10. Show that the following functions defined in $[0, 1]$ are invertible :

(*i*) $x \to \dfrac{x}{x+1}$, (*ii*) $x \to \dfrac{x}{x+1} + \dfrac{x}{x+2}$

Also describe the inverse functions.

11. Show that $f(x) = \dfrac{2x}{2x-1}$ is uniformly continuous on $[1, \infty[$.

[**Hint.** $|f(x) - f(y)| = \dfrac{2|x-y|}{(2x-1)(2y-1)} \quad \forall \ x, y \in [1, \infty[$

Since $x > 1$ and $y > 1$, $2x - 1 > 1$, $2y - 1 > 1$ and so
$$|f(x) - f(y)| < 2|x - y|.$$
For any $\varepsilon > 0$, choose $\delta = \varepsilon/2$.]

OBJECTIVE QUESTIONS

Multiple Choice Type Questions : *Select (a), (b), (c) or (d), whichever is correct.*

1. If $\lim_{x \to a} f(x) = l$ and $\lim_{x \to a} g(x) = m$, then $\lim_{x \to a} \dfrac{f(x)}{g(x)} = \dfrac{l}{m}$, if :

(*a*) $m = 0$ (*b*) $m \neq 0$ (*c*) $m = l$ (*d*) None of these. **(Kanpur, 2001)**

2. If $f(x) = (\sin x)/x$, then $f(0 - 0)$ is :

(*a*) 0 (*b*) 1 (*c*) -1 (*d*) None of these. **(Kanpur, 2001)**

3. The function $f(x) = 2^{1/x}$ is not continuous at :

(*a*) 0 (*b*) 1 (*c*) -1 (*d*) None of these. **(Kanpur, 2001)**

4. The function $f(x) = (x^2 - a^2)/(x - a)$ at $x = a$ is :
(a) Continuous (b) Not continuous
(c) Undecided (d) None of these. **(Kanpur, 2001)**

5. The function $f(x) = \sin(1/x)$ at $x = 0$ has a :
(a) discontinuity of first kind (b) discontinuity of second kind
(c) mixed continuity (d) removable discontinuity. **(Kanpur, 2001)**

6. The points at which $f(x) = [x]$ is discontinuous are (Here $[x]$ is the greatest integer less than or equal to x) :
(a) set of all rational numbers (b) set of all irrational numbers
(c) set of all integral points (d) set of all prime numbers. **(Kanpur, 2003, 03)**

7. Let $f(x) = \begin{cases} x, & \text{when } 0 \le x < 1/2 \\ 1, & \text{when } x = 1/2 \\ 1 - x, & \text{when } 1/2 < x < 1 \end{cases}$, then f is :
(a) continuous at $x = 1/2$ (b) not defined at $x = 1/2$
(c) discontinuous at $x = 1/2$ (d) continuous for all x, $0 \le x < 1$.

8. For function $f(x) = \sin(1/x)$, $x \ne 0$ at origin :
(a) left limit is -1 (b) right limit is $+1$
(c) limit is zero (d) None of these. **(Kanpur, 2003)**

9. *function* $f(x) = |x|/x$, $x \ne 0$ may be continuous at origin, if :
(a) $f(0) = 0$ (b) $f(0) = -1$ (c) $f(0) = 0$
(d) Cannot be continuous for any value of $f(0)$. **(Kanpur, 2003)**

10. If $f(x) = \dfrac{\sin[x]}{[x]}$, $[x] \ne 0$, $f(x) = 0$, $[x] = 0$, where $[x]$ denotes the greatest integer less than or equal to x, then $\lim_{x \to 0} f(x)$ is :
(a) 1 (b) 0 (c) -1 (d) None of these.
[Kanpur 2010]

11. If $f(x) = x^2$ for all $x \in \mathbf{R}$, f is :
(a) not continuous on **R** (b) uniformly continuous on **R**
(c) not uniformly continuous on **R** (d) None of these. **(Bharathiar, 2004)**

ANSWERS

1. (b) **2.** (b) **3.** (a) **4.** (b) **5.** (b) **6.** (c) **7.** (c) **8.** (d) **9.** (d)
10. (d) **11.** (c)

MISCELLANEOUS PROBLEMS ON CHAPTER 8

1. Describe the continuity of the following fuction aat $x = a$ $f(x) = \begin{cases} (x^3/a) - a^2, & \text{if } 0 < x < a \\ 0, & \text{if } x = a \\ a - (a^2/x), & \text{if } x > a \end{cases}$ **[Kanpur 2006]**

2. Show that the function $f(x) = x^2$ is uniformly continuous on the interval $[-1, 1]$
[Delhi B.Sc. (Prog) II 2011; Kanpur 2005]

3. Let $f : \mathbf{R} \to \mathbf{R}$ be continuous in $\mathbf{R}$ and suppose that $f(r) = 0$ for every rational number. Show that $f(x) = 0$ for all $x \in \mathbf{R}$. **[Delhi B.Sc. (Hons.) II 2018]**

4. Discuss the continuity of function f defined by $f(x)=\begin{cases} x^2, \text{for } x<-2 \\ 4, \text{for} -2\le x\le 2 \\ x^2, \text{for } x>2 \end{cases}$ **[Meerut 2006, 2014]**

5. The value of k for which makes $f(x)=\begin{cases} \sin(1/x), x\neq 0 \\ k, \ x=0 \end{cases}$ is continuous at $x = 0$:

(*a*) 8 (*b*) 1 (*c*) –1 (*d*) none of these **[Agra 2006]**

6. If the function $f(x)$ is continuous at $x = 0$ then which is true (*a*) $\lim_{x\to c} f(x)=f(c)$ (*b*) $\lim_{x\to c} f(x)\neq f(c)$ (*c*) $\lim_{x\to c+} f(x)=f(c)$ (*d*) $\lim_{x\to c-} f(x)=f(c)$ **[Agra 2005]**

7. (*a*) Show that equation $f(x)=xe^x-2=0$ has a root in [0, 1]

[Delhi B.Sc. (Hons) II 2018]

(*b*) Show that the equation $x=\cos x$ has a solution in $[0, \pi/2]$

[Delhi B.Sc. I(H) 2010]

8. Let f and g be continuous from R to R and $f(r) = g(r)$. $\forall\, r\in Q$. Is it true that $f(x)=g(x), \forall\, x\in R.$. Justify your answer. **[Delhi B.Sc. (Hons) II 2018]**

9. Define uniform continuity of a function over an interval. Show that the function $f(x)=\sin(1/x)$ is uniformly continuous in $[k, 1]$ for $0 < k < 1$ but not uniformly continuous on [0, 1]. **[Delhi Maths (G) 2006]**

10. Examine for continuity at $x = 0$, the function f defined as :

$$f(x)=\begin{cases} (e^{1/x}-e^{-1/x})/(e^{1/x}+e^{-1/x}), \text{ when } x\neq 0 \\ 0, \quad \text{when} \quad x=0 \end{cases}$$ **[Delhi Maths (G) 2006]**

11. Examine the continuity of the function : $f(x)=\sin(1/x), \quad x\neq 0; \quad f(0)=1$ at $x=0$. If it is discontinuous, then state the nature of discontinuity. **[Delhi Maths (H) 2006]**

12. Prove that a polynomial of odd degree has at least one real root.

[Delhi B.Sc. (Hons.) II, 2017]

13. *Discuss the continuity at $x = 3$ of the function f defined by $f(x) = x - [x]$, $\forall\, x \geq 0$, where $[x]$ is the greatest integer < x.* **[Delhi Maths (H) 2007]**

14. *If f is defined and continuous on closed interval $[a, b]$, prove that there exists a real number M such that $f(x)\le M, \ \forall x\in[a,b]$* **[Delhi Maths (Prog) 2007]**

15. *If a function f is continuous on $[0, 2]$ and assumes only rational values and $f(1) = 1$, then prove that $f(x) = 1$ for all $x\in[0,2]$.* **[Delhi Maths (H) 2007]**

16. *Prove that a function defined and continuous on a closed interval* $[a, b]$ *is bounded below and attains the infimum on* $[a, b]$. **[Delhi B.A. (Prog) III 2009; Delhi Maths (H) 2007]**

17. State intermediate theorem and show that if the continuity condition of the theorem is dropped, then its conclusion may fail to hold. **[Delhi B.Sc. II (Prog) 2007, 10]**

18. *Prove that function* $f(x) = (\sin x)/x, x \neq 0; f(0) = 0$ *is continuous at* $x = 0$.

[Kanpur 2007]

19. *If* $f : \mathbf{R} \to \mathbf{R}$ *is continuous and* $f(x+y) = f(x) + f(y)$ *for all* $x, y \in \mathbf{R}$, *then show that* $f(x) = xf(1)$ *for all* $x \in \mathbf{R}$ **[Kanpur 2010; I.A.S. 2008]**

20. Examine the curve $y = x^3 - 6x^2 + 11x - 6$ for concavity (convexity) and points of inflexion.

[Delhi Maths (Prog) 2008]

21. Consider the function f given by : $f(x) = x + \sin x, 0 \leq x \leq \pi/6$. Verify all the hypothesis of intermediate value theorem and use it to show that f has at least one zero in the interval $[0, \pi/6]$. Is the function f bounded ? Justify. **[Delhi Maths (Prog) 2008]**

22. Prove that the function f defined by $f(x) = 1/x, 0 < x \leq 4$ is not uniformly continuous on $]0, 4]$. Is the function uniformly continuous on [1, 4]. Justify. **[Delhi Maths (Prog) 2008]**

23. Verify the continuity of the following function at $x = 0$: $f(x) = e^x \operatorname{sgn}(x + [x]), x \neq 0, f(0) = 0$ where sgn denotes the signum function and $[x]$ is the greatest integer $\leq x$

[Delhi Maths (Prog) 2008]

[Hint : Use the following definition of signum function :

$\operatorname{sgn}(x) = 1$, if $x > 0$; $\operatorname{sgn}(x) = 0$, if $x = 0$; $\operatorname{sgn}(x) = -1$, if $x < 0$**]**

24. Prove that a function continuous on a closed interval $[a. b]$ attains its bounds on the interval **[Delhi Maths (Prog) 2008]**

25. Define uniform continuity of a function f on an interval I and prove that $f(x) = \sqrt{x}$ is uniformly continuous on [1, 3] **[Delhi Maths (Prog) 2008]**

26. State intermediate theorem. Prove that a continuous function $f(x)$ on $[a, b]$ assumes its supremum and infimum on $[a, b]$. **[Delhi Maths (Prof) 2008]**

27. Define a uniformly continuous function. Show that $f(x) = x^2, 0 \leq x < \infty$ is not uniformly continuous. **[Delhi Maths (Prog) 2008]**

28. Prove that function defined by $f(x) = x \log \sin^2 x, x \neq 0, f(x) = 0$ for $x = 0$ is continuous at $x = 0$. **[Agra 2004]**

29. Show that $f(x) = 2x$ is continuous at each point of **R**. **[Agra 2007]**

30. Prove that the image at a Cauchy sequence under a uniformly continuous mapping is itself a Cauchy sequence. **[Agra 2006, 08]**

31. Show that $f(x) = (1/x) \times \sin^{-1} x$ for $x \neq 0, f(0) = 1$ is continuous at the origin. **[Delhi 2008]**

32. Show that the function f defined by $f(x) = e^{1/x}, x = 0;\ f(x) = 0$ when $x = 0$ is discontinous at $x = 0$ and the discontinuity is of second kind. **[K.U. BCA (II) 2008]**

33. Show that the function f defined by $f(x) = x^2 + 3x + 2$ is uniformly continuous on [1,2]. **[G.N.D.U. Amritsar 2010]**

34. Is the function $f(x) = x^3$ uniformly continuous in [–2, 2]. **[Agra 2008]**

35. Show that if $f : [a, b] \to \mathbf{R}$ is continuous function then $f([a, b]) = [c, d]$ for some real numbers c and d, where $c < d$. **[I.A.S. 2009]**

Hint Use theorem V, page 8.39

36. Show that if a function in continuous on a closed and bounded interval, then it is bounded above and attain its supremum in that interval **[Delhi 2009]**

37. Let f and g be two functions defined on some nbd of a real number c such that $\lim_{x\to c} f(x) = l$ and $\lim_{x\to c} g(x) = m$. Show that $\lim_{x\to c}(fg)(x)$ exists and is equal to lm. **[Delhi 2008]**

38. State Cauchy's definition of continuity **[Kanpur 2008]**

39. Show that the function $f(x) = 1/(1 - e^{1/x})$ has a discontinuity of the first kind at $x = 0$. **[KANPUR 2008]**

40. Discuss the continuity and discontinuity of the function :
$f(x) = \{(xe^{1/x}) / (1 + e^{1/x})\} + \sin(1/x)$ when $x \neq 0$ and $f(x) = 0$ when $x = 0$ **[Kanpur 2008]**

41. Fill up the gap : If $f(a + 0)$ and $f(a - 0)$ both exist and $f(a + 0) \neq f(a - 0)$, then f is said to have discontinuity of .... **[Agra 2010]**

42. The function $f(x) = |x|/x$ is not defined at

(*a*) $x = \infty$ (*b*) $x = 1$ (*c*) $x = -1$ (*d*) $x = 0$ **[Agra 2009]**

43. Prove that $\lim_{x\to\infty} (1 + 1/x)^x = e$ **[G.N.D.U. Amritsar 2010]**

44. Let the function f be defined as
$$f(x) = \begin{cases} 2x + 3, & \text{if } x \le 4 \\ 7 + (16/x), & \text{if } x > 4 \end{cases}$$
Find the value of x (if any) at which f is not continuous. **[Delhi B.Sc. (Prog) I 2011]**

45. Find a value of k that will make the function f continuous everywhere, given that
$$f(x) = \begin{cases} kx^2, & \text{if } x \le 2 \\ 2x + k, & \text{if } x > 2 \end{cases}$$
[Delhi B.Sc. (Prog) I 2011]

46. If f is non-decreasing on $(a, ,b)$ and $c \in (a, b)$, prove that $\lim_{x\to c^+} f(x)$ and $\lim_{x\to c^-} f(x)$ both exist. **[Channai 2011]**

47. If f is non-decreasing function on the bounded interval $[a, b]$ and f is bounded above on $[a, b]$, prove that $\lim_{x\to b^-} f(x)$ exists. If f is bounded below on $[a, b]$, prove that $\lim_{x\to a^+} f(x)$ exists. **[Chennai 2011]**

48. Examine the nature of disticontinuity of the function $f(x) = [x] - [-x]$ at $x = 0$. *(Kanpur 2010)*

[**Sol**. Here $[x]$ denotes the greatest integer function. Hence, by definition

$$f(0) = [0] - [0] = 0, \; f(0+h) = [h] - [-h] \text{ and } f(0-h) = [-h] - [h].$$

Therefore, $\lim_{h \to 0} f(0+h) = 0 - (-1) = 1$ and $\lim_{h \to 0} f(0-h) = -1 - 0 = -1$.

Thus, $\lim_{h \to 0} f(0+h) \neq \lim_{h \to 0} f(0-h)$. Therefore, f is discontinuous at x = 0 and the discontinuity is of the first kind]

49. If f : $A \to \mathbf{R}$ is uniformly continuous on a subset A of $\boldsymbol{R}$, show that f is continuous on A. Show by an example that continuous function on A may not be unifomly continous. **[Delhi B.Sc. (Hons) I 2011]**

50. Use the definition of limit to show that $\lim_{x \to 2}(x^2 + 4x) = 12$ **[Delhi B.Sc. (Hons) 2011]**

51. Use (ε, δ) definition to prove that

(*a*) $\lim_{x \to 2}(3x - 5) = 1$ (b) $\lim_{x \to \infty} \{x/(x+1)\} = 1$ **[Delhi B.Sc. (Prog) II 2011]**

52. Determine condition on $|x - 4|$ that will assume that $|\sqrt{x} - 2| < 1/2$.

[Delhi B.Sc. (Hons.) I 2011]

ANSWERS

1. Continuous at $x = a$ **4.** Continuous on [–2, 2] **5.** (*d*) **6.** (*a*) **7.** (*a*)

10. Discontinuous at $x = 0$ **11.** Discontinuous at $x = 0$ and the discontinuity is of second kind. **37.** 1. **40.** Continuous at $x = 0$ **41.** First kind **42.** (d)

44. f is continuous $\forall x \in \boldsymbol{R}$ **45.** 4/3 **52.** $|x - 4| <$ |

Chapter

9

Real Functions. The Derivative

9.1. DERIVABILITY OF A FUNCTION

Let f be a real function with an interval $[a, b]$ as its domain.

Derivability at an interior point. Let c be a interior point of $[a, b]$ so that $a < c < b$.

Consider $$\frac{f(x)-f(c)}{x-c},\ x \neq c$$

The function f is said to be derivable at c, if

$$\lim_{x \to c} \frac{f(x)-f(c)}{x-c}$$

exists and the limit, called the derivative of f at c, is denoted by the symbol

$$f'(c) \quad \text{or} \quad \text{by } Df(c).$$

Right-hand derivative.

$$\lim_{x \to (c+0)} \frac{f(x)-f(c)}{x-c} \quad \text{or} \quad \lim_{h \to 0+} \frac{f(c+h)-f(c)}{h}$$

if it exists, is called the right-hand derivative of f at c and is denoted by

$$f'(c+0) \quad \text{or} \quad f'(c+) \quad \text{or} \quad Rf'(c)$$

Left-hand derivative.

$$\lim_{x \to (c-0)} \frac{f(x)-f(c)}{x-c} \quad \text{or} \quad \lim_{h \to 0+} \frac{f(c-h)-f(c)}{-h}$$

if it exists, is called the left-hand derivative of f at c and is denoted by

$$f'(c-0) \quad \text{or} \quad f'(c-) \quad \text{or} \quad Lf'(c)$$

It may be seen that

$f'(c)$ *exists* $\Leftrightarrow$ $f'(c+0)$ *and* $f'(c-0)$ *both exist and are equal.*

Note. In practice, we also use the following method for finding $Rf'(c)$ and $Lf'(c)$. Let $h > 0$. Then

$$Rf'(c) = \lim_{h \to 0} \frac{f(c+h)-f(c)}{h} \quad \text{and} \quad Lf'(c) = \lim_{h \to 0} \frac{f(c-h)-f(c)}{-h}$$

Derivatives at the end points of [*a*, *b*]. A function f is said to be derivable at the left end point a if $f'(a+0)$ exists and by $f'(a)$ would then be meant $f'(a+0)$. Similarly, f is said to be derivable at the right end point b if $f'(b-0)$ exists and by $f'(b)$ would then be meant $f'(b-0)$.

Derivability in the closed interval [*a*, *b*]. **[Meerut 2010]**

The function f is said to be derivable in the interval $[a, b]$, if it is derivable at every point thereof.

Derivatives. Let f be a function whose domain is an interval I. Let I_1 be the set of all those points x of I at which $f'(x)$ exists. If $I_1 \neq \phi$, we obtain another function $f'(x)$ with domain I_1 and $f'(x)$ is known as the derivative of f. Let $a \in I_1$ and let the derivative of f exist at a. Then the derivative of f' at a is called the second derivative of f at $x = a$ and is denoted by $f''(a)$ or $D^2 f(a)$ or $(f')'(a)$. Let I_2 be the set of all those points at which $f''(a)$ exists. If $I_2 \neq \phi$, we obtain another function $f''(x)$ with domain I_2. Then $f''(x)$ is known as the second derivative of f.

Proceeding likewise the nth derivative $f^n(a)$ of f at a point a and the function f^n may be defined.

9.2. A NECESSARY CONDITION FOR THE EXISTENCE OF A FINITE DERIVATIVE

Theorem. *Continuity is a necessary but not a sufficient condition for the existence of a finite derivative.* **[Agra, 1998, 99; Bharathiar, 2004; Delhi B.Sc. (Prog) II 2011; Garhwal, 2001; Kanpur, 2001, 04, 06, 07, 08, 11, 12; Meerut, 2006, 10; Kumaon, 1992; Patna, 2003 Delhi B.Sc. I (H) 2010; Purvanchal, 1994; Rohilkhand, 1993**

Proof. Let f be derivable at the point c so that

$$\lim_{x \to c} \frac{f(x) - f(c)}{x - c}$$

exists. We have

$$f(x) - f(c) = \left[\frac{f(x) - f(c)}{x - c}\right](x - c);\ x \neq c$$

$$\Rightarrow \quad \lim_{x \to c} [f(x) - f(c)] = \lim_{x \to c} \frac{f(x) - f(c)}{x - c} \lim_{x \to c} (x - c)$$

$$\Rightarrow \quad \lim_{x \to c} [f(x) - f(c)] = f'(c) \cdot 0 = 0$$

$$\Rightarrow \quad \lim_{x \to c} f(x) = f(c)$$

$\Rightarrow$ f is continuous at c.

Thus continuity is a necessary condition for differentiability but it is not a sufficient condition for the existence of a finite derivative as shown in the following example. Thus, the converse may not be true. **[Agra 2010; Kanpur 2005]**

Let f be defined by $f(x) = |x|$

We shall show that while f is continuous at 0, it is not derivable at 0. We have

$$\frac{f(x) - f(0)}{x - 0} = \frac{|x|}{x} = \begin{cases} 1 \ \forall\ x > 0, \\ -1 \ \forall\ x < 0. \end{cases}$$

Thus, $$\lim_{x \to (0+0)} \frac{f(x) - f(0)}{x - 0} = 1 \neq -1 = \lim_{x \to (0-0)} \frac{f(x) - f(0)}{x - 0}$$

$\Rightarrow$ $\lim_{x \to 0} \dfrac{f(x) - f(0)}{x - 0}$ does not exist.

$\Rightarrow$ f is *not* derivable at 0.

To examine the continuity at 0, take any $\varepsilon > 0$.

We have $|f(x) - f(0)| = |x| < \varepsilon$ when $|x| \leq \delta$

δ being any positive number less than ε. Thus f is continuous at 0.

Thus $f(x)$ is not differentiable at $x = 0$, though it is continuous there.

Note. While f, defined by $f(x) = |x|$, is not derivable for $x = 0$, it is derivable for every non-zero x. In fact, it may be easily seen that

$$f'(x) = \begin{cases} 1 \ \forall \ x > 0, \\ -1 \ \forall \ x < 0. \end{cases}$$

Thus, there is only one point where this function ceases to be derivable while remaining continuous.

9.3. ALGEBRA OF DERIVATIVES

Theorem I. *If and g be two functions which are defined on* $[a, b]$ *and derivable at any point* $c \in [a, b]$, *then their sum* $f + g$ *is also derivable at* $x = c$ *and* $(f+g)'(c) = f'(c) + g'(c)$.

Proof. Since f and g are derivable at c, therefore

$$\lim_{x \to c} \frac{f(x) - f(c)}{x - c} = f'(c) \text{ and } \lim_{x \to c} \frac{g(x) - g(c)}{x - c} = g'(c) \qquad ...(1)$$

Consider
$$\lim_{x \to c} \frac{(f+g)(x) - (f+g)(c)}{x - c} = \lim_{x \to c} \frac{[f(x) + g(x)] - [f(c) + g(c)]}{x - c}$$

$$= \lim_{x \to c} \frac{f(x) - f(c)}{x - c} + \lim_{x \to c} \frac{g(x) - g(c)}{x - c}$$

$$= f'(c) + g'(c), \text{ using (1)}$$

Hence, $f + g$ is derivable at c and $(f+g)'(c) = f'(c) + g'(c)$.

Theorem II. *If f and g are two derivable functions at* $x = c$, *then fg is also derivable at* $x = c$ *and* $(fg)'(c) = f(c)\, g'(c) + g(c)\, f'(c)$.

[*Kanpur, 2004, 07, 09; Delhi Maths (H), 2004, 06, 08*]

Proof. Since f and g are derivable at $x = c$, therefore

$$f'(c) = \lim_{x \to c} \frac{f(x) - f(c)}{x - c} \text{ and } g'(c) = \lim_{x \to c} \frac{g(x) - g(c)}{x - c} \qquad ...(1)$$

Since f and g are derivable at $x = c$, so f and g are continuous at $x = c$. Thus,

$$\lim_{x \to c} f(x) = f(c), \quad \lim_{x \to c} g(x) = g(c) \qquad ...(2)$$

Now,
$$\lim_{x \to c} \frac{(fg)(x) - (fg)(c)}{x - c} = \lim_{x \to c} \frac{f(x)\, g(x) - f(c)\, g(c)}{x - c}$$

$$= \lim_{x \to c} \frac{f(x)[g(x) - g(c)] + g(c)[(f(x) - f(c)]}{x - c}$$

$$= \lim_{x \to c} f(x) \cdot \lim_{x \to c} \frac{g(x) - g(c)}{x - c} + g(c) \cdot \lim_{x \to c} \frac{f(x) - f(c)}{x - c}$$

$$= f(c)\, g(c) + g(c)\, f'(c), \text{ using (1) and (2)}$$

Hence, fg is derivable at c and $(fg)'(c) = f(c)\, g'(c) + g(c)\, f'(c)$.

Theorem III. (*i*) *If a function f is derivable at c and if* $f(c) \neq 0$, then $1/f$ *is also derivable at c and*

$$\left(\frac{1}{f}\right)(c) = \frac{-f'(c)}{[f(c)]^2}$$

[*Delhi Maths (H), 1994, 2000*]

(*ii*) *If f and g be two functions on* $[a, b]$ *and derivable at any point* $c \in [a, b]$ *and if* $g(c) \neq 0$, *then f/g is also derivable at c and*

$$\left(\frac{f}{g}\right)(c) = \frac{g(c)\, f'(c) - f(c)\, g'(c)}{[g(c)]^2}$$

Proof. (*i*) Since f is derivable at c, so f is continuous at c.

Thus, $$\lim_{x \to c} f(x) = f(c)$$

Also, $$f'(c) = \lim_{x \to c} \frac{f(x) - f(c)}{x - c}$$

Now, $$\frac{\dfrac{1}{f(x)} - \dfrac{1}{f(c)}}{x - c} = -\frac{f(x) - f(c)}{x - c} \cdot \frac{1}{f(x)} \cdot \frac{1}{f(c)}$$

$\therefore$ $$\lim_{x \to c} \frac{\dfrac{1}{f(x)} - \dfrac{1}{f(c)}}{x - c} = -\lim_{x \to c} \frac{f(x) f(c)}{x - c} \cdot \lim_{x \to c} \cdot \frac{1}{f(x)} \cdot \frac{1}{f(c)}$$

$$= -f'(c) \cdot \frac{1}{f(c)} \cdot \frac{1}{f(c)} = \frac{-f'(c)}{[f(c)]^2}$$

Hence, $1/f$ is derivable at c and $$\left(\frac{1}{f}\right)'(c) = -\frac{f'(c)}{[f(c)]^2} \quad \text{...(1)}$$

(*ii*) Since g is derivable at c and $g(c) \neq 0$. So, $1/g$ is derivable at c.

Now, $$\left(\frac{f}{g}\right)(c) = f'(c) \cdot \frac{1}{g(c)} + f(c) \cdot \left(\frac{1}{g}\right)'(c)$$

$$= f'(c) \cdot \frac{1}{g(c)} - f(c) \cdot \frac{g'(c)}{[g(c)]^2}, \text{ using (1)}$$

Hence, $$\left(\frac{f}{g}\right)'(c) = \frac{g(c) f'(c) - f(c) g'(c)}{[g(c)]^2}.$$

Theorem IV. *If a function is differentiate at a point c and k is any real number, then the function* kf *is also differentiable at c and* $(kf)'(c) = kf'(c)$.

Proof. Left as an exercise.

Theorem V. (Chain rule). *Let f and g be two functions such that (i) range of* $f \subset$ *domain of g (ii) f is derivable at c and (iii) g is derivable at* $f(c)$, *then the composite function gof is derivable at c and* $(gof)'(c) = g'(f(c)) \cdot f'(c)$. ***(Agra, 1999; Bharathiar, 2004)***

Proof. Let $y = f(x)$ and $\alpha = f(c)$. ...(1)

Since f is differentiable at c, we have

$$\lim_{x \to c} \frac{f(x) - f(c)}{x - c} = f'(c)$$

$$\Rightarrow f(x) - f(c) = (x - c)[f'(c) + \varepsilon(x)], \quad \text{...(2)}$$

where $\varepsilon(x)$ depends on x and is such that

$$\varepsilon(x) \to 0 \text{ as } x \to c \quad \text{...(3)}$$

Again, since g is differentiable at $f(c)$, *i.e.*, α, we have

$$\lim_{y \to \alpha} \frac{g(y) - g(\alpha)}{y - \alpha} = g'(\alpha)$$

$$\Rightarrow g(y) - g(\alpha) = (y - \alpha)[g'(\alpha) + \varepsilon'(y)], \quad \text{...(4)}$$

where $\varepsilon'(y)$ depends on y and is such that

$$\varepsilon'(y) \to 0 \text{ as } y \to c \quad \text{...(5)}$$

Now, $(gof)(x) - (gof)(c)$

$$= g(f(x)) - g(f(c)) = g(y) - g(\alpha)$$
$$= (y - \alpha)[g'(\alpha) + \varepsilon'(y)], \text{ using (4)}$$
$$= [f(x) - f(c)][g'(\alpha) + \varepsilon'(y)], \text{ using (1)}$$
$$= (x - c)[f'(c) + \varepsilon(x)][g'(\alpha) + \varepsilon'(y)], \text{ using (2)}$$

Hence if $x \neq c$, we obtain

$$\frac{(gof)(x) - (gof)(c)}{x - c} = [f'(c) + \varepsilon(x)][g'(\alpha) + \varepsilon'(y)] \quad \text{...(6)}$$

Now, $\quad f$ is derivable at $c \Rightarrow f$ is continuous at c

Hence $\quad x \to c \Rightarrow f(x) \to f(c) \Rightarrow y \to \alpha$

So from (3) and (5), we obtain

$$\varepsilon(x) \to 0 \quad \text{and} \quad \varepsilon'(y) \to 0 \quad \text{as} \quad x \to c \quad \text{...(7)}$$

Taking the limits on both sides of (6) as $x \to c$ and using (7), we obtain

$$\lim_{x \to c} \frac{(gof)(x) - (gof)(c)}{x - c} = f'(c)\, g'(\alpha)$$

or
$$(gof)'(c) = g'(f(c))\, f'(c), \text{ as } \alpha = f(c).$$

Theorem VI. Inverse function theorem for derivatives

Let f be continuous one-to-one function defined on an interval and f be derivable at c with $f'(c) \neq 0$. Then the inverse of the function f is derivable at f(c) and its derivative at f(c) is $1/f'(c)$. **[Delhi Maths (H), 1994, 2000]**

Proof. Let the domain and range of f be X and Y respectively. Let $g\ (= f^{-1})$ be the inverse of f. Hence the domain and range of g will be Y and X respectively such that

$$f(x) = y \Leftrightarrow g(y) = x \quad \text{...(1)}$$

Let $f(c) = \alpha$. Then $g(\alpha) = c$. Let $(\alpha + k) \in Y$ such that $(\alpha + k) \neq \alpha$. But f is one-to-one, so there exists a point $(c + h) \in X$ such that $(c + h) \neq c$ and $f(c + h) = \alpha + k$. Again, by definition of g, we have $g(\alpha + k) = c + h$.

$$\therefore \quad f(c) = \alpha \quad \text{and} \quad f(c + h) = \alpha + k \quad \text{...(2)}$$
$$g(\alpha) = c \quad \text{and} \quad g(\alpha + k) = c + h \quad \text{...(3)}$$

Also, $$k \neq 0 \Rightarrow h \neq 0 \quad \text{...(4)}$$

Now, f is differentiable at $c \Rightarrow f$ is continuous at c. So g is continuous at $f(c)$ *i.e.* α and hence we obtain

$$\lim_{k \to 0} [g(\alpha + k) - g(\alpha)] = 0 \quad \text{or} \quad \lim_{k \to 0} [(c + h) - c] = 0, \text{ by (3)}$$

$$\Rightarrow \lim_{k \to 0} h = 0 \Rightarrow h \to 0 \text{ as } k \to 0 \quad \text{...(5)}$$

Assuming that $k \neq 0$, we obtain

$$\frac{g(\alpha + k) - g(\alpha)}{k} = \frac{(c + h) - c}{k}, \text{ by (3)}$$

$$= \frac{h}{(\alpha + k) - \alpha} = \frac{h}{f(c + h) - f(c)}, \text{ by (2)}$$

Since $h \neq 0$ by –, the above equation can be rewritten as

$$\lim_{k \to 0} \frac{g(\alpha + k) - g(\alpha)}{k} = \lim_{k \to 0} \frac{1}{\{f(c + h) - f(c)\}/h}$$

or $$g'(\alpha) = \lim_{h \to 0} \frac{1}{\{f(c+h) - f(c)\}/h}, \text{ using (5)}$$

or $$g'(f(c)) = 1/f'(c)$$

which is meaningful as $f'(c) \neq 0$ by hypothesis.

9.4. GEOMETRICAL MEANING OF THE DERIVATIVE

Let $f: [a, b] \to \mathbf{R}$ be derivable at $c \in\,]a, b[$. Let RQ be the graphical representation of the function $y = f(x)$. We take two neighbouring points $P[c, f(c)]$ and $Q[c + h, f(c + h)]$ on the curve such that $a < c + h < b$. Join PQ and produce it to meet the x-axis at L. Let QL make an angle θ with x-axis. Draw the tangent PT at P which makes an angle ψ with x-axis. Draw PM and QN perpendiculars from P and Q on the x-axis. Also draw PH perpendicular from P on QN. Then, we have

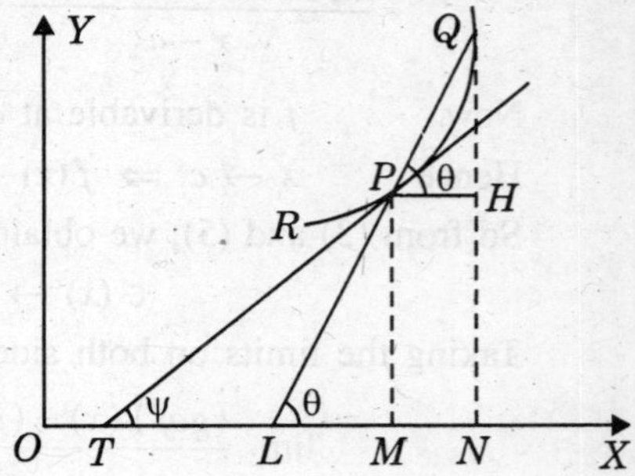

$$QH = QN - HN = QN - PM = f(c + h) - f(c)$$

and $$PH = MN = ON - OM = (c + h) - c = h$$

$$\therefore \quad \text{Slope of } PQ = \tan\theta = \frac{QH}{PH} = \frac{f(c+h) - f(c)}{h} \qquad ...(1)$$

From the figure, we see that as $h \to 0$, the point Q approaches the point P along the graph of f and, hence, the chord PQ approaches the tangent line PT as its limiting position. Thus, as $h \to 0$, $\theta \to \psi$.

$$\therefore \quad \tan\psi = \lim_{Q \to P} \tan\theta = \lim_{h \to 0} \frac{f(c+h) - f(c)}{h} = f'(c),$$

showing that $f'(c)$ is the slope of the tangent to the curve $y = f(x)$ at the point $(c, f(c))$.

EXAMPLES

Example 1. *Show that the function* $f(x) = \begin{cases} x \sin(1/x), & \text{if } x \neq 0 \\ 0, & \text{if } x = 0 \end{cases}$

is continuous at $x = 0$ *but not differentiable at* $x = 0$.

Solution. Test for continuity at $x = 0$. Let $h > 0$

$$\text{Left-hand limit} = \lim_{h \to 0} f(0 - h) = \lim_{h \to 0} (0 - h)\sin\left(\frac{1}{0 - h}\right) = \lim_{h \to 0} h \sin\frac{1}{h}$$

$$= 0 \times k, \text{ where } -1 \leq k \leq 1$$

$$= 0$$

$$\text{Right-hand limit} = \lim_{h \to 0} f(0 + h) = \lim_{h \to 0} (0 + h)\sin\left(\frac{1}{0 + h}\right) = \lim_{h \to 0} h \sin\frac{1}{h}$$

$$= 0, \text{ as before}$$

Also $f(0) = 0$. Thus, we have

$$\text{Left-hand limit} = \text{Right-hand limit} = f(0)$$

$$\Rightarrow \quad f(x) \text{ is continuous at } x = 0$$

Test for differentiability at $x = 0$

$$\text{Left-hand derivative} = \lim_{h \to 0} \frac{f(0 - h) - f(0)}{-h} = \lim_{h \to 0} \frac{h \sin(1/h)}{-h} = -\lim_{h \to 0} \sin(1/h),$$

which does not exist because as $h \to 0$, sin $(1/h)$ oscillates between -1 and 1 and does not tend to a unique and definite limit.

$$\text{Right-hand derivative} = \lim_{h \to 0} \frac{f(0+h) - f(0)}{h} = \lim_{h \to 0} \frac{h \sin(1/h)}{h},$$

which does not exist as before.

Since neither the left-hand derivative nor the right-hand derivative exists at $x = 0$, it follows that $f(x)$ is not differentiable at $x = 0$.

Example 2. *(a) Let* $f(x) = \begin{cases} x^p \sin(1/x), & x \neq 0 \\ 0, & x = 0 \end{cases}$

Obtain p such that (i) f(x) is continuous at x = 0 (ii) f(x) is differentiable at x = 0.

(I.A.S., 2002)

(b) Let $f(x) = \begin{cases} x^p \cos(1/x), & x \neq 0 \\ 0, & x = 0 \end{cases}$

Obtain p such that (i) f(x) is continuous at x = 0 (ii) f(x) is differentiable at x = 0.

Solution. *(a) (i)* Here $f(0) = 0$. Let $h > 0$. Also, we have

$$f(0+0) = \lim_{h \to 0} f(0+h) = \lim_{h \to 0} h^p \sin(1/h) \qquad \ldots(1)$$

and $$f(0-0) = \lim_{h \to 0} f(0-h) = \lim_{h \to 0} (-h)^p \sin(-1/h) = \lim_{h \to 0} (-1)^{p+1} h^p \sin(1/h) \qquad \ldots(2)$$

In order that f may be continuous at $x = 0$, we must have

$$f(0+0) = f(0-0) = f(0) \quad i.e. \quad f(0+0) = 0 \quad \text{and} \quad f(0-0) = 0,$$

i.e. the limits given by (1) and (2) both must tend to zero. This is possible only if we choose $p > 0$.

$$(ii)\ Rf'(0) = \lim_{h \to 0} \frac{f(0+h) - f(0)}{h} = \lim_{h \to 0} \frac{h^p \sin(1/h)}{h} = \lim_{h \to 0} h^{p-1} \sin(1/h) \qquad \ldots(3)$$

and $$Lf'(0) = \lim_{h \to 0} \frac{f(0-h) - f(0)}{-h} = \lim_{h \to 0} \frac{(-1)^{p+1} h^p \sin(1/h)}{-h}$$

$$= \lim_{h \to 0} (-1)^p h^{p-1} \sin(1/h) \qquad \ldots(4)$$

Now in order that f may be differentiable at $x = 0$, the limits (3) and (4) must be equal. This can happen only if $p > 1$, for in that case $Rf'(0) = 0 = Lf'(0)$.

(b) Try like Ex. 2 *(a)*. **Ans.** *(i)* $p > 0$ *(ii)* $p > 1$.

Example 3. *If* $f(x) = x^2 \sin(1/x)$ *when* $x \neq 0$ *and* $f(0) = 0$, *show that f is derivable for every value of x but the derivative is not continuous for x = 0.*

(Delhi B.Sc. (Prog) II 2011; Agra, 1998; Kanpur 2014; Patna, 2003)

Solution. Test for differentiability at $x = 0$. Let $h > 0$. We have

$$Rf'(0) = \lim_{h \to 0} \frac{f(0+h) - f(0)}{h} = \lim_{h \to 0} \frac{h^2 \sin(1/h)}{h} = \lim_{h \to 0} h \sin(1/h)$$

$$= 0 \times k = 0, \text{ where } -1 \leq k \leq 1$$

$$Lf'(0) = \lim_{h \to 0} \frac{f(0-h) - f(0)}{-h} = \lim_{h \to 0} \frac{(-h)^2 \sin(-1/h)}{-h} = \lim_{h \to 0} h \sin(1/h)$$

$$= 0, \text{ as before.}$$

Since $Rf'(0) = Lf'(0)$, so $f(x)$ is derivable at $x = 0$ and $f'(0) = 0$.

When $x \neq 0$, we easily see that $f(x)$ is also differentiable for such non-zero values of x. Thus $f(x)$ is derivable for every value of x.

To test for continuity of $f'(x)$ at $x = 0$

Here, $\quad f'(x) = 2x \sin(1/x) - \cos(1/x)$ at $x \neq 0$ and $f'(0) = 0$.

$$\text{Right-hand limit} = f'(0+0) = \lim_{h\to 0} f'(0+h) = \lim_{h\to 0} f'(h) = \lim_{h\to 0} \{2h\sin(1/h) - \cos(1/h)\}$$

$$= 2\lim_{h\to 0} h \times \lim_{h\to 0} \sin(1/h) - \lim_{h\to 0} \cos(1/h)$$

$$= 2\times 0\times k - \lim_{h\to 0} \cos(1/h) = -\lim_{h\to 0} \cos(1/h) \text{ where } -1 \le k \le 1$$

As $h \to 0$, $\cos(1/h)$ oscillates between -1 and 1 and so $\cos(1/h)$ does not tend to a unique and definite limit. Hence, it follows that right-hand limit does not exist. Similarly we easily see that left-hand limit also does not exist. Hence $f'(x)$ is not continuous at $x = 0$.

Example 4. *Let* $f(x) = x\dfrac{e^{1/x} - e^{-1/x}}{e^{1/x} + e^{-1/x}}$, $x \neq 0$; $f(0) = 0$. *Show that f is continuous but not differentiable at* $x = 0$. **[Agra, 2001; Delhi Maths (H), 1999, Kanpur 2008]**

Solution. We have $f(0) = 0$. Let $h > 0$. Also, here

$$f(0+0) = \lim_{h\to 0} f(0+h) = \lim_{h\to 0} f(h) = \lim_{h\to 0} h\frac{e^{1/h} - e^{-1/h}}{e^{1/h} + e^{-1/h}}$$

$$= \lim_{h\to 0} h\frac{1 - e^{-2/h}}{1 + e^{-2/h}} = 0 \times \frac{1-0}{1+0} = 0, \text{ as } \lim_{h\to 0} e^{-2/h} = 0$$

$$f(0-0) = \lim_{h\to 0} f(0-h) = \lim_{h\to 0} f(-h) = \lim_{h\to 0} (-h)\frac{e^{-1/h} - e^{1/h}}{e^{-1/h} + e^{1/h}}$$

$$= \lim_{h\to 0} (-h)\frac{e^{-2/h} - 1}{e^{-2/h} + 1} = 0 \times \frac{0-1}{0+1} = 0$$

Thus, $\quad f(0+0) = f(0-0) = f(0)$ and so $f(x)$ is continuous at $x = 0$.

Again, $\quad Rf'(0) = \lim_{h\to 0} \dfrac{f(0+h) - f(0)}{h} = \lim_{h\to 0} \dfrac{f(h)}{h}$

$$= \lim_{h\to 0} \frac{h(e^{1/h} - e^{-1/h})/(e^{1/h} + e^{-1/h})}{h} = \lim_{h\to 0} \frac{1 - e^{-2/h}}{1 + e^{-2/h}} = 1$$

and $Lf'(0) = \dfrac{f(0-h) - f(0)}{-h} = \lim_{h\to 0} \dfrac{f(-h)}{-h}$

$$= \lim_{h\to 0} \frac{(-h)(e^{-1/h} - e^{1/h})/(e^{-1/h} + e^{1/h})}{(-h)} = \lim_{h\to 0} \frac{e^{-2/h} - 1}{e^{-2/h} + 1} = -1$$

Thus $Rf'(0) \neq Lf'(0)$ and so $f(x)$ is not differentiable at $x = 0$.

Example 5. *Examine the following function for continuity and differentiability at* $x = 0$ *and* $x = 1$.

$$y = \begin{cases} x^2, & \text{for } x \le 0 \\ 1, & \text{for } 0 < x \le 1 \\ 1/x, & \text{for } x > 1 \end{cases}$$

Also draw graph of the function. **(Meerut, 2001, 02)**

Solution. Let $y = f(x)$. We first test $f(x)$ at $x = 0$. Let $h > 0$. Here $f(0) = 0$

Also, $$f(0+0) = \lim_{h\to 0} f(0+h) = \lim_{h\to 0} f(h) = \lim_{h\to 0} 1 = 1$$

$$f(0-0) = \lim_{h\to 0} f(0-h) = \lim_{h\to 0} f(-h) = \lim_{h\to 0} (-h)^2 = 0$$

Since $f(0) = f(0-0) \neq f(0+0)$, so $f(x)$ is not continuous at $x = 0$. Again, since $f(x)$ is discontinuous at $x = 0$, so it is not differentiable at $x = 0$.

Test for continuity and differentiability at $x = 1$. Here $f(1) = 1$

Here $$f(1+0) = \lim_{h\to 0} f(1+h) = \lim_{h\to 0} \{1/(1+h)\} = 1$$

and $$f(1-0) = \lim_{h\to 0} f(1-h) = \lim_{h\to 0} 1 = 1$$

Since $f(1+0) = f(1-0) = f(1)$, so $f(x)$ is continuous at $x = 1$

Again, $$Rf'(1) = \lim_{h\to 0} \frac{f(1+h)-f(1)}{h} = \lim_{h\to 0} \frac{\frac{1}{1+h}-1}{h} = \lim_{h\to 0} \frac{-1}{1+h} = -1$$

and $$Lf'(1) = \lim_{h\to 0} \frac{f(1-h)-f(1)}{-h} = \lim_{h\to 0} \frac{1-1}{-h} = \lim_{h\to 0} 0 = 0$$

Since $Rf'(1) \neq Lf'(1)$, so $f(x)$ is not differentiable at $x = 1$.

The graph of the function $y = f(x)$ is made up of the following three curves.

$y = x^2$ for $x \le 0$ (parabola AO)

$y = 1$ for $0 < x \le 1$ (straight line BC)

$y = 1/x$ for $x \ge 1$ (rectangular hyperbola CD)

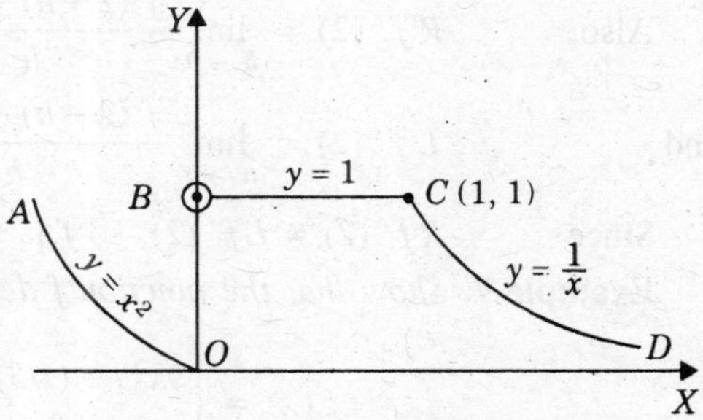

Example 6. *Draw the graph of the function $y = |x-1| + |x-2|$ in the interval $[0, 3]$ and discuss the continuity and differentiability of the function in this interval.*

(Agra. 2000, 04, 05, 09; Delhi Maths (H) 2007)

Solution. Let $y = f(x)$. According to the definition of f, we have

$$y = f(x) = \begin{cases} 1-x+2-x = 3-2x, & \text{for } 0 \le x \le 1 \\ x-1+2-x = 1, & \text{for } 1 \le x \le 2 \\ x-1+x-2 = 2x-3, & \text{for } 2 \le x \le 3 \end{cases}$$

The graph of the function $y = f(x)$ is made of the following three straight lines :

$y = 3 - 2x$ for $0 \le x \le 1$ shown by AB

$y = 1$ for $1 \le x \le 2$ shown by BC

$y = 2x - 3$ for $2 \le x \le 3$ shown by CD

Thus the graph of the given function in the interval $[0, 3]$ is as shown in the figure.

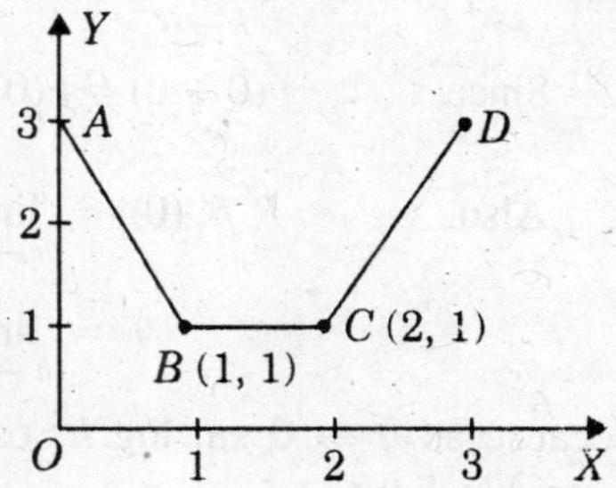

Since $f(x)$ is a linear function or a constant function over the various sub-intervals of $[0, 3]$, it follows that $f(x)$ is continuous as well as differentiable over each of corresponding sub-intervals except at two doubtful points, namely, the breaking points $x = 1$ and $x = 2$.

To test for continuity and differentiability at $x = 1$. Let $h > 0$. Here $f(1) = 1$.

$$f(1+0) = \lim_{h \to 0} f(1+h) = \lim_{h \to 0} 1 = 1$$

and
$$f(1-0) = \lim_{h \to 0} f(1-h) = \lim_{h \to 0} \{3 - 2(1-h)\} = 1$$

Since $f(1+0) = f(1-0) = f(1)$; so f is continuous at $x = 1$.

Also,
$$Rf'(1) = \lim_{h \to 0} \frac{f(1+h) - f(1)}{h} = \lim_{h \to 0} \frac{1-1}{h} = 0$$

and
$$Lf'(1) = \lim_{h \to 0} \frac{f(1-h) - f(1)}{-h} = \lim_{h \to 0} \frac{3-2(1-h)-1}{-h} = -2$$

Since $Rf'(1) \neq Lf'(1)$, so f is differentiable at $x = 1$.

To test for continuity and differentiability at $x = 2$. Here $f(2) = 1$.

$$f(2+0) = \lim_{h \to 0} f(2+h) = \lim_{h \to 0} \{2(2+h) - 3\} = 1$$

and
$$f(2-0) = \lim_{h \to 0} f(2-h) = \lim_{h \to 0} 1 = 1$$

Since $f(2+0) = f(2-0) = f(2)$; so $f(x)$ is continuous at $x = 2$.

Also,
$$Rf'(2) = \lim_{h \to 0} \frac{f(2+h) - f(2)}{h} = \lim_{h \to 0} \frac{2(2+h)-3-1}{h} = 2$$

and
$$Lf'(2) = \lim_{h \to 0} \frac{f(2-h) - f(2)}{-h} = \lim_{h \to 0} \frac{1-1}{-h} = 0$$

Since $Rf'(2) \neq Lf'(2)$, so f is not differentiate at $x = 2$.

Example 7. *Show that the function f defined by*

$$f(x) = \begin{cases} x[1 + (1/3) \times \sin(\log x^2)], & \text{for } x \neq 0 \\ 0, & \text{for } x = 0 \end{cases}$$

is continuous at = 0 but not differentiable at $x = 0$.

[Delhi Maths (H), 1996; Meerut, 2003, 10; Osmania, 2004]

Solution. Here $f(0) = 0$. Let $h > 0$. Then, we have

$$f(0+0) = \lim_{h \to 0} f(0+h) = \lim_{h \to 0} \{h + (h/3) \sin \log h^2\}$$
$$= 0 + 0 \times k = 0, \text{ where } -1 < k < 1$$

[$\because$ $\sin \log h^2$ oscillates finites between -1 and 1 as $h \to 0$]

and
$$f(0-0) = \lim_{h \to 0} f(0-h) = \lim_{h \to 0} \{-h - (h/3) \times \sin \log(-h)^2\} = 0, \text{ as before.}$$

Since $f(0+0) = f(0-0) = f(0)$, so $f(x)$, so $f(x)$ is continuous at $x = 0$

Also,
$$Rf'(0) = \lim_{h \to 0} \frac{f(0+h) - f(0)}{h} = \lim_{h \to 0} \frac{h\{1 + (1/3) \times \sin \log h^2\}}{h}$$
$$= \lim_{h \to 0} \{1 + (1/3) \times \sin \log h^2\}, \text{ which does not exist}$$

because as $h \to 0$ $\sin \log h^2$ oscillates finitely between -1 and 1 and so it does not tend to a unique limit.

By a similar argument, we easily see that $Lf'(0)$ also does not exist. This implies that $f(x)$ is not differentiable at $x = 0$

Example 8. *The function f defined by*

$$f(x) = \begin{cases} x^2 + 3x + a, & \text{if } x \le 1 \\ bx + 2, & \text{if } x > 1 \end{cases}$$

is given to be derivable for every x. Find a and b.

Solution. Since f is derivable for every x, f must be derivable at $x = 1$ and hence, f must be continuous at $x = 1$. Let $h > 0$. Then f is continuous at $x = 1$.

$$\Rightarrow f(1) = f(1+0) = f(1-0)$$

$$\Rightarrow 4 + a = \lim_{h \to 0} f(1+h) = \lim_{h \to 0} f(1-h)$$

$$\Rightarrow 4 + a = \lim_{h \to 0} \{b(1+h) + 2\} = \lim_{h \to 0} \{(1-h)^2 + 3(1-h) + a\}$$

$$\Rightarrow 4 + a = b + 2 = 4 + a \quad \Rightarrow \quad a - b = -2 \qquad \ldots(1)$$

Again, f is differentiable at $x = 1 \Rightarrow Rf'(1) = Lf'(1)$

$$\Rightarrow \lim_{h \to 0} \frac{f(1+h) - f(1)}{h} = \lim_{h \to 0} \frac{f(1-h) - f(1)}{-h}$$

$$\Rightarrow \lim_{h \to 0} \frac{b(1+h) + 2 - (4+a)}{h} = \lim_{h \to 0} \frac{(1-h)^2 + 3(1-h) + a - (4+a)}{-h}$$

$$\Rightarrow \lim_{h \to 0} \frac{b - a + bh - 2}{h} = \lim_{h \to 0} \frac{h^2 - 5h}{-h}$$

$$\Rightarrow \lim_{h \to 0} \frac{b - (b-2) + bh - 2}{h} = \lim_{h \to 0} (5 - h), \text{ using (1)}$$

$$\Rightarrow b = 5$$

Since $b = 5$, so (1) gives $a = b - 2 = 5 - 2 = 3$. Thus, $a = 3$, $b = 5$.

Example 9. *Show that* $f(x) = x \tan^{-1} (1/x)$ *for* $x \ne 0$ *and* $f(0) = 0$ *is not differentiable at* $x = 0$.

Solution. Let $h > 0$. Then we have

$$Rf'(0) = \lim_{h \to 0} \frac{f(0+h) - f(0)}{h} = \lim_{h \to 0} \frac{h \tan^{-1}(1/h)}{h} = \lim_{h \to 0} \tan^{-1} \frac{1}{h} = \frac{\pi}{2}$$

and $$Lf'(0) = \lim_{h \to 0} \frac{f(0-h) - f(0)}{-h} = \lim_{h \to 0} \frac{(-h)\tan^{-1}(-1/h)}{(-h)} = -\lim_{h \to 0} \tan^{-1} \frac{1}{h} = -\frac{\pi}{2}$$

Since $Rf'(0) \ne Lf'(0)$, so $f(x)$ is not differentiable at $x = 0$.

Example 10. *Let f and g be two functions having the same domain D. If (i)* $f + g$, *(ii)* $f - g$ *and (ii)* f/g *are derivable at* $c \in D$, *is it necessary that f and g are both derivable at c ?* **(Kanpur, 2004)**

Solution. (*i*) Let $f(x) = x \sin(1/x)$, $x \ne 0$ and $f(0) = 0$

and $g(x) = -f(x) \; \forall \; x \in \mathbf{R}$. Then we have

$$(f + g)(x) = f(x) + g(x) = 0 \; \forall \; x \in \mathbf{R}$$

Since $f + g$ is a constant function, so it is derivable at $x = 0$.

But we can show that f and g both are not derivable at $x = 0$.

(*ii*) Let $f(x) = g(x) = |x|$, $\forall \; x \in \mathbf{R}$. Then, we have

$$(f - g)(x) = f(x) - g(x) = 0, \; \forall \; x \in \mathbf{R}.$$

Since $f - g$ is a constant function, so it is derivable at $x = 0$.

But we can show that both f and g are not derivable at $x = 0$.

(*iii*) Let $f(x) = x \sin(1/x)$, if $x \neq 0$ and $f(0) = 0$

and $g(x) = 1/x$, $x \neq 0$ and $g(0) = 1$. Then, we have

$$\left(\frac{f}{g}\right)(x) = \begin{cases} x^2 \sin(1/x), & \text{if } x \neq 0 \\ 0, & \text{if } x = 0 \end{cases}$$

Then proceed as in example 3, to show that f/g is derivable at $x = 0$. Also we can show that $f(x)$ and $g(x)$ are not derivable at $x = 0$.

Example 11. (*a*) *Let f and g be two functions with domain D satisfying $g(x) = x f(x)\ \forall\ x \in D$. Show that if f be continuous at $x = 0 \in D$, then g is derivable at $x = 0$.*

(*b*) *Suppose the function f satisfies the conditions (i) $f(x + y) = f(x) f(y)\ \forall\ x, y$ (ii) $f(x) = 1 + x\, g(x)$, where $\lim_{x \to 0} g(x) = 1$. Show that the derivative of $f'(x)$ exists and $f'(x) = f(x)$ for all x.*

Solution. (*a*) Since f is continuous at $x = 0$, we have

$$\lim_{x \to 0} f(x) = f(0). \qquad ...(1)$$

Also given that $g(x) = x f(x)\ \forall\ x \in D$ ...(2)

Now, $$g'(0) = \lim_{x + 0} \frac{g(x) - g(0)}{x - 0} = \lim_{x \to 0} \frac{x f(x)}{x}, \text{ using (2)}$$

$$= \lim_{x \to 0} f(x) = f(0), \text{ using (1)}$$

$\Rightarrow$ $g(x)$ is derivable at $x = 0$.

(*b*) $$f'(x) = \lim_{h \to 0} \frac{f(x+h) - f(x)}{h} = \lim_{h \to 0} \frac{f(x) f(h) - f(x)}{h}, \text{ where } h > 0$$

[$\because$ by condition (*i*), $f(x + h) = f(x) f(h)$]

$$= f(x) \lim_{h \to 0} \frac{f(h) - 1}{h} = f(x) \lim_{h \to 0} \frac{1 + h\, g(h) - 1}{h}$$

[$\because$ by condition (*ii*), $f(h) = 1 + h\, g(h)$]

$$= f(x) \lim_{h \to 0} g(h) = f(x) \times 1 = f(x) \qquad \left[\because \text{ given that } \lim_{x \to 0} g(x) = 1\right]$$

Example 12. *If $f(x) = e^{-1/x^2} \sin(1/x) \neq 0$, and $f(0) = 0$, show that*

(*i*) *the function f has at every point a differential coefficient and this is continuous at $x = 0$;*

(*ii*) *the differential coefficient vanishes at $x = 0$ and at an infinite number of points in every neighbourhood of $x = 0$.*

Solution. We have

$$f'(x) = \frac{e^{-1/x^2}}{x^3}[2 \sin(1/x) - \cos(1/x)], \text{ when } x \neq 0$$

and $$f'(0) = \lim_{x \to 0} \frac{e^{-1/x^2} \sin(1/x) - 0}{x}.$$

Since $e^{-1/x^2} > 1/x^2\ \forall\ x \neq 0$, we see that

$$\left|\frac{e^{-1/x^2}\sin(1/x)}{x}\right| < \frac{x^2 \cdot 1}{|x|} = |x|, \ \forall\, x \neq 0.$$

$$\Rightarrow f'(0) = 0.$$

Also, when $x \neq 0$,

$$|f'(x) - f'(0)| = \left|\frac{e^{-1/x^2}}{x^3}[2\sin(1/x) - x\cos(1/x)]\right| \leq \frac{e^{-1/x^2}}{|x|^3}(2 + |x|).$$

Since $e^{1/x^4} > 1/2x^4 \ \forall\, x \neq 0$ we see that

$$|f'(x) - f'(0)| \leq 2|x|(2 + |x|)$$

$$\Rightarrow \quad [f'(x) - f'(0)] \to 0 \text{ as } x \to 0$$

$$\Rightarrow f' \text{ is continuous for } x = 0.$$

Also clearly f is continuous for every non-zero value of x.

We have now to show that f' vanishes at a point in *every* neighbourhood of $x = 0$

Let $\delta > 0$ be any number. There surely exists a positive integer n such that

$$0 < \frac{2}{(2n+1)\pi} < \frac{1}{n\pi} < \delta.$$

It is easy to see that,

for $x = 1/n\pi$, $f'(x)$ is –ve or +ve according as n is even or odd,

for $x = 2/(2n + 1)\pi$, $f'(x)$ is –ve or +ve according as n is odd or even.

Therefore, f', which is continuous, must vanish at least once between

$$2/(2n + 1)\pi \quad \text{and} \quad 1/n\pi.$$

We may similarly dispose of the left-handed neighbourhoods of 0.

EXERCISES

1. (*i*) Show that the function $f(x) = \begin{cases} x\cos(1/x), & \text{if } x \neq 0 \\ 0, & \text{if } x = 0 \end{cases}$

is continuous at $x = 0$ but not differentiable at $x = 0$.

(*ii*) If $f(x) = x\sin(1/x)$ when $x \neq 0$ and $f(0) = 0$, show that f is continuous but not derivable for $x = 0$. **(Rajasthan 2010; Osmania, 2004)**

2. Show that the following function are continuous at $x = 0$ but not differentiable at $x = 0$

(*i*) $f(x) = \begin{cases} x^2\sin(1/x), & x \neq 0 \\ 0, & x = 0 \end{cases}$ **(Delhi Maths (Prog) 2008; Calicut, 2004)**

(*ii*) $g(x) = \begin{cases} x^2\cos(1/x), & x \neq 0 \\ 0, & x = 0 \end{cases}$

3. Show that $f(x) = |x + 2|$ is continuous at $x = -2$ but not differentiable at this point. **(Kanpur, 2004)**

4. Are the following functions continuous and derivable ?

(*i*) $f(x) = 1 + x$ if $x < 2$ and $f(x) = 5 - x$ if $x \geq 2$ at the point $x = 2$

(*ii*) $f(x) = 2 + x$ if $x \geq 0$ and $f(x) = 2 - x$ if $x < 0$ at the origin

(*iii*) $f(x) = x$ if $0 \leq x < 1$ and $f(x) = 2 - x$ if $x \geq 1$ at the point $x = 1$ **(Meerut, 2003)**

5. Show that the function $f(x) = (x-1)^{1/3}$ has no finite derivative. **(Agra, 1999)**

6. Show that the $f(x) = |x| + |x-1|$ is continuous but not derivable at $x = 0$ and $x = 1$. **(Delhi B.Sc. (Prog) I 2010, 11; Agra, 1997, 99; Meerut 2011;**

7. Discuss the derivability of the function $f(x) = \begin{cases} 2x-3, & 0 \le x \le 2 \\ x^2-3, & 2 < x \le 4 \end{cases}$

at $x = 2, 4$. **[Delhi (Hons), 1997]**

8. Show that the function defined as $f(x) = x\dfrac{e^{1/x}-1}{e^{1/x}-1}$, if $x \ne 0$

$= 0,$ if $x = 0$

is continuous at $x = 0$ but not derivable at $x = 0$.

9. Show that the function defined as $f(x) = \dfrac{x}{1+e^{1/x}}$, if $x \ne 0$

$= 0,$ if $x = 0$

is continuous at $x = 0$ but not derivable at $x = 0$.

(Meerut, 2004; Srivenkateshwara, 2003)

10. Show that the function defined by $f(x) = |x-2| + |x+2|$ for all $x \in \mathbf{R}$ is derivable everywhere except at the points $x = -2$ and $x = 2$. **[Delhi Maths (Hons), 1998]**

11. Show that the function defined by $f(x) = |x-2| + |x| + |x+2|$ for all $x \in \mathbf{R}$ is derivable everywhere except at the points $x = -2$, $x = 0$ and $x = 2$.

[Delhi Maths (H), 2005]

12. Is the function $|\sin x - 1|$ differentiable at $x = \pi/2$? Write your answer with justification. **(Utkal, 2003)**

13. Is the function $|\cos x - (3/2)|$ differentiable at $x = 0$. Write your answer with justification.

14. Show that the function f defined on $\mathbf{R}$ as follows $f(x) = \begin{cases} x^2, & \text{if } x \text{ is rational} \\ 0, & \text{if } x \text{ is irrational} \end{cases}$

is derivable at $x = 0$ and $f'(0) = 0$. Further show that f is not differentiable for all $x \ne 0$.

15. If f and g are derivable at a and $f(a) \ne g(a)$, prove that each of the functions max $\{f, g\}$ and min $\{f, g\}$ is derivable at a. Discuss also the case $f(a) = g(a)$.

16. Show that $x/(1 + |x|)$ is differentiable in the open interval $(-\infty, \infty)$.

17. Let $f(x+y) = f(x) + f(y)\ \forall\ x, y \in \mathbf{R}$. Prove that f is derivable on $\mathbf{R}$ if it is derivable at one point of $\mathbf{R}$.

18. Let $f(x+y) = f(x)\, f(y)\ \forall\ x, y \in \mathbf{R}$. Prove that

(*i*) f is derivable on $\mathbf{R}$ if it is derivable at one point of $\mathbf{R}$

(*ii*) f is derivable at $x = 0$ if it is continuous at $x = 0$

19. Let $f(xy) = f(x) + f(y)\ \forall\ x, y \in \mathbf{R}^+$, where $\mathbf{R}^+$ denotes the set of all positive real numbers. Prove that f is derivable on $\mathbf{R}^+$ if it is derivable at one point of $\mathbf{R}^+$.

20. For all real numbers, $f(x)$ is given as $f(x) = \begin{cases} e^x + a\sin x, & \text{if } x < 0 \\ b(x-1)^2 + x - 2, & \text{if } x \ge 0 \end{cases}$

Find values of a and b for which f is differentiable at $x = 0$. **(I.A.S., 2003)**

21. Let n be a positive integer and let f be defined on $\mathbf{R}$ as

(*i*) $f(x) = x^{2n}\sin(1/x)$ if $x \ne 0$, and $f(0) = 0$. Prove that f^n exists $\forall\ x \in \mathbf{R}$, but f^n is not continuous at $x = 0$.

(*ii*) $f(x) = x^{2n+1} \sin(1/x)$ if $x \neq 0$, and $f(0) = 0$. Prove that f^n exists $\forall\ x \in \mathbf{R}$, f^n is continuous at $x = 0$, but f^n is not derivable at $x = 0$.

22. If a function f is defined by $\quad f(x) = (xe^{1/x})/(1 + e^{1/x}), \quad x \neq 0$

$$= 0, \qquad x = 0,$$

show that f is continuous but not derivable at $x = 0$. **(Meerut 2009)**

ANSWERS

4. (*i*) Continuous at $x = 2$, not derivable at $x = 2$ (*ii*) Continuous at $x = 0$, not derivable at $x = 0$ (*iii*) Continuous at $x = 1$, not derivable at $x = 1$

6. Differentiable at $x = 4$, not differentiable at $x = 2$

9.5. MEANING OF THE SIGN OF DERIVATIVE AT A POINT

Let c be an interior point of the domain of a function f. We suppose that $f'(c)$ exists and is positive. We have

$$\lim_{x \to c} \frac{f(x) - f(c)}{x - c} = f'(c) > 0.$$

Let ε be any positive number smaller than the positive number $f'(c)$. Then there exists $\delta > 0$ such that

$$\left| \frac{f(x) - f(c)}{x - c} - f'(c) \right| < \varepsilon \text{ when } 0 < |x - c| < \delta$$

$$\Rightarrow \quad f'(c) - \varepsilon < \frac{f(x) - f(c)}{x - c} < f'(c) + \varepsilon \ \forall\ x \in [c - \delta, c + \delta],\ x \neq c.$$

Now ε being smaller than $f'(c)$, we see that

$$\frac{f(x) - f(c)}{x - c} > 0 \ \forall\ x \in [c - \delta, c + \delta],\ x \neq c.$$

$$\Rightarrow \quad \begin{cases} f(x) - f(c) > 0 \Rightarrow f(x) > f(c) \text{ when } c < x \leq c + \delta. \\ f(x) - f(c) < 0 \Rightarrow f(x) < f(c) \text{ when } c - \delta \leq x < c. \end{cases}$$

Thus, we conclude that if $f'(c) > 0$, there exists a neighbourhood $[c - \delta, c + \delta]$ of c such that

$$f(x) > f(c) \ \forall\ x \in [c, c + \delta] \quad \text{and} \quad f(x) < f(c) \ \forall\ x \in [c - \delta, c].$$

The conclusion is often stated in the following manner :

If $f'(c) < 0$, then f is increasing at c.

Now suppose that $f'(c) < 0$. We define a function g as follows :

$$g(x) = -f(x)$$

$$\Rightarrow \quad g'(x) = -f'(c) > 0.$$

We conclude that if $f'(c) < 0$, there exists a neighbourhood $[c - \delta, c + \delta]$ of c such that

$$g(x) > g(c) \quad \Rightarrow \quad f(x) < f(c) \ \forall\ x \in\ ]c, c + \delta]$$

$$g(x) < g(c) \quad \Rightarrow \quad f(x) > f(c) \ \forall\ x \in\ ]c - \delta, c[$$

so that if $f'(c) < 0$, the function f is decreasing at c.

We now consider the end points. It may be shown that

(*a*) there exists an interval $]a, a + \delta]$ such that

$$f'(a) > 0 \quad \Rightarrow \quad f(x) > f(a) \ \forall\ x \in\ ]a, a + \delta],$$

$$f'(a) < 0 \quad \Rightarrow \quad f(x) < f(a) \ \forall\ x \in [a, a + \delta[.$$

(*b*) there exists an interval $]b-\delta, b[$ such that

$$f'(b) > 0 \Rightarrow f(x) < f(b) \ \forall \ x \in [b-\delta, b[,$$
$$f'(b) < 0 \Rightarrow f(x) > f(b) \ \forall \ x \in]b-\delta, b[.$$

9.6. DARBOUX'S THEOREM (OR INTERMEDIATE VALUE THEOREM FOR DERIVATIVES)

(Kanpur 2013)

If a function is derivable in a closed interval $[a, b]$ and $f'(a), f'(b)$ are of opposite signs, then there exists at least one point c of the open interval $]a, b[$ such that $f'(c) = 0$.

(Agra 2008, 09, 10; Srivenkateshwara, 2003; Kanpur, 2003, 09, 11, 13)

Proof. For the sake of definiteness, we suppose that $f'(a)$ is positive and $f'(b)$ negative. On this account there exist intervals $]a, a+h]$, $[b-h, b[$, $(h < 0)$, such that

$$x \in]a, a+h] \Rightarrow f(x) > f(a), \quad ...(i)$$
$$x \in [b-h, b[\Rightarrow f(x) > f(b) \quad ...(ii)$$

Again *f*, being derivable, is continuous in $[a, b[$. Therefore it is bounded and attains its bounds. Thus if *M* be the least upper bound of *f* in $[a, b]$ there exists $c \in [a, b]$ such that

$$f(c) = M.$$

From (*i*) and (*ii*), we see that the least upper bound is not attained at the end points *a* and *b* so that *c* is interior point of $[a, b]$.

If $f'(c)$ be positive, then there exists an interval $[c, c+\eta]$, $(\eta > 0)$ such that for every point *x* of this interval $f(x) > f'(c) = M$ and this is a contradiction.

If $f'(c)$ be negative, then there exists an interval $[c-\eta, c]$, $(\eta > 0)$ such that for every point *x* of this interval $f(x) > f(c) = M$ and this is, again, a contradiction.

Hence,
$$f'(c) = 0.$$

Cor. *(Darboux property).* *If f is derivable in a closed interval $[a, b]$ and $f'(a) \neq f'(b)$ and k is any number lying between $f'(a)$ and $f'(b)$, then there exists at least one point $c \in]a, b[$ such that $f'(c) = k$.*

Proof. Let
$$\varphi(x) = f(x) - kx.$$

The function φ is derivable in $[a, b]$. Here

$$\varphi'(a) = f'(a) - k \quad \text{and} \quad \varphi'(b) = f'(b) - k,$$

are of opposite signs. Therefore, there exists at least one point, *c*, of $]a, b[$ such that

$$\varphi(c) = 0 \Rightarrow f'(c) - k = 0 \Rightarrow f'(c) = k.$$

Note. We have seen that if $\varphi(a)$, $\varphi(b)$ are of opposite signs, then φ vanishes for at least one value, *c*, in $[a, b]$ if φ is continuous in $[a, b]$. In this context, the importance of the Darboux's theorem above lies in the fact that if *φ is the derivative of a function*, then the conclusion of the vanishing of φ remain valid even when *φ is not continuous.*

Example. *If f be derivable at a point c, then show that $|f|$ is also derivable at c, provided $f(c) \neq 0$. Show by means of an example that if $f(c) = 0$, then f may be derivable at c and $|f|$ may not be derivable at c.*

[Delhi Maths (Hons), 2003]

Solution. Since *f* is derivable at *c*, *f* is continuous at *c*.

Again,
$$f(c) \neq 0 \Rightarrow f(c) > 0 \text{ or } f(c) < 0.$$

Hence, there exist positive numbers δ_1 and δ_2 such that

$$f(x) > 0 \ \forall \ x \in]c-\delta_1, c+\delta_1[$$

or

$$f(x) < 0 \ \forall \ x \in]c-\delta_2, c+\delta_2[$$

Therefore,
$$|f(x)| = f(x) \ \forall \ x \in]c-\delta_1, c+\delta_1[\quad ...(1)$$

or

$$|f(x)| = -f(x) \ \forall \ x \in]c-\delta_2, c+\delta_2[\quad ...(2)$$

Since *f* is derivable at *c*, it follows from (1) and (2), that $|f|$ is also derivable at *c*.

We now give an example of a function $f(x)$ such that $f(c) = 0$ and $f(x)$ is derivable at $x = c$ but $|f|$ may not be derivable at $x = c$.

Let $f(x) = x \; \forall \; x \in \mathbf{R}$. Then $f(0) = 0$ and $|f(x)| = |x| \; \forall \; x \in \mathbf{R}$.

We can show that $|f(x)|$ is not derivable at $x = 0$ and $f(x)$ is derivable at $x = 0$.

EXERCISES

1. If f is defined and derivable on $[a, b]$, $f(a) = f(b) = 0$, and $f'(a)$ and $f'(b)$ are of the same sign, then prove that f must vanish at least once in $]a, b[$ **[Delhi Maths (H) 2007]**
2. If f is derivable on $[a, b]$, $f(a) = f(b) = 0$, and $f(x) \neq 0$ for any x in $]a, b[$, then prove that $f'(a)$ and $f'(b)$ must be of opposite signs.

MISCELLANEOUS EXERCISES ON CHAPTER 9

1. Given that f' is a strictly monotonic and derivable function with domain $[a, b]$, show that the inverse ϕ of f is also derivable and that $f'(x)\,\phi'(y) = 1$; x, y being the corresponding numbers. Hence find the derivative of the root function $x \to x^{1/n}$; $n \in \mathbf{N}$ and $x \geq 0$.
2. f and g are two derivable functions such that the range of f is a subset of the domain of g so that gof has a meaning. Show that gof is also derivable and $(gof)' = (g'of)\,f'$

   For the following pairs f, g of functions, find $(gof)'(x)$:

   (*i*) $f : x \to x + 1,\; g : x \to \sqrt{x}$,   (*ii*) $f : x \to x^2 + 2,\; g : x \to x^{1/3}$.
3. If $f(x) = \begin{cases} x^2 \sin(1/x), & \text{when } x \neq 0, \\ 0, & \text{when } x = 0 \end{cases}$ and $g(x) = x$, show that

   $\lim\limits_{x \to 0} [f'(x)/g(x)]$ does not exist but $\lim\limits_{x \to 0} [f(x)/g(x)]$ exists

   and is equal to $f'(0)/g'(0)$.
4. If $f(x) = |x|$, $g(x) = 2|x|$ show that $f'(0)$ and $g'(0)$ do not exist but $\lim [f(x)/g(x)]$ exists and is equal to $\lim [f'(x)/g'(x)]$ when $x \to 0$.
5. If
$$f(x) = \sqrt{x}\,[1 + x \sin(1/x)] \;\forall\; x > 0,$$
$$f(x) = -\sqrt{(-x)}\,[1 + x \sin(1/x)] \;\forall\; x < 0,$$
$$f(0) = 0$$
   show that f' exists everywhere and is finite except at $x = 0$, in the neighbourhood of which it oscillates between arbitrarily large positive and negative values.
6. Show that the function f given by $f(x) = 4 + 7x + x^2\,[8 + x \sin(1/x)]$, where $x \sin(1/x)$ is zero for $x = 0$ has a first dirivative but no second derivative at the origin.
7. If $f(x) = \sin x \sin(1/\sin x)$, when $0 < x < \pi < x < 2\pi$

   and $f(x) = 0$, when $x = 0, \pi, 2\pi$;

   show that f is continuous but not derivable for $x = 0, \pi, 2\pi$.
8. If $a_0, a_1, a_2, ..., a_n$ are real and $|a_0| + |a_1| + |a_2| + ... + |a_{n-1}| < a_n$, show that
$$u(x) = a_0 + a_1 \cos x + a_2 \cos 2x + ... + a_n \cos nx$$
   has at least $2n$ zeros in the interval $0 < x < 2\pi$. Show also that $u'(x)$ has at least $2n$ zeros in the interval $\alpha \leq x < 2\pi + \alpha$ for every real α.

   [$u(0)$, $u(\pi/n)$, $u(2\pi/n)$, ..., $u(2n\pi/n)$ have positive and negative signs alternatively.]

9. Determine the differential coefficient, if any, of the function f defined by

$$f(x) = (x-a)\frac{Ae^{1/(x-a)} + Be^{-1/(x-a)}}{e^{1/(x-a)} + e^{-1/(x-a)}} \sin\frac{\pi}{2(x-a)} \text{ when } x \neq a$$

$$f(a) = 0.$$

10. A function f is defined as follows :

$$f(x) = (x - w_1)(x - w_2)^2 (x - w_3)^3 \sin\frac{1}{(x-w_1)} \sin\frac{1}{(x-w_2)} \sin\frac{1}{(x-w_3)}$$

for all values of x except w_1, w_2, w_3, in the domain $[a, b]$ and $f(x) = 0$ when $x = w_1$ or w_2, w_3. Show that

(*i*) df/dx does not exist at the point $x = w_1$.

(*ii*) df/dx exists but has a discontinuity of the second kind at $x = w_2$.

(*iii*) df/dx exists and is continuous at $x = w_3$.

OBJECTIVE QUESTIONS

Multiple Choice Type Questions : *Select (a), (b), (c) or (d), whichever is correct.*

1. (*a*) Divergent sequences is not bounded

(*b*) A function which is not continuous at a point may be differentiable at that point

(*c*) Bounded sequence is always convergent

(*d*) Least upper bound of a bounded set is an element of the set. **(Kanpur, 2004)**

2. The function f defined by $f(x) = (x^2/a) - a$, if $0 < x < a$

$= a - (a^3/x^2)$, if $x \geq a$

(*a*) is not continuous on $]0, \infty[$ (*b*) is not differentiable on $]0, \infty[$

(*c*) is differentiable on $]0, \infty[$ (*d*) is differentiable on $]0, \infty[$ except at $x = a$.

[I.A.S. (Prel.), 1999]

3. The function $f(x) = |x|$ at $x = 0$ is :

(*a*) Continuous and differentiable (*b*) Continuous but not differentiable

(*c*) Not continuous but differentiable (*d*) Neither continuous nor differentiable.

(Kanpur, 2001, 02)

4. The function $f(x) = |x + 2|$ is not differentiable at a point :

(*a*) $x = 2$ (*b*) $x = -2$ (*c*) $x = -1$ (*d*) $x = 1$. **(Kanpur, 2003)**

5. Function $f(x) = |x|$ is defined on $[-2, 2]$. The points at which f is differentiable are :

(*a*) $-1, 0$ (*b*) $-1, 0, 1$

(*c*) $-1, 0, 1, 2$ (*d*) None of the above. **(Kanpur, 2004)**

6. On the interval $[-1, 1]$, $f(x)$ is such that $f'(x) = 0$ for the function $f(x) = x^2 + |x| + 2$. The values of x are :

(*a*) 2/3, – 1/3 (*b*) 1/4, – 1/3 (*c*) 1/3, – 1/2 (*d*) 1/2 – 1/2. **(Kanpur, 2004,**

7. A function f defined such that for all real x, y (*i*) $f(x + y) = f(x) \cdot f(y)$ (*ii*) $f(x) = 1 + x\, g(x)$ where $\lim_{x \to 0} g(x) = 1$. What is $df(x)/dx$ equal to ?

(*a*) $g(x)$ (*b*) $f(x)$ (*c*) $g'(x)$ (*d*) $g(x) + x\,g'(x)$. **[I.A.S. (Prel.), 2005]**

8. If $f'(a)$ exists, then $\lim_{h \to 0} \dfrac{f(a+h) - f(a-h)}{h}$ is equal to :

(*a*) $f'(a)/2$ (*b*) $f'(a)$ (*c*) $2f'(a)$ (*d*) None of these.

9. If $f(x+y) = f(x) \cdot f(y)$ for all x and y. Suppose that $f(3) = 3$ and $f'(0) = 11$. Then $f'(3)$ is equal to :

(*a*) 22 (*b*) 33 (*c*) 28 (*d*) None of these.

ANSWERS

1. (*a*) **2.** (*a*) **3.** (*b*) **4.** (*b*) **5.** (*d*) **6.** (*d*) **7.** (*d*)
8. (*b*) **9.** (*b*)

MISCELLANEOUS PROBLEMS ON CHAPTER 9

1. Give an example of a function which is continuous but not differentiable. **[Kanpur 2005]**

2. If $f + g$ is differentiable at a point, then will f and g both be differentiable at that point ? Justify your answer ? **[Kanpur 2005]**

3. Let $f(x+y) = f(x)f(y)$ for all x and y and $f(5) = -2$, $f'(0) = 3$. What is the value of $f'(5)$? (*a*) 3 (*b*) 1 (c) –6 (*d*) 6 **[IAS Prel. 2006]**

[Sol. Ans. (c). $f'(0) = 3 \Rightarrow \lim_{h \to 0} \dfrac{f(0+h) - f(0)}{h} = 3,\ h > 0$ ... (1)

Now, $f'(5) = \lim_{h \to 0} \dfrac{f(5+h) - f(5)}{h} = \lim_{h \to 0} \dfrac{f(5)f(h) - f(5)}{h}$

$$\Rightarrow f'(5) = -2 \lim_{h \to 0} \frac{f(h) - 1}{h} \qquad ...(2)$$

Also, $f(x+y) = f(x)f(y) \Rightarrow f(5+0) = f(5)f(0) \Rightarrow f(0) = 1$

$\therefore$ (2) $\Rightarrow f'(5) = -2 \lim_{h \to 0} \dfrac{f(h) - f(0)}{h} = (-2) \times 3 = -6$, by (1)**]**

4. Give an example so that the function $f(x) + g(x)$ is continous in closed interval [a, b] but not differentiable. **[Agra 2005]**

5. Let f be the function on R defined by $f(x) = 2x + |x|$. Is f derivable at $x = 0$?
Ans. Not derivable **[Delhi B.Sc. I (Hon) 2010]**

6. *If f defined and derivable on* $[a, b]$ *with* $f(a) = f(b) = 0$ *and* $f'(a)f'(b) > 0$, *then prove that f must vanish at least once in* $]a, b[$. **[Delhi Maths (H) 2007]**

7. *Examine the continuity and differentiability of the following function at* $x = 0$.

$$f(x) = \begin{cases} x \tan^{-1}(1/x), & x \neq 0 \\ 0, & x = 0 \end{cases}$$ **[Purvanchal 2006]**

8. Give an example so that the function $f(x) + g(x)$ is continuous in closed interval [a, b] but not differentiable. **[Agra 2005]**

9. Select (*a*), (*b*), (*c*), (*d*) whichever is correct

(*i*) The function f defined by $f(x) = x\sin(1/x),\ x \neq 0;\ f(x) = 0, x = 0$ is

(*a*) noı continuous (*b*) continuous but not differentiable

(*c*) differentiable but not continuous (*d*) differentiable and continuous **[Agra 2005]**

(*ii*) The function $f(x)=|x-1|$ is continuous but not differentiable at point.

(*a*) 0 (*b*) –1 (*c*) 1 (*d*) at each point. **[Agra 2007]**

[Ans. (i) (*b*); (*ii*) (*c*)]

10. Let *f* be defined in ***R*** by setting $f(x) = |x-1| + |x| + |x+1|$, for all $x \in \mathbf{R}$. Show that f in continuous but not derivable at $x = -1$, $x = 0$ and $x = 1$. **[Delhi Maths (H) 2009]**

11. Let f and g be real valued functions defined in a neighbourhood of a real number x_0, show that fg is derivable at x_0 and $(fg)'(x_0) = f'(x_0)\, g'(x_0)$ **[Delhi B.Sc. I (Hons) 2008]**

12. Let *f* be defined on an interval *I* and $c \in I$, then show that *f* is differentiable at $x = c$ if and only if there exists a function ϕ on I that is continuous at c and satisfies $f(x) - f(c) = \phi(x)\,(x-c),\ x \in I$. In this case $\phi(c) = f'(c)$. **[Delhi B.Sc. (Hons) II 2016, 2018]**

Note: This is known as coratheodory's theorem.

13. Let $I \subseteq \mathbf{R}$ be an interval and $f: I \to \mathbf{R}$. Let $c \in I$ and *f* is derivable at *c*. Show that if $f'(c) > 0$, there is a number $\delta > 0$ such that $f(x) > f(c)$ for $x \in I$ for $c < x < c + \delta$.

[Delhi B.Sc. I (Hons) 2010]

14. Show that the derivative of an even function is always an odd function.

[Kanpur 2010]

[Sol. Since *f* is differentiable, so by definition, we have

$$f'(x) = \lim_{h \to 0} \frac{f(x+h) - f(x)}{h} \quad \text{... (1)}$$

and $$f'(x) = \lim_{h \to 0} \frac{f(x+h) - f(x)}{-h} \quad \text{... (2)}$$

From (1), $$f'(-x) = \lim_{h \to 0} \frac{f(-x+h) - f(-x)}{h} = \lim_{h \to 0} \frac{f(-(x-h)) - f(-x)}{h} \quad \text{... (3)}$$

Let f be an even function. Then, by definition, we must have

$$f(-(x-h)) = f(x-h) \quad \text{and} \quad f(-x) = f(x) \quad \text{... (4)}$$

Now, using (4), (3) reduces

$$f'(-x) = \lim_{h \to 0} \frac{f(x-h) - f(x)}{h} = -\lim_{h \to 0} \frac{f(x-h) - f(x)}{-h} = -f'(x) \text{ using (2)}$$

Since $f'(-x) = -f'(x)$, by definiton $f'(x)$ is an odd function. Thus, derivative $f'(x)$ of an even function $f(x)$ is an odd function.

15. Show that the derivation of an odd function is always an even function.

[Hint. If $f(x)$ is odd, then $f(-x) = -f(x)$. Now, proceed as in above problem 14.]

16. Show that there exists a real continuous function on the real line which is nowhere differentiable. **[Himanchal 2010]**

[Sol. In 1872, Karl Weierstrass surprised mathematicans of the world by giving an example of function which is continuous everywhere but differentiable nowhere. This function, known as

Weierstrass's non-differentiable function, is given by $f(x) = \sum_{n=0}^{\infty} a^n \cos(b^n \pi x)$

where $0 < a < 1$ and b is an odd positive integer subject to the condition $ab > 1 + (3\pi/2)$

For example, the functions $f_1(x) = \sum_{n=0}^{\infty} (2/3)^n \cdot \cos(5^n \pi x)$ and $f_2(x) = \sum_{n=0}^{\infty} (5/6)^n \cdot \cos(7^n \pi x)$ are continuous for $\forall x \in \boldsymbol{R}$ but not differentiable for any $x \in \boldsymbol{R}$.

It can also be easily verified that the function $f(x) = \sum_{n=0}^{\infty} (1/2^n) \cdot \cos(3^n x)$ is continuous everywhere but differentiable nowhere.

17. *Evaluate $f'(0)$ where* $f(x) = \begin{cases} x, \text{ if } x \text{ is rational} \\ \sin x, \text{ if } x \text{ is irrational} \end{cases}$

Sol. Since $f(x) = x$, if x is rational and $f(x) = \sin x$, if x is irrational, we get

$$\lim_{x \to 0} \frac{f(x) - f(0)}{x - 0} = \lim_{x \to 0} \frac{x}{x} = 1, \text{ if } x \text{ takes only rational values}$$

and $$\lim_{x \to 0} \frac{f(x) - f(0)}{x - 0} = \lim_{x \to 0} \frac{\sin x}{x} = 1, \text{ if } x \text{ takes only irrational values}$$

Therefore, by definition, $$f'(0) = \lim_{x \to 0} \frac{f(x) - f(0)}{x - 0} = 1.$$

18. *A function $f : \boldsymbol{R} \to \boldsymbol{R}$, satisfies the equation $f(x + y) = f(x) + f(y)$ for all $x \in \boldsymbol{R}$. If f is differentiable at 0 and $f'(0) = 2$, show that $f'(x) = 2f(x)$ for all $x \in \boldsymbol{R}$.*

Sol. Given $f(x + y) = f(x) + f(y)$ for all $x \in \mathbf{R}$ ... (1)

$f'(0) = 2$ and $f(x) \neq 0$ for any $x \in \mathbf{R}$ so that $f(0) \neq 0$ ... (2)

By definition, $$f'(x) = \lim_{h \to 0} \frac{f(x+h) - f(x)}{h} = \lim_{h \to 0} \frac{f(x)\, f(h) - f(x)}{h}, \text{ using (1)}$$

Thus, $$f'(x) = f(x) \times \lim_{h \to 0} \frac{f(h) - 1}{h} \quad \text{... (3)}$$

Again, by definition, $$f'(0) = \lim_{h \to 0} \frac{f(0+h) - f(0)}{h} = \lim_{h \to 0} \frac{f(0)\, f(h) - f(0)}{h}, \text{ using (1)}$$

or $$2 = f(0) \lim_{h \to 0} \frac{f(h) - 1}{h}, \text{ since } f'(0) = 2 \quad \text{... (4)}$$

Setting $x = y = 0$ in (1) yields $f(0 + 0) = f(0)\, f(0)$ or $f(0)\{f(0) - 1\} = 0$

or $f(0) - 1 = 0$, since $f(0) \neq 0$ by (2). Hence, we have

$$f(0) = 1 \quad \text{... (5)}$$

Using (5), (4) reduces to $$\lim_{h \to 0} \frac{f(h) - 1}{h} = 2 \quad \text{... (6)}$$

Using (6), (3) reduces to $$f'(x) = 2f(x), \quad \text{for all } x \in \mathbf{R}.$$

19. *Examine the deriv ability of the function defined by the relation $f(x) = x^m \sin(1/x)$, if $x = 0$ and $f(x) = 0$, if $x = 0$. Determine m, when f' is continuous at $x = 0$.*

Sol. Given $\quad f(x) = x^m \sin(1/x)$, if $x \neq 0 \quad$ and $\quad f(0) = 0$ ... (1)

By definition, $Rf'(0) = \lim_{h \to 0} \dfrac{f(0+h) - f(0)}{h} = \lim_{h \to 0} \dfrac{h^m \sin(1/h)}{h} \to 0$, if $m > 1$ ... (2)

and $\quad Lf'(0) = \lim_{h \to 0} \dfrac{f(0-h) - f(0)}{-h} = \lim_{h \to 0} \dfrac{(-h)^m \sin(-1/h)}{(-h)} \to 0$, if $m > 1$ ... (3)

(2) and (3) $\Rightarrow$ $Rf'(0) = Lf'(0)$, if $m > 1$ $\Rightarrow$ f' exists if $m > 1$

From (1), $\quad f'(x) = mx^{m-1} \sin(1/x) - x^{m-2} \cos(1/x)$, if $x \neq 0$, $\quad f'(0) = 0$ ... (4)

Now, for continuity of f' at $x = 0$, we must have

$f'(0+0) = f'(0-0) = f'(0) \quad$ or $\quad f'(0+0) = f'(0-0) = 0$, $\quad$ as $f'(0) = 0$ ... (5)

Using (4), by definition, we have

$$f'(0+0) = \lim_{h \to 0} f'(0+h) = \lim_{h \to 0} \{mh^{m-1} \sin(1/h) - h^{m-2} \cos(1/h)\}$$

or
$$f'(0+0) = \lim_{h \to 0} h^{m-2} \{mh \sin(1/h) - \cos(1/h)\} \quad \text{... (6)}$$

and
$$f'(0-0) = \lim_{h \to 0} f'(0-h) = \lim_{h \to 0} \{m(-h)^{m-1} \sin(-1/h) - (-h)^{m-2} \cos(-1/h)\}$$

or
$$f'(0-0) = \lim_{h \to 0} (-1)^{m-2} h^{m-2} \{mh \sin(1/h) - \cos(1/h)\} \quad \text{... (7)}$$

When $m > 2$, (6) and (7) yield $f'(0+0) = 0$ and $f'(0-0) = 0$. Thus, (5) is satisfied. Hence f' is continuous when $m > 2$.

Chapter 10

Mean Value Theorems

10.1. *ROLLE'S THEOREM (Chennai 2011)

If a function f with domain $[a, b]$ *is such that it is*

(*i*) *continuous in the closed interval* $[a, b]$,

(*ii*) *derivable in the open interval* $]a, b[$,

(*iii*) $f(a) = f(b)$.

then there exists $c \in\ ]a, b[$ *such that* $f'(c) = 0$.

[Delhi Maths (Prog.) 2007, 08; Bharathiar, 2004; Calicut, 2004; Delhi B.Sc. (Prog) II 2011; Delhi B.A. (Prog.) II 2007, 08; Bharathiar 2004; Calicut 2004; Delhi (G) 2001, 03, 08 Kanpur, 2001; 06, 07, 09, 12; Kumaon, 2002, 04; Meerut, 2004, 05, 09; Patna, 2003; K.U. BCA II 2007, 08; Purvanchal, 2001, 06; Rohilkhand, 1997]

Proof. The function f, being continuous in the closed interval $[a, b]$, is bounded and attains its least upper bound and greatest lower bound. Let M, m be the least upper bound and the greatest lower bound of f respectively and let c, d be such that

$$f(c) = M, \quad f(d) = m.$$

Either $\quad M = m \quad$ or $\quad M \neq m$.

Now
$$M = m \Rightarrow f(x) = M \quad \forall\, x \in [a, b]$$
$$\Rightarrow f'(x) = 0 \quad \forall\, x \in [a, b].$$

Thus, the theorem is true in this case.

Now suppose that $M \neq m$. As $f(a) = f(b)$ and $M \neq m$ at least one of the numbers M and m must be different from $f(a)$ and $f(b)$.

Let M be different from each of $f(a)$ and $f(b)$. Thus, $\quad M = f(c), M \neq f(a), M \neq f(b)$.

Now, $\quad f(c) \neq f(a) \Rightarrow c \neq a \quad$ and $\quad f(c) \neq f(b) \Rightarrow c \neq b$.

Thus,
$$a < c < b.$$

The function is derivable at c. We shall show that $\quad f'(c) = 0$.

If $f'(c) > 0$, there exists $\delta > 0$ such that $\quad f(x) > f(c) = M \quad \forall\, x \in\ ]c, c + \delta]$.

But M, being the least upper bound, we have $\quad f(x) \leq M \quad \forall\, x \in [a, b]$.

Thus, we have a contradiction, so that we cannot have $\quad f'(c) > 0$.

Now suppose that $f'(c) < 0$ that there exists $\delta > 0$ such that

$$f(x) > f(c) = M \quad \forall\, x \in [c - \delta, c[.$$

This again is not possible. Thus, we cannot have $f'(c) < 0$.

We conclude that
$$f'(c) = 0.$$

Corollary. *Let f be a function defined on* $[a, a + h]$ *such that (i) f is continuous on* $[a, a + h]$ *(ii) f is derivable on* $]a, a + h[$ *and (iii)* $f(a) = f(a + h)$. *Then there exists a real number* θ, $0 < \theta < 1$, *such that* $f'(a + \theta h) = 0$.

* Rolle (1652-1719) was a British Mathematician.

Proof. Let $b = a + h$. Then, here $[a, b] = [a + 0 \cdot h, a + 1 \cdot h]$.

Therefore $c \in]a, b[$ will be of the form $c = a + \theta h$, for some θ satisfying $0 < \theta < 1$.

Thus, $$f'(c) = 0 \Rightarrow f'(a + \theta h) = 0, 0 < \theta < 1.$$

10.2. FAILURE OF ROLLE'S THEOREM

Rolle's theorem fails to hold good for a function which does not satisfy all the three conditions of the theorem. Hence Rolle's theorem will not hold good

(*i*) if $f(x)$ is discontinuous at some point in $[a, b]$

or (*ii*) if $f(x)$ is not derivable at some point in $]a, b[$

or (*iii*) if $f(a) \neq f(b)$.

Note. The conditions of Rolle's theorem are only sufficient but not necessary for $f'(x)$ to vanish at some point in $[a, b]$. For example, consider the function

$$f(x) = \begin{cases} 0, & \text{when } 0 \le x \le 1 \\ x + 1, & \text{when } 1 < x \le 2 \end{cases}$$

We can easily show that $f(x)$ is discontinuous and not derivable at $x = 1$. Thus $f(x)$ is not continuous on $[0, 2]$ and $f(x)$ is not derivable on $]0, 2[$. Also, $f(0) \neq f(2)$. But $f'(x) = 0$ $\forall x \in [0, 1] \subset [0, 2[$.

10.3. GEOMETRICAL INTERPRETATION OF ROLLE'S THEOREM

[Garhwal, 2001; Gorakhpur, 2001; Delhi Maths (G), 2003, 08 Meerut; 2004; Srivenkateshwara, 2003]

Let A and B be the points on the graph $AP_1P_2P_3B$ of the function $y = f(x)$ corresponding to $x = a$ and $x = b$ respectively. Then, geometrically, Rolle's theorem asserts that there is at least one point between $x = a$ and $x = b$, at which the tangent to the curve of the function, is parallel to x-axis. In the figure, we have shown the possibility of three points P_1, P_2 and P_3 where the tangent is parallel to x-axis.

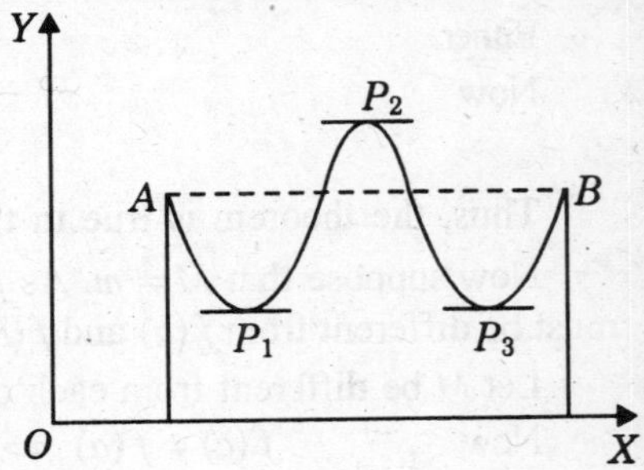

Note. Rolle's theorem tells us about the existence of at least one real number $c \in]a, b[$ such that $f'(c) = 0$. But it does not rule out the possibility of more than one such point like $x = c$. We have shown in the adjoining diagram the existence of three such points P_1, P_2 and P_3 where $f'(x) = 0$.

EXAMPLES

Example 1. *Verify whether the function $f(x) = \sin x$ in $[0, \pi]$ satisfies the conditions of Rolle's theorem and hence find c as prescribed by the theorem.* ***(Srivenkateshwara, 2003)***

Solution. Given $$f(x) = \sin x \text{ in } [0, \pi].$$

Here $f(0) = 0 = f(\pi)$. Also we know that $\sin x$ is continuous in $[0, \pi]$ and differentiable in $]0, \pi[$. Hence, f satisfies all the conditions of the Rolle's theorem. So there must exist at least one value of $x \in]0, \pi[$ such that $f'(x) = 0$.

Now, $$f'(x) = 0 \Rightarrow \cos x = 0 \Rightarrow x = \pi/2 \in [0, \pi].$$

Hence, in Rolle's theorem, $$c = \pi/2.$$

Example 2. *Verify Rolle's theorem for $f(x) = | x |$ in $[-1, 1]$.*

(Kanpur, 2002; Meerut, 2003)

Solution. Given $f(x) = |x|$ in $[-1, 1]$. Let $h > 0$. Then

$$Rf'(0) = \lim_{h \to 0} \frac{f(0+h) - f(0)}{h} = \lim_{h \to 0} \frac{|h|}{h} = \lim_{h \to 0} 1 = 1$$

and $$Lf'(0) = \lim_{h \to 0} \frac{f(0-h) - f(0)}{-h} = \lim_{h \to 0} \frac{|-h|}{-h} = \lim_{h \to 0} -1 = -1$$

Since $Rf'(0) \neq Lf'(0)$, $f(x)$ is not differentiable at $0 \in]-1, 1[$. Thus $f(x)$ is not derivable in $]-1, 1[$. Hence Rolle's theorem is not applicable to $f(x) = |x|$ in $[-1, 1]$.

Example 3. *Examine the validity of the hypothesis and the conclusion of Rolle's theorem for the function $f(x) = (x-a)^m (x-b)^n$, $x \in [a, b]$; m, n being positive integer.*

[Delhi B.Sc. (Prog) I 2011; K.U. BCA (II) 2008; Purvanchal 2006

Solution. Given $$f(x) = (x-a)^m (x-b)^n, x \in [a, b]. \quad ...(1)$$

Since m and n are positive integers, $(x-a)^m$ and $(x-b)^n$ can be expanded by binomial theorem. Then, $f(x)$ is a polynomial of degree of $m + n$. Hence $f(x)$ is continuous and differentiable on $[a, b]$. Again, from (1), $f(a) = f(b)$.

Thus all the three conditions of Rolle's theorem are satisfied. So there exists at least one value of x in $]a, b[$ such that $f'(x) = 0$.

Now, $$f'(x) = 0 \Rightarrow m(x-a)^{m-1}(x-b)^n + n(x-a)^m(x-b)^{n-1} = 0$$

or $$(x-a)^{m-1}(x-b)^{n-1}\{(m+n)x - (mb+na)\} = 0$$

$$\Rightarrow x = a, x = b \quad \text{and} \quad x = (mb+na)/(m+n).$$

Here $x = (mb+na)/(m+n)$ is point within $]a, b[$ because it divides a and b internally in the ratio $m : n$. We reject $x = a$, $x = b$ as they do not belong to $]a, b[$. Thus, Rolle's theorem is valid and $c = (mb+na)/(m+n)$.

Example 4. *Discuss the applicability of Rolle's theorem to $f(x) = 2 + (x-1)^{2/3}$ in $[0, 2]$.*

(K.U. BCA (II) 2003; Osmania, 2004)

Solution. Here $$f'(x) = (2/3) \times (x-1)^{-1/3} = 2/\{3(x-1)^{1/3}\},$$

which does not exist (*i.e.*, is not finite) at $x = 1 \in]0, 2[$. Hence the condition "$f(x)$ is derivable in $]0, 2[$" is not satisfied. Therefore, Rolle's theorem is not applicable to $f(x)$ in $[0, 2]$.

Example 5. *Discuss the applicability of Rolle's theorem to $f(x) = \log\{(x^2+ab)/(a+b)x\}$, in the interval $[a, b]$, $0 < a < b$.* **(Purvanchal, 2004)**

Solution. Here $f(a) = \log \dfrac{a^2+ab}{(a+b)a} = \log 1 = 0$ and $f(b) = \log \dfrac{b^2+ab}{(a+b)b} = \log 1 = 0$

Thus, we have $$f(a) = f(b).$$

Here $f(x) = \log(x^2+ab) - \log(a+b)x \Rightarrow f'(x) = \dfrac{2x}{x^2+ab} - \dfrac{1}{x} = \dfrac{x^2-ab}{x(x^2+ab)}$,

which does not become infinite or indeterminate for $a < x < b$ and hence $f(x)$ is derivable in $]a, b[$. Since $f(x)$ is derivable in $[a, b]$ also, so it is continuous in $[a, b]$.

Thus $f(x)$ satisfies all the three conditions of Rolle's theorem. Here $f'(x) = 0$ for at least one value of x in $]a, b]$.

Now, $f'(x) = 0 \Rightarrow (x^2-ab)/x(x^2+ab) = 0 \Rightarrow x^2 - ab = 0 \Rightarrow x = \pm\sqrt{ab}$.

Of these two values of x, clearly $x = \sqrt{ab}$ lies between a and b, being the geometric mean of a and b. So here $c = \sqrt{ab} \in]a, b[$. Thus Rolle's theorem is applicable to $f(x)$ in $[a, b]$.

Example 6. *Show that there is no real number k for which the equation $x^3 - 3x + k = 0$ has two distinct roots in [0, 1].* **[Delhi Maths (H), 2004; Delhi Maths (G), 1999]**

Solution. Let, on the contrary, that there is a real number k for which the given equation has two distinct root α and β in [0, 1], where $\alpha < \beta$. Obviously $[\alpha, \beta] \subset [0, 1]$ ...(1)

Consider $f(x) = x^3 - 3x + k$ in $[\alpha, \beta]$.

Since $f(\alpha) = f(\beta) = 0$, the conditions of Rolle's theorem are satisfied. Hence, there exists some $c \in]\alpha, \beta[$ such that

$$f'(c) = 0 \Rightarrow 3c^2 - 3 = 0 \Rightarrow c = \pm 1 \notin]0, 1[\Rightarrow c \notin]\alpha, \beta[, \text{ using (1)}$$

This is a contradiction to the condition that $c \in]\alpha, \beta[$. Hence, our assumption is wrong and therefore the result follows.

Example 7. *Prove that if P be any polynomial and P′ the derivative of P, then between any two consecutive zeros of P′, there lies at the most one zero of P.* **[Delhi, Maths (H), 2001]**

Solution. Let α and β be any two consecutive zeros of P'. Then

$$P'(\alpha) = 0 = P'(\beta) \quad \text{and} \quad P'(x) \neq 0 \quad \forall\, x \in]\alpha, \beta[. \qquad ...(1)$$

Let, if possible, there exist two zeros c and d of P, where $a < c < d < \beta$. Then $P(c) = 0 = P(d)$. Since P is a polynomial, P satisfies all the conditions of Rolle's theorem in $[c, d]$. Consequently, there exists some point $x_0 \in]c, d[$, (*i.e.*, $x_0 \in]\alpha, \beta[$) such that $P'(x_0) = 0$, which contradicts (1). Hence, our assumption is wrong and so there exists at the most one zero of P, which lies between two consecutive zeros of P'.

Example 8. *Show that between any two roots of $e^x \cos x = 1$, there exists at least one root of $e^x \sin x - 1 = 0$.* **[Kanpur 2012; I.A.S. 2009; Delhi Maths (G), 1997]**

Solution. Let α and β be any two distinct roots of $e^x \cos x = 1$

$$\therefore \quad e^\alpha \cos \alpha = 1 \quad \text{and} \quad e^\beta \cos \beta = 1 \qquad ...(1)$$

Define a function f as follows $\quad f(x) = e^{-x} - \cos x, \quad \forall\, x \in [\alpha, \beta]$ ...(2)

Obviously f is continuous in $[\alpha, \beta]$ and f is derivable in $]\alpha, \beta[$

Indeed, $\quad f'(x) = -e^{-x} + \sin x \quad \forall\, x \in]\alpha, \beta[$ ...(3)

From (2), $\quad f(\alpha) = e^{-\alpha} - \cos \alpha = \dfrac{1 - e^\alpha \cos \alpha}{e^\alpha} = 0$, by (1)

Similarly, $\quad f(\beta) = 0$ and so $f(\alpha) = f(\beta)$

Thus, f satisfies all the conditions of Rolle's theorem in $[\alpha, \beta]$ and so there exists some $\gamma \in]\alpha, \beta[$ such that $f'(\gamma) = 0$

Then (3) $\Rightarrow \sin \gamma - e^{-\gamma} = 0 \Rightarrow e^\gamma \sin \gamma - 1 = 0, \ \alpha < \gamma < \beta$

Hence, γ is a root of $e^x \sin x - 1 = 0$, $\alpha < \gamma < \beta$.

Example 9. *Prove that if $a_0, a_1, \ldots, a_n$ are real numbers such that*

$$\frac{a_0}{n+1} + \frac{a_1}{n} + \ldots + \frac{a_{n-1}}{2} + a_n = 0, \qquad ...(1)$$

then there exists at least one real number x between 0 and 1 such that

$a_0 x^n + a_1 x^{n-1} + \ldots + a_n = 0.$ **[Delhi Maths (G) 2006; Delhi Maths (H), 2004]**

Solution. Define a function f as follows :

$$f(x) = \frac{a_0}{n+1} x^{n+1} + \frac{a_1}{n} x^n + \ldots + \frac{a_{n-1}}{2} x^2 + a_n x, \ x \in [0, 1] \qquad ...(2)$$

Since f is a polynomial in x, therefore

(*i*) f is continuous in [0, 1],

(ii) f is derivable in $]0, 1[$,

(iii) $f(0) = f(1)$, since $f(0) = 0$ and $f(1) = 0$, by (1) and (2)

Thus, there exists some $x \in \,]0, 1[$ such that $f'(x) = 0$

or
$$\frac{a_0}{(n+1)} \cdot (n+1)\, x^n + \frac{a_1}{n} \cdot nx^{n-1} + \ldots.. + \frac{a_{n-1}}{2} \cdot 2x + a_n \cdot 1 = 0$$

Hence,
$$a_0 x^n + a_1 x^{n-1} + \ldots.. + a_{n-1} x + a_n = 0.$$

Example 10. *If f' is continuous on $[a, a+h]$ and derivable on $]a, a+h[$, then prove that there exists a real number c between a and $a + h$ such that*

$$f(a+h) = f(a) + h\, f'(a) + (h^2/2) \times f''(c).$$ **[Delhi Maths (H), 1999, 2001]**

Solution. Define a function ϕ on $[a, a+h]$ as follows :

$$\phi(x) = f(x) + (a+h-x)\, f'(x) + (1/2) \times (a+h-x)^2 A, \quad \text{...(1)}$$

where A is a constant to be determined by $\phi(a) = \phi(a+h)$

i.e., $$f(a) + h f'(a) + (h^2/2) \times A = f(a+h) \quad \text{...(2)}$$

Since f' is continuous on $[a, a+h]$, f and f' are continuous on $[a, a+h]$. Further $(a + h - x)$, $(a + h - x)^2$ are also continuous on $[a, a+h]$ and so by (1), ϕ is continuous on $[a, a+h]$. Since f' is derivable on $]a, a+h[$, so by (1), ϕ is derivable on $]a, a+h[$. Thus ϕ satisfies all the conditions of Rolle's theorem and so there exists some $c \in \,]a, a+h[$ such that

$$\phi'(c) = 0. \quad \text{...(3)}$$

From (1), $$\phi'(x) = f'(x) - f'(x) + (a+h-x) f''(x) - (a+h-x) A.$$

$\therefore$ $$0 = \phi'(c) = (a+h-c)\,[f''(c) - A] \quad \text{...(4)}$$

$\Rightarrow$ $f''(c) - A = 0$, since $c \in \,]a, a+h[$ means $(a+h-c) \neq 0$.

Putting $A = f''(c)$ in (2), we obtain

$$f(a+h) = f(a) + h f'(a) + (h^2/2) \times f''(c).$$

Example 11. *If a function f is twice differentiable on $[a, a+h]$, then show that*

$$f(a+h) = f(a) + hf'(a) + (h^2/2) \times f''(a + \theta h),\ 0 < \theta < 1.$$

Hint : Take $c = a + \theta h$, $0 < \theta < 1$ and do as in example 10.

Example 12. *If f' and g' exist for all $x \in [a, b]$ and if $g'(x) \neq 0\ \forall\ x \in \,]a, b[$, then prove that for some $c \in \,]a, b[$*

$$\frac{f(c) - f(a)}{g(b) - g(c)} = \frac{f'(c)}{g'(c)}.$$ **[I.A.S. 2005; Delhi Maths (G), 1993; Maths (H) 1998, 2002]**

Solution. Define a function ψ on $[a, b]$ as follows :

$$\psi(x) = f(x)\, g(x) - f(a)\, g(x) - g(b)\, f(x), \quad \forall\ x \in \,]a, b[\quad \text{...(1)}$$

Since f and g are derivable in $[a, b]$, so ψ is continuous in $[a, b]$ and derivable in $]a, b[$.

Also $$\psi(a) = \psi(b) = -f(a)\, g(b).$$

Since ψ satisfies all the conditions of Rolle's theorem, therefore, there exists some $c \in \,]a, b[$ such that $\psi'(c) = 0$. We have, from (1)

$$\psi'(x) = f'(x)\, g(x) + f(x)\, g'(x) - f(a)\, g'(x) - g(b)\, f'(x).$$

$\therefore$ $$\psi'(c) = 0 \quad \Rightarrow \quad f'(c)\, g(c) + f(c)\, g'(c) - f(a)\, g'(c) - g(b)\, f'(c) = 0$$

or $$g'(c)\,\{f(c) - f(a)\} = f'(c)\,\{g(b) - g(c)\}.$$

Hence, $$\frac{f(c) - f(a)}{g(b) - g(c)} = \frac{f'(c)}{g'(c)}.$$

Example 13. *If a function f is such that its derivative f' is continuous on $[a, b]$ and derivable on $]a, b[$, then show that there exists a number c between a and b such that*

$$f(b) = f(a) + (b-a) f'(a) + (1/2) \times (b-a)^2 f''(c).$$ **[Delhi Maths (H), 2005, 08]**

Solution. Define a function ϕ on $[a, b]$ as follows :

$$\phi(x) = f(b) - f(x)\,(b-x) f'(x) - (b-x)^2 A, \qquad ...(1)$$

where A is a constant to be determined by $\phi(a) = \phi(b)$ ...(2)

i.e., $f(b) - f(a) - (b-a) f'(a) - (b-a)^2 A = 0$

i.e., $f(b) = f(a) + (b-a) f'(a) + (b-a)^2 A.$ ...(3)

Since f' is continuous on $[a, b]$, so f is also continuous on $[a, b]$. Also $(b-x)$, $(b-x)^2$ are continuous on $[a, b]$. Then by (1), ϕ is continuous on $[a, b]$. Similarly, ϕ is derivable on $]a, b[$. Further $\phi(a) = \varphi(b)$, by (2).

Thus φ satisfies the conditions of Rolle's theorem and so there exists some point $c \in]a, b[$ such that $\phi'(c) = 0.$ ...(4)

From (1), $\phi'(x) = -f'(x) - \{-f'(x) + (b-x) f''(x)\} + 2(b-x) A$

or $\phi'(x) = -(b-x) f''(x) + 2(b-x) A.$ ...(5)

$\therefore$ $\phi'(c) = 0 \Rightarrow -(b-c) f''(c) + 2(b-c) A = 0$

Thus, $A = (1/2) \times f''(c)$, since $a < c < b \Rightarrow b - c \neq 0$.

Substituting this value of A in (3), the result is proved.

Example 14. *If a function f is such that f is continuous on $[a, b]$ and derivable in $[a, b]$, show that there exists a real number θ, $0 < \theta < 1$, such that*

$$f(b) = f(a) + (b-a) f'(a) + \frac{(b-a)^2}{2!} f''\{a + \theta(b-a)\}$$

Hint : Take $c = a + \theta(b-a)$ and do as in example 13.

Example 15. *If f'' be continuous on $[a, b]$ and derivable on $]a, b[$, then prove that*

$$f(b) - f(a) - \frac{1}{2}(b-a)[f'(a) + f'(b)] = -\frac{(b-a)^3}{12} f'''(d)$$

for some number d between a and b.

Solution. Define a function g on $[a, b]$ as follows :

$$g(x) = f(x) - f(a) - (1/2) \times (x-a)\{f'(a) + f'(x)\} + A(x-a)^3, \qquad ...(1)$$

where A is constant to be suitably chosen.

Since f'' is continuous on $[a, b]$, it follows that f and f' are continuous on $[a, b]$. Hence $g(x)$ given by (1) is continuous on $[a, b]$ and derivable on $[a, b]$. If A is so chosen that $g(a) = g(b)$, then $g(x)$ satisfies all the conditions of Rolle's theorem in $[a, b]$. Accordingly, there exists a number $c \in]a, b[$ such that $g'(c) = 0$.

From (1), $g'(x) = f'(x) - (1/2) \times \{f'(a) + f'(x)\} - (1/2) \times (x-a) f''(x) + 3A(x-a)^2$

$\Rightarrow$ $g'(c) = f'(c) - (1/2) \times \{f'(a) + f'(c)\} - (1/2) \times (c-a) f''(c) + 3A(c-a)^2$

$\therefore$ $g'(c) = 0 \Rightarrow (1/2) \times \{f'(c) - f'(a)\} - (1/2) \times (c-a) f''(c) + 3A(c-a)^2 = 0$ ...(2)

By hypothesis $g(a) = g(b)$ and $g(a) = 0$, by (1). So, we must have $g(b) = 0$

Putting $x = b$ in (1) and noting that $g(b) = 0$, we get

$$f(b) - f(a) - (1/2) \times (b-a)\{f'(a) + f'(b)\} + A(b-a)^3 = 0 \qquad ...(3)$$

Now, define a function h on $[a, c]$ as follows :

$$h(x) = (1/2) \times \{f'(x) - f'(a)\} - (1/2) \times (x-a) f''(x) + 3A(x-a)^2 \qquad ...(4)$$

Here h is continuous on $[a, c]$ and derivable on $]a, c[$. Again $h(c) = 0$ and $h(a) = 0$ by (2) and (4). So $h(a) = h(c)$. Hence h satisfies all the conditions of Rolle's theorem in $[a, c]$ and so there exists a number $d \in]a, c[$ such that $h'(d) = 0$.

From (4), $h'(x) = (1/2) \times f''(x) - (1/2) \times (x-a) f'''(x) - (1/2) \times f''(x) + 6A(x-a)$

Putting $x = d$ and using the fact that $f'(d) = 0$, we get

$$0 = (1/2) \times f''(d) - (1/2) \times (d-a) f'''(d) - (1/2) \times f''(d) + 6A(d-a)$$

or $6A(d-a) = (1/2) \times (d-a) f'''(d)$ or $A = (1/12) \times f'''(d)$, as $d \neq a$

Substituting this value of A in (3), we have

$$f(b) - f(a) - \frac{1}{2}(b-a)\{f'(a) + f'(b)\} = -\frac{1}{12}(b-a)^3 f'''(d).$$

Example 16. *A function f is such that its second derivative is continuous on $[a, a+h]$ and derivable on $]a, a+h[$, show that there exists a number θ between 0 and 1 such that*

$$f(a+h) - f(a) - \frac{1}{2}h\{f'(a) + f'(a+h)\} = -\frac{h^3}{12} f'''(a+\theta h).$$

Hint : Take $b = a + h$, $h = b - a$ and $d = a + \theta h \in]a, a+h[$ in example 15 to get the required result.

EXERCISES

1. Apply Rolle's theorem to show that tangent to the graph of the function $f(x) = 1/(x^2 + 1)$ is parallel to x-axis, at least one point between -3 and 3. **[Delhi Maths (G), 2002**

2. Verify Rolle's theorem for the function $f(x) = \sqrt{4 - x^2}$ in $[-2, 2]$

[Delhi B.Sc. (Prog) II 2011]

3. Verify Rolle's theorem for $f(x) = x(x+3)e^{-x/2}$, $-3 \le x \le 0$. **(Avadh, 2002)**

4. Show that Rolle's theorem is not applicable to the function $f(x) = 1 - (x-1)^{2/3}$ in $[0, 2]$. **[Delhi Maths (G), 1993]**

5. Verify Rolle's theorem for the following functions : **[Rajasthan 2010]**

(i) $f(x) = (x-a)^3 (x-b)^4$ on $[a, b]$ (ii) $f(x) = e^x \sin x$ on $[0, \pi]$

(iii) $f(x) = \log\{(x^2 + 3)/4x\}$ on $[1, 3]$ (iv) $f(x) = x(x-3)^2$ on $[0, 3]$

(v) $f(x) = e^{-x} \sin x$ on $[0, \pi]$ (vi) $f(x) = e^x(\sin x - \cos x)$ on $[\pi/4, 5\pi/4]$

6. Examine the validity of the hypotheses and the conclusion of Rolle's theorem for the following functions :

(i) $f(x) = \cos(1/x)$ on $[-1, 1]$ (ii) $f(x) = \tan x$ on $[0, \pi]$

(iii) $f(x) = (x-2)\sqrt{x}$ on $[0, 2]$

7. Does the function $f(x) = |x - 2|$ satisfy the conditions of Rolle's theorem in the interval $[1, 3]$? Justify your answer with correct reasoning.

8. The function f is defined on $[0, 1]$ as follows :

$$f(x) = 1 \text{ for } 0 \le x < 1/2$$
$$= 2 \text{ for } 1/2 \le x \le 1.$$

Show that $f(x)$ satisfies none of the conditions of Rolle's theorem, yet $f'(0) = 0$ for many points in $[0, 1]$.

9. If $a + b + c = 0$, then show that the quadratic equation $3ax^2 + 2bx + c = 0$ has at least one root in $]0, 1]$.

10. If $a_0 + a_1 + \ldots.. + a_n = 0$, where $a_0, a_1, \ldots.., a_n \in \mathbf{R}$, show that the equation $a_0 + 2a_1x + \ldots.. + (n+1)a_n x^n = 0$ has at least one real root in $(0, 1)$.

11. By considering the function $(x-4)\log x$, show that the equation $x\log x = 4 - x$ is satisfied by at least one value of x lying between 1 and 4.

12. Prove that if $p(x)$ is a polynomial, then between any two roots of $p(x) = 0$ lies a root of $p'(x) = 0$.

13. Prove that for any real number k, the polynomial given by $f(x) = x^3 + x + k$ has exactly one real root.

14. Show that $x = 0$ is the only real root of the equation $e^x = 1 + x$.

15. If $p(x)$ is a polynomial and $k \in \mathbf{R}$, prove that between any two real roots of $p(x) = 0$, there is a root of $p'(x) + k\,p(x) = 0$.
[**Hint.** Consider $f(x) = e^{kx} f(x)$, $x \in \mathbf{R}$. Apply Rolle's theorem to this new function $f(x)$].

16. If $f(x)$ and $g(x)$ are differentiable on (a, b) and continuous on $[a, b]$ and $f(a) = f(b) = 0$, then show that there exists a point $c \in (a, b)$ such that $f'(c) + f(c)\,g'(c) = 0$.
[**Hint :** Consider $F(x) = f(x)\,e^{g(x)}$ and apply Rolle's theorem to $F(x)$].

17. If $$f(x) = \begin{vmatrix} \sin x & \sin\alpha & \sin\beta \\ \cos x & \cos\alpha & \cos\beta \\ \tan x & \tan\alpha & \tan\beta \end{vmatrix}, \text{ where } 0 < \alpha < \beta < \pi/2$$
show that $f'(c) = 0$ where $\alpha < c < \beta$.

18. If $f(x)$, $g(x)$ and $h(x)$ have derivatives when $a \le x \le b$, show that there is a value of c in $]a, b[$ such that
$$\begin{vmatrix} f(a) & g(a) & h(a) \\ f(b) & g(b) & h(b) \\ f'(c) & g'(c) & h'(c) \end{vmatrix} = 0$$
(Garhwal, 1998; Delhi Maths (G) 2006 Kanpur 2010)

Hint : Consider $$F(x) = \begin{vmatrix} f(a) & g(a) & h(a) \\ f(b) & g(b) & h(b) \\ f(x) & g(x) & h(x) \end{vmatrix}$$
and apply Rolle's theorem to $F(x)$ on $[a, b]$.

19. Let f and g be two functions defined and continuous on $[a, b]$ and derivable on $]a, b[$. Show that there exists some $c \in\,]a, b[$ such that
$$\begin{vmatrix} f(a) & f(b) \\ g(a) & g(b) \end{vmatrix} = (b-a)\begin{vmatrix} f(a) & f'(c) \\ g(a) & g'(c) \end{vmatrix}.$$

Hint : Consider $$F(x) = \begin{vmatrix} f(a) & f(x) \\ g(a) & g(x) \end{vmatrix} - \left(\frac{x-a}{b-a}\right)\begin{vmatrix} f(a) & f(b) \\ g(a) & g(b) \end{vmatrix}$$
and apply Rolle's theorem to $F(x)$ on $[a, b]$.

20. If the functions f, g, h are continuous on $[a, b]$ and twice differentiable on $]a, b[$, prove that there exist $\xi, \eta \in [a, b]$ such that
$$\begin{vmatrix} f(a) & f(b) & f(c) \\ g(a) & g(b) & g(c) \\ h(a) & h(b) & h(c) \end{vmatrix} = \frac{1}{2}(b-c)(c-a)(a-b)\begin{vmatrix} f(a) & f'(\xi) & f''(\eta) \\ g(a) & g'(\xi) & g''(\eta) \\ h(a) & h'(\xi) & h''(\eta) \end{vmatrix}$$

Hint : Consider the function F defined as follows :
$$F(x) = \begin{vmatrix} f(a) & f(b) & f(x) \\ g(a) & g(b) & g(x) \\ h(a) & h(b) & h(x) \end{vmatrix} - \frac{(x-a)(x-b)}{(c-a)(c-b)}\begin{vmatrix} f(a) & f(b) & f(c) \\ g(a) & g(b) & g(c) \\ h(a) & h(b) & h(c) \end{vmatrix}$$
Now apply Rolle's theorem to $F(x)$.

21. If f' and g' are continuous and differentiable on $[a, b]$, show that for $a < c < b$

$$\frac{f(b) - f(a) - (b-a) f'(a)}{g(b) - g(a) - (b-a)(g'(a)} = \frac{f''(c)}{g''(c)}. \quad \textbf{[Delhi B.Sc. (H), 1991]}$$

Hint : Consider $F(x) = f(x) - (b-x) f'(x) + A\{g(x) + (b-x) g'(x)\}$ and apply Rolle's theorem to $F(x)$ with the assumption that A satisfies $F(a) = F(b)$.

22. Assuming that $f''(x)$ exists for all x in $[a, b]$, show that

$$f(c) - f(a)\frac{b-c}{b-a} - f(b)\frac{c-a}{b-a} - \frac{1}{2}(c-a)(c-b) f''(\xi) = 0,$$

where c and ξ both lie in $[a, b]$.

23. If f', g' are continuous on $[a-h, a+h]$ and derivable on $]a-h, a+h[$, then prove that

$$\frac{f(a+h) - 2f(a) + f(a-h)}{g(a+h) - 2g(a) + g(a-h)} = \frac{f''(d)}{g''(d)}$$

for some $d \in\,]a-h, a+h[$, provided $g(a+h) - 2g(a) + g(a-h) \neq 0$ and $g''(t) \neq 0$ for each $t \in\,]a-h, a+h[$.

ANSWERS

2. Yes; $c = 1/2$ **3.** Yes; $c = -2$

4. Not applicable as $f(x)$ is not differentiable at $x = 1 \in\,]0, 2[$

5. (*i*) Yes (*ii*) Yes (*iii*) Yes, $c = \sqrt{3}$ (*iv*) Yes, $c = 1$ (*v*) Yes, $c = \pi/4$ (*vi*) Yes, $c = \pi$

6. (*i*) No, as $f(x)$ is discontinuous at $x = 0 \in\,]-1, 1[$

(*ii*) No, as $f(x)$ is discontinuous at $x = \pi/2 \in\,]0, \pi]$ (*iii*) Yes, $c = 2/3$

7. No, as $f(x)$ is not derivable at $x = 2 \in\,]1, 3[$

10.4. LAGRANGE'S MEAN VALUE THEOREM OR FIRST MEAN VALUE THEOREM [AGRA 2011; DELHI B.Sc. (Prog) I, 2011]

If a function f with domain $[a, b]$ is such that it is

(*i*) *continuous in the closed interval $[a, b]$,* (*ii*) *derivable in the open interval $]a, b[$,*

then there exists $c \in\,]a, b[$ such that $\dfrac{f(b) - f(a)}{b-a} = f'(c)$.

[*Agra, 2012*; *Delhi B.Sc. (Hons.) II 2018*; *Delhi B.Sc. (Prog), II 2007*; *Garhwal, 2001*; *Kakatiya, 2003*; *Kanpur, 2011*; *Meerut, 2009*; *Patna, 2003*; *Purvanchal, 2006*; *Utkal, 2003*]

Proof. We consider a function ϕ defined as follows :

$$\phi(x) = f(x) + Ax,$$

where A is such that $\phi(a) = \phi(b)$

This requires $f(a) + Aa = f(b) + Ab$

$$\Rightarrow A = -[f(b) - f(a)]/(b-a) \qquad ...(1)$$

Now ϕ is (*i*) continuous in the closed interval $[a, b]$ (*ii*) derivable in the open interval $]a, b[$, and (*iii*) $\phi(a) = \phi(b)$ and as such, by the Rolle's theorem, there exists $c \in\,]a, b[$ such that

$$\phi'(c) = 0$$

Also $\phi'(x) = f'(x) + A$

$$\Rightarrow 0 = \phi'(c) = f'(c) + A \Rightarrow A = -f'(c) \qquad ...(2)$$

From (1) and (2), $[f(b) - f(a)]/(b-a) = f'(c)$.

* Lagrange (1736-1813) was a French Mathematician.

ANOTHER FORM OF LAGRANGE'S MEAN VALUE THEOREM

If a function f defined in $[a, a + h]$ *is such that*

(*i*) *f is continuous in* $[a, a + h]$ (*ii*) *f is derivable in* $]a, a + h[$

Then there exists some $\theta \in\,]0, 1[$ *such that*

$$f(a + h) - f(a) = h f'(a + \theta h) \qquad \textbf{[Delhi Maths (H), 2003]}$$

Proof. Take $b = a + h$ and $c \in\,]a, a + h[$ as $c = a + \theta h$, where $0 < \theta < 1$.

GEOMETRICAL INTERPRETATION OF THE MEAN VALUE THEOREM

[Delhi Maths (G), 2002; Garhwal, 2001; Kanpur, 2001; Kumaon, 1999; Meerut, 2001, 02; Purvanchal 2008]

Let A and B be the points on the graph of the function $y = f(x)$ corresponding to $x = a$ and $x = b$ respectively. Then the coordinates of A and B are $(a, f(a))$ and $(b, f(b))$ respectively.

Now, the slope of chord $AB = [f(b) - f(a)]/(b - a)$.

Since $f(x)$ is continuous on $[a, b]$ and differentiable on $]a, b[$, figures show that there is at least one point between A and B the tangent at which is parallel to the chord AB.

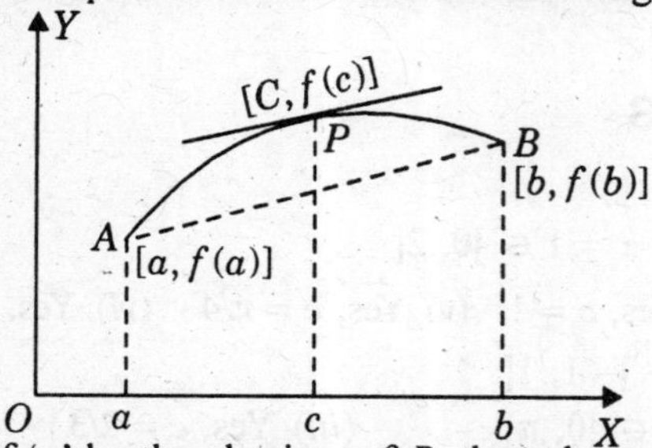

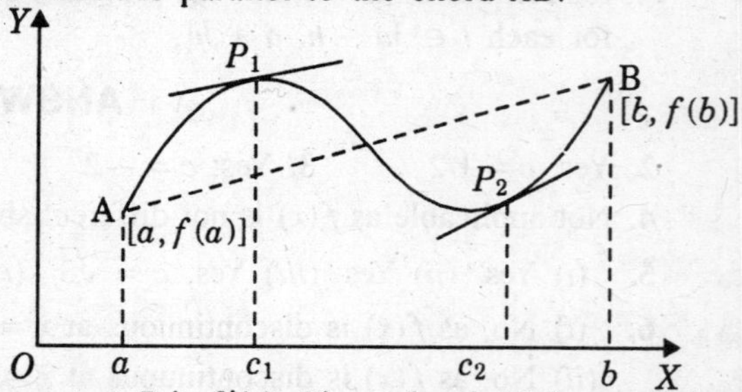

If 'c' be the abscissa of P, then slope of tangent at P is $f'(c)$. Since chord AB is parallel to the tangent at P, we have

$$[f(b) - f(a)]/(b - a) = f'(c).$$

Thus, interpreted geometrically, Lagrange's mean value theorem says that the tangent to the graph $y = f(x)$ at some suitable point between a and b is parallel to the chord joining the points on the graph with abscissae a and b.

10.5. INCREASING & DECREASING FUNCTIONS. MONOTONE FUNCTIONS

Increasing function. Definition. *A function f defined on an interval I is said to be increasing in I if* $f(x_1) \le f(x_2)$ *whenever* $x_1, x_2 \in I$ *and* $x_1 < x_2$. *Refer figure* (*i*).

Strictly increasing function. Definition. *A function f defined on an interval I is said to be strictly increasing in I if* $f(x_1) < f_1(x_2)$ *whenever* $x_1, x_2 \in I$ *and* $x_1 < x_2$. *Refer figure* (*ii*).

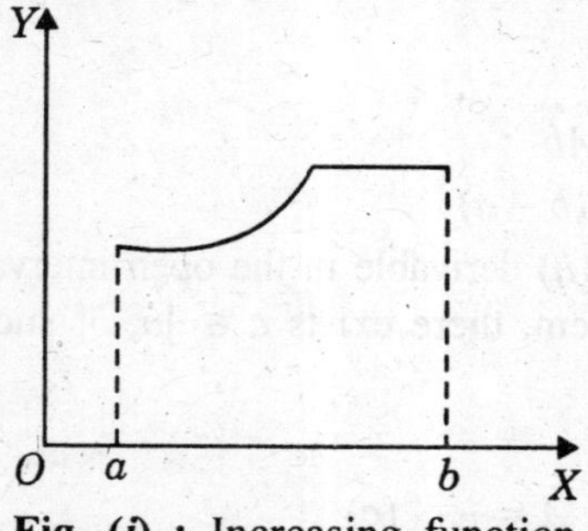

Fig. (*i*) : Increasing function

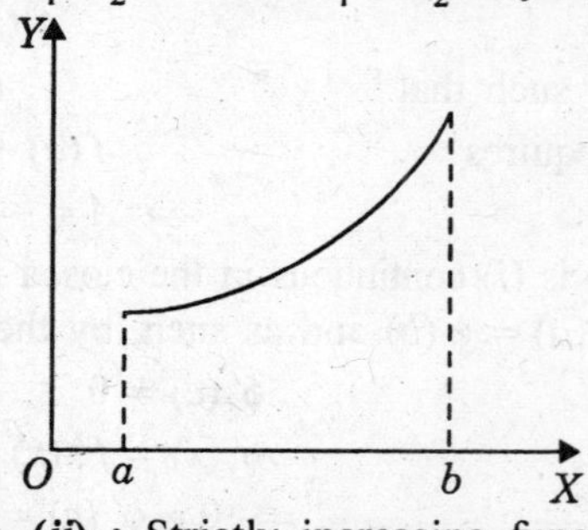

Fig. (*ii*) : Strictly increasing function

Decreasing function. Definition. *A function f defined on an interval I is said to be decreasing in I if* $f(x_1) \ge f(x_2)$ *whenever* $x_1, x_2 \in I$ *and* $x_1 < x_2$. *Refer figure* (*iii*).

Strictly decreasing function. Definition. *A function f defined on an interval I is said to be strictly decreasing in I if* $f(x_1) > f(x_2)$ *whenever* $x_1, x_2 \in I$ *and* $x_1 < x_2$. *Refer figure (iv).*

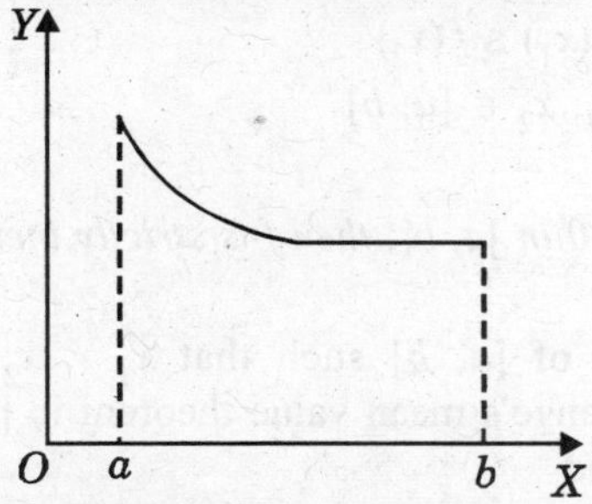

Fig. (*iii*) : Decreasing function

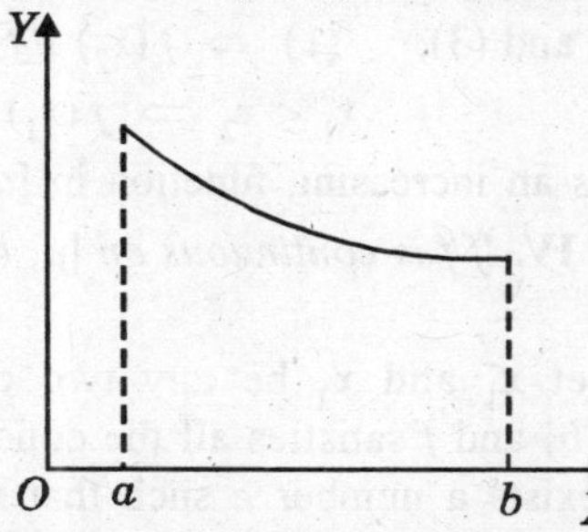

Fig. (*iv*) : Strictly decreasing function

Monotone function. Definition. A function is known as monotone in an interval I if it is either increasing in I or decreasing in I.

Strictly monotone function. Definition. A function is known as strictly monotone in an interval I if it is either strictly increasing or strictly decreasing in I.

10.6. SOME USEFUL DEDUCTIONS FROM THE MEAN VALUE THEOREM

Theorem I. *If f is defined and continuous on* $[a, b]$ *and is derivable on* $]a, b[$, *and if* $f'(x) = 0$ *for all x in* $]a, b[$, *then* $f(x)$ *has a constant value throughout* $[a, b]$.

[Delhi B.Sc. I (Hons) 2008]

Proof. Let c be any point of $]a, b[$. Then f is continuous on $[a, c]$ and derivable on $]a, c[$. Since f satisfies all the conditions of Lagrange's mean value theorem on $[a, c]$, it follows that there exists a real number d between a and c such that

$$f(c) - f(a) = (c - a) f'(d) \qquad ...(1)$$

But by hypothesis $f'(x) = 0 \ \forall \ x \in \ ]a, b[$. Hence $f'(d) = 0$ and so (1) reduces to $f(c) - f(a) = 0$ or $f(c) = f(a)$. Since c is an arbitrary point of $]a, b[$, it now follows that $f(x) = f(a)$ $\forall \ x \in [a, b]$. Therefore $f(x)$ has a constant value throughout $[a, b]$.

Theorem II. *If* $f(x)$ *and* $g(x)$ *are both defined and continuous on* $[a, b]$, *and are derivable on* $]a, b[$, *and if* $f'(x) = g'(x) \ \forall \ x \in \ ]a, b[$, *then* $f(x)$ *and* $g(x)$ *differ by a constant on* $[a, b]$.

Proof. Consider the function $\phi(x) = f(x) - g(x) \quad \forall \ x \in [a, b]$ ...(1)

Since $f(x)$ and $g(x)$ are defined and continuous on $[a, b]$, it follows that $\phi(x)$ is also defined and continuous on $[a, b]$. Again, since $f(x)$ and $g(x)$ are derivable $]a, b[$, it follows that $\phi(x)$ is also derivable on $]a, b[$.

Again, by (1) $\qquad \phi'(x) = f'(x) - g'(x)$ ...(2)

By hypothesis $\qquad f'(x) = g'(x) \quad \forall \ x \in \ ]a, b[$ ...(3)

From (2) and (3), we see that $\qquad \phi'(x) = 0 \quad \forall \ x \in \ ]a, b[$.

Hence from theorem I, we must have

$$\phi(x) = \text{constant} \quad \forall \ x \in [a, b]$$

$$\Rightarrow f(x) - g(x) = \text{constant} \quad \forall \ x \in [a, b], \text{ using (1)}$$

Thus $f(x)$ and $g(x)$ differ by a constant on $[a, b]$.

Theorem III. *If f is continuous on* $[a, b]$, *and* $f'(x) \geq 0$ *in* $]a, b[$, *then f is increasing in* $[a, b]$.

Proof. Let x_1 and x_2 be any two distinct points of $[a, b]$ such that $x_1 < x_2$. Then $[x_1, x_2] \subset [a, b]$ and f satisfies all the conditions of Lagrange's mean value theorem in $[x_1, x_2]$. Hence there exists a number c such that $x_1 < c < x_2$ and

$$f(x_2) - f(x_1) = (x_2 - x_1) f'(c). \qquad ...(1)$$

Now, $$x_1 < x_2 \Rightarrow x_2 - x_1 > 0 \quad ...(2)$$

Also, $$f'(x) \geq 0 \quad \forall\ x \in\ ]a, b[\quad \text{and} \quad x_1 < c < x_2 \Rightarrow f'(c) \geq 0 \quad ..(3)$$

Using (2) and (3), $$(1) \Rightarrow f(x_2) - f(x_1) \geq 0 \Rightarrow f(x_1) \leq f(x_2)$$

Thus $$x_1 < x_2 \Rightarrow f(x_1) \leq f(x_2) \quad \forall\ x_1, x_2 \in [a, b]$$

Hence f is an increasing function in $[a, b]$.

Theorem IV. *If f is continuous on* $[a, b]$ *and* $f'(x) > 0$ *in* $]a, b[$, *then f is strictly increasing in* $[a, b]$.

Proof. Let x_1 and x_2 be any two distinct points of $[a, b]$ such that $x_1 < x_2$. Then $[x_1, x_2] \subset [a, b]$ and f satisfies all the conditions of Lagrange's mean value theorem in $[x_1, x_2]$. Hence there exists a number c such that $x_1 < c < x_2$ and

$$f(x_2) - f(x_1) = (x_2 - x_1) f'(c) \quad ...(1)$$

Now, $$x_1 < x_2 \Rightarrow x_2 - x_1 > 0 \quad ...(2)$$

Also, $$f'(x) > 0 \quad \forall\ x \in\ ]a, b[\quad \text{and} \quad x_1 < c < x_2 \Rightarrow f'(c) > 0 \quad ..(3)$$

Using (2) and (3), $$(1) \Rightarrow f(x_2) - f(x_1) > 0 \Rightarrow f(x_1) < f(x_2).$$

Thus $$x_1 < x_2 \Rightarrow f(x_1) < f(x_2) \quad \forall\ x_1, x_2 \in [a, b]$$

Hence f is a strictly increasing function.

Theorem V. *If f is continuous on* $[a, b]$ *and* $f'(x) \leq 0$ *in* $]a, b[$, *then f is decreasing in* $[a, b]$.

Proof. Proceed as in theorem III yourself.

Theorem VI. *If f is continuous on* $[a, b]$ *and* $f'(x) < 0$ *in* $]a, b[$, *then f is strictly decreasing in* $[a, b]$.

Proof. Proceed as in theorem IV yourself.

Theorem VII. *If f' exists and is bounded on some interval I, then f is uniformly continuous on I.*

Proof. Since f' is bounded in the interval I, there exists a $k > 0$ such that

$$| f'(x) | \leq k \quad \forall\ x \in I. \quad ...(1)$$

Let x_1 and x_2 be any two distinct points of I such that $x_1 < x_2$. Then $[x_1, x_2] \subset I$ and f satisfies all the conditions of Lagrange's mean value theorem in $[x_1, x_2]$. Hence there exists a number c such that $x_1 < c < x_2$ and

$$f'(c) = \frac{f(x_2) - f(x_1)}{x_2 - x_1} \quad ...(2)$$

Then, from (1) and (2), $$\left| \frac{f(x_2) - f(x_1)}{x_2 - x_1} \right| < k$$

$$\Rightarrow | f(x_2) - f(x_1) | < k \, | x_2 - x_1 | \quad \forall\ x_1, x_2 \in I \quad ...(2)$$

Let $\varepsilon > 0$ be given. Then we can choose $\delta = \varepsilon/k$.

Let $x_1, x_2 \in I$ such that $| x_2 - x_1 | < \delta$. Then (3) gives

$$| f(x_2) - f(x_1) | < k\delta < \varepsilon, \text{ as } \delta = \varepsilon/k$$

Thus, $$| x_2 - x_1 | < \delta \Rightarrow | f(x_2) - f(x_1) | < \varepsilon \quad \forall\ x_1, x_2 \in I$$

Hence f is uniformly continuous on I.

EXAMPLES

Example 1. *Find the value of c of Lagrange's mean value theorem when*

$$f(x) = 2x^2 + 3x + 4 \text{ in } [1, 2]. \qquad \textbf{(Kanpur, 2002)}$$

Solution. Here $f(x) = 2x^2 + 3x + 4$ being a polynomial in x is continuous in the closed interval [1, 2].

Again, $f'(x) = 4x + 3$. Since $f'(x)$ is always unique and definite for $\forall\, x \in\,]1, 2[$. Hence $f(x)$ is derivable in]1, 2[.

Thus $f(x)$ satisfies conditions of Lagrange's mean value theorem and so there must exist $c \in\,]1, 2[$ such that

$$\frac{f(2) - f(1)}{2 - 1} = f'(c) \qquad \left[\because \frac{f(b) - f(a)}{b - a} = f'(c)\right]$$

or $$\frac{18 - 9}{2 - 1} = 4c + 3 \quad \text{so that} \quad c = \frac{3}{2}.$$

Since $c \in\,]1, 2[$, so required value of $c = 3/2$.

Example 2. *Verify Lagrange's Mean Value Theorem for the function*

$f(x) = x(x - 1)(x - 2)$ *in* $[0, 1/2]$. ***(Agra 2009; Kumaon, 1999; Rohilkhand, 1996)***

Solution. Since f is a polynomial, f is continuous in [0, 1/2] and derivable in]0, 1/2[. Thus, there exists $c \in\,]0, 1/2[$ such that

$$\frac{f(1/2) - f(0)}{(1/2) - 0} = f'(c). \qquad ...(1)$$

Now $$f'(x) = (x - 1)(x - 2) + x(x - 1) + x(x - 2)$$

or $$f'(x) = 3x^2 - 6x + 2. \text{ Now } f(0) = 0,\ f(1/2) = 3/8.$$

From (1), $$3/8 = (1/2) \times (3c^2 - 6c + 2)$$

or $$12c^2 - 24c + 5 = 0 \quad \text{or} \quad c = (6 \pm \sqrt{21})/6$$

Out of these two values of c only $1 - (\sqrt{21}/6)$ lies in]0, 1/2[. Hence we get $c = 1 - (\sqrt{21}/6)$ and the theorem is verified.

Example 3. *Test if Lagrange's Mean Value Theorem holds for the function* $f(x) = |x|$ *in the interval* $[-1, 1]$. ***(Kanpur 2010; Meerut, 2001)***

Solution. Here $f(x) = |x|$ is not differentiable at $x = 0$. Prove this for complete solution. Hence $f(x)$ is not differentiable in $]-1, 1[$ and so Lagrange's mean value theorem does not hold for $f(x) = |x|$ in $[-1, 1]$.

Example 4. *Prove that if f be defined for all x such that* $|f(x) - f(y)| < (x - y)^2$ *for all x and y, then f is constant.* ***[Delhi Maths (H), 2002, 03]***

Solution. Given $$|f(x) - f(y)| < (x - y)^2 \quad \forall\, x, y \in \mathbf{R} \qquad ...(1)$$

Let c be any real number. Then, for $x \neq c$, (1) gives

$$\left|\frac{f(x) - f(c)}{x - c}\right| \le |x - c| \qquad ...(2)$$

Let $\varepsilon > 0$ be given. Then, we choose $\delta = \varepsilon$ and then (2) gives

$$\left|\frac{f(x) - f(c)}{x - c} - 0\right| < \varepsilon \text{ whenever } |x - c| < \delta$$

$$\Rightarrow \qquad \lim_{x \to c} \frac{f(x) - f(c)}{x - c} = 0 \Rightarrow f'(c) = 0 \qquad ...(3)$$

Since c is any point of $\mathbf{R}$, (3) shows that $\quad f'(x) = 0 \quad \forall\, x \in \mathbf{R}$.

Hence by theorem I of Art. 10.6, f is constant.

Example 5. *Let f be defined and continuous on [a – h, a + h] and derivable on]a – h, a + h[. Prove that there is real number θ between 0 and 1 for which*

$$f(a+h) - 2f(a) + f(a-h) = h\,[f'(a+\theta h) - f'(a-\theta h)].$$

[Delhi B.Sc. (G), 2001; Delhi Maths (H), 2004]

Solution. Consider $\phi(x) = f(a+hx) + f(a-hx)$ on [0, 1].

Then, (*i*) ϕ is continuous in [0, 1], (*ii*) ϕ is derivable in]0, 1[. So by Lagrange's mean value theorem,

$$\frac{\phi(1) - \phi(0)}{1-0} = \phi'(\theta) \text{ for some } \theta \in \,]0, 1[$$

or $$\phi(1) - \phi(0) = \phi'(\theta), \text{ if } 0 < \theta < 1$$

or $$[f(a+h) + f(a-h)] - [f(a) + f(a)] = h\,[f'(a+\theta h) - f'(a-\theta h)], \text{ if } 0 < \theta < 1.$$

Hence, $$f(a+h) - 2f(a) + f(a-h) = h\,[f'(a+\theta h) - f'(a-\theta h)], \text{ if } 0 < \theta < 1.$$

Example 6. *Show that* $\dfrac{v-u}{1+v^2} < \tan^{-1} v - \tan^{-1} u < \dfrac{v-v}{1+u^2}$, *if* $0 < u < v$,

and deduce that $\dfrac{\pi}{4} + \dfrac{3}{25} < \tan^{-1}\dfrac{4}{3} < \dfrac{\pi}{4} + \dfrac{1}{6}$. **[Delhi Math (G) 2006; Nagpur 2010]**

Solution. Applying Lagrange's mean value theorem to the function

$f(x) = \tan^{-1} x$ in [u, v], we obtain

$$\frac{f(v) - f(u)}{v-u} = f'(c) \text{ for some } c \in \,]u, v[$$

or $$\frac{\tan^{-1} v - \tan^{-1} u}{v-u} = \frac{1}{1+c^2} \text{ for } u < c < v. \qquad ...(1)$$

Now $$c > u \Rightarrow 1 + c^2 > 1 + u^2 \Rightarrow 1/(1+c^2) < 1/(1+u^2) \qquad ...(2)$$

Again $$c < v \Rightarrow 1 + c^2 < 1 + v^2 \Rightarrow 1/(1+c^2) > 1/(1+v^2) \qquad ...(3)$$

From (1), (2), (3), we obtain

$$\frac{1}{1+v^2} < \frac{\tan^{-1} v - \tan^{-1} u}{v-u} < \frac{1}{1+u^2}$$

Hence, $$\frac{v-u}{1+v^2} < \tan^{-1} v - \tan^{-1} u < \frac{v-u}{1+u^2}, \text{ as } u < v \Rightarrow v - u > 0 \qquad ...(4)$$

Taking $u = 1$ and $v = 4/3$ in (4), we obtain

$$\frac{3}{25} < \tan^{-1}\frac{4}{3} - \tan^{-1} 1 < \frac{1}{6} \quad \text{or} \quad \frac{3}{25} < \tan^{-1}\frac{4}{3} - \frac{\pi}{4} < \frac{1}{6}$$

Hence, $$\frac{\pi}{4} + \frac{3}{25} < \tan^{-1}\frac{4}{3} < \frac{\pi}{4} + \frac{1}{6}.$$

Example 7. *Evaluate the value of θ that appears in Lagrange's mean value theorem for the function $x^2 - 2x + 3$ given that a = 1 and b = 1/2.*

Solution. Here $f(x) = x^2 - 2x + 3$ so that $f'(x) = 2x - 2$ ...(1)

Since *f* is a polynomial, *f* satisfies the conditions of Lagrange's mean value theorem and so there exist θ, 0 < θ < 1, satisfying

$$f(a+h) - f(a) = hf'(a+\theta h)$$

or $$f(1 + 1/2) - f(1) = (1/2) \times f'(1 + \theta/2), \text{ as } a = 1 \text{ and } h = 1/2$$

or $\quad (3/2)^2 - 2 \times (3/2) + 3 - (1 - 2 + 3) = (1/2) \times \{2\,(1 + \theta/2) - 2\}$, by (1)

or $\quad (9/4) - 2 = \theta/2 \quad$ so that $\quad \theta = 1/2.$

Example 8. *Using mean value theorem, show that*

$$\frac{x}{1+x} < \log(1+x) < x,\ x > 0.$$ **[Agra 2011]**

Solution. Let $f(x) = \log(1 + x)$ in $[0, x]$, so that $f'(x) = 1/(1+x)$. ...(1)

Since f is continuous in $[0, x]$ and derivable in $]0, x[$, so by Lagrange's mean value theorem, there exists some θ, $0 < \theta < 1$ such that

$$\frac{f(x) - f(0)}{x - 0} = f'(\theta x)$$

or $$\log(1+x) = x/(1+\theta x), \text{ using (1)} \qquad ...(2)$$

Now $$0 < \theta < 1 \text{ and } x > 0 \Rightarrow \theta x < x$$

$$\Rightarrow \quad 1 + \theta x < 1 + x \;\Rightarrow\; 1/(1+\theta x) > 1/(1+x)$$

$$\Rightarrow \quad \frac{x}{1+\theta x} > \frac{x}{1+x} \;\Rightarrow\; \frac{x}{1+x} < \frac{x}{1+\theta x}. \qquad ...(3)$$

Again $$0 < \theta < 1 \text{ and } x > 0 \;\Rightarrow\; 1 < 1 + \theta x$$

$$\Rightarrow \quad \frac{1}{1+\theta x} < 1 \;\Rightarrow\; \frac{x}{1+\theta x} < x. \qquad ...(4)$$

From (3) and (4)

$$\frac{x}{1+x} < \frac{x}{1+\theta x} < x. \qquad ...(5)$$

From (2) and (5), we obtain $\quad x/(1+x) < \log(1+x) < x.$

Example 9. *Applying Lagrange's mean value theorem to the function defined by*

$$f(x) = \log(1 + x) \text{ for all } x > 0,$$

show that $$0 < [\log(1 + x)]^{-1} - x^{-1} \text{ whenever } x > 0.$$

[Delhi Maths (H), 2000; Delhi Maths (G), 2000]

Solution. First proceed like Example 8 to obtain

$$\frac{x}{1+x} < \log(1+x) < x \;\Rightarrow\; \frac{1+x}{x} > \frac{1}{\log(1+x)} > \frac{1}{x}$$

$$\Rightarrow 1 > \frac{1}{\log(1+x)} - \frac{1}{x} > 0.$$

Hence, $$0 < [\log(1 + x)]^{-1} - x^{-1} < 1,\ x > 0.$$

Example 10. *Use the mean value theorem to prove* $\dfrac{x}{1+x^2} < \tan^{-1} x < x$, *if* $x > 0$.

(Bundelkhand, 1994; Kanpur, 1996; Kakatiya, 1992)

Solution. Let $\quad f(x) = \tan^{-1} x$ in $[0, x]$.

Then f satisfies the conditions of Lagrange's mean value theorem in $[0, x]$. Consequently, there exists some θ satisfying $0 < \theta < 1$ such that

$$\frac{f(x) - f(0)}{x - 0} = f'(\theta x) \qquad \left[\because\ f'(x) = \frac{1}{1+x^2} \text{ and } f(0) = 0\right]$$

or $$\tan^{-1} x = x/(1+\theta^2 x^2) \quad \ldots(1)$$

Now $0 < \theta < 1$ and $x > 0 \Rightarrow \theta^2 x^2 < x^2 \Rightarrow 1 + \theta^2 x^2 < 1 + x^2$

$$\Rightarrow \frac{x}{1+\theta^2 x^2} > \frac{x}{1+x^2} \Rightarrow \frac{x}{1+x^2} < \frac{x}{1+\theta^2 x^2} \quad \ldots(2)$$

Again $0 < \theta < 1$ and $x > 0 \Rightarrow 1 < 1 + \theta^2 x^2$

$$\Rightarrow \frac{1}{1+\theta^2 x^2} < 1 \Rightarrow \frac{x}{1+\theta^2 x^2} < x. \quad \ldots(3)$$

From (2) and (3), we obtain $$\frac{x}{1+x^2} < \frac{x}{1+\theta^2 x^2} < x,\ x > 0 \quad \ldots(4)$$

From (1) and (4), we obtain $x/(1+x^2) < \tan^{-1} x < x,\ x > 0.$

Example 11. *If $f''(x) > 0$ for all $x \in \mathbf{R}$, then show that*

$$f\left[\frac{1}{2}(x_1 + x_2)\right] \le \frac{1}{2}\{f(x_1) + f(x_2)\}.$$ **[*Delhi Maths (H), 1998, 2000, 01*]**

Solution. If $x_1 = x_2$, then the given relation becomes equality $f(x_1) = f(x_1)$ and hence the result is proved.

Now, let $x_1 \ne x_2$ and let us assume that $x_1 < x_2$.

Since $f''(x)$ is exists $\forall\ x \in \mathbf{R}$, it follows that f satisfies the conditions of Lagrange's mean value theorem in each of the intervals $[x_1, (x_1 + x_2)/2]$ and $[(x_1 + x_2)/2, x_2]$. Hence there exist numbers c_1 and c_2 such that

$$f[(x_1 + x_2)/2] - f(x_1) = \left[\frac{1}{2}(x_1 + x_2) - x_1\right] f'(c_1), \quad x_1 < c_1 < \frac{1}{2}(x_1 + x_2)$$

and $$f(x_2) - f[(x_1 + x_2)/2] = \left[x_2 - \frac{1}{2}(x_1 + x_2)\right] f'(c_2), \quad \frac{1}{2}(x_1 + x_2) < c_2 < x_2$$

On subtracting the corresponding sides of the above equations, we get

$$2f[(x_1 + x_2)/2] - f(x_1) - f(x_2) = [(x_2 - x_1)/2]\ [f'(c_1) - f'(c_2)] \quad \ldots(1)$$

Since f'' exists $\forall\ x \in \mathbf{R}$, it follows that f' satisfies the conditions of Lagrange's mean value theorem in the interval $[c_1, c_2]$ and hence there exists a number d such that

$$f'(c_2) - f'(c_1) = (c_2 - c_1) f''(d), \text{ when } d \in\ ]c_1, c_2[. \quad \ldots(2)$$

Since $c_2 - c_1 > 0$ and $f''(d) > 0$, hence (2) reduces to

$$f'(c_2) - f'(c_1) > 0 \quad \text{or} \quad f'(c_1) - f'(c_2) < 0 \quad \ldots(3)$$

Again, since $$x_1 < x_2 \Rightarrow (x_2 - x_1)/2 > 0 \quad \ldots(4)$$

Using (3) and (4), (1) reduces to

$$2f[(x_1 + x_2)/2] - f(x_1) - f(x_2) < 0 \quad \text{or} \quad f[(x_1 + x_2)/2] < (1/2) \times \{f(x_1) + f(x_2)\}.$$

Example 12. *If $a = -1$, $b \ge 1$ and $f(x) = 1/|x|$, show that the conditions of Lagrange's mean value theorem are not satisfied by the function f in $[a, b]$ but that the conclusion is true if and only if $b > 1 + \sqrt{2}$.*

Solution. We have $\forall\ x < 0, f(x) = 1/|x| = -1/x \Rightarrow f'(x) = 1/x^2;$

and $\forall\ x > 0, f(x) = 1/|x| = 1/x \Rightarrow f'(x) = -1/x^2;$

for $x = 0$, f is not derivable.

Thus, the conditions of the Lagrange's mean value theorem are not satisfied in as much as f is not derivable at the point '0' of the open interval $]a, b[$.

Again, $\dfrac{f(b)-f(a)}{b-a}=\dfrac{1/|b|-1/|a|}{b-a}=\dfrac{(1/b)-1}{b+1}=\dfrac{1-b}{b(b+1)}$ for $b \geq 1$ and $a=-1$.

As $[f(b)-f(a)]/(b-a)$ is negative, it cannot be equal to the derivative of the function f, for any negative value of x; the derivative being necessarily positive for such values of x. Let, if possible, there exists $c > 0$ such that

$$f'(c)=[f(b)-f(a)]/(b-a).$$

This requires $-\dfrac{1}{c^2}=\dfrac{1-b}{b(b+1)}=-\dfrac{b-1}{b(b+1)} \Rightarrow c^2=\dfrac{b(b+1)}{b-1}$.

For the conclusion of the theorem to be true, we must have $c < b$. This requires

$$\frac{b(b+1)}{b-1} < b^2 \Rightarrow b > 1+\sqrt{2}.$$

EXERCISES

1. *Verify Lagrange's mean value theorem for the following functions in the specified intervals :*

(*i*) $f(x)=2x^2-7x+10$ in $[2, 5]$ **(Meerut, 2003)**

(*ii*) $f(x)=\sqrt{x^2-4}$ in $[2, 4]$ **[Delhi Maths (G), 1994]**

(*iii*) $f(x)=x^2-2x+3$ in $[-2, 2]$ **(Kanpur, 2003)**

(*iv*) $f(x)=\cos x$ in $[0, \pi/2]$ **(Kakatiya, 2003)**

(*v*) $f(x)=x \sin x$ in $[0, \pi/2]$ **(Meerut, 2004)**

(*vi*) $f(x)=1/x$ in $[1, 4]$

(*vii*) $f(x)=x^n$, (n being a positive integer) in $[-1, 1]$

(*viii*) $f(x)=\log x$ in $[1, e]$ **(Kanpur, 2003; Manipur, 2002)**

(*ix*) $f(x)=x^3$ in $[a, b]$ **(Kanpur, 2001)**

2. *Examine the validity of the hypotheses and the conclusion of Lagrange's mean value theorem for the following functions in the specified intervals :*

(*i*) $f(x)=1/x$ in $[-1, 1]$ (*ii*) $f(x)=x^{1/3}$ in $[-1, 1]$

(*iii*) $f(x)=1+x^{2/3}$ in $[-8, 1]$

(*iv*) $f(x)=x \sin(1/x)$ in $[-1, 1]$; $f(0)=0$. **(Rohilkhand, 1993)**

3. *Find 'c' of Lagrange's mean value theorem of the following functions in the specified intervals :*

(*i*) $f(x)=(x-1)(x-2)(x-3)$ in $[0, 4]$ **(Meerut 2010; Gorakhpur, 2001)**

(*ii*) $f(x)=lx^2+mx+n$ in $[a, b]$

(*iii*) $f(x)=\log x$ in $[1, e]$ **(Gorakhpur, 1992)**

(*iv*) $f(x)=e^x$ in $[0, 1]$ **(Kanpur, 1994)**

(*v*) $f(x)=x^2-2x+3$ in $[-2, 2]$ **(Kanpur, 2003)**

(*vi*) $f(x)=x(x-1)$ in $[1, 2]$ **(Agra 2007, 10)**

4. Prove that (*i*) $|\sin x-\sin y| \leq |x-y| \ \forall \ x, y \in \mathbf{R}$ **[Delhi B.Sc. I (H) 2010, 11**

(*ii*) $|\tan^{-1} x-\tan^{-1} y| \leq |x-y| \ \forall \ x, y \in \mathbf{R}$ **[Delhi Maths (G), 1998]**

5. Verify that on the curve $y=px^2+qx+r$ (p, q, r being real numbers, $p \neq 0$), the chord joining the points for which $x=a$, $x=b$ is parallel to the tangent at the point given by $x=(a+b)/2$.

6. *Find the number (numbers) θ that appears in the conclusion of Lagrange's mean value theorem for the following functions for the specified values of a and h:*
(*i*) $f(x) = x^2$; $a = 1$, $h = 1/2$ (*ii*) $f(x) = \log_e x$; $a = 1$, $h = 1/10$
(*iii*) $f(x) = \log_e x$; $a = 1$, $h = 1/e$

7. Prove that for any quadratic $px^2 + qx + r$, the value of θ in Lagrange's theorem is always 1/2, whatever p, q, r, h may be.

8. Let f be defined and continuous in $[a - h, a + h]$ and derivable in $]a - h, a + h[$. Prove that there is a real number θ between 0 and 1 such that
$f(a + h) - f(a - h) = h\,[f'(a + \theta h) + f'(a - \theta h)]$. **[Delhi Maths (G), 1995]**

9. Use Lagrange's mean value theorem to prove that (*i*) $1 + x < e^x < 1 + xe^x \ \forall\ x > 0$.
(*ii*) $e^x > 1 + x,\ x > 0$ **[Delhi B.Sc. I (Hons) 2010]**

10. Show that the function f' if it exists in an interval, cannot have an ordinary or removable discontinuity in that interval.

11. A twice differentiable function f on $[a, b]$ is such that $f(a) = f(b) = 0$ and $f(c) > 0$ for $a < c < b$. Prove that there is at least one value ξ between a and b for which $f''(\xi) < 0$.

12. If $f(0) = 0$ and $f''(x)$ exists on $[0, \infty[$, show that
$$f'(x) - \frac{f(x)}{x} = \frac{1}{2} x f''(\xi),\ 0 < \xi < x,$$
and deduce that if $f''(x)$ is positive for positive values of x, then $f(x)/x$ strictly increases in $]0, \infty[$. **[Delhi Maths (H), 1995]**

13. If f'' be defined on $[a, b]$ and if $|f''(x)| \le M$ for all x in $[a, b]$, then prove that
$$\left| f(b) - f(a) - \frac{1}{2}(b - a)\{f'(a) + f'(b)\} \right| \le \frac{1}{2}(b - a)^2 M.$$

14. If $f''(x)$ exists $\forall\ x \in [a, b]$ and $\dfrac{f(c) - f(a)}{c - a} = \dfrac{f(b) - f(c)}{b - c}$, where $c \in\ ]a, b[$, then prove that there exists some $\xi \in\ ]a, b[$ such that $f''(\xi) = 0$. **[Delhi maths (H) 2007)**

15. Use the mean value theorem to show that
(*i*) $\sqrt{1 + x} < 1 + x/2$, if $-1 < x < 0$ or $x > 0$
(*ii*) $1 + \dfrac{x}{2\sqrt{1 + x}} < \sqrt{1 + x} < 1 + \dfrac{x}{2}$, for $-1 < x < 0$
(*iii*) $e^a(x - a) < e^x - e^a < e^x(x - a)$, if $a < x$

16. Examine the validity of the (*i*) hypothesis, and (*ii*) conclusion of the Lagrange's mean value theorem in the cases
(*a*) $f(x) = 1/x$, (*b*) $f(x) = x^{1/3}$, for the interval $[-1, 1]$.

17. Applying Lagrange's mean value theorem in turn to the functions $\log x$ and e^x, determine the corresponding values of θ in terms of a and h. Deduce that
(*i*) $0 < [\log(1 + x)^{-1} - x^{-1}] < 1$. (*ii*) $0 < x^{-1} \log[(e^x - 1)/x] < 1$.
[Delhi Maths (H) 2007]

ANSWERS

3. (*i*) $2 \pm 2\sqrt{3}/2$ (*ii*) $(a + b)/2$ (*iii*) $e - 1$ (*iv*) $\log(e - 1)$ (*v*) 3/2
6. (*i*) $(91/3)^{1/2} - 5$ (*ii*) $[\log_e(1{\cdot}1)]^{-1} - 10$ (*iii*) $[\log(1 + 1/e)]^{-1} - e$

10.7. INCREASING AND DECREASING FUNCTIONS AND THEIR APPLICATION IN ESTABLISHING SOME INEQUALITIES

Reader is advised to study Art. 10.5 and theorems III, IV, V and VI of Art. 10.6 carefully. For ready reference, we now present some useful results in the following table :

S. No.	*Value of $f'(x)$ in $[a, b]$*	*Nature of f in $[a, b]$*
1.	$f'(x) \geq 0$	f is increasing
2.	$f'(x) > 0$	f is strictly increasing
3.	$f'(x) \leq 0$	f is decreasing
4.	$f'(x) < 0$	f is strictly decreasing

EXAMPLES

Example 1. *Separate the intervals in which the function*

$$f(x) = 2x^3 - 15x^2 + 36x - 7 \quad \forall\ x \in \mathbf{R}$$

is increasing or decreasing. **[*Delhi Maths* (G), *2003*]**

Solution. Here $\qquad f(x) = 2x^3 - 15x^2 + 36x - 7$

From (1), $\qquad f'(x) = 6x^2 - 30x + 36 = 6(x^2 - 5x + 6) = 6(x-2)(x-3)$

We observe that

$f'(x) > 0$, whenever $x < 2$, $\qquad f'(x) = 0$, when $x = 2$,

$f'(x) < 0$, whenever $2 < x < 3$, $\qquad f'(x) = 0$, when $x = 3$,

and $\quad f'(x) > 0$, when $x > 3$.

Since $f'(x) > 0$ in each of the intervals $]-\infty, 2[$ and $]3, \infty[$ and since f is continuous everywhere, it follows that f is strictly increasing in both the intervals $]-\infty, 2]$ and $[3, \infty[$.

Again, since $f'(x) < 0$ in $]2, 3[$ and since f is continuous in $[2, 3]$, it follows that f is strictly decreasing in $[2, 3]$.

Example 2. *Show that $x^3 - 3x^2 + 3x + 2$ is monotonically increasing in every interval.*

Solution. Let $\qquad f(x) = x^3 - 3x^2 + 3x + 2 \qquad$ ...(1)

From (1), $\qquad f'(x) = 3x^2 - 6x + 3 = 3(x-1)^2$.

Thus, $\qquad f'(x) = 0$ when $x = 1 \quad$ and $\quad f'(x) > 0$ when $x \neq 1$.

Let c be any real number less than 1. Since $f'(x) > 0$ in $]c, 1[$ and since f is continuous in $[c, 1]$ it follows that f is increasing in $[c, 1]$.

Again, let d be any real number greater than 1. Since $f'(x) > 0$ in $]1, d[$ and f is continuous in $[1, d]$, it follows that f increasing in $[1, d]$.

Combining the above two results, we see that f is increasing in every interval.

Example 3. *Show that*

(*a*) $x - \dfrac{x^2}{2} < \log(1+x) < x - \dfrac{x^2}{2(1+x)}$, $x > 0$. **[Agra 2005]**

[*Meerut 2005; I.A.S., 2004; Delhi B.Sc.* (*G*), *1996; Delhi Maths* (*G*). *2002*]

(*b*) $\dfrac{x^2}{2(1+x)} < x - \log(1+x) < \dfrac{x^2}{2}$, $x > 0$. **[*Delhi Physics* (*H*), *1999*]**

(*c*) $\dfrac{x^2}{2} < x - \log(1+x) < \dfrac{x^2}{2(x+1)}$, $-1 < x < 0$. **[*Delhi, Maths* (*H*), *2005*]**

Solution. (*a*) Let $\qquad f(x) = \log(1+x) - (x - x^2/2);\ x > 0$

$$\therefore \qquad f'(x) = \frac{1}{1+x} - 1 + x = \frac{x^2}{1+x} > 0. \qquad (\because\ x > 0)$$

$\Rightarrow f(x)$ is an increasing function of x, for $x > 0$ ...(1)

$\Rightarrow f(x) > f(0)$, for $x > 0 \Rightarrow f(x) > 0$, as $f(0) = 0$

$\Rightarrow \quad \log(1+x) - (x - x^2/2) > 0$, for $x > 0$

$$\therefore \qquad x - x^2/2 < \log(1+x),\ x > 0 \qquad \text{...(2)}$$

$$\text{Let} \qquad g(x) = x - \frac{x^2}{2(1+x)} - \log(1+x),\ x > 0 \qquad \text{...(3)}$$

$$\therefore \qquad g'(x) = 1 - \frac{1}{2}\left\{\frac{2x + x^2}{(1+x)^2}\right\} - \frac{1}{1+x} = \frac{x^2}{2(1+x^2)^2} > 0 \text{ as } x > 0$$

$\Rightarrow g(x)$ is an increasing function, for $x > 0$

$\Rightarrow g(x) > g(0)$, for $x > 0 \quad \Rightarrow \quad g(x) > 0$ as $g(0) = 0$, by (3)

$$\Rightarrow x - \frac{x^2}{2(1+x)} - \log(1+x) > 0, \text{ for } x > 0, \text{ using (3)}$$

$$\therefore \qquad \log(1+x) < x - \frac{x^2}{2(1+x)}, \text{ for } x > 0 \qquad \text{...(4)}$$

Hence, from (2) and (4), $\qquad x - \dfrac{x^2}{2} < \log(1+x) < x - \dfrac{x^2}{2(1+x)}$, for $x > 0$.

(*b*) By part (*a*), we have

$$-x + \frac{x^2}{2} > -\log(1+x) > \frac{x^2}{2(1+x)} - x,\ x > 0 \text{ or } \frac{x^2}{2} > x - \log(1+x) > \frac{x^2}{2(1+x)},\ x > 0$$

$$\text{Hence,} \qquad \frac{x^2}{2(1+x)} < x - \log(1+x) < \frac{x^2}{2},\ x > 0.$$

(*c*) Left as an exercise for the reader.

Example 4. *Prove that* $\dfrac{\tan x}{x} > \dfrac{x}{\sin x}$, *whenever* $0 < x < \dfrac{\pi}{2}$. **[I.A.S. 2008]**

[*Delhi Maths (H), 1999, 2002; Delhi Maths (G), 1991; Delhi Maths (G), 1992*]

Solution. We have, $\qquad \dfrac{\tan x}{x} - \dfrac{x}{\sin x} = \dfrac{\tan x \sin x - x^2}{x \sin x}.$

Since $x \sin x > 0$ for all x in $]0, \pi/2[$, therefore, in order to prove the inequality we must prove that $\qquad \tan x \sin x - x^2 > 0, \quad \forall\ x \in\ ]0, \pi/2[$.

Let c be any real number in $]0, \pi/2[$.

Let $\qquad f(x) = \tan x \sin x - x^2 \quad \forall\ x \in [0, c]$.

Then f is continuous as well as derivable in $[0, c]$.

Now, $\quad f'(x) = \sec^2 x \sin x + \tan x \cos x - 2x = \sin x(\sec^2 x + 1) - 2x$.

Let $\qquad g(x) = f'(x)$ for all x in $[0, c]$.

Clearly, function g is continuous as well as derivable on $[0, c]$.

Now, $$g'(x) = \cos x\,(\sec^2 x + 1) + \sin x \cdot 2\sec^2 x \tan x - 2$$
$$= (\sqrt{\sec x} - \sqrt{\cos x})^2 + 2\sin^2 x \sec^3 x.$$

Since $g'(x) > 0$ for all $x \in]0, c[\Rightarrow g$ is strictly increasing in $[0, c] \Rightarrow g(x) > g(0)$, whenever $0 < x < c$. Since $g(0) = 0$, this means that $g(x) > 0$, whenever $0 < x < c \Rightarrow f'(x) > 0$, whenever $0 < x < c \Rightarrow f$ is strictly increasing in $[0, c] \Rightarrow f(c) > f(0) = 0 \Rightarrow f(c) > 0$

$$\Rightarrow \quad \tan c \sin c - c^2 > 0 \quad \forall\, c \in]0, \pi/2[$$

$$\Rightarrow \quad \frac{\tan c}{c} - \frac{c}{\sin c} > 0 \quad \forall\, c \in]0, \pi/2[.$$

Since c is any point of $]0, \pi/2[$, it follows that

$$\frac{\tan x}{2} > \frac{x}{\sin x}, \text{ whenever } 0 < x < \frac{\pi}{2}.$$

Example 5. *Prove that* $x < \sin^{-1} x < \dfrac{x}{\sqrt{1 - x^2}}$, *if* $0 < x < 1$. **[Delhi Maths (G), 2006]**

Solution. Let $$f(x) = \sin^{-1} x - x,\ 0 < x < 1$$

$$\therefore \quad f'(x) = \frac{1}{\sqrt{1-x^2}} - 1 = \frac{1 - \sqrt{1-x^2}}{\sqrt{1-x^2}}$$

$\Rightarrow f'(x) > 0$, for $0 < x < 1$ $\Rightarrow$ f is strictly increasing in $0 < x < 1$

$\Rightarrow f(x) > f(0)$, for $0 < x < 1$ $\Rightarrow$ $f(x) > 0$, for $0 < x < 1$.

$\Rightarrow \sin^{-1} x - x > 0$, for $0 < x < 1$ $\Rightarrow$ $\sin^{-1} x > x$, for $0 < x < 1$

Thus, $$x < \sin^{-1} x, \text{ for } 0 < x < 1. \quad \text{...(1)}$$

Let $$g(x) = \frac{x}{\sqrt{1-x^2}} - \sin^{-1} x,\ 0 < x < 1$$

$$\therefore \quad g'(x) = \frac{\sqrt{1-x^2} + \dfrac{x(-2x)}{2\sqrt{1-x^2}}}{(1-x^2)} - \frac{1}{\sqrt{1-x^2}} = \frac{1}{(1-x^2)\sqrt{1-x^2}} - \frac{1}{\sqrt{1-x^2}} = \frac{x^2}{(1-x^2)^{3/2}}$$

$\Rightarrow g'(x) > 0$, for $0 < x < 1$ $\Rightarrow$ g is strictly increasing, for $0 < x < 1$

$\Rightarrow g(x) > g(0)$, for $0 < x < 1$ $\Rightarrow$ $g(x) > 0$, for $0 < x < 1$

$$\Rightarrow \quad \frac{x}{\sqrt{1-x^2}} - \sin^{-1} x > 0, \text{ for } 0 < x < 1$$

Thus, $$\sin^{-1} x < x/\sqrt{1-x^2}, \text{ for } 0 < x < 1 \quad \text{...(2)}$$

From (1) and (2), $$x < \sin^{-1} x < x/\sqrt{1-x^2}, \text{ if } 0 < x < 1.$$

Example 6. *If* $0 < x < 1$, *show that* $2x < \log \dfrac{1+x}{1-x} < 2x\left(1 + \dfrac{1}{3} \cdot \dfrac{x^2}{1-x^2}\right)$.

Deduce that $$e < \left(1 + \frac{1}{n}\right)^{n + 1/2} < e \cdot e^{\frac{1}{12n(n+1)}}.$$ **[Delhi Maths (G), 2003, 04]**

Solution. Let $$f(x) = \log\frac{1+x}{1-x} - 2x,\ 0 \le x < 1 \qquad ...(1)$$

Let $c \in]0, 1[$. Then f is continuous and derivable in $[0, c]$. Again, from (1), we have

$$f'(x) = \frac{1-x}{1+x} \times \frac{(1-x)+(1+x)}{(1-x)^2} - 2 = \frac{2}{1-x^2} - 2 = \frac{2x^2}{1-x^2} \qquad ...(2)$$

From (2), we find that $f'(x) > 0$ for $0 \le x < c$

Hence f is strictly increasing in $[0, c]$.

In particular, $f(c) > f(0)$ if $c > 0$. Also $f(0) = 0$, by (2).

Now, $$f(c) > 0 \Rightarrow \log\frac{1+c}{1-c} - 2c > 0,\ 0 < c < 1 \Rightarrow 2c < \log\frac{1+c}{1-c}\ \forall\ c \in]0, 1[$$

Since c is any point of $]0, 1[$, it follows that

$$2x < \log\frac{1+x}{1-x},\ \text{for } 0 < x < 1 \qquad ...(3)$$

Let $$g(x) = 2x\left(1 + \frac{1}{3}\cdot\frac{x^2}{1-x^2}\right) - \log\frac{1+x}{1-x} \qquad ...(4)$$

From (4), $$g'(x) = 2 + \frac{2}{3}\cdot\frac{(1-x^2)\cdot 3x^2 + x^3\cdot 2x}{(1-x^2)^2} - \left(\frac{1}{1+x} + \frac{1}{1-x}\right)$$

$$g'(x) = 2 + \frac{2(3x^2 - x^4)}{3(1-x^2)^2} - \frac{2}{1-x^2} = \frac{4x^4}{3(1-x^2)^2} \qquad ...(5)$$

From (5), we find that $g'(x) > 0$ for $0 \le x < c$.

Hence f is strictly increasing in $[0, c]$. In particular, $g(c) > g(0)$ if $c > 0$. Also $g(0) = 0$, by (4).

Now, $$g(c) > 0 \Rightarrow 2c\left(1 + \frac{1}{3}\cdot\frac{c^2}{1-c^2}\right) - \log\frac{1+c}{1-c} > 0,\ \text{for } 0 < c < 1$$

Since c is any point of $]0, 1[$, it follows that

$$2x\left(1 + \frac{1}{3}\cdot\frac{x^2}{1-x^2}\right) - \log\frac{1+x}{1-x} > 0,\ \text{for } 0 < x < 1$$

or $$\log\frac{1+x}{1-x} < 2x\left(1 + \frac{1}{3}\cdot\frac{x^2}{1-x^2}\right)\ \text{for } 0 < x < 1 \qquad ...(6)$$

From (3) and (6), $$2x < \log\frac{1+x}{1-x} < 2x\left(1 + \frac{1}{3}\cdot\frac{x^2}{1-x^2}\right),\ 0 < x < 1 \qquad ...(7)$$

Deduction. Since $0 < x < 1$, we take $x = 1/(2n+1)$ in (7) and get

$$\frac{2}{2n+1} < \log\frac{1 + \frac{1}{2n+1}}{1 - \frac{1}{2n+1}} < \frac{2}{2n+1}\left[1 + \frac{1}{3}\,\frac{\left(\frac{1}{2n+1}\right)^2}{1 - \left(\frac{1}{2n+1}\right)^2}\right]$$

or $$\frac{2}{2n+1} < \log\frac{2n+2}{2n} < \frac{2}{2n+1}\left[1+\frac{1}{3}\frac{1}{(2n+1)^2-1}\right]$$

or $$\frac{2}{2n+1} < \log\frac{n+1}{n} < \frac{2}{2n+1}\left[1+\frac{1}{12n(n+1)}\right]$$

or $$1 < \left(n+\frac{1}{2}\right)\log\left(1+\frac{1}{n}\right) < 1+\frac{1}{12n(n+1)}$$

or $$e < \left(1+\frac{1}{n}\right)^{n+1/2} < e.e^{\frac{1}{12n(n+1)}}$$

EXERCISES

1. Show that the function f, defined on **R** by $f(x) = x^3 + 3x^2 + 3x - 8$, for all $x \in$ **R**, is increasing in every interval.

2. Separate the intervals in which the function f defined on **R** by $f(x) = x^3 - 6x^2 + 9x + 1$, $\forall\ x \in$ **R** is increasing or decreasing.

3. Determine the intervals in which the function $(x^4 + 6x^3 + 17x^2 + 32x + 32)\, e^{-x}$ is increasing or decreasing.

4. Prove that

(i) $\dfrac{x}{1+x} < \log(1+x) < x$, if $x > -1$, $x \neq 0$ **[I.A.S. 2008; Delhi Maths (H), 2001]**

(ii) $\dfrac{x^2}{2} < x - \log(1+x) < \dfrac{x^2}{2(1+x)}$, if $-1 < x < 0$ **[Delhi Maths (H), 1997]**

(iii) $x - \dfrac{x^2}{2} + \dfrac{x^3}{3(1+x)} < \log(1+x) < x - \dfrac{x^2}{2} + \dfrac{x^3}{3}$, if $x > 0$ **[Delhi Maths (H), 1996]**

(iv) $\tan x > x$, if $0 < x < \pi/2$ (v) $\dfrac{2}{\pi} < \dfrac{\sin x}{x} < 1$, if $0 < x < \dfrac{\pi}{2}$

[Kanpur 2011; Delhi Maths (H) 2009; Delhi Maths (G), 2005

(vi) $\log(1+x) > 2x/(1+x)$, if $x > 0$ (vii) $x - x^2/3 < \tan^{-1} x$, if $x > 0$

(viii) $x - x^3/6 < \sin x < x$, if $x > 0$

(ix) $\log(1+1/x) < 1/\sqrt{x^2+x}$, if $x > 0$ **[Delhi Maths (G), 2002]**

5. Prove that

(i) If $f'(x) > 0$ in $]a, \infty[$, then f is strictly increasing in $]a, \infty[$

(ii) If $f'(x) \geq 0$ in $]a, \infty[$, then f is increasing in $]a, \infty[$

(iii) If $f'(x) > 0$ in $]a, \infty[$ and f is continuous at a, then f is strictly increasing in $[a, \infty]$

(iv) If $f'(x) \geq 0$ in $]a, \infty[$ and f is continuous at a, then f is increasing in $[a, \infty]$.

6. Prove that

(i) If $f'(x) < 0$ in $]a, \infty[$, then f is strictly decreasing in $]a, \infty[$

(ii) If $f'(x) \leq 0$ in $]a, \infty[$, then f is decreasing in $]a, \infty[$

(iii) If $f'(x) < 0$ in $]a, \infty[$ and f is continuous at a, then f is strictly decreasing in $[a, \infty]$

(iv) If $f'(x) \geq 0$ in $]a, \infty[$ and f is continuous at a, then f is decreasing in $[a, \infty]$.

7. Define an increasing function. If f is increasing on $[a, b]$, prove that $f(c+)$ and $f(c-)$ both exist for each $c \in (a, b)$ and that $f(c-) \leq f(c) \leq f(c+)$. **(Bharathiar, 2004)**

ANSWERS

2. Strictly increasing in $]-\infty, 1]$ and $[3, \infty[$; strictly decreasing in $[1, 3]$
3. Strictly increasing in $[-2, -1]$, $[0, 1]$ and strictly decreasing in $]-\infty, -2]$, $[-1, 0]$ and $[1, \infty[$

10.8. CAUCHY'S MEAN VALUE THEOREM (OR SECOND MEAN VALUE THEOREM OF DIFFERENTIAL CALCULUS)

If two functions F and f are (i) continuous in the closed interval $[a, b]$; *(ii) derivable in the open interval* $]a, b[$; *and (iii)* $F'(x) \neq 0 \ \forall \ x \in \ ]a, b[$, *then there exists at least one point,* $c \in \ ]a, b[$ *such that*

$$\frac{f(b)-f(a)}{F(b)-F(a)} = \frac{f'(c)}{F'(c)}.$$

[Delhi B.Sc. I (H) 2009; Delhi Maths (Prog) 2007]

[Kanpur 2008, 2013; Gorakhpur, 2002; Nagpur 2010; Delhi Maths (G), 2001, 02, 04, 05; Manipur, 2002; Meerut, 2002, 03; Agra 2009; Delhi Maths (P), 2001, 03, 04, 05]

Proof. Let a function ϕ be defined by $\phi(x) = f(x) + AF(x)$, where, A, is a constant to be so determined that $\phi(a) = \phi(b)$.

This requires $[F(b) - F(a)]A = -[f(b) - f(a)]$. ...(1)

Now $[F(b) - F(a) \neq 0$, for, if it were 0, then the function F would satisfy all the conditions of Rolle's theorem, and its derivative would, therefore, vanish at least once in $]a, b[$ and the condition *(iii)* would be violated. On this account, we have from (1),

$$A = -[f(b) - f(a)]/[F(b) - F(a)]. \quad ...(2)$$

The function ϕ is continuous in the closed interval $[a, b]$, derivable in the open interval $]a, b[$, and $\varphi(a) = \varphi(b)$. Hence by Rolle's theorem, there exists at least one point $c \in \ ]a, b[$ such that $\varphi'(c) = 0$.

But $\varphi'(x) = f'(x) + AF'(x)$

$\Rightarrow$ $0 = \varphi'(c) = f'(c) + AF'(c)$

$\Rightarrow$ $\dfrac{f'(c)}{F'(c)} = -A = \dfrac{f(b)-f(a)}{F(b)-F(a)}$, using (2) $\quad [\because \ F'(c) \neq 0]$

Another form statement. *If two functions f, F are continuous in a closed interval* $[a, a+h]$, *derivable in the open interval* $]a, a+h[$ *and* $F'(x) \neq 0$ *for any* $x \in \ ]a, a+h[$, *then there exists at least one number* θ, *between 0 and 1 such that*

$$\frac{f(a+h)-f(a)}{F(a+h)-F(a)} = \frac{f'(a+\theta h)}{F'(a+\theta h)}; \ \theta \in \ ]0, 1[.$$

Note. Taking $F(x) = x$, we may be see that the Lagrange's mean value theorem is only a particular case of the Cauchy's theorem. **[Delhi Maths (G), 2002]**

Geometrical interpretation of Cauchy's mean value theorem. Rewriting Cauchy's mean value theorem, we have

$$\frac{f(b)-f(a)}{F(b)-F(a)} \times F'(c) = f'(c),$$

showing that there is an ordinate $x = c$ between $x = a$ and $x = b$ such that the tangents at the points where $x = c$ cuts the graphs of the functions $f(x)$ and $\dfrac{f(b)-f(a)}{F(b)-F(a)} F(x)$ are mutually parallel.

10.9. GENERALISED MEAN VALUE THEOREM

If the three function f, g and h defined on [a, b] are such that

(i) f, g and h are continuous on [a, b] *(ii) f, g and h are derivable on]a, b[.*

Then there exists a number c in]a, b[such that

$$\begin{vmatrix} f'(c) & g'(c) & h'(c) \\ f(a) & g(a) & h(a) \\ f(b) & g(b) & h(b) \end{vmatrix} = 0. \qquad \text{(Bharathiar, 2004; I.A.S., 1993)}$$

Proof. Define a function $\phi(x)$ defined on $[a, b]$ as follows :

$$\phi(x) = \begin{vmatrix} f(x) & g(x) & h(x) \\ f(a) & g(a) & h(a) \\ f(b) & g(b) & h(b) \end{vmatrix} \qquad ...(1)$$

Rewriting (1), we have $\qquad \phi(x) = A f(x) + Bg(x) + Ch(x), \qquad$ ...(2)

where A, B, C are constants since each of f, g and h is continuous on $[a, b]$ and derivable on $]a, b[$, (2) shows that $\phi(x)$ is also continuous on $[a, b]$ and derivable on $]a, b[$.

From (1), we see that two rows in $\phi(x)$ become identical by putting $x = a$ or $x = b$, hence $\phi(a) = 0$ and $\phi(b) = 0$. Hence, $\qquad \phi(a) = \phi(b)$

Thus ϕ satisfies all the three conditions of Rolle's theorem on $[a, b]$. Hence there exists a number c in $]a, b[$ such that $\qquad \phi'(c) = 0.$

But $\phi'(x) = \begin{vmatrix} f'(x) & g'(x) & h'(x) \\ f(a) & g(a) & h(a) \\ f(b) & g(b) & h(b) \end{vmatrix}$ so $\phi'(c) = \begin{vmatrix} f'(c) & g'(c) & h'(c) \\ f(a) & g(a) & h(a) \\ f(b) & g(b) & h(b) \end{vmatrix}$

Since $\phi'(c) = 0$, we have $\begin{vmatrix} f'(c) & g'(c) & h'(c) \\ f(a) & g(a) & h(a) \\ f(b) & g(b) & h(b) \end{vmatrix} = 0.$

Note 1. Taking $g(x) = x$ and $h(x) = 1 \ \forall \ x \in [a, b]$ in the above generalised mean value theorem, we get

$$\begin{vmatrix} f'(c) & 1 & 0 \\ f(a) & a & 1 \\ f(b) & b & 1 \end{vmatrix} = 0 \quad \text{or} \quad \begin{vmatrix} f'(c) & 1 & 0 \\ f(a) & a & 1 \\ f(b) - f(a) & b - a & 0 \end{vmatrix} = 0$$

or $\quad f'(c)\cdot(b-a) - \{f(b) - f(a)\} = 0 \quad$ or $\quad \dfrac{f(b) - f(a)}{b - a} = f'(c),$

which is Lagrange's mean value theorem.

Note 2. Taking $g(x) = F(x)$ and $h(x) = 1 \ \forall \ x \in [a, b]$ in the above generalised mean value theorem, we get

$$\begin{vmatrix} f'(c) & F'(c) & 0 \\ f(a) & F(a) & 1 \\ f(b) & F(b) & 1 \end{vmatrix} = 0 \quad \text{or} \quad \begin{vmatrix} f'(c) & F'(c) & 0 \\ f(a) & F(a) & 1 \\ f(b) - f(a) & F(b) - F(a) & 0 \end{vmatrix} = 0$$

or $\qquad -f'(c)\{F(b) - F(a)\} + F'(c)\{f(b) - f(a)\} = 0$

giving $\qquad \dfrac{f(b) - f(a)}{F(b) - F(a)} = \dfrac{f'(c)}{F'(c)},$

which is Cauchy's mean value theorem.

EXAMPLES

Example 1. *Verify the Cauchy's mean value theorem for the functions x^2 and x^3 in [1, 2] and also find 'c' of this theorem.* **(Agra 2005; Meerut, 2002, 06; Utkal, 2003)**

Solution. Let $f(x) = x^2$ and $F(x) = x^3$. Since $f(x)$ and $g(x)$ are both polynomial functions, so they are continuous on [1, 2] and derivable on [1, 2]. Also $F'(x) = 3x^2 \neq 0$ for any point in]1, 2[. Hence by Cauchy's mean value theorem there exists at least one number c in]1, 2[such that

$$\frac{f(2)-f(1)}{F(2)-F(1)} = \frac{f'(c)}{F'(c)} \quad \text{or} \quad \frac{4-1}{8-1} = \frac{2c}{3c^2}, \text{ as } f'(x) = 2x \quad \text{and} \quad F'(x) = 3x^2$$

or $$3/7 = 2/3c \quad \text{or} \quad c = 14/9,$$

which lies in]1, 2[. Hence Cauchy's mean value theorem is verified.

Example 2. *If, in the Cauchy's mean value theorem, we write*

(*i*) *$f(x) = 1/x^2$ and $F(x) = 1/x$, then c is the harmonic mean between a and b*

(*ii*) *$f(x) = x$ and $F(x) = x$, then c is the arithmetic mean between a and b*

(*iii*) *$f(x) = \sqrt{x}$ and $F(x) = 1/\sqrt{x}$, then c is the geometric mean between a and b.*

(Agra, 2001)

Solution. (*i*) Let $f(x) = 1/x^2$ and $F(x) = 1/x \Rightarrow f'(x) = -2/x^3$ and $F'(x) = -1/x^2$

$\therefore \; f(a) = 1/a^2, F(a) = 1/a, f(b) = 1/b^2, F(b) = 1/b, f'(c) = -2/c^3, F'(c) = -1/c^2$

Hence by Cauchy's theorem there exists a point $c \in\,]a, b[$ such that

$$\frac{f(b)-f(a)}{F(b)-F(a)} = \frac{f'(c)}{F'(c)} \quad \text{or} \quad \frac{(1/b^2)-(1/a^2)}{(1/b)-(1/a)} = \frac{(-2/c^3)}{(-1/c^2)}$$

or $$(1/b)+(1/c) = 2/c \quad \text{or} \quad c = (2ab)/(a+b),$$

showing c is the harmonic mean between a and b.

Parts (*ii*) and (*iii*). Left as exercises for the reader.

Example 3. *Show that* $\dfrac{\sin\alpha - \sin\beta}{\cos\beta - \cos\alpha} = \cot\theta$, *where* $0 < \alpha < \theta < \beta < \pi/2$.

[Delhi Maths (G), 1998; Delhi Maths (H), 2005, 06; Meerut, 1996]

Solution. Let $f(x) = \sin x,\ F(x) = \cos x$ so that $f'(x) = \cos x,\ F'(x) = -\sin x$.

Here $f(x)$ and $F(x)$ are continuous on [α, β] and derivable in]α, β[. Also, $F'(x) \neq 0$ for any $x \in\,]\alpha, \beta[$. Hence by Cauchy's mean value theorem, there exists at least one number $\theta \in\,]\alpha, \beta[$ such that

$$\frac{f(\beta)-f(\alpha)}{F(\beta)-F(\alpha)} = \frac{f'(\theta)}{F'(\theta)} \quad \text{or} \quad \frac{\sin\beta-\sin\alpha}{\cos\beta-\cos\alpha} = \frac{\cos\theta}{-\sin\theta}$$

or $$\frac{\sin\alpha-\sin\beta}{\cos\beta-\cos\alpha} = \cot\theta, \text{ where } 0 < \alpha < \theta < \beta < \pi/2.$$

EXERCISES

1. Find 'c' of Cauchy's mean value theorem for the following pairs of functions :

(*i*) $f(x) = e^x, F(x) = e^{-x}$ in $[a, b]$

(*ii*) $f(x) = \sin x, F(x) = \cos x$ in $[-\pi/2, 0]$ **[Agra 2005]**

2. Use Cauchy's mean value theorem to show that $\lim_{x \to 1} \dfrac{\cos(\pi x/2)}{\log(1/x)} = \dfrac{\pi}{2}$.

3. Let the function f be continuous in $[a, b]$ and derivable in $]a, b[$. Show that there exists a number c in $]a, b[$ such that $2c\,[f(a) - f(b)] = f'(c)\,(a^2 - b^2)$.

4. If $f(x) = x^2$, $g(x) = \cos x$, then find the point $c \in [0, \pi/2]$ which gives the result of Cauchy's mean value theorem in the interval $[0, \pi/2]$ for the functions $f(x)$ and $g(x)$.

ANSWERS

1. (*i*) $(a + b)/2$ (*ii*) $-\pi/4$

10.10. HIGHER DERIVATIVES

Let f be a function such that f' exists in a certain neighbourhood of c. This implies that f is defined and continuous in a neighbourhood of c. If the function f' has derivative at c, then this derivative is called the second derivative of f at c, and is denoted by $f''(c)$. In this case f' is necessarily continuous at c.

In general, if $f^{(n-1)}(x)$ exists in a certain neighbourhood of c, then the derivative of $f^{(n-1)}$ at c, in case it exists, is called the nth derivative of f at c and is denoted by $f^{(n)}(c)$.

10.11. *TAYLOR'S THEOREM WITH SCHLOMILCH AND ROCHE FORM OF REMAINDER (Meerut 2009; Agra, 2000; 2011)

If a function f is such that

(*i*) *the* $(n - 1)$*th derivative* $f^{(n-1)}$ *is continuous in a closed interval* $[a, a + h]$,

(*ii*) *the nth derivative* $f^{(n)}$ *exists in the open interval* $]a, a + h[$,

and (*iii*) *p is a given positive integer,*

then there exists at least one number, θ *between 0 and 1 such that*

$$f(a+h) = f(a) + hf'(a) + \frac{h^2}{2!}f''(a) + \dots + \frac{h^{n-1}}{(n-1)!}f^{(n-1)}(a) + \frac{h^n(1-\theta)^{n-p}}{(n-1)!\,p}f^{(n)}(a+\theta h). \quad ...(1)$$

The condition (*i*) implies the continuity of each of $f, f', f'', \ldots, f^{(n-2)}$ in the closed interval $[a, a + h]$. Let a function ϕ be defined by

$$\phi(x) = f(x) + (a+h-x)f'(x) + \frac{(a+h-x)^2}{2!}f''(x) + \dots + \frac{(a+h-x)^{n-1}}{(n-1)!}f^{(n-1)}(x) + A(a+h-x)^p$$

Here A is a constant to be determined such that $\phi(a) = \phi(a + h)$.

Thus A is given by

$$f(a+h) = f(a) + hf'(a) + \frac{h^2}{2!}f''(a) + \dots + \frac{h^{n-1}}{(n-1)!}f^{(n-1)}(a) + Ah^p \quad ...(2)$$

The function ϕ is continuous in the closed interval $[a, a + h]$, derivable in the open interval $]a, a + h[$ and $\phi(a) = \phi(a + h)$.

Hence, by Rolle's theorem, there exists at least one number, θ, between 0 and 1 such that

$$\phi'(a + \theta h) = 0.$$

* Taylor (1685-1731) was a British Mathematician.

But $$\phi'(x) = \frac{(a+h-x)^{n-1}}{(n-1)!} f^{(n)}(x) - pA(a+h-x)^{p-1}$$

$$\therefore \quad 0 = \phi'(a+\theta h) = \frac{h^{n-1}(1-\theta)^{n-1}}{(n-1)!} f^{(n)}(a+\theta h) - pA(1-\theta)^{p-1} h^{p-1}$$

$$\Rightarrow \quad A = \frac{h^{n-p}(1-\theta)^{n-p}}{p(n-1)!} \cdot f^{(n)}(a+\theta h), \text{ for } (1-\theta) \neq 0 \text{ and } h \neq 0.$$

Substituting the value of A in (2), we get the required result (1).

Remainder after *n* terms. The term $$R_n = \frac{h^n (1-\theta)^{n-p}}{p \cdot (n-1)!} f^n(a+\theta h),$$

is known as the *Schlomilch and Roche form* of remainder R_n after n terms and is due to *Schlomilch* and *Roche.*

TWO PARTICULAR CASES

10.11A. Taylor's Theorem with Cauchy's Form of Remainder [Kanpur 2019; Ranchi 2010; Delhi B.A. (Prog.) III, 2007; Meerut, 2011]

If a function f is such that

(i) the (n – 1)th derivative $f^{(n-1)}$ *is continuous in a closed interval* $[a, a+h]$

(ii) the nth derivative $f^{(n)}$ *exists in the open interval* $[a, a+h[$,

then there exists at least one number, θ *between 0 and 1, such that*

$$f(a+h) = f(a) + hf'(a) + \frac{h^2}{2!} f''(a) + \ldots + \frac{h^{n-1}}{(n-1)!} f^{(n-1)}(a) + \frac{h^n (1-\theta)^{n-1}}{(n-1)!} f^{(n)}(a+\theta h) \quad \ldots(2)$$

Proof. Taking $p = 1$ in the proof of Art. 10.11, we easily get (2).

Remainder after *n* terms. The term $$R_n = \frac{h^n (1-\theta)^{n-1}}{(n-1)!} f^{(n)}(a+\theta h)$$

is known as the *Cauchy's form of remainder* R_n after n terms and is due to *Cauchy.*

10.11B. Taylor's Theorem with Lagrange's Form of Remainder [Delhi B.Sc. (Hons.) II 2009; Kanpur, 2013; Meerut, 2003; Agra 2012; Osmania, 2004]

If a function f is such that

(i) the (n – 1)th derivative $f^{(n-1)}$ *is continuous in a closed interval* $[a, a+h]$

(ii) the nth derivative $f^{(n)}$ *exists in the open interval* $]a, a+h[$,

then there exists at least one number, θ *between 0 and 1, such that*

$$f(a+h) = f(a) + hf'(a) + \frac{h^2}{2!} f''(a) + \ldots + \frac{h^{n-1}}{(n-1)!} f^{(n-1)}(a) + \frac{h^n}{n!} f^{(n)}(a+\theta h) \quad \ldots(3)$$

Proof. Taking $p = n$ in the proof of Art. 10.11, we easily get (3).

Remainder after *n* terms. The term $$R_n = \frac{h^n}{n!} f^{(n)}(a+\theta h)$$

is known as the *Lagrange's form of remainder* R_n after n terms and is due to *Lagrange.*

Cor. 1. Let x be a point of the closed interval $[a, a + h]$.

Let f satisfy the conditions of Taylor's theorem in the interval $[a, a + h]$. Then it satisfies the conditions in the interval $[a, x]$ also.

Changing $a + h$ to x, *i.e.*, h to $x - a$, in (1), we obtain

$$f(x) = f(a) + (x-a) f'(a) + \frac{(x-a)^2}{2!} f''(a) + \frac{(x-a)^3}{3!} f'''(a) + \ldots\ldots$$

$$+ \frac{(x-a)^{n-1}}{(n-1)!} f^{(n-1)}(a) + \frac{(x-a)^n (1-\theta)^{n-p}}{p\,(n-1)!} f^{(n)}[a + \theta(x-a)],\ 0 < \theta < 1.$$

This result holds for each $x \in [a, a + h]$. Of course, θ may be different for different points x.

Cor. 2. *Maclaurin's Theorem **(Meerut, 2004)**

Putting $a = 0$ and $h = x$ in (1), (2) and (3), we get three different forms of the so-called Maclaurin's theorem.

(I) Maclaurin's theorem with Schlomilch and Roche form of remainder

If a function f is such that

(*i*) $f^{(n-1)}$ is continuous in $[0, x]$ (*ii*) $f^{(n)}$ exists in $]0, x[$ (*iii*) p is a given positive integer,

then there exists at least one number, θ between 0 and 1, such that

$$f(x) = f(0) + x f'(0) + \frac{x^2}{2!} f''(0) + \ldots\ldots + \frac{x^{n-1}}{(n-1)!} f^{(n-1)}(0) + \frac{x^n (1-\theta)^{n-p}}{p\,(n-1)!} f^{(n)}(\theta x)$$

(II) Maclaurin's theorem with Cauchy's form of remainder

If a function f is such that

(*i*) $f^{(n-1)}$ is continuous in $[0, x]$ (*ii*) $f^{(n)}$ exists in $]0, x[$,

then there exists at least one number, θ between 0 and 1, such that

$$f(x) = f(0) + x f'(0) + \frac{x^2}{2!} f''(0) + \ldots\ldots + \frac{x^{n-1}}{(n-1)!} f^{(n-1)}(0) + \frac{x^n}{(n-1)!} (1-\theta)^{n-1} f^{(n)}(\theta x)$$

(III) Maclaurin's theorem with Lagrange's form of remainder

If a function f is such that

(*i*) $f^{(n-1)}$ is continuous in $[0, x]$ (*ii*) $f^{(n)}$ exists in $]0, x[$

then there exists at least one number, θ between 0 and 1, such that

$$f(x) = f(0) + x f'(0) + \frac{x^2}{2!} f''(0) + \ldots\ldots + \frac{x^{n-1}}{(n-1)!} f^{(n-1)}(0) + \frac{x^n}{n!} f^{(n)}(\theta x)$$

11.12. POWER SERIES REPRESENTATION OF FUNCTIONS

Taylor's Infinite Series. Suppose that a given function f possesses a continuous derivative of every order in $[a, a + h]$.

Then for any natural number n, there exists an equality of the form

$$f(a+h) = f(a) + hf'(a) + \ldots\ldots + \frac{h^{n-1}}{(n-1)!} f^{(n-1)}(a) + R_n,$$

where R_n denotes Taylor's remainder after n terms.

* Maclaurin (1698-1746) was a British Mathematician.

We write $$S_n = f(a) + hf'(a) + ... + \frac{h^{n-1}}{(n-1)!} f^{(n-1)}(a),$$

$$\Rightarrow \quad f(a+h) = S_n + R_n.$$

If $R_n \to 0$, as $n \to \infty$, we have $\lim S_n = f(a+h)$ so that the infinite series

$$f(a) + hf'(a) + ... + \frac{h^{n-1}}{(n-1)!} f^{(n-1)}(a) + \frac{h^n}{n!} f^{(n)}(a) + ...$$

converges and its sum is equal to $f(a+h)$.

Thus we have proved that if

(i) *a function f possesses continuous derivative of every order in the closed interval* $[a + a + h]$

and (ii) *Taylor's remainder* $R_n \to 0$ *as* $n \to \infty$,

then $$f(a+h) = f(a) + hf'(a) + ... + \frac{h^n}{n!} f^{(n)}(a) + ...$$

The infinite series $$f(a) + hf'(a) + ... + \frac{h^n}{n!} f^{(n)}(a) + ...$$

is known as *Taylor's series.*

It should be clearly understood that *the mere convergence of this series does not mean that its sum is equal to* $f(a+h)$.

10.13. MACLAURIN'S INFINITE SERIES

From Art. 10.12, we deduce that *if a function f possesses continuous derivatives of every order in the closed interval* $[0, h]$ *and* $R_n \to 0$ *as* $n \to \infty$, *then for each* $x \in [0, h]$,

$$f(x) = f(0) + xf'(0) + ... + \frac{x^n}{n!} f^{(n)}(0) + ...$$

Note. The student should note that there is no fundamental distinction between the Taylor's and the Maclaurin's infinite series each of which seeks to express the value of a function at point in terms of the various derivatives of the function at any other point and the distance between the two points.

In this connection, it is also interesting to see 'a priori' why Taylor's or Maclaurin's expansions are not valid for any arbitrary function. Clearly the system of derivatives of a function at a point takes account of the function in only a neighbourhood of the point and accordingly the value of an arbitrary function at a point cannot possibly be given by those in a neighbourhood of another point. The conditions imposed on the function for validity of the Taylor's or the Maclaurin's expansions provide, so to say, *linking up* of the values of the function at the two points.

10.14. SOME STANDARD RESULTS

We prove the following two results :

I. $\lim_{n\to\infty} \frac{x^n}{n!} = 0$ for all real numbers x.

II. $\lim_{n\to\infty} \frac{m(m-1).....(m-n+1)}{(n-1)!} x^n = 0$ when $|x| < 1$ and m is any real number.

The proof of these two results will depend upon the fact that if a series Σu_n converges, then $\lim u_n = 0$.

Proof. I. Consider the series $\Sigma x^n / n!$

viz., $1 + x + \dfrac{x^2}{2!} + \dfrac{x^3}{3!} + \ldots + \dfrac{x^n}{n!} + \ldots$

and the series $$1 + |x| + \frac{|x|^2}{2!} + \frac{|x|^3}{3!} + \ldots + \frac{|x|^n}{n!} + \ldots$$

We have $$u_n = \frac{|x|^n}{n!}, \quad u_{n+1} = \frac{|x|^{n+1}}{(n+1)!}$$

so that $$\lim_{n\to\infty} \frac{u_{n+1}}{u_n} = \lim_{n\to\infty} \frac{|x|}{n+1} = 0 \quad \forall\, x.$$

It follows by D'Alembert's ratio test that Σu_n is a convergent series for all x.

This shows that the given series $\Sigma x^n / n!$ is absolutely convergent and, therefore, also convergent for all x.

Since the series $\Sigma x^n / n!$ is convergent for all x, it follows that the limit of the general term

i.e., $$\lim_{n\to\infty} x^n / n! = 0 \quad \forall\, x.$$

II. Consider the series $$\sum \frac{m(m-1)\ldots(m-n+1)}{(n-1)!} x^n$$

and the series $$\sum \left| \frac{m(m-1)\ldots(m-n+1)}{(n-1)!} x^n \right|$$

We write

$$u_n = \left| \frac{m(m-1)\ldots(m-n+1)}{(n-1)!} \right| |x|^n, \quad u_{n+1} = \left| \frac{m(m-1)\ldots(m-n)}{n!} \right| |x|^{n+1}$$

so that $$\frac{u_{n+1}}{u_n} = \left| \frac{m-n}{n} \right| |x| = \left| \frac{m}{n} - 1 \right| |x|$$

implying that $$\lim_{n\to\infty} \frac{u_{n+1}}{u_n} = |x|.$$

Thus the series Σu_n converges if $|x| < 1$ implying that the given series also converges if $|x| < 1$.

Thus the limit of the general term, *viz.*,

$$\lim_{n\to\infty} \frac{m(m-1)\ldots(m-n+1)}{(n-1)!} x^n = 0 \text{ if } |x| < 1.$$

10.15. POWERS SERIES EXPANSIONS OF SOME STANDARD FUNCTIONS

To obtain Maclaurin's series expansion of

(*i*) e^x **[Delhi B.Sc. (Hons) I 2011; Chennai 2011; Meerut, 2004; Delhi B.Sc. (Prog) III 2011]**

(*ii*) $\sin x$ **[Delhi Maths (H), 2004; Delhi B.Sc. (Hons) I 2011]** (*iii*) cos x **[*Delhi B.A. (Prog.) II, 2008*]**

(*iv*) log $(1 + x), -1 < x \le 1$. **[Delhi B.Sc. (Hons.) II, 2007; Meerut, 2003]**

(*v*) $(1 + x)^m, -1 < x < 1$, m being any real number.

[Delhi Maths (H), 1998; Delhi Physics (H), 2001; Delhi Maths (G), 1999]

(*i*) **Expansion of e^x**

We have $\qquad f(x) = e^x \qquad$ so that $\qquad f^{(n)}(x) = e^x$

$\Rightarrow$ f possesses derivatives of every order for every value of x.

Taking Lagrange's form of remainder, we have

$$R_n = \frac{x^n}{n!} f^{(n)}(\theta x) = \frac{x^n}{n!} e^{\theta x}.$$

We know that when $n \to \infty$, $x^n/n! \to 0$ whatever value x may have.

[Refer result I of Art. 10.14]

It follows that for all $x \qquad R_n \to 0$ as $n \to \infty$.

Conditions for Maclaurin's series being satisfied, we get

$$e^x = 1 + x + \frac{x^2}{2!} + \frac{x^3}{3!} + \ldots.. + \frac{x^n}{n!} + \ldots..$$

which is valid for every value of x.

(*ii*) **Expansion of sin x**

We have $\qquad f(x) = \sin x \qquad$ so that $\qquad f^{(n)}(x) = \sin(x + n\pi/2)$

$\Rightarrow$ f possesses derivatives of every order for every value of x.

We have $$R_n = \frac{x^n}{n!} f^{(n)}(\theta x) = \frac{x^n}{n!} \sin \theta x + \frac{n\pi}{2}$$

$$| R_n | \quad \left| \frac{x^n}{n!} \right| \left| \sin \quad x \quad \frac{n\pi}{2} \right| \quad \left| \frac{x^n}{n!} \right|$$

$$\Rightarrow \quad R_n \to 0 \text{ as } n \to \infty \text{ for every value of } x.$$

Thus the conditions for Maclaurin's infinite expansion are satisfied.

Now $$f^n(0) = \sin \frac{n\pi}{2}$$

Making these substitutions in Maclaurin's series, we get

$$\sin x = x - \frac{x^3}{3!} + \frac{x^5}{5!} - \frac{x^7}{7!} + \ldots..$$

which is valid for every value of x.

(*iii*) **Expansion of cos x.** As above, it may easily be shown that

$$\cos x = 1 - \frac{x^2}{2!} + \frac{x^4}{4!} - \frac{x^6}{6!} + \ldots..$$

for every value of x.

(*iv*) **Expansion of log $(1 + x)$.**

Let $$f(x) = \log(1 + x).$$

We know that log $(1 + x)$ possesses derivatives of every order for $(1 + x) > 0$, *i.e.*, for > -1.

Moreover $$f^{(n)}(x) = \frac{(-1)^{n-1}(n-1)!}{(1+x)^n}, \text{ for } x > -1$$

If R_n denotes the Lagrange's form of remainder, we have

$$R_n = \frac{x^n}{n!} f^{(n)}(\theta x) = (-1)^{n-1} \cdot \frac{1}{n}\left(\frac{x}{1+\theta x}\right)^n.$$

Two cases arise :

(*a*) Let $0 \le x \le 1$, so that $x/(1 + \theta x)$ and, therefore, also $[x/(1 + \theta x)]^n$ is positive and less than 1, whatever value n may have. Since, also, $1/n \to 0$, as $n \to \infty$, we see that $R_n \to 0$.

Thus $$R_n \to 0 \text{ as } n \to \infty \text{ when } 0 \le x \le 1.$$

(*b*) Let $-1 < x < 0$. In this case $x/(1 + \theta x)$ may not be numerically less than unity so that we fail to draw any definite conclusions from the Lagrange's form of R_n.

Taking Cauchy's form of remainder, we have

$$R_n = \frac{x^n}{(n-1)}(1-\theta)^{n-1} f^{(n)}(\theta x) = (-1)^{n-1} x^n \cdot \frac{1}{1+\theta x}\left(\frac{1-\theta}{1+\theta x}\right)^{n-1}.$$

Here $(1 - \theta)/(1 + \theta x)$ is positive and less than 1 and

$$\frac{1}{1+\theta x} < \frac{1}{1-|x|}.$$

Also $$x^n \to 0 \text{ as } n \to \infty \Rightarrow R_n \to 0 \text{ as } n \to \infty.$$

Hence, the conditions for Maclaurin's theorem are satisfied for $-1 < x \le 1$.

Also $$f^{(n)}(0) = (-1)^{n-1}(n-1)!.$$

Making these substitutions in the Maclaurin's series, we get

$$\mathbf{\log(1+x) = x - \frac{x^2}{2} + \frac{x^3}{3} - \frac{x^4}{4} + \ldots\ldots} \text{ for } -1 < x \le 1, \textit{ i.e., } \forall\, x \in\,]-1, 1].$$

(*v*) **Expansion of $(1 + x)^m$; (*m* is any real number)**

Let $$f(x) = (1+x)^m.$$

When m is any real number, $(1 + x)^m$ possesses continuous derivatives of every order only when $1 + x > 0$, *i.e.*, when $x > -1$. Now

$$f^{(n)}(x) = m(m-1)(m-2)\ldots\ldots(m-n+1)(1+x)^{m-n}.$$

We notice that if m is any positive integer, the derivatives of $f(x)$ of order higher than the mth vanish identically and thus, for $n > m$, R_n identically vanishes, so that $(1 + x)^m$ is expanded as a finite series consisting of $(m + 1)$ terms.

If m be not a positive integer, then no derivative vanishes identically so that we have to examine the case still further.

If R_n denotes the Cauchy's form of remainder, we get

$$\begin{aligned} R_n &= \frac{x^n}{(n-1)!}(1-\theta)^{n-1} f^{(n)}(\theta x) \\ &= \frac{x^n}{(n-1)!}(1-\theta)^{n-1} m(m-1)\ldots\ldots(m-n+1)(1+\theta x)^{m-n} \\ &= x^n \frac{m(n-1)\ldots\ldots(m-n+1)}{(n-1)!}\left(\frac{1-\theta}{1+\theta x}\right)^{n-1}(1+\theta x)^{m-1}. \end{aligned}$$

Let $-1 < x < 1$, *i.e.*, $|x| < 1$.

Now, $$0 < \theta < 1 \;\Rightarrow\; 0 < 1 - \theta < 1 + \theta x$$

$$\Rightarrow 0 < \frac{1-\theta}{1+\theta x} < 1 \;\Rightarrow\; 0 < \left(\frac{1-\theta}{1+\theta x}\right)^{n-1} < 1.$$

Let $(m-1)$ be positive.

Now, $$0 < 1 + \theta x < 1 + 1 = 2 \;\Rightarrow\; 0 < (1+\theta x)^{m-1} < 2^{m-1}.$$

Let, now $(m-1)$ be negative.

Now, $$\theta x \geq -|x| \;\Rightarrow\; 1 + \theta x \geq 1 - |x|,$$
$$\Rightarrow (1+\theta x)^{m-1} \geq (1 - |x|)^{m-1}.$$

Also we know that $$\frac{m(m-1)\ldots\ldots(m-n+1)}{(n-1)!}x^n \to 0 \text{ when } |x| < 1$$

(Refer result II of Art. 10.14)

Thus, $$R_n \to 0 \text{ as } n \to \infty \text{ when } |x| < 1$$

The conditions for Maclaurin's expansion are, therefore, satisfied.

Now $$f^{(n)}(0) = m(m-1)\ldots\ldots(m-n+1).$$

Making substitutions, we get

$$(1+x)^m = 1 + mx + \frac{m(m-1)}{2!}x^2 + \frac{m(m-1)(m-2)}{3!}x^3 + \ldots\ldots$$

when $-1 < -x < 1$ *i.e.*, $\forall\, x \in\,]-1, 1[$.

EXAMPLES

Example 1. *Show that* $1 - \frac{x^2}{2} \leq \cos x \leq 1 - \frac{x^2}{2} + \frac{x^4}{24}, \forall\, x \in \mathbf{R}$.

[*Delhi B.A. (Prog.) III, 2007; Delhi B.Sc. (Hons.) II, 2018*]

Solution. Case I. Let $x = 0$. Then there is nothing to prove because each of the expressions

$$1 - \frac{x^2}{2}, \cos x, 1 - \frac{x^2}{2} + \frac{x^4}{24} \text{ has the value 1.}$$

Case II. Let $x > 0$. Applying Taylor's theorem : $f(x) = f(0) + xf'(0) + \frac{x^2}{2}f''(\theta x)$, $0 < \theta < 1$, to the function $f(x) = \cos x$ in $[0, x]$, we obtain

$$\cos x = 1 - \frac{x^2}{2}\cos(\theta x),\; 0 < \theta < 1. \qquad \ldots(1)$$

We know that $\cos(\theta x) \leq 1, \forall\, \theta x \;\Rightarrow\; -\frac{x^2}{2}\cos(\theta x) \geq -\frac{x^2}{2}$

$$\Rightarrow \quad 1 - \frac{x^2}{2}\cos(\theta x) \geq 1 - \frac{x^2}{2} \;\Rightarrow\; \cos x \geq 1 - \frac{x^2}{2}, \text{ by (1)}$$

$$\therefore \quad 1 - \frac{x^2}{2} \leq \cos x. \qquad \ldots(2)$$

On applying Taylor's theorem to the function $f(x) = \cos x$ in $[0, x]$ with remainder after four terms, we get

$$f(x) = f(0) + f'(0) + \frac{x^2}{2!}f''(0) + \frac{x^3}{3!}f'''(0) + \frac{x^4}{4!}f^{iv}(\theta' x),\; 0 < \theta' < 1$$

or $$\cos x = 1-\frac{x^2}{2}+\frac{x^4}{24}(\cos\theta' x),\ 0<\theta'<1. \quad ...(3)$$

Since $\cos(\theta' x) \le 1$, for $0<\theta'<1$ and $x>0$,

$$\therefore \quad 1-\frac{x^2}{2}+\frac{x^4}{24}\cos(\theta' x) \le 1-\frac{x^2}{2}+\frac{x^4}{24} \Rightarrow \cos x \le 1-\frac{x^2}{2}+\frac{x^4}{24}, \text{ using (3).} \quad ...(4)$$

Hence, (2) and (4) $\Rightarrow 1-\dfrac{x^2}{2} \le \cos x \le 1-\dfrac{x^2}{2}+\dfrac{x^4}{24}$, $x>0$

Case III. Let $x<0$. Then $-x>0$ or $y>0$, where $y=-x$.

By case II, $$1-\frac{y^2}{2} \le \cos y \le 1-\frac{y^2}{2}+\frac{y^4}{24}. \quad (\because\ y>0)$$

Putting $y=-x$, we obtain $$1-\frac{x^2}{2} \le \cos(-x) \le 1-\frac{x^2}{2}+\frac{x^4}{24}.$$

Hence, $$1-\frac{x^2}{2} \le \cos x \le 1-\frac{x^2}{2}+\frac{x^4}{24},\ \forall\ x\in \mathbf{R}.$$

Example 2. *Show that sin x lies between* $x-\dfrac{x^3}{6}$ *and* $x-\dfrac{x^3}{6}+\dfrac{x^5}{120}$.

[*Delhi Maths (P), 2001; Delhi Maths (H). 1994, 2005*

Solution. If $x=0$, then there is nothing to prove because each of the expressions

$$x-\frac{x^3}{6},\ \sin x \text{ and } x-\frac{x^3}{6}+\frac{x^5}{120} \text{ has the value 0.}$$

Case I. Let $x>0$. On applying Taylor's theorem to the function $f(x)=\sin x$ in $[0, x]$ with remainder after three terms, we have

$$f(x) = f(0)+xf'(0)+\frac{x^2}{2}f''(0)+\frac{x^3}{3!}f'''(\theta x),\ 0<\theta<1$$

or $$\sin x = x-\frac{x^3}{6}\cos(\theta x), \text{ where } 0<\theta<1 \quad ...(1)$$

We know $\cos(\theta x)\le 1$, whatever θx may be. Therefore, for $x>0$,

$$\frac{x^3}{6}\cos(\theta x) \ge -\frac{x^3}{6} \Rightarrow x-\frac{x^3}{6}\cos(\theta x) \ge x-\frac{x^3}{6}$$

$$\Rightarrow \sin x \ge x-x^3/6, \text{ using (1)}$$

$$\therefore \quad x-x^3/6 \le \sin x. \quad ...(2$$

Again applying Taylor's theorem to the function $f(x)=\sin x$ in $[0, x]$ with remainder after five terms, we have

$$\sin x = x-\frac{x^3}{6}+\frac{x^5}{120}\cos(\theta_1 x), \text{ where } 0<\theta_1<1. \quad ...(3)$$

We know $\cos(\theta_1 x)\le 1$, whatever $\theta_1 x$ may be. For $x>0$, we get

$$x-\frac{x^3}{6}+\frac{x^5}{120}\cos(\theta_1 x) \le x-\frac{x^3}{6}+\frac{x^5}{120}$$

Using (3), $$\sin x \le x-\frac{x^3}{6}+\frac{x^5}{120} \quad ...(4)$$

From (2) and (4), $x - \frac{x^3}{6} \le \sin x \le x - \frac{x^3}{6} + \frac{x^5}{120},\ x > 0$ ...(5)

Case II. Let $x < 0$, so that $-x > 0$. Putting $y = -x > 0$ in (5), we get

$$y - \frac{y^3}{6} \le \sin y \le y - \frac{y^3}{6} + \frac{y^5}{120} \quad ...(6)$$

Putting $y = -x > 0$, we get

$$-\left(x - \frac{x^3}{6}\right) \le -\sin x \le -\left(x - \frac{x^3}{6} + \frac{x^5}{120}\right)$$

$$\Rightarrow \quad x - \frac{x^3}{6} \ge \sin x \ge x - \frac{x^3}{6} + \frac{x^5}{120},\ x < 0. \quad ...(7)$$

From (5) and (7), we find that $\sin x$ lies between $x - x^3/6$ and $x - x^3/6 + x^5/120$, $\forall\, x \in \mathbf{R}$.

Example 3. *Use Taylor's theorem to prove that*

$$1 + x + \frac{x^2}{2} < e^x < 1 + x + \frac{x^2}{2} e^x;\ x > 0.$$

[Delhi B.Sc. Physics (H), 1997, 2001, 05]

Solution. Let $f(x) = e^x,\ x > 0.$ ...(1)

By applying Taylor's theorem to f in $[0, x]$ and writing remainder after two terms. we obtain

$$f(x) = f(0) + x f'(0) + \frac{x^2}{2!} f''(\theta x), \text{ where } 0 < \theta < 1 \quad ...(2)$$

From (1), $f'(x) = f''(x) = e^x$. So $f(0) = 1, f'(0) = 1$ and $f''(\theta x) = e^{\theta x}$

So by (2), $e^x = 1 + x + \frac{x^2}{2} e^{\theta x}$ ...(3)

Now, $0 < \theta < 1$ and $x > 0 \Rightarrow 0 < \theta x < 0$

$\Rightarrow \quad e^0 < e^{\theta x} < e^x$, as e^x is an increasing function

$$\Rightarrow \quad \frac{x^2}{2} < \frac{x^2}{2} e^{\theta x} < \frac{x^2}{2} e^x \quad \Rightarrow \quad 1 + x + \frac{x^2}{2} < 1 + x + \frac{x^2}{2} e^{\theta x} < 1 + x + \frac{x^2}{2} e^x$$

$$\Rightarrow \quad 1 + x + \frac{x^2}{2} < e^x < 1 + x + \frac{x^2}{2} e^x, \text{ using (3).}$$

Example 4. *Expanding the following by Maclaurin's theorem with Lagrange's remainder, show that*

(i) $\sin ax = ax - \frac{a^3x^3}{3!} + \frac{a^5x^5}{5!} - \ldots + \frac{a^{n-1}x^{n-1}}{(n-1)!} \sin\left(\frac{n-1}{2}\pi\right) + \frac{a^n x^n}{n!} \sin\left(a\theta x + \frac{n\pi}{2}\right)$

(G.N.D.U. Amritsar 2010; Meerut, 2000; Kumaon, 1999)

(ii) $e^x = 1 + x + \frac{x^2}{2!} + \frac{x^3}{3!} + \ldots + \frac{x^{n-1}}{(n-1)!} + \frac{x^n}{n!} e^{\theta x}$ ***(Rohilkhand, 1996)***

(iii) $\log(1 + x) = x - \frac{x^2}{2} + \frac{x^3}{3} + \ldots + (-1)^{n-2} \frac{x^{n-1}}{n-1} + (-1)^{n-1} \frac{x^n}{n(1 + \theta x)^n}$, *for* $n > -1$

(iv) $\cos x = 1 - \frac{x^2}{2!} + \frac{x^4}{4!} - \ldots + \frac{x^{n-1}}{(n-1)!} \cos\left(\frac{x-1}{2}\pi\right) + \frac{x^n}{n!} \cos\left(\theta x + \frac{n\pi}{2}\right)$

(Srivenkateshwara, 2003)

Solution. (*i*) By Maclaurin's theorem with Lagrange's remainder,

$$f(x) = f(0) + xf'(0) + \frac{x^2}{2!} f''(0) + ... + \frac{x^{n-1}}{(n-1)!} f^{(n-1)}(0) + \frac{x^n}{n!} f^{(n)}(\theta x) \quad ...(1)$$

Here $\quad f(x) = \sin ax \quad$ and $\quad f^{(n)}(x) = a^n \sin(ax + x\pi/2) \quad ...(2)$

So $\quad f^n(0) = \sin(n\pi/2) \quad$ and $\quad f^{(n)}(\theta x) = a^n \sin(a\theta x + x\pi/2) \quad ...(3)$

From (2) and (3), we have

$$f(0) = 0, f'(0) = a, f''(0) = 0, f'''(0) = -a^3, f^{iv}(x) = 0, f^v(0) = a^5, ...$$

Hence (1) reduces to

$$\sin ax = xa - \frac{x^3 a^3}{3!} + ... + \frac{x^{n-1} a^{n-1}}{(n-1)!} \sin\left(\frac{n-1}{2}\pi\right) + \frac{x^n a^n}{n!} \sin\left(a\theta x + \frac{x\pi}{2}\right)$$

Parts (*ii*) to (*iv*). Left as an exercise for the reader.

Example 5 (*a*). *If* $f(a+h) = f(a) + hf'(a) + \frac{h^2}{2!} f''(a) + \frac{h^3}{3!} f'''(a+\theta h)$ *and if* $f^{iv}(x)$ *is continuous and non-zero at* $x = a$, *show that* $\lim_{h \to 0} \theta = 1/4$. **[*Agra 2002, 03, 10*]**

Solution. Given that

$$f(a+h) = f(a) + hf'(a) + \frac{h^2}{2!} f''(a) + \frac{h^3}{3!} f'''(a+\theta h), \; 0 < \theta < 1 \quad ...(1)$$

Since $f^{iv}(x)$ is continuous and non-zero at $x = a$, so $f^{iv}(a)$ exists. By Taylor's theorem with Lagrange's form of remainder after four terms, we have

$$f(a+h) = f(a) + hf'(a) + \frac{h^2}{2!} f''(a) + \frac{h^3}{3!} f'''(a) + \frac{h^4}{4!} f^{iv}(a+\theta' h), \quad ...(2)$$

where $0 < \theta' < 1$.

By subtracting, (1) and (2) give

$$\frac{h^3}{3!} f'''(a) + \frac{h^4}{4!} f^{iv}(a+\theta' h) = \frac{h^3}{3!} f'''(a+\theta h)$$

or $\quad (h/4) \times f^{iv}(a+\theta' h) = f'''(a+\theta h) - f'''(a)$

or $\quad (h/4) \times f^{iv}(a+\theta' h) = (a+\theta h - a) f^{iv}(a+\theta'' \theta h)$, where $0 < \theta'' < 1$

(By Lagrange's mean value theorem)

$$\therefore \quad \lim_{h \to 0} \theta = \lim_{h \to 0} \frac{1}{4} \times \frac{f^{iv}(a+\theta' h)}{f^{iv}(a+\theta'' \theta h)} = \frac{1}{4}.$$

Example 5 (*b*). *Show that 'θ' which occurs in the Lagrange's mean value theorem tends to the limit 1/2 as* $h \to 0$, *provided* f''' *is continuous.* **[*Delhi B.Sc. (Hons.) II, 2007*]**

Hint. Proceed as in Example 5 (*a*) yourself.

Example 6. *If* $f(x) = f(0) + x f(0) + \frac{x^2}{2!} f''(\theta x)$, *find the value of* θ *as x tends to 1,* $f(x)$ *being* $(1-x)^{5/2}$. **[*Agra 2015; Manipur 2011*]**

Solution. Here $\quad f(x) = (1-x)^{5/2}$.

So $\quad f'(x) = -(5/2) \times (1-x)^{3/2} \quad$ and $\quad f''(x) = (15/4) \times (1-x)^{1/2}$

$\therefore \qquad f(0) = 1, f'(0) = -5/2$ and $f'(\theta x) = (15/4) \times (1 - \theta x)^{1/2}$

Substituting these values in the given relation, we get

$$(1 - x)^{5/2} = 1 - (5/2) \times x + (15/8) \times x^2 (1 - \theta x)^{1/2}$$

Taking limit as $x \to 1$ on both sides, we have

$$0 = 1 - \frac{5}{2} + \frac{15}{8}(1-\theta)^{1/2} \quad \text{or} \quad (1-\theta)^{1/2} = \frac{4}{5} \quad \text{or} \quad 1-\theta = \frac{16}{25}$$

Thus, $\qquad \theta \to 9/25$ as $x \to 1$.

Example 7. *Assuming the validity of expansion, show that*

$$\log(1 + \sin x) = x - \frac{x^2}{2} + \frac{x^3}{6} - \frac{x^4}{12} + \ldots.. \qquad \textit{[Delhi Maths (H), 2003]}$$

Solution. Here $\quad f(x) = \log(1 + \sin x)$ and so $\quad f(0) = 0 \qquad$ ...(1)

Maclaurin's expansion for $f(x)$ is given by

$$f(x) = f(0) + x f'(0) + \frac{x^2}{2!} f''(0) + \frac{x^3}{3!} f'''(0) + \frac{x^4}{4!} f^{iv}(0) + \ldots.. \qquad \ldots(2)$$

From (1), $\qquad f'(x) = \dfrac{\cos x}{1 + \sin x}$ so that $\quad f'(0) = 1$

Again, $\quad f''(x) = \dfrac{-\sin x(1 + \sin x) - \cos^2 x}{(1 + \sin x)^2} = -\dfrac{1 + \sin x}{(1 + \sin x)^2} = -\dfrac{1}{1 + \sin x}$

and so $\qquad f''(0) = -1$

Next, $\qquad f'''(x) = \dfrac{\cos x}{(1 + \sin x)^2}$ so that $\quad f'''(0) = 1$

and $\quad f^{iv}(x) = \dfrac{-\sin x(1 + \sin x)^2 - 2(1 + \sin x)\cdot\cos^2 x}{(1 + \sin x)^4}$ so $\quad f^{iv}(0) = -2.$

Substituting the above values of $f(0), f'(0), \ldots.., f^{iv}(0)$ in (1), we get

$$\log(1 + \sin x) = x - \frac{x^2}{2} + \frac{x^3}{6} - \frac{x^4}{12} + \frac{x^5}{24} + \ldots...$$

Example 8. *Expand log sin (x + h) in powers of h by Taylor's theorem.*

(Meerut, 1994, 2001)

Solution. Expanding $f(x + h)$ by Taylor's theorem in powers of h, we obtain

$$f(x+h) = f(x) + hf'(x) + \frac{h^2}{2!} f''(x) + \frac{h^3}{3!} f'''(x) + \ldots.. \qquad \ldots(1)$$

Here $\qquad f(x + h) = \log \sin(x + h) \Rightarrow f(x) = \log \sin x \qquad$ ...(2)

From (2), $\quad f'(x) = \cot x, f''(x) = -\operatorname{cosec}^2 x, f'''(x) = -2 \operatorname{cosec}^2 x \cot x$ etc.

Substituting these values in (1), we get

$\log \sin(x + h) = \log \sin x + h \cot x - (h^2/2!) \operatorname{cosec}^2 x + (2h^3/3!) \operatorname{cosec}^2 x \cot x + \ldots..$

Example 9. *Assuming the validity of expansion, show that*

$$\tan^{-1} x = \tan^{-1}\frac{\pi}{4} + \frac{x - \pi/4}{1 + \pi^2/16} - \frac{\pi(x - \pi/4)^2}{4(1 + \pi^2/16)^2} + \ldots..$$

[Delhi Maths (H), 2000]

Solution. Assuming the validity of expansion, by Taylor's theorem, we have

$$f(a+h)=f(a)+hf'(a)+(h^2/2\,!)f''(a)+\ldots.. \qquad \ldots(1)$$

Rewriting, $\tan^{-1}x=\tan^{-1}\left(\frac{\pi}{4}+x-\frac{\pi}{4}\right)=\tan^{-1}(a+h)$

$$\Rightarrow\; a=\pi/4,\; h=x-\pi/4 \;\text{ and }\; a+h=x$$

Now, $f(x)=\tan^{-1}x \;\Rightarrow\; f'(x)=\frac{1}{1+x^2},\; f''(x)=-\frac{2x}{(1+x^2)^2}$ etc.

$\therefore\quad f(a)=f(\pi/4)=\tan^{-1}(\pi/4)$

$f'(a)=f'(\pi/4)=\frac{1}{1+\pi^2/16},\; f''(a)=f''(\pi/4)=-\frac{2\times(\pi/4)}{(1+\pi^2/16)^2}$ and so on.

Substituting these values in (1), we obtain

$$\tan^{-1}x=\tan^{-1}\frac{\pi}{4}+\frac{x-\pi/4}{1+\pi^2/16}-\frac{\pi(x-\pi/4)^2}{4(1+\pi^2/16)^2}+\ldots\ldots$$

Example 10. *A function f is twice derivable and satisfies* $\forall\; x>a$ *the inequalities*

$$|f(x)|<A,\;|f''(x)|<B,$$

where A and B are constants. Prove that $\forall\; x>a,\;|f'(x)|<2\sqrt{(AB)}$.

Solution. Let $x>a$ and h be any positive number. Now there exists a number θ, $(0<\theta<1)$ such that

$$f(x+h)=f(x)+hf'(x)+\frac{h^2}{2!}f''(x+\theta h)$$

$$\Rightarrow\quad |hf'(x)|=\left|\,f(x+h)-f(x)-\frac{h^2}{2!}f''(x+\theta h)\right|$$

$$\le|f(x+h)|+|f(x)|+\frac{h^2}{2}|f''(x+\theta h)|$$

$$<2A+\frac{Bh^2}{2}\quad\forall\; x>a$$

$$\Rightarrow\quad |f(x)|<\frac{2A}{h}+\frac{Bh}{2}\quad\forall\; x>a. \qquad \ldots(i)$$

Now $|f'(x)|$ is independent of h and also, by (*i*), it is less than $(2A/h+Bh/2)$ $\forall\; h>0$. Thus $|f'(x)|$ must be less than the least value of $(2A/h+Bh/2)$. Also

$$\frac{2A}{h}+\frac{Bh}{2}=\left(\sqrt{\frac{2A}{h}}-\sqrt{\frac{Bh}{2}}\right)^2+2\sqrt{AB}$$

$$\Rightarrow\quad 2\sqrt{AB}\le\frac{2A}{h}+\frac{Bh}{2}\quad\forall\; h>0. \qquad \ldots(iii)$$

From (*i*) and (*ii*), we see that $\forall\; x>a$

$$|f'(x)|<2\sqrt{(AB)}.$$

Example 11. *Show that* θ *which occurs in the Lagrange's form of remainder, viz.,* $(h^n/n\,!)f^{(n)}(a+\theta h)$, *tends to the limit,* $1/(n+1)$, *when* $h\to 0$, *provided that* $f^{(n+1)}$ *is continuous at a and* $f^{(n+1)}(a)\neq 0$. **[Delhi B.Sc. (Hons.) II, 2003]**

Solution. Since $f^{(n+1)}$ is continuous at a, there exists an interval $[a-\delta, a+\delta]$ at every point x of which $f^{(n+1)}(x)$ exists. Also, therefore, $f(x), f'(x), \ldots, f^{(n)}(x)$ are all continuous in $[a-\delta, a+\delta]$.

If $(a+h)$ be any point of this interval, the necessary conditions being satisfied, by Taylor's theorem with Lagrange's form of remainder after n terms and $(n+1)$ terms successively, we obtain

$$f(a+h) = f(a) + hf'(a) + \ldots + \frac{h^n}{n!} f^{(n)}(a+\theta h),\ 0<\theta<1$$

and
$$f(a+h) = f(a) + hf'(a) + \ldots + \frac{h^n}{n!} f^{(n)}(a) + \frac{h^{n+1}}{(n+1)} f^{(n+1)}(a+\theta' h),\ 0<\theta'<1.$$

On subtracting, we get
$$f^{(n)}(a+\theta h) - f^{(n)}(a) = \frac{h}{n+1} f^{(n+1)}(a+\theta' h).$$

Again applying Lagrange's mean value theorem to the expression on the left, we see that

$$\theta h f^{(n+1)}(a+\theta\theta'' h) = \frac{h}{n+1} f^{(n+1)}(a+\theta' h), \text{ where } 0<\theta''<1$$

$$\Rightarrow \quad \theta f^{(n+1)}(a+\theta\theta'' h) = \frac{1}{n+1} f^{(n+1)}(a+\theta' h).$$

Let $h \to 0$. Then $f^{(n+1)}$ being continuous at a, we obtain

$$\lim \theta f^{(n+1)}(a) = \frac{1}{n+1} f^{(n+1)}(a) \quad \text{so that} \quad \lim \theta = \frac{1}{n+1}, \text{ for } f^{(n+1)}(a) \neq 0.$$

EXERCISES

1. Use Taylor's theorem to establish the following inequalities :

(*i*) $\cos x \geq 1 - x^2/2$, for all $x \in \mathbf{R}$

(*ii*) $x - x^3/6 < \sin x < x$, if $x > 0$ **[Delhi B.A. (Prog.) II, 2005]**

(*iii*) $1 + x < e^x < 1 + xe^x$, if $x > 0$ (*iv*) $1 - x < e^{-x} < 1 - x + x^2/2$, if $x > 0$

(*v*) $0 \leq \sin x - (x - x^3/3! + x^5/5! - x^7/7!) \leq x^9/9!$, for $x > 0$

(*vi*) $1 + x/2 - x^3/8 < \sqrt{1+x} < 1 + x/2$, if $x > 0$

2. Assuming the validity of expansion, show that

(*i*) $\tan^{-1} x = x - x^3/3 + x^5/5 - \ldots$ (*ii*) $\log \sec x = x^2/2 + x^4/12 + \ldots$

(*iii*) $e^x \cos x = 1 + x - \dfrac{2x^3}{3!} - \dfrac{2^2 x^4}{4!} - \dfrac{2^2 x^5}{5!} + \ldots$

(*iv*) $\sin x = 1 - \dfrac{(x-\pi/2)^2}{2!} + \dfrac{(x-\pi/2)^4}{4!} + \ldots$

(*v*) $e^{\sin x} = 1 + x + x^2/2 - x^4/8 + \ldots$

(*vi*) $\sin(\theta + \pi/4) = (1/\sqrt{2}) \times (1 + \theta - \theta^2/2! - \theta^3/3! + \ldots)$

(*vii*) $a^x = 1 + \dfrac{x \log a}{1!} + \dfrac{x^2 (\log a)^2}{2!} + \dfrac{x^3 (\log a)^3}{3!} + \ldots$

(*viii*) $\sin x = \dfrac{1}{\sqrt{2}} \left[1 + \dfrac{(x-\pi/4)}{1!} - \dfrac{(x-\pi/4)^2}{2!} - \dfrac{(x-\pi/4)^3}{3!} + \ldots \right]$

(*ix*) $\sin(e^x - 1) = x + \frac{x^2}{2!} + \frac{5x^4}{4!} + \ldots$ **[Delhi Maths (H), 2002**

(*x*) $$f(x) = f(a) + 2\left\{\frac{x-a}{2} f'\left(\frac{x+a}{2}\right) + \frac{(x-a)^3}{8 \times (3!)} f'''\left(\frac{x+a}{2}\right) + \frac{(x-a)^5}{32 \times (5!)} f^{(5)}\left(\frac{x+a}{2}\right) + \ldots\right\}$$

3. If f'' is continuous on $[a - \delta, a + \delta]$ for some $\delta > 0$, prove that

$$\lim_{h \to 0} \frac{f(a+h) - 2f(a) + f(a-h)}{h^2} = f(a).$$

4. If $0 < x < 2$, then prove that

$$\log x = x - 1 - (x-1)^2/2 + (x-1)^3/3 - (x-1)^4/4 + \ldots$$

5. Show that $\sin(\alpha + h)$ differs from $\sin\alpha + h\cos\alpha$ by not more than $h^2/2$.

6. If $f(x) = e^x(x^2 - x + 2) - (x^2 + x + 2)$, show that when $x > 0$, $f(x)$ increases as x increases.

7. Show that $e^x - 1 > (1 + x)\log(1 + x)$, if $x > 0$.

8. If $f(x + h) = f(x) + hf'(x) + (h^2/2!)\, f''(x + \theta h)$, find the value of θ as $x \to a$ if $f(x) = (x - a)^{5/2}$. **[Ans.** 64/225**]**

9. Prove that, if $f''(c)$ exists and $\neq 0$, then, as $h \to 0$

$$\lim \frac{f(c+h) + f(c-h) - 2f(c)}{h} \quad \text{and} \quad \lim \frac{f(c+h) - f(c)}{f(c-h) - f(c)}$$

exist and find their values. State the theorem on limits you require in the proof. Examine the existence of these limits if $f(x) = |x|$.

10. If $f(0) = 0$ and $f''(x)$ exists in $[0, \infty[$, show that

$$f'(x) - \frac{f(x)}{x} = \frac{1}{2} x f''(\xi),\ 0 < \xi < x$$

and deduce that if $f'(x)$ is positive for positive values of x, then $f(x)/x$ strictly increases as x increases.

11. Show that

(*i*) $x^2 > (1 + x)[\log(1 + x)]^2\ \forall\ x > 0$.

(*ii*) $x < \log[1/(1 - x)] < x/(1 - x)$, when $0 < x < 1$.

12. Prove that, if $0 \le x \le 1$, then $|\log(1 + x) - x + x^2/2| \ge x^3/3$.

13. Show that $a^x > x^a$ if $x > a \ge e$.

[Hint. Let $f(x) = x\log a - a\log x$. Show that $f(a) = 0$ and $f'(x) > 0$].

14. Show by means of an example that the mere convergence of

$$f(x) + f'(x) + \frac{h^2}{2!} f''(x) + \ldots$$

does not mean that it converges to $f(x + h)$.

15. Assuming f'' to be continuous in $[a, b]$, show that

$$f(c) - f(a)\frac{b-c}{b-a} - f(b)\frac{c-a}{b-a} = \frac{1}{2}(c-a)(c-b) f''(\xi),$$

where c and ξ both lie in $[a, b]$.

16. If $f^{(n)}(x) = 0$ for every x in $[a, b]$, then there are numbers $a_0, a_1, a_2, ..., a_{n-1}$ such that

$$f(x) = \sum_{r=0}^{n-1} a_r x^r, \quad \forall\, x \in [a, b].$$

17. Show that the derivative of the sum of the infinite series

$$\sum_{n=1}^{\infty} \left[\frac{nx}{1+n^2x^2} - \frac{(n-1)\,x}{1+(n-1)^2 x^2}\right],$$

is not equal to the sum of the derivative for $x = 0$.

10.16. CONVEX FUNCTION [Delhi B.Sc. (Hons.) II 2017; Delhi B.Sc. (Hons.) I 2011]

Definition. Let $I \subseteq \boldsymbol{R}$ be an interval and let x_1, x_2 be two arbitrary points in I. Then, a function $f: I \to \boldsymbol{R}$ is said to be a *convex function* on I if for any t satisfying $0 \le t \le 1$, we have

$$f((1-t)x_1 + t\,x_2) \le (1-t)\,f(x_1) + t\,f(x_2)$$

Remark. Let $x_1 < x_2$. Then as t ranges from 0 to 1, the point $(1-t)\,x_1 + tx_2$ traverses the interval from x_1 to x_2. Therefore, as shown in the figure, if f is convex on I and if x_1, $x_2 \in I$, then the chord P_1P_2 joining any two points $P(x_1, f(x_1))$ and $P_2(x_2, f(x_2))$ on the graph of f lies above the graph P_1PP_2 of f.

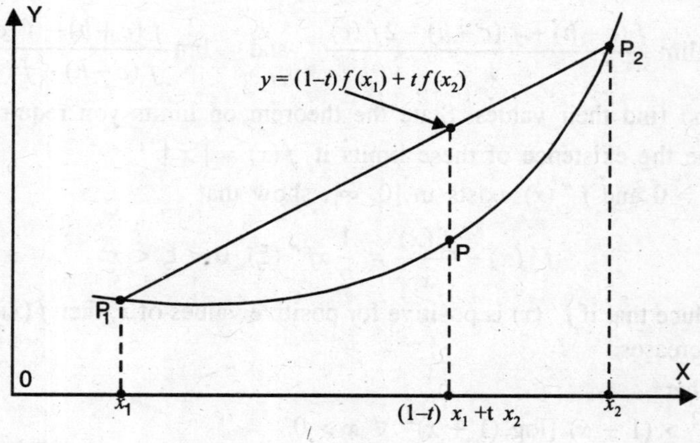

Theorem. Let I be an open interval and let $f: I \to \boldsymbol{R}$ have a second derivative on I. Then f is convex on I if and only if $f''(x) \ge 0\ \forall\ x \in I.$ **[Delhi B.Sc. (Hons) I 2011]**

Proof. Suppose that f is convex on I and a is any point in I. Let h be such that $a+h \in I$ and $a-h \in I.$ Then $a = (1/2)\times\{(a+h)-(a-h)\}$, and since f is convex on I, by definition, it follows that $$f(a) = f((1/2)\times(a+h)+(1/2)\times(a-h)) \le (1/2)\times f(a+h)+(1/2)\times f(a-h)$$

so that $$f(a+h) - 2f(a) + f(a-h) \ge 0 \qquad ... (1)$$

Since $f'(x)$ and $f''(x)$ exist at $x = a$, they also exist in the neighbourhood of $]\,a-h,\ a+h[$. Hence, on applying Taylor's theorem for the intervals $]a-h,\ a[$ and $]a,\ a+h[$, we have

$$f(a-h) = f(a) - h\,f''(a) + (h^2/2!)\times f''(a-\theta_1 h), \quad 0 < \theta_1 < 1 \qquad ... (2)$$

and $$f(a+h)=f(a)+hf'(a)+(h^2/2!)\times f''(a+\theta_2 h), \quad 0<\theta_2<1 \qquad \text{... (3)}$$

Adding (1) and (2), $f(a-h)+f(a+h)=2f(a)+(h^2/2!)\times\{f''(a+\theta_1 h)+f''(a+\theta_2 h)\}$

Hence $$\lim_{h\to 0}\frac{f(a-h)-2f(a)+f(a+h)}{h^2}=\lim_{h\to 0}\frac{f''(a-\theta_1 h)+f''(a+\theta_2 h)}{2}=f''(a) \qquad \text{... (4)}$$

Since $h^2>0$ for all $h\neq 0$, it follows that the limit in (4) is non-negative. Hence, from (4), it follows that $f''(a)\geq 0\ \forall\ a\in I$.

Conversely, let $f''(a)\geq 0\ \forall\ a\in I$. Let x_1 and x_2 be two arbitrary points of I. Also, let $0<t<1$ and $x_0=(1-t)x_1+tx_2$. Applying Taylor's theorem to f at x_0, we obtain a point c_1 between x_0 and x_1 such that

$$f(x_1)=f(x_0)+f'(x_0)(x_1-x_0)+(1/2)\times f''(c_1)(x_1-x_0)^2 \qquad \text{... (5)}$$

and a point c_2 between x_0 and x_2 such that

$$f(x_2)=f(x_0)+f'(x_0)(x_2-x_0)+(1/2)\times f''(c_2)(x_2-x_0)^2 \qquad \text{... (6)}$$

Since f'' is nonnegative on I and $0<t<1$, clearly the term

$$R=(1/2)\times(1-t)f''(c_1)(x_1-x_0)^2+(1/2)\times t\,f''(c_2)(x_2-x_0)^2\geq 0 \qquad \text{... (7)}$$

Multiplying both sides of (5) and (6) by $(1-t)$ and t respectively and then adding the corresponding sides of the resulting equations, we have.

$$(1-t)f(x_1)+t\,f(x_2)=f(x_0)+f'(x_0)((1-t)x_1+tx_2-x_0)+(1/2)\times(1-t)f''(c_1)(x_1-x_0)^2$$
$$+(1/2)\times t\,f''(c_2)(x_2-x_0)^2$$

Since $R\geq 0$ by (7), we have $\quad (1-t)f(x_1)+tf(x_2)\geq f(x_0)$

or $(1-t)f(x_1)+t(x_2)\geq f((1-t)x_1+tx_2)$ or $f((1-t)x_1+tx_2)\leq (1-t)f(x_1)+f(x_2)$

showing that f must be a convex function.

MISCELLANEOUS PROBLEMS ON CHAPTER 10

1. State and prove Rolle's theorem and verify for the function $f(x)=x^2-3x+2$ on the interval [1, 2] **[Meerut 2005]**

2. The function $f(x)=x-x^3$ satisfies the Lagrange's, mean valve theorem in the interval [–2, 1], Find the valve of c. **[Ans. c = –1]** **[Agra 2005]**

3. Find 'c' of Lagrange's mean valve theorem for $f(x)=(x+1)(x-2)(x+3), \forall\, x\in[0,1]$

Ans. $c=(\sqrt{13}-2)/2$ **[Kanpur 2006]**

4. Expand log sin x in powers of $(x-2)$ by Taylor's theorem. **[Kanpur 2006]**

5. (*i*) *If* $4a+2b+c=0$, *the equation* $3ax^2+2bx+c=0$ *has the one real root between which of the following* : (*a*) 0 *and* 1 (*b*) 1 and 2 (*c*) 0 and 2 (*d*) None of these.

(ii) What is abscissa of the point at which the tangent to the curve $y = e^x$ *is parallel to the chord joining the extrimities of the curve in the interval* [0, 1].

(*a*) 1/2 (*b*) *ln* (1/*e*) (*c*) *ln* (*e* – 1) (*d*) 1/*e*

(iii) If the tangent to the curve $f(x) = x^2$ *at any point* $(c, f(c))$ *is parallel to the line joining* $(a, f(a))$ *and* $(b, f(b))$ *on the curve, then which one of the following is true ? (a) a, c, b in A.P (b) a, c, b in G.P. (c) a, c, b in H.P. (d) a, c, b do not follow definite sequence.*

Ans. (*i*) (*c*); (*ii*) (*c*); (*iii*) (*a*)

6. *If* f'' *is continuous at* $x = a$ *with* $f''(a) \neq 0$, *then prove that* θ, *which occurs in Lagrange's mean value theorem approaches the limit* 1/2 *as* $h \to 0$

Hint: Refer Ex. 5(*b*), page 10.37.

7. *If a function f defined on* $[a, a + x]$ *is such that (i)* $f^{(n)}$ *is continuous on* $[a, a+ x]$ *(ii)* $f^{(n)}$ exists on $[a, a + x]$, *then prove that there exists a real number* θ *between* 0 *and* 1 *such that*

$$f(a+x) = f(a) + xf'(a) + ... + \frac{x^{n-1}}{(n-1)!} f^{(n-1)}(a) + \frac{x^n(1-\theta)^{n-1}}{(n-1)!} f^{(n)}(a+\theta x)$$

8. *Prove that* $e^x = 1 + x + x^2/2! + x^3/3! + ... + x^n/n! + ..., x \in \mathbf{R}$

9. *Prove that* $\log(1 + x) = x - x^2/2 + x^3/3 - ....$ for $-1/2 \leq x \leq 0$

10. *Using Taylor's theorem, prove that* $\cos x \geq 1 - (x^2/2), x \in \mathbf{R}$

11. State Rolle's Theorem. Prove that there is no real number k for which the equation $x^2 - 7x + k = 0$ has two distinct real roots in [0, 3]

[**Hint :** Proceed as in Ex.6, page 10.4]

12. Show that $x/(1+x^2) < \tan^{-1} x < x, x > 0$.

13. Derive the expansion of cos x in terms of power series.

[Delhi B.A. (Prog.) III, 2008, 12]

14. Find '*c*' in Cauchy's mean value theorem if $f(x) = x(x-1)(x-2)$ and $g(x) = x(x-2)(x-3)$ in $[0, 1/2]$ **[Agra 2008]**

15. Using Lagrange's mean value theorem, prove the following inequality for x > 0

$$0 < (1/x) \times \log_e \{(e^x - 1)/x\} < 1$$

16. State Lagrange's mean value theorem and prove that if $f'(x) > 0$, $a < x < b$, then *f*(*x*), monotonically increasing in [*a*, *b*]

17. Suppose that f'' is continuous on [1, 2] and that *f* has three zeros in the interval (1, 2). Show that f'' has at least one zero in the interval (1, 2).

18. Determine the expansion of e^x in terms of power series. **[Delhi B.Sc. (Prog.) III, 2010]**

19. Find the Maclaurin's series for a^{x+1} assuming that $\lim_{n\to\infty} R_n(x) = 0$ **[Delhi 2010]**

20. Find the Taylor's expansion for e^x. **[Delhi B.Sc. (Hons.) I 2010]**

21. Let $f:[a, b] \to \boldsymbol{R}$ be such that f and its derivatives $f', f'', \ldots, f^{(n)}$ are continuous on $[a, b]$ and $f^{(n+1)}$ exists on $[a, b]$. If $x_0 \in [a, b]$ and x is any point in $[a, b]$, then show that there exists a point c between x and x_0 such that

$$f(x) = f(x_0) + f'(x_0)(x - x_0) + \frac{f''(x_0)}{2!}(x - x_0)^2 + \ldots + \frac{f^{(n)}(x_0)}{n!}(x - x_0)^n + \frac{f^{(n+1)}(c)}{(n+1)!}(x - x_0)^{n+1}$$

[Delhi B.Sc. I (Hons) 2010]

22. Write the expansion of $f(x) = e^x$ in terms of a power series.

[Delhi B.Sc. (Prog) III 2011]

23. Suppose that $f:[0,2] \to \boldsymbol{R}$ is continuous on [0, 2] and differentiable on (0, 2) and that $f(0) = 0, f(1) = 1, f(2) = 1$

(*a*) Show that there exists $c_1 \in (0,1)$ such that $f'(c_1) = 1$

(*b*) Show that there exists $c_2 \in (1,2)$ such that $f'(c_2) = 0$

(*c*) Show that there exists $c_3 \in (0,2)$ such that $f'(c_3) = 1/3$

[Delhi B.Sc. (Hons) I 2011]

Hint. For parts (*a*) and (*b*), apply the mean value theorem For part (c), apply Darboux's theorem to the results of (*a*) and (*b*).

24. *If $f''(x)$ exists for all $x \in [a, b]$ and*

$$[f(c) - f(a)]/(c - a) = [f(b) - f(c)]/(b - c), \quad \text{where} \quad c \in (a, b),$$

then show that there exists a $x_0 \in (a, b)$ such that $f''(x_0) = 0$.

Sol. Since $f''(x)$ exists for all $x \in [a, b]$, it follows that f is continuous on $[a, b]$ and differentiable in (a, b). Hence, applying Lagrange's mean value theorem to $f(x)$ on $[a, c]$ and then on $[c, b]$, we obtain two points $x_1 \in (a, c)$ and $x_2 \in (c, b)$ such that

$$\{f(c) - f(a)\}/(c - a) = f'(x_1) \quad \text{and} \quad \{f(b) - f(c)\}/(b - c) = f'(x_2) \quad \ldots (1)$$

Using the given relation in the equation (1) yields $\quad f'(x_1) = f'(x_2) \quad \ldots (2)$

Since $f''(x)$ exists in $[a, b]$ and $[x_1, x_2] \subset [a, b]$, it follows that f' is continuous in $[x_1, x_2]$ and derivable in (x_1, x_2). Also, f' satisfies (2). Thus, f' satisfies all the three conditions of Rolle's theorem in $[x_1, x_2]$. Now, applying Rolle's theorem (see page 10.1) to f' on $[x_1, x_2]$, we get a point $x_0 \in (x_1, x_2) \subset (a, b)$ such that $f''(x) = 0$.

25. *A twice differentiable function f on $[a, b]$ is such that $f(a) = f(b) = 0$ and $f(c) > 0$ for $a < c < b$. Prove that there is at least one value x_0 between a and b in which $f''(x_0) < 0$.*

Sol. Since f is twice differentiable on $[a, b]$, hence f, f' exist on $[a, b]$. Therefore, f, f' are derivable on $[a, b]$ and hence continuous on $[a, b]$.

Since $a < c < b$, applying Larange's mean value theorem to $f(x)$ on $[a, c]$ and then on $[c, b]$, we obtain two points $x_1 \in (a, c)$ and $x_2 \in (c, b)$ such that

$$\{f(c) - f(a)\}/(c - a) = f'(x_1) \quad \text{and} \quad \{f(b) - f(c)\}/(b - c) = f'(x_2) \quad \ldots (1)$$

Since by hypothesis, $f(a) = f(b) = 0$, hence equations in (1) reduce to

$$f'(x_1) = f(c)/(c - a) \quad \text{and} \quad f'(x_2) = -\{f(c)/(b - c)\}, \text{ where } a < x_1 < c < x_2 < b \quad \ldots (2)$$

Since $(x_1, x_2) \subset (a, b)$ and f' is derivable and continuous on $[a, b]$, it follows that f' is continuous on $[x_1, x_2]$ and derivable on (a, b). Applying Lagrange's mean value theorem to f' on $[x_1, x_2]$, we have

$$\{f'(x_2) - f'(x_1)\}/(x_2 - x_1) = f''(x_0), \qquad \text{where} \qquad x_1 < x_0 < x_2 \qquad \text{... (3)}$$

Substituting the values of $f'(x_1)$ and $f'(x_2)$ given by (2) in (3) yields

$$f''(x_0) = \frac{1}{x_2 - x_1}\left[-\frac{f(c)}{b-c} - \frac{f(c)}{c-a}\right] = -\frac{(b-a)\, f(c)}{(x_2 - x_1)(b-c)(c-a)} \qquad \text{... (4)}$$

Since $a < x_1 < c < x_2 < b$ and $f(c) > 0$ by hypothesis, (4) shows that there exists at least one value x_0 between a and b such that $f''(x_0) < 0$.

Chapter

11

Maxima and Minima

11.1. INTRODUCTION

Application of Taylor's Theorem to the problem of extreme values of a function will be taken up in this chapter.

If c be any interior point of the domain $[a, b]$ of a function f, then

(*i*) $f(c)$ is said to be a *maximum* value of the function f, if there exists some neighbourhood $]c-\delta, c+\delta[$ of c, such that $\quad f(c) > f(x) \quad \forall\ x \in\ ]c-\delta, c+\delta[$ other than c.

(*ii*) $f(c)$ is said to be a *minimum* value of the function f, if there exists some neighbourhood $]c-\delta, c+\delta[$ of c, such that $\quad f(c) < f(x) \quad \forall\ x \in\ ]c-\delta, c+\delta[$ other than c.

(*iii*) $f(c)$ is said to be an *extreme* value of f, if it is either a maximum or a minimum value.

For $f(c)$ to be an extreme value, $f(c) - f(x)$ must keep the same sign for every point x, other than c, in some neighbourhood of c.

Note. Extreme value is also known as an *extremum* or a *turning value*.

11.2. A NECESSARY CONDITION FOR THE EXISTENCE OF EXTREME VALUES

If $f(c)$ be an extreme value of a function f, then $f'(c)$, in case it exists, is zero.

[*Bharathiar, 2004*; *Delhi Maths (H), 2003, 07*]

Proof. Since $f(c)$ is an extreme value of $f(x)$, it follows that $f(x)$ has either a maximum value or a minimum value at $x = c$.

Since $f'(c)$ exists, we have $\qquad Lf'(c) = Rf'(c) = f'(c) \qquad$ ...(1)

Let $f(x)$ have a maximum value at $x = c$. So, there exists a $\delta > 0$, such that

$$c - \delta < x < c \Rightarrow f(x) < f(c) \qquad \text{...(2)}$$

and

$$c < x < c + \delta \Rightarrow f(x) < f(c) \qquad \text{...(3)}$$

From (2), $\qquad f(x) - f(c) < 0 \quad$ and $\quad x - c < 0 \quad$ whenever $c - \delta < x < c$

$$\Rightarrow \frac{f(x) - f(c)}{x - c} > 0 \quad \text{whenever } c - \delta < x < c \qquad \text{...(4)}$$

Taking limits as $x \to c$, $\qquad$ (4) $\Rightarrow Lf'(c) \geq 0 \qquad$ ...(5)

Again, from (3), $\qquad f(x) - f(c) < 0 \quad$ and $\quad x - c > 0 \quad$ whenever $c < x < c + \delta$

$$\Rightarrow \frac{f(x) - f(c)}{x - c} < 0 \quad \text{whenever } c < x < c + \delta \qquad \text{...(6)}$$

Taking limits as $x \to c$, $\qquad$ (6) $\Rightarrow Rf'(c) \leq 0 \qquad$ ...(7)

From (1), (5) and (7), we have $\qquad f'(c) = 0$

Again, if $f(x)$ has a minimum value at $x = c$, by similar argument, we have $f'(c) = 0$.

Note 1. *The above condition is a necessary but not a sufficient condition for $f(x)$ to have a maximum or minimum at $x = c$ as shown below :*

(*i*) Consider the function $f(x) = x^3, x \in \mathbf{R}$. Here $f'(0) = 0$ but f does not have an extreme value at $x = 0$ as shown in the following figure. Thus, $f'(c) = 0$ may not imply that $f(x)$ has an extreme value at $x = c$.

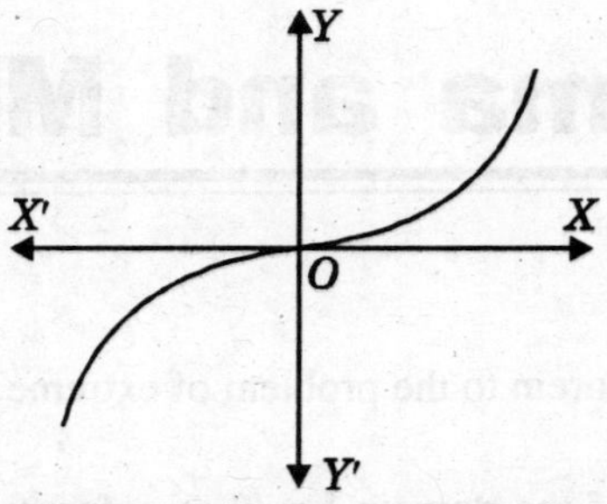

(*ii*) A function may have an extreme value at a point without having derivative at that point. For example, if $f(x) = |x|\ \forall\ x \in \mathbf{R}$. Then $f(x)$ is not derivable at $x = 0$ but it has a minimum value 0 at $x = 0$. as shown in the following figure.

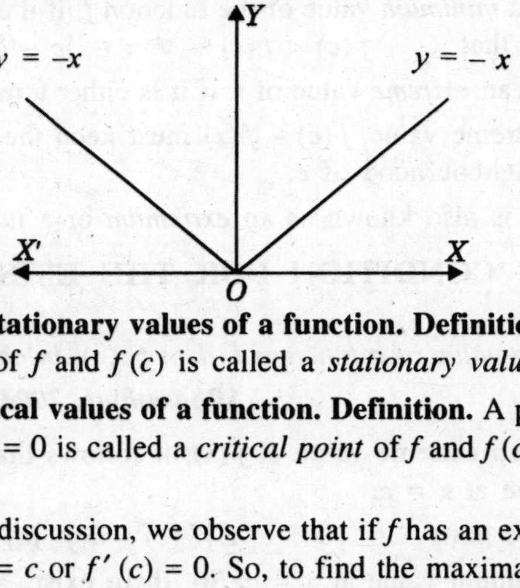

Stationary points and stationary values of a function. Definition. If $f'(c) = 0$, then $x = c$ is called a *stationary point* of f and $f(c)$ is called a *stationary value* of f.

Critical points and critical values of a function. Definition. A point $x = c$ such that either $f'(c)$ does not exist or $f'(c) = 0$ is called a *critical point* of f and $f(c)$ is called a *critical value* of f.

Note 2. From the above discussion, we observe that if f has an extreme value at $x = c$, then either f is not derivable at $x = c$ or $f'(c) = 0$. So, to find the maxima or minima of a function we first find the values of x for which either $f'(x)$ does not exist or if $f'(x)$ exists, then it vanishes thereat.

11.3. SUFFICIENT CRITERIA FOR THE EXISTENCE OF EXTREME VALUES

In what follows, we shall present two sets of sufficient conditions for the existence of extreme values.

Theorem I (First derivative test). *Let a function f be derivable in a neighbourhood of c, where f has an extreme value at c. Then f(c) is a maximum value if the sign of f′ changes from plus to minus and f(c) is a minimum if the sign of f′ changes from minus to plus as x passes through c.*

Proof. Suppose f' changes sign from plus to minus as x passes through c. Then there exists a $\delta > 0$ such that

$$f'(x) > 0 \text{ in }]c - \delta, c[\quad \text{and} \quad f'(x) < 0 \text{ in }]c, c + \delta[$$

$\Rightarrow$ f is strictly increasing in $]c - \delta, c[$ and strictly decreasing in $]c, c + \delta[$

$\Rightarrow$ $f(x) < f(c) \quad \forall\ x \in\ ]c - \delta, c[\quad$ and $\quad f(x) < f(c) \quad \forall\ x \in\ ]c, c + \delta[$

$\Rightarrow \quad f(x) < f(c) \quad \forall\ x\ ]c-\delta, c+\delta[,\ x \neq c$

$\Rightarrow \quad f$ has a maximum value at $x = c$.

If f' changes sign from minus to plus as x passes through c, by similar argument, we can show that f has a minimum value at $x = c$.

Note. If f' does not change sign, f' keeps the same sign in a neighbourhood of $x = c$, say $]c-\delta,\ c+\delta[$, then $f(x)$ is neither maximum nor minimum at $x = c$.

Illustration : *Find the values of x for which the function f, defined by*

$$f(x) = x^5 - 5x^4 + 5x^3 - 1, \quad \forall\ x \in \mathbf{R}$$

is maximum or minimum.

Solution. Given $\qquad f(x) = x^5 - 5x^4 + 5x^3 - 1 \qquad$ ...(1)

From (1), $\qquad f'(x) = 5x^4 - 20x^3 + 15x = 5x^2(x-1)(x-3)$

$f'(x) = 0 \Rightarrow x = 0, 1, 3$, which are critical points.

(*i*) Now, for x slightly $< 0, f'(x) > 0$ and for x slightly $> 0, f'(x) > 0$

Since $f'(x)$ does not change sign as x passes through $x = 0$, f is neither maximum nor minimum at $x = 0$.

(*ii*) Next, for x slightly $< 1, f'(x) > 0$ and for x slightly $> 1, f'(x) < 0$

Since $f'(x)$ changes sign from plus to minus as x passes through 1, f is maximum at $x = 1$.

(*iii*) Finally, for x slightly $< 3, f'(x) < 0$ and for x slightly $> 3, f'(x) > 0$

Since $f'(x)$ changes sign from minus to plus as x passes through 3, f is minimum at $x = 3$.

Theorem II (Sufficient criteria for extreme values) (General test)

Let c be an interior point of the domain [a, b] of a function f. Let

(i) $f^{(n)}(c)$ exist and be not zero, and (ii) $f'(c) = f''(c) = f'''(c) = \ldots\ldots = f^{(n-1)}(c) = 0$; then if n is odd, f (c) is not an extreme value and if n is even, f (c) is a maximum or a minimum value according as $f^{(n)}(c)$ is negative or positive.

Proof. The given condition (*i*) implies that

$f', f'', \ldots\ldots, f^{(n-1)}$ all exist in a certain neighbourhood $]c-\delta_1, c+\delta_1[$ of c. ...(1)

As $f^{(n)}(c)$ exists and $\neq 0$, there exists a neighbourhood $]c-\delta, c+\delta[$, of c, where $(0 < \delta < \delta_1)$ such that

$$\left.\begin{aligned} c-\delta < x < c &\Rightarrow f^{(n-1)}(x) < f^{(n-1)}(c) = 0, \\ c < x < c+\delta &\Rightarrow f^{(n-1)}(x) > f^{(n-1)}(c) = 0. \end{aligned}\right\} \qquad \text{...(2)}$$

[Using results of Art. 9.5, Chapter 9]

in case $f^{(n)}(c)$ is positive;

and

$$\left.\begin{aligned} c-\delta < x < c &\Rightarrow f^{(n-1)}(x) > f^{(n-1)}(c) = 0, \\ c < x < c+\delta &\Rightarrow f^{(n-1)}(x) < f^{(n-1)}(c) = 0. \end{aligned}\right\} \qquad \text{...(3)}$$

[Using results of Art. 9.5, Chapter 9]

in case $f^{(n)}(c)$ is negative.

Because of (*i*), we have by Taylor's theorem, when $|h| < \delta$,

$$f(c+h) = f(c) + hf'(c) + \frac{h^2}{2!} f''(c) + \ldots\ldots + \frac{h^{n-1}}{(n-1)!} f^{(n-1)}(c+\theta h)$$

which, by virtue of the given condition (*ii*), gives

$$f(c+h)-f(c)=\frac{h^{n-1}}{(n-1)!}f^{(n-1)}(c+\theta h), \quad ...(4)$$

where $c + \theta h \in\ ]c-\delta, c+\delta[$.

Let n be even. From (2) and (4), we deduce that if $f^{(n)}(c)$ be > 0, then for every point $x = c + h \in\ ]c - \delta, c + \delta[$, other than c, $f(c + h) > f(c)$, implying that $f(c)$ is a minimum.

From (3) and (4), it may similarly be shown that $f(c)$ is a maximum if $f^{(n)}(c)$ be < 0.

Let n be odd. From (2) and (4), we deduce that if $f^{(n)}(c) > 0$, then

$$f(c+h) > f(c), \text{ when } c < x = c + h \le c + \delta,$$

and

$$f(c+h) < f(c), \text{ when } c - \delta \le x = c + h < c$$

so that $f(c)$ is *not* an extreme value.

It may similarly be shown that $f(c)$ is not an extreme value when $f^{(n)}(c) < 0$.

Particular case (When $n = 2$). If a function f is such that $f'(c) = 0$ and $f''(c)$ exists and is non-zero, then

(*i*) $f(c)$ is a maximum if $f''(c) < 0$ (*ii*) $f(c)$ is a minimum if $f''(c) > 0$.

ILLUSTRATIONS

1. *Find the maximum value of the function $f(x) = x^2e^{-x}, x > 0$.*

[Delhi B.A. (Prog.) III, 2003; Kanpur 2009]

Solution. Let $y = f(x) = x^2e^{-x}, x > 0$ ...(1)

$\therefore$ $dy/dx = 2xe^{-x} - x^2e^{-x} = (2x - x^2)\, e^{-x}$ ...(2)

For maximum and minimum values of y, we have

$$dy/dx = 0, \text{ i.e., } x\,(2 - x)\, e^{-x} = 0 \text{ giving } x = 0, x = 2.$$

Also, $d^2y/dx^2 = (2 - 2x)\, e^{-x} + (2x - x^2)(-e^{-x}) = (2 - 4x + x^2)\, e^{-x}$.

(*i*) For $x = 0$, $d^2y/dx^2 = 2 > 0$ and so y is minimum at $x = 0$.

(*ii*) For $x = 2$, $d^2y/dx^2 = (2 - 8 + 4)\, e^{-2} = -2e^{-2} < 0$.

So, y is maximum at $x = 2$ and maximum value $= 4e^{-2}$, by (1).

2. *Suppose $f(x)$ is a function satisfying the conditions : (i) $f(0) = 2, f(1) = 1$ (ii) f has a minimum value at $x = 5/2$ (iii) $f'(x) = 2ax + b$ for all x. Determine the constants a, b and the function $f(x)$.*

Solution. Given that $f'(x) = 2ax + b$ for all x ...(1)

Since $f(x)$ has a minimum at $x = 5/2$, $f'(5/2) = 0$

$\therefore$ (1) gives $(2a) \times (5/2) + b = 0$ or $b = -5a$ ...(2)

So (1) reduces to $f'(x) = 2ax - 5a$

Integrating it, $f(x) = ax^2 - 5ax + x$, ...(3)

where c is an arbitrary constant of integration.

Putting $x = 0$ and $x = 1$ in succession in (3) and using the given facts $f(0) = 2$ and $f(1) = 1$, we have $2 = c$ and $1 = c - 4a$.

These give $c = 2$ and $a = 1/4$. Therefore, (2) $\Rightarrow$ $b = -5/4$.

So the required values of a and b are $a = 1/4$ and $b = -5/4$ and from (3), the required function is given by $f(x) = (1/4)\, x^2 - (5/4)\, x + 2$.

EXAMPLES

Example 1. *Find points of maxima and minima of $f(x) = 3\cos^2 x + \sin^6 x, -\pi/2 < x < \pi/2$.* **[Delhi Maths (P), 2003]**

Solution. Let $y = 3\cos^2 x + \sin^6 x, -\pi/2 < x < \pi/2$ ...(1)

Then $dy/dx = -6\cos x \sin x + 6\sin^5 x \cos x$ ...(2)

For maximum and minimum values of y, we have

$$dy/dx = 0 \quad \text{or} \quad 6\sin x \cos x(\sin^4 x - 1) = 0 \qquad ...(3)$$

But for $-\pi/2 < x < \pi/2$, $\cos x \neq 0$ and $(\sin^4 x - 1) \neq 0$

Hence, (3) $\Rightarrow$ $\sin x = 0$ or $x = 0$ for $-\pi/2 < x < \pi/2$.

Now, from (2), $d^2y/dx^2 = -6(\cos^2 x - \sin^2 x) + 6(5\sin^4 x \cos^2 x - \sin^6 x)$

$\therefore$ At $x = 0$, $d^2y/dx^2 = -6 < 0$, showing that y is maximum when $x = 0$.

Example 2. *Prove that the function f defined by $f(x) = |x-2| \cdot |x-3|\ \forall x \in \mathbf{R}$ has a minimum value 0 at 2, 3 and a maximum value 1/4 at 5/2.* **[Delhi B.Sc. (H), 2000]**

Solution. Let $y = f(x)$. Then, we have

$$y = |x-2| \cdot |x-3| = \begin{cases} (2-x)(3-x) \text{ if } x \leq 2 \\ (x-2)(3-x) \text{ if } 2 \leq x \leq 3 \\ (x-2)(x-3) \text{ if } x \geq 3 \end{cases}$$

or

$$y = f(x) = \begin{cases} 6 - 5x + x^2, \text{ if } x \leq 2 \\ -6 + 5x - x^2, \text{ if } 2 \leq x \leq 3 \\ x^2 - 5x + 6, \text{ if } x \geq 3 \end{cases} \qquad ...(1)$$

Clearly $f(x)$ is continuous everywhere. Also, $f(2) = f(3) = 0$.

Here $$Lf'(2) = \lim_{h \to 0} \frac{f(2-h) - f(2)}{-h} = \lim_{h \to 0} \frac{6 - 5(2-h) + (2-h)^2 - 0}{-h} = -1$$

and $$Rf'(2) = \lim_{h \to 0} \frac{f(2+h) - f(2)}{h} = \lim_{h \to 0} \frac{-6 + 5(2+h) - (2+h)^2 - 0}{h} = 1$$

Since $Lf'(2) \neq Rf'(2)$ so $f'(x)$ does not exist at $x = 2 \Rightarrow x = 2$ is a critical point of y.

Again, $Lf'(2) < 0$ and $Rf'(2) > 0 \Rightarrow f(x)$ is minimum at $x = 2$.

Further, minimum value at $x = 2$ is given by $f(2) = 0$.

Next, $$Lf'(3) = \lim_{h \to 0} \frac{f(3+h) - f(3)}{-h} = \lim_{h \to 0} \frac{-6 + 5(3-h) - (3-h)^2 - 0}{-h} = -1$$

and $$Rf'(3) = \lim_{h \to 0} \frac{f(3+h) - f(3)}{h} = \lim_{h \to 0} \frac{(3+h)^2 - 5(3+h) + 6 - 0}{h} = 1$$

Since $Lf'(3) \neq Rf'(3)$ so $f'(x)$ does not exist at $x = 3 \Rightarrow x = 3$ is a critical point of y.

Also, $Lf'(3) < 0$ and $Rf'(3) > 0 \Rightarrow f(x)$ is minimum at $x = 3$.

Further, minimum value at $x = 3$ is given by $f(3) = 0$.

Now, for $2 \leq x \leq 3$, $y = -6 + 5x - x^2$

$\therefore$ $dy/dx = 5 - 2x$ and $d^2y/dx^2 = -2$.

Since d^2y/dx^2 is negative, it follows that the value of x for which y is maximum is given by

$$dy/dx = 0 \quad \text{or} \quad 5 - 2x = 0 \quad \text{or} \quad x = 5/2.$$

Further, maximum value at $x = 5/2$ is given by $f(5/2) = -6 + 25/2 - 25/4 = 1/4$.

Example 3. *Find the maximum value of the function*

$$f(x) = |3 - x| + |2 + x| + |5 - x|.$$ **[Delhi Maths (H), 2003]**

Solution. Here $$f(x) = |3 - x| + |2 + x| + |5 - x| \quad ...(1)$$

From (1),

$$f(x) = \begin{cases} 3 - x - (2 + x) + 5 - x = 6 - 3x, \text{ if } x \le -2 \\ 3 - x + 2 + x + 5 - x = 10 - x, \text{ if } -2 \le x \le 3 \\ x - 3 + 2 + x + 5 - x = 4 + x, \text{ if } 3 \le x \le 5 \\ x - 3 + 2 + x + x - 5 = -6 + 3x, \text{ if } x \ge 5. \end{cases} \quad ...(2)$$

Here, from (2) $f(-2) = 12, f(3) = 7$ and $f(5) = 9$.

Now, $$Lf'(-2) = \lim_{h \to 0} \frac{f(-2-h) - f(-2)}{-h} = \lim_{h \to 0} \frac{6 - 3(-2-h) - 12}{-h} = -3$$

and $$Rf'(-2) = \lim_{h \to 0} \frac{f(-2+h) - f(-2)}{h} = \lim_{h \to 0} \frac{10 - (-2+h) - 12}{h} = -1$$

Since $Lf'(-2) \ne Rf'(-2)$, so $f'(x)$ does not exist at $x = -2$. Hence $x = -2$ is a critical point. But $Lf'(-2) < 0$ and $Rf'(-2) < 0$ show that there is neither a maxima nor minima of $f(x)$ at $x = -2$.

Next, $$Lf'(3) = \lim_{h \to 0} \frac{f(3-h) - f(3)}{-h} = \lim_{h \to 0} \frac{10 - (-3-h) - 7}{-h} = -1$$

and $$Rf'(3) = \lim_{h \to 0} \frac{f(3+h) - f(3)}{h} = \lim_{h \to 0} \frac{4 + 3 + h - 7}{h} = 1$$

Since $Lf'(3) \ne Rf'(3)$, so $f'(x)$ does not exist at $x = 3$. Hence $x = 3$ is a critical point.

But $Lf'(3) < 0$ and $Rf'(3) > 0$ and hence $f(x)$ is minimum at $x = 3$ and the required minimum value $= f(3) = 7$.

Finally, $$Lf'(5) = \lim_{h \to 0} \frac{f(5-h) - f(5)}{-h} = \lim_{h \to 0} \frac{4 + 5 - h - 9}{-h} = 1$$

and $$Rf'(5) = \lim_{h \to 0} \frac{f(5+h) - f(5)}{h} = \lim_{h \to 0} \frac{-6 + 3(5+h) - 9}{h} = 3$$

Since $Lf'(5) \ne Rf'(5)$, so $f'(x)$ does not exist at $x = 5$. Hence $x = 5$ is a critical point. But $Lf'(5)$ and $Rf'(5)$ are of the same sign, so there is neither a maxima nor a minima at $x = 5$.

Example 4. *Find the maxima and minima of the function*

$$f(x) = \sin x + (1/2) \sin 2x + (1/3) \sin 3x, \text{ for all } x \in [0, \pi].$$

[Delhi Maths (G), 2004; Delhi Maths (H), 1999, 2004]

Solution. $$f'(x) = \cos x + \cos 2x + \cos 3x = \cos 2x + 2 \cos 2x \cos x$$

Thus, $$f'(x) = \cos 2x\,(1 + 2\cos x),\ x \in [0, \pi].$$

$\therefore$ $f'(x) = 0 \Rightarrow \cos 2x = 0 \Rightarrow 2x = \pi/2, 3\pi/2 \Rightarrow x = \pi/4, 3\pi/4$, for $0 \le x \le \pi$

and $1 + 2\cos x = 0 \Rightarrow \cos x = -1/2 \Rightarrow x = 2\pi/3$, for $0 \le x \le \pi$

Now $$f''(x) = -\sin x - 2\sin 2x - 3\sin 3x.$$

$\therefore$ $f''(\pi/4) = -1/\sqrt{2} - 2 - 3/\sqrt{2} < 0 \Rightarrow f(x)$ has a maxima at $x = \pi/4$.

$f''(3\pi/4) = -1/\sqrt{2} + 2 - 3/\sqrt{2} = 2 - 2\sqrt{2} < 0 \Rightarrow f(x)$ has a maxima at $x = 3\pi/4$

and $f''(2\pi/3) = -\sqrt{3}/2 + 2 \times (\sqrt{3}/2) - 0 = \sqrt{3}/2 > 0 \Rightarrow f(x)$ has a minima at $x = 2\pi/3$.

$$\text{Maximum value at } x = \frac{\pi}{4} \text{ is } \sin\frac{\pi}{4} + \frac{1}{2}\sin\frac{\pi}{2} + \frac{1}{3}\sin\frac{3\pi}{4} = \frac{1}{\sqrt{2}} + \frac{1}{2} + \frac{1}{3\sqrt{2}} = \frac{4\sqrt{4}+3}{6}$$

$$\text{Maximum value at } x = \frac{3\pi}{4} \text{ is } \sin\frac{3\pi}{4} + \frac{1}{2}\sin\frac{3\pi}{2} + \frac{1}{3}\sin\frac{9\pi}{4} = \frac{1}{\sqrt{2}} - \frac{1}{2} + \frac{1}{3\sqrt{2}} = \frac{4\sqrt{2}-3}{6}$$

$$\text{Minimum value at } x = \frac{2\pi}{3} \text{ is } \sin\frac{2\pi}{3} + \frac{1}{2}\sin\frac{4\pi}{3} + \frac{1}{3}\sin 2\pi = \frac{\sqrt{3}}{2} - \frac{1}{2} \times \frac{\sqrt{3}}{2} = \frac{\sqrt{3}}{4}.$$

Example 5. *Given n real numbers $a_1, a_2, \ldots, a_n$, find the value of x for which the sun $\sum_{=1}^{n} (x - a_i)^2$ is a minimum.* **[Delhi B.A. 2009; Delhi Maths (G), 1990]**

Solution. Let
$$f(x) = \sum_{i=1}^{n} (x - a_i)^2$$

$$\therefore \quad f'(x) = \sum_{i=1}^{n} 2(x - a_i) = 2nx - 2(a_1 + a_2 + \ldots + a_n).$$

The arithmetic mean $(\bar{a})$ of $a_1, a_2, \ldots, a_n$ is given by

$$\bar{a} = (a_1 + a_2 + \ldots + a_n)/n \quad \text{so that} \quad n\bar{a} = (a_1 + a_2 + \ldots + a_n).$$

$$\therefore \quad f'(x) = 2nx - 2n\bar{a} \quad \text{and} \quad f''(x) = 2n > 0.$$

The extreme values of f are given by $f'(x) = 0 \Rightarrow 2nx - 2n\bar{a} = 0 \Rightarrow x = \bar{a}$.

Since $f''(x) = 2n > 0$, f has a minimum at $x = \bar{a}$.

Example 6. *Find the maximum value of $(1/x^2)^{2x^2}$, $x > 0$.* **[I.A.S. (Prel.), 2002]**

Solution. Let $f(x) = (1/x^2)^{2x^2}$, then $\log f(x) = (2x^2) \log (1/x^2) = -4x^2 \log x$

$$\therefore \quad f'(x)/f(x) = -4\{2x \log x + x^2 \times (1/x)\} = -4x(2\log x + 1)$$

$$\Rightarrow \quad f'(x) = -4x\, f(x)(2\log x + 1) \qquad \ldots(1)$$

For maxima and minima of $f(x)$, $f'(x) = 0$

i.e., $-4x\, f(x)\,(2\log x + 1) = 0 \Rightarrow 2\log x + 1 = 0$ as $x \neq 0$ and $f(x) \neq 0$

Now, $2\log x + 1 = 0 \Rightarrow \log x = -1/2 \Rightarrow x = e^{-1/2} = 1/\sqrt{e}$

From (2), $f''(x) = -4f(x)\,(2\log x + 1) - 4x f'(x)\,(2\log x + 1) - 4x\, f(x) \times (2/x)$

or $f''(x) = -4f(x) \cdot (2\log x + 3) - 4x f'(x)\,(2\log x + 1)$

Also $f''(1/\sqrt{e}) = -4f(1/\sqrt{e}) \cdot (2\log e^{-1/2} + 3) = -8f(1/\sqrt{e}) < 0$

$$[\because \ f(1/\sqrt{e}) = e^{2/e} > 0]$$

So f is maximum at $x = 1/\sqrt{e}$ and maximum value of $f(x)$ at $x = 1/\sqrt{e}$ is $f(1/\sqrt{e}) = e^{2/e}$

11.4. APPLICATIONS TO PROBLEMS

We now proceed to discuss problems in which the quantity whose maximum or minimum value is required is not directly given as a function of one variable. In such problems we shall write down a functional relation from the given problem and then proceed as before. Very often the quantity whose maximum or minimum value is required can be expressed as a function of two variables and these two variables can be connected by a relation with the help of the given problem. With help of this relation the quantity can be expressed in terms of one variable.

EXAMPLES

Example 1. *If the sum of the lengths of the hypotenuse and another side of a right-angled triangle be given, show that the area of the triangle is a maximum when the angle between the hypotenuse and the given side is* $\pi/3$. **[Delhi Maths (P), 2002]**

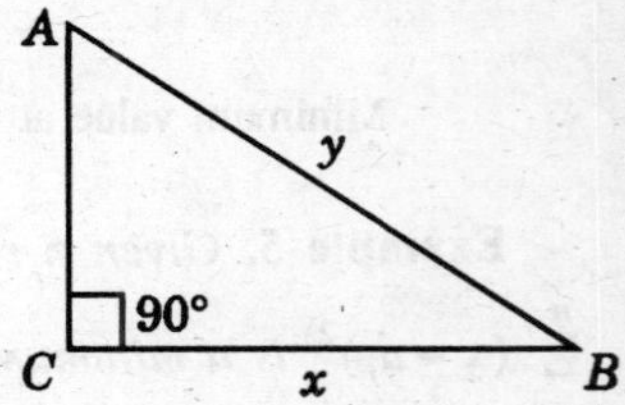

Solution. Let ABC be the given triangle right-angled at C. Let $AB = y$. Let $BC = x$ be the another side under consideration.

Given that $x + y =$ constant $= k$ (say) $\Rightarrow y = k - x$.

Then $\quad S =$ area of the triangle $= \dfrac{1}{2} \times AC \times BC$

$$S = (1/2) \times \sqrt{y^2 - x^2} \times x$$

$$S^2 = (1/4) \times x^2 (y^2 - x^2) = (1/4) \times x^2 \times \{(k - x)^2 - x^2\} = (1/4) \times (k^2x^2 - 2kx^3)$$

Now if S is maximum, S^2 will also be maximum. So even if we consider the maxima or minima of S^2 there is no loss of generality. Thus denoting S^2 by y, we have

$$y = S^2 = (1/4) \times (k^2x^2 - 2kx^3) \quad ...(1)$$

$$\therefore \quad dy/dx = (1/4) \times (2k^2x - 6kx^2) \quad ...(2)$$

For a maximum or a minimum of y, we must have

$$dy/dx = 0 \quad \text{or} \quad (1/4) \times (2k^2x - 6kx^2) = 0 \quad \text{or} \quad x = 0 \quad \text{or} \quad x = k/3.$$

The solution $x = 0$ is not proper for in this case $BC = 0$ and so the triangle will reduce to a straight line. So we shall consider $x = k/3$ only.

From (2), $\quad d^2y/dx^2 = (1/4) \times (2k^2 - 12kx)$

When $x = k/3$, $\quad d^2y/dx^2 = (1/4) \times (2k^2 - 4k^2) = -(k^2/2) = -$ ve,

showing that y and therefore S is maximum, when $x = k/3$.

Now $\quad x + y + k$ and $x = k/3 \Rightarrow y = 2k/3$.

Then, from ΔABC, $\quad \cos B = \dfrac{BC}{BA} = \dfrac{x}{y} = \dfrac{k/3}{2k/3} = \dfrac{1}{2} \Rightarrow B = \pi/3$.

Example 2. *Show that the height of the cylinder of maximum volume that can be inscribed in a sphere of radius a is* $2a/\sqrt{3}$.

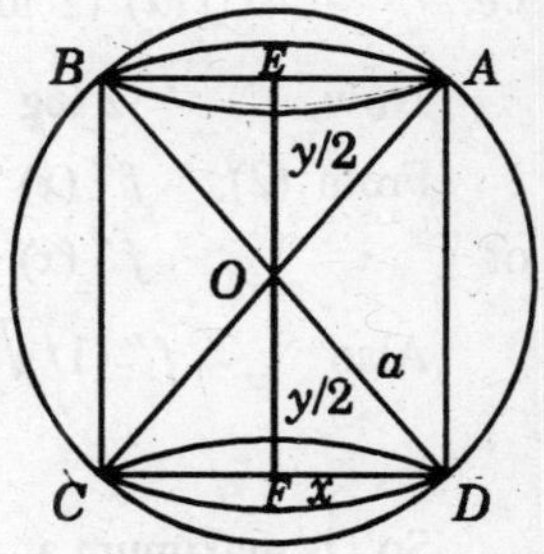

Solution. Let $ABCD$ be the given sphere of radius a so that $OD = a$. Let $ABCD$ be a right circular cylinder of radius FD $(= x)$ and height EF $(= y)$ which can be inscribed in the given sphere.

From ΔOFD, $\quad x^2 + y^2/4 = a^2 \quad$ or $\quad x^2 = a^2 - y^2/4$.

Note that if O is the centre of the sphere then it will also be the middle point of the height of the cylinder. Let V be the volume of the cylinder. Then, we have

$$V = \pi x^2 y = \pi (a^2 - y^2/4)\, y = \pi (a^2y - y^3/4) \quad ...(1)$$

$$\therefore \quad dV/dy = \pi (a^2 - 3y^2/4) \quad ...(2)$$

For maximum or minimum value of V, $dV/dy = 0$, so that $\pi (a^2 - 3y^2/4) = 0$ or $y = 2a/\sqrt{3}$, y being + ve only.

Again, from (2), $d^2y/dx^2 = -(3\pi y/2)$, which is – ve when $y = 2a/\sqrt{3}$. Therefore $y = 2a/\sqrt{3}$ g es the cylinder of maximum volume inscribed in a given sphere.

Example 3. *Find the dimensions of a right circular cone of maximum volume which can be circumscribed about a sphere of radius a.* **(I.A.S., 1999)**

Solution. Let cone OAB is circumscribed to a given sphere EMD with centre C. Let $OC = x$. Then, here

$$\angle ODC = \angle OEC = 90°.$$

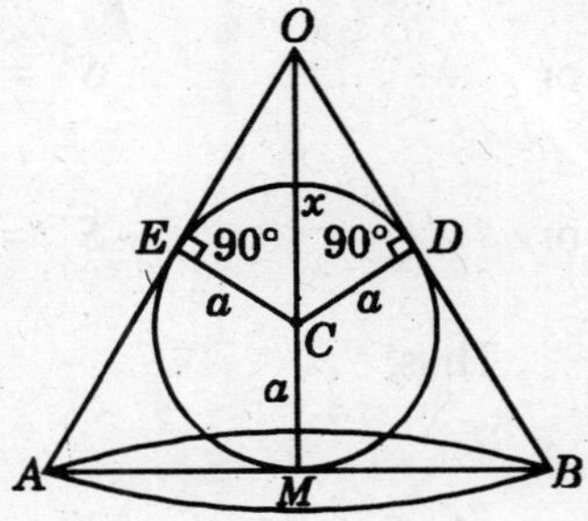

From $\Delta\, ODC$, we have $\quad OD = \sqrt{x^2 - a^2}$.

Also, $\quad OM = a + x$.

Then, volume V of the cone is given by

$$V = \frac{1}{3}\pi BM^2 \cdot OM = \frac{\pi}{3}(a + x)\, BM^2.$$

Since triangles OMB and OCD are similar, therefore

$$\frac{BM}{OM} = \frac{CD}{OD} \quad \text{or} \quad \frac{BM}{a + x} = \frac{a}{\sqrt{x^2 - a^2}}$$

or
$$BM = a\,(a + x)/(x^2 - a^2)^{1/2}$$

$$\therefore \quad V = \frac{1}{3}\pi (a + x) \times \frac{a^2 (a + x)^2}{(x^2 - a^2)} = \frac{\pi a^2 (a + x)^2}{3(x - a)}. \quad \ldots(1)$$

From (1),
$$\frac{dV}{dx} = \frac{\pi a^2}{3} \times \frac{2(a + x)\cdot(x - a) - (a + x)^2}{(x - a)^2} = \frac{\pi a^2 (a + x)(x - 3a)}{3(x - a)^2} \quad \ldots(2)$$

For a maximum or minimum value of V, we must have $dV/dx = 0$ and hence $x = 3a$, as $x = -a$ is not a proper solution.

Now, when x is slightly less than $3a$, dV/dx is negative and when x is slightly greater than $3a$, dV/dx is positive.

Hence there is a change of sign of dV/dx, namely, from negative to positive as x passes through the point $x = 3a$. So $x = 3a$ gives a minimum value of V.

When $x = 3a$, $\quad OM = a + x = 4a$

and
$$BM = \frac{a(a + x)}{\sqrt{x^2 - a^2}} = \frac{a \times 4a}{\sqrt{9a^2 - a^2}} = \frac{a \times 4a}{2\sqrt{2}\,a} = a\sqrt{2}.$$

Thus dimensions of the required cone are given by : radius $= BM = a\sqrt{2}$ and height $OM = 4a$.

Example 4. *A thin closed rectangular box is to have one edge n times the length of another edge and the volume of the box is given to be V. Prove that the least surface S is given by*

$$nS^3 = 54\,(n + 1)^2\, V^2 \qquad \textbf{(I.A.S., 1998)}$$

Solution. Let the lengths of the edges of rectangular box be x, nx and y. Then

$$V = nx^2 y \quad \text{so that} \quad y = V/nx^2 \quad \ldots(1)$$

and
$$S = 2\,(nx^2 + xy + nxy) \quad \ldots(2)$$

Substituting the value of y from (1) in (2), we have

$$S = 2\,(nx^2 + V/nx + V/x) \quad \ldots(3)$$

$$\therefore \quad dS/dx = 2\,(2nx - V/nx^2 - V/x^2). \quad \ldots(4)$$

For maximum or minimum value of S, $dS/dx = 0$

$$\therefore \quad 2nx - \frac{V}{nx^2} - \frac{V}{x^2} = 0 \quad \text{so that} \quad x^3 = \frac{V(n + 1)}{2n^2} \quad \ldots(5)$$

Also, from (4), $\quad d^2S/dx^2 = 2\,(2n + 2V/nx^3 + 2V/x^3)$,

which is positive for x given by (5). Hence S is least when x is given by (5).

Now, from (3), we have

$$S^3 = (8/n^3x^3) \times \{n^2x^3 + (n+1)V\}^3$$

or $$S^3 = \frac{8}{n^3} \times \frac{2n^2}{V(n+1)} \left\{ n^2 \times \frac{V(n+1)}{2n^2} + (n+1)V \right\}^3, \text{ using (5)}$$

or $$S^3 = \frac{8}{n^3} \times \frac{2n^2}{V(n+1)} \times \frac{27V^3(n+1)^3}{8} \quad \text{or} \quad S^3 = \frac{54V^2(n+1)^2}{n}$$

Thus, $$nS^3 = 54(n+1)^2 V^2.$$

EXERCISES

1. *Examine the following functions for maximum and minimum values :*
   (*i*) $(x-3)^2(x+3) \ \forall\ x \in \mathbf{R}$ **[Delhi Maths (P), 1995, 97]**
   (*ii*) $x^4 + 4x^3 - 2x^2 - 12x + 7 \ \forall\ x \in \mathbf{R}$ (*iii*) $12x^5 - 45x^4 + 40x^3 + 6 \ \forall\ x \in \mathbf{R}$
   (*iv*) $(x-3)^5(x+1)^4 \ \forall\ x \in \mathbf{R}$
   (*v*) $4x^{-1} - (x-1)^{-1} \ \forall\ x \in \mathbf{R} \sim \{0, 1\}$ **[Delhi Maths (G), 1995]**
   (*vi*) $(x-1)(x-2)(x-3) \ \forall\ x \in \mathbf{R}$ (*vii*) $x^4/(x-1)(x-3)^3$
   (*viii*) $x^{1/x}$ **(Calicut, 2004)** (*ix*) x^2/e^{x^2} **[Kanpur 2006]**
2. Show that the function f defined by $f(x) = (ax+b)/(cx+d)$, for all $x \in \mathbf{R} \sim \{-d/c\}$, does not possess an extreme value unless $ad - bc = 0$. **[Delhi Maths (G), 2005]**
3. Investigate the extreme values of the function f, defined by $f(x) = x^3 + 3px + q$, for all $x \in \mathbf{R}$, p, q being fixed real numbers.
   [Delhi Maths (G), 2004; Delhi Maths (H), 2004, 09]
4. Show that the function $f(x) = \sin x\,(1 + \cos x)$ has a maximum value when $x = \pi/3$.
   [Delhi Maths (G), 1997, 2005; Delhi Maths (H), 1996, 2005; Kumaon, 1998]
5. Find the maximum and minimum as well as the greatest and the least value of $x^3 - 12x^2 + 45x$ in the interval $[0, 7]$.
6. If $(x-a)^{2n}(x-b)^{2m+1}$, where m and n are positive integers, is the derivative of a function f, then show that $x = b$ gives a minimum but $x = a$ gives neither a maximum nor a minimum. **[Delhi Maths (H) 2006]**
7. Show that the function f, defined by $f(x) = \sin(x - \pi/3)\sin(x + \pi/3)$ for all $x \in \mathbf{R}$ has a minima at $x = \pi/6$ and maxima at $x = \pi/2$ and $x = -\pi/6$.
8. Prove that the function $f(x) = x^4(x - \pi/2)^2 + \sin^4 x \ \forall\ x \in \mathbf{R}$, has extreme values at $x = 0$ and $x = \pi/2$ and determine whether they are maxima or minima.
9. Show that the function $f(x) = \sin^m x \sin mx + \cos^m x \cos mx$, for all $x \in \mathbf{R}$, has a minimum at $x = \pi/4$ when $m = 2$ and a maximum at $x = \pi/4$ when $m = 4$ or 6.
10. Prove that the function f defined by $f(x) = 3|x| + 4|x-1|$, for all $x \in \mathbf{R}$, has a minimum value, and that this value is 3. **[Delhi Maths (G), 1996,**
11. Prove that $f(x) = 2|x-2| + 5|x-3|$ for all $x \in \mathbf{R}$, has a minimum value 2 at $x = 3$.
12. Prove that the function $f(x) = 4|x| + 5|x-1|$ is not derivable at $x = 0, 1$. Also show that the minimum value is 4 when $x = 1$
13. (*a*) Show that the function f, defined by $f(x) = x^p(1-x)^q$ for all $x \in \mathbf{R}$, where p and q are positive integers, has a maximum value, whatever the values of p and q may be. **[Delhi Maths (G), 1994]**

(*b*) Show that the function *f*, defined by

$$f(x) = |x|^p \cdot |x-1|^q, \text{ for all } x \in \mathbf{R}$$

has a maximum value $p^p q^q/(p+q)^{p+q}$, *p* and *q* being positive integers.

[Delhi Maths (G), 1999; Delhi Maths (H), 2001, 02]

Hint. Re-write the given function as

$$f(x) = \begin{cases} (-1)^m x^m (1-x)^n & , \text{ for } x < 0 \\ 0 & , \text{ for } x = 0 \\ x^m (1-x)^n & , \text{ for } 0 < x < 1 \\ 0 & , \text{ for } x = 1 \\ x^m (x-1)^n & , \text{ for } x > 1 \end{cases}$$

14. Show that the height of closed cylinder of given volume and least surface is equal to its diameter. **[Delhi Maths (H), 1995]**

15. Given the perimeter of a rectangle, show that its area is maximum when it is a square. **[Delhi Maths (G), 1998]**

16. Show that of all rectangles of given area, the square has the smallest perimeter.

17. Show that the rectangle of maximum area that can be inscribed in a circle is a square.

18. Show that the semi-vertical angle of a cone of maximum volume and of a given slant height in $\tan^{-1}\sqrt{2}$.

19. (*a*) Show that the right circular cylinder of the given surface and maximum volume is such that its height is equal to the diameter of the base.

(*b*) Show that the height of an open cylinder of given surface and greatest volume is equal to the radius of the base. **(G.N.D.U. Amritsar, 2004)**

20. Show that the volume of the greatest cylinder which can be inscribed in a cone of height *h* and semi-vertical angle α is $(4/27)\ \pi h^3 \tan^2 \alpha$. **(Garhwal, 2003)**

21. Prove that a conical tent of a given capacity will require the least amount of canvas when the height is $\sqrt{2}$ times the radius of the base.

22. Divide 15 into two parts such that the square of one multiplied with the cube of the other is a maximum. **(G.N.D.U. Amritsar, 2004)**

23. A given quantity of metal is to be cast into a half cylinder, *i.e.*, with a rectangular base and semi-circular ends. Show that in order that the total surface area may be minimum, the ratio of the length of the cylinder to the diameter of its semi-circular ends is $\pi : \pi + 2$.

24. A lane runs at right angles out of a road '*a*' metre wide. Find how many metre wide the lane must be if it is just possible to carry a pole '*b*' metre long ($b > a$) from the road into the lane, keeping it horizontal.

25. A rectangular sheet of metal has four equal square portions removed at the corners, and the sides are then turned up so as to form an open rectangular box. Show that when the volume contained in the box is a maximum, the depth will be

$$\{(a+b) - \sqrt{a^2 - ab + b^2}\}/6$$

where *a*, *b* are the sides of the original rectangle.

26. Show that $x^5 - 5x^4 + 5x^2 - 1$ has a maximum value when $x = 1$ and a minimum value when $x = 3$ and neither when $x = 0$. **(Kumaun, 2003)**

27. Investigate for maximum and minimum values of function given by $y = \sin x + \cos 2x$. **(Garhwal, 1997; G.N.D.U. Amritsar, 2004)**

28. Find the maximum and minimum of the radii vectors of the curve

$$c^4/r^2 = a^2/\sin^2\theta + b^2/\cos^2\theta.$$ **(Garhwal, 1999)**

29. Show that $x/(1 + x \tan x)$ has maximum value when $x = \cos x$.

30. Show that the area of the greatest isosceles triangle that can be inscribed in a given ellipse $x^2/a^2 + y^2/b^2 = 1$, the triangle having its vertex coincident with one extremity of the major axis, is $(3\sqrt{3}\,ab)/4$.

31. A perpendicular is let fall from the centre to a tangent to the ellipse $x^2/a^2 + y^2/b^2 = 1$. Show that the greatest value of the intercept between the point of contact and the foot of the perpendicular is $(a - b)$.

32. Tangents are drawn to the ellipse $x^2/a^2 + y^2/b^2 = 1$ and the circle $x^2 + y^2 = a$ at the point where a common ordinate cuts them. Show that if θ be the greatest inclination of the tangents, then $\tan\theta = (a-b)/2\sqrt{ab}$.

33. Normal is drawn at a variable point P of the ellipse $x^2/a^2 + y^2/b^2 = 1$. Show that the maximum distance of the normal from the centre is $(a - b)$.

34. A person being in a boat 'a' metres from the nearest point of the beach wishes to reach as quickly as possible a point 'b' metres from the point along the shore. The ratio of his rate of walking to his rate of rowing is $\sec\alpha$. Prove that he should land at a distance $(b - a\cot\alpha)$ from the place to be reached.

ANSWERS

1. (*i*) Max. value is 32 at $x = 1$; Min. value is 0 at $x = 3$

(*ii*) Max. value is 14 at $x = -1$; Min. value is -2 at $x = 1, -3$

(*iii*) Max. value is 13 at $x = 1$; Min. value is -10 at $x = 2$; Neither max. nor min. at $x = 0$

(*iv*) Maxima at $x = -1$; Minima at $x = 7/9$; Neither maxima nor minima at $x = 3$

(*v*) Max. value is 1 at $x = 2$; Min. value is 9 at $x = 2/3$

(*vi*) Max. value $= 2/3\sqrt{3}$; Min. value $= -2/3\sqrt{3}$;

(*vii*) Maxima at $x = 6/5$; Minima at $x = 0$

(*viii*) Max. value $= e^{1/e}$ when $x = e$.

3. When $p \not< 0$, f has no extreme value; if $p < 0$, then there is maxima at $x = -\sqrt{-p}$ and minima at $x = \sqrt{-p}$

5. Max. value = 54; Min. value = 50; Greatest value = 70 and least value = 0

7. Minima at $x = 0$ and $x = \pi/2$ **22.** 6 and 9 **24.** $(b^{2/3} - a^{2/3})^{3/2}$

27. y is max. for $x = \sin^{-1}(1/4)$, $\pi - \sin^{-1}(1/4)$ and min. for $x = \pi/2, 3\pi/2$

28. Maximum value of $r = c^2/(a + b)^2$

OBJECTIVE QUESTIONS

Multiple Choice Type Questions : *Select (a), (b), (c) or (d), whichever is correct.*

1. The maximum value of $(\log x)/x$ is :

(*a*) 1 (*b*) e (*c*) $2/e$ (*d*) $1/e$. **[I.A.S. (Prel.), 2004]**

2. The function defined by $f(x) = x^{1/x}$ has a maximum at :

(*a*) 1 (*b*) e (*c*) $\log_2 e$ (*d*) 2. **[I.A.S. (Prel.), 1998]**

3. The maximum value of $(1/x)^x$ is equal to :

(*a*) e (*b*) 1 (*c*) $e^{1/e}$ (*d*) $(1/e)^e$. **[I.A.S. (Prel.), 2001]**

4. The maximum value of $(1/x^2)^{2x^2}$, $x > 0$ is equal to :

(*a*) e (*b*) $e^{2/e}$ (*c*) $e\sqrt{e}$ (*d*) $1/e$. **[I.A.S. (Prel.), 2004]**

5. Let $f(x) = x^2 - 4x + 3$. The following statements are associated with f:

1. f is increasing in $(2, \infty)$ 2. f is increasing in $(-\infty, -2)$ 3. f has a stationary point at $x = 2$. Which of these statements are correct ?

(*a*) 1 and 2 (*b*) 1 and 3 (*c*) 2 and 3 (*d*) 1, 2 and 3.

[I.A.S. (Prel.), 2003]

6. The difference between the maximum and minimum values of the function $a \sin x + b \cos x$ is :

(*a*) $2\sqrt{a^2+b^2}$ (*b*) $2(a^2+b^2)$ (*c*) $\sqrt{a^2+b^2}$ (*d*) $-\sqrt{a^2+b^2}$.

[I.A.S. (Prel.), 1995]

7. Which one of the following statements is correct for the function $f(x) = x^3$?

(*a*) $f(x)$ has a maximum value at $x = 0$ (*b*) $f(x)$ has a minimum value at $x = 0$

(*c*) $f(x)$ has neither a maximum nor a minimum at $x = 0$

(*d*) $f(x)$ has no point of inflexion. **[I.A.S. (Prel.), 1997]**

ANSWERS

Multiple Choice Type Questions :

1. (*d*) **2.** (*b*) **3.** (*c*) **4.** (*b*) **5.** (*d*) **6.** (*a*) **7.** (*c*)

MISCELLANEOUS PROBLEMS ON CHAPTER 11

1. The function $f(x) = \sin^3 x - m \sin x$ is defined on the open interval $(-\pi/2, \pi/2)$ and it assumes only one maximum value and only one minimum value on this interval. Then, which one of the following must be correct

(*a*) $0 < m < 3$ (*b*) $-3 < m < 0$ (*c*) $m = 0$ (*d*) $m = 3$ **[I.A.S. Prel. 2006]**

2. Let A and B be fixed points with co-ordinates $(0, a)$ and $(0, b)$ respectively and P is a variable point $(x, 0)$ referred to rectangular axes. When is the angle APB extremum ?

(*a*) $x^2 = \sqrt{ab}$ (*b*) $x^2 = ab$ (*c*) $x^2 = a + b$ (*d*) $x^2 = b - a$ **[I.A.S. Prel. 2006]**

3. Find the maximum and minimum values of the function $f(x) = x^2 e^{-2x}$, $x \geq 0$

[Delhi Maths (G) 2006]

4. *Find the maximum and minimum value of the function* $f(x) = x^4 + 4x^3 - 2x^2 - 12x + 7$, $x \in \mathbf{R}$. **[Delhi Maths (Prog) 2007]**

[**Ans.** max. at $x = -1$, value $= 14$; Min at $x = -3$, value $= -2$; min. at $x = 1$; value $= -2$]

5. *If* $f(c)$ *is minimum value of the function, then prove that* $f'(c) = 0$ *whenever* $f'(c)$ *exists. What happens when* $f'(c)$ *does not exist ? Justify.* **[Delhi Maths (H) 2007]**

6. *Find the maximum and minimum values of the function*

$f(x) = \cos(x - \pi/6) \cos x \cos(x + \pi/6)$ *for all* $x \in [0, \pi]$ **[Delhi Maths (H) 2007]**

7. (*i*) What is the maximum value of $y = \sin^3 x \cos x,\ 0 < x < \pi$

(*a*) $-3\sqrt{3}/16$ (*b*) $3\sqrt{3}/4$ (*c*) $-3/16$ (*d*) $3\sqrt{6}/16$ **[I.A.S. Prel. 2007]**

(*ii*) *What is the maximum area of the rectangle whose sides pass through the angular points of a given rectangle of sides a and b.*

(*a*) $(a+b)^2/2$ (*b*) $(a+b)^2$ (*c*) $(a^2+b^2)/2$ (*d*) a^2+b^2 **[I.A.S. Prel. 2007]**

(*iii*) *Match list I with list II and select the correct answer using the code given below the lists:*

List – I	List – II
A. *The function* $x^3 - 6x^2 - 36x + 7$ *is increasing when*	1 $x = -2$
B. *The function* $x^3 - 6x^2 - 36x + 7$ *is maximum at*	2 $x = 6$
C. *The function* $x^3 - 6x^2 - 36x + 7$ *is minimum at*	3 $x < -2$ or $x > 6$
D. *The function* $x^3 - 6x^2 - 36x + 7$ *decreases when*	4 $-2 < x < 6$

Code:		A	B	C	D
	(*a*)	4	2	1	3
	(*b*)	3	1	2	4
	(*c*)	3	2	1	4
	(*d*)	4	1	2	3

[I.A.S. Prel 2007]

8. What is the point on the curve $y^2 = 4x$ which is nearest to the point (2,1)?

(*a*) (1, 2) (*b*) (1, –2) (*c*) (0, 0) (*d*) None of the above **(I.A.S. Prel 2009]**

9. What are the points of the extrema of the function $y = \int_0^x \frac{\sin t}{t}\, dt, x > 0$

(*a*) $0, \pm n\pi$ (*b*) $\pm n\pi$ only (*c*) $n\pi$ only (*d*) $0, n\pi$ only **(I.A.S. (Prel.) 2009]**

10. The function $f(x) = \int_0^{x^2} \left(\frac{t^2 - 5t + 4}{2 + e^t}\right) dt$ has (*a*) two maxima and two minima points

(*b*) two maxima and three minima points (*c*) three maxima and two minima points

(*d*) one maximum point and one minimum point **[I.A.S. (Prel) 2009]**

11. Define absolute maximum point and absolute minimum point of a function $f: A \to \mathbf{R}$. Is the absolute maximum point of f always unique ? Justify your answer.

[Delhi B.Sc. I (Hons) 2010]

ANSWERS

1. (a) **2.** (b) **3.** Minimum value = 0 and Maximum value = e^{-2} **7.** (*i*) (*d*); (*ii*) (*a*) (*iii*) (*a*) **8.** (*c*) **9.** (*c*) **10.** (*b*)

Chapter

12

Indeterminate Forms

12.1. INTRODUCTION

In this chapter we shall deal with applications of Taylor's theorem to the evaluation of certain limits, known as *indeterminate forms*.

We know that (refer Chapter 9)

$$\lim_{x \to a} \frac{f(x)}{g(x)} = \frac{\lim_{x \to a} f(x)}{\lim_{x \to a} g(x)},$$

provided $\lim_{x \to a} f(x)$ and $\lim_{x \to a} g(x)$ both exist and $\lim_{x \to a} g(x) \neq 0$.

In what follows we shall discuss a method (known as L' Hospital's rule) to evaluate $\lim_{x \to a} \{f(x)/g(x)\}$ even when $\lim_{x \to a} f(x) = \lim_{x \to a} g(x) = 0$. In such a situation $\lim_{x \to a} \{f(x)/g(x)\}$ is said to take the form 0/0.

There are various kinds of indeterminate forms $0/0$, ∞/∞, $0 \times \infty$, $(\infty - \infty)$, 1^{∞}, 0^0 and ∞^0.

The forms 0/0 and ∞/∞ are known as fundamental forms because the remaining forms can be easily converted into one of these forms.

12.2. THE INDETERMINATE FORM (0/0) (Gujarat, 2004)

Theorem I. *Let f, g be two functions such that*

(i) $\lim_{x \to c} f(x) = 0, \; \lim_{x \to c} g(x) = 0$ ...(*i*)

and (ii) $f'(c), g'(c)$ *both exist and* $g'(c) \neq 0$.

Then $$\lim_{x \to c} \frac{f(x)}{g(x)} = \frac{f'(c)}{g'(c)}$$

Proof. Since the functions *f*, *g* are derivable at the point *c*, therefore, they are continuous at *c* and accordingly, by the given condition (*i*), $f(c) = g(c) = 0$.

We have

$$f'(c) = \lim_{h \to 0} \frac{f(c+h) - f(c)}{h} = \lim_{h \to 0} \frac{f(c+h)}{h},$$

and $$g'(c) = \lim_{h \to 0} \frac{g(c+h) - g(c)}{h} \lim_{h \to 0} \frac{g(c+h)}{(h)}.$$

$$\therefore \quad \frac{f'(c)}{g'(c)} = \lim_{h \to 0} \frac{[f(c+h)/h]}{[g(c+h)/h]} = \lim_{h \to 0} \frac{f(c+h)}{g(c+h)} = \lim_{x \to c} \frac{f(x)}{g(x)}$$

Note. This result may also be stated as follows :

If $f(c) = g(c) = 0$, and $f'(c)$, $g'(c)$ exist but $g'(c) \neq 0$, then

$$\lim [f(x)/g(x)] = f'(c)/g'(c) \text{ when } x \to c.$$

***L' Hospital's Theorem II.** *Let f, g be two functions such that*

(*i*) $\lim_{x \to c} f(x) = 0, \ \lim_{x \to c} g(x) = 0$

and (*ii*) $\lim [f'(x)/g'(x)] = l$ *when* $x \to c$.

Then $\lim [f(x)/g(x)] = l$ *when* $x \to c$.

The condition (*ii*) implies that $f'(x)$ and $g'(x)$ exist and $g'(x) \neq 0$ at every point x, other than c, of a certain neighbourhood $]c - \delta, c + \delta[$ of c.

We *suppose* that $f(c) = g(c) = 0$ for this change in the definitions of f and g influences neither the hypothesis nor the conclusion of the theorem.

If $x \in [c - \delta, c + \delta]$, we have, by Cauchy's theorem,

$$\frac{f(x)}{g(x)} = \frac{f(x) - f(c)}{g(x) - g(c)} = \frac{f'(\xi)}{g'(\xi)},$$

where ξ lies between c and x and also depends upon x. By virtue of the given condition (*ii*)

$$f'(\xi)/g'(\xi) \to l \text{ as } x \to c \ \Rightarrow \ f(x)/g(x) \to l \text{ as } x \to c.$$

Note. The theorem may also be stated in another form as follows :

$$\lim_{x \to c} f(x) = 0 = \lim_{x \to c} g(x),$$

$$\Rightarrow \qquad \lim_{x \to c} \frac{f(x)}{g(x)} = \lim \frac{f'(x)}{g'(x)},$$

provided that the limit on the right exists.

It will later on be seen by means of an example that $\lim [f(x)/g(x)]$ may exist without $\lim [f'(x)/g'(x)]$ existing.

Note. The reader will find it useful to compare the hypothesis and the conclusions, of the preceding two theorems. In theorem I the derivability of f and g is assumed at $x = c$ *only*, whereas in theorem II while assuming the derivability of f, g in a neighbourhood at $x = c$, it is exactly at $x = c$ where we have not needed it.

Theorem III. *If f, g be two functions such that when* $x \to c$,

$$(i) \quad \begin{cases} \lim f(x) = \lim f'(x) = \dots = \lim f^{n-1}(x) = 0 \\ \lim g(x) = \lim g'(x) = \dots = \lim g^{n-1}(x) = 0 \end{cases}$$

and $\lim [f^n(x)/g^n(x)] = l$, when $x \to c$

then $\lim [f(x)/g(x)] = l$, *when* $x \to c$.

This may be proved through repeated applications of the preceding theorem.

Theorem IV. *Let f, g be two functions such that when* $x \to c$,

$$(i) \quad \begin{cases} \lim f(x) = \lim f'(x) = \dots = \lim f^{n-1}(x) = 0 \\ \lim g(x) = \lim g'(x) = \dots = \lim g^{n-1}(x) = 0 \end{cases}$$

and (*ii*) $f^n(c)$, $g^n(c)$ *exist and* $g^n(c) \neq 0$

then $\lim [f(x)/g(x)] = f^n(c)/g^n(c)$, *when* $x \to c$.

Proof. By virtue of the given condition (*ii*), on applying the theorem I, we see that

$$\lim \frac{f^{n-1}(x)}{g^{n-1}(x)} = \frac{f^n(c)}{g^n(c)}.$$

* L' Hospital (1661-1704) was a French Mathematician.

Also from theorem III, changing n to $(n-1)$, we see that

$$\lim \frac{f(x)}{g(x)} = \lim \frac{f^{n-1}(x)}{g^{n-1}(x)} = \frac{f^n(c)}{g^n(c)}.$$

Theorem V. *If f, g be two functions such that*

(*i*) $\lim\limits_{x \to \infty} f(x) = 0$, $\lim\limits_{x \to \infty} g(x) = 0$,

and (*ii*) $\lim\limits_{x \to \infty} [f'(x)/g'(x)] = l$,

then $$\lim_{x \to \infty} [f(x)/g(x)] = l.$$

We write $z = 1/x$ so that $x \to +\infty \Rightarrow z \to 0+0$.

Let $$F(z) = f(1/z),\ G(z) = g(1/z)$$

so that $$\lim_{z \to (0+0)} F(z) = 0, \quad \lim_{z \to (0+0)} G(z) = 0. \quad ...(1)$$

We have $$F'(z) = -f'\left(\frac{1}{z}\right)\cdot\frac{1}{z^2}, \quad G'(z) = -g'\left(\frac{1}{z}\right)\cdot\frac{1}{z^2}$$

$$\Rightarrow \frac{F'(z)}{G'(z)} = \frac{f'(1/z)}{g'(1/z)},$$

so that, by the given condition (*ii*),

$$F'(z)/G'(z) \to l, \text{ when } z \to (0+0). \quad ...(2)$$

Hence, from (1) and (2), $$\lim_{z \to (0+0)} [F(z)/G(z)] = l,$$

$$\Rightarrow \lim_{x \to \infty} [f(x)/g(x)] = l.$$

We may similarly see the truth of the result when $x \to -\infty$.

12.3. THE INDETERMINATE FORM (∞/∞)

Theorem I. *If f, g be two functions such that when $x \to c$,*

(*i*) *$\lim [g(x)] = \infty$ and (ii) $\lim [f'(x)/g'(x)] = 0$*

then $$\lim [f(x)/g(x)] = 0, \text{ when } x \to c.$$

Proof. The given condition (*ii*) implies that there exists an interval $[c-\delta, c+\delta]$ around c such that for every point $x \neq c$ of this interval, $f'(x)$, $g'(x)$ exist and $g'(x) = 0$.

From the Darboux's theorem (refer Art. 9.6), it follows $g'(x)$ keeps the same sign, positive or negative $\forall\ x \in\ ]c, c+\delta[$, and the same is true for $[c-\delta, c[$ also.

Firstly we consider the interval $]c, c+\delta[$. Let $g'(x)$ remain positive in it.

Let $\varepsilon > 0$ be any number. Then, by virtue of (*ii*), there exists a positive number $\delta_1 > \delta$ such that $\forall\ x \in\ ]c, c+\delta_1[$

$$\left| \frac{f'(x)}{g'(x)} - 0 \right| < \frac{\varepsilon}{2},$$

$$\Rightarrow \quad |f'(x)| < \frac{1}{2}\varepsilon\, |g'(x)| = \frac{1}{2}\varepsilon\, g'(x)$$

$$\Rightarrow \quad -\frac{1}{2}\varepsilon\, g'(x) < f'(x) < \frac{1}{2}\varepsilon\, g'(x).$$

By the Art. 10.6, we see that $\forall\ x \in\ ]c, c+\delta_1[$

$$-\frac{1}{2}\varepsilon [g(c+\delta) - g(x)] < [f(c+\delta) - f(x)] < \frac{1}{2}\varepsilon [g(c+\delta) - g(x)]$$

$$\Rightarrow \quad |f(c+\delta)-f(x)| < \frac{1}{2}\varepsilon |g(c+\delta)-g(x)|$$

$$\Rightarrow \quad |f(x)|-|f(c+\delta)| < \frac{1}{2}\varepsilon |g(c+\delta)|+\frac{1}{2}\varepsilon |g(x)|$$

$$\Rightarrow \quad |f(x)| < \frac{1}{2}\varepsilon |g(x)|+\frac{1}{2}\varepsilon |g(c+\delta)|+|f(c+\delta)|$$

$$\Rightarrow \quad |f(x)| < \frac{1}{2}\varepsilon |g(x)|+k,$$

$$\Rightarrow \quad \left|\frac{f(x)}{g(x)}\right| < \frac{1}{2}\varepsilon+\frac{k}{|g(x)|}.$$

Here $\quad k = (1/2)\times\varepsilon |g(c+\delta)|+|f(c+\delta)|$ is free of x.

Again $\lim_{x\to c} |g(x)| = \infty$ implies that there exists a positive number $\delta_2 < \delta_1$ such that $\forall\ x \in\ ]c, c+\delta_2[$, we have $\quad k/|g(x)| < \varepsilon/2.$

Thus, we see that to every $\varepsilon > 0$, there corresponds $\delta_2 > 0$ such that $\forall\ x \in\ ]c, c+\delta_2[$,.

$$|f(x)/g(x)| < \varepsilon$$

$$\Rightarrow \quad \lim_{x\to(c+0)} \frac{f(x)}{g(x)} = 0.$$

It may similarly be shown that $\quad \lim_{x\to(c-0)} \frac{f(x)}{g(x)} = 0.$

Thus, $\quad f(x)/g(x) \to 0$ as $x \to c$.

Theorem II. *If f, g be two functions such that when $x \to c$,*

(i) lim $|g(x)| = \infty$, (ii) lim $[f'(x)/g'(x)] = l$,

then $\quad$ *lim $[f(x)/g(x)] = l$, when $x \to c$.*

Proof. We write $\quad \phi(x) = f(x) - l\,g(x)$

We have $\quad \lim_{x\to 0} \frac{\phi'(x)}{g'(x)} = \lim_{x\to 0}\left[\frac{f'(x)}{g'(x)} - l\right] = 0$, by the given condition (*ii*)

$$\Rightarrow \quad \lim_{x\to c} \frac{\phi(x)}{g(x)} = 0,$$

$$\Rightarrow \quad \lim_{x\to c}\left[\frac{f(x)}{g(x)} - l\right] = 0$$

$$\Rightarrow \quad \lim \frac{f(x)}{g(x)} = l.$$

Note 1. As in theorem V of Art. 12.2, it may be shown that the result remains true even when $x \to \infty$.

Note 2. It should be specially noted that in the preceding theorem nothing whatsoever has been said about the limit of $f(x)$ as $x \to c$ so that the result holds good independently of the behaviour of $f(x)$ as $x \to c$. In particular, therefore, the result holds good when $f(x) \to \infty$ as $x \to c$; this being the form in which the theorem is usually stated.

12.4. SOME USEFUL RESULTS

I. The following standard infinite series expansion of functions can be used to compute limits of indeterminate forms.

(*i*) $e^x = 1 + x + \frac{x^2}{2!} + \frac{x^3}{3!} + \ldots \text{ ad. inf}$

(*ii*) $\sin x = x - \frac{x^3}{3!} + \frac{x^5}{5!} - \frac{x^7}{7!} + \ldots \text{ ad. inf}$ $\qquad \sinh x = x + \frac{x^3}{3!} + \frac{x^5}{5!} + \frac{x^7}{7!} + \ldots \text{ ad. inf}$

(*iii*) $\cos x = 1 - \frac{x^2}{2!} + \frac{x^4}{4!} - \frac{x^6}{6!} + \ldots \text{ ad. inf}$ $\qquad \cosh x = 1 + \frac{x^2}{2!} + \frac{x^4}{4!} + \frac{x^6}{6!} + \ldots \text{ ad. inf}$

(*iv*) $\tan x = x + \frac{x^3}{3} + \frac{2x^5}{15} + \ldots \text{ ad. inf}$ $\qquad \tan^{-1} x = x - \frac{x^3}{3} + \frac{x^5}{5} - \frac{x^7}{7} + \ldots \text{ ad. inf}$

(*v*) $\log(1 + x) = x - \frac{x^2}{2} + \frac{x^3}{3} - \frac{x^4}{4} + \ldots \text{ ad. inf}, -1 < x \le 1$

$\log(1 - x) = -x - \frac{x^2}{2} - \frac{x^3}{3} - \frac{x^4}{4} - \ldots \text{ ad. inf}, -1 \le x < 1$

II. The following standard limits can be used directly while finding limits of indeterminate forms.

(*i*) $\lim_{x \to 0} \frac{\sin x}{x} = 1, \quad \lim_{x \to 0} \frac{x}{\sin x} = 1, \quad \lim_{x \to 0} \frac{\tan x}{x} = 1, \quad \lim_{x \to 0} \frac{x}{\tan x} = 1$

(*ii*) $\lim_{x \to 0} (1 + x)^{1/x} = e, \quad \lim_{x \to 0} \left(1 + \frac{1}{x}\right)^x = e, \quad \lim_{x \to 0} \frac{\log(1 + x)}{x} = 1$

Note. The application of the above results either in the beginning or at an appropriate stage in some problems will shorten the whole process of evaluation of given limit.

12.5. WORKING RULE TO EVALUATE LIMIT IN INDETERMINATE FORM $\frac{0}{0}$, NAMELY, $\lim_{x \to a} \frac{f(x)}{g(x)}$ WHERE $f(a) = g(a) = 0$.

Step 1. Differentiate the numerator and denominator separately.

Step 2. Find $\lim_{x \to a} [f'(x) / g'(x)]$.

Step 3. If the indeterminate form 0/0 still persists, then repeat the process. Sometimes more than one application of L' Hospital's rule will be necessary.

Note. Before applying L' Hospital's rule at any stage, make sure that the expression to which the rule in being applied is an indeterminate form. For example, the following solution of evaluation of limit, namely,

$$\lim_{x \to 1} \frac{x^3 + 3x - 4}{2x^2 + x - 3} = \lim_{x \to 1} \frac{3x^2 + 3}{4x + 1} = \lim_{x \to 1} \frac{6x}{4} = \frac{3}{2}$$

is wrong because the expression $(3x^2 + 3)/(4x + 1)$ is not of the form 0/0 as $x \to 1$. Hence we cannot apply L' Hospital's rule to evaluate $\lim_{x \to 1} \{(3x^2 + 3)/(4x + 1)\}$. Actually, we have

$$\lim_{x \to 1} \frac{x^3 + 3x - 4}{2x^2 + x - 3} = \lim_{x \to 1} \frac{3x^2 + 3}{4x + 1} = \frac{\lim_{x \to 1} (3x^2 + 3)}{\lim_{x \to 1} (4x + 1)} = \frac{6}{5}.$$

EXAMPLES

Example 1. *Evaluate* $\lim_{x \to 0} \frac{x \cot x - 1}{x^2}$. **[*Delhi Maths (G), 1998, 2000*]**

Solution. We have

$$\lim_{x \to 0} \frac{x \cot x - 1}{x^2} = \lim_{x \to 0} \frac{x \cos x - \sin x}{x^2 \sin x} \qquad \left[\text{Form } \frac{0}{0}\right]$$

$$= \lim_{x \to 0} \frac{\cos x - x \sin x - \cos x}{2x \sin x + x^2 \cos x}$$

$$= \lim_{x \to 0} \frac{-\sin x}{2 \sin x + x \cos x} \qquad \left[\text{Form } \frac{0}{0}\right]$$

$$= \lim_{x \to 0} \frac{-\cos x}{2 \cos x + \cos x - x \sin x} = -\frac{1}{3}.$$

Example 2. *Evaluate* $\lim_{x \to 0} \frac{\tan x - x}{x^2 \tan x}$.

[*Delhi Maths (H), 2003; Delhi Maths (P), 1998; Manipur, 2002*]

Solution. We have

$$\lim_{x \to 0} \frac{\tan x - x}{x^2 \tan x} = \lim_{x \to 0} \left(\frac{\tan x - x}{x^3} \cdot \frac{x}{\tan x} \right) = \lim_{x \to 0} \frac{\tan x - x}{x^3} \times \lim_{x \to 0} \frac{x}{\tan x}$$

$$= \lim_{x \to 0} \frac{\tan x - x}{x^3} \times 1, \text{ as } \lim_{x \to 0} \frac{x}{\tan x} = 1 \qquad \left[\text{Form } \frac{0}{0}\right]$$

$$= \lim_{x \to 0} \frac{\sec^2 x - 1}{3x^2} \qquad \left[\text{Form } \frac{0}{0}\right]$$

$$= \lim_{x \to 0} \frac{2 \sec^2 x \tan x}{6x} = \frac{1}{3} \lim_{x \to 0} \sec^2 x \times \lim_{x \to 0} \frac{\tan x}{x}$$

$$= (1/3) \times 1 \times 1 = 1/3.$$

Example 3. *Evaluate* $\lim_{x \to 0} \frac{1 - \cos x^2}{x^2 \sin x^2}$. **[*Delhi Maths (H), 2004*]**

Solution. We have

$$\lim_{x \to 0} \frac{1 - \cos x^2}{x^2 \sin x^2} = \lim_{x \to 0} \left(\frac{1 - \cos x^2}{x^4} \cdot \frac{x^2}{\sin x^2} \right).$$

$$= \lim_{x \to 0} \frac{1 - \cos x^2}{x^4} \times 1 \qquad \left[\text{Form } \frac{0}{0}\right]$$

$$= \lim_{x \to 0} \frac{2x \sin x^2}{4x^3} = \frac{1}{2} \lim_{x \to 0} \frac{\sin x^2}{x^2} = \frac{1}{2}.$$

Example 4. *Evaluate* $\lim_{x \to 0} \frac{\sinh x - \sin x}{x \sin^2 x}$.

Solution. We have

$$\lim_{x\to 0}\frac{\sinh x-\sin x}{x\sin^2 x}=\lim_{x\to 0}\left\{\frac{\sinh x-\sin x}{x^3}\cdot\left(\frac{x}{\sin x}\right)^2\right\}$$

$$=\lim_{x\to 0}\frac{\sinh x-\sin x}{x^3}\times 1 \qquad \left[\text{Form }\frac{0}{0}\right]$$

$$=\lim_{x\to 0}\frac{\cosh x-\cos x}{3x^2} \qquad \left[\text{Form }\frac{0}{0}\right]$$

$$=\lim_{x\to 0}\frac{\sinh x+\sin x}{6x} \qquad \left[\text{Form }\frac{0}{0}\right]$$

$$=\lim_{x\to 0}\frac{\cosh x+\cos x}{6}=\frac{1+1}{6}=\frac{1}{3}.$$

Example 5. *Evaluate* $\lim_{x\to 0}\dfrac{\sin x-x+x^3/6}{x^5}$.

[*Agra, 2005*; *Delhi Maths (G), 2002*; *Kanpur, 2002*;]

Solution. We have

$$\lim_{x\to 0}\frac{\sin x-x+x^3/6}{x^5}=\lim_{x\to 0}\frac{\left(x-\dfrac{x^3}{3!}+\dfrac{x^5}{5!}-\dfrac{x^7}{7!}+\ldots\text{ ad. inf}\right)-x+\dfrac{x^3}{6}}{x^5}$$

$$=\lim_{x\to 0}\frac{x^5/5!-x^7/7!+\ldots}{x^5}=\lim_{x\to 0}\left(\frac{1}{5!}-\frac{x^2}{7!}+\ldots\right)=\frac{1}{120}.$$

Example 6. *Prove that*

(*i*) $\lim_{x\to 0}\dfrac{(1+x)^{1/x}-e}{x}=-\dfrac{e}{2}$. **(*Kanpur 2013*; *Avadh, 2000*; *Kumaon, 2004*)**

(*ii*) $\lim_{x\to 0}\dfrac{(1+x)^{1/x}-x+ex/2}{x^2}=\dfrac{11e}{24}$. **[*Delhi B.Sc.I (H) 2008*, Meerut 2011, *2006*]**

Solution. Since $\lim_{x\to 0}(1+x)^{1/x}=e$, it follows that the given limit is of the form 0/0. Let us first find the expansion of $(1+x)^{1/x}$ as follows :

Let $\qquad y=(1+x)^{1/2}$ so that

$$\log y=\frac{1}{x}\log(1+x)=\frac{1}{x}\left(x-\frac{x^2}{2}+\frac{x^3}{3}-\frac{x^4}{4}+\ldots..\right)$$

or $$\log y=1+\left(-\frac{x}{2}+\frac{x^2}{3}-\ldots..\right)=1+z,\text{ where } z=-\frac{x}{2}+\frac{x^2}{3}-\ldots..$$

$\therefore$ $$y=e^{1+z}=e\cdot e^z=e\,(1+z+z^2/2!+\ldots..)$$

or $$y=e\left\{1+\left(-\frac{x}{2}+\frac{x^2}{3}-\ldots..\right)+\frac{1}{2}\left(-\frac{x}{2}+\frac{x^2}{3}-\ldots..\right)^2-\ldots..\right\}$$

So $(1+x)^{1/x} = e\left\{1-\frac{1}{2}x+\left(\frac{1}{3}+\frac{1}{8}\right)x^2+.....\right\} = e\left(1-\frac{1}{2}x+\frac{11x^2}{24}+.....\right)$

(*i*) We have

$$\lim_{x\to 0}\frac{(1+x)^{1/x}-e}{x} = \lim_{x\to 0}\frac{e\left(1-\frac{x}{2}+\frac{11x^2}{24}+.....\right)-e}{x}$$

$$= \lim_{x\to 0}\left(-\frac{e}{2}+\text{terms containing } x\right) = -\frac{e}{2}$$

(*ii*) We have

$$\lim_{x\to 0}\frac{(1+x)^{1/x}-e+\frac{ex}{2}}{x^2} = \lim_{x\to 0}\frac{e\left(1-\frac{x}{2}+\frac{11x^2}{24}+.....\right)-e+\frac{ex}{2}}{x^2}$$

$$= \lim_{x\to 0}\left(\frac{11e}{24}+\text{terms containing } x\right) = \frac{11e}{24}.$$

Example 7. *For what value of a does* $(\sin 2x + a\sin x)/x^3$ *tend to a finite limit as* $x\to 0$*? Find this limit.* **[Delhi Maths (G), 2000; G.N.D.U. Amritsar, 2004; I.A.S. (Prel.), 2001]**

Solution. Here $(\sin 2x + a\sin x)/x^3$ is of the form 0/0 as $x\to 0$ whatever value a may have. So, by L' Hopital's rule,

$$\lim_{x\to 0}\frac{\sin 2x + a\sin x}{x^3} = \lim_{x\to 0}\frac{2\cos 2x + a\cos x}{3x^2} \quad ...(1)$$

Since the denominator of the expression on the R.H.S. of (1) is 0 when $x = 0$, it follows that in order that the limit of the expression on R.H.S. of (1) be finite as $x\to 0$, the numerator of the expression on R.H.S. of (1) should also be zero when $x = 0$. This will happen if we choose

$$2 + a = 0, \quad i.e., \quad a = -2.$$

Substituting this value of a on R.H.S. of (1), we have

$$\lim_{x\to 0}\frac{\sin 2x + a\sin x}{x^3} = \lim_{x\to 0}\frac{2\cos 2x - 2\cos x}{3x^2} \quad \left[\text{Form } \frac{0}{0}\right]$$

$$= \lim_{x\to 0}\frac{-4\sin 2x + 2\sin x}{6x} \quad \left[\text{Form } \frac{0}{0}\right]$$

$$= \lim_{x\to 0}\frac{-8\cos 2x + 2\sin x}{6} = -1.$$

Example 8. *Determine the values of A and B for which* $\lim_{x\to 0}\frac{\sin 3x + A\sin 2x + B\sin x}{x^5}$ *exists and find the limit.* **[Delhi Maths (G), 2003]**

Solution. Let L be the value of given limit.

$$\therefore \quad L = \lim_{x\to 0}\frac{\sin 3x + A\sin 2x + B\sin x}{x^5} \quad \left[\text{Form } \frac{0}{0}\right]$$

or

$$L = \lim_{x\to 0}\frac{3\cos 3x + 2A\cos 2x + B\cos x}{5x^4} \quad ...(1)$$

Here the denominator of (1) has the value 0 when $x = 0$. So in order that the limit (1) may exist, it is necessary that the numerator of (1) has the value 0 at $x = 0$. Thus, we must have

$$3 + 2A + B = 0 \quad \text{or} \quad 2A + B = -3 \quad ...(2)$$

Now, if the relation (2) holds, then (1) is of the form 0/0 as $x \to 0$, and hence its limit as $x \to 0$ can be obtained by L' Hopital's rule. Then, (1) yelds

$$L = \lim_{x \to 0} \frac{-9 \sin 3x - 4A \sin 2x - B \sin x}{20x^3} \quad \left[\text{Form } \frac{0}{0}\right]$$

or
$$L = \lim_{x \to 0} \frac{-27 \cos 3x - 8A \cos 2x - B \cos x}{60x^2} \quad ...(3)$$

Again we find that the denominator of (3) has the value 0 when $x = 0$. So in order that the limit (3) may exist, it is necessary that the numerator of (3) has the value 0 at $x = 0$. Thus, we must have

$$-27 - 8A - B = 0 \quad \text{or} \quad 8A + B = -27 \quad ...(4)$$

Solving (2) and (4), we have $A = -4$, $B = 5$. With these values, (3) reduces to

$$L = \lim_{x \to 0} \frac{-27 \cos 3x + 32 \cos 2x - 5 \cos x}{60x^2} \quad \left[\text{Form } \frac{0}{0}\right]$$

$$= \lim_{x \to 0} \frac{81 \sin 3x - 64 \sin 2x + 5 \sin x}{120x} \quad \left[\text{Form } \frac{0}{0}\right]$$

$$= \lim_{x \to 0} \frac{243 \cos 3x - 128 \cos 2x + 5 \cos x}{120} = 1.$$

12.6. APPLICATION OF L'!HOSPITAL'S RULE FOR ∞/∞ FORM

If $\lim_{x \to a} f(x) = \infty$ and $\lim_{x \to a} g(x) = \infty$, then $\lim_{x \to a} \frac{f(x)}{g(x)} = \lim_{x \to a} \frac{f'(x)}{g'(x)}$.

Note 1. Sometimes more than one application of L' Hospital's rule will be necessary to evaluate an indeterminate form ∞/∞.

Note 2. The above result is also true when $x \to \infty$.

Note 3. In some problems of the form ∞/∞, it is necessary to convert it into the form 0/0 at the proper stage, otherwise the process of application of L' Hospital's rule will never end.

EXAMPLES

Example 1. *Evaluate* $\lim_{x \to \pi/2} \frac{\tan 3x}{\tan x}$. **[*Delhi Maths (H)*, *1994*]**

Solution. We have

$$\lim_{x \to \pi/2} \frac{\tan 3x}{\tan x} \quad \left[\text{Form } \frac{\infty}{\infty}\right]$$

$$= \lim_{x \to \pi/2} \frac{3 \sec^2 3x}{\sec^2 x} \quad \left[\text{Form } \frac{\infty}{\infty}\right]$$

$$= \lim_{x \to \pi/2} \frac{3 \cos^2 x}{\cos^2 3x} \quad \left[\text{Form } \frac{0}{0}\right]$$

$$= \lim_{x \to \pi/2} \frac{-6 \cos x \sin x}{-6 \cos 3x \sin 3x} = \lim_{x \to \pi/2} \frac{\sin 2x}{\sin 6x} \quad \left[\text{Form } \frac{0}{0}\right]$$

$$= \lim_{x \to \pi/2} \frac{2 \cos 2x}{6 \cos 6x} = \frac{-2}{-6} = \frac{1}{3}.$$

Example 2. *Evaluate* $\lim\limits_{x \to \pi - 0} \dfrac{\log \sin x}{\log \sin 2x}$.

Solution. We have

$$\lim_{x \to \pi - 0} \frac{\log \sin x}{\log \sin 2x} \qquad \left[\text{Form } \frac{\infty}{\infty}\right]$$

$$= \lim_{x \to \pi - 0} \frac{\cot x}{2 \cot 2x} \qquad \left[\text{Form } \frac{\infty}{\infty}\right]$$

$$= \lim_{x \to \pi - 0} \frac{-\operatorname{cosec}^2 x}{-4 \operatorname{cosec}^2 2x} = \lim_{x \to \pi - 0} \cos^2 x = 1.$$

Example 3. *Evaluate* $\lim\limits_{x \to \infty} (x^n / e^x)$, *n being a positive integer.*

Solution. We have

$$\lim_{x \to \infty} \frac{x^n}{e^x} \qquad \left[\text{Form } \frac{\infty}{\infty}\right]$$

$$= \lim_{x \to \infty} \frac{n(n-1)x^{n-2}}{e^x} \qquad \left[\text{Form } \frac{\infty}{\infty}\right]$$

....

$$= \lim_{x \to \infty} \frac{n(n-1)(n-2) \ldots 3 \cdot 2 \cdot 1}{e^x}, \text{ after applying L' Hopital's rule } (n-2) \text{ times more}$$

$$= \lim_{x \to \infty} \frac{n!}{e^x} = 0.$$

Example 4. *Evaluate* $\lim\limits_{x \to 0+} \log_{\tan x} \tan 2x$. ***(Gujarat, 2004)***

Solution. We have

$$\lim_{x \to 0+} \log_{\tan x} \tan 2x = \lim_{x \to 0+} \frac{\log \tan 2x}{\log \tan x} \qquad \left[\text{Form } \frac{\infty}{\infty}\right]$$

$$= \lim_{x \to 0+} \frac{\dfrac{1}{\tan 2x} \cdot 2 \sec^2 2x}{\dfrac{1}{\tan x} \cdot \sec^2 x} = \lim_{x \to 0+} \frac{2 \sin 2x}{\sin 4x} \qquad \left[\text{Form } \frac{0}{0}\right]$$

$$= \lim_{x \to 0+} \frac{4 \cos 2x}{4 \cos 4x} = \frac{4}{4} = 1.$$

12.7. THE INDETERMINATE FORM $0 \times \infty$

If $\lim\limits_{x \to a} f(x) = 0$ and $\lim\limits_{x \to a} g(x) = \infty$, then $f(x) \times g(x)$ takes the form $0 \times \infty$. In such cases, $f(x) \times g(x)$ is expressed as

$$\frac{f(x)}{1/g(x)} \quad \text{or} \quad \frac{g(x)}{1/f(x)}$$

which has respectively 0/0 or ∞/∞ forms. Their limits can then be evaluated by using L' Hospital's rules explained in Art. 12.5 and 12.6 respectively.

EXAMPLES

Example 1. *Evaluate* $\lim_{x \to 1} (1 - x) \tan (\pi x / 2)$. **[*Delhi (G), 1998*]**

Solution. $\lim_{x \to 1} (1 - x) \tan (\pi x / 2)$ [Form $0 \times \infty$]

$$= \lim_{x \to 1} \frac{1 - x}{\cot (\pi x / 2)} \qquad \left[\text{Form } \frac{0}{0}\right]$$

$$= \lim_{x \to 1} \frac{-1}{(-\pi / 2) \operatorname{cosec}^2 (\pi x / 2)} = \frac{2}{\pi}.$$

Example 2. *Evaluate* $\lim_{x \to 0+0} (x^k \log x)$, *for each positive real number k.*

[*Delhi Maths (H), 2002*]

Solution. $\lim_{x \to 0+0} (x^k \log x)$ [Form $0 \times \infty$]

$$= \lim_{x \to 0+0} \frac{\log x}{x^{-k}} \qquad \left[\text{Form } \frac{\infty}{\infty}\right]$$

$$= \lim_{x \to 0+0} \frac{1/x}{(-k)\, x^{-k-1}} = \lim_{x \to 0+0} \left(\frac{x^k}{-k}\right) = 0.$$

12.8. THE INDETERMINATE FORM ∞ – ∞

If $\lim_{x \to a} f(x) = \infty$ and $\lim_{x \to a} g(x) = \infty$, then $f(x) - g(x)$ takes the form $\infty - \infty$. In such cases, $f(x) - g(x)$ is expressed in the form 0/0 as follows :

$$\lim_{x \to a} [f(x) - g(x)] \qquad [\text{Form } \infty - \infty]$$

$$= \lim_{x \to a} \left[\frac{1}{1/f(x)} - \frac{1}{1/g(x)}\right] = \lim_{x \to a} \frac{1/g(x) - 1/f(x)}{1/\{f(x)\, g(x)\}}, \qquad \left[\text{Form } \frac{0}{0}\right]$$

which is evaluated by L' Hospital's rule as explained in Art. 12.5.

EXAMPLES

Example 1. *Evaluate* $\lim_{x \to 0} (1/x - \cot x)$.

[*Delhi Maths (G), 1996; Delhi B.A. (P), 2003, 04, 09*]

Solution. $\lim_{x \to 0} (1/x - \cot x)$ [Form $\infty - \infty$]

$$= \lim_{x \to 0} \left(\frac{1}{x} - \frac{\cos x}{\sin x}\right) = \lim_{x \to 0} \frac{\sin x - x \cos x}{x \sin x} \qquad \left[\text{Form } \frac{0}{0}\right]$$

$$= \lim_{x \to 0} \frac{\sin x - x \cos x}{x^2} \cdot \frac{x}{\sin x} = \lim_{x \to 0} \frac{\sin x - x \cos x}{x^2} \qquad \left[\text{Form } \frac{0}{0}\right]$$

$$= \lim_{x \to 0} \frac{\cos x - (\cos x - x \sin x)}{2x} = \lim_{x \to 0} \frac{\sin x}{2} = 0.$$

Example 2. *Evaluate* $\lim_{x \to 0} \left(\frac{1}{e^x - 1} - \frac{1}{x}\right)$.

[*Delhi Maths (H), 2000; Delhi Maths (G), 2002; Delhi Maths (P), 2002*]

Solution. We have

$$\lim_{x\to 0}\left(\frac{1}{e^x-1}-\frac{1}{x}\right) \qquad [\text{Form } \infty-\infty]$$

$$= \lim_{x\to 0}\frac{x-e^x+1}{x(e^x-1)} \qquad \left[\text{Form } \frac{0}{0}\right]$$

$$= \lim_{x\to 0}\frac{1-e^x}{e^x-1+xe^x} \qquad \left[\text{Form } \frac{0}{0}\right]$$

$$= \lim_{x\to 0}\frac{-e^x}{e^x+e^x+xe^x} = -\frac{1}{2}.$$

Example 3. *Evaluate* $\lim_{x\to 0}\left[\frac{1}{x}-\frac{1}{x^2}\log(1+x)\right]$. **[*Delhi Maths* (*H*), *2003*]**

Solution. Since $\lim_{x\to 0}\{\log(1+x)\}/x = 1$, we have

$$\lim_{x\to 0}\left[\frac{1}{x}-\frac{1}{x^2}\log(1+x)\right] \qquad [\text{Form } \infty-\infty]$$

$$= \lim_{x\to 0}\frac{x-\log(1+x)}{x^2} \qquad \left[\text{Form } \frac{0}{0}\right]$$

$$= \lim_{x\to 0}\frac{1-1/(1+x)}{2x} = \lim_{x\to 0}\frac{1}{2(1+x)} = \frac{1}{2}.$$

12.9. THE INDETERMINATE FORMS 0^0, ∞^0 AND 1^∞

Let
$$l = \lim_{x\to a}\{f(x)\}^{g(x)}, \qquad ...(1)$$
where the R.H.S. assumes one of the above three indeterminate forms.

Taking logarithms on both sides, (1) reduces to
$$\log l = \lim_{x\to a} g(x)\log f(x) \qquad ...(2)$$

The limit on R.H.S. of (2) can be evaluated by one of the previous methods. If k is the value of the limit on L.H.S. of (2), then

$$\log l = k \Rightarrow l = e^k, \text{ which is the required value of the limit.}$$

EXAMPLES

Example 1. *Evaluate* $\lim_{x\to 0+0} x^x$. **[Agra 2006, Delhi Maths (P) 2008]**

Solution. Let
$$l = \lim_{x\to 0+0} x^x \qquad [\text{Form } 0^0]$$

$$\therefore \quad \log l = \lim_{x\to 0+0} x\log x \qquad [\text{Form } 0\times\infty]$$

$$= \lim_{x\to 0+0}\frac{\log x}{1/x} \qquad [\text{Form } \infty/\infty]$$

$$= \lim_{x\to 0+0}\frac{1/x}{-1/x^2} = \lim_{x\to 0+0}(-x) = 0$$

Thus $\log l = 0$ so that $l = e^0 = 1.$

Example 2. *Evaluate* $\lim_{x \to 1} (1 - x^2)^{1/\log(1-x)}$.

Solution. Let $$l = \lim_{x \to 1} (1 - x^2)^{1/\log(1-x)} \quad [\text{Form } 0^0]$$

$$\therefore \quad \log l = \lim_{x \to 1} \frac{\log(1 - x^2)}{\log(1 - x)} \quad \left[\text{Form } \frac{\infty}{\infty}\right]$$

$$= \lim_{x \to 1} \frac{(-2x)/(1 - x^2)}{(-1)/(1 - x)} = \lim_{x \to 1} \frac{2x}{1 + x} = 1.$$

Example 3. *Evaluate* $\lim_{x \to 0+0} (\cot x)^{\sin x}$

[*Bangalore, 2004*; *Delhi Maths (G), 2000, 04*; *Delhi Maths (H), 2006*; *Delhi Maths (P), 1998*]

Solution. Let $$l = \lim_{x \to 0+0} (\cot x)^{\sin x} \quad [\text{Form } \infty^0]$$

$$\therefore \quad \log l = \lim_{x \to 0+0} \sin x \log \cot x \quad [\text{Form } 0 \times \infty]$$

$$= \lim_{x \to 0+0} \frac{\log \cot x}{\operatorname{cosec} x} \quad \left[\text{Form } \frac{\infty}{\infty}\right]$$

$$= \lim_{x \to 0+0} \frac{-(1/\cot x) \times (-\operatorname{cosec}^2 x)}{-\operatorname{cosec} x \cot x} = \lim_{x \to 0+0} \sin x \sec^2 x = 0$$

Thus, $\log l = 0$ so that $l = e^0 = 1.$

Example 4. *Evaluate* $\lim_{x \to 0+0} (1/x)^{\tan x}$. **[*Delhi Maths (G), 1994*]**

Solution. Let $$l = \lim_{x \to 0+0} (1/x)^{\tan x} \quad [\text{Form } \infty^0]$$

$$\therefore \quad \log l = \lim_{x \to 0+0} \frac{\log(1/x)}{\cot x} \quad \left[\text{Form } \frac{\infty}{\infty}\right]$$

$$= \lim_{x \to 0+0} \frac{x \times (-1/x^2)}{-\operatorname{cosec}^2 x} = \lim_{x \to 0} \left(\frac{\sin x}{x} \times \sin x\right) = 1 \times 0 = 1$$

Thus $\log l = 0$ so that $l = e^0 = 1.$

Example 5. *Evaluate* (*i*) $\lim_{x \to 0} (\cos x)^{1/x^2}$

[*Delhi Maths (P), 1998, 2003*; *M.D.U. Rohtak, 2000*]

(*ii*) $\lim_{x \to 0+0} (\cos x)^{1/x^3}$ **[*Delhi Maths (H), 2003*]**

Solution. (*i*) Let $$l = \lim_{x \to 0} (\cos x)^{1/x^2} \quad [\text{Form } 1^\infty]$$

$$\therefore \quad \log l = \lim_{x \to 0} \frac{\log \cos x}{x^2} \quad \left[\text{Form } \frac{0}{0}\right]$$

$$= \lim_{x \to 0} \frac{-\tan x}{2x} = -\frac{1}{2} \lim_{x \to 0} \frac{\tan x}{x} = -\frac{1}{2}.$$

Thus, $\log l = -1/2$ so that $l = e^{-1/2} = 1/\sqrt{e}$.

(*ii*) Let $$l = \lim_{x \to 0+0} (\cos x)^{1/x^3} \qquad \text{[Form } 1^\infty]$$

$\therefore$ $$\log l = \lim_{x \to 0+0} \frac{\log \cos x}{x^3} \qquad \left[\text{Form } \frac{0}{0}\right]$$

$$= \lim_{x \to 0+0} \frac{-\tan x}{3x^2} = -\frac{1}{3}\left(\lim_{x \to 0+0} \frac{\tan x}{x} \times \lim_{x \to 0+0} \frac{1}{x}\right)$$

Thus, $\log l = (-1/3) \times 1 \times \infty = -\infty$ so that $l = e^{-\infty} = 0$.

Example 6. *Evaluate* $\lim_{x \to 0} \left(\frac{\tan x}{x}\right)^{1/x^2}$. ***[Agra 2010; Delhi Math 2007; Bangalore, 2004; Delhi Maths (H), 2000, 09; Garhwal, 2001; G.N.D.U., Amritsar, 2004; Kumaon, 2002]***

Solution. Let $$l = \lim_{x \to 0} \left(\frac{\tan x}{x}\right)^{1/x^2} \qquad \text{[Form } 1^\infty]$$

$$\left[\text{Note that } \lim_{x \to 0} \frac{\tan x}{x} = 1\right]$$

$\therefore$ $$\log l = \lim_{x \to 0} \frac{\log \{(\tan x)/x\}}{x^2} \qquad \left[\text{Form } \frac{0}{0}\right]$$

$$= \lim_{x \to 0} \frac{\dfrac{x}{\tan x}\left(\dfrac{x \sec^2 x - \tan x}{x^2}\right)}{2x}$$

$$= \lim_{x \to 0} \frac{x \sec^2 x - \tan x}{2x^2 \tan x} \qquad \left[\text{Form } \frac{0}{0}\right]$$

$$= \lim_{x \to 0} \frac{\sec^2 x + 2x \sec^2 x \tan x - \sec^2 x}{4x \tan x + 2x^2 \sec^2 x}$$

$$= \lim_{x \to 0} \frac{\sec^2 x \tan x}{2 \tan x + x \sec^2 x} \qquad \left[\text{Form } \frac{0}{0}\right]$$

$$= \lim_{x \to 0} \frac{\sec^4 x + 2 \sec x (\sec x \tan x) \tan x}{2 \sec^2 x + \sec^2 x + 2x \sec^2 x \tan x} = \frac{1}{3}.$$

Thus, $\log l = 1/3$ so that $l = e^{1/3}$.

Example 7. *Evaluate* $\lim_{x \to a} (2 - x/a)^{\tan(\pi x/2a)}$. ***(Rohilkhand, 2002)***

Solution. Let $$l = \lim_{x \to a} (2 - x/a)^{\tan(\pi x/2a)} \qquad \text{[Form } 1^\infty]$$

$\therefore$ $$\log l = \lim_{x \to a} \tan \frac{\pi x}{2a} \log\left(2 - \frac{x}{a}\right) \qquad [\text{Form } 0 \times \infty]$$

$$= \lim_{x \to a} \frac{\log (2 - x/a)}{\cot (\pi x/2a)} \qquad \left[\text{Form } \frac{0}{0}\right]$$

$$= \lim_{x \to a} \frac{(-1/a) \times \{1/(2 - x/a)\}}{-(\pi/2a) \operatorname{cosec}^2 (\pi x/2a)} = \frac{2}{\pi}.$$

Thus, $\log l = 2/\pi$ so that $l = e^{2/\pi}$.

EXERCISES

1. *Evaluate the following limits :*

(*a*) $\lim\limits_{x\to 0} \dfrac{1-\cos x}{x^2}$ (*b*) $\lim\limits_{x\to \pi/2} \dfrac{\cos x}{\pi/2 - x}$ (*c*) $\lim\limits_{x\to 0} \dfrac{\sin ax}{\sin bx}$

(*d*) $\lim\limits_{x\to 0} \dfrac{\tan x - x}{\sin x - 1}$ (*e*) $\lim\limits_{x\to 0} \dfrac{1-\cos x^3}{x^2 \sin x^3}$ (*f*) $\lim\limits_{x\to 0} \dfrac{x-\sin x}{\tan^3 x}$

(*g*) $\lim\limits_{x\to 0} \dfrac{x-\sin x}{x^3}$ **(Agra 2011)** (*h*) $\lim\limits_{x\to 0} \dfrac{x-\tan x}{x^3}$ (*i*) $\lim\limits_{x\to 0} \dfrac{a^x - 1}{b^x - 1}, b \neq 1$

(*j*) $\lim\limits_{x\to 0} \dfrac{\log(1+x) - x}{1-\cos x}$ (*k*) $\lim\limits_{x\to 0} \dfrac{\tan^2 x - x^2}{x^2 \tan^2 x}$ (*l*) $\lim\limits_{x\to 0} \dfrac{1-\cos x}{x \log(1+x)}$

(*m*) $\lim\limits_{x\to 0} \dfrac{x\cos x - \log(1+x)}{x^2}$ **[Delhi Maths (G), 2005; M.D.U. Rohtak, 2000, Meerut 2005, 09]**

(*n*) $\lim\limits_{x\to 0} \dfrac{\log(1-x^2)}{\log\cos x}$ **(Kumaon, 1998, 99)**

(*o*) $\lim\limits_{x\to 0} \dfrac{e^{ax} - e^{-ax}}{\log(1+bx)}$ (*p*) $\lim\limits_{x\to 0} \dfrac{e^x \sin x - x - x^2}{x^2 + x\log(1-x)}$ **(Delhi Maths (H) 2009)**

(*q*) $\lim\limits_{x\to 0} \dfrac{\cosh x - \cos x}{x \sin x}$ **[Meerut 2006]** (*r*) $\lim\limits_{x\to 0} \dfrac{x^{1/2}\tan x}{(e^x - 1)^{1/2}}$

(*s*) $\lim\limits_{x\to 0} \dfrac{e^x - e^{-x} - 2\log(1+x)}{x\sin x}$ **(G.N.D.U. Amritsar, 2004)**

(*t*) $\lim\limits_{x\to 0} \dfrac{a^x - b^x}{x}$ **(Rohilkhand, 2001)**

(*u*) $\lim\limits_{x\to 0} \dfrac{e^x - e^{-x} - 2\sin x}{x^3}$ **(Calicut, 2004)**

(*v*) $\lim\limits_{x\to 0} \dfrac{\sin 2x + 2\sin^2 x - 2\sin x}{\cos x - \cos^2 x}$ (*w*) $\lim\limits_{x\to 0} \dfrac{xe^x - \log(1+x)}{x}$ **(Agra 2010)**

2. (*a*) If $\lim\limits_{x\to 0} \dfrac{\sin 3x - a\sin x}{x^3}$ is finite, find the value of *a* and the limit.

(*b*) Determine the values of *p* and *q* for which $\lim\limits_{x\to 0} \dfrac{x(1+p\cos x) - q\sin x}{x^3}$ exists and equals 1. **[Delhi Maths (G) 2006; Avadh, 1999; Delhi Maths (P), 2001; Delhi Maths (H), 1997, 2005; Rohilkhand, 2000]**

(*c*) Find the values of *a* and *b* in order that $\lim\limits_{x\to 0} \dfrac{x(1-a\cos x) + b\sin x}{x^3} = \dfrac{1}{3}$.

[Delhi Maths (Prog) 2007; Delhi Maths (G), 2001, 04; Delhi Maths (H), 1995]

(*d*) Find the values of *a*, *b* and *c*, so that $\lim_{x \to 0} \dfrac{ae^x - b\cos x + ce^{-x}}{x \sin x} = 2.$

(*e*) Find the values of *a*, *b* and *c* so that $\lim_{x \to 0} \dfrac{a + b\cos x + c\sin x}{x^2}$ exists and equals 1/2.

(*f*) If $\lim_{x \to 0} \dfrac{xe^x - q\cos x + pe^{-x}}{x \tan x} = 3$, find the values of *p*, *q* and *r*.

(*g*) Find the values of *a*, *b* and *c* such that $\lim_{x \to 0} \dfrac{x(a + b - \cos x) - c\sin x}{x^5} = 1.$

(Agra, 2001; Garhwal, 1998; Kanpur, 2001)

3. *Evaluate the following limits :*

(*a*) $\lim_{x \to 0+0} \dfrac{\cot x}{\log \sin x}$ **[Delhi Maths (G) 2006]**

(*b*) $\lim_{x \to 1-0} \dfrac{\log(1-x)}{\cot \pi x}$ **[Agra 2003]**

(*c*) $\lim_{x \to \infty} \dfrac{\log x}{x^n}, n > 0$

(*d*) $\lim_{x \to \pi/2} \dfrac{\log(x - \pi/2)}{\tan x}$

(*e*) $\lim_{x \to a} \dfrac{\log(x-a)}{\log(e^x - e^a)}$

(*f*) $\lim_{x \to \infty} \log_x \tan x$

(*g*) $\lim_{x \to \infty} x^n e^{-x}$ where *n* is a positive integer.

(*h*) $\lim_{x \to 0} \dfrac{\log x}{\cot x}$ **[Agra 2006]**

4. *Evaluate the following limits :*

(*a*) $\lim_{x \to 1} (1-x)\tan\left(\dfrac{1}{2}\pi x\right)$

(*b*) $\lim_{x \to \pi/2} (1 - \sin x)\tan x$

(*c*) $\lim_{x \to 0} \sin x \ln x^2$

(*d*) $\lim_{x \to 1} \sec\left(\dfrac{1}{2}\pi x\right) \ln\left(\dfrac{1}{x}\right)$

(*e*) $\lim_{x \to \infty} 2^x \sin(a/2^x)$

(*f*) $\lim_{x \to \infty} (a^{1/x} - 1)x$

(*g*) $\lim_{x \to 0} x^m (\log x)^n$, where *m*, *n* are positive integers.

(*h*) $\lim_{x \to 0} \sin x \log x$ **(Delhi 2009)**

(*i*) $\lim_{x \to 1} \sec(\pi/2x) \log x$ **(Lucknow 2010)**

5. *Evaluate the following limits :*

(*a*) $\lim_{x \to 0} \left(\dfrac{1}{x} - \dfrac{1}{\sin x}\right)$ **[Delhi Maths (G), 1995, 2005; Delhi Maths (H), 2006; Delhi Maths (P) 2008]**

(*b*) $\lim_{x \to 0} \left(\dfrac{1}{x^2} - \dfrac{1}{\sin^2 x}\right)$ **[Kanpur 2011; Nagpur 2010; Delhi B.Sc. I (H) 2008; Garhwal, 2000; G.N.D.U. Amritsar, 2004; Delhi Maths (G), 1997]**

(*c*) $\lim_{x \to \pi/2} \left(x \tan x - \dfrac{\pi}{2} \sec x\right)$ **(Agra, 1999)**

(*d*) $\lim_{x \to 0} (\cot^2 x - 1/x^2)$ **[Delhi Maths (G), 1997; Delhi Maths (G), 2002; Garhwal, 1998; Kumaon, 1997; G.N.D.U. Amritsar, 2004]**

(*e*) $\lim_{x \to 0} (\operatorname{cosec} x - \cot x)$ **[Delhi Maths (P), 1995, 2002]**

(*f*) $\lim_{x \to 1} \left(\dfrac{x}{x-1} - \dfrac{1}{\log x}\right)$ **[Delhi Maths (G), 2001]**

(g) $\lim\limits_{x \to \pi/2} (\sec x - \tan x)$

(h) $\lim\limits_{x \to 4} \left\{ \dfrac{1}{\ln (x-3)} - \dfrac{1}{x-4} \right\}$

(i) $\lim\limits_{x \to \pi/2} \left(\sec x - \dfrac{1}{1 - \sin x} \right)$ **(Purvanchal 2006)**

(j) $\lim\limits_{x \to 0} \left\{ \dfrac{\pi}{4x} - \dfrac{\pi}{2x\,(e^{\pi x} + 1)} \right\}$ **(Agra 2011)**

(k) $\lim\limits_{x \to 0} \left\{ \dfrac{1}{x\,(1+x)} - \dfrac{\log (1+x)}{x^2} \right\}$

(l) $\lim\limits_{x \to 1} \left\{ \dfrac{x}{x-1} - \dfrac{1}{\log x} \right\}$

6. *Evaluate the following limits :*

(a) $\lim\limits_{x \to \pi/2} (\sin x)^{\tan x}$ **[Calicut, 2004; Delhi B.A. (P), 2000]**

(b) $\lim\limits_{x \to \pi/2} (\sin x)^{\tan^2 x}$ **[Delhi B.A. (P), 1996]**

(c) $\lim\limits_{x \to 0} (1 + x)^{1/x}$ **[Delhi B.Sc. (G), 1995]**

(d) $\lim\limits_{x \to \pi/2 - 0} (\tan x)^{\sin 2x}$ **[Delhi Maths (H), 1999]**

(e) $\lim\limits_{x \to 0+0} x^{1/\ln x}$

(f) $\lim\limits_{x \to 0+0} (\tan x)^{\sin 2x}$

(g) $\lim\limits_{x \to 0} (\cos x)^{1/x}$

(h) $\lim\limits_{x \to \pi/2 - 0} (\cos x)^{\cos x}$

(i) $\lim\limits_{x \to 1} x^{1/(x-1)}$

(j) $\lim\limits_{x \to 0+0} (\cot x)^{1/\ln x}$

(k) $\lim\limits_{x \to 0} (1 + \sin x)^{\cot x}$

(l) $\lim\limits_{x \to 0+0} \{\ln (1/x)\}^x$

(m) $\lim\limits_{x \to \pi/2 - 0} \left(\dfrac{\pi}{2} - x \right)^{\tan x}$

(n) $\lim\limits_{x \to 1} (2 - x)^{\tan (\pi x/2)}$

(o) $\lim\limits_{x \to 0} \left(\dfrac{\sin x}{x} \right)^{1/x}$

(p) $\lim\limits_{x \to 0} \left(\dfrac{\sin x}{x} \right)^{1/x^2}$ **(Kumaun, 2003; Manipur, 2002; M.D.U. Rohtak, 1999)**

(q) $\lim\limits_{x \to 0} \left(\dfrac{\tan x}{x} \right)^{1/x}$ **(Garhwal, 2003; Patna, 2003)**

(r) $\lim\limits_{x \to 0-0} \left(\dfrac{\tan x}{x} \right)^{1/x^3}$ **(Avadh, 2001; Kanpur, 2000)**

(s) $\lim\limits_{x \to 0} \left(\dfrac{\sinh x}{x} \right)^{1/x^2}$ **(Patna, 2000)**

7. If $f''(x)$ exists and is continuous in a neighbourhood of $x = a$, then show that

$$\lim_{h \to 0} \frac{f(a+h) - 2f(a) + f(a-h)}{h^2} = f''(a).$$

8. Show that

(i) $\lim\limits_{x \to 0} \dfrac{e^x - e^{x \cos x}}{x - \sin x} = 3$

(ii) $\lim\limits_{x \to 0} \dfrac{e^x - e^{\sin x}}{x - \sin x} = 1$ **(Patna, 2003)**

(iii) $\lim\limits_{x \to b} \dfrac{x^b - b^x}{x^x - b^b} = \dfrac{1 - \log b}{1 + \log b}$ (iv) $\lim\limits_{x \to 0} \dfrac{e^x + \log\left(\dfrac{1-x}{e}\right)}{\tan x - x} = -\dfrac{1}{2}$

(v) $\lim\limits_{x \to a} \dfrac{a^x - x^a}{x^x - a^a} = \log(a/e)\log(ae)$ **(Rohilkhand, 2003)**

(vi) $\lim\limits_{a \to b} \dfrac{a^b - b^a}{a^a - b^b} = \log(e/b)\log(be)$ **(M.D.U. Rohtak, 1998)**

(vii) $\lim\limits_{x \to 1} \dfrac{x^x - x}{x - 1 - \log x} = 2$ **(M.D.U. Rohtak, 2001)**

ANSWERS

1. (a) 1/2 (b) 1 (c) a/b (d) –2 (e) 0
(f) 1/6 (g) 1/6 (h) – 1/3 (i) $\log_b a$ (j) – 1
(k) – 2/3 (l) 1/2 (m) 1/2 (n) 2 (o) $2a/b$
(p) – 2/3 (q) 1 (r) 1 (s) 1 (t) $\log_e (a/b)$
(u) 2/3 (v) 4 (w) 0

2. (a) $a = 3$, limit = – 4 (b) $p = -5/2$, $q = -3/2$
(c) $a = 1/2$, $b = -1/2$ (d) $a = 1$, $b = 2$, $c = 1$
(e) $a = 1$, $b = -1$, $c = 0$ (f) $p = 3/2$, $q = 3$, $r = 3/2$
(g) $a = 120$, $b = 60$, $c = 180$

3. (a) $-\infty$ (b) 0 (c) 0 (d) 0 (e) 1
(f) 1 (g) 0 (h) 0

4. (a) $2/\pi$ (b) 0 (c) 0 (d) $2/\pi$ (e) a
(f) log a (g) 0 (h) 0 (i) $-(2/\pi)$

5. (a) 0 (b) – 1/3 (c) – 2 (d) – 2/3 (e) 0
(f) 1/2 (g) 0 (h) 1/2 (i) ∞ (j) $\pi^2/8$
(k) – 1/2 (l) 1/2

6. (a) 1 (b) $e^{-1/2}$ (c) e (d) 1 (e) e
(f) 1 (g) $e^{-1/2}$ (h) 1 (i) e (j) $1/e$
(k) e (l) 1 (m) 0 (n) $e^{-2/\pi}$ (o) 1
(p) $e^{-1/6}$ (q) 1 (r) 0 (s) $e^{1/6}$

OBJECTIVE QUESTIONS

Multiple Choice Type Questions : *Select (a), (b), (c) or (d), whichever is correct.*

1. Which of the following is not an indeterminate form ?

(a) $\infty + \infty$ (b) $\infty - \infty$ (c) ∞/∞ (d) $0 \times \infty$. **(Agra, 2001; Kanpur, 2000)**

2. $\lim\limits_{x \to 0} \dfrac{a^x - b^x}{x}$ is equal to :

(a) 0 (b) ∞ (c) $\log(a/b)$ (d) $\log(a-b)$. **(Agra, 2001, Kanpur, 2001)**

3. What is the value of of $\lim_{x \to 0} \frac{e^x - x - 1}{x^2}$?

(*a*) 0 (*b*) 1/2 (*c*) 2 (*d*) e **(I.A.S. Prel. 2009)**

4. The value of $\lim_{x \to 0} \frac{1 - \cos x}{3x^2}$ is :

(*a*) 3 (*b*) 1/3 (*c*) 1/6 (*d*) 1/9. **(Garhwal, 2002)**

5. The value of $\lim_{x \to 0} (1 + x)^{1/x}$ is :

(*a*) 1 (*b*) – 1 (*c*) *e* (*d*) 1/*e*. **(Agra, 2003; Kanpur, 2003)**

6. $\lim_{x \to 0} \frac{e^x + e^{-x} - 2}{x^2}$ is equal to :

(*a*) 1 (*b*) – 1 (*c*) 1/2 (*d*) – 1/2. **[I.A.S. (Prel.), 2002]**

7. $\lim_{x \to \infty} \frac{\log x}{x} =$ (*a*) 0 (*b*) ∞ (*c*) $-\infty$ (d) 1 **[Agra 2008, 10]**

8. What is the value of $\lim_{n \to \infty} (2^{n+1} + 3^{n+1})/(2^n + 3^n)$?

(*a*) 1/.3 (*b*) 1 (*c*) 3 (*d*) ∞ **(I.A.S. Prel. 2009)**

ANSWERS

1. (*a*) 2. (*c*) 3. (*b*) 4. (*c*) 5. (*c*) 6. (*a*) 7. (*a*) 8. (*c*)

MISCELLANEOUS PROBLEMS ON CHAPTER 12

1. Evaluate (*i*) $\lim_{x \to 1} \frac{\log x}{x - 1}$ **[Meerut 2006]** (*ii*) $\lim_{x \to 0} \frac{xe^x - \log(1 + x)}{x^2}$

[Agra 2006; Meerut 2010]

2. $\lim_{x \to 0} \frac{\log(1 + x)}{\sin x}$ is (*a*) 1 (*b*) 0 (*c*) 2 (*d*) None of these **[Agra 2006]**

3. $\lim_{x \to \pi/2} (\sin x)^{\tan x} = l$, the value of l is

(*a*) 0 (*b*) 1 (*c*) ∞ (*d*) e **[Agra 2005]**

4. Evaluate $\lim_{x \to \infty} x\,(a^{1/x} - 1),\ a > 0$ **[Delhi Maths (Prog) 2007]**

5. Evaluate $\lim_{x \to 0} (\operatorname{cosec} x)^{1/\log_e x}$ **[Purvanchal 2006]**

6. Evaluate $\lim_{x \to 0} \left(\frac{1}{\log(x + \sqrt{1 + x^2})} - \frac{1}{\log(1 - x)} \right)$ **[Delhi Maths (H) 2007]**

7. The value of $\lim_{x \to 0} \left(\frac{\tan x}{x} \right)^{1/x^2}$ is

(*a*) $e^{-1/3}$ (*b*) $e^{1/3}$ (*c*) e (*d*) 1 **[Agra 2010]**

8. Evaluate $\lim_{x\to 1}\left[\frac{2}{x^2-1}-\frac{1}{x-1}\right]$ **[Agra 2008, 10]**

9. Evaluate : (*i*) $\lim_{x\to 0}\frac{\operatorname{cosec} x}{\log x}$ **(Agra 2009)** (*ii*) $\lim_{x\to 0}\frac{\sin x\sin^{-1}x}{x^2}$ **(Meerut 2009)**

10. $\lim_{x\to 0}(\tan x-x)/(x-\sin x)$ is (*a*) 0 (*b*) 1 (*c*) 3 (*d*) 2 **(Agra 2009)**

11. $\lim_{n\to 0}(e^{1/n}-1)/(e^{1/n}+1)$ is (*a*) 1 (*b*) –1 (*c*) 0 (*d*) 2 **(Agra 2010)**

12. $\lim_{x\to\infty}(\log x)/x$ is (*a*) 0 (*b*) ∞ (*c*) $-\infty$ (*d*) 1 **(Agra 2010)**

13. Determine constants a and b so that $\lim_{x\to 0}\frac{a\cos x+bx\sin x-5}{x^4}$ exists and is finite

(G.N.D.U. Amritsar 2010)

ANSWERS 1. (*i*) 1 (*ii*) 3/2 **2.** (*a*) **3.** (*b*) **4.** $\log_e a$ **5.** –1
6. –(1/2) **7.** (*b*) **8.** –(1/2) **9.** (*i*) ∞ (*ii*) 1 **10.** (*d*) **11.** (*a*) **12.** (*a*)
13. $a = 5,\ b = 5/2$

Chapter

13

Riemann Integrability

13.1 INTRODUCTION

In elementary treatment, the subject of Integration is treated from the point of view of the inverse of Differentiation so that a function φ is called an integral of a given function f if $\phi'(x) = f(x)$ for all x belonging to the domain of the function f.

Historically, however, the subject arose in connection with the determination of areas of plane regions and was based on the notion of the limit of a type of sum when the number of terms in the sum tends to infinity each terms tending to zero.

In fact the name Integral Calculus has had its origin in this process of summation and the words *'To integrate'* literally means *'To give the sum of'*. It was only afterwards that it was seen that the subject of integration can also be viewed from the point of view of the inverse of differentiation.

In elementary works the reference to integration from summation point of view is always associated with intuitively perceived geometrical concepts.

Consistent with our general purpose, we shall, in this book, give a purely arithmetic treatement of the subject and the same will, moreover, basically be independent of the notion of differentiation. A function ϕ such that $\phi'(x) = f(x)$ will be called a *primitive* of f and the relation between the *Integrals and primitives* will be given later on by means of what is known as the *Fundamental Theorem of Integral Calculus*.

The first satisfactory rigorous arithmetic treatment, free from geometrical notions, of *Integration* was given by Riemann (1782-1867) and we shall be following the definition given by him. While earlier Cauchy (1782-1857) had confined the scope of his theory of integration to continuous functions only, Riemann succeeded in bringing some classes of discontinuous functions also within the purview of his theory of integration. Many refinements and generalisations of the subject have appeared since then. The most noteworthy of these is the theory of Integration by **Lebesgue* in 1902.

13.2 PARTITIONS AND RIEMANN (OR *DARBOUX) SUMS

(*i*) Partition (or dessection or net) of a closed interval. Let $I = [a, b]$ be a finite closed interval. If $a = x_0 < x_1 < x_2 < \dots < x_n = b$, then the finite ordered set $P = \{x_0, x_1, x_2 \dots, x_n\}$ is known as a partition of I. The n closed intervals $I_1 = [x_0, x_1]$, $I_2 = [x_1, x_2]$, $I_r = [x_{r-1}, x_r] \dots$, $I_n = [x_{n-1}, x_n]$ determined by P are known as the segments of the partition P. The length of the rth sub-interval $I_r = [x_{r-1}, x_r]$ is denoted by δ_r. Thus, $\delta_r = x_r - x_{r-1}$.

By chaning the partition points, the partition can be changed and therefore, we can obtain an infinite number of partitions of the interval $[a, b]$. The set (or family) of all partitions of $[a, b]$ is denoted by $P[a, b]$.

Note : $\sum_{r=1}^{n} \delta_r = \delta_1 + \dots + \delta_n = (x_1 - x_0) + (x_2 - x_1) + \dots + (x_n - x_0) = x_n - x_0 = b - a$

* **Darboux :** 1842-1917, and Lebesgue (1875-1941) were French mathematicians.

(*ii*) **Norm of a partition.** **[Kanpur 2018]**

The length of the greatest of all the intervals $[x_{r-1}, x_r]$ of the portition P will be called its norm and is denoted by $\| P \|$ or $\mu(P)$. Thus

$$\| P \| = \mu(P) = \max\{\delta_r : 1 \le r \le n\}$$

(*iii*) **Refiniment of a partition.** **[Meerut 2018]**

If P_1, P_2 be two partitions of $[a, b]$ such that $P_2 \supset P_1$, *i.e.*, every point of the partition P_1 is as well as point of P_2, we shall say that P_2 is finer than P_1 (or P_2 is a refinement of P_1).

If P_1, P_2 are two partition of $[a, b]$, then $P_1 \cup P_2$ is called a common refinement of P_1 and P_2.

(*iv*) **Upper and lower (Riemann or Darboux) sums.** **[Delhi B.Sc. (Prog.) II, 2018]**

The function f which is defined in $[a, b]$ is also necessarily bounded in each sub-interval I_r. Let M_r, m_r be the bounds (*i.e.*, the supremum and the infimum of f in I_r, *i.e.*, in $[x_{r-1}, x_r]$.

Set up the two sums as follows :

$$U(P, f) = M_1\delta_1 + M_2\delta_2 + \cdots + M_n\delta_n = \sum_{r=1}^{r=n} M_r\delta_r$$

and $$L(P, f) = m_1\delta_1 + m_2\delta_2 + \cdots + m_n\delta_n = \sum_{r=1}^{r=n} m_r\delta_r$$

$U(P, f)$ and $L(P, f)$ are called the upper Darboux sum (or upper Riemann sum) and the lower Darboux sum (or lower Riemann sum) of f corresponding to the partition P.

Note : For every portion P of $[a, b]$, we have

$$L(P, f) \le U(P, f),$$ (Srivenkatawara 2003)

since for $\forall r,\ m_r \le M_r \Rightarrow m_r\delta_r \le M_r\delta_r \Rightarrow \sum_{r=1}^{n} m_r\delta_r \le \sum_{r=1}^{n} M_r\delta_r \Rightarrow L(P, f) \le U(P, f)$

(*v*) **Oscillatory sum.** We have

$$U(P, f) - L(P, f) = \Sigma M_r\delta_r - \Sigma m_r\delta_r = \Sigma(M_r - m_r)\delta_r = \Sigma O_r\delta_r,$$

where O_r denotes the oscillation of the function in subinterval $[x_{r-1}, x_r]$. The sum $\Sigma O_r\delta_r$ is called the *oscillatory sum* and is denoted by $w(P, f)$. As $O_r \ge 0\ \forall r$, each oscillatory sum consists of the sum of a finite number of non-negative terms.

13.3 SOME PROPERTIES OF DARBOUX SUMS

Theorem I. *Let f be a bounded function defined on* $[a, b]$ *and let m, M be the infimum and supremium of f on* $[a, b]$. *Then for any partition P of* $[a, b]$, *we have* **[Agra 2009;**

$$m(b - a) \le L(P, f) \le U(P, f) \le M(b - a)$$ **Delhi B.Sc. (Prog.) II 2018; Meerut 2012]**

Proof. Let $P = \{a = x_0, x_1, x_2, \ldots x_{r-1}, x_r \ldots, x_n = b\}$ be any portion of $[a, b]$. Since f is bounded on $[a, b]$, so f is bounded on each sub-interval $[x_{r-1}, x_r]$, $r = 1, 2, \ldots$, n. Let m_r and M_r be infimum and supremum of f on $[x_{r-1}, x_r]$. Then for every value of r, we have

$$m \le m_r \le M_r \le M$$

$$\Rightarrow \quad m\delta_r \le m_r\delta_r \le M_r\delta_r \le M\delta_r$$

$$\Rightarrow \quad \sum_{r=1}^{n} m\delta_r \le \sum_{r=1}^{n} m_r\delta_r \le \sum_{r=1}^{n} M_r\delta_r \le \sum_{r=1}^{n} M\delta_r$$

$$\Rightarrow \quad m\sum_{r=1}^{n} \delta_r \le L(P, f) \le U(P, f) \le M\sum_{r=1}^{n} \delta_r$$

$$\Rightarrow \quad m(b - a) \le L(P, f) \le U(P, f) \le M(b - a) \qquad \ldots(1)$$

Note : A pair of Darboux sums corresponds to each division of the interval $[a, b]$ and from (1), we see that these sums obtained by considering all possible partitions of $[a, b]$ are bounded.

Theorem II. *Let f be a bounded function on* $[a, b]$ *and let P be a portion of* $[a, b]$. *It* P' *is a refinement of P, then*

(i) $L(P, f) \le L(P', f)$ **(Bhopal 2004; Meerut 2009)**

(ii) $U(P, f) \ge U(P', f)$ **(Bhopal 2004; Meerut 2009)**

(iii) $L(P, f) \le L(P', f) \le U(P', f) \le U(P, f)$

(iv) $U(P, f) - L(P, f) \ge U(P', f) - L(P', f)$

(v) $w(P, f) \ge w(P', f)$.

Proof : Let $P = \{a = x_0, x_1, x_2 ..., x_{r-1}, x_r, ..., x_n = b\}$

and $P' = \{a = x_0, x_1, x_2, ..., x_{r-1}, \xi, x_r, ..., x_n = b\}$

Clearly P' is refinement of P because it contains an additional point ξ such that $x_{r-1} < \xi < x_r$.

Let m_r', m_r'' and m_r be the infimum of f in the intervals $[x_{r-1}, \xi]$, $[\xi, x_r]$ and $[x_{r-1}, x_r]$ respectively. Again, let M_r', M_r'' and M_r be the supremum of f in the intervals $[x_{r-1}, \xi]$, $[\xi, x_r]$ and $[x_{r-1}, x_r]$ respectively. Then, we have

$$m_r \le m_r',\ m_r \le m_r'' \quad \text{and} \quad M_r \ge M_r',\ M_r \ge M_r'',$$

(i) From our construction of partitions P and P', it follows than the contribution to $L(P, f)$ and $L(P', f)$ of each sub-interval except $[x_{r-1}, x_r]$ is the same.

Now, contribution of the sub-interval $[x_{r-1}, x_r]$ to $L(P, f)$ is $m_r(x_r - x_{r-1})$.

Since the subinterval $[x_{r-1}, x_r]$ is split into two sub-interval $[x_{r-1}, \xi]$ and $[\xi, x_r]$, it follows that the contribution of the sub-interval $[x_{r-1}, x_r]$ to $L(P', f)$ is $m_r'(\xi - x_{r-1}) + m_r''(x_r - \xi)$. Hence, we have

$$\therefore\ L(P', f) - L(P, f) = m_r'(\xi - x_{r-1}) + m_r''(x_r - \xi) - m_r(x_r - x_{r-1})$$

$$= m_r'(\xi - x_{r-1}) + m_r''(x_r - \xi) - m_r\{(x_r - \xi) + (\xi - x_{r-1})\}$$

$$= (m_r' - m_r)(\xi - x_{r-1}) + (m_r'' - m_r)(x_r - \xi)$$

$$\ge 0, \text{ as } m_r' \ge m_r,\ m_r'' \ge m_r \text{ and } x_{r-1} < \xi < x_r$$

Thus, $L(P', f) - L(P, f) \ge 0 \Rightarrow L(P, f) \le L(P', f)$

If P' contains p points more than P, then repeating the above argument p times, we have

$$L(P, f) \le L(P', f) \quad ...(1)$$

that is, any refinement of P does not lower the lower Darboux sum

(ii) Proceed exactly an in part (i) to prove that

$$U(P, f) \ge U(P', f) \quad ...(2)$$

i.e., any refinement of P does not raise the upper Darboux sum

(iii) From (1) and (2), $L(P, f) \le L(P', f)$ and $U(P', f) \le U(P, f)$ ...(3)

Also, we know that $L(P', f) \le U(P', f)$ ...(4)

From (3) and (4), $L(P, f) \le L(P', f) \le U(P', f) \le U(P, f)$

(iv) We have $U(P, f) \ge U(P', f)$ ...(5)

Also, $L(P, f) \le L(P', f) \Rightarrow -L(P, f) \ge -L(P', f)$ ...(6)

Adding (5) and (6), $U(P, f) - L(P, f) \ge U(P', f) - L(P', f)$ ...(7)

(v) Using definition of $w(P, f)$ and $w(P', f)$, we have

$$w(P, f) = U(P, f) - L(P, f) \quad \text{and} \quad w(P', f) = U(P', f) - L(P', f).$$

So, from (7), $w(P, f) \ge w(P', f)$

Theorem III. *If P_1 and P_2 are any two partitions of* $[a, b]$.*Then* $U(P_2, f) \geq L(P_1, f)$.

i.e., no lower Darboux *sum can exceed any Darboux upper sum.* **(Meerut 2004)**

Proof : Let $P = P_1 \cup P_2$ be a common refinement of P_1 and P_2

Since any refinement does not lower the lower Darboux sum and does not raise the upper Darboux sum, it follows that

$$L(P_1, f) \leq L(P, f) \quad \text{and} \quad U(P, f) \leq U(P_2, f) \qquad ...(1)$$

Also, we have $$L(P, f) \leq U(P, f) \qquad ...(2)$$

Combining (1) and (2), $L(P_1, f) \leq L(P, f) \leq U(P, f) \leq U(P_2, f)$

Thus, $L(P_1, f) \leq U(P_2, f)$ or $U(P_2, f) \geq L(P_1, f)$

Theorem IV. *If f is a bounded function defined on* $[a, b]$ *and P be and partion of* $[a, b]$, *then*

$$U(P, -f) = -L(P, f) \quad \text{and} \quad L(P, -f) = -U(P, f)$$

(Meer ut 2012; Srivenkateswere 2003)

Proof : Since f is bounded on $[a, b]$, so $-f$ is also bounded on $[a, b]$. Again, it M_r, m_r are supremum and infimum of f on $[x_{r-1}, x_r]$, then $-M_r$, $-m_r$ are infimum and supremum of $-f$ on $[x_{r-1}, x_r]$. By definition of upper and lower Darboux sums, we have

$$U(P, -f) = \sum_{r=1}^{n} (-m_r)\delta_r = -\sum_{r=1}^{n} m_r \delta_r = -L(P, f)$$

and $$L(P, -f) = \sum_{r=1}^{n} (-M_r)\delta_r = -\sum_{r=1}^{n} M_r \delta_r = -U(P, f).$$

Theorem V. *Estimation of the difference between Darboux sums*

If P' is a refinement of P containing p points more than P and $|f(x)| \leq k \ \forall \ x \in [a, b]$, then

(*i*) $U(P', f) \geq U(P, f) - 2p\,k\,\delta$ (*ii*) $L(P', f) \leq L(P, f) + 2p\,k\,\delta$

[G.N.D.U. Amritsar 2010]

(*iii*) $w(P, f) - w(P', f) \leq 4p\,k\,\delta$, where $\|P\| = \text{norm of P} \leq \delta$

Proof : Let $P = \{a = x_0, x_1, x_2, ..., x_{r-1}, x_r, ..., x_n = b\}$

and $P' = \{a = x_0, x_1, x_2, ..., x_{r-1}, \xi, x_r, ..., x_n = b\}$

where $x_{r-1} < \xi < x_r$. Then P' is a refinement of P. Here P' contains just one point ξ (say) more than P.

Let m_r', m_r'' and m_r be the infimum of f in the intervals $[x_{r-1}, \xi]$, $[\xi, x_r]$ and $[x_{r-1}, x_r]$ respectively. Again, let M_r', M_r'' and M_r be the supremum of f in the intervals $[x_{r-1}, \xi]$, $[\xi, x_r]$ and $[x_{r-1}, x_r]$ respectively.

Given that $|f(x)| \leq k, \forall \ x \in [a, b]$ so that $-k \leq f(x) \leq k \ \forall \ x \in [a, b]$

$$\therefore \quad -k \leq M_r' \leq M_r \leq k, \qquad -k \leq M_r'' \leq M_r \leq k \qquad ...(1)$$

and $$-k \leq m_r \leq m_r' \leq k, \qquad -k \leq m_r \leq m_r'' \leq k \qquad ...(2)$$

$$(1) \Rightarrow \quad 0 \leq M_r - M_r' \leq 2k \quad \text{and} \quad 0 \leq M_r - M_r'' \leq 2k \qquad ...(3)$$

and $$(2) \Rightarrow \quad 0 \leq m_r' - m_r \leq 2k \quad \text{and} \quad 0 \leq m_r'' - m_r \leq 2k \qquad ...(4)$$

(*i*) The contribution to $U(P, f)$ and $U(P', f)$ of each sub-interval except $[x_{r-1}, x_r]$ is the same. Here the contribution of the sub-interval $[x_{r-1}, x_r]$ to $U(P, f)$ is $M_r(x_r - x_{r-1})$ whereas the contribution of $[x_{r-1}, x_r]$ (which is broken into two sub-intervals $[x_{r-1}, \xi]$ and $[\xi, x_r]$ in partition P') to $U(P', f)$ is $M_r'(\xi - x_{r-1}) + M_r''(x_r - \xi)$. Hence, we have

$$U(P, f) - U(P', f) = M_r (x_r - x_{r-1}) - [M_r'(\xi - x_{r-1}) + M_r''(x_r - \xi)]$$
$$= M_r[(x_r - \xi) + (\xi - x_{r-1})] - M_r'(\xi - x_{r-1}) - M_r''(x_r - \xi)$$
$$= (M_r - M_r'')(x_r - \xi) + (M_r - M_r')(\xi - x_{r-1})$$
$$\leq 2k(x_r - \xi) + 2k(\xi - x_{r-1}), \text{ using (3)}$$
$$= 2k(x_r - x_{r-1}) = 2k\delta_r, \text{ as } \delta_r \leq \|P\| < \delta$$

Thus, $$U(P, f) - U(P', f) \leq 2k\delta$$

$$\Rightarrow \qquad U(P', f) \geq U(P, f) - 2k\delta \qquad \text{...(5)}$$

If P' contains p additional points, then we can go on adjoining these additional point one by one. So, proceeding as above p times, we finally obtain

$$U(P', f) \geq U(P, f) - 2pk\delta \qquad \text{...(6)}$$

(*ii*) Proceed as in part (*i*) and use (4) in place of (3). Then

$$L(P', f) \leq L(P, f) + 2pk\delta \qquad \text{...(7)}$$

(*iii*) From parts (*i*) and (*ii*), we have

$$U(P, f) - U(P', f) \leq 2pk\delta \qquad \text{...(8)}$$

and $$L(P', f) - L(P, f) \leq 2pk\delta \qquad \text{...(9)}$$

Adding (8) and (9), $\{U(P, f) - L(P, f)\} - \{U(P', f) - L(P', f)\} \leq 4pk\delta$

or $$w(P, f) - w(P', f) \leq 4pk\delta$$

Theorem VI. Let f, g be bounded functions defined on $[a, b]$ and let P be any partition of $[a, b]$. Then

$L(P, f+g) \geq L(P, f) + L(P, g)$, $U(P, f+g) \leq U(P, f) + U(P, g)$ **(Bhopal 2004 : Garhwal 1998)**

Proof Let $P\{a = x_0, x_1, ..., x_n = b\}$ be any partition of $[a, b]$

Since f and g are bounded in $[a, b]$, therefore $f + g$ is also bounded on $[a, b]$ Let $I_r = [x_{r-1}, x_r]$ be rth sub-interval.

Let m_r', M_r' be infimum and supremum of f on I_r,

m_r'', M_r'' be infimum and supremum of g on I_r

and m_r, M_r be infimum and supremum of $f + g$ on I_r.

Now, m_r', m_r'' are infimum of f, g on I_r. So by definition

$$f(x) \geq m_r' \text{ and } g(x) \geq m_r'' \quad \forall\, x \in I_r$$
$$\Rightarrow \quad f(x) + g(x) \geq m_r' + m_r'' \quad \forall\, x \in I_r$$
$$\Rightarrow \quad (f+g)(x) \geq m_r' + m_r'' \quad \forall\, x \in I_r$$
$$\Rightarrow \quad m_r' + m_r'' \text{ is a lower bound of } f + g \text{ on } I_r$$

But, by hypothesis, m_r is the greatest lower bound (*i.e.* infimum) of $f + g$ on I_r. Hence, we have

$$m_r \geq m_r' + m_r''$$
$$\Rightarrow \quad m_r\delta_r \geq m_r'\delta_r + m_r''\delta_r$$
$$\Rightarrow \quad \sum_{r=1}^{n} m_r\delta_r \geq \sum_{r=1}^{n} m_r'\delta_r + \sum_{r=1}^{n} m_r''\delta_r$$
$$\Rightarrow \quad L(P, f+g) \geq L(P, f) + L(P, g)$$

(*ii*) Proof of the second part in left as an exercise

13.4. UPPER AND LOWER RIEMANN INTEGRALS. RIEMANN INTEGRAL

[Purvanchal 2012; Lucknow 2010; Agra 2010; Delhi B.A. (Prog) III, 2014; Meerut 2005]

Let f be defined on $[a, b]$. Then for every partition P of $[a, b]$, we have

$$m(b-a) \leq L(P, f) \leq U(P, f) \leq M(b-a), \qquad ...(1)$$

where m and M are the infimum and supremum of f on $[a, b]$

From (1), it follows that the set $\{L(P, f) : P \text{ in a partition of } [a, b]\}$ of lower sums is bounded above by $M(b-a)$ and, consequently, has the least upper bound. Similarly, from (1), it also follows that the set $\{U(P, f) : P \text{ in a partition of } [a, b]\}$ is bounded below by $m(b-a)$ and, consquently, has the greatest lower bound.

Upper and lower Integrals Def. The infimum *of the set of the upper sums* $U(P, f)$ *is called the upper integral of f over* $[a, b]$ *and is denoted by*

$$\overline{\int_a^b} f(x)\, dx.$$

The supremum *of the set of the lower sums,* $L(P, f)$ *is called the lower integral of f over* $[a, b]$ *and is denoted by*

$$\underline{\int_a^b} f(x)\, dx$$

Thus, $$\overline{\int_a^b} f(x)\, dx = \inf\{U(P, f) : P \text{ is a partion of } (a, b)\}$$

and $$\underline{\int_a^b} f(x)\, dx = \sup\{L(P, f) : P \text{ is a partion of } (a, b)\}$$

A bounded function f is said to be Riemann integrable, or simply integrable over $[a,b]$, if its upper and lower integrals are equal; the common value being called the *Riemann integral* or simply the *integral* denoted by the symbol.

$$\int_a^b f(x)\, dx.$$

The family (or class) of all bounded functions which are Riemann integrable on $[a, b]$ is denoted by $R[a, b]$. If f is R-integrable on $[a, b]$ we express it by writing $f \in R[a, b]$ or R, simply.

Notes 1. The numbers a, b are respectively called the *lower* and the *upper limits of integration.*

2. The definition of integrability as given above is based on the *notion of bounds*. Another equivalent approach based on the notion of limits is given in Art. 13.5.

3. If should be clearly understood that *every bounded function is not integrable i.e.,* there exists bounded functions with unequal corresponding upper and lower integrals. (Refer example 5, page 13.12).

The necessary and sufficient condition for the integrability of a bounded function f over $[a,b]$ is obtained in Art. 13.5.

4. The statement that the integral,

$$\int_a^b f(x)\, dx,$$

exists is equivalent to saying that the function f is bounded and integrable over $[a,b]$.

5. The concept of integrability of a function over an interval, as introduced here, is subject to two very important limitations, *viz.* (*i*) *the function is bounded* (*ii*) *the interval is finite* so that neither of the end points is infinite.

In his later studies, the reader will see how these limitations can be removed and the concept gereralised so as to be applicable sometimes even to cases where the function is not bounded or where one or both the limits of integration are infinite.

6. Since $L(P, -f) = -U(P, f)$ and $U(P, -f) = -L(P, f)$,

we have $\underline{\int_a^b} (-f(x))\, dx = -\overline{\int_a^b} f(x)\, dx$ and $\overline{\int_a^b} (-f(x))\, dx = -\underline{\int_a^b} f(x)\, dx$

7. $\underline{\int_a^b} f(x)\, dx$ is also written as $\underline{\int_a^b} f\, dx$ or $\underline{\int_a^b} f$.

Likewise $\overline{\int_a^b} f(x)\, dx$ is written as $\overline{\int_a^b} f\, dx$ or $\overline{\int_a^b} f$

and $\int_a^b f(x)\, dx$ is written as $\int_a^b f\, dx$ or $\int_a^b f$

Theorem I *The lower R-integral cannot exceed the upper R-integral i.e.,*

$$\underline{\int_a^b} f(x)\, dx \le \overline{\int_a^b} f(x)\, dx$$

[Rohitkhand 2011; Agra 2009; Purvanchal 2007, 2009]

Proof. Let $P[a, b]$ denote the set of all partions of $[a, b]$

Let $P_1, P_2 \in P[a, b]$. Since no lower sum can exceed any upper sum, we have

$$L(P_1, f) \le U(P_2, f), \quad \text{...(1)}$$

(1) in true for each $P_1 \in P[a, b]$. Keeping P_2 fixed, we see that the set $\{L(P_1, f) : P_1 \in [a, b]\}$ has an upper bound $U(P_2, f)$. Again, we know that

$$\underline{\int_a^b} f(x) dx = \sup\{L(P_1, f) : P_1 \in [a, b]\}$$

But supremum $\le$ any upper bound. Hence, we get

$$\underline{\int_a^b} f(x) dx \le U(P_2, f), \quad \text{...(2)}$$

which is true for each $P_2 \in P[a, b]$. From (2), it follows that the set $\{U(P_2, f) : P_2 \in P[a, b]\}$ has a lower bound $\underline{\int_a^b} f(x) dx$. Also we know that

$$\overline{\int_a^b} f(x) dx = \inf\{U(P_2, f) : P_2 \in [a, b]\}$$

Bus any lower bound $\le$ infimum. Hence, we get

$$\underline{\int_a^b} f(x)\, dx \le \overline{\int_a^b} f(x)\, dx$$

Theorem II (Darboux theorem) *Let f be a bounded function defined on* $[a, b]$. *Then, to every* $\varepsilon \le 0$, *there corresponds* $\delta > 0$ *such that*

(*i*) $U(P, f) < \overline{\int_a^b} f(x)\, dx + \varepsilon$ (*ii*) $L(P, f) > \underline{\int_a^b} f(x)\, dx - \varepsilon$

for all partitions P such that $\| P \| \le \delta$, $\| P \|$ being norm of the partition P.

[K.U. BCA (II) 2008; Purvanchal 2006; Agra 2006, 07, 08]

Proof (*i*) As f is bounded, there exists $k > 0$, such that

$$|f(x)| \le k \ \forall\, x \in [a,b].$$

Since $\overline{\int_a^b} f(x)\,dx$ is the least upper bound of the set of sums $U(P, f)$, there exists a partition

$$P_1 = \{a = x_0, x_1, x_2, ..., x_{n-1}, x_n = b\}$$

such that
$$U(P_1, f) < \overline{\int_a^b} f(x)\,dx + \frac{1}{2}\varepsilon. \qquad ...(1)$$

The points of P_1 are $(n + 1)$ in number.

Let δ be the positive number such that

$$2k(n-1)\,\delta = \varepsilon/2$$

Let P be any partition with norm less than or equal to δ.

We write
$$P_2 = P \cup P_2$$

so that P_2 is a partition finer than P and consisting of at the most $(n - 1)$ additional points. Also P_2 is finer than P_1. Thus we have

$$U(P, f) - 2\,(n-1)\,k\delta \le U(P_2, f) \le U(P_1, f)$$

$$\Rightarrow \quad U(P, f) \le 2(n-1)\,k\delta + U(P_1, f)$$

$$< \frac{1}{2}\varepsilon + \overline{\int_a^b} f(x)\,dx + \frac{1}{2}\varepsilon = \overline{\int_a^b} f(x)\,dx + \varepsilon, \text{ by (1) and (2)}$$

Hence, the result.

(*ii*) Proof is similar to that of the corresponding result on the upper integral.

Corollary. *If f is bounded and P is a partition of* $[a, b]$, *then*

(*i*) $\lim\limits_{\|P\|\to 0} U(P, f) = \overline{\int_a^b} f(x)\,dx$ (*ii*) $\lim\limits_{\|P\|\to 0} L(P, f) = \underline{\int_a^b} f(x)\,dx$

Proof (*i*) For each $\varepsilon > 0$, there exists $\delta > 0$ such that

$$U(P, f) < \overline{\int_a^b} f(x)\,dx + \varepsilon \quad \text{or} \quad U(P, f) - \overline{\int_a^b} f(x)\,dx < \varepsilon \qquad ..(1)$$

for all partitions P with $\|P\| \le \delta$. Now from (1), we have

$$-\varepsilon < 0 \le U(P, f) - \overline{\int_a^b} f(x)\,dx < \varepsilon \quad \text{or} \quad \left|U(P, f) - \overline{\int_a^b} f(x)\,dx\right| < \varepsilon$$

By definition of limit, this gives

$$\lim_{\|P\|\to 0} U(P, f) = \overline{\int_a^b} f(x)\,dx$$

(*ii*) Left as an exercise

Note. When $\|P\| \to 0$, the number n of sub-intervals become large in such a manner that $n \to \infty$ as $\|P\| \to 0$. Hence results (*i*) and (*ii*) may be re-written as

$$\lim_{n\to\infty} U(P,f) = \overline{\int_a^b} f(x)\,dx \quad \text{and} \quad \lim_{n\to\infty} L(P,f) = \underline{\int_a^b} f(x)\,dx$$

The above results in the limit form will be employed in solving problems on integrability of functions.

Carollary II. $w(P, f) < \int_a^{-b} f(x)\,dx - \int_{-a}^{b} f(x) + \varepsilon$, for any partition P with $\| P \| \le \delta$.

Hint Use definition, $w(P, f) = U(P, f) - L(P, f)$ and Darboux theorem to prove this corollary.

EXAMPLES

Example 1 *Compute $L(P, f)$ and $U(P, f)$ if*

(i) $f(x) = x^2$ on $[0, 1]$ and $P = \{0, 1/4, 2/4, 3/4, 1\}$ be a partition of $[0, 1]$

(ii) $f(x) = x$ on $[0, 1]$ and $P = \{0, 1/3, 2/3, 1\}$ be a partition of $[0, 1]$

[Kanpur 2018; Meerut 2010; Pune 2010]

(iii) $f(x) = x$ on $[0, 1]$ and $P = \{0, 1/4, 2/4, 3/4, 1\}$ on $[0, 1]$ **(Meerut 2012)**

Solution. (*i*) Here partition P divides $[0, 1]$ into sub-intervals

$$I_1 = [0, 1/4], I_2 = [1/4, 2/4], I_3 = [2/4, 3/4] \text{ and } I_4 = [3/4, 1].$$

The length of these intervals are given by $\delta_1 = \delta_2 = \delta_3 = \delta_4 = 1/4$

Let M_r and m_r be the supremum and infimum of f in interval I_r, $r = 1, 2, 3, 4$,. Since $f(x) = x^2$ is increasing on $[0, 1]$, we have

$$m_1 = 0, M_1 = 1/16;\ m_2 = 1/16, M_2 = 4/16;\ m_3 = 4/16, M_3 = 9/16;\ m_4 = 9/16, M_4 = 1$$

$$\therefore\quad L(P,f) = \sum_{r=1}^{4} m_r\delta_r = m_1\delta_1 + m_2\delta_2 + m_3\delta_3 + m_4\delta_4$$

$$= 0 \times (1/4) + (1/16) \times (1/4) + (4/16) \times (1/4) + (9/16) \times (1/4) = 7/32$$

$$U(P,f) = \sum_{r=1}^{4} M_r\delta_r = M_1\delta_1 + M_2\delta_2 + M_3\delta_3 + M_4\delta_4$$

$$= (1/16) \times (1/4) + (4/16) \times (1/4) + (9/16) \times (1/4) + 1 \times (1/4) = 15/32$$

(ii). Ans. $L(P, f) = 1/3$; $U(P, f) = 2/3$ **(*iii*). Ans.** $L(P, f) = 3/8$, $U(P, f) = 5/8$

Example 2 *Show that a constant function is integrable and* $\int_a^b k\,dx = k(b-a)$.

[Delhi B.Sc. (Prog.) II 2018; Agra 2002, 06; Calicut 2004; Kanpur 2018]

Solution. Let $f(x) = k,\ \forall\, x \in [a, b]$. ...(1)

be a constant function, k being a constant. f is a bounded function. Let $P = \{a = x_0, x_1, x_2, ..., x_n = b\}$ be any partition of $[a, b]$.

If M_r and m_r be respectively the supremum and infimum of f in $(x_{r-1}, x_r]$. Then

$$M_r = k \text{ and } m_r = k \text{ as } f(x) = k\ \forall\, x \in [a, b]$$

$$\therefore\quad U(P,f) = \sum_{r=1}^{n} M_r\delta_r = \sum_{r=1}^{n} k(x_r - x_{r-1}) = k\sum_{r=1}^{n}(x_r - x_{r-1})$$

$$= k[(x_1 - x_2) + (x_2 - x_1) + ... + (x_n - x_{n-1})]$$

$$= k(x_n - x_0) = k(b - a) = \text{constant}$$

and $$L(P,f) = \sum_{r=1}^{n} m_r\delta_r = \sum_{r=1}^{n} k(x_r - x_{r-1}) = k(b-a) = \text{constant, as before}$$

$$\therefore \quad \overline{\int_a^b} f(x)\,dx = \inf. U(P, f) = \inf\{k(b-a)\} = k(b-a)$$

and $$\underline{\int_a^b} f(x)\,dx = \sup L(P, f) = \sup\,\{k(b-c)\} = k(b-a)$$

Since $$\overline{\int_a^b} f(x)\,dx = \underline{\int_a^b} f(x)\,dx,$$

f is Riemann integrable and $$\int_a^b f(x)\,dx = k(b-a)$$

Example 3 *If $f(x) = x^3$ is defined on $[0, a]$, show that $f \in R[a, 0]$ and* $\int_0^a f(x)\,dx = \frac{a^4}{4}$

(Agra 2000; Meerut 2002; Purvanchal 1998)

Solution Let $P = \left\{0, \frac{a}{n}, \frac{2a}{n}, ..., \frac{(r-1)a}{n}, \frac{ra}{n}, ..., \frac{na}{n} = a\right\}$ be any partition of $[0, a]$

Then, its rth sub-interval $= I_r = \left[\frac{(r-1)a}{n}, \frac{ra}{n}\right], r = 1, 2, ..., n$

If δ_r be the length of I_r, then

$$\delta_r = \frac{ra}{n} - \frac{(r-1)a}{n} = \frac{a}{n}.$$

Let M_r and m_r be respectively the supremum and infimum of f in I_r. Since f is increasing on $[0, a]$, we have

$$M_r = \frac{r^3a^3}{n^3} \text{ and } m_r = \frac{(r-1)^3a^3}{n^3}, \text{ as } f(x) = x^3\ \forall\, x \in [0, a]$$

$$\therefore \quad U(P, f) = \sum_{r=1}^{n} M_r\delta_r = \sum_{r=1}^{n} \frac{r^3a^3}{n^3}\cdot\frac{a}{n} = \frac{a^4}{n^4}\sum_{r=1}^{n} r^3$$

$$= \frac{a^4}{n^4}(1^3 + 2^3 + \cdots + n^3) = \frac{a^4}{n^4}\cdot\frac{n^2(n+1)^2}{4} = \frac{a^4}{4}\left(1+\frac{1}{n}\right)^2$$

and $$L(P, f) = \sum_{r=1}^{n} m_r\delta_r = \sum_{r=1}^{n} \frac{(r-1)^3}{n^3}\cdot\frac{a}{n} = \frac{a^4}{n^4}\sum_{r=1}^{n}(r-1)^3$$

$$= \frac{a^4}{n^2}\{(1^3 + 2^3 + \cdots + (n-1)^3\} = \frac{a^4}{n^2}\cdot\frac{(n-1)^2n^2}{4} = \frac{a^4}{4}\left(1-\frac{1}{n}\right)^2$$

$$\left[\text{Since } 1^3 + 2^3 + ... + n^3 = \frac{n^2(n+1)^2}{4} \Rightarrow 1^3 + 2^3 + ... + (n-1)^3 = \frac{(n-1)^2(n-1+1)^2}{4}\right]$$

$$\therefore \quad \overline{\int_a^b} f(x)\,dx = \lim_{n\to\infty} U(P, f) = \lim_{n\to\infty} \frac{a^4}{4}\left(1+\frac{1}{n}\right)^2 = \frac{a^4}{4}$$

and $$\underline{\int_a^b} f(x)\,dx = \lim_{n\to\infty} L(P, f) = \lim_{n\to\infty} \frac{a^4}{4}\left(1-\frac{1}{n}\right)^2 = \frac{a^4}{4}$$

Since $$\overline{\int_0^a} f(x)\,dx = \underline{\int_0^a} f(x)\,dx,$$

f is Riemann integrable and $$\int_0^a f(x)\,dx = \frac{a^4}{4}.$$

Example 4 *Show that f(x) = sin x is integrable on [0, π/2] and* $\int_0^{\pi/2} \sin x\,dx = 1$.

(Meerut 1996)

Solution Let any partition P of $[0, \pi/2]$ be given by

$$P = \left\{0, \frac{\pi}{2n}, \frac{2\pi}{2n}, \ldots, \frac{(r-1)\pi}{2n}, \frac{r\pi}{2n}, \ldots, \frac{n\pi}{2n} = \frac{\pi}{2}\right\}$$

Then, its rth subinterval $= I_r = \left[\frac{(r-1)\pi}{2n}, \frac{r\pi}{2n}\right]$, $r = 1, 2, \ldots, n$

and $\delta_r =$ length of $I_r = (r\pi)/n - (r-1)\pi/2n = \pi/2n$

Let M_r and m_r be respectively the supremum and infimum of f in I_r. Since f is increasing on $(0, \pi/2)$, we have

$$M_r = \sin\frac{r\pi}{2n},\ m_r = \sin\frac{(r-1)\pi}{2n}, \text{ as } f(x) = \sin x \text{ on } \left[0, \frac{\pi}{2}\right]$$

In what follows, we shall use the following result of Trigonometry :

$$\sin\alpha + \sin(\alpha+\beta) + \sin(\alpha+2\beta) + \ldots \text{ to } n \text{ terms} = \frac{\sin\left(\alpha + \frac{n-1}{2}\beta\right)\sin\frac{n\beta}{2}}{\sin(\beta/2)} \qquad \ldots(1)$$

Now, $$U(P, f) = \sum_{r=1}^{n} M_r\delta_r = \sum_{r=1}^{n} \sin\frac{r\pi}{2n}\cdot\frac{\pi}{2n} = \frac{\pi}{2n}\sum_{r=1}^{n}\sin\frac{r\pi}{2n}$$

$$= \frac{\pi}{2n}\left(\sin\frac{\pi}{2n} + \sin\frac{2\pi}{2n} + \ldots + \sin\frac{n\pi}{2n}\right)$$

$$= \frac{\pi}{2n} \times \frac{\sin\left(\frac{\pi}{2n} + \frac{n-1}{2}\cdot\frac{\pi}{2n}\right)\sin\left(\frac{n}{2}\cdot\frac{\pi}{2n}\right)}{\sin(\pi/4n)}, \text{ using (1)}$$

$$= \frac{\pi}{2n} \times \frac{\sin\frac{(n+1)\pi}{4}\sin\frac{\pi}{4}}{\sin(\pi/4n)} = \frac{\pi}{2\sqrt{2}n} \times \frac{\sin(\pi/4 + \pi/4n)}{\sin(\pi/4n)}$$

$$= \frac{\pi}{2\sqrt{2}n} \times \frac{\sin(\pi/4)\cos(\pi/4n) + \cos(\pi/4)\sin(\pi/4n)}{\sin(\pi/4n)}$$

$$= \frac{\pi}{2\sqrt{2}n} \times \frac{1}{\sqrt{2}}\,\frac{\cos(\pi/4n) + \sin(\pi/4n)}{\sin(\pi/4n)} = \frac{\pi}{4n}\left(\cot\frac{\pi}{4n} + 1\right)$$

and $$L(P, f) = \sum_{r=1}^{n} m_r\delta_r = \sum_{r=1}^{n} \sin\frac{(r-1)\pi}{2n}\cdot\frac{\pi}{2n} = \frac{\pi}{2n}\sum_{r=1}^{n}\sin\frac{(r-1)\pi}{2n}$$

$$=\frac{\pi}{2n}\left\{\sin\frac{\pi}{2n}+\sin\frac{2\pi}{2n}+...+\frac{\sin(n-1)\pi}{2n}\right\}$$

$$=\frac{\pi}{2n}\times\frac{\sin\left(\frac{\pi}{2n}+\frac{n-2}{2}\cdot\frac{\pi}{2n}\right)\sin\left(\frac{n-1}{2}\cdot\frac{\pi}{2n}\right)}{\sin(\pi/4n)}, \text{ using (1)}$$

$$=\frac{\pi}{2n}\times\frac{\sin\frac{\pi}{4}\sin\left(\frac{\pi}{4}-\frac{\pi}{4n}\right)}{\sin(\pi/4n)}=\frac{\pi}{2\sqrt{2}n}\times\frac{\sin(\pi/4-\pi/4\pi)}{\sin(\pi/4n)}$$

$$=\frac{\pi}{2\sqrt{2}n}\times\frac{\sin(\pi/4)\cos(\pi/4n)-\cos(\pi/4)\sin(\pi/4n)}{\sin(\pi/4n)}$$

$$=\frac{\pi}{2\sqrt{2}n}\times\frac{1}{\sqrt{2}}\frac{\cos(\pi/4n)-\sin(\pi/4n)}{\sin(\pi/4n)}=\frac{\pi}{4n}\left(\cot\frac{\pi}{4n}-1\right)$$

$$\therefore \quad \overline{\int_0^{\pi/2}} f(x)\,dx=\lim_{n\to\infty}U(P,f)=\lim_{n\to\infty}\frac{\pi}{4n}\left(\cot\frac{\pi}{4n}+1\right)=\lim_{n\to\infty}\left\{\frac{(\pi/4n)}{\tan(\pi/4n)}+\frac{\pi}{4n}\right\}=1$$

and
$$\underline{\int_0^{\pi/2}} f(x)\,dx=\lim_{n\to\infty}L(P,f)=\lim_{n\to\infty}\frac{\pi}{4n}\left(\cot\frac{\pi}{4n}-1\right)=\lim_{n\to\infty}\left\{\frac{(\pi/4n)}{\tan(\pi/4n)}-\frac{\pi}{4n}\right\}=1$$

Since
$$\overline{\int_0^{\pi/2}} f(x)\,dx=\underline{\int_0^{\pi/2}} f(x)\,dx,$$

f in integrable and
$$\int_0^{\pi/2} f(x)\,dx=1$$

Example 5. *Show by an example that every bounded function need not be Riemonn integrable.* **(Delhi B.A. (Prog.) III 2014; Kanpur 2008)**

OR

Let $f(x)$ be defind on $[a, b]$ as follows

$$f(x)=\begin{cases}0, \text{ when } x \text{ is rational}\\ 1, \text{ when } x \text{ is irrational}\end{cases}$$

Show that f is not integrable on $[a, b]$. **[Agra 2007, 2011; Delhi Maths (Prog) 2008 Himachal 2015; Calicut 2004; Purvanchal 2008, 09; Rohilkhand 2011; Nagpur 2003]**

Solution. Clearly $f(x)$ is bounded on $[a, b]$ because $0 \le f(x) \le 1 \ \forall x \in [a, b]$

Let $P = \{a = x_0, x_1, x_2, ..., x_n = b\}$ be any partition of $[a, b]$ and let its sub-intervals be $I_r = [x_{r-1}, x_r]$, for $r = 1, 2, ... n$.

Here $\qquad \delta_r$ = the length of $I_r = x_r - x_{r-1}$.

Let M_r and m_r be respectively the supremum and infimum of the function f in I_r. Since rational and irrational points are everywhere dense so every sub-interval I_r will contain rational and irrational numbers. Hence, by definition of $f(x)$, it follows that

$$m_r = 0 \text{ and } M_r = 1, \text{ for } r = 1, 2, ..., n.$$

$$\therefore \quad U(P,f) = \sum_{r=1}^{n} M_r \delta_r = \sum_{r=1}^{n} 1.\delta_r = \sum_{r=1}^{n} \delta_r = \sum_{r=1}^{n} (x_r - x_{r-1})$$

$$= (x_1 - x_0) + (x_2 - x_1) + \ldots + (x_n - x_{n-1}) = x_n - x_0 = b - a$$

and $$L(P,f) = \sum_{r=1}^{n} m_r \delta_r = \sum_{r=1}^{n} 0.\delta_r = 0$$

$$\therefore \quad \overline{\int_a^b} f(x)\,dx = \lim_{n\to\infty} U(P,f) = 1 \quad \text{and} \quad \underline{\int_a^b} f(x)\,dx = \lim_{n\to\infty} L(P,f) = 0$$

Since $\overline{\int_a^b} f(x)\,dx \neq \underline{\int_a^b} f(x)\,dx$, so f is not R-integrable.

Example 6. *If* $f(x)$ *be defined on* [0, 2] *as follows,*

$$f(x) = x + x^2, \text{ when } x \text{ is rational.}$$
$$= x^2 + x^3, \text{ when } x \text{ is irrational}$$

then evaluate the upper and lower Riemann integrals of f over [0, 2] *and show that f is not R-integrable over* [0, 2]. **(Purvanchal 2007; Agra 2001; Meerut 2010, 2019)**

Solution. Here $(x + x^2) - (x^2 + x^3) = x - x^3 = x(1 - x^2)$, so

$(x + x^2) - (x^2 + x^3) > 0$ when $0 < x < 1$.

and $(x + x^2) - (x^2 + x^3) < 0$ when $1 < x < 2$.

Let M_r and m_r be the supremum and infimum of the given function $f(x)$ in I_r where I_r is the rth usual sub-interval of any partition. Then for all values of n we have

$$M_r = x + x^2, \text{ if } 0 < x < 1$$
$$= x^2 + x^3, \text{ if } 1 < x < 2$$

and
$$m_r = x^2 + x^3, \text{ if } 0 < x < 1$$
$$= x + x^2, \text{ if } 1 < x < 2$$

Now, by definition,

the upper Riemann integral $= \overline{\int_0^2} f(x)\,dx = \int_0^1 (x + x^2)\,dx + \int_1^2 (x^2 + x^3)\,dx = \dfrac{83}{12}$

and the lower Riemann integral $= \underline{\int_0^2} f(x)\,dx = \int_0^1 (x^2 + x^3)\,dx + \int_1^2 (x + x^2)\,dx = \dfrac{53}{12}$

Since $\overline{\int_0^2} f(x)\,dx \neq \underline{\int_0^2} f(x)\,dx$, so f is not R-integrable on [0, 2]

Example 7. *Find the upper and lower Riemann integrals for the function f defined on* [0, 1] *as follows*

$$f(x) = (1 - x^2)^{1/2}, \text{ if } x \text{ is rational.}$$

and $$f(x) = 1 - x, \text{ if } x \text{ is irrational.}$$

Hence show that f is not Riemann integral on [0, 1]

(Meerut 2009, 2020; Agra 2007)

Solution Here $f(x) = \sqrt{(1-x^2)} = \sqrt{(1-x)(1+x)}$, *if x is rational.*

and $$f(x) = 1 - x = \sqrt{(1-x)(1-x)}, \text{ if } x \text{ is irrational}$$

So, $$\sqrt{(1-x^2)} = \sqrt{(1-x)(1+x)} > \sqrt{(1-x)(1-x)} \text{ for } 0 < x < 1$$

Thus $$(1-x^2)^{1/2} > 1 - x, \text{ for } 0 < x < 1.$$

Let M_r and m_r be the supremum and infimum of the given function f in I_r where I_r is the rth usual sub-interval of any partition P of [0 1]. Then, for all values of r, we have

$$M_r = (1-x^2)^{1/2} \quad \text{and} \quad m_r = 1 - x$$

Now, by definition, the upper Riemann integral

$$= \overline{\int_0^1} f(x)\,dx = \int_0^1 \sqrt{(1-x^2)}\,dx = \left[\frac{1}{2}x\sqrt{(1-x^2)} + \frac{1}{2}\sin^{-1} x\right]_0^1 = \frac{\pi}{4}$$

and the lower Riemonn integral

$$= \underline{\int_0^1} f(x)\,dx = \int_0^1 (1-x)dx = \left[x - \frac{1}{2}x^2\right]_0^1 = \frac{1}{2}$$

Since upper and lower Riemann integrals are not equal, hence the given function f is not R-integrable on [0, 1]

EXERCISES

1. If f is defined on [0, 1] by $f(x) = x \ \forall\ x \in [0, 1]$, then prove that f is Riemann integrable on [0, 1] and $\int_0^1 f(x)\,dx = \frac{1}{2}$. *(Rajasthan 2010; Agra 2003, Garhwal 1997)*

2. If f is defined on $[0, a]$, $a > 0$ by $f(x) = x^2 \ \forall\ x \in [0, a]$ then prove that f is Riemann integrable on $[0, a]$ and $\int_0^a f(x)\,dx = \frac{a^3}{3}$

   (Agra 1999; Delhi Physics (H) 2000; Garhwal 2001; Meerut 2001)

3. Give an example to show that every bounded function need not be Riemann integrable.
   Hine Refer solved example 5, page 13.12.

4. Show that $(3x + 1)$ is R-integrable on [1, 2] and $\int_1^2 (3x+1)\,dx = \frac{11}{2}$.

   (G.N.D.U. Amritsar 2010; Delhi Physics (H) 1993)

5. Show that the function f : [1, 2] $\to$ **R** defined by $f(x) = \alpha x + \beta$, $x \in [1, 2]$, where α, β are constants, is integrable on [1, 2]. *(Delhi Physics(H) 2001)*

6. Show that the Dirichlet function f : [0, 1] $\to$ **R** defined by

   $$f(x) = 1, \text{ when } x \text{ is rational.}$$
   $$= 0, \text{ when } x \text{ is irrational}$$

   is not integrable on [0, 1]. *(Delhi B.A. (Prog.) III 2013; Purvanchal 1994, 96)*

7. (*a*) If $f(x)$ be defined on [0, 1] as follows :
$f(x) = 1$, when x is rational and $f(x) = -1$, when x is irrational , then prove that f is not R-integrable over [0, 1].
(*b*) If a real valued function f is defined on $[a, b]$ by $f(x) = -1$, if x is rational and $f(x) = 1$, if x is irrational then show that f is not R-integrable on $[a, b]$ **[Garhwal 1995]**

8. If $f(x)$ be a function defined on $[0, \pi/4]$ by $f(x) = \cos x$, if x is rational and $f(x) = \sin x$, if x is irrational, then prove that f is not Riemann-integrable over $[0, \pi/4]$. **[Agra 2008]**

9. Given $a > 0, b > 0$ and a function $f(x)$ defined on interval $[a, b]$ such that $f(x) = 1$, for rational x and $f(x) = 2$, for irrational x, then find $\underline{\int_a^b} f(x), dx$ and $\overline{\int_a^b} f(x)\, dx$. Also show that f is not Riemann integrable. *(Garhwal 1996, Purvanchal 98)*

10. Let f be a function defined on [0, 1] as follows $f(x) = \begin{cases} 1, \text{ if } x \neq 1/2 \\ 0, \text{ if } x = 1/2 \end{cases}$

Show that f is *R*-integrable on [0, 1] and $\int_0^1 f(x)\, dx = 1$.

11. Calculate the values of upper and lower integrals for the function f defined on [0, 2] as follows :
$f(x) = x^2$ when x is rational and $f(x) = x^3$; when x is irrational. **[Ans.** 49/12; 31/12**]**

12. If $f(x) = 1 + x$, when x is rational
$= x + x^2$, when x is irrational,

then show that $\overline{\int_0^2} f(x)\, dx = \dfrac{16}{3}$ and $\underline{\int_0^2} f(x)\, dx = \dfrac{10}{3}$ *(Garhwal 1999)*

13. Show that $f(x) = x$ is integrable on $[a, b]$ and $\int_a^b f(x)\, dx = \frac{1}{2}(b^2 - a^2)$.

14. Prove that if $f \in R\,[a, b]$, then the value of the integral is uniqually determined. *(Calicut 2004)*

15. A function f is defined on [0, 1/2] by $f(x) = \begin{cases} x, \text{ if } x \text{ is rational} \\ 1 - x, \text{ if } x \text{ is irrational} \end{cases}$

Prove that $\overline{\int_0^{1/2}} f = \dfrac{3}{8}, \underline{\int_0^{1/2}} f = \dfrac{1}{8}$ and f is not integrable.

16. Let $f : [0, 1] \to$ R be defined by $f(x) = 1$, if x is rational; $f(x) = 2$, if x is irrational

Show that $\int_0^1 f(x)\, dx$ does not exist. *(Garhwal 1994, Purvanchal 98)*

17. Using only the definition of integral, show that
$f(x) = \sin(1/x)$, x irrational, $0 \leq x \leq 1$
$= 0$, otherwise, $0 \leq x \leq 1$
is not Riemann integrable. *(Delhi Maths (H) 1995)*

18. Show that lower Reimann integral of $\int_0^1 (x^2 + 1)\, dx$ is 4/3. **(Delhi B.Sc. III (Prog.) 2010)**

18. Let $$h(x)=\begin{cases} x+1 \text{ for rational } x \text{ in } [0,1] \\ 0 \text{ for irrational } x \text{ in } [0,1] \end{cases}$$

Prove that h is not Riemann integrable (*Calicut 2004*)

13.5 ANOTHER EQUIVALENT DEFINITION OF INTEGRABILITY AND INTEGRAL

We shall firstly state and prove a theorem which will suggest an alternative definition of the integrability and the integral of a function. Let f be a funtion defined in $[a, b]$.

Let $$P=\{a=x_0, x_1, x_2, ..., x_{r-1}, x_r, ..., x_n=b\}$$

be a partition of $[a,b]$ and ξ_r be any arbitrary point of $I_r=[x_{r-1}, x_r]$. Let $\delta_r=x_r-x_{r-1}$

Form the sum

$$\sum_{r=1}^{r=n} f(\xi_r)\,\delta_r = f(\xi_1)\,\delta_1+...+f(\xi_r)\,\delta_r+...f(\xi_n)\,\delta_n.$$

Theorem *If f is bounded and integrable over $[a,b]$ then to every $\varepsilon>0$ there corresponds $\delta>0$, such that for every partition $P=\{a=x_0, x_1,...x_{r-1}, x_r, ..., x_n=b\}$ of norm $\le\delta$, and for every arbitrary choice of $\xi_r\in[x_{r-1}, x]$,*

$$\left|\sum_{r=1}^{r=n} f(\xi_r)\,\delta_r-\int_a^b f(x)\,dx\right|<\varepsilon. \text{ where } \delta_r=x_r-x_{r-1}$$ (*Delhi Physics (H) 1998*)

Proof. Since f is a bounded and integrable function, therefore

$$\overline{\int_a^b} f(x)\,dx=\underline{\int_a^b} f(x)\,dx=\int_a^b f(x)\,dx.$$

Let ε be any positive number.

By Darboux's theorem, there exists $\delta>0$ such that for every partition P whose norm is $\le\delta$

$$U(P,f)<\overline{\int_a^b} f(x)\,dx+\varepsilon=\int_a^b f(x)\,dx+\varepsilon, \quad ...(1)$$

and

$$L(P,f)>\underline{\int_a^b} f(x)\,dx-\varepsilon=\int_a^b f(x)\,dx-\varepsilon. \quad ...(2)$$

If ξ_r be *any* point of the interval I_r of P, we have

$$L(P,f)\le\sum_{r=1}^{r=n} f(\xi_r)\,\delta_r\le U(P,f) \quad ...(3)$$

From (1), (2) and (3) we deduce that for every partition P whose norm is $\le\delta$,

$$\int_a^b f(x)\,dx-\varepsilon<\sum_{r=1}^{r=n} f(\xi_r)\,\delta_r<\int_a^b f(x)\,dx+\varepsilon$$

$$\Rightarrow \left|\sum_{r=1}^{r=n} f(\xi_r)\,\delta_r-\int_a^b f(x)\,dx\right|<\varepsilon.$$

Hence the theorem.

In a more concise but less precise manner, the result may be stated a little differently as follows : *if a function f is bounded and integrable then*

$$\lim_{\|P\|\to 0} \sum_{r=1}^{r=n} f(\xi_r)\,\delta_r, \qquad \text{... (4)}$$

exists and is the integral of f over $[a,b]$.

If $S(P,f)=\sum_{r=1}^{n} f(\xi_r)(x_r-x_{r-1})=\sum_{r=1}^{n} f(\xi_r)\delta_r$, where $x_{r-1}\le \xi \le x_r$, for $r = 1, 2, ..., n$, then $S(P, f)$ is called a Riemann sum of f over $[a, b]$ with respect to partition P of $[a, b]$. Thus, (4) may be re-written as :

$$\int_a^b f(x)\,dx = \lim_{\|P\|\to 0} S(P,f) = \lim_{n\to\infty} S(P,f).$$

13.6 SECOND DEFINITION OF RIEMANN INTEGRABILITY

We shall now prove that the definition of integrability as given in Art. 13.4 is eqivalent to the following :

A function f is said to be integrable over $[a, b]$, *if there exists a number I such that to every* $\varepsilon > 0$ *there corresponds* $\delta > 0$ *and that for every partition* $\{a = x_0, x_1, ..., x_{r-1}, x_r, ..., x_n = b\}$ of norm $\le \delta$ and $\forall\, \xi_r \in I_r$ where $I_r = [x_{r-1}, x_r]$,

$$\left|\sum_{r=1}^{r=n} f(\xi_r)(x_r - x_{r-1}) - I\right| < \varepsilon.$$

Also then I is said to be the integral of f over $[a,b]$.

In a more concise but less precise manner, this definition may be stated as follows : *A function f is integrable, if*

$$\lim_{\|P\|\to 0} \sum_{r=1}^{r=n} f(\xi_r)\,\delta_r$$

exists and is independent of the choice of the interval I_r *and of the point* ξ_r of I_r; *the limit I, if it exists, is called the integral of f over* $[a, b]$.

The equivalence of the two definitions will bow be established.

As a consequence of the theorem proved in Art. 13.5 above, we see that if a function f be integrable according to the former definition, then it is so according to the latter also.

Now, Let f be integrable according to the latter definition so that

$$\lim \Sigma f(\xi_r)\,\delta_r,$$

exists as the norm $\|P\| \to 0$.

It will firstly be deduced that f is bounded in $[a,b]$.

If possible let f be not bounded.

There exists a partition P such that for every choice of ξ_r in I_r *i.e.*, $[x_{r-1}, x_r]$

$$\left|\Sigma f(\xi_r)\,\delta_r - \text{I}\right| < 1.$$

$$\Rightarrow \qquad \left|\Sigma f(\xi_r)\,\delta r\right| < |\text{I}| + 1$$

As f is not bounded in $[a,b]$, it must also not be bounded in at least one I_r say in I_m.

We take $\xi_r = x_r$, when $r \neq m$ so that every number ξ_r, except ξ_m, is fixed and, accordingly, every term of $\Sigma f(\xi_r)\,\delta_r$ except the term $f(\xi_m)\,\delta_m$ is also fixed. Since f is not bounded in I_m, we can now choose a point ξ_m in I_m such that

$$|\Sigma f(\xi_r)\,\delta_r| > |I| + 1,$$

and thus we arrive at a contradiction. Hence f is bounded in $[a,b]$.

Now, let ε be any positive number. There exists a $\delta > 0$ such that for every partition whose norm is $\leq \delta$, we have

$$|\Sigma f(\xi_r)\,\delta_r - I| < \varepsilon/2$$

$$\Rightarrow \qquad I - \varepsilon/2 < \Sigma f(\xi_r)\,\delta_r < I + \varepsilon/2 \qquad \text{... (1)}$$

for every choice of a point ξ_r in I_r.

If M_r, m_r be the bounds of f in I_r, there exist points α_r, β_r of I_r such that

$$f(\alpha_r) > M_r - \varepsilon/2\,(b-a), \qquad f(\beta_r) > m_r + \varepsilon/2\,(b-a),$$

From these we deduce that

$$\Sigma f(\alpha_r)\,\delta_r > U(P,f) - \varepsilon/2 \Rightarrow U(P,f) < \Sigma f(\alpha_r)\,\delta_r + \varepsilon/2 \qquad \text{... (2)}$$

$$\Sigma f(\beta_r)\,\delta_r > L(P,f) + \varepsilon/2 \Rightarrow L(P,f) > \Sigma f(\beta_r)\,\delta_r - \varepsilon/2 \qquad \text{... (3)}$$

From (1), (2) and (3), we deduce taking $\xi_r = \alpha_r$ and β_r, that

$$I - \varepsilon < L(P,f) \leq U(P,f) < I + \varepsilon \qquad \text{... (4)}$$

for every partition whose norm is $\leq \delta$.

Also we know that

$$L(P,f) \leq \underline{\int_a^b} f(x)\,dx \leq \overline{\int_a^b} f(x)\,dx \leq U(P,f)$$

From (4) and (5), we have

$$I - \varepsilon < \underline{\int_a^b} f(x)\,dx \leq \overline{\int_a^b} f(x)\,dx < I + \varepsilon. \qquad \text{... (6)}$$

$$\Rightarrow \qquad 0 \leq \overline{\int_a^b} f(x)\,dx - \underline{\int_a^b} f(x)\,dx < 2\,\varepsilon.$$

so that the non-negative number

$$\overline{\int_a^b} f(x)\,dx - \underline{\int_a^b} f(x)\,dx$$

is less than every positive number; ε being arbitrary. Thus it follows that

$$\overline{\int_a^b} f(x)\,dx - \underline{\int_a^b} f(x)\,dx = 0 \qquad \text{... (7)}$$

implying that the function is integrable according to the former definition also.

From (6) and (7), we have

$$I-\varepsilon<\int_a^b f(x)\,dx<I+\varepsilon$$

and since ε is arbitrary, this gives

$$I=\int_a^b f(x)\,dx.$$

Thus the equivalence is completely established.

An important Resutl : f is integrable on $[a, b]$ if lim S(P,f) exist as $\|P\| \to 0$, *i.e.*,

$$\lim_{\|P\|\to 0} \sum_{r=1}^{n} f(\xi_r)\,\delta_r \text{ exists, where } \xi_r \in [x_{r-1}, x_r]$$

and $$\int_a^b f(x)\,dx=\lim_{\|P\|\to 0}\sum_{r=1}^{n} f(\xi_r)\delta_r=\lim_{\|P\|\to 0} S(P,f)$$ *(Delhi Maths (H) 2002)*

Example Compute $\int_{-1}^{1} f(x)\,dx$, where $f(x)=|x|$ *(Delhi Physics (H) 1994)*

Solution. Here $f(x)$ is bounded and continuous on $[-1, 1]$

Now, we have $f(x) = x$, when $x \geq 0$

$= -x$, when $x \leq 0$

Consider a partition P of $[-1, 1]$ into 2n equal sub-intervals each of length $1/n$ given by

$$P=\left\{-1=x_0,\ x_1=-1+\frac{1}{n},\ x_2=-1+\frac{2}{n},\ldots,\ x_n=0=y_0,\ y_1=\frac{1}{n},\ y_2=\frac{2}{n},\ldots,\ y_n=1\right\}$$

Here, norm of $P = \|P\| = 1/n$ so that $\|P\| \to 0$ as $n\to\infty$.

Let $\xi_i=x_i\in[x_{i-1},x_i]$ *and* $\xi_i'=y_i\in[y_{i-1},y_i]$, for $i = 1, 2\ldots, n$.

Then For each i, $x_i = -1 + i/n$ and $y_i = i/n$.

Now, $$S(P,f)=\sum_{i=1}^{n} f(\xi_i)\delta x_i+\sum_{i=1}^{n} f(\xi_i')\delta y_i=\sum_{i=1}^{n} f(x_i).1/n+\sum_{i=1}^{n} f(y_i)\cdot(1/n)$$

$$=\frac{1}{n}\sum_{i=1}^{n}(-x_i)+\frac{1}{n}\sum_{i=1}^{n} y_i=-\frac{1}{n}\sum_{i=1}^{n}\left(-1+\frac{i}{n}\right)+\frac{1}{n}\sum_{i=1}^{n}\frac{i}{n}$$

$[\because x_i\leq 0 \text{ and } y_i\geq 0,\ \forall\, i=1,2,\ldots,n]$

$$=\sum_{i=1}^{n}\left[\frac{1}{n}-\frac{i}{n^2}+\frac{i}{n^2}\right]=\sum_{i=1}^{n}\frac{1}{n}=n\times\frac{1}{n}=1$$

$\therefore$ $$\lim_{\|P\|\to 0} S(P,f)=1$$

Since the limit exists, the function is integrable and $\int_{-1}^{1}|x|\,dx=\lim_{\|P\|\to 0} S(P,f)=1.$

Summation of series. Theorem

(*i*) *If f is Riemann integrable an* $[a, b]$,

$$\int_a^b f(x)\,dx=\lim_{n\to\infty}\sum_{r=1}^{n} hf(a+rh),\ \text{where } h=\frac{b-a}{n}$$

(*ii*) If f is integrable on $[0, 1]$, then

$$\int_0^1 f(x)dx=\lim_{n\to\infty}\frac{1}{n}\sum_{r=1}^{n} f\left(\frac{r}{n}\right)$$

(*iii*) If f is integrable on $[a, b]$, then

$$\int_a^b f(x)dx = \lim_{n\to\infty} \sum_{r=1}^{n} (ah^r - ah^{r-1}) f(ah^r), \text{ where } h = \left(\frac{b}{a}\right)^{1/n}$$

Proof (*i*) Let $P = \{a, a+h, a+2h, ..., a+nh = b\}$ be a partition of $[a, b]$. This partition divides $[a, b]$ into n equal sub-intervals, each of length $h = (b-a)/n$.

Here, $\|P\| = (b-a)/n$. Hence $\|P\| \to 0$ as $n \to \infty$

The rth sub-intervals $= I_r = [a + (r-1)h, a + rh]$

Then, δ_r = length of $I_r = a + rh - \{a + (r-1)h\} = h, \ \forall r = 1, 2, ..., r.$

Let ξ_r be such that $a + (r-1)h \le \xi_r \le a + rh$, *for each* $r = 1, 2, ..., n$

Then $$\int_a^b f(x)\, dx = \lim_{\|P\|\to 0} \sum_{r=1}^{n} f(\xi_r)\,\delta_r$$

$$= \lim_{n\to\infty} \sum_{r=1}^{n} f(a+rh)\cdot(h), \text{ taking } \xi_r = a + rh$$

(*ii*) Let $P = \{0, 1/n, 2/n, ..., n/n = 1\}$ be a partition of $[0, 1]$ so that the partition divides $[0, 1]$ into n equal sub-intervals, each of length $1/n$. Then $\delta_r = 1/r$, for $r = 1, 2, ..., n$.

So $\|P\| = 1/n$. Also $\|P\| \to 0 \Rightarrow n \to \infty$

The rth sub-interval is $[(r-1)/n, r/n]$

Let $\xi_r \in [(r-1)/n, r/n], \forall r = 1, 2, ..., n.$

$$\therefore \int_0^1 f(x)\, dx = \lim_{\|P\|\to 0} \sum_{r=1}^{n} f(\xi_r)\delta_r = \lim_{n\to\infty} \sum_{r=1}^{n} f\left(\frac{r}{n}\right)\cdot\frac{1}{n} \qquad (\text{Taking } \xi_r = r/n)$$

or $$\int_0^1 f(x)\, dx = \lim_{n\to\infty} \frac{1}{n} \sum_{r=1}^{n} f\left(\frac{r}{n}\right).$$

Note. To evaluate the limit of a sum, proceed as follows.

Step 1. Re-write the limit of sum in the form $\lim_{n\to\infty} \frac{1}{n} \sum_{r=1}^{n} f\left(\frac{r}{n}\right)$.

Step 2. Replace r/n by x and $1/n$ by dx

Step 3. Replace $\lim_{n\to\infty} \frac{1}{n} \sum_{r=1}^{n}$ by $\int_0^1$

[Here the limits of integration are the values of lim r/n for the first and the last term as $n \to \infty$]

(*iii*) Let $P = \{a, ah, ah^2, ..., ah^n = b\}$ be a partition of $[a, b]$ so that $h^n = b/a$ or $h = (b/a)^n$.

Then, I_r = the rth sub-interval $= [ah^{r-1}, ah^r]$

and $\delta_r = ah^r - ah^{r-1}$. As $\|P\| \to 0$, $h \to 1$ and $n \to \infty$

Let $\xi_r \in [ah^{r-1}, ah^r]$ for $r = 1, 2, 3, ..., n$.

Then, $$\int_a^b f(x)\, dx = \lim_{\|P\|\to 0} \sum_{r=1}^{n} f(\xi_r)\,\delta_r$$

$$= \lim_{n\to\infty} \sum_{r=1}^{n} f(ah^r)\cdot(ah^r - ah^{r-1}), \text{ taking } \xi_r = ah^r$$

$$= \lim_{n\to\infty} \sum_{r=1}^{n} (ah^r - ah^{r-1}) f(ah^r)$$

Example 1. Show that

$$(i)\ \lim_{n\to\infty}\left[\frac{1}{n}+\frac{n^2}{(n+1)^3}+\frac{n^2}{(n+2)^3}+\ldots+\frac{1}{8n}\right]=\frac{3}{8} \quad (ii)\ \lim_{n\to\infty}\left(\frac{n^n}{n!}\right)^{1/n}=e \quad \textbf{(Agra 2008)}$$

Solution. (*i*) $\lim_{n\to\infty}\left[\frac{1}{n}+\frac{n^2}{(n+1)^3}+\frac{n^2}{(n+2)^3}+\ldots+\frac{1}{8n}\right]$

$$= \lim_{n\to\infty}\left[\frac{n^2}{(n+0)^2}+\frac{n^2}{(n+1)^3}+\ldots+\frac{n^2}{(n+n)^3}\right] = \lim_{n\to\infty}\sum_{r=0}^{n}\frac{n^2}{(n+r)^3} = \lim_{n\to\infty}\sum_{r=0}^{n}\frac{1/n}{(1+r/n)^3}$$

$$= \int_0^1 \frac{dx}{(1+x)^3}, \text{ replacing } \frac{r}{n} \text{ by } x \text{ and } \frac{1}{n} \text{ by } dx$$

$$= \left[\frac{-1}{2(1+x)^2}\right]_0^1 = \frac{3}{8}.$$

(*ii*) Let $\qquad L = \lim_{n\to\infty}\left(\frac{n^n}{n!}\right)^{1/n} = \lim_{n\to\infty}\left(\frac{n}{1}\cdot\frac{n}{2}\cdot\frac{n}{3}\cdots\frac{n}{n}\right)^{1/n}$

Taking logarithm of both sides, we have

$$\log L = \lim_{n\to\infty}\frac{1}{n}\left(\log\frac{n}{1}+\log\frac{n}{2}+\ldots+\log\frac{n}{n}\right)$$

$$= \lim_{n\to\infty}\frac{1}{n}\sum_{r=1}^{n}\log\frac{n}{r} = \lim_{n\to\infty}\frac{1}{n}\sum_{r=1}^{n}\log\frac{1}{(r/n)}$$

$$= \int_0^1 \log\frac{1}{x}\,dx = -\int_0^1 \log x\,dx = -[x\log x + x]_0^1 = 1$$

Thus, $\qquad \text{Log } L = 1 \quad$ so that $\quad L = e \quad$ or $\quad \lim_{n\to\infty}\left(\frac{n^n}{n!}\right)^{1/n} = e$

Example 2. Show that

$$(i)\ \lim_{n\to\infty}\frac{1}{n}\left[\sin\frac{\pi}{n}+\sin\frac{2\pi}{n}+\ldots+\sin\frac{n\pi}{n}\right]=\frac{2}{\pi} \quad (ii)\ \lim_{n\to\infty}\left\{\left(1+\frac{1}{n}\right)\left(1+\frac{2}{n}\right)\ldots\left(1+\frac{4n}{n}\right)\right\}^{1/n}=\frac{5^5}{e^4}$$

Solution. Left as an exercise

Example 3. *Using the definition of integral as the limit of a sum show that*

$$\int_0^a \sin x\,dx = 1-\cos a \qquad \textit{(Osmanic 2004)}$$

Sol. Since $f(x) = \sin x$ is bounded and continuous on $[0, a]$ therefore, f in integrable on $[0, a]$.

Consider a partition $P = \{0 = x_0, x_1, x_2, \ldots, x_n = a\}$ of $[0, a]$ dividing it into n equal sub-intervals, each of length a/n. Then $\|P\| \to 0$ as $n\to\infty$. Here $x_r = 0 + ra/n = ra/n$ and δ_r a/n, $r = 1, 2, \ldots, n$.

Using the definition of integral as the limit of a sum, we have

$$\int_0^a f(x)\,dx = \lim_{\|P\|\to 0} \sum_{r=1}^{n} f(\xi_r)\delta_r = \lim_{n\to\infty} \sum_{r=1}^{n} f(x_r)\delta_r, \text{ taking } \xi_r = x_r$$

$$= \lim_{n\to\infty} \sum_{r=1}^{n} f\left(\frac{ra}{n}\right)\cdot\frac{a}{n} = \lim_{n\to\infty} \frac{a}{n}\sin\frac{ra}{n} = \lim_{n\to\infty} \frac{a}{n}\left[\sin\frac{a}{n} + \sin\frac{2a}{n} + \ldots + \sin\frac{na}{n}\right]$$

$$= \lim_{n\to\infty} \frac{a}{n} \frac{\sin\left(\frac{a}{n} + \frac{n-1}{2}\cdot\frac{a}{n}\right)\sin\left(\frac{n}{2}\cdot\frac{a}{n}\right)}{\sin(a/2n)}$$

$$\left[\because \sin\alpha + \sin(\alpha+\beta) + \sin(\alpha+2\beta) + \ldots \text{ to } n \text{ term} = \frac{\sin\{\alpha + (n-1)\beta/2\}\sin(n\beta/2)}{\sin(\beta/2)}\right]$$

$$= \lim_{n\to\infty} 2\times\frac{a/2n}{\sin(a/2n)}\times\sin\left\{\frac{a}{2}\left(1+\frac{1}{n}\right)\right\}\cdot\sin\frac{a}{2} = 2 \times 1 \times \sin(a/2) \times \sin(a/2) = 1 - \cos a$$

$$\left[\because \lim_{n\to\infty}\frac{a/2n}{\sin(a/2n)} = \lim_{(a/2n)\to 0}\frac{a/2n}{\sin(a/2n)} = 1\right]$$

Ex. 4. From definition, prove that

(i) $\int_0^1 (2x^2 - 3x + 5)dx = \frac{25}{3}$ *(ii)* $\int_{-1}^2 f(x)dx = \frac{5}{2}$, where $f(x) = |x|$

(iii) $\int_0^a \cos x\, dx = \sin a$ *(iv)* $\int_0^{\pi/2} \cos x\, dx = 1$

(v) $\int_0^{\pi/2} \sin x\, dx = 1$

13.7 NECESSARY AND SUFFICIENT CONDITION FOR INTEGRABILITY

I. First form. *A necessary and sufficient condition for the integrability of a bounded function f is, that to every* $\varepsilon > 0$, *there coresponds* $\delta > 0$ *such that for every partition P, whose norm is* $\leq \delta$, *the oscillartory sum w (P, f) is* $< \varepsilon$, *i.e.,* $U(P, f) - L(P, f) < \varepsilon$.

(Delhi B.Sc.(Prog) III 2009, 11; Purvanchal 2006; Agra 2006; Meerut 2006; Meerut 2010; (Delhi B.A.. (Prog.) III 2012)

Proof. The condition is necessary. The bounded function f being integrable.

$$\overline{\int_a^b} f(x)\,dx = \underline{\int_a^b} f(x)\,dx = \int_a^b f(x)\,dx.$$

Let ε be any positive numbger. By Darboux's theorem, there exists $\delta > 0$ such that for every partition P whose norm is $\leq \delta$,

$$U(P, f) < \overline{\int_a^b} f(x)\,dx + \frac{\varepsilon}{2} = \int_a^b f(x)\,dx + \frac{\varepsilon}{2},$$

and
$$L(P, f) > \underline{\int_a^b} f(x)\,dx - \frac{\varepsilon}{2} = \int_a^b f(x)\,dx - \frac{\varepsilon}{2}$$

implying

$$\int_a^b f(x)\,dx - \frac{\varepsilon}{2} < L(P,f) \le U(P,f) < \int_a^b f(x)\,dx - \frac{\varepsilon}{2}$$

$$\Rightarrow \qquad w(P,f) = U(P,f) - L(P,f) < \varepsilon$$

for every partition P whose norm is $\le \delta$.

The condition is sufficient. Let ε be any positive number. There exists a partition P such that

$$U(P,f) - L(P,f) = \left[U(P,f) - \overline{\int_a^b} f(x)\,dx \right]$$

$$+ \left[\overline{\int_a^b} f(x)\,dx - \underline{\int_a^b} f(x)\,dx \right] + \left[\underline{\int_a^b} f(x)\,dx - L(P,f) \right] < \varepsilon.$$

Each one of the three numbers

$$U(P,f) - \overline{\int_a^b} f(x)\,dx, \quad \overline{\int_a^b} f(x)\,dx - \underline{\int_a^b} f(x)\,dx, \quad \underline{\int_a^b} f(x)\,dx - L(P,f).$$

being non-negative, it follows that

$$0 < \overline{\int_a^b} f(x)\,dx - \underline{\int_a^b} f(x)\,dx < \varepsilon.$$

As ε is an arbitrary positive number, we see that the non-negative number

$$\overline{\int_a^b} f(x)\,dx - \underline{\int_a^b} f(x)\,dx,$$

is less than every positive number, and hence

$$\overline{\int_a^b} f(x)\,dx - \underline{\int_a^b} f(x)\,dx = 0 \Rightarrow \overline{\int_a^b} f(x)\,dx = \underline{\int_a^b} f(x)\,dx$$

so that f is integrable.

Thus the theorem is established.

Note. This theorem is sometimes stated differently as follows :

A necessary and sufficient condition for a bounded function f to be integrable in [a,b] is that

$$\lim w(P,f) = 0,$$

when $\| P \|$, *the norm of the partition P, tends to 0.*

II. The condition of integrability. Second form.

A necessary and sufficient condition for the integrability of a bounded function f is that to every $\varepsilon > 0$ *there corresponds a partition P such that the corrresponding oscillatory sum* $w(P,f) < \varepsilon$, *i.e.,* $U(P,f) - L(P,f) < \varepsilon$. ***[Rohilkhand 2009; Agra 2010; Kanpur 2018 Delhi B.Sc. (Prog.) II, 2017; Delhi B.A. (Prog.) III, 2017; Chennai 2011; Meerut 2020; Delhi B.Sc. (Hons.) II 2011, 2018)***

Proof. The condition is necessary. Let f be integrable on $[a, b]$

Hence $$\int_{\underline{a}}^{b} f(x)\,dx = \int_{a}^{\overline{b}} f(x)\,dx = \int_{a}^{b} f(x)\,dx \qquad ...(i)$$

Let $\varepsilon > 0$ be given. Since

$$\int_{a}^{\overline{b}} f(x)\,dx = \inf \{ U(P,f) : P \text{ is a partition of } [a, b]\}$$

and $$\int_{\underline{a}}^{b} f(x)\,dx = \sup \{L(P,f) : P \text{ is a partition of } [a, b]\}$$

therefore, there exists partitions P_1 and P_2 of $[a, b]$ such that

$$U(P_1,f) < \int_{a}^{\overline{b}} f(x)\,dx + \frac{1}{2}\varepsilon \qquad ...(2)$$

and $$L(P_2,f) > \int_{\underline{a}}^{b} f(x)\,dx - \frac{1}{2}\varepsilon \qquad ..(3)$$

Let $P = P_1 \cup P_2$ so that P is a common refinement of P_1 and P_2.

$\therefore \quad U(P, f) \le U(P_1, f) \quad$ and $\quad L(P, f) \ge L(P_2, f) \qquad ...(4)$

From (2) and (4), we have

$$U(P, f) < \int_{a}^{\overline{b}} f(x)\,dx + \frac{1}{2}\varepsilon \quad \text{or} \quad U(P,f) < \int_{a}^{b} f(x)\,dx + \frac{1}{2}\varepsilon, \text{ by (1)} \qquad ...(5)$$

Similarly, from (3) and (4), we have

$$L(P, f) > \int_{\underline{a}}^{b} f(x)\,dx - \frac{1}{2}\varepsilon \quad \text{or} \quad L(P, f) > \int_{a}^{b} f(x)\,dx - \frac{1}{2}\varepsilon, \text{ by (1)} \qquad ...(6)$$

Re-writing (6), $$-L(P,f) < -\int_{a}^{b} f(x)dx + \frac{1}{2}\varepsilon \qquad ...(7)$$

Adding the corresponding sides of (5) and (7), we get

$$U(P, f) - L(P, f) \le \varepsilon \quad \text{or} \quad W(P, f) < \varepsilon.$$

The condition is sufficient. Let $\in > 0$ be given and let P be a partition of $[a, b]$ such that

$$w(P, f) < \varepsilon, \quad \text{i.e.,} \quad U(P, f) - L(P, f) < \varepsilon \qquad ...(8)$$

Since $$L(P,f) \le \int_{\underline{a}}^{b} f(x)\,dx \le \int_{a}^{\overline{b}} f(x)\,dx \le U(P,f), \qquad ...(9)$$

we have $$\int_{a}^{\overline{b}} f(x)\,dx \le U(P,f) \quad \text{and} \quad -\int_{\underline{a}}^{b} f(x)\,dx \le -L(P,f)$$

$$\Rightarrow \quad \int_{a}^{\overline{b}} f(x)\,dx - \int_{\underline{a}}^{b} f(x)\,dx \le U(P,f) - L(P,f) < \varepsilon, \quad \text{by (8)}$$

From the above relation which holds for every positive $\in$, we deduce that

$$\overline{\int_a^b} f(x)\,dx - \underline{\int_a^b} f(x)\,dx = 0 \quad \text{or} \quad \overline{\int_a^b} f(x)\,dx = \underline{\int_a^b} f(x)\,dx$$

Hence $f(x)$ is integrable

Note. On comparing the two forms of conditions, the reader will easily see that, from the point of view of necessity, the first form is stronger than the second but, from the point of view of sufficiency, the second form is stronger than the first.

13.8. PARTICULAR CLASSES OF BOUNDED INTEGRABLE FUNCTIONS

Theorem I. *Every continuous function is integrable.* **[Kanpur 2016, 17]**

[Rohilkhand 2011; Rajasthan 2010; Meerut 2006; Agra 2007, 10; 2012; Delhi B.Sc. (Prog.) II 2017; Delhi B.A. (Prog) III 2011, 15; Purvanchal 2007, 08, 09, 15]

Proof. Suppose that f is continuous in $[a,b]$.

Since f is continuous, it is bounded. Also it is uniformly continuouss.

Let ε be any positive number.

We divide $[a, b]$ into a finite number of sub-intervals, say I_r, such that the oscillation of f in each of these sub-intervals is $< \varepsilon/(b-a)$. Let δ_r be the length of I_r.

We have for this partition,

$$\begin{aligned} w(P,f) &= \Sigma\,(M_r - m_r]\,\delta_r \\ &< \Sigma\,[\varepsilon/(b-a)]\,\delta_r \\ &= [\varepsilon/(b-a)]\,\Sigma\,\delta_r = [\varepsilon/(b-a)]\,.\,(b-a) = \varepsilon \end{aligned}$$

Thus, $\qquad w(P,f) < \varepsilon.$

Hence f is integrable in $[a,b]$ [Refer Art.13.7, second form].

Theorem II. *A bounded function f is integrable in [a,b], if the set of its points of discontinuity is finite.* ***(G.N.D.U. Amritsar 2010; Meerut 2016, 2017)***

Proof. Let $\{a_1, a_2, a_3, ..., a_p\}$ be the finite set of points of discontinuity of f in $[a,b]$.

Let ε be any positive number.

We enclose the points $a_1, a_2, ..., a_p$, respectively in p non-overlapping intervals

$$[a_1', a_1''], [a_2', a_2''], ..., [a_p', a_p'']$$

such that the sum of their lengths is $< \varepsilon/2\,(M-m)$; M, m being the bounds of f in $[a,b]$. The oscillation of f in each of these intervals being $\le (M-m)$, their total contribution to the oscillatory sum is

$$< [\varepsilon/2\,(M-m)].\,(M-m) = \varepsilon/2.$$

Now, f is continuous in each of the $(p + 1)$ sub-intervals

$$[a, a_1'], [a_1'', a_2'], ..., [a_p'', b].$$

As in Theorem I, above, each of these $(p + 1)$ sub-intervals can be further sub-divided so that the contribution to the oscillatory sum of the sub-intervals of each of them separately is

$$< \varepsilon/2\,(p+1).$$

Thus there exists a partition of $[a,b]$ such that, the corresponding oscillatory sum is

$$<\frac{\varepsilon}{2}+\frac{\varepsilon}{2(p+1)}(p+1)=\varepsilon$$

Hence the function f is integrable in $[a,b]$ [See Art. 13.7, second form].

Theorem III. *A bounded function f is integrable in [a,b] if the set of its points of discontinuity has a finite number of limit points.* **(Meerut 2020; Patna 2003)**

Let $\{a_1, a_2, a_3,..., a_p\}$ be the finite set of limit points of the set of the points of discontinuity of f in $[a,b]$.

We enclose them in, p, non-overlapping intervals

$$[a_1', a_1''], [a_2', a_2''], ..., [a_p', a_p''],$$

such that the sum of their lengths is $<\varepsilon/2(M-m)$; M, m being the bounds of f in $[a,b]$. The oscillation of f in each of these intervals being $\leq (M-m)$, their total contribution to the oscillatory sum of these intervals is $< \{\varepsilon/2(M-m)\}\times(M-m)=\varepsilon/2$.

Only a *finite* number of points of discontinuity of f can lie in each of the $(p+1)$ intervals

$$[a, a_1'], [a_1'', a_2'], ..., [a_p'', b]$$

so that, as in theorem II each of them can be so sub-divided that the contribution to the oscillatory sum of the sub-intervals of each of these $(p+1)$ intervals is separately $<\varepsilon/2(p+1)$.

Thus there exists a partition of $[a,b]$ such that corresponding oscillatory sum is

$$<\frac{\varepsilon}{2}+\frac{\varepsilon}{2(p+1)}(p+1)=\varepsilon.$$

Hence f is integrable in $[a,b]$. [See Art. 13.7 second form].

Theorem IV. *If f is monotonic in [a,b], then it is integrable in [a,b].*

(Delhi B.Sc. (Hons.) II 2018; Agra 2010; Delhi B.A. (Prog) III, 2011; Garhwal 2001; Osmania 2004; Meerut 2009, 2020; Tirupati 2003; Purvanchal 2008, 12)

Proof. Clearly f is bounded and $f(a), f(b)$ are its two bounds.

Let ε be any positive number.

For the sake of definiteness, we suppose that f is monotonically increasing.

Let $\qquad P=\{a=x_0, x_1, x_2, ..., x_{r-1}, x_r, ..., x_n=b\}$

be any partition of $[a,b]$ such that the length of each sub-interval is $<\varepsilon/[f(b)-f(a)+1]$.

Let $\delta_r = x_r - x_{r-1}$. Here $M_r = f(x_r), m_r = f(x_{r-1})$.

We have

$$w(P,f)=\Sigma(M_r-m_r)\delta_r$$

$$=\Sigma[f(x_r)-f(x_{r-1})]\delta_r$$

$$<\frac{\varepsilon}{f(b)-f(a)+1}\Sigma[f(x_r)-f(x_{r-1})]$$

$$=\frac{\varepsilon}{f(b)-f(a)+1}[f(b)-f(a)]<\varepsilon$$

Hence f is integrable in $[a,b]$.

Note : 1. If we had taken $[f(b)-f(a)]$ instead of $[f(b)-f(a)+1]$. that proof would not have been valid for the case when $f(b)-f(a) = 0$, *i.e.*, when f is a constant function. The artifice of taking, $[f(b)-f(a)+k]$ where k is positive, or in particular $f(b)-f(a)+1$, serves to make the proof applicable even in this case.

Note : 2. The above theorem I, II, III and IV give sufficient conditions for the Riemann integral to exist. Once we know that an integral exists, we may calculate it by a specific method.

Note : 3. The conditions given in theorem I, II, III and IV are not necessary for the Riemann integral to exist. Thus there exist functions which are integrable but do not satisfy conditions stated in the above mentioned theorems. This will be shown in the set of solved examples, which follow.

SOLVED EXAMPLES

Example 1 *Give an example of a Riemann integrable function on* $[a, b]$ *which is not nonotonic* $[a, b]$. [*Delhi Maths (H) 2005, 09*]

Solution. Let $f(x) = |x - 1/2|\ \forall\ x \in [0, 1]$

Then $f(x)$ in monotonic decreasing in the interval $[0, 1/2]$ and monotonic increasing in $[1/2, 1]$. Hence f in not monotonic in $[0, 1]$. Since $f(x)$ is continuous on $[0, 1]$, so by theorem I, Art 13.8, $f(x)$ in Riemann integrable on $[0, 1]$. Thus f is integrable on $[0, 1]$ although it is not monotonic on $[0, 1]$.

Note : This example shows that the condition of theorem IV of Art 13.8 is not necessarv for Riamann integrability.

Example 2 (*a*). Show that the function f defined as follows :

$$f(x) = 1/2^n \quad \text{when} \quad 1/2^{n+1} < x \le 1/2^n : (n = 0, 1, 2, 3, \ldots),\ f(0) = 0$$

is integrable in $[0, 1]$, although it has an infinite number of points of discontinuity.

[Garhwal 1996 ; Kumaun 1999]

Solution. We have, as given

$$f(0) = 0$$

$$f(x) = 1 \text{ when } 1/2 < x \le 1,$$

$$f(x) = \frac{1}{2}, \text{ when } \left(\frac{1}{2}\right)^2 < x \le \frac{1}{2},$$

$$\ldots\ldots\ldots\ldots\ldots\ldots\ldots\ldots\ldots\ldots\ldots\ldots\ldots\ldots$$

$$f(x) = \left(\frac{1}{2}\right)^{n-1}, \text{ when } \left(\frac{1}{2}\right)^n < x \le \left(\frac{1}{2}\right)^{n-1}$$

$$\ldots\ldots\ldots\ldots\ldots\ldots\ldots\ldots\ldots\ldots\ldots\ldots\ldots\ldots$$

Since f is bounded and monotonically increasing in $[0, 1]$, it is integrable [Refer theorem IV of Art. 13.8]

Alternative solution. For n = 1, 2, 3, ..., we have

$$f\left(\frac{1}{2^n}+0\right) = \frac{1}{2^{n-1}} \quad \text{and} \quad f\left(\frac{1}{2^n}-0\right) = \frac{1}{2^n},$$

showing that f is discontinuous at $x = 1/2^n$, $n = 1, 2, 3 \ldots$,

Again, $f(1-0) = 1 = f(1) \Rightarrow f(x)$ is continuous at $x = 1$.

Clearly, $f(x)$ is not continuous at $x = 0$

Thus, we notice that f is bounded and continuous in $[0, 1]$ except at the set of points

$$0, 1/2, 1/2^2, 1/2^3, \ldots, 1/2^n, \ldots$$

which has only one limit point, namely 0, and hence f is integrable (see theorem III of Art. 13.8).

Note. The above alternative solution shows that a function can be integrable on $[a, b]$ without being continuous on $[a, b]$. This shows that the condition given in theorem I of Art. 13.8 is only sufficient and not necessary.

Example. 2 (*b*) *Give an example of a Riemann integrable function having infinite number of points of discontinuity. Also find the value of the integral.* (*Delhi Meths (H) 2005, 09*)

Solution. Refer Ex. 2(*a*) for complete solution.

Example. 3 A function f is defined in $[0, 1]$ as follows :

$f(x) = 1/q$ *when x is any non-zero rational number p/q in its lowest terms, and*

$f(x) = 0$, *when x is irrational or* 0.

Show that f is integrable in $[0,1]$ and the value of the integral is 0. **(Lucknow 2010)**

Solution. Let, ε, be any positive number.

There exists only a finite number of integers q such that

$$1/q > \varepsilon/2 \quad \Leftrightarrow \quad q < 2/\varepsilon.$$

We call points p/q in $[0, 1]$ for which $1/q > \varepsilon/2$ exceptional points. Surely they are *finite* in number. Also we know that every interval encloses rational as well as irrational points.

Clearly the oscillation of the funtion f in an interval which includes no exceptional point is $< \varepsilon/2$ and that in an interval which includes an exceptional point is at the most 1.

We now enclose the exceptional points of $[0, 1]$ which are finite in number in intervals the sum of whose lengths is $< \varepsilon/2$ so that the part of the oscillatory sum arising from these is $\varepsilon/2$. Also howsoever, we may divide the remaining part of $[0, 1]$, the part of the oscillatory sum arising from the same will be $< \varepsilon/2$.

Thus we have a partition of $[0, 1]$ such that the corresponding oscillatory sum is less than a given pre-assigned positive number ε. Hence, by second form of Art. 13.7, the function is integrable,

Also since for every partition P the lower Darboux sum $L(P, f) = 0$, we have

$$\underline{\int_0^1} f(x)\,dx = \int_0^1 f(x)\,dx = 0.$$

Note. It can be shown that the function f is discontinuous for every non-zero rational value of x so that the set of the limit points of the set of points of discontinuity of f is the entire interval $[0, 1]$. This shows that the condition obtained for integrability in Theorem III of Art. 13.8 is only sufficient and not necessary.

Example 4. *If the function f be defined on [a, b] as follows :*

$$f(x) = 1/q^2, \text{ when } x = p/q.$$
$$= 1/q^3, \text{ when } \quad x = \sqrt{(p/q)},$$

where p and q are relatively prime integers and f (x) = 0 for all other values of x, then show that the function f is Riemann-integrable on [a, b]. [*I.A.S. 1996, 2001*]

Sol. We can easily show that the function f is discontinuous for rational values and continuous for irrational values.

Also, we are given that if p/q is any rational number in $[a, b]$, then

$$f(p/q) = 1/q^2$$

We know that in every interval, however small it may be, there are an infinite number of irrational numbers, so every interval around p/q contains irrational number x, for which

$$|f(x)-f(p/q)| = \left|0-\frac{1}{q^2}\right| = \frac{1}{q^2}$$

Hence function f is discontinuous for all rational points in $[a, b]$.

Again let y be an irrational points in $[a, b]$, then we have $f(y) = 0$ or $1/q^2$ according as

$$y \neq \sqrt{(p/q)} \quad \text{or} \quad y = \sqrt{(p/q)}.$$

∴ If ε be any given positive number, then there will be only a finite number of fractions $1/q^2$ for which $1/q^2 > \varepsilon$ or $1/q > \sqrt{\varepsilon}$ or $q < 1/\sqrt{\varepsilon}$. Thus an interval around y can be found which does not contain any rational number p/q for which $q < 1/\sqrt{\varepsilon}$.

∴ For any irrational point x in this interval, we get

$$|f(x)-f(y)| = 0 \text{ or } 1/q^3$$

$$\Rightarrow \qquad |f(x)-f(y)| < \varepsilon.$$

Similarly, for any rational point $x = p/q$ in the same interval.

$$|f(x)-f(y)| = \left(\frac{1}{q^2}\right)-\left(\frac{1}{q^3}\right) \text{ or } \frac{1}{q^2}.$$

Thus in each case $f(x)-f(y) < \varepsilon$, hence $f(x)$ is continuous at $x = y$, where y is any rational point in $[a, b]$.

Hence all the irrational points in $[a, b]$ are points of continuity and all the rational points in $[a, b]$ are points of discontinuity of the function f and so all points of discontinuity being rational points in $[a, b]$ form a set of measure zero.

Hence the function f is R-integrable in $[a, b]$.

13.9 PROPERTIES OF INTEGRABLE FUNCTIONS

I. *If a bounded function f is integrable in [a, b], then it is also integrable in [a, c] and [c, b], where c is a point of [a, b].*

Conversely, if f is bounded and integrable in [a, c], [c, b], then it is also integrable is [a, b].

Also $$\int_a^b f(x)\,dx = \int_a^c f(x)\,dx + \int_c^b f(x)\,dx,\ a \le c \le b.$$

(*Garhwal 1999, Kumaun 1998, Meerut 2000*)

Proof. Suppose that f is bounded and integrable in $[a, b]$.

Let a be any positive number.

There exists $\delta > 0$ such that for each partition of $[a, b]$ whose norm is $\le \delta$, the oscillatory sum is $< \varepsilon$. Let P be a partition of $[a, b]$ such that $c \in P$ and $\|P\| \le \delta$.

The oscillatory sum for the partition P breaks itself into two parts, respectively consisting of terms arising from the sub-intervals of $[a, c]$ and $[c, b]$. Since the terms of an oscillatory sum are all positive each part must itself be $< \varepsilon$. Hence f is integrable both in $[a,c]$ and $[c, b]$.

Let now f be bounded and integrable in $[a, c]$ and $[c, b]$.

Let ε be any positive number. There exist partitions of $[a, c]$ and $[c, b]$ such that each of the corresponding oscillatory sums is $< \varepsilon/2$. The partitions of $[a, c]$ and $[c, b]$ give rise to a partition of $[a, b]$ for which the oscillatory sum is $< (\varepsilon/2 + \varepsilon/2) = \varepsilon$. Hence f is integrable in $[a, b]$.

As f is simultaneously integrable in $[a, c]$, $[c, b]$ and $[a, b]$ there exists $\delta > 0$ such that for partitions of norm $\leq \delta$, and of which c is a point, we have

$$\left| \sum_{[a,c]} f(\xi_r)\, \delta_r - \int_a^c f(x)\, dx \right| < \frac{\varepsilon}{3}, \quad \left| \sum_{[c,b]} f(\xi_r)\, \delta_r - \int_c^b f(x)\, dx \right| < \frac{\varepsilon}{3},$$

and
$$\left| \sum_{[a,b]} f(\xi_r)\, \delta_r - \int_a^b f(x)\, dx \right| < \frac{\varepsilon}{3},$$

where the meanings of the symbols $\sum_{[a,c]} f(\xi_r)\, \delta_r$ etc., are obvious.

Since
$$\sum_{[a,c]} f(\xi_r)\, \delta_r + \sum_{[c,b]} f(\xi_r)\, \delta_r = \sum_{[a,b]} f(\xi_r)\, \delta_r$$

we deduce that

$$\left| \int_a^b f(x)\, dx - \int_a^c f(x)\, dx - \int_c^b f(x)\, dx \right| < \varepsilon$$

$$\Rightarrow \quad \int_a^b f(x)\, dx - \int_a^c f(x)\, dx - \int_c^b f(x)\, dx = 0,$$

ε being an arbitrary positive number. Thus, $\int_a^b f(x)\, dx = \int_a^c f(x) dx + \int_c^b f(x) dx$

Cor. *If f is bounded and integrale in $[a, b]$, it is also bounded and integrable in $[\alpha, \beta]$ where $a \leq \alpha < \beta \leq b$.*

Proof : Now f is integrable in $[a, b]$ $\Rightarrow$ f is integrable in $[a, \beta]$

$\Rightarrow$ f is integrable in $[\alpha, \beta]$.

13.10 INTEGRABILITY OF THE SUM, DIFFERENCE, PRODUCT AND QUOTIENT OF INTEGRABLE FUNCTIONS

Before, taking up the main question, we state and prove a simple lemma.

Lemma. *The oscillation of a bounded function f in a interval $[a, b]$ is the supremum of the set*

$$\{|f(\alpha) - f(\beta)| : \alpha, \beta \in [a, b]\}$$

of numbers.

Proof : Let m, M be the bounds of f in $[a, b]$, Now

$$m \leq f(a), f(\beta) \leq M; \forall\, \alpha, \beta \in [a, b]$$

$$\Rightarrow \quad |f(\alpha) - f(\beta)| \leq M - m, \forall\, \alpha, \beta \in [a, b] \qquad \text{... (1)}$$

$\Rightarrow$ $M - m$ is an upper bound of the set in question.

Let $\varepsilon > 0$ be given. Now

M is the supremum of f $\Rightarrow$ there exists $\alpha_1 \in [a, b]$ such that

$$f(\alpha_1) > M - \varepsilon/2. \qquad \text{... (2)}$$

m is the infimum of $f \Rightarrow$ there exists $\beta_1 \in [a, b]$ such that

$$f(\beta_1) < m + \varepsilon/2. \qquad \text{... (3)}$$

Now (2) and (3), imply that there exist $\alpha_1, \beta_1 \in [a, b]$ such that

$$f(\alpha_1) - f(\beta_1) > M - m - \varepsilon \;\; \Rightarrow \;\; |f(\alpha_1) - f(\beta_1)| > M - m - \varepsilon.$$

There exists, therefore a pair of numbers $\alpha_1, \beta_1 \in [a, b]$ such that

$$|f(\alpha_1) - f(\beta_1)| > M - m - \varepsilon \quad \text{... (4)}$$

where $\varepsilon >$ is arbitrary, so that no number less than $M - m$ is an upper bound of the set in question.

It follows from (1) and (4) that

$$M - m = \sup\{|f(\alpha) - f(\beta)|; \alpha, \beta \in [a, b]\}.$$

I. Integrability of sum and difference. If f and g are two functions both bounded and integrable in $[a, b]$, then $f \pm g$ are also bounded and integrable in $[a, b]$ and

$$\int_a^b [f(x) \pm g(x)[dx = \int_a^b f(x)\,dx \pm \int_a^b g(x)\,dx.$$

[Purvanchal 2006; Kanpur 2005; Pune 2010; Himanchal 2009]

Proof : Let $P = \{a = x_0, x_1, ..., x_{-1}, ..., x_n = b\}$ be any partition of $[a, b]$.

Let $M'_r, m'_r; M''_r, m''_r; M_r, m_r$ be the bounds of f, g and $f + g$ in $I_r = [x_{r-1}, x_r]$. If α_1, α_2 be any two points of I_r. Then, we have

$$|[f(\alpha_2) + g(\alpha_2)] - [f(\alpha_1) + g(\alpha_1)]|$$

$$\leq |f(\alpha_2) - f(\alpha_1)| + |g(\alpha_2) - g(\alpha_1)|$$

$$\leq (M'_r - m'_r) + (M''_r - m'_r).$$

$$\Rightarrow \quad M_r - m_r \leq (M'_r - m'_r) + (M''_r - m''_r) \quad \text{... (1)}$$

Let $\varepsilon > 0$ be any given number.

Since f, g are integrable, there exists $\delta > 0$ such that for every partition of norm $\leq \delta$, the corresponding oscillatory sums for f and g are both less than $\varepsilon/2$.

Now for any partition with norm $\leq \delta$, we have, as implied by (1).

$$\Sigma (M_r - m_r)\,\delta_r \leq \Sigma (M'_r - m'_r)\,\delta_r + \Sigma (M''_r - m''_r)\,\delta_r$$

$$< \varepsilon/2 + \varepsilon/2 = \varepsilon$$

Thus $f + g$ is integrable in $[a, b]$.

Now we prove that

$$\int_a^b [f(x) + g(x)]\,dx = \int_a^b f(x)\,dx + \int_a^b g(x)\,dx.$$

Let ε be any positive number.

Since f, g are integrable, there exists $\delta > 0$ such that for every partition of norm $\leq \delta$ and for every $\xi_r \in I_r$

$$\left|\Sigma f(\xi_r)\,\delta_r - \int_a^b f(x)\,dx\right| < \frac{\varepsilon}{2} \quad \text{and} \quad \left|\Sigma g(\xi_r)\,\delta_r - \int_a^b g(x)\,dx\right| < \frac{\varepsilon}{2}$$

$$\Rightarrow \quad \left|\Sigma\left[f(\xi_r) + g(\xi_r)\right]\delta_r - \left[\int_a^b f(x)\,dx + \int_a^b g(x)\,dx\right]\right| < \varepsilon.$$

Thus we obtain

$$\int_a^b [f(x) + g(x)]\,dx = \int_a^b f(x)\,dx + \int_a^b g(x)\,dx, \text{ refer Art. 13.5}$$

The case of difference may be similarly discussed.

II. Integrability of product. *If f, g are two functions, both bounded and integrable in [a, b], then their product fg is also bounded and integrable in [a, b].* **[Rajasthan 2010; Bundelkhand 2011; Delhi Maths (H) 2000, 2001, 2005, Meerut 2003, 06, 07, 11; Purvanchal 2010]**

Show by means of an example that a product of two nonintegrable functions may be integrable. **[Delhi Maths (H) 2005]**

Proof. Since f and g are bounded, there exists k, such that $\forall\, x \in [a, b]$,

$$|f(x)| \le k, |g(x)| \le k$$

$$\Rightarrow \quad |f(x)\, g(x)| \le k^2,\ \forall\, x \in [a, b]$$

$\Rightarrow$ fg is bounded.

Let $P = \{a = x_0, x_1, x_2, \ldots x_{r-1}, x_r \ldots x_n = b\}$ be any partition of $[a, b]$.

Let $M'_r, m'_r; M''_r, m''_r; M_r, m_r$ be the bounds of f, g and fg respectively in $I_r = [x_{r-1}, x_r]$.

We have, $\forall\, \alpha_1, \alpha_2 \in I_r$

$$f(\alpha_2)\, g(\alpha_2) - f(\alpha_1)\, g(\alpha_1) = g(\alpha_2)[f(\alpha_2) - f(\alpha_1)] + f(\alpha_1)[g(\alpha_2) - g(\alpha_1)]$$

$$\Rightarrow \quad |f(\alpha_2)\, g(\alpha_2) - f(\alpha_1)\, g(\alpha_1)| \le |g(\alpha_2)||f(\alpha_2) - f(\alpha_1)| + |f(\alpha_1)||g(\alpha_2) - g(\alpha_1)|$$

$$\le k\,(M'_r - m'_r) + k\,(M''_r - m''_r)$$

$$\Rightarrow \quad (M_r - m_r) \le k\,(M'_r - m'_r) + k\,(M''_r - m''_r) \qquad \ldots (1)$$

Now let ε be any positive number.

Since f, g are integrable, there exists $\delta > 0$ such that for every partition of norm $\le \delta$, the corresponding oscillatory sums for f and g are both $< \varepsilon / 2k$.

Now for every partition P of norm $\le \delta$, we have,

$$\Sigma\,(M_r - m_r) < k\,\Sigma\,(M'_r - m'_r)\delta_r + k\,\Sigma(M''_r - m'_r)\,\delta_r$$

$$< k\,(\varepsilon/2k) + k(\varepsilon/2k) = \varepsilon,$$

implying that fg is integrable in $[a, b]$.

We now give an example to show that even though f and g are not integrable on $[a, b]$, fg may be integrable on $[a, b]$

Consider f : $[a, b] \to R$ and g : $[a, b] \to R$ defined by

$$f(x) = \begin{cases} 0, & x \in Q \\ 1, & x \in R - Q \end{cases} \quad \text{and} \quad g(x) = \begin{cases} 1, & x \in Q \\ 0, & x \in R - Q \end{cases}$$

where R and Q are sets of all real and rational numbers respectively.

Then f and g are not integrable on $[a, b]$ (prove it) but $(fg)\,(x) = f(x)\, g(x) = 0\ \forall\, x \in [0, 1]$. Since fg is a constant function, it is integrable.

III. Integrability of Quotient. *If f, g are two functions both bounded and integrable in [a, b] and there exists a number $t > 0$ such that $|g(x)| \ge t\ \forall\, x \in [a, b]$, then f/g is bounded and integrable in [a, b].*

Proof. Now there exist positive numbers k and t such that

$$|f(x)| \le k,\ |g(x)| \le k, |g(x) \ge t,\ \forall\, x \in [a, b]$$

$\Rightarrow \qquad |f(x)/g(x)| \le k/t \ \forall x \in [a,b] \Rightarrow f/g$ is bounded.

Let $\qquad P = \{a = x_0, x_1, ..., x_{r-1}, x_r, ..., x_n = b\}$

by any partition of [*a, b*] and let $M'_r, m'_r; M''_r, m''_r; M_r, m_r$ be the bounds of *f*, *g*, *f/g* in $I_r = [x_{r-1}, x_r]$.

Now, $\forall \alpha_1, \alpha_2 \in I_r$, we have

$$\left|\frac{f(\alpha_2)}{g(\alpha_2)} - \frac{f(\alpha_1)}{g(\alpha_1)}\right| = \left|\frac{g(\alpha_1)[f(\alpha_2) - f(\alpha_1)] - f(\alpha_1)[g(\alpha_2) - g(\alpha_1)]}{g(\alpha_1)\, g(\alpha_2)}\right|$$

$$\le (k/t^2)\,|f(\alpha_2) - f(\alpha_1)| + (k/t^2)\,|g(\alpha_2) - g(\alpha_1)|$$

$$\le (k/t^2)(M'_r - m'_r) + (k/t^2)(M''_r - m''_r)$$

$$\Rightarrow \qquad (M_r - m_r) \le (k/t^2)(M'_r - m'_r) + (k/t^2)(M''_r - m''_r) \qquad ...(1)$$

Let, now ε be any positive number.

Now *f*, *g* are integrable, implies that there exists a number $\delta > 0$ such that for every partition *P* of norm $\le \delta$, the corresponding oscillatory sums for *f*, *g* are both less than $t^2\varepsilon/2k$.

Now for any partition *P* of norm $\le \delta$, we have, from (1)

$$\Sigma(M_r - m_r)\,\delta_r \le (k/t^2)\,\Sigma(M'_r - m'_r)\,\delta_r + (k/t^2)\,\Sigma(M''_r - m''_r)\,\delta_r$$

$$\le (k/t^2)(t^2\varepsilon/2k) + (k/t^2)(t^2\varepsilon/2k) = \varepsilon,$$

implying that *f/g* is bounded and integrable in [*a, b*].

13.11 INTEGRABILITY OF THE MODULUS OF A BOUNDED INTEGRABLE FUNCTION

If f is bounded and integrable in [a, b], then |f| is also bounded and integrable in [a,b] and

$$\left|\int_a^b f(x)\,dx\right| \le \int_a^b |f(x)|\,dx$$

(Chennai 2011; Agra 2000, 10; Calicut 2004; Delhi B.Sc. (Hons.) II 2018, 2019; Meerut 2020; Himanchal 2009; Kanpur 2014; Patna 2003; Purvanchal 2006, 08)

Proof. For a bounded function *f*(*x*), there exists a positive number *k* such that

$$|f(x)| \le k, \; x \in [a,b],$$

$\Rightarrow$ the function $|f|$ is bounded.

Let ε be any positive number.

Now *f* is integrable implies that there exists a partition $P = \{a = x_0, x_1, x_2, ..., x_{r-1}, x_r, ..., x_n = b\}$, such that the corresponding oscillatory sum for *f* is $< \varepsilon$.

Let $M'_r, m'_r; M_r, m_r$ be the bounds of *f* and $|f|$ in $I_r = [x_{r-1}, x_r]$

Now, $\forall\ \alpha_1, \alpha_2 \in I_r$, we have

$$|[|f(\alpha_2)| - |f(\alpha_1)|]| \le |f(\alpha_2) - f(\alpha_1)| \le M'_r - m'_r$$

$$\Rightarrow \quad M_r - m_r \le M'_r - m'_r.$$

$$\Rightarrow \quad \Sigma (M_r - m_r)\,\delta_r \le (M'_r - m'_r)\,\delta_r < \varepsilon.$$

Hence $|f|$ is integrable in $[a, b]$.

Since $|f| = \max\{f, -f\}$

$\therefore \; f(x) \le |f(x)| = |f|(x)$ and $-f(x) \le |f(x)| = |f|(x),\ \forall\, x \in [a, b]$

$$\Rightarrow \quad \int_a^b f(x)\,dx \le \int_a^b |f|(x)\,dx \text{ and } \int_a^b \{-f(x)\}\,dx \le \int_a^b |f|(x)\,dx$$

$$\Rightarrow \quad \int_a^b f(x)\,dx \le \int_a^b |f|(x)\,dx \text{ and } -\int_a^b f(x)\,dx \le \int_a^b |f|(x)\,dx$$

$$\Rightarrow \quad \left|\int_a^b f(x)\,dx\right| \le \int_a^b |f|(x)\,dx \qquad \Rightarrow \quad \left|\int_a^b f(x)\,dx\right| \le \int_a^b |f(x)|\,dx$$

Remarks. The converse of this theorem is not always true, that is, if $|f|$ is integrable, then f may not the integrable. **[Delhi B.Sc. (Hons.) II, 2017, 2019; Himachal 2009]**

To this end, consider the following function : $f(x) = \begin{cases} 1, & \text{when } x \text{ is rational} \\ -1, & \text{when } x \text{ is irrational.} \end{cases}$

Let $P = \{a = x_0, x_1, x_2, ..., x_n = b\}$ be any partition of $[a, b]$

Let m_r and M_r be the infimum and supremum of f on $I_r = [x_{r-1}, x_r]$,

then $\qquad m_r = -1 \quad$ and $\quad M_r = 1, \quad r = 1, 2, 3, ...n$

$$\therefore \quad L(P, f) = \sum_{r=1}^{n} m_r \delta_r = \sum_{r=1}^{n} (-1)\cdot\delta_r = -(b-a) = a - b$$

and
$$U(P, f) = \sum_{r=1}^{n} M_r \delta_r = \sum_{r=1}^{n} (1)\cdot\delta_r = b - a$$

Now, $\displaystyle \underline{\int_a^b} f(x)\,dx$ = sub $\{L(P, f)$: P is a partion of $[a, b]\}$ = sub$\{(a - b)\} = (a - b)$

and $\displaystyle \overline{\int_a^b} f(x)\,dx$ = inf $\{U(P, f)$: P is a partition of $[a, b]\}$ = inf $\{(b - a)\} = b - a$

Since $\displaystyle \underline{\int_a^b} f(x)\,dx \ne \overline{\int_a^b} f(x)\,dx$, $\quad R$ is not Riemann integrable

But $|f|(x) = |f(x)| = 1,\ \forall\, x \in [a, b]$. Since $|f|$ is a constant function, $|f|$ is integrable.

INTEGRABILITY OF THE SQUARE OF A BOUNDED INTEGRABLE FUNCTION.

Theorem. *If f is integrable on $[a, b]$, then f^2 is also integrable.*

Proof. Since f is bounded on $[a, b]$, so there exists a positive number k such that

$$|f(x)| \le k,\ \forall\, x \in [a, b] \qquad ...(1)$$

Now, $\quad f$ is integrable on $[a, b] \quad \Rightarrow \quad |f|$ is integrable on $[a, b]$.

Since $|f|$ is integrable on $[a, b]$, so for a given $\varepsilon > 0$ there exists a partition P of $[a, b]$ such that
$$U(P, |f|) - L(P, |f|) < \in /2k \qquad ...(2)$$

Now, $\quad |f^2(x)| = |f(x)|^2 < k^2 \Rightarrow f^2$ is bounded on $[a, b]$

Let M_r and m_r be the respectively the supremum and infimum of $|f|$ in rth sub-interval $[x_{r-1}, x_r]$ of partition of $[a, b]$. Then, clearly, M_r^2, m_r^2 will be respectively the supremum and infimum of f^2 in $[x_{r-1}, x_r]$, as $f^2(x) = |f(x)|^2$. Let δ_r be the length of the rth sub-interval $[x_{r-1}, x_r]$.

$$\therefore \quad U(P, f^2) - L(P, f^2) = \sum_{r=1}^{n} (M_r^2 - m_r^2)\delta_r$$

$$= \sum_{r=1}^{n} (M_r - m_r)(M_r + m_r)\delta_r$$

$$\leq 2k \sum_{r=1}^{n} (M_r - m_r)\delta r_r, \text{ using (1)}$$

$$= 2k\{U(P, |f|) - L(P, |f|)\}$$

$$< 2k \times (\varepsilon / 2k) = \varepsilon \text{, using (2)}$$

Hence f^2 in Riemann integrable

EXERCISES

1. A function f is bounded in $[a, b]$, show that

(*i*) $\overline{\int_a^b} kf(x)\,dx = k \overline{\int_a^b} f(x)\,dx$, $\underline{\int_a^b} kf(x)\,dx = k \underline{\int_a^b} f(x)\,dx$, where k is a positive constant.

(*ii*) $\overline{\int_a^b} kf(x)\,dx = k \underline{\int_a^b} f(x)\,dx$, $\underline{\int_a^b} kf(x)\,dx = k \overline{\int_a^b} (fx)\,dx$, where k is a negative constant.

Deduce that if f is bounded and integrable over $[a, b]$, then so is kf, where k is any constant

and
$$\int_a^b k\, f(x)\,dx = k \int_a^b f(x)\,dx$$

Hint. If M_r, m_r be the bounds of f in I_r then $k\,M_r$, $k\,m_r$, $(k\,m_r,\ k\,M_r)$ are the bounds of kf in I_r where k is positive, (k is negative).

2. If f and g are bounded integrable functions on $[a, b]$ and p and q be any constants. Then prove that $p\,f + qg$ is integrable on $[a, b]$ and

$$\int_a^b (pf + qg) = p \int_a^b f + q \int_c^b g \qquad \textit{(Delhi Maths (H) 1995)}$$

3. Does the integrability of $|f|$ implier that of f. Justify your answer. ***(Himanchal 2009***
[**Hint**. For solution, refer remarks on page 13.34.]

13.12 DEFINITION OF $\int_a^b f(x)\,dx$, if $b \leq a$

If f be bounded and integrable in $[b, a]$ where $a > b$, then by definition, we have

$$\int_a^b f(x)\,dx = -\int_b^a f(x)\,dx$$

Also, by definition, $$\int_a^a f(x)\,dx = 0$$

It is easy to show that the results about integrals obtained in Art. 13.10 hold true when the upper limit is less than or equal to the lower limit.

Note. The reader may carefully note that the statement $\int_a^b f(x)\,dx$ exists, is equivalent to the statement that f is bounded and integrable in $[a, b]$.

13.13 INEQUALITIES FOR AN INTEGRAL

Theorem I. *If f is bounded in [a, b] and M, m are the infimum and supremum of f in [a, b], then*

$$m(b-a) \le \underline{\int_a^b} f(x)\,dx \le \overline{\int_a^b} f(x)\,dx \le M(b-a)$$

Proof. For every partition P of $[a, b]$, we have

$$m(b-a) \le L(P,f) \le U(P,f) \le M(b-a) \qquad \text{...(1)}$$

Now, $\sup\{L(P, f) : \text{P is a partition of } [a, b]\} = \underline{\int_a^b} f(x)\,dx$

$$\Rightarrow \quad L(P, f) \le \underline{\int_a^b} f(x)\,dx \qquad \text{...(2)}$$

and $\inf\{U(P, f) : P \text{ is a partition of } [a, b]\} = \overline{\int_a^b} f(x)\,dx$

$$\Rightarrow \quad \overline{\int_a^b} f(x)\,dx \le U(P, f) \qquad \text{...(3)}$$

Again, we know that $\underline{\int_a^b} f(x)\,dx \le \overline{\int_a^b} f(x)\,dx$ ...(4)

Then, from (1), (2), (3) and (4), we obtain

$$m(b-a) \le L(P,f) \le \underline{\int_a^b} f(x)\,dx \le \overline{\int_a^b} f(x)\,dx \le U(P,f) \le M(b-a)$$

$$\Rightarrow \quad m(b-a) \le \underline{\int_a^b} f(x)\,dx \le \overline{\int_a^b} f(x)\,dx \le M(b-a)$$

Theorem II. *If f bounded and integrable in [a, b] and M, m are the bounds of f in [a, b], then*

$$m(b-a) \le \int_a^b f(x)\,dx \le M(b-a), \text{ if } b \ge a.$$

and $m(b-a) \ge \int_a^b f(x)\,dx \ge M(b-a)$, if $b \le a$.

(Purvanchal 2008; Delhi Maths (H) 2000, 01)

Proof. For $a = b$, the result is trivial.
Since f is integrable on $[a, b]$, we have

$$\underline{\int_a^b} f(x)\,dx = \overline{\int_a^b} f(x)\,dx = \int_a^b f(x)\,dx \qquad \text{...(1)}$$

If $b > a$, then for any partition P of $[a, b]$, we have

$$m(b-a) \le L(P,f) \le U(P,f) \le M(b-a) \qquad \text{...(2)}$$

Now, $\sup\{L(P, f) : P \text{ is a partition of } [a, b]\} = \underline{\int_a^b} f(x)\,dx$

$$\Rightarrow \quad L(P,f) \le \underline{\int_a^b} f(x)\,dx \quad \Rightarrow \quad L(P,f) \le \int_a^b f(x)\,dx, \text{ by (1)} \qquad \text{...(3)}$$

and $\inf\{U(P,f) : \text{P is a partion of } [a, b]\} = \overline{\int_a^b} f(x)\,dx$

$$\Rightarrow \quad \overline{\int_a^b} f(x)\,dx \le U(P,f) \quad \Rightarrow \quad \int_a^b f(x)\,dx \le U(P,f), \text{ by (1)} \qquad \text{...(4)}$$

Then from (2), (3) and (4), we have

$$m(b-a) \le L(P,f) \le \int_a^b f(x)\,dx \le U(P,f) \le M(b-a)$$

$$\Rightarrow \quad m(b-a) \le \int_a^b f(x)\,dx \le M(b-a) \qquad \text{...(5)}$$

If $b < a$ so that $a > b$, then, as proved above is (5), we have

$$m(a-b) \le \int_b^a f(x)\,dx \le M(a-b)$$

$$\Rightarrow \quad -m(b-a) \le -\int_a^b f(x)\,dx \le -M(b-a)$$

$$\Rightarrow \quad m(b-a) \ge \int_a^b f(x)\,dx \ge M(b-a)$$

Theorem III (First mean value theorem) **(Agra 2011; Meerut 2009, 10)**
If f is bounded and integrable in $[a, b]$, then there exists a number, μ, lying between the bounds of f such that $\int_a^b f(x)\,dx = \mu(b-a)$ **(Purvanchal 2011, 12**

(Kanpur 2005; Meerut 2006; Agra 2001, 06, 08, 10; Patna 2003, Delhi Maths (H) 2000)

Proof. Let M, m be the bounds of f in $[a, b]$ so that

$$m \le \mu \le M$$

$$\Rightarrow \quad m(b-a) \le \mu(b-a) \le M(b-a), \text{ if } b > a \qquad \text{...(1)}$$

But we know that $\quad m(b-a) \le \int_a^b f(x)\,dx \le M(b-a) \qquad$ ...(2)

From (1) and (2), we have $\int_a^b f(x)\,dx = \mu(b-a)$, where in $m \le \mu \le M$

Corollary. *If f is continuous in [a, b], then there exists a number c, lying between a and b such that* $\int_a^b f(x)\,dx = (b-a)f(c)$ *(Pune 2010; Delhi Maths (H) 2000, Garhwal 1998)*

Proof. Since f is continuous, f is Riemann integrable. Hence by Theoram III, there exists a number μ lying between the bounds m, M of f in $[a, b]$, such that

$$\int_a^b f(x)\,dx = \mu(b-a)$$

Since f is continuous, it will take every value between m and M. In particular, f will take its value μ where $m < \mu < M$. Therefore, there exists a point c, lying between a and b such that $f(c) = \mu$. Then (2) reduces to

$$\int_a^b f(x)\,dx = (b-a)\,f(c)$$

Theorem IV. *If f is bounded and integrable in $[a, b]$, and k is a number such that $x \in [a,b]$, $|f(x)| \le k$, then*

$$\left|\int_a^b f(x)\,dx\right| \le k\,|b-a|.$$ *(Delhi Maths (H) 2000, Garhwai 1997)*

Proof. For $a = b$, the result in trivial.

Let M, m be the bounds of f in $[a, b]$. Let $b > a$.

We have, $\forall\, x \in [a,b]$, $|f(x)| \le k$

$$\Rightarrow \quad -k \le f(x) \le k \quad \Rightarrow \quad -k \le m \le f(x) \le M \le k \qquad \ldots (i)$$

$$\Rightarrow \quad -k(b-a) \le m(b-a) \le \int_a^b f(x)\,dx \le M(b-a) \le k(b-a)$$

$$\Rightarrow \quad \left|\int_a^b f(x)\,dx\right| \le k\,|b-a|.$$

If $b < a$ so that $a > b$, we have

$$\left|\int_b^a f(x)\,dx\right| \le k\,|a-b| \quad \Rightarrow \quad \left|\int_a^b f(x)\,dx\right| \le k\,|b-a|.$$

Theorem V. *If f is bounded and integrable in $[a, b]$ and $f(x) \ge 0$, $\forall\, x \in [a,b]$, then*

$$\int_a^b f(x)\,dx = \begin{cases} \ge 0, & \text{when } b \ge a; \\ \le 0, & \text{when } b \le a. \end{cases}$$

Proof. Since $f(x) \ge 0$, $\forall\, x \in [a, b]$ it follows that $m = \inf f \ge 0$. By theorem II, we have

$$m(b-a) \le \int_a^b f(x)\,dx$$

$$\Rightarrow \quad \int_a^b f(x)\,dx \ge m(b-a) \ge 0, \text{ if } b \ge a$$

Again, since $M = \sup f \ge 0$, we have

$$\int_a^b f(x)\,dx \le M(b-a) \le 0, \text{ if } b \le a$$

Theorem VI. *If f, g are bounded and integrable, then*

$$f \ge g \quad \Rightarrow \quad \begin{cases} \displaystyle\int_a^b f(x)\,dx \ge \int_a^b g(x)\,dx, & \text{when } b \ge a, \\ \displaystyle\int_a^b f(x)\,dx \le \int_a^b g(x)\,dx, & \text{when } b \le a. \end{cases}$$

[Delhi B.Sc. (Hons) I 2011; Delhi B.Sc. (Hons) II 2011]

Proof. Now $f \geq g \Rightarrow [f(x) - g(x)] \geq 0, \ \forall \ x \in [a, b]$

$\Rightarrow \quad \int_a^b [f(x) - g(x)] \, dx \geq 0$ or ≤ 0 according as $b \geq a$ or $b \leq a$, using theorem V.

$\Rightarrow \quad \left[\int_a^b f(x) \, dx - \int_a^b g(x) \, dx \right] \geq 0$ or ≤ 0 according as $b \geq a$ or $b \geq a$.

Hence the result.

13.14 FUNCTIONS DEFINED BY DEFINITE INTEGRALS

Let f be bounded and integrable in $[a, b]$. We write $\quad \phi(t) = \int_a^t f(x) \, dx, \ t \in [a, b]$

It is now proposed to study the properties of the function ϕ with domain $[a, b]$ in relation to its continuity and derivability.

The function ϕ may be called the *Integral function* of the function f.

Property I. Continuity of the integral function.

Theorem I. The integral function of an integrable function is continuous.

(*Kanpur 2011; Agra 2009; Himanchal 2006; Garhwal 2001, Meerut 2000*)

Proof : We have $\quad \phi(t) = \int_a^t f(x) \, dx.$

Now there exists a number k such that $\quad |f(x)| \leq k, \ \forall \ x \in [a, b].$

Let c, be any point of $[a, b]$ and let ε be *any* positive number.

We have $\quad \phi(c) = \int_a^c f(x) \, dx, \quad \phi(c+h) = \int_a^{c+h} f(x) \, dx.$

$\Rightarrow \quad |\phi(c+h) - \phi(c)| = \left| \int_c^{c+h} f(x) \, dx \right| \leq |h| k$, using theorem IV of Art. 13.13.

Thus, $\quad |\phi(c+h) - \phi(c)| < \varepsilon, \ $ if $|h| < \varepsilon / k.$

Hence ϕ is continuous at any point $c \in [a, b]$ and so in the interval $[a, b]$.

Exercise. Let $f \in R[a, b]$. Then prove that the function F defined on $[a, b]$ by $F(x) = \int_a^x f(t) \, dt$ is continuous in $[a, b]$. [*Kanpur 2005, Delhi Maths (H) 2009*]

Proof : Left as an exercise.

Propeıty II. Derivability of the integral function.

Theorem II. *The integral function ϕ of a continuous function f is continuous and $\phi' = f$.*

or

Let f be continuous of on $[a, b]$ and let $\phi(t) = \int_a^t f(x) dx \ \forall \ t \in [a, b]$. Then f is continuous in $[a, b]$ and $\phi' = f$. **(Meerut 2010)**

Proof. For the first part, refer theorem I. We now that $\phi' = f$

Let $c \in [a, b]$. Then, we have

$$\phi(c+h) - \phi(c) = \int_c^{c+h} f(x) \, dx = hf(c + \theta h) \text{ where } 0 \leq \theta \leq 1 \qquad \text{... (i)}$$

[Using corollary of theorem III of Art. 13.13]

Now, f is continuous at $c \Rightarrow \lim_{h \to 0} f(c+\theta h) = f(c)$ ... (*i*)

$$\therefore \phi'(c) = \lim_{h\to 0} \frac{\phi(c+h)-\phi(c)}{h} = \lim_{h\to 0} f(c+\theta h) = f(c), by\ (i)\ and\ (ii)$$

As c, is any point of $[a, b]$, we have $\forall\ t \in [a,b]$,

$$\phi'(t) = f(t) \Rightarrow \phi' = f.$$

Primitive. Def. *A derivable function* φ, *if it exists, such that its derivative* φ' *is equal to a given function f is called a primitive of f.*

The theorem above shows that a sufficient condition for a function to admit of a primitive is that it is continuous so that every continuous function f possesses a primitive viz. $\int_a^b f(x)\,dx$.

$$\int_a f(x)\,dx.$$

Note. We now give a counter example to show that a function admitting of a primitive may not necessarily be continuous. This is equivalent to saying that the derivative of a function may not necessarily to continuous.

We consider ϕ defined in [0, 1] as follows :

$$\phi(x) = \begin{cases} x^2 \sin(1/x) & \text{for } x \neq 0. \\ 0 & \text{for } x = 0. \end{cases}$$

We have, as may easily be shown.

$$\phi'(x) = \begin{cases} 2x\sin(1/x) - \cos(1/x) & \text{for } x \neq 0 \\ 0 & \text{for } x = 0. \end{cases}$$

We now define f as ϕ' so that

$$f(x) = \begin{cases} 2x\sin(1/x) - \cos(1/x) & \text{for } x \neq 0 \\ 0 & \text{for } x = 0. \end{cases}$$

The function f admits of a primitive *viz.* ϕ but fails to be continuous because of the failure of its continuity at 0 so that we have an instance of a function which admits of a primitive without being continuous.

13.15 FUNDAMENTAL THEOREM OF INTEGRAL CALCULUS *[Kanpur 2004; Punjab 2003]*

If f is bounded, integrable and admits of a primitive ϕ *in [a, b], then*

$$\int_a^b f(x)\,dx = \phi(b) - \phi(a).$$

[G.N.D.U. Amritsar 2000, 10; Gwalior 2003; Rohtak 2004; Ravishankar 2006; Agra 2006, 11, 12; Bhopal 2004, Calicut 2004, Delhi B.A. (Prog) III 2001, Garhwal 2001, Himachal 2014, 2015, Kumaun 1999, Meerut 2009, 10, 12, 13; Rohilkhand 2011, 12; Kanpur 2014)

Proof. Let ε be any positive number. Since $\phi' = f$ is bounded and integrable in $[a, b]$, there exists a partition $P = (a = x_0, x_1, x_2, \ldots, x_{r-1}, x_r, \ldots x_n = b)$ such that

$$\left| \sum_{r=1}^{r=n} \phi'(\xi_r)\,\delta_r - \int_a^b \phi'(x)\,dx \right| < \varepsilon. \qquad \ldots (1)$$

We particularise the *arbitrary* point $\xi_r \in I_r = [x_{r-1}, x_r]$ in the following manner :

By the Lagrange's mean value theorem of Differential Calculus there exists $\xi_r \in I_r$ such that

$$\phi(x_r) - \phi(x_{r-1}) = \phi'(\xi_r)\,\delta_r$$

$$\Rightarrow \qquad \Sigma\,\phi'(\xi_r)\,\delta_r = \Sigma\,[\phi(x_r) - \phi(x_{r-1})] = \phi(b) - \phi(a) \qquad \text{... (2)}$$

From (1) and (2), it follows that

$$\left|\phi(b) - \phi(a) - \int_a^b \phi'(x)\,dx\right| < \varepsilon.$$

As ε is an arbitrary positive number, we conclude that

$$\phi(b) - \phi(a) - \int_a^b \phi'(x)\,dx = 0$$

or

$$\int_a^b f(x)\,dx = \phi(b) - \phi(a), \quad \text{as} \quad \phi' = f.$$

Hence the result.

Alternative form of statement of fundamental theorem of Integral Calculus

Let f be the Riemann integrable and bounded on [a, b]. If there is a differentiable function ϕ on [a, b] such that $\phi' = f$, then, $\int_a^b f(x)\,dx = \phi(b) - \phi(a)$. **[Himachal 2014, 2015]**

13.16 GENERALIZED FIRST MEAN VALUE THEOREM OR SECOND MEAN VALUE THEOREM

If $\int_a^b f(x)\,dx$ and $\int_a^b \phi(x)\,dx$ both exist and $\phi(x)$ keeps the same sign, positive or negative, throughout the interval of integration, then there exists a number μ, lying between the bounds of f such that $\int_a^b f(x)\,\phi(x)\,dx = \mu \int_a^b \phi(x)\,dx$. **[Delhi 2011; Meerut 2011; Purvanchal 2012]**

Proof. First suppose that $\phi(x)$ is positive $\forall\, x \in [a, b]$. If M, m be the bounds of f, we have

$$m \le f(x) \le M$$

$$\Rightarrow \qquad m\,\phi(x) \le f(x)\,\phi(x) \le M\,\phi(x), \qquad \text{for} \quad \phi(x) \ge 0, \quad \forall x \in [a, b]$$

$$\Rightarrow \qquad m \int_a^b \phi(x)\,dx \le \int_a^b f(x)\,\phi(x)\,dx \le M \int_a^b \phi(x)\,dx, \quad \text{if } b \ge a$$

and

$$m \int_a^b \phi(x)\,dx \ge \int_a^b f(x)\,\phi(x) \ge M \int_a^b \phi(x)\,dx, \qquad \text{if } b \le a$$

In either case we see that there exists a number μ, lying between M and m, such that

$$\int_a^b f(x)\,\phi(x)\,dx = \mu \int_a^b \phi(x)\,dx \qquad \text{... (1)}$$

Hence the result.

The case when ϕ is negative may be similarly disposed of

Corollary. In addition to the conditions of the theorem, if f is continuous also, then there exists a number, ξ, belonging to the domain of integration such that

$$\int_a^b f(x)\,\phi(x)\,dx = f(\xi) \int_a^b \phi(x)\,dx.$$

Proof. Since f is continuous on $[a, b]$, therefore f will assume every value lying between the bounds m and M of f.

Now $m \le \mu \le M \Rightarrow$ there exists a number $\xi \in [a,b]$ such that $f(\xi) = \mu$.

Substituting this value of μ in (1), we obtain the required result.

13.17 ABEL'S LEMMA.

If (i) $a_1, a_2, ..., a_n$ is a monotonically decreasing set of n positive numbers,

(ii) $v_1, v_2, ..., v_n$ is a set of any, n, numbers, and

(iii) k, K are two numbers such that $\quad k < v_1 + v_2 + ... + v_p < K$, *for* $1 \le p \le n$,

then $$a_1 k < \sum_{r=1}^{r=n} a_r v_r < a_1 K.$$

Proof. We write $$S_p = v_1 + v_2 + ... + v_p.$$

We have

$$\sum_{r=1}^{r=n} a_r v_r = a_1 S_1 + a_2 (S_2 - S_1) + ... + a_r (S_r - S_{r-1}) + ... + a_n (S_n - S_{n-1})$$

Now by condition (*i*), $\quad (a_1 - a_2), (a_2 - a_3), ..., (a_{n-1} - a_n), a_n$

are all positive. Also by condition (*iii*) $\quad k < S_p < K \ \forall \ p \le n.$

$$\therefore \quad \sum_{r=1}^{r=n} a_r v_r < (a_1 - a_2)K + (a_2 - a_3)K + ... + (a_{n-1} - a_n)K + a_n K = a_1 K,$$

and $$\sum_{r=1}^{r=n} a_r v_r > (a_1 - a_2)k + (a_2 - a_3)k + ... + (a_{n-1} - a_n)k + a_n k = a_1 k.$$

Hence the lemma.

13.18 SECOND MEAN VALUE THEOREM (WEIERSTRASS FORM) (Meerut 2011)

If $$\int_a^b f(x)\, dx \quad \textit{and} \quad \int_a^b \varphi(x)\, dx$$

both exist and φ is monotonic in [a, b], then there exists $\xi \in [a,b]$ such that

$$\int_a^b f(x)\, \phi(x)\, dx = \phi(a) \int_a^\xi f(x)\, dx + \phi(b) \int_\xi^b f(x)\, dx.$$

(*Agra 1999, Delhi Maths (H) 1998*)

Proof. Firstly, we prove the following result, known as as Bonnet's theorem or second mean value theorem (Bonnet form).

If $$\int_a^b f(x)\, dx \text{ and } \int_a^b \psi(x)\, dx$$

both exist, ψ is monotonically decresing and positive in [*a, b*], *then there exists a point* $\xi \in [a,b]$ *such that*

$$\int_a^b f(x)\, \psi(x)\, dx = \psi(a) \int_a^\xi f(x)\, dx.$$ (*Purvanchal 2006; Delhi Maths (H) 1999, 2000*)

Let $\quad P = \{a = x_0, x_1, ... x_{r-1}, x_r, ..., x_n = b\}$

be any partition of $[a, b]$. Let M_r, m_r be the bounds of f in $I_r = [x_{r-1}, x_r]$.

Let $\quad \xi_1 = a$ and ξ_r, where $r \ne 1$, be any point of i_r

We have, $$m_r\delta_r \le \int_{x_{r-1}}^{x_r} f(x)\,dx \le M_r\delta_r,\quad m_r\delta_r \le f(\xi_r)\delta_r \le M_r\delta_r.$$

Putting $r = 1, 2, 3, \ldots, p$ where $p \le n$, and adding, we obtain

$$\sum_{r=1}^{r=p} m_r\delta_r \le \int_a^{x_p} f(x)\,dx \le \sum_{r=1}^{r=p} M_r\delta_r,$$

and $$\sum_{r=1}^{r=p} m_r\delta_r \le \sum_{r=1}^{r=p} f(\xi_r)\delta_r \le \sum_{r=1}^{r=p} M_r\delta_r.$$

Thus, we have

$$\left|\int_a^{x_p} f(x)\,dx - \sum_{r=1}^{r=p} f(\xi_r)\delta_r\right| \le \sum_{r=1}^{r=p} (M_r - m_r)\delta_r \le \sum_{r=1}^{r=n} (M_r - m_r)\delta_r,$$

$$\Rightarrow \qquad \int_a^{x_p} f(x)\,dx - \sum_{r=1}^{r=n} O_r\delta_r \le \sum_{r=1}^{r=p} f(\xi_r)\delta_r \le \int_a^{x_p} f(x)\,dx + \sum_{r=1}^{r=n} O_r\delta_r,$$

where $O_r = (M_r - m_r)$ is the oscillation of f in I_r.

Now, $\int_a^t f(x)\,dx$, being a continuous function with t as variable, is bounded. Let C, D be its bounds. Therefore we have

$$C - \sum_{r=1}^{r=n} O_r\delta_r \le \sum_{r=1}^{r=p} f(\xi_r)\delta_r \le D + \sum_{r=1}^{r=n} O_r\delta_r.$$

Using the Abel's lemma, we put, as is justifiable,

$v_r = f(\xi_r)\delta_r,\ a_r = \psi(\xi_r);\qquad k = C - \Sigma O_r\delta_r,\ K = D + \Sigma O_r\delta_r,$

and obtain

$$\psi(a)\left[C - \sum_{r=1}^{r=n} O_r\delta_r\right] \le \sum_{r=1}^{r=n} f(\xi_r)\psi(\xi_r)\delta_r \le \psi(a)\left[D + \sum_{r=1}^{r=n} O_r\delta_r\right].$$

Let the norm of the partition tend to 0. We then obtain, in the limit,

$$C\psi(a) \le \int_a^b f(x)\psi(x)\,dx \le D\psi(a)$$

$$\Rightarrow \qquad \int_a^b f(x)\psi(x)\,dx = \mu\psi(a),$$

where μ is some number between C and D.

The continuous function

$$\int_a^t f(x)\,dx$$

must assume, for some $\xi \in [a, b]$ the value μ which lies between its bounds C, D. Thus we obtain

$$\int_a^b f(x)\psi(x)\,dx = \psi(a)\int_a^\xi f(x)\,dx.$$

We now turn to the theorem proper.

Let φ be monotonically decreasing so that the function ψ where

$$\psi(x) = \varphi(x) - \varphi(b)$$

is monotonically decreasing and positive.

There exists, therefore, a number, ξ, between a and b, such that

$$\int_a^b f(x)\,[\varphi(x)-\varphi(b)]\,dx = [\varphi(a)-\varphi(b)]\int_a^\xi f(x)\,dx$$

$$\Rightarrow \qquad \int_a^b f(x)\,\varphi(x)\,dx = \varphi(a)\int_a^\xi f(x)\,dx + \varphi(b)\left\{\int_a^b f(x)\,dx - \int_a^\xi f(x)\,dx\right\}$$

$$= \varphi(a)\int_a^\xi f(x)\,dx + \varphi(b)\int_\xi^b f(x)\,dx.$$

Let φ be monotonically increasing so that, $-\varphi$, is monotonically decreasing.

There exists, therefore, by the preceding, a number ξ between a and b, such that

$$\int_a^b f(x)\,[-\varphi(x)]\,dx = -\varphi(a)\int_a^\xi f(x)\,dx - \varphi(b)\int_\xi^b f(x)\,dx,$$

$$\Rightarrow \qquad \int_a^b f(x)\,\varphi(x)\,dx = \varphi(a)\int_a^\xi f(x)\,dx + \varphi(b)\int_\xi^b f(x)\,dx.$$

Thus we have completely established the second mean value theorem.

Note. The reader may easily show that the theorem holds good even if $a > b$.

13.19 CHANGE OF VARIABLE IN AN INTEGRAL

If (i) $\int_a^b f(x)\,dx$ *exists.*

(ii) φ *is a derivable function with domain* $[\alpha, \beta]$ *such that* $\varphi(\alpha) = a$, $\varphi(\beta) = b$,

(iii) φ' *is bounded and integrable and* $\varphi(t) \neq 0\ \forall\, t \in [\alpha, \beta]$,

(iv) $f \circ \varphi$ *is bounded and integrable in* $[\alpha, \beta]$.

then
$$\int_a^b f(x)\,dx = \int_\alpha^\beta f[\varphi(t)]\,\varphi'(t)\,dt.$$

Proof. Since $\varphi'(t) \neq 0$ for any $t \in [\alpha, \beta]$ it follows by Darboux's theorem that $\varphi'(t)$ must always have the same sign and therefore φ must be strictly monotonic in $[\alpha, \beta]$. Also therefore $[a, b]$ is the range of the function f.

Let
$$P = \{\alpha = t_0, t_1, t_2, \dots, t_{r-1}, t_r, \dots, t_{n-1}, t_n = \beta\}$$

be a partition of $[\alpha, \beta]$ and let

$$P' = \{\alpha = x_0, x_1, x_2, \dots, x_{r-1}, x_r, \dots, x_{n-1}, x_n = \beta\}$$

be the corresponding partition of $[a, b]$; $\varphi(t_r)$ being equal to x_r.

By the Lagrange's mean value theorem, we have

$$x_r - x_{r-1} = \varphi(t_r) - \varphi(t_{r-1}) = (t_r - t_{r-1})\,\varphi'(\eta_r),$$

where η_r, lies between t_{r-1} and t_r.

Let $\varphi(\eta_r) = \xi_r$. Then, we have

$$\sum_{r=1}^{r=n} f(\xi_r)(x_r - x_{r-1}) = \sum_{r=1}^{r=n} f\big[\varphi(\eta_r)\big]\,\varphi'(\eta_r)\,(t_r - t_{r-1}) \qquad \dots (1)$$

Now f is integrable in $[a, b]$.

Also $f \,o\, \varphi$, and φ' are integrable in $[\alpha, \beta]$ so that $(f \,o\, \varphi)\, \varphi'$ is integrable in $[\alpha, \beta]$.

When the norm of the partition $\to 0$, then the norm of P' also $\to 0$.

From (1), therefore we obtain in the limit

$$\int_a^b f(x)\, dx = \int_\alpha^\beta [(f \,o\, \varphi)\, \varphi'](t)\, dt = \int_\alpha^\beta f\, [\varphi(t)]\, \varphi'(t)\, dt.$$

Note. The theorem holds even if $\varphi'(t) = 0$ for a finite number of values of $t \in [\alpha, \beta]$. In this case we can divide the interval $[\alpha, \beta]$ into a finite number of sub-intervals in each of which φ is strictly increasing or decreasing and repeat the argument for each interval in turn and add the results.

13.20 INTEGRATION BY PARTS

If $\int_a^b f(x)\, dx, \int_a^b g(x)\, dx$ both exist and

$$F(x) = A + \int_a^x f(x)\, dx, \quad G(x) = B + \int_a^x g(x)\, dx$$

where A, B are two constants, then

$$\int_a^b F(x)\, g(x)\, dx = \left|F(x)\, G(x)\right|_a^b - \int_a^b G(x)\, f(x)\, dx.$$

[Here $\mid F(x)\, G(x) \mid_a^b$ denotes the difference $[F(b)\, G(b) - F(a)\, G(a)]$.]

Proof. Let $P = \{a = x_0, x_1, x_2, x_3, \ldots x_{r-1}, x_r, \ldots, x_n = b\}$ be a partition of $[a, b]$.

We have $\left|F(x)\, G(x)\right|_a^b = \sum_{r=1}^{r=n} \left[F(x_r)\, G(x_r) - F(x_{r-1})\, G(x_{r-1})\right]$.

$$= \Sigma\, F(x_r)\left[G(x_r) - G(x_{r-1})\right] + \Sigma\, G(x_{r-1})\left[F(x_r) - F(x_{r-1})\right]$$

$$= \Sigma\, F(x_r) \int_{x_{r-1}}^{x_r} g(x)\, dx + \Sigma\, G(x_{r-1}) \int_{x_{r-1}}^{x_r} f(x)\, dx \quad \ldots (1)$$

Let M_r, m_r, O_r denote the bounds and the oscillation of f and M'_r, m'_r, O'_r those of g in $I_r = [x_{r-1}, x_r]$. Now $\forall\, x \in I_r$, we have

$$\left|g(x) - g(x_r)\right| \le O'_r, \quad \left|f(x) - f(x_{r-1})\right| \le O_r,$$

$$\Rightarrow \begin{cases} g(x_r) - O'_r \le g(x) \le g(x_r) + O'_r; \\ f(x_{r-1}) - O_r \le f(x) \le f(x_{r-1}) + O_r, \end{cases}$$

It follows that

$$\left.\begin{aligned} [g(x_r) - O'_r]\, \delta_r &\le \int_{x_{r-1}}^{x_r} g(x) \le [g(x_r) + O'_r]\, \delta_r; \\ [f(x_{r-1}) - O_r]\, \delta_r &\le \int_{x_{r-1}}^{x_r} f(x)\, dx \le [f(x_{r-1}) + O_r]\, \delta_r \end{aligned}\right\} \quad \ldots (2)$$

These give

$$\int_{x_{r-1}}^{x_r} g(x)\, dx = [g(x_r) + \theta'_r\, O'_r]\, \delta_r \quad \text{and} \quad \int_{x_{r-1}}^{x_r} f(x)\, dx = [f(x_{r-1}) + \theta_r O_r]\, \delta_r \quad \ldots (3)$$

where

$$-1 \le \theta_r,\, \theta'_r \le 1.$$

From (1), (2) and (3), we obtain

$$\left|f(x)\,G(x)\right|_a^b = \Sigma\, F(x_r)\, g(x_r)\,\delta_r + \Sigma\, G(x_{r-1})\, f(x_{r-1})\,\delta_r + \sigma \quad \text{... (4)}$$

where
$$\sigma = \Sigma\left[F(x_r)\,\theta'_r\, O'_r + G(x_{r-1})\,\theta_r\, O_r\right]\delta_r.$$

Since F, G are continuous, therefore, they are bounded. Let k be a number such that $\forall\, x \in [a, b]$, $|F(x)| \le k, |G(x)| \le k$,

$$\therefore \qquad |\sigma| \le k\,(\Sigma O_r + \Sigma O'_r)\,\delta_r.$$

Let the norm of the partition $P \to 0$. Then $\sigma \to 0$.

From (4), we now obtain

$$\left|F(x)\,G(x)\right|_a^b = \int_a^b F(x)\, g(x)\, dx + \int_a^b G(x)\, f(x)\, dx.$$

Hence the result.

Cor. If a function g is bounded and integrable in $[a, b]$ and a function f is derivable in $[a, b]$ and its derivative f' is bounded and integrable, then

$$\int_a^b f(x)\, g(x)\, dx = \left| f(x) \int_a^x g(x)\, dx \right|_a^x - \int_a^b \left\{ f'(x) \int_a^x g(x)\, dx \right\} dx$$

$$= f(b) \int_a^b g(x)\, dx - \int_a^b \left\{ f'(x) \int_a^x g(x)\, dx \right\} dx.$$

Example. *Show that the second mean value theorem does not hold good in the interval* $[-1, 1]$ for $f(x) = \phi(x) = x^2$. *What about the validity of the generalised first mean value theorem in this case.*

Solution. Since $f(x)$ and $\phi(x)$ are both continuous on $[-1, 1]$, so they are both integrable on $[-1, 1]$. Here $\phi(x) = x^2$ is not monotonic in $[-1, 1]$ because $\phi(x)$ is monotonically decreasing on $[-1, 0]$ and monotonically increasing on $[0, 1]$.

If possible, suppose the second mean value theorem holds for $f(x) = x^2$ and $\phi(x) = x^2$. Then there exist same $\xi \in [-1, 1]$ such that

$$\int_{-1}^1 f(x)\,\phi(x)\, dx = \phi(-1) \int_{-1}^{\xi} f(x)\, dx + \phi(1) \int_{\xi}^1 f(x)\, dx$$

or
$$\int_{-1}^1 x^4 dx = \int_{-1}^{\xi} x^2\, dx + \int_{\xi}^1 x^2 dx$$

or
$$\frac{2}{5} = \frac{1}{3}\left(\xi^3 + 1\right) + \frac{1}{3}\left(1 - \xi^3\right) \qquad \text{so that} \qquad \frac{2}{5} = \frac{2}{3},$$

which is absurd. Hence the second mean value theorem in not true.

We now examine the validity of the generalised first mean value theorem. Here $f(x)$ and $\phi(x)$ are integrable, as before. Also, $\phi(x) = x^2 \Rightarrow \phi(x) \ge 0\ \forall\, x \in [-1, 1]$ *i.e.*, $\phi(x)$ keeps the same sign in $[-1, 1]$. If possible, suppose the generalised first mean value theorem is true for $f(x) = x^2$ and $\phi(x) = x^2$. Then there must exist a number, μ, lying between the bounds 0 and 1 of $f(x)$ such that

$$\int_{-1}^1 f(x)\,\phi(x)\, dx = \mu \int_{-1}^1 \phi(x)\, dx$$

i.e., $\int_{-1}^{1} x^4\,dx = \mu \int_{-1}^{1} x^2\,dx$ or $\frac{2}{5} = \mu \times \frac{2}{3}$ or $\mu = \frac{3}{5}$.

Since $0 < 3/5 < 1$, it follows that the generalized first mean value theorem holds for the given functions defined in $[-1, 1]$.

EXERCISES

1. Taking $f(x) = x$ and $\phi(x) = e^x$, verify second mean value theorems for the interval $[-1, 1]$.

[Meerut 2011; Delhi Maths (H) 2006]

2. Varity first mean value theorem (generalized form) for the function $f(x) = \sin x$ and $\phi(x) = e^x, x \in [0, 1]$

[Meerut 2011]

3. Show that the Bonnet's mean value theorem holds on $[-1, 1]$, for $f(x) = e^x$, $\phi(x) = x$.

4. Show that the Bonnet's theorem does not hold on $[-1, 1]$ for $f(x) = \phi(x) = x^2$

5. Show that for the validity of the second mean value theorem of Integral calculary ϕ must be necessarily monotonic by showing that the theorem does not hold it $\phi(x) = \cos x$, $f(x) = x^2$

6. Prove Bonnet's form of the second mean value theorem that if f' is continuous and of constant sign and $f(b)$ has the same sign as $f(b) - f(a)$, then

$$\int_a^b f(x)\,\phi(x)\,dx = f(a) \int_a^{\xi} \phi(x)\,dx, \text{ when } \xi \text{ lies between } a \text{ and } b.$$

7. Text the validity of the second mean value theorem in $[-1, 1]$ for the functions $f(x) = e$ $g(x) = x$.

[Delhi Maths (H) 2009]

EXAMPLES

Example 1. *If f is non-negative continous function an $[a, b]$ such that $\int_a^b f(x)\,dx > 0$. then show that $f(x) = 0$ $\forall x \in [a, b]$* (*Delhi Maths (H) 1999, 2004, Nagpur 2003*)

Solution. Let, if possible, for some $c \in [a, b]$, $f(c) > 0$

Let $\epsilon = (1/2) \times f(c) > 0$. Since f is continuous at c, so for given $\epsilon > 0$, there exists $\delta > 0$ such that

$$|f(x) - f(c)| < \epsilon \quad \forall x \in]c - \delta, c + \delta[$$

$$\Rightarrow \quad f(c) - \epsilon < f(x) < f(c) + \epsilon \text{ if } c - \delta < x < c + \delta$$

$$\Rightarrow \quad f(x) > f(c) - \epsilon \text{ if } c - \delta < x < c + \delta$$

$$\Rightarrow \quad f(x) > (1/2) \times f(c), \text{ if } c - \delta < x < c + \delta$$

Now f is continuous an $[a, b] \Rightarrow f$ is integrable on $[a, b]$.

$$\therefore \quad \int_a^b f(x)\,dx = \int_a^{c-\delta} f(x)\,dx + \int_{c-\delta}^{c+\delta} f(x)\,dx + \int_{c+\delta}^{b} f(x)\,dx$$

$$\geq \int_{c-\delta}^{c+\delta} f(x)\,dx, \text{ as } f(x) \geq 0 \ \forall x \in [a, b]$$

$$> \frac{1}{2} f(c) \int_{c-\delta}^{c+\delta} dx = \delta\, f(c) > 0$$

Thus, $\int_a^b f(x)\,dx > 0$, which controdicts the given hypothesis $\int_a^b f(x)\,dx = 0$

Hence $f(c) > 0$ cannot hold. Similarly, we can show that $f(c) < 0$ cannot hold. Hence $f(x) = 0$ $\forall x \in [a, b]$.

Example 2 (a) *Prove that the function f defined on* [0, 4] *by* $f(x) = [x]$, *where* $[x]$ *denotes the greatest integer not greater then x, is integralble on* [0, 4] *and* $\int_0^4 f(x)\,dx = 6$

[Kanpur 2018, 19; Agra 2009; Delhi B.Sc. (Hons.) II, 2004)

(*b*) Evaluate $\int_0^2 x\,[2x]\,dx$, *where* $[x]$ *denotes the greatest integer function.*

[Delhi B.Sc. (Hons.) II, 2002]

Solution. (*a*) We have, by definition of the function $[x]$

$$\begin{aligned} f(x) = [x] &= 0 \text{ if } 0 \le x < 1 \\ &= 1 \text{ if } 1 \le x < 2 \\ &= 2 \text{ if } 2 \le x < 3 \\ &= 3 \text{ if } 3 \le x < 4 \end{aligned}$$

Here $f(x)$ is bounded and has four points of discontinuity at x = 1, 2, 3 and 4. Since the points of discontinity of f on [0, 4] are finite in number, so f is integrable on [0, 4] and

$$\int_0^4 f(x)\,dx = \int_0^1 f(x)\,dx + \int_1^2 f(x)\,dx + \int_2^3 f(x)\,dx + \int_3^4 f(x)\,dx$$

$$= \int_0^1 0\,dx + \int_1^2 1\,dx + \int_2^3 2\,dx + \int_3^4 3\,dx = 6$$

(*b*)Here, we have

$$\begin{aligned} f(x) = x\,[2x] &= x \times 0 \text{ if } 0 \le x < 1/2 \\ &= x \times 1 \text{ if } 1/2 \le x < 1 \\ &= x \times 2 \text{ if } 1 \le x < 3/2 \\ &= x \times 3 \text{ if } 3/2 \le x < 2 \end{aligned}$$

As in part (*a*), $f(x)$ has only finite (3 here) number of points of discontinuity at x = 1/2, 1, 3/2 and so $f(x)$ is integrable and

$$\int_0^2 f(x)\,dx = \int_0^{1/2} f(x)\,dx + \int_{1/2}^1 f(x)\,dx + \int_1^{3/2} f(x)\,dx + \int_{3/2}^2 f(x)\,dx$$

$$= \int_0^{1/2} 0\,dx + \int_{1/2}^1 x\,dx + \int_1^{3/2} 2x\,dx + \int_{3/2}^2 3x\,dx = \frac{17}{4}$$

Example 3 *Show that the function f defined by*

$$f(x) = 1/2^n, \text{ when } 1/2^{n+1} < x \le 1/2^n, \ (n = 0, 1, 2, \ldots)$$
$$f(0) = 0$$

in integrable on [0, 1], *although it has an infinite number of points of discontinuities. Also show that* $\int_0^1 f(x)\,dx = \frac{2}{3}$ **[Delhi B.Sc. (Prog.) II, 2017; Delhi B.A. (Prog.) III 2013, 16]**

Solution. Here, $f(x)$ = 1, when $1/2 < x \le 1$

$= 1/2$, when $1/2^2 < x \le 1/2$

$= 1/2^2$, when $1/2^3 < x \le 1/2^2$

....

$= 1/2^{n-1}$, when $1/2^n < x \le 1/2^{n-1}$

....

$= 0$, when $x = 0$

Observe that f is bounded and continuous on [0, 1] except at the points 0, 1/2, $1/2^2$, ... which are infinite in number. The set of points has only one limit point, namely, 0 and hence f is integrable in [0, 1]. Now, we have

$$\int_{1/2^n}^{1} f(x)\,dx = \int_{1/2}^{1} f(x)\,dx + \int_{1/2^2}^{1/2} f(x)\,dx + \int_{1/2^3}^{1/2^2} f(x)\,dx + \ldots + \int_{1/2^n}^{1/2^{n-1}} f(x)\,dx$$

$$= \int_{1/2}^{1} 1dx + \int_{1/2^2}^{1/2} \frac{1}{2}dx + \int_{1/2^3}^{1/2^2} \frac{1}{2^2}dx + \ldots + \int_{1/2^n}^{1/2^{n-1}} \frac{1}{2^{n-1}}dx$$

$$= 1 - \frac{1}{2} + \frac{1}{2}\left(\frac{1}{2} - \frac{1}{2^2}\right) + \frac{1}{2^2}\left(\frac{1}{2^2} - \frac{1}{2^3}\right) + \ldots + \frac{1}{2^{n-1}}\left(\frac{1}{2^{n-1}} - \frac{1}{2^n}\right)$$

$$= \frac{1}{2} + \frac{1}{2}\cdot\frac{1}{2^2} + \frac{1}{2^2}\cdot\frac{1}{2^3} + \ldots + \frac{1}{2^{n-1}}\cdot\frac{1}{2^n} = \frac{1}{2}\left[1 + \frac{1}{2^2} + \left(\frac{1}{2^2}\right)^2 + \ldots + \left(\frac{1}{2^2}\right)^{n-1}\right]$$

$$= \frac{1}{2} \times \frac{1-(1/2^2)^n}{1-(1/2^2)} = \frac{2}{3}\left(1 - \frac{1}{4^n}\right)$$

Proceeding to the limit when $n \to \infty$, we get

$$\int_0^1 f(x)\,dx = \lim_{n\to\infty} \int_{1/2^n}^{1} f(x)\,dx = \lim_{n\to\infty} \frac{2}{3}\left(1 - \frac{1}{4^n}\right) = \frac{2}{3}$$

Example 4. *Show that the function f defined by*

$$f(x) = \begin{cases} 0, \text{ if } x \text{ is an integer} \\ 1, \text{otherwise} \end{cases}$$

is integrable on [0, *m*], *m* being an integer.

Solution. $f(x) = \begin{cases} 0, \text{if } x = 0, 1, 2, \ldots, m \\ 1, \text{if } r-1 < x < r, \textit{ for } r = 1, 2, 3, \ldots \end{cases}$

Clearly f is bounded and is continuous everywhere except at $(m + 1)$ points $x = 0, 1, 2, 3, \ldots, m$. Since the points of discontinuity are finite, so f is integrable as $[0, m]$ and

$$\int_0^m f(x)\,dx = \sum_{r=1}^{m} \int_{r-1}^{r} f(x)\,dx = \sum_{r=1}^{m} \int_{r-1}^{r} 1\,dx = \sum_{r=1}^{n} \{r-(r-1)\} = \sum_{r=1}^{n} 1 = m$$

Example 5. Show that $$\frac{1}{\pi} \le \int_0^1 \frac{\sin \pi x}{1+x^2}dx \le \frac{2}{\pi}$$

Solution. Let $f(x) = 1/(1+x^2)$ and $\phi(x) = \sin \pi x$. Then f and ϕ are continuous on [0, 1] and hence integrable on [0, 1]. Also, $\phi(x) = \sin \pi x \ge 0$ on [0, 1].

Since f is decreasing on [0, 1], $\inf f = f(1) = 1/2$ and $\sup f = f(0) = 1$. Hence by the generalised first mean value theorem, there exists $\mu \in [1/2, 1]$ such that

$$\int_0^1 f(x)\,\phi(x)\,dx = \mu \int_0^1 \phi(x)\,dx \quad \text{or} \quad \int_0^1 \frac{\sin \pi x}{1+x^2}dx = \mu \int_0^1 \sin \pi x\,dx$$

or $$\int_0^1 \frac{\sin \pi x}{1+x^2}dx = \frac{2\mu}{\pi} \quad \text{or} \quad \mu = \frac{\pi}{2}\int_0^1 \frac{\sin \pi x}{1+x^2}dx \qquad \ldots(1)$$

Since f is continuous on [0, 1], it attains every value between its bounds 1/2 and 1. Since $1/2 < \mu < 1$, so there exists a number $c \in [a, b]$ such that $f(c) = \mu$.

Then (1) becomes $$f(c) = \frac{\pi}{2}\int_0^1 \frac{\sin \pi x}{1+x^2}\,dx$$

Now, $0 \le c \le 1$ and f is decreasing on [0, 1]

$$\Rightarrow \quad f(0) \ge f(c) \ge f(1) \quad \Rightarrow \quad f(1) \le f(c) \le f(0)$$

$$\Rightarrow \quad \frac{1}{2} \le \frac{\pi}{2}\int_0^1 \frac{\sin \pi x}{1+x^2}\,dx \le 1 \;\Rightarrow\; \frac{1}{\pi} \le \int_0^{\pi} \frac{\sin \pi x}{1+x^2}\,dx \le \frac{2}{\pi}$$

Example 6. *By an example, prove that the equation* $\int_a^b f'(x)\,dx = f(b) - f(a)$ *is not always true.*

Solution. Consider the followin function :

$$f(x) = x^2 \sin(1/x^2), \text{ if } 0 < x \le 1$$
$$= 0, \text{ if } x = 0.$$

Then it can easily be proved that f is differentiable on [0, 1] and its derivative is given by

$$f'(x) = 2x \sin(1/x^2) - (2/x) \times \cos(1/x^2), \text{ if } 0 < x \le 1$$
$$= 0 \text{ if } x = 0$$

Since f' is not bounded, so f' is not Riemann integrable, *i.e.*, $\int_0^1 f'(x)\,dx$ does not exist and hence the given equation fails to hold.

Example 7. *If* $$G(x, \xi) = \begin{cases} x(\xi - 1), \text{ when } x \le \xi, \\ \xi(x - 1), \text{ when } \xi < x, \end{cases}$$

and if f is a continuous function of x in [0, 1] *and if*

$$g(x) = \int_0^1 f(\xi)\, G(x, \xi)\, d\xi,$$

show that $g''(x) = f(x),\ \forall\, x \in [0, 1]$ and find g (0) and g (1).

Sol. We have $$g(x) = \int_0^x f(\xi)\, \xi\, (x-1)\, d\xi + \int_x^1 f(\xi)\, x(\xi - 1)\, d\xi$$

$$= (x-1)\int_0^x \xi f(\xi)\, d\xi + x\int_x^1 (\xi - 1)\, f(\xi)\, d\xi$$

$$= x\int_0^1 \xi f(\xi)\, d\xi - \int_0^x \xi\, f(\xi)\, d\xi - x\int_x^1 f(\xi)\, d\xi$$

Hence $$g'(x) = \int_0^1 \xi\, f(\xi)\, d\xi - x\, f(x) + x\, f(x) - \int_x^1 f(\xi)\, d\xi.$$

and so $$g''(x) = f(x).$$

We may easily see that g (0) and g (1) are both zero.

Example 8. *Prove that if the functions f and* ϕ *are bounded and integrable in* $[a, b]$, *then*

$$\left[\int_a^b f(x)\,\phi(x)\,dx\right]^2 \le \int_a^b [f(x)]^2\,dx \int_a^b \left[\phi(x)^2\right] dx.$$

Under what conditions does the sign of equality hold ?

Sol. We have $$\left[\int_a^b f(x)\,\phi(x)\,dx\right]^2 = \left[\lim \Sigma\,(x_r - x_{r-1})\,f(\xi_r)\,\phi(\xi_r)\right]^2$$

$$\int_a^b [f(x)]^2\,dx = \lim \Sigma\left[\sqrt{(x_r - x_{r-1})}\,f(\xi_r)\right]^2$$

and $$\int_a^b [\phi(x)]^2\,dx = \lim\left[\Sigma\sqrt{(x_r - x_{r-1})}\,\phi(\xi_r)\right]^2.$$

Now putting $a_r = \sqrt{(x_r - x_{r-1})}\,f(\xi_r),\ b_r = \sqrt{(x_r - x_{r-1})}\,\phi(\xi_r)$, in the Cauchy's inequality

$$(\Sigma\, a_r b_r)^2 \le \Sigma\, a_r^{\,2}\,\Sigma\, b_r^{\,2}$$

we get the required result.

The sign of equality holds only when

$$\frac{a_1}{b_1} = \frac{a_2}{b_2} = \dots \quad \Leftrightarrow \quad \frac{f(\xi_1)}{\phi(\xi_1)} = \frac{f(\xi_2)}{\phi(\xi_2)} = \dots \Leftrightarrow \quad f, \phi \text{ are both constant functions.}$$

Example 9. *If f is positive and monotonically decreasing in* $[1, \infty[$, *show that the sequence* $\{A_n\}$, *where* $A_n = \left\{f(1) + f(2) + \dots + f(n) - \int_1^n f(x)\,dx\right\}$, *is convergent.*

Deduce the convergence of $\{1 + 1/2 + 1/3 + \dots + 1/n - \log n\}$.

Sol. We have

$$A_n = \left[f(1) - \int_1^2 f(x)\,dx\right] + \left[f(2) - \int_2^3 f(x)\,dx\right] + \dots + \dots + \left[f(n-1) - \int_{n-1}^n f(x)\,dx\right] + f(n).$$

Now, because of the monotonic character of the function f, each of the expressions within brackets is positive. Also $f(n)$ is positive. Thus A_n is positive $\forall\ n$. Again $\forall\ n$

$$A_{n+1} - A_n = f(n+1) - \int_1^{n+1} f(x)\,dx + \int_1^n f(x)\,dx$$

$$= f(n+1) - \int_n^{n+1} f(x)\,dx < 0$$

$$\Rightarrow \quad A_{n+1} < A_n \ \forall\, n.$$

Thus $\{A_n\}$ is monotonically decreasing. Also being positive, A_n is bounded below for each $n \in N$. Hence $\{A_n\}$ is convergent.

Taking $f(x) = 1/x$, we can now deduce that

$$\lim_{n \to \infty} \left(1 + 1/2 + 1/3 + \dots + 1/n - \log n\right) \text{ exists.} \qquad \dots (1)$$

Note. (1) is known as *Euler's constant* whose numerical valve is 0.577215

Example 10. *Show that the function F defined in the interval* [0, 1] *by the condition that if r is a positive integer*

$$F(x) = 2rx \text{ when } 1/(r+1) < x < 1/r \text{ for each } r \in N..$$

is integrable over [0, 1] *and that*

$$\int_0^1 F(x)\,dx = \frac{\pi^2}{6}. \qquad \textit{(Agra 2004, 08, 10; Rohilkhand 2010; Patna 2003)}$$

Sol. The function F, as given, is not defined at the set of points $\{0, 1, 1/2, \ldots\ldots, 1/r, \ldots\}$ ... (1)

We may, however, define F at these points in any manner we please provided F remains bounded.

Now, the only points of discontinuity of F are those given above in (1). The set formed by these points is infinite having only one limit point *viz*, 0. Thus the function is integrable.

Consider $\psi(\varepsilon) = \int_\varepsilon^1 F(x)\,dx.$

We know that ϕ is a continuous function so that

$$\psi(0) = \int_0^1 F(x)\,dx = \lim_{\varepsilon \to 0} \psi(\varepsilon).$$

We, now, find $\psi(\varepsilon)$. We take $\varepsilon = 1/n$ so that $\varepsilon \to 0$ as $n \to \infty$. we have

$$\int_{1/n}^1 F(x)\,dx = \int_{1/2}^1 F(x)\,dx + \int_{1/3}^{1/2} F(x)\,dx + \ldots \int_{1/(r+1)}^{1/r} F(x)\,dx + \ldots + \int_{1/n}^{1/(n-1)} F(x)\,dx. \quad \ldots (2)$$

Now, $$\int_{1/(r+1)}^{1/r} F(x)\,dx = \int_{1/(r+1)}^{1/r} 2r\,x\,dx = 2r\left[\frac{x^2}{2}\right]_{1/(r+1)}^{1/r} = r\left[\frac{1}{r^2} - \frac{1}{(r+1)^2}\right] = \frac{2r+1}{r(r+1)^2}$$

$\therefore$ from (2), $$\int_{1/n}^1 F(x)\,dx = \sum_{r=1}^{n-1} \frac{2r+1}{r(r+1)^2} = \sum_{r=1}^{n-1}\left(\frac{1}{r} - \frac{1}{r+1} + \frac{1}{(r+1)^2}\right)$$

$$= \sum_{r=1}^{n-1}\left(\frac{1}{r} - \frac{1}{r+1}\right) + \sum_{r=1}^{n-1} \frac{1}{(r+1)^2} = 1 - \frac{1}{n} + \sum_{r=1}^{n-1} \frac{1}{(r+1)^2}.$$

* Now the series $1/1^2 + 1/2^2 + 1/3^2 + 1/4^2 + \ldots$ is convergent and its sum is $\pi^2/6$

$$\therefore \quad \lim_{n\to\infty} \int_{1/n}^1 F(x)\,dx = 1 + \frac{\pi^2}{6} - 1 = \frac{\pi^2}{6} \quad \text{and so} \quad \int_0^1 F(x)\,dx = \frac{\pi^2}{6}$$

Example 11. *Show that when* $-1 < x \le 1$, $\lim_{m\to\infty} \int_0^x \frac{t^m}{t+1}\,dt = 0$

Sol. Now $\forall\, x \in [0,1]$, we have

$$0 \le \int_0^x \frac{t^m\,dt}{1+t} \le \int_0^x t^m\,dt = \frac{x^{m+1}}{m+1} < \frac{1}{m+1}$$

Let $-1 < x < 0$. Putting $t = -u$, we obtain

$$\left|\int_0^x \frac{t^m}{1+t}\,dt\right| = \left|\int_0^{-x} \frac{u^m}{1-u}\,du\right| < \left|\frac{1}{1+x}\int_0^{-x} u^m\,du\right| < \frac{1}{(m+1)(x+1)}.$$

Hence the result.

Example 12. *Show that, when* $|x| < 1$.

$$\int_0^x \frac{dt}{1+t^4} = x - \frac{1}{5}x^5 + \frac{1}{9}x^9 - \frac{1}{13}x^{13} + \ldots$$

* We assume that $\pi^2/6 = 1/1^2 + 1/2^2 + 1/3^2 + 1/4^2 + \ldots$

Sol. We have

$$\frac{1}{1+t^4} = 1 - t^4 + t^8 - t^{12} + \ldots + (-1)^{n-1}\, t^{4n-4} + \frac{(-1)^n t^{4n}}{1+t^4}$$

$$\Rightarrow \quad \int_0^x \frac{dt}{1+t^4} = x - \frac{1}{5}x^5 + \frac{1}{9}x^9 - \frac{1}{13}x^{13} + \ldots + \frac{(-1)^{n-1}\; x^{4n-3}}{4n-3} + (-1)^n \int_0^x \frac{t^{4n}}{1+t^4}\, dt.$$

Now, $\forall\, x \in [-1, 1]$, we have

$$0 \le \left| \int_0^x \frac{t^{4n}}{1+t^4}\, dt \right| < \left| \int_0^x t^{4n}\, dt \right| = \left| \frac{x^{4n+1}}{4n+1} \right| < \frac{1}{4n+1},$$

$$\lim_{n\to\infty} \int_0^x \frac{t^{4n}}{1+t^n}\, dt = 0.$$

Hence the result.

Example 13. *If a function f is continuous in* [0, 1], *show that*

$$\lim_{n\to\infty} \int_0 \frac{nf\,(x)}{1+n^2\, x^2}\, dx = \frac{\pi}{2}\, f(0)$$ **[I.A.S. 2008; Purvanchal 2006; Delhi Maths (H) 2003, 07]**

Sol. We write
$$\int_0^1 \frac{nf\,(x)}{1+n^2\, x^2}\, dx = \int_0^{1/\sqrt{n}} \frac{nf\,(x)}{1+n^2\, x^2}\, dx + \int_{1/\sqrt{n}}^1 \frac{nf\,(x)}{1+n^2\, x^2}\, dx.$$

By the first mean value theorem, we have

$$\int_0^{1/\sqrt{n}} \frac{nf\,(x)}{1+n^2\, x^2}\, dx = f(\alpha_n) \int_0^{1/\sqrt{n}} \frac{ndx}{1+n^2\, x^2}, \text{ where } 0 \le \alpha_n \le 1/\sqrt{n}$$

$$= f(\alpha_n) \tan^{-1} \sqrt{n}, \text{ which} \to f(0).\frac{\pi}{2} \text{ as } n \to \infty.$$

Again,
$$\left| \int_{1/\sqrt{n}}^1 \frac{nf\,(x)}{1+n^2\, x^2}\, dx \right| = \left| f(\beta_n) \int_{1/\sqrt{n}}^1 \frac{ndx}{1+n^2\, x^2} \right|, \text{ where } 1/\sqrt{n} \le \beta_n \le 1$$

$$= f(\beta_n)\,(\tan^{-1} n - \tan^{-1} \sqrt{n})$$

$$\le M\,(\tan^{-1} n - \tan^{-1} \sqrt{n}), \text{ which} \to 0 \text{ as} \to \infty,$$

M, being the supremum of $|\,f(x)\,|$. Hence the result.

Example 14. *If f is bounded and integrable in the interval* [*a*, *b*], *show that*

$$\lim_{n\to\infty} \int_a^b f(x) \cos nx\; dx = 0.$$

Sol. We write
$$I_n = \int_a^b f(x) \cos nx\; dx.$$

Let ε be a positive number. Since f is bounded and integrable in [a, b], there exists a partition

$$P = \{a = x_0 < x_1 < x_2 < \ldots < x_{r-1} < x_r < \ldots < x_p = b\}$$

such that the corresponding oscillatory sum

$$\Sigma\,(x_r - x_{r-1})\, O_r < \varepsilon/2;$$

O_r being the oscillation of f in $[x_{r-1}, x_r]$.

We have, $I_n = \Sigma \int_{x_{r-1}}^{x_r} f(x) \cos nx \, dx$

$$= \Sigma f(x_{r-1}) \int_{x_{r-1}}^{x_r} \cos nx \, dx + \Sigma \int_{x_{r-1}}^{x_r} \left[f(x) - f(x_{r-1})\right] \cos nx \, dx,$$

$$\Rightarrow \quad |I_n| \le \Sigma |f(x_{r-1})| \left| \int_{x_{r-1}}^{x_r} \cos nx \, dx \right| + \Sigma \left| \int_{x_{r-1}}^{x_r} \{f(x) - f(x_{r-1})\} \cos nx \, dx \right|.$$

We have $\forall \, x \in [x_{r-1}, x_r]$

$$\left|f(x) - f(x_{r-1})\right| \le O_r$$

$$\Rightarrow \quad \left|\left[f(x) - f(x_{r-1})\right] \cos nx\right| \le O_r.$$

Also $\left| \int_{x_{r-1}}^{x_r} \cos nx \, dx \right| \le \frac{1}{n} \{|\sin nx_r|\} + |\sin nx_{r-1}|\} < \frac{2}{n}.$

It follows that

$$|I_n| \le \frac{2}{n} \Sigma \left|f(x_{r-1})\right| + \Sigma (x_r - x_{r-1}) O_r \le \frac{2}{n} \Sigma \left|f(x_{r-1})\right| + \frac{\varepsilon}{2}.$$

Keeping the partition P fixed, we see that $\Sigma \left|f(x_{r-1})\right|$ is fixed. We now choose a positive integer m such that $\forall \, n \ge m$

$$\frac{2}{n} \Sigma \left|f(x_{r-1})\right| < \frac{\varepsilon}{2}.$$

Thus $\forall \, n \ge m$ we have $|I_n| < \varepsilon$.
Hence the result.
It may similarly be shown that

$$\lim_{n \to \infty} \int_a^b f(x) \sin nx \, dx = 0. \qquad \textit{(Delhi B. Sc. (H) 2004)}$$

Example 15. *Show that* $\lim \{I_n\}$, *where*

$$I_n = \int_0^h \frac{\sin nx}{x} \, dx, n \in 1$$

exists and that the limit is equal to $\pi/2$.

Sol. The integrand becomes continuous for every value of x, if we assign to it the value n for $x = 0$. The result will be proved in *three* steps.

I. Firstly, it will be proved that $\{I_n\}$ is convergent. Putting $nx = t$, we have

$$I_n = \int_0^{nh} \frac{\sin t}{t} \, dt.$$

$$\Rightarrow \quad \left|I_{n+p} - I_n\right| = \left| \int_{nh}^{(n+p)h} \frac{\sin t}{t} \, dt \right|.$$

As $1/t$ is positive and monotnicaly decreasing when $t \in \,]\, nh, (n+p)\, h \,[$, we have, by Bonnett's form of the second mean value theorem,

$$\left|I_{n+p} - I_n\right| = \frac{1}{nh} \left| \int_{nh}^{\alpha} \sin t \, dt \right| \le \frac{2}{nh} < \varepsilon \;\; \forall \, n > 2/\varepsilon h.$$

Hence, by Cauchy's principle of convergence, $\{I_n\}$ converges.

II. It will now be proved that, when $n \to \infty$,

$$\lim I_n = \lim \int_0^{\pi/2} \frac{\sin nx}{\sin x}\, dx.$$

We write $\displaystyle\int_0^{\pi/2} \frac{\sin nx}{x}\, dx = \int_0^{h} \frac{\sin nx}{x}\, dx + \int_h^{\pi/2} \frac{\sin nx}{x}\, dx$

As proved in the preceding example,

$$\int_0^{\pi/2} \frac{\sin nx}{x}\, dx \to 0 \text{ as } n \to \infty.$$

$$\lim I_n = \lim \int_0^{\pi/2} \frac{\sin nx}{x}\, dx.$$

Again, taking $f(x) = (1/x - 1/\sin x)$ in the preceding example,

$$\lim \int_0^{\pi/2} \left(\frac{1}{x} - \frac{1}{\sin x}\right) \sin nx\, dx = 0,$$

for f is continuous in $[0, \pi/2]$, if we set $f(0) = 0$.

It follows that

$$\lim \int_0^{\pi/2} \frac{\sin nx}{x}\, dx = \lim \int_0^{\pi/2} \frac{\sin nx}{\sin x}\, dx$$

$$\lim I_n = \lim \int_0^{\pi/2} \frac{\sin nx}{x}\, dx = \lim \int_0^{\pi/2} \frac{\sin nx}{\sin x}\, dx.$$

To determine the actual value of the limit, we proceed by making $n \to \infty$ through odd integer values.

III. We have, as may be easily shown,

$$\frac{\sin (2n+1)\, x}{\sin x} = 2\left[\frac{1}{2} + \cos 2x + \cos 4\, x + \ldots + \cos 2nx\right]$$

so that $\displaystyle\int_0^{\pi/2} \frac{\sin (2n+1)\, x}{\sin x}\, dx = \frac{\pi}{2}.$

Hence the result.

EXERCISES

1. Give example of a function when upper Riemann integral is not equal to lower Riemann integral. Justify your answer. *(Delhi Maths (H) 2001)*
2. Give example of Riemann integral function defined on [0, 2] and having discontinuity only at 1 and 2. *(Delhi Maths (H) 1999)*
3. Show that the function f defined by

$$f(x) = \lim_{n\to\infty} \frac{\log(2+x) - x^{2n} \sin x}{1 + x^{2n}}$$

in integrable on $[0, \pi/2]$ *(Delhi Maths (H) 2000)*

4. Let f be a real valued function defined on $[a, b]$. Define f^+ and f^- by $f^+(x) = \max\{f(x), 0\}$, $f^-(x) = \max\{-f(x), 0\}$. If $f \in R[a, b]$ then show that $f^+, f^- \in R[a, b]$ and that

$$\int_a^b f(x)\,dx = \int_a^b f^+(x)\,dx - \int_a^b f^-(x)\,dx.$$ *(Delhi Maths (H) 2004)*

5. (*a*) Give example of Riemann integrable function defined on [0, 1] having infinite number of discontinuties. *(Delhi Maths (H) 1999, 2004)*

   (*b*) Give example of an integrable function which has an infinite set of points of discontinuties having only finitely many limit points. *(Delhi Maths (H) 1998)*

6. State the class of Riemann integrable function and prove the result for one of them. *(Delhi Maths (H) 2003)*

7. A function f is defined on [0, 1] by

   $f(x) = 1/x$ for $1/n > x \geq 1/(n+1)$, $n = 1, 2, 3, \ldots$

   Prove that f is integrable on [0, 1] and $\int_0^1 f(x)\,dx = (\pi^2/6) - 1$

8. Show that the Oscillation $(M-m)$ of a bounded function f defined on an interval $[a, b]$ is $\sup\{|f(x) - f(t)| : x, t \in [a, b]\}$ *(Delhi Maths (H) 2004)*

9. If a function f is bounded and integrable on $[a, b]$ prove that f^2 is also bounded and integrable. Is the converse of this result true ? *(Delhi Physice (H) 1999)*

10. Prove that $|f| \in R[a, b] \Rightarrow f \in R[a, b]$ *(Garhwal 1999)*

11. Let $f \in R[a, b]$. Put $F(x) = \int_a^x f(t)\,dt$, $a \leq x \leq b$. Prove that F is continuous on $[a, b]$. Also, if f is continuous at a point x_0 of $[a, b]$ then prove that F in differentiable at x_0 and $F'(x_0) = f(x_0)$. *(Nagpur 2003)*

12. If f is Riemann integrable over every interval of finite length and $f(x + y) = f(x) + f(y)$ for every pair of real numbers, show that $f(x) = cx$, where $c = f(1)$ *(IAS 1999)*

13. If f is Riemann integrable on $[a, b]$, show that the indefinite integral F of f given by

    $F(x) = \int_a^x f(t)\,dt$ for all $x \in [a, b]$ is uniformly continuous. *(Delhi Maths (H) 1998)*

14. If $f(y, x) = 1 + 2x$ for y rational and $f(y, x) = 0$ for y irrational, calculate

    $$F(y) = \int_0^1 f(y, x)\,dx.$$

15. Show that $\int_0^2 f(x)\,dx = 2$,

    where $f(x) = 0$, when $x = n/(n+1), (n+1)/n, (n = 1, 2, 3, \ldots)$
    $f(x) = 1$, elsewhere.

    Examine for continuity the function f so defined at the point $x = 1$

    [Delhi Maths (H) 2006]

16. A function f is defined for $x \geq 0$ by

    $$f(x) = \int_{-1}^1 \frac{dt}{\sqrt{(1 - 2tx + x^2)}}.$$

    Prove that if $0 \leq x \leq 1$, $f(x) = 2$. What is the value of f for $x > 1$? Has the function f a differential coefficient for $x = 1$?

    [For $x > 1$, $f(x) = 2/x$: f is not derivable for $x = 1$ even though it is continuous threat]

17. If for $x \geq 0$, φ is defined as

$$\varphi(x) = \lim \frac{x^n + 2}{x^n + 1} \text{ as } n \to \infty \text{ and } f(x) = \int_0^x \varphi(t)\, dt,$$

prove that f is continuous but not differentiable for $x = 1$.

18. If $f(x, y) = xy^2 e^{-xy} + x^2 y/(1 + y)$ and a, b are positive, show that

$$\lim_{y \to \infty} \int_a^b f(x, y)\, dx = \int_a^b \left[\lim_{y \to \infty} f(x, y) \right] dx.$$

Also show that the equality does not hold for $a = 0$.

19. f is bounded and integrable in $[a, b]$; show that

$$\int_a^b [f(x)]^2\, dx = 0$$

if, and only if, $f(c) = 0$ at every point, c, of continuity of f.

20. If $a > 0$, $n > 0$, show that $\quad 0 < n \displaystyle\int_0^1 \frac{x^{n-1}}{(1 + x^{-2})} a\, dx < 1.$

21. The functions f and g are bounded and integrable in $[a, b]$. If further

$$F(x) = \int_a^x f(t)\, dt \text{ and } H(x) = \int_a^x f(t)\, g(t)\, dt,$$

where $a \leq x \leq b$ and if $F' = f$ and g is continuous, show that

$$H' = f(x)\, g(x)$$

22. A function f is integrable in $[a - c, a + c]$ and $|f(x)| \leq M \;\; \forall\, x \in [a - c, a + c]$.

$$\int_{a-c}^{a+c} f(x)\, dt = 0 \text{ and } F(x) = \int_{a-c}^{a} f(t)\, dt.$$

Prove that $\quad \left| \displaystyle\int_{a-c}^{a+c} f(x)\, dx \right| \leq Mc^2.$

23. If f be defined in the interval $[0, 1]$ by the condition that if, r, is a positive integer,

$$f(x) = (-1)^{r-1}, \text{ when } 1/(r + 1) < x < 1/r.$$

prove that $\quad \displaystyle\int_0^1 f(x)\, dx = \log 4 - 1.$ *(Agra 2012)*

24. A function f is defined in $[0, 1]$ as follows :

$$f(x) = \frac{1}{a^r - 1} \text{ when } \frac{1}{a^r} < x \leq \frac{1}{a^r - 1} \text{ for } r = 1, 2, 3, \ldots$$

where a is an integer greater than 2. Show that

$\displaystyle\int_0^1 f(x)\, dx$ exists and is equal to $\dfrac{a}{a + 1}$. *(Agra 2008; Delhi Maths (H) 1997)*

25. If $f(x) = 0 \;\; \forall\, x \in [0, 1]$ except at the set of points $\{x_1, x_2, \ldots, x_r, \ldots, x_n\}$, $n \in N$

and $f(x_n) = 1/\sqrt{n}$; show that f is integrable in $[0, 1]$. *(Delhi Maths (H) 2002)*

26. By repeadtly employing the method of integration by parts to the integral $\int_0^x e^{-t}\, dt$.

show that $$e^x = 1 + x + \frac{x^2}{2!} + \frac{x^3}{3!} + ... + \frac{x^n}{n!} + e^x \int_0^x \frac{t^n}{n!} e^{-t}\, dt,$$

and deduce the Maclaurin's infinite series for e^x.

27. Obtain by integration, from the identity

$$\frac{1}{1+t} = 1 - t + t^2 - t^3 + ... + (-1)^{n-1} t^n + (-1)^n \frac{t^n}{1+t},$$

the Maclaurin's infinite series for log $(1 + x)$ in $[-1, 1]$.

28. By applying the mean value theorem of Integral Calculus, show that

(*i*) $\frac{\pi^2}{9} \le \int_{\pi/6}^{\pi/2} \frac{x}{\sin x}\, dx \le \frac{2\pi^2}{9}$ (*ii*) $\frac{1}{3\sqrt{2}} \le \int_0^1 \frac{x^2}{\sqrt{1-x^2}}\, dx \le \frac{1}{3}$

(*iii*) $\frac{\pi}{6} < \int_0^{1/3} \frac{dx}{\sqrt{[(1-x^2)\ (1-k^2x^2)]}} \le \frac{\pi}{6} \cdot \frac{1}{\sqrt{1-(k^2/2)}}$.

29. If f is Riemann integrable on [a, b], show that the indefinite integral $F(x) = \int_a^x f(t)\, dt$ $\forall\, x \in [a, b]$ is differentiable from the right at each x_0 such that $a \le x_0 \le b$ for which f is continuous from the right. *(Delhi Maths (H) 1998)*

30. By an example, show that the continuity assumption cannot be dropped in the first mean value theorem.

31. Let $f(x) \ge g(x)$ $\forall\, x \in [a, b]$ and f and g are both bounded and Riemann integrable on $[a, b]$. At a point c $\in$ $[a, b]$, let f and g be continuous and $f(c) > g(c)$, then prove that $\int_a^b f(x)\, dx > \int_a^b g(x)\, dx$ and hence show that $-\frac{1}{2} < \int_a^b \frac{x^3 \cos 5x}{2+x^2} dx < \frac{1}{2}$

(Delhi Maths (H) 2004)

32. It f is Riemann integrable function as $[a, b]$, show that the function F defined by $F(x) = \int_a^x f(t)\, dt$ for all x in $[a, b]$ is uniformly continuous. *(Delhi Maths (H) 2005, 09)*

OBJECTIVE QUESTIONS

Multiple Choice Questions. *Choose the corrent answer from the given alternatives in the following questions.*

1. If f be a bounded function defined on $[a, b]$ and P_1, P_2 be two partitions of $[a, b]$ such that P_2 is refinement of P_1, then

(*A*) $L(P_2, f) \le L(P_1, f)$ (*B*) $L(P_2, f) \ge U(P_1, f)$

(C) $U(P_2, f) \ge U(P_1, f)$ (*D*) None of these **(Meerut 2013)**

2. For Riemann integrability, condition of continuity is

(*A*) necessary (*B*) sufficient (*C*) necessary and sufficent (*D*) None of these

3. If f is Riemann integrable on $[a, b]$, then

(A) $\left|\int_a^b f(x)\,dx\right| \le \int_a^b |f(x)|\,dx$ (B) $\left|\int_a^b f(x)\,dx\right| \ge \int_a^b |f(x)|\,dx$

(C) $\left|\int_a^b f(x)\,dx\right| = \int_a^b |f(x)|\,dx$ (D) None of these **(Meerut 2013)**

4. If the function $f(x)$ is bounded and integrable on $[a, b]$ such that $f(x) \ge 0 \ \forall x \in [a, b]$, where $b \ge a$, then $\int_a^b f(x)\,dx$.

(A) ≤ 0 (B) $=0$ (C) ≥ 0 (D) None of these

5. If $f(x) = x \ \forall x \in [0, 3]$ and $P = \{0, 1, 2, 3\}$ be a partition of P, then $L(P, f)$ and $U(P, f)$ are respectively

(A) 3, 6 (B) 2, 8 (C) 1, 4 (D) 6, 3

ANSWERS

1. (B) 2. (B) 3. (A) 4. (C) 5. (A)

MISCELLANEOUS PROBELMS ON CHAPTER 13

1. Let $f(x) = \sin(1/x),\ 0 < x \le 1, f(0) = 7$. Is f Riemann integrable on [0, 1] Justify **(Pune 2010]**

[Hint: 0 is the only point of discontinuity of f in [0, 1] and continuous elsewhere in [0,1]. Hence f is integrable by theorem II, page 13.25.

2. Prove that the function of defined as

$$f(x) = \begin{cases} x & \text{when } x \text{ is rational} \\ -x & \text{when } x \text{ is irrational} \end{cases}$$

is not integrable on [0, 1] **[Kanpur 2005]**

3. Let f be defined and bounded over an interval $[a, b]$ and P be a partition of $[a, b]$. If $\lim_{\|P\| \to 0} S(P, f)$ exits, then prove that f is integrable is $[a, b]$ and $\lim_{\|P\| \to 0} S(P, f) = \int_a^b f(x)dx$

[Delhi Maths (H) 2006]

4. Prove that the function f defined on [0, 1] as $f(x) = \begin{cases} 2n, & \text{if } x = 1/n, n = 1, 2, 3, \ldots \\ 0, & \text{other wise} \end{cases}$

is not Riemann integrable on [0, 1]. **[G.N.D.U. Amritsar 2010]**

Hint: $\lim_{x \to 0} f(x) = \infty$, f is not bounded above and hence not Riemann integrable on [0,1]

5. By using the generalised first mean value theorem, prove that $\dfrac{\pi^2}{24} \le \int_0^{\pi} \dfrac{x^2\,dx}{5 + 3\cos x} \le \dfrac{x^3}{6}$

[Meerut 2006]

6. *Let f be the function defined on* [0, 1] *by* $f(x) = \begin{cases} 1, & \textit{when } x \textit{ is rational} \\ 0, & \textit{when } x \textit{ is irrational} \end{cases}$

Then calculate $\underline{\int_0^1} f$ *and* $\overline{\int_0^1} f$ *and hence show that* $f \notin R[0,1]$ **[Meerut 2007]**

7. *If* $f(x) = x,\ x \in [0, 3]$ *and* $P = [0, 1, 2, 3]$ *is the partition of* $[0, 3]$, *then show that* $U(P, f) = 6, L(P, f) = 3.$ **[Meerut 2007]**

8. *Show that the function defined by* $f(x) = \begin{cases} 0, & \text{when } x \text{ is rational} \\ 2, & \text{when } x \text{ is irrational} \end{cases}$

is not Reimann integrable on $[1, 2]$. **[Purvanchal 2006]**

9. *Let* $f: [a, b] \to R$ *be a continuous function on interval* $[a, b]$, *then prove that* $\overline{\int_a^b} f = \underline{\int_a^b} f$ **[Meerut 2007]**

10. *If* f_1, f_2 *are two R-integrable functions on* $[a, b]$ *then for* $k_1,\ k \in R$, prove that $k_1 f_1 + k_2 f_2$ *is also R-integrable.* **[Meerut 2007]**

11. *State and prove Darboux theorem involving upper Riemann integral.* **[Purvanchal 2006]**

12. *Show that a continuous function on a closed interval* $[a, b]$ *is integrable. Give example of a function which is integrable but not continuous on* $[a, b]$ **[Delhi Maths (Prog) 2007]**

13. State two classes of Riemann integrable functions. Give an example of a function which lies in neither of two of the classes stated above but still Riemann integrable. Justify your answer in the example with your mathematical argument. **[Delhi maths (H) 2008]**

14. Suppose f is Riemann integrable on the closed and bounded interval $[a, b]$. Define $f^+, f^- : [a,b] \to \boldsymbol{R}$ by $f^+(x) = \max\{f(x), 0\}$ and $f^-(x) = \max\{-f(x), 0\}$. Show that f^+ and f^- are integrable on $[a, b]$ and $\int_a^b f = \int_a^b f^+ - \int_a^b f^-$. **[Delhi Maths (H) 2008]**

15. If a function f is bounded aand integrable on $[a, b]$, show that $|f|$ is also bounded and integrable on $[a, b]$. **[Delhi maths (H) 2008]**

16. Evaluate the integral $\int_0^1 x\,dx$ using Riemann concept of integration. **[Agra 2006]**

17. Evaluate the integral $\int_a^b x^2 dx$ by using the concept of Riemann integration. **[Agra 2007]**

18. Find the lower Riemann integral of $\int_0^1 (x^2 + 1)\,dx$. **[Delhi B.Sc. (Prog) III 2008]**

[Ans. 4/3**]**

19. Let $f(x)$ be a function defined on [0, 1] as follows : $f(x) = \begin{cases} 0, & \text{when } x \text{ is rational} \\ 1, & \text{when } x \text{ is irrational} \end{cases}$

Calculate $\underline{\int_0^1} f$ and $\overline{\int_0^1} f$ and show that $f \notin R[a,b]$. **[Agra 2007]**

20. Let $f(x) = x$ on closed interval [0,1]. Calculate $\underline{\int_0^1} x\,dx$ and $\overline{\int_0^1} x\,dx$ by dividing [0, 1] into n equal parts. Then prove that $f \in R[a,b]$. **[Agra 2003, 08]**

21. Prove the (*i*) $\frac{1}{3\sqrt{2}} \le \int_0^1 \frac{x^2}{\sqrt{1+x}}\,dx \le \frac{1}{3}$ **(Pune 2010)**

(*ii*) $2\sqrt{2} \le \int_1^3 \sqrt{1+x^3}\,dx \le 2\sqrt{28}$ **(G.N.D.U. Amritsar 2010)**

22. Let $P = \{0, 1, 2, 4\}$ be a partition of the interval [0, 4]. Let $f(x) = x^2$. Find

(*i*) norm P (*ii*) $U(P, f)$ (*iii*) $L(P, f)$. **[Delhi B.Sc III (Prog) 2009]**

Ans (*i*) Norm $P = 2$ (*ii*) $U(P, f) = 37$ (*iii*) $L(P, f) = 9$

23. (a) Show that the function f(x) defined on the interval [0, 5] as $f(x) = x[x]$ is integrable State clearly any result that you will use. $[x]$ denotes greatest integer less than or equal to x.

[Delhi B.A III (Prog) 2009]

(*b*) Show that the function defined by $f(x) = x[x], \forall\, x \in [0,3]$ is Riemann integrable on [0, 3], $[x]$ being the greatest integer function. Also evaluate $\int_0^3 f(x)dx$. **[Delhi B.A. (Prog) III 2010]**

Sol. (*a*) Since the given function is bounded and has only five points of discontinuity (namely $x = 1, 2, 3, 4, 5$), hence it is integrable (refer theorem II, Art. 13.8) and

$$\int_0^5 x[x]\,\text{dx} = \int_0^1 x[x]\,\text{dx} + \int_1^2 x[x]\,\text{dx} + \int_2^3 x[x]\,\text{dx} + \int_3^4 x[x]\,\text{dx} + \int_4^5 x(x)\,\text{dx}$$

$$= 0 + \int_1^2 x\,dx + 2\int_2^3 x\,dx + 3\int_3^4 x\,dx + 4\int_4^5 x\,dx$$

$$= [x^2/2]_1^2 + 2[x^2/2]_2^3 + 3[x^2/2]_3^4 + 4[x^2/2]_4^5 = 35, \text{ on simplification}$$

(*b*) **Hint.** Proceed as in part (a). The value of integral is 13/2.

24. If $f : [1, 2] \to \boldsymbol{R}$ is a non-negative Riemann integrable function such that $\int_1^2 \frac{f(x)}{\sqrt{x}}dx = k\int_1^2 f(x)dx \ne 0$, then k belongs to the

(*a*) [0, 1/3] (*b*) (1/3, 2/3] (*c*) (2/3, 1] (*d*) (1, 4/3] **(GATE 2010)**

25. Using fundamental theorem of Integral Calculus, compute $\int_0^{\pi/3} \cos x\,dx$ **[Meerut 2011]**

26. Let f be a monotonically increasing function on a closed and bounded interval $[a, b]$. Prove that f is Riemann integrable on $[a, b]$. **[Delhi B.A. (Prog.) III 2011]**

27. Let f be bounded function on $[a, b]$ so that there exists $B > 0$ such that $|f(x)| \le B \;\; \forall\, x \in [a,b]$. Show that $U(P, f^2) - L(P, f^2) < 2B\,[U(P, f) - L(P, f)]$ for all partitions. Hence show that if f is integrable then f^2 is also integrable. **[Delhi B.Sc. (Hons) II 2011]**

Hint. Refer theorem on page 13.34.

28. Suppose the functions f and g are integrable on $[a, b]$ and suppose that $f(x) \le g(x)$ for all $x \in [a,b]$. Show that $\int_a^b f \le \int_a^b g$. **[Delhi B.Sc. (Hons) II 2011]**

[**Sol.** $g(x) \geq f(x) \Rightarrow |g(x) - f(x)| \geq 0$, for all $x \in [a, b]$

$\Rightarrow \int_a^b [g(x) - f(x)]dx \geq 0$, if $b \geq 0$ by theorem V, page 13.38

$$\Rightarrow \int_a^b g(x)dx - \int_a^b f(x)dx \geq 0 \quad \Rightarrow \quad \int_a^b f(x)dx \leq \int_a^b g(x)dx \quad \Rightarrow \quad \int_a^b f \leq \int_a^b g$$

29. *Find a point c in [0, π/2] such that* $\int_0^1 \frac{x}{1+x^2}\, dx = 1.$

Sol. Let $f(x) = x$ and $\phi(x) = 1/(1 + x^2)$. Since f and ϕ are continuous in [0, 1], hence $\int_0^1 f(x)\, dx$ and $\int_0^1 \phi(x)\, dx$ exist. Also, $\phi(x) > 0 \ \forall \ x \in [0, 1]$. Since 0 and 1 are lower and upper bounds of $f(x)$ in [0, 1], hence by second mean value theorem (see page 13.41), there exists a number $c \in [0, 1]$ such that

$$\int_0^1 f(x)\, \phi(x)\, dx = c \int_0^1 \phi(x)\, dx \quad \text{or} \quad \int_0^1 \frac{x}{1+x^2} dx = c \int_0^1 \frac{1}{1+x^2} dx \quad \text{or} \quad \frac{1}{2}[\log(1+x^2)]_0^1 = c[\tan^{-1} x]_0^1$$

Thus. $(1/2) \times \log 2 = c \times (\pi/4)$ so that $c = (2/\pi) \times \log 2 \in [0, \pi/2]$

Chapter

14

The Riemann – Stieltjes Integral

14.1 INTRODUCTION

In the previous chapter on Riemann integration we have seen that Riemann was the first mathematician who gave a rigorous arithmetic treatment of integration. This concept was later generalised by Thomas Joaner Stieltjes and we now propose to study this generalisation, known as Riemann-Stieltjes integral. Riemann integral is its special case.

14.2 Partition, refinement of a partition and norm of a partition. Lower and Upper Riemann-Stieltjes sums (or simply the lower and upper sums)

(*i*) **Partition of a closed interval-** Let $I = [a,b]$ be a finite closed interval. If $a = x_0 < x_1 < x_2 < \ldots < x_n = b$, then the finite ordered set $P = \{x_0, x_1, x_2, \ldots, x_n\}$ in known as a partition of I. The n closed intervals $I_1 = [x_0,x_1]$, $I_2 = [x_1,x_2], \ldots, I_r = [x_{r-1},x_r], \ldots, I_n = [x_{n-1},x_n]$ determined by P are known the segments of the partition P. The length of the rth sub-interval $I_r = [x_{r-1},x_r]$ is denoted by δ_r.

Thus, $\delta_r = x_r - x_{r-1}$.

By changing the partition points, the partition can be changed and therefore, we can obtain an infinite number of partitions of $[a,b]$. The set (or family) of all partitions of $[a,b]$ is denoted by $P[a,b]$

(*ii*) **Norm of a partition-** The length of the greatest of all the intervals I_r of the partition P will be called its norm and denoted by $\| P \|$ or $\mu(P)$. Thus,

$$\| P \| = \mu(P) = \max\{\delta_r : 1 \le r \le n\}$$

(*iii*) **Refinement of a partition-** If P_1, P_2 be two partitions of $[a,b]$ such that $P_2 > P_1$, *i.e.,* every point of the partition P_1 in as well as point of P_2, we shall say that P_2 is finer than P_1 (or P_2 is a refinement of P_1).

If P_1 and P_2 are two partitions of [a,b], then $P_1 \cup P_2$ is called a common refinement of P_1 and P_2.

(*iv*) **Upper and lower sums-** Let $f(x)$ and $\alpha(x)$ be bounded functions on $[a,b]$ and α be monotonic increasing on $[a,b]$. Also, let $b \ge a$. We shall write

$$\delta\alpha_r = \alpha(x_r) - \alpha(x_{r-1}),\ r = 1,2,\ldots,n \qquad \ldots (1)$$

corresponding to any partition P of $[a,b]$. Since α is monotomic increasing on $[a,b]$, we have

$$\delta\alpha_r \ge 0\ \forall\, r = 1,2,\ldots,n \qquad \ldots (2)$$

The function f which is defined in $[a,b]$ is also necessarily bounded in each sub-interval I_r. Let M_r, m_r be respectively, the supremum and the infimum of f in I_r, *i.e.,* in $[x_{r-1}, x_r]$. Set up the two sums as follows.

$$U(P, f, \alpha) = \sum_{r=1}^{n} M_r \delta\alpha_r \quad \text{and} \quad L(P, f, \alpha) = \sum_{r=1}^{n} m_r \delta\alpha_r \qquad \text{... (3)}$$

$U(P, f, \alpha)$ and $L(P, f, \alpha)$ are called upper Riemann-Stieltjes sum (or simply upper sum) and lower Riemann Stieltjes sum (or simply lower sum) respectively, of f with respect to α and corresponding to the partition P.

Note 1. Since $\forall r, m_r \le M_r$, so $m_r \delta\alpha_r \le M_r \delta\alpha_r$

$$\Rightarrow \sum_{r=1}^{n} m_r \delta\alpha_r \le \sum_{r=1}^{n} M_r \delta\alpha_r \Rightarrow L(P, f, \alpha) \le U(P, f, \alpha) \qquad \text{... (4)}$$

Note 2. As a particular case, if $\alpha(x) = x$, then the so called Riemann-Stieltjes sums reduce to the Riemann sums which have been defined in the previous chapter. So by taking $\alpha(x) = x$, the Riemann integral can be easily seen to be a special case of Riemann-Stieltjes integral.

Note 3. In this chapter, unless otherwise stated, all functions will be taken as bounded and the function α will be always monotonic increasing.

Note 4. Some authors write $g(x)$ instead of $\alpha(x)$.

Theorem I. *If P^* is a refinement of P, then prove that*

(*i*) $L(P, f, \alpha) \le L(P,{}^* f, \alpha)$ (*ii*) $U(P^*, f, \alpha) \le U(P, f, \alpha)$ **[Bhopal 2004; Jabalpur 2004; Agra 2012; Himachal 2015; Jiwaji 1999; Kanpur 2001,05; Ravishankar 2003]**

Proof: Let $P = \{a = x_0, x_1, x_2, ..., x_{r-1}, x_r, ..., x_n = b\}$

and $P^* = \{a = x_0, x_1, x_2, ..., x_{r-1}, \xi, x_r, ..., x_n = b\}$.

Clearly, P^* is refinement of P because it contains an additional point ξ, where $x_{r-1} < \xi < x_r$.

Let m'_r, m''_r and m_r be the infimum of f in the intervals $[x_{r-1}, \xi], [\xi, x_r]$ and $[x_{r-1}, x_r]$ respectively. Again, let M'_r, M''_r and M_r be the supremum of f in the intervals $[x_{r-1}, \xi], [\xi, x_r]$ and $[x_{r-1}, x_r]$ respectively. Then, we have

$$m_r \le m'_r, m_r \le m''_r \text{ and } M_r \ge M'_r, M_r \ge M''_r \qquad \text{... (1)}$$

(*i*) We now prove that $L(P, f, \alpha) \le L(P^* f, \alpha)$.

From our construction of partitions P and P^*, it follows that the contribution to $L(P, f, \alpha)$ and $L(P^* f, \alpha)$ of each sub-interval except $[x_{r-1}, x_r]$ is the same. Now, the contribution of the sub-interval $[x_{r-1}, x_r]$ to $L(P, f, \alpha)$ is $m_r \{\alpha(x_r) - \alpha(x_{r-1})\}$.

Since the sub-interval $[x_{r-1}, x_r]$ in spilt into two sub-intervals $[x_{r-1}, \xi]$ and $[\xi, x_r]$ it follows that the contribution of the sub-interval $[x_{r-1}, x_r]$ to $L(P^* f, \alpha)$ is

$$m'_r \{\alpha(\xi) - \alpha(x_{r-1})\} + m''_r \{\alpha(x_r) - \alpha(\xi)\}.$$

Hence, we have

$$\begin{aligned}
&L(P^* f, \alpha) - L(P, f, \alpha) \\
&= m'_r \{\alpha(\xi) - \alpha(x_{r-1})\} + m''_r \{\alpha(x_r) - \alpha(\xi)\} - m_r \{\alpha(x_r) - \alpha(x_{r-1})\} \\
&= m'_r \{\alpha(\xi) - \alpha(x_{r-1})\} + m''_r \{\alpha(x_r) - \alpha(\xi)\} - m_r [\{\alpha(x_r) - \alpha(\xi)\} + \{\alpha(\xi) - \alpha(x_{r-1})\}] \\
&= (m'_r - m_r) \{\alpha(\xi) - \alpha(x_{r-1})\} + (m''_r - m_r) \{\alpha(x_r) - \alpha(\xi)\} \\
&\ge 0,
\end{aligned}$$

because $m'_r \geq m_r, m''_r \geq m_r$ by (1) and $\alpha(\xi_r) > \alpha(x_{r-1})$ and $\alpha(x_r) > \alpha(\xi)$ since α is monotonic increasing function on $[x_{r-1}, x_r]$ and $x_{r-1} < \xi < x_r$.

Thus, $L(P^*, \alpha, f) - L(P, f, \alpha) \geq 0$ so that $L(P, f, \alpha) \leq L(P^*, f, \alpha)$.

If P^* contains p points more than P, then repeating the above argument p times we obtain

$$L(P, f, \alpha) \leq L(P^*, f, \alpha)$$

(*ii*) Proceed exactly as in part (*i*)

Theorem II. *If P_1 and P_2 be any two partitions of $[a,b]$, then $U(P_2, f, \alpha) \geq L(P_1, f, \alpha)$, that is no lower sum can exceed any upper sum.*

Proof. Let $P^* = P_1 \cup P_2$ be a common refinement of P_1 and P_2. Then, by theorem I, we have,

$$L(P_1, f, \alpha) \leq L(P^*, f, \alpha) \quad \text{... (1)}$$

and $$U(P^*, f, \alpha) \leq U(P_2, f, \alpha) \quad \text{... (2)}$$

Also, we have $$L(P^*, f, \alpha) \leq U(P^*, f, \alpha) \quad \text{... (3)}$$

From (1), (2) and (3), we have

$$L(P_1, f, \alpha) \leq L(P^*, f, \alpha) \leq U(P^*, f, \alpha) \leq U(P_2, f, \alpha)$$

$$\Rightarrow \quad L(P_1, f, \alpha) \leq U(P_2, f, \alpha)$$

$$\Rightarrow \quad U(P_2, f, \alpha) \geq L(P_1, f, \alpha)$$

Theorem III. *Let m, M be the infimum and supremum of f on $[a,b]$. Then, for any partition P of $[a,b]$*

$$m\{\alpha(b) - \alpha(a)\} \leq L(P, f, \alpha) \leq U(P, f, \alpha) \leq M\{\alpha(b) - \alpha(a)\}$$ **[Meerut 2010]**

Proof. Let $P = \{a = x_0, x_1, x_2, ..., x_{r-1}, x_r, ..., x_n = b\}$ be any partition of $[a,b]$. Since f is bounded on $[a, b]$, so f is bounded on each sub-interval $[x_{r-1}, x_r]$ $r = 1, 2, ..., n$. Let m_r and M_r be the infimum and supremum of f on $[x_{r-1}, x_r]$. Then, for every value of r, we have

$$m \leq m_r \leq M_r \leq M$$

$$\Rightarrow m\delta\alpha_r \leq m_r\delta\alpha_r \leq M_r\delta\alpha_r \leq M\delta\alpha_r \text{ as } \delta\alpha_r \geq 0$$

$$\Rightarrow \sum_{r=1}^{n} m\delta\alpha_r \leq \sum_{r=1}^{n} m_r\delta\alpha_r \leq \sum_{r=1}^{n} M_r\delta\alpha_r \leq \sum_{r=1}^{n} M\delta\alpha_r$$

$$\Rightarrow m\sum_{r=1}^{n} \delta\alpha_r \leq L(P, f, \alpha) \leq U(P, f, \alpha) \leq M\sum_{r=1}^{n} \delta\alpha_r$$

$$\Rightarrow m\{\alpha(b) - \alpha(a)\} \leq L(P, f, \alpha) \leq U(P, f, \alpha) \leq M\{\alpha(b) - \alpha(a)\}$$

$$\left[\because \sum_{r=1}^{n} \delta\alpha_r = \sum_{r=1}^{n} \{\alpha(x_r) - \alpha(x_{r-1})\}\right.$$

$$= \alpha(x_1) - \alpha(x_0) + \alpha(x_2) - \alpha(x_1) + ... + \alpha(x_n) - \alpha(x_{n-1})$$

$$\left.= \alpha(x_n) - \alpha(x_0) = \alpha(b) - \alpha(a), \text{ as } x_0 = a, x_n = b\right]$$

Note. The above result shows that the set $\{L(P, f, \alpha): P$ in a partition of $[a, b]\}$ of lower sums in bounded above by $M(b-a)$ and hence, has the least upper bound. Similarly, we see that the set $\{U(P, f, \alpha): P$ is a partition of $[a, b]\}$ is bounded below by $m(b-a)$ and, hence, has the greatest lower bound.

14.3 THE LOWER AND UPPER RIEMANN-STIELTJES INTEGRALS. DEFINATIONS

(Agra 2011)

Let $f(x)$ and $\alpha(x)$ be bounded functions on $[a,b]$ and α be monotonic increasing on $[a,b]$, where $b \geq a$.

The lower Riemann-Stieltjes integral of f relative to α over $[a,b]$ is the supremum (or the least upper bound) of $L(P, f, \alpha)$ over all partition P of $[a,b]$ and is denoted by

$$\underline{\int_a^b} f \, d\alpha \qquad \text{or} \qquad \underline{\int_a^b} f(x) \, d\alpha(x)$$

Thus, $$\underline{\int_a^b} f \, d\alpha = \sup \{L(P, f, \alpha): P \text{ is a partition of } [a, b]\}$$

The upper Riemann-Stieltjes integral of f relative to α over $[a, b]$ is the infimum (or the greatest lower bound) of $U(P, f, \alpha)$ over all partition of P of $[a, b]$ and is denoted by

$$\overline{\int_a^b} f \, d\alpha \qquad \text{or} \qquad \overline{\int_a^b} f(x) \, d\alpha(x)$$

Thus, $$\overline{\int_a^b} f \, d\alpha = \inf \{U(P, f, \alpha): P \text{ is a partition of } [a,b]\}$$

Theorem I. *The lower Riemann-Stieltjes integral cannot exceed the upper Riemann-Stieltjes integral i.e.,*

$$\underline{\int_a^b} f \, d\alpha \leq \overline{\int_a^b} f \, d\alpha$$ ***(G.N.D.U. Amritsar 2001, 03, 05; Agra 2011; Jiwaji 2000; Himachal 2013, 14)***

Proof. Let $P[a, b]$ denote the set of all partitions of $[a, b]$. Let $P_1, P_2 \in P[a, b]$. Since no lower sum can exceed any upper sum, we have

$$L(P_1, f, \alpha) \leq U(P_2, f, \alpha) \qquad \text{... (1)}$$

(1) is true for each $P_1 \in P[a, b]$. Keeping P_2 fixed we see that the set $\{L(P_1, f, \alpha): P_1 \in [a,b]\}$ has an upper bound $U(P_2, f, \alpha)$. Again, we know that

$$\underline{\int_a^b} f \, d\alpha = \sup \{L(P_1, f, \alpha): P_1 \in [a,b]\}$$

But supremum $\leq$ any upper bound. Hence, we get

$$\underline{\int_a^b} f \, d\alpha \leq U(P_2, f, \alpha), \qquad \text{... (2)}$$

which is true for each $P_2 \in P[a, b]$. From (2), it follows that the set $\{U(P_2, f, \alpha) : P_2 \in P[a, b]\}$

has a lower bound $\underline{\int_a^b} f\, d\alpha$. Also, we know that

$$\overline{\int_a^b} f\, d\alpha = \inf\{U(P_2, f, \alpha) : P_2 \in P[a, b]\}.$$

But any lower bound $\leq$ infimum. Hence, we get

$$\underline{\int_a^b} f\, d\alpha \leq \overline{\int_a^b} f\, d\alpha.$$

Theorem II. Let f and α be bounded functions on $[a, b]$ and α be monotonic increasing on $[a, b]$. Then, for any partition P of $[a, b]$,

$$L(P, f, \alpha) \leq \underline{\int_a^b} f\, d\alpha \leq \overline{\int_a^b} f\, d\alpha \leq U(P, f, \alpha)$$

Proof. From theorem I, we have

$$\underline{\int_a^b} f\, d\alpha \leq \overline{\int_a^b} f\, d\alpha \qquad \text{... (1)}$$

By the definition of upper and lower integrals, we have

$$L(P, f, \alpha) \leq \underline{\int_a^b} f\, d\alpha \quad \text{and} \quad \overline{\int_a^b} f\, d\alpha \leq U(P, f, \alpha) \qquad \text{... (2)}$$

From (1) and (2), $\quad L(P, f, \alpha) \leq \underline{\int_a^b} f\, d\alpha \leq \overline{\int_a^b} f\, d\alpha \leq U(P, f, \alpha).$

14.4 THE RIEMANN-STIELTJES INTEGRAL. (Meerut 2004, 05, 06, 08; Ravishankar 2004]

Definition. Let f and α be bounded functions on $[a, b]$ and α be monotonic increasing on $[a,b]$. Also, let $b \geq a$. Then f in called Riemann-Stieltjes integrable (or simply or RS- integrable) on $[a,b]$ relative to α if and only if $\quad \underline{\int_a^b} f\, d\alpha = \overline{\int_a^b} f\, d\alpha$

and their common value is called the Riemann-Stieltjes integral, or simply the Stieltjes integral of f on $[a, b]$ relative to α and is denoted by

$$\int_a^b f\, d\alpha \quad \text{or} \quad \int_a^b f(x)\, d\alpha\, (x)$$

The function f is called the integrand and α is called the integrator. The class of all R-S integrable functions on $[a,b]$ relative to α is denoted by

$$R(\alpha) \quad \text{or} \quad R\text{–}S(\alpha) \quad \text{or} \quad RS([a,b], \alpha) \quad \text{or} \quad R_\alpha\, [a,b]$$

Note. When $b < a$, we define $\int_a^b f\, d\alpha = -\int_b^a f\, d\alpha$ Also, we define $\int_a^a f\, d\alpha = 0$

NECESSARY AND SUFFICIENT CONDITION FOR RS- INTEGRABILITY

Theorem *Let f and α be bounded functions on $[a,b]$ and α be monotonic increasing on $[a,b]$. Then the function f is integrable with respect to α on $[a,b]$ if and only if for every $\varepsilon > 0$ there exists a partition P of $[a,b]$ such that* $\quad U(P, f, \alpha) - L(P, f, \alpha) < \varepsilon$

[*Agra 2000, 03, 04, Himachal 2007, 2010, 2015; Bhopal 1998; Jiwaji 2001; Kanpur 2002,11; Meerut 2004, 06, 08; Nagpur 2003; Punjab 2003, 05; Ravinshankar 2004, 06*]

Proof. The condition is necessary. Let f be RS-integrable relative to α on $[a,b]$ so that

$$\underline{\int_a^b} f\,d\alpha = \overline{\int_a^b} f\,d\alpha = \int_a^b f\,d\alpha.$$

Let $\varepsilon > 0$ be any real number. Since the upper and lower RS-integrals are the infimum and the supremum, respectively, of the upper and lower RS- sums, hence there exist partition P_1 and P_2 of $[a, b]$ such that

$$U(P_1, f, \alpha) < \overline{\int_a^b} f\,d\alpha + \frac{1}{2}\varepsilon \quad \text{or} \quad U(P_1, f, \alpha) < \int_a^b f\,d\alpha + \frac{1}{2}\varepsilon \qquad \text{... (1)}$$

and $$L(P_2, f, \alpha) > \overline{\int_a^b} f\,d\alpha - \frac{1}{2}\varepsilon \quad \text{or} \quad \int_a^b f\,d\alpha < L(P_2, f, \alpha) + \frac{1}{2}\varepsilon \qquad \text{... (2)}$$

Let $P = P_1 \cup P_2$ be the common refinement of P_1 and P_2.

Then $$\begin{aligned} U(P, f, \alpha) &\le U(P_1, f, \alpha) \\ &< \int_a^b f\,d\alpha + \frac{1}{2}\varepsilon, \text{ by (1)} \\ &< L(P_2, f, \alpha) + \varepsilon, \text{ by (2)} \\ &< L(P, f, \alpha) + \varepsilon \end{aligned}$$

Thus, $$U(P, f, \alpha) - L(P, f, \alpha) < \varepsilon,$$

showing that the condition is necessary.

The condition is sufficient. Let for every $\varepsilon > 0$, there exists a partition P such that

$$U(P, f, \alpha) - L(P, f, \alpha) < \varepsilon \qquad \text{... (3)}$$

By the definition of upper and lower RS-integrals, we have

$$\overline{\int_a^b} f\,d\alpha \le U(P, f, \alpha) \quad \text{and} \quad \underline{\int_a^b} f\,d\alpha \ge L(P, f, \alpha)$$

$$\Rightarrow \quad \overline{\int_a^b} f\,d\alpha \le U(P, f, \alpha) \qquad \text{... (4)}$$

and $$-\underline{\int_a^b} f\,d\alpha \le -L(P, f, \alpha) \qquad \text{... (5)}$$

Adding (4) and (5), $$\overline{\int_a^b} f\,d\alpha - \underline{\int_a^b} f\,d\alpha \le U(P, f, \alpha) - L(P, f, \alpha) \qquad \text{... (6)}$$

From (3) and (6), $$0 < \overline{\int_a^b} f\,d\alpha - \underline{\int_a^b} f\,d\alpha < \varepsilon, \text{ for every } \varepsilon > 0$$

showing that a non-negative number is less than every positive number (namely $\in$) and it can happen only if the that non-negative number is itself zero, *i.e.*, if

$$\overline{\int_a^b} f\,d\alpha - \underline{\int_a^b} f\,d\alpha = 0 \quad \text{or} \quad \overline{\int_a^b} f\,d\alpha = \underline{\int_a^b} f\,d\alpha,$$

showing that the function f is RS-integrable with respect to α on $[a,b]$ and so the condition is sufficient.

14.5 THE RIEMANN STIELTJES SUM AS A LIMIT OF SUM [Kanpur 2006]

Riemann-Stieltjes sum (or RS-sum) Definition: Let f be a bounded function and α be a monotonic increasing function on $[a,b]$. Let $P = \{a = x_0, x_1, ..., x_{r-1}, x_r, ..., x_n = b\}$ be a partition of $[a,b]$. Let $\xi_r \in [x_{r-1}, x_r]$, for $r = 1, 2, ..., n$. Then, the sum

$$S(P, f, \alpha) = \sum_{r=1}^{n} f(\xi_r)\,\delta\alpha_r$$

is called the Riemann-Stieltjes sum of f relative to α on $[a, b]$ corresponding to the partition P.

The sum $S(P, f, \alpha)$ is said to converge to l as $\mu(P) \to 0$, if for every $\varepsilon > 0$, there exists $\delta > 0$ such that
$$|S(P, f, \alpha) - l| < \varepsilon \quad \forall \|P\| < \delta$$

Theorem I. *If* $\lim\limits_{\|P\|\to 0} S(P, f, \alpha)$ *exists, then f is Riemann Stieltjes integrable on $[a,b]$ and*

$$\lim_{\|P\|\to 0} S(P, f, \alpha) = \int_a^b f\,d\alpha$$

(*Himachal 2002, 03; Agra 1999, 2000; Khurkshetra 2005; Rohtak 2004*)

Proof. Assume that $\lim S(P, f, \alpha)$ exists as $\|P\| \to 0$ and
$$\lim_{\|P\|\to 0} S(P, f, \alpha) = l$$

Hence for $\varepsilon > 0$, there exists $\delta > 0$ such that for every partition P of $[a,b]$ with $\|P\| < \delta$, we have
$$|S(P, f, \alpha) - l| < \varepsilon/2$$

i.e.,
$$l - \varepsilon/2 < S(P, f, \alpha) < l + \varepsilon/2 \qquad \text{... (1)}$$

Let P be one such partition. Letting the points ξ_r range over the sub-interval $[x_{r-1}, x_r]$ and taking the infimum and the supremum of the sums $S(P, f, \alpha)$ so obtained, (1) gives
$$l - \varepsilon/2 < L(P, f, \alpha) \le U(P, f, \alpha) < l + \varepsilon/2 \qquad \text{... (2)}$$

$$\Rightarrow \quad U(P, f, \alpha) < l + \varepsilon/2 \quad \text{and} \quad -L(P, f, \alpha) < -l + \varepsilon/2$$

$$\Rightarrow \quad U(P, f, \alpha) - L(P, f, \alpha) < \varepsilon \quad \Rightarrow \quad f \text{ is RS- integrable over } [a, b]$$

Since f is RS-integrable, we have

$$\underline{\int_a^b} f\,d\alpha = \overline{\int_a^b} f\,d\alpha = \int_a^b f\,d\alpha \qquad \text{... (3)}$$

We know that

$$L(P, f, \alpha) \le \underline{\int_a^b} f\,d\alpha \le \overline{\int_a^b} f\,d\alpha \le U(P, f, \alpha) \qquad \text{... (4)}$$

From (3) and (4),
$$L(P, f, \alpha) \le \int_a^b f\,d\alpha \le U(P, f, \alpha) \qquad \text{... (5)}$$

Then, from (2) and (5), we obtain

$$l-\varepsilon/2<\int_a^b f\,d\alpha<l+\varepsilon/2 \qquad \text{... (6)}$$

From (2) and (6), we have

$$l-\varepsilon/2<\int_a^b f\,d\alpha<l+\varepsilon/2,\quad i.e.,\quad |l-\int_a^b f\,d\alpha|<\varepsilon/2<\varepsilon \qquad \text{... (7)}$$

Since a non-negative number can be less than every positive number, if it is itself zero, hence (7) yields

$$\therefore \qquad l=\int_a^b f\,d\alpha,\qquad i.e.,\qquad \lim_{\|P\|\to 0} S(P,f,\alpha)=\int_a^b f\,d\alpha.$$

Note. We have shown that a sufficient condition for the existence of RS-integral is the existance of the $\lim S(P,f,\alpha)$ as $\|P\|\to 0$. But it can be shown that the existence of this limit is not a necessary condition for the existence of RS-integral.

Theorem II. *If f in continuous on [a,b], then f is RS-integrable*

and $$\lim_{\|P\|\to 0} S(P,f,\alpha)=\int_a^b f\,d\alpha.$$ **(Himanchal 2013)**

Proof. Let $\varepsilon>0$ be given. Then, we can choose $\eta>0$ such that

$$\eta\,[\alpha(b)-\alpha(a)]<\varepsilon \qquad \text{... (1)}$$

Since f is continuous on the closed interval $[a, b]$, so f is uniformly continuous on $[a, b]$. Hence for $\eta>0$, there points $\delta>0$ such that

$$x', x''\in[a,b], [x'-x'']<\delta \Rightarrow |f(x')-f(x'')|<\eta \qquad \text{... (2)}$$

Let $P=\{a=x_0, x_1, x_2,\ldots, x_{r-1}, x_r,\ldots, x_n=b\}$ be a partition of $[a,b]$ with $\|P\|<\delta$.

Let m_r, M_r be the infimum and supremum of f in $[x_{r-1}, x_r]$. Since f is continuous on $[a,b]$, so it is continuous on each sub-interval $[x_{r-1}, x_r]$ and hence f will attain its bounds m_r, M_r on $[x_{r-1}, x_r]$. Therefore, there exist $x', x''\in[x_{r-1}, x_r]$ such that $f(x')=M_r$ and $f(x'')=m_r$. Then, from (2)

$$|f(x')-f(x'')|<\eta \qquad \Rightarrow \qquad M_r-m_r<\eta, \text{ where } r=1, 2,\ldots n \qquad \text{... (3)}$$

$$\therefore \qquad U(P,f,\alpha)-L(P,f,\alpha)=\sum_{r=1}^{n}(M_r-m_r)\,\delta\alpha_r<\eta\sum_{r=1}^{n}\alpha\delta_r,\ \text{using (3)}$$

$$=\eta\sum_{r=1}^{n}\{\alpha(x_r)-\alpha(x_{r-1})\}$$

$$=\eta\,[\alpha(x_1)-\alpha(x_0)+\alpha(x_2)-\alpha(x_1)+\ldots+\alpha(x_n)-\alpha(x_{n-1})\}$$

$$=\eta\,\{\alpha(x_n)-\alpha(x_0)\}=\mu\,\{\alpha(b)-\alpha(a)\}<\varepsilon,\ \text{by (1)}$$

Thus, $\qquad U(P,f,\alpha)-L(P,f,\alpha)<\varepsilon \quad\Rightarrow\quad f\varepsilon\, RS(\alpha)$ over $[a,b]$.

Now, since $f\in RS(\alpha)$, so for $\varepsilon>0$, there exists $\delta>0$ such that for all partitions P with $\|P\|<\delta$, we have

$$|U(P,f,\alpha)-L(P,f,\alpha)|<\varepsilon$$

We know that $S(P,f,\alpha)$ and $\int_a^b f\,d\alpha$ lie between $U(P,f,\alpha)$ and $L(P,f,\alpha)$ for all partitions P with $\|P\|<\delta$ and for every choice of ξ_r in $[x_{r-1}, x_r]$. Hence, we have

$$|S(P,f,\alpha)-\int_a^b f\,d\alpha| < U(P,f,\alpha)-L(P,f,\alpha)<\varepsilon$$

$$\therefore \quad \lim_{\|P\|\to 0} S(P,f,\alpha)=\int_a^b f\,d\alpha, \quad \text{where} \quad S(P,f,\alpha)=\sum_{r=1}^{n} f(\xi_r)\,\delta\alpha_r.$$

Note. From the above theorem, it follows that continuity is sufficient condition for RS-integrabity of a function. But it is not necessary condition, *i.e.,* there exist functions which are integrable but for which function is not continuous.

Theorem III. *If f is monotonic on* $[a,b]$, *and if* α *is continuous and monotonic increasing on* $[a,b]$, *then f is RS-integrable.* ***(Agra 2000, G.N.D.U. 1997, Himachal 2010, 2014; Kanpur 2007, 11, 14; Indoor 1997; Jiwaji 1999; Ravishankar 1997; Sagar 1998)***

Proof. Let $\in\,>0$ be given. Since α is continuous and monotonic increasing on $[a,b]$, so α attains every value between its bounds $\alpha(a)$ and $\alpha(b)$. Consequently, for a chosen positive number n, we can choose a partition $P=\{a=x_0, x_1,\ldots, x_{r-1}, x_r,\ldots, x_n=b\}$ of $[a,b]$ such that

$$\delta\alpha_r=\{\alpha(b)-\alpha(a)\}/n, \text{ for } r=1, 2, \ldots, n \qquad \ldots (1)$$

Suppose that f is monotonic increasing on $[a,b]$. Then its lower and the upper bounds m_r, M_r in $[x_{r-1}, x_r]$ are given by

$$m_r=f(x_{r-1}) \text{ and } M_r=f(x_r), \text{ for } r=1, 2, 3,\ldots, n \qquad \ldots (2)$$

Now, $U(P,f,\alpha)-L(P,f,\alpha)$

$$=\sum_{r=1}^{n}(M_r-m_r)\,\delta\alpha_r=\frac{\alpha(b)-\alpha(\alpha)}{n}\sum_{r=1}^{n}\{f(x_r)-f(x_{r-1})\}, \text{ by (1)}$$

$$=\frac{\alpha(b)-\alpha(a)}{n}[f(x_1)-f(x_0)+f(x_2)-f(x_1)+\ldots+f(x_n)-f(x_{n-1})]$$

$$=\frac{\alpha(b)-\alpha(a)}{n}\{f(x_n)-f(x_0)\}=\frac{\alpha(b)-\alpha(a)}{n}\{f(b)-f(a)\}$$

Hence, for large values of n, we have

$$U(P,f,\alpha)-L(P,f,\alpha)<\varepsilon \quad \Rightarrow \quad f \text{ is RS-integrable on } [a, b].$$

EXAMPLES

Example 1. *If f be bounded in* $[a,b]$ *and c be a constant, then evaluate* $\int_a^b f(x)\,dc.$

[*Himanchel 2006, 08, 09; Delhi Maths (H) 2004*]

Solution. Let $P=\{a=x_0, x_1, x_2,\ldots, x_n=b\}$ be any partition of $[a,b]$ and $\xi_r\in[x_{r-1}, x_r]$.

Riemann-Stieltjes sum is given by (noting that $\alpha(x)=c$ in our problem)

$$S(P,f,\alpha)=\sum_{r=1}^{n} f(\xi_r)\delta\alpha_r=\sum_{r=1}^{n} f(\xi_r)[\alpha(x_r)-\alpha(x_{r-1})]$$

$$= \sum_{r=1}^{n} f(\xi_r), [c-c] = 0, \text{ as } \alpha(x) = c$$

$\therefore$ lim $(S, P, \alpha) = 0$ as $\| P \| \to 0$. Since lim $S(P, f, \alpha)$ exists as $\| P \| \to 0$, f is RS-integrable

and $$\int_a^b f(x)\, dc = \lim_{\|P\| \to 0} S(P, f, \alpha) = 0.$$

Example 2. Let $f(x) = k$ be a constant function defined on $[a,b]$ and α a monotonically non-decreasing function on $[a,b]$. Prove that f is RS-integrable. Also prove that

(*i*) $\int_a^b k\, d\alpha = k\{\alpha(b) - \alpha(a)\}$ (*ii*) $\int_a^b k\, d\alpha = k \int_a^b d\alpha$.

[Agra 2003, 09; Kanpur 2002, 06; Meerut 2004; 05, 11; Rewa 1998]

Solution. Since $f(x)$ is a constant function, it is also bounded on $[a,b]$. Let $P = \{a = x_0, x_1, ..., x_n = b\}$ be any partition of $[a,b]$. Let m_r and M_r be the infimum and the supremum of f in sub-interval $[x_{r-1}, x_r]$. Then $m_r = M_r = k$

$$\therefore \quad U(P, f, \alpha) = \sum_{r=1}^{n} M_r \,\delta\alpha_r = k \sum_{r=1}^{n} \{\alpha(x_r) - \alpha(x_{r-1})\}$$

$$= k\{\alpha(x_1) - \alpha(x_0) + \alpha(x_2) - \alpha(x_1) + ... + \alpha(x_n) - \alpha(x_{n-1})\}$$

$$= k\{\alpha(x_n) - \alpha(x_1)\} = k\{\alpha(b) - \alpha(a)\}$$

Similarly, $$L(P, f, \alpha) = \sum_{r=1}^{n} m_r \,\delta\alpha_r = k\{\alpha(b) - \alpha(a)\}$$

Now, $$\overline{\int_a^b} f\, d\alpha = \inf \{U(P, f, \alpha) : P \text{ is a partition of } [a,b]\}$$

$$= \inf\, [k\{\alpha(b) - \alpha(a)\}] = k\, \{\alpha(b) - \alpha(a)\}$$

and $$\underline{\int_a^b} f\, d\alpha = \sup \{L(P, f, \alpha) : P \text{ is a partition of } [a, b]\}$$

$$= \sup\, [k\{\alpha(b) - \alpha(a)\}] = k\, \{\alpha(b) - \alpha(a)\}$$

Since $\underline{\int_a^b} f\, d\alpha = \overline{\int_a^b} f\, d\alpha$, so f is RS-integrable and

(*i*) $\int_a^b f\, d\alpha = k\{\alpha(b) - \alpha(a)\}$ or $\int_a^b k\, d\alpha = k\{\alpha(b) - \alpha(a)\}$ ... (1)

(*ii*) For $k = 1$, (1) $\Rightarrow$ $\int_a^b d\alpha = \alpha(b) - \alpha(a)$

so that $$k \int_a^b d\alpha = k\{\alpha(b) - \alpha(a)\} \quad ... (2)$$

From (1) and (2), $$\int_a^b k\, d\alpha = k \int_a^b d\alpha.$$

Example 3. (*i*) Let $f(x) = x$ and $\alpha(x) = x^2$. Does $\int_0^1 f\, d\alpha$ exist ? If it exists, find its value.

[*Meerut 2010*]

(*ii*) Evaluate $\int_0^1 x\, dx^2$, by using only the definition of the integral. [*Kanpur 2007, 2009*]

or evaluate (*RS*) $\int_x^1 x dx^2$ [*Himachal 2002, 2006, Delhi Maths (H) 1995, 2003*]

Solution. (*i*) since f is a continuous function and α is a non-decreasing on $[0,1]$, hence is RS-integrable (refer theorem II, Part 14.5) and

$$\int_0^1 f\, d\alpha = \lim_{\|P\|\to 0} S(P,f,\alpha) \qquad \text{... (1)}$$

Consider a partition P of $[0,1]$ as follows :

$$P = \left\{0, \frac{1}{n}, \frac{2}{n}, \ldots, \frac{r-1}{n}, \frac{r}{n}, \ldots, \frac{n}{n} = 1\right\}.$$

Here rth sub-interval $I_r = [x_{r-1}, x_r] = \left[\frac{r-1}{n}, \frac{r}{n}\right]$, where $r = 1, 2, \ldots, n$

Suppose that $\xi_r = \frac{r}{n} \in \left[\frac{r-1}{n}, \frac{r}{n}\right]$, for $r = 1, 2, \ldots, n$

$$\therefore \quad S(P,f,\alpha) = \sum_{r=1}^{n} f(\xi_r)\,\delta\alpha_r = \sum_{r=1}^{n} f(\xi_r)[\alpha(x_r) - \alpha(x_{r-1})]$$

$$= \sum_{r=1}^{n} f\left(\frac{r}{n}\right)\left[\alpha\left(\frac{r}{n}\right) - \alpha\left(\frac{r-1}{n}\right)\right]$$

$$= \sum_{r=1}^{n} \frac{r}{n}\left[\frac{r^2}{n^2} - \frac{(r-1)^2}{n^2}\right], \text{ as } f(x) = x, \alpha(x) = x^2$$

$$= \frac{1}{n^3}\sum_{r=1}^{n}(2r^2 - 1) = \frac{2}{n^3}\sum_{r=1}^{n} r^2 - \frac{1}{n^3}\sum_{r=1}^{n} r$$

$$= \frac{2}{n^3}\cdot\frac{n(n+1)(2n+1)}{6} - \frac{1}{n^3}\cdot\frac{n(n+1)}{2}$$

$$= \frac{1}{6n^2}\cdot\{2(2n^2+3n+1) - 3n - 3\} = \frac{1}{6}\left(4 + \frac{1}{2n} - \frac{1}{6n^2}\right)$$

$$\therefore \text{ By (1),} \quad \int_0^1 f\, d\alpha = \lim_{\|P\|\to 0}\frac{1}{6}\left(4 + \frac{1}{2n} - \frac{1}{6n^2}\right) = \lim_{n\to\infty}\frac{1}{6}\left(4 + \frac{1}{2n} - \frac{1}{6n^2}\right)$$

$$= 2/3 \qquad [\because \text{ as } \|P\| \to 0, n \to \infty]$$

(*ii*) Take $f(x) = x$ and $\alpha(x) = x^2$. Now proceed as in part (*i*).

Example 4. *If* $f(x) = x$ *and* $\alpha(x) = \sin(\pi x/2)$, *find* $U(P,f,\alpha)$ *and* $L(P,f,\alpha)$, *where* $P = [0, 2, 4]$. *Prove that, refinements of P need not decrease upper sums or increase lower sums.*

(*Delhi Maths (H) 1995*)

Solution. Here partition of $[0, 4]$ is given as $[0 = x_0, 2 = x_1, 4 = x_2]$ and so there are two sub-intervals. $I_1 = [x_0, x_1] = [0, 2]$ and $I_2 = [x_1, x_2] = [2, 4]$. Then, $m_1 = \inf f$ in $[0, 2] = f(0) = 0$, $M_1 = \sup f$ in $[0, 2] = f(2) = 2$ and $m_2 = \inf f$ in $[2, 4] = f(2) = 2$, $M_2 = \sup f$ in $[2, 4] = f(4) = 4$

$$\therefore \quad U(P,f,\alpha) = \sum_{r=1}^{2} M_r\,\delta\alpha_r = \sum_{r=1}^{2} M_r\{\alpha(x_r) - \alpha(x_{r-1})\}$$

$$= M_1\{\alpha(x_1)-\alpha(x_0)\} + M_2\{\alpha(x_2)-\alpha(x_1)\}$$
$$= M_1\{\alpha(2)-\alpha(0)\} + M_2\{\alpha(4)-\alpha(2)\}$$
$$= 2\,(0-0) + 4\,(0-0), \text{ as } \alpha(x) = \sin(\pi x/2)$$

Thus, $$U(P, f, \alpha) = 0 \qquad \text{... (1)}$$

Similarly, $$L(P, f, \alpha) = \sum_{r=1}^{2} m_r\,\delta\alpha_r = \sum_{r=1}^{2} m_r\{\alpha(x_r)-\alpha(x_{r-1})\}$$
$$= m_1\{\alpha(x_1)-\alpha(x_0)\} + m_2\{\alpha(x_2)-\alpha(x_1)\}$$
$$= m_1\{\alpha(2)-\alpha(0)\} + m_2\{\alpha(4)-\alpha(2)\}$$
$$= 0\,(0-0) + 2\,(0-0) = 0, \text{ as } \alpha(x) = \sin(\pi x/2)$$

Thus, $$L(P, f, \alpha) = 0 \qquad \text{... (2)}$$

Let $P^* = \{0, 1, 2, 3, 4\}$ be refinement of P. Then, as before

$$m_1 = \inf f \text{ in } [0,1] = f(0) = 0,\ M_1 = \sup f \text{ in } [0,1] = f(1) = 1$$
$$m_2 = \inf f \text{ in } [1,2] = f(1) = 1,\ M_2 = \sup f \text{ in } [1,2] = f(2) = 2$$
$$m_3 = \inf f \text{ in } [2,3] = f(2) = 2,\ M_3 = \sup f \text{ in } [2,3] = f(3) = 3$$
and $$m_4 = \inf f \text{ in } [3,4] = f(3) = 3,\ M_4 = \sup f \text{ in } [3,4] = f(4) = 4$$

$\therefore$ $$U(P^*, f, \alpha) = M_1\{\alpha(1)-\alpha(0)\} + M_2\{\alpha(2)-\alpha(1)\}$$
$$+ M_3\{\alpha(3)-\alpha(2)\} + M_4\{\alpha(4)-\alpha(3)\} \qquad \text{... (3)}$$

Now, $$\alpha(x) = \sin(x/2) \Rightarrow \alpha(0) = \alpha(1) = \alpha(2) = \alpha(4) = 0 \text{ and } \alpha(3) = -1$$

$\therefore$ $$U(P^*, f, \alpha) = 1\times(1-0) + 2\times(0-1) + 3\times(-1-0) + 4\times\{0-(-1)\} = 0 \qquad \text{... (4)}$$

From (1) and (4), $U(P^*, f, \alpha) = U(P, f, \alpha)$, showing that refinement need not decrease upper sums.

Again, $$L(P^*, f, \alpha) = m_1\{\alpha(1)-\alpha(0)\} + m_2\{\alpha(2)-\alpha(1)\}$$
$$+ m_3\{\alpha(3)-\alpha(2)\} + m_4\{\alpha(4)-\alpha(3)\}$$
$$= 0\times(1-0) + 1\times(0-1) + 2\times(-1-0) + 3\times\{0-(-1)\} = 0 \qquad \text{.... (5)}$$

From (2) and (5), $L(P^*, f, \alpha) = L(P, f, \alpha)$, showing that refinement need not increase lower sums.

Example 5. *Evaluate* $\int_0^3 x\, d[x]$, *where* $[x]$ *is the greatest integer not exceeding* x.

[Meerut 2004; Delhi Maths (H) 2006, 09]

Solution. Here $\alpha(x) = [x]$ is monotonic increasing function on [0, 3] given by

$$\alpha(x) = 0, \text{ for } 0 \le x < 1$$
$$\alpha(x) = 1, \text{ for } 1 \le x < 2$$
$$\alpha(x) = 2, \text{ for } 2 \le x < 3$$
and $$\alpha(x) = 3, \text{ for } x = 3$$

Consider any partition of P of [0,3] as follows.

$$P = \{0 = x_0, x_1, \ldots, x_{l-1}, x_l = 1, x_{l+1}, \ldots, x_{m-1}, x_m = 2, x_{m+1}, \ldots, x_{n-1}, x_n = 3\}$$

Then by definition, we have

$$U(P,f,\alpha)=\sum_{r=1}^{l-1} M_r\,\delta\alpha_r + M_l\delta\alpha_l + \sum_{r=l+1}^{m-1} M_r\,\delta\alpha_r + M_m\,\delta\alpha_m + \sum_{r=m+1}^{n-1} M_r\,\delta\alpha_r + M_n\,\delta\alpha_n$$

$$=\sum_{r=1}^{l-1} M_r\,(0-0)+1\times(1-0)+\sum_{r=1}^{m-1} m_r\,(1-1)+2\times(2-0)+\sum_{r=m}^{n-1} M_r\,(2-2)+3(3-2)=6$$

$$\therefore \quad \overline{\int_0^3} x\,d[x]=\inf\{U(P,f,\alpha): P \text{ is a partition of } [0,3]\}=6$$

Similarly, $\underline{\int_0^3} xd\,[x]=\sup\{L(P,f,\alpha): P \text{ is a partition of } [0,3]\}=6$

Since $\overline{\int_0^3} x\,d[x]=\underline{\int_0^3} xd\,[x]$, hence $xd\,[x]$ is RS-integrable and $\int_0^3 xd\,[x]=6.$

Example 6. *A function* α *increases on* $[a,\ b]$ *and is continuous at* x_0 *where* $a\le x_0\le b$. *Another function* f *is such that* $f(x_0)=1$, *and* $f(x)=0$, *for* $x\ne x_0$. *Prove that* $f\in R(a)$ *over* $a,b]$ *and* $\int_a^b f\,d\alpha=0.$ *(Himachal 2004, 05, 07, Agra 1997, Delhi Maths (H) 2004*

Solution. Let $P=\{a=x_0, x_1, x_2,\ldots, x_{r-1}, x_r,\ldots, x_n=b\}$ be a partition of $[a,b]$ and let $x_0\in[x_{r-1},x_r]$ so that $x_{r-1}\le x_0\le x_r$.

Since α in continuous at x_0, for given $\in>0$, there exists $\delta>0$ such that

$$|\alpha(x)-\alpha(x_0)|<\in/2 \quad \text{wherever} \quad |x-x_0|<\delta \qquad \ldots(1)$$

Since α lis an increasing function, (1) gives

$$\alpha(x)-\alpha(x_0)<\in/2 \quad \text{wherever} \quad 0<x-x_0<\delta \qquad \ldots(2)$$

and

$$\alpha(x_0)-\alpha(x)<\in/2 \quad \text{wherever} \quad 0<x_0-x<\delta \qquad \ldots(3)$$

Suppose that $\|P\|<\delta$ Now, we have

$$\delta\alpha_r=\alpha(x_r)-\alpha(x_{r-1})=\{\alpha(x_r)-\alpha(x_0)\}+\{\alpha(x_0)-\alpha(x_{r-1})\}$$

$$=\in/2+\in/2, \text{ using (2) and (3) and noting that } x_{r-1}\le x_0\le x_r$$

Thus, $\delta\alpha_r<\in$ when $\|P\|<\delta$ ... (4)

Now, $S(P,f,\alpha)=\sum_{r=1}^{n} f(\xi_r)\,\delta\alpha_r=f(\xi_r)\,\delta\alpha_r$

$$=\begin{cases}0, & if\ \ \xi_r\ne x_0\\ \delta\alpha_r, & if\ \ \xi_r=x_0\end{cases}$$

$\therefore$ Using (4), $S(P,f,\alpha)=0,\ \ if\ \ \xi_r\ne x_0$

and $S(P,f,\alpha)<\in,\ \ if\ \ \xi_r=x_0$

Hence in both cases, $\lim\limits_{\|P\|\to 0} S(P,f,\alpha)=0$ and so f is RS- integrable over $[a, b]$ and

$$\int_a^b f\,d\alpha = \lim_{\|P\|\to 0} S(P,f,\alpha)=0.$$

Example 7. *If $f(x)$ is bounded on* $[-1, 1]$ *and*

$$\beta(x) = 0, \text{ when } x<0$$
$$= 1, \text{ when } x=0$$
$$= 2, \text{ when } x>0$$

then show that $f(x)\in R(\beta(x))$ *on* $[-1, 1]$ *iff $f(x)$ is continuous at* $x=0$.

(*Delhi Maths (H) 2002*)

Solution. Let $P=\{-1=x_0, x_1,\ldots,x_{r-2}, 0=x_{r-1}, x_r,\ldots, x_n=1\}$ be a partition of $[-1,1]$ and let $\xi_i\in[x_{i-1},x_i]$. Then, we have

$$S(P,f,\beta)=\sum_{i=1}^{n} f(\xi_i)\{\beta(x_i)-\beta(x_{i-1})\}$$

Then, $$S(P,f,\beta)=f(\xi_{r-1})(1-0)+f(\xi_r)(2-1) \quad \ldots (1)$$

In particular, when $\xi_{r-1}=\xi_r=0$, we have

$$S(P,f,\beta)=2f(0) \quad \ldots (2)$$

Now, when $\|P\|\to 0$, then $x_{r-1}\to 0-0$ and $x_r\to 0+0$. From this, it follows that

$$\lim_{\|P\|\to 0} S(P,f,\beta) \text{ exists, if}$$

$$\lim_{x_{r-1}\to 0-0} f(\xi_{r-1}) = \lim_{x_r\to 0+0} f(\xi_r)=f(0),$$

which follows from (1) and (2) by taking $\xi_{r-1}=0=\xi_r$ in (1)

Hence, $\lim\limits_{\|P\|\to 0} S(P,f,\beta)$ exists and equals $f(0)$, if

$$f(0-0)=f(0+0)=f(0), \text{ i.e., } f \text{ is continuous at } x=0$$

Thus, if f is continuous at 0, then $f\in R(\beta)$ on $[-1,1]$.

Conversely, if $f\in R(\beta)$ on $[-1, 1]$ then $\lim\limits_{\|P\|\to 1} S(P,f,\beta)$ must exist and hence we easily see that f must be continuous at $x=0$.

Therefore, $$f\in R(\beta) \Leftrightarrow f \text{ is continuous at } x=0$$

Also, then by (2), $$\int_{-1}^{1} f\,d\beta = 2f(0).$$

EXERCISE

1. Using only the definition of the integral, show that $\int_0^1 x^3\,d\,[x]=1$.

(*Delhi B.Sc. Maths (H) 1995, 2003*)

2. Define Riemann-Stieltjes integral $\int_a^b f(x)\,d(\alpha(x))$ and use it to show that $\int_0^2 [x]\,dx^2=3$, where $[x]$ denotes greatest integer function. (*Delhi Maths (H) 1997, 2002*)

3. If $\beta_1(x) = 0$, when $x \le 0$ $\quad$ and $\quad$ $\beta_2(x) = 0$, when $x \le 0$
$= 1$, when $x > 0$ $\quad\quad$ $= 1$, when $x > 1$

then prove that $\beta_2 \in R(\beta_1)$ on $[-1, 1]$ even though $\lim\limits_{\mu(P)\to 0} S(P, \beta_2, \beta_1)$ does not exists.

4. Compute $\int_0^2 f\, d\alpha$, when $f(x) = \begin{cases} 0, & 0 \le x \le 1 \\ 1, & 1 < x \le 2 \end{cases}$ and $\alpha(x) = \begin{cases} 0, & 0 \le x \le 1 \\ 1, & 1 \le x \le 2 \end{cases}$

(Delhi Maths (H) 2000)

5. If f is bounded on $[-1, 1]$ and
$\beta(x) = 0$, when $x < 0$
$= 1/2$, when $x = 0$
$= 1$, when $x > 0$

then show that $f \in R(\beta)$ if f is continuous at $x = 0$.

6. If f is bounded on $[-1, 1]$ and

$$\beta_1(x) = \begin{cases} 0, & x \le 0 \\ 1, & x > 0 \end{cases} \quad \text{and} \quad \beta_2 = \begin{cases} 0, & x < 0 \\ 1, & x \ge 0 \end{cases}$$

Then prove that

(*i*) $f \in R(\beta_1)$ iff $f(0+) = f(0)$ and then $\int_{-1}^{1} f\, d\beta_1 = f(0)$

(*ii*) $f \in R(\beta_2)$ iff $f(0-) = f(0)$ and then $\int_{-1}^{1} f\, d\beta_2 = f(0)$

7. Let $f(x) = \begin{cases} 0, & \text{if } 0 \le x \le 1 \\ 1, & \text{if } 1 < x < 2 \end{cases}$ and $\alpha(x) = \begin{cases} 0, & \text{if } 0 \le x < 1 \\ 1, & \text{if } 1 \le x \le 2 \end{cases}$

Is $f \in R(\alpha)$? Is so, compute $\int_0^2 f\, d\alpha$

8. If f is bounded on $[a, b]$ and $\alpha(x) = \begin{cases} 0, & \text{when } x < b \\ 1, & \text{when } x = b \end{cases}$,

prove that $f \in R(\alpha)$ on $[a, b]$ iff $f(b - a) = f(b)$, and then $\int_a^b f\, d\alpha = f(b)$

9. If f is bounded on $[a,b]$ and $\alpha(x) = \begin{cases} 0, & \text{if } x = a \\ 1, & \text{if } x > a \end{cases}$

prove that $f \in R(\alpha)$ iff $f(a + 0) = f(a)$, and then $\int_a^b f\, d\alpha = f(a)$

10. Let α be monotonically non-decreasing on $[a, b]$ and $S(P, f, \alpha) = \sum\limits_{r=1}^{n} f(\xi_r)\, \delta\alpha_r$, where $\xi_r \in [x_{r-1}, x_r]$, $r = 1, 2, \ldots, n$ Prove that if $\lim S(P, f, \alpha)$ exists as $\| P \| \to 0$, then $f \in R(\alpha)$ on $[a,b]$ and $\lim\limits_{\|P\|\to 0} S(P, f, \alpha) = \int_a^b f\, d\alpha$

11. If f is bounded on $[a, b]$ and there are only finitely many points at which f is discontinuous, then prove that $f \in R(\alpha)$

14.6 SOME USEFUL INEQUALITIES RELATED TO RS-INTEGRALS

Theorem I *If* $f \in R(\alpha)$ *on* $[a,b]$ *and m, M be the bounds of f on* $[a,b]$, *then,*

$$m\{\alpha(b)-\alpha(a)\} \le \int_a^b f\, d\alpha \le M\{\alpha(b)-\alpha(a)\}$$

[*Agra 2016; Kanpur 2014, 16; Meerut 2014, 17; Ravishankar 2000*]

Proof. Let $P = \{a = x_0, x_1, ..., x_n = b\}$ be any partition of $[a, b]$ and let m_r and M_r be the infimum and supremum of $f(x)$ in $[x_{r-1}, x_r]$, then, for each r, we have

$$m \le m_r \le M_r \le M$$

$$\Rightarrow \quad \sum_{r=1}^{n} m\delta\alpha_r \le \sum_{r=1}^{n} m_r \delta\alpha_r \le \sum_{r=1}^{n} M_r \delta\alpha_r \le \sum_{r=1}^{n} M\delta\alpha_r$$

$$\Rightarrow \quad m\sum_{r=1}^{n} \delta\alpha_r \le L(P,f,\alpha) \le U(P,f,\alpha) \le M\sum_{r=1}^{n} \delta\alpha_r \qquad \text{... (1)}$$

But
$$\sum_{r=1}^{n} \delta\alpha_r = \sum_{r=1}^{n} \{\alpha(x_r) - \alpha(x_{r-1})\}$$
$$= \alpha(x_1) - \alpha(x_0) + \alpha(x_2) - \alpha(x_1) + ... + \alpha(x_n) - \alpha(x_{r-1})$$
$$= \alpha(x_n) - \alpha(x_0) = \alpha(b) - \alpha(a)$$

$$\therefore (1) \Rightarrow \quad m\{\alpha(b)-\alpha(a)\} \le L(P,f,\alpha) \le U(P,f,\alpha) \le M\{\alpha(b)-\alpha(a)\} \qquad \text{... (2)}$$

But
$$L(P,f,\alpha) \le \underline{\int_a^b} f\, d\alpha = \overline{\int_a^b} f\, d\alpha \le U(P,f,\alpha) \qquad \text{... (3)}$$

$$\therefore (2) \text{ and } (3) \Rightarrow \quad m\{\alpha(b)-\alpha(a)\} \le \underline{\int_a^b} f\, d\alpha \le \overline{\int_a^b} f\, d\alpha \le M\{\alpha(b)-\alpha(a)\} \qquad \text{... (4)}$$

Now,
$$f \in R(\alpha) \Rightarrow \underline{\int_a^b} f\, d\alpha = \overline{\int_a^b} f\, d\alpha = \int_a^b f\, d\alpha \qquad \text{... (5)}$$

$$\therefore (4) \text{ and } (5) \Rightarrow \quad m\{\alpha(b)-\alpha(a)\} \le \int_a^b f\, d\alpha \le M\{\alpha(b)-\alpha(a)\} \qquad \text{... (6)}$$

Corollary 1. *Let* $f \in R(\alpha)$. *Then there exists a number* μ *lying between the bounds m and M such that*

$$\int_a^b f\, d\alpha = \mu\,\{\alpha(b)-\alpha(a)\}$$

Proof. From (6), it follows that there exists a number μ, where $m \le \mu \le M$ such that

$$\int_a^b f\, d\alpha = \mu\,\{\alpha(b)-\alpha(a)\} \qquad \text{... (7)}$$

Corollary 2. First mean value theorem. *If f is continuous function on* $[a,b]$ *and* α *is monotonic increasing on* $[a,b]$, *there exists a number* ξ *in* $[a,b]$ *such that*

$$\int_a^b f\, d\alpha = f(\xi)\,\{f(b) - f(a)\}$$ **[*Himachal 2008, 2010, 2013; G.N.D.U Amritsar 2005; Kanpur 2019; Meerut 2009, 2011; Rohtak 2004*]**

Proof. Since α is continuous and α is monotonic, so $f \in R(\alpha)$. Then as in corollary 1 above, we have

$$m\{\alpha(b)-\alpha(a)\} \leq \int_a^b f\, d\alpha \leq M\{\alpha(b)-\alpha(a)\},$$

where m, M are the infimum and supremum of f in $[a,b]$. Hence there exists a number μ, satisfying

$m \leq \mu \leq M$ such that $$\int_a^b f\, d\alpha = \mu\, \{\alpha(b)-\alpha(a)\} \qquad \text{... (8)}$$

Since f is continuous on $[a,b]$, there exists a number $\xi \in [a,b]$ such that $f(\xi)=\mu$. Then, (8) reduce to

$$\int_a^b f\, d\alpha = f(\xi)\, \{\alpha(b)-\alpha(a)\}. \qquad \text{... (9)}$$

Note. The following example shows that we may fail to choose ξ such that $a<\xi<b$, and so the condition $\xi \in [a,b]$ cannot be replaced by $c \in]a,b[$

Consider $\alpha(x)=0$, if $x=a$ and $\alpha(x)=1$, if $a<x\leq b$.
Then for a continuous function f, we obtain

$$\int_a^b f\, d\alpha = f(a) = f(a)\, \{\alpha(b)-\alpha(a)\}.$$

Corollary 3. If $f \in R(\alpha)$ on $[a, b]$ and $|f(x)| \leq k$ on $[a,b]$, then $\left|\int_a^b f\, d\alpha\right| \leq k\{\alpha(b)-\alpha(a)\}$

(Agra 2006; Kanpur 2002; Himachal 2001)

Solution. $|f(x)| \leq k \;\forall\; x \in [a,b]$

$\Rightarrow \quad -k \leq f(x) \leq k$

$\Rightarrow \quad -k \leq m \leq f(x) \leq M \leq k$

$\Rightarrow \quad -k \leq m \leq m_r \leq f(x) \leq M_r \leq M \leq k$

$\Rightarrow \quad -k \leq m_r \leq M_r \leq k$

$\Rightarrow \quad -\sum_{r=1}^{n} k\delta\alpha_r \leq \sum_{r=1}^{n} m_r\delta\alpha_r \leq \sum_{r=1}^{n} M_r\delta\alpha_r \leq \sum_{r=1}^{n} k\delta\alpha_r$

$\Rightarrow \quad -k\sum_{r=1}^{n} \delta\alpha_r \leq L(P,f,\alpha) \leq U(P,f,\alpha) \leq k\sum_{r=1}^{n} \delta\alpha_r$

$\Rightarrow \quad -k\,\{\alpha(b)-\alpha(a)\} \leq L(P,f,\alpha) \leq U(P,f,\alpha) \leq k\{\alpha(b)-\alpha(a)\}$ ... (10)

$\left[\because \sum_{r=1}^{n} \delta\alpha_r = \alpha(b)-\alpha(a)\right.$, *as* in Therorem I

But $$L(P,f,\alpha) \leq \underline{\int_a^b} f\, d\alpha \leq \overline{\int_a^b} f\, d\alpha \leq U(P,f,\alpha) \qquad \text{... (11)}$$

From (10) and (11), $$-k\{\alpha(b)-\alpha(a)\} \leq \underline{\int_a^b} f\, d\alpha \leq \overline{\int_a^b} f\, d\alpha \leq k\{\alpha(b)-\alpha(a)\} \qquad \text{... (12)}$$

Now, $$f \in R(\alpha) \Rightarrow \underline{\int_a^b} f\, d\alpha = \overline{\int_a^b} f\, d\alpha = \int_a^b f\, d\alpha \qquad \text{... (13)}$$

From (12) and (13), $$-k\{\alpha(b)-\alpha(a)\} \leq \int_a^b f\, d\alpha \leq k\{\alpha(b)-\alpha(a)\}$$

$$\Rightarrow \quad \left|\int_a^b f\, d\alpha\right| \leq k\{\alpha(b)-\alpha(a)\} \qquad \text{... (14)}$$

Corollary 4. *If* $f \in R(\alpha)$ and $f(x) \geq 0 \; \forall \; x \in [a,b]$, *then*

$$\int_a^b f \, d\alpha \geq 0.$$

Proof. Here, $$f(x) \geq 0 \;\; \forall \; x \in [a,b] \Rightarrow m \geq 0 \qquad \text{... (15)}$$

Also, since α is monotonic increasing on $[a, b]$, so

$$\alpha(b) - \alpha(a) \geq 0 \qquad \text{... (16)}$$

$\therefore$ (1) and (2) $\Rightarrow$ $$m\{\alpha(b) - \alpha(a)\} \geq 0 \qquad \text{... (17)}$$

Now, from (16) $$\int_a^b f d\alpha \geq m\{\alpha(b) - \alpha(a)\} \qquad \text{... (18)}$$

$\therefore$ (17) and (18) yield $$\int_a^b f \, d\alpha \geq 0 \qquad \text{... (19)}$$

Corollary 5. *Let* $f_1, f_2 \in R(\alpha)$ *on* $[a,b]$ *and* $f_1(x) \leq f_2(x)$ *for all* $x \in [a,b]$, *then*

$$\int_a^b f_1 \, d\alpha \leq \int_a^b f_2 \, d\alpha.$$

[Kanpur 2013; Jiwaji 2001; Meerut 1993; Ravishankar 1997]

Proof. Here, $f_1(x) \leq f_2(x) \;\forall\; x \in [a,b]$

$\Rightarrow \quad f_2(x) - f_1(x) \geq 0, \;\forall\; x \in [a,b]$

$\Rightarrow \quad (f_2 - f_1)(x) \geq 0, \;\forall\; x \in [a,b]$

$\Rightarrow \quad \int_a^b (f_2 - f_1) \, d\alpha \geq 0$, by corollary 4

$\Rightarrow \quad \int_a^b f_2 \, d\alpha - \int_a^b f_1 \, d\alpha \geq 0 \quad \Rightarrow \quad \int_a^b f_1 \, d\alpha \leq \int_a^b f_2 \, d\alpha$

14.7 ALGEBRA OF RS-INTEGRABLE FUNCTIONS

Theorem I. *If* $f \in R(\alpha)$ *on [a,b] and c is a contant, then* $cf \in R(\alpha)$ *on [a,b] and*

$$\int_a^b cf \, d\alpha = c \int_a^b f \, d\alpha. \quad \textit{(Kanpur 2009, 2010; Meerut 2009; Himachal 2005, 2008)}$$

Proof. If $c = 0$, we have done. So, let $c \neq 0$. To begin with, we further assume that $c > 0$.

Now, $f \in R(\alpha)$ on $[a,b]$, so for a given $\varepsilon > 0$, there exists a partition $P = \{a = x_0, x_1, ..., x_n = b\}$ such that

$$U(P, f, \alpha) - L(P, f, \alpha) < \varepsilon / c \qquad \text{... (1)}$$

Let m_r and M_r be the infimum and the supremum of f in $[x_{r-1}, x_r]$, then cm_r and cM_r will be the infimum and the supremum of cf in $[x_{r-1}, x_r]$. Hence for the same partition P of $[a,b]$, we obtain

$$U(P, cf, \alpha) = c\, U(P, f, \alpha) \qquad \text{... (2)}$$

and $$L(P, cf, \alpha) = c\, L(P, f, \alpha) \qquad \text{... (3)}$$

$\therefore$ $$U(P, cf, \alpha) - L(P, cf, \alpha) = c\{U(P, f, \alpha) - L(P, f, \alpha)\}$$

or $$U(P, cf, \alpha) - L(P, cf, \alpha) < c \times (\varepsilon / c) = \varepsilon, \text{ by (1)} \qquad \text{... (4)}$$

If $c < 0$, we take $c = -c'$, where $c' > 0$. Then, we have

$$U(P, -c'f, \alpha) = -c' L(P, f, \alpha) \qquad \text{... (5)}$$

and $$L(P, -c'f, \alpha) = -c' U(P, f, \alpha) \qquad \text{... (6)}$$

Using relations (5) and (6), we see as before that the inequality (4) holds for $c < 0$. Hence for $\varepsilon > 0$, there exist a partition P of $[a,b]$ such that

$$U(P, cf, \alpha) - L(P, cf, \alpha) < \in$$

Hence for every constant c, $f \in R(\alpha)$ on $[a, b]$

$$\therefore \qquad \underline{\int_a^b} cf\, d\alpha = \overline{\int_a^b} cf\, d\alpha = \int_a^b cf\, d\alpha \qquad \text{... (7)}$$

Since $f \in R(\alpha)$, we have

$$\underline{\int_a^b} f\, d\alpha = \overline{\int_a^b} f\, d\alpha = \int_a^b f\, d\alpha \qquad \text{... (8)}$$

When $c > 0$, taking the infimum of both sides of (2) over all partition P of $[a,b]$, we have

$$\overline{\int_a^b} cf\, d\alpha = c \overline{\int_a^b} f\, d\alpha$$

or

$$\int_a^b cf\, d\alpha = c \int_a^b f\, d\alpha, \text{ by (7) and (8)} \qquad \text{... (9)}$$

If $c < 0$, then take $c = -c'$ where $c' > 0$.

Now,

$$\int_a^b cf\, d\alpha = \overline{\int_a^b} cf\, d\alpha = \overline{\int_a^b} (-c'f)\, d\alpha$$

$$= \inf \{U(P, -c'f, \alpha)\} = -\inf \{-c'L(P, f, \alpha)\}$$

$$= -c' \sup L(P, f, \alpha) = c \underline{\int_a^b} f\, d\alpha = c \int_a^b f\, d\alpha$$

Thus, for every constant, we have

$$\int_a^b cf\, d\alpha = c \int_a^b f\, d\alpha.$$

Corollary. If $f \in R(\alpha)$, then so is $-f$ and

$$\int_a^b (-f)\, d\alpha = -\int_a^b f\, d\alpha.$$

Theorem II. If $f_1, f_2 \in R(\alpha)$ on $[a, b]$ then $f_1 + f_2 \in R(\alpha)$

and

$$\int_a^b (f_1 + f_2)\, d\alpha = \int_a^b f_1\, d\alpha + \int_a^b f_2\, d\alpha$$

(Meerut 2013; Agra 2012)

(Himachal 2000, 04, 07, 08, 09: Bharathiar 2004; Kanpur 2003; Bhopal 2004)

Proof. Since $f_1, f_2 \in R(\alpha)$ on $[a,b]$, hence for a given $\varepsilon > 0$, there exist partitions P_1 and P_2 of of $[a, b]$ such that

$$U(P_1, f_1, \alpha) - L(P_1, f_1, \alpha) < \varepsilon/2 \qquad \text{... (1)}$$

and

$$U(P_2, f_2, \alpha) - L(P_2, f_2, \alpha) < \varepsilon/2 \qquad \text{... (2)}$$

Let $P = P_1 \cup P_2$ so that P is a common refinement of P_1 and P_2. Hence, (1) and (2), yield

$$U(P, f_1, \alpha) - L(P, f_1, \alpha) < \varepsilon/2, \; U(P, f_1, \alpha) - L(P, f_1, \alpha) < \varepsilon/2, \qquad \text{... (3)}$$

Adding these inequalities, we have

$$\{U(P, f_1, \alpha) + U(P, f_2, \alpha)\} - \{L(P, f_1, \alpha) + L(P, f_2, \alpha)\} < \varepsilon \qquad \text{... (4)}$$

Let $f_1 + f_2 = f$. Corresponding to the partition $P = \{a = x_0, x_1, ..., x_n = b\}$ of $[a,b]$, let m'_r, M'_r, m''_r, M''_r and m_r, M_r be the infimum and the supremum of f_1, f_2 and f respectively on the sub-interval $[x_{r-1}, x_r]$. Then, we have

$$m'_r \le f_1(x) \le M'_r,\ \ m''_r \le f_2(x) \le M''_r \text{ and } m_r \le f(x) \le M_r$$

$$\Rightarrow \qquad m'_r + m''_r \le f_1 + f_2 \le M'_r + M''_r \text{ and } m_r \le f_1 + f_2 \le M_r$$

$$\Rightarrow \qquad m'_r + m''_r \le m_r \le M_r \le M'_r + M''_r \qquad \text{... (5)}$$

Now, $U(P, f, \alpha) = \sum_{r=1}^{n} M_r \delta\alpha_r \le \sum_{r=1}^{n} (M'_r + M''_r)\, \delta\alpha_r$, by (5)

$$\therefore U(P, f, \alpha) \le \sum_{r=1}^{n} M'_r \delta\alpha_r + \sum_{r=1}^{n} M''_r \delta\alpha_r$$

Thus, $$U(P, f, \alpha) \le U(P, f_1, \alpha) + U(P, f_2, \alpha) \qquad \text{... (6)}$$

Again, $L(P, f, \alpha) = \sum_{r=1}^{n} m_r \delta\alpha_r \ge \sum_{r=1}^{n} (m'_r + m''_r)\, \delta\alpha_r$, by (5)

$$\therefore L(P, f, \alpha) \ge \sum_{r=1}^{n} m'_r \delta\alpha + \sum_{r=1}^{n} m''_r \delta\alpha_r \qquad \text{or} \qquad L(P, f, \alpha) \ge L(P, f_1, \alpha) + L(P, f_2, \alpha)$$

Thus, $$-L(P, f, \alpha) \le -L(P, f_1, \alpha) - L(P, f_2, \alpha) \qquad \text{... (7)}$$

From (6) and (7), we have

$$U(P, f, \alpha) - L(P, f, \alpha) \le U(P, f_1, \alpha) + U(P, f_2, \alpha) - L(P, f_1, \alpha) - L(P, f_2, \alpha)$$

or $$U(P, f, \alpha) - L(P, f, \alpha) < \varepsilon, \text{ by (4)}$$

Hence $$f \in R(\alpha),\ i.e.,\ f_1 + f_2 \in R(\alpha)$$

Since the upper RS-integral is the infimum of the upper sums, hence there exist partitions P_1, P_2 such that

$$U(P_1, f_1, \alpha) < \int_a^b f_1\, d\alpha + \frac{\varepsilon}{2} \qquad \text{and} \qquad U(P_2, f_2, \alpha) < \int_a^b f_2\, d\alpha + \frac{\varepsilon}{2} \qquad \text{... (8)}$$

If $P = P_1 \cup P_2$, then (8) reduces to

$$U(P, f_1, \alpha) < \int_a^b f_1\, d\alpha + \frac{\varepsilon}{2} \qquad \text{and} \qquad U(P, f_2, \alpha) < \int_a^b f_2\, d\alpha + \frac{\varepsilon}{2} \qquad \text{... (9)}$$

Corresponding to the partition P, we have

$$\int_a^b f\, d\alpha = \overline{\int_a^b} f\, d\alpha \le U(P, f, \alpha) \le U(P, f_1, \alpha) + U(P, f_2, \alpha), \text{ by (6)}$$

$$< \int_a^b f_1\, d\alpha + \int_a^b f_2\, d\alpha + \varepsilon, \text{ by (9)}$$

Since $\varepsilon\ (> 0)$ is arbitrary, it follows that

$$\int_a^b f\, d\alpha \le \int_a^b f_1 d\alpha + \int_a^b f_2\, d\alpha \qquad \text{... (10)}$$

Since f_1, f_2 and $f \in R(\alpha)$, so $-f_1, -f_2$ and $-f \in R(\alpha)$ and using result (10) for $-f_1$, $-f_2$ and $-f$, we obtain

$$\int_a^b (-f)\, d\alpha \le \int_a^b (-f_1)\, d\alpha + \int_a^b (-f_2)\, d\alpha$$

or $$-\int_a^b f\, d\alpha \le -\left\{ \int_a^b f_1\, d\alpha + \int_a^b f_2\, d\alpha \right\}$$

or $$\int_a^b f d\alpha \ge \int_a^b f_1\, d\alpha + \int_a^b f_2\, d\alpha \qquad \text{... (11)}$$

From (10) and (11), $$\int_a^b f\, d\alpha = \int_a^b f_1 d\alpha + \int_a^b f_2 d\alpha$$

or $$\int_a^b (f_1 + f_2)\, d\alpha = \int_a^b f_1\, d\alpha + \int_a^b f_2\, d\alpha$$

Corollary If $f_1, f_2 \in R(\alpha)$ on $[a,b]$, then so is $c_1 f_1 + c_2 f_2$, where c_1, c_2 are constants and

$$\int_a^b (c_1 f_1 + c_2 f_2)\, d\alpha = c_1 \int_a^b f_1\, d\alpha + c_2 \int_a^b f_2\, d\alpha. \qquad \textit{(Kanpur 2005)}$$

Proof. Use theorems I and II.

Note. As a particular case, if $c_1 = 1$ and $c_2 = -1$, then $f_1, f_2 \in R(\alpha) \Rightarrow (f_1 - f_2) \in R(\alpha)$ and

$$\int_a^b (f_1 - f_2)\, d\alpha = \int_a^b f_1\, d\alpha - \int_a^b f_2\, d\alpha$$

Theorem III. *If $f \in R(\alpha_1)$ and $f \in R(\alpha_2)$, then $f \in R(\alpha_1 + \alpha_2)$*

and $$\int_a^b f\, d(\alpha_1 + \alpha_2) = \int_a^b f\, d\alpha_1 + \int_a^b f\, d\alpha_2$$ *[Delhi Maths (H) 2003, Nagpur 2003)*

Gwalior 2004; Kanpur 2005]

Proof. Since $f \in R(\alpha_1)$ and $f \in R(\alpha_2)$, hence for a given $\varepsilon > 0$, there exist partitions P_1 and P_2 of $[a, b]$ such that

$$U(P_1, f, \alpha_1) - L(P_1, f, \alpha_1) < \varepsilon/2 \quad \text{and} \quad U(P_2, f, \alpha_2) - L(P_2, f, \alpha_2) < \varepsilon/2$$

Let $P = P_1 \cup P_2$ so that P is a common refinement of P_1 and P_2. Then the above results hold good and so $U(P, f, \alpha_1) - L(P, f, \alpha_1) < \varepsilon/2$, $U(P, f, \alpha_2) - L(P, f, \alpha_2) < \varepsilon/2$

Adding these inequalities, we have

$$U(P, f, \alpha_1) + U(P, f, \alpha_2) - L(P, f, \alpha_1) - L(P, f, \alpha_2) < \varepsilon \qquad \text{... (1)}$$

Let the partition P be $\{a = x_0, x_1, x_2, \ldots, x_n = b\}$ and m_r and M_r be respectively the infimum and the supremum of f in the sub-interval $[x_{r-1}, x_r]$. Now, by definition, we have

$$U(P, f, \alpha) = \sum_{r=1}^{n} M_r \{\alpha(x_r) - \alpha(x_{r-1})\}, \text{ where } \alpha = \alpha_1 + \alpha_2$$

$$= \sum_{r=1}^{n} M_r \{(\alpha_1 + \alpha_2)(x_r) - (\alpha_1 + \alpha_2)(x_{r-1})\}$$

$$= \sum_{r=1}^{n} M_r \{\alpha_1(x_r) + \alpha_2(x_r) - \alpha_1(x_{r-1}) - \alpha_2(x_{r-1})\}$$

$$= \sum_{r=1}^{n} M_r \{\alpha_1(x_r) - \alpha_1(x_{r-1})\} + \sum_{r=1}^{n} M_r \{\alpha_2(x_r) - \alpha_2(x_{r-1})\}$$

Thus, $$U(P, f, \alpha) = U(P, f, \alpha_1) + U(P, f, \alpha_2) \qquad \text{... (2)}$$

Similarly, $$L(P, f, \alpha) = L(P, f, \alpha_1) + L(P, f, \alpha_2) \qquad \text{... (3)}$$

Subtraction (3) from (2), we have

$$U(P, f, \alpha) - L(P, f, \alpha) = U(P, f, \alpha_1) + U(P, f, \alpha_2) - L(P, f, \alpha_1) - L(P, f, \alpha_2)$$

or $$U(P, f, \alpha) - L(P, f, \alpha) < \varepsilon, \text{ using (1)}$$

$\Rightarrow$ $f \in R(\alpha)$ *i.e.*, $f \in R(\alpha_1 + \alpha_2)$, as $\alpha = \alpha_1 + \alpha_2$.

Now, from (2), we obtain

$$\inf U(P, f, \alpha) = \inf \{U(P, f, \alpha_1) + U(P. f, \alpha_2)\}$$

or $$\inf U(P, f, \alpha) \geq \inf U(P, f, \alpha_1) + \inf U(P, f, \alpha_2)$$

or $$\overline{\int_a^b} f\,d\alpha \geq \overline{\int_a^b} f\,d\alpha_1 + \overline{\int_a^b} f\,d\alpha_2 \qquad \text{... (4)}$$

But $f \in R(\alpha)$, $f \in R(\alpha_1)$ and $f \in R(\alpha_2)$ and so

$$\overline{\int_a^b} f\,d\alpha = \int_a^b f\,d\alpha,\ \overline{\int_a^b} f\,d\alpha_1 = \int_a^b f\,d\alpha_1,\ \overline{\int_a^b} f\,d\alpha_2 = \int_a^b f\,d\alpha_2$$

Hence (4) reduces to

$$\int_a^b f\,d\alpha \geq \int_a^b f\,d\alpha_1 + \int_a^b +f\,d\,\alpha_2 \qquad \text{... (5)}$$

Similarly, taking supremum of both sides of (3) and proceeding as above, we obtain

$$\int_a^b f\,d\alpha \leq \int_a^b f\,d\alpha_1 + \int_a^b f\,d\,\alpha_2 \qquad \text{... (6)}$$

From (5) and (6), $$\int_a^b f\,d\alpha = \int_a^b f\,d\alpha_1 + \int_a^b f\,d\alpha_2$$

i.e., $$\int_a^b f\,d(\alpha_1+\alpha_2) = \int_a^b f\,d\alpha_1 + \int_a^b f\,d\alpha_2$$

Theorem IV *If $f \in R(\alpha)$ on $[a, b]$, then $|f| \in R(\alpha)$ on $[a, b]$ and* $\left|\int_a^b f\,d\alpha\right| \leq \int_a^b |f|\,d\alpha$

(G.N.D.U. Amritsar 2004; Kanpur 2001, 02, 04, 07; Ravishankar 2001; Sagar 1995; Rewa 1998; Jiwaji 1996; Himachal 2014, 2015)

Proof. Since $f \in R(\alpha)$, therefore for a given $\varepsilon > 0$, there exist a partition $P = \{a = x_0, x_1, ..., x_n = b\}$ of $[a,b]$ such that

$$U(P,f,\alpha) - L(P,f,\alpha) < \varepsilon \qquad \text{... (1)}$$

Let M_r, m_r and M'_r, m'_r be respectively the supremum and infimum of f and $|f|$ respectively in the sub-interval $[x_{r-1}, x_r]$.

Now, if ξ_1 and ξ_2 be lany two points in $[x_{r-1}, x_r]$, then we have

$$\left|\{|f(\xi_1)| - |f(\xi_2)|\}\right| \leq \left|f(\xi_1) - f(\xi_2)\right| \qquad \text{... (2)}$$

The above relation (2) suggests that, we must have

$$M'_r - m'_r \leq M_r - m_r$$

$$\Rightarrow \quad \sum_{r=1}^{n} (M'_r - m'_r)\,\delta\alpha_r \leq \sum_{r=1}^{n} (M_r - m_r)\,\delta\alpha_r$$

$$\Rightarrow \quad U(P,|f|,\alpha) - L(P,|f|,\alpha)| \leq U(P,f,\alpha) - L(P,f,\alpha)$$

$$\Rightarrow \quad U(P,|f|,\alpha) - L(P,|f|,\alpha) < \varepsilon, \text{ by (1)}$$

$$\Rightarrow \quad |f| \in R(\alpha) \text{ on } [a,b]$$

Again, we know that

$$-|f(x)| \leq f(x) \leq |f(x)| \ \forall\, x \in [a,b] \qquad \text{... (3)}$$

Further, we know that if $f_1, f_2 \in R(\alpha)$ and $f_1(x) \leq f_2(x)$ on $[a,b]$, then

$$\int_a^b f_1\,d\alpha \leq \int_a^b f_2\,d\,\alpha \qquad \text{... (4)}$$

Now, $f \in R(\alpha)$, $|f| \in R(\alpha)$ so $-|f| \in R(\alpha)$, Hence, using result (4), (3) yields

$$-\int_a^b |f|\,d\alpha \leq \int_a^b f\,d\alpha \leq \int_a^b |f|\,d\alpha \quad \Rightarrow \quad \left|\int_a^b f\,d\alpha\right| \leq \int_a^b |f|\,d\alpha$$

Note. The converse of the theorem is not true. For example, consider $f(x) = 1$, if x is rational and $f(x) = -1$, if x is irrational. Then $|f| \in R(\alpha)$ whereas $f \notin R(\alpha)$ by taking $\alpha(x) = x$.

[Himanchal 2007]

Theorem V. *If $f \in R(\alpha)$ on $[a,b]$, then $f^2 \in R(\alpha)$, on $[a,b]$* *(Himachal 2014)*

Proof. Since f is bounded on $[a,b]$, there exist $M > 0$ such that

$$|f(x)| \le M \quad \forall \ x \in [a,b] \qquad \text{... (1)}$$

Since $f \in R(\alpha)$, therefore for a given $\varepsilon > 0$, there exist a partition $P = \{a = x_0, x_1, ..., x_n = b\}$ of $[a,b]$ such that

$$U(P,f,\alpha) - L(P,f,\alpha) < \varepsilon / 2M \qquad \text{... (2)}$$

Let M_r, m_r and M'_r, m'_r be respectively the supremum and infimum lof f and f^2 respectively on the sub-interval $[x_{r-1}, x_r]$. Then, for any two points $\xi_1, \xi_2 \in [x_{r-1}, x_r]$, we have

$$\left|f^2(\xi_1) - f^2(\xi_2)\right| = \left|f(\xi_1) + f(\xi_2)\right| \times \left|f(\xi_1) - f(\xi_2)\right|$$

$$\le 2M \left|f(\xi_1) - f(\xi_2)\right|, \text{ by (1)}$$

Hence $$M'_r - m'_r \le 2M\,(M_r - m_r)$$

$$\Rightarrow \quad \sum_{r=1}^{n} (M'_r - m'_r)\,\delta\alpha_r \le 2M \sum_{r=1}^{n} (M_r - m_r)\,\delta\alpha_r$$

$$\Rightarrow \quad U(P,f^2,\alpha) - L(P,f^2,\alpha) \le 2M\,\{U(P,f,\alpha) - L(P,f,\alpha)\}$$

$$\Rightarrow \quad U(P,f^2,\alpha) - L(P,f^2,\alpha) < \varepsilon, \text{ by (2)}$$

$$\Rightarrow \quad f^2 \in R(\alpha)$$

Theorem VI. *If $f, g \in R(\alpha)$ on $[a,b]$, then $f\,g \in R(\alpha)$ on $[a,b]$ (Himachal 2014; Kanpur 2012)*

Proof. $f, g \in R(\alpha)$ on $[a,b]$

$\Rightarrow \quad (f+g)$ and $(f-g) \in R(\alpha)$

$\Rightarrow \quad (f+g)^2$ and $(f-g)^2 \in R(\alpha)$

$\Rightarrow \quad (f+g)^2 - (f-g)^2 \in R(\alpha)$

$\Rightarrow \quad (1/4) \times \{(f+g)^2 - (f-g)^2\} \in R(\alpha)$

$\Rightarrow \quad f\,g \in R(\alpha)$

Theorem VII. *If $f \in R(\alpha)$ on $[a,b]$ and c be a point such that $a < c < b$, then $f \in R(\alpha)$ on $[a, c]$ as well as on $[c, b]$ and* $\displaystyle\int_a^b f\,d\alpha = \int_a^c f\,d\alpha + \int_c^b f\,d\alpha$ *(Himanchal 2008)*

Proof. Since $f \in R(\alpha)$ on $[a,b]$, for a given $\varepsilon > 0$, there exists a partition P of $[a,b]$ such that

$$U(P,f,\alpha) - L(P,f,\alpha) < \varepsilon \qquad \text{... (1)}$$

Let $P^* = P \cup \{c\}$ then, from (1), we have

$$U(P^*, f,\alpha) - L(P^*, f,\alpha) < \varepsilon \qquad \text{... (2)}$$

Let P_1 and P_2 be the set of those points of P^* which constitute the partitions of $[a, c]$ and $[c, b]$ respectively, then

$$U(P_1, f, \alpha) + U(P_2, f, \alpha) = U(P^*, f, \alpha) \quad \text{... (3)}$$

and
$$L(P_1, f, \alpha) + L(P_2, f, \alpha) = L(P^*, f, \alpha) \quad \text{... (4)}$$

Subtracting (4) from (3), we have

$$U(P_1, f, \alpha) - L(P_1, f, \alpha) + U(P_2, f, \alpha) - L(P_2, f, \alpha) = U(P^*, f, \alpha) - L(P^*, f, \alpha)$$

$$\Rightarrow \quad \{U(P_1, f, \alpha) - L(P_1, f, \alpha)\} + \{U(P_2, f, \alpha) - L(P_2, f, \alpha)\} < \varepsilon, \text{ by (2)}$$

Since both terms on the R.H.S. are positive, so each term is less than ε

i.e., $U(P_1, f, \alpha) - L(P_1, f, \alpha) < \varepsilon$ and $U(P_2, f, \alpha) - L(P_2, f, \alpha) < \varepsilon$

$\Rightarrow$ $f \in R(\alpha)$ on $[a, c]$ and $f \in R(\alpha)$ on $[c, b]$

Now, $f \in R(\alpha)$ on $[a, c]$ and $f \in R(\alpha)$ on $[c, b]$ and $f \in R(\alpha)$ on $[a, b]$

$$\Rightarrow \quad \underline{\int_a^c} f\, d\alpha = \overline{\int_a^c} f\, d\alpha = \int_a^c f\, d\alpha, \quad \underline{\int_c^b} f\, d\alpha = \overline{\int_c^b} f\, d\alpha = \int_c^b f\, d\alpha \quad \text{... (5)}$$

and
$$\underline{\int_a^b} f\, d\alpha = \overline{\int_a^b} f\, d\alpha = \int_a^b f\, d\alpha \quad \text{... (6)}$$

Now, (3) $\Rightarrow$
$$U(P_1, f, \alpha) + U(P_2, f, \alpha) = U(P^*, f, \alpha) \geq \overline{\int_a^b} f\, d\alpha$$

$$\Rightarrow \quad U(P_1, f, \alpha) + U(P_2, f, \alpha) \geq \int_a^b f\, d\alpha, \text{ by (6)} \quad \text{... (7)}$$

Keeping P_1 fixed and taking the supremum over all partitions P_2, (7) yields

$$U(P_1, f, \alpha) + \overline{\int_c^b} f\, d\alpha \geq \int_a^b f\, d\alpha$$

or
$$U(P_1, f, \alpha) + \int_c^b f\, d\alpha \geq \int_a^b f\, d\alpha, \text{ by (5)} \quad \text{... (8)}$$

Now, taking the supremum over all partitions P_1, (8) yields

$$\overline{\int_a^c} f\, d\alpha + \int_c^b f\, d\alpha \geq \int_a^b f\, d\alpha$$

or
$$\int_a^c f\, d\alpha + \int_c^b f\, d\alpha \geq \int_a^b f\, d\alpha, \text{ by (5)} \quad \text{... (9)}$$

Similarly, (4) $\Rightarrow$ $L(P_1, f, \alpha) + L(P_2, f, \alpha) = L(P^*, f, \alpha) \leq \underline{\int_a^b} f\, d\alpha$

or
$$L(P_1, f, \alpha) + L(P_2, f, \alpha) \leq \int_a^b f\, d\alpha, \text{ by (6)} \quad \text{... (10)}$$

Keeping P_1 fixed and taking the infimum over all partitions P_2, (10) yields

$$L(P_1, f, \alpha) + \underline{\int_c^b} f\, d\alpha \leq \int_a^b f\, d\alpha$$

or
$$L(P_1, f, \alpha) + \int_c^b f\, d\alpha \leq \int_a^b f\, d\alpha, \text{ by (5)} \quad \text{... (11)}$$

Now, taking the infimum over all partitions P_1, (11) yields

$$\underline{\int_a^c} f\,d\alpha + \underline{\int_c^b} f\,d\alpha \le \underline{\int_a^b} f\,d\alpha \qquad \text{or} \qquad \int_a^c f\,d\alpha + \int_c^b f\,d\alpha \le \int_a^b f\,d\alpha \qquad \text{... (12)}$$

From (9) and (12), $$\int_a^b f\,d\alpha = \int_a^c f\,d\alpha + \int_c^b f\,d\alpha$$

Note. The converse of the above theorem is not always true. For example, let f and α be defined on [0, 2] as follows

$$f(x) = \begin{cases} 0, & \text{if } 0 \le x < 1 \\ 1, & \text{if } 1 \le x \le 2 \end{cases} \qquad \text{and} \qquad \alpha(x) = \begin{cases} 0, & \text{if } 0 \le x < 1 \\ 1, & \text{if } 1 < x \le 2 \end{cases}$$

Then it can be shown that $\int_0^1 f\,d\alpha$ and $\int_1^2 f\,d\alpha$ both exist, but the $\int_0^2 f\,d\alpha$ does not exist.

Theorem VIII. *Let* $f \in R(\alpha)$ *on* $[a,b]$, *let* $m = \inf f$ *and* $M = \sup f$ *on* $[a,b]$ *and let* $\phi : [m, M] \to \mathbf{R}$ *be continuous. Then the composite function* $h(x) = \phi(f(x)) \in R(\alpha)$ *on* $[a,b]$

(*Gwalior. 2004; Meerut 1997; Ravishankar 2005; Himachal 2003, I.A.S. 1997*)

Proof. Since ϕ is continuous on $[m, M]$, so it is uniformly continuous on $[m, M]$. Therefore, for $\varepsilon > 0$, there exist $\delta > 0$ such that

$$s, t \in [m, M], |s - t| < \delta \Rightarrow |\phi(s) - \phi(t)| < \varepsilon, \qquad \text{...(1)}$$

where, without any loss of gernerality, assume that

$$\delta < \varepsilon. \qquad \text{... (2)}$$

Since $f \in R(\alpha)$, for $\delta^2 > 0$, there exists a partition $P = \{a = x_0, x_1, x_2, \ldots, x_n = b\}$ such that

$$U(P, f, \alpha) - L(P, f, \alpha) < \delta^2 \qquad \text{... (3)}$$

Let $m_r = \inf f$, $M_r = \sup f$ on sub-interval $[x_{r-1}, x_r]$ and $m_r{}^* = \inf h$, $M_r{}^* = \sup h$ on $[x_{r-1}, x_r]$. We now divide the set $\{r : r = 1, 2, \ldots, n\}$ into two sets A and B such that

$$A = \{r : M_r - m_r < \delta\} \quad \text{and} \quad B = \{r : M_r - m_r \ge \delta\} \qquad \text{... (4)}$$

Then $$r \in A \text{ and } x_{r-1} < s < t \le x_r$$

$$\Rightarrow \quad |f(s) - f(t)| \le M_r - m_r < \delta$$

$$\Rightarrow \quad |\phi(f(s)) - \phi(f(t))| < \varepsilon, \text{ using (1)}$$

$$\Rightarrow \quad |h(s) - h(t)| < \varepsilon \Rightarrow |M_r^* - m_r^*| \le \varepsilon$$

$$\Rightarrow \quad M_r^* - m_r^* < \varepsilon \qquad \text{... (3)}$$

From (3), $$\sum_{r \in A} (M_r^* - m_r^*)\,\delta\alpha_r \le \sum_{r \in A} \varepsilon\,\delta\alpha_r \le \varepsilon \sum_{r=1}^{n} \delta\alpha_r = \varepsilon\,\{\alpha(b) - \alpha(a)\} \qquad \text{...(4)}$$

Now, $$r \in B \Rightarrow M_r - m_r \ge \delta, \text{ by (4)}$$

$$\therefore \quad \delta \sum_{r \in B} \delta\alpha_r \le \sum_{r \in B} (M_r - m_r)\,\delta\alpha_r$$

$$\le \sum_{r=1}^{n} (M_r - m_r)\,\delta\alpha_r$$

$$= U(P, f, \alpha) - L(P, f, \alpha) < \delta^2, \text{ by (3)}$$

Thus, $$\sum_{r \in B} \delta\alpha_r < \delta \quad \text{or} \quad \sum_{r \in B} \delta\alpha_r < \varepsilon, \text{ by (2)} \qquad \text{... (5)}$$

Since ϕ is continuous on $[m, M]$, it follows that it is bounded on $[m, M]$ and hence exists $k \geq 0$ such that $|\phi| \leq \lambda$ on $[a,b]$ so that

$$M_r^* - m_r^* \leq 2k \qquad \text{... (6)}$$

$$\therefore \quad U(P,h,\alpha) - L(P,h,\alpha) = \sum_{r\in A} (M_r^* - m_r^*)\, \delta\alpha_r + \sum_{r\in B} (M_r^* - m_r^*)\, \delta\alpha_r$$

$$\leq \in \{\alpha(b) - \alpha(a)\} + \sum_{r\in B} 2k\, \delta\alpha_r, \text{ by (4) and (5)}$$

$$< \in \{\alpha(b) - \alpha(a)\} + 2k\, \varepsilon, \text{ by (5)}$$

Thus, $\quad U(P,h,\alpha) - L(P,h,\alpha) < \in \{\alpha(b) - \alpha(\alpha) + 2k\}$ ... (7)

Since $\{\alpha(b) - \alpha(a) + 2k\}$ is a constant and $\varepsilon > 0$ is arbitrary, (7) implies that

$$h = \phi \circ f \in R(\alpha) \text{ on } [a,b].$$

14.8 REDUCTION OF RIEMANN-STIELTJES INTEGRAL INTO RIEMANN INTEGRAL.

We have already seen that Riemann integral is a special case of Riemann-Stieltjes integral when $\alpha(x) = x$. In the following theorem, we shall present conditions under which R-S integral reduces to R-integral.

Theorem. *If $f \in R[a,b]$ and α be a monotonic non-decreasing function on $[a, b]$ such that $\alpha' \in R\,[a, b]$. Then $f \in R(\alpha)$ and $\int_a^b f\, d\alpha = \int_a^b f(x)\, \alpha'(x)\, dx$*

(Bhopal 2000, 04; Jiwaji 1996; G.N.D.U. 1998; Kanpur 2000; 03; Punjab 2002; Agra 2002, 2012; Himachal 2001, 05, 07, 08, 09; G.N.D.U. Amritsar 1998)

Proof. Since $f \in R[a,b]$, f is bounded on $[a,b]$ and hence there exists $k > 0$ such that

$$|f(x)| \leq k \quad \forall\, x \in [a,b] \qquad \text{... (1)}$$

Let $\varepsilon > 0$ be given. Since $f, \alpha' \in R\,[a,b]$, hence $f\alpha' \in R\,[a,b]$. Therefore, there exist $\delta_1 > 0$ such that

$$\left| \sum_{r=1}^{n} f(\xi_r)\, \alpha'(\xi_r)\, \delta x_r - \int_a^b f\, \alpha' dx \right| < \varepsilon/2, \qquad \text{... (2)}$$

for $\|P\| < \delta_1$ and all $\xi_r \in I_r$, where $I_r = [x_{r-1}, x_r]$ is the rth sub-interval of the partition $P = \{a = x_0, x_1, \ldots, x_{r-1}, x_r, \ldots, x_n = b\}$ of $[a, b]$.

Again, since $\alpha' \in R\,[a,b]$, so there exists $\delta_2 > 0$ such that

$$\left| \sum_{r=1}^{n} \alpha'(\xi_r)\, \delta x_r - \int_a^b \alpha'\, dx \right| < \varepsilon/4k \qquad \text{... (3)}$$

for $\|P\| < \delta_2$ and all $\xi_2 \in I_r$.

Now, for $\|P\| < \delta_2$ and all $\eta_r \in I_r$, (2) yields

$$\left| \sum_{r=1}^{n} \alpha'(\eta_r)\, \delta x_r - \int_a^b \alpha' dx \right| < \varepsilon/4k \qquad \text{... (4)}$$

Now, we have

$$\left|\sum_{r=1}^{n} \{\alpha'(\xi_r)-\alpha'(\eta_r)\}\,\delta x_r\right| = \left|\sum_{r=1}^{n}\left\{\alpha'(\xi_r)\,\delta x_r - \int_a^b \alpha' dx + \int_a^b \alpha' dx - \alpha'(\eta_r)\,\delta x_r\right\}\right|$$

$$\leq \left|\sum_{r=1}^{n} \alpha'(\xi_r)\,\delta x_r - \int_a^b \alpha' dx\right| + \left|\int_a^b \alpha' dx - \alpha'(\eta_r)\,\delta x_r\right|$$

$$< \varepsilon/4k + \varepsilon/4k, \text{ using (3) and (4).}$$

Thus,
$$\sum_{r=1}^{n} |\alpha'(\xi_r)-\alpha'(\eta_r)|\,\delta x_r < \varepsilon/2k, \qquad \text{... (5)}$$

for $\|P\| < \delta_2$ and all $\xi_r, \eta_r \in I_r$.

Let $\delta = \min\{\delta_1, \delta_2\}$. Let P be a partition with $\|P\| < \delta$ and $\xi_r \in [x_{r-1}, x_r]$. Then, by Lagrange's mean value theorem, there are points $\eta_r \in [x_{r-1}, x_r]$ such that

$$\alpha'(\eta_r) = \frac{\alpha(x_r)-\alpha(x_{r-1})}{x_r - x_{r-1}} = \frac{\delta\alpha_r}{\delta x_r}$$

so that
$$\delta\alpha_r = \alpha'(\eta_r)\,\delta x_r \qquad \text{... (6)}$$

Now, we have

$$\left|\sum_{r=1}^{n} f(\xi_r)\,\delta\alpha_r - \int_a^b f\alpha'\,dx\right| = \left|\sum_{r=1}^{n} f(\xi_r)\,\alpha'(\eta_r)\,\delta x_r - \int_a^b f\alpha'\,dx\right|, \text{ using (6)}$$

$$= \left|\sum_{r=1}^{n} f(\xi_r)\,\alpha'(\xi_r)\,\delta x_r - \int_a^b f\,\alpha'\,dx + \sum_{r=1}^{n} f(\xi_r)\{\alpha'(\eta_r)-\alpha'(\xi_r)\}\,\delta x_r\right|$$

$$\leq \left|\sum_{r=1}^{n} f(\xi_r)\,\alpha'(\xi_r)\delta x_r - \int_a^b f\alpha'\,dx\right| + \sum_{r=1}^{n} \left|f(\xi_r)\right|\left|\alpha'(\eta_r)-\alpha'(\xi_r)\right|\delta x_r,$$

$$< \varepsilon/2 + k \times (\varepsilon/2k), \text{ using (1), (2) and (5)}$$

Thus, for any $\varepsilon > 0$, there exists $\delta > 0$ such that for all partition with $\|P\| < \delta$

$$\left|\sum_{r=1}^{n} f(\xi_r)\,\delta\alpha_r - \int_a^b f\alpha'\,dx\right| < \varepsilon$$

$\Rightarrow$ $\lim_{\|P\|\to 0} \sum_{r=1}^{n} f(\xi_r)\,\delta\alpha_r$ exists and is equal to $\int_a^b f\,\alpha'\,dx$

$\Rightarrow$ $f \in R(\alpha)$ and $\lim_{\|P\|\to 0} \sum_{r=1}^{n} f(\xi_r)\,\delta\alpha_r = \int_a^b f\alpha'\,dx,$

Since $\lim_{\|P\|\to 0} \sum_{r=1}^{n} f(\xi_r)\,\delta\alpha_r = \int_a^b f\,d\alpha,$ we have

$$\int_a^b f\,d\alpha = \int_a^b f\,\alpha'\,dx$$

Theorem II *If f is continuous on* $[a,b]$ *and* α *has a continuous derivatives on* $[a,b]$, *then*

$$\int_a^b f\,d\alpha = \int_a^b f\,\alpha'\,dx$$ **[Himanchal 2013, 2015]**

Proof. Left as an exercise for the reader.

Note. The result of the theorem is a particular case of theorem I.

Example. Evaluate (*i*) $\int_0^1 x\, d(e^{2x})$ (*Himachal 2004, 09; Delhi Maths (H) 1994*)

(*ii*) $\int_0^2 [x]\, dx^2$ (*Himachal 2007, 08; Delhi Maths (H) 2008*)

Solution. (*i*) $\int_0^1 x\, d(e^{2x}) = \int_0^1 x \cdot 2e^{2x}\, dx = [x \cdot e^{2x}]_0^1 - \int_0^1 1 \cdot e^{2x}\, dx$

$$= e^2 - (1/2) \times [e^{2x}]_0^1 = e^2 - (1/2) \times (e^2 - 1) = (1 + e^2)/2$$

(*ii*) $\int_0^2 [x]\, dx^2 = \int_0^2 [x]\, 2x\, dx = \int_0^1 [x]\, 2x\, dx + \int_1^2 [x]\, 2x\, dx$

$$= \int_0^1 (0 \times 2 \times x) + \int_1^2 (1 \times 2 \times x)\, dx = 0 + [x^2]_1^2 = 4 - 1 = 3$$

EXERCISE

1. Show that

(*i*) $\int_0^2 x^2 dx^2 = 8$ **(Meerut 2007)** **(Himachal 2005)**

(*ii*) $\int_\pi^{2\pi} \sin x\, d(\cos x) = -\frac{\pi}{2}$ **(Kanpur 2013)**

(*iii*) $\int_0^3 \sqrt{x}\, dx^3 = \frac{6}{7} \times 3^{7/2}$

(*iv*) $\int_0^{\pi/2} x\, d(\sin x) = \frac{\pi}{2} - 1$

(*v*) $\int_0^3 [x]\, d(e^x) = 2e^3 - e^2 - e$ **(Kanpur 2013)**

(*vi*) $\int_1^4 (x - [x])\, dx^2 = 6$ **(Himachal 2014)**

(*vii*) $\int_0^{\pi/2} \sin x\, d(\cos x) = -(\pi/4)$

2. Show that $\int_0^1 f\, d\alpha = \frac{3}{5}$, where $f(x) = x^2$ and $\alpha(x) = x^3$ **(Himachal 1997)**

3. Evaluate $\int_0^1 x\, dx^2$, by using

(*i*) definition of Riemann-Stieltjes integral.

(*ii*) a property of Riemann- Stieltjes integral. (*Delhi Maths (H) 2005*)

Solution. Left as an exercise. [Ans. 2/3]

14.9 SOME USEFUL THEOREMS

Theorem I. Partial integration formula. *If f is continuous and* α *is monotonic on* $[a, b]$, *then*

$$\int_a^b f\, d\alpha = [f(x)\, \alpha(x)]_a^b - \int_a^b \alpha\, df,$$

where $$[f(x)\, \alpha(x)]_a^b = f(b)\, \alpha(b) - f(a)\, \alpha(a).$$ **(Himachal 2014)**

Proof. Since f is continuous and α is monotonic, both integrals $\int_a^b f\, d\alpha$ and $\int_a^b \alpha\, df$ exist.

For a partition $P = \{a = x_0, x_1, \ldots, x_{r-1}, x_r, \ldots, x_n = b\}$ of $[a,b]$, choose ξ_r such that $\xi_r \in [x_{r-1}, x_r], \forall\, r = 1, 2, \ldots, n$. Also assume that $\xi_0 = a$ and $\xi_{n+1} = b$. Then, clearly, $Q = \{a = \xi_0, \xi_1, \xi_2, \ldots, \xi_n, \xi_{n+1} = b\}$ is also a partition of $[a,b]$. Then, we have

$$S(P, f, \alpha) = \sum_{r=1}^{n} f(\xi_r)\, \delta\alpha_r$$

$$= f(\xi_1)\{\alpha(x_1) - \alpha(x_0)\} + f(\xi_2)\{\alpha(x_2) - \alpha(x_1)\} + \ldots + f(\xi_n)\}\{\alpha(x_n) - \alpha(x_{n-1})\}$$

$$= -\alpha(x_0) f(\xi_1) - \alpha(x_1)\{f(\xi_2) - f(\xi_1)\} + \alpha(x_2)\{f(\xi_3) - f(\xi_2)\} + \ldots$$

$$+ \alpha(x_{n-1})\{f(\xi_n) - f(\xi_{n-1})\} + \alpha(x_n)\, f(\xi_n) \quad \ldots (1)$$

On adding and subtracting $\alpha(x_0)\, f(\xi_0) + \alpha(x_n)\, f(\xi_{n+1})$ on R.H.S. of (1), we get

$$S(P, f, \alpha) = \alpha(x_n)\, f(\xi_{n+1}) - \alpha(x_0)\, f(\xi_0) - \sum_{r=1}^{n} \alpha(x_r)\, \{f(\xi_{r+1}) - f(\xi_r)\}$$
$$= f(b)\, \alpha(b) - f(a)\, \alpha(a) - S(Q, \alpha, f), \quad \text{... (2)}$$

where $\xi_r \le x_r \le \xi_{r+1}$. Now, if $\| P \| \to 0$, then $\| Q \| \to 0$.

Again, $$\lim_{\|P\|\to 0} S(P, f, \alpha) = \int_a^b f\, d\alpha$$

and hence $\lim_{\|Q\|\to 0} S(Q, \alpha, f)$ also exists and equals to $\int_a^b \alpha\, df$

Hence taking limit of both sides of (2) as $\| P \| \to 0$ (and hence $\| Q \| \to 0$) we obtain

$$\int_a^b f\, d\alpha = f(b)\, \alpha(b) - f(a)\, \alpha(a) - \int_a^b \alpha\, d\, f$$

or $$\int_a^b f\, d\alpha = \left[f(x)\, \alpha(x) \right]_a^b - \int_a^b \alpha\, d\, f$$

Note. The above theorem is similar to the well known theorem "Integration of parts" for Riemann integration. Also observe that this result holds whenever out of the two functions f and α, one continuous and the other is monotonic.

Theorem II The second mean value theorem. *Let f be continuous and* α *be continuous and monotone on* $[a,b]$, *then there exist* $\xi \in [a,b]$ *such that*

$$\int_a^b f\, d\alpha = f(a)\, \{\alpha(\xi) - \alpha(a)\} + f(b)\, \{\alpha(b) - \alpha(\xi)\}$$ (*Himachal 2013, 2014; Meerut 2007; Kanpur 2003, 2012; Kurukshetra 1996*)

Proof. By theorem I, we have

$$\int_a^b f\, d\alpha = f(b)\, \alpha(b) - f(a)\, \alpha(a) - \int_a^b \alpha\, df \quad \text{... (1)}$$

Since α is continuous, so by first mean value theorem, we have

$$\int_a^b \alpha\, df = \alpha(\xi)\, \{f(b) - f(a)\} \quad \text{... (2)}$$

$\therefore$ From (1) and (2), we obtain

$$\int_a^b f\, d\alpha = f(b)\, \alpha(b) - f(a)\, \alpha(a) - \alpha(\xi)\, \{f(b) - f(a)\}$$

or $$\int_a^b f\, d\alpha = f(a)\, \{\alpha(\xi) - \alpha(a))\} + f(b)\, \{\alpha(b) - \alpha(\xi)\}$$

Theorem III (The change of variable) *If f is continuous on* $[a, b]$ *and* ϕ *is continuous and strictly monotone on* $[a, b]$ *then*

$$\int_a^b f(x)\, dx = \int_{\phi(a)}^{\phi(b)} f(\psi(y))\, d\psi(y)$$ (*Kanpur 2004; Bharathiar 2004*)

Proof. Let $P = \{a = x_0, x_1, ..., x_n = b\}$ be a partition of $[a,b]$. Let $y_r = \phi(x_r),\ r = 0, 1, 2, ..., n$. Then $Q = \{\phi(a) = y_0, y_1, y_2, ..., y_n = \phi(b)\}$ is the corresponding partition of $[\phi(a), \phi(b)]$.

Let us write $g(y) = f(\psi(y))$

Now, $\phi^{-1} = \psi \Rightarrow y_r = \phi(x_r) = \psi^{-1}(x_r) \Rightarrow x_r = \psi(y_r)$. Then, we get

$$\sum_{r=1}^{n} f(x_r)(x_r - x_{r-1}) = \sum_{r=1}^{n} f(\psi(y_r))\{\psi(y_r) - \psi(y_{r-1})\}$$

or
$$\sum_{r=1}^{n} f(x_r)(x_r - x_{r-1}) = \sum_{r=1}^{n} g(y_r)\{\psi(y_r) - \psi(y_{r-1})\} \qquad \dots (1)$$

Now, ϕ is continuous on $[a,b] \Rightarrow \phi$ is uniformly continuous on $[a,b]$

Hence $\qquad \|Q\| \to 0$ as $\|P\| \to 0$

Continuity of $f \Rightarrow \sum_{r=1}^{n} f(x_r)(x_r - x_{r-1})$ tends to $\int_a^b f\, dx$ as $\|P\| \to 0$.

ϕ is continuous $\Rightarrow \phi^{-1}$ is continuous *i.e.*, ψ is continuous.

Also f and ψ are continuous $\Rightarrow g$ is continuous *i.e.*, $fo\psi$, is continuous.

continuity of $g \Rightarrow \sum_{r=1}^{n} g(y_r)\{\psi(y_r) - \psi(x_{r-1})\}$ tends to $\int_{\phi(a)}^{\phi(b)} g(y)\, d\psi(y)$ when $\|Q\| \to 0$

Hence proceeding to limits when $\|P\| \to 0$ (and hence $\|Q\| \to 0$), (1) reduces to

$$\int_a^b f(x)\, dx = \int_{\phi(a)}^{\phi(b)} f(\psi(y))\, d\psi(y)$$

EXAMPLES

Example 1. Evaluate $\int_0^3 x^2\, d([x] - x)$ *(Kanpur 2008; Himachal 2014)*

Solution. $\int_0^3 x^2\, d([x]-x) = \left[x^2([x]-x)\right]_0^3 - \int_0^3 2x([x]-x)\, dx$, by theorem I, page 14.28

$$= 0 - 2\int_0^3 x[x]dx + 2\int_0^3 x^2 dx$$

$$= -2\left[\int_0^1 x[x]\, dx + \int_1^2 x[x]\, dx + \int_2^3 x[x]\, dx\right] + 2\cdot\left[\frac{x^3}{3}\right]_0^3$$

$$= -2\left[\int_0^1 0\, dx + \int_1^2 x\, dx + \int_2^3 2x\, dx\right] + 18$$

$$= -2\left[\left[x^2/2\right]_1^2 + \left[x^2\right]_2^3\right] + 18 = -2[3/2 + 5] + 18 = 5$$

Example 2. Evaluate $\int_{-1}^2 x^5 d(|x|^3)$. *(Himachal 2002, 09)*

Solution. $\int_{-1}^2 x^5 d(|x|^3) = \left[x^5|x|^3\right]_{-1}^2 = \int_{-1}^2 |x|^3 \cdot 5x^4\, dx$, b y theorem I, page 14.28

$$= (32\times 8) - (-1)^5 \times |-1|^3 - 5\int_{-1}^2 x^6 |x|\, dx, \text{ as } |x|^2 = x^2$$

$$= 257 - 5\left[\int_{-1}^0 x^6|x|\, dx + \int_0^2 x^6 |x|\, dx\right]$$

$$= 257 - 5\left[-\int_{-1}^0 x^7 dx + \int_0^2 x^7\, dx\right] = 257 - 5\left[-\left[\frac{x^8}{8}\right]_{-1}^0 + \left[\frac{x^8}{8}\right]_0^2\right]$$

$$= 257 - 5(1/8 + 2^8/8) = 771/8$$

EXERCISE

1. Show that

(i) $\int_{-1}^{1} x\,d|x| = 1$ **(*Himachal 2004, 07*)** (ii) $\int_0^2 x\,d([x]-x) = 1$

(*Himachal 2010*)

(iii) $\int_0^3 x\,d([x]-x) = \frac{3}{2}$ **(*Himachal 2015*)**

(iv) $\int_{-\pi}^{\pi} x^2\,d|\sin x| = -4\pi$ (v) $\int_0^x d[t] = [x]\ \forall\ x \in \boldsymbol{R}$

(vi) $\int_2^5 [5-x]\,d(\log[x]) = \log 3$ (vii) $\int_0^4 x d[x] = 10$ **(*Himachal 2006, 07, 08*)**

2. Show that $\int_0^1 f\,d\alpha = \frac{29}{2}$, where $f(x) = x^3 + 7$ and $\alpha(x) = \begin{cases} 1+x^3, & if\ \ 0 \le x \le 1 \\ 0, & if\ \ x = 0 \end{cases}$

3. Evaluate the following integrals : **(*Himachal 2013*)**

(i) $\int_0^3 f(x)\,d([x]+x)$, where $f(x) = \begin{cases} [x], & if\ \ 0 \le x < 3/2 \\ e^x & if\ \ 3/2 \le x \le 3 \end{cases}$

(ii) $\int_0^2 x\,d\alpha(x)$, where $\alpha(x) = \begin{cases} x, & if\ \ 0 \le x \le 1 \\ 2+x, & if\ \ 1 < x \le 2 \end{cases}$

(*Himachal 2014*)

4. Evaluate the Riemann-Stieltjes integral of $f(x) = x^2$ on $[0,3]$ with respect to $\alpha(x) = [x] - x$, where $[x]$ is the greatest integer function. (*Delhi Maths (H) 1998*)

Hint : Refer solved example 2 on page 14.30.

MISCELLANEOUS PROBLEMS ON CHAPTER 14

1. Let f be continuous and ∞ monotonically increasing on $[a, b]$, then prove that $f \in R(\alpha)$ on $[a,b]$ **[Himanchal 2008, 09; Bhopal 1996; Jiwaji 2001; Kanpur 2003]**

2. Functions f and α are defined on $[-1, 1]$ as follows :

$f(x) = \begin{cases} 0, \forall\ x < 0 \\ 1, \forall\ x \ge 0 \end{cases}$ and $\alpha(x) = \begin{cases} 0, \forall\ x \le 0 \\ 1, \forall\ x > 0 \end{cases}$

Prove that $f \in R(\alpha)$ although $\lim_{\|P\| \to 0} S(R, f, \alpha)$ does not exist. **[Delhi Maths (H) 2006]**

3. Define $R-S$ sum. If f is continuous and g is monotonically non-decreasing on $[a, b]$, then $f \in RS(g)$. Moreover for every $\varepsilon > 0$, there corresponds a $\delta > 0$ such that $\left| \sum_{r=1}^{n} f(\xi_r)\,\delta g_r - \int_a^b f\,dg \right| < \varepsilon$ every partition $P = \{x_0 = a, x_1, x_2, \ldots, x_n = b\}$ with $\| P \| < \delta$ and for every intermediate partition $Q = \{\xi_1, \xi, .. \xi_n\}$ of P, that is, $\lim_{\|P\| \to 0} RS\,(P, Q, f, g) = \int_a^b f dg.$ **(Kanpur 2006)**

4. Let $f(x) = x$, $g(x) = x^2$. Does $RS \int_0^1 f\,dg$ exist ? If it exists, find its value. **(Agra 2007)**

5. If f be monotonic and g be continuous and monotonic non-decreasing on $[a, b]$, then prove that f is $(R-S)$ integrable on $[a, b]$. **[Agra 2007]**

6. Evaluate the Riemann Stieltjes integral $\int_0^3 f\,d\alpha$ where $f(x) = x^2, 0 \le x \le 3$ and $\alpha(x) = x - [x]$ where $[x]$ denotes the greatest integer $\le x$ **[Delhi Maths (H) 2008]**

7. Write the name of integral of f relative to g over $[a, b]$ if $\underline{\int_a^b} f\,dg = \overline{\int_a^b} f dg$ where f is bounded function and g is monotonic increasing function. **[Agra 2007]**

8. Functions f and α are defined on $[-1, 1]$ as follows:

$$f(x) = \begin{cases} 0, & x < 0 \\ 1, & x \geq 0 \end{cases} \quad \text{and} \quad \alpha(x) = \begin{cases} 0, & \text{if } x \leq 0 \\ 1, & \text{if } x > 0 \end{cases}$$

Show that f is RS – integrable with respect to α on $(-1, 1)$ by definition. **[Delhi 2009**

9. State and prove first mean value theorem for RS – integrable. Find ξ for theorem it

$f(x) = x^2 + 2x - 5, \alpha(x) = x^2$, $a = 1, b = 3$. **[Himanchal 2005, 08]**

10. If $f \in R(\alpha)$ and c is a positive constant, show that $f \in R(c\alpha)$ and $\int_a^b fd(c\alpha) = c\int_a^d f\,d\,\alpha$ **(Himanchal 2004, 05)**

11. If $f \in R(\alpha)$ on $[a, b]$, show that $f^2 \in R(\alpha)$. What can you say about its converse ? Prove or disprove your assertion. **(Himanchal 2008)**

12. State and prove mean value theorem for R-S integral. **[Meerut 2011**

Chapter

15

Uniform Convergence of Sequences and series of Functions

15.1 INTRODUCTION

In this chapter we shall consider sequences whose members are real valued functions defined in a set S and distinguish between two types of convergence of such sequences.

Let f_n be a real valued function defined in a set S for each n.

Point-wise convergence. **(Kanpur 2010)**

To each $c \in S,$ there corresponds a real number sequence $\{f_n(c)\}$ with values

$$f_1(c), f_2(c), ..., f_n(c) ...$$

We suppose that this sequence is convergent. In fact, we suppose that each of the sequences arising for different members of S is convergent. Thus we define in a natural way a real valued function say f, with domain S such that its value $f(c)$ for $c \in S$ is $\lim \{f_n(c)\}$.

The function f, thus defined, is referred to as the *Pointwise* limit of the sequence $\{f_n\}$ of functions. Also in this case, we say that the sequence is point-wise convergent.

Thus if a function f is the point-wise limit of the point-wise convergent sequence $\{f_n\}$ of functions defined in S, to each $c \in S$ and to each $\varepsilon > 0$ there corresponds an integer m such that

$$\forall\, n \geq m, \quad \left|f_n(c) - f(c)\right| < \varepsilon.$$

Of course, if we fix, the choice of m may depend upon the choice of c.

Uniform convergence. **[Kanpur 2010; Chennai 2011; Meerut 2005, 13]**

We say that a sequence $\{f_n\}$ of real valued functions with domain S is uniformly convergent, if there exists a real valued function ϕ, with domain S and to each $\varepsilon > 0$ there corresponds an integer m such that

$$\forall\, x \in S \quad \text{and} \quad \forall\, n \geq m. \quad \left|f_n(x) - \phi(x)\right| < \varepsilon.$$

Also in this case we say that the function ϕ is the uniform limit of the sequence $\{f_n\}$.

It may be easily seen that

Uniform convergence $\Rightarrow$ *point-wise convergence*

and that in the event of uniform convergence

Uniform limit = Point-wise limit.

As a result, we shall denote the uniform limit of $\{f_n\}$ by f instead of ϕ.

It should, however, be remembered that every point-wise convergent sequence is not uniformly convergent as is illustrated by the following counter-exmaple.

Let $$f_n(x) = \frac{nx}{1 + n^2 x^2},\ x \in R.$$

Here $\lim_{n\to\infty} f_n(x) = \lim_{n\to\infty} \frac{nx}{1+n^2x^2} = \lim_{n\to\infty} \frac{x}{nx^2+1/n} = 0, \ \forall\, x \in R,$

showing that the sequence $\{f_n\}$, is point-wise convergent with point-wise limit f such that

$$f(x) = 0 \qquad \forall\, x \in R.$$

We shall now show that the convergence is *not* uniform in any interval $[a, b]$ with, 0, as an interior point. **[Agra 1998, 2004, Delhi B.Sc. Maths (H) 2002]**

Suppose that $\{f_n\}$ is uniformly convergent in $[a, b]$ so that the point-wise limit f is also the uniform limit.

Let $\varepsilon > 0$ be given. Then there exists m such that $\forall\, x \in [a,b]$ and $\forall\, n \geq m$.

$$\left|\frac{nx}{1+n^2x^2} - 0\right| < \varepsilon.$$

We take $\varepsilon = 1/4$. Now there exists an integer k such that $k \geq m$ and $1/k \in [a,b]$.

Taking $n = k$ and $x = 1/k$, we have $(nx)/(1+x^2x^2) < 1/2$,
which is not less that 1/4.

Thus we arrive at a contradiction and as such see that the sequence is *not* uniformly convergent in any interval $[a, b]$ with, 0, as an interior point even though it is point-wise convergent there.

Exercise. Show that the sequence $<f_n>$ where $f_n(x) = (nx)/(1+n^2x^2)$ does not converge uniformly on **R**. **[Meerut 2010]**

15.2 CAUCHY'S GENERAL PRINCIPLE OF UNIFORM CONVERGENCE
(Necessary and sufficient condition for uniform convergence) [Meerut 2011]

Theorem. *A necessary and sufficient condition for a sequence $\{f_n\}$ of functions defined in a set S to be uniformly convergent is that to each $\varepsilon > 0$, there corresponds m such that $\forall\, n \geq m, \forall\, p \geq 1$ and $\forall\, x \in S$,* $\left|f_{n+p}(x) - f_n(x)\right| < \varepsilon$. **[Himachal 2013**

Purvanchal 2006; Kanpur 2005, 13; Kanpur 2013; Delhi B.Sc. (Hons.) II 2018]

Proof. The condition is necessary. Let the sequence $\{f_n\}$ be uniformly convergent with f as its uniform limit.

Let $\varepsilon > 0$ be given. Then there exists m such that $\forall\, n \geq m$ and $\forall\, x \in S$,

$$\left|f_n(x) - f(x)\right| < \varepsilon/2. \qquad \text{... (1)}$$

Also, since $\forall\, n \geq m$, $\forall\, p \geq 1$ and $\forall\, x \in S$,

$$\left|f_{n+p}(x) - f(x)\right| < \varepsilon/2, \qquad \text{... (2)}$$

it follows that $\forall\, n \geq m$, $\forall\, p \geq 1$ and $\forall\, x \in S$

$$\begin{aligned}\left|f_{n+p}(x) - f_n(x)\right| &= \left|f_{n+p}(x) - f(x) + f(x) - f_n(x)\right| \\ &\leq \left|f_{n+p}(x) - f(x)\right| + \left|f(x) - f_n(x)\right| \\ &< \varepsilon/2 + \varepsilon/2 = \varepsilon, \text{ using (1) and (2).}\end{aligned}$$

The condition is sufficient. From Cauchy's principle of convergence as proved in theorem III of Art. 5.9 of chapter 5 it follows that the sequence is pointwise convergent. All that we have now to show is that the convergence is uniform. Let f be the point-wise limit of the sequence $\{f_n\}$. Let $\varepsilon > 0$ be given. Then there exists m such that $\forall\, n \geq m, \forall\, p \geq 1,\ \forall\, x \in S$,

$$\left| f_{n+p}(x) - f_n(x) \right| < \frac{1}{2}\varepsilon \quad \Rightarrow \quad f_n(x) - \frac{1}{2}\varepsilon < f_{n+p}(x) < f_n(x) + \frac{1}{2}\varepsilon.$$

Keeping n fixed and letting p tent to infinity, we see that $\forall\, n \geq m$ and $\forall\, x \in S$

$$f_n(x) - \varepsilon/2 \leq f(x) \leq f_n(x) + \varepsilon/2 \quad \Rightarrow \quad \left| f_n(x) - f(x) \right| \leq \varepsilon/2 < \varepsilon.$$

Thus the convergence is uniform.

EXAMPLES

Ex. 1. *Show that the sequence $\{f_n\}$ of functions where $f_n(x) = n/(x+n)$, is uniformly convergent in $[0, k]$ whatever k may be, but not uniformly convergent in $[0, \infty[$.*

(Himachal 2013; Delhi B.A. (Prog.) III, 2014, 15; Delhi B.Sc. (Prog.) II, 2015)

Sol. The sequence $\{f_n\}$ is point-wise convergent $\forall\, x \geq 0$ and the point-wise limit f is given by

$$f(x) = 1 \ \forall\, x \geq 0$$

Let $\varepsilon > 0$ be given. We have

$$\left| f_n(x) - f(x) \right| = \frac{x}{x+n} < \varepsilon \quad \text{if} \quad n > x\left(\frac{1}{\varepsilon} - 1\right).$$

Let $m(\varepsilon, x)$ denote the integer just greater than $x(1/\varepsilon - 1)$. Obviously $m(\varepsilon, x)$ increases as x increases and $\to \infty$ as $x \to \infty$ so that it is not possible to choose any number m such that $\forall\ n \geq m$ and $\forall\, x \geq 0$,

$$\left| f_n(x) - f(x) \right| < \varepsilon$$

so that the convergence is *not* uniform in $[0, \infty[$.

Now consider the interval $[0, k]$.

Let m be a integer greater than $k(1/\varepsilon - 1)$. We then see that $\forall\, n \geq m$ and $\forall\ x \in [0, k]$,

$$\left| f_n(x) - f(x) \right| < \varepsilon$$

so that the convergence is uniform in $[0, k]$.

Ex. 2. *Show that the sequence $\{f_n\}$ where $f_n(x) = x^n$ is uniformly convergent in $[0, k]$ where k is a number less than 1 and only point-wise convergent in $[0, 1]$.* **(Himachal 2013)**

[Delhi B.Sc. (Prog) 2008; Agra 2003; Delhi B.A. (Prog.) III 2014; Purvanchal 2006 Bhopal 2000; Rewa 2001; Ravishankar 2001; Agra 2003, Delhi B.Sc. Physics (H) 2004]

Sol. The given sequence is point-wise convergent in $[0, 1]$ and the point- wise limit f is given by

$$f(x) = \begin{cases} 0, & \text{when } 0 \leq x < 1 \\ 1, & \text{when } x = 1 \end{cases}$$

Let $\varepsilon > 0$ be given. Then for $0 < x \leq k < 1$, we have

$$\left| f_n(x) - f(x) \right| = x^n < \varepsilon$$

$$\Rightarrow \quad (1/x)^n > (1/\varepsilon) \quad \Rightarrow \quad n \log(1/x) > \log(1/\varepsilon), x \neq 0 \quad \Rightarrow \quad n > (\log 1/\varepsilon) \div \log(1/x).$$

Also if $x = 0$, $\left|f_n(x) - f(x)\right| = 0 < \varepsilon \;\; \forall\, n \geq 1$.

Let $m(\varepsilon, x)$ denote the integer next greater than 1 and $\log(1/\varepsilon) \div \log(1/x)$.

Now, $m(\varepsilon, x)$ increases and $\to \infty$ as $x \to 1$, so that there does not exist an m such that

$$\forall\, n \geq m \text{ and } \forall\, x \in [0,1], \qquad \left|f_n(x) - f(x)\right| < \varepsilon\,;$$

so that the convergence is *not* uniform in [0, 1].

Now suppose that k is a number such that $0 \leq k < 1$.

We see that in [0, k], the greatest value of $\log(1/\varepsilon) \div \log(1/x)$ is $((\log(1/\varepsilon)) \div (\log 1/k))$.

Let any integer greater than this value be denoted by m. Then we see that $\forall\, n \geq m$ and

$$\forall\, x \in [0, k], \qquad \left|f_n(x) - f(x)\right| < \varepsilon$$

so that the convergence is uniform in [0, k].

Note : A point, like $x = 1$ which is such that the sequence is not uniformaly conversent in any interval containing $x = 1$, is known as *a point of non-uniform convergence.*

Ex. 3. *Show that if* $f_n(x) = nxe^{-nx^2}$, *the sequence* $\{f_n\}$ *is point-wise, but not uniformly, convergent in* $[0, k[, k > 0$. ***[Delhi B.A. (Prog.) III, 2017; Kanpur 2017; Pune 2010; Delhi B.Sc. (Prog), II, 2010, 11; Meerut 2007, 11]***

Sol. It may be easily seen that the sequence is point-wise convergent and that the point-wise limit is the function f such that $\forall\, x, f(x) = 0$.

If possible, let the sequence be uniformly convergent in $[0, k[$, so that, $\varepsilon > 0$ being given, there exists m such that $\forall\, n \geq m$ and $\forall\, x \geq 0$

$$\left|f_n(x) - f(x)\right| = nxe^{-nx^2} < \varepsilon. \qquad \text{... (1)}$$

Let m_0 be an integer greater than m and $e^2\varepsilon^2$ and let $x = 1/m_0$. Then the inequality (1) holds for $x = 1/\sqrt{m_0}$ and $n = m_0$. These give $\sqrt{m_0}/e < \varepsilon \Leftrightarrow m_0 < e^2\varepsilon^2$ so that we arrive at a contradiction.

Thus the convergence is *not* uniform in $[0, \infty[$., 0 being *a* point of non-uniform convergence of $< f_n(x) >$.

EXERCISES

1. Show that the sequence f_n where $f_n(x) = e^{-nx}$ is point-wise but not uniformly convergent in $[0, \infty]$. Also show that the convergence is uniform in $[k, \infty[$; k being any positive number. ***[Agra 2000]***
2. Show that the sequence $\{e^{-nx}\}$ is uniformaly convergent in any interval $[a, b]$, where a and b are positive numbers but only point-wise in $[a, b]$.
3. Show that the sequence $\{f_n(x)\}$, where $f_n(x) = \tan^{-1} nx$, $x \geq 0$ is uniformaly convergent in any interval $[a, b]$, $a > 0$ but is only pointwise convergent in $[0, b]$ ***[Himachal 2008; Delhi B.Sc. (Hons.) II, 2009; Jiwaji 2001)***
4. Show that the sequence $< f_n >$ defined as $f_n(x) = x^n/n$ on
   (*i*) $]-\infty, \infty[$ is not uniformly convergent. ***[Kanpur 2018]***
   (*ii*) [0, 1] converges uniformly to 0
5. Show that the sequence $< f_n >$, where $f_n(x) = (n^2 x) / (1 + n^2 x^2)$ is not uniformly convergent on [0, 1].

6. Show that the sequence $< f_n >$, where $f_n(x) = (nx) / (nx + 1)$ is uniformly convergent on $[a, b]$, $a > 0$ but in only pointwise convergent on [0, b].

7. Show that $< x_n >$ where $f_n(x) = (n^2x)/1 + n^4x^2)$ is not uniformly convergent on [0, 1]

15.3 A TEST FOR UNIFORM CONVERGENCE OF SEQUENCE OF FUNCTIONS

In order to test whether a given sequence $< f_n(x)$ in uniformly convergent or not in a given interval, so far we have been using the definition of uniform convergence. Accordingly, we tried to get $m \in N$, independent of x, which is not easy in practice. This method can be replaced by an easy method given in the following theorem.

Theorem (M_n-test). *Let $< f_n >$ be a sequence of functions defined on an interval I such that* $\lim_{n\to\infty} f_n(x) = f(x) \forall\, x \in [a, b]$ *and let* $M_n = \sup\{ | f_n(x) - f(x) | : x \in [a, b]\}$.
Then $< f_n >$ converges uniformly on $[a, b]$ if and only if $M_n \to 0$ as $n \to \infty$.

[Delhi B.Sc. (Prog) III 2008; Kanpur 2005, 06, 10; Bhopal 2004; Himachal 2013, 2014, Delhi B.Sc. (Prog.) III 2008, 2012; Meerut 2012]

Proof. The condition is necessary. Let $< f_n >$ converge uniformly to f on $[a, b]$. Then, for a given $\varepsilon > 0$, there exists a positive integer m such that

$$| f_n(x) - f(x) | < \varepsilon \;\forall\, x \geq m \text{ and } \forall\, x \in [a, b]$$

$$\Rightarrow \quad \sup \{| f_n(x) - f(x) | : x \in [a, b]\} < \varepsilon \;\forall\; n \geq m$$

$$\Rightarrow M_n < \varepsilon \;\forall\; n \geq m \quad \Rightarrow |M_n - 0| < \varepsilon \;\forall\; n \geq m \quad \Rightarrow M_n \to 0, \text{ as } n \to \infty$$

The condition is sufficient. Let $M_n \to 0$ and $n \to \infty$. Then for a given $\varepsilon > 0$, there exists a positive integer m such that

$$| M_n - 0 | < \varepsilon \;\forall\, n \geq m \quad \Rightarrow \quad M_n < \varepsilon \;\forall\, n \geq m$$

$$\Rightarrow \quad \sup \{| f_n(x) - f(x) | : x \in [a, b]\} < \varepsilon \;\forall\, n \geq m$$

$$\Rightarrow \quad | f_n(x) - f(x) | < \varepsilon \;\forall\, n \geq m, \forall\, x \in [a, b]$$

$\Rightarrow \quad < f_n >$ converges uniformly to f on $[a, b]$

EXAMPLES

Remark. If M_n does not tend to zero as $n \to \infty$, then the sequence $< f_n >$ is not uniformly convergent.

Ex. 1. *Prove that the sequence $< f_n >$, where $f_n(x) = x / (1 + nx^2)$ converges uniformly on any closed intervals I.* **[Bhopal 1998; Indore 2000; Meerut 2001, 04, 08, 13; Delhi B.Sc. (Hons.) II 2002, 03; 06, 07; Kanpur 2003, 2007, 2015]**

Sol. Here pointwise limit $= f(x) = \lim_{n\to\infty} f_n(x) = 0, \;\forall\, x \in \text{I}$

$$\therefore \quad | f_n(x) - f(x) | = \left| \frac{x}{1 + nx^2} - 0 \right| = \left| \frac{x}{1 + nx^2} \right| = | y |, \text{ say} \qquad ...(1)$$

where $$y = x / (1+nx^2) \qquad ...(2)$$

From (2), $$\frac{dy}{dx} = \frac{(1+nx^2)\cdot 1 - x \cdot 2nx}{(1+nx^2)^2} = \frac{1 - nx^2}{(1+nx^2)^2} \qquad ...(3)$$

For maximum and minimum value of y, we have

$$dy/dx = 0 \quad \Rightarrow \quad 1 - nx^2 = 0 \quad \Rightarrow \quad x = 1/\sqrt{n} \in \text{I}$$

From (3), $$\frac{d^2y}{dx^2} = \frac{(1+nx^2)^2 \cdot (-2nx) - (1 - nx^2) \cdot 2(1+nx^2) \cdot 2nx}{(1+nx^2)^4}$$

or
$$\frac{d^2y}{dx^2} = \frac{-2nx\,(1+nx^2) - 4nx\,(1-nx^2)}{(1+nx^2)^3}$$

$$\text{When } x = 1/\sqrt{n}, \quad d^2y/dx^2 = -(\sqrt{n}/2) < 0,$$

showing that y is maximum when $x = 1/\sqrt{n}$ and from (2) the maximum value of $y = 1/2\sqrt{n}$

$\therefore$ $M_n = \sup\{\,|f_n(x) - f(x)| : x \in \mathrm{I}\,\} = \sup\{|\,y\,| : x \in \mathrm{I}\} = 1/2\,\sqrt{n}$

Since $M_n \to 0$ as $n \to \infty$, $<f_n>$ is uniformly convergent on any closed interval I.

Ex. 2. *Show that the sequence* $<f_n>$, *where* $f_n(x) = nx\,(1-x)^n$ *is not uniformly convergent on* [0, 1].*(Pune 2010; Rajasthan 2010; Bhopal 2003; Kanpur 2004; Meerut 08, 09; Agra 2000, 01)*

Sol. When $x = 0, f_n(x) = 0\ \forall\, n \in N$; when $x = 1, f_n(x) = 0\ \forall\, n \in N$

Hence $f(x) = \lim_{n\to\infty} f_n(x) = 0$ when $x = 0$ and $x = 1$.

Again, for $0 < x < 1$, we have

$$f(x) = \lim_{n\to\infty} f_n(x) = \lim_{n\to\infty} nx(1-x)^n = \lim_{n\to\infty} \frac{nx}{(1-x)^{-n}} \qquad \left[\text{Form } \frac{\infty}{\infty}\right]$$

$$= \lim_{n\to\infty} \frac{x}{-(1-x)^{-n}\log(1-x)} = \lim_{n\to\infty} \frac{-x(1-x)^n}{\log(1-x)} \qquad \text{[using L' Hopital's rule]}$$

$= 0$ as $0 < x < 1$ so that $(1-x)^n \to 0$ as $n \to \infty$

Thus, we have $f(x) = 0\ \forall\, x \in [0, 1]$

$\therefore$ $|f_n(x) - f(x)| = |\,nx\,(1-x)^n - 0\,| = nx(1-x)^n = y$, say ...(1)

where $y = nx\,(1-x)^n$ ...(2)

$\therefore$ $dy/dx = n(1-x)^n - n^2x(1-x)^n = n\,(1-x)^{n-1}\,\{1 - (n+1)\,x\}$ ...(3)

For maximum and minimum value of y, we have

$dy/dx = 0 \Rightarrow x = 1/(n+1)$

From (3), $d^2y/dx^2 = -n(n-1)\,(1-x)^{n-2}\,\{1-(n+1)x\} - n\,(n+1)\,(1-x)^{n-1}$.

$$\text{When } x = \frac{1}{n+1}, \ \frac{d^2y}{dx^2} = -n(n+1)\cdot\left(\frac{n}{n+1}\right)^{n-1} < 0,$$

showing that y is maximum when $x = 1/(n+1)$ and from (2),

$$\text{the maximum value of } y = \frac{n}{n+1}\left(1 - \frac{1}{n+1}\right) = \left(1 - \frac{1}{n+1}\right)^{n+1} \qquad \text{...(4)}$$

$$\therefore M_n = \sup\{\,|f_n(x) - f(x)| : x \in [0, 1]\} = \sup\{\,|\,y\,| : x \in [0, 1]\} = \left(1 - \frac{1}{n+1}\right)^{n+1}, \text{ by (4)}$$

$\therefore M_n \to e^{-1}$ as $n \to \infty$. Since M_n does not tend to 0 as $n \to \infty$, the sequence $<f_n>$ is not uniformly anvergent on [0, 1]. Here 0 is a point of non-uniform conversence because $x = 1/(n+1) \to 0$ as $n \to \infty$.

Ex. 3. *Show that the sequence* $<f_n>$ *where* $f_n(x) = (\sin nx)/\sqrt{n}$ *is uniformly convergent on* $[0, \pi]$. ***(Himachal 2014; Delhi B.A. (Prog) III 2011; Kanpur 2005; Agra 1998)***

Sol. Here $f(x) = \lim_{n\to\infty} f_n(x) = \lim_{n\to\infty} \frac{1}{\sqrt{n}} \cdot \sin nx = 0 \quad \forall\, x \in [0, \pi]$

$$\therefore \qquad |f_n(x) - f(x)| = \left|\frac{\sin nx}{\sqrt{n}} - 0\right| = |y|, \text{ say} \qquad ...(1)$$

where
$$y = \frac{\sin nx}{\sqrt{n}} \text{ so that } \frac{dy}{dx} = \sqrt{n} \cos nx \qquad ...(2)$$

For maximum and minimum value of y, we have

$$dy/dx = 0 \Rightarrow \cos nx = 0 \Rightarrow nx = \pi/2 \quad \text{or} \quad x = \pi/2n \in [0, \pi]$$

Again, $\quad d^2y/dx^2 = -n^{3/2} \sin nx$

$\Rightarrow \quad$ When $x = \pi/2n$, $d^2y/dx^2 = -n^{3/2} \sin(\pi/2) = -n^{3/2} < 0$,

showing that y is maximum when $x = \pi/2n$ and

the maximum value of $y = (1/\sqrt{n}) \times \sin(\pi/2) = 1/\sqrt{n}$. ...(3)

Moreover $\quad x = \pi/2n \to 0$ as $n \to \infty$

From (1), (2) and (3) $\quad M_n = \sup\{|f_n(x) - f(x)| : x \in [0, \pi]\} = 1/\sqrt{n}$

Since $M_n \to 0$ and $n \to \infty$, $< f_n >$ is uniformly convergent on $[0, \pi]$.

EXERCISE

1. Show that the sequence $< f_n >$ where $f_n(x) = nx/(1+n^2x^2)$ is not uniformly convergent on any interval containing zero. *(G.N.D.U. Amristar 2002; Jabalpur 2000; Agra 2008, 10; Kanpur 2007, Himanchal 2014; Meerut 2000; 02, 05, 06; Delhi B.Sc. (Prog.) II, 2016)*
2. Show that the sequence $< f_n >$ where $f_n(x) = nx\, e^{-nx^2}$, is not uniformally convergent on (*i*) $[0, 1]$ (*ii*) $[0, k]$, $k > 0$ *(Himanchal Pradesh 2002)*
3. Show that if $f_n(x) = n^2x/(1+x^4x^2)$, then $< f_n >$ converges non-uniformly on $[0, 1]$.
4. Show that the sequence $< f_n >$, where $f_n(x) =$
   (*i*) $n^2x / (1+n^3x^2)$ is not uniformly convergent on $[0, 1]$ ***(Himachal 2014)***
   (*ii*) $nx / (1 + n^3x^2)$ converges uniformly on any closed interval $[a, b]$
   (*iii*) $n^2x / (1 + n^2x^2)$ is not uniformly convergent on $[0, 1]$
   (*iv*) $(1/n) \times e^{-nx}$ converges uniformly to 0 on $[0, \infty[$ ***(Kanpur 2011)***
5. Show that the sequence $< f_n >$ is uniformly convergent for all $x \geq 0$, when
   (*i*) $f_n(x) = x / (n + x^2)^2$ (*ii*) $x/n(1 + nx^2)$
6. Show that the sequence $< x^{n-1}(1 - x) >$ is uniformaly convergent on $[0, 1]$ *(Sagar 1995)*
7. If a sequence $< f_n(x) >$ conveges uniformly to $f(x)$ on $[a, b]$ and x_0 is a point of $[a, b]$ such that $\lim\limits_{n \to x_0} f_n(x) = a_n$ $n = 1, 2, 3, ...$, then prove that
   (*i*) $< a_n >$ converges (*ii*) $\lim\limits_{x \to x_0} f(x) = \lim\limits_{n \to \infty} a_n$. *(Delhi Maths (H) 2000, 2005)*

15.4 CONTINUITY OF THE UNIFORM LIMIT OF A UNIFORMLY CONVERGENT SEQUENCE OF CONTINUOUS FUNCTIONS. ***[Delhi B.Sc. (Prog.) II, 2016]***

Theorem. *If $< f_n >$ is a sequence of continuous functions on an interval $[a, b]$ and if $f_n \to f$ uniformly on $[a, b]$, then f is continuous on $[a, b]$.* ***[Bilaspur 2000; Kanpur 2003, 04; Ravishankar 2000; G.N.D.U. Amritsar 2000, 02, Jiwaji 2001; Kurukshetra 2003; Himachal 2013, 2014, 2015; Meerut 2005; Delhi B.Sc. (Hons) II, 2011]***

Proof. Consider a uniformly convergent sequence $\{f_n\}$ of continuous functions defined in $[a, b]$. Let the function f be the limit which is, of course, the uniform limit of the sequence. It will now be proved that f is a continuous function. Let c be any point of $[a, b]$.

We have $\forall\ x$ and $\forall\ m \in N$,

$$|f(x) - f(c)| = |f(x) - f_m(x) + f_m(x) - f_m(c) + f_m(c) - f(c)|$$
$$\leq |f(x) - f_m(x)| + |f_m(x) - f_m(c)| + |f_m(c) - f(c)|. \qquad ...(1)$$

Let $\varepsilon > 0$ be given. As $\{f_n\}$ converges uniformly to f, there exists $m \in N$, such that $\forall\, x \in [a, b]$ and $\forall\, n \geq m$, we have

$$|f(x) - f_n(x)| < \varepsilon/3$$

In particular, we see that $\forall\, x \in [a,b]$,

$$|f(x) - f_m(x)| < \varepsilon/3 \quad \Rightarrow \quad |f(c) - f_m(c)| < \varepsilon/3 \qquad \text{... (2)}$$

As f_m is continuous at c, there exists $\delta > 0$ such that $\forall\, x \in]c-\delta, c+\delta[$

$$|f_m(x) - f_m(c)| < \varepsilon/3 \qquad \text{... (3)}$$

From (1), (2) and (3), we have

$$|f(x) - f(c)| < \varepsilon/3 + \varepsilon/3 + \varepsilon/3 \quad \text{or} \quad |f(x) - f(c)| < \varepsilon.$$

Thus we see that there exists $\delta > 0$ such that $\forall\, x \in]c-\delta, c+\delta[$

$$|f(x) - f(c)| < \varepsilon,$$

and as such f is continuous at c and, therefore, also at every point of the domain.

Note1. Uniform convergence of the sequence $<f_n>$ is only a sufficient condition but not a necessary for the continuity of the limit function f, i.e. if the limit function f is continuous on $[a, b]$, then it is not necessary that the sequence $<f_n>$ is uniformly convergent on $[a, b]$.

Note 2. From the above theorem, it follows that if the limit function f is discontinuous on $[a, b]$, then the sequence $<f_n>$ of continuous function cannot be uniformly convergent on $[a, b]$. Therefore, the above theorem provides us an easy method to prove that a certain sequence is not uniformly convergent.

Example 1. *Show that the sequence* $<f_n>$ *where* $f_n(x) = \tan^{-1} nx$ *is not uniformaly convergent on* [0, 1]. *(Delhi Maths (H) 1998, 2000)*

Solution. The limit function f is given by

$$f(x) = \lim_{n\to\infty} f_n(x) = \lim_{n\to\infty} \tan^{-1} nx = \pi/2 \text{ for } 0 < x \leq 1$$

When $x = 0$, the sequence $<f_n>$ converges to 0

Thus, $$f(x) = \begin{cases} \pi/2, \text{ if } 0 < x \leq 1 \\ 0, \text{ if } x = 0 \end{cases}$$

Clearly, f is discontinuous at $x = 0$ and so f is discontinuous on [0, 1].

Also, $f_n(x) = \tan^{-1} nx$, $0 \leq x \leq 1$ is continous on [0, 1] $\forall\, n \in N$

Thus, $<f_n>$ is a sequence of continuous functions and its limit function f is discontinuous on [0, 1]. Hence the $<f_n>$ is not uniformly conversent on [0, 1].

Example 2. Let f_n be defined by $f_n(x) = nx/(1+x^2x^2)$, in any domain $[a, b]$ with 0 as an interior point. If has been seen on page 15.2 that the sequence is point wise convergent but not uniformly convergent. Also the point wise limit f is given by

$$f(x) = 0 \;\forall\, x \in [a, b]$$

Thus we see that while convergence is not uniform, the point wise limit itself is a continuous function.

Example 3. *Let* f_n *be defined by* $f_n(x) = 1 - |1 - x^2|^n$ *in the domain*

$$\{x : |1 - x^2| \leq 1\} = [-\sqrt{2}, \sqrt{2}].$$

The given sequence $\{f_n\}$ is point-wise convergent in $\left[-\sqrt{2}, \sqrt{2}\right]$ with the point-wise limit f, where

$$f(x) = \begin{cases} 1 \text{ when } \left|1-x^2\right| < 1 \\ 0 \text{ when } \left|1-x^2\right| = 1 \end{cases} \quad \textit{i.e.}, \text{ for } x = 0, \pm\sqrt{2}.$$

Thus we see that the point-wise limit is not continuous for $x = 0$ even though each f_n is continuous there at.

Thus we may conclude that the sequence cannot be uniformly convergent in $\left[-\sqrt{2}, \sqrt{2}\right]$. We may also see this fact directly as follows :

Suppose that the sequence is uniformly convergent in $\left[-\sqrt{2}, \sqrt{2}\right]$ so that f is the uniform limit.

Take $\varepsilon = 1/2$. Then there exists $m \in N$ such that

$$\left|f_n(x) - f(x)\right| < 1/2, \quad \forall\, n \geq m \quad \text{and} \quad \forall\, x \in \left[-\sqrt{2}, \sqrt{2}\right].$$

In particular $\quad \forall\, x \in \left[-\sqrt{2}, \sqrt{2}\right], \quad \left|f_m(x) - f(x)\right| < 1/2.$

Now, $$\left|f_m(x) - f(x)\right| = \begin{cases} \left|1-x^2\right|^m & \text{when } \left|1-x^2\right| < 1 \\ 0 & \text{when } \left|1-x^2\right| = 1 \end{cases}$$

Since $\lim\limits_{x \to 0} \left|1-x^2\right|^m = 1$, there exists a neighbourhood of 0 for every x of which $\left|1-x^2\right|^m$ belongs to a neighbourhood

$$\left]1-\frac{1}{4}, 1+\frac{1}{4}\right[= \left]\frac{3}{4}, \frac{5}{4}\right[$$

of 1, and is as such greater than 1/2.

Thus, we arrive at a contradiction and as such we see directly that the convergence is *not* uniform in $\left[-\sqrt{2}, \sqrt{2}\right]$.

EXERCISE

1. Show that the sequence $<f_n>$, where $f_n(x) = x^n$, $0 \leq x \leq 1$ is not uniformly convergent on [0, 1]. **[Delhi B.Sc. (Hons) II 2011]**
2. Show that the sequence $<f_n>$, where $f_n(x) = 1/(1+nx)$, $0 \leq x \leq 1$ is not uniformly convergent on [0, 1] **[Kanpur 2013]**

15.5 INTEGRABILITY OF THE UNIFORM LIMIT OF A UNIFORMLY CONVERGENT SEQUENCE OF INTEGRABLE FUNCTIONS

Theorem. Let $\{f_n\}$ be a uniformly convergent sequence with uniform limit f on $[a, b]$ and let f_n be integrable on $[a, b]$ $\forall\, n \in N$. Then the limit f is itself integrable and

$$\int_a^b f(x)\, dx = \lim_{n \to \infty} \int_a^b f_n(x)\, dx.$$ **[Mumbai 2010; Meerut 2001, 03, 04]**

[Himachal 2002, Agra 2001, Delhi B.Sc. (Hons.) II, 2017, 2019]

Proof. Let $\varepsilon > 0$ be given so that there exists $m \in N$ such that $\forall\, x \in [a,b]$ and $\forall\, n \geq m$,

$$\left| f_n(x) - f(x) \right| < \varepsilon / 4\,(b-a).$$

In particular, we have $\forall\, x \in [a,b]$, $\quad \left| f_m(x) - f(x) \right| < \varepsilon / 4\,(b-a).$

We write $\quad f(x) - f_m(x) = R_m(x),$

so that we have defined a new function R_m.

As the function f_m is integrable, there exists a partition, say P, of $[a, b]$ such that the oscillatory sum $w(P, f_m)$ for the function f_m corresponding to the partition P is $< \varepsilon / 2$.

Also since $\quad \left| R_m(x) \right| < \varepsilon / 4\,(b-a)\ \forall\, x \in [a,b],$

the oscillation of R_m in each sub-interval is $< \varepsilon / 2\,(b-a)$.

implying that $\quad w(P, R_m) < \dfrac{\varepsilon}{2\,(b-a)}(b-a) = \dfrac{\varepsilon}{2}.$

Also $\quad f = f_m + R_m \Rightarrow w(P, f) \leq w(P, f_m) + w(P, R_m).$

Thus if $\varepsilon > 0$ be given, there exists a partition P of $[a, b]$ such that $w(P, f) < \varepsilon$, and accordingly f is integrable.

Now, we prove that $\quad \displaystyle\int_a^b f(x)\, dx = \lim_{n \to \infty} \int_a^b f_n(x)\, dx.$

Let $\varepsilon' > 0$ be given so that there exists $m' \in N$ such that $\forall\, x \in [a,b]$ and $\forall\, n \geq m'$

$$\left| f(x) - f_n(x) \right| < \varepsilon' / (b-a)$$

$$\Rightarrow \quad -\varepsilon' / (b-a) < f(x) - f_n(x) < \varepsilon' / (b-a).$$

$$\Rightarrow \quad -\varepsilon' / (b-a) < (f - f_n)(x) < \varepsilon' / (b-a).$$

Now f_n is integrable and f has also been proved integrable. Thus $f - f_n$ is integrable. We have

$$\left| \int_a^b (f - f_n)\,(x)\, dx \right| < \varepsilon'.$$

Thus if $\varepsilon' > 0$ is given, there exists $m' \in N$ such that $\forall\, n \geq m'$

$$\left| \int_a^b f(x)\, dx - \int_a^b f_n(x)\, dx \right| < \varepsilon' \Rightarrow \lim_{n \to \infty} \int_a^b f_n(x)\, dx = \int_a^b f(x)\, dx.$$

Cor. *It may be proved without difficulty that the sequence*

$$\left\{ \int_a^t f_n(x)\, dx \right\}$$

of functions is uniformly convergent with limit

$$\int_0^t f(x)\, dx,\ t \in [a, b].$$

Note 1. The converse of the above theorem may not be true, *i.e.*, a sequence may converge to an integrable limit without being uniformly convergent.

Note 2. If $<f_n>$ is a sequence of integrable function converging to f on $[a, b]$ and if

$$\int_a^b f(x)\,dx \neq \lim_{n\to\infty}\int_a^b f_n(x)$$

then $<f_n>$ cannot converge uniformly to f.

Example 1. *Show that the sequence* $<f_n>$, where $f_n(x) = nx\,e^{-nx^2}$ is not uniformly convergent on [0, 1]

Solution. Here $f(x) = \lim_{n\to\infty} f_n(x) = \lim_{n\to\infty}\dfrac{nx}{e^{nx^2}} = \lim_{n\to\infty}\dfrac{nx}{1+\dfrac{nx^2}{11.}+\dfrac{n^2x^4}{21.}} = 0 \;\forall\, x\in[0,1]$

Thus, $$\int_0^1 f(x)\,dx = \int_0^1 0\,dx = 0 \qquad ...(1)$$

Also, $$\int_0^1 f_n(x)\,dx = \int_0^1 nx\,e^{-nx^2}\,dx = \frac{1}{2}\int_0^n e^{-t}\,dt, \text{ on putting } t = nx^2$$

$$= (1 - e^{-n})/2$$

$\therefore$ $$\lim_{n\to\infty}\int_0^1 f_n(x)\,dx = \lim_{n\to\infty}\frac{1}{2}(1-e^{-n}) = \frac{1}{2} \qquad ...(2)$$

From (1) and (2), we have

$$\lim_{n\to\infty}\int_0^1 f_n(x)dx = \frac{1}{2} \neq 0 = \int_0^1 f(x)\,dx = \int_0^1 \lim_{n\to\infty} f_n(x)$$

and as such the integral of the point wise limit is not equal to the limit of the sequence of the integrals.

It follows from theorem of Art. 15.5 that the sequence $<f_n>$ is not uniformly convergent in [0, 1]

Example 2. Let f_n be defined by

$$f_n(x) = \frac{nx}{1+n^2x^2},\; x\in[0,1].$$

It has been shown on page 15.2 that while the sequence $\{f_n\}$ is point-wise convergent, it is not uniformly convergent in [0, 1]. In spite of this however, it will be seen that

$$\lim \int_0^1 f_n(x)\,dx = \int_0^1 \lim f_n(x)\,dx.$$

We have $$\int_0^1 f_n(x)\,dx = \int_0^1 \frac{nx}{1+n^2x^2}\,dx = \frac{1}{2n}\log(1+n^2)$$

$\Rightarrow$ $$\lim \int_0^1 f_n(x)\,dx = 0.$$

Also since the point-wise limit f is such that $f(x) = 0\;\forall\, x$, we have

$$\int_0^1 \lim f_n(x) = 0.$$

Thus we have the equality, in spite of the absence of uniform convergence

Example 3. (a) *If* $<g_n>$ and $<f_n>$ *are two sequences of functions defined for* $0 \le x \le 1$ *by* $g_n(x) = nx/(1+n^3x^2)$ *and* $f_n(x) = \{\log(1 + n^3x^2)\}/2n^2$ *Prove that* $<g_n>$ *converges uniformly to zero on* [0, 1] *and hence obtain the uniform convergence of* $<f_n>$ (Delhi Maths (H) 1996) (*b*) *Show that the sequence* $<f_n(x)>$, *where* $(f_n(x) = \{\log(1+n^3x^2)\}/n^2$ *is uniformly convergent on* [0, 1] (*Delhi B.Sc. Maths (H) 1995, 1999, 2000*)

Solution. (*a*) For $x \neq 0$, $\lim_{n\to\infty} g_n(x) = \lim_{n\to\infty} \frac{nx}{1+n^3x^2} = \lim_{n\to\infty} \frac{x/n^2}{(1/n)^3 + x^2} = 0$

Also, when $x = 0, g_n(0) = 0 \;\; \forall\, n \in N$

$\therefore \qquad \lim_{n\to\infty} g_n(0) = 0 \;\; \forall\, n \in N$

Thus, $\qquad g(x) = \lim_{n\to\infty} g_n(x) = 0, \; \forall\, x \in [0,1]$

$$\therefore \qquad \left|g_n(x) - g(x)\right| = \left|\frac{nx}{1+n^3x^2} - 0\right| = \frac{nx}{1+n^3x^2} = y, \text{ say} \qquad ...(1)$$

$$\therefore \qquad \frac{dy}{dx} = \frac{n(1+n^3x^2) - nx\cdot 2n^3x}{(1+n^3x^2)^2} = \frac{n(1-n^3x^2)}{(1+n^3x^2)^2} \qquad ...(2)$$

For max and min of y, $dy/dx = 0 \;\Rightarrow\; 1 - n^3x^2 = 0 \;\Rightarrow\; x = 1/n^{3/2}$

From (2), $$\frac{d^2y}{dx^2} = n \times \frac{-2n^3x(1+n^3x^2) - (1-n^3x^2)\cdot 2(1+n^3x^2)\cdot 2n^3x}{(1+n^3x^2)^4}$$

or $\frac{d^2y}{dx^2} = -\frac{2n^4x(3-n^3x^2)}{(1+n^3x^2)^3} \;\Rightarrow\;$ value of $\frac{d^2y}{dx^2}$ at $x = \frac{1}{n^{3/2}}$ is $-n^{5/2} < 0$,

showing that y is max. when $x = 1/n^{3/2}$ and the maximum value of $y = \frac{n\times(1/n^{3/2})}{1+1} = \frac{1}{2n^{1/2}}$

$$\therefore \qquad M_n = \sup\{|g_n(x) - g(x)| : x \in [0,1]\} = 1/n^{3/2}$$

Since $M_n \to 0$ as $n \to \infty$, so $<g_n>$ converges uniformly to $g(x) = 0 \;\forall\, x \in [0,1]$

We now proceed to show that $<f_n>$ is uniformly convergent, if

$$f_n(x) = \{\log(1+n^3x^2)\}/2n^2 \qquad ...(3)$$

From (3), $$f_n'(x) = \frac{1}{2n^2}\cdot\frac{2n^3x}{1+n^3x^2} = \frac{nx}{1+n^3x^2} = g_n(x) \qquad ...(4)$$

Since $<g_n>$ converges uniformly to $g(x)$, where $g(x) = 0$ on [0, 1] and since each $g_n(x)$ is continuous on [0, 1] and hence integrable on [0, 1], it follows that the sequence

$$\left\{\int_0^x g_n(x)\,dx\right\},\; i.e., \left\{\int_0^1 f_n'(x)\,dx\right\},\; i.e., <f_n(x)> \text{ is uniformly convergent}$$

(*b*) Left as an exercise

EXERCISE

1. If a sequence $<f_n>$ uniformly converges to f and each function f_n is integrable then f is integrable on $[a, b]$ and the sequence $<\int_a^x f_n\,dt>$ converges uniformly to $\int_0^x f\,dt$ on $[a, b]$.

(*Delhi Maths (H) 1997*)

15.6 DERIVABILITY OF THE POINT-WISE LIMIT OF A SEQUENCE OF DERIVABLE FUNCTIONS IF THE DERIVATIVES ARE CONTINUOUS AND THE SEQUENCE OF DERIVATIVES IS UNIFORMLY CONVERGENT

Theorem. *Let $\{f_n\}$ be a sequence of derivable functions with point-wise limit f. Let f'_n be continuous on $[a, b]$ $\forall\, n \in N$ and let the sequence $\{f'_n\}$ be uniformly convergent with ϕ as uniform limit on $[a, b]$. Then ϕ is derivable and derivative is equal to f, i.e.,*

$$\phi' = \left(\lim_{n\to\infty} \{f_n\}\right)' = \lim_{n\to\infty} \{f'_n\}$$

i.e., the derivative of the limit of the sequence is equal to the limit of the sequence of derivatives. *[G.N.D.U. Amritsar 2003; Jiwaji 2001; Ravishankar 2003]*

Proof. As $\{f'_n\}$ is a uniformly convergent sequence of continuous functions, it follows that its uniform limit ϕ is a cootinuous function. It follows that

$$\int_a^t \phi\,(x)\,dx = \lim \int_a^t f'_n\,(x)\,dx = \lim \{f_n\,(t) - f_n(a)\}$$

$$= \lim \{f_n(t)\} - \lim \{f_n(a)\} = f(t) - f(a).$$

Also ϕ being continuous, we have,

$$\phi'(t) = f(t)\ \forall\, t \in [a,b] \Rightarrow \phi' = f.$$

Note. The above theorem provides a good negative test for the uniform convergence of $< f_n' >$. Accordingly, if the derivative of the limit of the sequence is not equal to the limit of the sequence of the derivates, then the sequence $< f_n'(x) >$ cannot be uniformly convergent.

Example. *Show that the sequence $< f_n >$ of functions where $f_n(x) = nx/(1+n^2x^2)$ converges to f where $f(x) = 0$ for all x and that the equation $f'(x) = \lim_{n\to\infty} f_n'(x)$ is true for all $x \neq 0$ but is false if $x = 0$. What you say about the uniform convergence of the sequence $< f_n' >$ in an interval containing zero* *(Delhi Maths (H) 2001)*

Sol. By examnle on page 15.1, the sequence $< f_n >$ converges pointwise to $f(x) = 0\ \forall\, x \in [0, 1]$

Now, $\quad f(x) = 0 \quad \Rightarrow \quad f'(x) = 0\ \forall\, x \in [0, 1]$ ...(1)

For $x \neq 0$,
$$f_n'(x) = \frac{n(1+n^2x^2) - 2n^2x \cdot nx}{(1+n^2x^2)^2} = \frac{n(1-n^2x^2)}{(1+n^2x^2)^2} = \frac{1/n^2 - x^2}{n(1/n^2 + x^2)}$$

$$\Rightarrow \quad \lim_{n\to\infty} f_n'(x) = 0 \text{ for all } x \neq 0. \quad \text{... (2)}$$

From (1) and (2), $\quad f'(x) = \lim_{n\to\infty} f_n'(x)$ is true for all $x \neq 0$

Now, for $x = 0$, we have

$$f_n'(0) = \lim_{x\to 0} \frac{f_n(x) - f_n(0)}{x - 0} = \lim_{x\to 0} \frac{(nx)/(1+n^2x^2)}{x} = \lim_{x\to 0} \frac{n}{1+n^2x^2} = n$$

$$\Rightarrow \quad f_n'(0) \to \infty \text{ as } n \to \infty$$

Thus at $x = 0$, $f'(x) = \lim_{n\to\infty} f_n'(x)$, is false.

Thus, $\left(\lim_{n\to\infty} < f_n >\right)' \neq \lim_{n\to\infty}\left\{f_n'\right\}$ in any interval containing zero, It follows, that $< f_n' >$ cannot converge uniformly in any interval containing zero.

EXERCISES

1. Show that the sequence $< f_n >$, where $f_n(x) = x/(1+nx^2)$ converges uniformly to a function f on [0, 1], and that the equation $f_n' = \lim_{n\to\infty} f_n'(x)$ is true if $x \neq 0$ and false if $x = 0$. Why so ? *[Rohilkhand 1996, 97. Agra 1998]*
2. Show that the sequence $< f_n >$ of functions where $f_n(x) = nx/(1 + n^3x^2)$ converges uniformly to f on [0, 1] when $f(x) = 0 \;\forall\; x \in [0, 1]$. Check whether the equation $f'(x) = \lim_{n\to\infty} f_n'(x)$ holds on [0, 1]. *(Delhi Maths (H) 2002)*
3. (*a*) Show that the sequence $< f_n >$, where $f_n(x) = x - x^n/n$, converges uniformly on [0, 1]. Show that the sequence $< f_n' >$ of differentails does not converge uniformaly on [0, 1]. *(Delhi Maths (H) 1999)*

   (*b*) Decide whether or not the sequence $< f_n' >$ converges uniformly on [0, 1], where $f_n(x) = x - x^n/n$. *(Delhi B.Sc. Physics (H) 1999)*

15.7 INFINITE SERIES OF FUNCTIONS *(Himachal 2007; Kanpur 2007)*

We now consider series whose terms are functions defined in some set S. Let

$$\sum_{n=1}^{\infty} f_n(x)$$

be such a series. We write

$$S_n(x) = \sum_{r=1}^{n} f_r(x)$$

so that $\{S_n(x)\}$ is a sequence of functions.

We say that the series is point-wise convergent if the sequence $\{S_n(x)\}$ is point-wise convergent. Also the series is said to be uniformly convergent, if the sequence $\{S_n(x)\}$ is uniformly convergent. Also, the point-wise limit or the uniform limit of $\{S_n(x)\}$ as the case may be is said to be the point-wise sum or the uniform sum of the series and is denoted by $S(x)$. $S(x)$ is also known as sum function or limit function. Thus, $S(x) = \lim_{n\to\infty} S_n(x) = \sum_{n=1}^{\infty} f_n(x)$

15.8 TEST FOR THE UNIFORM CONVERGENCE OF A SERIES

15.8.1 Cauchy's general principle of uniform convergence for series.

The necessary and sufficient condition for the uniform convergence in [*a*, *b*] of a series Σf_n is that to every positive number, ε, there corresponds a positive integer m such that $\forall\, n \geq m$; $\forall\, p \geq 1$ and $\forall\, x \in [a, b]$, $\left|f_{n+1}(x) + f_{n+2}(x) + ... + f_{n+p}(x)\right| < \varepsilon$. ***(Himachal 2014)***

This result is an immediate consequence of the corresponding result for sequences proved in Art. 15.2.

15.8.2. Weierstrass's M-test for Uniform Convergence *(Himanchal 2007, 08, 10)*

(Meerut 2006, 09, 13; Kanpur 2007, 09; Agra 2007, 09, 10, 12; Delhi B.Sc. (Hons) II 2011, Delhi B.Sc. (Prog.) II, 2015; Delhi B.A. (Prog.) III 2012, 2013)

Theorem. A series Σf_n will converge uniformly in [*a*, *b*], if there exists a convergent series ΣM_n of positive numbers such that $\forall\; x \in [a, b]$,

$$\left|f_n(x)\right| \le M_n$$ *(Kanpur 2009, 10; Meerut 2010)*

Proof. Let ε be a positive number. Since ΣM_n is convergent, there exists a positive integer m such that

$$\left|M_{n+1} + M_{n+1} + \ldots + M_{n+p}\right| < \varepsilon,\ \forall\, n \ge m \text{ and } \forall\, p \ge 1. \qquad \ldots (i)$$

Also, given that $$|f_n(x)| \le M_n\ \forall\ x \in [a,b] \qquad \ldots (ii)$$

From (*i*) and (*ii*), we see that

$$\forall\, x \in [a,b],\ \forall\, n \ge m \text{ and } \forall\, p \ge 1,$$

$$\left|f_{n+1}(x) + f_{n+2}(x) + \ldots + f_{n+p}(x)\right| \le \left[M_{n+1} + M_{n+2} + \ldots + M_{n+p}\right] < \varepsilon.$$

Hence Σf_n is uniformly convergent in $[a, b]$

Remark. Note carefully that M_n must be independent of x for every $n = 1, 2, 3, \ldots\ldots$

EXAMPLES

Ex. 1. *Show that the following series are uniformly convergent for real values of x :*

(*i*) $\sum_{n=1}^{\infty} \frac{\sin nx}{n^p}$, $p > 1$ *(Himanchal 2008, 09);* *(Kanpur 2012)* $\sum_{n=1}^{\infty} \frac{\cos nx}{n^4}, x \in R$ *(Delhi B.Sc. (Prog) II 2008)*

(*ii*) $\sum_{n=1}^{\infty} \frac{\cos nx}{n^2}$ *(Kanpur 2005; Meerut 2009)* *(Delhi B.A. (Prog.) III 2012)*   (*iii*) $\sum_{n=1}^{\infty} \frac{\sin nx}{n^2}$ *(Delhi Maths (H) 2001)*

(*iv*) $\sum_{n=1}^{\infty} \frac{\sin(x^2 + nx^2)}{n(n+1)}$ *(Himanchal 2004, 05)*   *(Delhi Maths (H) 2001)*

Sol. (*i*) Here, $f_n(x) = (\sin nx) / n^p$ and so

$$|f_n(x)| = \left|\frac{\sin nx}{n^p}\right| = \frac{|\sin nx|}{n^p} \le \frac{1}{n^p} = M_n, \text{ say } \forall\ x \in R$$

Since $\sum_{n=1}^{\infty} M_n = \sum_{n=1}^{\infty} \frac{1}{n^p}$ is convergent for p > 1, so by Weierstrass's *M* test, the given series converges uniformly for all real values of x.

(*ii*) and (iii). Left as exercises for the reader.

(*iv*) Here $f_n(x) = \{\sin(x^2+nx^2)\}/n(n+1)$

$$\therefore\ |f_n(x)| = \left|\frac{\sin(x^2 + nx^2)}{n(n+1)}\right| = \frac{|\sin(x^2 + nx^2)|}{n^2(1 + 1/n^2)} \le \frac{1}{n^2} = M_n \text{ for } \forall\ x \in R$$

Since $\sum_{n=1}^{\infty} M_n = \sum_{n=1}^{\infty} \frac{1}{n^2}$ is convergent, so by Weierstrass's M-test, the given series is uniformly convergent for all real values of x.

Ex. 2. *Show that the series* $\sum_{n=1}^{\infty} \frac{x}{n(1+nx^2)}$ *converges uniformly.* *(Kanpur 2007, 14)*

(Agra 2001, 09; Delhi Maths (H) 2000; Delhi B.Sc. Physics (H) 1997, 98)

Sol. Here $$f_n(x) = x/n(1+nx^2) \qquad \ldots(1)$$

$$\therefore\ \frac{df_n(x)}{dx} = \frac{1}{n}\cdot\frac{(1+nx^2)\cdot 1 - x\cdot 2nx}{(1+nx^2)^2} = \frac{1-nx^2}{n(1+nx^2)^2} \qquad \ldots(2)$$

For max, or min, $d\,f_n(x)/dx = 0 \quad \Rightarrow \quad 1 - nx^2 = 0 \quad \Rightarrow \quad x = 1/\sqrt{n}$

From (2), $\dfrac{d^2 f_n(x)}{dx^2} = \dfrac{1}{n} \times \dfrac{(1+nx^2)^2 \cdot (-2nx) - (1-nx^2) \cdot 2(1+nx^2) \cdot 2nx}{(1+nx^2)^4}$

$\therefore$ $\dfrac{d^2 f_n(x)}{dx^2} = -\dfrac{2x\{(1+nx^2)+2(1-nx^2)\}}{(1+nx^2)^3}$ and $\left[\dfrac{d^2 f_n(x)}{dx^2}\right]_{x=1/\sqrt{n}} = -\dfrac{1}{2\sqrt{n}} < 0,$

showing that $f_n(x)$ is maximum at $x = 1\sqrt{n}$ and the maximum value of $f_n(x)$ from (1) is

$$\frac{1/\sqrt{n}}{n(1+1)}, \quad \textit{i.e.,} \quad \frac{1}{2n^{3/2}}.$$

Henc $\quad | f_n(x) | \le \dfrac{1}{2n^{3/2}} < \dfrac{1}{n^{3/2}} = M_n$, say

Since $\sum\limits_{n=1}^{\infty} M_n = \sum\limits_{n=1}^{\infty} \dfrac{1}{n^{3/2}}$ is conversent, so by Weierstrass's M-test, the given series is uniformly convergent for all real values of x.

Ex. 3. *Show that* $\sum\limits_{n=1}^{\infty} n^2 x^n$ *is uniformly convergent in* $[-\alpha, \alpha]$, *when* $0 < \alpha < 1$.

(Delhi Maths (H) 2001)

Sol. Here $f_n(x) = n^2 x^n$ so $| f_n(x)| = n^2 \; |x|^n < \alpha^n n^2 = M_n$ as $| x | < \alpha$

Here $\quad \lim\limits_{n\to\infty} \dfrac{M_{n+1}}{M_n} = \lim\limits_{n\to\infty} \dfrac{(n+1)^2 \alpha^{r+1}}{n^2 \alpha^n} = \lim\limits_{n\to\infty} \alpha \left(1 + \dfrac{1}{n}\right)^2 = \alpha < 1$

Hence, by *D'* Alembert's ratio test, ΣM_n converges and so by Weierstrass's M-test, the given series is uniformly convergent in $[-\alpha, \alpha]$.

Ex. 4. *If* (x) *denotes the positive or negative excess of x over the nearest integer and if x is midway between two integers, let x be zero. Test the uniform convergence of the series* $\Sigma \dfrac{(nx)}{n^2}$.

Sol. We have

$$(x) = \begin{cases} x, & \text{for } -(1/2) < x < 1/2 \\ 0, & \text{for } x = 1/2 \\ x-1, & \text{for } 1/2 < x < 3/2 \\ 0, & \text{for } x = 3/2 \\ x-2, & \text{for } 3/2 < x < 5/2 \text{ and so on} \end{cases}$$

Similarly, $\quad (2x) = \begin{cases} 2x, & \text{for } -(1/4) < x < 1/4 \\ 0, & \text{for } x = 1/4 \\ 2x-1, & \text{for } 1/4 < x < 3/4 \\ 0, & \text{for } x = 3/4 \text{ and so on.} \end{cases}$

Similarly, we can write expressions for (3x), (4x), etc. It is obvious from above that

$$|(nx)| < 1/2.$$

We then have $\quad |u_n(x)| = \left|\dfrac{(nx)}{n^2}\right| < \dfrac{1}{2n^2} < \dfrac{1}{n^2}$

But $\Sigma \dfrac{1}{n^2}$ is convergent. Hence the series $\Sigma \dfrac{(nx)}{n^2}$ is uniformly convergent.

Ex. 5. *Show that the series* $\sum_{n=1}^{\infty} \frac{x}{n^p + n^q x^2}$ *is uniformly convergent for all real x if p + q > 2.*

(Himachal 2015; Patna 2003)

Sol. Here $$f_n(x) = x/(n^p + n^q x^2) \qquad ...(1)$$

$$\therefore \quad \frac{df_n(x)}{dx} = \frac{(n^p + n^q x^2)\cdot 1 - x\cdot 2n^q x}{(n^b + n^q x^2)^2} = \frac{n^p - n^q x^2}{(n^p + n^q x^2)^2} \qquad ...(2)$$

For max. or min, $df_n(x)/dx = 0 \;\Rightarrow\; n^p - n^q x^2 = 0 \;\Rightarrow\; x = n^{(p-q)/2}$

From (2), $$\frac{d^2 f_n(x)}{dx^2} = \frac{(n^b + n^q x^2)^2\cdot(2n^q x) - (n^p - n^q x^2)\cdot 2(n^p + n^q x^2)\cdot 2n^2 x}{(n^p + n^q x^2)^4}$$

$$= -\frac{2n^q x\{n^p + n^q x^2 + 2(n^p - n^q x^2)\}}{(n^p + n^q x^2)^3}$$

When $x = n^{(p-q)/2}$, then $$\frac{d^2 f_n(x)}{dx^2} = -\frac{2n^q \cdot n^{(p-q)/2}\cdot 2n^p}{(2n^p)^3} = -\frac{1}{2} n^{(q-3p)/2} < 0,$$

showing that $f_n(x)$ is maximum at $x = n^{(p-q)/2}$ and so

maximum value of $f_n(x)$, from (1) is $\frac{n^{(p-q)/2}}{2n^p}$, *i.e.*, $\frac{1}{2n^{(p+q)/2}}$

Hence $$|f_n(x)| = \left|\frac{x}{n^p + n^q x^2}\right| \le \frac{1}{2n^{(p+q)/2}} < \frac{1}{n^{(p+q)/2}} = M_n, \text{ say}$$

Since $\sum_{n=1}^{\infty} M_n = \sum_{n=1}^{\infty} \frac{1}{n^{(p+q)/2}}$ is convergent when $(p + q)/2 > 1$ i.e., when $p + q > 2$, hence, by Weierstrass's M-test, the given series is convergent for all real x, when $p + q > 2$

EXERCISES

1. Show that if $0 < r < 1$, then each of the following series is uniformly convergent $\forall x \in \mathbf{R}$

[Delhi B.Sc. (Prog.) II, 2015]

(*i*) $\Sigma\, r^n \cos nx$ (*ii*) $\Sigma\, r^n \sin nx$ (*iii*) $\Sigma\, r^n \cos n^2 x$ (*iv*) $\Sigma\, r^n \sin a^n x$

(*v*) $\Sigma \frac{\cos(x^2 + nx)}{n(n^2+2)}$ (*vi*) $\Sigma \frac{x}{(n + x^2)^2}$ (*vii*) $\Sigma \frac{\cos nx}{n^p}, p > 1$

2. Show that each of the following series is uniformly convergent for all values of x.

(*i*) $\Sigma \frac{1}{n^4 + n^2 x^2}$ **[Delhi B.Sc. (Prog.) II, 2016]** (*ii*) $\Sigma \frac{1}{n^2 + n^4 x^2}$ (*iii*) $\Sigma \frac{1}{n^3 + n^4 x^2}$

3. Let Σa_n converge absolutely. Then prove that each of the following series is uniformly convergent for all real values of x.

(*i*) $\Sigma a_n \cos nx$ (*ii*) $\Sigma a_n \sin nx$ (*iii*) $\Sigma \frac{a_n x^{2n}}{1 + x^{2n}}$ (*iv*) $\Sigma \frac{a_n x^n}{1 + x^{2n}}$

4. Show that $\Sigma\, 1/(n^p + n^q x^2)$ is uniformly convergent for all real x and $p > 1$.

5. (*a*) Show that the series $\Sigma \frac{(-1)^n}{n^p}\cdot\frac{x^{2n}}{1 + x^{2n}}$ is absolutely and uniformly convergent for all real x, if $p > 1$.

[Himachal 2010]

(*b*) Show that $\sum_{n=0}^{\infty}\{(-1)^n/(x^2 + n^2)\}$ is uniformly convergent on **R**.

[Delhi B.Sc. (Hons) II 2011]

(*c*) Find the sum function of $\sum_{n=0}^{\infty} x(1-x)^n$ on [0, 1] **(Pune 2010)** **Ans.** $\begin{cases} 0, \text{if } x = 0 \text{ or } 1 \\ 1, \text{ if } 0 < x < 1 \end{cases}$

6. Show that each of the following series is uniformly covnergent is $[-\delta, \delta]$, if $0 < \delta < 1$:

$$(i)\ \Sigma x^n \quad (ii)\ \Sigma (n+1)x^n$$

7. (*a*) Show that the series $\Sigma \frac{x}{1+n^2 x}$ is not uniformly convergent on [0, 1]. while it is so on [1/2, 1] *(Delhi Maths (H) 1996)*

(*b*) Show that $\sum_{n=1}^{\infty} \frac{x}{1+n^2 x}$ is uniformly convergent in $[\delta, 1]$ for any $\delta > 0$ but is not uniformly convergent in [0, 1]. *(Delhi Maths (H) 2000)*

8. Show that the series $\frac{1}{a} - \frac{2a\cos\theta}{a^2-1^2} + \frac{2a\cos 2\theta}{a^2-2^2}$.... is uniformly convergent with respect to θ in any finite interval. *(Agra 1999, Delhi Maths (H) 1998)*

9. Show that the series $1 + x + x^2/2! + x^3/3! + ... + x^n/n! + ...$ is uniformly convergent in $[-1, 1]$.

10. Show that the series $\frac{1}{(1+x)^3} + \frac{2}{(2+x)^3} + \frac{3}{(3+x)^3} + \cdots$ is uniformly convergent for all $x \geq 0$

11. Show that $\sum_{n=1}^{\infty} \frac{1}{1+n^2+n^4x^2}$ converges uniformly for all values of x.

15.9 ABEL'S TEST AND DIRICHLET'S TEST

Since Weierstrass M-tests can be used for those series which are absolutely convergent as weil, so we proceed to find two important tests.

Theorem I (Abel's test). *[Himachal 2007, 2009, 2013, 2015]*

If (i) $\Sigma f_n(x)$ is uniformly convergent in [a, b]

(ii) The sequence $< g_n(x) >$ is monotonic decreasing $\forall\, x \in [a, b]$

(iii) There exists a positive number k such that $|g_n(x)| < k\ \forall\, x \in [a, b]$ and $\forall\, n \in \mathbf{N}$, then the series $\Sigma f_n(x) g_n(x)$ is uniformly convergent on [a, b].

Proof. Since $\Sigma f_n(x)$ converges uniformly on [*a*, *b*], so by Cauchy's general principle of uniform convergence of a series (see Art. 15.8.1), for each $\varepsilon > 0$ and $\forall\, x \in [a, b]$, there exist a positive integer *m* (depending only on ε) such that

$$|f_{n+1}(x) + f_{n+2}(x) + \ldots + f_{n+p}(x)| < \varepsilon / k \quad \forall\, n \geq m, \forall p \geq 1$$

i.e.,
$$\left| \sum_{r=n+1}^{n+p} f_r(x) \right| < \varepsilon / k \quad \forall \geq m, \forall p \geq 1$$

From the given facts (*ii*) and (*iii*), we see that $< g_n(x) >$ is monotonic decreasing on [*a, b*] such that
$$|g_n(x)| < k \quad \forall\, x \in [a, b]$$

Hence by *Abel's lemma, we obtain

$$\left| \sum_{r=n+1}^{n+p} f_r(x)\, g_r(x) \right| < (\varepsilon / k) \times k = \varepsilon \quad \forall\, n \geq m, \forall\, p \geq 1 \text{ and}, \forall\, x \in [a, b]$$

***Abel's Lemma** Let $k < \sum_{r=m}^{p} u_r < K$, for $p = m, m+1, ..., n$ and $< a_n >$ be a positive monotonic decreasing sequence, then
$$a_m k \leq \sum_{r=m}^{n} a_r u_r \leq a_m K$$

$\Rightarrow$ $\quad |f_{n+1}(x)\, g_{n+1}(x) \ldots + f_{n+p}(x)\, g_{n+p}(x)| < \varepsilon \;\forall\; n \ge m, \forall p \ge 1$ and $\forall\; x \in [a, b]$

$\Rightarrow$ $\quad f_n(x) g_n(x)$ converges uniformly on $[a, b]$

[using Cauchy's general principle of uniform convergences]

Theorem II. (DIRICHLET'S TEST)

(Himachal 2015)

If (i) there exists a positive real number k such that

$$|S_n(x)| = |\sum_{r=1}^{n} f_r(x)| < k \;\forall\; x \in [a, b], \forall n \in \boldsymbol{N},$$

and $\quad$ *(ii)* $< g_n(x) >$ *is positive monotonic decreasing sequence converging uniformly to zero on* $[a, b]$ *then the series* $\Sigma f_n(x)\, g_n(x)$ *is uniformly convergent on* $[a, b]$.

Proof. Given $\quad |S_n(x)| < \text{k} \;\forall\; x \in [a, b], \forall\; n \in \boldsymbol{N}$ ...(1)

Now, $\forall\; x \in [a, b]$, $\forall\; n \ge m$ and $\forall\; p \ge 1$, we have

$$|S_{n+p}(x) - S_n(x)| \le |S_{n+p}(x)| + |S_n(x)| < k + k = 2k, \text{ using (1)}$$

Thus, $\quad \left|\sum_{r=n+1}^{n+p} f_r(x)\right| < 2k \;\forall\; n \ge m,\; p \ge 1, \forall x \in [a, b]$ ...(2)

From given fact *(ii)*, $< g_n(n) >$ is in positive monotone decreasing sequence on $[a, b]$. Hence, by Abel's lemma,, we obtain

$$\left|\sum_{r=n+1}^{n+p} f_r(x)\, g_r(x)\right| < 2k\, g_{n+1}(x) \;\forall\; n \ge m_1, \;\; \forall\, p \ge 1, \forall x \in [a, b] \text{ and } \forall\, m_1 \in \boldsymbol{N} \qquad \text{...(3)}$$

Since $< g_n(x) >$ converges uniformly to zero on $[a, b]$, so for a given $\varepsilon > 0$, there exists a positive number m_2 such that

$$|g_n(x) - 0| < \varepsilon / 2k \;\forall\; n \ge m_2, \quad \text{i.e.,} \quad |g_n(x)| < \varepsilon / 2k \;\forall\; n \ge m_2 \qquad \text{...(4)}$$

Let m = max $\{m_1, m_2\}$. Since $g_n(x)$ is positive $\forall\; n \in \boldsymbol{N}$, so from (3) and (4), we obtain

$$\left|\sum_{r=n+1}^{n+p} f_r(x)\, g_r(x)\right| < 2k\; g_{n+1}(x) \;\forall\; n \ge m, \forall p \ge 1 \text{ and } \forall\; x \in [a, b] \qquad \text{...(5)}$$

and $$g_{n+1}(x) < \varepsilon / 2k \;\forall\; n \ge m \qquad \text{...(6)}$$

From (5) and (6), $\quad \left|\sum_{r=n+1}^{n+p} f_r(x) g_r(x)\right| < 2k \times (\varepsilon / 2k) < \varepsilon,$

which hold for $\forall\; n \ge m$, $\forall\, p \ge 1$ and $\forall\; x \in [a, b]$ Hence by Cauchy's general principle of uniform convergence, $\Sigma f_n(x)\, g_n(x)$ is uniformly convergent on $[a, b]$.

EXAMPLES

Example 1. *Show that the series* $\Sigma \dfrac{(-1)^{n-1}}{n} |x|^n$ *is uniformly convergent in* $[-1, 1]$.

(Himachal 2013, 14; G.N.D.U. Amritsar1998)

Solution. Let $\quad f_n(x) = (-1)^{n-1}/n$ and $g_n(x) = |x|^n$

Here $\Sigma f_n(x)$ is convergent by Leibnitz' test. Since each $f_n(x)$ is independent of x, hence $\Sigma f_n(x)$ is uniformly convergent

Again, for $-1 \le x \le 1$, $\;|x|^n \le 1 \;\forall\; n \in \boldsymbol{N}$ and $\;|g_n(x)| = |x|^n \le 1$

Hence $< g_n(x) >$ is monotonic decreasing and bounded on $[-1, 1]$ $\forall\; n \in \boldsymbol{N}$. Consequently, by

$$\Sigma f_n(x) g_n(x), \ i.e., \ \Sigma \frac{(-1)^{n-1}}{n} |x|^n$$

is uniformly convergent on [–1, 1].

Example 2 *Show that the series* $\Sigma \dfrac{(-1)^{n-1}}{n+x^2}$ *is uniformly convergent for all values of x.*

Solution. Let $f_n(x) = (-1)^{n-1}$ and $g_n(x) = 1/(n+x^2)$

Here $S_n(x) = \sum_{r=1}^{n} f_r(x) = 1 + -1 + 1 - 1 + ... + (-1)^{n-1} = \begin{cases} 1, \text{ if } n \text{ is odd} \\ 0, \text{ if } n \text{ is even} \end{cases}$

Hence $S_n(x)$ is bounded $\forall \ n \in N$ and for all x.

Now, $1/(n+x^2) < 1/n, \forall x \in R$ and $g_{n+1}(x) \le g_n(x) \ \forall x \in \mathbf{R}$

Therefore, $< g_n(x) >$ is a positive monotonic decreasing sequence converging to 0 for all values of x. Hence by Dirichlet's test the series

$$\Sigma f_n(x) g_n(x), \qquad i.e., \qquad \Sigma \frac{(-1)^{n-1}}{n+x^2}$$

is uniformly convergent for all x.

Example 3. *Prove that* $\Sigma \ a_n \ n^{-x}$ *is uniformly convergent on* [0, 1] *if* $\Sigma \ a_n$ *converges uniformly in* [0, 1].

Solution. Take $g_n(x) = n^{-x} = 1/n^x$ and $f_n(x) = a_n$. The sequence $< n^{-x} >$ is monotonic decreasing on [0, 1].

$$\therefore \qquad 1/n^x \le 1/n^0 = 1 \ \forall \ n \in \mathbf{N} \text{ and } \forall \ x \in [0,1].$$

$$\therefore \qquad |g_n(x)| = |n^{-x}| \le 1 \ \forall \ n \in \mathbf{N} \text{ and } \forall \ x \in [0,1].$$

Thus $< g_n(x) >$ is uniformly bounded and monotonic decreasing sequence on [0, 1].

Also $\Sigma \ f_n(x) = \Sigma \ a_n$ is uniformly convergent on [0, 1].

Hence, by Abel's test, the series $\Sigma \ f_n(x) \ g_n(x)$, i.e., $\Sigma \ a_n \ n^{-x}$ is uniformly convergent on [0, 1].

EXERCISES

1. If Σa_n is convergent, series of positive constants, then show the following series are all uniformly convergent in [0, 1].

   (i) $\Sigma \ a_n \ x^n$ ***(Kanpur 2010)*** (ii) $\Sigma a^n / n^x$ (iii) $\Sigma \dfrac{a_n x^n}{1+x^n}$

   (iv) $\Sigma \dfrac{a_n x^n}{1+x^{2n}}$ (v) $\Sigma \dfrac{nx^n(1-x)}{1+x^n}$ (vi) $\Sigma \dfrac{2na_n x^n(1-x)}{1+x^{2n}}$

2. Show that the series $\dfrac{1}{1+x^2} - \dfrac{1}{2+x^2} + \dfrac{1}{3+x^2} - \dfrac{1}{4+x^2} + ...$ is uniformly convergent is any interval. ***(Kanpur 2010, 2011)***

3. Prove that the series $\Sigma(-1)^n \dfrac{x^2+n}{n^2}$ converges uniformly in every bounded interval, but does not converge absolutely for any value of x. ***(Himachal 2014, 15)***

4. Show that the series $1 + \dfrac{e^{-2x}}{2^2-1} - \dfrac{e^{-4x}}{4^2-1} + \dfrac{e^{-6x}}{6^2-1} ...$ is uniformly convergent for all real $x \ge 0$. ***[Kanpur 2018]***

5. Prove that the series cos $x + (1/2) \times \cos 2x + (1/3) \times \cos 3x + \ldots$ converges uniformly in $[0, 2\pi]$. *(Meerut 2007)*

6. Prove that both the series $\Sigma \frac{\sin n x}{n^2}$ and $\Sigma \frac{\cos n x}{n^2}$ are uniformly convergent in $[0, 2\pi]$.

7. Show that the following series are uniformly convergent in $[\alpha, 2\pi \quad \alpha]$ where $0 < \alpha < \pi$.

(i) $\Sigma \frac{\sin nx}{n}$ (ii) $\Sigma \frac{\cos nx}{n}$ (iii) $\Sigma \frac{\sin nx}{n^p}, p > 0$ (iv) $\Sigma \frac{\cos nx}{n^p}, p > 0$

(Kanpur 2006, 2011)

8. Prove that if δ is any fixed positive number less than unity, the series $\Sigma x^n/(n+1)$ is uniformly convergent in $(-\delta, \delta)$.

9. Show that the series $\Sigma \{\log(n+1)\}^{-x} \cos nx$ is uniformly convergent in $[\alpha, \beta]$, where $0 < \alpha \le x \le \beta < 2\pi$.

15.10 PROPERTIES OF UNIFORMLY CONVERGENT SERIES OF FUNCTIONS

It has been seen in Chapters 8, 9 and 13 that the sum of a *finite* number of continuous (derivable, integrable) functions is continuous (derivable, integrable). These results, however, may not be true in relation to the point-wise limits of infinite series. The following three results, however, are direct consequences of those proved in Art. 15.4, 15.5 and 15.6, in relation to uniformly convergent sequences. For complete proofs, refer Appendix A.

In all results, we shall suppose that each of the function f_n has $[a, b]$ as its domain.

I. The sum of a uniformly convergent series of continuous functions is continuous. *[Rajasthan 2010; Kanpur 2005]*

II. The sum of a uniformly convergent series of integrable functions is integrable and the integral of the sum is equal to the sum of the series of integrals of the functions.

Thus if Σf_n be a uniformly convergent series of integrable functions in $[a, b]$, and $S(x)$ denotes it sum, then $S(x)$ is integrable in $[a, b]$ and

$$\int_a^b \Sigma f_n(x)\, dx = \int_a^b S(x)\, dx = \Sigma \int_a^b f_n(x)\, dx.$$

i.e., the series is term by term integrable. *(Kanpur 2005)*

III. If Σf_n is a point-wise convergent series of derivable functions with continuous derivatives and the series $\Sigma f'_n$ of derivatives is uniformly convergent, then

$$\frac{d}{dx}(\Sigma f_n) = \Sigma f'_n,$$

i.e., the derivative of the sum is equal to the sum of the derivatives. Thus, the term-by-term differentiation of the series is valid.

Note 1. The converse of result I may not be true, *i.e.*, series of continuous functions exist which have a continuous sum but are not uniformly convergent. Thus the condition of result I is only sufficient, not necessary.

The result I also shows that if the sum function of the series is discontinuous, then the series Σf_n of continuous function cannot be uniformly convergent. This fact is often used to prove that Σf_n is not uniformly convergent.

Note 2. For result II, uniform convergence of the series Σf_n is only sufficient but not a necessary condition for the validity of term by term integration.

Note 3. For result III, uniform convergence of the series Σf_n is only sufficient but not a necessary condition for the validity of term-by term differentiation.

Note 4. If $S_n(x) = f_1(x) + ... + f_n(x)$ and $S(x) = \lim\limits_{n\to\infty} S_n(x)$, then

(*i*) result II can be re-written as $\lim\limits_{n\to\infty} \int_a^b S_n(x)\,dx = \int_a^b \lim\limits_{n\to\infty}\{S_n(x)\}\,dx$

(*ii*) result III can be re-written as $S'(x) = \lim\limits_{n\to\infty} S_n'(x)$ *i.e.*, $\dfrac{d}{dx}\lim\limits_{n\to\infty} S_n(x) = \lim\limits_{n\to\infty}\dfrac{d\,S_n(x)}{dx}$

EXAMPLES

Example 1. *Discuss the uniform convergence of*

(*i*) $x^4 + \dfrac{x^4}{1+x^4} + \dfrac{x^4}{(1+x^4)^2} + \dfrac{x^4}{(1+x^4)^3} + ...$ *on [0, 1]* [*Kanpur 2010; Delhi B.Sc. (Prog.) II, 2016*]

[*Himanchal 2007*]

(*ii*) $x^2 + x^2/(1+x^2) + x^2/(1+x^2)^2 + x^2/(1+x^2)^3 + ...$

Solution. Here $S_n(x) = n$th partial sum $= f_1(x) + f_2(x) + ... + f_n(x)$

$$= x^4 + \frac{x^4}{1+x^4} + \frac{x^4}{(1+x^4)^2} + ... + \frac{x^4}{(1+x^4)^{n\ 1}}$$

$$= \frac{x^4[1 \quad \{1/(1+x^4)^n\}]}{1 \quad \{1/(1+x^4)\}} = 1 + x^4 \quad \frac{1}{(1+x^4)^{n\ 1}}$$

$\therefore$ Sum function $S(x) = \lim\limits_{n\to\infty} S_n(x) = \begin{cases} 1+x^4, x \neq 0 \\ 0, x = 0 \end{cases}$

Since the sum function $S(x)$ is discontinuous at $x = 0 \in [0, 1]$ the given series is not uniformly convergent on [0, 1]

Example 2 *Show that the series* $\sum\limits_{n=1}^{\infty} \dfrac{x}{(nx+1)\{(n \quad 1)\,x+1\}}$ *is uniformly convergent on any interval* [*a*, *b*] *where* $0 < a < b$, *but only pointwise convergent on* [0, *b*]

(*Mumbai 2010; Delhi B.Sc. (Prog) III 2009, 2012; Agra 1998, Delhi Maths (H) 2004*)

Solution. Here $f_n(x) = \dfrac{x}{(nx+1)\{(n \quad 1)x+1\}} = \dfrac{1}{(n \quad 1)x+1} \quad \dfrac{1}{nx+1}$

$\therefore$ nth partial sum $= S_n(x) = f_1(x) + f_2(x) + ... + f_n(x)$

$$= \left(1 \quad \frac{1}{x+1}\right) + \left(\frac{1}{x+1} \quad \frac{1}{2x+1}\right) + ... + \left\{\frac{1}{(n \quad 1)\,x+1} \quad \frac{1}{nx+1}\right\} = 1 \quad \frac{1}{nx+1}$$

$\therefore$ Sum function $= S(x) = \lim\limits_{n\to\infty} S_n(x) = \begin{cases} 1, \text{ if } x \neq 0 \\ 0, \text{ if } x = 0 \end{cases}$

Since the sum function $S(x)$ is discontinous at x = $0 \in [0, b]$ the given series is not uniformly convergent on [0, *b*]

We now test Σf_n for uniform convergence in [*a*, *b*], when $0 < a < b$. We have for $0 < a < b$ and for a given $\varepsilon > 0$,

$$|S_n(x) - S(x)| < \varepsilon \Rightarrow \left|1 \quad \frac{1}{nx+1} \quad 1\right| < \varepsilon \quad \Rightarrow \frac{1}{nx+1} < \varepsilon$$

$$\Rightarrow \quad nx + 1 > \frac{1}{\varepsilon} \quad \Rightarrow \quad n > \frac{1}{x}\left(\frac{1}{\varepsilon} \quad 1\right)$$

Now $\frac{1}{x}\left(\frac{1}{\varepsilon}-1\right)$ increases as x decreases, its maximum value being $\frac{1}{a}\left(\frac{1}{\varepsilon}-1\right)$ in $[a, b]$, where $a > 0$. Let $m \geq \frac{1}{a}\left(\frac{1}{\varepsilon}-1\right)$.

Then, for given $\varepsilon > 0$, we have $|S_n(x) - S(x)| < \varepsilon \ \forall \ n \geq m$ and $\forall \ x \in [a, b]$, where $a > 0$.

Hence the given series is uniformly convergent

Example 3. *Test for uniform convergence and term by term integration of the series* $\sum_{n=1}^{\infty} \frac{x}{(n+x^2)^2}$. *Also prove that* $\int_0^1 \left(\sum_{n=1}^{\infty} \frac{x}{(n+x^2)^2}\right) dx = \frac{1}{2}$. *(Kanpur 2007; Agra 2004, 10)*

[Bilaspur 1998; Meerut 2005; Ravishankar 1999; Rewa 1999; Sager 2001]

Solution. Here $f_n(x) = x/(n+x^2)^2$ ...(1)

$$\therefore \quad \frac{df_n(x)}{dx} = \frac{(n+x^2)^2 \cdot 1 - x \cdot 2(n+x^2) \cdot 2x}{(n+x^2)^4} = \frac{n-3x^2}{(n+x^2)^3} \quad ...(2)$$

For max. or min., $d f_n(x)/dx \Rightarrow n - 3x^2 = 0 \Rightarrow x = (n/3)^{1/2}$

$$\text{From (2),} \quad \frac{d^2 f_n(x)}{dx^2} = \frac{(n+x^2)^3 \cdot (-6x) - (n-3x^2) \cdot 3(n+x^2) \cdot 2x}{(n+x^2)^6} = \frac{-6x(n+x^2+n-3x^2)}{(n+x^2)^4}$$

So when $x = (n/3)^{1/2}$, $\frac{d^2 f_n(x)}{dx^2} = \frac{-6(n/3)^{1/2}(n+n/3)}{(n+n/3)^4} = -\frac{27\sqrt{3}}{32n^{5/2}} < 0$,

showing that $f_n(x)$ is maximum when $x = (n/3)^{1/2}$ and from (1),

$$\text{the maximum value of } f_n(x) = \frac{(n/3)^{1/2}}{(n+n/3)^2} = \frac{3\sqrt{3}}{16n^{3/2}}$$

$$\text{Hence} \quad |f_n(x)| \leq \frac{3\sqrt{3}}{16n^{3/2}} < \frac{1}{n^{3/2}} = M_n \text{ say}$$

Since $\sum_{n=1}^{\infty} M_n = \sum_{n=1}^{\infty} \frac{1}{n^{3/2}}$ is convergent, by Weirstrass M-test, the given series is uniformly convergent for all values of x.

Also, $\forall \ n \in N$, f_n is integrable on [0, 1]. Hence the series can be integrated term by term.

$$\therefore \quad \int_0^1 \sum_{n=1}^{\infty} \frac{x}{(n+x^2)^2} dx = \sum_{n=1}^{\infty} \int_0^1 x(n+x^2)^{-2} dx = \sum_{n=1}^{\infty} \left[\frac{(n+x^2)^{-1}}{-2}\right]_0^1 = \sum_{n=1}^{\infty} \frac{1}{2}\left(\frac{1}{n} - \frac{1}{n+1}\right)$$

$$= \lim_{n\to\infty} \sum_{n=1}^{n} \frac{1}{2}\left(\frac{1}{n} - \frac{1}{n+1}\right) = \frac{1}{2} \lim_{n\to\infty} \left[1 - \frac{1}{2} + \frac{1}{2} - \frac{1}{3} + ... + \frac{1}{n} - \frac{1}{n+1}\right]$$

$$= \frac{1}{2} \lim_{n\to\infty} \left(1 - \frac{1}{n+1}\right) = \frac{1}{2}$$

Example 4. *Show that the series for which* $S_n(x) = nx(1-x)^n$ *can be integrated term by term on* [0, 1], *though it is not uniformly convergent on* [0, 1]. *(Patna 2003)*

Solution. Given, nth partial sum $= S_n(x) = nx(1-x)^n$.

For $0 < x < 1$, we have $\lim_{n\to\infty} S_n(x) = \lim_{n\to\infty} nx(1-x)^n = \lim_{n\to\infty} \frac{nx}{(1-x)^{-n}} \quad \left[\text{form } \frac{\infty}{\infty}\right]$

$$= \lim_{n\to\infty} \frac{x}{-(1-x)^{-n} \log(1-x)} = 0$$

Again, for $x = 0$ or $x = 1$, we have $\quad S_n(x) = 0 \Rightarrow \lim_{n\to\infty} S_n(x) = 0$

Thus, $\quad S(x) = \text{sum function} = 0 \ \forall\, x \in [0, 1]$

Here $$\int_0^1 S(x)\,dx = \int_0^1 0\,dx = 0$$

and $$\int_0^1 S_n(x)\,dx = \int_0^1 nx(1-x)^n\,dx = n\int_0^1 (1-x)x^n\,dx$$

$$= n\left[\frac{x^{n+1}}{n+1} - \frac{x^{n+2}}{n+2}\right]_0^1 = n\left[\frac{1}{n+1} - \frac{1}{n+2}\right] = \frac{n}{(n+1)(n+2)}$$

$$\therefore \quad \lim_{n\to\infty}\int_0^1 S_n(x)\,dx = \lim_{n\to\infty}\frac{1}{(1+1/n)(1+2/n)n} = 0$$

Since $\lim_{n\to\infty}\int_0^1 S_n(x)\,dx = \int_0^1 \lim_{n\to\infty}[S_n(x)]\,dx,$

it follows that the given series is integrable term by term.

Proceed as in solved example 2, Art. 15.3, to show that 0 is a point of non-uniform convergence of the serves.

Thus the given series is integrable term by term on [0, 1] although $x = 0$ is a point of non-uniform convergence of the series.

Ex. 5. *Examine for term by term integration the series for which* $S_n(x) = nxe^{-nx^2}$ *indicating the interval over which your conclusion holds.* *(Meerut 2004)*

Sol. Here $S(x) = \text{sum function} = \lim_{n\to\infty} S_n(x) = \lim_{n\to\infty}\frac{nx}{e^{nx^2}}$

$$= \lim_{n\to\infty}\frac{nx}{1 + nx^2 + \dfrac{n^2x^4}{2!} + \ldots} = 0 \text{ for all finite values of } x.$$

Consider the interval $0 \le x \le 1$.

$$\int_0^1 S(x)\,dx = \int_0^1 0.dx = 0$$

and $$\int_0^1 S_n(x)\,dx = \int_0^1 nxe^{-nx^2}\,dx = \left[-\frac{1}{2}e^{-nx^2}\right]_0^1 = \frac{1}{2}\left[1 - e^{-n}\right] = \frac{1}{2} \text{ as } n \to \infty.$$

Hence term by term integration over the interval $0 \le x \le 1$ is now justified. In fact zero is a point of non-uniform convergence as we have already seen. However, term by term integration is justified over the interval $[c, 1]$ where $0 < c < 1$. For we have

$$\int_c^1 S_n(x)\,dx = \int_c^1 nxe^{-nx^2}\,dx = \frac{1}{2}\left[e^{-nc^2} - e^{-n}\right] = 0 \text{ as } n \to 0.$$

Hence $$\int_c^1 S(x)\,dx = \lim_{n\to\infty}\int_c^1 S_n(x)\,dx.$$

Hence $[c, 1]$ is the interval in which term by term integration holds.

Ex. 6. Show that near $x = 0$, the series $u_1(x) + u_2(x) + \ldots$, where $u_1(x) = x$, $u_n(x) = x^{1/(2n-1)} - x^{1/(2n-3)}$ and real values of x are concerned, is discontinuous and non-uniformly convergent. Can the series be integrated term by term ?

Sol. Here

$$u_1(x) = x$$
$$u_2(x) = x^{1/3} - x$$
$$u_3(x) = x^{1/5} - x^{1/3},$$
$$\ldots\ldots\ldots\ldots\ldots\ldots\ldots\ldots$$
$$u_n(x) = x^{1/(2n-1)} - x^{1/(2n-3)}$$

Hence, $S_n(x) = n$th partial sum $= x^{1/(2n-1)}$

Let $\qquad S(x) = \lim_{n\to\infty} S_n(x)$

$\therefore \qquad S(x) = 0$ for $x = 0$

and $\qquad S(x) = 1$ for other values of x.

Hence S is discontinuous at $x = 0$ and consequently zero is a point of non-uniform convergence of the series.

Now for $0 \le x \le c < \infty$, we have $\displaystyle\int_0^c S(x)\,dx = \int_0^c dx = c$

and $$\int_0^c S_n(x)\,dx = \int_0^c x^{1/(2n-1)}\,dx = \frac{2n-1}{2n}\cdot c^{2n/(2n-1)} \to c \text{ as } n\to\infty.$$

Hence the series is term by term integrable in the above interval although 0 is a point of non-uniform convergence of the series.

Ex. 7. Given the series $\Sigma\, u_n(x)$ for which $\qquad S_n(x) = (1/2n^2) \times \log(1+n^4x^2)$.

Show that the series $\Sigma\, u_n'(x)$ does not converge uniformly, but that the given series can be differentiated term by term. *(Kanpur 2007,*

Sol. We have

$$\text{Sum function} = S(x) = \lim_{n\to\infty} S_n(x) = \lim_{n\to\infty} \frac{\log(1+n^4x^2)}{2n^2} \qquad \left[\text{From } \frac{\infty}{\infty}\right]$$

$$= \lim_{n\to\infty} \frac{\dfrac{4n^3x^2}{1+n^4x^2}}{4n} = \lim_{n\to\infty} \frac{n^2x^2}{1+n^4x^2} = 0 \text{ for } 0 \le x \le 1$$

Hence $\qquad S'(x) = 0. \qquad$ and $\qquad \displaystyle\lim_{n\to\infty} S_n'(x) = \frac{n^2x}{1+n^4x^2} = 0$ for $0 \le x \le 1$

$$\therefore \qquad S'(x) = \lim_{n\to\infty} S_n'(x),$$

so that term by term differentiation holds. The series $\Sigma\, u_n'(x)$ is not uniformly convergent in $0 \le x \le 1$ since the sequence $\{S'_n(x)\}$ has 0 as a point of non-uniform convergence.

Ex. 8. *Let* $u_n(x) = x^2\,(x^{1/(2n-1)} - x^{1/(2n-3)}) \sin(1/x)$ *for* $x \ge 0$

$u_n(0) = 0$, *for any positive integer greater than unity and*

$u_1(x) = x^3 \sin(1/x)$ *for* $x \lessgtr 0$, $u_1(0) = 0$.

show that $\sum_{n=1}^{\infty} u_n(x)$ *converges for all values of x to S (x), where* $S(x) = x^2 \sin(1/x)$ *for* $x \gtrless 0$ *and* $S(0) = 0$. *Also show that f is discontinuous at* $x = 0$, *that* $\sum_{n=1}^{\infty} u_n'(x)$ *is not uniformly convergent in any interval including the origin, and that* $S'(x) = \sum_{n=1}^{\infty} u_n'(x)$ *for all values of x.*

Sol. We have, when $x \neq 0$

$$u_1(x) = x^3 \sin(1/x) = x^2 (x - 0) \sin(1/x)$$

$$u_2(x) = x^2 \left(x^{1/3} - x\right) \sin(1/x),$$

$$u_3(x) = x^2 \left(x^{1/5} - x^{1/3}\right) \sin(1/x),$$

....

$$u_n(x) = x^2 \left(x^{1/(2n-1)} - x^{1/(2n-3)}\right) \sin(1/x),$$

then $$S_n(x) = \sum_{n=1}^{n} u_n(x) = x^2 \sin(1/x) \cdot x^{1/(2n-1)}.$$

$\therefore$ $$S(x) = \lim_{n\to\infty} S_n(x) = x^2 \sin(1/x) \text{ when } x \neq 0;$$

and $$S(x) = 0 \text{ when } x = 0.$$

Now, when $x \neq 0$, we have

$$S'(x) = 2x \sin(1/x) - \cos(1/x)$$

Since $\lim\limits_{n\to\infty} \cos(1/x)$ does not exist, $S(x)$ is discontinuous at $x = 0$,

$$\sum_1^\infty u_n'(x) = \lim_{n\to\infty} \sum_1^n u_n'(x) = \lim_{n\to\infty} \left[\frac{d}{dx}\left\{x^2 \sin\frac{1}{x} - x^{1/(2n-1)}\right\}\right]$$

$$= \lim_{n\to\infty} \left[\frac{4n-1}{2n-1} x^{2n/(2n-1)} \sin\frac{1}{x} - x^{1/(2n-1)} \cos\frac{1}{x}\right]$$

$$= 2x \sin(1/x) - \cos(1/x)$$

and $$S'(x) = 2x \sin(1/x) - \cos(1/x)$$

Hence $$S'(x) = \sum_1^\infty u_n'(x) \text{ when } x \neq 0.$$

Obviously, $$S'(0) = \sum_1^\infty u_n'(0)$$

Since the sum function $\{2x \sin(1/x) - \cos(1/x)\}$ of the series $\Sigma u_n'(x)$ is discontinuous at $x = 0$, the series $\Sigma u_n'(x)$ is non-uniformly convergent near $x = 0$.

Example 9. *Show that* $\Sigma \dfrac{1}{n^p + n^q x^2}$, $p > 1$ *is uniformly convergent for all values of x and can be differentiated term by term if* $q < 3p - 2$. *(Agra 1998: I.A.S. 1998)*

Solution. Let $$f_n(x) = 1/(n^p + n^q x^2) \qquad ...(1)$$

From (1), $$f_n'(x) = \frac{d f_n(x)}{dx} = (-2n^q x)/(n^p + n^q x^2)^{-2} \qquad ...(2)$$

For max. or min., $$d f_n(x)/dx = 0 \quad \Rightarrow \quad x = 0$$

From (2), $$d^2 f_n(x)/dx^2 = -2n^q(n^p + n^q x^2)^{-2} + 8n^{2q} x^2 (n^p + n^q x^2)^{-3} \qquad ...(3)$$

When $x = 0$, $$d^2 f_n(x)/dx^2 = -n^q (x^p)^{-2} < 0,$$

showing that $f_n(x)$ is maximum when $x = 0$ and from (1), the maximum value of $f_n(x) = 1/n^p$.

Hence $$|f_n(x)| = |1/(n^p + n^q x^2)| \le 1/n^p = M_n, \text{ say}$$

Since $\sum_{n=1}^{\infty} M_n = \sum_{n=1}^{\infty} \frac{1}{n^p}$ is convergent for $p > 1$, so by Weirstrass's test, the given series is uniformly convergent for all real values of x.

From (3), $d^2 f_n(x)/dx^2 = 2n^q(n^p + n^q x^2)^{-3} \{4n^q x^2 - (n^p + n^q x^2)\}$

or $$df_n'/dx = 2n^q(n^p + n^q x^2)^{-3}(3n^q x^2 - n^p) \qquad \text{... (4)}$$

For max. or min. of $f_n'(x)$, $d f_n'(x)/dx = 0$ giving $x = \pm(1/\sqrt{3})\, n^{(p-q)/2}$

From (4), $d^2 f_n'/dx^2 = 2n^q(n^p + n^q x^2)^{-3}.(6n^q x) - 6x^q(n^p + n^q x^2)^{-4}.(2n^q x).(3n^q x^2 - n^p)$

When $x = -(1/\sqrt{3})\, n^{(p-q)/2}$ *i.e.*, $x^2 = n^p/3n^q$,

the value of $d^2 f_n'(x)/dx^2 = 2n^q(n^p + n^p/3)^{-3}.\{-6\, n^q\ (1/\sqrt{3})\, n^{(p-q)/2}\} < 0$,

showing that $f_n'(x)$ is maximum when $x = -(1/\sqrt{3}).n^{(p-q)/2}$ and from (2), the maximum value of $f_n'(x)$.

$$= \frac{2n^q \times (1/\sqrt{3})n^{(p-q)/2}}{(n^p + n^p/3)^2} = \frac{3\sqrt{3}}{8} \cdot \frac{1}{n^{(3p-q)/2}} \qquad \text{... (5)}$$

$\therefore$ (2) and (5) $\Rightarrow \left|f_n'(x)\right| = \left|-\frac{2n^p x}{(n^p + n^q x^2)^2}\right| \le \frac{3\sqrt{3}}{8} \cdot \frac{1}{n^{(3p-q)/2}} < \frac{1}{n^{(3p-q)/2}} = M_n$, say

Since $\sum_{n=1}^{\infty} M_n = \sum_{n=1}^{\infty} \frac{1}{n^{(3p-q)/2}}$ is convergent if $(3p-q)/2 > 1$ *i.e.*, if $q < 3p - 2$, therefore, by Weierstrass's M-test, the series $\sum_{n=1}^{\infty} f_n'(x)$ is convergent if $q < 3p - 2$.

Hence term by term differentiation is possible if $q < 3p - 2$.

EXERCISES

1. Show that the series $\sum_{n=0}^{\infty} (1-x)x^n$ is not uniformly convergsent on [0, 1].

2. Show that the series $\sum_{n=0}^{\infty} \frac{x^2}{(1+x^2)^n}$ is not uniformly convergent for $x \ge 0$. *(Gwalior 2003)*

3. Prove that $\int_0^1 \left(\sum_{n=1}^{\infty} \frac{x^n}{n^2}\right) dx = \sum_{n=1}^{\infty} \frac{1}{n^2(n+1)}$ **(Kanpur 2010)**

4. Show that the series the sum of whose first n terms is $n^2 x(1-x)^n$, $0 \le x \le 1$ cannot be integrated term by term. *[Indore 2000; Jabalpur 1999, Vikram 2001]*

5. Show that the series for which $S_n(x) = 1/(1+nx)$ can be integrated term by term on [0, 1], though it is not uniformly convergent on [0, 1]. *[Meerut 2010, 12; Delhi 2009]*

6. Show that the series $\sum_{n=0}^{\infty} (-1)^n x^n$, $0 \le x \le 1$ admits of term by term integration on [0, 1], though it is not uniformly convergent on [0, 1].

7(*a*) Show that the following series can be integrated term by term on [0, 1], although it is not uniformly convergent on [0, 1]. $\sum_{n=1}^{\infty} \left[\frac{nx}{1+n^2x^2} - \frac{(n-1)x}{1+(n-1)^2x^2}\right]$

(*b*) Test for uniform convergence and term by term integration of the series $\sum_{n=1}^{\infty} \{x/(n+x)^2\}$

(Agra 2004, 2005, 2010)

8. Let $<f_n>$ be a sequence of real valued continuous functions on $\mathbf{R}$ converging uniformly to a continuous function f on $\mathbf{R}$. Prove that $\lim_{n\to\infty} f_n(x+1/n) = f(x)$ for every $x \in \mathbf{R}$. *(Mumbai 2010)*

9. Show that the function represented by $\Sigma\,(\sin n\,x)/n^3$ is differentiable for every x and its derivative is $\Sigma\,(\cos n\,x)/n^2$. *(Kanpur 1999, 2012)*

10. Given $S(x) = \Sigma\, 1/(n^2 + n^4 x^2)$, justify the validity of $S'(x) = -2x\,\Sigma \dfrac{1}{n^2(1+nx^2)^2}$

11. Prove that $\Sigma\, 1/(n^3 + x^4x^2)$ is uniformly convergent for all real x and that it may be differentiated term by term. *(Meerut 2013; Kanpur 1999, 2001)*

12. (*a*) Show that the series for which $S_n(x) = (nx)/(1 + n^2x^2)$, $0 \le x \le 1$, cannot be differentiated term by term at $x = 0$. *(Kanpur 2009, 10; Meerut 2010; Agra 2010)*

(*b*) Show that the series for which $S_n(x) = 1/(n + n^3x^2)$ can be differentiated term by term on $[0, 1]$.

13. Show that the series $\dfrac{x}{x+1} + \dfrac{x}{(x+1)\,(2x+1)} + \dfrac{x}{(2x+1)\,(3x+1)} + \ldots$

is uniformly convergent in $]\,k, \infty\,[$, where k is a positive number. Show also that the series is non-uniformly convergent in $[\,0, \infty\,[$.

14. Show that the series $(1-x)^2 + x\,(1-x)^2 + x^2(1-x)^2 + \ldots$ is not uniformly convergent in $[0, 1]$. *(Delhi Maths (H) 2005)*

15. Show that the series $\Sigma\,[x\,/\,n(n+1)]$ is uniformly convergent in $[0, k]$ where k is any positive number whatsoever but that it does converge uniformly in $]\,0, \infty\,[$.

16. Show that $\sum\limits_{n=-\infty}^{\infty} e^{-(x-n)^2}$ converges uniformly in any fixed interval $[a, b]$.

17. Show that $S(x) = \sum\limits_{n=0}^{\infty} \dfrac{1}{1+n^2+n^4x^2}$ converges uniformly for all values of x; examine whether $S'(0)$ can be found by term by term differentiation.

18. Show that the series $\Sigma\, n^{-x}$ is uniformly convergent in $]\,1+\delta, \infty\,[$ where δ is a positive number. Show also that term by term differentiation is valid in the same interval.

19. Show that the following series converge uniformly in the intervals indicated :

(*i*) $x - x^2 + x^3 - x^4 + x^5 - x^6 + \ldots,$ $\quad [-3/4, 3/4]$

(*ii*) $e^x + e^{2x} + e^{3x} + e^{4x} + \ldots,$ $\quad [-2, -1/2]$

20. Show that $\sum\limits_{n=1}^{\infty} \dfrac{x}{1+n^2x^2}$ is uniformly convergent in $[\delta, 1]$ for any $\delta > 0$ but is not uniformly convergent in $[0, 1]$.

21. Show that the series $\sum\limits_{n=1}^{\infty} \left[e^{-n|x|} - e^{-2n|x|}\right]$ though convergent for all values of x, is not uniformity convergent in any interval which contains $x = 0$.

22. Show that, in the interval, $[-1, 1]$, each of the functions $\dfrac{nx}{1+nx^2}, \dfrac{nx}{1+n^2x^2}, \dfrac{1}{1+n^3x^3}$ tends to a limit as $n \to \infty$ and discuss the uniformly of convergence.

23. Discuss the uniform convergence of $1 + \dfrac{e^{-2x}}{2^2-1} + \dfrac{e^{-4x}}{4^2-1} + \dfrac{e^{-6x}}{6^2-1} \ldots$ in $[\,0, \infty\,[$.

24. Show that the sequences $< f_n >$ and $< s_n >$ defined on [0, 1] by $f_n(x) = 1/x + 1/n$ and $s_n = 1/n$ are uniformly convergent but their product $< f_n s_n >$ is not. *(Agra 1999)*

25. Examine (*i*) absolute convergence (*ii*) uniform convergence of the series $(1 - x) + x(1 - x) + x^2(1 - x) + \ldots$ in $[c, 1]$ where $0 < c < 1$. *(I.A.S. 1994)*

26. Let $< f_n >, < g_n >$ be sequences of functions which converge to f, g respectively on an interval I. Show that if f and g are bounded on I, then the sequence $< f_n g_x >$ must converge uniformly to fg on I. *(Delhi Maths (H) 1996)*

27. (*a*) If a sequence of continuous function $< f_n >$ defined on $[a, b]$ is monotonic increasing, and converges (point wise) to a continuous function f, then prove that the convergence is uniform on $[a, b]$. *(Agra 1997, 99)*

(*b*) If the sum function of a series Σf_n with non-negative continuous terms defined on an interval $[a, b]$, is continuous on $[a, b]$, then prove that the series is uniformly convergent on the interval

[This is known as Dini's Theorem on uniform convergence].

28. Test the point wise and uniform convergence of $\Sigma\, 1/(n^4 + n^2x^2)$ *(Utkal 2003)*

15.11 THE WEIERSTRASS APPROXIMATION THEOREM *(Himachal 2013)*

The great German mathematician Weierstrass stated and proved remarkable and interesting theorem, which we shall now state and prove.

Every continuous function, defined in an interval, can be approximated to uniformly, by a sequence of polynomials. *(G.N.D.U. Amritsar 2000, 01, 02, 03; Gwalior 20002, 03, Kanpur 2004; Meerut 2001, 00, 06, 07, 09; Rajasthan 2006; Delhi Maths (H) 1997, 2003)*

Proof. Let a real valued continuous function f be defined on $[a, b]$. We shall prove that there exists a sequence $\{P_n\}$ of polynomials which converges uniformly to f.

Without any loss of generality we suppose that $[a, b] = [0, 1]$ and that $f(0) = 0 = f(1)$ and when $x \in R \sim [0, 1]$.

If the theorem is proved for this case, we consider.

$$g(x) = f(x) - f(0) - x[-f(1) - f(0)]$$

Clearly $g(0) = 0 = g(1)$. If g can be obtained as the limit of a uniformly convergent sequence of polynomials, then the same is true for $f, f - g$ being a polynomial.

I. We write $Q_n(x) = C_n(1 - x^2)^n;\ n \in N$, where C_n is such that $\int_{-1}^{1} Q_n(x)\,dx = 1;\ n \in N$.

II. We now obtain an estimate of C_n. This result will be made to depend upon the inequality

$$(1 - x^2)^n > (1 - nx^2).$$

To see the truth of this inequality, we consider the expression

$$(1 - x^2)^n - 1 + nx^2$$

which is 0 for $x = 0$, and whose derivative $2nx\{1 - (1 - x^2)^n\}$ is positive for $x \in [0, 1]$.

Thus, $$(1 - x^2)^n - 1 + nx^2 > 0 \text{ for } x \in [0,1].$$

We have $\int_{-1}^{1}(1 - x^2)^n\,dx = 2\int_{0}^{1}(1 - x^2)^n\,dx$

$$\geq \int_{0}^{1/\sqrt{n}} 2(1 - x^2)^n\,dx \geq 2\int_{0}^{1/\sqrt{n}} (1 - nx^2)\,dx = \frac{4}{3}\sqrt{n} > \frac{1}{\sqrt{n}}$$

Thus, $\int_{-1}^{1} C_n (1-x^2)^n \, dx = 1 \Rightarrow C_n > \sqrt{n}.$

It follows that $Q_n(x) \le \sqrt{n}\,(1-\delta^2)^n$ where $\delta \le |x| \le 1$

and as such $Q_n \to 0$ uniformly in $\delta \le |x| \le |$.

We now write $P_n(x) = \int f(x+t)\, Q_n(t)\, dt.$

We have $P_n(x) = \int_{-x}^{1-x} f(x+t)\, Q_n(t)\, dt.$

on the basis of the fact that f is zero for all x outside [0, 1].

Also we have

$$P_n = \int_{-x}^{1-x} f(x+t)\, Q_n(t)\, dt = \int_0^1 f(t)\, Q_n(t-x)\, dt$$

so that P_n is a polynomial in x.

Now f being continuous, corresponding to $\varepsilon > 0$ there exists $\delta > 0$ such that

$$|f(y) - f(x)| < \epsilon/2 \text{ whenever } |y - x| < \delta.$$

Let $M = \sup. f(x)$. Then, we have

$$|P_n(x) - f(x)| = \int_{-1}^{+1} |[f(x+t) - f(x)]\, Q_n(t)|\, dt$$

$$\le \int_{-1}^{+1} |f(x+t) - f(x)|\, |Q_n(t)|\, dt$$

$$\le 2M \int_{-1}^{-\delta} Q_n(t)\, dt + \frac{\varepsilon}{2} \int_{-\delta}^{\delta} Q_n(t)\, dt + 2M \int_{\delta}^{1} Q_n(t)\, dt$$

$$\le 4M\sqrt{n}\,(1-\delta^2)^n + (\varepsilon/2) < \varepsilon.$$

for sufficiently large vales of x.

Thus the theorem has been proved.

Note. Weierstrass Approximation theorem may also be stated as follows :

If f is a real valued continuous function defined on a closed interval $[a, b]$, then there exists a sequence of real polynomials $\{P_n(x)\}$ which converge uniformly to $f(x)$ on $[a, b]$.

Corollary. *For any interval $[-a, a]$, there is a sequence of real polynomials P_n such that* $P_n(0) = 0$ and that $\lim_{n\to\infty} P_n(x) = |x|$, uniformly on $[-a, a]$. *(Himachal 2014; Delhi 2003)*

Proof. Here $f(x)$ $(= |x|)$ is real valued continuous function defined on the closed interval $[-a, a]$, hence by Weierstrass approximation theorem, there exists a sequence $\{Q_n(x)\}$ of real polynomials which converge uniformly to $f(x)$, i.e., $|x|$ on $[-a, a]$

As a particular case, we have $Q_n(0) \to 0$ as $n \to \infty$

We write $P_n(x) = Q_n(x) - Q_n(0)$, where $n \in \boldsymbol{N}$.

Then the sequence $\{P_n(x)\}$ exists such that $P_n(0) = 0$ and $\lim_{n\to\infty} P_n(x) = |x|$.

Example 1. *If f is continuous on* [0, 1] *such that* $\int_0^1 x^n f(x)\, dx = 0$, for $n = 0, 1, 2, 3, \ldots$ *then show that* $f(x) = 0$ on [0, 1]. *(GATE 2004; Himachal 2010, 2014, 2015)*

Solution. Given that $\int_0^1 x^n f(x)\, dx = 0$, for $n = 0, 1, 2, 3, \ldots$ ...(1)

Since f is continuous on [0, 1], hence by Weierstrass approximation theorem, there exists a sequence $\{P_n\}$ of polynomials such that

$P_n \to f$ uniformly on [0, 1] and therefore, $P_n f \to f^2$ uniformly on [0, 1]

Since f is continuous on [0, 1], so it is bounded on [0, 1]. Hence, by Art. 15.5, we get

$$\int_0^1 [f(x)]^2\, dx = \lim_{n\to\infty} \int_0^1 P_n(x)\, f(x)\, dx \qquad \text{... (2)}$$

Let any real polynomial be given by $\qquad g(x) = c_0 + c_1 x + ... + c_c x^n = \sum_{r=0}^{n} c_r x^r$

Then, $\quad \int_0^1 g(x)\, f(x)\, dx = \int_0^1 f(x)\left[\sum_{r=0}^{n} c_r x^r\right] dx = \sum_{r=0}^{n} c_r \int_0^1 x^r f(x)\, dx = 0,$ using (1)

Thus, the integral of the product of f with any polynomial is zero. Hence, (2) yields

$$\int_0^1 [f(x)]^2\, dx = 0 \quad \Rightarrow \quad [f(x)]^2 = 0 \text{ on } [0, 1] \quad \Rightarrow \quad f(x) = 0 \text{ on } [0, 1]$$

OBJECTIVE QUESTIONS

1. Which of the following sequence $<f_n>$) of functions does not converge on [0, 1]:

(*a*) $f_n(x) = e^{-x}/n$ (*b*) $f_n(x) = (1 - x)^n$

(*c*) $f_n(x) = (x^2 + nx)$ (*d*) $f_n(x) = (1/n) \times \sin(nx + n)$ **(GATE 2009)**

Sol. Ans. (*c*) Left as an exercise

2. Let $f_n(x) = x^n/(1 + x)$ and $g_n(x) = x^n/(1 + nx)$ for $x \in [0,1]$ and $n \in N$. Then on [0, 1] :

(*a*) both $<f_n>$ and $<g_n>$ converge uniformly (*b*) neither $<f_n>$ nor $<g_n>$ converge uniformly

(*c*) $<f_n>$ converges uniformly but $<g_n>$ does not converge uniformly

(*d*) $<g_n>$ converges uniformly but $<f_n>$ does not converge uniformly. **(GATE 2010)**

Sol. Ans. (d). Let as an exercise.

3. Let $<f_n>$ be a sequence of real valued differential functions on [*a, b*] such that $f_n(x) \to f(x)$ as $n \to \infty$ for every $x \in [a,b]$ and for some Riemann integrable function $f : [a,b] \to \mathbf{R}$. Consider the statements :

$P_1 : <f_n>$ converges uniformly $\qquad P_2 : <f_n'>$ converges uniformly

$P_3 : \int_a^b f_n(x)dx \to \int_a^b f(x)\, dx \qquad P_4 : f$ is differentiable

Then, which one of the following need not be true

(*a*) P_1 implies P_3 (*b*) P_2 implies P_1

(*c*) P_2 implies P_4 (*d*) P_3 implies P_1 (*GATE 2010*)

Sol. Ans. (*c*). Try yourself

4. Let $f_n(x) = x / \{(n-1)x + 1\}\ (nx + 1)$ and $s_n(x) = \sum_{J=1}^{n} f_j(x)$ for $x \in [0,1]$. The sequence $\{s_n\}$

(*a*) converger on [0, 1] (*b*) converge pointwise on [0,1] but not uniformly

(*c*) converge pointwise for $x = 0$ but not for $x \in (0,1]$

(*d*) does not coverge for $x \in [0,1]$. (*GATE 2011*)

Hint : Ans. (*b*) See Ex.2, page 15.22.

MISCELLONEOUS PROBLEMS ON CHAPTER 15

1. If $\sum_{k=1}^{\infty} u_k(x)$ is a series of Riemann functions on $[a, b]$ which converges uniformuly to $f(x)$ on $[a, b]$, then prove that f is Riemann integrable on $[a, b]$ and $\int_a^b f(x)\,dx = \sum_{k=1}^{\infty} \int_a^b u_k(x)\,dx$.

Hence show that if $\sum_{n=0}^{\infty} |a_n| < \infty$, *then* $\int_0^1 [\sum_{n=0}^{\infty} a_n x^n]\,dx = \sum_{n=0}^{\infty} \frac{a_n}{n+1}$ *[Pune 2010]*

. Prove that the sequence $<f_n>$ where $f_n(x) = x^{n-1}(1-x)$ converges uniformly in the interva (0, 1). *(Kanpur 2014; Agra 2009;*

3. Examine the series $\Sigma\, f_n,\ f_n = x^2/(1+x^2)^n$ for uniform convergence in [0,1].
[Hint : Refer Ex, 2, page 15.27] *(Delhi Maths (H) 2006)*

4. Show that the series $\sin x + (1/2)\times \sin 2x + (1/3)\times \sin 3x + \ldots\ldots$ convergen uniformly in $0 < a < x < b < \pi$. *(Kanpur 2006, 09)*

5. Examine for continuity of the sum function and for term by term integration the series whose nth term is $n^2 x e^{-n^2x^2} - (n-1)^2 x\, e^{-(n-1)^2x^2}$, having all values in the interval [0, 1] *(Kanpur 2006)*

[Sol. Let $\qquad u_n(x) = n^2 x e^{-n^2x^2} - (n-1)^2 x\, e^{-(n-1)^2x^2}$

Hence
$$u_1(x) = x e^{-x^2} - 0$$
$$u_2(x) = 2^2 x\, e^{-2^2x^2} - x\, e^{-x^2}$$
$$u_3(x) = 3^2 x\, e^{-3^2x^2} - 2^2 x\, e^{-2^2x^2}$$
$$\text{-------------------------------------}$$
$$u_{n-1}(x) = (n-1)^2 x\ \ e^{-(n-1)^2x^2} - (n-2)^2 x\, e^{-(n-2)^2x^2}$$
$$u_n(x) = n^2 x\, e^{-n^2x^2} - (n-1)^2 x\, e^{-(n-1)^2x^2}$$

Then,
$$S_n(x) = \sum_{n=1}^{n} u_n(x) = n^2 x\, e^{-x^2x^2}$$

Hence,
$$S(x) = \lim_{n\to\infty} S_n(x) = 0 \quad \text{for} \quad 0 \le x \le 1,$$
showing that the sum function $S(x)$ is continuous for all x in [0, 1].

Now,
$$\int_0^1 S(x)dx = \int_0^1 (0)\,dx = 0 \ \text{ and}$$
$$\int_0^1 S_n(x)dx = \int_0^1 n^2 x\, e^{-n^2x^2}\,dx = \frac{1}{2}\int_0^{n^2} e^{-t}\,dt, \text{ putting } n^2x^2 = t \text{ and } n^2 x dx = (1/2)\times dt$$
$$= (1/2)\times[-e^{-t}]_0^{n^2} = (1/2)\times(1-e^{-n^2})$$

$\therefore$
$$\lim_{n\to\infty} \int_0^1 S_n(x)\,dx = \lim_{n\to\infty} (1/2)\times(1-e^{-n^2}) = 1/2$$

Thus,
$$\int_0^1 S(x)dx \neq \lim_{n\to\infty} \int_0^1 S_n(x)\,dx,$$
showing the given series cannot be integrated term by term in [0,1]

6. Show that the series
$$\frac{x^2}{1+x} + \left(\frac{2x^2}{1+2x} - \frac{x^2}{1+x}\right) + \ldots + \left(\frac{nx^2}{1+nx} - \frac{(n-1)x^2}{1+(n-1)x}\right) + \ldots\ldots$$
converges uniformly on [0, 1]

[Sol. Let the given series be denoted by $\sum_{n=1}^{\infty} u_n$

Then $u_1(x) = x^2/(1+x)$

$$u_2(x) = 2x^2/(1+2x) - x^2/(1+x)$$

....

$$u_n(x) = \frac{nx^2}{1+nx} - \frac{(n-1)x^2}{1+(n-1)x}$$

$\therefore\ S_n(x) = \sum_{n=1}^{\infty} u_n(x) = \dfrac{nx^2}{1+nx}$ and so $S(x) = \lim_{n\to\infty} S_n(x) = \lim_{n\to\infty} \dfrac{nx^2}{1+nx} = x$

Let $\varepsilon > 0$ be given. Then, we have $|S_n(x) - S(x)| < \varepsilon$

or $\left|nx^2/(1+nx) - x\right| < \varepsilon$ or $x/(1+nx) < \varepsilon$ or $1+nx > x/\varepsilon$

or $n > (1/\varepsilon) - (1/x)$. ... (1)

(1) holds for all $n \geq m$, where m is positive integer just greater than $1/\varepsilon$. In particular for $x = 0$, $m = 1$. Hence the series converges uniformly on [0, 1]

7. Show that the sequence $<f_n>$ defined by $f_n(x) = 1/(x+n)$ is uniformly convergent in (0, 1). **[Himanchal 2009]**

8. Show that the series $\sum_{n=1}^{\infty} x^n e^{-nx}$ is unifirmly convergent on [0, a], $a > 0$

[Pune 2010; Mumbai 2010]

9. Find the limit function of the sequence $\{f_n(x)\}_{n=1}^{\infty}$ where $f_n(x) = x^n/(1+x^n), 0 \leq x \leq 1$

Ans. Required limit = $\begin{cases} 0, \text{if } 0 \leq x < 1 \\ 1/2, \text{if } x = 1 \end{cases}$ **[Pune 2010]**

10. *If $b_n(x)$ is a monotonic function of n for each fixed value of x in [a, b] and $b_n(x)$ tends uniformly to zero for $a \leq x \leq b$ and if there is a number $k > 0$ independent of x and n, such that for all values of x in [a, b], $\left|\sum_{r=1}^{\infty} u_r(x)\right| \leq k, \forall n,$ then prove that the series $\Sigma b_n(x)\, u_n(x)$ is uniformly convergent on [a, b].* **[Delhi Maths (H) 2007]**

11. *Prove that uniform limit f of the sequence of functions $< f_n >$ is integrable whenever the each function f_n is integrable on [a, b].* **[Purvanchal 2006]**

12. Suppose the series $\sum_{n=1}^{\infty} f_n$ converges uniformly to the function f on [a, b] where each f_n is continuous function on [a, b]. Prove that f is integrable on [a. b] and that the series $\sum_{n=1}^{\infty}\left(\int_a^x f_n(t)dt\right)$ converges uniformly to $\int_a^x f(t)dt\ \forall x$ in $[a,b]$. **[Delhi Maths (H) 2008]**

13. Examine the uniform convergence of the following series :

(i) $\dfrac{2x}{1+x^2} + \dfrac{4x^3}{1+x^4} + \dfrac{8x^7}{1+x^8} + \ldots, -\dfrac{1}{2} < x < \dfrac{1}{2}$ (ii) $\dfrac{\cos\theta}{1^p} - \dfrac{\cos 2\theta}{2^p} + \dfrac{\cos 3\theta}{3^p} - \ldots, \theta$ real

[Delhi Maths (H) 2008]

14. Let $<f_n>$ be a sequence of functions such that $\lim_{n\to\infty} f_n(x) = f(x),\ x \in [a,b]$ and let $M_n = \sup_{x\in[a,b]} |f_n(x) - f(x)|,$ then show that $f_n \to f$ uniformly on $[a, b]$ if and only if $M_n \to 0$ *as* $n \to \infty$.

[Delhi Maths (Prog) 2008]

15. Test the uniform convergence of series $\sum_{n=1}^{\infty} x\, e^{-nx}$ in the interval [0,1] **[Agra 2008]**

16. If the series $1 - x + x^2 - x^3 + \ldots.$ converges uniformly to $1/(1+x)$ on $[0, y]$, $y \in (0,1)$, then show that $y - y^2/2 + y^3/3 - y^4/4 + \ldots\ldots = \log(1+y)$. **(Pune 2010)**

17. Show that $\lim_{x\to 1} \sum_{n=1}^{\infty} \frac{n^2x^2}{n^4+x^4} = \sum_{n=1}^{\infty} \frac{n^2}{n^4+1}$. Justify all steps of your answer by quoting the theorem you are using **[I.A.S. 2009]**

[Hint: To prove the required result, we use the following result : "The limit of the sum function of a series = the sum of the series of limits of functions", i.e., $\lim_{x\to x_0} \sum_{n=1}^{\infty} f_n(x) = \sum_{n=1}^{\infty} \lim_{x\to x_0} f_n(x),$ where $\sum_{n=1}^{\infty} f_n$ converges uniformly in $[a, b]$ and x_0 is a point in $[a, b]$. Here take $x_0 = 1$]

18. If a sequence $<f_n(x)>$ converges uniformly in $[a, b]$ and x_0 is a point of $[a, b]$ such that $\lim_{x\to x_0} f_n(x) = a_n$, $n = 1, 2, 3, \ldots$ then prove that (i) sequence $<a_n>$ is convergent (ii) $\lim_{x\to x_0} f(x)$ exists and is equal to $\lim_{n\to\infty} a_n$. **[Delhi Maths (H), 2009)]**

19. Examine for convergence and continuity of the limit function of the sequence $<f_n>$ where $f_n(x) = (nx)/(1+n^2x^2),\ 0 < x < 1$. **[Delhi B.Sc. (Prog) III 2010]**

20. Test for pointwise and uniform convergence for the sequence $<f_n>$ defined by $f_n(x) = [\log(1+n^4x^2)]/(2n^2),\ 0 \le x \le 1$. State clearly the results you are using.

[Delhi B.A. (Prog) III 2010]

21. When a sequence is said to be uniformly convergent ? Show that uniform convergence implies point-wise convergence but converse is not true. **[Himachal 2008, 09]**

22. Let $f_n(x) = (\sin nx)/n,\ 0 \le x \le 1$. Find $m \in N$ such that $|f_n(x) - 0| < 1/10$ for $n > m$ for all $x \in [0,1]$. **[Pune 2010]** **Ans.** $m = 11$

23. Show that, if f is continuous on R, there exists a sequence $\{P_n\}$ of polynomials converging uniformly to f on each bounded subset of **R**.

24. State weierstrass approximation theorem. Show that there does not exist a sequence of polynomials converging uniformly on **R** to $f(x) = e^x$. **[Himachal 2013]**

Chapter 16

Improper Integrals

16.1 PROPER AND IMPROPER INTEGRALS

(Agra 2010)

The theory of Riemann integration, as developed in chapter 13, requires that the domain of integration is finite and that the integrand is bounded in the domain.

It is possible, however, so to extend the theory that the symbol

$$\int_a^b f(x)\,dx$$

may sometimes have a meaning (*i.e.,* denote a number), even when f is not bounded or when either a or b or both are infinite. In case f is unbounded or the limits a or b are infinite, the symbol

$$\int_a^b f(x)\,dx$$

is called an Improper or Generalised or Infinite integral. Thus

$$\int_0^1 \frac{dx}{x^3},\ \int_1^2 \frac{dx}{(1-x)(2-x)},\ \int_{-\infty}^{\infty} \frac{dx}{1+x^2},\ \int_0^{\infty} \frac{1}{\sqrt{x}}\,dx$$

are examples of improper integrals.

An improper integral with finite interval of integration $[a,b]$ with its integrand having a finite number of points of infinite discontinuity is known as the *improper integral* of *the first kind.* Again, the improper integral with infinite interval of integration and bounded integrand is known as *the improper integral of the second kind.* Finally, an improper integral with infinite interval of integration with its integrand having a finite number of points of infinite discontinuity is known as the *improper integral of the mixed kind.*

For the sake of distinction an integral which is not improper will be called a proper integral.

It can be *proved that if a function f is not bounded in a finite interval $[a,b]$, then there exists *at least one point, c, of the interval such that in no neighbourhood of c, f is bounded.* Such a point is known as a point of *Infinite discontinuity* of the function f. It will always be assumed that the function f is such that the set of its points of infinite discontinuity in any interval, finite or infinite, are finite in number; the consideration of functions having an infinite number of points of infinite discontinuity being beyond the scope of the book.

In a finite interval which encloses no point of infinite discontinuity, the function is always bounded and we assume, once for all, in order to avoid tedious repetition, that it is also integrable in such an interval.

16.2 CONVERGENCE OF IMPROPER INTEGRALS OF THE FIRST KIND. DEFINITIONS.

Convergence at the Left-End Let a, be the only point of infinite discontinuity of a function f in a finite interval $[a,b]$.

We write $$\varphi(\varepsilon) = \int_{a+\varepsilon}^{b} f(x)\,dx,\ 0 < \varepsilon \le b-a,$$

* For the proof, use may be made of the completeness of order of the field of real numbers.

the integral on the right existing according to the assumption made in the last paragraph. Thus we have defined a function φ with domain $]\,0, b-a\,[$.

If, when $\varepsilon \to (0+0)$, the function φ tends to a finite limit, say l, we say that the improper

integral $$\int_a^b f(x)\,dx, \qquad ...(1)$$

exists, or converges at a, and use the symbol (1) to denote the number l.

Thus, $$\int_a^b f(x)\,dx = \lim_{\varepsilon \to 0} \int_{a+\varepsilon}^b f(x)\,dx,$$

provided that the limit on the right exists. In case the limit does not exist, we say that the improper integral $\int_a^b f(x)\,dx$ does not exists or that it does not converge.

Ex. Show that if $\int_a^b f(x)\,dx$ converges at a, then $\int_a^c f(x)\,dx$ $(a < c < b)$ also converges at a.

Ex. The improper integral $\int_a^b f(x)\,dx$ converges at a, and k is any constant: show that $\int_a^b kf(x)\,dx$ converges at a, and conversely.

Convergence at the right-end. *Let b be the only point of infinite discontinuity of f in a finite interval* $[a, b]$. If then the proper integral

$$\int_a^{b-\varepsilon} f(x)\,dx, \; 0 < \varepsilon < (b-a) \qquad ...(2)$$

tends to a finite limit, as $\varepsilon \to (0+0)$, we say that the improper integral

$$\int_a^b f(x)\,dx \qquad ...(3)$$

exists or converges at b and use the symbol (3) to denote the limit of (2).

Ex. 1. Examine the convergence of the improper integrals :

(*i*) $\displaystyle\int_0^1 \frac{dx}{x^2}$. (*ii*) $\displaystyle\int_0^1 \frac{dx}{(1-x^2)^{1/2}}$

Sol. (*i*) Now 0 is the only point of infinite discontinuity of the integrand. We have, when $0 < \varepsilon \le 1$.

$$\varphi(\varepsilon) = \int_\varepsilon^1 \frac{dx}{x^2} = \left[-\frac{1}{x}\right]_\varepsilon^1 = \frac{1}{\varepsilon} - 1, \text{ which } \to +\infty \text{ as } \varepsilon \to (0+0).$$

Thus we see that the improper integral, in question, does not converge.

(*ii*) Now 1, is the only point of infinite discontinuity of the integrand in [0, 1]. We have

$$\varphi(\varepsilon) = \int_0^{1-\varepsilon} \frac{dx}{(1-x^2)^{1/2}} = \sin^{-1}(1-\varepsilon), \text{ which } \to \pi/2 \text{ as } \varepsilon \to (0+0)$$

Thus the improper integral converges with value $\pi/2$

Ex. 2. *Show that* $\int_0^{\pi/2} \sin x \log \sin x$ *is convergent with value* log (2/e).

(*Agra 2007; Delhi Maths (H) 2002*)

Sol. Here 0 is the only point of infinite discontinuity of the integrand in $[0, \pi/2]$. Here we have

$$\int_{\epsilon}^{\pi/2} \sin x \log \sin x\, dx = [-\cos x \log \sin x]_{\varepsilon}^{\pi/2} + \int_{\varepsilon}^{\pi/2} \frac{\cos^2 x}{\sin x} dx, \text{ integrating by parts}$$

$$= \cos\varepsilon \log \sin\varepsilon + \int_{\varepsilon}^{\pi/2} (\operatorname{cosec} x - \sin x)\, dx = \cos\varepsilon \log\sin\varepsilon + [\log\tan(x/2) + \cos x]_{\varepsilon}^{\pi/2}$$

$$= \cos\varepsilon \log\sin\varepsilon - \log\tan(\varepsilon/2) - \cos\varepsilon$$

$$= \cos\varepsilon \log\{2\sin(\varepsilon/2)\cos(\varepsilon/2)\} - \log\{\sin(\varepsilon/2)/\cos(\varepsilon/2)\} - \cos\varepsilon$$

$$= \cos\varepsilon \log(2\cos\varepsilon/2) + \cos\varepsilon \log\sin(\varepsilon/2) - \log\sin\varepsilon/2 + \log\cos(\varepsilon/2) - \cos\varepsilon$$

$$= \cos\varepsilon \log(2\cos\varepsilon/2) + \log\cos(\varepsilon/2) - \cos\varepsilon - (1 - \cos\varepsilon)\log\sin(\varepsilon/2)$$

Now, $\int_0^{\pi/2} \sin \log \sin x\, dx = \lim_{\varepsilon \to 0+} \int_{\varepsilon}^{\pi/2} \sin \log \sin x\, dx$, by definition

$$\lim_{\varepsilon\to 0+} \{\cos\varepsilon \log(2\cos\varepsilon/2) + \log\cos(\varepsilon/2) - \cos\varepsilon - 2\sin^2(\varepsilon/2)\log\sin\varepsilon/2\}$$

[using (1)]

$$= \log 2 + 0 - 1 - 0 = \log 2 - \log e = \log(2/e),$$

as $\lim_{\varepsilon\to 0+} 2\sin^2(\varepsilon/2)\log\sin\varepsilon/2 = \lim_{t\to 0+} 2t^2 \log t$, if $t = \sin(\varepsilon/2)$

$$= 2\lim_{t\to 0+} \frac{\log t}{t^{-2}} \qquad \left[\text{From } \frac{\infty}{\infty}\right]$$

$$= 2\lim_{t\to 0+} \frac{(1/t)}{(-2)\times t^{-3}} = 0$$

Ex. 3. Define $f: [0,1] \to R$ where $f(x) = 1/\sqrt{x}$. Show that the improper integral of *f* on *R* exists. (*Calicut 2004*)

Convergence at both the end points

Let the end points *a* and *b* be the only points of infinite discontinuity of *f*. We take any point, *c*, such that $a < c < b$.

If the improper integrals $\int_a^c f(x)\, dx$ and $\int_c^b f(x)\, dx$

both converge at *a* and *b* respectively, we say that the improper integral

$$\int_a^b f(x)\, dx \text{ converges}$$

and write

$$\int_a^b f(x)\, dx = \int_a^c f(x)\, dx + \int_c^b f(x)\, dx.$$

It is easy to show that the fact of convergence and the value of the improper integral is independent of the choice of *c*.

If *d* be any point of] *a,b* [, we have

$$\int_{a+\varepsilon}^{d} f(x)\, dx = \int_{a+\varepsilon}^{c} f(x)\, dx + \int_c^d f(x)\, dx,$$

and see that

$$\lim_{\varepsilon \to 0} \int_{a+\varepsilon}^{d} f(x)\, dx \text{ exists finitely} \Rightarrow \lim_{\varepsilon \to 0} \int_{a+\varepsilon}^{c} f(x)\, dx \text{ exists finitely.}$$

It follows that

$$\int_a^d f(x)\, dx \text{ converges} \Leftrightarrow \int_a^c f(x)\, dx \text{ converges.}$$

Also, in case they converge finitely, we have

$$\int_a^d f(x)\, dx = \int_a^c f(x)\, dx + \int_c^d f(x)\, dx. \quad \text{... (1)}$$

It may similarly be shown that

$$\int_d^b f(x)\, dx \text{ converges} \Leftrightarrow \int_c^b f(x)\, dx \text{ converges,}$$

and in case they converge, we have

$$\int_d^b f(x)\, dx = \int_d^c f(x)\, dx + \int_c^b f(x)\, dx. \quad \text{... (2)}$$

Adding (1) and (2), we get

$$\int_a^d f(x)\, dx + \int_d^b f(x)\, dx = \int_a^c f(x)\, dx + \int_c^b f(x)\, dx = \int_a^b f(x)\, dx.$$

Ex. Examine the convergence of the improper integrals:

(i) $\int_0^1 \frac{dx}{\sqrt{(x - x^2)}}$, *(ii)* $\int_0^2 \frac{dx}{x(2-x)}$, *(iii)* $\int_0^\pi \frac{dx}{\sin x}$.

General Case. Any finite number of points of infinite discontinuity. Let $c_1, c_2, c_3, \ldots c_n$ be any finite number of points of discontinuity of f in $[a,b]$ where $a \le c_1 < c_2 < \ldots < c_{n-1} < c_n \le b$.

If the improper integrals

$$\int_a^{c_1} f(x)\, dx, \quad \int_{c_1}^{c_2} f(x)\, dx, \ldots, \quad \int_{c_{n-1}}^{c_n} f(x)\, dx, \quad \int_{c_n}^{b} f(x)$$

all exist in accordance with the definitions given above and we regard

$$\int_a^{c_1} f(x)\, dx = 0 \text{ if } c_1 = a \qquad \text{and} \qquad \int_{c_n}^{b} f(x)\, dx = 0 \text{ if } b = c_n,$$

then we say that $\int_a^b f(x)\, dx$ exists and write

$$\int_a^b f(x)\, dx = \int_a^{c_1} f(x)\, dx + \int_{c_1}^{c_2} f(x)\, dx + \ldots + \int_{c_n}^{b} f(x)\, dx.$$

Note. In the examples given above, the improper integrals are such that the integrands admit of primitives in terms of elementary functions. In such cases the examination of the existance is generally easy, but more advanced methods become necessary when the integrand does not possess a primitive in terms of elementary functions.

16.3. TEST FOR CONVERGENCE AT '*a*'. POSITIVE INTEGRAND

Just as in the case of infinite series with positive terms it is comparatively easier to consider the convergence of integrals with positive integrands. We now proceed to consider the case of positive intergrands. Let a be the only point of infinite discontinuity of f in $[a,b]$. The case where the integrand f is positive in a certain neighbourhood $]\, a, c\, [$ of, a, is particularly simple and important.

Since
$$\int_{a+\varepsilon}^{b} f(x)\,dx = \int_{a+\varepsilon}^{c} f(x)\,dx + \int_{c}^{b} f(x)\,dx,$$

it follows that
$$\int_{a}^{c} f(x)\,dx, \int_{a}^{b} f(x)\,dx$$

are either both convergent at, a or both non-convergent. It is, therefore, no loss of generality to suppose that f is positive in $[a,b]$.

The question of the existance of the integral is, in such a case, decided by comparison with another suitably integral whose existance or otherwise is already known.

16.4 THE NECESSARY AND SUFFICIENT CONDITION FOR THE CONVERGENCE OF THE IMPROPER INTEGRAL

$$\int_{a}^{b} f(x)\,dx$$

at a, where $f(x)$ is positive when $x \in\,]\,a,b\,]$, is that there exists $k > 0$, such that

$$\int_{a+\varepsilon}^{b} f(x)\,dx < k \;\forall\; \varepsilon \in\,]\,0, b-a\,].$$

The proof follows from the fact that, since $f(x)$ is positive when $x \in\,]a,b]$, the integral

$$\int_{a+\varepsilon}^{b} f(x)\,dx$$

monotonically increases as ε decreases and will, therefore, tend to a finite limit if, and only if, it is bounded above.

Note. In case $\varphi(\varepsilon) = \int_{a+\varepsilon}^{b} f(x)\,dx$ is not bounded above, then $\varphi(\varepsilon) \to +\infty$ as $\varepsilon \to (0+0)$

and we say that the improper integral $\int_{a}^{b} f(x)\,dx$ diverges to ∞.

16.5 COMPARISON OF TWO INTEGRALS

Let f and φ be two functions such that $\forall\; x \in\,]a,b]$, $f(x) \geq 0, \varphi(x) \geq 0$.

Also let $f(x) \leq \varphi(x)$. Then

(*i*) $\int_{a}^{b} \varphi(x)\,dx$ converges $\Rightarrow \int_{a}^{b} f(x)\,dx$ converges

(*ii*) $\int_{a}^{b} f(x)\,dx$ does not converge $\Rightarrow \int_{a}^{b} \varphi(x)\,dx$ does not converges. (*Lucknow 93, 95*)

It is assumed that f and φ are both bounded and integrable in $[a+\varepsilon, b]$, $0 < \varepsilon \leq (b-a)$, and a, is the only point of infinite discontinuity $[a,b]$. We have $\forall\; \varepsilon \in\,]0, b-a]$.

$$\int_{a+\varepsilon}^{b} f(x)\,dx \leq \int_{a+\varepsilon}^{b} \varphi(x)\,dx \qquad ...(1)$$

Let $\int_{a}^{b} \varphi(x)\,dx$ converge so that there exists a number k such that $\forall\; \varepsilon \in\,]\,0, b-a\,]$.

$$\int_{a+\varepsilon}^{b} \varphi(x)\, dx < k \qquad \text{... (2)}$$

From (1) and (2), we have $\forall\, \varepsilon \in]\, 0, b-a]$,

$$\int_{a+\varepsilon}^{b} f(x)\, dx < k \Rightarrow \int_{a}^{b} f(x)\, dx \text{ converges at } a.$$

For the second part, we see that if $\int_{a}^{b} f(x)\, dx$ does not converge at a, then $\int_{a+\varepsilon}^{b} f(x)\, dx$ is not bounded above and consequently, from (1), $\int_{a+\varepsilon}^{b} \varphi(x)\, dx$ is also not bounded above so that $\int_{a}^{b} \varphi(x)\, dx$ does not converge.

16.5A. PRACTICAL COMPARISON TEST

If $f(x)/\varphi(x) \to l$ when $x \to a$, and l, is neither 0 nor infinite, then the two integrals

$$\int_{a}^{b} f(x)\, dx \qquad \text{and} \qquad \int_{a}^{b} \varphi(x)\, dx$$

either both converge or both do not converge.

Since $f(x)/\varphi(x)$ is positive $\forall\, x$, limit $f(x)/\varphi(x) = l$ cannot be negative.

Let d be a positive number less than l so that there exists a number c $(a < c < b)$, such that

$$l - \delta < f(x)/\varphi(x) < l + \delta$$

when $a < x \leq c$, *i.e.*, $\forall\, x \in]\, a, c]$. We have $\forall\, x \in]a, c]$,

$$(l-\delta)\, \varphi(x) < f(x) \qquad \text{and} \qquad f(x) < (l+\delta)\, \varphi(x). \qquad \text{... } (i)$$

Now $\int_{a}^{b} f(x)\, dx$ converges $\Rightarrow \int_{a}^{c} f(x)\, dx$ converges

$\Rightarrow (l-\delta) \int_{a}^{c} f(x)\, dx$ converges $\Rightarrow \int_{a}^{c} \varphi(x)\, dx$ converges $\Rightarrow \int_{a}^{b} \varphi(x)\, dx$ converges.

It may similarly be shown that

$$\int_{a}^{b} \varphi(x)\, dx \text{ does not converge} \Rightarrow \int_{a}^{b} f(x)\, dx \text{ does not converge.}$$

Also, from (i), we may prove that

$$\int_{a}^{b} \varphi(x)\, dx \text{ converges} \Rightarrow \int_{a}^{b} f(x)\, dx \text{ converges,}$$

$$\int_{a}^{b} f(x)\, dx \text{ does not converges} \Rightarrow \int_{a}^{b} \varphi(x)\, dx \text{ does not converge.}$$

Ex. Prove that if $f(x)/\varphi(x) \to 0$.

$$\int_{a}^{b} \varphi(x)\, dx \text{ converges} \Rightarrow \int_{a}^{b} f(x)\, dx \text{ converges}$$

and that if $f(x)/\varphi(x) \to \infty$,

$$\int_{a}^{b} f(x)\, dx \text{ converges} \Rightarrow \int_{a}^{b} \varphi(x)\, dx \text{ converges.}$$

16.6 USEFUL COMPARISON INTEGRALS

(*i*) The improper integral $\int_a^b \frac{dx}{(x-a)^n}$ converges if and only if $n<1$. **[*Kanpur 2005, 2017*]**

[*Purvanchal 2011; Himanchal 2004; Meerut 2009; 2015*]

We have, if $n \neq 1$,

$$\int_{a+\varepsilon}^{b} \frac{dx}{(x-a)^n} = \left| \frac{1}{(1-n)(x-a)^{n-1}} \right|_{a+\varepsilon}^{b} = \frac{1}{(1-n)} \left[\frac{1}{(b-a)^{n-1}} - \frac{1}{\varepsilon^{n-1}} \right],$$

which tends to $1/(1-n)(b-a)^{n-1}$ or $+\infty$ according as $n<1$ or >1.

Again, if $n=1$,

$$\int_{a+\varepsilon}^{b} \frac{dx}{x-a} = \log(b-a) - \log\varepsilon, \text{ which } \to +\infty \text{ as } \varepsilon \to (0+0).$$

Hence the result.

(*ii*) The improper integral $\int_a^b \frac{dx}{(b-x)^n}$ is convergent if and only if $n<1$. **[*Kanpur 2018*]**

Proof. Proceed as in part (*i*), yourself.

16.7 TWO USEFUL TESTS

With the help of article 16.4, 16.5 and 16.6, we deduce two important practical tests for convergence at, *a*, of the integral of *f* by comparison with the integral of $\varphi(x) = 1/(x-a)^n$

Practical test I.

Let *f* be positive in $[a,b]$. Then the integral of *f* converges at *a*, if there exists a number $n>0$ and less than 1 and a fixed positive number *k* such that $f(x) \le k/(x-a)^n \ \forall\, x \in\,]a,b]$.

Also, the integral of *f* does not converge, if, there exists a number $a \ge 1$ and a fixed positive number *k* such that $f(x) \ge k/(x-a)^n \ \ \forall\, x \in\,]a,b]$.

Practical test II. The μ - test [*Meerut 2010*]

(*i*) *If a is the only point of infinite discontinuity of f on [a, b] and* $\lim_{x \to a+0} (x-a)^{\mu} f(x)$ *exists and is non-zero finite, then* $\int_a^b f(x)\,dx$ *converges if and only if* $\mu < 1$.

(*ii*) *If b is the only point of infinite discontinuity of f on [a, b] and* $\lim_{x \to b-0} (b-x)^{\mu} f(x)$ *exists and is non-zero finite, then* $\int_a^b f(x)\,dx$ *converges if and only if* $\mu < 1$

EXAMPLES

Ex. 1. Examnine the convergence of (*i*) $\int_0^1 \frac{dx}{x^{1/3}(1+x^2)}$ **[*Kanpur 2016; Meerut 2016*]**

(*ii*) $\int_0^1 \frac{dx}{x^2(1+x)^2}$ (Purvanchal 2006) (*iii*) $\int_0^1 \frac{dx}{x^{1/2}(1-x)^{1/2}}$

Sol. The integrands are all positive.

(*i*) Here 0, is the only point of infinite discontinuity of the integrand.

We have, $f(x)=\dfrac{1}{x^{1/3}(1+x^2)}$ Take $\varphi(x)=\dfrac{1}{x^{1/3}}$

Now $$\lim_{x\to 0}\frac{f(x)}{\varphi(x)}=\lim_{x\to 0}\frac{1}{1+x^2}=1.$$

Hence $\int_0^1 f(x)\,dx$ and $\int_0^1 \varphi(x)\,dx$

have identical behaviours. But $n = 1/3$ being less than 1, the latter integral converges by Art. 16.5. Hence the given integral also converges.

(*ii*) Here, 0, is the only point of infinite discontinuity of the given integrand. We have

$$f(x)=\frac{1}{x^2(1+x)^2} \qquad \text{Take} \qquad \varphi(x)=\frac{1}{x^2}.$$

Now $$\lim_{x\to 0}\frac{f(x)}{\varphi(x)}=\lim_{x\to 0}\frac{1}{(1+x)^2}=1$$

Thus, $\int_0^1 f(x)\,dx$ and $\int_0^1 \varphi(x)\,dx$

have identical behaviours.

But $n = 2$ being greater than 1, the latter integral does not converge by Art. 16.5. Hence the given integral also does not converge.

(*iii*) Here, 0 and 1, are the two points of infinite discontinuity of the integrand. We have

$$f(x)=\frac{1}{x^{1/2}(1-x)^{1/3}}$$

We take any number between 0 and 1, say 1/2, and examine the convergence of the two improper integrals.

$$\int_0^{1/2} f(x)\,dx \quad \text{and} \quad \int_{1/2}^1 f(x)$$

at 0, and 1, respectively.

To examine the convergence of the former integral, we take $\varphi(x)=1/x^{1/2}$

so that $$f(x)/\varphi(x)\to 1 \text{ as } x\to 0.$$

$\therefore$ By Art. 16.5, $\int_0^{1/2} f(x)\,dx$ converges.

For the latter, we take $\phi(x)=1/(1-x)^{1/3}$

so that $f(x)/\varphi(x)\to 1$ as $x\to 1$. So by Art. 16.5,

$$\int_{1/2}^1 f(x)\,dx \text{ converges. Hence } \int_0^1 f(x)\,dx \text{ converges.}$$

Ex. 2. *Show that the Beta function* $\int_0^1 x^{m-1}(1-x)^{n-1}\,dx$ *converges if and only if* $m>0$ *and* $n>0$.

[Kanpur 2005, 12; Delhi B.A. (Prog) III 2013, 15; G.N.D. U 2010;l Himachal 2003, 04; Utkal 2003, Delhi B.Sc. (Prog) III 2010, 2011; Purvanchal 2007, 12, Agra 2009]

[Delhi B.Sc. (Hons.) II, 2018, 2019]

Sol. The integral is proper if $m \geq 1$ and $n \geq 1$.

The number, 0 is a point of infinite discontinuity if $m < 1$ and the number, 1, is a point of infinite discontinuity if $n < 1$.

Let $$m < 1 \text{ and } n < 1$$

We take any number, say, 1/2 between 0 and 1 and examine the convergence of the two improper integrals

$$\int_0^{1/2} x^{m-1}(1-x)^{n-1}\,dx, \qquad \int_{1/2}^{1} x^{m-1}(1-x)^{n-1}\,dx$$

at 0, 1 respectively.

Convergence at 0. We write

$$f(x) = x^{m-1}(1-x)^{n-1} = (1-x)^{n-1}/x^{1-m}.$$

and take $\varphi(x) = 1/x^{1-m}$, so that

$$f(x)/\varphi(x) \to 1 \text{ as } x \to 0.$$

As $$\int_0^{1/2} \varphi(x)\,dx = \int_0^{1/2} \frac{1}{x^{1-m}}\,dx$$

is convergent at, 0, if and only if $1-m<1 \Leftrightarrow 0<m$, we deduce that the integral

$$\int_0^{1/2} x^{m-1}(1-x)^{n-1}\,dx$$

is convergent at 0, if and only if, $m > 0$.

Convergence at 1. We write

$$f(x) = x^{m-1}(1-x)^{n-1} = x^{m-1}/(1-x)^{1-n}$$

and take $\varphi(x) = 1/(1-x)^{1/n}$, so that

$$f(x)/\varphi(x) \to 1 \text{ as } x \to 1.$$

As $$\int_{1/2}^{1} \varphi(x)\,dx = \int_{1/2}^{1} \frac{1}{(1-x)^{1/n}}\,dx$$

is convergent, if and only if, $1-n<1 \Leftrightarrow 0<n$, we deduce that the integral

$$\int_{1/2}^{1} x^{m-1}(1-x)^{n-1}\,dx$$

converges if, and only if, $n > 0$.

Thus $\int_0^1 x^{m-1}(1-x)^{n-1}\,dx$ exists for positive values of m, n only. It is a function of two variables called Beta function denoted by $B(m, n)$ and defined for all positive values of m and n.

Ex.3 (*a*) Show that $$\int_0^{\pi/2} x^m \operatorname{cosec}^n x\,dx \qquad \ldots (i)$$

exists if and only if $n < (m + 1)$ *(Delhi Maths (Prog) 2007; Pune 2010, I.A.S. 2001)*

Sol. (*a*) Writing $f(x) = \dfrac{x^m}{(\sin x)^n} = \left(\dfrac{x}{\sin x}\right)^n \cdot x^{m-n} = \left(\dfrac{x}{\sin x}\right)^n \dfrac{1}{x^{n-m}}$

we see that $$\lim_{x\to 0} f(x) = \begin{cases} 0 & \text{if } m-n>0 \\ 1 & \text{if } m-n=0 \\ \infty & \text{if } m-n<0 \end{cases}$$

Thus (*i*) is a proper integral if $(m-n) \geq 0$; and improper if $(m-n) < 0$; 0 being the only point of infinite discontinuity of the integrand in this case,

Let $\quad (m-n) < 0 \quad$ so that $\quad (n-m) > 0$

Take $\quad \varphi(x) = 1/x^{n-m}$.

Then $\quad f(x)/\varphi(x) \to 1$ as $x \to 0$.

Also
$$\int_0^{\pi/2} \varphi(x)\, dx = \int_0^{\pi/2} \frac{1}{x^{n-m}}\, dx$$

converges, if only if, $(n-m) < 1 \Leftrightarrow n < (m+1)$.

Therefore the integral converges if and only if $n < (m+1)$, which also includes the case $n \leq m$ when the integral is proper.

3 (*b*) Show that $\displaystyle\int_0^{\pi/2} \frac{\sin^m x}{x^n}\, dx$ exists iff $n < m+1$

[Delhi B.Sc. (Hons) II 2003, 07]

Sol. Try Yourself.

Ex.4. *Examine the convergence of*

$$\int_0^1 x^{n-1} \log x\, dx.$$ **[Purvanchal 2006, 09; Meerut 2019]**

The integrand is negative in the interval] 0, 1] and we therefore, consider

$$\int_0^1 -x^{n-1} \log x\, dx = \int_0^1 x^{n-1} \log\left(\frac{1}{x}\right) dx.$$

The integrand is proper if $(n-1) > 0$, inasmuch as the integrand, in that case $\to 0$ as $x \to 0$ and accordingly, 0, is not a point of infinite discontinuity in this case.

Let $(n-1) \leq 0$ so that we have now to examine the convergence at 0. Let m be a positive number such that $(m+n-1) > 0$

We have
$$\lim_{x \to 0}\left[-x^{m+n-1} \log x\right] = 0,$$

so that ε being a given positive number.

$$-x^{m+n-1} \log x < \varepsilon \Rightarrow -x^{n-1} \log x < \varepsilon / x^m$$

for values of x, sufficiently near 0.

Now the integral of ε / x^m converges at 0 if and only if $m < 1$.

It is possible to choose a number $m < 1$ such that

$$(m+n-1) = [(m-1)+n] > 0$$

if and only if $n > 0$. Thus the integral converges if $n > 0$

When $n = 0$, the integrand becomes $x^{-1} \log x$. We have

$$\int_\varepsilon^1 x^{-1} \log x\, dx = -\frac{(\log \varepsilon)^2}{2} \text{ which} \to -\infty \text{ as} \to 0.$$

When $n < 0$, we have $x^{-1} < x^{n-1} \quad \forall\, x \in [0,1]$,
so that in this case also the integral does not converge.

Thus the given integral converges if and only if $n > 0$.

Ex. 5. Show that $\int_0^1 x^{n-1}\, e^{-x}\, dx$ is convergent if $x > 0$ ***(Rohilkhand 2010, 11; Kanpur 2008; Purvanchal 2010, 11; Delhi B.Sc. (Prog) III 2011)***

Sol. If $n \geq 1$, then the given integral I (say) is a proper integral because its integrand $f(x) = x^{n-1} e^{-x}$ is bounded in [0, 1]. So I is convergent when $n \geq 1$

If $0 < n < 1$, then the integrand of I is unbounded at $x = 0$. Take $\phi(x) = x^{n-1}$. Then, we have

$\lim_{x \to 0} \dfrac{f(x)}{\phi(x)} = \lim_{x \to 0} e^{-x} = 1$, which is finite and non-zero.

So by comparison test, $\int_0^1 f(x)\, dx$ and $\int_0^1 \phi(x)\, dx$ either both converge or both diverge.

Now $$\int_0^1 \phi(x)\, dx = \lim_{\varepsilon \to 0} \int_\varepsilon^1 x^{n-1}\, dx = \lim_{\varepsilon \to 0} \left[\frac{x^n}{n}\right]_\varepsilon^1 = \lim_{\epsilon \to 0} \left[\frac{1}{n} - \frac{\varepsilon^n}{n}\right] = \frac{1}{n},$$
which is finite number

$\therefore \int_0^1 \phi(x)\, dx$ is convergent and hence $\int_0^1 x^{n-1}\, e^{-x}\, dx$ is also convergent when $n > 0$.

Ex. 6. Test the convergence of $\int_0^2 \dfrac{\log x}{\sqrt{2-x}}\, dx$

(Agra 2012; Delhi Maths (H) 2004, Delhi B.Sc. Physics (H) 1989, Kumann 1999)

Sol. Let $f(x) = (\log x)/\sqrt{2-x}$. Clearly, f is unbounded at $x = 0$ and $x = 2$. Let $0 < c < 2$. Then, we have

$$I = \int_0^2 \frac{\log x}{\sqrt{2-x}} = \int_0^c \frac{\log x}{\sqrt{2-x}}\, dx + \int_c^2 \frac{\log x}{\sqrt{2-x}} = I_1 + I_2, \text{ say} \qquad \text{... (1)}$$

For I_1, 0 is the only point of infinite discontinuity in [0, c].

Now, $$\lim_{x \to 0} x^\mu f(x) = \lim_{x \to 0} x^\mu \frac{\log x}{\sqrt{2-x}} = 0, \text{ if } \mu > 0$$

Hence, if we choose $0 < \mu < 1$, then by μ-test, I_1 is convergent

Again, for I_2, 2 is the only point of infinite discontinuity in $[c, 2]$. Now choosing $\mu = 1/2$, we have

$$\lim_{x+2-0} (2-x)^\mu\, f(x) = \lim_{x+2-0} (2-x)^{1/2} \frac{\log x}{(2-x)^{1/2}} = \log 2$$

Hence by μ test, I_2 is convergent because $0 < \mu < 1$.

Hence, by (1), the given integral I is also convergent because it is sum of two convergent integrals I_1 and I_2.

Ex. 7. *Find the values of m and n for which the integral* $\int_0^1 e^{-mx}\, x^n\, dx$ converges

[*G.N.D.U. Amritsar 1997*]

Sol. For all values of m, when $n \geq 0$, the given integral is a proper integral and hence convergent. When $n < 0$ and m is any number, 0 is the only point of infinite discontinuity.

Let $$f(x) = e^{mx}\, x^n \text{ and take } \varphi(x) = x^n = 1/x^{-n}.$$

$$\therefore \quad \lim_{x \to 0+} \frac{f(x)}{\varphi(x)} = \lim_{x \to 0+} e^{-mx} = 1, \text{ which is non-zero and finite}$$

But $$\int_0^1 \varphi(x)\, dx = \int_0^1 \frac{dx}{x^{-n}} \text{ converges if } -n < 1, \text{ i.e., if } n > -1.$$

$\therefore$ By comparison test, $\int_0^1 f(x)\, dx$ also converges if $n > -1$.

Ex. 8. (*a*) Show that $\int_0^{\pi/2} \log \sin x \, dx$ is convergent and hence evaluate it

(*Rohilkhand 2012; Agra 2011; Meerut 2012; Delhi B.A. (Prog.) III 2013*)

(*b*) Test the convergence of $\int_0^1 \frac{\log x}{\sqrt{1-x^2}}\, dx.$

Sol. (*a*) Let $f(x) = \log \sin x$. Here 0 is the only point of infinite discontinuity of f in $[0, \pi/2]$. Since f is negative in $[0, \pi/2]$, we consider $-f$ for testing convergence of the integral.

Take $$\varphi(x) = 1/x^n, \text{ where } 0 < n < 1. \text{ Then}$$

$$\lim_{x \to 0+} \frac{-f(x)}{\varphi(x)} = \lim_{x \to 0+} \left(-\frac{\log \sin x}{1/x^n} \right) \qquad \left[\text{From } \frac{\infty}{\infty}\right]$$

$$= \lim_{x \to 0+} \frac{\cos x}{n / x^{n+1}} = \lim_{x \to 0+} \frac{x^n}{n} \cdot \frac{x}{\tan x} = 0$$

Also, $\int_0^{\pi/2} \varphi(x)\, dx$ convergent and hence by comparison test $\int_0^{\pi/2} \{-f(x)\} dx$ is convergent.

Consequently, $\int_0^{\pi/2} f(x)\, dx$, *i.e.*, $\int_0^{\pi/2} \log \sin dx$ is convergent

To evaluate the integral, let $$I = \int_0^{\pi/2} \log \sin dx \qquad \ldots (1)$$

From (1), $$I = \int_0^{\pi/2} \log \sin (\pi/2 - x) dx = \int_0^{\pi/2} \log \cos dx \qquad \ldots (2)$$

$$\left[\because \int_0^a f(x)\, dx = \int_0^a f(a-x)\, dx\right]$$

Adding (1) and (2), $$2I = \int_0^{\pi/2} \log \sin dx + \int_0^{\pi/2} \log \cos x \, dx$$

$$\therefore \quad 2I = \int_0^{\pi/2} \log (\sin x \cos x) = \int_0^{\pi/2} \log \{(\sin 2x)/2\}\, dx$$

or $$2I = \int_0^{\pi/2} \log \sin 2x \, dx - \int_0^{\pi/2} \log 2 \, dx = \int_0^{\pi/2} \log \sin 2x \, dx - \frac{\pi}{2} \log 2 \quad \text{... (3)}$$

Now, $$\int_0^{\pi/2} \log \sin 2x \, dx = \frac{1}{2} \int_0^{\pi/2} \log \sin t \, dt, \text{ putting } 2x = t$$

$$= \frac{1}{2} \times 2 \int_0^{\pi/2} \log \sin t \, dt = \int_0^{\pi/2} \log \sin dx = I, \text{ using (1)}$$

Then (3) reduces to $\quad 2I = I - (\pi/2) \log 2 \quad$ or $\quad I = -(\pi/2) \log 2.$

(*b*) Put $x = \sin y$ so that $dx = \cos y \, dy$. Then

$$\int_0^1 \frac{\log x}{\sqrt{1-x^2}} dx = \int_0^{\pi/2} \log \sin y \, dy,$$

Now proceed as in part (*a*).

Ex. 9. Test for convergence $\displaystyle\int_0^1 \frac{dx}{\sqrt{1-x^3}}$ ***(Delhi B.Sc. (Prog) III 2010; Calicut 2004)***

Sol. Hence $$\frac{1}{\sqrt{1-x^3}} = \frac{1}{(1-x)^{1/2}} \times \frac{1}{(1+x+x^2)^{1/2}},$$

where $1/(1+x+x^2)^{1/2}$ is a bounded function. If M be its upper bound then we have

$$1/(1-x^3)^{1/2} \le M/(1-x)^{1/2}$$

Since $\displaystyle\int_0^1 \frac{dx}{(1-x)^{1/2}}$ is convergent, so by comparison test, the given integral is also convergent.

EXERCISES

1. *Test the convergence of the following integrals :*

(*a*) $\displaystyle\int_0^{\pi/4} \frac{dx}{\sqrt{\tan x}}$ ***(Agra 2012; Himachal 2003; Kanpur 2001; Meerut 2020)***

(*b*) $\displaystyle\int_0^1 \frac{dx}{x^3(1+x^2)}$ *(Meerut 2003)* (*Delhi B.Sc. (Prog) III 2009*) (*c*) $\displaystyle\int_1^3 \frac{dx}{\sqrt{x(3-x)}}$ ***(Himachal 2002)***

(*d*) $\displaystyle\int_0^{\pi/2} \frac{\cos x}{x^n} dx$ *(Meerut 2003, 04)* (*e*) $\displaystyle\int_0^{\pi/2} \frac{\sin x}{x^n} dx$ ***[Delhi B.A. (Prog) III 2010; Delhi B.Sc. (Prog.) III 2012]***

(*f*) $\displaystyle\int_0^1 \frac{\operatorname{cosec} x}{x} dx, \quad \int_0^1 \frac{\sec x}{x} dx$ ***(Pune 2010; Meerut 2007)***

(*g*) $\displaystyle\int_0^1 \frac{\log x}{\sqrt{x}} dx$ ***(Delhi B.Sc. (Hons.) II, 2008; Kanpur 2004)***

(*h*) $\displaystyle\int_1^2 \frac{\sqrt{x}}{\log x} dx$

(*i*) $\displaystyle\int_0^{\pi/2} \frac{\log x}{x^a} dx, a < 1$ (*j*) $\displaystyle\int_0^{\pi} \frac{\sqrt{x}}{\sin x} dx$

(*k*) $\int_0^{\pi/2} \frac{\sin x}{x^p}\, dx$ (*Delhi 2009, Kanpur 2012*) (*l*) $\int_0^1 \frac{x^n}{1-x}\, dx$ (*m*) $\int_0^1 \frac{x^n}{1+x}\, dx$

2. Show that $\int_0^1 \frac{\operatorname{cosec} x}{x^n}$ is divergent if $n \geq 1$.

3. Show that $\int_0^{\pi/2} \sin^{m-1} x \cos^{n-1} x\, dx$ is convergent if $m > 0$, $n > 0$.

4. (*a*) Show that $\int_0^1 x^{m-1}(1-x)^{n-1} \log x$ is convergent if $m > 0$, $n > -1$. (*Himachal 2002*)

(*b*) Show that $\int_0^1 x^{m-1}(1-x)^{n-1} \log(1/x)\, dx$ is convergent if $m > 0$, $n > -1$.

(*Delhi Maths (H) 2002*)

5. Test for convergence :

(*a*) $\int_0^1 \frac{x^n \log x}{(1+x)^2}\, dx$ (*Himachal 2003; Delhi Math (H) 2006, 09*) (*b*) $\int_0^1 \frac{x^n \log x}{1+x^2}\, dx$

(*c*) $\int_0^{\pi/2} \frac{\log \sin x}{(\sin x)^n}\, dx, n < 1$ (*Kanpur 2010, 12*) (*d*) $\int_0^{\pi/2} \cos 2n\, x \, \log \sin x\, dx$

(*e*) $\int_0^1 \frac{(x^p + x^{-p}) \log(1+x)}{x}\, dx$ (*f*) $\int_0^1 \left(\log \frac{1}{x}\right)^m dx$ (*g*) $\int_0^{\pi/2} \frac{x^{a-1} dx}{1-x}$ (*Dehli Maths (H) 1997*)

(*h*) $\int_1^2 \frac{dx}{x \log x}$ (*i*) $\int_1^2 \frac{dx}{2x - x^2}$ (*j*) $\int_0^1 x^b \left(\log \frac{1}{x}\right)^b dx$ (*G.N.D.U. Amritsar 2010*)

ANSWERS

1. (*a*) convergent (*b*) divergent (*c*) convergent (*d*) convergent, if $0 < n < 1$
(*e*) convergent if $n < 2$ (*f*) both divergent (*g*) convergent
(*h*) divergent (*i*) convergent (*j*) divergent
(*k*) convergent if $p < 2$, divergent it $p \geq 2$. (*l*) divergent for all $n \in \boldsymbol{R}$
(*m*) convergent if $n > -1$, divergent if $n \leq -1$.

5. (*a*) convergent if $n > -1$ (*b*) convergent if $n > -1$
(*c*) Convergent (*d*) convergent for $\forall\, n \in \boldsymbol{R}$
(*e*) convergent if $-1 < p < 1$ (*f*) convergent if $-1 < m < 0$
(*h*) Divergent (*i*) Divergent

16.8 $f(x)$ NOT NECESSARILY POSITIVE

(General) Test for Convergence. We now obtain a general test for convergence at, *a*, of the infinite integral

$$\int_a^b f(x)\, dx. \quad \text{... (1)}$$

Cauchy's Test. *The necessary and sufficient condition for the convergence of the improper integral (1) at, a is that to every given* $\eta > 0$ *there corresponds* $\delta > 0$ *such that*

$$\left| \int_{a+\varepsilon_1}^{a+\varepsilon_2} f(x)\, dx \right| < \eta,$$

$\forall$ *positive numbers* $\varepsilon_1, \varepsilon_2$ *less than or equal to* δ. (*Delhi Maths (H) 2004*)

Proof. We write $\varphi(\varepsilon) = \int_{a+\varepsilon}^{b} f(x)\, dx.$

The necessary and sufficient condition for $\lim \varphi(\varepsilon)$ to exist finitely is that to every $\eta > 0$ there corresponds $\delta > 0$ such that $\forall\, 0 < \varepsilon_1, \varepsilon_2 < \delta$

$$\Rightarrow \quad |\varphi(\varepsilon_1) - \varphi(\varepsilon_2)| < \eta$$

$$\Leftrightarrow \quad \left| \int_{a+\varepsilon_1}^{b} f(x)\, dx - \int_{a+\varepsilon_2}^{b} f(x)\, dx \right| < \eta$$

$$\Leftrightarrow \quad \left| \int_{a+\varepsilon_1}^{a+\varepsilon_2} f(x)\, dx \right| < \eta.$$

16.9 ABSOLUTE AND CONDITIONALLY CONVERGENCE OF IMPROPER INTEGRALS OF THE SECOND KIND

Definitions. The improper integral $\int_a^b f(x)\, dx$ is said to be *aboslutely convergent* if $\int_a^b |f(x)|\, dx$ is convergent,

A convergent integral which is not absolutely convergent is said to be a *conditionally* c[illegible]*gent integral.*

Theorem. *Every absolutely convergent integral is convergent.*

(Delhi Maths (H) 2004)

i.e., $\int_a^b |f(x)|\, dx$ exists $\Rightarrow$ $\int_a^b f(x)\, dx$ exists

Proof. Since $\int_a^b |f(x)|\, dx$ exists, so by Cauchy's test for every $\eta > 0$, there corresponds $\delta > 0$ such that

$$\left| \int_{a+\epsilon_1}^{a+\epsilon_2} |f(x)|\, dx \right| < \eta, \text{ for } 0 < \varepsilon_1, \varepsilon_2 < \delta \qquad \dots (1)$$

Also, we know that (refer Art 13.11, Chapter 13) that

$$\left| \int_{a+\varepsilon_1}^{a+\varepsilon_2} f(x)\, dx \right| < \int_{a+\varepsilon_1}^{a+\varepsilon_2} |f(x)|\, dx \qquad \dots (2)$$

From (1) and (2), $\left| \int_{a+\varepsilon_1}^{a+\varepsilon_2} f(x)\, dx \right| < \eta$, for $0 < \varepsilon_1, \varepsilon_2 < \delta$

$$\Rightarrow \quad \int_a^b f(x)\, dx \text{ exists, by Cauchy's test}$$

Note. Since $|f(x)|$ is always positive, the comparison test can be used to examine the convergence of $\int_a^b |f(x)|\, dx$ *i.e.,* absolute convergence of $\int_a^b f(x)\, dx$

Note 2. The converse of the above theorem is not always true, *i.e.,* every convergent integral need not always be absolutely convergent

Note 3. The above theorem provides us with sufficient test for convergence.

EXAMPLES

Ex.1. Test the convergence of $\int_0^1 \frac{\sin(1/x)}{\sqrt{x}}\, dx.$ (*I.A.S. 2001*)

Sol. Let $f(x) = \sin(1/x)/\sqrt{x}$. Here there is no neighbourhood of the point 0, in which $f(x)$ constantly keeps the same sign.

Now $\forall\, x \in\,]0,1]$, we have

$$\left|\frac{\sin(1/x)}{\sqrt{x}}\right| = \frac{|(\sin 1/x)|}{\sqrt{x}} \le \frac{1}{\sqrt{x}}.$$

Also $\int_0^1 \frac{1}{\sqrt{x}}\, dx$ is convergent. It follows that

$$\int_0^1 \left|\frac{\sin(1/x)}{\sqrt{x}}\right| dx \text{ is convergent} \Rightarrow \int_0^1 \frac{\sin(1/x)}{\sqrt{x}}\, dx \text{ is absolutely convergent.}$$

Ex. 2. Show that $\int_0^1 \frac{\sin(1/x)}{x^n}\, dx,\ x > 0,$ converges absolutely, if $n < 1$.

Sol. Left as an exercise.

16.10 CONVERGENCE OF IMPROPER INTEGRALS OF THE SECOND KIND.

Infinite Range of Integration. Convergence at ∞

Let f be bounded and integrable in $[a, t]$ where t is a number $\ge a$ so that the proper integral

$$\int_a^t f(x)\, dx \text{ exists.}$$

We write $$\varphi(t) = \int_a^t f(x)\, dx$$

so that φ is a function with domain $[a, \infty[$. If φ tends to a finite limit as $t \to \infty$, we say that the improper integral

$$\int_a^\infty f(x)\, dx \qquad \text{... (1)}$$

exists or that it converges at ∞ and regard the symbol (1) as denoting the limit. Thus by definition

$$\int_a^\infty f(x)\, dx = \lim_{t \to \infty} \int_a^t f(x)\, dx, \qquad \text{...(2)}$$

provided the limit exists.

Convergence at $-\infty$

If f be bounded and integrable in $[t, b]$ where $t \le b$, and

$$\int_t^b f(x)\, dx \qquad \text{... (3)}$$

tends to a finite limit at $t \to -\infty$, we say that

$$\int_{-\infty}^a f(x)\, dx \qquad \text{...(4)}$$

converges or exists and regard the symbol (4) as denoting the limit of (3).

Let the Range of Integration be $]-\infty, \infty[$.

If c, is any number and

$$\int_{-\infty}^{c} f(x)\,dx, \quad \int_{c}^{\infty} f(x)\,dx,$$

both exist in accordance with the definitions already given, then we say that

$$\int_{-\infty}^{\infty} f(x)\,dx$$

exists and write

$$\int_{-\infty}^{\infty} f(x)\,dx = \int_{-\infty}^{c} f(x)\,dx + \int_{c}^{\infty} f(x)\,dx.$$

It is easy to show that the existence and the value of

$$\int_{-\infty}^{\infty} f(x)\,dx$$

is independent of the choice of c.

General Case

If an infinite range of integration includes a finite number of points of infinite discontinuity, then we arbitrarily consider an interval $[a,b]$ which embraces all the points of infinite discontinuity and examine the existence of the three improper integrals

$$\int_{-\infty}^{a} f(x)\,dx, \int_{a}^{b} f(x)\,dx, \int_{b}^{\infty} f(x)\,dx,$$

in accordance with the definitions given above, and in case they all exist, we say that $\int_{-\infty}^{\infty} f(x)\,dx$ exists and write

$$\int_{-\infty}^{\infty} f(x)\,dx = \int_{-\infty}^{a} f(x)\,dx + \int_{a}^{b} f(x)\,dx + \int_{b}^{\infty} f(x)\,dx.$$

Example. *Examine the existence of the following improper integrals and evaluate those which exist ?*

(*i*) $\int_{2}^{\infty} \frac{2x^2}{x^4-1}\,dx$ (*ii*) $\int_{-\infty}^{\infty} \frac{dx}{(1+x^2)^2}$ **(Kanpur 2003)**

(*iii*) $\int_{1}^{\infty} \frac{dx}{x}$ (*iv*) $\int_{1}^{\infty} \frac{dx}{x(1+x)}$ (*v*) $\int_{0}^{\infty} \frac{dx}{(1+x)\sqrt{x}}$

(*vi*) $\int_{-1}^{\infty} \frac{dx}{\sqrt{x(1-x)}}$ (*vii*) $\int_{-\infty}^{\infty} \frac{dx}{x(1+x^2)}$ (*viii*) $\int_{0}^{\infty} x^2 e^{-x}\,dx$

Solution. (*i*) By definition, we have

$$\int_{2}^{\infty} \frac{2x^2 dx}{x^4-1} = \lim_{t\to\infty} \int_{2}^{t} \frac{2x^2 dx}{x^4-1} = \lim_{t\to\infty} \int_{2}^{t} \frac{(x^2+1)+(x^2-1)}{(x^2+1)(x^2-1)}\,dx$$

$$= \lim_{t\to\infty} \int_{2}^{t} \left(\frac{1}{x^2-1} + \frac{1}{x^2+1} \right) dx = \lim_{t\to\infty} \left[\frac{1}{2} \log \frac{x-1}{x+1} + \tan^{-1} x \right]_{2}^{t}$$

$$= \lim_{t\to\infty} \left[\frac{1}{2} \log \frac{t-1}{t+1} + \tan^{-1} t - \frac{1}{2} \log \frac{1}{3} - \tan^{-1} 2 \right]$$

$$=\frac{1}{2}\lim_{t\to\infty}\log\frac{1-(1/t)}{1+(1/t)}+\frac{\pi}{2}+\frac{1}{2}\log 3-\tan^{-1}2$$

$$=\pi/2+(1/2)\times\log 3-\tan^{-1}2, \text{ which is finite}$$

Hence the given integral is convergent and its value is $\pi/2+(1/2)\times\log 3-\tan^{-1}2$.

(*ii*) By definition, we have

$$\int_{-\infty}^{\infty}\frac{dx}{(1+x^2)^2}=\int_{-\infty}^{0}\frac{dx}{(1+x^2)^2}+\int_{0}^{\infty}\frac{dx}{(1+x^2)^2}=2\int_{0}^{\infty}\frac{dx}{(1+x^2)^2}=\lim_{t\to\infty}\int_{0}^{t}\frac{dx}{(1+x^2)^2} \quad \text{...(1)}$$

Now, $\displaystyle\int\frac{dx}{(1+x^2)^2}=\int\frac{\sec^2\theta\, d\theta}{(1+\tan^2\theta)^2}=\frac{1}{2}\int(1+\cos 2\theta)\,d\theta$, putting $x=\tan\theta$

$$=\frac{1}{2}\theta+\frac{1}{4}\sin 2\theta=\frac{1}{2}\theta+\frac{1}{2}\cdot\frac{\tan\theta}{1+\tan^2\theta}=\frac{1}{2}\tan^{-1}x+\frac{x}{2(1+x^2)}$$

$\therefore$ (1) reduces to

$$\int_{-\infty}^{\infty}\frac{dx}{(1+x^2)^2}=2\lim_{t\to\infty}\left[\frac{1}{2}\tan^{-1}x+\frac{x}{2(1+x^2)}\right]_0^t=\lim_{t\to\infty}\left[\tan^{-1}t+\frac{t}{(1+t^2)}\right]$$

$$=\lim_{t\to\infty}\left[\tan^{-1}t+\frac{(1/t)}{1+(1/t^2)}\right]=\pi/2, \text{ which is finite}$$

Hence the given integral in convergent and its value is $=\pi/2$,

(*iii*) By definition, we have

$$\int_1^{\infty}\frac{dx}{x}dx=\lim_{t\to\infty}\int_1^t\frac{dx}{x}=\lim_{t\to\infty}[\log x]_1^t=\lim_{t\to\infty}\log t=\infty$$

Since the limit does not exist finitely, so the given integral is divergent, *i.e.*, it does not exist.

(*iv*) – (*viii*) - Left as exercises for the reader.

16.11 CONVERGENCE AT ∞. THE INTEGRAND BEING POSITIVE

Theorem. *Let $f(x)$ be positive in $[a,t]$. The necessary and sufficient condition for* $\int_a^{\infty}f(x)\,dx$ *to be convergent is that there exists a positive number k, independent of t, such that* $\int_a^t f(x)\,dx\le k\ \forall\, t\ge a.$ *(Delhi Maths (H) 1999)*

Proof. Let $$\phi(t)=\int_a^t f(x)\,dx$$

Since f is positive in $[a,t]$, the function $\varphi(t)$ monotonically increases with t and will therefore tend to a finite limit if and only if it is bounded above, *i.e.*, there exists a positive number k, independent of t, such that $\varphi(t)\le k\ \ \forall\, t\ge a$. Hence, we have

$$\int_a^t f(x)<k\ \ \forall\, t\ge a$$

Note. If no such number k exists, then the monotonic increasing function $\varphi(t)$ is unbounded above and so it tends to ∞ as $t\to\infty$. Hence $\int_a^{\infty} f(x)\,dx$ diverges to ∞

16.12 COMPARISON OF TWO INTEGRALS (COMPARISON TEST) (Meerut 2012)

If f and φ are two functions such that $0 < f(x) \le \varphi(x)\ \forall\, x \in [a, \infty[$. Then,

(*i*) $\int_a^\infty \varphi(x)\, dx$ is convergent $\Rightarrow$ $\int_a^\infty f(x)$ is convergent

(*ii*) $\int_a^\infty f(x)\, dx$ is divergent $\Rightarrow$ $\int_a^\infty \varphi(x)$ is divergent

Proof. Since f and φ are both positive and $f(x) \le \varphi(x)\ \forall \in [a, t]$, we have

$$\int_a^t f(x)\, dx \le \int_a^t \varphi(x)\, dx \qquad \text{... (1)}$$

(*i*) Let $\int_a^\infty \varphi(x)\, dx$ be convergent. Then there exists a positive number k such that

$$\int_a^t \varphi(x)\, dx < k \ \ \forall\, t \ge a \qquad \text{... (2)}$$

From (1) and (2), $\int_a^t f(x)\, dx < k\ \forall\, t \ge a$

$\Rightarrow$ $\int_a^\infty f(x)\, dx$ is convergent

(*ii*) Let $\int_a^\infty f(x)\, dx$ be divergent. Then $\int_a^t f(x)\, dx$ is unbounded above. Hence (1) shows that $\int_a^t \varphi(x)\, dx$ is unbounded above and therefore $\int_a^\infty \varphi(x)\, dx$ is divergent.

Practical Comparison test. *If f and φ are positive on* $[a, \infty[$ and $f(x)/\varphi(x) \to l$ when $x \to \infty$, then

(*i*) *if l is non-zero finite, then the two integrals.*

$\int_a^\infty f(x)\, dx$ and $\int_a^\infty \varphi(x)\, dx$ converge or diverge together.

(*ii*) if $l = 0$ and $\int_a^\infty \varphi(x)\, dx$ converges, then $\int_a^\infty f(x)\, dx$ converges

(*iii*) if $l = \infty$ and $\int_a^\infty \varphi(x)\, dx$ diverges, then $\int_a^\infty f(x)\, dx$ diverges

Proof. Left as an exercise for the reader.

16.13 A USEFUL COMPARISON INTEGRAL

To prove that

$$\int_a^\infty \frac{dx}{x^n}, (a > 0) \qquad \text{... (1)}$$

converges if and only if $n > 1$

[Kanpur 2009; Agra 2007; 11; Meerut 2005, 07, 11; Himanchal 2008; Delhi B.A. (Prog) III 2010]

Proof: We have, if $n \ne 1$.

$$\int_a^t \frac{dx}{x^n} = \frac{1}{1-n}\left[\frac{1}{t^{n-1}} - \frac{1}{a^{n-1}}\right]$$

so that when $t \to \infty$, the integral tends to $1/(n-1)\, a^{n-1}$ or ∞ according as $n > 1$ or $n < 1$.

For $n = 1$, we have $\int_a^t \frac{dx}{x} = \log \frac{t}{a}$ which $\to \infty$ as $t \to \infty$.

Hence the result.

Note. Adopting (1) as the comparison integral and employing the test of Art. 16.11, we may now easily obtain the following practical tests for convergence at ∞.

Test I. *If $f(x)$ is positive in $[a, \infty[$ then the integral converges, if there exists a positive number n greater than 1 and a fixed positive number k such that*

$$f(x) \le k/x^n \;\forall\; x \ge a.$$

Again, the integral does not converge, if there exists a positive number $n \le 1$ and a fixed positive number k such that

$$f(x) \ge k/x^n \;\;\forall\; x \ge a.$$

Test II (μ-test). *If $f(x)$ be bounded and integrable in $[a, \infty[$, which $a > 0$. Then*

(i) if there exists a number $\mu > 1$ such that $\lim_{x \to \infty} x^{\mu} f(x)$ is finite, then $\int_a^{\infty} f(x)\, dx$ is convergent.

(ii) if there exists a number $\mu \le 1$, such that $\lim_{x \to \infty} x^{\mu} f(x)$ exists and is non-zero, then $\int_a^{\infty} f(x)\, dx$ is divergent. The same still holds if $\lim_{x \to \infty} x^{\mu} f(x)$ is $+\infty$ or $-\infty$.

Note. Let m be the highest power of x in the denominater and n be the highest power of x in the numberator of the integrand, then choose $\mu = m - n$.

EXAMPLES

1. Examine the convergence of

(i) $\int_0^{\infty} \frac{x\, dx}{(1+x)^3}$ *(Himachal 2003; Kanpur 2006)* ***[Purvanchel 2011; Rohilkhand 2011]***

(ii) $\int_1^{\infty} \frac{dx}{(1+x)\sqrt{x}}$

(iii) $\int_1^{\infty} \frac{dx}{x^{1/3}(1+x)^{1/2}}$ *(Meerut 2009)*

(iv) $\int_0^{\infty} \frac{\sin^2 x}{x^2}\, dx$. ***(Kanpur 2014)***

Sol. (*i*) Let $f(x) = x/(1+x)^3$. We take $\varphi(x) = x/x^3 = 1/x^2$,

so that
$$\lim_{x \to \infty} \frac{f(x)}{\varphi(x)} = \lim_{x \to \infty} \frac{x^3}{(1+x)^3} = 1.$$

Thus the two integrals
$$\int_1^{\infty} \frac{x\, dx}{(1+x)^3} \quad \text{and} \quad \int_1^{\infty} \frac{1}{x^2}\, dx$$

have identical behaviour for convergence at ∞.

By Art. 16.13 the latter integral is convergent. Accordingly, the given integral is also convergent.

(*ii*) Let $f(x) = 1/(1+x)\sqrt{x}$. We take $\varphi(x) = 1/x\sqrt{x} = 1/x^{3/2}$

so that
$$\lim_{x \to \infty} \frac{f(x)}{\varphi(x)} = \lim_{x \to \infty} \frac{x}{1+x} = \lim_{x \to \infty} \frac{1}{(1/x)+1} = 1$$

Also $\int_1^\infty \frac{1}{x^{3/2}}\,dx$ is convergent. Thus

$$\int_1^\infty f(x)\,dx \text{ i.e., } \int_1^\infty \frac{dx}{(1+x)\sqrt{x}} \text{ is convergent.}$$

(*iii*) Let $f(x) = 1/x^{1/3}(1+x)^{1/2}$. We take $\varphi(x) = 1/x^{1/3}\,x^{1/2} = 1/x^{5/6}$.

We have $\lim \frac{f(x)}{\varphi(x)} = 1$, when $x \to \infty$, and $\int_1^\infty \varphi(x)$ is not convergent.

So $\int_1^\infty f(x)\,dx$ *i.e.,* $\int_1^\infty \frac{dx}{x^{1/3}(1+x)^{1/2}}$ is divergent

(*iv*) Taking $a > 0$, we have

$$\int_0^\infty \frac{\sin^2 x}{x^2}\,dx = \int_0^a \frac{\sin^2 x}{x^2}\,dx + \int_a^\infty \frac{\sin^2 x}{x^2}\,dx \qquad \text{... (1)}$$

Here $\int_0^\infty \frac{\sin^2 x}{x^2}\,dx$ is a proper integral for $(\sin^2 x)/x^2 \to 1$ as $x \to 0$, so that, 0 is not a point of infinite discontinuity. So $\int_0^a \frac{\sin^2 x}{x^2}\,dx$ is convergent.

We now test $\int_a^\infty \frac{\sin^2 x}{x^2}\,dx$ for its convergence.

Let $f(x) = (\sin^2 x)/x^2$ and take $\varphi(x) = 1/x^2$. Then clearly $(\sin^2 x)/x^2 \le 1/x^2$ as $\sin^2 x \le 1$.

Again, we know that $\int_a^\infty \frac{dx}{x^2}$ is convergent. So by the comparison test, $\int_a^\infty \frac{\sin^2 x}{x^2}\,dx$ is also convergent.

Since, both the integrals on R.H.S. of (1) are convergent, it follows that the integral on L.H.S. of (1), *i.e.,* the given integral is also convergent.

Ex. 2. *Show that the Gamma function* $\int_0^\infty x^{n-1}e^{-x}\,dx$ *is convergent if, and only, it* $n > 0$.

(Delhi B.Sc. (Hons.) II, 2017; Delhi B.A. (Prog) III 2011, 12; Himachal 2019; Agra 2009; Purvanchal 2006; M.D. U 2001; Kanpur 2006, 2019)

Sol. Let $f(x) = x^{n-1}e^{-x}$.

Now 0, is a point of infinite discontinuity of the function f, if $n < 1$. Thus we have to examine the convergence at ∞ as well as at 0.

We consider any positive number > 0, say 1, and examine the convergence of the two integrals

$$\int_0^1 f(x)\,dx \qquad \text{and} \qquad \int_1^\infty f(x)\,dx,$$

at 0, and ∞, respectively.

(*i*) **Convergence at 0**. Let $a < 1$. We take

$$\phi(x) = \frac{1}{x^{1-n}} \qquad \text{so that} \qquad \lim_{x\to 0} \frac{f(x)}{\phi(x)} = 1.$$

Also $$\int_0^1 \frac{1}{x^{1-n}}\,dx$$

converges, if and only if $(1-n) < 1 \Leftrightarrow 0 < n$.

(*ii*) **Convergence at** ∞. Now given n we have for sufficiently large values of x

$$e^x > x^{n+1} \Leftrightarrow x^{n-1}\,e^{-x} < 1/x^2.$$

But $$\int_1^\infty \frac{1}{x^2}\,dx \text{ converges.}$$

Therefore $$\int_1^\infty x^{n-1}\,e^{-x}dx \text{ also converges } \forall\, n.$$

Thus $$\int_0^\infty x^{n-1}\,e^{-x}\,dx \text{ converges if, and only if, } n > 0.$$

Ex. 3. Show that $\int_0^\infty \left(\frac{1}{x}-\frac{1}{\sinh x}\right)\frac{dx}{x}$ is convergent.

(*Agra 2006, 11; Delhi B.Sc. (H) Maths 2008, Gerhwal 1997, Himanchal Pradesh 2003*)

Sol. The point 0, is not a point of infinite discontinuity of the integrand inasmuch as the integrand $\to 1/6$ as $x \to 0$. We have, therefore, to examine convergence at ∞ only. We have

$$\left(\frac{1}{x}-\frac{1}{\sinh x}\right)\frac{1}{x} = \frac{e^x - e^{-x} - 2x}{x^2\,(e^x - e^{-x})} < \frac{e^x}{x^2\,(e^x - e^{-x})},\ \forall\, x > 0.$$

Thus, $$\left(\frac{1}{x}-\frac{1}{\sinh x}\right)\frac{1}{x} < \frac{e^{2x}}{e^{2x}-1}\cdot\frac{1}{x^2}.$$

Since $e^{2x}/(e^{2x}-1) \to 1$ as $x \to \infty$, there exists a number k such that

$$e^{2x}/(e^{2x}-1) < 3/2,\ \forall\, x \geq k.$$

Thus, $$\forall\, x \geq k,\ \left(\frac{1}{x}-\frac{1}{\sinh x}\right)\frac{1}{x} < \frac{3}{2}\cdot\frac{1}{x^2}.$$

Since $\int_k^\infty \left(\frac{1}{x}-\frac{1}{\sinh x}\right)\frac{1}{x}\,dx$ is convergent, the given integral is also convergent.

Ex. 4. *Examine the convergence of* $\int_1^\infty \frac{\log x}{x^2}\,dx$ (*Delhi Maths (H) 2003*)

Sol. Here $f(x) = (\log x)/x^2$. Taking $\mu = 3/2$, we have

$$\lim_{x\to\infty} x^\mu f(x) = \lim_{x\to\infty} \frac{x^{3/2}\log x}{x^2} = \lim_{x\to\infty}\frac{\log x}{x^{1/2}} \quad \left[\text{From } \frac{\infty}{\infty}\right]$$

$$= \lim_{x\to\infty} \frac{1/x}{(1/2)\,x^{-1/2}} = \lim_{x\to\infty}\frac{2}{\sqrt{x}} = 0, \text{ which is finite.}$$

Since $\mu = 3/2 > 1$, it follows by the μ- test that the given integral in convergent.

Ex. 5. *Test the convergence of* $\int_0^\infty \frac{x^{2m}}{1+x^{2n}}\,dx$, *m and n being positive integers.*

(*Rohilkhand 2012; Purvanchal 2007; Garhwal 1999*)

Sol. Let $a > 0$. Then, we have

$$\int_0^{\infty} \frac{x^{2m}}{1+x^{2n}}\,dx = \int_0^{a} \frac{x^{2m}}{1+x^{2n}}\,dx + \int_a^{\infty} \frac{x^{2m}}{1+x^{2n}}\,dx = I_1 + I_2, \text{ say} \quad \text{... (1)}$$

Since I_1 is a proper integral, so it is convergent. We now test I_2, by μ-test. Take $\mu = 2n - 2m$. Then, we have

$$\lim_{x\to\infty} x^{\mu} f(x) = \lim_{x\to\infty} x^{2n-2m} \frac{x^{2m}}{1+x^{2n}} = \lim_{x\to\infty} \frac{1}{1+(1/x^{2n})} = 1,$$

which is finite. So, by μ-test I_2 is convergent if $\mu > 1$ *i.e.*, $2n - 2m > 1$, which is possible if $n > m$ because m and n are given to be positive integers. Again, by μ test I_2 is divergent if $\mu \le 1$, *i.e.*, $2n - 2m < 1$, which is possible if $n \le m$.

Ex. 6. Show that $\int_0^1 \left(\frac{1}{1+x} - e^{-x}\right)\frac{dx}{x}$ is convergent. (*Delhi Maths (H) 2001, 04, 05*)

Sol. Let
$$f(x) = \left(\frac{1}{1+x} - e^{-x}\right)\frac{1}{x} = \frac{e^x - 1 - x}{x(1+x)e^x} > 0, \ \forall\, x > 0$$

Now,
$$\int_0^{\infty} f(x)\,dx = \int_0^1 f(x)\,dx + \int_1^{\infty} f(x)\,dx = I_1 + I_2, \text{ say} \quad \text{... (1)}$$

Here 0 is not a point of infinite discontinuity for the integrand tends to 0 as $x \to 0$. Hence $\int_0^1 f(x)\,dx$ is a proper integral and so it is convergent.

To test I_2 for convergence at ∞, we take $\varphi(x) = 1/x^2$.

Then,
$$\lim_{x\to\infty} \frac{f(x)}{\varphi(x)} = \lim_{x\to\infty} \frac{e^x - (1+x)}{e^x}\cdot\frac{x}{1+x} = \lim_{x\to\infty}\left(1 - \frac{1+x}{e^x}\right) \times \lim_{x\to\infty} \frac{1}{1+1/x}$$
$$= \left(1 - \lim_{x\to\infty} \frac{1+x}{e^x}\right) \times 1 = 1 - 0 = 1$$

Also, $\int_1^{\infty} \varphi(x)\,dx = \int_1^{\infty} \frac{1}{x^2}\,dx$ is convergent. So by the comparison test, I_2 is also convergent.

Since I_1 and I_2 are both convergent, so from (1), the integral on L.H.S., *i.e.*, the given integral is convergent.

EXERCISES

1. *Discuss the convergence of the following.*

(*a*) $\int_0^{\infty} \frac{x^{n-1}}{1+x}\,dx$ [*Agra 2007, 11*] (*a*) $\int_1^{\infty} \frac{x^{a-1}\log x}{1+x}\,dx$ (*c*) $\int_1^{\infty} \frac{\tan^{-1} x}{x^2}\,dx$

(Delhi Maths (H) 2009, Pune 2010)

(*d*) $\int_0^{\infty} \frac{x \tan^{-1} x}{(1+x^4)^{1/3}}\,dx$ (*e*) $\int_{e^2}^{\infty} \frac{dx}{x \log\log x}$ (*f*) $\int_1^{\infty} x^m e^{-nx}\,dx$

(*g*) $\int_0^{\infty} \frac{\cos mx}{x^2 + a^2}\,dx$, $\int_0^{\infty} \frac{\cos x}{1+x^2}\,dx$ (*Rohilkhand 2011*)

(h) $\int_0^{\infty} e^{-\lambda x}\, x^{a-1}\, dx$ *(Delhi B.A. (Prog) III 2010; Delhi Maths (H) 1995, 2004)*

(i) $\int_1^{\infty} \frac{dx}{x\sqrt{x^2+1}}$ *(Lucknow 2010)*

2. *Prove that* $\int_0^{\infty} \frac{x \log x}{(1+x^2)^2}\, dx$ *converges to 0.* *[Delhi B.Sc.(Hons) III 2011]*

3. *Discuss the convergence of the following :*

(a) $\int_0^{\infty} \frac{x^m (1+x^n)}{1+x^p}\, dx$ *(Delhi Maths Prog 2007)* (b) $\int_0^{\infty} x^m (\log x)^n\, dx$

(c) $\int_0^{\infty} \frac{x^{m-1}+x^{n-1}}{1-x}\, dx$ (d) $\int_0^{\infty} \frac{x^a}{(1+x)^b\,[1+(\log x)^2]}\, dx$

4. *Show that the improper integral* $\int_0^{\infty} \log(1+2\,\text{sech}\, x)\, dx$. *converges.*

[Hint : Since $\log(1+x) < x \;\; \forall\, x > 0$

so $\log(1+2\,\text{sec h}\, x) < 2\,\text{sec h}\, x = \frac{4}{e^x+e^{-x}} < \frac{4}{e^x} = 4e^{-x}$ **]**

5. *Show that* $\int_0^{\infty} \frac{\cosh bt}{\cosh at}\, dt \quad a > 0,\, b > 0$

converges if, and only if $b < a$.

[Hint : $b < a \quad \frac{\cosh bt}{\cosh at} \quad \frac{e^{bt}+e^{-bt}}{e^{at}+e^{-at}} \quad \frac{e^{bt}+e^{bt}}{e^{at}} \quad 2e^{-(a-b)t}$

$$b > a \Rightarrow \frac{\cosh bt}{\cosh at} = \frac{e^{bt}+e^{-bt}}{e^{at}+e^{-at}} > \frac{e^{bt}}{e^{at}+e^{at}} = \frac{1}{2}e^{(b-a)t}$$ **]**

6. *Show that* $\int_0^{\infty} \frac{\sinh bx}{\sinh ax}\, dx,\; a > 0,\, b > 0,$

converges if and only if, $a > b$.

[Hint : $a > b \Rightarrow \frac{\sinh bx}{\sinh ax} = \frac{e^{bx}-e^{-bx}}{e^{ax}-e^{-ax}} < \frac{e^{bx}}{e^{ax}-1};$

$$a < b \Rightarrow \frac{\sinh bx}{\sinh ax} = \frac{e^{bx}-e^{-bx}}{e^{ax}-e^{-ax}} < \frac{e^{bx}-1}{e^{ax}}.$$ **]**

7. *Discuss the convergence of the improper integral*

$$\int_0^{\infty} \left(\frac{1}{e^x-1} - \frac{1}{x} + \frac{1}{2}\right)\frac{e^{-kx}}{x}\, dx.$$

8. *Show that the integrals* $\int_{-\infty}^{\infty} e^{-x^2}\, dx$ *[Delhi BA (Prog) 2009]*

$\int_0^{\infty} e^{-x^2}\, dx$ and $\int_{-\infty}^{\infty} e^{-(x-a/x)^2}\, dx$ converge. ***(Meerut 2012; Rohilkhand 2010)***

9. *Discuss the convergence of* $\displaystyle\int_0^{\infty} \frac{x^{\alpha}+x^{-\alpha}}{1+x^2}$

10. *Show that* $\displaystyle\int_0^{\infty} x^{2n+1}\, e^{-x^2}\, dx$ *is convergent for all positive integral values of n.*

11. *If* $a > 1$, *show that* $\displaystyle\int_a^{\infty} \frac{dx}{x(\log x)^{n+1}}$ *is convergent if* $n > 0$ *and divergent if* $n \le 0$.

ANSWERS

1. (*a*) convergent if $0 < n < 1$ (*b*) convergent if $a < 0$
(*c*) convergent (*d*) divergent
(*e*) divergent (*f*) convergent if $n > 0$
(*g*) both convergent (*h*) convergent if $p > -1,\ q > -1$ (*i*) convergent

5. convergent.

16.14 GENERAL TEST. CONVERGENCE AT ∞, THE INTEGRAND BEING NOT NECESSARILY POSITIVE

Cauchy's test for convergence. *The necessary and sufficient condition for the convergence of* $\int_a^{\infty} f(x)\, dx$ *at* ∞ *is that corresponding to every* $\eta > 0$ *there exists a number k, such that*

$$\left|\int_{t_1}^{t_2} f(x)\, dx\right| < \eta,\ \forall\, t_1, t_2 \ge k.$$ (*Delhi B.Sc. (H) 2000, 03*)

Proof. We write $\varphi(t) = \int_a^t f(x)\, dx.$

The necessary and sufficient condition for $\lim \varphi(x)$ to exist finitely is that to every positive number, η, there corresponds a number k, such that $\forall\, t_1, t_2 > k$

$$|\varphi(t_2) - \varphi(t_2)| < \eta \iff \left|\int_a^{t_2} f(x)\, dx - \int_a^{t_1} f(x)\, dx\right| < \eta \iff \left|\int_{t_1}^{t_2} f(x)\, dx\right| < \eta.$$

16.15 ABSOLUTE AND CONDITIONALLY CONVERGENCE OF IMPROPER INTEGRALS OF THE SECOND KIND [DELHI B.Sc. (HONS.) II 2011]

Definitions. The improper integral $\int_a^{\infty} f(x)\, dx$ is said to be *absolutely convergent* if $\int_a^{\infty} |f(x)|\, dx$ is convergent.

A convergent integral which is not absolutely convergent is said to be a *conditionally convergent integral.*

Theorem. *Every absolutely convergtent integral is convergent.*

i.e., $\int_a^{\infty} |f(x)|\, dx$ converges $\Rightarrow$ $\int_a^{\infty} f(x)\, dx$ converges

Proof. Let η be a positive number. Since $\int_a^{\infty} |f(x)|\, dx$ converges, there exists a number k such that $\forall\, t_1, t_2 \ge k$

$$\left|\int_{t_1}^{t_2} |f(x)|\, dx\right| < \eta \qquad \text{... (1)}$$

Also, $$\left|\int_{t_1}^{t_2} f(x)\, dx\right| \le \left|\int_{t_1}^{t_2} |f(x)|\, dx\right|. \qquad \text{... (2)}$$

From (1) and (2) the result follows.

Note. The converse of the above theorem is not true. Every convergent integral need not always be absolutely convergent.

16.16 TEST FOR THE ABSOLUTE CONVERGENCE OF THE INTEGRAL OF A PRODUCT

Let φ *be bounded in* $[a, \infty[$ *and integrable in* $[a, t]\ \forall\ t \ge a$.

Let $\int_a^\infty f(x)\, dx$ *converge absolutely at* ∞. *Then* $\int_a^\infty f(x)\, \varphi(x)\, dx$, *is absolutely onvergent.*

(Delhi Maths (H) 2003)

Proof. Since φ is bounded in $[a, \infty[$, there exists a number A such that

$$|\varphi(x)| \le A\ \forall\ x \ge a. \qquad \text{... (1)}$$

Since the improper integral $\int_a^\infty |f(x)|\, dx$ with positive integrand is convergent, there exists a number B, such that

$$\int_a^t |f(x)|\, dx \le B\ \forall\ t \ge a. \qquad \text{... (2)}$$

We have from (1) and (2)

$|f(x)\, \varphi(x)| \le A|f(x)|\ \forall\ x \ge a.$

$$\Rightarrow \quad \int_a^t |f(x)\, \varphi(x)|\, dx \le A\int_a^t |f(x)|\, dx \le AB\ \forall\ t \ge a.$$

$$\Rightarrow \quad \int_a^t |f(x)\, \varphi(x)|\, dx \text{ is bounded above } \forall\ t \ge a.$$

$$\Rightarrow \quad \int_a^\infty |f(x)\, \varphi(x)|\, dx \text{ is convergent}$$

$$\Rightarrow \quad \int_a^\infty f(x)\, \varphi(x)\, dx \text{ is absolutely convergent.}$$

Ex. Discuss the convergence of the following integrals :

(*i*) $\int_1^\infty \frac{\sin x}{x^2}\, dx.$ (*ii*) $\int_0^\infty e^{-ax} \cos x\, dx.$

16.17. ABEL'S TEST

(Meerut 2010, 2012

If $\int_a^\infty f(x)\, dx$ is convergent at ∞ and $\varphi(x)$ is bounded and monotonic for $x \ge a$, then $\int_a^\infty f(x)\, \varphi(x)\, dx$ is convergent at ∞. *(Delhi Maths (H) 2001, 04, 08; Lucknow 1995 Kanpur 2012; Rohilkhand 2008, 11; Purvanchal 2007, 10, 11; M.D.U. Rohtak 2001*

Proof. The bounded function φ, which is monotonic in $[a, \infty[$, is integrable in $[a, t]\ \forall\ t \ge a$. Applying the second mean value theorem, we have

$$\int_{t_1}^{t_2} f(x)\,\varphi(x)\,dx = \varphi(t_1)\int_{t_1}^{\xi} f(x)\,dx + \varphi(t_2)\int_{\xi}^{t_2} f(x)\,dx, \qquad \text{... (1)}$$

where $a < t_1 \le \xi \le t_2$.

Let η be a positive number.

Since φ is bounded in $[a, \infty[$, there exists a positive number A such that

$$|\varphi(x)| \le a \ \forall\ x \ge a.$$

In particular it follows that

$$|\varphi(t_1)| \le A, |\varphi(t_2)| \le A. \qquad \text{... (2)}$$

Also, since $\int_a^{\infty} f(x)\,dx$ is convergent, there exists by Art. 16.14, a number k such that $\forall\ t_1, t_2 > k$

$$\left|\int_{t_1}^{t_2} f(x)\,dx\right| < \frac{\eta}{2A}.$$

We now suppose that t_1, t_2 are numbers $\ge k$ so that ξ which lies between t_1 and t_2 is also $\ge k$

$$\therefore \quad \left|\int_{t_1}^{\xi} f(x)\,dx\right| < \frac{t_1}{2A} \quad \text{and} \quad \left|\int_{\xi}^{t_2} f(x)\,dx\right| < \frac{\eta}{2A} \qquad \text{...(3)}$$

From (1), (2) and (3), we deduce that there exists a number k such that $\forall\ t_1, t_2 \ge k$

$$\left|\int_{t_1}^{t_2} f(x)\,\varphi(x)\,dx\right| \le |\varphi(t_1)|\left|\int_{t_1}^{\xi} f(x)\,dx\right| + \varphi(t_2)\left|\int_{\xi}^{t_2} f(x)\,dx\right|$$

$$< A.\frac{\eta}{2A} + A.\frac{\eta}{2A} = \eta,$$

where η is a positive number assigned arbitrarily.

Hence $\int_a^{\infty} f(x)\,\varphi(x)\,dx$ converges at ∞.

16.18 DIRICHLET TEST. (*Meerut 2007, 10, 2012*)

Let φ be bounded and monotonic in $[a, \infty[$ and let $\varphi(x) \to 0$, when $x \to \infty$. Also let $\int_a^t f(x)\,dx$ be bounded $\forall\ t \ge a$. Then $\int_a^{\infty} f(x)\,\varphi(x)\,dx$ converges. (*Purvanchal 2012*)

(*K.U. Kurukshetra 2000, M.D.U. Rohtak 1997, Delhi Maths (H) 1999, 2000, 01, 02, 05, 07*)

Proof. The function φ, which is monotonic in $[a, \infty[$, is integrable in $[a, t]\ \forall\ t \ge a$. Applying the second mean value theorem, we have

$$\int_{t_1}^{t_2} f(x)\,\varphi(x)\,dx = \varphi(t_1)\int_{t_1}^{\xi} f(x)\,dx + \varphi(t_2)\int_{\xi}^{t_2} f(x)\,dx, \qquad \text{... (1)}$$

where $a < t_1 \le \xi \le t_2$.

Since $\int_a^t f(x)\,dx$ is bounded $\forall\, t \geq a$ there exists a number A such that

$$\left|\int_a^t f(x)\,dx\right| \leq A\ \forall\, t \geq a, \qquad \text{... (2)}$$

$$\therefore \quad \left|\int_{t_1}^{\xi} f(x)\,dx\right| = \left|\int_a^{\xi} f(x)\,dx - \int_a^{t_1} f(x)\,dx\right|$$

$$\leq \left|\int_a^{\xi} f(x)\,dx\right| + \left|\int_a^{t_1} f(x)\,dx\right|$$

$$\leq A + A = 2A.$$

Similarly $$\left|\int_{\xi}^{t_2} f(x)\,dx\right| \leq 2A. \qquad \text{... (3)}$$

Let η be a positive number.

Since $\varphi(x) \to 0$ as $x \to \infty$, there exists a number k such that

$$|\varphi(x)| < \eta/4A\ \forall\, x \geq k. \qquad \text{... (4)}$$

We now suppose that t_2, t_2 are any two numbers $\leq k$, so that from (4) $\forall\, t_1, t_2 \geq k$

$$|\varphi(t_2)| < \eta/4A, \qquad |\varphi(t_2)| < n/4A \qquad \text{... (5)}$$

From (1), (2), (3) and (5), we deduce that there exists a number k such that $\forall\, t_1, t_2 \geq k$

$$\left|\int_{t_1}^{t_2} f(x)\,\varphi(x)\,dx\right| \leq |\varphi(t_1)|\left|\int_{t_1}^{\xi} f(x)\,dx\right| + |\varphi(t_2)|\left|\int_{\xi}^{t_2} f(x)\,dx\right|$$

$$\leq (\eta/4A)\times 2A + (\eta/4A)\times 2A = \eta$$

where η is a positive number arbitratily assigned.

$$\Rightarrow \quad \int_a^{\infty} f(x)\,\varphi(x)\,dx \text{ converges at } \infty.$$

EXAMPLES

Ex. 1. *Show that* $\int_0^{\infty} \sin x^2\,dx$ *is convergent* (*Delhi B.Sc. (Prog) III 2009*)

(*Agra 2012; Delhi Maths (H) 2002. Garhwal 1994. M.D.U. Rohtak 1997*)

Sol. We have $$\int_0^{\infty} \sin x^2\,dx = \int_0^1 \sin x^2\,dx + \int_1^{\infty} \sin x^2\,dx \qquad \text{... (1)}$$

But $\int_0^1 \sin x^2\,dx$ is a proper integral and so it is convergent.

We now test $\int_1^{\infty} \sin x^2\,dx$ for convergence at ∞. We have

$$\int_1^{\infty} \sin x^2\,dx = \int_1^{\infty} (2x \sin x^2)\cdot \frac{1}{2x}\,dx \qquad \text{... (2)}$$

Let $$f(x) = 2x \sin x^2 \text{ and } \varphi(x) = 1/2x. \quad \text{... (3)}$$

Now, $\varphi(x)$ is monotonic and $\to 0$ as $x \to \infty$. Also,

$$\left|\int_1^t f(x)\,dx\right| = \left|\int_1^t 2x \sin x^2 dx\right| = \left|\left[-\cos x^2\right]_1^t\right| = \left|\cos 1 - \cos t^2\right| \le |\cos 1| + |\cos t^2| \le 2,$$

showing that $\int_1^t f(x)\,dx$ is bounded for all $t \ge 1$.

$\therefore$ By Dirichlet's test, $\int_1^\infty f(x)\,\varphi(x)\,dx$, *i.e.*, $\int_1^\infty \sin x^2 dx$ is convergent. Hence, from (1), the given integral is convergent.

Ex. 2. (*a*) Show that $\int_0^\infty \frac{\sin x}{x}\,dx$ converges but not absolutely. ***(Delhi B.Sc. (Prog) II 2009)***

(Calicut 2003; Delhi B.Sc. (Hons.) II 2019; Himachal 2004, 05; Purvanchel 2010, 12; G.N.D.U. Amritsar 2010)

(*b*) Show that $\int_0^\infty \frac{\sin x}{x}\,dx$ is conditionally convergent.

(Calicut 2003, I.A.S. 1993, Delhi Maths (H) 1998, 2005, 06; Himachal Pradesh 1999)

Sol. (*a*) We have $$\int_0^\infty \frac{\sin x}{x}\,dx = \int_0^1 \frac{\sin x}{x}\,dx + \int_1^\infty \frac{\sin x}{x}\,dx \quad \text{... (1)}$$

Consider $\int_0^1 \frac{\sin x}{x}\,dx$. Here, $\lim_{x\to 0} \frac{\sin x}{x} = 1 = \text{finite}$, showing thereby that (sin *x*)/*x* is bounded in (0, 1). Hence, $\int_0^1 \frac{\sin x}{x}\,dx$ is a proper integral and therefore it is convergent.

We now test $\int_1^\infty \sin x \cdot \frac{1}{x}\,dx$ for its convergence.

Let $f(x) = \sin x$ and $\varphi(x) = 1/x$.

Now, $\varphi(x)$ is monotonic and $\to 0$ as $x \to \infty$. Also,

$$\left|\int_1^t f(x)dx\right| = \left|\int_1^t \sin x\,dx\right| = \left|\left[-\cos x\right]_1^t\right| = \left|\cos 1 - \cos t\right| \le \left|\cos 1\right| + \left|\cos t\right| \le 2,$$

showing that $\int_1^t f(x)\,dx$ is bounded $\forall\, t \ge 1$. Hence, by Dirichlet's test, $\int_1^\infty f(x)\,\varphi(x)\,dx$, *i.e.*, $\int_1^\infty \frac{\sin x}{x}\,dx$ is convergent. Therefore, from (1), the given integral is convergent.

In order to show that the given integral is not absolutely convergent, we must show that $\int_0^\infty \frac{|\sin x|}{x}\,dx$ is not convergent. Consider the proper integral $\int_0^{n\pi} \frac{|\sin x|}{x}\,dx$ where *n* is a positive integer. We have

$$\int_0^{n\pi} \frac{|\sin x|}{x}\,dx = \sum_{r=1}^{n} \int_{(r-1)\pi}^{r\pi} \frac{|\sin x|}{x}\,dx.$$

We put $x=(r-1)\pi+y$ so that y varies in $[0, \pi]$. We have

$$\left|\sin\left[(r-1)x+y\right]\right| = \left|(-1)^{r-1}\sin y\right| = \sin y.$$

$$\therefore \quad \int_{(r-1)\pi}^{r\pi} \frac{|\sin x|}{x}\,dx = \int_0^{\pi} \frac{\sin y}{(r-1)\pi+y}\,dy.$$

Since, $r\pi$ is the maximum value of $[(r-1)\pi+y]$ when $y\in[0,\pi]$, we have

$$\int_0^{\pi} \frac{\sin y}{(r-1)\pi+y}\,dy \geq \frac{1}{r\pi}\int_0^{\pi}\sin y\,dy = \frac{2}{r\pi}$$

$$\therefore \quad \int_0^{n\pi} \frac{|\sin x|}{x}\,dx \geq \sum_1^n \frac{2}{r\pi} = \frac{2}{\pi}\sum_1^n \frac{1}{r}.$$

Since $\sum_1^n \frac{1}{r} \to \infty$ as $n\to\infty$, we see that

$$\int_0^{n\pi} \frac{|\sin x|}{x}\,dx \to \infty \text{ as } n\to\infty.$$

Let, now, t be a real number. There exists a positive integer n such that $n\pi \leq t < (n+1)\pi$.

We have $$\int_0^t \frac{|\sin x|}{x}\,dx \geq \int_0^{n\pi} \frac{|\sin x|}{x}\,dx.$$

Let $t\to\infty$ so that n also $\to\infty$. Thus we see that

$$\int_0^t \frac{|\sin x|}{x}\,dx \to \infty \Rightarrow \int_0^{\infty} \frac{|\sin x|}{x}\,dx \text{ does not converge.}$$

Hence $\int_0^{\infty} \frac{\sin x}{x}\,dx$ is not absolutely convergent

(b) From part (*a*), we see that the given integral is convergent while $\int_0^{\infty}\left|\frac{\sin x}{x}\right|dx$ is not absolutely convergent. Hence, by definition, the given integral is conditionally convergent.

Ex. 3. Show that $\int_0^{\infty} e^{-ax}\frac{\sin x}{x}\,dx,\ a\geq 0$ is convergent.

(Kanpur 2009, 11; Delhi Maths (H) 2001, 04, I.A.S. 1998, Kurukshetra 2000)

Sol. Let $f(x)=(\sin x)/x$ and $\varphi(x)=e^{-ax}, a\geq 0$. Proceed as in first part of Ex. 2, to show that $\int_0^{\infty} f(x)\,dx$, *i.e.*, $\int_0^{\infty}\frac{\sin x}{x}\,dx$ is convergent. Also, $\varphi(x)$ is a bounded and monotonically decreasing function for $x>0$.

So by Abel's test, $\int_0^{\infty} f(x)\,\varphi(x)\,dx$ *i.e.*, $\int_0^{\infty} e^{-ax}\frac{\sin x}{x}\,dx$ is convergent.

Ex. 4. *If* $a\neq 0$, *then prove that* $\int_0^{\infty} e^{-a^2x^2}\sin bx\,dx$ *is absolutely convergent.*

Sol. We have $$\int_0^{\infty}\left|e^{-a^2x^2}\sin bx\right|dx = \int_0^1\left|e^{-a^2x^2}\sin bx\right|dx + \int_1^{\infty}\left|e^{-a^2x^2}\sin bx\right|dx \quad ...(1)$$

We have, $$\int_0^t \left|e^{-a^2x^2} \sin bx\right| dx \le \int_0^t e^{-a^2x^2} dx \quad \forall\ t>0$$

Also, for $a \ne 0, \lim\limits_{x\to 0} \dfrac{e^{-a^2x^2}}{1/x^2} = 0.$ and $\int_1^\infty \dfrac{dx}{x^2}$ is convergent

$\therefore$ $\int_1^\infty e^{-a^2x^2}\, dx$ is convergent and so $\int_1^\infty e^{-a^2x^2} \sin bx\, dx$ converges absolutely.

Here $\int_0^1 e^{-a^2x^2} \sin bx\, dx$ is a proper integral and so it is absolutely convergent, since f is R-integrable $\Rightarrow |f|$ is also R -integrable. Hence, from (1), it follows that the given integral is absolutely convergent, if $a \ne 0$.

Ex. 5. Show that the improper integral $\int_0^\infty \dfrac{x\, dx}{1+x^6 \sin^2 x}$ is convergent.

Sol. The integral is positive for positive values of ξ but the test obtained in Art. 16.11 does not enable us to establish the convergence. In order to show that the integral converges, we proceed as follows. Consider the proper integral

$$\int_0^{n\pi} \frac{x\, dx}{1+x^6 \sin^2 x},$$

and write
$$\int_0^{n\pi} \frac{x\, dx}{1+x^6 \sin^2 x} = \sum_{r=1}^{n} \int_{(r-1)\pi}^{r\pi} \frac{x\, dx}{1+x^6 \sin^2 x}.$$

Now $\forall\ x \in [(r-1)\pi, r\pi]$, we have

$$\frac{x}{1+x^6 \sin^2 x} \le \frac{r\pi}{1+(r-1)^6 \pi^6 \sin^2 x}.$$

$$\therefore \quad \int_{(r-1)\pi}^{r\pi} \frac{x\, dx}{1+x^6 \sin^2 x} \le \int_{(r-1)\pi}^{r\pi} \frac{r\pi\, dx}{1+(r-1)^6 \pi^6 \sin^2 x} = a_r, \text{ say}$$

Putting $x = (r-1)\pi + y$, we see that

$$a_r = \int_0^{\pi} \frac{r\pi\, dy}{1+(r-1)^6 \pi^6 \sin^2 y} = 2\int_0^{\pi/2} \frac{r\pi\, dy}{1+(r-1)^6\pi^6 \sin^2 y}$$

If, now, A is a positive number, we have

$$\int_0^{\pi/2} \frac{dy}{1+A\sin^2 y} = \int_0^{\pi/2} \frac{\operatorname{cosec}^2 y\, dy}{A+1+\cot^2 y} = -\frac{1}{\sqrt{(A+1)}}\left[\tan^{-1} \frac{\cot y}{\sqrt{(A+1)}}\right]_0^{\pi/2} = \frac{\pi}{2} \cdot \frac{1}{\sqrt{(A+1)}}$$

$$\therefore \quad a_r = 2r\pi \times \left(\frac{\pi}{2}\right) \times \frac{1}{\sqrt{[(r-1)^6 \pi^6 + 1]}} = \frac{r\pi^2}{\sqrt{[(r-1)^6\pi^6 + 1]}}$$

$$\therefore \quad \int_{(r-1)\pi}^{r\pi} \frac{x}{1+x^6 \sin^2 x}\, dx \le \frac{r\pi^2}{\sqrt{[1+(r-1)^6 \pi^6]}} < \frac{r}{(r-1)^3} \cdot \frac{1}{\pi}, \text{ where } r \ne 1.$$

Since $$\frac{r}{(r-1)^3}=\frac{1}{(r-1)^2}+\frac{1}{(r-1)^3},$$

the two infinite series

$$\sum_{r=2}^{\infty}\frac{1}{(r-1)^2} \quad \text{and} \quad \sum_{r=2}^{\infty}\frac{1}{(r-1)^3}$$

are both convergent. It follows that

$$\sum_{r=2}^{\infty}\frac{r}{(r-1)^3}\cdot\frac{1}{\pi} \text{ is a convergent series.}$$

$$\therefore \quad \int_0^{n\pi}\frac{x\,dx}{1+x^6\sin^2 x}\to \text{a finite limit as } n\to\infty.$$

Hence the given integral converges

Ex. 6. Show that $\int_0^{\infty}\frac{x\,dx}{1+x^4\sin^2 x}$ is divergent. (*I.A.S. 2003*)

Sol. We write

$$\int_0^{n\pi}\frac{x\,dx}{1+x^4\sin^2 x}=\sum_{r=1}^{n}\int_{(r-1)\pi}^{r\pi}\frac{x\,dx}{1+x^4\sin^2 x}$$

Now $\forall\, x\in[(r-1)\pi,\, r\pi]$, we have

$$\frac{(r-1)\pi}{1+(r\pi)^4\sin^2 x}\le\frac{x}{1+x^4\sin^2 x}$$

$$\therefore \quad \int_{(r-1)\pi}^{r\pi}\frac{(r-1)\pi}{1+(r\pi)^4\sin^2 x}\,dx\le\int_{(r-1)\pi}^{r\pi}\frac{x}{1+x^4\sin^2 x}\,dx.$$

Now, $$\int_{(r-1)\pi}^{r\pi}\frac{(r-1)\pi}{1+(r\pi)^4\sin^2 x}\,dx=(r-1)\pi\int_0^{\pi}\frac{dy}{1+(r\pi)^4\sin^2 y}$$

$$=2(r-1)\pi\int_0^{\pi/2}\frac{dy}{1+(r\pi)^4\sin^2 y}$$

$$=2(r-1)\pi\times\frac{\pi}{2}\times\frac{1}{\sqrt{[1+(r\pi)^4]}}=\frac{(r-1)\pi^2}{\sqrt{[1+(r\pi)^4]}}.$$

The infinite series $\Sigma\frac{(r-1)\pi^2}{\sqrt{[1+(r\pi)^4]}}$ diverges, as we may see on comparison with the series $\Sigma(1/r)$. Hence the given integral does not converge.

EXERCISE

1. *Examine the convergence of*

(*a*) $\int_0^{\infty}\frac{\sin x}{\sqrt{x}}\,dx$ (*Delhi Math (H) 2004*)

(*b*) $\int_1^{\infty}\frac{\sin x}{\sqrt{x}}\,dx$ (*G.N.D.U. Amritsar 1996; Himachal 2002*)

(c) $\int_0^\infty e^{-a^2x^2} \cos bx \, dx$ (Kumaun 1998)

(d) $\int_0^\infty \sin x \, dx$ (Delhi Maths (H) 2004, Himachal Pradesh 2003)

(e) $\int_0^\infty \cos^2 x \, dx$ (Delhi B.Sc. Physics (H) 1998)

(f) $\int_0^\infty \frac{\cos x}{\sqrt{x+x^2}}$ (Delhi Maths (H) 1996, 2008)

(g) $\int_1^\infty \sin x^p \, dx, p > 1$ (h) $\int_1^\infty \frac{\sin x}{x^p} \, dx, p > 0$ (Meerut 2010)

(i) $\int_1^\infty (1-e^{-x}) \frac{\cos x}{x} \, dx$ **(Rohilkhand 2009)** (j) $\int_a^\infty \frac{\sin x \log x}{x} \, dx$ (Himanchal 2004

(k) $\int_1^\infty f(x) \, dx$, where $f(x) = \begin{cases} 1/x^2, & \text{if } x \text{ is rational} \\ -(1/x^2), & \text{if } x \text{ is irrational} \end{cases}$

(l) $\int_0^\infty \frac{\sin x^m}{x^n} \, dx$ (Delhi Maths (H) 2002, 05)

2. Show that $\int_0^\infty \frac{\sin mx}{a^2+x^2} \, dx$ converges absolutely. (Kanpur 2002)

3. Show that $\int_1^\infty \frac{\sin x}{x^p} \, dx$ converges absolutely if $p > 1$

4. (a) Show that $\int_2^\infty \frac{\cos x}{\log x} \, dx$ is conditionally convergent.

(Delhi Maths (H) 1999; Himachal 2003, 04)

(b) Show that $\int_2^\infty \frac{\sin x}{\log x} \, dx$ converges, but not absolutely. (Delhi Maths (H) 2007)

5. Examine the convergence of

(a) $\int_0^\infty \frac{\cos ax \cos bx}{x} \, dx$ (b) $\int_0^\infty \frac{\sin(x+x^2)}{x^n} \, dx$

(c) $\int_0^\infty \frac{x^p \sin^2 x}{1+x^2} \, dx$ (Delhi Maths (H) 2002)

(d) $\int_0^\infty \frac{x^m \cos ax}{1+x^n} \, dx$ (Delhi Maths (H) 2002, 07)

6. Examine the convergence of

(a) $\int_0^\infty \frac{x \, dx}{1+x^4 \cos^2 x}$ (b) $\int_0^\infty \frac{dx}{1+x^4 \cos^2 x}$

7. If $\alpha \geq 0, \beta \geq 0$, then show that $\int_0^{\infty} \frac{x^{\beta} dx}{1+x^{\alpha} \sin^2 x}$ converges for $\alpha > 2(\beta+1)$ and diverges for $\alpha \leq 2(\beta+1)$ *(I.A.S. 1992)*

8. Show that $\int_0^{\infty} \frac{dx}{1+x^2 \sin^2 x}$ is divergent **(I.A.S. 2003)**

9. Show that $\int_a^{\infty} \frac{\cos \alpha x - \cos \beta x}{x} dx$ is convergent if $a > 0$ **(Rohilkhand 2011)**

10. Examine whether the integral $\int_1^{\infty} \frac{\sin x}{1+x^2} dx$ is convergent. **(Pune 2010)**

ANSWERS

1. (*a*) convergent (*b*) convergent (*c*) convergent (*d*) convergent (*e*) convergetnt (*f*) convergent (*g*) convergent (*h*) convergent (*i*) convergent (*j*) convergent (*k*) convergent

(*m*) convergent for $1 - m < n < 1 + m$ and $\forall$ *m*, + ve or – ve

5. (*a*) Divergent (*b*) convergent if $-1 < n < 2$ (*c*) convergent for $-3 < p < 1$

(*d*) convergent for $n > 0, -1 < m < n$, or $n < 0, 0 > m > n - 1$

6. (*a*) convergent (*b*) convergent **10.** convergent

OBJECTIVE QUESTIONS

Choose the correct answers for each question: $\frac{\cos x}{(x^2+x)^{1/2}}dx$

1. If $m > 0, n > 0$ then integral $\int_0^1 x^{m-1}(1-x)^{n-1} dx$ is

(*a*) convergent (*b*) divergent (*c*) oscillatory (*d*) None of these *(Agra 2009)*

2. The integral $\int_0^{\pi/2} \sin^{2m-1} x \cos^{2n-1} x \, dx$ is convergent iff

(*a*) $m > 0, n > 0$ (*b*) $m < 0, n < 0$ (*c*) $m \neq n$ (*d*) None of these *(Meerut 2011)*

3. $\int_{-\infty}^{0} e^x dx$ is

(*a*) convergent (*b*) divergent (*c*) oscillatory (*d*) None of these *(Kanpur 2001)*

4. $\int_0^1 x^{m-1}(1-x)^{n-1} dx$ is convergent when

(*a*) $m > 0$ (*b*) $n > 0$ (*c*) $m > 0, n > 0$ (*d*) $m > 1, n > 1$

5. $\int_0^{\infty} x^{n-1} e^{-x} dx$ is convergent when

(*a*) $n > 0$ (*b*) $n = 0$ (*c*) $n < 0$ (*d*) None of these *(Meerut 2011)*

6. $\int_a^{\infty} \frac{\sin x}{\sqrt{x}} dx$, where $a > 0$ is :

(*a*) convergent (*b*) divergent (*c*) oscillatory (*d*) proper.

7. $\int_0^\infty \frac{\sin mx}{a^2+x^2}\,dx$:

(*a*) converges but not absolutely (*b*) absolutely convergent
(*c*) divergent (*d*) None of these

8. Integral $\int_0^{\pi/4} \frac{dx}{\sqrt{\tan x}}$:

(*a*) convergent but not absolutely (*b*) absolutely convergent
(*c*) divergent (*d*) proper

9. If $\int_a^\infty |f(x)|\,dx$ is convergent then integral $\int_a^\infty f(x)\,dx$

(*a*) converges (*b*) diverges
(*c*) oscillates (*d*) none of these **(Meerut 2013)**

10. If, when $x \to \infty$. $\lim [f(x)\, x^n]$ exists finitely, the limit being neither 0 nor infinite, then the integral $\int_a^\infty \frac{dx}{x^n}$; $(a > 0)$ converges if

(*a*) $n < 1$ (*b*) $n > 1$ (*c*) $n = 1$ (*d*) $n = 0$ **(Meerut 2013)**

11. $\int_\pi^\infty \frac{\sin^2 x}{x^2}\,dx$ is :

(*a*) convergent (*b*) divergent (*c*) oscillatory (*d*) proper.

12. Integral $\int_0^\infty \frac{x^{2n}}{1+x^{2m}}\,dx$ is convergent if

(*a*) $n < m$ (*b*) $n > m$ (*c*) $n = m$ (*d*) None of these

ANSWERS

1. (*a*) **2.** (*a*) **3.** (*a*) **4.** (c) **5.** (*a*) **6.** (*a*)
7. (*b*) **8.** (*a*) **9.** (*a*) **10.** (*b*) **11.** (*a*) **12.** (*b*)

MISCELLENEOUS PROBLEMS ON CHAPTER 16

1. Is it true or false : $\int_0^1 \frac{dx}{1-x}$ is convergent **[Ans.** True] [*Meerut 2005*]

2. Test the convergence of

(*i*) $\int_0^\infty \frac{dx}{(x+1)^2 x^{1/2} (x-1)^{1/3}}$ [*Delhi Maths (H) 2006*]

(*ii*) $\int_0^\infty \frac{dx}{1+x^2}$ **[Ans.** Conv.] [*Meerut 2006*]

3. Write the statement of Abel's test for the convergence of first and second kind of integral of the product of two functions. [*Meerut 2005*]

4. State are prove Abel's test for the improper integral of a product of two functions. [*Meerut 2010; Delhi Maths (H) 2006*]

5. Test the convergence of the following

(*i*) $\int_{-\infty}^{\infty} \frac{dx}{1+x^2}$ **(Kanpur 2009)** (*ii*) $\int_0^1 \frac{dx}{x^{1/2}(1-x)^{1/3}}$ **[Purvanchal 2006]**

(*iii*) $\int_0^{\infty} \frac{dx}{x^{1/3}(1+x^{1/2})}$ **[Kanpur 2007]** (*iv*) $\int_0^{\infty} x^3 e^{-x^2}\, dx$ **[Delhi B.Sc. (Hons.) II 2007]**

(*v*) $\int_0^{2a} \frac{dx}{(x-a)^2}$ **[(Kanpur 2011]** (*vi*) $\int_0^{\pi/2} \frac{\log \sin x}{(\sin x)^n} dx,\ n<1$ **[Kanpur 2010]**

[Ans. (*i*) Conv. (*ii*) Conv. (*iii*) Div. (*iv*) Conv. (*v*) Div. (*vi*) conv.]

6. *Show that the integral* $\int_a^{\infty} \frac{dx}{x^n}$ *converges when* $n > 1$ *and diverges when* $n \le 1$.

[Meerut 2007]

7. *State Dirichlet test for the convergence of integral of product of two functions and hence test of convergence of* $\int_a^{\infty} \frac{\sin x}{x^n} dx,\ x > 0$ **[Meerut 2007]**

8. Examine the convergence of $\int_0^{\pi/2} \frac{\sin^3 x}{x^{28}} dx$ **[Delhi B.Sc. (Hons.) II 2008]**

[Hint. Ans. Convergent. Use Ex. 3(b) page 16.10. Note that here n = 28 and m = 3]

9. Discuss the convergence of $\int_0^{\infty} e^{-\lambda x} x^{a-1} dx,\ \lambda > 0,\ a < 0$ **[Delhi B.Sc. (Prog) II 2008]**

10. Show that $\int_0^{\infty} \frac{\sin x}{x^4} dx$ is absolutely convergent. **[Agra 2008]**

11. Discuss the convergence of (*i*) $\int_0^{\infty} \cos (x^2 + x) dx$ **[Delhi B.Sc. (Hons.) II 2009]**

(*ii*) $\int_1^{\infty} \frac{\sin x}{x^p} dx$, where $p > 0$ **[Delhi B.Sc. (Hons.) II 2009]**

[Ans (*i*) Convergent (*ii*) Convergent]

12. Show that $\int_1^2 \frac{x dx}{\sqrt{x-1}}$ is convergent **[Delhi B.Sc. (Prog) II 2010]**

13. Examine the convergence of $\int_0^1 x^p \left(\log \frac{1}{x}\right)^q dx$ **[Kanpur 2011]**

14. Show that $\int_1^{\infty} \frac{\cos t}{t} dt$ is not absolutely convergent. **[Delhi B.Sc. (Hons) II 2011]**

15. Show that $\int_0^{\infty} \frac{\cos x}{(x^2 + x)^{1/2}} dx$ converges **[Delhi B.Sc. (Hons) III 2011]**

16. Show that $\int_{\pi}^{\infty} \frac{\sin x}{x^2} dx$ converges **[Delhi B.Sc. (Hons.) II 2018]**

[Bundelkhand 2011; Rohilkhand 2012; Purvanchal 2010]

Chapter

17

Power Series

17.1 Introduction

In this chapter we propose to study the theory of power series which is very useful tool in the study of analysis, Theory of power series is employed to deal with "integration in series" in study of differential equations. The study of power series is very useful due to two reasons, namely (*i*) power series behave fairly well and (*ii*) a set of values of the variable power series converge to functions, which are infinitely differentiable. **[*Kanpur 2007*]**

17.2 POWER SERIES

A series of the form $$\sum_{n=0}^{\infty} a_n x^n \qquad \text{... (1)}$$

is known as real infinite power series where a_0, a_1, ... a_n, are real coefficients free from x, and x is the real variable. More generally $\sum_{n=0}^{\infty} a_n (x-x_0)^n$ is taken to represent a general power series. Since with a shift of origin to x_0 *i.e.*, with change of variable $x - x_0$ to x this precisely reduces to the form (1), hence without any loss of generally our studies shall be confined to the form (1).

For the sake of brevity we shall write $\Sigma a_n x^n$ instead of $\sum_{n=0}^{\infty} a_n x^n$

17.3 Some important facts aboutr the power series $\Sigma a_n x^n$.

(*i*) For all values of the coefficients, every power series converges for $x = 0$. Hence if a power series converges for no value other than $x = 0$, we say that the given power series is *nowhere convergent*. For example, the power series $\Sigma n^n x^n$ is nowhere convergent.

(*ii*) If a given series converges for all values of x, we say that the given power series is everywhere convergent. For example, the power series $\Sigma(x^n / n!)$ is everywhere convergent.

(*iii*) If the given power series converges for some value of x and diverges for other values of x, then the set of all values of x for which it is convergent is known as its *region of convergence*.

For example, the series $\Sigma(x/r)^n$ converges for $|x| < r$ and diverges for $|x| \geq r$, since for every fixed x, the given power series is reduced to a geometric series of common ratio x/r, which converges if $|x/r| < 1$ and diverges if $|x/r| \geq 1$

17.4 Radius of convergence and interval of convergence

[*Delhi B.Sc. (Prog.) II, 2016; Kanpur 2005, 07; Meerut 2008, 2012; Delhi B.Sc. (Hons.) II 2011*]

If a given power series does not converge everywhere or nowhere, then a definite positive number R exists such that the given power series converges (indeed absolutely) for every $|x| < R$ and diverges for every $|x| > R$. Such a number R is known as the *radius of convergence* and the interval $]-R, R[$, the *interval of convergence,* of the given power series.

17.5 Formulas for determining the radius of convergence

Theorem 1. *If the power series* $\Sigma a_n x^n$ *is such that* $a_n \neq 0$ *for all n and* $\lim_{n\to\infty} |a_{n+1}/a_n| = 1/R$, *then* $\Sigma a_n x^n$ *is convergent (indeed absolutely) for* $|x| < R$ *and divergent for* $|x| > R$.

[Himanchal 2009; Delhi Maths (H) 2006]

Proof. Let $u_n = a_n x^n$ so that $u_{n+1} = a_{n+1} x^{n+1}$. Then, we have

$$\lim_{n\to\infty}\left|\frac{u_{n+1}}{u_n}\right| = \lim_{n\to\infty}\left|\frac{a_{n+1}\ x}{a_n}\right| = |x| \lim_{n\to\infty}\left|\frac{a_{n+1}}{a_n}\right| = \frac{|x|}{R} \qquad \text{... (1)}$$

$\therefore$ By D' Alembert's ratio test, $\Sigma a_n x^n$ converges absolutely if $|x|/R < 1$, *i.e.*, $|x| < R$. Also, $\Sigma a_n x^n$ diverges if $|x| > R$.

Theorem II. *If the power series* $\Sigma a_n x^n$ *is such that* $a_n \neq 0$ *for all n and* $\lim_{n\to\infty} |a_n|^{1/n} = 1/R$, *then* $\Sigma a_n x^n$ *is convergent (indeed absolutely) for* $|x| < R$ *and divergent for* $|x| > R$.

Proof. According to Cauchy's second theorem on limits, if $<|a_n|>$ is a sequence of positive constants, then

$$\lim_{n\to\infty} |a_n|^{1/n} = \lim_{n\to\infty}\left|\frac{a_{n+1}}{a_n}\right|, \qquad \text{... (1)}$$

provided the limit on the right side of (1) exists, wherther finite or infinite. Also given that

$$\lim_{n\to\infty} |a_n|^{1/n} = 1/R \qquad \text{... (2)}$$

$$\therefore \quad \text{(1) and (2)} \quad \Rightarrow \quad \lim_{n\to\infty} |a_{n+1}/a_n| = 1/R \qquad \text{... (3)}$$

Using (3), the result of the theorem follows from theorem I.

In view of the above discussion, the radius of convergence R of the power series $\Sigma a_n x^n$ can be determined as follows :

$$R = \lim_{n\to\infty}\left|\frac{a_n}{a_{n+1}}\right| \quad \text{or} \quad 1/R = \lim_{n\to\infty} |a_n|^{1/n} \qquad \text{... (*)}$$

$$R = 0 \quad \text{if} \quad \lim_{n\to\infty}\left|\frac{a_n}{a_{n+1}}\right| = 0 \quad \text{or} \quad \lim_{n\to\infty} |a_n|^{1/n} = \infty$$

$$R = \infty, \quad \text{if} \quad \lim_{n\to\infty}\left|\frac{a_n}{a_{n+1}}\right| = \infty \quad \text{or} \quad \lim_{n\to\infty} |a_n|^{1/n} = 0$$

Cauchy-Hadamard theorem

The radius of convergence of the power series $\Sigma a_n x^n$ *is R, where* $R = 1\Big/\lim_{n\to\infty} |a_n|^{1/n}$

Proof. Refer theorem II of Art. 17.5.

Note 1. Every power series converges absolutely within its interval of convergence.

Note 2. Observe that formula (*) is derived with the supposition of existence of the finite limit $\lim |a_n / a_{n+1}|$, that is, with the supposition that the power series $\Sigma a_n x^n$ contains all powers of x. Indeed for the power series $\Sigma\{(2x)^{2n+1}/(2n+1)\}$, the coefficients of even powers of x are equal to zero, $a_{2m} = 0$ and hence $\lim_{n\to\infty}(a_{2n+1}/a_{2n}) = \infty$ and $\lim_{n\to\infty}(a_{2n+2}/a_{2n+1}) = 0$. This shows that we cannot apply the formula (*) to the given power series. However, a direct application of D' Alembert's ratio test leads to the desired result:

Here, let $u_n = (2x)^{2n+1}/(2n+1)$ so that $u_{n+1} = (2x)^{2n+3}/(2n+3)$

$$\therefore \quad \lim_{n\to\infty}\left|\frac{u_{n+1}}{u_n}\right| = \lim_{n\to\infty}\left|\frac{(2x)^{2n+3}}{(2x)^{2n+1}} \times \frac{2n+1}{2n+3}\right| = 4|x|^2 \lim_{n\to\infty}\left|\frac{2+1/n}{2+3/n}\right| = 4|x|^2$$

Therefore, by D' Alembert's ratio test, the given power series converges absolutely if

$4|x|^2 < 1$ or $|x|^2 < 1/4$ or $|x| < 1/2$.

Hence, the radius of conversence of the given series = R = $1/\sqrt{2}$.

Note. If the given power series is given in general form $\Sigma a_n (x - x_0)^n$, then formula (*) can be used to find the radius of convergence R. In such a case, we say that the given power series converges if $|x - x_0| < R$ and diverges if $|x - x_0| > R$. The interval of convergence is given by $]x_0 - R, x_0 + R[$.

17.5A. Solved examples based on art. 17.5

Ex. 1. *Find the radius of convergence of the following series*

(*i*) $\dfrac{x}{2} + \dfrac{1\cdot 3}{2\cdot 5}x^2 + \dfrac{1\cdot 3\cdot 5}{2\cdot 5\cdot 8}x^3 + \ldots$ **[Himachal 2014; Kanpur 2005, 10; Delhi Maths (H) 2006]**

(*ii*) $1 + \dfrac{a\cdot b}{1\cdot c} + \dfrac{a(a+1)\,b(b+1)}{1\cdot 2c\,(c+1)} + \ldots.$ **[Delhi Maths (H) 2003]**

Sol. (*i*) Let the give series be denoted by $\sum_{n=1}^{\infty} a_n x^n$.

Then, here $a_n = \dfrac{1\cdot 3\cdot 5\ldots(2n-1)}{2\cdot 5\cdot 8\ldots(3n-1)}$ and $a_{n+1} = \dfrac{1\cdot 3\cdot 5\ldots(2n-1)\,(2n+1)}{2\cdot 5\cdot 8\ldots(3n-1)\,(3n+2)}$

$$\therefore \quad \text{Radius of convergence} = \lim_{n\to\infty}\left|\frac{a_n}{a_{n+1}}\right| = \lim_{n\to\infty}\left|\frac{3n+2}{2n+1}\right| = \lim_{n\to\infty}\left|\frac{3+2/n}{2+1/n}\right| = \frac{3}{2}$$

(*ii*) Omitting the first term, let the given series be denoted by $\Sigma a_n x^n$. Then, here we have

$$a_n = \frac{a(a+1)\ldots(a+n-1)\,b\,(b+1)\ldots(b+n-1)}{1\cdot 2\ldots n\,c(c+1)\ldots(c+n-1)}$$

and $$a_{n+1} = \frac{a(a+1)\ldots(a+n-1)\,(a+n)\,b(b+1)\ldots(b+n-1)\,(b+n)}{1.2\ldots n(n+1)c(c+1)\ldots(c+n-1)\,(c+n)}$$

$$\text{Radius of convergence} = \lim_{n\to\infty}\left|\frac{a_n}{a_{n+1}}\right| = \lim_{n\to\infty}\left|\frac{(n+1)(c+n)}{(a+n)(b+n)}\right| = \lim_{n\to\infty}\left|\frac{(1+1/n)(1+c/n)}{(1+a/n)(1+b/n)}\right| = 1$$

Ex. 2. *Find the radius of convergence and the exact interval of convergence of the power series* $\Sigma\dfrac{(n+1)}{(n+2)(n+3)}x^n$. **[Delhi B.A. (Prog.) III, 2014]**

Sol. Let the given series be denoted by $\Sigma a_n x^n$ or Σu_n. Then, we have

$a_n = (n+1)/\{(n+2)(n+3)]$ and $a_{n+1} = (n+2)/\{(n+3)(n+4)\}$

$$\therefore\quad R = \text{radius of convergence} = \lim_{n\to\infty}\left|\frac{a_n}{a_{n+1}}\right| = \lim_{n\to\infty}\frac{(n+1)(n+4)}{(n+2)^2} = 1$$

Hence the given converges for $|x| < 1$ and diverges for $|x| > 1$. We now investigate the nature of the given power series when $|x| = 1$, *i.e.*, when $x = 1$ and $x = -1$.

For $x = 1$, $$u_n = \frac{(n+1)}{(n+2)(n+3)} = \frac{1}{n}\times\frac{(1+1/n)}{(1+2/n)(1+3/n)} \qquad \text{... (1)}$$

Let the companion series Σv_n be such that $v_n = 1/n$.

Then, $\lim_{n\to\infty}(u_n/v_n) = 1$, which is finite and non zero.

Again, $\Sigma v_n = \Sigma(1/n)$ is a divergent series. So by comparison test Σu_n diverges for $x = 1$.

Next, for $x = -1$, the given series is an alternating series for which $u_{n+1} < u_n$ for each natural number n and $\lim_{n\to\infty} u_n = 0$, by (1). Hence, by Leibnitz's test the given series converges for $x = -1$.

Hence the exact interval of convergence is $[-1, 1[$.

Ex. 3(*i*). *Determine the interval of convergence of the power series* $\Sigma\{(1/n)\times(-1)^{n+1}(x-1)^n\}$. **[Delhi B.Sc. (Prog) III 2010; 2012]**

(*ii*) Find the radius of convergence of $\sum\{(1/n)\times(-1)^{n-1}x^n\}$, *i.e.*, $x - (x^2/2) + (x^3/3) - \ldots\ldots$ **[Himanchal 2009]**

Sol. (*i*) Let the given series be denoted by $\Sigma a_n(x-1)^n$ or Σu_n. Then, we have

$a_n = (-1)^{n+1}/n$ and $a_{n+1} = (-1)^{n+2}/(n+1)$.

$$\therefore\quad R = \text{radius of convergence} = \lim_{n\to\infty}\left|\frac{a_n}{a_{n+1}}\right| = \lim_{n\to\infty}\left|-\frac{n+1}{n}\right| = 1$$

Since the given power series is about the point $x = x_0 = 1$, the interval of convergence is

$x_0 - R < x < x_0 + R$, *i.e.*, $-1 + 1 < x < 1 + 1$, *i.e.*, $0 < x < 2$.

For $x = 2$, the given series reduces to the alternating series $\Sigma(-1)^{n-1}/n\,(= \Sigma(-1)^{n-1}u_n$, say) for which $u_{n+1} < u_n$ for each natural number n and $\lim_{n\to\infty} u_n = \lim_{n\to\infty}(1/n) = 0$. Hence by Leibnitz's test the given series is convergent when $x = 2$.

Next, for $x = 0$, clearly the given series diverges. Hence the exact interval of convergence is $]0, 2]$.

(*ii*). Do like part (*i*) **Ans.** Radius of convergence = 1

EXERCISE 17 (A)

Determine the radius of convergence and the exact interval of convergence of each of the following power series.

1. (*i*) $\Sigma \dfrac{nx^n}{(n+1)^2}$ (*ii*) $\Sigma \dfrac{2^n x^n}{n!}$ (*iii*) $\Sigma \dfrac{x^n}{n^3}$ (*iv*) $\Sigma \dfrac{x^n}{n^n}$ (*v*) $\Sigma \dfrac{n!}{n^n} x^n$

[*Himachal 2013*] [*Meerut 2011, 13, 15*]

2. (*i*) $\Sigma \dfrac{(2n)!x^{2n}}{(n!)^2}$ (*ii*) $\Sigma \dfrac{(-1)^n x^{2n}}{(n!)^2 2^{2n}}$ (*iii*) $\sum_{n=1}^{\infty} 3^{-n} x^{3n}$ (*iv*) $\Sigma \dfrac{(n!)^2 x^{2n}}{(2n)!}$

(*Kanpur 2017*) (*Delhi 2014*)

(*v*) $\Sigma(-1)^n \dfrac{x^{2n+1}}{(2n+1)}$ [*Delhi B.Sc. III (Prog) 2009*] (*vi*) $\sum_{n=1}^{\infty} \dfrac{(-1)^n}{n^x} x^n$ (*Himachal 2007*)

3. (*i*) $\sum_{n=1}^{\infty} \dfrac{(x-1)^n}{2^n}$ (*ii*) $\Sigma \dfrac{(-1)^n (x-1)^n}{2^n (3n-1)}$ (*iii*) $\Sigma \dfrac{n!(x+2)^n}{n^n}$ [*Delhi B.Sc.(Prog.) II, 2015, 16*]

(*iv*) $\sum_{n=1}^{\infty} \dfrac{(x-1)^n}{n^n}$ (*v*) $\sum_{n=1}^{\infty} \{x^n / \log n\}$ (*vi*) $\sum_{n=1}^{\infty} \dfrac{x^n}{n!}$ [*Delhi B.A. (Prog.) III, 2011, 13*]

(*Meerut 2017*)

4. *If the power series* $\Sigma a_n x^n$ *has radius of convergence R, then prove that, for any positive integer m,* $\Sigma a_n x^{mn}$ *has radius of convergence* $R^{1/m}$.

ANSWERS

1. (*i*) $R = 1, [-1, 1[$; (*ii*) $R = \infty, R$; (*iii*) $R = 1, [-1, 1]$ (*iv*) $R = \infty, R$ (*v*) $R = e$

2. (*i*) $R = 1/2, [-1/2, 1/2]$; (*ii*) $R = \infty, R$; (*iii*) $3^{1/3}, [-3^{1/3}, 3^{1/3}]$ (*iv*) $R = 2,]-2, 2[$

(*v*) $R = 1, [-1, 1]$ (*vi*) $1,]-1, 1]$

3. (*i*) $R = 2,]-1, 3[$, (*ii*) $R = 2,]-1, 3]$ (*iii*) $R = e,]-2, -e, -2 + e[$ (*iv*) $R = \infty,]-\infty, \infty[$

(*v*) Radius of convergence = 1. (*vi*) Radius of convergence $= \infty$

17.6 Some basic theorems

Theorem I. *If a power series* $\Sigma a_n x^n$ *converges for* $x = x_0$, *then it is absolutely convergent for every* $x = x_1$, *where* $|x_1| < |x_0|$ **[Delhi 2005; Kanpur 2013; Himachal 2014, 2015]**

Proof. Since $\Sigma a_n x^n$ converges for $x = x_0$, we have

$$\Sigma a_n x_0^n \text{ is convergent} \Rightarrow \lim_{n \to \infty} a_n x_0^n = 0$$

Hence, for $\varepsilon = 1/2 (say)$, there exists a positive integer m such that

$$|a_n x_0^n - 0| < 1/2 \quad \forall n \geq m \qquad \text{or} \qquad |a_n| < 1/2 |x_0|^n, \forall n \geq m \quad \ldots (1)$$

Now, $|a_n x_1^n| = |a_n||x_1|^n < |x_1|^n / 2|x_0|^n$ or $|a_n x_1^n| < (1/2) \times |x_1 / x_0|^n \ \forall n \geq m$ ... (2)

Given that $|x_1| < |x_0|$. Hence $\Sigma |x_1 / x_0|^n$ is a geometric series with common ratio $|x_1 / x_0| < 1$. Hence, by comparison test, $\Sigma a_n x_1^n$ is convergent for $|x_1| < |x_0|$.

Therefore, $\Sigma a_n x^n$ is absolute convergent for every $x = x_1$ where $|x_1| < |x_0|$.

Remark 1. The above theorem says that if a power series converges for $x = x_0$ then it converges absolutely for all those x for which $|x| < |x_0|$. Later on we shall study problems wherein a power series will be absolutely convergent within the interval of convergence while it will be just convergent at one or both the end points of the interval of convergence.

Theorem II. *If a power series diverges for* $x = x'$, *then it diverges for every* $x = x''$ *where* $|x''| > |x'|$. **[Himachal 2013, 2014; Delhi B.Sc. (Hons.) II 2004, 05]**

Proof. Given that the power series $\Sigma a_n x^n$ diverges for $x = x'$. Suppose, if possible, that the series was convergent for $x = x''$. Then, by theorem I, it must converge for every x where $|x| < |x''|$, and in particular at $x = x'$, which contradicts the hypothesis.

Hence, if a power series diverges for $x = x'$, then it must diverge for every $x = x''$ where $|x''| > |x'|$.

Remark 2. From the above theorems I and II, it follows that the radius of convergence is supremum (least upper bound) of all the numbers $|x|$ for which $\Sigma |a_n x^n|$ converges. Thus, we have following result.

Theorem III. *A power series is absolutely convergent within its interval convergence and divergent outside it.*

17.7 Properties of functions expressible as power series.

In what follows we propose to study some properties of functions which can be expressed in terms of power series, *i.e.*, the function of the form $f(x) = \Sigma a_n x^n$. In the interval of convergence, the power series has a definite sum $f(x)$ for each x, and, in general, different sum for a different x. To exhibit this dependence on x, we write $f(x) = \Sigma a_n x^n$ and call $f(x)$ as the sum function of the series.

It is very important to note the difference between the concepts of intervals of absolute and of uniform convergence. Remember that an interval of uniform convergence must include its end points whereas the interval of absolute convergence may or may not include its end points. Therefore, if a power series convergence absolutely, and uniformly for $|x| < R$, then we say that it converges in $]-R, R[$ and uniformly in $[-R+\varepsilon, R-\varepsilon]$, no matter which $\varepsilon > 0$ is chosen (or alternatively, we say that it converges uniformly in $[-R', R']$, where $R' < R$.)

Theorem I. *If a power series $\Sigma a_n x^n$ converges for $|x| < R$, and let us define a function $f(x)$, $f(x) = \Sigma a_n x^n$, $|x| < R$, then $\Sigma a_n x^n$ converges uniformly on $[-R+\varepsilon, R-\varepsilon]$, no matter which $\varepsilon > 0$ in chosen, and that the function f is continuous and differentiable on $]-R, R[$ and $f'(x) = \Sigma n a_n x^{n-1}, |x| < R$.* **[Mumbai 2010; Delhi B.Sc. (Prog) II 2018; Delhi B.Sc. (Hons.) II 2019; Kanpur 2004]**

Proof. Let $\varepsilon > 0$ be any given number. Then, we have

$$|x| \le R - \varepsilon \quad \Rightarrow \quad |a_n x^n| \le |a_n| (R-\varepsilon)^n \qquad \text{... (1)}$$

Since every power series converges absolutely within its interval of convergence, it follows that $\Sigma a_n (R-\varepsilon)^n$ converges absolutely. Hence by Weierstrass's M-test it follows that the series $\Sigma a_n x^n$ converges uniformly on $[-R+\varepsilon, R-\varepsilon]$.

Since each term of the series $\Sigma a_n x^n$ is continuous on $]-R, R[$ and $\Sigma a_n x^n$ is uniformly convergent on $[-R+\varepsilon, R-\varepsilon]$, hence the sum function $f(x)$ of $\Sigma a_n x^n$ is also continuous.

Again, since every term of the power series $\Sigma a_n x^n$ is differentiable on $]-R, R[$ and $\Sigma a_n x^n$ is uniformly convergent on $[-R+\varepsilon, R-\varepsilon]$, it follows that its sum function f is also differentiable on $]-R, R[$,

Furthermore, $\limsup_{n\to\infty} |na_n|^{1/n} = \limsup_{n\to\infty} (n^{1/n}) \times |a_n|^{1/n} = 1/R,$

[$\because$ R is radius of convergence of $\Sigma a_n x^n \Rightarrow \limsup_{n\to\infty} |a_n|^{1/n} = 1/R$ and $\lim_{n\to\infty} n^{1/n} = 1$]

Thus, the differentiated series $\Sigma na_n x^{n-1}$ is also power series and has the same radius of convergence R as the given power series $\Sigma a_n x^n$. Hence $\Sigma na_n x^{n-1}$ is uniformly convergent in $[-R+\varepsilon, R-\varepsilon]$. Also, we have $f'(x) = \Sigma na_n x^{n-1}, |x| < R$

Theorem II. *If a power se' .es $\Sigma a_n x^n$ converge for $|x| < R$, and let us define a function $f(x)$, $f(x) = \Sigma a_n x^n$, $|x| < R$, then $\Sigma a_n x^n$ converges uniformly on $[-R+\varepsilon, R-\varepsilon]$, no matter which $\varepsilon > 0$ is chosen and that the function f is continuous and integrable on $[-R, R[$ and*

$$\int f(x)\,dx = \Sigma\{a_n/(n+1)\}\, x^{n+1}$$

Proof. Refer theorem I for proof for uniform convergence of $\Sigma a_n x^n$ and continuity of function f.

Last part. Since every term of the power series $\Sigma a_n x^n$ is integrable on $]-R, R[$ and $\Sigma a_n x^n$ is uniformly convergent on $[-R+\varepsilon, R-\varepsilon]$, it follows that its sum function is also integrable on $]-R, R[$

Furthermore, $\limsup_{n\to\infty} \left|\frac{a_n}{n+1}\right|^{1/n} = \limsup_{n\to\infty} \frac{1}{(n+1)^{1/n}} \times |a_n|^{1/n} = \frac{1}{R},$

[$\because$ R is radius of convergence of $\Sigma a_n x^n \Rightarrow \limsup_{n\to\infty} |a_n|^{1/n} = 1/R$ and $\lim_{n\to\infty} (n+1)^{1/n} = 1$]

Thus, the integrated series $\Sigma\{a_n/(n+1)\}x^{n+1}$ is also a power series and has the same radius of convergence R as the given power series $\Sigma a_n x^n$. Hence $\Sigma\{a_n/(n+1)\}x^{n+1}$ is uniformly convergent in $[-R+\varepsilon, R-\varepsilon]$. Also, we have $\int f(x)dx = \Sigma\{a_n/(n+1)\}x^{n+1}$

Corollary. *Under the hypothesis of theorem I, f has derivates of all orders in $]-R, R[$, which are given by*

$$f^{(k)}(x) = \sum_{n=k}^{\infty} n(n-1).....(n-k+1)a_n x^{n-k}$$

and in particular,

$$f^{(k)}(0) = k!\, a_k \qquad (k = 0, 1, 2, ...)$$

[Here $f^{(k)}$ is the kth derivative f for $k = 1, 2, 3$, and $f^{(0)} = f$]

Proof. Applying theorem 1. successively to f, for $|x| < R$, we get

$$f(x) = \sum_{n=0}^{\infty} a_n x^n, \qquad f'(x) = \sum_{n=1}^{\infty} n\, a_n x^{n-1}. \qquad f''(x) = \sum_{n=2}^{\infty} n(n-1)a_n x^{n-2}, ...$$

$$f^{(k)}(x) = \sum_{n=k}^{\infty} n(n-1)....(n-k+1)a_n x^{n-k}. \quad ...(1)$$

Putting $x = 0$ in (1), we have $\qquad f^{(k)}(0) = k(k-1)...3\cdot 2\cdot 1\cdot a_k = k!\,a_k \qquad ...(2)$

From (2), we have $\qquad a_k = (1/k!)\times f^k(0),\ k = 0, 1, 2, ... \qquad ...(3)$

Remark : Formula (3) shows that the coeffcients of the power series expansion of f can be obtained in terms of the values at the origin of f and that of its derivatives.

Theorem III. *The series obtained by integrating and differentiating power series term by term has the same radius of convergence as the original series.*

[Delhi B.Sc. (Hons.) II 2017, 2018]

Proof. Let R be the radius of convergence of the given power series $\sum_{n=0}^{\infty} a_n x^n$. ... (1)

On integrating (1) term by term, we get $\qquad \sum_{n=0}^{\infty} \dfrac{a_n x^{n+1}}{n+1} \qquad$ ... (2)

Let R' be the radius of convergence of (2). Then, we have

$$R = \lim_{n\to\infty} \frac{1}{|a_n|^{1/n}} \quad \text{and} \quad R' = \lim_{n\to\infty} \frac{(n+1)^{1/n}}{|a_n|^{1/n}} \quad ...(3)$$

Let $l = \lim_{n\to\infty} (n+1)^{1/n}$ so that $\log l = \lim_{n\to\infty} \dfrac{\log(1+n)}{n} = \lim_{n\to\infty} \dfrac{1/(n+1)}{1}$, by L' Hospital rule.

$$\Rightarrow \quad \log l = 0 \quad \Rightarrow \quad l = e^0 = 1 \quad \Rightarrow \quad \lim_{n\to\infty} (n+1)^{1/n} = 1 \quad ...(4)$$

Using (4), $\qquad$ (3) $\Rightarrow$ $\qquad R' = R$

Next, differentiating (1) term by term, we get $\qquad \sum_{n=1}^{\infty} na_n x^{n-1} \qquad$ ... (5)

Let R'' be the radius of convergence of (5). Then

$$R'' = \lim_{n\to\infty} \frac{1}{n^{1/n}\,|a_n|^{1/n}} = \lim_{n\to\infty} \frac{1}{n^{1/n}} \times \lim_{n\to\infty} \frac{1}{|a_n|^{1/n}} \quad ...(6)$$

Let $m = \lim_{n\to\infty} n^{1/n}$ so that $\log m = \lim_{n\to\infty} \dfrac{\log n}{n} = \lim_{n\to\infty} \dfrac{(1/n)}{1}$, by L' Hospital's rule

$$\Rightarrow \quad \log m = 0 \quad \Rightarrow \quad m = e^0 = 1 \quad \Rightarrow \quad \lim_{n\to\infty} n^{1/n} = 1 \quad ...(7)$$

From (3), (6) and (7), we have $\qquad R'' = R.$

Exercise. *Show that both the power series* $\sum_{n=0}^{\infty} a_n x^n$ *and corresponding series of derivatives* $\sum_{n=1}^{\infty} na_n x^{n-1}$ *have the same radius of convergence.* **[Delhi Maths (H) 2001]**

[Hint. Refer the second part of the above theorem III.]

17.8 Abel's theorem

In this article we propose to show that for a power series with a given radius of connergence and convergent at an end point of the interval, the interval of uniform convergence extends upto and includes that end point. Furthermore, in such a situation the sum function f is continuous not only in open interval $]-R, R[$, but also at the end point.

Here two forms of Abel's theorem are given. The first is regarding uniform convergence and the second is regarding continuity.

Theorem I (First form of Abel's theorem) *If a power series* $\Sigma a_n x^n$ *converges at the end point* $x = R$ *of the interval of convergence* $[-R, R[$; *then it is uniformly convergent in the closed interval* $[0, R]$. **[Delhi Maths (H) 2002, 08; Ravishankar 2005]**

Proof. In order to prove the theorem, we proceed to prove that under the hypothesis of the theorem, Cauchy's criterion for uniform convergence is satisfied on the interval $[0, R]$. From this it would follow that $\Sigma a_n x^n$ is uniformly convergent on $[0, R]$

Let $$S_{n,p} = a_{n+1}R^{n+1} + a_{n+2}R^{n+2} + ... + a_{n+p}R^{n+p}, \quad p = 1, 2, 3,...$$

Then, it is easy to verify that

$$\left.\begin{aligned} a_{n+1}\, R^{n+1} &= S_{n,1} \\ a_{n+2}\, R^{n+2} &= S_{n,2} - S_{n,1} \\ &\cdots\cdots\cdots \\ a_{n+p}\, R^{n+p} &= S_{n,p} - S_{n,\,p-1} \end{aligned}\right\} \qquad ...(1)$$

Let $\varepsilon > 0$ be given. By the hypothesis of the theorem, the number series $\Sigma a_n R^n$ is convergent. Therefore, by Cauchy's general principle of convergence, there exists a positive integer m such that

$$|S_{n,p}| < \varepsilon \qquad \forall n \geq m, \quad \text{and} \quad \forall p \geq 1 \qquad ...(2)$$

We have $$(x/R)^{n+p} \leq (x/R)^{n+p-1} \leq ... \leq (x/R)^{n+1} \leq 1, \text{ for } 0 \leq x \leq R \qquad ...(3)$$

Using (1) and (3), for $\forall n \geq m$, we have

$$\left|a_{n+1}x^{n+1} + a_{n+2}x^{n+2} + ... + a_{n+p}x^{n+p}\right|$$

$$= \left|a_{n+1}R^{n+1}(x/R)^{n+1} + a_{n+2}R^{n+2}(x/R)^{n+2} + ... + a_{n+p}R^{n+p}(x/R)^{n+p}\right|$$

$$= \Big|S_{n+1}\left\{(x/R)^{n+1} - (x/R)^{n+2}\right\} + S_{n+2}\left\{(x/R)^{n+2} - (x/R)^{n+3}\right\} + ...$$
$$+ S_{n+p-1}\left\{(x/R)^{n+p-1} - (x/R)^{n+p}\right\} + S_{n+p}(x/R)^{n+p}\Big|$$

$$\leq |S_{n+1}|\left\{(x/R)^{n+1} - (x/R)^{n+2}\right\} + |S_{n+2}|\left\{(x/R)^{r+2} - (x/R)^{n+3}\right\} + ...$$
$$+ \left|S_{n+p-1}\right|\left\{(x/R)^{n+p-1} - (x/R)^{n+p}\right\} + \left|S_{n+p}\right|(x/R)^{n+p}$$

$$\leq \varepsilon\left\{(x/R)^{n+1} - (x/R)^{n+2} + (x/R)^{n+2} - (x/R)^{n+3} + ... - (x/R)^{n+p} + (x/R)^{n+p}\right\}, \text{ using (2)}$$

$$= \varepsilon \times (x/R)^{n+1} \leq \varepsilon$$

Thus, $$\left|a_{n+1}x^{n+1} + ... + a_{n+p}x^{n+p}\right| \leq \varepsilon \qquad \forall n \geq m, \quad \forall p \geq 1 \quad \text{and} \quad \forall x \in [0, R]$$

Therefore, by Cauchy's principle for uniform convergence the power series $\Sigma a_n x^n$ converges uniformly on $[0, R]$

Remark. We know that a series of continuous functions that converges uniformly in a given interval, defines a continuous function in that interval. Hence if a power series (having R as its radius of convergence) which converges at the end point R, defines a continuous function in $[-R+\varepsilon, R]$, no matter which $\varepsilon > 0$ is chosen.

Corollary 1. *If power series with interval of convergene* $]-R, R[$ *converges at the left end* $x = -R$, *or at both the ends, then in the former situation the series is uniformly convergent on* $[-R, 0]$ *and in the latter situation on* $[-R, R]$.

Proof. Left as an exercise.

Corollary 2. *If a power series with interval of convergence* $]-R, R[$, *diverges at the end point* $x = R$, *it cannot be uniformly convergent on* $[0, R]$.

Proof. If possible, let the series be uniformly convergent on [0, R]. Then, it must converge at x = R as well, which contradicts the hypothesis of the corollary.

Theorem II. (Second form of Abel's theorem)

If $\sum_{n=0}^{\infty} a_n x^n$ *be a power series with finite radius of convergence R, and let* $f(x) = \sum_{n=0}^{\infty} a_n x^n$, $-R < x < R$. *If the series* $\sum_{n=0}^{\infty} a_n R^n$ *converges, then* $\lim_{x \to R-0} f(x) = \Sigma a_n R^n$

[Delhi Maths (H) 1999, 2000, 05, 09]

Proof. We first prove that there is no loss of generality in taking $R = 1$. Accordingly, putting $x = Ry$, we have

$$\sum_{n=0}^{\infty} a_n x^n = \sum_{n=0}^{\infty} a_n R^n y^n = \sum_{n=0}^{\infty} b_n y^n, \qquad \text{where} \qquad b_n = a_n R^n$$

Now $\Sigma b_n y^n$ is a power series. Let its radius of convergence be R'. Then we have

$$R' = \frac{1}{\limsup_{n\to\infty} |b_n|^{1/n}} = \frac{1}{\limsup_{n\to\infty} |a_n R^n|^{1/n}} = \frac{1}{\limsup_{n\to\infty} R|a_n|^{1/n}} = 1,$$

$$[\because \quad R = \text{Radius of convergence of } \Sigma a_n = 1/\left(\limsup_{n\to\infty} |a_n|^{1/n}\right)]$$

Thus, we see that it suffices to prove the following modified form of the statement of the above theorem.

Let $\sum_{n=0}^{\infty} a_n x^n$ *be a power series with unit radius of convergence and let* $f(x) = \sum_{n=1}^{\infty} a_n x^n, -1 < x < 1$. *If the series* Σa_n *converges, then* $\lim_{x \to 1-0} f(x) = \sum_{n=0}^{\infty} a_n$.

[Bhopal 2002; G.N.D.U. Amritsar 2000, 02, 03, 04, 05; Meerut 2004, 05]

We now prove the above form of Abel's theorem

Let $\qquad S_n = a_0 + a_1 + a_2 + \ldots + a_n \qquad$ and $\qquad S_{-1} = 0 \qquad \ldots (1)$

Then, $$\sum_{n=0}^{m} a_n x^n = \sum_{n=0}^{m} (S_n - S_{n-1}) x^n = \sum_{n=0}^{m-1} S_n x^n + S_m x^m - \sum_{n=0}^{m} S_{n-1} x^n$$

$$= \sum_{n=0}^{m-1} S_n x^n - x \sum_{n=0}^{m} S_{n-1} x^{n-1} + S_m x^m$$

$$= \sum_{n=0}^{m-1} S_n x^n - x \sum_{n=0}^{m-1} S_n x^n + S_m x^m, \quad \text{as from (1)}, \quad S_{-1} = 0$$

Thus, $$\sum_{n=0}^{m} a_n x^n = (1-x) \sum_{n=0}^{m-1} S_n x^n + S_m x^m \qquad \text{... (2)}$$

For $|x| < 1$, we let $m \to \infty$ in (2). Note that $x^m \to 0$ and $\sum_{n=0}^{m} a_n x^n = \sum_{n=0}^{\infty} a_n x^n = f(x)$

Thus, (2) becomes $$f(x) = (1-x) \sum_{n=0}^{\infty} S_n x^n, \quad \text{for } 0 < x < 1 \qquad \text{... (3)}$$

Suppose $S = \lim_{n \to \infty} S_n$. Let $\varepsilon > 0$ be given. Choose $m \in N$ such that

$$|S_n - S| < \varepsilon/2 \quad \text{for all } n \geq m \qquad \text{... (4)}$$

Again, we have $$(1-x) \sum_{n=0}^{\infty} x^n = 1 \quad \text{for } |x| < 1 \qquad \text{... (5)}$$

Therefore for $n \geq m$ and $0 < x < 1$, we have

$$|f(x) - S| = \left|(1-x) \sum_{n=0}^{\infty} S_n x^n - S\right|, \text{ using (3)}$$

$$= \left|(1-x) \sum_{n=0}^{\infty} S_n x^n - S(1-x) \sum_{n=0}^{\infty} x^n\right|, \text{ using (5)}$$

$$= \left|(1-x) \sum_{n=0}^{\infty} (S_n - S) x^n\right|$$

$$\leq (1-x) \sum_{n=0}^{m} |S_n - S| x^n + \frac{\varepsilon}{2}(1-x) \sum_{n=m+1}^{\infty} x^n, \text{ using (4)}$$

Thus, $$|f(x) - S| \leq (1-x) \sum_{n=0}^{m} |S_n - S| x^n + \frac{\varepsilon}{2} \qquad \text{... (6)}$$

Now, for a fixed m, $(1-x) \sum_{n=0}^{m} |S_n - S| x^n$ is a positive continuous function of x, having zero value at $x = 1$. Therefore, there exists $\delta > 0$, such that for $1 - \delta < x < 1$, we have

$$(1-x) \sum_{n=0}^{m} |S_n - S| x^n < \varepsilon/2 \qquad \text{... (7)}$$

Then, (6) and (7) $\Rightarrow$ $$|f(x) - S| < \varepsilon/2 + \varepsilon/2 = \varepsilon, \quad \text{where} \quad 1 - \delta < x < 1$$

Therefore, $$\lim_{x\to 1-0} f(x) = S = \sum_{n=0}^{\infty} a_n$$

Corollary 1. *If the series* $\Sigma(-1)^n\ a_n$ *converges, then* $\lim_{x\to -1+0} f(x) = \Sigma(-1)^n a_n$

Proof. Putting $y = -x$ and on taking $b_n = (-1)^n a_n$, we get

$$\lim_{x\to -1+0} f(x) = \lim_{x\to -1+0} \Sigma a_n x^n = \lim_{x\to -1+0} \Sigma(-1)^n a_n(-x)^n = \lim_{y\to 1-0} \Sigma b_n y^n = \Sigma b_n$$

Remark 1. Let R be the radius of convergence of $\Sigma a_n x^n$. Then $\Sigma a_n R^n$ is convergent. Hence, by Abel' test for uniform convergence, it follows that the series $\Sigma a_n x^n$ converges uniformly in $[-R+\varepsilon, R]$, showing that the interval of uniform convergence extends upto and includes the end point R.

Corollary 2. *Sum function f is continuous at the end point R of the interval of convergence of the power series.*

Proof. We have shown that the power series converges uniformly in $[-R+\varepsilon, R-\varepsilon]$ and each term of the series is continuous. Hence the sum function f is also continuous on $[-R+\varepsilon, R-\varepsilon]$. Again, by the above theorem, f is continuous at $x = R$. Therefore f is continuous (actually uniformly) on $[-R+\varepsilon, R-\varepsilon]$.

An application of Abel's theorem:

Theorem. *If* $\sum_{n=0}^{\infty} a_n$, $\sum_{n=0}^{\infty} b_n$, $\sum_{n=0}^{\infty} c_n$, *converge to A, B and C respectively and if* $c_n = a_0 b_n + a_1 b_{n-1} + \dots + a_n b_0$, *then* $C = AB$

Proof. Let $f(x) = \sum_{n=0}^{\infty} a_n x^n$, $\quad g(x) = \sum_{n=0}^{\infty} b_n x^n$, $\quad h(x) = \sum_{n=0}^{\infty} c_n x^n$, for $0 \le x \le 1$

For $|x| < 1$, the above three series converge absolutely and so $\Sigma c_n x^n$ is the Cauchy product of $\Sigma a_n x^n$ and $\Sigma b_n x^n$. When the multiplication is carried out, we see that

$$f(x) \times g(x) = h(x) \qquad \text{for} \qquad 0 \le x \le 1 \qquad \dots (1)$$

Again, using Able's theorem, we have

$$\left.\begin{aligned} &\lim_{x\to 1-0} f(x) = \sum_{n=0}^{\infty} a_n \Rightarrow f(x) \to A \quad \text{as} \quad x \to 1-0 \\ \text{Similarly} \quad &g(x) \to B \quad \text{and} \quad h(x) \to C \quad \text{as} \quad x \to 1-0 \end{aligned}\right\} \qquad \dots (2)$$

Equations (1) and (2) $\Rightarrow$ $AB = C$

17.9 Same theorems on power series

Theorem I. *Given a double sequence* $< a_{ij} >, i = 1, 2, 3, \dots,\ j = 1, 2, 3, \dots$ *suppose that* $\sum_{j=1}^{\infty} |a_{ij}| = b_i$, $i = 1, 2, 3, \dots$ *and* $\sum_{i=1}^{\infty} b_i$ *converges. Then* $\sum_{i=1}^{\infty}\sum_{j=1}^{\infty} a_{ij} = \sum_{j=1}^{\infty}\sum_{i=1}^{\infty} a_{ij}$

[Bangalore 2002; Kurukshetra 2003, 05; Nagpur 2003, 05]

Proof. Let E be a countable set, consisting of the points $x_0, x_1, x_2, \dots$, and suppose $x_n \to x_0$ as $n \to \infty$. Define

$$f_i(x_0) = \sum_{j=1}^{\infty} a_{ij}, \qquad i = 1, 2, 3, ... \qquad ...(1)$$

$$f_i(x_n) = \sum_{j=1}^{n} a_{ij}, \qquad i = 1, 2, 3, ..., \qquad n = 1, 2, 3, ... \qquad ...(2)$$

and $$g(x) = \sum_{i=1}^{\infty} f_i(x), \qquad \text{where } x \in E \qquad ...(3)$$

Also, given that $$\sum_{j=1}^{\infty} |a_{ij}| = b_i, \qquad i = 1, 2, 3, ... \qquad ...(4)$$

Now, (1), (2) and (4) together show that each f_i is continuous at x_0. Since $|f_i(x)| \le b_i$ for $x \in E$, (3) converges uniformly, so that g is continuous at x_0. It follows that

$$\sum_{i=1}^{\infty} \sum_{j=1}^{\infty} a_{ij} = \sum_{i=1}^{\infty} f_i(x_0) = g(x_0) = \lim_{n \to \infty} g(x_n) = \lim_{n \to \infty} \sum_{i=1}^{\infty} f_i(x_n)$$

$$= \lim_{n \to \infty} \sum_{i=1}^{\infty} \sum_{j=1}^{n} a_{ij} = \lim_{n \to \infty} \sum_{j=1}^{n} \sum_{i=1}^{\infty} a_{ij} = \sum_{j=1}^{\infty} \sum_{i=1}^{\infty} a_{ij}$$

Theorem II (Taylor's theorem) for power series. *Suppose* $f(x) = \sum_{n=0}^{\infty} c_n x^n$, *the series converging in* $|x| < R$. *If* $-R < a < R$, *then f can be expanded in a power series about the point* $x = a$ *which converges in* $|x-a| < R - |a|$, *and* $f(x) = \sum_{n=0}^{\infty} \frac{f^{(n)}(a)}{n!}(x-a)^n$, *where* $|x-a| < R - |a|$ **[G.N.D.U. Amristar 2005]**

Proof. Here $|x-a| < R - |a| \Rightarrow \quad |x| \le |x-a| + |a| < R \Rightarrow \Sigma c_n x^n$ converges.

Now, we have

$$f(x) = \sum_{n=0}^{\infty} c_n x^n = \sum_{n=0}^{\infty} c_n \{(x-a)+a\}^n = \sum_{n=0}^{\infty} c_n \sum_{m=0}^{n} {}^nc_m \, a^{n-m} (x-a)^m$$

[using binomial theorem for positive integral values]

Thus, $$f(x) = \sum_{m=0}^{\infty} \left[\sum_{n=m}^{\infty} {}^nc_m c_n \, a^{n-m} \right] (x-a)^n \qquad ...(1)$$

(1) is the desired expansion about $x = a$. To prove its validity, we have to justify the change which was made in the order of summation. Theorem I shows that this is permissible provided

$$\sum_{n=0}^{\infty} \sum_{m=0}^{n} \left| c_n \, {}^nc_m a^{n-m} (x-a)^m \right| \qquad ...(2)$$

converges. But (2) is the same as

$$\sum_{n=0}^{\infty} |c_n| \times (|x-a| + |a|)^n, \qquad ...(3)$$

and (3) converges if $$|x-a| + |a| < R$$

We now proceed to re-write the coefficients in (1) in terms of the values of f and its derivatives at $x = a$.

Now, $f(x)=\sum_{n=0}^{\infty} c_n x^n \quad \Rightarrow \quad f^{(m)}(x)=\sum_{n=m}^{\infty} n(n-1)....(n-m+1)\, c_n x^{n-m}$

Replacing x by a, we obtain

$$f^{(m)}(a)=\sum_{n=m}^{\infty} n(n-1)....(n-m+1)c_n a^{n-m}=\sum_{n=m}^{\infty}\frac{n!}{(n-m)!}c_n a^{n-m}$$

$$\Rightarrow \quad \frac{f^{(m)}(a)}{m!}=\sum_{n=m}^{\infty}\frac{n!}{m!(n-m)!}c_n a^{n-m} \Rightarrow \frac{f^{m}(a)}{m!}=\sum_{n=m}^{\infty} {}^{n}c_m\, c_n\, a^{n-m} \quad ... (4)$$

Using (4), (1) $\Rightarrow \quad f(x)=\sum_{m=0}^{\infty}\frac{f^{(m)}(a)}{m!}(x-a)^m$

Thus, we have $\quad f(x)=\sum_{n=0}^{\infty}\frac{f^{(n)}(a)}{n!}(x-a)^n$, where $|x-a|<R-|a|$

Remark. If should be noted that (5) may actually converge in a larger interval than one given by $|x-a|<R-|a|$.

17.10. Solved examples

Ex. 1. *Show that* $\log(1+x)=x-x^2/2+x^3/3-x^4/4+...,-1<x\le 1$ and deduce that $\log 2=1-1/2+1/3-1/4+....$ **[Delhi B.Sc. (Prog) III 2010, 2012; Delhi Maths (H) 2003, 08]**

or Show by integrating the series for $1/(1+x)$ that if $|x|<|$, then $\log(1+x)=\sum_{n=1}^{\infty}\{(-1)^{n-1}/n\}x^n$

[Delhi B.Sc. (Hons) II 2011)

Sol. We know that $\quad \log(1+x)^{-1}=1-x+x^2-x^3+...,-1<x<1 \quad ... (1)$

The radius of convergence of the power series on R.H.S. of (1) is 1. Hence it converges absolutely in]–1, 1 [and uniformly in [–*a*, *a*] where $|a|<1$. The integrated series also converges absolutely in]–1, 1 [and uniformly in [–*a*, *a*].

Integrating (1), $\quad \log(1+x)=c+x-x^2/2+x^3/3-....,-1<x<1 \quad ... (2)$

Putting $x=0$ in (2), we get $c=0$. Hence, (2) reduces to

$$\log(1+x)=x-x^2/2+x^3/3-x^4/4+....,-1<x<1 \quad ... (3)$$

Since the power series on R.H.S of (3) converges at $x=1$ also, hence, by Abel's theorem,

$$\log(1+x)=x-x^2/2+x^3/3-x^4/4+...,-1<x\le 1 \quad ... (4)$$

Deduction Using the second form of Abel's theorem, at $x=1$, (4) yields

$$\log 2=\lim_{x\to 1-0}\log(1+x)=1-1/2+1/3+1/4+.....$$

Ex. 2. *Show that* $\frac{1}{2}\{\log(1+x)\}^2=\frac{x^2}{2}-\frac{x^3}{3}\left(1+\frac{1}{2}\right)+\frac{x^4}{4}\left(1+\frac{1}{2}+\frac{1}{3}\right)....,-1<x\le 1$

Sol. Proceeding as in Ex. 1, we have

$$\log(1+x)=x-x^2/2+x^3/3-x^4/4+.....,-1<x\le 1 \quad ... (1)$$

Also $\quad (1+x)^{-1}=1-x+x^2-x^3+x^4-.....,\ -1<x<1 \quad ... (2)$

Since both the both the above series are absolutely convergent in]–1, 1 [, hence their Cauchy product will also converge to $(1+x)^{-1}\log(1+x)$

$$\therefore \quad \frac{\log(1+x)}{1+x} = x - x^2\left(1+\frac{1}{2}\right) + x^3\left(1+\frac{1}{2}+\frac{1}{3}\right)......, -1 < x < 1 \quad ...(3)$$

Integrating (3), we have

$$\frac{1}{2}\{\log(1+x)\}^2 = c + \frac{x^2}{2} - \frac{x^3}{3}\left(1+\frac{1}{2}\right) + \frac{x^4}{4}\left(1+\frac{1}{2}+\frac{1}{3}\right) - ...., -1 < x < 1, \quad ...(4)$$

where c is constant of integration. Putting $x = 0$ in (4), we get $c = 0$.

$$\therefore \quad \frac{1}{2}\{\log(1+x)\}^2 = \frac{x^2}{2} - \frac{x^3}{3}\left(1+\frac{1}{2}\right) + \frac{x^4}{4}\left(1+\frac{1}{2}+\frac{1}{3}\right) - ....., -1 < x < 1, \quad ...(5)$$

Since the power series on R.H.S. of (5) convergent at $x = 1$ also, hence by Abel's theorem,

$$\frac{1}{2}\{\log(1+x)\}^2 = \frac{x^2}{2} - \frac{x^3}{3}\left(1+\frac{1}{2}\right) + \frac{x^4}{4}\left(1+\frac{1}{2}+\frac{1}{3}\right) - ....., -1 < x \le 1,$$

Ex. 3. *Show that (i)* $\tan^{-1} x = x - x^3/3 + x^5/5 - x^7/7 + ..., -1 \le x \le 1$

[Himachal 2009; Delhi Maths (H) 2000, 04]

(ii) $\pi/4 = 1 - 1/3 + 1/5 - 1/7 + ...$ **[Himachal 2008, 2014; Delhi Maths (H)2004]**

Sol. *(i)* We know that $(1+x^2)^{-1} = 1 - x^2 + x^4 - x^6 + ..., -1 < x < 1$ ... (1)

The radius of convergence of the power series on R.H.S. of (1) is 1. Hence it converges absolutely in $]-1, 1[$, and uniformly in $[-a, a]$ where $|a| < 1$. The integrated series also converges absolutely in $]-1, 1[$ and uniformly in $[-a, a]$.

Integrating (1), $\tan^{-1} x = c + x - x^3/3 + x^5/5 - x^7/7 + ..., -1 < x < 1$ ... (2)

where c is a constant of integration. Putting $x = 0$ in (2), we get $c = 0$

$$\therefore \quad \tan^{-1} x = x - x^3/3 + x^5/5 - x^7/7 + ..., -1 < x < 1 \quad ...(3)$$

Since the power series on R.H.S. of (3) converges at $x = 1$ also, hence, by Abel's theorem it is uniformly convergent in $[-1, 1]$ and therefore, we have

$$\tan^{-1} x = x - x^3/3 + x^5/5 - x^7/7 + ..., -1 \le x \le 1 \quad ...(4)$$

(ii) Using the second form of Abel's theorem, at $x = 1$, (4) yields

$$\pi/4 = 1 - 1/3 + 1/5 - 1/7 + ...$$

Ex. 4. *Show that* $\frac{1}{2}(\tan^{-1} x)^2 = \frac{x^2}{2} - \frac{x^4}{4}\left(1+\frac{1}{3}\right) + \frac{x^6}{6}\left(1+\frac{1}{3}+\frac{1}{5}\right) - ...., -1 < x \le 1$

[Delhi Maths (H) 2000]

Sol. Proceeding as in Ex. 1, we have $\tan^{-1} x = x - x^3/3 + x^5/5 - ...., -1 \le x \le 1$ ... (1)

Also, $(1+x^2)^{-1} = 1 - x^2 + x^4 - x^6 + ..., -1 < x < 1$ ... (2)

Since both the above series are absolutely convergent in $[-1, 1[$, hence their Cauchy product will also converge to $(1+x^2)^{-1}\tan^{-1}x$ in $]-1, 1[$.

$$\therefore \quad \frac{\tan^{-1} x}{1+x^2} = x - \left(1+\frac{1}{3}\right)x^3 + \left(1+\frac{1}{3}+\frac{1}{5}\right)x^5 - ...., -1 < x < 1 \quad ...(4)$$

Integrating, $\frac{1}{2}(\tan^{-1} x)^2 = c + \frac{x^2}{2} - \frac{x^4}{4}\left(1+\frac{1}{3}\right) + \frac{x^6}{6}\left(1+\frac{1}{3}+\frac{1}{5}\right) - \ldots, \; -1 < x < 1,$ ... (5)

where c is the constant of integration. Putting $x = 0$ in (5), we get $c = 0$

$\therefore$ From (5), $\frac{1}{2}(\tan^{-1} x)^2 = \frac{x^2}{2} - \frac{x^4}{4}\left(1+\frac{1}{3}\right) + \frac{x^6}{6}\left(1+\frac{1}{3}+\frac{1}{5}\right) - \ldots, \; -1 < x < 1$ ... (6)

Since the series on the R.H.S. of (6) converges at $x = 1$ also, hence by Abel's theorem, we get

$$\frac{1}{2}(\tan^{-1} x)^2 = \frac{x^2}{2} - \frac{x^4}{4}\left(1+\frac{1}{3}\right) + \frac{x^6}{6}\left(1+\frac{1}{3}+\frac{1}{5}\right) - \ldots, -1 < x \le 1$$

Ex. 5. Show that

(*i*) $\sin^{-1} x = x + \frac{1}{2}\frac{x^3}{3} + \frac{1\times 3}{2\times 4}\frac{x^5}{5} + \frac{1\times 3\times 5}{2\times 4\times 6}\frac{x^7}{7} + \ldots, -1 < x \le 1$

and $\frac{\pi}{2} = 1 + \frac{1}{2}\times\frac{1}{3} + \frac{1\times 3}{2\times 4}\times\frac{1}{5} + \frac{1\times 3\times 5}{2\times 4\times 6}\times\frac{1}{7} + \ldots$ **[Delhi Maths (H) 2005; Delhi B.Sc. (Prog) III 2009]**

(*ii*) $\frac{1}{2}(\sin^{-1} x)^2 = \frac{x^2}{2} + \frac{2}{3}\times\frac{x^4}{4} + \frac{2\times 4}{3\times 5}\times\frac{x^6}{6} + \ldots, -1 < x \le 1$

(*iii*) Find similar expressions for $\cos^{-1}x$, $\sinh^{-1}x$, $\cosh^{-1}x$ and $\tanh^{-1}x$

Sol. (*i*) We know that, for $-1 < x < 1$, we have

$$(1-x^2)^{-1/2} = 1 + \left(-\frac{1}{2}\right)\times(-x^2) + \frac{(-1/2)\times(-3/2)}{2!}\times(-x^2)^2 + \frac{(-1/2)\times(-3/2)\times(-5/2)}{3!}\times(-x^2)^3 + \ldots$$

[Using binomial theorem for any rational index]

or $$\frac{1}{\sqrt{1-x^2}} = 1 + \frac{x^2}{2} + \frac{1\times 3}{2\times 4}x^4 + \frac{1\times 3\times 5}{2\times 4\times 6}x^6 + \ldots \quad \text{for } -1 < x < 1$$

Integrating, $\sin^{-1} = c + x + \frac{1}{2}\frac{x^3}{3} + \frac{1\times 3}{2\times 4}\frac{x^5}{5} + \frac{1\times 3\times 5}{2\times 4\times 6}\frac{x^7}{7} + \ldots, \; -1 < x < 1$ ... (1)

where c is the constant of integration. Putting $x = 0$ in (1), we get $c = 0$. The power series on the R.H.S. of (1) converges at $x = 1$ also, Hence, by Abel's theorem, we obtain

$$\sin^{-1} x = x + \frac{1}{2}\frac{x^3}{3} + \frac{1\times 3}{2\times 4}\frac{x^5}{5} + \frac{1\times 3\times 5}{2\times 4\times 6}\frac{x^7}{7} + \ldots, -1 < x \le 1 \quad \ldots (2)$$

At $x = 1$, by Abel's theorem, (2) yields

$$\frac{\pi}{2} = 1 + \frac{1}{2}\times\frac{1}{3} + \frac{1\times 3}{2\times 4}\times\frac{1}{5} + \frac{1\times 3\times 5}{2\times 4\times 6}\times\frac{1}{7} + \ldots$$

(*ii*) Since both the series (1) and (2) are absolutely in $]-1, 1[$, hence their Cauchy product will converge to $(\sin^{-1} x)/\sqrt{1-x^2}$.

$$\therefore \qquad (\sin^{-1} x)/\sqrt{1-x^2} = x+\frac{2}{3}x^3+\frac{2\times 4}{3\times 5}x^5+\ldots.., \; -1<x<1$$

Integrating, $$\frac{1}{2}(\sin^{-1} x)^2 = c'+\frac{x^2}{2}+\frac{2}{3}\times\frac{x^4}{4}+\frac{2\times 4}{3\times 5}\times\frac{x^6}{6}+\ldots., -1<x<1 \qquad \ldots (3)$$

where c' is the constant of integration. Putting $x = 0$ in (3), we get $c' = 0$. The power series on the R.H.S. of (3) converges at x = 1 also. Hence, by Abel's theorem, we obtain.

$$\frac{1}{2}(\sin^{-1} x)^2 = \frac{x^2}{2}+\frac{2}{3}\times\frac{x^4}{4}+\frac{2\times 4}{3\times 5}\times\frac{x^6}{6}+\ldots., \; -1<x\le 1$$

EXERCISES 17 (B)

1. Examine the uniform convergence with respect to x of the series $\sum_{n=1}^{\infty} (x^n/n^n)$.

2. Show that (*i*) $\log(1-x) = -x-x^2/2-x^3/3-\ldots., -1\le x<1$ **[Delhi B.Sc. (Hons.) II 2007]**

(*ii*) $\log 2 = 1-1/2+1/3-1/4+\ldots..$

(*iii*) $\frac{1}{2}\{\log(1-x)\}^2 = \frac{x^2}{2}+\left(1+\frac{1}{2}\right)\frac{x^3}{3}+\left(1+\frac{1}{2}+\frac{1}{3}\right)\frac{x^4}{4}+\ldots., \; -1\le x<1$ **[Delhi B.Sc. (Hons.) II 2007]**

3. Show that the series $x-x^3/3+x^5/5-x^7/7+\ldots.$ is uniformly convergent in $-1\le x\le 1$ and that its sum is $\tan^{-1}x$. **[Delhi B.Sc. (Hons.) II 2001]**

4. Prove the uniform convergence of the following series:

(*i*) $x+x^2/2^2+x^3/3^2+x^4/4^2+\ldots.., \; -1\le x\le 1$

(*ii*) $1+x/2+x^2/3+x^3/4+\ldots., \; -1\le x\le a,$ where $0<a<1$

(*iii*) $x+\frac{1}{2}\frac{x^3}{3}+\frac{1\times 3}{2\times 4}\frac{x^5}{5}+\frac{1\times 3\times 5}{2\times 4\times 6}\times\frac{x^7}{7}+\ldots, \; -1\le x\le 1$

5. Show that $\int_0^x \frac{dt}{1+t^n} = x-\frac{x^{n+1}}{n+1}+\frac{x^{2n+1}}{2n+1}-\ldots.., \; -1<x\le 1, \; n>0$

17.11 Elementary functions

It is proposed in this article to define the elementary functions $\sin x$, $\cos x$, $\log x$, e^x, a^x and develop their basic properties. The present treatment will be dependent on the set of real numbers as a complete ordered field and the properties of continuity, derivability and integrabality of functions as based thereon.

17.11A. Trigonometric functions

Theorem. *The two series* $\sum_{n=0}^{\infty} \frac{(-1)^n x^{n+1}}{(2n+1)!}$ *and* $\sum_{n=0}^{\infty} \frac{(-1)^n x^{2n}}{(2n)!}$

are uniformly convergent in every interval $[a, b]$.

Proof. Let M be any positive number greater than $|a|$ as well as $|b|$. Then, $\forall\, x \in [a, b]$,

$$\left|\frac{(-1)^n x^{2n+1}}{(2n+1)!}\right| \le \frac{M^{2n+1}}{(2n+1)!} \quad \text{and} \quad \left|\frac{(-1)^n x^{2n}}{(2n)!}\right| \le \frac{M^{2n}}{(2n)!}$$

Now, as may be easily shown, the two series $\sum \frac{M^{2n+1}}{(2n+1)!}$ and $\sum \frac{M^{2n}}{(2n)!}$ are convergent.

Therefore, by Weierstrass M-test, we prove that the two series are uniformly convergent in $[a, b]$

Definitions. This theorem justifies the following two definitions :

$$\sin x = x - \frac{x^3}{3!} + \frac{x^5}{5!} - .... + \frac{(-1)^n x^{2n+1}}{(2n+1)!} + ..., \forall x \in \boldsymbol{R} \qquad ... (1)$$

and

$$\cos x = 1 - \frac{x^2}{2!} + \frac{x^4}{4!} - .... + \frac{(-1)^n x^{2n}}{(2n)!} + ..., \forall x \in \boldsymbol{R} \qquad ... (2)$$

17.11B. The number π. *To find the smallest positive root of equation* $\cos x = 0$.

Theorem. *To prove that there exists a positive number π such that* $\cos(\pi/2) = 0$ *and* $\cos x > 0$, *for* $0 < x < \pi$. **[Delhi Maths (H) 2001]**

Proof. Consider the interval [0, 2]. We know that cos 0 (= 1), is positive and we will now show that cos 2 is negative, we have

$$\cos 2 = 1 - \frac{2^2}{2!} + \frac{2^4}{4!} - \frac{2^6}{6!} + ... = 1 - \frac{2^2}{2!}\left(1 - \frac{2^2}{3\times 4}\right) - \frac{2^6}{6!}\left(1 - \frac{2^2}{7\times 8}\right) + ...$$

Since the breakets are all positive, we have

$$\cos 2 < 1 - \frac{2^2}{2!}\left(1 - \frac{2^2}{3\times 4}\right) \quad \text{or} \quad \cos 2 < -\frac{1}{3} \quad \text{so that} \quad \cos 2 \text{ is negative.}$$

This there exists at least, one number, c between 0 and 2, such that cos c = 0. Also there cannot exist more than one such value; for, if possible, let k be another so that

$\cos c = 0$ and $\cos k = 0$, where $0 < c < 2$, $0 < k < 2$

By Rolle's theorem there exist, at least, one number between c and k such that the derivative, $-\sin x$ of $\cos x$ vanishes for $x = 1$, so that $\sin \lambda = 0$, where $0 < \lambda < 2$.

But $\sin \lambda = \frac{\lambda}{1!}\left(1 - \frac{\lambda^2}{2\times 3}\right) + \frac{\lambda^5}{5!}\left(1 - \frac{\lambda^2}{6\times 7}\right) + ...$, which is clearly positive.

Thus there exists one and only one root of the equation $\cos x = 0$ lying between 0 and 2. Denoting this value by π, we see that $\pi/2$ is the least positive root of the equation $\cos x = 0$. Also, therefore, we have $\cos x > 0$ when $0 \le x < \pi/2$

17.11C. The function tan *x*

Tan x is defined by the relation $\tan x = \sin x/\cos x$. Clearly $\tan x$ is defined, continuous and derivable for all values of x except those for which $\cos x = 0$, *i.e.*, when $x = (2n+1)\pi/2$, n being any integer.

To show that $\lim_{x\to\pi/2-0} \tan x = \infty$ *and* $\lim_{x\to\pi/2+0} \tan x = -\infty$ **[Delhi Maths (H) 2002, 04]**

Proof. Let k be a positive number.

As $\lim_{x\to\pi/2} \sin x = 1$, there exists $\delta_1 > 0$ such that (taking $\varepsilon = 1/2$)

$$1/2 < \sin x, \qquad \forall x \in [\pi/2 - \delta_1, \pi/2 + \delta_1] \qquad \text{... (1)}$$

Again, since $\lim_{x\to\pi/2} \cos x = 0$, there exists $\delta_2 > 0$ such that

$$-(1/2k) < \cos x < 1/2k, \quad \forall x \in [\pi/2 - \delta_2, \pi/2 + \delta_2]$$

As cos x is positive when $x \in [0, \pi/2]$ and negative when $x \in]\pi/2, \pi]$, we have

$$0 < \cos x < 1/2\,k, \qquad \forall x \in [\pi/2 - \delta_2, \pi/2] \qquad \text{... (2)}$$

and
$$-(1/2k) < \cos x < 0, \qquad \forall x \in]\pi/2, \pi/2 + \delta_2] \qquad \text{... (3)}$$

From (1) and (2), we have if $\delta = \min(\delta_1, \delta_2)$

$$\tan x = \sin x / \cos x > k, \qquad \forall x \in [\pi/2 - \delta, \pi/2[$$

and from (1) and (3),
$$\tan x = \sin x / \cos x < -k, \quad \forall x \in]\pi/2, \pi/2 + \delta]$$

Hence the result.

17.11D. Logarithmic function The integral $\int_1^x \frac{dt}{t}$ exists if and only if $x > 0$ as in that case the integrand is bounded. We write $\log x = \int_1^x \frac{dt}{t}$ so that log x has a meaning if and only if $x > 0$.

Theorem. *The function* log x *is strictly monotonically increasing with range* $]-\infty, \infty[$

Proof We have
$$\log x = \int_1^x \frac{dt}{t}$$

The integrand being positive we see that log x is strictly monotonically increasing. In fact, if $b > a$, we have
$$\log b - \log a = \int_1^b \frac{dt}{t} - \int_1^a \frac{dt}{t} = \int_a^b \frac{dt}{t} > 0$$

Thus, the function is strictly monotonically increasing.

17.11E. Exponential function. Since, as has been seen, $\log x$ is a strict monotonically increasing continuous function with domain $]0, \infty[$ and range $]-\infty, \infty[$, it admits of continuous strictly, monotonically increasing inverse function with domain $]-\infty, \infty[$ and range $]0, \infty[$. This inverse function is called the exponential function and we write $y = \log x \Leftrightarrow x = E(y)$ or $x = \exp(y)$

Then, exp (x) is also defined by
$$\exp(x) = \sum_{n=0}^{\infty} (x^n / n!), \forall x \in \boldsymbol{R} \qquad \text{... (1)}$$

Here the series converges $\forall x \in \boldsymbol{R}$, as its radius of convergence is ∞.

The number *e*. We denote by e the number $E(1)$ so that e is a number such that $\log e = 1$.

We have, by definition, $e^x = E(x \log e) = E(x)$, so that what we have so far denoted by $E(x)$ or exp (x) can as is usual be denoted by e^x.

Theorem. e^x *is a monotonically increasing functio on* $\boldsymbol{R}$.

Proof. From definition (1) of exp (x) by comparison of terms, it follows that

$$0 < x < y \Rightarrow \exp(x) < \exp(y) \quad \text{and so,} \quad -y < -x < 0 \Rightarrow \exp(-y) < \exp(-x)$$

Hence, exp (x) is monotonically increasing on $\boldsymbol{R}$.

17.12. Abel's theorem.

If the series Σa_n is convergent and has the sum s, then the series $\Sigma a_n x^n$ is uniformly convergent for $0 \le x \le 1$ and $\lim_{x \to 1} \Sigma a_n x^n = s$. **[Meerut 2011, 2012, 1013; Kanpur 2007, 08, 10]**

Proof. Since the series Σa_n is convergent, for $n \ge m$ and $\varepsilon > 0$

$$|a_{n+1} + a_{n+2} + \ldots + a_{n+p}| < \varepsilon, p \, \varepsilon \, N$$

Again, since the sequence $<x^n>$ is monotonic decreasing $\forall$ $x \in [0,1]$, we have from Abel's inequality
$$|a_n x^n + a_{n+1} x^{n+1} + \ldots + a_{n+p} x^{n+p}| \le \varepsilon x^n \le \varepsilon, \quad \forall x \in [0,1],$$

showing that $\Sigma a_n x^n$ is continuous function of x in [0, 1] and therefore $\lim_{x \to 1} \Sigma a_n x^n = \Sigma a_n = s$.

MISCELLANEOUS EXAMPLES ON CHAPTER 17

Example 1. *Find the exact interval of absolute convergence and of uniform convergence of each of the following series* : (*i*) $x - \frac{x^3}{3} + \frac{x^5}{5} - \ldots$ (*ii*) $\frac{1}{2.3} + \frac{2}{3.4} x + \frac{3}{4.5} x^2 + \ldots$ **[Delhi 2009]**

Sol. (i) Proceed as in example 6, page 7.6. The given series is absolutely convergent if $-1 < x < 1$. Again, the given series in convergent if $x = 1$ and $x = -1$. Hence, by corollary I of Abel' Theorem on page 17.10, it follows that the given series converges uniformly for $-1 \le x \le 1$.

(ii) Try yourself. The series is absolutely convergent if $-1 < x < 1$ and uniformly convergent if $-1 \le x < 1$.

2. Find the interval of absobute convergence for the following series.

(*i*) $x + (1/2^2) \times x^2 + (2!/3^3) \times x^3 + (3!/4^4) \times x^4 + \ldots$ **(Himachal 2013, 2015)**

(*ii*) $x + x^2/2^2 + x^3/3^3 + x^4/4^4 + \ldots$ **(Delhi 2008)**

Ans. (*i*) $]-e, e[$ (*ii*) $]-\infty, \infty[$

3. Prove that a power series having *R* as radius of convergence convergences absolutely and uniformly in $[-R + \varepsilon, R - \varepsilon]$ with $0 < \varepsilon < R$ **[Mumbai 2010; Delhi Maths (Prog) 2008]**

4. Show that if a function *f* is represented by a power series, then the power series is necessarily its Taylor's series. **(Himanchal 2008)**

5. Let $E(x) = \sum_{n=0}^{\infty} (x^n/n!)$ for $x \in \mathbf{R}$. Show that for x, y $\in \mathbf{R}$ (*i*) E is differentiable over **R** and $E'(x) = E(x)$ (*ii*) $E(x+y) = E(x) \, E(y)$ *and* $E(-x) = 1/E(x)$ (*iii*) E' is strictly increasing. **(Mumbai 2010)**

6. Show that the exponential function is strictly increasing on ***R*** and has range equal to $\boldsymbol{R}^+ = \{y \in \boldsymbol{R} : y > 0\}$. **[Delhi B.Sc. (Hons) II 2011]**

Chapter 18

Double Sequences and Series

18.1 Double sequence. Definition

A double sequence is a collection of numbers having correspondence with each pair of natural numbers. Thus, a double sequence is the range of a function from $N \times N$ to R. A double sequence is usually denoted by $< s_{m,n} >$ where each of m and n varies over R

18.2 Convergence of a double sequence. Definition

A double sequence $<s_{m,n}>$ is said to converge to $s \in R$ if for every $\varepsilon > 0$, there exists $n_0 \in N$ such that

$$|s_{m,n} - s| < \varepsilon, \quad m, n \geq n_0 \qquad \text{... (1)}$$

The number s is known as the limit of the double sequence $< s_{m,n} >$. We can easily verify that s is always unique. Symbolically (1) is written as

$$\lim_{m,n\to\infty} s_{m,n} = s, \quad \text{or} \quad \text{simply as} \quad \lim s_{m,n} = s \quad \text{or} \quad s_{m,n} \to s$$

Such sequences are called *convergent sequences.*

From the above definition, it follows that $< s_{m,n} >$ tends to s when m and n tend to infinity independent of each other.

The concepts of divergence of sequence $< s_{m,n} >$ to ∞, or $-\infty$, or of its oscillation are easily associated.

18.3 Cauchy general principle of convergence for double sequences.

Theorem. *A necessary and sufficient condition for the convergence of double sequence* $< s_{m,n} >$ *is that of every* $\varepsilon > 0$, *there exist* $n_0 \in N$ *such that*

$$| s_{m,n} - s_{p,q} | < \varepsilon \qquad \forall\, m > p \geq n_0 \qquad \text{and} \qquad n > q \geq n_0$$

Proof. Left as an exercise.

18.4 Repeated double limits. Definition

Let $< s_{m,n} >$ be a given sequence. Then,

$$\lim_{m\to\infty}\left(\lim_{n\to\infty} s_{m,n}\right) \quad \text{and} \quad \lim_{n\to\infty}\left(\lim_{m\to\infty} s_{m,n}\right) \text{ are called repeated limits}$$

Theorem : (Pringsheim) *If* $\lim\limits_{m,n\to\infty} s_{m,n}$ *exists and also* $\lim\limits_{n\to\infty} s_{m,n}$ *and* $\lim\limits_{m\to\infty} s_{m,n}$ *exist, then the repeated limits* $\lim\limits_{m\to\infty}\left(\lim\limits_{n\to\infty} s_{m,n}\right)$ *and* $\lim\limits_{n\to\infty}\left(\lim\limits_{m\to\infty} s_{m,n}\right)$ *exist and all the three are equal.*

Proof. Let s denote the double limit of $< s_{m,n} >$, *i.e.*, $\lim\limits_{m,n} s_{m,n} = s$. Then for any arbitrary $\varepsilon > 0$, we have $n_0(\varepsilon) \in N$ such that

$$\left|s_{m,n}-s\right|<\varepsilon/2, \quad \forall\, m, n\geq n_0$$

$$\Rightarrow \qquad s-\varepsilon/2<s_{m,n}<s+\varepsilon/2, \quad \forall\, m, n\geq n_0$$

Letting $n\to\infty$, the above inequalities yield

$$s-\varepsilon/2\leq \lim_{n\to\infty} s_{m,n}\leq s+\varepsilon/2, \forall\, m\geq n_0, \text{ since repeated limits exist}$$

Again, letting $m\to\infty$, we have

$$s-\varepsilon/2\leq \lim_{m\to\infty}\lim_{n\to\infty} s_{m,n}\leq s+\varepsilon/2, \text{ since repeated limit exist}$$

Thus, $$\left|\lim_{m}\lim_{n} s_{m,n}-s\right|\leq\varepsilon/2<\varepsilon \qquad \text{... (1)}$$

Since ε is arbitrary, (1) $\Rightarrow$ $\lim\limits_{m}\lim\limits_{n} s_{m,n}=s$

Likewise, we have $\lim\limits_{n}\lim\limits_{m} s_{m,n}=s$

Thus, we have $\lim\limits_{m}\lim\limits_{n} s_{m,n}=\lim\limits_{n}\lim\limits_{m} s_{m,n}=\lim\limits_{m,n} s_{m,n}$

Remarks. Double limits may exist even though repeated limits do not exist.

The mere existence and even equality of the repeated limits do not guarantee the existence of the double limit. We consider the following examples to support the above remarks.

Example. *Show that the sequence* $<(mn^2)/(m^2+n^4)>$ *has its repeated limits but not the double limit*

Sol. We write $$s_{m,n}=(mn^2)/(m^2+n^4)$$

$$\therefore \quad \lim_{m\to\infty}\lim_{n\to\infty} s_{m,n}=\lim_{m\to\infty}\lim_{n\to\infty}\frac{m/n^2}{(m^2/n^2+1)}=\lim_{m\to\infty}\frac{0}{1}=0$$

and $$\lim_{n\to\infty}\lim_{m\to\infty} s_{m,n}=\lim_{n\to\infty}\lim_{m\to\infty}\frac{n^2/m}{1+n^4/m^2}=\lim_{n\to\infty}\frac{0}{1}=0$$

Thus, the repeated limits exist and are equal.

Now, if $m=n^2$, then as $n\to\infty$, $m\to\infty$ and we have

$$\lim_{m,n\to\infty} s_{m,n}=\lim_{n\to\infty}\frac{n^2\times n^2}{n^4+n^4}=\lim_{n\to\infty}\frac{1}{2}=\frac{1}{2}$$

Again, if $m=2n^2$, then as $n\to\infty$, $m\to\infty$ and we have

$$\lim_{m,n\to\infty} s_{m,n}=\lim_{n\to\infty}\frac{(2n^2)\times n^2}{4n^4+n^4}=\lim_{n\to\infty}\frac{2}{5}=\frac{2}{5}$$

Hence double limit of $<(mn^2)/(m^2+n^4)>$ does not exist.

18.5 Double series. Definition

If $<a_{m,n}>$ is a double sequence of real numbers, then a double series is defined as a double summation extended over all terms $a_{m,n}$ which are generally given for all $m, n\in N$. Unless stated otherwise, the terms are to be summed up in an arbitrary manner.

A double series is denoted by $\sum_{m,n=1}^{\infty} a_{m,n}$ or more briefly by $\sum_{m,n} a_{m,n}$ or $\Sigma a_{m,n}$

Since in its summation no term $a_{m,\,n}$ with finite suffixes m, n escapes addition, before $(m^2+n^2)\to\infty$, the sums like (repeated series) $\sum_{m=1}^{\infty}\left(\sum_{n=1}^{\infty} a_{m,n}\right)$ and $\sum_{n=1}^{\infty}\left(\sum_{m=1}^{\infty} a_{m,n}\right)$ are not treated as double series.

A double series $\Sigma a_{m,n}$ can be written as

$$\begin{array}{l} a_{11} + a_{12} + a_{13} + a_{14} + \ldots\ldots + a_{1n} + \ldots\ldots \\ + a_{21} + a_{22} + a_{23} + a_{24} + \ldots\ldots + a_{2n} + \ldots\ldots \\ + a_{31} + a_{32} + a_{33} + a_{34} + \ldots\ldots + a_{3n} + \ldots\ldots \\ \text{— — — — — — — — — — — — —} \\ + a_{m1} + a_{m2} + a_{m3} + a_{m4} + \ldots\ldots + a_{mn} + \ldots\ldots \\ \text{— — — — — — — — — — — — —} \\ \text{— — — — — — — — — — — — —} \end{array}$$

The sum of terms contained in the finite rectangle of the first m rows and n columns is (m, n)th partial sum, which is denoted by $s_{m,\,n}$. Thus, we have

$$s_{m,n} = a_{11} + a_{12} + \ldots + a_{1n} + a_{21} + a_{22} + \ldots + a_{2n} + \ldots + a_{m1} + a_{m2} + \ldots + a_{mn}$$

If m and n are finite, we can sum up terms in $s_{m,\,n}$ in any manner. However, if $m, n \to \infty$, we must take into account the manner in which the terms of $s_{m,\,n}$ are summed up. We shall use the following notations:

$\lim_{m,n} s_{m,n}$ denotes sum by rectangles, $\lim_{n} s_{n,n}$ denotes sum by squares,

$\lim_{m}\left(\lim_{n} s_{m,n}\right)$ denotes sum by rows, $\lim_{n}\left(\lim_{m} s_{m,n}\right)$ denotes sum by columns.

There are variety of manners of summing up the terms of non-negative numbers $a_{m,\,n}$. For example,

$\sum_{n=1}^{\infty}\left(\sum_{m=1}^{n} a_{m,n-m+1}\right) = a_{11} + (a_{21}+a_{12}) + (a_{31}+a_{22}+a_{13}) + \ldots$ is the sum by triangles, or diagonals.

All the above sums consisting of non-negative numbers only either converage or diverage to ∞.

The rectangular array is said to converge or diverge by rectangles, squares, rows, columns, or triangles according as the corresponding limit converge or diverge to ∞.

Consider dots as the terms of the double infinite reactangular array and the corresponding line as an illustration of the region within which the terms are to be summed up. Then, we have

I. First mode of summation

Sum of rectangles

$\lim_{m,n} s_{m,\,n}$

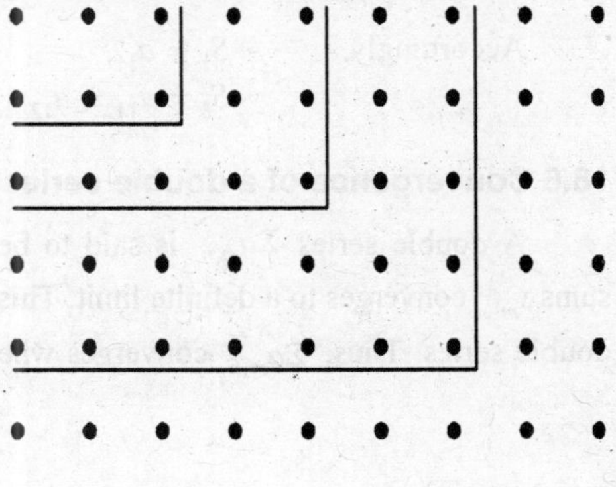

II. Second mode of summation

Sum by squares

$\lim_{n} s_{n,n}$

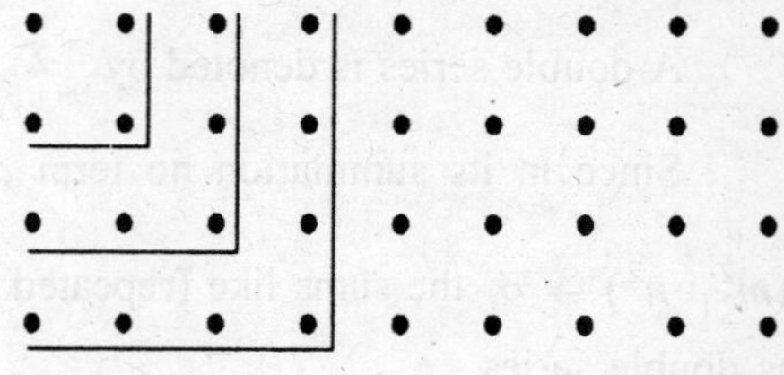

III. Third mode of summation

Sum by triangles

$\sum_{n=1}^{\infty} \sum_{m=1}^{n} a_{m,n-m+1}$

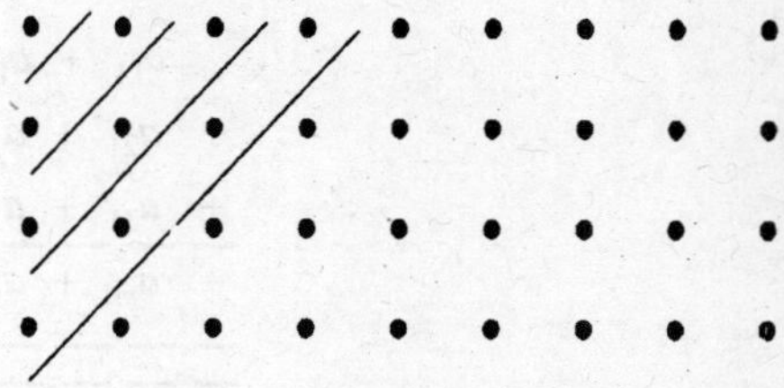

IV. Forth mode of summation

Sum by columns

$\sum_{n=1}^{\infty} \sum_{m=1}^{\infty} a_{m,n}$

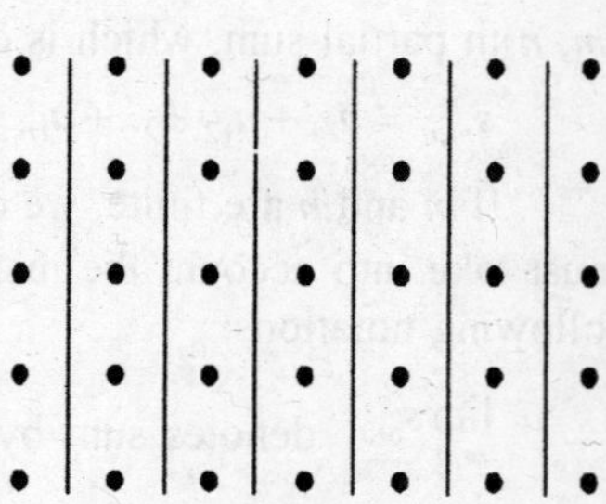

V. Fifth mode of summation

Sum by rows

$\sum_{m=1}^{\infty} \sum_{n=1}^{\infty} a_{m,n}$

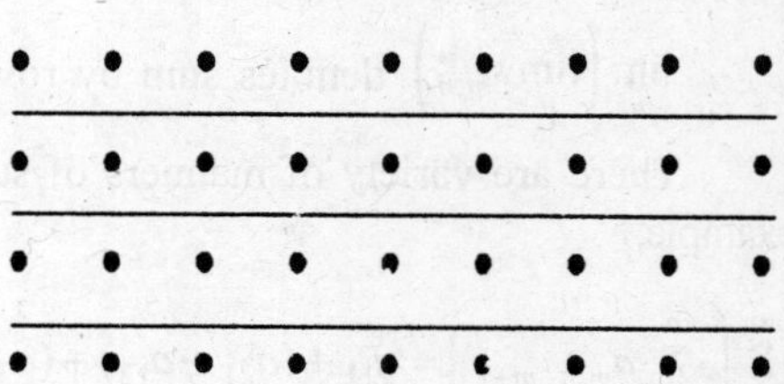

Remark : We have $a_{n,n} = s_{m,n} - s_{m-1,n} - s_{m,n-1} + s_{m-1,n-1}$ for $m, n \geq 2$

and $a_{n,n} = s_{n,n} - s_{n-1,n} - s_{n,n-1} + s_{n-1,n-1}$ for $n \geq 2$

This gives the sum contained in the first n rows and n columns.

Notation. $\sum_{p \leq n} \sum_{q \leq n} a_{m,n}$ is denoted by S_n, *i.e.*, $S_n = s_{n,n}$.

Accordingly, $S_1 = a_{11}$, $S_2 = a_{11} + a_{12} + a_{21} + a_{22}$,

$S_3 = a_{11} + a_{12} + a_{13} + a_{21} + a_{22} + a_{23} + a_{31} + a_{32} + a_{33}$ and so on.

18.6 Convergence of a double series. Definition.

A double series $\Sigma a_{m,n}$ is said to be convergent if the double sequence of $(m, n)^{\text{th}}$ partial sums $s_{m,n}$ converges to a definite limit. This definite limit is known as the (Pringsheim's) sum of the double series. Thus, $\Sigma a_{m,n}$ converges whenever $\sum_{m,n} s_{m,n}$ exists finitely.

In case, if $< s_{m,n} >$ diverges to ∞, or $-\infty$ or oscillates, the series $\Sigma a_{m,n}$ is said to diverge to ∞, or $-\infty$ or oscillate.

The limits $\lim_{m} \lim_{n} s_{m,n}$ and $\lim_{n} \lim_{m} s_{m,n}$ are known as the repeated sums of $\Sigma a_{m,n}$. Observe that the repeated sum $\lim_{m} \lim_{n} s_{m,n}$ is the sum of the sequences of row totals, *i.e.*, if all the rows are summed up and then these sums are added, Thus, we arrive at $\lim_{m} \lim_{n} s_{m,n}$. Likewise $\lim_{n} \lim_{m} s_{m,n}$ is the sum of the sequences of column totals

Example 1. *The double series $\Sigma a_{m,n}$ whose (m, n)th partial sum $s_{m,n} = 1/m + 1/n$ is convergent, as* $\lim_{m,n}(1/m+1/n)=0$. *Hence the sum of $\Sigma a_{m,n}$ is zero.*

Example 2. The double series $\Sigma a_{m,n}$ whose (m, n) th partial sum $s_{m,n} = m + n$ is divergent, as $\lim_{m,n}(m+n)=\infty$.

Example 3. The double series $\Sigma a_{m,n}$ whose (m, n)th partial sum $s_{m,n} = (-1)^{m+n}$ is oscillatory, as $\limsup_{m,n} s_{m,n} = 1$ and $\liminf_{m,n} s_{m,n} = -1$

Cauchy's criterian of convergence of a double series. Theorem

A double series $\Sigma a_{m,n}$ converges if and only if for every $\varepsilon > 0$, there exists $n_0 \in N$ such that

$$\left|s_{m+p,n+q} - s_{m,n}\right| < \varepsilon, \ \forall \ m, n \geq n_0 \ \ p \geq 0 \text{ and } q \geq 0$$

and

$$s_{m,n} = \sum_{\alpha=1}^{m} \sum_{\beta=1}^{n} a_{\alpha,\beta}$$

Proof: The proof follows from the corresponding Cauchy's theorem on double sequence (see Art 18.3) when applied to $s_{m,n}$ *i.e.*, the (m, n)th partial sum of $\Sigma a_{m,n}$

Corollary. If $\Sigma a_{m,n}$ is convergent, then $\lim_{m,n} a_{m,n} = 0$

18.7 Double series of positive terms

Theorem I. *A double series $\Sigma a_{m,n}$ of positive terms is convergent if and only if $< S_n >$ is convergent.*

Proof. *The condition is necessary.* Let the double series $\Sigma a_{m,n}$ of positive terms be convergent. Then, by definition $< S_n >$ is convergent.

The condition is sufficient. Let $< S_n >$ be convergent and let it converge to s. Then, for $\varepsilon > 0$, there exists $n_0 \in N$ such that

$$| S_n - s | < \varepsilon, \text{ if } n \geq n_0 \qquad i.e., \qquad s - \varepsilon < S_n < s + \varepsilon \quad \forall \ n \geq n_0$$

Again, $\qquad m, n \geq n_0 \quad \Rightarrow \quad S_{m+n} \geq s_{m,n} \geq S_{n_0} > s - \varepsilon$

Hence, $\qquad s - \varepsilon < s_{m,n} < s + \varepsilon \qquad \forall \ m, n \geq n_0$

Since ε is arbitrary, we have $\lim_{m,n} s_{m,n} = s$ and so $\Sigma a_{m,n}$ is convergent.

Remark 1. By Dirichlet's theorem (refer Art. 7.8 chapter 7), every re-arrangement of a positive term convergent series is convergent and the sum also does not change. It follows that a double series of positive terms remains convergent even if the order of the terms is changed in any way. This is also true for a double series diverging to ∞.

Remark 2. A double series of positive terms can either converge or diverge to ∞.

Theorem II. *If a double series of positive terms is convergent, then so also is each row and each column and then the repeated sums, i.e., sum of the row totals and the sum of the column totals are each equal to the sum of the double series.*

Proof. Let $\Sigma a_{m,n}$ be a double series of positive terms and let it converge to s. Let $s_{m,n}$ be the (m, n)th partial sum. Then, we have $s_{m,n} \le s.$

Now, for a fixed m, $\sum\limits_{\beta=1}^{n} a_{m,\beta}$ is a monotone increasing sequence and bounded above by s. Therefore, it is convergent. Likewise, for a fixed n, $\sum\limits_{\alpha=1}^{m} a_{\alpha,n}$ is a monotone increasing and bounded above by s. Therefore it is also convergent.

Note that the partial sums of the rows totals being a monotone increasing sequence bounded above by s is convergent and so also is the case of partial sums of column totals. Hence the repeated sums exist.

Since $< s_{m,n} >$ is convergent, for $\varepsilon > 0$, there exists $n_0(\varepsilon) \in N$ such that

$$|s_{m,n} - s| < \varepsilon/2, \forall\, m, n \ge n_0, \quad i.e., \quad s - \varepsilon/2 < s_{m,n} < s + \varepsilon/2 \quad \text{... (1)}$$

Taking limit as $n \to \infty$, (1) gives

$$s - \varepsilon/2 \le \lim_{n} s_{m,n} \le s + \varepsilon/2, \forall\, m \ge n_0 \quad \text{... (2)}$$

Next, taking limit as $m \to \infty$, (2) gives

$$s - \varepsilon/2 \le \lim_{m} \lim_{n} s_{m,n} \le s + \varepsilon/2, \quad \text{or} \quad |\lim_{m} \lim_{n} s_{m,n} - s| \le \varepsilon/2 < \varepsilon \quad \text{... (3)}$$

Since ε is arbitrary, (3) $\Rightarrow$ $\lim\limits_{m} \lim\limits_{n} s_{m,n} = s$ i.e., $\sum\limits_{n}\sum\limits_{m} a_{m,n} = s$

Similarly, $\sum\limits_{m}\sum\limits_{n} a_{m,n} = s$

Theorem III. *If any one of two repeated sums of a double series of positive terms is finite then so is the other and then the double series is also convergent and has the same sum.*

Proof. Left as an exercise.

18.8 Some results and tests for convergence of a double series of positive terms.

Result I Every particular manner of summation of a double series either converges or diverges to ∞.

Result II. The double series converges to the sum s iff any of its particular manner of summatioin converges to the sum s.

Result III. The double diverges to ∞ iff any of its particular manner of summation diverges to ∞.

Test I. *Let* $\Sigma a_{m,n}$ *and* $\Sigma b_{m,n}$ *be two double series of positive terms. Then*

(*i*) $\Sigma a_{m,n}$ *converges if* $\Sigma b_{m,n}$ *is convergent and* $a_{m,n} \le b_{m,n}$ *for all but finitely many m and n.*

(*ii*) $\Sigma a_{m,n}$ *diverges to* ∞ *if* $\Sigma b_{m,n}$ *diverges to* ∞ *and* $a_{m,n} \ge b_{m,n}$ *for infinitely many m or n.*

Proof. Left as an exercise.

Test II. Let c_r denote a function of r, *e.g.*, $c_m = m^3$ or $c_n = 1/n + 2$. Then the double series $\Sigma a_{m,n}$ of positive terms is convergent if $a_{m,n} = 1/c_m c_n$ and series $\sum_r 1/c_r$ is convergent, where $c_r > 0$

Proof. Left as an exercise.

Test III. *The double series* $\Sigma a_{m,n}$ *of positive terms is convergent or divergent according as* $\sum_p (1/c_p)$ *is convergent or divergent, where* $a_{m,n} = 1/(pc_p)$ *and* $p = m + n$.

Proof. Left as an exercise

Example 1. *Show that* $\Sigma\, 1/(m^\alpha n^\beta)$ *converges if* $\alpha > 1, \beta > 1$

Sol. Since $\Sigma(1/m^\alpha n^\beta) = \Sigma\{(1/m^\alpha) \times (1/n^\beta)\}$, so by test I, if $\Sigma 1/m^\alpha$ converges, then $\Sigma(1/m^\alpha n^\beta)$ also converges. Thus, if $\alpha > 1, \beta > 1$, the series $\Sigma(1/m^\alpha)$ and $\Sigma(1/n^\beta)$ converge and hence $\Sigma(1/m^\alpha n^\beta)$ converges.

Example 2. *Show that* $\Sigma\left\{1/(m+n)^\alpha\right\}$ *converges if* $\alpha > 2$ *and diverges if* $\alpha < 2$.

Solution. Here $\Sigma \dfrac{1}{(m+n)^\alpha} = \Sigma \dfrac{1}{(m+n)\,(m+n)^{-\alpha}}$ and so by test III, $\Sigma \dfrac{1}{(m+n)^\alpha}$ converges (diverges) if $\Sigma \dfrac{1}{(m+n)^{\alpha-1}}$ converges (diverges).

Again, we know that $\Sigma\left\{1/(m+n)^{\alpha-1}\right\}$ converges or diverges according as $\alpha - 1 > 1$ or $\alpha - 1 \le 1$. Thus, the given double series converges if $\alpha > 2$ and diverges if $\alpha \le 2$.

18.9 Mixed series Definition

If the terms of a double series are arbitrarily positive or negative, the series is known as a *mixed series.*

Absolute convergence. Definition. A mixed double series $\Sigma a_{m,n}$ is said to be absolutely convergent if the double series $\Sigma |a_{m,n}|$ of absolute values of the terms of $\Sigma a_{m,n}$ is convergent.

Theorem I. *Every absolutely convergent double series is convergent.*

Proof. Use Cauchy's criterion of convergence to prove the theorem.

Theorem II. (Jordon). *The sum of a non-absolutely convergent double series may have different values according to the mode of summation but for an absolutely convergent series, the sum remains unchanged.*

Proof. Left as an exercise.

Remark. The converse of theorem I is not true, *i.e.*, a double series may be convergent but it may not be absolutely convergent. The following example proves this assertion.

Example. *Prove that the double series* $\Sigma\{(-1)^{m+n}/mn\}$ *converges but not absolutely.*

Sol. We write $a_{m,n} = (-1)^{m+n}/mn$. Let $s_{m,n}$ be the (m, n)th partial sum of $\Sigma a_{m,n}$.

$$\therefore \qquad s_{m,n} = \left\{-1 + 1/2 - 1/3 + \ldots + (-1)^m/m\right\} \times \left\{-1 + 1/2 - 1/3 + \ldots + (-1)^n/n\right\}$$

$$= \{1-1/2+1/3-...-(-1)^m/m\}\times\{1-1/2+1/3-...-(-1)^n/n\}$$

$$\Rightarrow \quad \lim_{m,n} s_{m,n} = \lim_{m\to\infty, n\to\infty}\left[\{1-1/2+1/3-...-(-1)^m/m\}\times\{1-1/2+1/3-...-(-1)^n/n\}\right]$$

$$= \lim_{m\to\infty}\{1-1/2+1/3-...-(-1)^m/m\}\times\lim_{n\to\infty}\{1-1/2+1/3-...-(-1)^n/n\}$$

$$= (\log 2)\times(\log 2) = (\log 2)^2$$

$$\Rightarrow \quad \Sigma\{(-1)^{m+n}/mn\} \text{ is convergent.}$$

Again, $\Sigma|a_{m,n}| = \Sigma(1/mn)$ diverges by test II (see Art. 18.8)

[$\because$ $\Sigma 1/m$ and $\Sigma 1/n$ are divergent]

18.10 Solved examples

Ex. 1. *Show that the double series shown in margin has the sum* 1 *if summed by rows and* -1 *if summed by columns.*

$$\begin{array}{ccccccccccc} & 0 &+& 1 &+& 0 &+& 0 &+& 0 &+ ......\\ - & 1 &+& 0 &+& 1 &+& 0 &+& 0 &+ ......\\ + & 0 &-& 1 &+& 0 &+& 1 &+& 0 &+ ......\\ + & 0 &+& 0 &-& 1 &+& 0 &+& 1 &+ ......\\ + & \multicolumn{10}{l}{..............................} \end{array}$$

Sol. Let the given series be denoted by $\Sigma a_{m,n}$.

Let R_i denote the ith row total and C_j denote the jth column total

Then, $R_1 = 1$ and $R_i = 0,\ \forall\, i \ge 2$ and $C_1 = -1$ and $C_j = 0,\ \forall\, j \ge 2$

$\therefore$ Sum by rows $= \sum_m\sum_n a_{m,n} = \sum_m R_m = (1+0+0+...) = 1$

and sum by columns $= \sum_n\sum_m a_{m,n} = \sum_n C_n = (-1+0+0+...) = -1$

Ex. 2. *For* $|x|<1$, *convert the series* $x/(1+x^2)+x^2/(1+x^4)+x^3/(1+x^6)+....$ *into a double series and prove that it is equal to* $x/(1-x)-x^3/(1-x^3)+x^5/(1-x^5)-....$

[Delhi Maths (H) 1999, 2007]

Or *Convert the series* $x/(1+x^2)+x^2/(1+x^4)+x^3/(1+x^6)+ ....$ *into a double series and examine it for absolute convergence by rows and columns.* **[Delhi Maths (H) 2006]**

Sol. Re-writing the terms of the given series, we have

$$\begin{array}{l|llllllll} x/(1+x^2) = & x & - & x^3 & + & x^5 & - & x^7 & + \;....\\ x^2/(1+x^4) = & +\;x^2 & - & x^6 & + & x^{10} & - & x^{14} & + \;....\\ x^3/(1+x^6) = & +\;x^3 & - & x^9 & + & x^{15} & - & x^{21} & + \;....\\ .... & \multicolumn{8}{l}{..} \end{array} \quad ...(1)$$

The given series is $x/(1+x^2)+x^2/(1+x^4)+x^3/(1+x^6)+.....$ ... (2)

The other series is $x/(1-x)-x^3/(1-x^3)+x^5/(1-x^5)-....$ ... (3)

From (1), it follows that the series (2) and (3) represent respectively the sum by rows and the sum by columns of the double series (1) and that these two sums will be equal provided (1) converges absolutely. We now proceed to prove the absolute convergence of (1).

Replacing x by $|x|$ and taking all the terms with positive sign, the sum by rows of the double series, so amended

$$= y/(1-y^2)+y^2/(1-y^4)+y^3/(1-y^6)-...., \text{ when } y = |x| < 1 \quad ...(4)$$

Let the series (4) be denoted by Σu_n. Then, we have

$$u_n = y^n/(1-y^{2n}) \qquad \text{and} \qquad u_{n+1} = y^{n+1}/(1-y^{2n+2})$$

$$\therefore \quad \lim_{n\to\infty} \frac{u_n}{u_{n+1}} = \lim_{n\to\infty} \frac{1}{y} \times \frac{1-y^{2n+2}}{1-y^{2n}} = \frac{1}{y} = \frac{1}{|x|} > 1, \text{ as } |x| < 1$$

Hence by D'Alembert's ratio test, (4) is convergent and so (1) is absolutely convergent. Thus, we see that (1) converges by rows and it will behave similarly by columns as well and the two sums will be equal. Therefore, the sum of (2) is equal to the sum of (3). Hence the result.

Ex. 3. *Convert the series* $x/(1+x) - 2x^2/(1+x^2) + 3x^3/(1+x^3) \; - 4x^4/(1+x^4) + \ldots, |x| < 1$ *into a double series and show that it is equal to*

$x/(1+x^2)^2 - x^2/(1+x^2)^2 + x^3/(1+x^2)^2 - x^4/(1+x^2)^2 + \ldots$ **[Delhi Maths (H) 2004, 11]**

Or If $|x| < 1$, show that $x/(1+x) - 2x^2/(1+x^2) + 3x^3/(1+x^3) - \ldots\ldots$

$= x/(1+x^2)^2 - x^2/(1+x^2)^2 + x^3/(1+x^2)^2 - \ldots\ldots$ **[Delhi Maths (H) 1998]**

Sol. Re-writing the terms of $x/(1+x) - 2x^2/(1+x^2) + 3x^3/(1+x^3) - \ldots\ldots$ we get

$$\begin{array}{rc|l}
x/(1+x) & = & x \; - \; x^2 \; + \; x^3 \; - \; \ldots\ldots \\
-2x^2/(1+x^2) & = & -\; 2x^2 \; + \; 2x^4 \; - \; 2x^6 \; - \; \ldots\ldots \\
3x^3/(1+x^3) & = & 3x^3 \; - \; 3x^6 \; + \; 3x^9 \; - \; \ldots\ldots \\
\ldots\ldots\ldots\ldots & & \ldots\ldots\ldots\ldots\ldots\ldots
\end{array} \qquad \ldots (1)$$

The given series is $\quad x/(1+x) - 2x^2/(1+x^2) + 3x^3/(1+x^3) - \ldots\ldots \quad$ ... (2)

The other series is $\quad x/(1+x^2)^2 - x^2/(1+x^2)^2 + x^3/(1+x^2)^2 - \ldots\ldots \quad$ ... (3)

From (1), it follows that the series (2) and (3) represent respectively the sum by rows and the sum by columns of the double series (1) and that these two sums will be equal provided (1) converges absolutely. We now proceed to prove the absolute convergence of (1).

Replacing x by $|x|$ and taking all the terms with positive sign, the sum by rows of the double series, so amended

$$= y/(1-y) + 2y^2/(1-y^2) + 3y^3/(1-y^3) + \ldots, \text{ where } y = |x| < 1 \qquad \ldots (4)$$

Let the series (4) be denoted by Σu_n. Then, we have

$$u_n = ny^n/(1-y^n) \qquad \text{and} \qquad u_{n+1} = (n+1)y^{n+1}/(1-y^{n+1})$$

$$\therefore \quad \lim_{n\to\infty} \frac{u_n}{u_{n+1}} = \lim_{n\to\infty} \frac{ny^{-1}}{n+1} \; \frac{1-y^{n+1}}{1-y^n} = \lim_{n\to\infty} \frac{y^{-1}}{1+(1/n)} \; \frac{1-y^{n+1}}{1-y^n} = \frac{1}{y} = \frac{1}{|x|} > 1$$

Hence, series (4) is convergent, by D'Alembert's ratio test. Therefore, (1) is absolutely convergent by rows and consequently it will do so by columns as well and the two sums will be equal.

Ex. 4. *Show that for* $|x| < 1$, $x/(1+x^2) + x^3/(1+x^6) + x^5/(1+x^{10}) - \ldots = x/(1-x^2) - x^3/(1-x^6) + x^5/(1-x^{10}) - \ldots$ **[Delhi Maths (H) 2000, 03, 08]**

Or *Convert the series* $x/(1+x^2) + x^3/(1+x^6) + x^5/(1+x^{10}) - \ldots\ldots$ into a double series and show that it is equal to $x/(1-x^2) - x^3/(1-x^6) + x^5/(1-x^{10}) - \ldots$, for $|x| < 1$ **[Delhi Maths (H) 2001]**

Sol. Re-writing the terms of $x/(1+x^2) + x^3/(1+x^6) + x^5/(1+x^{10}) - \ldots$, we get

$$\begin{array}{rcl|lllllll} x/(1+x^2) & = & x & - & x^3 & + & x^5 & - & \ldots\ldots \\ x^3/(1+x^6) & = & x^3 & - & x^9 & + & x^{15} & + & \ldots\ldots \\ x^5/(1+x^{10}) & = & x^5 & - & x^{15} & + & x^{25} & - & \ldots\ldots \end{array} \quad \ldots(1)$$

The given series is $\qquad x/(1+x^2)+x^3/(1+x^6)+x^5/(1+x^{10})-\ldots\ldots$ ... (2)

Other series is $\qquad x/(1-x^2)-x^3/(1-x^6)+x^5/(1-x^{10})-\ldots\ldots$ ... (3)

From (1), it follows that the series (2) and (3) represent respectively the sum by rows and the sum by columns of the double series (1) and that these two sums will be equal provided (1) converges absolutely. We now proceed to prove the aboslute convergence of (1).

Replacing x by $|x|$ and taking all the terms with positive sign, the sum by rows of the double series, so amended

$$= y/(1-y^2)+y^3/(1-y^6)+y^5/(1-y^{10})+\ldots., \text{ where } y=|x|<1 \qquad \ldots(4)$$

Let the series (4) be denoted by $\sum u_n$. Clearly nth term of sequence $y, y^3, y^5, \ldots$ is y^{2n-1}

Let S_n and T_n denote the sum of first n term and nth term of the series $2+6+10+\ldots$. Then,

$$S_n = 2 + 6 + 10 + \ldots + T_{n-1} + T_n \qquad \ldots(5)$$

Again, $$S_n = \quad 2 + 6 + \ldots + T_{n-1} + T_n \qquad \ldots(6)$$

Subtracting (6) from (5), we get $\quad 0 = 2 + \{4+4+\ldots (n-1) \text{ terms}\} - T_n$

$\Rightarrow \quad T_n = 2+4(n-1) \quad \Rightarrow \quad T_n = 4n-2$

$\therefore \; u_n = y^{2n-1}/(1-y^{4n-2})$ and so $\quad u_{n+1} = y^{2n+1}/(1-y^{4n+2})$

$$\therefore \quad \lim_{n\to\infty} \frac{u_n}{u_{n+1}} = \lim_{n\to\infty} \frac{1}{y^2}\,\frac{1-y^{4n+2}}{1-y^{4n-2}} = \frac{1}{y^2} = \frac{1}{|x|^2} > 1, \text{ as } |x|<1$$

$\Rightarrow$ Series (4) is convergent by D'Alembert's ratio test.

Therefore, (1) is absolutely convergent by rows and hence it will do so by columns as well and the two sums will be equal.

EXERCISES

1. *Prove that for a double series of positive terms, either the sum by rectangles, the sum by squares and the sum by triangles are equal or else all the three do not exist.* **[Delhi Maths (H) 2005]**

2. *Prove that a positive term double series $\Sigma a_{m,n}$ is convergent (by rectangles) iff it converges by squares and then both the sums are equal.* **[Delhi Maths (H) 2002]**

3. *Let $\Sigma a_{m,n}$ be a double series which converges by rows, columns and rectangles. Show that their sums are equal i.e., $R(\Sigma a_{m,n}) = C(\Sigma a_{m,n}) = \Sigma a_{m,n}$* **[Delhi Maths (H) 2001, 08]**

4. *If in a double series the sum by rows (column) exist and the sum by rectangles exist, then prove that the two are equal.* **[Delhi Maths (H) 2000]**

5. *Prove that* $\displaystyle \sum_{m=1}^{\infty}\sum_{n=1}^{\infty} \frac{(-1)^n}{(m+n^2)(m+n^2-1)} = -\frac{\pi^2}{12}$ **[Delhi Maths (H) 1999]**

6. *Prove that the double series* $\sum\limits_{m,n=1}^{\infty} \dfrac{1}{(m+n^2)(m+n^2-1)}$ *converges to the sum* $\dfrac{\pi^2}{6}$.

7. *Show that* $\dfrac{x}{1+x^2} - \dfrac{x^2}{1+x^4} + \dfrac{x^3}{1+x^6} - \dots = \dfrac{x}{1+x} - \dfrac{x^3}{1+x^3} + \dfrac{x^5}{1+x^5} - \dots, \ |x|<1$

$$= \frac{1}{1+x} - \frac{1}{1+x^3} + \frac{1}{1+x^5} - \dots, \ |x|>1$$

[Hint: To solve the second part, replace x by $1/x$]

8. *Show that the double series*

$$\begin{array}{l} 3 + 0 - 1 + 0 + 0 + 0 + \dots \\ + 0 + 2 + 0 - 1 + 0 + 0 + \dots \\ - 1 + 0 + 2 + 0 - 1 + 0 + \dots \\ + 0 - 1 + 0 + 2 + 0 - 1 + \dots \\ \hline \end{array}$$

has the sum 2 *by rows and by columns and by diagonals, but the series does not converge.*

9. *If* $|x| < 1$, *prove that* $\dfrac{x}{1-x} + \dfrac{x^2}{1-x^2} + \dfrac{x^3}{1-x^3} + \dots = x\dfrac{1+x}{1-x} + x^4\dfrac{1+x^2}{1-x^2} + x^9\dfrac{1+x^3}{1+x^3} + \dots$

[Delhi Maths (H) 2009]

10. *If* $a > 0$, $b > 0$, *prove that* $\sum\limits_{m,n=1}^{\infty} \dfrac{1}{(am+bn)^p}$, *converges if* $p > 2$.

11. *Show that* $\Sigma(1/m^2n^2)$ *is convergent. Show also that the repeated sum exist and are equal.*

18.11. TAYLOR'S THEOREM FOR POWER SERIES **(Himanchal 2010]**

Let $\sum\limits_{n=0}^{\infty} a_n x^n$ *be a power series with radius of convergence R and let* $f(x) = \sum\limits_{n=0}^{\infty} a_n x^n$, $|x| < R$.

Then for any $a \in]-R, R[$, *prove that f can be expanded in a power series about 'a' wihich converges for* $|x-a| < R - |a|$ *and*

$$f(x) = \sum_{n=0}^{\infty} \frac{f^{(n)}(a)}{n!}(x-a)^n, \qquad \text{where} \qquad |x-a| < R - |a|$$

Proof. Assume that $|x-a| < R-|a|$. Then, $|x| \le |x-a| - |a| < R$ and so $\sum a_n x^n$ converges.

Now, $f(x) = \sum\limits_{n=0}^{\infty} a_n x^n = \sum\limits_{n=0}^{\infty} a_n \{(x-a)+a\}^n = \sum\limits_{n=0}^{\infty} a_n \sum\limits_{m=0}^{n} {}^nC_m\, a^{n-m} (x-a)^m$ ...(1)

We now proceed to change the order of summation on R.H.S. of (1). For proving its validity observe that it is the summation by rows of a double series (refer Art. 18.7), which if convergent absolutely, will converge by columns a well to the same sum.

Replace all quanties by their moduli and take all terms with positive sign on R.H.S. of (1). Then,

$$\text{R.H.S. of (1)} = \sum_{n=0}^{\infty} |a_n| \sum_{m=0}^{n} {}^nC_m |a_n|^{n-m} |x-a|^m = \sum_{n=0}^{\infty} |a_n| (|x-a|+|a|)^n,$$

which is a power series . Since $|x-a|+|a|<R$, this series is convergent. It follows that change of order of summation is valid for $|x-a|<R-|a|$ on R.H.S. of (1) and therefore, we get

$$f(x) = \sum_{m=0}^{\infty} \left\{ \sum_{n=m}^{\infty} {}^nC_m\, a_n\, a^{n-m} \right\} (x-a)^m \qquad \text{... (2)}$$

Given, $\quad f(x) = a_0 + a_1x + a_2x^2 + ... + a_mx^m + ... + a_nx^n + ...$ ... (3)

From (3), $\quad f'(x) = a_1 + 2a_2x + 3a_3x^2 + ... + ma_mx^{m-1} + ... + na_nx^{n-1} + ...$ ... (4)

From (4), $\quad f''(x) = 2!\, a_2 + 3\cdot 2\, a_3x + ... + m(m-1)a_mx^{m-2} + ... + n(n-1)a_nx^{n-2} + ...$

Proceeding likewise , we obtain

$$f^{(m)}(x) = m!a_m + (m+1)\,m\,(m-1) \cdots 3\cdot 2a_{m+1}x + \cdots + n(n-1).... (n-m+1)a_nx^{n-m} + ...$$

or $\quad f^{(m)}(x) = m!(a_m + {}^{m+1}C_m\, a_{m+1}x + {}^{m+2}C_m a_{m+2}x^2 + ...) = m! \sum_{n=m}^{\infty} {}^nC_m\, a_n\, x^{n-m}$

Hence, $\quad f^{(m)}(a) = m! \sum_{n=m}^{\infty} {}^nC_m\, a_n\, x^{n-m} \quad$ *so that* $\quad \sum_{n=m}^{\infty} {}^nC_m\, a_n\, a^{n-m} = \left\{ f^{(m)}(a) \right\} / m!$ ... (5)

Using (5), (2) yields $\quad f(x) = \sum_{m=0}^{\infty} \frac{f^{(m)}(a)}{m!}(x-a)^m \quad$ so that $\quad f(x) = \sum_{n=0}^{\infty} \frac{f^{(n)}(a)}{n!}(x-a)^n$

Chapter

19

Metric Spaces

19.1 INTRODUCTION.

During our study of theory of functions of real variables, the limit process has been our main concern. Again, we have seen that the notion of distance has played an important role in defining convergence, continuity and differentiability. In 1906, M. Frechet gereralized the notion of distance and extended it to arbitrary sets which includes the real line **R** as a particular case.

In due course of study of this chapter we shall observe that the most important notions of metric spaces are the generalization of concepts of real line. Reader is, therefore, advised to study chapter 3, 5 and 8 because the notions introduced in these chapters will provide necessary motivation for generalizing real line to metric spaces.

19.2 METRIC SPACE. *(Meerut 2014; M.S. Univ. 2006; Agra 2006; Kanpur 2015)*

Let X be any set. A metric for X is a function d with domain $X \times X$ and range $[0, \infty[$ such that

(*i*) $d(x,y) \geq 0 \;\; \forall\, x, y \in X$ (Non- negative property)

(*ii*) $d(x,y) = 0 \Leftrightarrow x = y$

(*iii*) $d(x,y) = d(y,x) \;\; \forall\, x, y \in X$ (Symmetry)

(*iv*) $d(x,y) \leq d(x,z) + d(z,y) \;\; \forall\, x, y, z \in X$ (Triangle inequality)

The set X with the metric d is called a metric space and it is denoted by (X, d). When there is no confusion about the metric d on X, we shall say that X is a metric space in place of saying that (X, d) is metric space.

Remark. If $x_1, x_2, \cdots, x_n \in X$, then triangle inequality can be generalised as follows:

$$d(x_1, x_2) + d(x_2, x_3) + \cdots + d(x_{n-1}, x_n) \leq d(x_1, x_n)$$

19.3 PSEUDOMETRIC SPACE. *(Meerut 2003, 11, 14)*

A mapping d of $X \times X$ into $[0, \infty[$ is called a pseudometric for X iff d satisfies the axioms (*i*), (*iii*) and (*iv*) of the above definition of metric space on X and the axiom

(*ii*)′ $d(x,x) = 0 \;\; \forall\, x \in X$

Clearly every metric space is a pseudometric but every pseudometic need not be a metric space.

19.4 SOME IMPORTANT RESULTS FOR DIRECT APPLICATIONS

(*i*) The absolute value function satisfies the following properties

$$|x| \geq 0, \quad |x| = 0 \Leftrightarrow x = 0, \quad |x| = |-x|, |x+y| \leq |x| + |y| \;\; \forall\; x, y \in \boldsymbol{R}$$

(*ii*) If u and v are any two complex numbers, then

$$|u+v| \leq |u| + |v|; \qquad \frac{|u+v|}{1+|u+v|} \leq \frac{|u|}{1+|u|} + \frac{|v|}{1+|v|}$$

(*iii*) **Cauchy- Schwarz inequality** Let $u = (u_1, u_2, \ldots, u_n)$ and $v = (v_1, v_2, \ldots, v_n)$ be any two n-triples of complex numbers, then
$$\sum_{i=1}^{n} |u_i\, v_i| \leq \left(\sum_{i=1}^{n} |u_i|^2 \right)^{1/2} \left(\sum_{i=1}^{n} |v_i|^2 \right)^{1/2}$$

(iv) *Minkowski's inequality.* Let u = $(u_1, u_2,..., u_n)$ and $v = (v_1,v_2,...v_n)$ be any two n -triple of complex numbers, then

$$\left(\sum_{i=1}^{n} |u_i + v_i|^p\right)^{1/p} \le \left(\sum_{i=1}^{n} |u_i|^p\right)^{1/p} + \left(\sum_{i=1}^{n} |v_i|^p\right)^{1/p}, \text{ if } p \ge 1$$

19.5 EXAMPLES OF METRIC SPACES

(M.S Univ. T.N. 2006)

Ex. 1. *Let* $\boldsymbol{R}$ *be the set of real numbers. Show that the function* $d : \boldsymbol{R}\times\boldsymbol{R} \to \boldsymbol{R}$ *defined by* $d(x, y) = |x - y|, \ \forall \ x, y \in \boldsymbol{R}$ *is a metric on* $\boldsymbol{R}$. *(Kanpur 2016; Meerut 2006, 2018)*

Sol. Here (i) $|x - y| \ge 0 \ \forall \ x, y \in \boldsymbol{R} \Rightarrow d(x, y) \ge 0 \ \forall \ x, y \in \boldsymbol{R}$

(ii) $|x - y| = 0 \Leftrightarrow x - y = 0 \Leftrightarrow x = y$ so that $d(x, y) = 0 \Leftrightarrow x = y$

(iii) $|x - y| = |y - x| \ \forall \ x, y \in \boldsymbol{R} \Rightarrow d(x, y) = d(y, x) \ \forall \ x, y \in \boldsymbol{R}$

(iv) $|x - y| = |(x - z) + (z - y)| \le |x - z| + |z - y| \ \forall \ x, y, z \in \boldsymbol{R}$

$$\Rightarrow \quad d(x, y) \le d(x, z) + d(z, y) \ \forall \ x, y, z \in \boldsymbol{R}$$

From (i) – (iv), we see that d is a metric on $\boldsymbol{R}$.

Note. The above metric d is known as the *usual metric* on $\boldsymbol{R}$ and the metric space ($\boldsymbol{R}$, d) is known as the *real line* (*or usual metric space*)

Ex.2. Discrete metric space. *(Agra 2009; Chennai 2011; Meerut 2011, 2015]*

Let X be a non-empty set and define a mapping $d : X \times X \to \boldsymbol{R}$ *as follows*

$$d(x, y) = \begin{cases} 0, & \text{when } x = y \\ 1, & \text{when } x \ne y \end{cases}, \quad \forall \ x, y \in X$$

Then show that d is a metric on X.

[(*X, d*) so defined is known as *discrete (or trival) metric space*]

Sol. We have (i) $d(x, y) \ge 0$, by definition of d.

(ii) $d(x, y) = 0 \Leftrightarrow x = y$, by definition

(iii) If $x = y$, then $d(x, y) = 0 = d(y, x)$ and if $x \ne y$, then $d(x, y) = 1 = d(y, x)$.

Hence $\quad d(x, y) = d(y, x) \ \forall \ x, y \in R$

(iv) Let x, y, z be any three element in X.

If $x = y$, then $d(x, y) = 0$. Also, $d(x, z) \ge 0$ and $d(z, y) \ge 0$

Hence $\quad d(x, y) \le d(x, z) + d(z, y)$.

If $x \ne y$, then either $x \ne y \ne z$ or $x \ne y = z$.

i.e., either $\quad d(x, y) = d(x, z) = d(z, y) = 1$

or $\quad d(x, y) = d(x, z) = 1$ and $d(z, y) = 0$

Hence in both these situations, $\quad d(x, y) \le d(x, z) + d(z, y)$.

Thus, $\quad d(x, y) \le d(x, z) + d(z, y) \ \forall \ x, y, z \in X$

Hence d is a metric on X and (X, d) is a metric space.

Ex. 3. *Let* $\boldsymbol{R}^2$ *be the set of all ordered pairs of real numbers and let* $d : \boldsymbol{R}^2 \times \boldsymbol{R}^2 \to \boldsymbol{R}$ *be defined by* $d(x, y) = \left\{(x_1 - y_1)^2 + (x_2 - y_2)^2\right\}^{1/2}$ *where* $x = (x_1, y_2)$ and $y = (y_1, y_2)$. Show that d is a metric on $\mathbf{R}^2$. *(Bhopal 2014; Meerut 2017)*

Sol. Here (*i*) since $(x_1 - y_1)^2$ and $(x_2 - y_2)^2$ are non-negative real numbers, we have

$$\{(x_1 - y_1)^2 + (x_2 - y_2)^2\}^{1/2} \geq 0 \quad \text{so that} \quad d(x, y) \geq 0$$

(*ii*) $d(x, y) = 0 \Leftrightarrow \{(x_1 - y_1)^2 + (x_2 - y_2)^2\}^{1/2} = 0$

$\Leftrightarrow (x_1 - y_1)^2 = 0$ and $(x_2 - y_2)^2 = 0 \Leftrightarrow x_1 = y_1$ and $x_2 = y_2$

$\Leftrightarrow (x_1, x_2) = (y_1, y_2) \Leftrightarrow x = y$

(*iii*) For all $x, y \in \boldsymbol{R}^2$, we have

$$d(x, y) = \{(x_1 - y_1)^2 + (x_2 - y_2)^2\}^{1/2} = \{(y_1 - x_1)^2 + (y_2 - x_2)^2\}^{1/2} = d(y, x)$$

(*iv*) Let $x = (x_1, x_2)$, $y (y_1, y_2)$ and $z = (z_1, z_2)$ be any three elements of $\boldsymbol{R}^2$. Then

$$d(x,y) = \{(x_1 - y_1)^2 + (x_2 - y_2)^2\}^{1/2}$$
$$= [\{(x_1 - z_1) + (z_1 - y_1)\}^2 + \{(x_2 - z_2) + (z_2 - y_2)\}^2]^{1/2}$$
$$\leq \{(x_1 - z_1)^2 + (x_2 - z_2)^2\}^{1/2} + \{(z_1 - y_1)^2 + (z_2 - y_2)^2\}^{1/2}$$

[Using Minkowski inequality, see Art 19.4]

$$\Rightarrow \quad d(x, y) \leq d(x, z) + d(z, y) \quad \forall\ x, y, z \in \boldsymbol{R}^2$$

Hence d is metric on $\boldsymbol{R}^2$ and $(\boldsymbol{R}^2, d)$ is metric space (known as the *Euclideam Plane*).

Ex. 4. Let $\boldsymbol{R}^n$ be the set of all ordered n-tuples of real numbers, and let $d : \boldsymbol{R}^n \times \boldsymbol{R}^n \to \boldsymbol{R}$ be defined by

$$d(x, y) = \{(x_1 - y_1)^2 + (x_2 - y_2)^2 + \ldots + (x_n - y_n)^2\}^{1/2} \quad \forall\ x, y, \in \boldsymbol{R}^n,$$

where $x = (x_1, x_2, \ldots, x_n)$ and $y = (y_1, y_2, \ldots, y_n)$ (*Meerut 2011; Delhi Maths (H) 200*)

Sol. (*i*) We have, $(x_i - y_i)^2 \geq 0 \ \forall\ i = 1, 2, \ldots, n$

$$\Rightarrow \sum_{i=1}^{n} (x_i - y_i)^2 \geq 0 \Rightarrow \left[\sum_{i=1}^{n} (x_i - y_i)^2\right]^{1/2} \geq 0 \Rightarrow d(x, y) \geq 0,\ \forall\ x, y \in \boldsymbol{R}^n$$

(*ii*) $d(x, y) = 0 \Leftrightarrow \left[\sum_{i=1}^{n} (x_i - y_i)^2\right]^{1/2} = 0$

$\Leftrightarrow (x_i - y_i)^2 = 0 \ \forall\ i = 1, 2, \ldots, n \Leftrightarrow x_i = y_i \ \forall\ i = 1, 2, \ldots, n$

$\Leftrightarrow (x_1, x_2, \ldots, x_n) = (y_1, y_2, \ldots, y_n) \Leftrightarrow x = y$

(*iii*) For all $x, y \in \boldsymbol{R}^n$, we have

$$d(x, y) = \left[\sum_{i=1}^{n} (x_i - y_i)^2\right]^{1/2} = \left[\sum_{i=1}^{n} (y_i - x_i)^2\right]^{1/2} = d(y, x)$$

(*iv*) Let $x = (x_1, x_2, \ldots, x_n)$, $y = (y_1, y_2, \ldots, y_n)$, $z = (z_1, z_2, \ldots, z_n)$ be any three elements of $\boldsymbol{R}^n$. Then, we have

$$d(x, y) = \{(x_1 - y_1)^2 + (x_2 - y_2)^2 + \ldots + (x_n - y_n)^2\}^{1/2}$$
$$= [\{(x_1 - z_1) + (z_1 - y_1)\}^2 + \{(x_2 - z_2) + (z_2 - y_2)\}^2 + \ldots + \{(x_n - z_n) + (z_n - y_n)\}^2]^{1/2}$$
$$\leq [\{(x_1 - z_1)^2 + (x_2 - z_2)\}^2 + \ldots + (x_n - z_n)^2\}^{1/2} + \{(z_1 - y_1)^2 + (z_2 - y_2)^2 + \ldots + (z_n - y_n)^2]^{1/2}$$

[Using Minkowski's inequality, refer Art. 19.4]

$$\Rightarrow \quad d(x, y) \leq d(x, z) + d(z, y) \quad \forall\ x, y, z \in \boldsymbol{R}^n$$

Hence d is metric on $\boldsymbol{R}^n$ and $(\boldsymbol{R}^n, d)$ is a metric space (known as *n-dimensional Euclidean space*)

Ex. 5. Let $\boldsymbol{R}$ be the set of real number. Show that the function $d: \boldsymbol{R}\times\boldsymbol{R}\to\boldsymbol{R}$ defined by $d(x,y)=|x^2-y^2|\ \forall\ x,y\in\boldsymbol{R}$ is pseudo-metric on $\boldsymbol{R}$ which is not a metric on R.

(*Kanpur 2014; Agra 2009, 2010, 2011, 2012; Meerut 2007, 2012*)

Sol. Here (*i*) $|x^2-y^2|\geq 0\ \forall\ x,y\in\boldsymbol{R}\Rightarrow d(x,y)\geq 0\ \forall x,y\in\boldsymbol{R}$

(*ii*) $d(x,x)=|x^2-x^2|=0\ \forall\, x\in\boldsymbol{R}$

(*iii*) $d(x,y)=|x^2-y^2|=|y^2-x^2|=d(y,x)\ \forall\ x,y\in\boldsymbol{R}$

(*iv*) $d(x,y)=|x^2-y^2|=|(x^2-z^2)+(z^2-y^2)|\leq|x^2-z^2|+|z^2-y^2|$

$\Rightarrow \qquad d(x,y)\leq d(x,z)+d(z,y)\ \forall\ x,y,z\in\boldsymbol{R}$

Hence d is a pseudo-metric on $\boldsymbol{R}$.

We now show that d is not a metric on R. Here

$$d(x,y)=0 \Rightarrow |x^2-y^2|=0 \Leftrightarrow x^2-y^2=0 \Leftrightarrow x=\pm y,$$

showing that $d(x,y)=0$ does not always imply only $x=y$. For example, we see that

$$d(1,-1)=|1^2-(-1)^2|=0, \quad \text{where} \quad 1\neq -1.$$

Hence, the function d is not a metric on $\boldsymbol{R}$.

19.6 DISTANCE BETWEEN TWO SETS. DIAMETER OF A SET

Distance of a point from a given set. Definition

Let (X, d) be a metric space and let A be any non-empty subset of X. The distance between a point $p\in X$ and A is denoted by $d(p, A)$ and defined by

$$d(p,A)=\inf\{d(p,x):\ x\in A\}$$

Distance between two subsets of a metric space. Definition.

Let (X,d) be a metric space and let A and B be any two non-empty subsets of X. The distance between the sets A and B is denoted by $d(A, B)$ and is defined as

$$d(A,B)=\inf\{d(x,y): x\in A,\ y\in B\}$$

Diameter of a subset of a metric space. Definition

Let (X, d) be a metric space and let A be a non-empty subset of X. Then the diameter of A is denoted by $\delta(A)$ and is defined by

$$\delta(A)=\sup\{d(x,y): x,y\in A\}.$$

The diameter of a set A is also denoted by $d(A)$. If $\delta(A)<\infty$, then the diameter of A is said to be finite otherwise infinite. By convention, $\delta(\phi)=-\infty$.

19.7 BOUNDED AND UNBOUNDED METRIC SPACES. DEFINITION (Meerut 2012)

A given metric space (X, d) is said to be *bounded* if there exists a real number $k>0$ such that

$$d(x,y)\leq k\ \forall\, x,y\in X,\ \text{i.e.,}\ \delta(X)\leq k.$$

A metric space which is not bounded is said to be *unbounded.*

EXERCISE

1. Let $d(x,y)=\min\{2,|x-y|\}$. Show that d is usual metric on $\boldsymbol{R}$. (*Meerut* 2002)
2. For a metirc space (X, d) if $d^*(x,y)=3d(x,y)$, then prove that (X, d^*) is also a metric space. (*Meerut* 2000)

3. Let R^2 be the set of all ordered pairs of real numbers are let $d\,;\, R^2 \times R^2 \to R$ be defined by $d(x, y) = \max\{[x_1 - y_1], [x_2 - y_2]\}$ where $x = (x_1, x_2)$ and $y = (y_1, y_2)$. Then prove that (R^2, d) is a metric space.

4. Let C be the set of all complex number and let $d = C \times C = R$ be defined by $d(z_1, z_1) = |z_1 - z_1|$, $\forall\, z_1, z_2 \in C$. Prove that (C, d) is a metric space.

5. Let R^n be the set of all ordered n-tuples of real numbers, and let $d : R^n \times R^n \to R$ be defined by $\quad d(x, y) = \max\{|x_i - y_i| : i = 1, 2, ..., n\}$ where $x = (x_1, x_2, ..., x_n)$, $y = (y_1, y_2, ..., y_n) \in R^n$. Prove that (R^n, d) is a metric space. *(Agra 2000, Delhi (H) 2002)*

6. Let P and Q be two bounded sets in a metric space (X, d). Show that $P \cup Q$ is also bounded. *(Rajasthan 2010)*

7. Give an example of pseudo-metric which is not a metric. *(Agra 2009; Meerut 2007)*

8. Prove that if x, y be any two point of a set A, then $|d(x, A) - d(y, A)| \le d(x, y)$.

9. Let A and B be any two subsets of X. Prove that

   (a) $\delta(A) = 0 \Leftrightarrow A$ contains at the most one point

   (b) $A \subseteq B \Rightarrow \delta(A) \le \delta(B)$

   (c) $A \cap B \ne \phi \Rightarrow \delta(A \cup B) \le \delta(A) + \delta(B)$

19.8 OPEN AND CLOSED SPHERES (OR BALLS). DEFINITION

Let (X, d) be any metric space and $p \in X$. Then for any $r > 0$, the set

$$S(p, r) = \{x \in X : d(x, p) < r\}$$

is known as an *open sphere* of radius r centred at p.

The notations $S_r(p)$ or $B_r(p)$ or $B(p, r)$ are also used to denote $S(p, r)$.

Similarly, a *closed sphere* is denoted and defined by

$$S[p, r] = \{x \in X : d(x, p) \le r\}.$$

Example. For the usual metric space (R, d), the open sphere $S(p, r)$ is the open interval $]\,p - r, p + r\,[$ while the closed sphere $S[p, r]$ is the closed interval $[p - r, p + r]$, where $p \in R$ and $r > 0$.

19.9 NEIGHBOURHOOD OF A POINT. DEFINITION *(Meerut 2007)*

Let (X, d) be a metric space. A set $N \subset X$ is called a *nbd* of a point $p \in X$ if there exists some $r > 0$ such that $S(p, r) \subset N$.

Examples. On the real line, an open interval $]\,a, b\,[$ is a *nbd* of each of its points. Refer Art. 3.2 for more examples in chapter 3. *(Meerut 2006)*

19.10 OPEN SET. DEFINITION

In a metric space (X, d), a set $G \subset X$ is said to be open if it is a *nbd* of each of its points. In other words, $G \subset X$ is said to be open in (X, d) if for each $p \in G$, there exists $r > 0$ such that

$$S(p, r) \subseteq G.$$

Illustrations : For the usual metric space (R, d), the open interval is an open set. For more examples, refer Art. 3.8 in Chapter 3.

19.11 PROPERTIES OF OPEN SETS

Theorem I. *Show that every open sphere is an open set but the converse is not true.*

Agra 2008, 10, 12; M.S. Univ T.N. 2006; Meerut 2006, 11, Delhi Maths (H) 1998, Kanpur 1996)

Sol. Let $S(p, r)$ be an arbitrary open sphere in metric space (X, d)

Now, $S(p, r)$ being non- empty, let x be an arbitrary point of $S(p, r)$ so that

$$x \in S(p,r) \Rightarrow d(p,x) < r.$$

Let $r' = r - d(p, x) > 0$, as $d(p, x) < r$ ... (1)

We shall now prove that $S(x,r') \subseteq S(p,r)$

Let y be an arbitrary element of $S(x, r')$ so that $d(x, y) < r'$.

Then $d(p,y) \le d(p,x) + d(x,y) < d(p, x) + r' = r$, using (1)

Thus, $d(x, y) < r$ and hence $y \in S(p,r)$

Since $y \in S(x,r') \Rightarrow y \in S(p,r)$, we have $S(x,r') \subseteq S(p,r)$,

showing that each point of $S(p, r)$ is the centre of an open sphere contained in it and so $S(p, r)$ is an open set.

Its converse is not true. For example in the usual metric space $(\mathbf{R}, d)$,]1, 2 [U] 2, 3[is an open set but it is not an open sphere. *(Agra 2006)*

Theorem II. *Let (X, d) be a metric space. A subset of X is open if and only if it is a union of open spheres.* *(Mumbai 2010; Delhi Maths (H) 2002)*

Proof. Refer Ex. 10, Page 19.57

Theorem III. *Let (X, d) be a metric space. Then*

(i) φ *is open* *(ii) X is open*

(iii) the union of an arbitrary collection open sets is open.

(Agra 2006, 07, 08; Delhi Maths (H) 1997, 98, Meerut 2001, 02, 09, 11)

(iv) the intersection of a finite number of open sets is open.

(Kanpur 1997; Agra 2006. 07, 10; Meerut 2010, 2012; M.S. Univ. T.N. 2006:)

Proof. *(i)* To prove that φ is open, we have to show that each point in φ is the centre of an open sphere contained in φ. Since there is no point in φ, this requirement is automatically satisfied. Hence φ is an open set.

(ii) Again, if $p \in X$, every open sphere $S(p, r)$ is contained in X. Hence X is an open set.

(iii) Let $\{G_\lambda : \lambda \in \Delta\}$ be and arbitrary collection of open sets in a metric space (X, d). Let x be an arbitrary point of $U\{G_\lambda : \lambda \in \Delta\}$ so that $x \in G_\lambda$ for at least one index $\lambda \in \Delta$. Since G_λ is open, therefore, there exists $r > 0$ such that $S(x,r) \subseteq G_\lambda$ and hence $S(x,r) \le U\{G_\lambda : \lambda \in \Delta\}$. Thus corresponding to each point $x \in U\{G_\lambda : \lambda \in \Delta\}$ there exists an open sphere $S(x, r)$ contained in it. Hence $U\{G_\lambda : \lambda \in \Delta\}$ is an open set.

(iv) Let (X, d) be a metric space and let $\{G_i : i = 1,2,...,n\}$ be a finite collection of open subsets of X. We wish to prove that $\cap\{G_i : i = 1,2,...,n\}$ is an open set.

Now, if $\cap\{G_i : i = 1,2,...,n\} = \varphi$, then it is open by result *(i)*.

Again, if $\cap\{G_i : i = 1,2,...,n\} \ne \varphi$, let x be an arbitrary point of $\cap\{G_i : i = 1,2,...,n\}$ so that $x \in G_i$ for each $i = 1, 2,..., n$. Since each G_i is open, there exists $r_i > 0$ such that

$$S(x, r_i) \subseteq G_i : \forall\ i = 1, 2, ..., n \quad ...(1)$$

Let $r = \min\{r_1, r_2, ..., r_n\}$. Then, from (1), we have

$$S(x, r) \subseteq S(x, r_i)\ \ \forall\ i = 1, 2, ..., n$$

$\Rightarrow$ $S(x, r) \subseteq G_i : \forall\ i = 1, 2, ..., n$, using (1)

$\Rightarrow$ $S(x, r) \subseteq \cap\{G_i = i = 1, 2, ..., n\}$,

showing each point of $\cap\{G_i = 1, 2, ..., n\}$ is an open sphere contained in it.

Hence $\cap\{G_i = i = 1, 2, ..., n\}$ is an open set.

Note : The intersection of an arbitrary family of open sets may not be an open set as shown below. *(Delhi Maths (H)* 1997, 98)

Let $I_n =]-1/n,\ 1/n\ [,\ n \in N$. Then $\{I_n : n \in N\}$ is an infinite family of open sets. Also, $\cap\{I_n : n \in N\} = \{0\}$, which being a non-empty finite set is not an open set.

Theorem IV (Housdroff) *In a metric space (X, d), for every pair of distinct points* $x, y \in X$, *there exists' disjoint open sets G and H such that* $x \in G$ *and* $y \in H$.

Proof. Left as an exercise for the reader.

19.12 EQUIVALENT METRICS

(Agra 2006, 2011

Two metrics d and d' on the same set X are said to be equivalent, if every open set in (X, d) is open in (X, d') and vice versa.

Ex.1. Let (X, d) be a metric space and ρ be a function on $X \times X$, defined by $\rho(x, y) = \min\{1, d(x, y)\}\ \ \forall\ x, y \in X$.

Then *(a)* (X, ρ) is a bounded metric space. (b) ρ is equivalent to d.

(Kanpur 2014; Delhi Maths (H) 1997, 2003, 2005,07 Meerut 2013

Sol. Given that $\rho(x, y) = \min\{1, d(x, y)\}$ ...(1)

(i) $d(x, y) \geq 0 \Rightarrow \min\{1, d(x, y)\} \geq 0 \Rightarrow \rho(x, y) \geq 0$

(ii) $\rho(x, y) = 0 \Rightarrow \min\{1, d(x, y)\} = 0 \Rightarrow d(x, y) = 0 \Leftrightarrow x = y$

(iii) $\rho(x, y) = \min\{1, d(x, y)\} = \min\{1, d(y, x)\} = \rho(y, x)\ \ \forall\ x, y \in X$

(iv) Let x, y, z be any three points of X. If at least one of $d(x, y)$ and $d(y, z)$ is ≥ 1, say

$$d(x, y) \geq 1,$$

then $\min\{1, d(x, y)\} = 1$ so that $\rho(x, y) = 1$, by (1)

$\therefore$ $\rho(x, y) + \rho(y, z) \geq 1 \geq \rho(x, z)$

Also, in case when $d(x, y) < 1$ and $d(y, z) < 1$, we have

$$\rho(x, y) = \min\{d(x, y), 1\} = d(x, y) \text{ and } \rho(y, z) = \min\{d(y, z), 1\} = d(y, z)$$

$\therefore$ $\rho(x, y) + \rho(y, z) = d(x, y) + d(y, z)$

$\geq d(x, z)$, by triangle inequality

$\geq \rho(x, z)$, since $\rho(x, z) \leq d(x, z)$ by (1)

Thus $\rho(x, y) + \rho(y, z) \geq \rho(x, z) \;\forall\; x, y, z \in X$

Hence ρ is a metric on X and (X, ρ) is a metric space.

From (1), $\rho(x, y) \leq 1 \;\forall\; x, y \in X$ Hence d (X, ρ) is a bounded metric space.

(*b*) **To show that ρ is equivalent to *d*.**

For this purpose we must show that every open set is (X, ρ) is open in (X, d) and vise-versa. Let G be any open subset of X in (X, ρ) and let x be any point of G. Then there exists an open set

$$\{y \in X : \rho(x, y) < r\} \subseteq G. \quad ...(2)$$

From (1), $\rho(x, y) \leq d(x, y) \;\forall\; x, y \in X$. Hence, using (2), we get

$$\{y \in X : d(x, y) < r\} \subseteq \{y \in X : \rho(x, y) < r\} \subseteq G.$$

showing that each point of G is the centre of an open sphere in (X, d) contained in G. Hence G is also open is (X, d) and therefore, every open set is (X, ρ) is open in (X, d).

Next, let H be an open set in (X, d). Then for each $x \in H$ Here exists an open sphere

$$\{y \in X : d(x, y) < r\} \subseteq H.$$

Let $r' = \min\{1, r\}$ so that $r' \leq r$. Then we have

$$\{y \in X : \rho(x, y) < r'\} \subseteq \{y \in X : d(x, y) < r\} \subseteq H,$$

showing that each point of H is the centre of a open sphere in (X, ρ) contained in H. Hence H is also open in (X, ρ), and, therefore every open set in (X, d) is open in (X, ρ).

Hence d and ρ are equivalent metrics.

Ex. 2. *Let (X, d) be a metric space, and let*

$$d^*(x, y) = \frac{d(x, y)}{1 + d(x, y)}, \text{ for all } x, y \text{ in } X.$$

Then prove that d^* is a bounded metric on X which is equivalent to *d*.

(*Meerut 2010, 2013; Agra 2006, 07, 10; Delhi Maths (H) 1997, 98, 2003, 04, 06*)

Sol. Given $d^*(x, y) = d(x, y) / \{1 + d(x, y)\}, \;\forall\; x, y \in X$ ... (1)

(*i*) Since $d(x, y) \geq 0$, (1) shows that $d^*(x, y) \geq 0$

(*ii*) $d^*(x, y) = 0 \Leftrightarrow d(x, y) = 0$, by (1)

$\Leftrightarrow x = y$, since d is metric on X.

(*iii*) $$d^*(x, y) = \frac{d(x, y)}{1 + d(x, y)} = \frac{d(y, x)}{1 + d(y, x)} = d^*(y, x)$$

[$\because$ d is metric on $X \Rightarrow d(x, y) = d(y, x)$]

(*iv*) For all x, y, z in X, we have

$$d^*(x, y) + d^*(y, z) = \frac{d(x, y)}{1 + d(x, y)} + \frac{d(y, z)}{1 + d(y, z)}$$

$$\geq \frac{d(x, y)}{1 + d(x, y) + d(y, z)} + \frac{d(y, z)}{1 + d(x, y) + d(y, z)}$$

[$\because$ d is metric $\Rightarrow$ $d(x, y) \geq 0$ and $d(y, z) \geq 0$]

$$= \frac{d(x,y)+d(y,z)}{1+d(x,y)+d(y,z)} = 1 - \frac{1}{1+d(x,y)+d(y,z)} \quad \text{... (2)}$$

Since d is metric on X, $d(x,y)+d(y,z) \geq d(x,z)$

$$\Rightarrow \quad 1+d(x,y)+d(y,z) \geq 1+d(x,z)$$

$$\Rightarrow \quad \frac{1}{1+d(x,y)+d(y,z)} \leq \frac{1}{1+d(x,z)}$$

$$\Rightarrow \quad 1-\frac{1}{1+d(x,y)+d(y,z)} \geq 1-\frac{1}{1+d(x,z)} = \frac{d(x,z)}{1+d(x,z)} = d^*(x,z) \quad \text{... (3)}$$

[Using (1)]

Then (2) and (3) $\Rightarrow$ $d^*(x,y)+d^*(y,z) \geq d^*(x,z)$

Hence d^* is a metric on X. Since $d(x,y) > 0$, (1) shows that $d^*(x,y) \leq 1 \ \forall \ x, y \in X$. Hence (X, d^*) is a bounded metric space.

We now show that d and d^* are equivalent. For this purpose it is sufficient to show that every d- open sphere centred at $x \in X$ contains a d^*-open sphere centred at $x \in X$ and conversely.

Let $S(x, r)$, $r > 0$ be any d-open sphere centered at $x \in X$ and $S(x, \rho)$ be d^*-open sphere centered at x such that $\rho = r/(1+r)$, then we shall show that

$$S(x,\rho) \subseteq S(x,r) \quad \text{...(4)}$$

Now, $\quad y \in S(x,\rho) \Rightarrow d^*(x,y) < \rho \Rightarrow \dfrac{d(x,y)}{1+d(x,y)} < \dfrac{r}{1+r}$

$$\Rightarrow \quad d(x,y) + r\,d(x,y) < r + r d(x,y) \Rightarrow d(x,y) < r \Rightarrow y \in S(x,r)$$

Thus, $y \in S(x,\rho) \Rightarrow y \in S(x,r)$. Hence as desired, (4) is true.

Next, let $S(x, \rho)$, $\rho > 0$ be any d^*-open sphere at $x \in X$. Since $d^*(x,y) \leq 1 \ \forall \ x, y \in X$, without any loss of generality, we take $0 < \rho < 1$. Let $S(x, r)$ be d-open sphere, where

$$r = \rho/(1-\rho)$$

We shall now show that

$$S(x,r) \subseteq S(x,\rho) \quad \text{... (5)}$$

Now, $\quad y \in S(x,r) \Rightarrow d(x,y) < r \Rightarrow \dfrac{d^*(x,y)}{1-d^*(x,y)} < \dfrac{\rho}{1-\rho} \quad \text{... (6)}$

$$\left[\because \text{ from (1), } d^*(x,y) = \frac{d(x,y)}{1+d(x,y)} \Rightarrow d(x,y) = \frac{d^*(x,y)}{1-d^*(x,y)}\right]$$

(6) $\Rightarrow$ $d^*(x,y) - \rho\, d^*(x,y) < \rho - \rho\, d^*(x,y) \Rightarrow d^*(x,y) < \rho \Rightarrow y \in S(x,\rho)$.

Thus $y \in S(x,r) \Rightarrow y \in S(x,\rho)$. Hence as desired, (5) is true.

Thus we have shown that every d-open sphere is a d^*-open sphere and conversely. Hence every d-open set is d^*-open and conversely. Consequently, the metrics d and d^* are equivalent.

EXERCISE

1. Let (X_1, d_1), (X_2, d_2) be two metric spaces and d^* be the function defined on $X_1 \times X_2$ by setting

$$d^*(x,y) = d_1(x_1, y_1) + d_2(x_2, y_2).$$

Then show that d^* is a metric on $X_1 \times X_2$ such that d^* is equivalent to the product metric space.

2. Let (X, d) be a metric space and $d^*(x, y) = \{M\, d(x, y)\} / \{1 + d(x, y)\}$ then prove that (X, d^*) is bounded metric space with $\delta(X) \le M$. (*Rajasthan 2010; Meerut* 2002)

3. Show that the metric $\sigma(x, y) = |x_1 - y_1| + |x_2 - y_2|$

and $\rho(x, y) = \max\{|x_1 - y_1|, |x_2 - y_2|\}$

where $x = (x_1, x_2)$ and $y = (y_1, y_2)$ are equivalent. (*Calicut* 2004)

4. Let mapping $d : \boldsymbol{R} \times \boldsymbol{R} \to \boldsymbol{R}$ be defined as $d(x, y) = |x - y| / \{1 + |x - y|\}$ prove that d is metric on $\boldsymbol{R}$. (*Agra 2010*)

5. Define a discrete metric space and describe open sphere in it.

6. Let (X, d) be a metric space and p, q two different points in X. Show that there exists two disjoint open spheres with centres p and q.

19.13 LIMIT POINT (OR AN ACCUMULATION POINT OR A CLUSTER POINT OR A CONDENSATION POINT) DEFINITION (*Meerut* 2007, 09)

Let (X, d) be a metric space. A point $p \in X$ is said to be a limit point of a set $A \subset X$ if every *nbd* of p contains a point of A different from p.

In other words, a point of $p \in X$ is said to be a limit point of a set $A \subset X$ if for each $r > 0$, the open sphere $S(p, r)$ contains a point of A other that p.

Derived set. Definition. The set of all limit points of a set $A \subset X$ is known as the derived set and is denoted by A' or $D(A)$.

Illustrations. On the real line with usual metric space $(\boldsymbol{R}, d)$, refer illustrations given in Art. 3.4.

19.14 CLOSED SET. DEFINITION. (*Agra 2011; Chennai 2011; Meerut 2006*)

A subset A of a metric space (X, d) is said to be closed if A contains all its limit points. For illustrations on usual metric space, refer Art. 3.9 of chapter 3.

19.15 PROPERTIES OF CLOSED SETS

Theorem I. *In a metric space (X, d), a subset $F \subset X$ is closed if and only if its complement is open.* (*Nagpur 2003, Bhopal 2004, Meerut 2004*)

Proof. Let F be a closed set. We shall prove that its complement F^C is open. If $F^c = \varphi$, then F^C is open as φ is an open set. Next, assume that $F^C \neq \varphi$. Let x be an arbitrary element of F^C. Then $x \notin F$. Now, we have

$$\begin{aligned} x \notin F \text{ and } F \text{ is closed } &\Rightarrow x \text{ is not a limit point of } F \\ &\Rightarrow \text{there exists } r > 0 \text{ such that } S(x, r) \cap F = \varphi \\ &\Rightarrow x \in S(x, r) \subset F^C \\ &\Rightarrow F^C \text{ is open} \end{aligned}$$

Now, assume that F^C is open and show that F is closed. If $x \in F^C$, then there exists $r > 0$ such that

$$S(x, r) \subseteq F^C \text{ so that } S(x, r) \cap F = \varphi$$

$$\Rightarrow \quad x \text{ cannot be a limit point of } F$$

$$\Rightarrow \quad F \text{ contains all its limit points} \Rightarrow F \text{ is closed}$$

Theorem II. *In a metric space (X, d), each closed sphere is closed set.*

(Agra 2000, 07, 11; Delhi Maths (H) 2000, 09, G.N.D.U. 1997)

Proof. Let $S[x, r]$ be a closed sphere in X. Let $G = X - S[x, r]$ denotes the complement of $S[x, r]$. If $G = \varphi$, then G is open and so $S[x, r]$ is closed by theorem I.

Next, let $G \neq \varphi$. Let y be an arbitrary point of G. Since $d(y, x) > r$, let $r' = d(y, x) - r$, which is a positive real number. In order to prove that G is open, we must prove that $S(y, r') \subset G$. If $z \in S(y, r')$, we have

$$d(z, y) < r' \Rightarrow d(z, y) < d(y, x) - r$$

$\Rightarrow$ $r < d(y, x) - d(z, y) \Rightarrow r < d(x, y)$, using triangle inequality

Thus, $z \in S(y, r') \subset X - S[x, r]$

$\Rightarrow$ $X - S[x, r]$ is open $\Rightarrow$ $S[x, r]$ is closed.

Theorem III. Let (X, d) be a metric space. Then

(i) φ is closed *(ii)* X is closed

(iii) The intersection of an arbitrary family of closed sets is closed.

(Meerut 2013; Delhi Maths (H) 2003; Agra 2007)

(iv) The union of a finite family of closed sets is closed.

Proof. Proceed exactly as in Art 3.6 of chapter 3.

19.16 SUBSPACES. DEFINITION.

Let (X, d) be a metric space and let $Y \subset X$. Then the mapping d_Y on Y defined by $d_Y(x, y) = d(x, y)$, $\forall\ x, y \in Y$ is called the metric induced on Y and the metric space (Y, d_Y) is known as a subspace of (X, d)

Example Let $(\boldsymbol{R}, d)$ be the usual metric space. Here $\boldsymbol{Q} \subset \boldsymbol{R}$ and define metric ρ on ` by $\rho(x, y) = |x - y|\ \forall\ x\ y \in \boldsymbol{Q}$. Then $(\boldsymbol{Q}, \rho)$ is a subspace of the real line.

Theorem I. *Let (X, d) be a metric space and $Y \subseteq X$, then a set A of Y is open in (Y, d_Y) if and only if there exists a set G open in (X, d) such that $A = G \cap Y$.(Mumbai 2010; Nagpur 2003)*

Proof. We first assume that A is open in (Y, d_Y). Then for each $a \in A$, there exists $r_a > 0$ such that

$$S^*(a, r_a) = \{y \in Y : d(a, y) < r_a\} \subseteq A$$

$\Rightarrow$ $A = U\{S^*(a, r_a) : a \in A\}$

Let $S(a, r_a) = \{y \in X : d(a, y) < r_a\}$

Clearly, $S^*(a, r_a) = Y \cap S(a, r_a)\ \forall\ a \in A$

$\Rightarrow$ $\cup\{S^*(a, r_a) : a \in A\} = Y \cap [\cup\{S(a, r_a) : a \in A\}]$

$\Rightarrow$ $A = Y \cap G$, where $G = U\{S(a, r_a) : a \in A\}$

But G is open in (X, d) because it is the union of open spheres in (X, d).

Thus, $A = Y \cap G$, where G is open in (X, d)

Conversely, suppose that there exists an open G in (X, d) such that $G \cap Y = A$. Then we shall prove that A is open in (Y, d_Y). Let a be an arbitrary element of A. Then, $a \in G$ and so there exists $r > 0$ such that

$$S(a, r) \subseteq G$$

$$\Rightarrow \quad S^*(a,r) = S(a,r) \cap Y \subseteq G \cap Y \Rightarrow S^*(a,r) \subseteq A,$$

showing that each point of A is the centre of an open sphere in (Y, d_Y) contained in A. Hence A is open in (Y, d_Y).

Theorem II. *Let (X, d) be a metric space and $Y \subseteq X$, then a set B of Y is closed in (Y, d_Y) if and only it there exists a set H closed in (X, d) such that $B = H \cap Y$.* *(Mumbai 2011)*

Proof. Left as an exercise.

19.17 ADHERENT POINT (OR A CONTACT POINT.) DEFINITION. *(Meerut 2011)*

Let (X, d) be a metric space and let $S \subset X \cdot A$ point $p \in X$ is called an adherent point of S if every *nbd* of p contains a point of S.

Equivalently, a point $p \in X$ is called an adherent point of S if each open sphere centered at p contains at least one point of S.

Closure of a set. Definition. *(Agra 2006)*

The set of all adherent point of S, called the closure of S, is denoted by Cl(S) or $\overline{S}$.

Illustrations. One usual metric, refer Art 3.14 of Chapter 3.

19.18 PROPERTIES OF CLOSURE OF A SET *(Lucknow 2010)*

Theorem I. *Let (X, d) be a metric space. If $S \subset X$ then*

$$\overline{S} = S \cup S' \qquad \textit{(Meerut 2001)}$$

Proof.l Refer Art. 3.14 of Chapter 3.

Theorem II. *Let (X, d) be a metric space and let A and B be arbitrary subsets of X.* Then

(*i*) $A \subseteq B \Rightarrow \overline{A} \subseteq \overline{B}$ (*ii*) $\overline{A \cap B} \subseteq \overline{A} \cap \overline{B}$ *(Meerut 2001, Delhi 2008)*

(*iii*) $\overline{A \cup B} = \overline{A} \cup \overline{B}$ *(Delhi Maths (H) 2008, G.N.D.U. 1997)*

Proof. (*i*) Let $x \in \overline{A}$. Then, by definition of $\overline{A}$, we get

$$S(x,r) \cap A \neq \varphi \;\; \forall \; r > 0$$

$$\Rightarrow \quad S(x,r) \cap B \neq \varphi \;\; \forall \; r > 0, \text{ as } A \subseteq B \text{ (given)}$$

$$\Rightarrow \quad x \in \overline{B}.$$

Thus, $x \in \overline{A} \Rightarrow x \in \overline{B}$ so that $\overline{A} \subset \overline{B}$.

(*ii*) Here $A \cap B \subseteq A$ and $A \cap B \subseteq B$

$\Rightarrow$ $\overline{A \cap B} \subseteq \overline{A}$ and $\overline{A \cap B} \subseteq \overline{B}$, by part (*i*)

$\Rightarrow$ $\overline{A \cap B} \subseteq \overline{A} \cap \overline{B}$

(*iii*) Here $A \subseteq A \cup B$ and $B \subseteq A \cup B$

$\Rightarrow$ $\overline{A} \subseteq \overline{A \cup B}$ and $\overline{B} \subseteq \overline{A \cup B}$, by part (*i*)

$\Rightarrow$ $\overline{A} \cup \overline{B} \subseteq \overline{A \cup B}$ ... (1)

Now, we shall show that $\overline{A \cup B} \subseteq \overline{A} \cup \overline{B}$ ... (2)

Then if possible, $x \in \overline{A \cup B}$ and $x \notin \overline{A} \cup \overline{B}$.

Then $x \notin \overline{A} \cap \overline{B}$

$\Rightarrow$ x is neither an adherent point of A nor that of B

$\Rightarrow$ there exist open spheres $S(x, x_1)$ and $S(x, r_2)$ containing no points of A and B respectively.

Let $r = \min\{r_1, r_2\}$. Then, $S(x, r)$ contains no point of A as well as that of B and so open sphere $S(x, r)$ contains no point of $A \cup B$ and hence x cannot be an adherent point of $A \cup B$, *i.e.*, $x \notin \overline{A \cup B}$. Thus we arrive at a contradiction. Hence

$$x \in \overline{A \cup B} \Rightarrow x \in \overline{A} \cup \overline{B} \text{ so that (2) is true.}$$

From (1), and (2), $$\overline{A \cup B} = \overline{A} \cup \overline{B}$$

Theorem III *Let A be a subset of a metric space (X, d), then*

(i) $\overline{A}$ ***is the smallest closed superset of A*** ***(Lucknow 2010; Delhi Maths (H) 2006)***

(ii) A is closed if and only if $\overline{A} = A$

(iii) $\overline{A}$ *is the intersection of all closed supersets of A*

(Delhi Maths (H) 2002, G.N.D.U. 1997)

Proof. Since $\overline{A} = A \cup A'$, it follows that $\overline{A}$ is a superset of A. In order to prove that $\overline{A}$ is closed, we shall show that its complement $(\overline{A})^C$ is open. If $(\overline{A})^C = \varphi$, then it is open.

If $(\overline{A})^C \neq \varphi$, let x be any point of $(\overline{A})^C$, then $x \notin \overline{A}$ *i.e.*, x is not an adherent point of A. Hence there exists an open sphere $S(x, r)$ such that

$$S(x, r) \cap A = \varphi.$$

We proceed to prove that $S(x, r) \cap \overline{A} = \varphi$. Let y be any point of $S(x, r)$ so that $d(y, x) < r$. Let $r' = r - d(y, x)$. Then $r' > 0$.

Now $S(y, r') \subseteq S(x, r)$

$\Rightarrow S(y, r') \cap A = \varphi$ for at least one $r' > 0$ $\quad [\because S(y, r') \cap A \subseteq S(x, r) \cap A]$

$\Rightarrow y \notin \overline{A}$

Since y is an arbitrary point of $S(x, r)$, so $\quad S(x, r) \subseteq (\overline{A})^C$,

showing that each point of $(\overline{A})^C$ is the centre of an open sphere contained in it. Hence $(\overline{A})^C$ is open and so $\overline{A}$ is closed.

(ii) If $A = \overline{A}$, then from *(i)* $\overline{A}$ is closed, and so A is closed.

Conversely, let us now assume that A is closed. Then it being the smallest closed superset of A must coincide with $\overline{A}$, *i.e.*, in this case $A = \overline{A}$ by *(i)*

(iii) Let B be the intersection of all closed sets containing A. Then B is closed.

Now, $\quad A \subset B \Rightarrow \overline{A} \subseteq \overline{B} \Rightarrow \overline{A} \subseteq B$, as $\overline{B} = B$, by *(ii)*

Hence every closed set which contains A, contain $\overline{A}$. Now $\overline{A}$ is a closed set containing A and B, being the intersection of all closed sets containing A, is contained in $\overline{A}$. Hence $\overline{A} = B$, so that $\overline{A}$ is the intersection of all closed sets containing A.

19.19 INTERIOR POINT OF A SET. DEFINITION. *(Meerut 2009, 11; Agra 2006, 2011*

Let (X, d) be a metric space and let $A \subset X$. A point $p \in A$ is said to be an *interior point* of A, if A is a *nbd* of p.

Equivalently, let (X, d) be a metric space and let $A \subset X$. A point $p \in A$ is said to be an interior point of A if here exists a number $r > 0$ such that $S(p, r) \subset A$.

Interior of a set. Definition.

The set of all *interior points* of A is called the *interior* of A and is denoted by int A or $A°$.

Illustrations. On the real line with usual metric space (R, d), refer Art. 3.10 for examples.

19.20 PROPERTIES OF INTERIOR OF A SET.

Theorem I. *In a metric space (X,d), the interior of a set $A \subset X$ is the largest open subset of A.*

(Delhi Maths (H) 1998, 2000, 2005, G.N.D.U. 1998)

Proof. By definition, it follows that $A° \subset A$. Now, if $A° = \varphi$, then it is open. So let $A° \neq \varphi$, and let x be any point of $A°$ so that x is an interior point of A. So there exists a real number $r > 0$ such that $S(x,r) \subseteq A$.

Now each open sphere being an open set, each point of $S(x, r)$ is the centre of an open sphere contained in $S(x, r)$ and so in A. Hence, each point of $S(x, r)$ is an interior point of A, *i.e.*, $S(x,r) \subseteq A°$. In other words, each point of $A°$ is the centre of an open sphere contained in it. Hence $A°$ is a an open subset of A.

Now it only remains to be shown that if G is any open set contained in A, then G is contained in $A°$. Since G is an open set, so each point of G is the centre of an open sphere contained in G and so in A. Hence each point of G is an interior point of A, that is, $G \subset A°$. Also $A°$ is open subset of A. Hence $A°$ must be the largest open subset of A.

Corollary I. *In a metric space (X,d), if $A \subset X$. Then*

(*i*) $\phi° = \phi$ (*ii*) $X° = X$ (*iii*) A is open $\Leftrightarrow$ $A° = A$

Proof. Left as an exercise.

Theorem II. *If A and B are two sebsets of a metric space (X, d), then*

(*i*) $A \subseteq B \Rightarrow A° \subseteq B°$ *(Meerut 2006)* (*ii*) $(A \cup B)° \supseteq A° \cup B°$ *(Delhi Maths (H) 2007)*

(*iii*) $(A \cap B)° = A° \cap B°$. *(Delhi Maths (H) 2004)*

Proof. (*i*) Let $x \in A°$. Then there exists $r > 0$ such that $S(x,r) \subseteq A$

Now $S(x,r) \subseteq A$ and $A \subseteq B \Rightarrow S(x,r) \subseteq B \Rightarrow x \in B°$

Thus $x \in A° \Rightarrow x \in B°$ so that $A° \subset B°$.

(*ii*) Using result (*i*), we have

$$A \subseteq A \cup B \quad \text{and} \quad B \subseteq A \cup B$$

$$\Rightarrow \quad A° \subseteq (A \cup B)° \text{ and } B° \subseteq (A \cup B)° \Rightarrow A° \cup B° \subseteq (A \cup B)°$$

Thus, $$(A \cup B)° \supseteq A° \cup B°$$

(*iii*) Using result (*i*), we have

$$A \cap B \subseteq A \quad \text{and} \quad A \cap B \subseteq B$$

$$\Rightarrow \quad (A \cap B)° \subseteq A° \quad \text{and} \quad (A \cap B)° \subseteq B° \Rightarrow (A \cap B)° \subseteq (A° \cap B°) \quad \text{... (1)}$$

To prove the required result, we now prove $A° \cap B° \subseteq (A \cap B)°$ ... (2)

Let x be an arbitrary point of $A° \cap B°$.

$\Rightarrow$ x is an interior point of A as well as that of B

$\Rightarrow$ there exists real numbers $r_1 > 0$ and $r_2 > 0$ such that

$$S(x,r_1) \subseteq A \quad \text{and} \quad S(x,r_2) \subseteq B \quad \text{... (3)}$$

Let $r = \min\{r_1, r_2\}$. Then (3), shows that

$$S(x,r) \subseteq A \text{ and } S(x,r) \subseteq B \text{ so that } S(x,r) \subseteq A \cap B$$

$\Rightarrow$ x is an interior point of $A \cap B$, *i.e.,* $x \in (A \cap B)^\circ$.

Thus, $x \in A^\circ \cap B^\circ \Rightarrow x \in (A \cap B)^\circ$ and so (2) is true.

From (1) and (2), $(A \cap B)^\circ = A^\circ \cap B^\circ$.

Note. For any two subsets A and B of a metric space (X, d), the result *(ii)* cannot always be written as $(A \cup B)^\circ = A^\circ \cup B^\circ$. For example, for real line, let $A = [1, 2]$ and $B = [2, 3]$. Then $A^\circ \cup B^\circ =]1, 2[\cup]2, 3[$ while $(A \cup B)^\circ =]1, 3[$ and so $A^\circ \cup B^\circ \neq (A \cup B)^\circ$.

Theorem III. *If A is a subset of a metric space (X, d), then* $X \sim A^\circ = \overline{(X \sim A)}$

(*Mumbai 2010; Delhi Maths (H)* 2005)

Proof. $x \in X \sim A^\circ$

$\Leftrightarrow$ $x \notin A^\circ$

$\Leftrightarrow$ x is not a *nbd* of x

$\Leftrightarrow$ There does not exist any $\varepsilon > 0$ such that $S(x, \varepsilon) \subset A$

$\Leftrightarrow$ for every $\varepsilon > 0$, $S(x, \varepsilon) \cap (X \sim A) \neq \phi$

$\Leftrightarrow$ $x \in \overline{(X - A)}$

Hence $X \sim A^\circ = \overline{(X - A)}$ ... (1)

Corollary I. $A^\circ = X \sim \overline{(X - A)}$

Taking complements of both sides of (1), we get

$$A^\circ = X \sim \overline{(X \sim A)} \qquad \text{... (2)}$$

Corollary II. $\overline{A} = X - (X \sim A)^\circ$

Proof. Replacing A by $X \sim A$ in (1), we have

$$X \sim (X \sim A)^\circ = \overline{A} \qquad \text{... (3)}$$

Corollary III. $X \sim \overline{A} = (X \sim A)^\circ$ (*Mumbai 2010*)

Proof. Taking complements of both sides of (3), we get

$$(X \sim A)^\circ = X \sim \overline{A}.$$

19.21 DENSE (OR EVERYWHERE DENSE), DENSE IN ITSELF, NOWHERE DENSE (OR NON-DENSE) AND PERFECT SETS. DEFINITIONS.

Let (X, d) be metric space and let $S \subseteq X$. Then, replace R by X in all definitions of Art. 3. 16.

For examples, for real line with usual metric, refer Art 3.16 in Chapter 3.

19.22 SEPARABLE SPACE DEFINITION. (Meerut 2011, 2012)

A metric space (X,d) is said to be *separable* if it possesses a countable dense subset.

Examples The real line with usual metric space is separable because the set $\boldsymbol{Q}$ of rational numbers is a countable dense subset of $\boldsymbol{R}$.

EXERCISE

1. Show that a discrete metric space is separable iff if is countable.

2. Show that the complex number space C is separable.
3. Define a function $d : \boldsymbol{R}^2 \times \boldsymbol{R}^2 \to \boldsymbol{R}$ such that
$$d((x_1, y_1), (x_2, y_2)) = \max (|x_1 - x_2|, |y_1 - y_2|).$$
Show tnat d is a metric for $\boldsymbol{R}^2$. Describe geometrically the open ball with centre (0, 0) and radius 2 in $(\boldsymbol{R}^2, d)$. Further show that (R^2, d) is a separable metric space. (*Delhi Hons.* 1996)
4. Show that the Euclidean space $\boldsymbol{R}^n$ is separable.
5. Let $\boldsymbol{R}$ be the set of real number and let d be the metric on $\boldsymbol{R}$ defined by $d(x, y) = |x - y| / \{1 + |x - y|\}, \ \forall\, x, y \in R$. Is the space (R, d)j separable? Justify your answer.

19.23 EXTERIOR, FRONTIER AND BOUNDARY POINTS. DEFINITIONS. (Agra 2006)

Let (X, d) be a metric space and $A \subseteq X$. Then a point $x \in X$ is said to be an *exterior point* of A, if is an interior point of the complement of A, i.e., there exists an open sphere $S(x, r) \subseteq A^c$, or $S(x, r) \cap A = \varphi$. Here A^c stands for the complement of A.

The set of all exterior points of A is called the *exterior of A* and is denoted by ext A or $e(A)$.

A point $x \in X$ is known as a frontier point of A if it is neither an interior nor an exterior point of A. If the frontier point belongs to A it is said to be a *boundary point of A*.

The set of all frontier points and boundary points are denoted by $Fr\ (A)$ and $Bd\ (A)$ respectively.

For examples on real line refer Art 3.11 in Chapter 3.

19.24 PRODUCT OF METRIC SPACES.

Let (X_1, d_1) and (X_2, d_2) be two metric spaces, then for any pair of points $x = (x_1, x_2)$ and $y = (y_1, y_2)$ in $X_1 \times X_2$, consider the function $d : X_1 \times X_2 \to \boldsymbol{R}$ defined by
$$d\,(x, y) = \max \{d_1\,(x_1, y_1), d_2\,(x_2, y_2)\} \qquad \text{... (1)}$$
Using the facts that d_1 and d_2 are metrics on X_1 and X_2 respectively, we shall prove that d is a metric on $X_1 \times X_2$.

(*i*) $d(x, y) = \max\{d_1(x_1, y_1), d_2(x_2, y_2)\} \geq 0$, as $d_1(x_1, y_1) \geq 0, d_2\,(x_2, y_2) \geq 0$

(*ii*) $d(x, y) = 0 \;\Leftrightarrow\; \max\{d_1(x_1, y_1), d_2(x_2, y_2)\} = 0$
$$\Leftrightarrow \quad d_1(x_1, y_1) = 0 \text{ and } d_2(x_2, y_2) = 0$$
$$\Leftrightarrow \quad x_1 = y_1 \text{ and } x_2 = y_2$$
$$\Leftrightarrow \quad (x_1, x_2) = (y_1, y_2) \Leftrightarrow x = y$$

(*iii*) $d\,(x, y) = \max\,\{d_1\,(x_1, y_1), d_2\,(x_2, y_2)\}$
$$= \max\,\{d_1\,(y_1, x_1), d_2\,(y_2, x_2)\} = d\,(y, x)$$

(*iv*) For all $x = (x_1, x_2), y = (y_1, y_2), z = (z_1, z_2)$ in $X_1 \times X_2$, we have
$$d\,(x, z) + d\,(z, y) = \max\,\{d_1\,(x_1, z_1), d_2\,(x_2, z_2)\} + \max\,\{d_1\,(z_1, y_1), d_2\,(z_2, y_2)\}$$
$$\geq \max\,\{d_1\,(x_1, z_1) + d_1\,(z_1, y_1), d_2\,(x_2, z_2) + d_2\,(z_2, y_2)\}$$
$$\geq \text{man}\,\{d_1\,(x_1, y_1), d_2\,(x_2, y_2)\} = d\,(x, y)$$

Thus, $\quad d(x, y) \leq d(x, z) + d(z, y) \ \forall\ x, y, z \in X_1 \times X_2$

Hence d is a metric on $X_1 \times X_2$ and the metric space $(X_1 \times X_2, d)$ is called the *product of metric spaces* (X_1, d_1) and (X_2, d_2).

Note. Unless stated otherwise, in product spaces, the metric will be defined as above. However, there exist other metrics on $X_1 \times X_2$ as given in the following theorem.

Theorem. Let (X_1, d_1) and (X_2, d_2) be two metric spaces and let ρ and σ be two functions defined on $X_1 \times X_2$ as follows

$$\rho(x, y) = \{d_1^2(x_1, y_1) + d_2^2(x_2, y_2)\}^{1/2}$$

and
$$\sigma(x, y) = d_1(x_1, y_1) + d_2(x_2, y_2)$$

Then ρ and σ are metrics on $X_1 \times X_2$ equivalent to the product metrics on $X_1 \times X_2$, where $x = (x_1, x_2)$ and $y = (y_1, y_2)$

Proof. Left as an exercise for the reader.

19.25 CONTINUOUS FUNCTIONS ON METRIC SPACES.

(Meerut 2011; Delhi 2012; Agra 2009)

Let (X_1, d_1) and (X_2, d_2) be any two metric spaces. A function $f : X \to Y$ is said to be *continuous at a point 'a' of X,* if for given $\varepsilon > 0$, there exists a $\delta > 0$ such that

$$d_1(x, a) < \delta \quad \Rightarrow \quad d_2(f(x), f(a)) < \varepsilon$$

Equivalently, for each open sphere $S(f(a), \varepsilon)$ centered at $f(a)$, there exists an open sphere $S(a, \delta)$ centred at a such that $f(S(a, \delta)) \subseteq S(f(a), \varepsilon)$

If f is continuous at every point of X, then it is said to be *continuous on X*.

19.26 PROPERTIES OF CONTINUOUS FUNCTIONS.

Theorem I. *Let (X, d_1) and (Y, d_2) be two metric spaces. A function $f : X \to Y$ is continuous on X iff for each open set $G \subset Y$, f^{-1} (G) is an open subset of X.* *(Agra 2009)*

(M.S. Univ. T.N. 2006; Bharathiar 2004, Mumbai 2010;, Meerut 2000, 01, 613)

Proof. The condition is necessary. Let f be continuous function and let G be any open subset of Y. Then we are to show that $f^{-1}(G)$ is open in X. If $f^{-1}(G) = \varphi$, then it is open. If $f^{-1}(G) \neq \varphi$, let x be any point of $f^{-1}(G)$ so that $f(x) \in G$. Since G is an open set containing $f(x)$, there exists an open sphere $S(f(x), \varepsilon)$ for some $\varepsilon > 0$ such that

$$S(f(x), \varepsilon) \subseteq G. \qquad \text{... (1)}$$

Now f being continuous at x we can find a $\delta > 0$ such that

$$f(S(x, \delta)) \subseteq S(f(x), \varepsilon) \qquad \text{... (2)}$$

From (1) and (2) $\quad f(S(x, \delta)) \subseteq G \quad$ so that $\quad S(x, \delta) \subseteq f^{-1}(G)$,
showing that each point of $f^{-1}(G)$ is the centre of an open sphere contained in it and hence $f^{-1}(G)$ is an open set in X.

Contidition is sufficient. Assume that the inverse image under f of every open subset of Y is an open subset of X. Then in order to prove that f is continuous, it is sufficient to show that it is continuous at an arbitrary point $x \in X$. Let $\varepsilon > 0$. Then,

$$S(f(x), \varepsilon) \text{ is open in } Y \quad \Rightarrow \quad f^{-1}(S(f(x), \varepsilon)) \text{ is open in } X.$$

Hence there exists $\delta > 0$ such that

$$S(x, \delta) \subseteq f^{-1}(S(f(x), \varepsilon)) \quad \text{so that } f(S(x, \delta)) \subseteq S(f(x), \varepsilon)$$

$\Rightarrow$ f is continuous at x.

Since x is any point of X, so f is continuous on X.

Theorem II. *Let (X, d_1) and (Y, d_2) be two metric spaces. A function $f : X \to Y$ is continuous on X iff for each closed set $H \subset Y$, $f^{-1}(H)$ is closed subset of X.*

(*M.S. Univ. T.N. 2006; Meerut 2000, 01*)

Proof. Condition is necessary. Let f be a continuous function and let H be any closed subset of Y. Then $Y - F$ is open in Y. So by theorem I, $f^{-1}(Y - F)$ is an open subset of X. But

$$f^{-1}(Y-F) = f^{-1}(Y) - f^{-1}(F) = X - f^{-1}(F).$$

Thus, $X - f^{-1}(F)$ is open and so $f^{-1}(F)$ is closed in X.

Condition is sufficient. Let the inverse image under f of every closed subset of Y be a closed subset of X. Let G be any open subset of Y so that $Y - G$ is closed and so $f^{-1}(Y - G)$ is closed in X. But

$$f^{-1}(Y-G) = f^{-1}(Y) - f^{-1}(G) = X - f^{-1}(G)$$

$\Rightarrow$ $X - f^{-1}(G)$ is closed and hence $f^{-1}(G)$ is open.

Thus, the inverse image under f of every open subset of Y is an open subset of X. Hence by theorem I, f is continuous on X.

Theorem III *A function f from a metric space X into metric space Y is continuous if $f(\bar{A}) \subseteq \overline{f(A)}$, for every subset A of X.*(*Meerut 2010; Agra 2006, 08; Delhi Maths (H) 2004*

Proof. Condition is necessary. Let f be continuous and let $A \subset X$. Since $\overline{f(A)}$ is closed in Y, so $f^{-1}\,\overline{f(A)}$ is closed in X. Now, we have

$$f(A) \subseteq \overline{f(A)} \Rightarrow A \subseteq f^{-1}\,\overline{f(A)} \Rightarrow \bar{A} \subseteq \overline{f^{-1}\,\overline{f(A)}} = f^{-1}\,\overline{f(A)}$$

[$\because B$ is closed $\Rightarrow \bar{B} = B$]

Thus, $\bar{A} \subseteq f^{-1}\,\overline{f(A)}$ so that $f(\bar{A}) \subseteq \overline{f(A)}$.

Condition is sufficient. Let $f(\bar{A}) \subseteq \overline{f(A)}$, ... (1)

be true for every subset A of X. Let H be any closed subset of Y. Then $f^{-1}(H)$ is a subset of X. Hence, by result (1), we have

$$f\,\overline{f^{-1}(H)} \subseteq \overline{f(f^{-1}(H))} = \bar{H} = H, \text{ since } H \text{ is closed} \Rightarrow \bar{H} = H$$

Thus, $f\,\overline{f^{-1}(H)} \subseteq H$ so that $\overline{f^{-1}(H)} \subseteq f^{-1}(H)$,

showing that $f^{-1}(H)$ is closed in X. Thus the inverse image under f of every closed set in Y is a closed subset of X. Hence by theorem II, f is continuous on X.

19.27 UNIFORM CONTINUITY. DEFINITION.

(*Agra 2006, 2011*)

Let (X, d_1) and (Y, d_2) be metric spaces. A function $f : X \to Y$ is said to be uniformly continuous on X if for every $\varepsilon > 0$, there exists a δ depending only on ε such that

$$d_1(x, y) < \delta \Rightarrow d_2(f(x), f(y)) < \varepsilon \ \forall\ x, y \in X$$

Note. From the above definition, we see that if a function is uniformly continuous on X, then it is continuous at each point of X and hence it is continuous on X. But the converse need not be true. For example, consider a function $f :]\,0,1\,] \to \mathbf{R}$ defined by $f(x) = 1/x$. Then it is continuous but not uniformly continuous on] 0, 1] as shown in Theorem I of Art 8. 19 of Chapter 8. (*Meerut 2000*)

Note 2. For examples on uniform condinuity refer examples based on Art 8.19 for real line with usual metric space.

Theorem I *Every real valued function defined and continuous on a closed and bounded interval is uniformly continuous.*

Proof. Refer theorem II, Art 8. 19 of Chapter 8.

EXERCISE

1. If (X, d) be a metric space and S be a non-empty subset of X, then prove that the function $D: X \to R$ defined by $D(x) = d(x, S)\ \forall\, x \in X$, is uniformly continuous function from (X, d) to the real line. *(Calicut 2004, Delhi Maths (H) 1995, 2001)*
2. Let (X, d) be a metric space and let A and B be disjoint non-empty closed subsets of X. Show that there exists a real valued continuous function $f: X \to [0,1]$ such that $f(A) = \{0\}$ and $f(B) = \{0\} \cdot \{[0,1]\}$ is equipped with the usual metric space.
[*It is known as Urysohn's lemma for metric spaces*] *(Delhi Maths (H) 1999)*
3. For any non-empty subset A of a metric space (X, d), prove that the mapping $x \to d(x, A)$ is uniformly continuous. Also show that $f(x) = 0 \Leftrightarrow x \in \bar{A}$.
4. Let (X, d) be a metric space, then show that any disjoint pair of closed sets in X can be separated by disjoint open sets in X.
5. Let (X, d) be a metric space and let x_0 be a fixed point of X. Show that the real valued function $f_{x_0}(x) = d(x, x_0)$ is continuous. *(Meerut 2000, 01)*

19.28 HOMEOMORPHISM.DEFINITION. (Agra 2009)

Let (X, d_1) and (Y, d_2) be two metric spaces. A function $f: X \to Y$ is said to be homeomorphism if *(i)* f is both one-one and onto *(ii)* f and f^{-1} are both continuous.

Homeomorphic spaces. Definition. A metric space (X, d_1) is said to homeomorphic to another metric space (Y, d_2) if there exists a homeomorphism of X onto Y. *(Delhi Maths (H) 2007)*

Theorem I. Let (X, d_1) and (Y, d_2) be metric spaces and f a mapping from X onto Y. *Then* f is homeomorphism from X to Y iff the following conditions hold :

(i) f is one-one onto *(ii)* f is continuous
(iii) for each open set G in X, $f(G)$ is open in Y

Theorem II. Let (X, d_1) and (Y, d_2) be metric spaces and f a mapping from X onto Y. *Then* f is homeomorphism from X to Y iff the following conditions hold :

(i) f is one-one onto *(ii)* f is continuous
(iii) for each closed set H in X, $f(H)$ is closed in Y.

Theorem III. Show that homeomorphism is an equivalence relation in the collection of al metric spaces. *(Agra 2006, 08, 11)*

Proof. Refer Ex. 25, page 19.62

Example 1. Let (X, d) and (Y, ρ) be two metric spaces. Prove that a necessary and sufficient condition for the bijection $f: X \to Y$ to be a homeomorphism is that $f(\bar{A}) = \overline{f(A)}$ for every subset A of X. *(Delhi Maths (H) 2003)*

Example 2. Show that the open interval $]-1, 1[$ is homeomorphic to $]a, b[$.

Example 3. Show that the mapping $f: \mathbf{R} \to]-1, 1[$ defined by $f(x) = x/(1+|x|), \forall\, x \in \mathbf{R}$, is a homeomorphism of the real line onto the open interval $]-1, 1[$ with the usual metric. *(Delhi Maths (H) 1998)*

Example 4. Show that the intervals $]a, b[$ and $]c, d[$ in $(\mathbf{R}, d)$ are homeomorphic, d being the usual metric. *(Agra 2010; Delhi Maths (H) 2005)*

19.29 ISOMETRY. DEFINITION. (Delhi Maths (H) 2007)

Let (X, d_1) and (Y, d_2) be metric spaces. A function $f: X \to Y$ is called an isometry if

$$d_1(x, y) = d_2(f(x), f(y)),\ \forall\ x, y \in X$$

Note. From above definition, we see that each isometry is always one-one function and uniformly continuous.

Example 1. Let (X, d) be a metric space. Then identity mapping of X to itself is an isometry.

Example 2. Show that the mapping $f : \boldsymbol{R}^2 \to \boldsymbol{R}^2$ defined by $f(x, y) = (x_1, y_1)$ where $x_1 = x\cos\theta - y\sin\theta$, $y_1 = x\sin\theta + y\cos\theta$, where θ is constant real number is an isometry of the plane onto itself. Give a geometrical interpretation of this isometry.

Example 3. Show that $f : \boldsymbol{R} \to \boldsymbol{R}$ defined by $f(x) = 3x + 4$ is not an isometry.

Example 4. Show that isometry is an equivalence relation in the collection of all metric spaces.

Theorem *Every isometry is a homeomorphism* *(Delhi Maths (H) 1995, 2001)*

Proof. Let f be isometry from metric space (X, d_1) to (Y, d_2)

Since each isometry is one-one onto, so f is one-one onto.

Let a be an arbitrary element of X and take an $\varepsilon > 0$.

Then f is an isometry $\Rightarrow$ $d_1(x,a) = d(f(x), f(a))\ \forall\, x \in X$

Taking $\delta = \varepsilon$, we have

$$d_1(x,a) < \delta \Rightarrow d_2(f(x), f(a)) < \varepsilon,$$

showing that f is continuous at a. Since a is an arbitrary point of X, it follows that f is continuous on X.

Now, f is isometry $\Rightarrow f^{1}$ is isometry. $\Rightarrow f^{1}$ is continuous, as before.

Thus, f is one-one onto mapping such that f and f^{1} are both continuous and hence f is a homeorphism.

Note. We have seen that every isometry is a homeomorphism but the converse need not be true. For example,

The mapping $f : \boldsymbol{R} \to]-1, 1[$ *defined by* $f(x) = x/(1/|x|)$ *is a homeomorphism but is not an isometry.* *(Delhi Maths (H) 2001)*

Sol. Left as an exercise.

19.30 SEQUENCES IN A METRIC SPACES.

Let (X, d) be a metric space. A function whose domain is the set of natural numbers $\boldsymbol{N}$ and range a subset of X is called a sequence.

When $X = \boldsymbol{R}$, the set of real numbers, then the sequence is known as a real sequence. Refer chapter 5 for notation, range and illustrations of sequences.

Convergent sequence. Definition.

A sequence $<x_n>$ in a metric space (X,d) is said to converge to $x \in X$, if given $\varepsilon > 0$, we can find a positive integer m (depending on ε) such that $d(x_n, x) < \varepsilon$, whenever $n \geq m$.

Equivalently, the sequence $< x_n >$ is said to converge to $x \varepsilon X$ if given $\varepsilon > 0$, there exist a positive integer m (depending on ε) such that $x_n\ \varepsilon\ S(x, \varepsilon)$ for all $n \geq m$

If $< x_n >$ converges to x, we say that x is a limit of the sequence and we write

$$\lim_{x\to\infty} x_n = n \quad \text{or} \quad x_n \to x \quad \text{as} \quad n \to \infty \quad \text{or} \quad d(x_n, x) \to 0 \quad \text{as} \quad n \to \infty$$

Theorem I *Limit of a sequence, if it exists is unique.*

Proof. Let $< x_n >$ be a sequence in metric space which converges to $x \in X$. Let, if possible, $< x_n >$ converges to another point y distinct from x. Let $d(x, y) = \varepsilon$. Then the open spheres $S(x, \varepsilon/4)$ and $S(y, \varepsilon/4)$ are disjoint.

Now, $x_n \to x$, so there exists a positive integer m_1, such that

$$\forall\ n \geq m_1 \quad \Rightarrow \quad x_n \,\varepsilon\, S(x, \varepsilon/4) \qquad \text{... (1)}$$

Again, $x_n \to y$, so there exists a positive integer m_2, such that

$$\forall\ n \geq m_2 \quad \Rightarrow \quad x_n \,\varepsilon\, S(y, \varepsilon/4). \qquad \text{... (2)}$$

Let $m = \max\{m_1, m_2\}$. Then, by (1) and (2), we see that $x_m \varepsilon S(x, \varepsilon/4)$ and $x_m\ \varepsilon\ S(y, \varepsilon/4)$ so that $S(x, \varepsilon/4)$ and $S(y, \varepsilon/4)$ are not disjoint. This is a contradiction. Hence the limit of $< x_n >$ is unique, if it is exists.

Theorem II. Sequential Characterization of Continuity

Let (X, d_1) and (Y, d_2) be two metric spaces. A function $f : X \to Y$ is continuous at $a \in X$ iff for each sequence $< a_n >$ in X converging to $a \in X$, the sequence $<f(a_n)>$ converges to f (a). (*Mumbai 2010; G.N.D.U. 1998, Meerut 1997, Delhi B.Sc. (Hons) III 2012*)

Proof. The condition is necessary. Let f be continuous at $a \in X$ and let $< a_n >$ be a sequence is X such that $a_n \to x$.

Let $\varepsilon > 0$. Since f is continuous at a, so there exists $\delta > 0$ such that

$$d_1(x, a) < \delta \Rightarrow d_2(f(x), f(a)) < \varepsilon \qquad \text{... (1)}$$

Since $a_n \to a$ as $n \to \infty$, there exists $m \in N$ such that

$$d_1(a_n, a) < \delta, \forall\ n \geq m \qquad \text{... (2)}$$

Putting $x = a_n$ in (1), we have

$$d_1(a_n, a) < \delta \Rightarrow d_2(f(a_n), f(a)) < \varepsilon \qquad \text{.. (3)}$$

From (2) and (3), $\quad d_2(f(a_n), f(a)) < \varepsilon\ \forall\ n \geq m$

$\Rightarrow \quad < f(a_n) >$ converges to $f(a)$.

The condition is sufficient. Let, if possible, f be not continuous. Then we shall show that there exists a sequence $< a_n >$ converging to 'a' but the sequence $< f(a_n)$ does not converge to $f(a)$.

Since f is not continuous, so there must exist at least one $\varepsilon > 0$ such that for every $\delta > 0$ and for some $x \in X$,

$$d_1(x, a) < \delta \quad \text{but} \quad d_2(f(x), f(a)) \geq \varepsilon \qquad \text{... (4)}$$

Takingl $\delta = 1/n$ in (4), we see that for each $n\ \varepsilon\ N$, there exists $a_n \varepsilon X$ such that

$$d_1(x, a) < 1/n \quad \text{but} \quad d_2(f(x), f(a)) \geq \varepsilon$$

$\Rightarrow \quad < f(a_n) >$ cannot converge to $f(a)$.

Theorem III *If x and y are any two points is a metric space (X, d) and if $< y_n >$ is a sequence converging to y, then $< d(x, y_n) >$ converge to d (x, y).*

Proof. Let as an exercise.

Theorem IV *If $<x_n>$ and $<y_n>$ converge to points x and y respectively in a metric space (X, d), then the sequence $< d(x_n, y_n) >$ converge to d (x, y).* **(Rajasthan 2010; Meerut 2011)**

Proof. Left as an exercise

Theorem V. *Let (X, d_1) and (Y, d_2) be two metric spaces. A sequence $< (x_n, y_n) >$ in the product space converges to (x, y) iff $x_n \to x$ and $y_n \to y$.* *(Delhi Maths (H) 2000)*

Proof. The product metric d on $X_1 \times X_2$ is defined by

$$d((x_1, x_2), (y_1, y_2)) = \max\{d_1(x_1, y_1), d_2(x_2, y_2)\} \quad \dots (1)$$

The condition is necessary. Let $(x_n, y_n) \to (x, y)$. Then, for a given $\varepsilon > 0$, we can find a positive integer m such that

$$d((x_n, y_n), (x, y)) < \varepsilon, \ \forall n \geq m$$

$$\Rightarrow \quad \max\{d_1(x_n, x), d_2(y_n, y)\} < \varepsilon, \ \forall n \geq m, \text{ using (1)}$$

$$\Rightarrow \quad d_1(x_n, x) < \varepsilon \text{ and } d_2(y_n, y) < \varepsilon \ \forall n \geq m$$

$$\Rightarrow \quad x_n \to x \text{ and } y_n \to y$$

The condition is sufficient. Let $x_n \to x$ and $y_n \to y$. Then for a given $\varepsilon > 0$, there exists $m_1, m_2 \in N$ such that

$$d_1(x_n, x) < \varepsilon \ \forall n \geq m_1 \text{ and } d_2(y_n, y) < \varepsilon \ \forall n \geq m_2.$$

Let $m = m_1 + m_2$ then we have

$$d_1(x_n, x) < \varepsilon \text{ and } d_2(y_n, y) < \varepsilon \ \forall n \geq m$$

$$\Rightarrow \quad \max\{d_1(x_n, x), d_2(y_n, y)\} < \varepsilon \ \forall n \geq m$$

$$\Rightarrow \quad d((x_n, y_n), (x, y)) < \varepsilon \ \forall n \geq m, \text{ by (1)}$$

$$\Rightarrow \quad (x_n, y_n) \to (x, y).$$

Theorem VI. *A point p in a metric space (X, d) is a limit point of a set $A \subset X$ iff there is a sequence of distinct points in A converging to p.* *(Mumbai 2010; Delhi B.Sc. (H) 2005)*

Proof. The condition is necessary. Let p be a limit point of A. Since each *nbd* of a limit point of a set in a metric space contains infinitely many points of A, hence for each $n \in N$, $S(p, 1/n)$ must contain infinitely many points of A. As a particular case, $S(p, 1) \cap A$ is an infinite set. Let us select $p_1 \in S(p, 1) \cap A$. As before $S(p, 1/2) \cap A$ is an infinite set. Now, select

$$p_2 \in S(p, 1/2) \cap (A - \{p_1\})$$

In this manner, in general, $\forall n > 1$, we can select

$$p_n \in S(p, 1/n) \cap (A - \{p_1, p_2, \dots, p_{n-1}\})$$

Hence the sequence $< p_n >$ is a sequence of distinct points in A converging to p. Also, we have

$$d(p_n, p) < 1/n \ \forall n \Rightarrow p_n \to p$$

The condition in sufficient. Assume that there exists a sequence $< p_n >$ of distinct points in A converging to p. Hence, for a given $\in > 0$, we can select $m \in \mathbf{N}$ such that $d(p_n, p) < \in \ \forall n \geq m$. Then $S(p, \in)$ will contain infinitely many points of A, namely, $p_n \ \forall n \geq m$. Hence p is a limit point of A.

Theorem VII. *A point p in a metric space (X, d) is an adherent point of a set $A \subset X$ if and only if there is a sequence in A converging to p.*

Proof. Let as an exercise.

Theorem VIII. *Let $< x_n >$ be a sequence in a metric space (X, d). If $x_n \to x$, then every subsequence of $< x_n >$ also converges to x.*

Proof. Left as an exercise.

19.31 CAUCHY SEQUENCE. DEFINITION. *(Rajasthan 2010)*

A sequence $< x_n >$ in a metric space (X, d) is said to be a Cauchy sequence iff for each $\varepsilon > 0$ there exists a positive integer p such that $d(x_n, x_m) < \varepsilon \;\forall\; n, m \ge p$

Theorem I. *Every convergent sequence is a Cauchy sequence but the converse is not necessarily true.* *(Rajasthan 2010; Agra 2009, 10, 12; Meerut 2011, 12)*

Proof. Let $< x_n >$ be a convergent sequence in a metric space (X, d) and let $x_n \to x$. Hence given $\varepsilon > 0$, there exists a positive integer p such that

$$d(x_n, x) < \varepsilon/2 \quad \text{and} \quad d(x_m, x) < \varepsilon/2 \;\forall\; m, n > p \qquad \text{... (1)}$$

Hence for $\forall\; m, n \ge p$, using triangle inequality, we have

$$d(x_n, x_m) \le d(x_n, x) + d(x, x_m)$$
$$= d(x_n, x) + d(x_m, x) < \varepsilon/2 + \varepsilon/2, \text{ by (1)}$$

Thus, $$d(x_n, x_m) < \varepsilon \;\forall\; n, m \ge p$$

Hence $< x_n >$ is a Cauchy sequence.

To show that the converse is not always true, consider the space $X =]\,0, 1]$ of the real line with usual metric. Let $x_n = 1/n \;\forall\; n \in N$. Then, $< x_n >$ is a Cauchy sequence, for, if for a giver $\in > 0$, we choose a positive integer $p > 2/\varepsilon$, then

$$\forall\; m, n \ge p \;\Rightarrow\; d(x_n, x_m) = |x_n - x_m| \le |x_n| + |x_m| = 1/m + 1/n$$

Thus, $$\forall\; m, n \ge p \;\Rightarrow\; d(x_n, x_m) < \varepsilon/2 + \varepsilon/2 = \varepsilon.$$

But $\lim\limits_{x \to \infty} (1/n) = 0$ which is not a point of $]\,0, 1]$. Hence $< x_n >$ is not a convergent sequence in $]0,1]$.

Thus, $< x_n >$ is a Cauchy sequence but it is not a convergent sequence.

Theorem II. *The image of a Cauchy sequen/ce under a uniformly continuous function is again a Cauchy sequence.* *(Delhi Maths (H) 2007)*

Proof. Left as an exercise.

Theorem III. Let $< x_n >$ be a Cauchy sequence in metric space (X, d). If $< x_n >$ possesses a convergent subsequence $< x_{n_k} >$ converging to x, then $< x_n >$ converges to x.

EXERCISE

1. Prove that a sequence in $\boldsymbol{R}^n$ is convergent iff it is a Cauchy sequence. *(Delhi Hons. 1997)*
2. Prove that (l_2, d), where l_2 is the set of all real sequence $< x_n >$ such that $\sum_{i=1}^{n} x_i^2$ converges and d defined by $d(x, y) = \left\{ \sum_{i=1}^{n} (x_i - y_i)^2 \right\}^{1/2}$, $\forall\; x, y \in l_2$, is a metric space and also show that it is separable. *(Delhi Maths Hons. 1995, 2001)*
3. Let (X, d) be a metric space and $A \subset X$. Show that $p \in \overline{A}$ iff there is a sequence $\{x_n\}$ in A such that $x_n \to p$. *[Mumbai 2010]*

19.32 COMPLETE METRIC SPACE. DEFINITION.

(Meerut 2012; Agra 2009, 10; Chennai 2011; Lucknow 2010; Rajasthan 2010)

A metric space is said to be *complete* if every Cauchy sequence in X converges.

A metric space which is not complete is said to be *incomplete*.

Example 1. *Any set X with the discrete metric forms a complete metric space.*

[Meerut 2013; Delhi Maths (H) 2008]

Solution. Let $< x_n >$ be a Cauchy sequence in the discrete metric space X. Then $d(x_n, x_m) = 0$ if $x_n = x_m$ and $d(x_n, x_m) = 1$ if $x_n \neq x_m$. On taking $\varepsilon = 1/2$ in the definition of Cauchy sequence, we obtain

$$d(x_n, x_m) < 1/2 \;\; \forall \; n, m \geq p \qquad \text{... (1)}$$

Using the definition of the discrete metric space, (1) yields

$$d(x_n, x_p) = 0 \quad \text{for each } n$$

$$\Rightarrow \quad x_n \to x_p \quad \text{as } n \to \infty,$$

showing that every Cauchy sequence converges to a point of X which is also a term of the sequence. Hence discrete metric space X is complete.

Example 2. *Let* $X =]0, 1[$ *and let* $d(x, y) = |x - y| \;\forall\; x, y \in X$, *then show that (X, d) is an incomplete metric space.* *(Delhi Maths (H) 2001)*

Sol. Let $< x_n >$ be sequence in X defined by $x_n = 1/n \;\forall\; n \in N$. Then as shown in theorem I, Art 19. 31, $< x_n >$ is a Cauchy sequence in X. We now show that $< x_n >$ cannot converge to any point $x \in X$.

Let x be an arbitrary point of X. Then, by the Archimedean property of real number, there exist $n \in \boldsymbol{N}$ such that

$$n \leq x < n + 1, \qquad i.e., \qquad 1/(n+1) < x \leq 1/n.$$

Case I. Let $x = 1/n$. Then $]\, 1/(n-1), 1/(n+1)\, [$ is a *nbd* of x in X which contains only one member of $< 1/n >$.

Case II. Let $x \neq 1/n$. Then $]\, 1/(n+1), 1/n\, [$ is a *nbd* of x in X which does not contains any member of $< 1/n >$.

Hence in both cases, $< 1/n >$ cannot converge to x. Since x is an arbitrary point of X, it follows that $< 1/n >$ cannot converge to any point of X and so (X, d) is an incomplete metric space.

Example 3 *Show that the real line is a complete metric space.*

(Agra 2006, 09, 10; Kanpur 2006; Delhi Maths (H) 1995, Meerut 2010)

Solution. Let $< x_n >$ be an arbitrary Cauchy sequence of real numbers. Let us define the sequence $< n_k >$ of positive integers as follows : n_{k+1} is the smallest integer $\geq n_k$ such that

$$|x_n - x_m| < 1/2^{k+1}, \quad \forall \; n, m \geq n_k \qquad \text{... (1)}$$

Let $\quad I_k = \left[x_{n_k} - 2^{-k}, x_{n_k} + 2^{-k} \right]$ so that each I_k is a closed interval.

Since $\quad \left| x_{n_k} - x_{n_{k+1}} \right| < 1/2^{k+1}$, we have $I_{k+1} \subset I_k$.

Here the length of $I_k = 2/2^k = 1/2^{k-1}$ and so the length of $I_k \to 0$ as $k \to \infty$. Hence by Cantor nested interval theorem (refer Art. 3.18), $\bigcap_{k=1}^{\infty} I_k$ consists of exactly one point, say $x \in \boldsymbol{R}$. Since $x \in I_k \;\forall\; k \in \boldsymbol{N}$, we have

$$\left|x - x_{n_k}\right| < 1/2^k \ \forall \ k \in N. \qquad \text{... (2)}$$

Also, for $n \geq n_k$, (1) yields

$$\left|x_{n_k} - x_n\right| < 1/2^{k+1} < 1/2^k \qquad \text{... (3)}$$

Thus for all $n \geq n_k$, we have

$$|x - x_n| = |(x - x_{n_k}) + (x_{n_k} - x_n)| \leq |x - x_{n_k}| + |x_{n_k} - x_n|$$

$$< (1/2^k) + (1/2^k) = 1/2^{k-1}, \text{ using (2) and (3)} \qquad \text{... (4)}$$

For a given $\varepsilon > 0$, choose $k \in N$ such that $1/2^{k-1} < \varepsilon$

Hence (4) gives $\qquad |x - x_n| < \varepsilon \ \forall \ n \geq n_k$

$$\Rightarrow \qquad x_n \to x \text{ as } n \to \infty$$

Thus every Cauchy sequence in ***R*** is convergent and hence the real line in complete.

Example 4. *Prove that the space of C [0, 1] of all continuous real valued function on [0, 1] with metric d defined by*

$$d(f, g) = \sup\{|f(x) - g(x)| : x \in [0,1]\} \text{ is a complete metric space.}$$

***(Delhi Maths (H)* 1999, 2002, 04, 07)**

Solution. (*i*) $d(f, g) = \sup\{|f(x) - g(x)| : x \in [0,1]\} \geq 0$

(*ii*) $d(f, g) = 0 \Leftrightarrow \sup\{|f(x) - g(x)| : x \in [0,1]\} = 0$

$$\Leftrightarrow f(x) - g(x) = 0 \ \forall \ x \in [0,1] \quad \Leftrightarrow \quad f(x) = g(x), \ \forall \ x \in [0,1]$$

$$\Leftrightarrow f = g$$

(*iii*) $d(f, g) = \sup\{|f(x) - g(x)| : x \in [0,1]\} = \sup\{|g(x) - f(x)| : x \in [0,1]\} = d(g, f)$

(*iv*) For any three function $f(x)$, $g(x)$ and $h(x)$, we have

$$d(f, g) = \sup\{|f(x) - g(x)| : x \in [0,1]\}$$

$$= \sup\{|f(x) - h(x) + h(x) - g(x)| : x \in [0,1]\}$$

$$\leq \sup\{|f(x) - h(x)| : x \in [0,1]\} + \sup\{|h(x) - g(x)| : x \in [0,1]\}$$

$$= d(f, h) + d(h, g)$$

Hence $(C[0,1], d)$ is a metric space.

Let $< f_n >$ be a Cauchy sequence in $C[0,1]$. Let $\varepsilon > 0$ be given. Then there exists a positive integer p such that

$$d(f_n, f_m) < \varepsilon, \ \forall \ n, m \geq p$$

$$\Rightarrow \qquad \sup\{|f_n(x) - f_m(x)| : x \in [0,1]\} < \varepsilon \ \forall \ n, m \geq p$$

$$\Rightarrow \qquad |f_n(x) - f_m(x)| < \varepsilon \ \forall \ n, m \geq p \text{ and } \forall \ x \in [0,1]$$

$$\Rightarrow \qquad < f_n > \text{ converges uniformly on } [0,1]$$

[Using Cauchy's condition for convergence]

If $f_n \to f$, then f is also continuous on [0, 1]. Hence the Cauchy sequence $< f_n >$ converges to a point of $C[0,1]$. Thus, every Cauchy sequence in ***R*** is convergent and hence the real line is complete.

EXERCISE

1. Prove that $(\boldsymbol{R}^n, d_1)$ is a complete metre space where $d_1(x, y) = \sum_{1=1}^{n} | x_i - y_i |$ for

$x = (x_1, ..., x_n)$ and $y = (y_1, ... y_n)$ *[Mumbai 2010]*

2. Show that the set $\boldsymbol{C}$ of complex numbers with usual metric is a complete metric space. *(Meerut 1993)*

3. Let $\boldsymbol{R}$ be the set of real numbers. Let $X = \boldsymbol{R} \sim \{o\}$ and let $d(x, y) = | x - y |,\ \forall\, x, y \in X$. Prove that (X, d) is an incomplete metric space. *[Delhi Maths (H) 2009]*

4. Show that the n-dimensional Euclidean metric space (R^n, d) where $d : \boldsymbol{R}^n \times \boldsymbol{R}^n \to \boldsymbol{R}$ is defined by $d(x, y) = \left\{ \sum_{i=1}^{n} (x_i - y_i)^2 \right\}^{1/2}$, $x = (x_1, x_2, ... x_n)$, $y = (y_1, y_2, ..., y_n) \in R$ is a complete metric space. *(Delhi Maths (H) 1993)*

Hint : Refer Ex. 36 (b), page 19.68

5. Prove that the metric space (l_2, d) where l_2 is the set of all real sequences $< x_n >$ with Σx_n^2 convergent and metric d defined by $d(x_m, x_n) = \left\{ \sum_{i=1}^{n} \left(x_{m_i} - x_{n_i} \right)^2 \right\}^{1/2}$, $x_n = \{x_{n_1}, x_{n_2} ....\}$ is a complete metric space. *(Delhi Maths (H) 1994)*

6. Define a complete metric space. Using the definition, show that $A = \{x, y) \in \boldsymbol{R}^2 \mid x^2 + y^2 < 1\}$ is not complete (distance being Euclidean) *(Mumbai 2010)*

7. Let X be the set of all continuous real-valued functions defined on [0,1], and let

$$d(x, y) = \int_0^1 | x(t) - y(t) |\, dt,\ \forall\ x, y \in X.$$

Show that (X, d) is not complete.

8. Show that the space $\mathbf{C}$ of all convergent sequences of real numbers with metric d defined as

$$d(< x_n >, < y_n >) = \sup \{| x_n - y_n |\} \text{ is complete.}$$

19.33 PROPERTIES OF COMPLETE METRIC SPACES

Theorem I. *Completeness in preservered under isometries.* (*Delhi Maths (M) 2000, 2005, 07*)

Proof. Let (X, d_1) and (Y, d_2) be metric spaces. Let $f : X \to Y$ be an isometry from (X, d_1) onto (Y, d_2) and let (X, d_1) be complete.

Let $< y_n >$ be any sequence in Y. Since every isometry is a one-one onto mapping, hence for each $n \in N$, there exists $x_n \in X$ such that $f(x_n) = y_n$. Since f is an isometry, we obtain.

$$d_1(x_n, x_m) = d_2\,(f(x_n), f(x_m)) = d_2(y_n, y_m),\ \forall\, n, m \in N \quad ... (1)$$

Since $< y_n >$ is a Cauchy sequence, hence for a given $\varepsilon > 0$, there exists $p \in N$ such that

$$d_2(y_n, y_m) < \varepsilon,\ \forall\, m, n \in N \quad ... (2)$$

From (1) and (2), $\quad d_1(x_n, x_m) < \varepsilon,\ \forall\, m, n \in N$

$\Rightarrow \quad < x_n >$ is a Cauchy sequence in X.

Since (X, d) is complete and $< x_n >$ is a Cauchy sequence, hence $< x_n >$ is a convergent sequence. Hence there exists $x \in X$ such that $x_n \to x$. Hence for a given $\varepsilon > 0$, there exists $q \in N$ such that

$$d_1(x_n, x) < \varepsilon, \ \forall \, n \geq q \quad \text{... (3)}$$

Since f is an isometry, we have

$$d_1(x_n, x) = d_2(f(x_n), f(x)) = d_2(y_n, f(x)) \quad \text{... (4)}$$

From (3) and (4), $\quad d_2(y_n, f(x)) < \varepsilon \ \forall \, n \geq q,$

showing that $y_n \to f(x)$. Thus every Cauchy sequences in Y is a convergent sequence and hence (Y, d_2) is complete

Note. *Homeomorphism need not preserve completeness* *(Delhi Maths (H) 2000)*

We know that the red line is complete and space $X =]\,0, 1[$ with usual metric is not complete (see examples, 2 and 3 of Art 19.32). It can be easily shown that homeomorphic image of $\boldsymbol{R}$ is X. But $\boldsymbol{R}$ is complete whereas X is incompelte.

Theorem II. *Every closed subspace of a complete metric space is complete.*

(Meerut 2010; Agra 2008, 11; Delhi B.Sc. (H) 2005, 06, 08)

Proof. Let (X, d) be a complete metric space and let Y be a closed subset of X. Then we are to prove that the subspace (Y, d) is complete

Let $< x_n >$ be an arbitrary Cauchy sequence in Y. The $< x_n >$ is a Cauchy sequence in X. Since X is complete, $< x_n >$ must converge to a point x in X. Thus there is a sequence $< x_n >$ in Y converging to x in X. Hence x is a limit point of Y. Since Y is closed, $x \in A$. Thus every Cauchy's sequence is Y converges to a point x in Y. Hence, Y is complete.

Note. Every open subspace of a complete metric space need not be complete as shown in the following example. *(Delhi B.Sc. (H) 2005, 08)*

We know that the real line is complete and space $X =]\,0, 1[$ with usual metric is not complete. Thus, open subspace $]\,0, 1[$ of $\boldsymbol{R}$ is not complete.

Theorem III. *A complete subspace of a metric space is closed.* *(Meerut 2010)*

Proof. Let (Y, d) be complete subspace of (X, d). In order to show that Y is closed, we shall prove that each limit point of Y belongs to Y. Let x be an arbitrary limit point of Y. Then for each $n \in N$, the open sphere $S(x, 1/n)$ contains at least one point x_n of Y, other than x. Thus, we get a sequence $< x_n >$ in Y such that

$$d(x_n, x) < 1/n \quad \text{and so} \quad x_n \to x, \text{ as } n \to \infty,$$

showing that $< x_n >$ is a convergent sequence in Y. Since every convergent sequence is a Cauchy sequence, it follows that $< x_n >$ is a Cauchy sequence in Y. Since (Y, d) is complete, the limit x of the Cauchy sequence $< x_n >$ must therefore be in Y. Thus every limit point of Y is contained in it. Hence Y is closed.

Theorem IV. Let (X, d) be a complete metric space and Y be subspace of X. Then Y is complete iff Y is closed. *(Meerut 2010)*

Proof. Combine results of theorems II & III for complete proof.

Exercise. Show that completeness is preserved under isometrics. What happens if the mapping is only continuous. Justify your answer. *(Delhi (Maths) (H) 2005)*

19.34 CANTOR'S INTERSECTION THEOREM *(Meerut 2011, 2012)*

In this article we propose to re-formulate the well known 'Cantor nested interval theorem' of Art 3.18 of chapter 3 in terms of metric spaces.

Nested sequence. Definition.

A sequence $< F_n >$ of sets is said to be nested if $F_{n+1} \subset F_n \ \forall \ n \in N$

Cantor intersection theorem (First form). *If $< F_n >$ is a nested sequence of non-empty closed subsets of metric space (X, d) such that $\delta(F_n) \to 0$ as $n \to \infty$, then X is complete if and only if $\bigcap_{n=1}^{\infty} F_n$ consists of exactly one point, where $\delta(F_n)$ denoted the diameter of F_n.*

(Mumbai 2010; M.S. Univ. T.N. 2006; Delhi Maths (H) 2008; Meerut 2011, 12)

Proof. The condition is necessary. Let (*X*, *d*) be a complete metric space and let $< F_n >$ be a nested sequence of non-empty closed sets with $\delta(F_n) \to 0$ as $n \to \infty$. We shall prove that $\bigcap_{n=1}^{\infty} F_n$ contains precisely one point.

Since $F_n \neq \varphi \ \forall \ n \in N$, we can construct a sequence $< x_n >$ by chooseing $x_n \in F_n$, $n = 1, 2, 3, \ldots$. We now prove that $< x_n >$ is a Cauchy sequence. Since $\delta(F_n) \to 0$ as $n \to \infty$, for a given $\varepsilon > 0$, there exists $m \in N$ such that

$$\delta(F_n) < \varepsilon, \ \forall \ n \geq m \qquad \ldots (1)$$

Since $< F_n >$ is a nested sequence so that $F_n \subset F_m \ \ \forall \ n \geq m$, hence we obtain,

$$x_n \in F_m, \ \forall \ n \geq m \qquad \ldots (2)$$

Using (2) and the definition of diameter of a bounded set, we have $d(x_n, x_m) \leq \delta(F_m) \ \forall \ n \geq m$

or $$d(x_n, x_m) < \varepsilon, \ \forall \ n \geq m, \text{ using (1)}$$

showing that $< x_n >$ is a Cauchy sequence in *X*. Since (*X*, *d*) is complete, there exists a point $x \in X$ such that $x_n \to x$ as $n \to \infty$.

Let us now prove that $x \in \bigcap_{n=1}^{\infty} F_n$. Let, if possible, $x \notin \bigcap_{n=1}^{\infty} F_n$ then there exists a positive integer *k* such that $x \notin F_k$.

Now F_k is closed and $x \notin F_k \Rightarrow d(x, F_k) > 0$

Let $d(x, F_k) = \delta$. Then $$d(x, y) \geq \delta, \ \forall \ y \in F_k$$

Therefore, F_k and the open sphere $S(x, \delta/2)$ are disjoint and hence

$$n > k \ \Rightarrow \ F_n \subset F_k \ \Rightarrow \ x_n \in F_k \ \Rightarrow \ x_n \notin S(x, \delta/2),$$

which is impossible since $x_n \to x$. Hence $x \in \bigcap_{n=1}^{\infty} F_n$, which proves that the intersection $\bigcap_{n=1}^{\infty} F_n$ is non-empty.

To prove that $\bigcap_{n=1}^{\infty} F_n$ contains exactly one point, suppose it contains two distinct points *x* and *y*. Then $\delta(F_n) > d(x, y)$ for all *n* and hence $\delta(F_n)$ cannot tend to zero which is a contradiction.

Therefore $\bigcap_{n=1}^{\infty} F_n$ contains exactly one point.

The condition is sufficient. Let every nested sequence of closed sets with diameters tending to zero has non-empty intersection. We shall prove that (X, d) is complete.

Let $< x_n >$ be an arbitrary Cauchy sequence in X.

For a given $\varepsilon = 1/2$, choose $n_1 \in N$ such that

$$d(x_m, x_n) < 1/2, \ \forall\, m, n \geq n_1$$

so that
$$d(x_{n_1}, x_n) < 1/2, \ \forall\, n \geq n_1$$

Similarly, for a given $\varepsilon = 1/2^2$, choose $n_2 \in N$ such that $n_2 > n_1$ and

$$d(x_m, x_n) < 1/2^2, \ \forall\, m, n \geq n_2$$

so that
$$d(x_{n_2}, x_n) < 1/2^2, \ \forall\, n \geq n_2$$

Proceeding likewise, we obtain an increasing sequence $< n_1, n_2, ... >$ of positive integers such that

$$d(x_{n_1}, x_n) < 1/2, \ \forall\, n \geq n_1, \qquad d(x_{n_2}, x_n) < 1/2^2, \ \forall\, n \geq n_2$$

$$d(x_{n_3}, x_n) < 1/2^3, \ \forall\, n \geq n_3, \ .... \ d(x_{n_k}, x_n) < 1/2^k, \ \forall\, n \geq n_k$$

....

and so on.

Let
$$F_k = S\,[x_{n_k}, 1/2^{k-1}], \ k = 1, 2, 3, ...$$

Hence, $< F_k >$ is sequence of closed sets (noting that the closed sphere $S[x_{n_k}, 1/2^{k-1}]$ is a closed set) such that $\delta(F_k) \to 0$ as $k \to \infty$. We shall now show that $< F_n >$ is nested, *i.e.*, $F_{k+1} \subset F_k \ \forall\, k \geq 1$. Let x be an arbitrary point of F_{k+1}. Then we have

$$d\left(x_{n_k}, y\right) \leq d\left(x_{n_k}, x_{n_{k+1}}\right) + d\left(x_{n_{k+1}}, y\right)$$
$$< 1/2^k + 1/2^k = 1/2^{k-1}$$

$$\Rightarrow \qquad y \in S\left(x_{n_k}, 1/2^{k-1}\right) \subset F_k \Rightarrow F_{k+1} \subset F_k,$$

where $S\left(x_{n_k}, 1/2^{k-1}\right)$ is an open sphere.

Hence $< F_n >$ is a nested sequence of non-empty closed sets with diameters tending to zero and by hypothesis we must have $x \in X$ such that $x \in F_k \ \forall\, k \in N$.

Now $< x_{n_1}, x_{n_2}, x_{n_3}, ... >$ is a subsequence of $< x_n >$. Since $x_{n_k} \in F_k$ and $x \in F_k$, we have

$$d\left(x_{n_k}, x\right) < 1/2^k \quad \forall\, k \in \mathbf{N},$$

showing that the sequence $< x_{n_1}, x_{n_2} ... >$ converges to x. But $< x_n >$ is a Cauchy sequence in X and $< x_{n_1}, x_{n_2} ... >$ is a subsequence of $< x_n >$ converging to x, hence $x_n \to x$ as $n \to \infty$.

Thus, every Cauchy sequence in X converges and hence (X, d) is complete.

Cantor intersection theorem (Second form)

A metric space (X, d) is complete iff the intersection of every nested sequence of closed spheres with radii tending to zero consists of exactly one point. (*Delhi Maths (H) 2004*)

Proof. Left as an exercise for the reader.

Note. The following examples show that in the statement of Cantor's intersection theorem, neither of the following conditions is redundant : (*i*) $\delta(F_n) \to 0$ as $n \to \infty$ (*ii*) the sets F_n are closed.

[Delhi Maths (H) 2008]

Example 1. Let X be the real line $\boldsymbol{R}$ with usual metric and let $F_n =]0, 1/n]$. Then $X = \boldsymbol{R}$ is complete and $< F_n >$ is nested sequence such that $F_1 \supset F_2 \supset F_3 ....$ and $\lim \delta(F_n) \to 0$ as $n \to \infty$. But $\bigcap\limits_{n=1}^{\infty} F_n = \varphi$. Note that this has happended because F_n's are not closed.

Example 2. Let X be the real line $\boldsymbol{R}$ with usual metric and let $F_n = [n, \infty[$. Then $X = \boldsymbol{R}$ is complete and $< F_n >$ is nested sequence of closed sets F_n's and F_n are closed sets. But $\bigcap\limits_{n=1}^{\infty} F_n = \varphi$. Note that this has happened because $\delta(F_n)$ does not tend to zero as $n \to \infty$.

19.35 CONTRACTION MAPPING PRINCIPLE

Fixed point of mapping Definition.

Let X be a non-empty set and let $T : X \to X$ be a mapping on X to itself. A point $x \in X$ is called a *fixed point* of T if $T(x) = x$.

Note. In this article we shall write T_x instead of $T(x)$ and use T^2 instead of $T.T$, T^3 instead of $T.T^2$ and in general T^n instead of $T.\ T^{n-1}$.

Contraction mapping Definition.

A mapping T from a metric space (X, d) to itself is said to be a *contraction mapping* if for some real number α such that $0 < \alpha < 1$,

$$d(T_x, T_y) \leq \alpha\, d(x, y),\ \ \forall\ x, y \in X$$

Note. The mapping T is contraction mapping in the sense that the distance between the images of any two points is less than that between their pre-images.

Theorem I. *If T is a contraction on X, then T is continuous on X.* (*Meerut 2005*)

Proof. Let a be any point of X. Let $\varepsilon > 0$. Hence if $d(x, a) < \varepsilon$, then from the definition of contraction, we have $d(T_x, T_a) < \alpha\,\varepsilon < \varepsilon$ so that T is continuous at a.

Since a is any point of X, so T is continuous at each of X and hence T is continuous on X.

Exercise. Show that every contraction mapping is uniformly continuous but not conversely.

(*Delhi Maths (H)* 2005)

Sol. Try Yourself.

Banach's fixed point theorem (or simply fixed point theorem)

Every contraction mapping T on a complete metric space (X, d) has a unique fixed point.

(*Bharathior 2004. Delhi B.Sc. (Hons.) III 2012; Meerut 20013*)

Proof. Let T be a contraction mapping in a complete metric space (X, d). Then for some $\alpha, 0 < \alpha < 1$, we have

$$d\,(T_x, T_y)\ \leq\ \alpha\ d(x, y),\ \ \forall x, y \in X \qquad ...(1)$$

Let x_0 be an arbitrary point in X. We shall now construct a Cauchy sequence depending on x_0.

Let $$x_1 = T_{x_0}, x_2 = T_{x_1} = T^2_{x_0}, x_3 = T_{x_2} = T^3_{x_0}, ..., x_n = T_{x_{n-1}} = T^n_{x_0} \quad ... (2)$$

We shall now prove that $< x_n >$ is a Cauchy sequence.

Using (1), $$d(T^2_x, T^2_y) \le \alpha \, d(T_x, T_y) \le \alpha^2 \, d(x, y) \; \forall \; x, y \in X$$

and hence proceeding likewise, we have

$$d(T^n_x, T^n_y) \le \alpha^n \, d(x, y) \; \forall \; x, y \in X \text{ and } \forall n \in N \quad ... (3)$$

If m, n are positive integers and $m > n$, let $m = n + p$, when p is a positive integer. Then, by triangle inequality, we have

$$\begin{aligned} d(x_n, x_m) &= d(x_n, x_{n+p}) \\ &\le d(x_n, x_{n+1}) + d(x_{n+1}, x_{n+2}) + ... + d(x_{n+p-1}, x_{n+p}) \\ &= d(T^n_{x_0}, T^n_{x_1}) + d(T^{n+1}_{x_0}, T^{n+1}_{x_1}) + ... + d(T^{n+p-1}_{x_o}, T^{n+p-1}_{x_1}) \\ &\le \alpha^n d(x_0, x_1) + \alpha^{n+1} d(x_0, x_1) + ... + \alpha^{n+p-1} d(x_0, x_1), \text{ using (3)} \\ &\le \alpha^n d(x_0, x_1) [1 + \alpha + \alpha^2 + ...]. \quad ... (4) \end{aligned}$$

Summing up the infinite geometric series in the square breaket, (4) gives

$$d(x_n, x_m) \le \{\alpha^n \, d(x_0, x_1)\} / (1 - \alpha). \quad ... (5)$$

Since $\alpha^n \to 0$ as $n \to \infty$, $d(x_n, x_m)$ can be made less than any pre-assigned positive real number by taking n and therefore m, sufficiently large. Hence $< x_n >$ is a Cauchy sequence. Since X is complete, there exists $x \in X$ such that $x_n \to x$ as $n \to \infty$. We shall now prove that $T_x = x$. Since T is a continuous, so

$$T_x = \lim_{n \to \infty} \left(T_{x_n} \right) = \lim_{n \to \infty} x_{n+1} = x,$$

showing that x is a fixed point of the mapping T.

Now, in order to prove the uniqueness of the point, if possible, let $y \in X$ such that $y \ne x$ and $T_y = y$. Then, we have

$$d(x, y) = d(T_x, T_y) \le \alpha d(x, y) \Rightarrow \quad 1 \le \alpha \Rightarrow \alpha \ge 1,$$
$$[\because \; x \ne y \Rightarrow d(x, y) \ne 0]$$

which is a contradiction. Hence $T_y \ne y$. Therefore x is the only point in X such that $f(x) = x$.

19.36 A SUBSET OF FIRST CATEGORY AND SECOND CATEGORY.

Definitions. A subset of a metric space is said to be of the *first category*, if it is the union of a countable family of nowhere dense sets; otherwise it is said to be of the *second category*.

Illustrations. (*i*) Any countable set A in $\boldsymbol{R}$ is of first category.

(*ii*) The set $\boldsymbol{R}$ is of second category.

Baire Category theorem. **[Agra 2007]**

Every complete metric space is of second category. (*M.S. Univ. T.N. 2006; Meerut 2013*)

Proof. Let (X,d) be a complete metric space. Let, if possible, X be not of second category. Then X is of first category so that X is the union of countable family of nowhere dense sets given by the sequence $< A_n >$, where $A_n \subset X \ \forall\, n \in \mathbf{N}$.

Let A be any nowhere dense subset of X. Then, by definition $\overline{ext\, A} = X$. Thus, each point $x \in X$ is an adherent point of $ext\, A$ and so open sphere $S(x, r)$ contains a point $y \in ext\, A$. Again, since $ext\, A$ is an open set there exists an open sphere $S(y,r') \subset S(y,r)$ such that $S(y,r') \subset ext\, A$. Hence $S(y,r')$ contains no point of A and hence if $0 < r'' < r'$, then closed sphere $S[y,r''] \subset S(x,r)$ such that $S[y,r''] \cap A = \varphi$.

As discussed above, A_1 is nowhere dense set implies that there exists a closed sphere K_1 with radius $r_1 < 1$ such that $K_1 \cap A_1 = \varphi$.

Let S_1 denote the open sphere having the same centre and radius as K_1. As before, in S_1, we can choose a closed sphere K_2, of radius $r_2 < 1/2$ such that $K_2 \cap A_2 = \varphi$ and so on. Proceeding like wise, we obtain a nested sequence $< K_n >$ of closed spheres having the following two properties.

(*i*) $\delta(K_n) \to 0$ as $n \to \infty$

(*ii*) $\forall\, n \in \mathbf{N}$, K_n is disjoint from $A_1, A_2, \ldots, A_n$

Since (X, d) is complete, by Cantor's intersection theorem it follows that $\bigcap_{n=1}^{\infty} K_n$ is non-empty. Let $x \in \bigcap_{n=1}^{\infty} K_n$. Then $x \in K_n$, $\forall\, n \in \mathbf{N}$ and hence x will not belong to any of the nowhere dense sets A_n, by condition (*ii*). But this is not possible because by hypothesis X is union of this family. Hence X cannot be of the first category and therefore X is of the second category.

EXERCISES

1. If A and B are sets of first category, then prove that $A \cup B$ is also of first category.
2. Prove that $\boldsymbol{R}$ with usual metric is of second category.
3. Prove that the set of rational numbers with usual metric is of second category.
4. Prove that any non-empty open interval of $\boldsymbol{R}$ is of the second category.

19.37 COMPACT METRIC SPACES **(Meerut 2011)**

We have already discussed definition of cover (or covering) and finite sub-cover etc. in Art 3.19 in relation to $\boldsymbol{R}$. The same definitions should be used by replacing $\boldsymbol{R}$ by X in this chapter.

Compact metric space. Definition. **(Rajastan 2010; Delhi Maths (H) 2009)**

A metric space (X, d) is said to be compact if every open cover of X admits of a finite subcover.

Also, a subset Y of a metric space (X, d) is said to be compact if the subspace (Y, d) is compact in its own right.

Note. Read Art 3.20 and Art 3.21 and examples of compact sets given just after Art 3.21.

Properties of compact metric spaces

Theorem I. *Every closed subset of a compact metric space is compact.*

(Meerut 2010, 11; Rajasthan 2010; M.S. Univ. T.N. 2006; Delhi Maths (H), 2002)

Proof. Let (X, d) be a compact metric space and let F be a closed subset of X. Let $C = \{G_\lambda : \lambda \in \Delta\}$ be any open cover of F, then $C^* = \{G_\lambda : \lambda \in \Delta\} \cup (X - F)$ is an open cover of X. Since X is compact, hence it has a finite sub-subcover. say $G_{\lambda_1}, G_{\lambda_2}, ..., G_{\lambda_n}, X - F$ so that

$$G_{\lambda_1} \cup G_{\lambda_2} \cup ... G_{\lambda_n} \cup (X - F) = X \quad \text{and so} \quad F \subseteq \cup \{G_{\lambda_i} : i = 1, 2, ..., n\}$$

$\Rightarrow$ $\{G_{\lambda_i} : i = 1, 2, ..., n\}$ is a finite subcover of F. Hence F is compact.

Theorem II. *Every compact subset of a metric space is closed.*

(M.S. Univ. T.N. 2006; Delhi Maths (H) 1998, 2002, 08; I.A.S. 1999

Proof. Let $x \in X$ be an arbitrary limit point of A. Then by theorem VI of Art 19.30, there exists a sequence $< x_n >$ in A converging to x. In order to prove that A is closed we must show that $x \in A$.

Since $< x_n >$ is a convergent sequence in A, it is a Cauchy sequence in A. Since A is compact, it is complete. Hence $< x_n >$ converges to a point in A and this point is clearly x. Thus $x \in A$. Since x is any limit point of A, so all limit points of A belong to A and hence A is closed.

Theorem III A continuous image of a compact metric space is compact. *(Mumbai 2010)*

(Bharathiar 2004; Chennai 2011; Nagpur 2003; Delhi B.Sc. (Hons) III 2012

Proof. Let f be a continuous mapping of a compact metric space X into a metric space Y. Then we are to show that image $f(X)$ of X is also compact.

Let $C = \{G_\lambda : \lambda \in A\}$ be an open cover of $f(X)$

$$\therefore \qquad f(X) \subset \cup \{G_\lambda : \lambda \in \Delta\} \qquad ... (1)$$

Since f is continuous, $f^{-1}(G_\lambda)$ is open subset of X for every open set G_λ in C. From (1),

$$X \subset f^{-1}[f(X)] \subset f^{-1}[\cup G_\lambda : \lambda \in \Delta]$$

$$\Rightarrow \qquad X \subset \cup \{f^{-1}(G_\lambda) : \lambda \in \Delta\},$$

showing that $C^* = \{f^{-1}(G_\lambda) : \lambda \in \Delta\}$ is an open cover of X. Since X is compact, $C^{**} = \{f^{-1}(G_{\lambda_1}), f^{-1}(G_{\lambda_2}), ..., f^{-1}(G_{\lambda_n})\}$ forms a finite subcover of X, so that

$$X \subseteq \cup \{f^{-1}(G_{\lambda_i}) : i = 1, 2, ..., n\}$$

$$\Rightarrow \qquad f(X) \subset f(\cup \{f^{-1}(G_{\lambda_i}) : i = 1, 2, ..., n\})$$

$$\Rightarrow \qquad f(X) \subset \cup \{f[f^{-1}(G_{\lambda_i})] : i = 1, 2, ..., n\}$$

$$\Rightarrow \qquad f(X) \subset \cup \{G_{\lambda_i} : i = 1, 2, ..., n\} \text{ as } f[f^{-1}(G_{\lambda_i})] \subset G_{\lambda_i}$$

$\Rightarrow$ The open cover C of $f(X)$ can be reduced to a finite sub-cover and so $f(X)$ is compact.

Corollary I. *Compactness is preserved under isometries.* *(Delhi Maths (H) 2001)*

Proof. Since every isometry is a continuous mapping, result follws by the above theorem III.

Corollary II. *Compactness is preserved under homeomorphism.*

Hint : Since every homeomorphism is a continuous mapping so result follows by theorem III.

Finite intersection property. Definition.

A family, of subsets of a non-empty sets is said to have a *finite intersection property* (*FIP*) if every finite sub-family of the same has a non-empty intersection.

Theorem IV. *The metric space (X, d) is compact iff every family of closed sets in (X, d) with the finite intersection property (FIP) has non-empty intersection.* *(M.S. Univ. T.N. 2006)*

Proof. Let (X, d) be compact and let $\{F_\lambda : \lambda \in \Delta\}$ be any family of closed sets in (X, d) with *F.I.P.* Let, if possible, $\cap\{F_\lambda : \lambda \in \Delta\} = \varphi$.

On taking complements in X, this $\Rightarrow \ \cup\{F_\lambda^c : \lambda \in \Delta\} = X$ ... (1)

Since F_λ is closed, so its complement F_λ^c is open. Hence (1) slhows that $C = \{F_\lambda^c : \lambda \in \Delta\}$ is an open cover of X. Since X is compact C is reducible to a finite subcover $\{F_{\lambda_i}^c : i = 1, 2, ..., n\}$ such that

$$\cup\{F_{\lambda_i}^c : i = 1, 2, ..., n\} = X$$

$$\Rightarrow \quad \cap\{F_{\lambda_i} : i = 1, 2, ..., n\} = \varphi, \text{ on taking complements}$$

This is a contradiction of the fact that $\{F_\lambda : \lambda \in \Delta\}$ has the *F.I.P.* Since the contradiction arises by assuming that $\cap\{F_\lambda : \lambda \in \Delta\} = \varphi$, it follows that $\cap\{F_\lambda : \lambda \in \Delta\} \neq \varphi$.

Conversely, let every family of closed sets in (X, d) with *FIP* has non-empty intersection. We are to show that (X, d) is compact. Let $\{G_\lambda : \lambda \in \Delta\}$ be an open cover of X.

$$\therefore \quad \cup\{G_\lambda : \lambda \in \Delta\} = X \text{ so that } \cap\{G_\lambda^c : \lambda \in \Delta\} = \varphi,$$

[on taking complements on both sides]

showing that $\{G_\lambda^c : \lambda \in \Delta\}$ is a family of closed sets in X, whose intersection is empty. Hence, by hypothesis this family cannot have the *FIP*. Consequently there exists a subfamily $\{G_{\lambda_i}^c : i = 1, 2, ..., n\}$ such that

$$\cap\{G_{\lambda_i}^c : i = 1, 2, ..., n\} = \varphi$$

$$\Rightarrow \left[\cap\{G_{\lambda_i}^c : i = 1, 2, ..., n\}\right]^c = \varphi^c \quad \Rightarrow \quad \cup\{G_{\lambda_i} : i = 1, 2, ..., n\} = X, \text{ by De-Morgan's law}$$

showing that there exists a finite subcover $\{G_{\lambda_i} : i = 1, 2, ..., n\}$ of open cover $\{G_\lambda : \lambda \in \Delta\}$ and hence (X, d) is compact.

Bolzano- Weierstrass property (*BWP*),. Definition

A metric space (X, d) is said to have the Bolzano-Weierstarn property if every infinite subset of X has a limit point.

Sequentially compact metric space. Definition.

A metric space (X, d) is said to be sequentially compact if every sequence in X has a convergent subsequence.

Countably Compact metric space. Definition.

A metric space is said to be countably compact if every countable open cover of the space admits of a finite sub-cover.

Theorem V. *A metric space is sequentially compact iff it has the Bolzano - Weierstrass property.* *(Delhi Maths (H) 1995, 96)*

Proof. Let (X, d) be sequentially compact metric space and let A be an infinite subset of X. We shall show that A has a limit point.

Since A is infinite, we can construct a sequence $< a_n >$ of distinct points of A. Since $< a_n >$ is a sequence in X and X is sequentially compact, there exists a convergent subsequence $< a_{n_k} >$ converging to a point $a \in X$.

We now show that a is limit point of A. Since $a_{n_k} \to a$, there exists $\varepsilon > 0$ and $m \in N$ such that $S(a_{n_k}, a) < \varepsilon$, $\forall\, k \geq m$. Then open sphere $S(a, \varepsilon)$ contain infinitely many terms of $< a_n >$ and therefore of A. Hence a is a limit point of A. Hence X has the BWP

Conversely, let us suppose that X has the BWP and let $< x_n >$ be an arbitrary sequence in X such that $B = \{x_n : n \in N\}$. We now have the following two possibilities :

Case (*i*) When B is finite. Then there exists some $x \in X$ such $x_n = x$ for infinitely many values of n. Hence, $< x_n >$ has a constant subsequence and therefore convergent.

Case (*ii*) When B is infinite. Then B has a limit point, say b. Since b is limit point of B, so open sphere $S\ (b\ ,1)$ contains infinitely many points of B. Hence, there exists $n_1 \in N$ such that $d(x_{n_1}, b) < 1$.

As before, $S\ (b,\ 1/2)$ also contains infinitely many points of B. Hence we have $n_2 \in N$ such that $d(x_{n_2}, b) < 1/2$ for $n_2 > n_1$. Proceeding likewise we construct a sequence $< x_{n_k} >$ such that $d(x_{n_k}, b) < 1/k$ for all $k \in N$.

Now $< x_{n_k} >$ is a subsequence of $< x_n >$ that converges to b. Hence $< x_n >$ has a convergent subsequence and so X is sequentially compact.

Theorem VI. *Every compact metric space is sequentially compact*

Proof. Let (X, d) be a compact metric space. Then in order to show that X is sequentially compact, it is sufficient to show that it has the BWP (see theorem V). Let, if possible, an infinite subset A of X has no limit point in X. Then corresponding to each $x \in X$, there exists an open sphere $S\ (x,\ r_x)$ containing no point of A, other than possibly, the point x.

Now, the collection $\{S(x, r_x) : x \in A\} \cup (X - A)$ is an open cover of X and this cover cannot be reduced to a finite subcover of X and so X is not compact. This contradicts the hypothesis of compactness of X. This contradiction arises by assuming that an infinite subset A has no limit point. Hence X has the BWP and therefore it is sequentially compact.

19.37A. ε-net and total boundedness. Definition. **(Delhi Maths (H) 2008)**

Let (X, d) be a metric space.

Let $\varepsilon > 0$ and $A \subset X$. A set B of points of X is said to be an $\in$-net for A if for each $x \in A$ we find some $y \in B$ such that $d(x, y) < \varepsilon$.

A metric space (X, d) is said to be *totally bounded* if given any $\varepsilon > 0$, we can find a finite ε-net, *i.e.*, we can find a finite set of points $\{x_1, x_2, ..., x_n\}$ in X such that

$$\cup\{S(x_i, \varepsilon): i = 1, 2, \ldots n\} = X.$$

Note. Clearly, X is totally bounded $\Rightarrow$ X is bounded. **(Delhi Maths (H) 2007)**

Theorem VII. *Every sequentially compact metric space is totally bounded.*

Proof. Let (X, d) be a sequently compact metric space and let $\varepsilon > 0$ be given. Let x_1 be an arbitrary point of X. If $S(x_1, \varepsilon) = X$, then the singleton set $\{x_1\}$ is an ε-net and so the theorem is proved. However, if there exists $x_2 \in X$ such that $x_2 \notin S(x_1, \varepsilon)$. and $S(x_1, \varepsilon) \cup S(x_2, \varepsilon) = X$, then $\{x_1, x_2\}$ is the $\in$-net and the theorem is proved If it is not so, we proceed likewise to obtain.

$$S(x_1, \varepsilon) \cup S(x_2, \varepsilon) \cup \ldots \cup S(x_n, \varepsilon) = X,$$

where note that the process terminates after n stepts because if the same process could be continued indefinitely, then the sequence $< x_n >$ would be a sequence containing no convergent sub-sequence. This Contradicts the hypothesis that X is sequentially compact. It follows that $\{x_1, x_2, \ldots, x_n\}$ is an ε-net and hence X is totally bounded.

Theorem VIII. *A metric space (X, d) is totally bounded iff each sequence in X admits of a Cauchy subsequence.*

Proof. Left as an exercise for the reader.

Theorem IX. *Every compact metric space is countably compact.*

Proof. Let (X, d) be a compact metric space. Then every open cover of X is reducible to a finite sub-cover. In particular, every countable open cover of X is reducible to a finite sub-cover. Hence X is countably compact.

Theorem X. *A metric space is countably compact iff every countable family of non-empty closed sets with finite intersection property has non-empty intersection.*

Proof. The proof of the theorem is the same as that of theorem IV except that you have to consider countable cover instead of any arbitrary cover

Theorem XI. *A metric space is compact iff it is countably compact.*

Proof. Proof follows from theorems IV and X.

19.37B LEBESGUE NUMBER FOR COVER. DEFINITION.

Let (X, d) be a metric space and let $\{G_\lambda : \lambda \in \Delta\}$ be an open cover of X. A real number $l > 0$ is called a Lebesgue number for the open cover if for each subset A of X with diameter $\delta\ (A) < l$ there is at least one G_λ which contains A.

Theorem XII. Lebesgue covering Lemma.

Every open cover of a sequentially metric space has a Lebesgue number.

Proof. Let (X, d) be a sequentially compact metric space and let $C = \{G_\lambda : \lambda \in \Delta\}$ be an open cover of X. Let, if possible X, has no Lebesgue number then, for each $n \in N$, there is a non-empty subset A_n of X such that

$$\delta(A_n) < 1/n \text{ and } A_n \not\subset G_\lambda, \forall \lambda \in \Delta. \qquad \ldots (1)$$

Choose a point $a_n \in A_n$ for each $n \in N$. Since X is sequentially compact, the sequence $< a_n >$ must contain a subsequence, say $< a_{n_k} >$, which converges to $a \in X$.

Since C is cover of X, there exists an open set $G_{\lambda_0} \in C$ such that $a \in G_{\lambda_0}$. Since G_{λ_0} is open, there exists $\varepsilon > 0$ such that $S(a,\varepsilon) \subset G_{\lambda_0}$ ... (2)

For the above $\varepsilon > 0$, $a_{n_k} \to a$ and $\delta(A_{n_k}) \to 0$ as $k \to \infty$. Hence there exists $k_0 \in N$ such that

$$d(a_{n_{k_0}}, a) < \varepsilon/2 \quad \text{and} \quad \delta(A_{n_{k_0}}) < \varepsilon/2 \quad \text{... (3)}$$

Let a' be an arbitrary element of $A_{n_{k_0}}$ Now using (3) and triangle inequality, we have

$$d(a', a) \le d(a', a_{n_{k_0}}) + d(a_{n_{k_0}}, a)$$

$$\le \delta(A_{n_{k_0}}) + d(a_{n_{k_0}}, a), \text{ as } a_{n_{k_0}} \in A_{n_{k_0}}$$

$$< \varepsilon/2 + \varepsilon/2 = \varepsilon$$

$$\Rightarrow \quad a' \,\varepsilon\, S(a,\varepsilon) \le G_{\lambda_0} \Rightarrow A_{n_{k_0}} \subseteq G_{\lambda_0},$$

which contradicts (1). Hencde C has a Lebesgue number.

Theorem XIII. *Every sequentially compact metric space is compact.*

Proof. Let (X, d) be a sequentially compact metric space and let $C = \{G_\lambda : \lambda \in \Delta\}$ be an arbitrary open cover of X. Since X is sequentially compact, by Lebesgue covering lemma C has a Lebesgue number, say $l > 0$.

We take $\varepsilon = l/3$ so that $\varepsilon > 0$. ... (1)

Since (X, d) is sequentially compact, it is totally bounded (by theorem VII) and hence it has ε-net, say $\{x_1, x_2, \ldots x_n\}$. Then

$$X = U\{S(x_k, \varepsilon) : k = 1, 2, \ldots, n\} \quad \text{... (2)}$$

Now, $\quad d(S(x_k, \varepsilon)) \le 2\varepsilon = 2l/3 < l, \ \forall\, k = 1, 2, \ldots, n$, using (1)

Thus for each k = 1, 2, ...,*n*, $\delta\,(S(x_k, \varepsilon)) < l$ and hence by definition of Lebesgue number there exist at least one G_{λ_k} such that

$$S(x_k, \varepsilon) \subseteq G_{\lambda_k}, \ \forall\ k = 1, 2, \ldots, n$$

$$\Rightarrow \quad \cup\{(S(x_k, \varepsilon) : k = 1, 2, .., n\} \subseteq \cup\{G_{\lambda_k} : k = 1, 2, \ldots, n\}$$

$$\Rightarrow \quad X \subset \cup\{G_{\lambda_k} : k = 1, 2, \ldots, n\}, \text{ using (2)}$$

$$\Rightarrow \quad [G_{\lambda_k} : k = 1, 2, \ldots, n] \text{ is a finite subcover of } C$$

Hence (X, d) is compact.

Theorem XIV. *A metric space is compact iff it is sequentially compact.*

Proof. Refer theorems VI and IX.

Theorem XV. *A metric space is compact iff it has BWP.* (*Delhi Maths (H) 2000, 01*)

Proof. Prove with help of theorems V and X.

Theorem XVI. *A metric space is sequentially compact iff it is complete and totally bounded.*

Proof. Let (X, d) be sequentially compact. Then by theorem VII, it is totally bounded. We now prove that X is complete.

Since X is sequentially compact, so every Cauchy sequence $< x_n >$ in X must have a convergent sequences and therefore it must itself converge. Hence X is complete.

Conversely, Let X be complete and totally bounded. Let $< x_n >$ be any sequence in X. Then, by theorem VIII, totally boundedness of X implies that $< x_n >$ has a Cauchy subsequence say, $< x_{n_k} >$. Then X being complete, $< x_{n_k} >$ must converge and therefore X is sequentially compact.

Theorem XVII. *A metric space X is compact iff it is compelete and totally bounded.*

(Meerut 2010; Delhi Maths (H) 1993, 2005

Proof. Refer theorems XIV and XVI.

Theorem XVIII. *Every continuous function on a compact metric space is uniformly continuous.* *(M.S. Univ. T.N. 2006; Delhi Maths (H) 2000, 2003, 06, 07; I.A.S. 1998)*

Proof. Let $f:(X,d_1) \to (Y,d_2)$ be continuous and let (X, d) be compact. Let $\varepsilon > 0$ be given. Let x be any point of X. Since f is continuous at x, hence there exists a δ_x (depending on δ and x) such that for $y \in X$, we have

$$d_1(x,y) < \delta_x \Rightarrow d_2(f(x), f(y)) < \varepsilon/2 \qquad \text{... (1)}$$

Then the set $\{S(x, \delta_x/2) : x \in X\}$ of open spheres is an open cover of X. Since X is compact, there exists a finite subcover $\{S(x_k, \delta_x/2) : k = 1, 2, ..., n\}$ such that

$$X = \cup\{S(x_k, \delta_{x_k}/2) : k = 1, 2, ..., n\} \qquad \text{... (2)}$$

Let $\quad \delta = \min\{\delta_{x_1}/2, \delta_{x_2}/2, ..., \delta_{x_n}/2\}$ so that $\delta > 0$.

Let x, y be arbitrary points of X with $d_1(x,y) < \delta$. Then (2) implies that $x \in S(x_k, \delta_{x_k}/2)$ for some $x_k \in X$

Since $\quad d_1(x, x_k) < \delta_{x_k}/2 < \delta_{x_k}$, (1) yields

$$d_2(f(x), f(x_k)) < \varepsilon/2 \qquad \text{... (3)}$$

Using triangle inequality, we have

$$d_1(y, x_k) \le d_1(y, x) + d_1(x, x_k)$$

$$< \delta + \delta_{x_k}/2 < \delta_{x_k}/2 + \delta_{x_k}/2 = \delta_{x_k} \qquad \text{... (4)}$$

Hence, $\quad (1) \Rightarrow d_2(f(y), f(x_k)) < \varepsilon/2$

Now, $\quad d_2(f(x), f(y)) \le d_2(f(x), f(x_k)) + d_2(f(x_k), f(y))$

$$< \varepsilon/2 + \varepsilon/2 = \varepsilon, \text{ by (3) and (4)}$$

Thus, $\quad d_1(x,y) < \delta \Rightarrow d_2(f(x), f(y)) < \varepsilon \;\forall\, x, y \in X$

Hence f is continuous.

19.37C LOCALLY COMPACT SPACES. DEFINITION.

A metric space (X, d) is said to be locally compact iff every point in X has at least one neighbourhood whose closure is compact.

Example. *Show by means of an example that a locally compact space need not be compact.* *(Meerut 2001)*

Solution. Consider the discrete metric space (X, d) such that X is infinite. Then, for each $x \in X$, the set $\{x\}$ is open as well as closed. But X is not compact because the collection of all singleton subsets of X is an infinite open cover of X which has no finite subcover.

Now, we show that X is locally compact. Let x be an arbitrary point of X. Then $\{ x \}$ is a *nbd* of x whose closure is $\{ x \}$. Also $\{ x \}$ is compact subset of X, being finite. Thus every point of X has a *nbd* whose closure in compact. Hence X is locally compact.

Theorem XIX. *Every compact metric space is locally compact.*

Proof. Let (X, d) be a compact metric space and let x be an arbitrary point of X. Then X is an open set and hence X is a *nbd* of x. Also $\overline{X} = X$. But X is given to be compact. Therefore every point of X has at least one *nbd* (the whole space) whose closure is compact. Hence X is locally compact.

Theorem XX. *Every closed subset of a locally compact space is locally compact.*

Proof. Left as an exercise for the reader.

Theorem XXI. *Let (X, d) be a metric space. Show that the following conditions are equivalent (i) X is compact (ii) X has the B.W.P. (iii) X is countably compact (iv) X is sequentlly compact (V) X is complete and totally bounded.*

Proof. Left as an exercise for the reader.

EXERCISES

1(*a*). Let (X, d) be a compact metric space and let d^* be defined on X by setting $d^*(x, y) = \min \{1, d(x, y)\}, \ \forall \ x, y \in X$. Show that (X, d^*) is compact. *(Delhi 1997)*

(*b*) Let $\boldsymbol{R}$ denote the set of real numbers, and let $d(x, y) = |x - y|/(1 + |x - y|)$, for all $x, y \in \boldsymbol{R}$. show that $(\boldsymbol{R}, d)$ is not compact.

2. Let A and B be two compact subsets of a metric space (X, d). Prove that $A \cup B$ is a compact subset of X. *(Mumbai 2010; Delhi Maths (H) 2000)*

3. Prove that a subset of the real line is compact if and only if it is closed and bounded. *(Meerut 2013; Delhi Maths (H) 2000, 2004, 06)*

4. Define a compact metric space. Show that the real line $\boldsymbol{R}$ is not compact. *(Delhi Maths (H) 1988, 2000)*

5. (*a*) Show that closed interval $[0, 1]$ is compact. *(Delhi Maths (H) 2004)*

(*b*) Show that every closed interval $[a, b]$ is compact. *(Delhi Maths (H) 2000)*

6. Prove that every continuous function defined on a compact metric space is uniformly continuous. Show also that this may not be true if the space is not compact. *(Delhi Maths Hons. 1996, 2000)*

7. If (X, d) is compact metric space then show that X has a countable dense subset. *(Delhi Maths Hons 2003)*

8. Let f be a continuous mapping from a compact metric space (X, d) onto a metric space (Y, d). Then prove that $f(A)$ is a closed subset of Y for every closed subset A of X. Hence conclude that every continuous bijection defined on a complete metric space is another homeomorphism. *(Delhi (Maths) Hons. 2004)*

9. If A is an infinite subset of a compact metric space X, then prove that A has a limit point in X. *(Meerut 2010)*

10. If F is closed and K is compact, then prove that $F \cap K$ is compact. *(Nagpur 2003)*

11. If f is a continuous mapping of a compact metric space into $\boldsymbol{R}^k$, then prove that $f(X)$ is closed and bounded. Further, suppose f is continuous real function on a compact metric space X and $M = \sup_{p \in X} f(p)$ and $m = \inf_{p \in X} f(p)$, then prove that there exist points $p, q \in X$ such that $f(p) = M$ and $f(q) = m$. *(Nagpur 2003)*

12. Show that continuous image of a sequentially compact metric space is sequentially compact.

13. Show that a metric space (X, d) is sequentially compact iff each sequence in X has a cluster point.

14. Show that a metric space (X, d) is countably compact iff each sequence in X has a cluster point.

15. Show that a metric space has the Bolzano-Weierstran property iff every sequence in X has a cluster point.

16. Show that a metric space has the Bolzana Weierstran property iff every infinite open cover of the space admits of a proper subcover. (*Delhi Maths Hons. 1999*)

17. Prove that the product of two metric spaces is compact iff each factor space is compact.

18. Let X be an infinite set with the discrete metric. Show that (X, d) is not compact.

19. Prove that every compact subset A of a metric space is bounded.

20. Let A be a non-empty compact subset of a metric space (X, d) and let B be a closed subset of X such that $A \cap B = \phi$, then prove that $d(A, B) > 0$.

19.38 CONNECTED METRIC SPACES. SEPARATED SETS. DEFINITIONS

Let (X, d) be a metric space. Two non-empty subsets A and B of X are said to be separated if

$$A \cap \overline{B} = \varphi \quad \text{and} \quad \overline{A} \cap B = \varphi.$$

In other words non-empty subsets A and B of X are said to be separated if A and B are disjoint and neither A nor B contains any limit point of the other.

Note. The separated sets are always disjoint, but two disjoint sets are not always separated as shown below.

Example 1. Consider intervals $A =]\,0, 1\,[$ and $B = [\,1, 2\,[$ which are disjoint sets on the real line. Then $\overline{A} = [0,1]$ and $\overline{B} = [1,2]$ so that $A \cap \overline{B} = \varphi$ and $\overline{A} \cap B = \{1\} \neq \varphi$. Hence A and B are disjoint but not separated.

Example 2. Sets $A =]\,0, 1\,[$ and $B =]\,1, 2\,[$ are disjoint and also separated because $\overline{A} \cap B = [0,1] \cap]\,1,2\,[= \varphi$ and $A \cap \overline{B} =]\,0,1\,[\cap [\,1,2\,] = \varphi$

Theorem I. *Let A and B be separated subsets of a metric space (X, d) and let P and Q be non-empty subsets of X such that $P \subset A$ and $Q \subset B$. Then P and Q are also separated.*

Proof. Since A and B are separated, we have

$$A \cap \overline{B} = \varphi \quad \text{and} \quad \overline{A} \cap B = \varphi \qquad \dots (1)$$

$$\text{Now,} \quad P \subset A \quad \text{and} \quad Q \subset B \qquad \dots (2)$$

$$\Rightarrow \quad \overline{P} \subset \overline{A} \quad \text{and} \quad \overline{Q} \subset \overline{B}$$

$$\Rightarrow \quad \overline{P} \cap Q \subset \overline{A} \cap Q \quad \text{and} \quad P \cap \overline{Q} \subset P \cap \overline{B}$$

$$\Rightarrow \quad \overline{P} \cap Q \subset \overline{A} \cap B \quad \text{and} \quad P \cap \overline{Q} \subset A \cap \overline{B}, \text{ by (2)}$$

$$\Rightarrow \quad \overline{P} \cap Q = \varphi \quad \text{and} \quad P \cap \overline{Q} = \varphi, \text{ by (1)}$$

$\Rightarrow$ P and Q are separated.

Connected and disconnected sets. Definition

A subset E of a metric space (X, d) is said to be separated if it can be experessed as the union of two non-empty separated sets. If E is not conneced, then it is said be disconnected.

When E is disconnected, we have two non-empty subsets A and B of X such that $A \cap \overline{B} = \phi$, $\overline{A} \cap B = \phi$ and $E = A \cup B$. The pair (A, B) is called a *separation (disconnection)* of X.

Examples. Since an empty set φ cannot be expressed as the union of two non-empty sets which are separated, so φ is a connected set. Again the singleton $\{ x \}$ of a metric space is also connected because it cannot be expressed as union of two non-empty disjoint sets.

Properties of connected and disconnected sets

Theorem II. *Let (X, d) be a metric space and let E be a connected subset of X such that $E \subset A \cup B$ when A and B are separated subsets of X. Prove that either $E \subset A$ or $E \subset B$.*

(Delhi Maths (Hons) 2001, 2003)

Proof. A and B are separated $\Rightarrow A \cap \bar{B} = \varphi, \bar{A} \cap B = \varphi$ ... (1)

Again, since $E \subset A \cup B$, we have

$$E = E \cap (A \cup B) = (E \cap A) \cup (E \cap B) \quad \text{... (2)}$$

We show now prove that at least one of the sets $E \cap A$ and $E \cap B$ is empty. For, if possible, suppose none of these sets is empty, *i.e.*, $E \cap A \neq \varphi$ and $E \cap B \neq \varphi$. Then, we obtain

$$(E \cap A) \cap \overline{(E \cap B)} \subset (E \cap A) \cap (\bar{E} \cap \bar{B}), \text{ as } \overline{E \cap B} \subset \bar{E} \cap \bar{B}$$

$$= (E \cap \bar{E}) \cap (A \cap \bar{B}) = (E \cap \bar{E}) \cap \varphi = \varphi, \text{ by (1)}$$

$$\Rightarrow \quad (E \cap A) \cap \overline{(E \cap B)} = \varphi \quad \text{... (3)}$$

Similarly, $\overline{(E \cap A)} \cap (E \cap B) = \varphi$ ... (4)

From (2), (3) and (4), we see that E can be expressed as the union of two non-empty disjoint separated sets and hence E is disconnected. This contractičts the connectedness of E. Hence at least one of the sets $E \cap A$ and $E \cap B$ must be empty.

If $E \cap A = \varphi$, then (2) $\Rightarrow$ $E = E \cap B \Rightarrow E \subset B$

Similarly, if $E \cap B = \varphi$, then (2) $\Rightarrow$ $E = E \cap A \Rightarrow E \subset A$

Hence either $E \subset A$ or $E \subset B$.

Theorem III. *Let (X, d) be a metric space and let E be a connected subset of X. If F is a subset of X such that $E \subset F \subset \bar{E}$, then F is connected. In particular, $\bar{E}$ is connected.*

(M.S. Univ. T.N. 2006; Delhi Maths (H) 2003, Meerut 2001)

Proof. Given that $E \subset F \subset \bar{E}$ ... (1)

We are to show that F is connected. Let, if possible, F be disconnected. Then there exist non-empty sets A and B such that

$$F = A \cup B, \quad \text{where } A \cap \bar{B} = \varphi \text{ and } \bar{A} \cap B = \varphi \quad \text{... (2)}$$

From (1) and (2), $E \subset F$ and $F = A \cup B \Rightarrow E \subset A \cup B$,

showing that the connected set E is contained in the union of two separated sets A and B. Hence, by theorem II, it follows either $E \subset A$ or $E \subset B$.

If $E \subset A$, then $\bar{E} \subset \bar{A} \Rightarrow \bar{E} \cap B \subset \bar{A} \cap B \Rightarrow \bar{E} \cap B \subset \varphi$, by (2)

Thus, $\bar{E} \cap B = \varphi$. ... (3)

From (1) and (2), $F \subset \bar{E}$ and $F = A \cup B$

$$\Rightarrow \quad B \subset \bar{E} \Rightarrow \bar{E} \cap B = B \quad \text{... (4)}$$

From (3) and (4), we have $B = \varphi$, which is a contradiction because B is non-empty. Hence F must be connected.

Particular Case Since $E \subseteq \bar{E} \subseteq \bar{E}$ and E is connected, so by the above theorem (taking $F = \bar{E}$), we see that $\bar{E}$ is connected.

Theorem IV. *The closure of a connected set is connected.*

(*M.S. Univ. T.N. 2006; Delhi Maths (H) 1995, 98, 2005, Meerut 2012*)

Proof. Let E be a connected subset of a metric space (X, d). We are to show that $\bar{E}$ is connected. Let, if possible, $\bar{E}$ be disconnected. Then there exist non-empty sets A and B such that

$$\bar{E} = A \cup B, \quad \text{where} \quad A \cap \bar{B} = \varphi \quad \text{and} \quad \bar{A} \cap B = \varphi \quad \text{... (1)}$$

Now, $$E \subset \bar{E} \text{ and so } E \subset A \cup B,$$

showing that the connected set E is contained in the union of two separated sets A and B. Hence, by theorem II, if follows that either $E \subset A$ or $E \subset B$.

If $E \subset A$, then $\bar{E} \subset \bar{A} \Rightarrow \bar{E} \cap B \subset \bar{A} \cap B \Rightarrow \bar{E} \cap B \in \varphi$, by (1)

Thus, $$\bar{E} \cap B = \varphi \quad \text{... (2)}$$

From (1), $$\bar{E} = A \cup B \Rightarrow B \subset \bar{E} \Rightarrow \bar{E} \cap B = B \quad \text{... (3)}$$

From (2) and (3), we have $B = \varphi$, which is a contradiction because B is non-empty. Hence $\bar{E}$ must be connected.

Theorem V. *Let (X, d) be a metric space. Show that (X, d) is disconnected if there exists a non-empty proper subset of X which is both open and closed.* (*Meerut 2010*)

Proof. Let A be a non-empty proper subset of X which is both open and closed. Let $B = X - A$. Then A being non-empty and open, it follows that B is non-empty and closed. Also, $A \cap B = \varphi$. Now, A and B are closed $\Rightarrow \bar{A} = A$ and $B = \bar{B}$.

$$\therefore \quad A \cap B = \varphi \Rightarrow \bar{A} \cap B = \varphi \quad \text{and} \quad A \cap \bar{B} = \varphi,$$

showing that $X = A \cup B$ and A and B are non-empty separated sets. Hence (X, d) is disconnected.

Conversely, let (X, d) be disconnected. Then there exist non-empty subsets A and B of X such that

$$A \cup B = X, \quad A \cap \bar{B} = \varphi \quad \text{and} \quad \bar{A} \cap B = \varphi$$

$$\Rightarrow \quad A \cup \bar{B} = X, \quad \bar{A} \cup B = X \quad \text{and} \quad A \cap B = \varphi$$

Now, $$A \cup B = X \quad \text{and} \quad A \cap B = \varphi \Rightarrow A = X - B \quad \text{... (1)}$$

$$A \cup \bar{B} = X \quad \text{and} \quad A \cap \bar{B} = \varphi \Rightarrow A = X - \bar{B} \quad \text{... (2)}$$

and $$\bar{A} \cup B = X \quad \text{and} \quad \bar{A} \cap B = \varphi \Rightarrow B = X - \bar{A} \quad \text{... (3)}$$

Since both A and B are non-empty, (1) shows that A is a non-empty proper subset X. Since $\bar{A}$ and $\bar{B}$ are closed, so (2) and (3) imply that their complements A and B are open. Now B being open, (1) imply that its complement A is closed. Thus there exists a non-empty proper subset A of X which is both open and closed.

Theorem VI. *A metric (X, d) is connected iff the only non-empty subset of X which is both open and closed is X itself.* (*M.S. Univ. T.N. 2006, G.N.D.U. 1997, Meerut 2000*)

Proof. Let (X, d) be connected. Then, by theorem V, there exists a non-empty subset A of X which is both open and closed. Now A is open and closed $\Rightarrow X - A$ is both open and closed.

Now, A and $X - A$ are closed $\Rightarrow \bar{A} = A$ and $\overline{X - A} = X - A$ ... (1)

$$\therefore \quad A \cap (X - A) = \varphi \quad \Rightarrow \quad A \cap (\overline{X - A}) = \varphi \text{ and } \bar{A} \cap (X - A) = \varphi, \text{ using (1).}$$

Let if possible $(X - A) \neq \varphi$. Then X can be expressed as union of two non-empty separated sets A and $(X - A)$ and so X is disconnected. But this is contradiction since by hypothesis X is connected. Hence we cannot assume that $(X - A) \neq \varphi$ and therefore we have $X - A = \varphi$ and so $A = X$, showing that the only non-empty subset of X which is both open and closed is X itself.

Conversely, let the only non-empty subset of X which is both open and closed is X itself. Hence, there cannot exist any non-empty proper subset of X which is both open and closed. Hence by theorem V, (X, d) cannot be disconnected. Therefore it is connected.

Theorem VII. (*i*) *A metric space (X, d) is disconnected iff X is the union of two non-empty disjoint open sets.*

(*ii*) *A metric space (X, d) is disconnected iff X is the union of two non-empty disjoint closed sets.*

Proof. (*i*) Let (X, d) be disconnected. Then by theorem V, there exists a non-empty proper subset A of X which is both open and closed. Let $B = X - A$. Then A being proper non-empty closed set of X, B is also non-empty open subset of X. Thus, $X = A \cup B$, where A and B are both non-empty disjoint open subsets of X.

Conversely, let $X = A \cup B$ and $A \cap B = \varphi$ where A and B are non-empty open subsets of X. Then $A = X - B \Rightarrow A$ is closed since B is open. Thus, there exists a non-empty proper subset A of X which is both open and closed. Hence by theorem V, (X, d) is disconnected.

(*ii*) Left as an exercise.

Theorem VIII (*i*) *A metric space (X, d) is connected iff X cannot be expressed as the union of two non-empty disjoint open sets.*

(*ii*) *A metric space (X, d) is connected iff X cannot be expressed as the union of two non-empty disjoint closed sets.*

Proof. Use theorem VII to prove this theorem.

Theorem IX. *A continuous image of a connected metric space is connected.*

(Rajasthan 2010; Bharathiar 2004, Delhi Maths (H) 2001, 2004, Meerut 2000)

Proof. Let (X, d_1) and (Y, d_2) be two metric spaces such that X is connected. Let $f : X \rightarrow Y$ be a continuous mapping from X onto Y. We are to prove that (Y, d_2) is also connected. If (Y, d_2) were not connected, then by theorem V, there would exist a non-empty proper subset A of Y such that A is both open and closed in (Y, d_2). Since f is continuous and onto, so $f^{-1}(A)$ would be a non-empty, proper subset of X which is both and closed in X. Hence (X, d) would be disconnected. This contradicts the hypothesis that X is connected. This contradiction proves that (Y, d_2) is connected.

Corollary 1. *Connectedness is preserved under homeomorphism.*

Corollary 2. *Connectedness is preserved under isometrics.*

Hint : Since f is homomesphlism (or isometry) $\Rightarrow f$ is continuous, the result of both corollaries follows from theorem IX.

Connected sets. Definition.

A subset E of a metric space (X, d) is said to be connected if the subspace (E, d_E) is connected.

A subset E of a metric space (X, d) is said to be disconnected if the subspace (E, d_E) is disconnected.

Theorem X. *Let (X, d) be a metric space and let $E \subset X$. The follwing statements are equivalent.*

(*i*) *E is disconnected.*

(*ii*) *E is expressibele as the union of two separated subsets of X.*

(*iii*) There exists d- closed sets A and B such that

$$E \subset A \cup B, E \cap A \neq \varphi, E \cap B \neq \varphi, E \cap A \cap B = \varphi.$$

(*iv*) There exists d-open sets G and H such that

$$E \subset G \cup H, E \cap G \neq \varphi, E \cap H \neq \varphi, E \cap G \cap H = \varphi$$

Proof. Left as an exercise.

Theorem XI. *A subset of the real line is connected iff it is an interval.*

(Mumbai 2010; Agra 1999, G.N.D. U. 1998, Delhi (Hons.) 2007; Meerut 2012)

Proof. The condition is necessary. Let E be a connected subset of $\boldsymbol{R}$ containing at least two points. If possible, let E be not an interval. Then there exist three real number a, b, c such that $a < c < b$, $a \in E, b \in E$ but $c \notin E$.

Let $G =]-\infty, c[$ and $H =]c, \infty[$. Then G and H are for disjoint open sets in $\boldsymbol{R}$ such that $a \in G$ and $b \in H$. Also $a, b \in E$. So $E \cap G \neq \varphi$ and $E \cap H \neq \varphi$.

Also, $E \cap G \cap H = \varphi$ and $E = (E \cap G) \cup (E \cap H) = E \cap (G \cup H) \Rightarrow E \subseteq G \cup H$.

Thus, there exists open set G and H in $\boldsymbol{R}$ such that

$$E \subset G \cup H, E \cap G \neq \varphi, E \cap H \neq \varphi \text{ and } E \cap G \cap H = \varphi$$

Hence, by theorem X, E is disconnected which contradicts the hypothesis that E is connected. Therefore E must be an interval.

The condition is sufficient. Let E be an interval. Then we are to show that E is connected. Let, if possible, E be disconnected. Then there exist non-empty subsets A and B such that

$$E = A \cup B, \ A \cap \overline{B} = \varphi \quad \text{and} \quad \overline{A} \cap B = \varphi$$

Since A and B are non-empty disjoint sets, so real numbers $a, b \in E$ such that $a \in A$ and $b \in B$ where $a \neq b$. Then either $a < b$ or $a > b$. We may assume without loss of generality, that $a < b$. Since E is an interval, we have $[a, b] \subset E$.

Let $\quad G = [a,b] \cap A \quad$ and $\quad H = [a,b] \cap B$

Since $G \subset A$ and $H \subset B$ and A and B are separated, so by theorem I, G and H are also separated.

Since G is non-empty and $G \subset [a,b]$, it follows that G is bounded above (by b) and hence it has a supremum, say c. Since c is the supremum of G and b is an upper bound of G, we have $c \leq b$. Again, since $a \in G$ and c is the supremum of G, we have $a \leq c$ Therefore, $a \leq c \leq b$.

We shall now prove that $c \notin G$. Now, we have

$$c \in G \Rightarrow c \text{ is the greatest element of } G.$$

$$\Rightarrow \quad]c, b[\subset H \Rightarrow [c, b] \subset \overline{H}$$

Then $\quad G \cap \overline{H} = \varphi \quad$ and $\quad c \in \overline{H} \Rightarrow c \notin G$.

Next, we shall show that $c \notin H$. Since c is the supremum of G, so for given $\varepsilon > 0$, there exists some $d \in G$ such that $d > c - \varepsilon$. We also have $d \leq c$. But $c \notin G$ hence there exists $d \neq c$ such that $]c - \varepsilon, c + \varepsilon[\cap G \neq \varphi$, showing there by that c is a limit point of G so that $c \in \overline{G}$. But $\overline{G} \cap H = \varphi$ and so $c \notin H$.

Now, $\quad c \notin G$ and $c \notin H \Rightarrow c \notin G \cup H \Rightarrow c \notin [a,b]$,

which contradicts (1). Hence E is connected.

Corollary. The real line $\boldsymbol{R}$ is connected. *[M.S. Univ. T.N. 2006]*

Proof. Since $\boldsymbol{R}$ is an interval $]-\infty, \infty[$, it follows that it is connected by the above theorem.

Theorem XII. *The real line **R** with usual metric is connected.*

(*Delhi Maths (H) 1995, 2000, Meerut 2001*)

Proof. Let, if possible, **R** be disconnected so that there exist two non-empty disjoint closed subsets G and H of **R** such that $\mathbf{R} = G \cup H$. Since $G \cap H = \varphi$, choose real numbers a and b such that $a \in G$ and $b \in H$ where $a \neq b$. Then either $a < b$ or $a > b$. Without loss of generality, we may assume that $a < b$. Also $[a, b] \subset \mathbf{R} = G \cup H$. Hence, every point of $[a, b]$ is either in G or in H.

Let $c = \sup ([a, b] \cap G)$ such that $c \in [a, b]$. Since c is supremum of $[a, b] \cap G$, for every $\varepsilon > 0$, there exists a real number $d \in [a, b] \cap G$ such that $c - \varepsilon < d < c$, showing thereby that each *nbd* of c contains a point of $[a, b] \cap G$ and hence a point of G. Hence $c \in G$ or c is a limit point of G. Since G is closed, so $c \in G$. Again, $G \cap H = \varphi \Rightarrow c \notin H$. But $b \in H$ and so $b \neq c$. Thus we get the strict inequality $c < b$.

Now, by definition of c, it follows that for each $\varepsilon > 0$, $c + \varepsilon \in H$ and $c + \varepsilon < b$. Thus each *nbd* of c contain a point of H other than c and hence it is a limit point of H. Since H is closed, we have $c \in H$.

Now,

$$c \in G \quad \text{and} \quad c \in H \Rightarrow c \in G \cap H,$$

which contradicts the fact that $G \cap H = \varphi$.

Since the contradiction arises by assuming that **R** is disconnected, it follows that **R** is connected.

Theorem XIII. *(i) The union of two connected sets, having non-empty intersection, is connected.* (*Mumbai 2010*)

(ii) An arbitrary union of connected sets, with non-empty intersection is connected.

Proof. Left as an exercise. (*Delhi Maths (H) 2006, 08*)

Theorem XIV. (Intermediate value theorem) *The range of a continuous real valued function defined on a connected metric space is an interval.* (*M.S. Univ. T.N. 2006*)

Proof. Let (X, d) be a connected metric space and let f be a continuous real valued function on X, so that the range $f(X)$ of f is a subset of **R**. Since the continuous image of a connected set is connected, so $f(X)$ is a connected subset of **R**. Hence by theorem *XI*, it follows that $f(X)$ is an interval.

Corollary I. *If f is a continuous real valued function defined on a connected metric space (X, d) and if x_0 is any real number between two values $f(x_1)$ and $f(x_2)$, then there exists some $x \in X$ such that $f(x) = x_0$.*

Corollary II. *If f is a continuous real valued function defined on a connected metric space (X, d) and if $x_1, x_2 \in X$ such that $f(x_1) f(x_2) < 0$, then there exists $x \in X$ such that $f(x) = 0$.*

Corollary III. *If $f : [a, b] \to \mathbf{R}$ be continuous and x_0 be any real number between $f(a)$ and $f(b)$, then there exists a real number $x \in \,]a, b[$ such that $f(x) = x_0$.*

Corollary IV. *If $f : [a, b] \to \mathbf{R}$ be continuous and $f(a) f(b) < 0$, then there exist a real number $x \in [a, b]$ such that $f(x) = 0$.*

Corollary V. *A continuous image of a compact connected metric space under a real valued function is closed and bounded interval.*

Proof. Prove the above corollaries yourself.

EXERCISE

1. Show that a metric space (X_1, d_1) is connected iff every continuous function from (X_1, d_1) into (X_2, d_2) is constant, where $X_2 = \{0, 1\}$ and d_2 is the discrete metric.

2. Let (X, d) be a metric space and let A and B be closed subsets of X such that $A \cup B$ and $A \cap B$ are connected. Show that each of the sets A and B are connected.
3. Let (X, d) be a connected metric space, P a connected subset of X, and let $P^C = A \cup B$, where P^C is complement of P and A and B are separated. Show that each of the sets $A \cup P$ and $B \cup P$ is connected.
4. Let X be an infinite set, and let d be the discrete metric on X. Show that every subset of X having more than one point is disconnected
5. Show that a line segment is connected subset of $\boldsymbol{R}$. (*Delhi Maths Hons.* 1998)
6. Consider a collection C of connected subsets of a metric space (X, d) such that $\cap\{A : A \in C\} = \phi$. Show that $\cup\{A : A \in C\}$ is connected in (X, d). Deduce that $\boldsymbol{R}^2$ with Euclidean metric is connected (connectedness of $\boldsymbol{R}$ can be assumed here). (*Delhi Maths Hons.* 1996)
7. Define connected sets. Give two examples of connected sets. (*Nagpur* 2003)
8. Give two examples to illustrate the separated sets and non-separated sets. (*Meerut* 2000)

19.39 COMPONENTS OF A METRIC SPACE. DEFINITION.

A subset A of a metric space (X, d) is said to be a component of X if it is the maximal connected subset of X.

Example. In discrete metric space (X, d), singletons are the only connected subsets of X because a subset A of X containing more than one element is not connected.

Properties of Components

Theorem I. *Every component of a metric space is a closed set.* (*G.N. D. U. 1995, 97*)

Proof. Let (X, d) be a metric space and let A be its component. Then by definition A is connected but not contained properly in any larger connected subset of X. Let, if possible, A be not closed. Then $A \subset \overline{A}$ Since A is closed, so $\overline{A}$ is also closed. Thus A is connected and is contained properly in larger connected subset $\overline{A}$ of X. This is a contradiction. Since the contradiction arises by assuming that A is not closed, hence A is closed.

Theorem II. *Components of a metric space are either identical or pairwise disjoint.*

Proof. Left as an exercise.

Theorem III. (*i*) *Every connected subset of a metric space (X, d) is contained in a component of X.*

(*ii*) *Each element of a metric space (X, d) is contained is exactly one component.*

(*iii*) *An non-empty connected subset of a metric space X is a component, if it is both open and closed.*

Proof. (*i*) Let A be an arbitrary connected subset of X. Let $\{A_\lambda : \lambda \in \Delta\}$ be a collection of connected subsets of X such that $A \subseteq A_\lambda \ \forall \lambda \in \Delta$. Let $B = \cup\{A_\lambda : \lambda \in \Delta\}$. Then B is a connected subset of X containing A, *i.e.,* $A \subseteq B$.

We now show that B is a maximal connected subset of X. Let, if possible, C be a connected subset of X such that $B \subseteq C$. Now $A \subseteq B$ ande $B \subseteq C \Rightarrow A \subseteq C$ and hence $C \in \{A_\lambda : \lambda \in \Delta\}$. Therefore $C \subseteq \cup\{A_\lambda : \lambda \in \Delta\}$, *i.e.,* $C \subseteq B$, Now $B \subseteq C$ and $C \subseteq B \Rightarrow C = B$, showing thereby that B is a maximal connected subset of X containing A. Thus B is the desired component of the X containing A. Since components are pairwise disjoint, it follows that B is unique component of X containing A.

Parts (*ii*) and (*iii*). Left as an exercise.

19.40 CONNECTEDNESS OF PRODUCT OF CONNECTED METRIC SPACES

Theorem I. *Product of two connected spaces is connected.* *(Meerut 1996, 1997)*

Proof. Let (X_1, d_1) and (X_2, d_2) be two connected metric spaces and let $(X_1 \times X_2, d)$ be their product space where d is defined by

$$d(x, y) = \max\{d_1(x_1, y_1), d_2(x_2, y_2)\} \ \forall \ x, y \in X_1 \times X_2$$

where $x = (x_1, x_2)$ and $y = (y_1, y_2)$.

We shall show that there exists a connected subset of $X_1 \times X_2$ that contains both (x_1, x_2) and (y_1, y_2).

Let $f : X_1 \to X_1 \times \{y_2\}$ be defined by $f(x_1) = (x_1, y_2) \ \forall \ x_1 \in X_1$. Then for $\forall \ t_1, s_1 \in X_1$, we have

$$d(f(t_1), f(s_1)) = d((t_1, y_2), (s_1, y_2)) = \max\{d_1(t_1, s_1), d_2(y_2, y_2)\}$$

Thus, $$d(f(t_1), f(s_1)) = d_1(t_1, s_1)$$

Now f is one-one and preserves the distance. Hence f is an isometry and therefore f is continuous. Since cotinuous image of a connected set is connected, it follows that $X_1 \times \{y_2\}$ is connected set conntaining (x_1, y_2) and (y_1, y_2). Similarly, $\{x_1\} \times X_2$ is also connected set containing (x_1, x_2) and (x_1, y_2).

The connected sets $\{x_1\} \times X_2$ and $X_1 \times \{y_2\}$ have a non-empty intersection because (x_1, y_1) is contained in both of these sets. Hence the set $A = (\{x_1\} \times X_2) \cup (X_1 \times \{y_2\})$ is connected set containing (x_1, x_2) and (y_1, y_2).

Thus for each pair of points (x_1, x_2) and (y_1, y_2) of $X_1 \times X_2$, there exists a connected subset A of $X_1 \times X_2$ such that $(x_1, x_2), (y_1, y_2) \in A$. Hence $X_1 \times X_2$ is connected.

Theorem II. (*i*) Let $\{(X_i, d_i) : i = 1, 2, ..., n\}$ *be a collection of connected metric spaces. Then their product metric space (X, d) is also connected.*

(*ii*) For each $n \in \boldsymbol{N}$, $\boldsymbol{R}^n$ is connected. *(Meerut 2001)*

Proof. Left as an exercise.

OBJECTIVE QUESTIONS

MULTIPLE CHOICE QUESTIONS

Indicate the correct answer by writing (A), (B), (C) or (D) whichever is correct.

1. *In the metric space ($\boldsymbol{R}$, d), where d is the usual metric on $\boldsymbol{R}$, if $A =]\,0, 1]$, then*

(*A*) $A =]\,0,1\,]$ (*B*) $\text{int}\,(A) =]\,0,1]$ (*C*) $\text{int}\,(A) =]\,0,1\,[$ (*D*) $A' =]\,0,1\,]$

2. *Let A and B be any two subsets of a metric space. Then*

(*A*) $(A \cap B)° = A° \cap B°$ (*B*) $(A \cup B)° = A° \cup B°$

(*C*) $\overline{A \cap B} = \bar{A} \cap \bar{B}$ (*D*) $(A \cap B)' = A' \cap B'$

3. *Which of the following statements is not correct*

(*A*)] 0, 1[and [1, 2[are separated. (*B*)] 0, 1[and] 1, 2[are separated.

(*C*) The closure of a connected set is connected.

(*D*) The real line is connected.

4. *Which of the following statements is not correct ?*

(*A*) The real line is a complete metric space.

(*B*) The complex plane with the usual metric is complete.
(*C*) The space of continuous functions on $[a,b]$ is complete.
(*D*) The space of rational numbers with the usual metric is complete.

5. *Which of the following statement is not correct.*
(*A*) Every closed interval is compact. (*B*) The real line ***R*** is compact.
(*C*) Every compact subset of a metric space is closed.
(*D*) Every finite metric space is sequentially compact.

(The following questions 6-25 were asked in M.S. Univ. T.N. 2006)

6. *The set {0, 1} in **R** with usual metric is*
(*A*) compact (*B*) open (C) closed (D) connected

7. Which of the following is not true in a metric space
(*A*) Finite union of open sets is open (B) finite intersection of open sets is open
(C) arbitrary union of open sets is open (D) arbitrary intersection of open sets is open

8. If $f:[a,b]\to \boldsymbol{R}$ is a continuous function, then
(A) *f* is bounded and uniformly continuous
(B) *f* is bounded but not uniformly continuous
(C) *f* is not bounded but uniformly continous
(D) *f* is neither bounded nor uniformly continuous

9. Let $f: M_1 \to M_2$ be a continous function, then which of the statement is false ?
(A) inverse image of open set is open (B) image of open set is open
(C) inverse image of closed set is closed (D) $<x_n> \to x \Rightarrow <f(x_n)> \to f(x)$

10. In a discrete metric space, the only connected subsets are
(A) all singleton subsets (B) all subsets
(C) all nonempty subsets (D) all finite subsets

11. Which of the following metric space is not complete under metric $d(x,y)=|x-y|$
(*A*) **Z** (*B*) **Q** (*C*) **R** (*D*) **C**

12. A subset in **R** is compact ⇔ it is
(A) both open and bounded (B) open and unbounded
(C) closed and unbounded (D) both closed and bounded

13. A subset of **R** is both compact and connected if it is
(A) any interval (B) any open interval
(C) any closed interval (D) any bounded interval

14. Every convergent sequence is
(A) bounded and Cauchy (B) bounded but not Cauchy
(C) not bounded but Cauchy (D) none of the above

15. If every sequence in metric space *M* has a convergent subsequence, then *M* is called
(A) compact (B) sequentially compact (C) connected (D) complete

16. In a discrete metric space ***R***, B (0, 2)
(A) {0} (B) ***R*** (C) {2} (D) (–2, 2)

17. D (***Q***) = (A) **Q** (B) ***R*** (C) **Z** (D) ϕ

18. Let $f: \boldsymbol{R}\to\boldsymbol{R}$ be a continuous map. Then $f^{-1}\{(0,1)\}$ is
(A) an open set (B) a closed set (C) both (D) none

19. Let (M_1, d_1) be a discrete metric space and (M_2, d_2) be any metric space. Then $f : M_1 \to M_2$ is *a*

(A) discrete function (B) continuous function
(C) uniformly continous function (D) discontinuous function

20. In ***R***, ***Q*** is
(A) dense (B) compact (C) connected (D) not compact

21. ***R*** is (A) complete (B) compact (C) connected (D) bounded

22. Let (*M, d*) be a metric space $A, B \subseteq M$. then

(A) $\text{int } M = \phi$ (B) $\text{int } (A \cup B) \subseteq \text{int } A \cup \text{int } B$

(C) $\overline{A \cap B} \subseteq \overline{A} \cap \overline{B}$ (D) $\overline{A} = A$

23. In ***R*** the diameter of the set {0, 1, 2, ..., 100} is
(A) 100 (B) 99 (C) 101 (D) 0

24. Let $f : \boldsymbol{R} \to \boldsymbol{R}$. Which is not uniformly continuous ?

(A) $f(x) = \sin x$ (B) $f(x) = x^3$ (C) $f(x) = 1/x$ (D) $f(x) = \cos x$

25. Any totally bounded metric space is
(A) separable (B) not bounded (C) not complete (D) not separable

TRUE OR FALSE.

Write 'T' for true and 'F' for false statement.

1. In metric space intersection of two open sets is open. *(Meerut 2004)*

2. If *A* is an open set in a metric space (*X, d*), then (*X* – *A*) = *A'* is closed. *(Meerut 2004)*

3. The mapping $f : \boldsymbol{R} \to \boldsymbol{R}$ defined by $f(x) = 2x + 5$ is an isometry.

4. The set ***I*** of all integers with usual metric is a complete metric space.

5. Continuous image of a compact metric space is compact.

6. $A = A \cup D(A)$ *(Meerut 2005)* **7.** $A^o = A - Fr(A)$ (Meerut 2005)

ANSWERS

Multiple Choice :

1. (*C*)	**2.** (*A*)	**3.** (*A*)	**4.** (*D*)	**5.** (*B*)
6. (*C*)	**7.** (*D*)	**8.** (*D*)	**9.** (*B*)	**10.** (*A*)
11. (*B*)	**12.** (*D*)	**13.** (*C*)	**14.** (*A*)	**15.** (*B*)
16. (*A*)	**17.** (*B*)	**18.** (*A*)	**19.** (*B*)	**20.** (*A*)
21. (*A*)	**22.** (*D*)	**23.** (*A*)	**24.** (*B*)	**25.** (*A*)

True or False :

1. T **2.** T **3.** F **4.** T **5.** T **6.** F

MISCELLANEOUS PROBLEMS ON CHAPTER 19

1. Prove that the metric induced on the set ***Z*** of all integers by the usual metric on ***R*** is the discrete metric. *(Delhi Maths (H) 2006)*

2. Give counter example for an open set which is not an open sphere *(Agra 2006)*

3. Prove that is a metric space (*X, d*), $D(A \cup B) = D(A) \cup D(B)$, when *A* and *B* are subsets of *X* and *D* denotes the derived set *(Meerut 2005)*

4. (*i*) Fill up gap : $\overline{A} \cup \overline{B}$......$\overline{A \cup B}$ (*Meerut 2005*)
(*ii*) Fill up gap : If *f* and *g* are continuous mapping, then *gof* is ..... (*Meerut 2005*)
Ans. (*i*) = (ii) continuous

5. If *f* and *g* are continuous mappings on a metric space (*X*, *d*), then show that *fg* is also continuous mapping. (*Meerut 2005*)

6. If *f* and *g* are continuous mappings on metric space then show that *f* + *g* and *fg* are continuous. (*Agra 2006*)

7. Give an example to show that a one-one, onto continuous mapping from one metric space to another may fail to be homeomorphism. (*Delhi Maths (H) 2006*)
[**Sol.** Let (***R***, *u*) be the usual metric space with *u* as usual metric on ***R*** and (***R***, *d*) be the discrete metric space with *d* as discrete metric on **R**. Consider the identity mapping $i: \boldsymbol{R} \to \boldsymbol{R}$. Then, clearly *i* is one-one onto mapping. Also *i* is (*d* ; *u*) continuous but is not (*u* , *d*) continuous.
This is a an example of a one-one onto continuous mapping *i* such that i^{-1} is not continuous. Hence i is not a homeomorphism.]

8. Define the completion of a metric space and obtain the completion of all rational numbers with usual metric. (*Delhi Maths (H) 2006*)

9. Let (*X*, *d*) be a complete metric space and let *Y* be a subspace of *X*, then show that *Y* is complete if and only if *Y* is closed. (*Meerut 2006, 10*)

10. Prove that every contraction mapping is continuous. (*Meerut 2005*)

11. Show that homeomophic image of a second countable space is second countable. (*Meerut 2005*)

12. Prove that every metric space is first countable. (*Meerut 2006*)

13. Let $\{A_\alpha : \alpha \in J\}$ be a family of connected subsets of a metric space *X* such that $\cap A_\alpha \neq \phi$. Prove that $\cup A_\alpha$ is connected. (*Delhi Maths (H) 2006, 08*]

[The following questions 14-26 were asked in M.S. Univ. T.N. 2006]

14. State and prove the intermediate value theorem

15. Prove that any compact subset *A* of a metric space is bounded.

16. Show that a closed sub-space of a compact metric space is compact.

17. Give an example of a complete metric space which is not compact.
[**Sol.** Let $< x_n >$ be a real sequence whose *n*th element is *n* and all other elements are zero. Again, let $A = \{x_n : n \in N\}$. Then, clearly A is a complete subset of the Hilbert space l_2 while *A* is not compact. Thus, a complete metric space need not always be a compact metric space.]

18. Let (M_1, d_1) and (M_2, d_2) be two metric spaces. Let $a \in M_1$. A function $f: M_1 \to M_2$ is continuous at *a* iff $x_n \to a \Rightarrow f(x_n) \to f(a)$. Prove.

19. Prove that a subspace of **R** is connected iff it is an interval.

20. *Let d_1 and d_2 be two metries on M. If $d(x, y) = d_1(x, y) + d_2(x, y)$, then prove that d is also a metric on M..*

21. *If f is continuous map from a compact metric space M_1 into anv metric space M_2 then prove that $f(M_1)$ is closed and bounded.*

22. *In a compact metric space M, prove that any infinite subset of M has a limit point ⇔ M is sequentially compact.*

23. *If f be a continuous function from the compact metric space M_1 into the metric space M_2, prove that $f(M_1)$ is also compact.* (*Chennai 2011; Mumbai 2010*)

24. *Prove that any totally bounded metric space is separable (ii) Is the converse true? Give example.*

25. *Prove that every coninuous mapping from a compact matric space onto any metric space is uniformly continuous.* **[Delhi Maths (H) 2006, 07]**

26. *Prove that every totally bounded metric space is bounded.* **[Delhi Maths (H) 2007]**

27. *Gives an example of pseudo–metric which is not a metric.* **[Meerut 2007]**

28. *Prove that every subset of the metric space is open iff it is nbd of each of its points.* **[Meerut 2007]**

29. *Prove that every interval is a connected subset of the real line.* **[Delhi Maths (H) 2007]**

30. *In a metric space (X, d) every nested sequence of closed sets whose diameters tend to zero has non-empty intersection. Show that (X, d) is complete.* **[Delhi Maths (H) 2007]**

31. *Show that the space C*[0, 1] *of all real valued continuous functions defined on* [0, 1] *with sub-metric is complete.* **[Delhi Maths (H) 2007]**

32. *When are two metric spaces said to be homeomorphic ? Show that the real line R is homeomorphic to the open interval*]–1, 1[*with usual metric.* **[Delhi Maths (H) 2007]**

33. *If S and T are non-empty subsets of a metric space (X, d). Show that (int S) U (int T)* $\subset$ *int (SUT)* but not conversely. Give example to justify. **[Delhi Maths (H) 2007]**

34. Let *d* and *d** be two metrics on $\boldsymbol{R}^2$ defined by $d(x,y)=\{(x_1-y_1)^2+(x_2-y_2)^2\}^{1/2}$ and $d^*(x,y)=\max\{|x_1-y_1|,|x_2-y_2|\}$, $\forall\, x=(x_1,x_2),\ y=(y_1,y_2)$ *in* $\boldsymbol{R}^2$. Prove that *d* and *d** are equivalent **[Delhi Maths (H) 2008]**

35. Let (*X, d*) be a metric space and let *S* and *T* be arbitrary subsets of *X*. Then prove that

(*i*) $\overline{S\cup T}=\overline{S}\cup\overline{T}$ and (*ii*) $\overline{S\cap T}\subseteq\overline{S}\cap\overline{T}$. Here $\overline{A}$ stands for the closure of the set A. Give an example to show that the inclusion in part (ii) can be strict. **[Delhi Maths (H) 2008]**

36. Let (*X, d*) be a metric space. Prove that the mapping $d: X\times X\to\boldsymbol{R}$ is uniformly continuous where $X\times X$ is given the product metric induced by the metric d and $\boldsymbol{R}$ is the set of all real numbers. **[Delhi Maths (H) 2008]**

37. Prove that every compact subspace of a metric space is closed. **[Delhi Maths (H) 2008]**

38. Prove the set of all rational numbers with usual metric is not connected. **[Delhi Maths (H) 2008]**

39. Prove that a function $f:(X,d)\to(Y,\rho)$ is continuous if $[f^{-1}(B)]^0\subset f^{-1}(B^0)$ where $B\subset Y$. **[Agra 2008, Mumbai 2010]**

40. Let $X=\boldsymbol{R}$ with metric $d\ (x,y)=|x-y|/(1+|x-y|)$. Then, metric space (*X, d*) is

(a) bounded but not compact (b) bounded but not complete

(c) complete but not bounded (d) compact but not complete **(GATE 2010)**

Hint. Ans (*a*). Since $d(x,y)<1,\ \forall\ x,y\in\boldsymbol{R}$, hence (*X, d*) is bounded. Again, from problem 1(a), page 19.39, (*x, d*) is not compact.

41. Prove that every Cauchy sequence in a metric space is bounded. **(Agra 2010)**

42. Let (*X, d*) be any metric space and let M be a positive number, then prove that there exists a metric *d** on *X* such that the metric space (*X, d**) is bounded with $\delta(X)\le M$. **(Meerut 2010)**

43. On the real line, show that the set Z of all integers has no limit point where the metric on it is the usual metric $d(x,y)=|x-y|,\forall\, x,y\in\boldsymbol{R}$. (Meerut 2010)

Hint. Refer illustration 4 on page 3.6

44. Show that 5/6 is not a limit point of A, where A = {1/*m* + 1/*n* | *m, n* ∈ N}, A C ***R***. (distance being usual) *(Mumbai 2010)*

45. Let d_1, d_2 be two metrics on X. Show that d_1 and d_2 are equivalent metrics on X iff $i:(X,d_1)\to(X,d_2)$ and $i:(X,d_2)\to(X,d_1)$ are continuous, i being the identity map.

46. Show that $A=\{(x,y)\in R^2 : xy>0\}$ is not a connected subset of $\mathbf{R}^2$, (distance being Euclidean). *(Mumbai 2010)*

47. Show that $A=\{(x,y)\in R^2 : x^2+y^2=1\}$ is a connected subject of $\boldsymbol{R}^2$ (distance being Euclidean). *(Mumbai 2010)*

48. Let $A=\{(x,y)\in R^2 : |x|+|y|\le 5\}$. Show that A is a compact subset of $\boldsymbol{R}^2$(distance being Euclidean). *(Mumbai 2010)*

49. Let (*X*, *d*) be a metric space such that no proper subset of X is dense in X. Show that every subset of X is an open set. *(Mumbai 2010)*

50. Let $\{f_n\}_{n=1}^{\infty}$ be a sequence of real valued functions on a metric space M which converges uniformly to the function *f* on *M*. If each $f_n(n\in N)$ is continuous at $a\in M$, Then prove that *f* is continuous at *a*. *(Pune 2010)*

Hint. Refer theorem on page 15.7 of chapter 15.

51. Let (*X*, *d*) be a metric space and $F\subset X$. Prove that the following statements are equivalent :

(*i*) X \ F is open (ii) F' contains all its limit points. *[Mumbai 2010]*

ADDITIONAL PROBLEMS ON CHAPTER 19

Ex. 1. *If (X, d) be a metric space and x, y, z be any three points of X, prove that $|d(x, z) - d(y, z)| \leq d(x, y)$.* **[Agra 2004, 08, 11; Meerut 2003, 04, 09]**

Sol. By triangle inequality, we have

$d(x, z) \leq d(x, y) + d(y, z)$ so that $d(x, z) - d(y, z) \leq d(x, y)$ ...(1)

Again, by triangle inequality, we have

$d(z, y) \leq d(z, x) + d(x, y)$ so that $d(y, z) \leq d(x, z) + d(x, y)$

[$\because$ d is metric on $X \Rightarrow d(z, y) = d(y, z)$ and $d(z, x) = d(x, z)$]

Thus, we have $d(y, z) - d(x, z) \leq d(x, y)$ ...(2)

(1) and (2) $\Rightarrow$ $|d(x, z) - d(y, z)| \leq d(x, y)$.

Ex. 2. *Let (X, d) be a metric space and let A be a non-empty subset of X. If x, y are any two points in A, then prove that $|d(x, A) - d(y, A)| \leq d(x, y)$*

Sol. For each $z \in A$, by triangle inequality, $d(x, z) \leq d(x, y) + d(y, z)$ ...(1)

Also, by definition of $d(x, A)$, we have $d(x, A) = \inf\{d(x, z) : z \in A\}$

or $d(x, A) \leq \inf\{d(x, y) + d(y, z) : z \in A\}$, using (1)

or $d(x, A) \leq d(x, y) + \inf\{d(y, z) : z \in A\}$

or $d(x, A) \leq d(x, y) + d(y, A)$, using definition of $d(y, A)$

or $d(x, A) - d(y, A) \leq d(x, y)$ ...(2)

Since (2) holds for all $x, y \in X$, so interchanging x and y, we have

$d(y, A) - d(x, A) \leq d(y, x)$ ...(3)

Since $d(y, x) = d(x, y)$, (3) gives $d(y, A) - d(x, A) \leq d(x, y)$ ... (4)

Now, (2) and (4) $\Rightarrow$ $|d(x, A) - d(y, A)| \leq d(x, y)$

Ex. 3. *Let X be a non-empty set. Then, a mapping d of $X \times X$ into **R** is a metric if and only if the following conditions are satisfied.*

$M-1: d(x, y) = 0$ if and only if $x = y$, $\forall$ x, y $\in$ X

$M-2: d(x, y) \leq d(x, z) + d(y, z)$, $\forall$ $x, y, z \in X$ **[Delhi B.Sc. (Hons.) III 2012; Meerut 2001, 02]**

[**Note** The above two conditions are known as an alternative set of conditions for a metric]

Sol. We know that d : X × X $\rightarrow$ **R** is a metric if the following four conditions are satisfied (refer Art. 19.2)

(*i*) $d(x, y) \geq 0$, $\forall$ $x, y \in X$ (*ii*) $d(x, y) = 0 \Leftrightarrow x = y$

(*iii*) $d(x, y) = d(y, x)$, $\forall$ $x, y \in X$ (*iv*) $d(x, y) \leq d(x, z) + d(z, y)$, $\forall$ $x, y, z \in X$

First, let us suppose that the given conditions M – 1 and M –2 hold, Then, in order to prove that d is a metric on X, we must show that the above mentioned four condition (*i*), (*ii*), (*iii*) and (*iv*) hold.

Applying condition $M-2$ for x, x and y, we get

$d(x, x) \leq d(x, y) + d(x, y)$ or $2d(x, y) \geq 0$, using $M-1$.

Thus, $d(x, y) \geq 0$, $\forall$ $x, y \in X$. Hence condition (*i*) is satisfied

Applying condition $M-2$ for x, y and x and using condition $M-1$ as before, we get

$d(x, y) \leq d(x, x) + d(y, x)$ or $d(x, y) \leq d(y, x)$ ...(1)

Applying condition $M-2$ for y, x and y and using condition $M-1$, we get

$d(y, x) \leq d(y, y) + d(x, y)$ or $d(y, x) \leq d(x, y)$ ...(2)

From (1) and (2), we get $d(x, y) = d(y, x)$, $\forall\, x, y \in X$ ...(3)

Thus, the condition (*iii*) is satisfied

Now, from (3), $d(y, z) = d(z, y), \quad \forall\ y, z \in X$ ...(4)

Making use of (4) in condition M–2, we have

$$d(x, y) \leq d(x, z) + d(z, y), \quad \forall\ x, y, z \in X$$

Thus, the condition (*iv*) is satisfied.

Finally, condition M–1 shows that the condition (*ii*) is satisfied.

Therefore, all the conditions (*i*), (*ii*), (*iii*) and (iv) are satisfied and hence d is a metric on X.

Conversely, let us suppose that d is a metric on X and so the conditions (*i*), (*ii*), (*iii*) and (*iv*) hold. We now proceed to prove conditions M–1 and M–2.

Clearly condition (*ii*) $\Rightarrow$ condition M–1.

Again, by condition (*iii*), $d(z, y) = d(y, z)$. Hence condition (*iv*) reduces to

$d(x, y) \leq d(x, z) + d(y, z)\ \forall\ x, y, z \in X$. So condition $M-2$ holds.

Ex. 4. ***(Unitary space)****. Let C^n be the set of all ordered n-tuples of complex numbers and let $d\colon C^n \times C^n \to \mathbf{R}$ be defined by setting* $d(\mathbf{x}, \mathbf{y}) = (\sum_{i=1}^{n} |x_i - y_i|^2)^{1/2}$,

where $\mathbf{x} = (x_1, x_2, \ldots, x_n)$ and $\mathbf{y} = (y_1, y_1, \ldots, y_n)$ are any two points of C^n. Prove that (C^n, d) is a metric space.

Sol. (*i*). $d(\mathbf{x}, \mathbf{y}) = (\sum_{i=1}^{n} |x_i - y_i|^2)^{1/2}$, which is clearly non-negative.

(*ii*) $d(\mathbf{x}, \mathbf{y}) = 0 \Leftrightarrow (\sum_{i=1}^{n} |x_i - y_i|^2)^{1/2} = 0 \Leftrightarrow x_i - y_i = 0,\ \forall\ \text{i} = 1, 2, \ldots, n$

$$\Leftrightarrow x_1 = y_1, x_2 = y_2, \ldots, \text{and } x_n = y_n$$
$$\Leftrightarrow (x_1, x_2, \ldots, x_n) = (y_1, y_2, \ldots, y_n) \Leftrightarrow \mathbf{x} = \mathbf{y}$$

(*iii*) $d(\mathbf{x}, \mathbf{y}) = (\sum_{i=1}^{n} |x_i - y_i|^2)^{1/2} = (\sum_{i=1}^{n} |y_i - x_i|^2)^{1/2}$

Thus, $d(\mathbf{x}, \mathbf{y}) = d(\mathbf{y}, \mathbf{x}),\ \forall\ \mathbf{x}, \mathbf{y} \in C^n$

(*iv*) Let $\mathbf{x} = (x_1, x_2, \ldots, x_n)$, $\mathbf{y} = (y_1, y_2, \ldots, y_n)$ and $\mathbf{z} = (z_1, z_2, \ldots, z_n)$ be any three elements of C^n. Then, we get

$$d(\mathbf{x}, \mathbf{y}) = (\sum_{i=1}^{n} |x_i - y_i|^2)^{1/2} = (\sum_{i=1}^{n} |(x_i - z_i) + (z_i - y_i)|^2)^{1/2}$$

$$\leq (\sum_{i=1}^{n} |x_i - z_i|^2)^{1/2} + (\sum_{i=1}^{n} |z_i - y_i|^2)^{1/2} \quad \text{by Minkowski's inequality.}$$

Thus, $d(\mathbf{x}, \mathbf{y}) \leq d(\mathbf{x}, \mathbf{z}) + d(\mathbf{z}, \mathbf{y}),\quad \forall \mathbf{x}, \mathbf{y}, \mathbf{z} \in C^n$

Hence d is a metric C^n. The metric space (C^n, d) is known as *unitary space.*

Ex. 5. (*a*) *State Holder and Minkowski's inequalities.*

(*b*) ***(l^p–spaces)****. For $1 < p < \infty$, we denote by l^p, the space of all sequences $< x_i >$ such that $\sum_{i=1}^{\infty} |x_i|^p < \infty$. Let $d\colon l^p \times l^p \to \mathbf{R}$ be defined by setting* $d(\mathbf{x}, \mathbf{y}) = \left(\sum_{i=1}^{\infty} |x_i - y_i|^p\right)^{1/p}$,

where $\mathbf{x} = < x_i > \in l^p$ and $\mathbf{y} = < y_i > \in l^p$. Then show that (l^p, d) is a metric space.

Sol. (*a*) **Holder's inequality**. For $1 < p < \infty$, we denote by l^p, the space of all sequences $<x_i>$ such that $\sum_{i=1}^{\infty} |x_i|^p < \infty$. If $< x_i > \in l^p$ and $< y_i > \in l^q$ with $1/p + 1/q = 1$, then

$$\sum_{i=1}^{\infty} |x_i y_i| \leq (\sum_{i=1}^{\infty} |x_i|^p)^{1/p} (\sum_{i=1}^{\infty} |y_i|^q)^{1/q}.$$

Particular case. (Cauchy inequality). For $p = q = 2$, we get $\sum_{i=1}^{\infty} |x_i y_i| \le (\sum_{i=1}^{\infty} x_i^2)^{1/2} (\sum_{i=1}^{\infty} y_i^2)^{1/2}$

Minkowski's inequality. For $1 < p < \infty$, let $< x_i > \in l^p$ and $< y_i > \in l^p$. Then, we have

$$(\sum_{i=1}^{\infty} |x_i + y_i|^p)^{1/p} \le (\sum_{i=1}^{\infty} |x_i|^p)^{1/p} + (\sum_{i=1}^{\infty} |y_i|^p)^{1/p}$$

Particular case. For $p = 2$, we have $\quad (\sum_{i=1}^{\infty} (x_i + y_i)^2)^{1/2} \le (\sum_{i=1}^{\infty} x_i^2)^{1/2} + (\sum_{i=1}^{\infty} y_i^2)^{1/2}$

(*b*). First, we shall show that d is well defined.

Now, $d(\boldsymbol{x}, \boldsymbol{y}) = (\sum_{i=1}^{\infty} |x_i + (-y_i)|^p)^{1/p} \le (\sum_{i=1}^{\infty} |x_i|^p)^{1/p} + (\sum_{i=1}^{\infty} |y_i|^p)^{1/p}$

[using Minkowski's inequality]

Hence, $\quad d(\boldsymbol{x}, \boldsymbol{y}) < \infty, \quad$ as $\quad <x_i>, <y_i> \in l^p \Rightarrow \sum_{i=1}^{\infty} |x_i|^p < \infty, \quad \sum_{i=1}^{\infty} |y_i|^p < \infty$

Thus, $d(\boldsymbol{x}, \boldsymbol{y}) < \infty$, and so d is well defined.

Now, by definition, $\quad d(\boldsymbol{x}, \boldsymbol{y}) = (\sum_{i=1}^{\infty} |x_i - y_i|^p)^{1/p}, \ \forall \ \boldsymbol{x}, \boldsymbol{y} \in l^p \quad$ ...(1)

(*i*). From (1), we find that $\quad d(\boldsymbol{x}, \boldsymbol{y}) \ge 0, \quad \forall \ \boldsymbol{x}, \boldsymbol{y} \in l^p$

(*ii*). $d(\boldsymbol{x}, \boldsymbol{y}) = 0 \Leftrightarrow |x_i - y_i|^p = 0, \quad \forall \ i = 1, 2, \ldots,$ using (1)

$\Rightarrow x_i = y_i, \quad \forall \ i = 1, 2, \ldots$

$\Leftrightarrow <x_i> = <y_i> \Leftrightarrow \boldsymbol{x} = \boldsymbol{v}$

(*iii*). $d(\boldsymbol{x}, \boldsymbol{y}) = (\sum_{i=1}^{\infty} |x_i - y_i|^p)^{1/p} = (\sum_{i=1}^{\infty} |y_i - x_i|^p)^{1/p} = d(\boldsymbol{y}, \boldsymbol{x}), \quad \forall \ \boldsymbol{x}, \boldsymbol{y} \in l^p$

(*iv*). Let $\boldsymbol{x} = < x_i >$, $\boldsymbol{y} = < y_i >$ and $\boldsymbol{z} = < z_i >$ belong to l^p. Then, using definition (1), we have

$$d(\boldsymbol{x}, \boldsymbol{y}) = (\sum_{i=1}^{\infty} |x_i - y_i|^p)^{1/p} = (\sum_{i=1}^{\infty} |(x_i - z_i) + (z_i - y_i)|^p)^{1/p}$$

$$\le (\sum_{i=1}^{\infty} |x_i - z_i|^p)^{1/p} + (\sum_{i=1}^{\infty} |z_i - y_i|^p)^{1/p}, \text{ by Minkowski's inequality.}$$

Thus, $\quad d(\boldsymbol{x}, \boldsymbol{y}) \le d(\boldsymbol{x}, \boldsymbol{z}) + d(\boldsymbol{z}, \boldsymbol{y})$, using (1).

From (*i*), (*ii*), (*iii*) and (*iv*), we find that (l^p, d) is a metric space.

Particular case : When $p = 2$, we have $d(x, y) = (\sum_{i=1}^{\infty} (x_i - y_i)^2)^{1/2}$. In this case the metric space (l^2, d) in known as *Hilbert space.*

Ex. 6 *Describe open spheres of unit radius about (0, 0) for each of the following matrics for* $\boldsymbol{R}^2$ $\quad$ (*i*) $d_1(z_1, z_2) = \{(x_1 - x_2)^2 + (y_1 - y_2)^2\}^{1/2}$

(*ii*) $d_2(z_1, z_2) = |x_1 - x_2| + |y_1 - y_2|$ $\quad$ (*ii*) $d_3(z_1, z_2) = \max(|x_1 - x_2|, |y_1 - y_2|)$

where $z_1 = (x_1, y_1)$ *and* $z_2 = (x_2, y_2)$ *are any two points of* $\boldsymbol{R}^2$

Sol. (*i*) For the usual metric space $(\boldsymbol{R}^2, d_1)$, the open sphere with centre (0, 0) and radius 1 is given by

$$S((0, 0), 1) = \{(x, y) \in \boldsymbol{R}^2 : x^2 + y^2 < 1\},$$

showing that the required open sphere consists of all points in the cartesian plane which lie within the circle $x^2 + y^2 = 1$ as shown in the adjoining figure

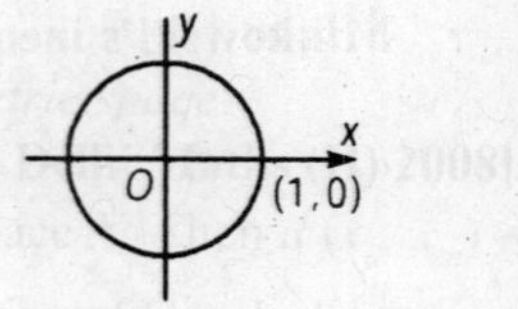

(*ii*) In this case the required open sphere is given by

$$S((0, 0), 1) = \{(x, y) \in \boldsymbol{R}^2 : |x| + |y| < 1\},$$

showing that the required open sphere is the square bounded by the four lines $x + y = 1, x - y = 1, -x + y = 1$ and $-x - y = 1$ as shown in the adjoining figure.

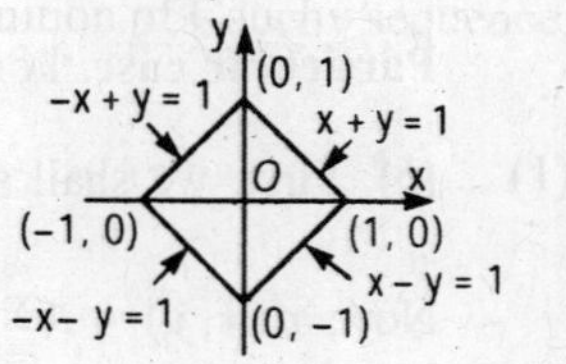

(*iii*) In this case the required open sphere is given by

$$S((0, 0), 1) = \{(x, y) \in \boldsymbol{R}^2 : \max\{|x|, |y|\} < 1\},$$

showing that in this case the required open sphere is the interior of the square bounded by the four straight lines $x = 1, x = -1, y = 1$ and $y = -1$ as shown in the adjoining figure.

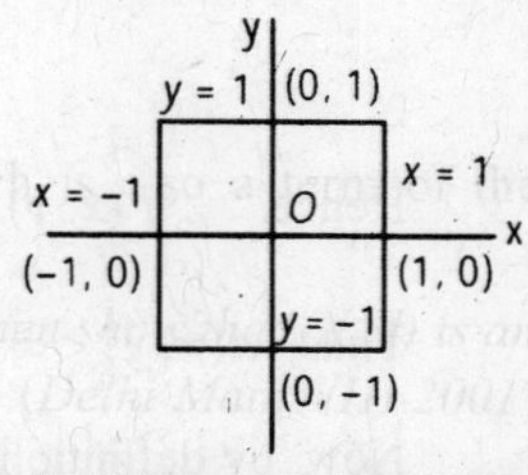

Ex. 7. *Prove that every closed sphere is a closed set in any metric space. Is the converse true? Justify.* **[Delhi Maths (H) 2009]**

Sol. For first part, refer theorem II, page 19.11.

We now give a counter example to show that the converse is not true.

Consider a subset A of a metric space (*X*, *d*) defined by

$$A = \{x \in X : d(x, p) \geq r, p \in X\} \quad \ldots(1)$$

where $r > 0$ is any real number. Now, by definition, we have

$$A' = \text{complement of } A = \{x \in X : d(x, p) < r, p \in X\}$$

Since A' is an open sphere, A' is also an open set. Hence *A* must be a closed set. But from (1), it follows that *A* is not a closed sphere (see Art 19.8). Thus, a closed set A need not imply that *A* is a closed sphere.

Ex. 8. (*a*) *Show that in a metric space every singleton set is closed More generally, a finite set is closed.* **[Agra 2012]**

(*b*) *Show that in a metric space, complement of every singleton set is open. More generally, the complement of a finite set is open.*

Sol. (*a*). Let $\{x\}$ be a singleton subset of a metric space (*X*, *d*). In order to prove that $\{x\}$ is closed, we must prove that its complement $\{x\}'$ is open. If $\{x\}' = \phi$, Then $\{x\}'$ is open because ϕ is an open set.

Now, if $\{x\}' \neq \phi$, then there exists $y \in X$ such that $y \in [x]'$ and $d(x, y) = r > 0$. If $r' \leq r$, then $d(x, y) = r \nless r'$ and hence the open sphere $S(y, r')$ does not contain *x* and is therefore contained in $\{x\}'$. Thus, for every $y \in \{x\}'$, there exist an open sphere $S(y, r')$ such that $S(y, r') \subseteq \{x\}'$. Hence $\{x\}'$ is open and therefore its complement $\{x\}$ is closed.

Again, let $A = \{x_1, x_2, \ldots, x_n\}$ be a finite subset of *X*. Then, A may treated as union of *n* singleton subsets of *X*, i.e., we may write $A = \{x_1\} \cup \{x_2\} \cup \ldots \cup \{x_n\}$

Hence, $A' = [\{x_1\} \cup \{x_2\} \cup \ldots \cup \{x_n\}]' = \{x_1\}' \cap \{x_2\}' \cap \ldots \cap \{x_n\}'$

Since each $\{x_1\}'$, $i = 1, 2, \ldots, n$ is open and the intersection of a finite collection of open sets is open, therefore A' is open and hence A is closed

(*b*). Refer part (*a*) for solution.

Ex. 9. *Prove that every subset of the discrete metric space is open as well as closed.*

[Delhi Maths (H) 2009]

Sol. Let (X, d) be a discrete metric space. Then, we have

$$d(x, y) = \begin{cases} 0, & \text{if } x \neq y \\ 1, & \text{if } x = y \end{cases}$$

Let A be any subset of X. If $A = \phi$, then A is open because ϕ is an open set. Now, let A be a non-empty subset of X. Let x be an arbitrary point of A and let r be a real number such that $r \leq 1$. Then, by definition, $S(x, r) = \{x\}$ and hence $S(x, r) \subset A$. Thus, for each $x \in A$, there exists $r > 0$ such that $S(x, r) \subset A$. Hence, by definition, A is open.

Since every subset of discrete space is open as shown above, it follows that X ~ A is also open and hence its complement A must be closed.

Thus, for a discrete space, every subset is open as well as closed.

Ex. 10. Show that a *subset of a metric space is open if and only if it is the union of a family of open spheres.* **[Mumbai 2010; Delhi Maths (H) 2002]**

Sol. Let A be an open subset of X. Then, for each $x \in A$, there exists a real number $r_x > 0$ such that

$$x \in S(x, r_x) \subset A$$

Taking the union of all such relations as x runs through A, we get

$$A \subset \cup\{S(x, r_x) : x \in A\} \subset A \qquad \text{so that} \qquad A = \cup\{S(x, r_x) : x \in A\}$$

showing that A is union of open spheres.

Conversely, let A be the union of open spheres. Let the collection of all these open spheres be denoted by T. Let x be an arbitrary element of A. Then, x must belong to some open sphere, say $S(p, r) \in T$. Since every open sphere is an open set, x is the centre of an open sphere $S(p, r')$ such that

$$S(x, r') \subset S(p, r) \qquad \ldots(1)$$

But, by hypothesis,

$$S(p, r) \subset \text{A} \qquad \ldots(2)$$

From (1) and (2), we have $S(x, r') \subset A$, showing that A is a *nbd* of x. Since x is an arbitrary point of A, it follows that A is a *nbd* of each of its points. Hence, by definition, A is an open set

Ex. 11. (Housdroff theorem). *In a metric space (X, d), show that for every pair of distinct points $x, y \in X$, there exists disjoint open sets G and H such that $x \in G$ and $y \in H$.*

Sol. Let x and y be two distinct points of X so that $d(x, y) > 0$. Let $r = (1/3) \times d(x, y)$, $G = S(x, r)$ and $H = S(y, r)$. Since G and H are open spheres, and since every open sphere is an open set, it follows that G and H are open sets.

We now show that G and H are are disjoint sets. Let, if possible, $G \cap H \neq \phi$ and let $z \in G \cap H$ so that $z \in G$ and $z \in H$.

Now, $z \in G$ and $z \in H \Rightarrow d(z, x) < r$ and $d(z, y) < r$ ...(1)

Now, by the triangle inequality, we have

$$d(x, y) \leq d(x, z) + d(z, y) = d(z, x) + d(z, y), \quad \text{as} \quad d(x, z) = d(z, x)$$

Hence, using (1), we get $d(x, y) < r + r$ or $d(x, y) < (2/3) \times d(x,y)$ or $1 < 2/3$, , which is impossible. Thus, we arrive at a contradiction. Since the contradiction arises by assuming that $G \cap H \neq \varphi$, it follows that we must have $G \cap H = \varphi$.

We have thus shown the existence of open subsets G and H of X such that $x \in G, y \in H$ and $G \cap H = \varphi$.

Ex. 12. *Show that a subset of a metric space is open if and only if it is nbd of each of its points.*

Sol. Let (X, d) be a metric space and let G be a subset of X.

First suppose that G is open and let x be an arbitrary point of G. Then, by definition G is a nbd of x because G is an open set such that $x \in G \subseteq G$. Thus, G is nbd of each of its points.

Conversely, let G be a *nbd* of each of its points and let x be any point of G. Since G is *nbd* of x, there exists an open set A_x such that $x \in A_x \subseteq G$. Let $A = \cup\{A_x : x \in G\}$. We now show that $A = G$,

If $x \in G$, then by definition of A_x, we have $x \in A_x$. Hence $x \in U\{A_x : x \in G\} = A$. Thus, $G \subseteq A$.. Again, if $y \in A$, then $y \in A_x$ for some $x \in G$. But $A_x \subseteq G$ and so $y \in G$. Thus, we have $A \subseteq G$.

Now, $G \subseteq A$ and $A \subseteq G \Rightarrow A = G$. But A is open, being the union of collection of open sets (see theorem III art 19.10), it follows that G is also open.

Ex. 13. *Let (X, d) be a metric space. Prove that a point $p \in X$ is a limit point of a set $G \subseteq X$ if and only if every nbd of p contains infinitely many points of G.* **[Delhi Maths (H) 2008]**

Sol. First, suppose that every *nbd* of p contains infinitely many points of G. This implies that every *nbd* of p contains a point of G different from p and hence, p is a limit point of G.

Next, suppose that there exists a *nbd*, say N, of p which contains only finitely many points $p_1, p_2, \ldots p_n$ of G.

Let $N \cap G \sim \{p\} = \{p_1, p_2 \cdots p_n\}$, $r = \min\{d(p_1, p), \ldots, d(p_n, p)\}$ and $M = N \cap S(p, r)$. Since N and $S(p, r)$ are both *nbds* of p, so their intersection is also a *nbd* of p. Thus, M is a *nbd* of p.

Now, $$M \cap G \sim \{p\} = N \cap S(p,r) \cap G \sim \{p\} = \phi,$$

showing that M is a *nbd* of p which contains no point of G other than p. Hence p cannot be a limit point of G. Therefore, if p is a limit point of G, then every *nbd* of p must contain infinitely many points of G.

Ex. 14. *Prove that every subset of a metric space is closed if and only if it contains all its limit points.* **[Meerut 2007]**

Sol. Let G be a subset of a given metric space and let $D(G)$ denote the set of all limit points of G. Then in order to prove the required result, we have to prove that

$$G \text{ is closed} \quad \Leftrightarrow \quad D(G) \subseteq G$$

First, let G be closed so that G' is open. If $D(G) = \phi$, then $D(G) \subseteq G$ holds. So let $D(G) \neq \phi$. Let $p \in D(G)$, i.e., let p be a limit point of G. Then, we shall prove that $p \in G$.

Let, if possible, $p \notin G$. Then $p \in G'$. Since G' is open, there exists $r > 0$ such that $S(p, r) \subseteq G'$. Hence $S(p,r) \cap G = \phi$, i.e., the open sphere $S(p, r)$, which is a *nbd* of p, contains a point of G. This contradicts the fact that p is a limit point of G. Hence, $p \in G$. Thus, if G is a closed set, then it must contain all its limit points.

Conversely, let $D(G) \subseteq G$, i.e., G contain all its limit points. Then, we are to show that G is closed. By definition G is closed whenever G' is open. Let $p \in G'$ so that $p \notin G$.

Therefore $p \notin D(G)$, since we have assumed that $D(G) \subseteq G$.

Now, $p \notin D(G)$

$\Rightarrow$ p is not a limit point of G

$\Rightarrow$ there exists $r > 0$ such that $S(p, r)$ contains no point of G other than p

$\Rightarrow$ there exists $r > 0$ such that $S(p, r)$ contains no point of G, because $p \notin G$

$\Rightarrow$ there exists $r > 0$ such that $S(p, r) \subset G'$

Thus, for each $p \in G'$, there exist $r > 0$ such that $S(p, r) \subseteq G'$ and so G' in open and hence G is closed.

• **Note.** In view of the characteristic property of closed set stated in the above Ex.14, some authors define closed set an follows :

Closed set. (Alternative definition). *A subset G of a metric space is said to be closed set if and only if G contains all its limit points, i.e.,* $D(G) \subseteq G$.

Using the above definition of closed set, the reader can prove the following result:

Ex. 15. *Let (X, d) be a metric space and G be a subset of X. Then, show that G is closed if and only if complement X ~ G is open.*

Ex. 16. Let A and B be any two subsets of a metric space (X, d). Then show that :

(*i*) $D(\phi) = \phi$ (*ii*) $A \subseteq B \Rightarrow D(A) \subseteq D(B)$

(*iii*) $D(A \cap B) \subseteq D(A) \cap D(B)$ (*iv*) $D(A \cup B) = D(A) \cup D(B)$,

Ex. 17. *Given an example to show that the inclusion in the result* $\overline{S \cap T} \subseteq \bar{S} \cap \bar{T}$ *can be strict.* **[Delhi Maths (H) 2008]**

Sol. Let $(\boldsymbol{R}, d)$ be the usual metric space. Let $S =]0, 1[$ and $T =]1, 2[$. Then, by definition, we have $\bar{S} = [0,1]$, $\bar{T} = [1,2]$, $S \cap T = \phi$ and $\overline{S \cap T} = \bar{\phi} = \phi$ and $\bar{S} \cap \bar{T} = \{1\}$. Thus, we have $\overline{S \cap T} \subset \bar{S} \cap \bar{T}$, showing that the given result can be strict, i.e., we may have $\overline{S \cap T} \neq \bar{S} \cap \bar{T}$.

Ex. 18. (*a*) *Let (X, d) be a given metric space and let* (Y, d_Y) *be a subspace of (X, d). Prove that every* d_Y*-closed subset of Y is d-closed if and only if Y is d-closed.*

[Delhi Maths (H) 2009)]

Sol. First, let every d_Y-closed subset of Y be d-closed. Then, since Y is d_Y-closed, Y must be d-closed.

Conversely, let Y be d-closed. Let G be an arbitrary d_Y-closed subset of Y so that $G = Y \cap H$ for some d-closed set H. Since intersection of two d-closed sets is again a d-closed set, so $Y \cap H$ is d–closed and consequently G must be d-closed.

Ex. 18(*b*) *Let (X, d) be a given metric space and let* (Y, d_Y) *be a subspace of (X, d). Prove that every* d_Y*-open subset of Y is d-open if and only if Y is d-open.*

Hint: Proceed an in Ex. 18 (*a*) by replacing word 'closed' by 'open'.

Ex. 19. *Let A be a subset of a metric space (X, d) and x be a point of X. Define d(X, A) and prove that d(X, A) > 0 if and only if* $x \in X \sim cl\, A$ **[Delhi Maths (H) 2009]**

Ex. 20. *Let* (X_1, d_1) *and* (X_2, d_2) *be two metric spaces. For any pair of points* $x = (x_1, x_2)$, $y = (y_1, y_2)$ *in* $X = X_1 \times X_2$, *define d by setting* $d(x, y) = d_1(x_1, y_1) + d_2(x_2, y_2)$. *Prove that d is a metric for X, i.e.,* $X_1 \times X_2$. **[M.S.Univ. T.N. 2006]**

Sol. Here $(x_1, x_2), (y_1, y_2), (z_1, z_2) \in X_1 \times X_2 \Rightarrow x_1, y_1, z_1 \in X_1$ and $x_2, y_2, z_2 \in X_2$.

Given $$d(x, y) = d_1(x_1, y_1) + d_2(x_2, y_2) \quad ...(1)$$

d_1 is a metric on X_1 and d_2 is a metric on X_2

$$\Rightarrow \quad d_1(x_1, y_1) \geq 0 \quad \text{and} \quad d_2(x_2, y_2) \geq 0 \quad ...(2)$$

$$d_1(x_1, y_1) = 0 \Rightarrow x_1 = y_1 \quad \text{and} \quad d_2(x_2, y_2) = 0 \Rightarrow x_2 = y_2 \quad ...(3)$$

$$d_1(x_1, y_1) = d_1(y_1, x_1) \quad \text{and} \quad d_2(x_2, y_2) = d_2(y_2, x_2) \quad ...(4)$$

and $$d_1(x_1, y_1) \leq d_1(x_1, z_1) + d_1(z_1, y_1) \quad \text{and} \quad d_2(x_2, y_2) \leq d_2(x_2, z_2) + d_2(z_2, y_2) \quad ...(5)$$

Now, we proceed to show that d is metric on $X_1 \times X_2$

(*i*). From (2), $d_1(x_1, y_1) + d_2(x_2, y_2) \geq 0$. Now, using (1), we get

$$d(x, y) \geq 0, \qquad \forall\ x, y \in X_1 \times X_2$$

(*ii*). From (3), $\quad d_1(x_1, y_1) + d_2(x_2, y_2) = 0 \quad \Rightarrow \quad x_1 = y_1$ and $x_2 = y_2$

Hence, using (1), $\quad d(x, y) = 0 \quad \Rightarrow \quad (x_1, x_2) = (y_1, y_2) \quad \Rightarrow \quad x = y$

(*iii*). From (4), $\quad d_1(x_1, y_1) + d_2(x_2, y_2) = d_1(y_1, x_1) + d_2(y_2, x_2)$

Hence, $\quad d(x, y) = d(y, x)\ \forall\ x, y \in X_1 \times X_2$, using (1)

(*iv*). Adding the corresponding sides of inequalities (5), we get

$$d_1(x_1, y_1) + d_2(x_2, y_2) \leq [d_1(x_1, z_1) + d_2(x_2, z_2)] + [d_1(z_1, y_1) + d(z_2, y_2)]$$

Hence, $\quad d(x, y) \leq d(x, z) + d(z, y), \quad \forall\ x, y, z \in X_1 \times X_2$, using (1)

From (*i*), (*ii*), (*iii*) and (*iv*), we see that $(X_1 \times X_2, d)$ is a metric space.

Ex. 21. *Let (X, d), (Y, ρ) and (Z, σ) be three metric spaces. Suppose $f : X \to Y$ and $g : Y \to Z$ be continuous functions. Then prove that gof, the composite of f and g, is also continuous.* **[Agra 2007; Meerut 2005]**

Sol. Let G be any σ-open set. Then, we have

$$(\text{gof})^{-1}(G) = (\text{f}^{-1}\text{og}^{-1})(G) = f^{-1}(g^{-1}(G)) \quad ...[1]$$

Since g is continuous and G is σ-open, hence $g^{-1}(G)$ is ρ-open. Again, since f is continuous and $g^{-1}(G)$ is ρ-open, it follows that $f^{-1}[g^{-1}(G)]$ is d-open. Thus, we find that G is σ-open $\Rightarrow$ $(\text{gof})^{-1}(G)$ is d-open. Hence, it follows that gof : $X \to Z$ is continuous.

Ex. 22. *Let (X, d) be a metric space and let x_0 be a fixed point of X. Show that the real valued function $f_{x_0}(x) = d(x, x_0)$ is continuous.* **[M.S. Univ. T.N 2006, Mumbai 2010]**

Sol. Let p be any point of X and let $\varepsilon > 0$ be given. We choose $\delta > 0$ such that $\delta \leq \in$. Then, for any point $x \in X$ and satisfying $d(x, p) < \delta$, we have

$$|f_{x_0}(x) - f_{x_0}(p)| = |d(x, x_0) - d(p, x_0)|, \text{ by definition of } f_{x_0}$$
$$\leq d(x, p), \text{ using result of Ex. 2}$$
$$< \delta, \text{ by hypothesis}$$
$$\leq \in, \text{ by hypothesis}$$

Thus, $\quad d(x, p) < \delta \quad \Rightarrow \quad |f_{x_0}(x) - f_{x_0}(p)| < \in,$

showing that f_{x_0} is continuous at $x = p$. Since p is an arbitrary point of X, it follows that f_{x_0} is a continuous function on X.

Ex. 23. (*a*) *Let (X, d) be a metric space and A be a non-empty subset of X, then prove that the function $f : X \to$ **R** given by $f(x) = d(x, A)$, $x \in A$ is uniformly continuous.*

[Calcutta 2004; Delhi Maths (H) 2001]

Sol. Since d is a metric on X, for each $a \in A$, by triangle inequality, we have

$$d(x,a) \le d(x,y) + d(y,a), \ \forall \ x, y \in X \quad \ldots(1)$$

By definition, $\quad d(x, A) = \inf \{d(x, a) : a \in A\}$

or $\quad d(x, A) \le \inf \{d(x, y) + d(y, a) : a \in A\}$, using (1)

or $\quad d(x, A) \le d(x, y) + \inf \{d(y, a) : a \in A\}$

or $\quad d(x, A) \le d(x, y) + d(y, A)$, by definition

Thus, $\quad d(x, A) - d(y, A) \le d(x, y) \quad \ldots(2)$

Since (2) is true for all $x, y \in X$, so interchanging x and y in (2), we obtain

$d(y, A) - d(x, A) \le d(y, x) \quad$ or $\quad d(y, A) - d(x, A) \le d(x, y) \quad \ldots(3)$

[$\because$ d is metric on $X \Rightarrow d(y, x) = d(x, y)$]

From (2) and (3), $\quad |d(x, A) - d(y, A)| \le d(x, y), \ \forall x, y \in X \quad \ldots(4)$

Let $\varepsilon > 0$ be given. We choose $\delta > 0$ such that $\delta \le \varepsilon$. Then, for $\forall x, y \in X$ and satisfying $d(x, y) < \delta$, we have

$$|f(x) - f(y)| = |d(x, A) - d(y, A)| \le d(x, y), \text{ using (4)}$$

or $\quad |f(x) - f(y)| < \delta$, since by hypothesis $d(x, y) < \delta$

or $\quad |f(x) - f(y)| < \varepsilon$, since by hypothesis $\delta \le \varepsilon$

Thus, $\quad d(x, y) < \delta \ \Rightarrow \ |f(x) - f(y)| < \varepsilon, \ \forall x, y \in X,$

showing that f is uniformly continuous on X.

Ex. 23. (*b*) *(Urysohn's Lemma for metric spaces).* *Let (X, d) be a metric space and let A and B be disjoint non-empty closed subsets of X. Show that there exists a real valued continuous function* $f : X \to [0,1]$ such that $f(A) = \{0\}$ and $f(B) = \{1\}$, *where ([0, 1]) is equipped with the usual metric.* **[Delhi Maths (H) 1999]**

Sol. Consider the function f given by

$$f(x) = d(x, A) / \{d(x, A) + d(x, B)\}, \ \forall \ x \in \mathbf{R} \quad \ldots(1)$$

Now, by definition, $d(x, A) \ge 0$ and $d(x, B) \ge 0$ for all $x \in X$. Again A and B being disjoint, $d(x, A)$ and $d(x, B)$ cannot vanish simultaneously and hence $d(x, A) + d(x, B) \neq 0$ for all $x \in X$. Therefore, f is well defined.

As shown in Ex. 23(*a*), $x \to d(x, A)$ and $x \to d(x, B)$ are uniformly continuous and hence these are continuous. Hence $x \to d(x, A) + d(x, B)$ is also continuous. Since $x \to d(x, A)$ and $x \to d(x, A) + d(x, B)$ are both continuous and hence their quotient is also continuous. Consequently, f is continuous.

Also, $\quad x \in A \Rightarrow d(x, A) = 0 \quad$ and $\quad$ so by (1), $\quad f(x) = 0, \quad$ if $x \in A$

and $\quad x \in B \Rightarrow d(x, B) = 0 \quad$ and $\quad$ so by (1), $\quad f(x) = 1, \quad$ if $x \in B$

Thus, there exists a continuous function defined by (1) such that $f(A) = \{0\}$ and $f(B) = \{1\}$

Ex. 24. *Prove that every contraction mapping on a metric space is uniformly continuous but not conversely.* **[Delhi Maths (H) 2005]**

Sol. Let T be a contraction mapping on a metric space (X, d) and let α be a real number such that $0 < \alpha < 1$. Then, by definition, we have

$$d(T_x, T_y) \le \alpha\ d(x, y), \qquad \forall\ x, y \in X \qquad \ldots(1)$$

Let $\varepsilon > 0$ be given. We choose $\delta > 0$ such that $\delta = \varepsilon / \alpha$. Then, for $x,\ y \in X$ and satisfying $d(x, y) < \delta$, (1) yields

$$d(T_x, T_y) < \alpha\ \delta < \varepsilon, \qquad \text{since by hypothesis} \qquad \delta = \varepsilon / \alpha.$$

Thus, $\quad d(x, y) < \delta \quad \Rightarrow \quad d(T_x, T_y) < \varepsilon, \qquad \forall\ x, y \in X,$

showing that T is uniformly continuous on X.

Ex. 25. *Show that homeomorphism is an equivalence relation in the collection of all metric spaces.* **[Meerut 2010; Agra 2006, 08, 11, 12]**

Sol. In what follows, we shall write $(X, d) \sim (Y, \rho)$ to indicate that (X, d) is homeomorphic to (Y, ρ).

Reflexivity Let (X, d) be any metric space. Consider the identity mapping $i : X \to X$ such that $i(x) = x,\ \forall\ x \in X$. Then, we have

(*a*) i is one-one onto mapping.

(*b*) Since $i^{-1}(G) = G$ for every open set G, hence i is continuous.

(*c*) For each open set G in X, $i(G)$ is open.

Hence (see theorem I, page 19.19), it follows that the identity mapping i is a homeomorphism and therefore $(X, d) \sim (X, d)$.

Symmetry. Let $(X, d) \sim (Y, \rho)$. Hence, there exists homeomorphism $f : X \to Y$, Then,

(*a*) f is homeomorphism $\Rightarrow$ f is one-one onto $\Rightarrow$ f^{-1} is one-one onto.

(*b*) f is homeomorphism $\Rightarrow$ f^{-1} in continuous.

(*c*) f is continuous and homeomorphism $\Rightarrow$ $(f^{-1})^{-1}$, i.e., f is continuous.

From (*a*), (*b*) and (*c*), we find that f^{-1} is a homomorphism.

Hence, $\qquad (X, d) \sim (Y, \rho) \Rightarrow (Y, \rho) \sim (X, d)$

Transitivity. Let (X, d), (Y, ρ) and (Z, σ) be three metric spaces such that $(X, d) \sim (Y, \rho)$ and $(Y, \rho) \sim (Z, \sigma)$. Now, we have

$(X, d) \sim (Y, \rho) \Rightarrow$ there exists homeomorphism $f : X \to Y$

and $\quad (Y, \rho) \sim (Z, \sigma) \Rightarrow$ there exists homeomorphism $g : Y \to Z$.

We now prove that the composite mapping gof : $X \to Z$ is homeomorphism.

(*a*) Since f and g are one-one onto mappings so gof is also one-one onto.

(*b*) Since f and g are continuous so g of is also continuous.

(*c*) Let G be any d-open set. Since f is homeomorphism from X to Y, $f(G)$ will be ρ-open in (Y, ρ). Again since g is homeomorphism from Y to Z and $f(G)$ is open in Y, so $g(f(G))$, i.e., gof(G) is σ-open in (Z, σ). Thus, for every d-open set G in (X, d), (gof)(G) is σ-open in (Z, σ).

Now, (*a*), (*b*) and (*c*) imply that gof is homeomorphism from X to Z. Hence $(X, d) \sim (Z, \sigma)$.

Thus, $\qquad (X, d) \sim (Y, \rho)$ and $(Y, \rho) \sim (Z, \sigma) \Rightarrow (X, d) \sim (Z, \sigma)$

Since the relation '~' is reflexive, symmetric and transitive, therefore it is an equivalence relation.

Ex.26. *Let (X, d), (Y, ρ) and (Z, σ) be metric spaces. If X is homeomorphic to Y and Y is homeomorphic to Z, then prove that X is homeomorphic to Z.*

Sol. Refer 'transitivity' part of Ex. 25.

Ex. 27. *Let (X, d) and (Y, ρ) be two metric spaces and let f : X → Y be one-one onto mapping. Then, show that the following statements are equivalent :*

(*i*) f^{-1} is continuous.

(*ii*) For each d-open set A, the set f(A) is ρ -open.

(*iii*) *For each d-closed set B, the set f(B) is ρ -closed.*

Sol. To show that (*i*) ⇒ (*ii*). Let f^{-1} be continuous. Hence, (by theorem I, page 19.17), if A is d-open set, then $(f^{-1})^{-1}(A)$, i.e., $f(A)$ must be ρ -open.

To show that (*ii*) ⇒ (*iii*). Let $f(A)$ be ρ -open whenever A is *d*-open. Let B be a d-closed set so that $X \sim B$ is d-open. Hence, by hypothesis, $f(X \sim B)$ is ρ -open. Since f is one-one onto mapping, we have $\qquad f(X \sim B) = f(X) \sim f(B) = Y \sim f(B)$

Since $Y \sim f(B)$ is ρ -open, hence $f(B)$ must be ρ -closed.

To show that (*iii*) ⇒ (*i*) Let $f(B)$ be ρ -closed whenever B is d-closed.

Now, $f(B) = (f^{-1})^{-1}(B)$ and hence $(f^{-1})^{-1}(B)$ is ρ -closed whenever B is d-closed.

Thus, The inverse image under f^{-1} of every d-closed set is ρ -closed. Hence, f^{-1} is continuous (refer theorem II, page 19.18).

Ex. 28. *Show that homeomorphism carries open sets and closed sets of one space onto open and closed sets of other respectively.* **[Agra 2008, 11]**

Sol. Let (X, d) and (Y, ρ) be two metric spaces. Let $f : X \to Y$ be homeomorphism from X to Y. Then, by definition, f is one-one onto, f is continuous and f^{-1} is continuous. (refer Art 19.28, page 19.19)

Let A be any d-open set in X. Since f^{-1} is continuous, so $(f^{-1})^{-1}(A)$, i.e., $f(A)$ is ρ -open in Y (refer theorem *I*, page 19.17).

Let B be any d-closed set in X. Since f^{-1} is continuous, so $(f^{-1})^{-1}(B)$, i.e., $f(B)$ is ρ -closed in Y (refer theorem II, page 19.18)

Thus, we find that homeomorphism carries open and closed sets of metric space (X, *d*) onto open and closed sets of metric space (*Y*, *d*) respectively.

Ex. 29 *Show that the mapping* $\boldsymbol{f : R \to]-1, 1[}$ *defined by setting* $f(x) = x/(1+|x|)$, $\forall\ x \in \boldsymbol{R}$ *is a homeomorphism of the real line onto the open interval]–1, 1[with usual metric* **[Delhi Maths (H) 2007]**

Sol. Given, $\qquad f(x) = x/(1+|x|),\ x \in \boldsymbol{R} \qquad \ldots(1)$

Since $1+|x| \neq 0$ for any $x \in \boldsymbol{R}$, hence the mapping f is well defined on $\boldsymbol{R}$. We, now show that f is one-one onto.

$$f(x_1) = f(x_2) \Rightarrow x_1/(1+|x_1|) = x_2/(1+|x_2|), \text{ using (1)}$$

$$\Rightarrow |x_1|/(1+|x_1|) = |x_2|/(1+|x_2|)$$

$$\Rightarrow |x_1| = |x_2| \Rightarrow 1+|x_1| = 1+|x_2|$$

$$\Rightarrow x_1 = x_2, \quad \text{as} \quad x_1/(1+|x_1|) = x_2/(1+|x_2|)$$

Thus, $f(x_1) = f(x_2) \Rightarrow x_1 = x_2$ and so f is one one mapping.

Next, since $|f(x)| = |x|/(1+|x|) < 1$, $\forall\ x \in \boldsymbol{R}$, so $f(x) \in]-1, 1[$, $\forall\ x \in \boldsymbol{R}$

We now show that f is an onto mapping. To, this end, let y be an arbitrary element of]–1, 1[. Let, if possible, x be its pre-image under f. Then, we have

$f(x) = y \Rightarrow x/(1+|x|) = y \Rightarrow |x|/(1+|x|) = |y|$ $[\because x > 0 \Leftrightarrow y > 0]$

$\Rightarrow 1-\{1/(1+|x|)\} = |y| \Rightarrow 1+|x| = 1/(1-|y|)$

$\Rightarrow 1/(1+|x|) = 1-|y| \Rightarrow x(1-|y|) = y$ $[\because x/(1+|x|) = y]$

Thus, $f(x) = y \Rightarrow x = y/(1-|y|)$, ...(2)

showing that for each $y \in]-1, 1[$, there exists an $x \in \boldsymbol{R}$ such that $f(x) = y$. Hence, f is onto.

Thus, f is one-one onto. Hence inverse mapping $f^{-1} :]-1, 1[\to \boldsymbol{R}$ exists and is defined by

$$f^{-1}(y) = y/(1-|y|) \qquad ...(3)$$

$[\because f(x) = y \Rightarrow f^{-1}(y) = x \Rightarrow f^{-1}(y) = y/(1-|y|)$, using (2)]

We now show that f is continuous. We know that function x is continuous. Also $1+|x|$ is a continuous function. Since the quotient of two continuous functions is also continuous (noting that $1+|x| \neq 0$ for any $x \in \boldsymbol{R}$), so the function f defined by (1) is continuous.

We now show that f^{1} is continuous. As before y and $1-|y|$ are both continuous and hence their quotient must be continuous. Therefore the function f^{1} defined by (3) is continuous.

Thus f is one-one onto such that f and f^{-1} are both continuous. Hence f in a homeomorphism.

Ex. 30. (*a*) *Let u be the usual metric on **R** and d the discrete metric on **R**. Let f be a function from (**R**, d) to (**R**, u) defined as f(x) = x, ∀ x ∈ **R**. Again let g be a function from (**R**, u) to (**R**, d) defined as g(x) = x, ∀ x ∈ **R**. Show that f is continuous whereas g is no. ontinuous.* **[GATE 2015; Mumbai 2010]**

(*b*) *Let u be the usual metric on **R*** and *d* the discrete metric on ***R***. *Let f be a function fron. (**R**, d) to (**R**, u) defined as f(x) = x, ∀ x ∈ **R**. Is the function f a homeomorphism.*

[Delhi Maths (H) 2009]

(*c*) *Give an example to show that a one-one onto continuous mapping from one metric space to another may fail to be homeomorphism.* **[Delhi Maths (H) 2006]**

Sol. (*a*). Let A be a u-open subset of $\boldsymbol{R}$, Now, by definition of f, $f^{-1}(A) = A$, which is a d– open subset of $\boldsymbol{R}$, since every subset of discrete space $(\boldsymbol{R}, d)$ is open as shown in Ex. 9. Thus, the inverse image under f of every u-open subset of $\boldsymbol{R}$ is d–open, hence f is continuous (by theorem *I*, page 19.17).

We now proceed to prove that g is not continuous. Consider a d-open subset B of $\boldsymbol{R}$ which is not u-open. By definition of g, $g^{-1}(B) = B$, which is not u-open by our mode of selection of subset B. Thus, we see the inverse image under g of a d-open set need not be u-open. Hence, as before, g cannot be continuous.

(*b*). First proceed as in first part of the above part (*a*) to show that f is continuous. Since f is one-one onto, its inverse mapping f^{-1} (= g, say) exists such that $f^{-1}(x) = g(x) = x$,

Now proceed as in second part of the above part (a) to show that g, i.e., f^{-1} is not continuous.

Thus f is one-one onto, continuous but f^{-1} in not continuous and so f is not a homeomorphism

(*c*). Give example given in part (*a*) and proceed as in part (*b*).

Ex. 31. *If (X, d) is a discrete metric space and (Y, ρ) is any metric space, then show tha every function $f : X \to Y$ is continuous on X.* **[M.S. Univ. T. N. 2006, Mumbai 1998]**

Sol. Let G be any ρ-open subset of Y. Since every subset of the discrete metric space (X, d) is d-open as shown in Ex. 9, it follows that $f^{-1}(G)$ in d-open whenever G is ρ-open. Hence f is continuous on X (refer theorem I, Art. 19.26).

Ex. 32. *Prove that the image of a Cauchy sequence under a uniformly continuous function is a Cauchy sequence. What happen if the function is only continuous.* **[Delhi Maths (H) 2007]**

Sol. Let (X, d) and (Y, ρ) be metric spaces and $f : X \to Y$ be a uniformly continuous function. Let $<x_n>$ be a Cauchy sequence in X, then we are to show that $< f(x_n) >$ is also a Cauchy sequence in Y.

Since f is uniformly continuous, for a given $\varepsilon > 0$, there exists $\delta > 0$ such that

$$d(x, x') < \delta, \quad x, x' \in X \quad \Rightarrow \quad \rho(f(x), f(x')) < \varepsilon$$

As a particular case of the above result, we have

$$d(x_m, x_n) < \delta \quad \Rightarrow \quad \rho(f(x_m), f(x_n)) < \varepsilon \qquad \ldots(1)$$

Since $<x_n>$ is a Cauchy sequence in X, given $\delta > 0$, there exists $p \in N$ such that

$$d(x_m, x_n) < \delta, \qquad \forall\ n, m \geq p \qquad \ldots(2)$$

Hence, from (1) and (2) it follows that $\quad \rho(f(x_m), f(x_n)) < \varepsilon, \quad \forall\, n, m \geq p,$

showing that $< f(x_n) >$ is a Cauchy sequence in Y.

We now present an example to show that the image of a Cauchy sequence under continuous function need not be a Cauchy sequence.

Let $X =]0, 1]$ and $Y = [1, \infty[$. Consider the usual metric on X and Y. Let $f : X \to Y$ defined by $f(x) = 1/x$, $\forall\ x \in X$. Then, f is continuous but it is not uniformly continuous (refer theorem I, Art 8.19 on page 8.43)

Consider a sequence $<x_n>$ where $x_n = 1/n$. Let $\varepsilon > 0$ be given and let $n > m$.

Now, $\quad |x_n - x_m| = |1/n - 1/m| = 1/m - 1/n, \quad$ as $\quad n > m \quad \Rightarrow \quad 1/n < 1/m$

Hence, $\quad | x_m - x_n | < 1/m < \varepsilon$, whenever $m > 1/\varepsilon$

Thus, for each $\varepsilon > 0$, there exists a positive integer m such that $| x_n - x_m | < \varepsilon, \forall n > m$.

Hence $<x_n> = <1/n>$ is a Cauchy sequence.

Now, consider $<f(x_n)>$ i.e. $< f(1/n)>$, i.e., $<n>$. Let $p = 2m + 1$ and $q = 2m$ so that $p, q > m$.

Now, $\qquad | f(x_p) - f(x_q) | = | p - q | = | (2m + 1) - 2m | = 1$

If $\varepsilon = 1/2$, it is not possible to find any positive integer m such that $| f(x_p) - f(x_q) | < \varepsilon$, $\forall\ p, q > m$. Hence $< f(x_n) >$ is not a Cauchy sequence.

Ex. 33 *Prove that every convergent sequence in a metric space is a Cauchy sequence ? Is the converse true ? If not can you give a sufficient condition for a Cauchy sequence to become convergent.* **[Delhi Maths (H) 2009]**

Sol. For the solution of the first part of the problem, refer theorem I, Art. 19.13.

Sufficient condition for a Cauchy sequence to become convergent (refer theorem III, Art 19.31).

We now state and prove this result:

Theorem. *Let $<x_n>$ be a Cauchy sequence in a metric space (X, a). If $<x_n>$ possesses a convergent subsequence $< x_{n_k} >$ converging to x, then $<x_n>$ converges to x.*

Proof of the above theorem. Let $<x_n>$ be a Cauchy sequence and let it possess $< x_{n_k} >$ as a convergent subsequence. This implies that $< x_{n_k} > = < x_{n_1}, x_{n_2}, \ldots >$ where $n_1 < n_2 < n_3 < \ldots$ are positive integers such that $x_{n_k} \to x$ as $n_k \to \infty$.

Let $\varepsilon > 0$ be given. Hence, there exists a positive integer p such that

$$d(x_{n_k}, x) < \varepsilon/2, \quad \forall \quad k > p \qquad \text{...(1)}$$

Since $<x_n>$ is a Cauchy sequence, there exists a positive integer q such that

$$d(x_m, x_n) < \varepsilon/2, \quad \forall \quad m, n > q \qquad \text{...(2)}$$

Let $k = p + q = r$, say. Then, as a particular case, (1) yields

$$d(x_{n_r}, x) < \varepsilon/2 \qquad \text{...(3)}$$

Also, note that $n_r \geq r \geq q$ and hence (2) yields $d(x_{n_r}, x_n) < \varepsilon/2, \quad \forall \quad n \geq q$ ...(4)

By triangle inequality, $\quad d(x_n, x) \leq d(x_n, x_{n_r}) + d(x_{n_r}, x)$

or $\quad d(x_n, x) < \varepsilon/2 + \varepsilon/2, \quad \forall\, n \geq q$, using (3) and (4).

Thus, $\quad d(x_n, x) < \varepsilon, \ \forall\, n \geq q \quad$ so that $\quad x_n \to x$.

Hence the Cauchy sequence $<x_n>$ converges to the limit of the convergent subsequence.

Remark. Combining the results of theorems I and III of Art 19.31, we have the following theorem IV

Theorem IV *A Cauchy sequence is convergent if and only if it has a convergent subsequence.*

Hint. If a Cauchy sequence $<x_n>$ in metric space (X, d) is convergent, then every subsequence of $<x_n>$ is convergent in (X, d). Now, refer to theorems I and III of Art. 19.31.

Ex. 34. *Let (X, d) be a metric space and Y be a subset of X, then show that Y in complete if and only if Y is closed.* **[Meerut 2005, 06]**

Hint Refer theorems II and III of Art. 19.33.

Ex. 35. (*a*) *Show that the set* ***Z*** *of integers with usual metric,* $d(x, y) = |x - y|, \ \forall \ x, y \in \boldsymbol{Z}$, *is a complete metric space.* **[Delhi Maths (H) 2009]**

Sol. Let $<x_n>$ be an arbitrary Cauchy sequence in $\boldsymbol{Z}$. Taking $\varepsilon = 1/2$ in the definition of Cauchy sequence, we have a positive integer p such that

$d(x_m, x_n) < \varepsilon, \ \forall \ m, n \geq p \quad$ or $\quad |x_m - x_n| < 1/2, \ \forall \ m, n \geq p$ ...(1)

Let N denote the set of all natural numbers.

Since $\quad x_n \in \boldsymbol{Z}, \quad \forall n \in N, \quad$ therefore $\quad |x_m - x_n| \geq 0, \ \forall \ m, n \in N$ ...(2)

Now, $\quad$ (1) and (2) $\Rightarrow \quad |x_m - x_n| = 0, \ \forall \ m, n \geq p$

Thus, $\quad x_m = x_n, \ \forall \ m, n \geq p \quad \Rightarrow \quad x_n = x_p, \ \forall \quad n \geq p,$

showing that $<x_n>$ is a convergent sequence and and hence it converges to $x_p \in \boldsymbol{Z}$.

Therefore, every sequence in $\boldsymbol{Z}$ converges to a point of $\boldsymbol{Z}$ and hence $(\mathbf{Z}, d)$ is a complete metric space.

Ex. 35. (*b*) *Show that set* $X = [0, 1]$ *equipped with the metric d defined by setting* $d(x, y) = |1/(1 - x) - 1/(1 - y)|, \ \forall \ x, y \in X$ *is a complete metric space.* **[Delhi Maths (H) 2009]**

Sol. Let $<x_n>$ be an arbitrary Cauchy sequence in metric space (X, d). Then, taking $\varepsilon = 1/2$ in the definition of Cauchy sequence, there exists $p \in N$ such that

$d(x_m, x_n) < 1/2, \ \forall \ n, m \geq p$

$\Rightarrow |1/(1 - x_m) - 1/(1 - x_n)| < 1/2, \ \forall \ n, m \geq p, \quad$ as $\quad d(x, y) = |1/(1 - x) - 1/(1 - y)|$

$\Rightarrow |x_m - x_n| < (1/2) \times |1 - x_m| \times |1 - x_n|, \ \forall \ n, m \geq p$

Thus, $|x_m - x_n| < 1/2, \ \forall \ n, m \geq p$ ...(1)

$[\because \ x_m, x_n \in [0, 1] \Rightarrow |1 - x_m| < 1 \text{ and } |1 - x_n| < 1]$

From (1), it follows that $<x_n>$ is a Cauchy sequence in X. Since the space X with the usual metric is complete, it follows that the Cauchy sequence $<x_n>$ converges. Since $<x_n>$ is any Cauchy sequence in X, it follows that every Cauchy sequence in X converges and hence the metric space (X, d) is a complete metric space.

Ex. 35(c). *Show that the space of rational numbers with usual metric is incomplete.*

[Meerut 2010; Delhi Maths (H) 2005]

Sol. Let $\boldsymbol{Q}$ be the set of rational numbers. Let d be the usual metric on $\boldsymbol{Q}$, defined by

$$d(x, y) = |x - y|, \qquad \forall \ x, y \in \boldsymbol{Q} \qquad \text{...(1)}$$

In order to show that $(\boldsymbol{Q}, d)$ is incomplete, we must show that there exists a sequence $<x_n>$ in $\boldsymbol{Q}$ which cannot converge to a point of $\boldsymbol{Q}$. To this end, consider a sequence, given by

$$x_1 = 1 \quad \text{and} \quad x_{n+1} = (4 + 3x_n) / (3 + 2x_n), \ \forall \ n \in \boldsymbol{N} \qquad \text{...(2)}$$

where $\boldsymbol{N}$ denotes the set of all natural numbers.

Let, if possible; $<x_n>$ converge to a limit, say x, in $\boldsymbol{Q}$, i.e., $x_n \to x$ as $n \to \infty$.

Then, $$x = \lim_{n\to\infty} x_{n+1} = \lim_{n\to\infty} \frac{4+3x_n}{3+2x_n} = \frac{4+3\lim x_n}{3+2\lim x_n} = \frac{4+3x}{3+2x}$$

or $x \times (3 + 2x) = 4 + 3x$ so that $x^2 = 2$

Since $x^2 = 2$ has no solution in $\boldsymbol{Q}$, so $<x_n>$ cannot converge to a limit in $\boldsymbol{Q}$.

We now proceed to prove that $x_n \geq 1, \forall n \in \boldsymbol{N}$. Note that $x_1 > 0$. Again $x_n > 0 \Rightarrow x_{n+1} > 0$. Hence, by the principle of finite induction, we have $x_n > 0, \forall n \in \boldsymbol{N}$.

From (1), $$x_n - 1 = \frac{4+3x_{n-1}}{3+2x_{n-1}} - 1 = \frac{x_{n-1}+1}{2x_{n-1}+3} \geq 0, \quad \forall n \geq 2$$

Thus, $x_n \geq 1, \ \forall \ n \geq 2$. Also, $x_1 = 1$. Hence, $x_n \geq 1, \quad \forall n \in \boldsymbol{N}$ ...(2)

We now proceed to prove that $x_n^2 \to 2$ as $n \to \infty$.

Now, from (1), $$x_n^2 - 2 = \left(\frac{4+3x_{n-1}}{3+2x_{n-1}}\right)^2 - 2 = \frac{x_{n-1}^2 - 2}{(3+2x_{n-1})^2}, \quad \forall n \geq 2$$

Also, $x_1^2 < 2$ and $x_{n-1}^2 - 2 < 0 \Rightarrow x_n^2 - 2 < 0,$

$\Rightarrow \ x_n^2 < 2, \forall \ n \in \boldsymbol{N}$, by the principle of finite induction.

Now, from (2), $x_n \geq 1, \quad \forall \ n \in \boldsymbol{N} \Rightarrow 1/(3+2x_{n-1})^2 < 5^{-2}, \quad \forall n \geq 2$

Hence, $$|x_n^2 - 2| = |x_{n-1}^2 - 2|/(3+2x_{n-1})^2 \leq |x_{n-1}^2 - 2| \times 5^{-2}$$

Thsu, $|x_n^2 - 2| < 5^{-(2n-2)}|x_1^2 - 2| \Rightarrow |x_n^2 - 2| < 5^{-(2n-2)}$ ...(3)

Taking $\varepsilon > 0$, we can choose $m \in \boldsymbol{N}$ such that $5^{-(2m-2)} < \varepsilon$. Then, we have

$$|x_n^2 - 2| < 5^{-(2n-2)} \leq 5^{-(2m-2)} < \varepsilon, \forall \ n \geq m,$$

showing that the sequence $< x_n^2 >$ converges to 2.

Since $<x_n^2>$ is a sequence in $\boldsymbol{Q}$ such that $x_n^2 \to 2 \in \boldsymbol{Q}$, so $<x_n^2>$ is a Cauchy sequence in $\boldsymbol{Q}$. Hence, by definition of Cauchy sequence, for a given $\varepsilon > 0$, there exists $n_0 \in \boldsymbol{N}$ such that

$$|x_p^2 - x_q^2| < \varepsilon, \quad \forall p, q \geq n_0 \qquad \ldots(4)$$

Now, $\quad |x_p + x_q| = x_p + x_q, \quad$ as $\quad x_p, x_q \geq 0, \; \forall \; p, q \in \boldsymbol{N}$

Thus, $\quad |x_p + x_q| \geq 2, \quad$ as $x_n \geq 1, \quad \forall \; n \in \boldsymbol{N} \qquad \ldots(5)$

Then, (4) yields $\quad |x_p - x_q| < \varepsilon / |x_p + x_q|, \quad \forall \quad p, q \geq n_0 \qquad \ldots(6)$

From (5) and (6), $\quad |x_p - x_q| \leq \varepsilon/2, \quad \forall \quad p, q \geq n_0,$

showing that $<x_n>$ is a Cauchy sequence in $\boldsymbol{Q}$. Thus, there exists a Cauchy sequence $<x_n>$ given by (1) in $\boldsymbol{Q}$, which does not converge in $\boldsymbol{Q}$. Hence, $(\boldsymbol{Q}, d)$ is not complete.

Ex. 36. (*a*) *Show that the Euclidean space $\boldsymbol{R}^2$ is complete.* **[Delhi Maths (H) 1993]**

Sol. Let $\boldsymbol{R}^2$ be the set of all ordered pairs of real numbers, and let $d : \boldsymbol{R}^2 \times \boldsymbol{R}^2 \to \boldsymbol{R}$ be defined by

$$d(\boldsymbol{x}, \boldsymbol{y}) = \{(x_1 - y_1)^2 + (x_2 - y_2)^2\}^{1/2}, \forall \; \boldsymbol{x}, \boldsymbol{y} \in \boldsymbol{R}^2, \qquad \ldots(1)$$

where $\boldsymbol{x} = (x_1, x_2)$ and $\boldsymbol{y} = (y_1, y_2)$. Then, $(\boldsymbol{R}^2, d)$ is a metric space (see Ex. 2, page 19.2).

Let $<x^{(n)}> \; = \; <x_1^{(n)}, x_2^{(n)}>$ be a Cauchy sequence in $\boldsymbol{R}^2$. Then, given $\varepsilon > 0$, there exists $n_0 \in \boldsymbol{N}$ such that

$$d(x^{(n)}, x^{(m)}) < \varepsilon, \qquad \forall \; m, n \geq n_0$$

$$\Rightarrow \quad (x_1^{(n)} - x_1^{(m)})^2 + (x_2^{(n)} - x_2^{(m)})^2 < \varepsilon^2, \text{ using (1)}$$

$$\Rightarrow \quad |x_1^{(n)} - x_1^{(m)}| < \varepsilon \quad \text{and} \quad |x_2^{(n)} - x_2^{(m)}| < \varepsilon, \; \forall \quad n, m \geq n_0$$

showing that $<x_1^{(n)}>$ and $<x_2^{(n)}>$ are Cauchy sequences of real number for $n = 1, 2, 3 \ldots$ Since $\boldsymbol{R}$ is complete, $x_1^{(n)} \to x_1$ and $x_2^{(n)} \to x_2$ as $n \to \infty$ in $\boldsymbol{R}$ with usual metric. Hence, for $\varepsilon > 0$

$$|x_1^{(n)} - x_1| < \varepsilon/\sqrt{2} \qquad \text{and} \qquad |x_2^{(n)} - x_2| < \varepsilon/\sqrt{2}, \qquad \forall \; n \geq n_0 \qquad \ldots(2)$$

Now, $\quad d(x^{(n)}, \boldsymbol{x}) = \{(x_1^{(n)} - x_1)^2 + (x_2^{(n)} - x_2)^2\}^{1/2}$, using (1)

$$< (\varepsilon^2/2 + \varepsilon^2/2)^{1/2}, \text{ using (2)}$$

Thus, $\quad d(x^{(n)}, \boldsymbol{x}) < \varepsilon, \quad \forall \; n \geq n_0 \quad$ where $\boldsymbol{x} = (x_1, x_2) \in \boldsymbol{R}^2$.

Hence every Cauchy sequence $<x^{(n)}>$ in $\boldsymbol{R}^2$ converges to a point of $\boldsymbol{R}^2$. Therefore, $(\boldsymbol{R}^2, d)$ is a complete metric space.

Ex. 36. (*b*). *Show that the n-dimensional Euclidean space $\boldsymbol{R}^n$* is complete. **[Delhi 1993]**

Sol. Let $\boldsymbol{R}^n$ be the set of n-tuples of real numbers, and let $d : \boldsymbol{R}^n \times \boldsymbol{R}^n \to \boldsymbol{R}$ be defined by

$$d(\boldsymbol{x}, \boldsymbol{y}) = \{(x_1 - y_1)^2 + (x_2 - y_2)^2 + \ldots + (x_n - y_n)^2\}^{1/2}, \; \forall \; \boldsymbol{x}, \boldsymbol{y} \in \boldsymbol{R}^n$$

where $\boldsymbol{x} = (x_1, x_2, \ldots, x_n)$ and $\boldsymbol{y} = (y_1, y_2, \ldots, y_n)$. Then, $(\boldsymbol{R}^n, d)$ is a metric space (see Ex. 3, page 19.3).

Let $< x^{(p)} > = < x_1^{(p)}, x_2^{(p)}, ..., x_n^{(p)} >$ be a Cauchy sequence in $\boldsymbol{R}^n$. Then given $\varepsilon > 0$, there exists $n_0 \in N$ such that

$$d(x^{(p)}, x^{(q)}) < \varepsilon, \qquad \forall \ p, q \geq n_0$$

$$\Rightarrow \qquad (x_1^{(p)} - x_1^{(q)})^2 + (x_2^{(p)} - x_2^{(q)})^2 + ... + (x_n^{(p)} - x_n^{(q)})^2 < \varepsilon^2, \text{ using (1)}$$

$$\Rightarrow \qquad | x_1^{(p)} - x_1^{(q)} | < \varepsilon, | x_2^{(p)} - x_2^{(q)} | < \varepsilon, \ ..., | x_n^{(p)} - x_n^{(q)} | < \varepsilon, \qquad \forall \ p, q \geq n_0$$

showing that $< x_1^{(p)} >, < x_2^{(p)} >, ..., < x_n^{(p)} >$ are Cauchy sequences of real numbers for $p = 1, 2, 3, ...$. Since $\boldsymbol{R}$ is complete, $x_1^{(p)} \to x_1, \ x_2^{(p)} \to x_2, ..., x_n^{(p)} \to x_n$ as $n \to \infty$ in $\boldsymbol{R}$ with usual metric. Hence, for $\varepsilon > 0$, we have

$$| x_1^{(p)} - x_1 | < \varepsilon / \sqrt{n}., \quad | x_2^{(p)} - x_2 | < \varepsilon / \sqrt{n}, ..., | x_n^{(p)} - x_n | < \varepsilon / \sqrt{n}, \ \forall \quad n \geq n_0 \qquad ...(2)$$

Now, $d(x^{(p)}, \boldsymbol{x}) = \{(x_1^{(p)} - x_1)^2 + (x_2^{(p)} - x_2)^2 + ... + (x_n^{(p)} - x_2)^2\}^{1/2}$, using (1)

$$< (\varepsilon^2 / n + \varepsilon^2 / n + ... + \varepsilon^2 / n)^{1/2}, \text{ using (2)}$$

Thus, $\qquad d(x^{(p)}, \boldsymbol{x}) < \varepsilon, \quad \forall \quad n \geq n_0$ where $\boldsymbol{x} = (x_1, x_2, ..., x_n) \in \boldsymbol{R}^n$.

Hence every Cauchy sequence $<x^{(p)}>$ *in* $\boldsymbol{R}^n$ converges to a point of $\boldsymbol{R}^n$. Therefore, $(\boldsymbol{R}^n, d)$ is a complete metric space.

Ex. 37. (*a*) *Write note on completion of a metric space*

(*b*) *State and prove the completion theorem*

Sol. (*a*). In Ex. 35, we have shown that the set of rational numbers $\boldsymbol{Q}$ with the usual metric in not a complete metric space. The set of rational number $\boldsymbol{Q}$ can be enlarged to the set $\boldsymbol{R}$ of real numbers such that $\boldsymbol{R}$ with usual metric is a complete metric space as shown in Ex. 3, Art 19.32. Again, in illustration 1 of Art. 3.17, we have shown that $\boldsymbol{Q}$ is dense in $\boldsymbol{R}$. This type of method can be employed to get a complete metric space from any incomplete metric space by the so called *completion theorem.*

(b). Completion theorem. *Every metric space is isometric to a dense subset of a complete metric space*

Proof We will complete the proof in the various steps:

Step I : Let (X, d) be any metric space and let S denote the set of all Cauchy sequences in X. Suppose that $<x_n>, <y_n> \in S$ are equivalent when $d(x_n, y_n) = 0$ as $n \to \infty$. We shall write $<x_n> \sim <y_n>$ to indicate that the sequences $<x_n>$ and $<y_n>$ are equivalent. Thus, we have

$$<x_n> \sim <y_n> \Leftrightarrow \lim_{n \to \infty} d(x_n, y_n) = 0 \qquad ...(1)$$

We now show that '$\sim$' is an equivalence relation.

Reflexivity. We have, $d(x_n, x_n) = 0, \ \forall \ n \in N$. So, $\lim_{n \to \infty} d(x_n, x_n) = 0 \Rightarrow < x_n > \sim < x_n >$, by (1)

Hence, the relation '$\sim$' is reflexive

Symmetry. Let $<x_n> \sim <y_n>$ This $\Rightarrow \ \lim_{n \to \infty} d(x_n, y_n) = 0 \ \Rightarrow \ \lim_{n \to \infty} d(y_n, x_n)$, using (1)

Thus, $<x_n> \sim <y_n> \Rightarrow \ <y_n> \sim <x_n>$, showing that '$\sim$' is a symmetric relation.

Transitivity. Let $\quad <x_n> \sim <y_n> \quad$ and $\quad <y_n> \sim <z_n>$.

Then, $\lim_{n\to\infty} d(x_n, y_n) = 0$ and $\lim_{n\to\infty} d(y_n, z_n) = 0$, by definition ...(2)

Now, by triangle inequality, $d(x_n, z_n) \le d(x_n, y_n) + d(y_n, z_n)$ ..(3)

Let $\varepsilon > 0$ be given. Then, results in (2) hold, if there exists $n_1, n_2 \in N$ such that

$$d(x_n, y_n) < \varepsilon/2, \quad \forall n \ge n_1 \quad \text{and} \quad d(y_n, z_n) < \varepsilon/2, \quad \forall n \ge n_2 \quad ...(4)$$

Let $n_0 = \max\{n_1, n_2\}$. Then results of (4) are true for $\forall n \ge n_0$. Using (4); (3) reduces to

$$d(x_n, z_n) < \varepsilon/2 + \varepsilon/2, \quad \forall n \ge n_0$$

Hence, $d(x_n, z_n) < \varepsilon, \ \forall n \ge n_0 \quad \Rightarrow \quad \lim_{n\to 0} d(x_n, z_n) = 0.$

Thus, $<x_n> \sim <y_n>$ and $<y_n> \sim <z_n> \Rightarrow <x_n> \sim <z_n>$

showing that the relation '~' is transitive.

Since the relation '~' is reflexive, symmetric and transitive, it follows that '~' is an equivalence relation. Hence this relation decomposes the set S into equivalence classes. Let X^* denote new space consisting of all these equivalence classes as its points. We now define a metric d* on X^* by setting

$$d^*(x^*, y^*) = \lim_{n\to\infty}(x_n, y_n), \quad \forall x^*, y^* \in X^*, \quad ...(5)$$

where for any two points x^* and y^*, we select any sequences $<x_n> \in x^*$ and $<y_n> \in y^*$.

We now proceed to prove that the limit on R.H.S. of (5) exists and is independent of the sequences chosen from x^* and y^*. We have, by triangle inequality,

$$d(x_n, y_n) \le d(x_n, x_m) + (x_m, y_m) + d(y_m, y_n)$$

so that $$d(x_n, y_n) - d(x_m, y_m) \le d(x_n, x_m) + d(y_m, y_n) \quad ...(6)$$

Interchanging m and n in (6), we have

$$d(x_m, y_m) - d(x_n, y_n) \le d(x_m, x_n) + d(y_n, y_m) \quad ...(7)$$

Since d is metrix on X, $d(x_m, x_n) = d(x_n, x_m)$ and $d(y_n, y_m) = d(y_m, y_n)$

Hence, (7) reduces to $d(x_m, y_m) - d(x_n, y_n) \le d(x_n, x_m) + d(y_m, y_n)$ ...(8)

Now, (7) and (8) $\Rightarrow$ $|d(x_m, y_m) - d(x_n, y_n)| \le d(x_n, x_m) + d(y_m, y_n)$ ..(9)

Since $<x_n>$ and $<y_n>$ are Cauchy sequences, the R.H.S of (9) can be made as small as we please. Thus, we find that the limit on R.H.S (5) exists because $\boldsymbol{R}$ is complete. Again. Let $<x_n> \sim <x'_n>$ and $<y_n> \sim <y'_n>$. Then, we have

$$|d(x_n, y_n) - d(x'_n, y'_n)| \le d(x_n\ x'_n) + d(y_n, y'_n) \to 0, \quad \text{as} \quad n \to \infty$$

Hence $$\lim_{n\to\infty} d(x_n, y_n) = \lim_{n\to\infty}(x'_n, y'_n)$$

showing that the limit on the R.H.S. of (5) is independent of the choice of the sequences in x^* and y^*.

Step 2: We now show that (X^*, d^*) is a metric space. We have

(*i*). $d^*(x^*, y^*) = \lim_{n\to\infty}(x_n, y_n) \ge 0, \ \forall x^*, y^* \in X^*$

(*ii*). $d^*(x^*, y^*) = 0 \Rightarrow \lim_{n\to\infty}(x_n, y_n) = 0 \Rightarrow <x_n> \sim <y_n> \Rightarrow x^* = y^*$

(*iii*). $d^*(x^*, y^*) = \lim_{n\to\infty}(x_n, y_n) = \lim_{n\to\infty} d(y_n, x_n) = d(y^*, x^*), \ \forall x^*, y^* \in X^*$

(*iv*). By triangle inequality, we have $d(x_n, y_n) \le d(x_n, z_n) + d(z_n, y_n)$ ...(10)

Now, $d^*(x^*, y^*) = \lim_{n\to\infty} d(x_n, y_n) \le \lim_{n\to\infty}[d(x_n, z_n) + d(z_n, y_n)]$, using (10)

Thus, $d^*(x^*, y^*) \le d^*(x^*, z^*) + d(z^*, y^*)$

From (*i*), (*ii*), (*iii*) and (*iv*), we find that (X^*, d^*) is a metric space.

Step 3: We now show that X is isometric to a sub space of X^*. To this end we define a mapping $f: X \to X^*$ which associates to an $x \in X$, the equivalence class of all Cauchy sequences each of which converges to x. Clearly, such a class is never empty because at least the constant sequence $<x, x, x, \ldots>$ belongs to this class. Now, by definition, we have

$$d^*(f(x), f(y)) = d(x, y), \ \forall \ x, y \in X, \text{ showing that } f \text{ is an isometry.}$$

Since f is one-one mapping and $f: X \to f(X)$ is onto mapping, it follows that X is isometric to a subspace $f(X)$ of X^*.

• **Step 4:** We now show that $f(X)$ is dense in X^*. Let $x^* \in X$ and $<x_n> \in x^*$. Since $<x_n>$ is a Cauchy sequence, for a given $\varepsilon > 0$, there exists $p \in N$ such that

$$d(x_n, x_m) < \varepsilon/2, \ \forall n, m \ge p \qquad \ldots(11)$$

In particular, for $m = p$, (11) $\Rightarrow$ $d(x_n, x_p) < \varepsilon/2$, $\forall n \ge p$.

Let us denote x_p by k. Then, $k \in X$ and the constant sequence $< k, k, k, \ldots> \in f(k)$ $(= k^*)$. Then, using definition of d^* given by (5), we get

$$d^*(x^*, k^*) = \lim_{n\to\infty}(x_n, \ k) \le \varepsilon/2 < \varepsilon \ \Rightarrow \ x^* \in \overline{f(X)}.$$

Thus, $X^* = \overline{f(X)}$, showing that $f(X)$ is dense in X^*.

Step 5: Finally, we now prove that (X^*, d^*) is a complete metric space.

Let $< x_n^* >$ be a Cauchy sequence in X^*. Since $\overline{f(X)} = X^*$, hence corresponding to each x_n^*, there must exist a $y_n^* \in f(X)$ such that

$$d^*(x_n^*, y_n^*) < 1/n \qquad \ldots(12)$$

Now, by triangle inequality in (X^*, d^*), we have

$$d^*(y_m^*, y_n^*) \le d^*(y_m^*, x_m^*) + d^*(x_m^*, x_n^*) + d^*(x_n^*, y_n^*)$$

Hence, $d^*(y_m^*, y_n^*) < 1/m + d^*(x_m^*, x_n^*) + 1/n$, using (12)

Thus, $d^*(y_m^*, y_n^*) < \varepsilon$, $\qquad \ldots(13)$

for sufficiently large values of m and n because $< x_n^* >$ is a Cauchy sequence.

Now, (13) $\Rightarrow$ $< y_n^* >$ in a Cauchy sequence in $f(X)$

Since f is one-one onto, we write $y_n = f^{-1}(y_n^*)$.

Since f is isometry, $<y_n>$ is a Cauchy sequence in X. Suppose that x* is the equivalence class in X such that $<y_n> \in x^*$. We now proceed to show that $< x_n^* >$ converges to $x^* \in X^*$

Now, $d^*(x_n^*, x^*) \le d^*(x_n^*, y_n^*) + d(y_n^*, x^*) < 1/n + d^*(y_n^*, x^*)$ $\qquad \ldots(14)$

Since $y_n^* \in f(X)$, hence the constant sequence $<y_n, y_n, y_n, \ldots> \in y_n^*$. Again, note that $< y_n > \in x^*$. Hence, (14) reduces tc

$$d^*(x_n^*, x^*) < 1/n + \lim_{n\to\infty} d(y_n, y_m) < \varepsilon, \qquad \ldots(15)$$

for sufficiently large value of n. Hence, (15) reduces to $\lim_{n\to\infty} x_n^* = x^* \in X^*$

Thus, each Cauchy sequence $< x_n^* >$ in X* converges to a point $x^* \in X^*$. Hence, (X^*, d^*) is a complete metric space.

Remark From the above theorem, it follows that every metric space has essentially one completion. The following theorem shows that if a metric space has more than one completion, then they must be isometric.

Theorem. *Completion of a metric space is unique upto isometry, i.e., any two completions of a metric space are isometric.*

Proof. Let (X, d) be a given metric space and let (X^*, d^*) and (X^{**}, d^{**}) be two completions of X. In particular, suppose that X^* and X^{**} both contain X as a dense subset and are complete. Define an isometry g between X^* and X^{**}. For each $x^* \in X^*$, since X is dense in X^*, there exists a sequence $<x_n>$ of points of X converging to x^*. Again, let $<x_n>$ be a Cauchy sequence in X^{**}. Since X^{**} is complete, $<x_n>$ converges to some point $x^{**} \in X^{**}$.

In view of the above construction, we define $f(x^*) = x^{**}$. We now prove that this construction is independent of the particular sequence $<x_n>$ converging to x^* and $f : X^* \to X^{**}$ is one-one onto mapping. Now, we have $f(x) = x, \forall x \in$ X. Again, let $x_n \to x^*$ in X^* and $x_n \to x^{**}$ in X^{**}; again, let $y_n \to y^*$ in X^* and $y_n \to y^{**}$ in X^{**}. Then, by definition,

$$d^*(x^*, y^*) = \lim_{n\to\infty} d(x_n, y_n) \quad \text{and} \quad d^{**}(x^{**}, y^{**}) = \lim_{n\to\infty} d(x_n, y_n).$$

$$\Rightarrow d^{**}(x^{**}, y^{**}) = d^*(x^*, y^*) \Rightarrow d^{**}(f(x^*), f(y^*)) = d^*(x^*, y^*),$$

showing that f is an isometry. Thus, any two completions of a metric space are isometric.

Ex. 38. (*a*) *Define the completion of a metric space and obtain the completion of all rational numbers with usual metric space.* **[Delhi Maths (H) 2006]**

(*b*) *Define the completion of an incomplete metric space. Explain the concept by an example* **[Delhi maths (H) 2008]**

Sol. (*a*). Completion of an incomplete metric space. Definition.

A metric space (X^*, d^*) is known as completion of a metric space (X, d) if

(*i*) (X^*, d^*) is complete

(*ii*) $X \subseteq X^*$ and $d^*(x, y) = d(x, y), \forall x, y \in X$

(*iii*) X is dense in X^*

Illustration by an example. Let $\boldsymbol{Q}$ and $\boldsymbol{R}$ be the set of all rational numbers and the set of all real numbers respectively. Let d and d* both denote usual metric on $\boldsymbol{Q}$ and $\boldsymbol{R}$. Then,

(*i*) $(\boldsymbol{R}, d^*)$ is a complete metric space (refer Ex. 3, Art 19.32)

(*ii*) $\boldsymbol{Q} \subset \boldsymbol{R}$ and $d^*(x, y) = |x - y| = d(x, y), \forall x, y \in \boldsymbol{Q}$

(*iii*) $\boldsymbol{Q}$ is dense in $\boldsymbol{R}$ as shown in illustration number 1 in Art 3.17.

(*b*) Refer part (*a*).

Ex. 39. *Prove that every compact metric space is totally bounded (using only the definition of compactness and totally boundedness of a metric space.). Is the converse true ?* **[Delhi Maths (H) 2009]**

Sol. Let (X, d) be a compact metric space. Then, by definition (refer Art. 19.37), every open cover of X admits of a finite subcover.

Consider the open cover $\{S(x,\varepsilon): x \in X\}$ of X. Since X is compact, by definition, this open cover must admit a finite subcover, say, $\{S(x_1, \varepsilon), S(x_2, \varepsilon), \ldots, S(x_n, \varepsilon)\}$ of X. Hence, $\cup\{S(x_n, \varepsilon): n = 1,2,\ldots,n\} = X$, showing that $\{x_1, x_2, \ldots, x_n\}$ is an ε-net.

Therefore, (X, d) is totally bounded (refer definition in Art. 19.37).

Converse of the given result is not always true. To this end we know that the metric space]0, 1[with usual metric is totally bounded but not complete. Hence, it is not compact (since every compact metric space is always complete)

Ex. 40. *Define a totally bounded metric space. Give an example that is not totally bounded with justification.* **[Delhi Maths (H) 2008]**

Sol. Totally bounded metric space Definition

Let (X, d) be a metric space. Then X is said to be totally bounded if given $\varepsilon > 0$, there exists a finite number of subsets $G_1, G_2, \ldots G_n$ of X such that $\delta(G_n) < \varepsilon$ $(n = 1, 2, \ldots n)$ and such that $X \subset \bigcup_{i=1}^{n} G_i$. For definition of $\delta(G_n)$ refer Art. 19.6, page 19.4.

Example of a metric space which is not totally bounded

Let X be an infinite set with discrete metric. Choose $\varepsilon = 1$. Then, the set $S(x, 1/2)$ is the singleton set $\{x\}$ because $d(x, y) = 1$ when $x \neq y$ for two points in X. Hence the only non-empty subsets of X with diameter less than 1 are the singleton subsets. Since X is infinite, X cannot be written as the union of a finite number of sets with diameters less than 1. Hence, the infinite set X with discrete metric cannot be totally bounded.

Ex. 40. (*b*) *Show that if (X, d) is a totally bounded metric space then X is bounded. Is the converse true?*

Sol. Let the given metric space be totally bounded. Hence, for a given $\varepsilon > 0$, we can find a finite set of points $\{x_1, x_2, \ldots, x_n\}$ in X such that

$$\cup\{S(x_i, \varepsilon): i = 1,2,3,\ldots n\} = X$$

In particular, choosing $\varepsilon = 1$, we have

$$\cup\{S(x_i, 1): i = 1,2,3,\ldots n\} = X \qquad \ldots(1)$$

Since finite union of bounded sets is finite, it follows the L.H.S. of (1) is a bounded set. Hence, from (1), we see that X is bounded.

The converse of the above result is not true in general as shown in the following example.

Let X be an infinite set with discrete metric d. Then, by definition of discrete metric on X, there exists a real number $1 > 0$ such that $d(x, y) \leq 1$, $\forall\, x, y \in X$. Hence (X, d) is a bounded metric space. But (X, d) is not totally bounded as shown in Ex 40 (*a*).

Ex. 41. (*a*). *Show that any bounded subset of **R** is totally vounded in the usual metric*

(*b*). *Prove that the interval [0, 1] with the metric d, defined as $d(x, y) = |x - y|$, $\forall\, x, y \in [0,1]$ is a totally bounded metric space.* **[Delhi Maths (H) 2009]**

(*c*). Prove that the interval]0, 1[with the metric d, defined as $d(x, y) = |x-y|$, $\forall\; x, y \in]0,1[$ is a totally bounded metric space.

Sol. (*a*). Let G be a bounded subset of ***R***. Let $<x_n>$ be an infinite sequence of G. Since each subset of a bounded set is bounded and G is bounded, it follows that $<x_n>$ is a bounded sequence. Again, since $<x_n>$ is bounded, it must contain a convergent subsequence $<x_{n_k}>$. Since every convergent sequence is a Cauchy sequence, it follows that $<x_{n_k}>$ is a Cauchy sequence.

Thus, each sequence $<x_n>$ in G admits of a Cauchy subsequence $< x_{n_k} >$. Hence G is totally bounded (refer theorem VIII, Art 19.37).

Remark Since a totally bounded set in any metric space is bounded, the above result shows that boundedness and total boundedness are equivalent in $\boldsymbol{R}$.

(*b*). Take G = [0, 1] so that G is a bounded subset of $\boldsymbol{R}$. Hence, by part (*a*), [0, 1] is totally bounded Here ($\boldsymbol{R}$, d) is usual metric space.

(*c*). Take G =]0, 1[and proceed as in part (*b*)

Ex. 42. *Prove that every totally bounded metric space is separable. Is the converse true? Justify.* **[Delhi Math (H) 2007]**

Sol. Let (X, d) be a totally bounded metric space. Then, for each $n \in \boldsymbol{N}, X$ has a finite $1/n$ – net, and we denote it by A_n. Let $A = \bigcup\limits_{n=1}^{\infty} A_n$.

We now show that A is countable dense subset of X. Since A is a union of countably many finite sets, therefore A must be countable. In order to show that A is dense subset of X, we proceed to show that $\overline{A} = X$.

Let $x \in X$ and $r > 0$. Choose $n \in \boldsymbol{N}$ such that $1/n < r$. Since A_n is finite, let $A = \{a_1, a_2, \ldots, a_m\}$. Then, we have

$$X = \bigcup_{i=1}^{m} S(a_i, 1/n) \qquad \ldots(1)$$

Let x be an arbitrary point of X. Then, in view of (1), $x \in S(a_i, 1/n)$, for some $i = 1, 2, \ldots, m$ and hence

$$d(x, a_i) < 1/n < r \quad \Rightarrow \quad a_i \in S(x, r) \quad \Rightarrow \quad S(x, r) \cap A \neq \phi, \quad \text{as} \quad a_i \in A_n \subset A$$

showing that every open sphere intersects A, and hence $\overline{A} = X$, i.e., A is dense subset of X. Since X has a countable dense subset, therefore (X, d) is separable.

The converse of the above result, in general, is not true. To this end, consider the following example :

The real line with usual metric space is separable because the set $\boldsymbol{Q}$ of rational numbers is a countable dense subset of $\boldsymbol{R}$. But $\boldsymbol{R}$ is not bounded and hence it is not totally bounded because boundedness is equivalent to total boundedness in Euclidean spaces.

Ex. 43. (*a*) Is the Euclidean space $\boldsymbol{R}^2$ separable **[Delhi Maths (H) 2009]**

Sol. Since the set of rational number $\boldsymbol{Q}$ is countable and since the cartesian product of two countable sets is countable (refer page 4.5), it follows that $\boldsymbol{Q} \times \boldsymbol{Q}$ is a countable subset of $\boldsymbol{R}^2$. We now show that $\boldsymbol{Q} \times \boldsymbol{Q}$ *i.e.*, $\boldsymbol{Q}^2$ is a dense subset of $\boldsymbol{R}^2$. Take a point $(p_1, p_2) \in \boldsymbol{R}^2 - \boldsymbol{Q}^2$ and any $\varepsilon > 0$. Since $\boldsymbol{Q}$ is dense subset of $\boldsymbol{R}$, hence we can find rational numbers r_1 and r_2 such that

$$|p_1 - r_1| < \varepsilon/\sqrt{2} \qquad \text{and} \qquad |p_2 - r_2| < \varepsilon/\sqrt{2}$$

Hence, $\{(p_1 - r_1)^2 + (p_2 - r_2)^2\}^{1/2} < \varepsilon \quad \Rightarrow \quad d(\boldsymbol{p}, \boldsymbol{r}) < \varepsilon,$ ...(1)

where d is the usual metric on $\boldsymbol{R}^2$ and $\boldsymbol{p} = (p_1, p_2)$ and $\boldsymbol{r} = (r_1, r_2)$.

(1) shows every open sphere centered at $\boldsymbol{p}$ contains a point of $\boldsymbol{Q}^2$. Therefore, $\boldsymbol{p} \in cl\, \boldsymbol{Q}^2$.

Since $\boldsymbol{p} \in \boldsymbol{R}^2 \sim \boldsymbol{Q}^2 \Rightarrow \boldsymbol{p} \in cl\, \boldsymbol{Q}^2$, it follows that $cl\, \boldsymbol{Q}^2 = \boldsymbol{R}^2$. Hence $\boldsymbol{Q}^2$ is a dense subset of $\boldsymbol{R}^2$.

Since there exists a countable subset of $\boldsymbol{R}^2$, hence $\boldsymbol{R}^2$ is separable.

Ex. 43. (*b*) *Show that the n - dimensional Euclidean space is separable.*

Hint : Prove as explained in Ex. 43 (*a*)

Ex. 44. *Is the discrete metric space connected.* **[Delhi Maths (H) 2009]**

Sol. Let (X, d) be discrete metric space where X contains more than one point. Let $p \in X$. Since (X, d) is discrete metric space, subset $\{p\}$ of X is both open and closed (See Ex. 9). Thus, there exists a proper subset $\{p\}$ of X which is both open and closed. Hence (X, d) is disconnected (refer theorem V, Art 19.38)

Ex. 45. (*a*) Prove that the set of all rational numbers with usual metric is not connected **[Delhi Maths (H) 2008]**

(*b*) *Is the set S of irrationals, a connected subset of the metric space* $(\mathbf{R}, d)$, *where* $\mathbf{R}$ *is the set of real numbers and d the usual metric on* $\mathbf{R}$? *Justify. If not find a disconnection of S and justify your answer* **[Delhi Maths [H] 2009]**

(*c*) *Show that the metric space* $X = \mathbf{R} - \{0\}$ *with usual metric is disconnected.*

Sol. (*a*). $\boldsymbol{Q}$, i.e., the set of all rational number is a subset of $\mathbf{R}$, i.e., the set of real numbers. But $\boldsymbol{Q}$ is not an interval by definition of an interval of a real line. Hence $\boldsymbol{O}$ is not connected (refer theorem X1, Art 19.38)

(*b*). Try yourself

(*c*). Since X is union of two non-empty open sets $(-\infty, 0)$ and $(0, \infty)$, it follows that X is disconnected (refer theorem VII, Art 19.38)

Ex. 46. *Let* $\{A_\alpha : \alpha \in \Delta\}$ *be a family of connected subsets of a metric space X and* $\cap \{A_\alpha : \alpha \in \Delta\} \neq \phi$. *Prove that* $\cup \{A_\alpha : \alpha \in \Delta\}$ *is connected* **[Delhi Maths (H) 2006, 08]**

Sol. Let
$$E = \cup\{A_\alpha : \alpha \in \Delta\} \qquad \ldots(1)$$

Let, if possible, E be disconnected. Then there exist two non-empty subsets A and B of X such that
$$E = A \cup B, \qquad A \cap \overline{B} = \phi, \qquad \overline{A} \cap B = \phi \quad \text{and} \quad A \cap B = \phi \qquad \ldots(2)$$

Given that
$$\cap \{A_\alpha : \alpha \in \Delta\} \neq \phi \qquad \ldots(3)$$

(3) implies that there exists at least one point p such that
$$p \in \cap\{A_\alpha : \alpha \in \Delta\} \qquad \ldots(4)$$

Clearly $p \in E$. But $E = A \cup B$ and $A \cap B = \phi$. Hence either $p \in A$ or $p \in B$. Without loss of generality, suppose that $p \in A$. Again, from (4), $p \in A_\alpha, \forall \alpha \in \Delta$.

Now, $\quad p \in A$ and $p \in A_\alpha, \;\forall\; \alpha \in \Delta \;\Rightarrow\; A_\alpha \cap A \neq \phi, \;\forall\; \alpha \in \Delta$

Since each A_α is connected subset of X such that $A_\alpha \subset E$, i.e., $A_\alpha \subset A \cup B$, where A and B are two non-empty separated subsets of X, hence (by theorem II, Art. 19.38), we have
$$A_\alpha \subset A \qquad \text{or} \qquad A_\alpha \subset B$$

Since A and B are disjoint and $A_\alpha \cap A \neq \phi$, $\forall \alpha \in \Delta$, we have
$$A_\alpha \subset A, \quad \forall \alpha \in \Delta \quad \Rightarrow \quad \cup\{A_\alpha : \alpha \in \Delta\} \subset A \quad \Rightarrow \quad E \subset A.$$

Thus, $A \cup B \subset A$ and hence $B = \phi$, which contradicts the hypothesis that B is non-empty.

Since this contradiction arises due to assuming that E is disconnected, hence E, i.e., $\cup\{A_\alpha : \alpha \in \Delta\}$ is connected.

Ex. 47(A) *Select (a), (b), (c) or (d) whichever is correct:*

In a metric space consider the following statements (i) The intersection of finitely many open spheres is an open sphere (ii) The intersection of countable many open sets is an open set. Then

(a) (i) and (ii) are both false *(b) (i) and (ii) both true*

(c) (i) is true (ii) is false *(d) (ii) is true (i) is false.* **[Agra 2005]**

Sol. Ans. (*c*). Use theorems III and (IV) of Art 19.11

Ex.47 (B). Two metrics d_1 and d_2 on the same set X are said to be equivalent metrics if they give rise to the same family of

(*a*) closed sets (*b*) compact sets (*c*) connected sets (*d*) open sets **(Agra 2009)**

Sol. Ans. (*d*). Refer Art 19.12, page 197

Ex. 47 (C). In a metric space (X, d) the distance between two non-empty subsets A and B of X, i.e., $d(A, B)$ is given by

(*a*) $\inf \{d(x, y) : x \in A, y \in A\}$ (*b*) $\sup \{d(x, y) : x \in A, y \in B\}$

(*c*) None of these (*Meerut 2011*)

Sol. Ans. (*a*) Refer Art. 19.6, page 19.4

Ex. 48. *State whether the following statement is true or false.*

Let A, B be nun-empty subsets of a metric space (X, d). The distance between A and B denoted by d(A, B) is defined as $d(A, B) = \inf [d(a, b) : a \in A, b \in B]$ **[Agra 2008]**

Sol. True. Refer Art. 19.6

Ex. 49. *Fill up gaps* :

(*i*) *Let* (X, d_1) *and* (Y, d_2) *be two metric spaces. A mapping f of X into Y is continuous on X iff* $f^{-1}(H)$ *is ... in X for every open H set in Y.* **[Agra 2004, 07]**

(*ii*) In a metric space (X, d) each open sphere is ... set.

(*iii*) If d is a metric on X, then the ordered pair (X, d) is called... **[Agra 2007]**

(*iv*) A metric space (X, d) is called ... iff there exists a real number k such that $d(x, y) \leq k$, for all $x, y \in X$. **[Agra 2008]**

Ans. (*i*) open (*ii*) open (*ii*)metric space (*iv*) bounded

50. *Define Hereditary property of a metric space.* (*Meerut 2011*)

Sol. A property of a metric space is said to be *hereditary* if and only if every subspace of that has that property.

Since every subspace of a separable metric space is separable, hence separability is a hereditary property on a metric space.

Chapter

20

Beta and Gamma Functions

20.1. INTRODUCTION

In this chapter we propose to discuss the Gamma and Beta functions. These functions arise in the solution of physical problems and are also of great importance in various branches of mathematical analysis. The reader in strongly advised to master important results of these functions in order to understand the topics coverved by this book.

20.2. Euler's integrals. Beta and Gamma functions

Beta function. Definition [Delhi B.Sc. (Hons) II 2011; Delhi B.A. (Prog) III, 2010; Agra 2012]

The definite integral $\int_0^1 x^{m-1}(1-x)^{n-1}\,dx,\quad \text{for } m>0,\ n>0$

is known as the *Beta function* and is denoted by $B(m, n)$ [read as "Beta m, n"]. Beta function is also called the *Eulerian integral of the first kind.*

Thus, $$B(m, n) = \int_0^1 x^{m-1}(1-x)^{n-1}\,dx,\quad m>0,\quad n>0. \qquad \text{... (1)}$$

Gamma function. Definition **[Delhi B.A. (Prog) III, 2010, 11, 15; Agra 2000, 2012] [Meerut 2015, 2018]**

The definite integral $\int_0^\infty e^{-x}x^{n-i}\,dx,\ \text{for } n>0$

is known as the *Gamma function* and is denoted by $\Gamma(n)$ [read as "Gamma n"]. Gamma function is also called the *Eulerain integral of the second kind.*

Thus, $$\Gamma(n) = \int_0^\infty e^{-x}x^{n-1}\,dx,\ \textit{for } n>0 \qquad \text{...(2)}$$

Remark. The integral (1) is valid only for $m>0$ and $n>0$ and the integral (2) is valid only for $n>0$, because it is for just these values of m and n that the above integrals are convergent.

20.3. Properties of Gamma function. [Agra 1999]

I. To show that $\Gamma(1) = 1$. **[Delhi B.A. (Prog) III 2011, 2012]**

Proof. By the definition of Gamma function, $$\Gamma(n) = \int_0^\infty e^{-x}x^{n-1}\,dx,\ n>0 \qquad \text{... (1)}$$

From (1), $$\Gamma(1) = \int_0^\infty e^{-x}x^{1-1}\,dx = \int_0^\infty e^{-x}\,dx = \left[-e^{-x}\right]_0^\infty = 1.$$

II. To show that $\Gamma(n+1) = n\Gamma(n),\ n>0$. **[Agra 2010; Lucknow 2010; Meerut 2016, 2019]**

Proof. We have from the definition of Gamma function,

$$\Gamma(n+1) = \int_0^\infty e^{-x}x^{n+1-1}\,dx = \int_0^\infty x^n e^{-x}\,dx$$

$$= \left[x^n(-e^{-x})\right]_0^{\infty} - \int_0^{\infty}(nx^{n-1})(-e^{-x})\,dx, \text{ on integrating by parts}$$

$$\therefore \qquad \Gamma(n+1) = -\lim_{x\to\infty}\frac{x^n}{e^x} + 0 + n\int_0^{\infty} e^{-x}x^{n-1}\,dx. \qquad ...(1)$$

$$\left(\because \lim_{x\to 0} x^n e^{-x} = 0 \text{ as } n > 0\right)$$

Now, we have

$$\lim_{x\to\infty}\frac{x^n}{e^x} = \lim_{x\to\infty}\frac{x^n}{1+\frac{x}{1!}+\frac{x^2}{2!}+...+\frac{x^n}{n!}+\frac{x^{n+1}}{(n+1)!}+...} = \lim_{x\to\infty}\frac{1}{\frac{1}{x^n}+\frac{1}{1!\,x^{n-1}}+...+\frac{1}{n!}+\frac{x}{(n+1)!}+...} = 0$$

Also, by definition $\qquad \Gamma(n) = \int_0^{\infty} e^{-x}x^{n-1}\,dx.$

Using the above facts (1) reduces to $\qquad \Gamma(n+1) = n\,\Gamma(n).$

III. If n is a non-negative integer, then $\Gamma(n+1) = n\,!$.

Proof. We known that for $n > 0$, we have (from property II)

$$\Gamma(n+1) = n\Gamma(n) = n\,\Gamma(n-1+1) = n\,(n-1)\Gamma(n-1), \text{ by property II again}$$

$$= n\,(n-1)\,(n-2)\,\Gamma\,(n-2), \text{ by property II again}$$

$$= n\,(n-1)\,(n-2)...3\cdot 2\cdot 1\,\Gamma\,(1)$$

(by repeated use of property II and the fact that n is positive integer)

$$= n\,!, \text{ as } \Gamma(1) = 1$$

Remark. *Gauss's Pi-function* is denoted by $\Pi\,(n)$ and is defined by $\Pi\,(n) = \Gamma(n+1)$. When n is +ve integer, $\Pi\,(n) = n\,!$.

20.4. Extension of definition of Gamma function $\Gamma(n)$ for n < 0.

When $n > 0$, we known that $\qquad \Gamma(n+1) = n\,\Gamma(n)$

so that $\qquad \Gamma(n) = \Gamma(n+1)/n. \qquad ...(1)$

Let $-1 < n < 0$. Then $-1 < n \Rightarrow n+1 > 0$ so that $\Gamma(n+1)$ is well defined by definition 20.2 and so R.H.S. of (1) is well defined. Thus $\Gamma(n)$ is defined for $-1 < n < 0$ by (1). Similarly, $\Gamma(n)$ is given by (1) for $-2 < n < -1$, $-3 < n < -2$ and so on. Thus (1) defines $\Gamma(n)$ for all values of n except $n = 0, -1, -2, -3, ...$

Property : To show that $\Gamma(n) = \infty$, if n is zero or a negative integer.

Proof. Putting $n = 0$ in (1), we get $\qquad \Gamma(0) = \Gamma(1)/0 \quad\Rightarrow\quad \Gamma(0) = \infty \qquad ...(2)$

Again, putting $n = -1$ in (1), we get $\qquad \Gamma(-1) = \frac{\Gamma(0)}{-1} = \infty$, by (2) $\qquad ...(3)$

Next putting $n = -2$ in (1) and using (3), we get $\qquad \Gamma(-2) = \frac{\Gamma(-1)}{-2} = \infty,$

and so on. Thus, we find that $\Gamma(n)$ is ∞ if n is zero or negative integer.

and so on. Thus, we find that $\Gamma(n)$ is ∞ if n is zero or negative integer.

20.5. Theorem. To show that $\Gamma(1/2)=\sqrt{\pi}$. **[Agra 2000, 05, 09; Meerut 2007]**
[Delhi B.A. (Prog) III, 2010, 2015]

Proof. From definition of gamma function, $\Gamma(n)=\int_0^\infty e^{-t}t^{n-1}dt,\ n>0$ ... (1)

Replacing n by 1/2 in (1), we have $\Gamma(1/2)=\int_0^\infty e^{-t}t^{-1/2}dt=2\int_0^\infty e^{-u^2}du$ ... (2)

[Putting $t=u^2$ so that $dt=2u\ du$]

$\therefore\ \Gamma(1/2)=2\int_0^\infty e^{-x^2}dx$ and $\Gamma(1/2)=2\int_0^\infty e^{-y^2}dy$. ... (3)

[Limits remaining the same, we can write x or y as the variable in the integrand of (2)].
Multiplying the corresponding sides of two equations of (3), we get

$$[\Gamma(1/2)]^2=\left(\int_0^\infty e^{-x^2}dx\right)\left(\int_0^\infty e^{-y^2}dy\right)=4\int_0^\infty\int_0^\infty e^{-(x^2+y^2)}dx\,dy=4\int_{\theta=0}^{\pi/2}\int_{r=0}^\infty e^{-r^2}r\,d\theta\,dr$$

(on changing the variables to polar co-ordinates (r,θ), where $x=r\cos\theta$, $y=r\sin\theta$ so that $x^2+y^2=r^2$ and $dx\,dy=r\,d\theta\,dr$. The area of integration is the positive quadrant of xy– plane).

$$\therefore\ [\Gamma(1/2)]^2=2\int_0^{\pi/2}\left\{\int_0^\infty 2e^{-r^2}r\,dr\right\}d\theta=2\int_0^{\pi/2}\left\{\int_0^\infty e^{-v}dv\right\}d\theta,\ \text{putting } r^2=v \text{ so that } 2r\,dr=dv$$

Hence, $[\Gamma(1/2)]^2=2\int_0^{\pi/2}\left[-e^{-v}\right]_0^\infty d\theta=2\int_0^{\pi/2}d\theta=2[\theta]_0^{\pi/2}=\pi$

Thus, $[\Gamma(1/2)]^2=\pi$ so that $\Gamma(1/2)=\sqrt{\pi}$. ... (4)

Remark. From (3) and (4), $2\int_0^\infty e^{-x^2}dx=\sqrt{\pi}$ **(Kanpur 2011)** ... (5)

20.6. Transformation of Gamma function.

Form I. To show that $\Gamma(n)=\dfrac{1}{n}\int_0^\infty e^{-x^{1/n}}dx,\ n>0$. **(Meerut 1996)**

Proof. By definition, $\Gamma(n)=\int_0^\infty e^{-x}x^{n-1}dx,\quad n>0$. ... (*i*)

Put $x^n=t$ so that $nx^{n-1}\ dx=dt$. Then (*i*) gives

$$\Gamma(n)=\frac{1}{n}\int_0^\infty e^{-t^{1/n}}dt \quad\text{or}\quad \Gamma(n)=\frac{1}{n}\int_0^\infty e^{-x^{1/n}}dx. \quad ...(ii)$$

Particular Case. Put $n=1/2$ in (*ii*). Then, $\Gamma(1/2)=2\int_0^\infty e^{-x^2}dx$. ... (*iii*)

Form II. Show that $\int_0^\infty e^{-kx}x^{n-1}dx=\dfrac{\Gamma(n)}{k^n},\ n>0,\ k>0$. **(Meerut 2011; Agra 2010)**

Proof. By definition, $\Gamma(n)=\int_0^\infty e^{-x}x^{n-1}dx,\quad n>0$. ... (*i*)

Put $x=kt$ so that $dx=k\ dt$, Then (*i*) gives

or $$\int_0^\infty e^{-kx} x^{n-1} dx = \frac{\Gamma(n)}{k^n}, \quad n > 0, \quad k > 0.$$

Form III. To show that $\Gamma(n) = \int_0^1 \left(\log \frac{1}{x} \right)^{n-1} dx, \quad n > 0.$ **[Meerut 1997, Kumaun 2000, Agra 2008; Garhwal 2003 Purvanchal 2005, Delhi Physics (H) 2001]**

Proof. By definition, $$\Gamma(n) = \int_0^\infty e^{-x} x^{n-1} dx. \quad \text{... } (i)$$

Put $e^{-x} = t$ so that $-e^{-x}\, dx = dt$. Then (i) gives

$$\Gamma(n) = -\int_1^0 \left(\log \frac{1}{t} \right)^{n-1} dx = \int_0^1 \left(\log \frac{1}{t} \right)^{n-1} dt, \quad \text{as} \quad e^{-x} = t \quad \Rightarrow \quad e^x = \frac{1}{t} \quad \Rightarrow \quad x = \log \frac{1}{t}$$

$$\therefore \quad \Gamma(n) = \int_0^1 \left(\log \frac{1}{x} \right)^{n-1} dx, \quad n > 0.$$

Form IV. To show that $\Gamma(n) = 2\int_0^\infty e^{-x^2} x^{2n-1} dx, \quad n > 0.$

Proof. By definition, $$\Gamma(n) = \int_0^\infty e^{-x} x^{n-1} dx. \quad \text{... } (i)$$

Put $x = t^2$ so that $dx = 2t\, dt$. Then (i) gives

$$\Gamma(n) = \int_0^x e^{-t^2} (t^2)^{n-1} 2t\, dt \quad \text{or} \quad \Gamma(n) = 2\int_0^\infty e^{-t^2} t^{2n-1} dt \quad \text{or} \quad \Gamma(n) = 2\int_0^\infty e^{-x^2} x^{2n-1} dx.$$

20.7. Solved examples based on Gamma function

Ex. 1. *Evaluate* (i) $\int_0^\infty x^4 e^{-x} dx$ $\qquad (ii)$ $\int_0^\infty x^6 e^{-2x} dx.$

Sol. (i) $\int_0^\infty x^4 e^{-x} dx = \int_0^\infty e^{-x} x^{5-1} dx = \Gamma(5) = 4! = 24$, by definition of Gamma function

(ii) Let $I = \int_0^\infty x^6 e^{-2x} dx$. Put $2x = t$ so that $dx = \frac{1}{2} dt$. Then, we have

$$I = \int_0^\infty \left(\frac{t}{2} \right)^6 e^{-t} \cdot \frac{1}{2} dt = \frac{1}{2^7} \int_0^\infty e^{-t} t^{7-1} dt = \frac{1}{2^7} \Gamma(7), \text{ by definition of Gamma function.}$$

$$= (1/2^7) \times 6! = 45/8$$

Ex. 2. *Compute* (i) $\Gamma\left(-\frac{1}{2}\right)$ **[Agra 2006, 10, 11]** $\quad (ii)$ $\Gamma\left(-\frac{3}{2}\right)$ $\quad (iii)$ $\Gamma\left(-\frac{5}{2}\right)$.

(iv) Prove that $\Gamma(-9/2) = -(32\sqrt{\pi})/945$ **[Kanpur 2004]**

Sol. We know that $$\Gamma(n) = \Gamma(1+n)/n \quad \text{... } (i)$$

Part (*i*). Putting $n = -\frac{1}{2}$ in (1), $\Gamma\left(-\frac{1}{2}\right) = \frac{\Gamma(1/2)}{(-1/2)} = \frac{\sqrt{\pi}}{(-1/2)} = -2\sqrt{\pi}$, as $\Gamma\left(\frac{1}{2}\right) = \sqrt{\pi}$

Part (*ii*). Putting $n = -3/2$ in (1), we have

$$\Gamma\left(-\frac{3}{2}\right) = \frac{\Gamma\left(-\frac{1}{2}\right)}{(-3/2)} = -\frac{2}{3}\Gamma\left(-\frac{1}{2}\right) = -\frac{2}{3}\left(-2\sqrt{\pi}\right) = \frac{4\sqrt{\pi}}{3}, \text{ using part } (i)$$

Part (*iii*). Putting $n = -\dfrac{5}{2}$ in (*i*), $\Gamma\left(-\dfrac{5}{2}\right) = \dfrac{\Gamma\left(-\dfrac{3}{2}\right)}{(-5/2)} = -\dfrac{2}{5}\left(\dfrac{4\sqrt{\pi}}{3}\right)$, using part (*ii*)

Part (*iv*) Left as an exercise

Ex. 3. *If n is a positive integer, prove that* $2^n\Gamma(1+1/2) = 1\cdot 3\cdot 5 ... (2n+1)\sqrt{\pi}$.

[Delhi Physics (H) 2002]

Sol. Using the formula $\Gamma(n+1) = n\Gamma(n), \quad n > 0.$ ... (1)

$$\Gamma\left(n+\frac{1}{2}\right) = \Gamma\left(n-\frac{1}{2}+1\right) = \left(n-\frac{1}{2}\right)\Gamma\left(n-\frac{1}{2}\right) = \left(n-\frac{1}{2}\right)\Gamma\left(n-\frac{3}{2}+1\right), \text{ using (1)}$$

$$= \left(n-\frac{1}{2}\right)\left(n-\frac{3}{2}\right)\Gamma\left(n-\frac{3}{2}\right) = \frac{2n-1}{2}\cdot\frac{2n-3}{2}\cdot\Gamma\left(\frac{2n-3}{2}\right) = \frac{2n-1}{2}\cdot\frac{2n-3}{2}\cdots\frac{5}{2}\cdot\frac{3}{2}\cdot\frac{1}{2}\,\Gamma\left(\frac{1}{2}\right)$$

[By repeated application of (1) and noting that $(2n-1), (2n-3), ...$ are all odd].

$$\therefore \quad \Gamma\left(n+\frac{1}{2}\right) = \frac{(2n-1)\,(2n-3)...5\cdot 3\cdot 1}{2^n}\sqrt{\pi}, \quad \text{as} \quad \Gamma(1/2) = \sqrt{\pi}$$

$$\therefore \quad 2^n\Gamma(n+1/2) = 1\cdot 3\cdot 5 ... (2n-1)\sqrt{\pi}$$

Ex. 4. *If n is a positive integer and m > –1, prove that* $\displaystyle\int_0^1 x^m(\log x)^n\,dx = \frac{(-1)^n\,n!}{(m+1)^{n+1}}$.

Sol. Let $I = \displaystyle\int_0^1 x^m(\log x)^n\,dx$. Put $\log x = -t$ so that $x = e^{-t}$ and $dx = -e^{-t}\,dt$.

$$\therefore \quad I = \int_\infty^0 (e^{-t})^m(-t)^n(-e^{-t}dt) \qquad [\because \log 0 = -\infty \text{ and } \log 1 = 0]$$

$$= (-1)^n\int_0^\infty e^{-(m+1)t}t^{(n+1)-1}dt = (-1)^n\cdot\frac{\Gamma(n+1)}{(m+1)^{n+1}}, \text{ provided } m+1>0 \text{ i.e., } m>-1$$

[using form II of Art. 20.6]

$$= \frac{(-1)^n\,n!}{(m+1)^{n+1}}. \qquad [\because \Gamma(n+1) = n!, \; n \text{ being the integer}]$$

Ex. 5 (*a*) *With certain limitations on the values of a, b, m and n, prove that*

$$\int_0^\infty\int_0^\infty e^{-(ax^2+by^2)}x^{2m-1}y^{2n-1}dx\,dy = \frac{\Gamma(m)\Gamma(n)}{4a^mb^n}.$$

Sol. Let $\quad I = \displaystyle\int_0^\infty\int_0^\infty e^{-(ax^2+by^2)}x^{2m-1}y^{2n-1}dx\,dy$ ... (1)

or $\quad I = \displaystyle\int_0^\infty e^{-ax^2}x^{2m-1}dx \times \int_0^\infty e^{-by^2}y^{2n-1}dy = I_1\times I_2$ ... (2)

where $\quad I_1 = \displaystyle\int_0^\infty e^{-ax^2}x^{2m-1}dx$ ... (3)

and $\quad I_2 = \displaystyle\int_0^\infty e^{-by^2}y^{2n-1}dy.$ ... (4)

Put $ax^2 = t$, *i.e.*, $x = (t/a)^{1/2}$ so that $dx = dt/2\sqrt{at}$. Then (3) becomes

$$I_1 = \int_0^\infty e^{-t}\left[\frac{t}{a}\right]^{(2m-1)/2}\frac{dt}{2\sqrt{at}} = \frac{1}{2a^m}\int_0^\infty e^{-t}t^{m-1}dt$$

$= \Gamma(m)/2a^m$, by definition of Gamma function, taking $m > 0,\ a > 0$

Similarly, $I_2 = \Gamma(n)/2b^n$, if $n > 0;\ b > 0$

$\therefore$ From (1) and (2), we obtain $I = I_1 \times I_2 = \dfrac{\Gamma(m)\,\Gamma(n)}{4a^m b^n}$.

Ex. 5 (*b*) *Show that* $\displaystyle\int_0^\infty e^{-ax^2}x^{2n-1}dx = \frac{\Gamma(n)}{2a^n}$ **(Purvanchal 2007)**

[**Hint.** Same as I_1 of Ex. 5(*a*)]

Ex. 6. Evaluate $\displaystyle\int_0^1 \frac{dx}{\sqrt{(-\log x)}}$ **[Meerut 1996, Agra 1998]**

Sol. Put $-\log x = t$ so that $x = e^{-t}$ and $dx = -e^{-t}\,dt$.

$$\therefore \quad \int_0^1 \frac{dx}{\sqrt{(-\log x)}} = \int_\infty^0 \frac{-e^{-t}dt}{\sqrt{t}} = \int_0^\infty e^{-t}t^{-1/2}dt = \int_0^\infty e^{-t}t^{(1/2)-1}dt = \Gamma(1/2) = \sqrt{\pi}$$

Ex. 7. Evaluate $\displaystyle\int_0^\infty t^{-3/2}(1-e^{-t})\,dt$

Sol. $$\int_0^\infty t^{-3/2}(1-e^{-t})\,dt = \left[(1-e^{-t})\left(\frac{t^{-1/2}}{-1/2}\right)\right]_0^\infty - \int_0^\infty (e^{-t})\left(\frac{t^{-1/2}}{-1/2}\right)dt$$

$$= 0 + 2\int_0^\infty e^{-t}t^{(1/2)-1}dt = 2\Gamma(1/2) = 2\sqrt{\pi}$$

Ex. 8. *Evaluate* (*i*) $\displaystyle\int_0^\infty x^m e^{-a\,x^n}dx$ (*ii*) $\displaystyle\int_0^\infty x^m e^{-x^n}dx.$

Sol. Part (*i*). Put $ax^n = t$ so that $x = t^{1/n}/a^{1/n} = (t/a)^{1/n}$, $dx = (1/a^{1/n}) \times (1/n) \times t^{(1/n)-1}\,dt$

$$\therefore \quad \int_0^\infty x^m e^{-ax^n}dx = \int_0^\infty \left(\frac{t}{a}\right)^{m/n} e^{-t}\,\frac{1}{a^{1/n}}\cdot\frac{1}{n}t^{(1/n)-1}\,dt = \int_0^\infty \frac{e^{-t}}{a^{(m/n)+(1/n)}}t^{(m/n)+(1/n)-1}\cdot\frac{1}{n}dt$$

$$= \frac{1}{n\,a^{(m+1)/n}}\int_0^\infty e^{-t}t^{[(m+1)/n]-1}dt = \frac{1}{n\,a^{(m+1)/n}}\Gamma\left(\frac{m+1}{n}\right),$$ by definition of Gamma function

Part (*ii*). Put $a = 1$ in part (*i*) **Ans.** $(1/2)\times\Gamma((m+1)/n)$

Ex. 9. *Show that* $\displaystyle\int_0^\infty \frac{x^c}{c^x}dx = \frac{\Gamma(c+1)}{(\log c)^{c+1}}, c > 0$ **(Agra 2010; Purvanchal 2006)**

Sol. $$\int_0^\infty \frac{x^c}{c^x}dx = \int_0^\infty x^c c^{-x}dx = \int_0^\infty x^c[e^{\log_e c}]^{-1}dx, \text{ as } c = e^{\log_e c}, \text{ if } c > 0$$

$$= \int_0^\infty x^{(c+1)-1}e^{-\log_e c}dx = \frac{\Gamma(c+1)}{(\log_e c)^{c+1}} \quad \left[\because \int_0^\infty x^{n-1}e^{-kx}dx = \frac{\Gamma(n)}{k^n}, n > 0, k > 0\right]$$

Ex. 10. *Show that* $[\Gamma(1/2)]^2 = 4\int_0^\infty \int_0^\infty e^{-(x^2+y^2)} dx\, dy = 4\int_0^{\pi/2} \int_0^\infty e^{-r^2} r\, dr\, d\theta.$

Sol. Let $$I = 4\int_0^\infty \int_0^\infty e^{-(x^2+y^2)} dx\, dy. \qquad ...(1)$$

$$\therefore \quad I = 4\int_0^\infty \int_0^\infty e^{-x^2} e^{-y} dx\, dy = \left(2\int_0^\infty e^{-x^2} dx\right)\left(2\int_0^x e^{-y^2} dy\right) = \Gamma\left(\frac{1}{2}\right)\cdot\Gamma\left(\frac{1}{2}\right), \text{ refer Art. 20.5}$$

$$\therefore \qquad I = [\Gamma(1/2)]^2 \qquad ...(2)$$

Again, put $x = r\cos\theta$, $y = r\sin\theta$ so that $x^2 + y^2 = r^2$ and $dxdy = r\, d\theta\, dr$. Furthermore, the region of integration in integral I is the first quadrant of xy plane and so in polar coordinates the corresponding limits will be $r = 0$ to $r = \infty$ and $\theta = 0$ to $\theta = \pi/2$ for the same first quadrant. Hence $$I = 4\int_0^{\pi/2} \int_0^\infty e^{-r^2} r\, d\theta\, dr. \qquad ...(3)$$

From (1), (2) and (3), the required result follows

Ex. 11. Show that $\int_0^\infty \exp(2ax - x^2)dx = \frac{1}{2}\sqrt{\pi}\exp a^2$, where $\exp k = e^k$.

Sol. $\int_0^\infty \exp(2ax - x^2)dx = \int_0^\infty e^{2ax-x^2} dx = \int_0^\infty e^{a^2-(x^2-2ax+a^2)} dx = \int_0^\infty e^{a^2-(x-a)^2} dx$

$= e^{a^2}\int_0^\infty e^{-(x-a)^2} dx = e^{a^2}\int_0^\infty e^{-t^2} dt$, on putting $x - a = t$ and $dx = dt$

$$\therefore \qquad \int_0^\infty \exp(2ax - x^2)dx = \exp a^2 \int_0^\infty e^{-t^2} dt. \qquad ...(1)$$

Now, $\Gamma(n) = \int_0^\infty e^{-u} u^{n-1} du. \quad \Rightarrow \quad \Gamma(1/2) = \int_0^\infty e^{-u} u^{-1/2} du. \qquad ...(2)$

Putting $u = t^2$ so that $du = 2t$ in (2), we get $\Gamma(1/2) = \int_0^\infty (e^{-t^2}\cdot t^{-1}\cdot 2t)\, dt$

or $\sqrt{\pi} = 2\int_0^\infty e^{-t^2} dt$ or $\int_0^\infty e^{-t^2} dt = \dfrac{\sqrt{\pi}}{2} \qquad ...(3)$

Using (3), (1) reduces to $$\int_0^\infty \exp(2ax - x^2)dx = \frac{1}{2}\sqrt{\pi}\exp a^2.$$

EXERCISE 20(A)

1. Prove that (*i*) $\int_0^\infty e^{-4x} x^{3/2} dx = \dfrac{3\sqrt{\pi}}{128}$ (*ii*) $\int_0^\infty e^{-x^2} x^2 dx = \dfrac{\sqrt{\pi}}{4}$

2. Show that $\Gamma\left(-\dfrac{15}{2}\right) = \dfrac{2^8\sqrt{\pi}}{1\cdot 3\cdot 5\cdot 7\cdot 9\cdot 11\cdot 13\cdot 15}$

3. Show that if n is a positive integer, then $\Gamma\left(-n + \dfrac{1}{2}\right) = \dfrac{(-1)^n 2^n\sqrt{\pi}}{1\cdot 3\cdot 5 \ldots (2n-1)}$

4. Prove that $\int_0^1 x^{n-1}\left(\log\dfrac{1}{x}\right)^{m-1} dx = \dfrac{\Gamma(m)}{n^m}$, $m > 0, n > 0$. **[Garhwal 2003]**

5. Prove that $\int_0^{\infty} \frac{e^{-st}}{\sqrt{t}}\, dt = \left(\frac{\pi}{s}\right)^{1/2}, s > 0$

6. Prove that (*i*) $\int_0^{\infty} \sqrt{x}\, e^{-x^3}\, dx = \frac{\sqrt{\pi}}{3}$ (*ii*) $\int_0^{\infty} 3^{-4x^2}\, dx = \frac{\sqrt{\pi}}{4\sqrt{(\log 3)}}$

7. Prove that $\int_0^{\infty} e^{-ax} x^n\, dx = \frac{\Gamma(n+1)}{a^{n+1}}, n > -1, a > 0.$

20.8. Symmetrical property of Beta function, i.e., B(m, n) = B(n, m).

Proof. By the definition of Beta function, we have

$$B(m, n) = \int_0^1 x^{m-1}(1-x)^{n-1}\, dx = \int_0^1 (1-x)^{m-1}[1-(1-x)]^{n-1}\, dx, \text{ as } \int_0^a f(x)\, dx = \int_0^a f(a-x)\, dx$$

$$= \int_0^1 (1-x)^{m-1} x^{n-1} dx = \int_0^1 x^{n-1}(1-x)^{m-1}\, dx = B(n, m), \text{ by the def. of Beta function.}$$

$$\therefore \qquad B(m, n) = B(n, m).$$

20.9. Evaluation of *B*(*m*, *n*) in an explicit form when *m* or *n* is a positive integer

By the definition of Beta function, $B(m, n) = \int_0^1 x^{m-1}(1-x)^{n-1}\, dx.$ ... (1)

The following three cases arise:

Case 1. When only *n* is positive integer. If $n = 1$, (1) gives

$$B(m, 1) = \int_0^1 x^{m-1}(1-x)^{1-1}\, dx = \int_0^1 x^{m-1} dx = \left[\frac{x^m}{m}\right]_0^1 = \frac{1}{m}, \qquad \text{... (2)}$$

showing that $B(m, n)$ can be evaluated when $n = 1$.

Now, let $n > 1$. Then from (1), we have

$$B(m, n) = \int_0^1 (1-x)^{n-1} x^{m-1} dx = \left[(1-x)^{n-1}\cdot\frac{x^m}{m}\right]_0^1 - \int_0^1 (n-1)\,(1-x)^{n-2}.(-1)\frac{x^m}{m}\, dx$$

$$= 0 + \frac{n-1}{m}\int_0^1 x^m (1-x)^{n-2}\, dx, \quad \text{as} \quad n > 1, \quad \text{so} \quad \lim_{x\to 0}(1-x)^{n-1}\frac{x^m}{m} = 0$$

$$= \frac{n-1}{m}\int_0^1 x^{(m+1)-1}(1-x)^{(n-1)-1}\, dx = \frac{n-1}{m} B(m+1, n-1), \text{ by def. of Beta function}$$

Thus, $B(m, n) = \frac{n-1}{m} B(m+1, n-1).$ ... (3)

Replacing m by $m + 1$ and n by $n - 1$ in (3), we get

$$B(m+1, n-1) = \frac{n-1-1}{m+1} B(m+2, n-2). \qquad \text{... (4)}$$

Using (4), (3) becomes $B(m, n) = \frac{n-1}{m}\cdot\frac{n-2}{m+1} B(m+2, n-2)$ ... (5)

Since n is a positive integer and $n > 1$, after applying the above process repeatedly, we get

$$B(m, n) = \frac{n-1}{m}\cdot\frac{n-2}{m+1}\cdot\frac{n-3}{m+1}\cdots\frac{1}{m+n-2} B(m+n-1, 1). \qquad \text{... (6)}$$

Replacing m by $m + n - 1$ is (2), we get $$B(m+n-1, 1) = \frac{1}{m+n-1} \quad ...(7)$$

Using (7), (6) becomes $$B(m, n) = \frac{n-1}{m} \cdot \frac{n-2}{m+1} \cdot \frac{n-3}{m+2} \cdots \frac{1}{m+n-2} \cdot \frac{1}{m+n-1}$$

or $$B(m, n) = \frac{(n-1)!}{m(m+1)\,(m+2)...(m+n-2)\,(m+n-1)} \quad ...(8)$$

Case II. When only *m* is a positive integer.

Since the Beta function is symmetrical in m and n, *i.e.*, $B(m, n) = B(n, m)$, hence from case *I*, interchanging m and n in (8), we get

$$B(m, n) = \frac{(m-1)!}{n(n+1)\,(n+2)...(n+m-2)\,(n+m-1)} \quad ...(9)$$

Case III. When both *m* and *n* are positive integers.

Since n is a positive integer, so by case *I*, we have

$$B(m, n) = \frac{(n-1)!}{m(m+1)\,(m+2)...(m+n-2)\,(m+n-1)}$$

$$= \frac{[1\,.\,2\,.\,3\,...\,(m-1)]\,(n-1)!}{1\,.\,2\,.\,3\,....\,(m-1)m(m+1)(m+2)...(m+n-2)\,(m+n-1)} = \frac{(m-1)!(n-1)!}{(m+n-1)!}$$

20.10. Transformation of Beta function

From I. To show that $B(m, n) = \int_0^\infty \frac{x^{n-1}dx}{(1+x)^{m+n}} = \int_0^\infty \frac{x^{m-1}dx}{(1+x)^{m+n}},\ m > 0,\ n > 0$ **(Meerut 2010)**

[Delhi B.A. (Prog.) III, 2015]

Proof. By definition, $$B(m, n) = \int_0^1 x^{m-1}(1-x)^{n-1}\,dx. \quad ...(1)$$

Put $x = 1/(1 + t)$, so that $dx = -dt/(1 + t)^2$. Then, from (1), we have

$$B(m, n) = -\int_\infty^0 \frac{1}{(t+1)^{m+1}}\left(1-\frac{1}{1+t}\right)^{n-1} \frac{dt}{(1+t)^2} = \int_0^\infty \frac{1}{(t+1)^{m+1}}\left(\frac{1}{1+t}\right)^{n-1} dt = \int_0^\infty \frac{t^{n-1}dt}{(1+t)^{m+n}}$$

or $$B(m, n) = \int_0^\infty \frac{x^{n-1}dx}{(1+x)^{m+n}} \quad ...(2)$$

Since m and n are interchangeable in Beta function, (2) gives

$$B(m, n) = \int_0^\infty \frac{x^{m-1}}{(1+x)^{m+n}}dx. \quad ...(3)$$

Thus, (2) and (3) $\Rightarrow$ $$B(m, n) = \int_0^\infty \frac{x^{n-1}dx}{(1+x)^{m+n}} = \int_0^\infty \frac{x^{m-1}dx}{(1+x)^{m+n}} \quad ...(4)$$

Deduction. Show that $B(m,n) = \frac{1}{2}\int_0^\infty \frac{x^{m-1}+x^{n-1}}{(1+x)^{m+n}}dx$ **[Delhi 2008; Kanpur 2004]**

Proof. Adding (3) and (4), we have

$$2B(m,n) = \int_0^\infty \frac{x^{m-1}+x^{n-1}}{(1+x)^{m+n}}dx \quad \Rightarrow \quad B(m,n) = \frac{1}{2}\int_0^\infty \frac{x^{m-1}+x^{n-1}}{(1+x)^{m+n}}dx$$

From II. To show that $B(m, n) = \int_0^1 \frac{x^{m-1} + x^{n-1}}{(1+x)^{m+n}} dx.$

[Delhi B.A. (Prog.) III, 2014; Kumaun 2002]

Proof. From form I, we have

$$B(m, n) = \int_0^\infty \frac{x^{m-1} dx}{(1+x)^{m+n}} = \int_0^1 \frac{x^{m-1} dx}{(1+x)^{m+n}} + \int_1^\infty \frac{x^{m-1} dx}{(1+x)^{m+n}}. \quad \text{... (1)}$$

Put $x = 1/t$, so that $dx = -1/t^2\, dt$. Then, we have

$$\int_1^\infty \frac{x^{m-1} dx}{(1+x)^{m+n}} = \int_1^0 \frac{(1/t)^{m-1}(-1/t^2)\, dt}{(1+1/t)^{m+n}} = \int_0^1 \frac{t^{n-1} dt}{(1+t)^{m+n}} = \int_0^1 \frac{x^{n-1} dx}{(1+x)^{m+n}}. \quad \text{... (2)}$$

Using (2), (1) reduces $\quad B(m, n) = \int_0^1 \frac{x^{m-1} dx}{(1+x)^{m+n}} + \int_0^1 \frac{x^{n-1} dx}{(1+x)^{m+n}} = \int_0^1 \frac{x^{m-1} + x^{n-1}}{(1+x)^{m+n}} dx.$

From III. To show that $\int_0^\infty \frac{x^{m-1} dx}{(ax+b)^{m+n}} = \frac{B(m, n)}{a^m b^n}$

[Delhi Maths (H) 2006; G.N.D.U. Amritsar 2010]

Proof. From form I, we have $\quad B(m, n) = \int_0^\infty \frac{x^{m-1}}{(1+x)^{m+n}} dx \quad \text{... (1)}$

Put $x = (at)/b$ so that $dx = (a\, dt)/b$. Then, (1) reduces to

$$B(m, n) = \int_0^\infty \frac{(at/b)^{m-1} \times (a/b)}{(1+at/b)^{m+n}} dt = a^m b^n \int_0^\infty \frac{t^{m-1} dt}{(at+b)^{m+n}} = a^m b^n \int_0^\infty \frac{x^{m-1} dx}{(ax+b)^{m+n}}.$$

$$\therefore \quad \int_0^\infty \frac{x^{m-1} dx}{(ax+b)^{m+n}} = \frac{B(m, n)}{a^m b^n}.$$

From IV. To show that $\int_0^{\pi/2} \frac{\sin^{2m-1}\theta \cos^{2n-1}\theta\, d\theta}{(a\sin^2\theta + b\cos^2\theta)^{m+n}} = \frac{B(m, n)}{2a^m b^n}$

Proof. Put $x = \tan^2\theta$ so that $dx = 2\tan\theta \sec^2\theta\, d\theta$. Then from form III above, we have

$$\frac{B(m, n)}{a^m b^n} = \int_0^{\pi/2} \frac{\tan^{2m-2}\theta \cdot 2\tan\theta \cdot \sec^2\theta\, d\theta}{(a\tan^2\theta + b)^{m+n}}$$

$$= 2\int_0^{\pi/2} \frac{\sin^{2m-2}\theta (\sin\theta/\cos\theta)(\cos\theta)^{2m+2n}\, d\theta}{\cos^{2m-2}\theta \cos^2\theta (a\sin^2\theta + b\cos^2\theta)^{m+n}} = 2\int_0^{\pi/2} \frac{\sin^{2m-1}\theta \cos^{2n-1}\theta\, d\theta}{(a\sin^2\theta + b\cos^2\theta)^{m+n}}$$

$$\therefore \quad \int_0^{\pi/2} \frac{\sin^{2m-1}\theta \cos^{2n-1}\theta\, d\theta}{(a\sin^2\theta + b\cos^2\theta)^{m+n}} = \frac{B(m,n)}{2a^m b^n}$$

Form V. To show that $\int_0^1 \frac{x^{m-1}(1-x)^{n-1}\, dx}{(x+a)^{m+n}} = \frac{B(m, n)}{a^n (1+a)^m}$

Proof. Put $\frac{x}{1+a} = \frac{t}{t+a}$ so that $dx = a(1+a)\frac{dt}{(t+a)^2}$. Then (1) reduces to

$$B(m,n)=\int_0^1 (1+a)^{m-1}\left(\frac{t}{t+a}\right)^{m-1} a^{n-1}\left(\frac{1-t}{1+t}\right)^{n-1}\frac{a(a+1)\,dt}{(t+a)^2}$$

$$=a^n(1+a)^m\int_0^1\frac{t^{m-1}(1-t)^{n-1}}{(t+a)^{m+n}}dt=a^n(1+a)^m\int_0^1\frac{x^{m-1}(1-x)^{n-1}}{(x+a)^{m+n}}dx$$

$$\therefore\quad \int_0^1\frac{x^{m-1}(1-x)^{n-1}}{(x+a)^{m+n}}dx=\frac{B(m,n)}{a^n(1+a)^m}.$$

Form VI. **To show that** $\int_b^a (x-b)^{m-1}(a-x)^{n-1}dx=(a-b)^{m+n-1}B(m,n)$, $m>0, n>0$ **[Delhi Maths (H) 2009]**

Proof. Put $x=\dfrac{t-b}{a-b}$ so that $dx=\dfrac{dt}{a-b}$. Then (1) gives

$$B(m,n)=\int_b^a\left(\frac{t-b}{a-b}\right)^{m-1}\left(\frac{a-t}{a-b}\right)^{n-1}\frac{dt}{a-b}$$

$$=\frac{1}{(a-b)^{m+n-1}}\int_b^a(t-b)^{m-1}(a-t)^{n-1}dt=\frac{1}{(a-b)^{m+n-1}}\int_b^a(x-b)^{m-1}(a-x)^{n-1}dx$$

$$\therefore\quad\int_b^a(x-b)^{m-1}(a-x)^{n-1}dx=(a-b)^{m+n-1}B(m,n)\qquad\text{... (1)}$$

Remark 1. By putting $a=1$, $b=-1$ in (1), we get

$$\int_{-1}^1(x+1)^{m+1}(1-x)^{n-1}dx=2^{m+n-1}B(m,n)=2^{m+n-1}\frac{(m)\,\Gamma(n)}{\Gamma(m+n)}$$ **[Delhi Maths(H) 2002]**

Remark 2. Putting $b=5$, $a=7$, $m=7$, $n=4$ in (1), we get

$$\int_5^7(x-5)^6(7-x)^3dx=2^{10}B(7,4).$$

Form VII. **To show that** (*i*) $\displaystyle\int_0^1\frac{x^{m-1}(1-x)^{n-1}dx}{\{a+(b-a)x\}^{m+n}}=\frac{1}{a^nb^m}B(m,n)$

(*ii*) $\displaystyle\int_0^1\frac{x^{m-1}(1-x)^{n-1}dx}{(b+cx)^{m+n}}=\frac{1}{(b+c)^mb^n}B(m,n), m>0,\ n>0.$ **[Delhi Maths (H) 2006, 09]**

Proof. By definition, $$B(m,n)=\int_0^1x^{m-1}(1-x)^{n-1}dx\qquad\text{... (1)}$$

Let $\dfrac{a}{y}-\dfrac{b}{x}=a-b$ so that $$x=\frac{by}{a+(b-a)y}\qquad\text{... (2)}$$

From (2), $$dx=\frac{b[a+(b-a)y]-by(b-a)}{\{a+(b-a)y\}^2}dy\quad\text{or}\quad dx=\frac{ab\,dy}{[a+(b-a)y]^2}\qquad\text{... (3)}$$

Again from (2), we see that when $x=1$, $y=1$ and when $x=0$, $y=0$. Again, using (2), we have

$$1-x=1-\frac{by}{a+bv-ay}=\frac{a(1-y)}{a+(b-a)}\qquad\text{... (4)}$$

Using (2), (3) and (4), (1) gives

$$B(m, n) = \int_0^1 \left\{\frac{by}{a+(b-a)y}\right\}^{m-1} \left\{\frac{a(1-y)}{a+(b-a)y}\right\}^{n-1} \frac{ab\, dy}{\{a+(b-a)y\}^2}$$

$$= a^n b^m \int_0^1 \frac{y^{m-1}(1-y)^{n-1}\, dy}{\{a+(b-a)y\}^{m+n}} = a^n b^m \int_0^1 \frac{x^{m-1}(1-x)^{n-1}\, dx}{\{a+(b-a)x\}^{m+n}}$$

$$\therefore \qquad \int_0^1 \frac{x^{m-1}(1-x)^{n-1}\, dx}{\{a+(b-a)x\}^{m+n}} = \frac{1}{a^n b^m} B(m, n). \qquad \text{... (5)}$$

Part (ii). Interchanging a and b in (5), $\displaystyle\int_0^1 \frac{x^{m-1}(1-x)^{n-1}\, dx}{\{b+(a-b)x\}^{m+n}} = \frac{1}{a^m b^n} B(m, n)$ ... (6)

Putting $a - b = c$, *i.e.*, $a = b + c$ in (6), we get

$$\int_0^1 \frac{x^{m-1}(1-x)^{n-1}\, dx}{(b+cx)^{m+n}} = \frac{1}{(b+c)^m b^n} B(m, n). \qquad \text{... (7)}$$

20.11. Relation between Beta and Gamma Functions. [Delhi B.A. (Prog) III, 2010]

$B(m, n) = \dfrac{\Gamma(m)\, \Gamma(n)}{\Gamma(m+n)}$, $m > 0, n > 0$. **[Agra 2009, 10, 12, Meerut 2004, 11; Delhi 2004]**

Proof. From form II of Art. 20.6, $\displaystyle\frac{\Gamma(m)}{z^m} = \int_0^\infty e^{-zx} x^{m-1} dx$ ... (1)

or $\displaystyle\Gamma(m) = \int_0^\infty z^m e^{-zx} x^{m-1} dx.$ ... (2)

Multiplying both sides of (2) by $e^{-z} z^{n-1}$, $\displaystyle\Gamma(m) e^{-z} z^{n-1} = \int_0^\infty e^{-z(1+x)} z^{m+n-1} x^{m-1} dx.$ ... (3)

Integrating both sides of (3) w.r.t. z from 0 to ∞, we have

$$\Gamma(m)\int_0^\infty e^{-z} z^{n-1} dz = \int_0^\infty \left\{\int_0^\infty e^{-z(1+x)} z^{m+n-1} x^{m-1} dx\right\} dz \text{ or } \Gamma(m)\Gamma(n) = \int_0^\infty \left\{\int_0^\infty e^{-z(1+x)} z^{m+n-1} dz\right\} x^{m-1}\, dx$$

[using definition of Gamma function on L.H.S. and interchanging the order of integration on R.H.S.]

or $\displaystyle\Gamma(m)\Gamma(n) = \int_0^\infty \frac{\Gamma(m+n)}{(1+x)^{m+n}} x^{m-1} dx$, by (1)

$$= \Gamma(m+n) \int_0^\infty \frac{x^{m-1} dx}{(1+x)^{m+n}} = \Gamma(m+n) B(m, n), \text{ by form I of Art. 20.10.}$$

$$\therefore \qquad B(m, n) = \frac{\Gamma(m)\Gamma(n)}{\Gamma(m+n)}.$$

Deduction IA. To show that $\Gamma(n)\, \Gamma(1-n) = \pi / \sin n\pi$, $0 < n < 1$

[Agra 1999, Delhi Maths (H) 2002, 08; Delhi Physics (H) 2000; Kanpur 2006]

Proof. We know that $\displaystyle B(m, n) = \int_0^\infty \frac{x^{n-1} dx}{(1+x)^{m+n}}$, $m > 0, n > 0$. ... (1)

The relation between Beta and Gamma is $B(m, n) = [\Gamma(m)\,\Gamma(n)]/\Gamma(m+n)$ ... (2)

From (1) and (2), $$\int_0^\infty \frac{x^{n-1}dx}{(1+x)^{m+n}} = \frac{\Gamma(m)\Gamma(n)}{\Gamma(m+n)}. \quad ..(3)$$

Taking $m + n = 1$ so that $m = 1 - n$, (3) reduces to

$$\int_0^\infty \frac{x^{n-1}dx}{1+x} = \frac{\Gamma(1-n)\Gamma(n)}{\Gamma(1)}, 0 < n < 1. \text{ as } m > 0 \Rightarrow 1 - n > 0 \Rightarrow n < 1; \text{ Also } n > 0$$

But we know that $\int_0^\infty \frac{x^{n-1}dx}{1+x} = \frac{\pi}{\sin n\pi}$ and $\Gamma(1) = 1$.

$$\therefore \qquad \pi/\sin n\pi = \Gamma(1-n)\Gamma(n), \quad 0 < n < 1$$

Deduction IB. To show that $\Gamma(1+n)\,\Gamma(1-n) = (n\pi)/\sin n\pi$.

Proof. L.H.S. $= n\,\Gamma(n)\,\Gamma(1-n)$, as $\Gamma(n+1) = n\,\Gamma(n)$

$= (n\pi)/\sin n\pi$, by deduction IA.

Deduction II. To show that $\Gamma(1/2) = \sqrt{\pi}$. **[Delhi B.A. (Prog) III, 2010]**

Proof. We have just proved that $\Gamma(n)\Gamma(1-n) = \pi/\sin n\pi$ ... (1)

Putting $n = 1/2$ in (1), we obtain

$$\Gamma(1/2)\Gamma(1-1/2) = \pi/\sin(\pi/2) \quad \text{or} \quad [\Gamma(1/2)]^2 = \pi \quad \text{or} \quad \Gamma(1/2) = \sqrt{\pi}$$

Deduction III. To show that

$$\int_0^\infty e^{-x^2}\,dx = \frac{\sqrt{\pi}}{2}.$$ **[Meerut 2009; Rohilkhand 1999, Delhi Maths (H) 2005]**

Proof. From the definition of Gamma function, we have

$$\Gamma(n) = \int_0^\infty e^{-x}x^{n-1}dx \quad \text{and so} \quad \Gamma\left(\frac{1}{2}\right) = \int_0^\infty e^{-x}x^{-1/2}dx. \quad ... (1)$$

Let $x = t^2$ so that $dx = 2t\,dt$. Then (1) becomes

$$\Gamma\left(\frac{1}{2}\right) = \int_0^\infty e^{-t^2}(t^2)^{-1/2}2t\,dt \quad \text{or} \quad \Gamma\left(\frac{1}{2}\right) = 2\int_0^\infty e^{-t^2}dt. \quad \text{or} \quad \Gamma\left(\frac{1}{2}\right) = 2\int_0^\infty e^{-x^2}dx. \quad ...(2)$$

Also $\Gamma(1/2) = \sqrt{\pi}$. ... (3)

From (2) and (3), we obtain $2\int_0^\infty e^{-x^2}dx = \sqrt{\pi}$ or $\int_0^\infty e^{-x^2}dx = \frac{\sqrt{\pi}}{2}$

Deduction IV. (*i*) $B(x+1, y) = \frac{x}{x+y}B(x, y)$ (*ii*) $B(x, y+1) = \frac{y}{x+y}B(x, y)$

[Delhi Maths (H) 2009]

Proof. (*i*) $$B(x+1, y) = \frac{\Gamma(x+1)\Gamma(y)}{\Gamma(x+1+y)} = \frac{x\Gamma(x)\,\Gamma(y)}{(x+y)\,\Gamma(x+y)}, \text{ as } \Gamma(n+1) = n\,\Gamma(n)$$

$$= \frac{x}{x+y}B(x, y).$$

(*ii*) Proceed as in part (*i*)

Deduction V. To show that for $m > 0, n > 0$, $B(m, n) = B(m+1, n) + B(m, n+1)$.

Proof. Using results (*i*) and (*ii*) of deduction IV, we have **[Agra 2006, 07]**

$$B(m+1, n) + B(m, n+1) = \frac{m}{m+n}B(m, n) + \frac{n}{m+n}B(m, n) = \frac{m+n}{m+n}B(m, n) = B(m, n)$$

Deduction VI. To show that

(i) $\int_0^{\pi/2} \sin^{2m-1}\theta\ \cos^{2n-1}\theta\, d\theta = \dfrac{\Gamma(m)\,\Gamma(n)}{2\Gamma(m+n)} = \dfrac{B(m,n)}{2}$, $m>0, n>0$

[Nagpur 2010; Delhi B.Sc. (Hons) III 2008, 11; Purvanchal 2005; Agra 2008]

(ii) $\int_0^{\pi/2} \sin^p\theta\cos^q\theta\, d\theta = \dfrac{\Gamma\left(\dfrac{p+1}{2}\right)\Gamma\left(\dfrac{q+1}{2}\right)}{2\Gamma\left(\dfrac{p+q+2}{2}\right)}$, $p>-1, q>-1$

[G.N.D.U. Amritsar 2010; Meerut 2018]

(iii) $\int_0^{\pi/2} \sin^p\theta\, d\theta = \int_0^{\pi/2} \cos^p\theta\, d\theta = \dfrac{1\cdot3\cdot5\ldots(p-1)}{2\cdot4\cdot6\ldots p}\dfrac{\pi}{2}$, if p is even +ve integer

$= \dfrac{2\cdot4\cdot6\ldots(p-1)}{1\cdot3\cdot5\ldots p}$, is p is odd +ve integer

(iv) $\int_0^{\pi/2} \sin^{p-1}\theta\cos^{q-1}\theta\, d\theta = \dfrac{\Gamma(p/2)\Gamma(q/2)}{2\Gamma\left(\dfrac{p+q}{2}\right)}$

(v) $\int_0^{\pi/2} \sin^{p-1}\theta\, d\theta = \int_0^{\pi/2} \cos^{p-1}\theta\, d\theta = \dfrac{\Gamma(p/2)\Gamma(1/2)}{2\Gamma\left(\dfrac{p+1}{2}\right)} = \dfrac{\sqrt{\pi}}{2}\dfrac{\Gamma(p/2)}{\Gamma\left(\dfrac{p+1}{2}\right)}$, as $\Gamma\left(\dfrac{1}{2}\right) = \sqrt{\pi}$

Proof. (*i*) By definition of Beta function, $B(m,n) = \int_0^1 x^{m-1}(1-x)^{n-1}dx$

Let $x=\sin^2\theta$ so that $dx = 2\sin\theta\cos\theta\, d\theta$. Then, we have

$B(m,n) = \int_0^{\pi/2} \sin^{2m-2}(1-\sin^2\theta)^{n-1}(2\sin\theta\cos\theta d\theta)$ or $\int_0^{\pi/2} \sin^{2m-1}\theta\cos^{2n-1}\theta\, d\theta = \dfrac{B(m,n)}{2}$.

$\therefore \int_0^{\pi/2} \sin^{2m-1}\theta\cos^{2n-1}\theta\, d\theta = \dfrac{\Gamma(m)\Gamma(n)}{2\Gamma(m+n)}$, as $B(m,n) = \dfrac{\Gamma(m)\Gamma(n)}{\Gamma(m+n)}$ ... (1)

Part (*ii*). Let $p = 2m-1$ and $q = 2n-1$, so that $m = (p+1)/2$ and $n = (q+1)/2$.

Then (1) becomes $\int_0^{\pi/2} \sin^p\theta\cos^q\theta\, d\theta = \dfrac{\Gamma\left(\dfrac{p+1}{2}\right)\Gamma\left(\dfrac{q+1}{2}\right)}{2\Gamma\left(\dfrac{p+q+2}{2}\right)}$ ... (2)

Part (*iii*) Replacing q by 0 in (2), $\int_0^{\pi/2} \sin^p\theta\, d\theta = \dfrac{\Gamma\left(\dfrac{p+1}{2}\right)\Gamma\left(\dfrac{1}{2}\right)}{2\Gamma\left(\dfrac{p+2}{2}\right)}$ ... (3)

Next, putting $p=0$ and $q=p$ in (2), $\int_0^{\pi/2} \cos^p\theta\, d\theta = \dfrac{\Gamma\left(\dfrac{p+1}{2}\right)\Gamma\left(\dfrac{1}{2}\right)}{2\Gamma\left(\dfrac{p+2}{2}\right)}$ ... (4)

Let p be even, say $p = 2r$. Then R.H.S. of (3) or (4)

$$= \frac{\Gamma\left(\frac{2r+1}{2}\right)\Gamma\left(\frac{1}{2}\right)}{2\Gamma\left(\frac{2r+2}{2}\right)} = \frac{\Gamma\left(r+\frac{1}{2}\right)\Gamma\left(\frac{1}{2}\right)}{2\Gamma(r+1)} = \frac{\left(r-\frac{1}{2}\right)\left(r-\frac{3}{2}\right)\ldots\frac{3}{2}\cdot\frac{1}{2}\cdot\Gamma\left(\frac{1}{2}\right)\Gamma\left(\frac{1}{2}\right)}{2\,.\,r\,(r-1)\,(2r-3)\ldots3\,.\,2\,.\,1}$$

$$= \frac{(2r-1)\,(2r-3)\ldots3\,.\,1}{(2r)\,(2r-2)\,(2r-4)\ldots6\,.\,4\,.\,2}\,\frac{\pi}{2} = \frac{1\,.\,3\,.\,5\ldots(2r-3)\,(2r-1)}{2\,.\,4\,.\,6\ldots(2r-2)\,(2r)}\,\frac{\pi}{2}, \text{ as } \Gamma\left(\frac{1}{2}\right) = \sqrt{\pi}$$

$$= \frac{1\,.\,3\,.\,5\ldots(p-3)\,(p-1)}{2\,.\,4\,.\,6\ldots(p-2)\,p}\,\frac{\pi}{2}, \text{ as } p = 2r \qquad \ldots(5)$$

Next, let $p = 2r + 1$ *i.e.*, odd +ve integer. Then R.H.S. of (3) and (4)

$$= \frac{\Gamma(r+1)\Gamma(1/2)}{2\Gamma\left(r+\frac{3}{2}\right)} = \frac{r(r-1)\ldots3\,.\,2\,.\,1\,\sqrt{\pi}}{2.\left(r+\frac{1}{2}\right)\left(r-\frac{1}{2}\right)\ldots\frac{3}{2}\cdot\frac{1}{2}\sqrt{\pi}} = \frac{2\,.\,4\,.\,6\ldots(2r-2)\,(2r)}{1\,.\,3\,.\,5\ldots(2r-1)\,(2r+1)} = \frac{2\,.\,4\,.\,6\ldots(p-1)}{1\,.\,3\,.\,5\ldots p}, \qquad \ldots(6)$$

since $2r + 1 = p$. Thus from (3), (4), (5) and (6) the required results follow.

Part (*iv*) Let $2m = p$ and $2n = q$ so that $m = p/2$ and $n = q/2$. Then (1) becomes

$$\int_0^{\pi/2} \sin^{p-1}\theta\cos^{q-1}\theta\,d\theta = \frac{\Gamma(p/2)\,\Gamma\,(q/2)}{2\Gamma\left(\frac{p+q}{2}\right)} \qquad \ldots(7)$$

Part (*v*) Replacing q by 1 in (7),

$$\int_0^{\pi/2} \sin^{p-1}\theta\,d\theta = \frac{\Gamma(p/2)\Gamma(1/2)}{2\Gamma\left(\frac{p+1}{2}\right)} \qquad \ldots(8)$$

Next, replacing p by 1 and q by p, in (7),

$$\int_0^{\pi/2} \cos^{p-1}\theta\,d\theta = \frac{\Gamma(1/2)\Gamma(p/2)}{2\Gamma\left(\frac{1+p}{2}\right)} \qquad \ldots(9)$$

From (8) and (9), the required results follow.

20.12. Solved Examples

Ex. 1. *Evaluate the following Integrals :*

(*i*) $\int_0^1 x^4(1-x)^2\,dx$ (*ii*) $\int_0^2 \frac{x^2\,dx}{\sqrt{(2-x)}}$,

(*iii*) $\int_0^a y^4\sqrt{(a^2-y^2)}\,dy$ (*iv*) $\int_0^2 x(8-x^3)^{1/3}\,dx.$ **[Agra 2000, 03]**

Sol. We know that

$$\int_0^1 x^{m-1}(1-x)^{n-1}\,dx = B(m,n) = \frac{\Gamma(m)\,\Gamma(n)}{\Gamma(m+n)} \qquad \ldots(1)$$

Part. (*i*). $\int_0^1 x^4(1-x)^2\,dx = \int_0^1 x^{5-1}(1-x)^{3-1}\,dx = \frac{\Gamma\,(5)\,\Gamma\,(3)}{\Gamma\,(5+3)} = \frac{4\,!\,2\,!}{7\,!} = \frac{4!\times2}{7\times5\times4!\times6} = \frac{1}{105}$

Part. (*ii*) Let $I = \int_0^2 x^2(2-x)^{-1/2}\,dx$. Let $x = 2t$, so that $dx = 2dt$, Then

$$I = \int_0^1 (2t)^2(2-2t)^{-1/2}(2dt) = 4\sqrt{2}\int_0^1 t^2(1-t)^{-1/2}\,dt = 4\sqrt{2}\int_0^1 t^{3-1}(1-t)^{1/2-1}\,dt$$

$$= 4\sqrt{2}\,\frac{\Gamma(3)\Gamma(1/2)}{\Gamma(3+1/2)} = 4\sqrt{2}\,\frac{2\,!\,\Gamma(1/2)}{\frac{5}{2}\cdot\frac{3}{2}\cdot\frac{1}{2}\Gamma(1/2)} = \frac{64\sqrt{2}}{15}$$

Part (*iii*) Let $I = \int_0^a y^4 \sqrt{(a^2 - y^2)}\, dy$. Let $y^2 = a^2 t$, so that $dy = \dfrac{a^2\,dt}{2y} = \dfrac{a\,dt}{2\sqrt{t}}$. Then

$$I = \int_0^1 (a^2 t)^2 \sqrt{(a^2 - a^2 t)}\,\frac{(a\,dt)}{2\sqrt{t}} = \frac{a^6}{2}\int_0^1 t^{3/2}(1-t)^{1/2}\,dt = \frac{a^6}{2}\int_0^1 t^{(5/2)-1}(1-t)^{(3/2)-1}\,dt$$

$$= \frac{a^6}{2} \times \frac{\Gamma(5/2)\Gamma(3/2)}{\Gamma(5/2+3/2)} = \frac{a^6}{2} \times \frac{\frac{3}{2}\cdot\frac{1}{2}\sqrt{\pi}\cdot\frac{1}{2}\sqrt{\pi}}{3\,!} = \frac{\pi a^6}{32}$$

Part (*iv*) Let $I = \int_0^2 x(8 - x^3)^{1/3}\,dx$. Put $x^3 = 8t$ or $x = 2t^{1/3}$ so that $dx = (2/3) \times t^{-2/3}$ dt

$$\therefore\; I = \int_0^1 (2t^{1/3})(8 - 8t)^{1/3}(2/3)t^{-2/3}\,dt = \frac{8}{3}\int_0^1 t^{-1/3}(1-t)^{1/3}\,dt = \frac{8}{3}\int_0^1 t^{(2/3)-1}(1-t)^{(4/3)-1}\,dt$$

$$= \frac{8}{3} \times \frac{\Gamma(2/3)\Gamma(4/3)}{\Gamma(2/3+4/3)} = \frac{8}{3} \times \frac{\Gamma(1-1/3)\Gamma(1+1/3)}{\Gamma(2)} = \frac{8}{3}\Gamma\left(1-\frac{1}{3}\right)\frac{1}{3}\Gamma\left(\frac{1}{3}\right), \text{ as } \quad \Gamma(n+1) = n\,\Gamma(n)$$

$$= \frac{8}{9} \times \frac{\pi}{\sin(\pi/3)} = \frac{16\pi}{2\sqrt{3}}, \text{ as } \Gamma(1-n)\Gamma(n) = \frac{\pi}{\sin n\pi}$$

Ex. 2. *Show that (a)* $\displaystyle\int_0^1 \frac{dx}{\sqrt{(1-x^n)}} = \frac{\Gamma(1/n)}{\Gamma(1/2+1/n)} \cdot \frac{\sqrt{\pi}}{n}$. **[Meerut 2004, Purvanchal 2006] Kanpur 2009**

(*b*) $\displaystyle\int_0^1 \frac{dx}{(1-x^4)^{1/2}} = \frac{\sqrt{\pi}}{4} \cdot \frac{\Gamma(1/4)}{\Gamma(1/2+1/4)}$ **[Meerut 2007]**

Sol. (*a*) Let $I = \displaystyle\int_0^1 \frac{dx}{(1-x^n)^{1/2}}$. Putting $x^n =$ so that $x = t^{1/n}$ and $dx = (1/n)t^{(1/n)-1}\,dt$, we get

$$I = \int_0^1 \frac{1}{(1-t)^{1/2}} \cdot \frac{1}{n} t^{(1/n)-1}\,dt = \frac{1}{n}\int_0^1 t^{(1/n)-1}(1-t)^{(1/2)-1}\,dt = \frac{1}{n}B\left(\frac{1}{n}, \frac{1}{2}\right)$$

$$= \frac{1}{n}\,\frac{\Gamma(1/n)\Gamma(1/2)}{\Gamma(1/n+1/2)} = \frac{\sqrt{\pi}\,\Gamma(1/n)}{n\,\Gamma(1/n+1/2)}.$$

(*b*) Taking $n = 4$ in part (*a*), we get the required result.

Ex. 3. *Show that* $\displaystyle\int_0^a \frac{dx}{(a^n - x^n)^{1/n}} = \frac{\pi}{n\sin(\pi/n)}$. **[Meerut 1998]**

Sol. Let $I = \displaystyle\int_0^a \frac{dx}{(a^n - x^n)^{1/n}}$. Putting $x^n = a^n\,t$ so that $x = at^{1/n}$ and $dx = a(1/n)\,t^{(1/n)-1}\,dt$, gives

$$I = \int_0^1 \frac{1}{(a^n - a^n t)^{1/n}}\,\frac{a}{n}\,t^{(1/n)-1}\,dt = \frac{1}{n}\int_0^1 t^{(1/n)-1}(1-t)^{-1/n}\,dt = \frac{1}{n}\int_0^1 t^{(1/n)-1}(1-t)^{(1-1/n)-1}\,dt$$

$$= \frac{1}{n} B\left(\frac{1}{n}, 1 - \frac{1}{n}\right), \ = \frac{1}{n} \cdot \frac{\Gamma\left(\frac{1}{n}\right)\Gamma\left(1-\frac{1}{n}\right)}{\Gamma\left(\frac{1}{n}+1-\frac{1}{n}\right)} = \frac{\Gamma\left(\frac{1}{n}\right)\Gamma\left(1-\frac{1}{n}\right)}{n} = \frac{1}{n} \cdot \frac{\pi}{\sin(\pi/n)}.$$

$[\because \ \Gamma(p)\Gamma(1-p) = \pi/\sin p\pi]$

Ex. 4. *Show that* $\displaystyle\int_0^\infty \frac{x\,dx}{1+x^6} = \frac{1}{6} B\left(\frac{1}{3}, \frac{2}{3}\right) = \frac{\pi}{3\sqrt{3}}$ **[Delhi Maths (H) 2007, 08]**

Sol. Let $I = \displaystyle\int_0^\infty \frac{x\,dx}{1+x^6}$ Putting $x^6 = t$ so that $x = t^{1/6}$ and $dx = (1/6)t^{-5/6}\,dt$, we get

$$I = \int_0^\infty \frac{t^{1/16} \cdot (1/6)t^{-5/6}}{1+t} = \frac{1}{6}\int_0^\infty \frac{t^{-2/3}dt}{1+t} = \frac{1}{6}\int_0^\infty \frac{t^{(1/3)-1}}{(1+t)^{1/3+2/3}} dt$$

$$= \frac{1}{6} B\left(\frac{1}{3}, \frac{2}{3}\right) \qquad \because \int_0^\infty \frac{x^{m-1}dx}{(1+x)^{m+n}} = B(m,n)$$

$$= \frac{1}{6} \times \frac{\Gamma(1/3)\Gamma(2/3)}{\Gamma(1/3+2/3)} = \frac{1}{6}\Gamma\left(\frac{1}{3}\right)\Gamma\left(1-\frac{1}{3}\right)$$

$$= \frac{1}{6} \times \frac{\pi}{\sin(\pi/3)} = \frac{\pi\sqrt{3}}{9} \qquad \because \Gamma(n)\Gamma(1-n) = \frac{\pi}{\sin n\pi}$$

$$= (\pi\sqrt{3})/9.$$

Ex. 5. *Evaluate (i)* $\displaystyle\int_0^\infty \frac{x^4(1+x^5)}{(1+x)^{15}} dx$ *(ii)* $\displaystyle\int_0^\infty \frac{x^{m-1} - x^{n-1}}{(1+x)^{m+1}} dx$

(iii) $\displaystyle\int_0^\infty \frac{x^8(1-x^6)}{(1+x)^{24}} dx.$ **[Garhwal 2000]**

Sol. Part (*i*) $\displaystyle\int_0^\infty \frac{x^4(1+x^5)}{(1+x)^{15}} dx = \int_0^\infty \frac{x^4 dx}{(1+x)^{15}} + \int_0^\infty \frac{x^9}{(1+x)^{15}} dx = \int_0^\infty \frac{x^{5-1}dx}{(1+x)^{5+10}} + \int_0^\infty \frac{x^{10-1}dx}{(1+x)^{10+5}}$

$$= B(5, 10) + B(10, 5), \text{ as } \int_0^\infty \frac{x^{m-1}}{(1+x)^{m+n}} dx = B(m,n)$$

$$= 2B(5, 10), \text{ as } B(5, 10) = B(10, 5)$$

$$= 2\frac{\Gamma(5)\,\Gamma(10)}{\Gamma(5+10)} = \frac{2 \times 4! \times 9!}{14!} = \frac{1}{5005}.$$

Part (*ii*) $\displaystyle\int_0^\infty \frac{x^{m-1} - x^{n-1}}{(1+x)^{m+n}} dx = \int_0^\infty \frac{x^{m-1}dx}{(1+x)^{m+n}} - \int_0^\infty \frac{x^{n-1}dx}{(1+x)^{m+n}}$ **(Meerut 2010,**

$= B(m, n) - B(n, m) = 0.$ $[\because B(m, n) = B(n, m)]$

Part (*iii*) $\displaystyle\int_0^\infty \frac{x^8(1-x^6)dx}{(1+x)^{24}} = \int_0^\infty \frac{x^8 dx}{(1+x)^{24}} - \int_0^\infty \frac{x^{14}dx}{(1+x)^{24}} = \int_0^\infty \frac{x^{9-1}dx}{(1+x)^{9+15}} - \int_0^\infty \frac{x^{15-1}}{(1+x)^{15+9}}$

$= B(9, 15) - B(15, 9) = 0$ $[\because B(9, 15) = B(15, 9)]$

Ex. 6. *Show by means of Beta function, that* $\displaystyle\int_t^z \frac{dx}{(z-x)^{1-\alpha}(x-t)^{-\alpha}} = \frac{\pi}{\sin \pi\alpha}, 0<\alpha<1$

Sol. Let $$I = \int_t^z \frac{dx}{(z-x)^{1-\alpha}(x-t)^{-\alpha}} \qquad \text{... (1)}$$

Putting $x - t = (z - t)y$ so that $x = t + (z - t)\, y$ and $dx = (z - t)dy$, (1) becomes

$$I = \int_0^1 \frac{(z-t)dy}{[z-t-(z-t)y]^{1-\alpha}[(z-t)y]^{-\alpha}} = \int_0^1 \frac{(z-t)dy}{(z-t)^{1-\alpha}(1-y)^{1-\alpha}(z-t)^{-\alpha}y^{-\alpha}}$$

$$= \int_0^1 (1-y)^{\alpha-1} y^{\alpha} dy = \int_0^1 y^{(1-\alpha)-1}(1-y)^{\alpha-1}dy = B(1-\alpha, \alpha), \text{ by definition of Beta function.}$$

$$= \frac{\Gamma(1-\alpha)\,\Gamma(\alpha)}{\Gamma(1-\alpha+\alpha)} = \frac{\pi}{\sin\pi\alpha}, \qquad \text{as} \qquad \Gamma(p)\Gamma(1-p) = \frac{\pi}{\sin \pi p}$$

Ex. 7. *Prove that (a)* $\displaystyle\int_0^{\pi/2} \sqrt{\tan\theta}\, d\theta = \frac{1}{2}\Gamma\left(\frac{1}{4}\right)\Gamma\left(\frac{3}{4}\right) = \frac{\pi\sqrt{2}}{2}$ **[Kumaun 2000 Meerut 2004]**

(b) $\displaystyle\int_0^{\pi/2} \tan^n x\, dx = \frac{\pi}{2}\sec\frac{n\pi}{2}, -1<n<1.$

Sol. Part (a) $\displaystyle\int_0^{\pi/2} \sqrt{\tan\theta}\, d\theta = \int_0^{\pi/2}\left(\frac{\sin\theta}{\cos\theta}\right)^{1/2} d\theta = \int_0^{\pi/2} \sin^{1/2}\theta\cos^{-1/2}\theta\, d\theta$

$$= \frac{\Gamma\left(\frac{1+1/2}{2}\right)\Gamma\left(\frac{1-1/2}{2}\right)}{2\Gamma\left(\frac{1/2-1/2+2}{2}\right)} \qquad \Big| \text{ Refer deduction VI (ii) of Art. 20.11.}$$

$$= \frac{\Gamma(3/4)\Gamma(1/4)}{2\Gamma(1)} = \frac{1}{2}\Gamma\left(\frac{3}{4}\right)\Gamma\left(\frac{1}{4}\right) = \frac{1}{2}\Gamma\left(\frac{1}{4}\right)\Gamma\left(1-\frac{1}{4}\right) = \frac{1}{2}\frac{\pi}{\sin(\pi/4)} = \frac{\pi\sqrt{2}}{2}$$

Part (b) $\displaystyle\int_0^{\pi/2} \tan^n x\, dx = \int_0^{\pi/2} \sin^n x \cos^{-n} x\, dx$

$$= \frac{\Gamma\left(\frac{1+n}{2}\right)\Gamma\left(\frac{1-n}{2}\right)}{2\Gamma\left(\frac{n-n+2}{2}\right)}$$

Refer deduction VI (ii) of Art. 20.11
Here $(1 + n)/2 > 0$ and $(1 - n)/2 > 0$
$\Rightarrow n > -1$ and $n < 1 \Rightarrow -1 < n < 1$.

$$= \frac{1}{2}\Gamma\left(\frac{1+n}{2}\right)\Gamma\left(1-\frac{1+n}{2}\right) = \frac{1}{2}\frac{\pi}{\sin\left(\frac{1+n}{2}\right)\pi} = \frac{\pi}{2\sin\left(\frac{\pi}{2}+\frac{n\pi}{2}\right)} = \frac{\pi}{2\cos\frac{n\pi}{2}}$$

$$= (\pi/2)\times\sec(n\pi/2), \text{ where } -1<n<1. \qquad [\because \Gamma(p)\Gamma(1-p) = \pi/\sin p\pi]$$

Ex. 8. *If* $p>0, q>0, m+1>0, n+1>0$, *prove* $\int_0^p x^m(p^q-x^q)^n\,dx = \frac{p^{nq+m+1}}{q}B\left(n+1, \frac{m+1}{q}\right)$

Sol. Let $$I = \int_0^p x^m(p^q-x^q)^n\,dx \qquad \text{... (1)}$$

Putting $x^q = p^q t$ so that $x = pt^{1/q}$ and $dx = (p/q)\, t^{(1/q)-1}\, dt$, (1) reduces to

$$I = \int_0^1 (pt^{t/q})^m (p^q - p^q t)^n (p/q) t^{(1/q)-1} dt = \frac{p^m \cdot p^{nq} \cdot p}{q}\int_0^1 t^{(m/q)+(1/q)-1}(1-t)^{(n+1)-1} dt$$

$$= \frac{p^{nq+m+1}}{q} B\left(\frac{m+1}{q}, n+1\right) = \frac{p^{nq+m+1}}{q} B\left(n+1, \frac{m+1}{q}\right)$$

Ex. 9. *Compute* $I = \int_0^\infty x^2 e^{-x^4} dx \cdot \int_0^\infty e^{-x^4} dx.$

Sol. Putting $x^4 = t$ so that $x = t^{1/4}$ and $dx = (1/4)\, t^{-3/4}\, dt$, we get

$$I = \int_0^\infty (t^{1/4})^2 e^{-t}\left(\frac{1}{4}t^{-3/4}\right) dt \cdot \int_0^\infty e^{-t}\frac{1}{4}t^{-3/4} dt = \frac{1}{16}\int_0^\infty e^{-t} t^{-1/4} dt \cdot \int_0^\infty e^{-t} e^{-3/4} dt$$

$$= \frac{1}{16}\int_0^\infty e^{-t} t^{(3/4)-1} \cdot \int_0^\infty e^{-t} t^{(1/4)-1} dt = \frac{1}{16}\Gamma\left(\frac{3}{4}\right)\Gamma\left(\frac{1}{4}\right) = \frac{1}{16}\Gamma\left(\frac{1}{4}\right)\Gamma\left(1-\frac{1}{4}\right)$$

$$= \frac{1}{16}\cdot\frac{\pi}{\sin(\pi/4)} = \frac{\pi\sqrt{2}}{16}. \qquad \left|\ \because\ \Gamma(n)\Gamma(1-n) = \frac{\pi}{\sin n\pi}\right.$$

Ex. 10. *Show that* $I = \int_0^{\pi/2}\sqrt{\sin\theta}\,d\theta.\int_0^{\pi/2}\frac{d\theta}{\sqrt{\sin\theta}} = \pi.$ **[Delhi B.Sc. (Prog.) 2009]**

[Garhwal 2002, Meerut 1998, Delhi B.A. (Prog.) III 2014, 15]

Sol. We know that $$\int_0^{\pi/2}\sin^p\theta d\theta = \frac{\Gamma\left(\frac{p+1}{2}\right)\sqrt{\pi}}{2\Gamma\left(\frac{p+2}{2}\right)} \qquad \text{... (1)}$$

$$I = \int_0^{\pi/2}\sin^{1/2}\theta\, d\theta \cdot \int_0^{\pi/2}\sin^{-1/2}\theta\, d\theta = \frac{\Gamma\left(\frac{1/2+1}{2}\right)\sqrt{\pi}}{2\Gamma\left(\frac{1/2+2}{2}\right)}\cdot\frac{\Gamma\left(\frac{-1/2+1}{2}\right)\sqrt{\pi}}{2\Gamma\left(\frac{-1/2+2}{2}\right)}, \text{ using (1)}$$

$$= \frac{\Gamma(3/4)\sqrt{\pi}}{2\Gamma(5/4)}\cdot\frac{\Gamma(1/4)\sqrt{\pi}}{2\Gamma(3/4)} = \frac{\pi\Gamma(1/4)}{4\Gamma(1+1/4)} = \frac{\pi\Gamma(1/4)}{4\times(1/4)\times\Gamma(1/4)} = \pi.$$

EXERCISE 20 B

1. Prove that (*a*) $\int_0^1 \frac{dx}{(1-x^3)^{1/3}} = \frac{2\pi}{3\sqrt{3}}$ (*b*) $\int_0^1 \frac{dx}{(1-x^n)^{1/n}} = \frac{\pi}{\sin(\pi/n)}$

(*c*) $\int_0^1 (1-x^n)^{1/n} dx = \frac{\{\Gamma(1/n)\}^2}{2n\Gamma(2/n)}$ (*d*) $\int_0^1 x^{m-1}(1-x^k)^n dx = \frac{1}{a}\frac{n!\Gamma(m/k)}{\Gamma(n+1+m/k)}$

[Nagpur 2010]

(e) $\int_0^1 x^{m-1}(1-x^2)^{n-1}\,dx = \frac{1}{2}B\left(\frac{1}{2}m, n\right)$ **[Garhwal 2003]** (f) $\int_0^1 \frac{dx}{\sqrt{(1+x^4)}} = \frac{\Gamma(1/4)\Gamma(1/2)}{4\sqrt{2}\,\Gamma(3/4)}$

(g) $\int_0^1 \frac{x^{n-1}dx}{\sqrt{(1-x^2)}} = \frac{\sqrt{\pi}\,\Gamma\{(n-1)/2\}}{2\Gamma(n/2)}$ (h) $\int_0^1 \frac{dx}{\sqrt{(1-x^4)}} = \frac{1}{6\sqrt{2\pi}}[\Gamma(1/4)]^2$

[Delhi Math (H) 2003, Meerut 2007]

2. Prove that (a) $\int_0^1 \left(\frac{1}{x}-1\right)^{1/4} dx = B\left(\frac{5}{4}, \frac{3}{4}\right) = \frac{\pi}{2\sqrt{2}}$ (b) $\int_0^\infty \frac{dt}{\sqrt{t}(1+t)} = B\left(\frac{1}{2}, \frac{1}{2}\right) = \pi.$

3. Show that $\int_0^1 \frac{x^n dx}{\sqrt{(1-x^2)}} = \frac{1.3.5\ldots(n-1)}{2.4.6\ldots n} \times \frac{\pi}{2}$ or $\frac{2.4.6\ldots(n-1)}{1.3.5\ldots n}$

according as n is even or odd positive integer.

4. Show that if p and q are positive, then

$B(p, q) = 2\int_0^{\pi/2} \cos^{2p-1}\theta \sin^{2q-1}\theta\, d\theta = \frac{\Gamma(p)\Gamma(q)}{\Gamma(p+q)}$ and deduce that $\int_0^\pi e^{-x^2} dx = \frac{\sqrt{\pi}}{2}.$

5. Prove that (i) $B(l, m)\, B(l + m, n) = B(m, n)\, B(m + n, l) = B(n, l)\, B(n + l, m)$.

(ii) $B(l, m)\, B(l + m, n)\, B(l + m + n, p) = \frac{\Gamma(l)\Gamma(m)\Gamma(n)\Gamma(p)}{\Gamma(l+m+n+p)}$

(iii) $l\, B(l, m + l) = mB\,(l + l, m)$. **[Agra 2005]**

6. Prove that $\int_{-a}^{a} (a+x)^{m-1}(a-x)^{n-1}\,dx = (2a)^{m+n-1} B(m, n)$

7. Prove $\int_{-1}^{1} (1+x)^{p-1}(1-x)^{q-1}\,dx = 2^{p+q-1} B(p, q)$

8. Using the integral $\int_0^\infty \frac{x^{n-1}}{1+x}dx = \frac{\pi}{\sin n\pi}, 0 < n < 1,$ prove that $\Gamma(n)\,\Gamma(1-n) = \pi/\sin n\pi,$ $0 < n < 1$ Hence obtain value of $\Gamma(1/2)$.

9. Prove that $\int_{-1}^{1} \frac{(x+1)^{a-1}(1-x)^{b-1}}{(x+2)^{a+b}}dx = \frac{2^{a+b-1}}{3a}B(a, b), a > 0, b > 0$

10. Show that the perimeter of a loop of the curve $r^n = a^n \cos n\theta$ can be expressed as

$$(a/n)\times 2^{(1/n)-1}\{\Gamma(1/2n)/\Gamma(1/n)\}$$

11. Show that the area enclosed by the curve $x^4 + y^4 = 1$ is $[\Gamma(1/4)]^2/2\sqrt{\pi}$.

12. With the help of double integral, prove that $\int_0^\infty e^{-x^2} dx = \sqrt{\pi}/2.$

[Delhi Maths (H) 2007]

20.13. Legendre-Duplication Formula. $\Gamma(n)\Gamma\left(n+\frac{1}{2}\right)=\frac{\sqrt{\pi}}{2^{2n-1}}\Gamma(2n),\ n>0.$

[Delhi B.Sc. (Hons) III 2005, 07, 11; Agra 2000, 01, 02, 03, 06, 08, 10; Meerut 1999]

Proof. We know that $B(m, n)=\frac{\Gamma(m)\Gamma(n)}{\Gamma(m+n)}$, where $m>0,\ n>0$. ... (1)

Putting $m=n$ in (1), we get $B(n, n)=[\Gamma(n)]^2/\Gamma(2n)$ ... (2)

By the definition of the Beta function, $B(n, n)=\int_0^1 x^{n-1}(1-x)^{n-1}dx.$ ... (3)

Putting $x=\sin^2\theta$ so that $dx=2\sin\theta\cos\theta\, d\theta$, (1) gives

$$B(n, n)=\int_0^{\pi/2}(\sin^2\theta)^{n-1}(1-\sin^2\theta)^{n-1}\cdot 2\sin\theta\cos\theta\, d\theta=2\int_0^{\pi/2}(\sin\theta\cos\theta)^{2n-1}d\theta$$

$$=2\int_0^{\pi/2}\left(\frac{\sin 2\theta}{2}\right)^{2n-1}d\theta=\frac{1}{2^{2n-2}}\int_0^{\pi/2}\sin^{2n-1}2\theta\, d\theta=\frac{1}{2^{2n-2}}\int_0^{\pi}\sin^{2n-1}\phi\frac{d\phi}{2}=\frac{1}{2^{2n-1}}\int_0^{\pi}\sin^{2n-1}\phi\, d\phi$$

[On putting $2\theta=\phi$ and $d\theta=(d\phi)/2$]

$$=\frac{1}{2^{2n-1}}\times 2\int_0^{\pi/2}\sin^{2n-1}\phi\, d\phi,\quad \text{as}\quad \int_0^{2a}f(x)dx=2\int_0^a f(x)\,dx\quad \text{when}\quad f(2a-x)=f(x)$$

$$=\frac{1}{2^{2n-2}}\int_0^{\pi/2}\sin^{2n-1}\phi\,(\cos\phi)^0\, d\phi=\frac{1}{2^{2n-2}}\frac{\Gamma\left(\frac{2n-1+1}{2}\right)\cdot\Gamma\left(\frac{0+1}{2}\right)}{2\Gamma\left(\frac{2n-1+0+2}{2}\right)}$$

$$\therefore \quad B(n, n)=\frac{1}{2^{2n-1}}\frac{\Gamma(n)\sqrt{\pi}}{\Gamma(n+1/2)},\ \text{as}\ \Gamma\left(\frac{1}{2}\right)=\sqrt{\pi} \quad ...(3)$$

Equating two values of $B(n, n)$ given by (2) and (3), we obtain

$$\frac{[\Gamma(n)]^2}{\Gamma(2n)}=\frac{1}{2^{2n-1}}\frac{\Gamma(n)\sqrt{\pi}}{\Gamma(n+1/2)}\quad \text{or}\quad \Gamma(n)\Gamma(n+1/2)=\frac{\sqrt{\pi}}{2^{2n-1}}\Gamma(2n). \quad ... (4)$$

Deduction 1. To show that $B(n, n)=\frac{\sqrt{\pi}\,\Gamma(n)}{2^{2n-1}\Gamma(n+1/2)},\ n>0$

Proof. From (3), we get the required result.

Deduction II For all positive real values of p, $2^p\Gamma\left(\frac{p+1}{2}\right)\Gamma\left(\frac{p+2}{2}\right)=\sqrt{\pi}\,\Gamma(p+1).$

Proof. Putting $2n-1=p$ so that $n=(p+1)/2$ in (4), we get

$$\Gamma\left(\frac{p+1}{2}\right)\Gamma\left(\frac{p+1}{2}+\frac{1}{2}\right)=\frac{\sqrt{\pi}}{2^p}\Gamma(p+1)\quad \text{or}\quad 2^p\Gamma\left(\frac{p+1}{2}\right)\Gamma\left(\frac{p+2}{2}\right)=\sqrt{\pi}\,\Gamma(p+1).$$

Deduction III. When n is positive integer, to show that

[Delhi Maths (H) 2003]

$$\Gamma\left(n+\frac{1}{2}\right)=\frac{(2n)!}{2^{2n}n!}\sqrt{\pi}.$$

Proof. Let n be positive integer, then $\frac{\Gamma(2n)}{\Gamma(n)} = \frac{(2n-1)!}{(n-1)} = \frac{(2n)\,(2n-1)!}{2 \cdot n\,(n-1)!} = \frac{(2n)!}{2 \cdot n!}$

Now, from the duplication formula (4) and the above result, we have

$$\Gamma\left(n+\frac{1}{2}\right) = \frac{\sqrt{\pi}}{2^{2n-1}} \cdot \frac{\Gamma(2n)}{\Gamma(n)} = \frac{\sqrt{\pi}}{2^{2n-1}} \cdot \frac{(2n)!}{2 \cdot n!} = \frac{(2n)!}{2^{2n}\, n!}\sqrt{\pi}.$$

20.14. SOLVED EXAMPLES.

Ex. 1. Express $\Gamma(1/6)$ in terms of $\Gamma(1/3)$.

Sol. From the duplication formula, $\Gamma(n)\Gamma\left(n+\frac{1}{2}\right) = \frac{\sqrt{\pi}}{2^{2n-1}}\Gamma(2n).$ ... (1)

Putting $n = 1/6$ in (1), we get

$$\Gamma\left(\frac{1}{6}\right)\Gamma\left(\frac{2}{3}\right) = \frac{\sqrt{\pi}\,\Gamma(1/3)}{2^{-2/3}} \quad \text{or} \quad \Gamma\left(\frac{1}{6}\right) = \frac{\sqrt{\pi}\,\Gamma(1/3)}{2^{-2/3}\,\Gamma(2/3)} \qquad ... (2)$$

Now, we know that $\Gamma(n)\Gamma(1-n) = \pi/\sin n\pi.$ ... (3)

Putting $n = 1/3$ in (3) we get

$$\Gamma\left(\frac{1}{3}\right)\Gamma\left(\frac{2}{3}\right) = \frac{\pi}{\sin(\pi/3)} = \frac{2\pi}{\sqrt{3}} \quad \text{or} \quad \Gamma\left(\frac{2}{3}\right) = \frac{2\pi}{\sqrt{3}\,\Gamma(1/3)} \qquad ... (4)$$

Substituting the value of $\Gamma(2/3)$ given by (4) in (2), we get

$$\Gamma\left(\frac{1}{6}\right) = \frac{\sqrt{\pi}\,\Gamma(1/3)}{2^{-2/3}} \cdot \frac{\sqrt{3}\,\Gamma(1/3)}{2\pi} \quad \text{or} \quad \Gamma\left(\frac{1}{6}\right) = \frac{\sqrt{3}}{2^{1/3}\sqrt{\pi}}\left[\Gamma\left(\frac{1}{3}\right)\right]^2$$

Ex. 2. Prove that $\Gamma(n)\Gamma\left(\frac{1-n}{2}\right) = \frac{\sqrt{\pi}\,\Gamma(n/2)}{2^{1-n}\cos(n\pi/2)}, 0 < n < 1.$

Sol. We know that $\Gamma(m)\,\Gamma(1-m) = \pi/\sin m\pi, 0 < m < 1$ ... (1)

and $\Gamma(m)\Gamma\left(m+\frac{1}{2}\right) = \frac{\sqrt{\pi}\,\Gamma(2m)}{2^{2m-1}}, m > 0,$ ... (2)

Putting $m = (n+1)/2$ in (1), we get

$$\Gamma\left(\frac{n+1}{2}\right)\Gamma\left(\frac{1-n}{2}\right) = \frac{\pi}{\sin\{(n+1)\pi/2\}} = \frac{\pi}{\sin(\pi/2+n\pi/2)} = \frac{\pi}{\cos(n\pi/2)} \qquad ... (3)$$

Putting $m = n/2$ in (2), we get $\Gamma\left(\frac{n}{2}\right)\Gamma\left(\frac{n+1}{2}\right) = \frac{\sqrt{\pi}\,\Gamma(n)}{2^{n-1}}$ ... (4)

Dividing the corresponding sides of (3) and (4), we get

$$\frac{\Gamma[(1-n)/2]}{\Gamma(n/2)} = \frac{\pi}{\cos(n\pi/2)} \times \frac{2^{n-1}}{\sqrt{\pi}\,\Gamma(n)} \quad \text{or} \quad \Gamma(n)\Gamma\left(\frac{1-n}{2}\right) - \frac{\sqrt{\pi}\,\Gamma(n/2)}{2^{1-n}\cos(n\pi/2)}$$

Ex. 3. Prove that $B(m, m)B(m+1/2, m+1/2) = (\pi m^{-1})/2^{4m-1}$

[Delhi B.A. (Prog) III 2010]

Ex. 4. By evaluating $I = \int_0^{\pi/2} \sin^{2p} x\, dx$ and $J = \int_0^{\pi/2} \sin^{2p} x\, dx$ derive the Legendre's duplication formula for gamma function.

[Kanpur 2006]

MISCELLANEOUS PROBLEMS BASED ON THIS CHAPTER

Ex. 1. *It 0 < n < 1, then* $\Gamma(n)\Gamma(1-n) =$

(a) $\pi/\sin n\pi$ *(b)* $\pi/\cos n\pi$ *(c) (n + 1)!* *(d) (n – 1)!* **[Agra 2008]**

Sol. Ans. (*a*). Refer deduction 1 *A*, page 20.12

Ex. 2. *B (l, m) is equal to*

(a) $\int_0^\infty \frac{x^{m-1}dx}{(1+x)^{l+m}}$ *(b)* $\int_0^\infty \frac{x^{m+1}dx}{(1+x)^{l+m}}$ *(c)* $\int_{-\infty}^0 \frac{x^{m-1}dx}{(1+x)^{l+m}}$ *(d) None of these*

Sol. Ans. (*a*). Refer form *I* of Art. 20.10, page 20.9 **[Agra 2008]**

Ex. 3. *If n is a positive integer, then* $\Gamma(n+1)$ *is equal to*

(a) n! *(b) (n + 1)!* *(c) (n – 1)!* *(d) n* **[Agra 2007]**

Sol. Ans. (*a*). Refer property III of Art. 20.3

Ex. 4. *B* (*m*, *n*) is equal to

(a) $\frac{\Gamma(m)\Gamma(n)}{\Gamma(m+n)}$ *(b)* $\frac{\Gamma(m-1)\Gamma(n-1)}{\Gamma(m+n)}$ *(c)* $\frac{\Gamma(m)\Gamma(n)}{\Gamma(m-n)}$ *(d)* $\frac{\Gamma(m)\Gamma(n)}{\Gamma(mn)}$

Sol. Ans. (*a*). Refer Art. 2011 **[Agra 2007, 09, 10**

Ex. 5. *B (m, n) =* *(a) B (m + 1, n) + B (m)* *(b) B (m + 1, n) + B (m, n + 1)*
(c) B (m + 1, n) + B (m) *(d) B (m – 1, m) + B (m, n – 1)* **[Agra 2006]**

Sol. Ans. (*b*). Refer deduction V, page 20.13

Ex. 6. *The value of nB (m + 1, n) equals*

(a) $\Gamma(m)$ *(b) mB (m, n + 1)* *(c) mB (m, n)* *(d) None of these* **[Agra 2005]**

Sol. Ans. (*b*). We have, *nB (m + 1, n)*

$$= n\frac{\Gamma(m+1)\Gamma(n)}{\Gamma(m+1+n)} = \frac{m\Gamma(m)\times n\Gamma(n)}{\Gamma(m+1+n)} = m\frac{\Gamma(m)\Gamma(n+1)}{\Gamma(m+n+1)} = mB(m,n+1)$$

Ex. 7. *Prove that* $\int_0^\infty e^{-x^2}x^2 dx = \sqrt{\pi}/4$ **[Agra 2008]**

Sol. As in Ex. 5 (*b*), page. 20.6, $\int_0^\infty e^{-ax^2}x^{2n-1}dx = \frac{\Gamma(n)}{2a^n}$ ...(1)

Setting $n = 3/2$ and $a = 1$ in (1), we obtain

$$\int_0^\infty e^{-x^2}x^2 dx = \frac{\Gamma(3/2)}{2} = \frac{(1/2)\times\Gamma(1/2)}{2} = \frac{\sqrt{\pi}/2}{2} = \frac{\sqrt{\pi}}{4}$$

Ex. 8. *Prove that* $\int_a^b (x-a)^p(b-x)^q dx = (b-a)^{p+q+1}\frac{\Gamma(p+1)\Gamma(q+1)}{\Gamma(p+q+2)}$ **[Agra 2007]**

Sol. Refer form VI on page 20.11 Thus, we have

$$\int_b^a (x-b)^{m-1}(a-x)^{n-1}dx = (a-b)^{m+n-1}B(m,n) \quad \text{...(1)}$$

Replacing *a*, *b*, *n* and m by *b*, *a*, *q* + 1 and *p* + 1 respectively, we have

$$\int_a^b (x-a)^p (b-x)^q\, dx = (b-a)^{p+q+1} B(p+1,q+1) = (b-a)^{p+q+1} \frac{\Gamma(p+1)\Gamma(q+1)}{\Gamma(p+q+2)}$$

Ex. 9. *Show that* $\displaystyle\int_0^1 \frac{x^2 dx}{\sqrt{1-x^4}} \times \int_0^1 \frac{dx}{\sqrt{1+x^4}} = \frac{\pi}{4\sqrt{2}}$

[Delhi B.Sc. (Hon) III 2008, 11; Delhi B.Sc. III (Prog) 2008]

Sol. Let $x^2 = \sin\theta$ so that $2x\, dx = \cos\theta\, d\theta$

Thus, $dx = (1/2) \times \cos\theta \times (\sin\theta)^{-1/2} d\theta$. Then, we have

$$\int_0^1 \frac{x^2 dx}{\sqrt{1-x^4}} = \frac{1}{2}\int_0^{\pi/2} \frac{\sin\theta}{\cos\theta} \cos\theta (\sin\theta)^{-1/2} d\theta = \frac{1}{2}\int_0^{\pi/2} \sin^{1/2}\theta \cos^{\circ}\theta\, d\theta$$

$$= \frac{1}{4} B\left(\frac{1/2+1}{2}, \frac{0+1}{2}\right) = \frac{1}{4} B(3/4, 1/2) \qquad \ldots(1)$$

$$\left[\because \int_0^{\pi/2} \sin^{2m-1}\theta \cos^{2n-1}\theta\, d\theta = \frac{1}{2} B(m,n)\right]$$

Again, let $x^2 = \tan\phi$ so that $2x\, dx = \sec^2\phi\, d\phi$

Thus, $dx = \left(\sec^2\phi / 2\sqrt{\tan\phi}\right) d\phi$. Then, we have

$$\int_0^1 \frac{dx}{\sqrt{1+x^4}} = \frac{1}{2}\int_0^{\pi/4} \left(\frac{1}{\sec\phi} \times \frac{\sec^2\phi}{\sqrt{\tan\phi}}\right) d\phi = \frac{1}{2}\int_0^{\pi/4} \frac{d\phi}{\sqrt{\sin\phi\cos\phi}} = \frac{1}{\sqrt{2}}\int_0^{\pi/4} \frac{d\phi}{\sqrt{\sin 2\phi}}$$

$$= \frac{1}{2\sqrt{2}}\int_0^{\pi/2} \sin^{-1/2} t \cos^{\circ} t\, dt, \text{ on putting } t = 2\phi \text{ and } d\phi = (dt)/2$$

$$= \frac{1}{4\sqrt{2}} B\left(\frac{-1/2+1}{2}, \frac{0+1}{2}\right) = \frac{1}{4\sqrt{2}} B(1/4, 1/2), \text{ as before} \qquad \ldots(2)$$

$$\therefore \int_0^1 \frac{x^2 dx}{\sqrt{1-x^4}} \times \int_0^1 \frac{dx}{\sqrt{1+x^4}} = \frac{1}{16\sqrt{2}} B(3/4, 1/2)\, B(1/4, 1/2)$$

$$= \frac{1}{16\sqrt{2}} \times \frac{\Gamma(3/4)\Gamma(1/2)}{\Gamma(5/4)} \times \frac{\Gamma(1/4)\Gamma(1/2)}{\Gamma(3/4)}, \text{ as } B(m,n) = \frac{\Gamma(m)\Gamma(n)}{\Gamma(m+n)}$$

$$= \frac{1}{16\sqrt{2}} \times \frac{\{\Gamma(1/2)\}^2 \times \Gamma(1/4)}{(1/4) \times \Gamma(1/4)}, \text{ as } \Gamma(n) = (n-1)\,\Gamma(n-1)$$

$$= \frac{1}{16\sqrt{2}} \times \frac{(\sqrt{\pi})^2}{(1/4)} = \frac{\pi}{4\sqrt{2}}$$

Ex. 10. *Show that* $\int_0^1 x^{-1/3}(1-x)^{-2/3}(1+2x)^{-1}\,dx = \dfrac{1}{3^{2/3}}B(2/3,1/3)$

[Delhi B.Sc. (Hons) III 2008, 11]

Sol. Let $$I = \int_0^1 x^{-1/3}(1-x)^{-2/3}(1+2x)^{-1}\,dx \qquad \dots(1)$$

Let $x/(1-x) = (at)/(1-t)$, *i.e.*, $x = (at)/\{1-(1-a)t\}$, ...(2)

where a is a constant to be so selected in order that I given by (1) reduces to Beta function.

From (2), $$dx = a\frac{1-(1-a)t-t\{-(1-a)\}}{\{1-(1-a)t\}^2}dt = \frac{a}{\{1-(1-a)t\}^2}dt$$

$$\therefore I = \int_0^1 \left\{\frac{at}{1-(1-a)t}\right\}^{-1/3}\left\{1-\frac{at}{1-(1-a)t}\right\}^{-2/3}\left\{1+\frac{2at}{1-(1-a)t}\right\}^{-1}\frac{a\,dt}{\{1-(1-a)t\}^2}$$

$$= \int_0^1 \frac{a^{-1/3}t^{-1/3}}{\{1-(1-a)t\}^{-1/3}}\cdot\frac{(1-t)^{-2/3}}{\{1-(1-a)t\}^{-2/3}}\cdot\frac{(1-t+3at)^{-1}}{\{1-(1-a)t\}^{-1}}\cdot\frac{a\,dt}{\{1-(1-a)t\}^2}$$

$$= a^{2/3}\int_0^1 \frac{t^{-1/3}(1-t)^{-2/3}}{1-t+3at}dt = \left(\frac{1}{3}\right)^{2/3}\int_0^1 t^{-1/3}(1-t)^{-2/3}\,dt, \text{ choosing } a = 1/3$$

$$= \left(\frac{1}{3}\right)^{2/3}\int_0^1 t^{(2/3)-1}(1-t)^{(1/3)-1} = \frac{1}{3^{2/3}}B(2/3,1/3)$$

$$B(m,n) = (\Gamma(m)\Gamma(n)/2\Gamma(m+n)$$

Ex. 11. *Show that* $B(m,m)\ B(m+1/2,\ m+1/2) = \pi\times m^{-1}\times 2^{1-4m}$ *where* $B(m,n)$ *denotes the Beta function of m and n.* **[Delhi B.Sc. (Prog) III 2010]**

Sol. We know that $$B(m,n) = [\Gamma(m)\Gamma(n)]/\Gamma(m+n)$$

Hence, L.H.S. $= B(m,m)\ B(m+1/2,\ m+1/2)$

$$= \frac{\Gamma(m)\,\Gamma(m)}{\Gamma\,(2m)}\times\frac{\Gamma(m+1/2)\Gamma(m+1/2)}{\Gamma(2m+1)}$$

$$= \left[\frac{\Gamma(m)\,\Gamma(m+1/2)}{\Gamma\,(2m)}\right]^2\times\frac{1}{2m}, \quad as \quad \Gamma(2m+1) = 2m\,\Gamma(2m)$$

$$= \left(\frac{\sqrt{\pi}}{2^{2m-1}}\right)^2\times\frac{1}{2m}, \text{ using duplication formula}$$

$$= \pi\times m^{-1}\times 2^{1-4m} = \text{R.H.S}$$

Ex. 12. Using Beta and Gamma function, show that $\displaystyle\int_0^\infty \frac{dt}{\sqrt{t}\,(1+t)} = \pi$

[Delhi B.A. (Prog) III 2010

Sol. We know that $$B(m,n) = \int_0^\infty \frac{t^{m-1}}{(1+t)^{m+n}}dt \qquad \dots(1)$$

Setting $m = n = 1/2$, in (1), we get

$$B(1/2,1/2) = \int_0^\infty \frac{t^{-1/2}}{(1+t)}dt \quad \text{or} \quad \int_0^\infty \frac{dt}{\sqrt{t}\,(1+t)} = \frac{\Gamma(1/2)\,\Gamma[1/2]}{\Gamma(1/2+1/2)} = \frac{(\sqrt{\pi})^2}{\Gamma(1)} = \pi$$

Ex. 13. The relation between Beta and Gemma function is

(*a*) $B(m,n) = (\Gamma(m)\Gamma(n)] / \Gamma(m,n)$

(*b*) $B(m,n) = (\Gamma(m+n) / [\Gamma(m)\Gamma(n)]$

(*c*) $B(m,n) = [(\Gamma(m+1)\Gamma(n+1)] / 2\Gamma(m+n)$

(*d*) $B(m,n) = (\Gamma(m)\Gamma(n) / 2\Gamma(m+n)$

Hint. Ans. (*a*). Refer Art. 20.11 **[Agra 2009, 10]**

Ex. 14. Evaluate $\int_0^1 x^4 \{\log(1/x)\}^3 \, dx.$ **(Delhi B.Sc. (Hons) III 2011)**

Chapter

21

Differentiation Under the Integral Sign

21.1 Introduction. If the integrand of a definite integral is a function of one or more parameters in addition to the variable of integration, then the given definite integral between the limits, which may be constants or functions of the parameters, is a function of these parameters as illustrated in the following examples:

$$\int_0^1 \cos(\alpha x)\, dx = [(1/\alpha) \times \sin(\alpha x)]_0^1 = (\sin \alpha)/\alpha$$

$$\int_0^1 (\cos \alpha x + \cos \beta x)\, dx = [(1/\alpha) \times \sin \alpha x + (1/\beta) \times \sin \beta x]_0^1 = (\sin \alpha)/\alpha + (\sin \beta)/\beta$$

Thus, in general, $\int_a^b f(x,\alpha)\, dx = F(\alpha)$ and $\int_a^b g(x,\alpha,\beta)\, dx = G(\alpha,\beta)$ ...(1)

where α and β are parameters and *a*, *b* are constants or functions of parameters.

In some problems functions $f(x, \alpha)$ and $g(x, \alpha, \beta)$ are such that the evaluation of the corresponding integrals given by (1) are either very complicated or impossible. In such problems sometimes the integrals

$$\int_a^b \frac{\partial f(x,\alpha)}{\partial \alpha}\, dx \quad \text{or} \quad \int_a^b \frac{\partial g(x,\alpha,\beta)}{\partial \alpha}\, dx \quad \text{or} \quad \int_a^b \frac{\partial g(x,\alpha,\beta)}{\partial \beta}\, dx$$

may be easily evaluated. In view of this fact, we propose to discuss the technique of differentiation under the integral sign.

21.2. Leibnitz's rule for differentiation under the integral sign. **(Kanpur 2011, 13)**

Theorem. *If* $f(x,\alpha)$ *and* $\partial f/\partial\alpha$ *are continuous functions of x and* α *for* $a \le x \le b$, $c \le \alpha \le d$, *a, b being independent of* α, *then*

$$\frac{d}{d\alpha}\int_a^b f(x,\alpha)\, dx = \int_a^b \frac{\partial f}{\partial \alpha}\, dx$$

Proof. Let $$F(\alpha) = \int_a^b f(x,\alpha)\, dx \quad \text{...(1)}$$

Let α change to $\alpha + \delta\alpha$ (α and $\alpha + \delta\alpha$ both lying in the closed interval [*c*, *d*]), then *a*, *b* and *x* being independent of α, remain unaltered and $F(\alpha)$ changes to $F(\alpha + \delta\alpha)$. Hence, we have

$$F(\alpha + \delta\alpha) = \int_a^b f(x, \alpha + \delta\alpha)\, dx. \quad \text{...(2)}$$

From (1) and (2), $$F(\alpha+\delta\alpha)-F(\alpha) = \int_a^b \{f(x,\alpha+\delta\alpha)-f(x,\alpha)\}\,dx \quad \ldots(3)$$

Using the Lagrange's mean value theorem for derivatives, we get

$$f(x,\alpha+\delta\alpha)-f(x,\alpha)=\delta\alpha\cdot\frac{\partial}{\partial\alpha}f(x,\alpha+\theta\,\delta\alpha),\ \text{where } 0<\theta<1 \quad \ldots(4)$$

Using (4), (3) reduces to

$$F(\alpha+\delta\alpha)-F(\alpha) = \int_a^b \delta\alpha\cdot\frac{\partial}{\partial\alpha}f(x,\alpha+\theta\,\delta\alpha)\,dx = \delta\alpha\int_a^b \frac{\partial}{\partial\alpha}f(x,\alpha+\theta\,\delta\,\alpha)\,dx$$

Thus, $$\frac{F(\alpha+\delta\alpha)-F(\alpha)}{\delta\alpha} = \int_a^b \frac{\partial}{\partial\alpha}f(x,\alpha+\theta\,\delta\alpha)\,dx \quad \ldots(5)$$

By definition, we have $$\lim_{\delta\alpha\to 0}\frac{F(\alpha+\delta\alpha)-F(\alpha)}{\delta\alpha} = \frac{dF(\alpha)}{d\alpha} \quad \ldots(6)$$

Taking limits as $\delta\alpha\to 0$ on both sides of (5) and using (6), we get

$$\frac{dF(\alpha)}{d\alpha} = \lim_{\delta\alpha\to 0}\int_a^b \frac{\partial}{\partial\alpha}f(x,\alpha+\theta\,\delta\alpha)\,dx = \int_a^b \lim_{\delta\alpha\to 0}\frac{\partial}{\partial\alpha}f(x,\alpha+\theta\,\delta\alpha)\,dx, \quad \ldots(7)$$

where we have assumed that the limit of integral is equal to the integral of limit. Again, since $\partial f/\partial\alpha$ is continuous, we have

$$\lim_{\delta x\to 0}\frac{\partial}{\partial\alpha}f(x,\alpha+\theta\,\delta\alpha) = \frac{\partial}{\partial\alpha}f(x,\alpha) \quad \ldots(8)$$

Using (8), (7) reduces to

$$\frac{dF(\alpha)}{d\alpha} = \int_a^b \frac{\partial f(x,\alpha)}{\partial\alpha}dx \qquad \text{or} \qquad \frac{d}{d\alpha}\int_a^b f(x,\alpha)\,dx = \int_a^b \frac{\partial f(x,\alpha)}{\partial\alpha}dx, \text{ using (1)}$$

Remark Let $$G(\alpha,\beta) = \int_a^b g(x,\alpha,\beta)\,dx, \quad \ldots(i)$$

where a and b are independent of parameters α and β. Then, proceeding as in the above theorem, we may show that

$$\frac{\partial G(\alpha,\beta)}{\partial\alpha} = \int_a^b \frac{\partial g(x,\alpha,\beta)}{\partial\alpha}dx \quad \text{and} \quad \frac{\partial G(\alpha,\beta)}{\partial\beta} = \int_a^b \frac{\partial g(x,\alpha,\beta)}{\partial\beta}dx \quad \ldots(ii)$$

While dealing with function $g(x,\alpha,\beta)$ of two parameters, we make a choice of appropriate parameter α or β in the above results (*ii*). The selected parameter must lead us to new integral which can be easily evaluated. For more discussion, refer Art. 21.5 A.

21.3. General form of Leibnitz's ruie of differentiation under the integral sign when the limits of integration are functions of the parameters.

Theorem I. *$f(x,\alpha)$ and $\partial f/\partial\alpha$ are contiinuous functions of x and α for $g(\alpha)\le x\le h(\alpha)$, $c\le\alpha\le d$ and $g(\alpha)$, $h(\alpha)$ are themselves functions of α, possessing continuous first order derivatives, then*

$$\frac{d}{d\alpha}\int_{g(\alpha)}^{h(\alpha)} f(x,\alpha)\,dx = \int_{g(\alpha)}^{h(\alpha)}\frac{\partial f(x,\alpha)}{\partial\alpha}dx + \frac{dh(\alpha)}{d\alpha}f(h(\alpha),\alpha) - \frac{dg(\alpha)}{d\alpha}f(g(\alpha),\alpha)$$

Proof. Let $$F(\alpha) = \int_{g(\alpha)}^{h(\alpha)} f(x,\alpha)\,dx \qquad \ldots(1)$$

Since $g(\alpha)$ and $h(\alpha)$ are functions of α, when α changes to $\alpha+\delta\alpha$, let $g(\alpha)$ change to $g(\alpha+\delta\alpha)$, $h(\alpha)$ change to $h(\alpha+\delta\alpha)$ and $F(\alpha)$ change to $F(\alpha+\delta\alpha)$. Hence, we get

$$F(\alpha+\delta\alpha) = \int_{g(\alpha+\delta\alpha)}^{h(\alpha+\delta\alpha)} f(x,\alpha+\delta\alpha)\,dx. \qquad \ldots(2)$$

$$(1) \text{ and } (2) \Rightarrow \quad F(\alpha+\delta\alpha)-F(\alpha) = \int_{g(\alpha+\delta\alpha)}^{h(\alpha+\delta\alpha)} f(x,\alpha+\delta\alpha)\,dx - \int_{g(\alpha)}^{h(\alpha)} f(x,\alpha)\,dx$$

$$= \int_{g(\alpha+\delta\alpha)}^{g(\alpha)} f(x,\alpha+\delta\alpha) + \int_{g(\alpha)}^{h(\alpha)} f(x,\alpha+\delta\alpha)\,dx + \int_{h(\alpha)}^{h(\alpha+\delta\alpha)} f(x,\alpha+\delta\alpha)\,dx - \int_{g(\alpha)}^{h(\alpha)} f(x,\alpha)\,dx$$

$$\text{Thus,} \quad F(\alpha+\delta\alpha)-F(\alpha) = \int_{g(\alpha)}^{h(\alpha)} \{f(x,\alpha+\delta\alpha) - f(x,\alpha)\}\,dx$$

$$+ \int_{h(\alpha)}^{h(\alpha+\delta\alpha)} f(x,\alpha+\delta\alpha)\,dx - \int_{g(\alpha)}^{g(\alpha+\delta\alpha)} f(x,\alpha+\delta\alpha)\,dx \qquad \ldots(3)$$

Using the Lagrange's mean theorem for derivatives, we get

$$f(x,\alpha+\delta\alpha) - f(x,\alpha) = \delta\alpha \frac{\partial}{\partial\alpha} f(x,\alpha+\theta\,\delta\alpha), \quad 0<\theta<1 \qquad \ldots(4)$$

Again, using the mean value theorem for integrals, we get

$$\int_{h(\alpha)}^{h(\alpha+\delta\alpha)} f(x,\alpha+\delta\alpha)\,dx = \{h(\alpha+\delta\alpha)-h(\alpha)\} \times f(\xi,\alpha+\delta\alpha) \qquad \ldots(5)$$

and $$\int_{g(\alpha)}^{g(\alpha+\delta\alpha)} f(x,\alpha+\delta\alpha)\,dx = \{g(\alpha+\delta\alpha)-g(\alpha)\} \times f(\eta,\alpha+\delta\alpha), \qquad \ldots(6)$$

where ξ lies between $h(\alpha)$ and $h(\alpha+\delta\alpha)$ and η lies between $g(\alpha)$ and $g(\alpha+\delta\alpha)$.

Using (4) (5) and (6), (3) reduces to

$$F(\alpha+\delta\alpha)-F(\alpha) = \int_{g(\alpha)}^{h(\alpha)} \delta\alpha \cdot \frac{\partial}{\partial\alpha} f(x,\alpha+\theta\delta\alpha)\,dx$$

$$+ \{h(\alpha+\delta\alpha)-h(\alpha)\} f(\xi,\alpha+\delta\alpha) - \{g(\alpha+\delta\alpha)-g(\alpha)\}\; f(\eta,\alpha+\delta\alpha)$$

or $$\frac{F(\alpha+\delta\alpha)-F(\alpha)}{\delta\alpha} = \int_{g(\alpha)}^{h(\alpha)} \frac{\partial}{\partial\alpha} f(x,\alpha+\theta\delta\alpha)\,dx$$

$$+ \frac{h(\alpha+\delta\alpha)-h(\alpha)}{\delta\alpha} f(\xi,\alpha+\delta\alpha) - \frac{g(\alpha+\delta\alpha)-g(\alpha)}{\delta\alpha} f(\eta,\alpha+\delta\alpha) \qquad \ldots(7)$$

$$\text{Now,} \quad \lim_{\delta\alpha\to 0} \frac{F(\alpha+\delta\alpha)-F(\alpha)}{\delta\alpha} = \frac{dF}{d\alpha}, \quad \lim_{\delta\alpha\to 0} \frac{h(\alpha+\delta\alpha)-h(\alpha)}{\delta\alpha} = \frac{dh}{d\alpha}, \quad \lim_{\delta\alpha\to 0} \frac{g(\alpha+\delta\alpha)-g(\alpha)}{\delta\alpha} = \frac{dg}{d\alpha}$$

Taking limit as $\delta\alpha \to 0$ on both sides of (7) and using the above three results, we obtain

$$\frac{dF}{d\alpha} = \int_{g(\alpha)}^{h(\alpha)} \frac{\partial f(x,\alpha)}{\partial \alpha}\, dx + \frac{dh}{d\alpha} f(h(\alpha),\alpha) - \frac{dg}{d\alpha} f(g(\alpha),\alpha)$$

[Since $h(\alpha) \le \xi \le h(\alpha+\delta\alpha)$ and $g(\alpha) \le \eta \le g(\alpha+\delta\alpha)$, hence

$\lim_{\delta\alpha\to 0} f(\xi, \alpha+\delta\alpha) = f(h(\alpha),\alpha)$ and $\lim_{\delta\alpha\to 0} f(\eta, \alpha+\delta\alpha) = f(g(\alpha),\alpha)$]

Thus, $$\frac{d}{d\alpha}\int_{g(\alpha)}^{h(\alpha)} f(x,\alpha)\, dx = \int_{g(\alpha)}^{h(\alpha)} \frac{\partial f(x,\alpha)}{\partial \alpha}\, dx + \frac{dh(\alpha)}{d\alpha} f(h(\alpha),\alpha) - \frac{dg(\alpha)}{d\alpha} f(g(\alpha),\alpha) \qquad \ldots(8)$$

Remark 1. If $g(\alpha) = a$ and $h(\alpha) = b$ are independent of α, then the last two terms in result (8) are zero and we get

$$\frac{d}{d\alpha}\int_a^b f(x,\alpha)\, dx = \int_a^b \frac{\partial(x,\alpha)}{\partial \alpha} d\alpha,$$

which is what we have already proved in Art 21.2

Remark 2. ***Differentiation under the integral sign in the case of improper integrals.***

The results obtained in Art. 21.2 and Art. 21.3 may not be applicable in the case of improper integrals, and the question of validity of the results to improper integral requires further investigation. However, in our discussion in this chapter we shall omit this investigation. Accordingly, whenever we shall deal with any improper integral, we shall assume that the necessary conditions for validity of the results are satisfied.

21.4A. Evaluation of integral $\int_a^b f(x,\alpha)\, dx$, where a and b are independent of parameter α. WORKING RULE.

Step 1: Let $$F(\alpha) = \int_a^b f(x,\alpha)\, dx \qquad \ldots(1)$$

Step 2: Differentiating both sides of (1) w.r.t. 'α' and using Leibnitz's rule for differentiation under the integral sign (refer Art 21.2), we have

$$\frac{dF}{d\alpha} = \int_a^b \frac{\partial f(x,\alpha)}{\partial \alpha}\, dx \qquad \ldots(2)$$

Step 3: Evaluate the integral on R.H.S of (2) as usual.

Step 4: Solve the resulting differential equation obtained in step 3. The solution so obtained will involve a constant of integration C.

Step 5: Using (1), compute the value of constant C (obtained in step 4) by giving a suitable value to the parameter. Substitute this value of C in the result of step 4 and thus get the value of the integral (1)

21.4 B. Solved examples of type 1 based on Art 21.4 A

Ex. 1. *Assuming the validity of differentiation under the integral sign, show that*

(*i*) $$\int_0^{\pi/2} \frac{\log(1+y\sin^2 x)}{\sin^2 x}\, dx = \pi(\sqrt{1+y}-1), \text{ where } y > -1$$

[Delhi B.Sc. (Hons) III 2008, 11; Kurukshetra B.C.A (II) 2007]

(*ii*) $\int_0^{\pi/2} \log(1-x^2\cos^2\theta)\,d\theta = \pi\left\{\log(1+\sqrt{1-x^2})-\log 2\right\}, \quad x^2 \le 1$

(*iii*) $\int_0^{\pi/2} \log(1-x^2\sin^2\theta)\,d\theta = \pi\left\{\log(1+\sqrt{1-x^2})-\log 2\right\}, \quad x^2 \le 1$

Sol. (*i*). Let $$F(y) = \int_0^{\pi/2} \frac{\log(1+y\sin^2 x)}{\sin^2 x}\,dx \qquad \text{...(1)}$$

Differentiating both sides of (1) w.r.t. '*y*' and using the Leibnitz's rule of differentiation under the integral sign, we get

$$\frac{dF}{dy} = \int_0^{\pi/2} \frac{\partial}{\partial y}\left[\frac{\log(1+y\sin^2 x)}{\sin^2 x}\right]dx = \int_0^{\pi/2} \frac{1}{\sin^2 x}\times\frac{1}{1+y\sin^2 x}\times \sin^2 x\,dx$$

$$= \int_0^{\pi/2} \frac{dx}{1+y\sin^2 x} = \int_0^{\pi/2} \frac{\sec^2 x}{\sec^2 x + y\tan^2 x}\,dx = \int_0^{\pi/2} \frac{\sec^2 x}{1+(1+y)\tan^2 x}\,dx$$

[on dividing the numerator and denominator by $\cos^2 x$]

$$= \int_0^{\infty} \frac{dt}{1+(1+y)t^2}, \text{ putting } \tan x = t \text{ and } \sec^2 x\,dx = dt$$

$$= \frac{1}{1+y}\int_0^{\infty} \frac{dt}{t^2+\{1/(1+y)\}} = \frac{1}{1+y}\times\frac{1}{(1/\sqrt{1+y})}\times\left[\tan^{-1}\frac{t}{(1/\sqrt{1+y})}\right]_0^{\infty}$$

$$= (1+y)^{-1/2}(\tan^{-1}\infty - \tan^{-1} 0) = (1+y)^{-1/2}\times(\pi/2-0) = (\pi/2)\times(1+y)^{-1/2}$$

Thus, $$dF = (\pi/2)\times(1+y)^{-1/2}\,dy \qquad \text{...(2)}$$

Integrating (2), $$F(y) = \pi(1+y)^{1/2} + C, \; C \text{ being an arbitrary constant} \qquad \text{...(3)}$$

Putting $y = 0$ in (1), we get $F(0) = 0$. Next, putting $y = 0$ and $F(0) = 0$ in (3), we get $C = -\pi$. Hence (3) reduces to

$$F(y) = \pi(1+y)^{1/2} - \pi \quad \text{or} \quad \int_0^{\pi/2} \frac{\log(1+y\sin^2 x)}{\sin^2 x}\,dx = \pi(\sqrt{1+y}-1), \text{ using (1)}$$

(*ii*) Let $$F(x) = \int_0^{\pi/2} \log(1-x^2\cos^2\theta)\,d\theta \qquad \text{...(1)}$$

Differntiating both sides of (1) w.r.t. '*x*' and using the Leibnitz's rule of differntiation under the integral sign, we have

$$\frac{dF}{dx} = \int_0^{\pi/2} \frac{\partial \log(1-x^2\cos^2\theta)}{\partial x}\,d\theta = \int_0^{\pi/2} \frac{1}{1-x^2\cos^2\theta}\times(-2x\cos^2\theta)\,d\theta$$

$$= \frac{2}{x}\int_0^{\pi/2} \frac{(-x^2\cos^2\theta)}{1-x^2\cos^2\theta}\,d\theta = \frac{2}{x}\int_0^{\pi/2} \frac{(1-x^2\cos^2\theta)-1}{1-x^2\cos^2\theta}\,d\theta$$

$$= \frac{2}{x}\left[\int_0^{\pi/2} d\theta - \int_0^{\pi/2} \frac{d\theta}{1-x^2\cos^2\theta}\right] = \frac{2}{x}\times\frac{\pi}{2} - \frac{2}{x}\int_0^{\pi/2}\frac{d\theta}{1-x^2\cos^2\theta}$$

$$= \frac{\pi}{x} - \frac{2}{x}\int_0^{\pi/2}\frac{\sec^2\theta}{\sec^2\theta - x^2}\,d\theta = \frac{\pi}{x} - \frac{2}{x}\int_0^{\pi/2}\frac{\sec^2\theta}{1+\tan^2\theta - x^2}\,d\theta$$

$$= \frac{\pi}{x} - \frac{2}{x}\int_0^{\infty}\frac{dt}{t^2+(1-x^2)}, \text{ putting } \quad \tan\theta = t \quad \text{and} \quad \sec^2\theta\, d\theta = dt$$

$$= \frac{\pi}{x} - \frac{2}{x}\left[\frac{1}{\sqrt{1-x^2}}\tan^{-1}\frac{t}{\sqrt{1-x^2}}\right]_0^{\infty} = \frac{\pi}{x} - \frac{2}{x\sqrt{1-x^2}}\times\frac{\pi}{2}$$

Thus, $$dF = \left(\frac{\pi}{x} - \frac{\pi}{x\sqrt{1-x^2}}\right)dx \qquad \ldots(2)$$

Integrating (2), $F(x) = \pi\log x - \pi\int\frac{dx}{x\sqrt{1-x^2}} + C$, C being an arbitrary constant ...(3)

Now, putting $x = 1/z$ and $dz = -(1/z^2)dz$, we have

$$\int\frac{dx}{x\sqrt{1-x^2}} = \int\frac{(-1/z^2)\,dz}{(1/z)\sqrt{1-(1/z)^2}} = -\int\frac{dz}{\sqrt{z^2-1}} = -\log\,(z+\sqrt{z^2-1})$$

$$= -\log\{1/x + \sqrt{(1/x)^2 - 1}\} = -\log\,\{(1+\sqrt{1-x^2})/x)\}$$

$$= -\{\log(1+\sqrt{1-x^2}) - \log x\} = \log x - \log(1+\sqrt{1-x^2})$$

Hence (3) yields, $$F(x) = \pi\log x - \pi\,\{\log x - \log(1+\sqrt{1-x^2})\} + C$$

or $F(x) = \pi\log(1+\sqrt{1-x^2}) + C$, C being an arbitrary constant ...(4)

Putting $x = 0$ in (1), we get $F(0) = 0$. Next, putting $x = 0$ and $F(0) = 0$ in (4), we get $0 = \pi\log 2 + C$ giving $C = -\pi\log 2$. Hence, (4) reduces to

$$F(x) = \pi\log(1+\sqrt{1-x^2}) - \pi\log 2$$

or $$\int_0^{\pi/2}\log(1-x^2\cos^2\theta)\,d\theta = \pi\{\log(1+\sqrt{1-x^2}) - \log 2\}, \text{ using (1)}$$

(*iii*). Using the property $\int_0^a f(x)\,dx = \int_0^a f(a-x)\,dx$ of definite integrals, we have

$$\int_0^{\pi/2}(1-x^2\sin^2\theta)\,d\theta = \int_0^{\pi/2}\{1-x^2\sin^2(\pi/2-\theta)\}\,d\theta = \int_0^{\pi/2}(1-x^2\cos^2\theta)\,d\theta$$

$$= \pi\{\log(1+\sqrt{1-x^2}) - \log 2)\}, \text{ by, part } (ii)$$

Ex. 2 *Assuming the validity of differentiation under the integral sign, show that*

(*i*) $\displaystyle\int_0^{\pi} \frac{\log(1+a\cos x)}{\cos x}\,dx = \pi\sin^{-1} a$, if $|a|<1$ **[Delhi Maths (H) 2005 [Pune 2010]**

(*ii*) $\displaystyle\int_0^{\pi} \frac{\log(1+\sin\alpha\cos x)}{\cos x}\,dx = \pi\alpha$

(*iii*) $\displaystyle\int_0^{\pi/2} \frac{\log(1+\cos\alpha\sin x)}{\cos x}\,dx = \frac{1}{2}\left(\frac{\pi^2}{4}-\alpha\right)$

(*iv*) $\displaystyle\int_{-\pi/2}^{\pi/2} \frac{\log(1+a\sin x)}{\sin x}\,dx = \pi\sin^{-1} a$, if $|a|<1$

Sol. (*i*). Let $\displaystyle F(a) = \int_0^{\pi} \frac{\log(1+a\cos x)}{\cos x}\,dx$ ...(1)

Differentiating both sides of (1) w.r.t 'a' and using the Leibnitz's rule of differentiation under the sign of integral, we get

$$\frac{dF}{da} = \int_0^{\pi} \frac{\partial}{\partial a}\left[\frac{\log(1+a\cos x)}{\cos x}\right]dx = \int_0^{\pi} \frac{1}{\cos x}\cdot\frac{\cos x}{1+a\cos x}\,dx = \int_0^{\pi} \frac{dx}{1+a\cos x} \quad ...(2)$$

From Integral Calculus, $\displaystyle\int \frac{dx}{a+b\cos x} = \frac{1}{\sqrt{a^2-b^2}}\cos^{-1}\frac{b+a\cos x}{a+b\cos x}$, if $b^2<a^2$ ...(3)

Here, given that $|a|<1$ so that $a^2<1$ Hence, using (3), (2) yields

$$\frac{dF}{da} = \frac{1}{\sqrt{1-a^2}}\left[\cos^{-1}\frac{a+\cos x}{1+a\cos x}\right]_0^{\pi} = \frac{1}{\sqrt{1-a^2}}\left[\cos^{-1}\frac{a-1}{1-a} - \cos^{-1}\frac{a+1}{1+a}\right]$$

$$= \{\cos^{-1}(-1)-\cos^{-1}(1)\}/(1-a^2)^{1/2} = (\pi-0)/(1-a^2)^{1/2} = \pi/(1-a^2)^{1/2}$$

Thus, $dF = \{\pi/(1-a^2)^{1/2}\}\,da$

Integrating it, $F(a) = \pi\sin^{-1} a + C$, C being an arbitrary constant ...(4)

Putting $a=0$ in (1), we get $F(0)=0$. Next, putting $a=0$ and $F(0)=0$ in (4), we get $C=0$. Hence, (4) reduces to

$F(a) = \pi\sin^{-1} a$ or $\displaystyle\int_0^{\pi} \frac{\log(1+a\cos x)}{\cos x}\,dx = \pi\sin^{-1} a$, using (1)

(*ii*). Taking $a=\sin\alpha$, proceed as in part (*i*). Also, note that the condition $|a|<1 \Rightarrow$ $|\sin\alpha|<1$, which is true. Thus, $\displaystyle\int_0^{\pi} \frac{\log(1+\sin\alpha\cos\alpha)}{\cos x}\,dx = \pi\sin^{-1}(\sin\alpha) = \pi\alpha$

(*iii*). Let $\displaystyle F(\alpha) = \int_0^{\pi/2} \frac{\log(1+\cos\alpha\cos x)}{\cos x}\,dx$ ...(1)

Differentiating both sides of (1) w.r.t. 'α' and using Leibnitz's rule of differentiation under the integral sign, we have

$$\frac{dF}{d\alpha} = \int_0^{\pi/2} \frac{\partial}{\partial\alpha}\left[\frac{\log(1+\cos\alpha\cos x)}{\cos x}\right]dx = -\sin\alpha\int_0^{\pi/2}\frac{dx}{1+\cos\alpha\cos x} \quad \ldots(2)$$

From Integral Calculus, $$\int\frac{dx}{a+b\cos x} = \frac{1}{\sqrt{a^2-b^2}}\cos^{-1}\frac{b+a\cos x}{a+b\cos x}, \text{ if } b^2 < a^2 \quad \ldots(3)$$

Here $\cos^2\alpha < 1$ and so using (3), (2) yields

$$\frac{dF}{d\alpha} = -\sin\alpha\left[\frac{1}{\sqrt{1-\cos^2\alpha}}\cos^{-1}\frac{\cos\alpha+\cos x}{1+\cos\alpha\cos x}\right]_0^{\pi/2} = -(\cos\cos^{-1}\alpha - \cos^{-1}1) = -\alpha$$

Thus, $$dF = -\alpha\, d\alpha$$

Integrating, $F(\alpha) = -(\alpha^2/2) + C, C$ being an arbitrary constant ...(4)

Putting $\alpha = \pi/2$ in (1), we get $F(\pi/2) = 0$. Next, putting $\alpha = \pi/2$ and $F(\pi/2) = 0$ in (4), we get $0 = -\pi^2/8 + C$ so that $C = \pi^2/8$. Hence (4) yields

$$F(\alpha) = -\frac{\alpha^2}{2}+\frac{\pi^2}{8} \quad \text{or} \quad \int_0^{\pi/2}\frac{\cos(1+\cos\alpha\cos x)}{\cos x}dx = \frac{1}{2}\left(\frac{\pi^2}{4}-\alpha^2\right), \text{ using (1)}$$

(iv). Let $$F(a) = \int_{-\pi/2}^{\pi/2}\frac{\log(1+a\sin x)}{\sin x}dx \quad \ldots(1)$$

Differentiating both sides of (1) w.r.t. 'a' and using Leibnitz's rule of differentiation under the integral sign, we obtain

$$\frac{dF}{da} = \int_{-\pi/2}^{\pi/2}\frac{\partial}{\partial a}\left[\frac{\log(1+a\sin x)}{\sin x}\right]dx = \int_{-\pi/2}^{\pi/2}\frac{dx}{1+a\sin x} \quad \ldots(2)$$

Putting $x = \pi/2 - t$ *and* $dx = -dt$, (2) reduces to

$$\frac{dF}{da} = \int_{\pi}^{0}\frac{(-dt)}{1+a\sin(\pi/2-t)} = \int_0^{\pi}\frac{dt}{1+a\cos t} \quad \ldots(3)$$

From Integral Calculus, $$\int\frac{dx}{a+b\cos x} = \frac{1}{\sqrt{a^2-b^2}}\cos^{-1}\frac{b+a\cos x}{a+b\cos x}, \ b^2 < a^2 \quad \ldots(4)$$

Here, given that $|a| < 1$ so that $a^2 < 1$. Hence, using (4), (3) yields

$$\frac{dF}{da} = \left[\frac{1}{\sqrt{1-a^2}}\cos^{-1}\frac{a+\cos t}{1+\cos t}\right]_0^{\pi} = \frac{1}{\sqrt{1-a^2}}\left[\cos^{-1}\frac{a-1}{1-a} - \cos^{-1}\frac{a+1}{1+a}\right]$$

$dF/da = \{[\cos^{-1}(-1) - \cos 1]\}/(1-a^2)^{1/2}$ or $dF = \{\pi/(1-a^2)^{1/2}\}\, da.$

Integrating, $F(a) = \pi\sin^{-1}a + C,\ C$ being an arbitrary constant ...(5)

Putting $a = 0$ in (1) yields F(0) = 0. Next, putting $a = 0$ and $F(0) = 0$ in (5) yields $C = 0$. Hence, (5) reduces to

$$F(a) = \pi \sin^{-1} a \qquad \text{or} \qquad \int_{-\pi/2}^{\pi/2} \frac{\log(1 + a\sin x)}{\sin x} dx = \pi \sin^{-1} a \text{, using (1)}$$

Ex. 3. *Assuming the validity of differentiation under the integral sign, show that*

$$\int_0^\infty \frac{\tan^{-1} ax}{x(1+x^2)} dx = \frac{\pi}{2} \log(1+a), a \geq 0.$$

Also find the value of integral if a < 0. **[Delhi B.Sc. (Hons) 2008, 11]**

Sol. Let
$$F(a) = \int_0^\infty \frac{\tan^{-1} ax}{x(1+x^2)} dx \qquad \ldots(1)$$

Differentiating both sides of (1) w.r.t. '*a*' and using Leilinitz's rule of differntiation under the integral sign, we obtain

$$\frac{dF}{da} = \int_0^\infty \frac{\partial}{\partial a}\left[\frac{\tan^{-1} ax}{x(1+x^2)}\right] dx = \int_0^\infty \frac{1}{x(1+x^2)} \times \frac{x}{1+a^2x^2} dx = \int_0^\infty \frac{dx}{(1+x^2)(1+a^2x^2)}$$

$$= \frac{1}{1-a^2}\int_0^\infty \left[\frac{1}{1+x^2} - \frac{a^2}{1+a^2x^2}\right], \text{ on resolving into partial fractions}$$

$$= \frac{1}{1-a^2}\left[\int_0^\infty \frac{dx}{1+x^2} - \int_0^\infty \frac{dx}{x^2+(1/a)^2}\right] = \frac{1}{1-a^2}\left[\tan^{-1} x - a\tan^{-1} ax\right]_0^\infty \qquad \ldots(2)$$

Case (*i*). Let $a \geq 0$. Then, (2) reduces to

$$\frac{dF}{da} = \frac{1}{1-a^2}\left[\frac{\pi}{2} - \frac{a\pi}{2}\right] = \frac{1}{(1-a)(1+a)} \times \frac{\pi(1-a)}{2} = \frac{\pi}{2(1+a)}$$

Thus,
$$dF = [\pi/2(1+a)]da$$

Integrating, $F(a) = (\pi/2) \times \log(1+a) + C$, C being an arbitrary constant ...(3)

Putting $a = 0$ in (1) yields $F(0) = 0$. Next, putting $a = 0$ and $F(0) = 0$ in (3) yields $C = 0$. Hence, (3) reduces to

$$F(a) = \frac{\pi}{2}(1+a) \qquad \text{or} \qquad \int_0^\infty \frac{\tan^{-1} ax}{x(1+x^2)} = \frac{\pi}{2}(1+a), \text{ using (1)} \qquad \ldots(4)$$

Case (*ii*). Let $a < 0$. Then, $a < 0 \Rightarrow \tan^{-1}(ax) = -\pi/2$. Therefore, (2) yields

$$\frac{dF}{da} = \frac{1}{1-a^2}\left[\frac{\pi}{2} + \frac{a\pi}{2}\right] = \frac{1}{(1-a)(1+a)} \times \frac{\pi(1+a)}{2} = \frac{\pi}{2(1-a)}$$

Then,
$$dF = [\pi/2(1-a)]da$$

Integrating, $F(a) = -(\pi/2) \times \log(1-a) + D$, D being an arbitrary constant ...(5)

Putting $a = 0$ in (1) yields $F(0) = 0$. Next, putting $a = 0$ and $F(0) = 0$ in (5) yields $D = 0$. Hence, (5) reduces to

$$F(a) = -\frac{\pi(1-a)}{2} \quad \text{or} \quad \int_0^\infty \frac{\tan^{-1} ax}{x(1+x^2)}\,dx = -\frac{\pi(1-a)}{2}, \text{ using (1)}$$

Ex. 4. If $|a| < 1$, prove that $\int_0^\pi \log(1+a\cos x)\,dx = \pi\log(1/2+\sqrt{1-a^2}/2)$ **[Delhi 2007]**

Sol. Let
$$F(a) = \int_0^\pi \log(1+a\cos x)\,dx \qquad \text{...(1)}$$

Differentiating boths sides of (1) w.r.t. 'a' and using Leibnitz's rule of differentiation under the integral sign, we get

$$\frac{dF}{da} = \int_0^\pi \frac{\partial}{\partial a}[\log(1+a\cos x)]\,dx = \int_0^\pi \frac{\cos x\,dx}{1+a\cos x} = \frac{1}{a}\int_0^\pi \frac{(1+a\cos x)-1}{1+a\cos x}\,dx$$

or
$$\frac{dF}{da} = \frac{1}{a}\int_0^\pi \left(1-\frac{1}{1+a\cos x}\right)dx = \frac{\pi}{a}-\frac{1}{a}\int_0^\pi \frac{dx}{1+a\cos x} \qquad \text{...(2)}$$

From Integral Calculus,
$$\int \frac{dx}{a+b\cos x} = \frac{1}{\sqrt{a^2-b^2}}\cos^{-1}\frac{b+a\cos x}{a+b\cos x}, \text{ if } b^2 < a^2$$

Here, given that $|a| < 1$ so that $a^2 < 1$. Hence, using the above formula, (2) reduces to

$$\frac{dF}{da} = \frac{\pi}{a}-\frac{1}{a}\left[\frac{1}{\sqrt{1-a^2}}\cos^{-1}\frac{a+\cos x}{1+a\cos x}\right]_0^\pi = \frac{\pi}{a}-\frac{1}{a\sqrt{1-a^2}}\left[\cos^{-1}\frac{a-1}{1-a}-\cos^{-1}\frac{a+1}{1+a}\right]$$

or
$$\frac{dF}{da} = \frac{\pi}{a}-\frac{\pi}{a\sqrt{1-a^2}} \quad \text{or} \quad dF = \left(\frac{\pi}{a}-\frac{\pi}{a\sqrt{1-a^2}}\right)da$$

[Using the fact that $\cos^{-1}(-1) = \pi$ *and* $\cos^{-1}1 = 0$]

Integrating,
$$F(a) = \pi\log a - \pi\int \frac{da}{a\sqrt{1-a^2}} + C, \text{ } C \text{ being an arbitrary constant} \qquad \text{...(3)}$$

Putting $a = 1/t$ and $da = -(1/t^2)dt$, we get

$$\int \frac{da}{a\sqrt{1-a^2}} = -\int \frac{(1/t^2)\,dt}{(1/t)\sqrt{1-(1/t^2)}} = -\int \frac{dt}{\sqrt{t^2-1}} = -\log(t+\sqrt{t^2-1})$$

$$= -\log\{1/a+\sqrt{1-(1/a^2)}\} = -\log\{(1+\sqrt{1-a^2})/a\} = \log a - \log(1+\sqrt{1-a^2})$$

Hence, (3) yields
$$F(a) = \pi\log a - \pi\{\log a - \log(1+\sqrt{1-a^2})\} + C$$

or
$$F(a) = \pi\log(1+\sqrt{1-a^2}) + C \qquad \text{...(4)}$$

Putting $a = 0$ in (1) yields $F(0) = 0$. Next, putting $a = 0$ and $F(0) = 0$ in (4) yields $0 = \pi\log 2 + C$ so that $C = -\pi\log 2$. Hence, (4) reduces to

$$F(a) = \pi\{\log(1+\sqrt{1-a^2}) - \log 2\} = \pi\log\{(1+\sqrt{1-a^2})/2\}$$

or
$$\int_0^\pi \log(1+a\cos x)\,dx = \pi\log\{1/2+(\sqrt{1-a^2})/2\}, \text{ using (1)}$$

Ex. 5. *Assuming the validity of differntiation under integral sign, show that*

$$\int_0^1 \log\left(\frac{1+ax}{1-ax}\right)\frac{dx}{x\sqrt{1-x^2}} = \pi \sin^{-1} a$$ **[Delhi Maths (II) 1994]**

Sol. Let $$F(a) = \int_0^1 \log\left(\frac{1+ax}{1-ax}\right)\frac{dx}{x\sqrt{1-x^2}}. \qquad \text{...(1)}$$

Differentiating boths sides of (1) w.r.t. 'a' and using Leibnitz's rule of differentiating under the integral sign, we get

$$\frac{dF}{da} = \int_0^1 \frac{1}{(1+ax)/(1-ax)} \times \frac{x(1-ax)-(-x)(1+ax)}{(1-ax)^2} \cdot \frac{dx}{x\sqrt{1-x^2}}$$

$$= 2\int_0^1 \frac{dx}{(1-a^2x^2)\sqrt{1-x^2}} = \int_0^1 \frac{(2/x^3)\,dx}{(1/x^2-a^2)\sqrt{(1/x^2)-1}}$$

$$= \int_\infty^0 \frac{(-2t)\,dt}{(t^2+1-a^2)\,t}, \quad \text{putting } \frac{1}{x^2}-1 = t^2 \quad \text{and} \quad -\frac{2}{x^3}dx = 2t\,dt$$

$$= 2\int_0^\infty \frac{dt}{t^2+(\sqrt{1-a^2})^2} = \frac{2}{\sqrt{1-a^2}}\left[\tan^{-1}\frac{t}{\sqrt{1-a^2}}\right]_0^\infty = \frac{2}{\sqrt{1-a^2}} \times \frac{\pi}{2}, \quad \text{if } a^2 < 1$$

Thus, $$F(a) = (\pi/\sqrt{1-a^2})\,da \qquad \text{...(2)}$$

Integrating (2), $$F(a) = \pi \sin^{-1} a + C, \; C \text{ being an arbitrary constant} \qquad \text{...(3)}$$

Putting $a = 0$ in (1) yields $F(0) = 0$. Next, putting $a = 0$ and $F(0) = 0$ in (3) yields $C = 0$. Hence (3) reduces to

$$F(a) = \pi \sin^{-1} a \quad \text{or} \quad \int_0^1 \log\left(\frac{1+ax}{1-ax}\right)\frac{dx}{x\sqrt{1-x^2}} = \pi \sin^{-1} a, \text{ using (1)}$$

Ex. 6. *If $y > 0$, snow that* $\displaystyle\int_0^\infty e^{-xy}\frac{\sin x}{x}\,dx = \cot^{-1} y = \pi/2 - \tan^{-1} y$

[Kurukshetra B.C.A (II) 2008]

Sol. Let $$F(y) = \int_0^\infty e^{-xy}\frac{\sin x}{x}\,dx \qquad \text{...(1)}$$

Differentiating boths sides of (1) w.r.t. 'y' and using Leibnitz's z rule of differentiation under integral sign, we get

$$\frac{dF}{dy} = \int_0^\infty \frac{\partial}{\partial y}\left(e^{-xy}\frac{\sin x}{x}\right)dx = \int_0^\infty \frac{\sin x}{x}e^{-xy}(-x)\,dx = -\int_0^\infty e^{-yx}\sin x\,dx$$

$$= -\left[\frac{e^{-yx}}{y^2+1}(-y\sin x - \cos x)\right]_0^\infty, \quad \text{as } \int e^{ax}\sin bx\,dx = \frac{e^{ax}(a\sin bx - b\cos bx)}{a^2+b^2}$$

or $\quad dF/dy = -[1/(1+y^2)] \quad$ or $\quad dF = -\{1/(1+y^2)\}\,dy$

Integrating, $F(y) = \cot^{-1} y + c$, c being an arbitrary constant . .(2)

Letting $y \to \infty$ in (1) $\Rightarrow$ $F(y) \to 0$ as $y \to \infty$. Next, letting $y \to \infty$ in (3) and using the fact that $F(y) \to 0$ as $y \to \infty$, we get $C = 0$. Hence, (2) reduces to

$$F(y) = \cot^{-1} y \qquad \text{or} \qquad \int_0^\infty e^{-xy} \frac{\sin x}{x} dx = \cot^{-1} y, \text{ using (1)}$$

or $$\int_0^\infty e^{-xy} \frac{\sin x}{x} dx = \pi/2 - \tan^{-1} y, \qquad \text{as} \qquad \tan^{-1} y + \cot^{-1} y = \pi/2$$

Ex.7. *Assuming the validity of differentiation under the integral sign, showt that*

(*i*) $\displaystyle\int_0^\infty e^{-x^2} \cos \alpha x\, dx = \frac{1}{2}\sqrt{\pi}\, e^{-\alpha^2/4}$ **[Delhi Physics. (H) 1994]**

(*ii*) $\displaystyle\int_0^\infty \exp(-x^2) \cos 2\alpha x\, dx = \frac{1}{2}\sqrt{\pi}\ \exp\ (-\alpha^2)$ **[Delhi Physics (H) 1998]**

Sol. (*i*) Let $$F(\alpha) = \int_0^\infty e^{-x^2} \cos \alpha x\, dx \qquad \text{...(1)}$$

Differentiating both sides of (1) w.r.t. 'α' and using Leibnitz's rule under the integral sign, we have

$$\frac{dF}{d\alpha} = \int_0^\infty \frac{\partial (e^{-x^2} \cos \alpha x)}{\partial \alpha} dx = \int_0^\infty e^{-x^2} (-x \sin \alpha x) dx = \frac{1}{2}\int_0^\infty \sin \alpha x \cdot (-2xe^{-x^2})\, dx \qquad \text{...(2)}$$

Putting $-x^2 = t$ and $-2x\, dx = dt$, we have

$$\int (-2x) e^{-x^2} dx = \int e^t\, dt = e^t = e^{-x^2} \qquad \text{as} \qquad t = -x^2 \qquad \text{...(3)}$$

Integrating by parts the integral on R.H.S. of (2) and using (3) while treating $(-2x e^{-x^2})$ as function to be integrated, (2) reduces to

$$\frac{dF}{d\alpha} = \frac{1}{2}\left\{\left[\sin \alpha x \cdot e^{-x^2}\right]_0^\infty - \int_0^\infty \alpha \cos \alpha x\, e^{-x^2} dx\right\} = -\frac{\alpha}{2}\int_0^\infty e^{-x^2} \cos \alpha x\, dx \qquad \text{...(4)}$$

Now, (1) and (4) $\Rightarrow$ $$\frac{dF}{d\alpha} = -\frac{\alpha}{2} F(\alpha) \qquad \text{or} \qquad \frac{F'(\alpha)}{F(\alpha)} d\alpha = -\frac{\alpha}{2}\, d\alpha$$

Integrating, $\log F(\alpha) = -(\alpha^2/4) + C$, C being an arbitrary constant ...(5)

Putting $\alpha = 0$ in (1) yields $$F(0) = \int_0^\infty e^{-x^2} dx \qquad \text{...(6)}$$

Putting $x^2 = u$, i.e., $x = u^{1/2}$ and $dx = (1/2) \times u^{-1/2} du$, (6) yields

$$F(0) = \frac{1}{2}\int_0^\infty e^{-u} u^{-1/2}\, du = \frac{1}{2}\int_0^\infty e^{-u} u^{(1/2)-1}\, du = (1/2) \times \Gamma(1/2)$$

[Since, by definition (see Art 20.2), $\int_0^\infty e^{-x} x^{n-1} dx = \Gamma(n)$]

or $F(0) = (1/2)\times\sqrt{\pi}$, as $\Gamma(1/2)=\sqrt{\pi}$, (see Art. 20.5)

Putting $\alpha = 0$ in (5) and using $F(0) = \sqrt{\pi}/2$, (5) yields $C = \log(\sqrt{\pi}/2)$. So, (5) gives

$$\log F(\alpha) = -(\alpha^2/4) + \log(\sqrt{\pi}/2) \qquad \text{or} \qquad F(\alpha) = (\sqrt{\pi}/2)\times e^{-\pi^2/4}$$

or
$$\int_0^\infty e^{-x^2}\cos\alpha x\, dx = (\sqrt{\pi}/2)\times e^{-\pi^2/4}, \text{ using (1)} \qquad \ldots(7)$$

(ii). Note that exp a stands for e^a. Hence, we are to show that

$$\int_0^\infty e^{-x^2}\cos 2\alpha x\, dx = (\sqrt{\pi}/2)\times e^{-\alpha^2} \qquad \ldots(8)$$

Replacing α by 2α in (7), we get the required result (8).

Ex. 8. Evaluate $\int_0^\infty (e^{-x}/x)\{a-(1/x)+(1/x)\times e^{-ax}\}dx$

Sol. Let
$$F(a) = \int_0^\infty (e^{-x}/x)\{a-(1/x)+(1/x)\times e^{-ax}\}dx \qquad \ldots(1)$$

Differentiating both sides of (1) w.r.t. 'a' and using Leibnitz's rule of differentiation under the integral sign, we have

$$\frac{dF}{da} = \int_0^\infty \frac{\partial}{\partial a}\left\{\frac{e^{-x}}{x}\left(a-\frac{1}{x}+\frac{1}{x}e^{-ax}\right)\right\}dx = \int_0^\infty \frac{e^{-x}}{x}(1-e^{-ax})dx \qquad \ldots(2)$$

Again, differentiating both sides of (2) w.r.t 'a' as before yields

$$\frac{d^2F}{da^2} = \int_0^\infty \frac{\partial}{\partial a}\left\{\frac{e^{-x}}{x}\left(1-e^{-ax}\right)\right\}dx = \int_0^\infty \frac{e^{-x}}{x}(xe^{-ax})dx = \int_0^\infty e^{-(1+a)x}dx$$

or
$$d^2F/da^2 = \left[e^{-(1+a)x}/\{-(1+a)\}\right]_0^\infty = 1/(1+a) \qquad \ldots(3)$$

Integrating (3), $dF/da = \log(1+a) + C_1$, C_1 being an arbitrary constant ...(4)

Putting $a = 0$ in (2), we get $dF/da = 0$. Next, putting $a = 0$ and $dF/da = 0$ in (4), we get $C_1 = 0$. Hence, (4) reduces to

$$dF/da = \log(1+a) \qquad \text{or} \qquad dF = \log(1+a)\,da \qquad \ldots(5)$$

Integrating (5), $F(a) = \int \log(a+1)da = a.\log(a+1) - \int\{a/(1+a)\}da$

or $F(a) = a\log(a+1) - \int\{1-1/(1+a)\}\,da = a\log(a+1) - a + \log(a+1) + C_2$

or $F(a) = (a+1)\log(a+1) - a + C_2$, C_2 being an arbitrary constant ... (6)

Putting $a = 0$ in (1), we get $F(a) = 0$. Next, putting $a = 0$ and $F(a) = 0$ in (6), we get $C_2 = 0$. Hence, (6) reduces to

$$F(a) = (a+1)\log(a+1) - a \qquad \ldots(7)$$

From (1) and (6),
$$\int_0^\infty (e^{-x}/x)\left\{a-(1/x)+(1/x)e^{-ax}\right\}dx = (a+1)\log(a+1) - a$$

EXERCISE 21 (A)

Assuming the validity of differentiation under the integral sign, show that

$$\int_0^1 \frac{x^\alpha - 1}{\log x}dx = \log(1+\alpha), \alpha > -1$$ **[Pune 2010; Delhi Maths (H) 2006]**

2. $\int_0^{\pi/2} \log(1-e^2\sin^2\theta)\,d\theta = \pi\log\{(1+\sqrt{1-e^2})/2\}$, if $e^2 < 1$

3. $\int_0^{\infty} \frac{1-\cos mx}{x} e^{-x}\,dx = \frac{1}{2}\log(1+m^2)$

4. $\int_0^{\infty} e^{-(x^2+b^2/x)}\,dx = (\sqrt{\pi}/2)\times e^{-2b},\ b \geq 0$

21.5A. Evaluation of integral $\int_a^b g(x,\alpha,\beta)\,dx$, where a and b are independent of parameters α and β. Working rule.

Step 1. Let
$$G(\alpha,\beta) = \int_a^b g(x,\alpha,\beta)\,dx \qquad ...(i)$$

Step 2. Read carefully remark of Art 21.2 for getting

$$\frac{\partial G(\alpha,\beta)}{\partial\alpha} = \int_a^b \frac{\partial g(x,\alpha,\beta)}{\partial\alpha}\,dx \quad \text{or} \quad \frac{\partial G(\alpha,\beta)}{\partial\beta} = \int_a^b \frac{\partial g(x,\alpha,\beta)}{\partial\beta}\,dx \qquad ...(ii)$$

Steps 3 to 5: Read Art 21.4 A with corresponding modifications. The process will be clear by the following solved examples given in Art 21.5 B.

21.5B. Solved examples of type 2 based on Art 21.5 A

Ex.1. *Show that* $\int_0^{\infty} e^{-\alpha x}\frac{\sin\beta x}{x}\,dx = \tan^{-1}\frac{\beta}{\alpha},\ \alpha \geq 0$

and deduce that $\int_0^{\infty} \frac{\sin\beta x}{x}\,dx = \begin{cases} \pi/2, & \text{if } \beta > 0 \\ 0, & \text{if } \beta = 0 \\ -\pi/2, & \text{if } \beta < 0 \end{cases}$ **[Kanpur 2001]**

Sol. First part. Here, the integrand $(e^{-\alpha x}\sin\beta x)/x$ contains two parameters α and β. In order to get rid of the factor $(1/x)$ in this integrad, we must treat only β as parameter. So, let

$$F(\beta) = \int_0^{\infty} e^{-\alpha x}\frac{\sin\beta x}{x}\,dx \qquad ...(1)$$

Differentiating both sides of (1) w.r.t. 'β' and using Leibnitz's rule of differentiation under the integral sign, we get

$$\frac{dF}{d\beta} = \int_0^{\infty}\frac{\partial}{\partial\beta}\left[\frac{e^{-\alpha x}\sin\beta x}{x}\right]dx = \int_0^{\infty}\frac{e^{-\alpha x}}{x}\times(x\cos\beta x)\,dx = \int_0^{\infty} e^{-\alpha x}\cos\beta\,dx$$

$$= \left[\frac{e^{-\alpha x}}{\alpha^2+\beta^2}(-\alpha\cos\beta x+\beta\sin\beta x)\right]_0^{\infty}, \text{ as } \int_0^{\infty} e^{ax}\cos bx\,dx = \frac{e^{ax}(a\cos bx+b\sin bx)}{a^2+b^2}$$

or $\quad \frac{dF}{d\beta} = \frac{\alpha}{\alpha^2+\beta^2} \quad$ or $\quad dF = \frac{\alpha}{\alpha^2+\beta^2}\,d\beta,\ \ \alpha > 0$

Integrating, $\quad F(\beta) = \tan^{-1}(\beta/\alpha) + C$, C being an arbitrary constant. $\quad ...(2)$

Putting $\beta = 0$ in (1) yields $F(0) = 0$. Next, putting $\beta = 0$ and $F(0) = 0$ in (2) yields $C = 0$. Hence, (2) reduces to

$$F(\beta) = \tan^{-1}\frac{\beta}{\alpha} \quad \text{or} \quad \int_0^\infty e^{-\alpha x}\frac{\sin\beta x}{x}\,dx = \tan^{-1}\frac{\beta}{\alpha}, \text{ using (1)}$$

Letting $\alpha \to 0$ on both sides of the above result, we get

$$\int_0^\infty \frac{\sin\beta x}{x}\,dx = \lim_{\alpha\to 0}\tan^{-1}\frac{\beta}{\alpha} = \begin{cases}\pi/2, \text{ if } \beta > 0\\ 0, \text{ if } \beta = 0\\ -\pi/2, \text{ if } \beta < 0\end{cases}$$

Ex. 2. *Assuming the validity of differentiation under the integral sign, show that*

(*i*) $\displaystyle\int_0^{\pi/2}\log\frac{a+b\sin\theta}{a-b\sin\theta}\frac{d\theta}{\sin\theta} = \pi\sin^{-1}\frac{b}{a}$, $a > b$. **[Delhi Maths (H) 2002, 03, 07, 09]**

(*ii*) $\displaystyle\int_0^{\pi/2}\log\frac{1+\lambda\sin\theta}{1-\lambda\sin\theta}\frac{d\theta}{\sin\theta} = \pi\sin^{-1}\lambda$, where $\lambda < 1$

Sol (*i*). Here, the integrand contains two parameters a and b. Because of the presence of $(1/\sin\theta)$ as a factor in the integrand, we treat only b as a parameter. So let

$$F(b) = \int_0^{\pi/2}\log\frac{a+b\sin\theta}{a-b\sin\theta}\frac{d\theta}{\sin\theta} \qquad \ldots(1)$$

Differentiating both sides of (1) w.r.t 'b' and using Leibnitz's rule of differentiation under the integral sign, we have

$$\frac{dF}{db} = \int_0^{\pi/2}\frac{\partial}{\partial b}\left[\{\log(a+b\sin\theta) - \log(a-b\sin\theta)\}\times\frac{1}{\sin\theta}\right]d\theta$$

$$= \int_0^{\pi/2}\left(\frac{\sin\theta}{a+b\sin\theta} + \frac{\sin\theta}{a-b\sin\theta}\right)\frac{d\theta}{\sin\theta} = 2a\int_0^{\pi/2}\frac{d\theta}{a^2 - b^2\sin^2\theta}$$

$$= 2a\int_0^{\pi/2}\frac{\sec^2\theta d\theta}{a^2\sec^2\theta - b^2\tan^2\theta} = 2a\int_0^{\pi/2}\frac{\sec^2\theta\, d\theta}{a^2(1+\tan^2\theta) - b^2\tan^2\theta}$$

$$= 2a\int_0^\infty\frac{dt}{a^2 + t^2(a^2-b^2)}, \text{ putting } \tan\theta = t \text{ and } \sec^2\theta\, d\theta = dt$$

$$= \frac{2a}{a^2-b^2}\int_0^\infty\frac{dt}{t^2 + (a/\sqrt{a^2-b^2})^2} = \left[\frac{2a}{a^2-b^2}\times\frac{1}{(a/\sqrt{a^2-b^2})}\tan^{-1}\frac{t}{a/\sqrt{a^2-b^2}}\right]_0^\infty$$

Thus, $\quad dF/db = (\pi/\sqrt{a^2-b^2}) \quad$ or $\quad dF = (\pi/\sqrt{a^2-b^2})\,db$.

Integrating, $\quad F(b) = \pi\sin^{-1}(b/a) + c$, c being an arbitrary constant $\qquad \ldots(2)$

Putting $b = 0$ in (1) yields $F(0) = 0$. Next, putting $b = 0$ and $F(0) = 0$ in (2), yields $C = 0$. Hence, (2) reduces to

$$F(b) = \pi\sin^{-1}\frac{b}{a} \quad \text{or} \quad \int_0^{\pi/2}\log\frac{a+b\sin\theta}{a-b\sin\theta}\frac{d\theta}{\sin\theta} = \pi\sin^{-1}\frac{b}{a}, \text{ using (1)} \qquad \ldots(4)$$

(*ii*). Re-writing (4), $$\int_0^{\pi/2} \log\frac{1+(b/a)\times\sin\theta}{1-(b/a)\times\sin\theta}\frac{d\theta}{\sin\theta} = \pi\sin^{-1}\frac{b}{a} \qquad \ldots(5)$$

(4) is true for $a > b$, i.e., $b < a$. Setting $b/a = \lambda$ in (5), we get

$$\int_0^{\pi/2} \log\frac{1+\lambda\sin\theta}{1-\lambda\sin\theta}\frac{d\theta}{\sin\theta} = \pi\sin^{-1}\lambda, \text{ where } \lambda < 1$$

Ex. 3. *Assuming the validity of differentiation under the integral sign, show that*

(*i*) $\int_0^{\pi/2} \log(\alpha\cos^2\theta+\beta\sin^2\theta)\,d\theta = \pi\log\{(\sqrt{\alpha}+\sqrt{\beta})/2\}$. **[Delhi-Maths (H) 2000,04]**

(*ii*) $\int_0^{\pi/2} \log(a^2\cos^2\theta+b^2\sin^2\theta)\,d\theta = \pi\log\{(a+b)/2\}$

Sol (*i*). Let $$G(\alpha,\beta) = \int_0^{\pi/2} (\alpha\cos^2\theta+\beta\sin^2\theta)d\theta \qquad \ldots(1)$$

Differentiating both sides of (1) partally w.r.t. 'α' and using Leibnitz's rule of differentiation under the integral sign, we get

$$\frac{\partial G}{\partial\alpha} = \int_0^{\pi/2}\frac{\cos^2\theta}{\alpha\cos^2\theta+\beta\sin^2\theta}d\theta = \int_0^{\pi/2}\frac{d\theta}{\alpha+\beta\tan^2\theta} \qquad \ldots(2)$$

Putting $\tan\theta = t$ so that $\sec^2\theta\, d\theta = dt$ i.e., $d\theta = (dt)/(1+\tan^2\theta)$ i.e., $d\theta = (dt)/(1+t^2)$, (1) reduces to

$$\frac{\partial G}{\partial\alpha} = \int_0^{\infty}\frac{dt}{(1+t^2)(\alpha+\beta t^2)} = \frac{1}{\alpha-\beta}\int_0^{\infty}\left(\frac{1}{1+t^2}-\frac{\beta}{\alpha+\beta t^2}\right)dt, \text{ if } \alpha\neq\beta$$

[on resolving into partial fractions]

$$= \frac{1}{\alpha-\beta}\int_0^{\infty}\left\{\frac{1}{1+t^2}-\frac{1}{t^2+(\alpha/\beta)}\right\}dt = \frac{1}{\alpha-\beta}\left[\tan^{-1}t-\frac{1}{\sqrt{\alpha/\beta}}\tan^{-1}\frac{t}{\sqrt{\alpha/\beta}}\right]_0^{\infty}$$

$$= \frac{1}{\alpha-\beta}\left[\frac{\pi}{2}-\frac{\sqrt{\beta}}{\sqrt{\alpha}}\times\frac{\pi}{2}\right] = \frac{\pi}{(\sqrt{\alpha})^2-(\sqrt{\beta})^2}\times\frac{\sqrt{\alpha}-\sqrt{\beta}}{2\sqrt{\alpha}} = \frac{\pi}{2\sqrt{\alpha}(\sqrt{\alpha}+\sqrt{\beta})} \qquad \ldots(3)$$

For $\beta = \alpha$, (2) reduces to

$$\frac{\partial G}{\partial\alpha} = \int_0^{\pi/2}\frac{\cos^2\theta}{\alpha}d\theta = \frac{1}{2\alpha}\int_0^{\pi/2}(1+\cos 2\theta)\,d\theta = \frac{1}{2\alpha}\left[\theta+\frac{\sin 2\theta}{2}\right]_0^{\pi/2}$$

Thus, $$\partial G/\partial\alpha = \pi/4\alpha. \qquad \ldots(4)$$

From (3) and (4), it follows that without exception, we have

$$\frac{\partial G}{\partial\alpha} = \frac{\pi}{2\sqrt{\alpha}(\sqrt{\alpha}+\sqrt{\beta})} = \pi\,\frac{(1/2\sqrt{\alpha})}{\sqrt{\alpha}+\sqrt{\beta}}$$

Integrating w.r.t. 'α', $$G(\alpha,\beta) = \pi\int\frac{d(\sqrt{\alpha}+\sqrt{\beta})/d\alpha}{\sqrt{\alpha}+\sqrt{\beta}}d\alpha = \pi\log(\sqrt{\alpha}+\sqrt{\beta})+C, \qquad \ldots(5)$$

where C is independent of α. Using the well known property $\int_0^a f(x)\,dx = \int_0^a f(a-x)\,dx$ of definite integrals, (1) yields

$$G(\alpha,\beta) = \int_0^{\pi/2} \log(\alpha\cos^2\theta+\beta\sin^2\theta)\,d\theta = \int_0^{\pi/2} \log\{\alpha\cos^2(\pi/2-\theta)+\beta\sin^2(\pi/2-\theta)\}\,d\theta$$

$$= \int_0^{\pi/2} \log(\beta\cos^2\theta+\alpha\sin^2\theta)\,d\theta = G(\beta,\alpha), \text{ using (1)}$$

In view of the relation $G(\alpha,\beta) = G(\beta,\alpha)$, it follows that C occruing in (5) is also independent of β. Hence, C is an absolute constant Now, putting $\alpha=\beta=1$ in (1), we get

$$G(1, 1) = \int_0^{\pi/2} \log(\cos^2\theta+\sin^2\theta)\,d\theta = \int_0^{\pi/2} (\log 1)\,d\theta = 0, \text{ as } \log 1 = 0$$

Putting $\alpha=\beta=1$ in (5) and using $G(1, 1) = 0$, we get $0 = \pi\log 2 + C$ so that $C = -\pi\log 2$. Hence, from (1) and (5), we get

$$\int_0^{\pi/2} \log(\alpha\cos^2\theta+\beta\sin^2\theta)\,d\theta = \pi\{\log(\sqrt{\alpha}+\sqrt{\beta})-\log 2\} = \pi\log\{(\sqrt{\alpha}+\sqrt{\beta})/2\} \quad \ldots(6)$$

(*ii*). Replacing α and β by a^2 and b^2 respectively in equation (6) of part (*i*), we get

$$\int_0^{\pi/2} \log(a^2\cos^2\theta+b^2\sin^2\theta)\,d\theta = \pi\log\{(a+b)/2\}$$

Ex. 4. *Show that* $\displaystyle\int_0^\infty \frac{\tan^{-1}\alpha x\tan^{-1}\beta x}{x^2}dx = \frac{\pi}{2}\log\left\{\frac{(\alpha+\beta)^{\alpha+\beta}}{\alpha^\alpha\beta^\beta}\right\}, \alpha>0, \beta>0$

Sol. Let
$$G(\alpha,\beta) = \int_0^\infty \frac{\tan^{-1}\alpha x\tan^{-1}\beta x}{x^2}dx \quad \ldots(1)$$

Differentiating both sides of (1) w.r.t. 'α' and using the Leibnitz's rule of differention under the integral sign, we have

$$\frac{\partial G}{\partial\alpha} = \int_0^\infty \frac{\partial}{\partial\alpha}\left\{\frac{\tan^{-1}\alpha x\tan^{-1}\beta x}{x^2}\right\}dx = \int_0^\infty \frac{\tan^{-1}\beta x}{x(1+\alpha^2x^2)}dx \quad \ldots(2)$$

Again, differentiating both sides of (2) w.r.t. 'β' and using the Leibnit's rule of differentiation under the integral sign, we have

$$\frac{\partial^2 G}{\partial\beta\partial\alpha} = \int_0^\infty \frac{\partial}{\partial\beta}\left\{\frac{\tan^{-1}\beta x}{x(1+\alpha^2x^2)}\right\}dx = \int_0^\infty \frac{dx}{(1+\alpha^2x^2)(1+\beta^2x^2)}$$

$$= \int_0^\infty \frac{1}{\alpha^2-\beta^2}\left(\frac{\alpha^2}{1+\alpha^2x^2}-\frac{\beta^2}{1+\beta^2x^2}\right)dx, \text{ on resolving into partial fractions}$$

$$= \frac{1}{\alpha^2-\beta^2}\int_0^\infty\left\{\frac{1}{(1/\alpha)^2+x^2}-\frac{1}{(1/\beta)^2+x^2}\right\}dx = \frac{1}{\alpha^2-\beta^2}\Big[\alpha\tan^{-1}\alpha x-\beta\tan^{-1}\beta x\Big]_0^\infty, \alpha\neq\beta$$

or $$\frac{\partial^2 G}{\partial\alpha\partial\beta} = \frac{1}{\alpha^2-\beta^2}\left(\alpha\times\frac{\pi}{2}-\beta\times\frac{\pi}{2}\right) = \frac{1}{(\alpha-\beta)(\alpha+\beta)}\cdot\frac{\pi(\alpha-\beta)}{2} = \frac{\pi}{2(\alpha+\beta)}, \alpha>0, \beta>0 \quad ...(3)$$

It is easy to show that (3) remains valid even for $\alpha = \beta$.

Now, integrating (3) w.r.t. 'β', we have

$$\partial G/\partial\alpha = (\pi/2)\times\log(\alpha+\beta)+f(\alpha),\ f(\alpha), \text{ being an arbitrary function of } \alpha \quad ...(4)$$

Putting $\beta = 0$ in (2) yields $\partial G/\partial\alpha = 0$. Next, putting $\beta = 0$ and $\partial G/\partial\alpha = 0$ in (4) yields $0 = (\pi/2)\times\log\alpha + f(\alpha)$ so that $f(\alpha) = -(\pi/2)\times\log\alpha$. Hence, (4) reduces to

$$\partial G/\partial\alpha = (\pi/2)\times\log(\alpha+\beta)-(\pi/2)\times\log\alpha \quad ...(5)$$

Integrating (5) w.r.t. 'α', we get

$$G(\alpha,\beta) = \frac{\pi}{2}\int\log(\alpha+\beta)\cdot 1\,d\alpha - \frac{\pi}{2}\int\log\alpha\cdot 1\,d\alpha = \frac{\pi}{2}\left\{\log(\alpha+\beta)\cdot\alpha - \int\frac{\alpha d\alpha}{\alpha+\beta}\right\} - \frac{\pi}{2}\left\{\log\alpha\cdot\alpha - \int\frac{1}{\alpha}\cdot\alpha d\alpha\right\}$$

or $$G(\alpha,\beta) = \frac{\pi}{2}\left\{\alpha\log(\alpha+\beta) - \int\left(1-\frac{\beta}{\alpha+\beta}\right)d\alpha\right\} - \frac{\pi}{2}(\alpha\log\alpha - \alpha)$$

$$G(\alpha,\beta) = (\pi/2)\{\alpha\log(\alpha+\beta)-\alpha+\beta\log(\alpha+\beta)\} - (\pi/2)\times(\alpha\log-\alpha)+g(\beta),$$

where $g(\beta)$ in an arbitrary function of β.

Thus. $$G(\alpha,\beta) = (\pi/2)\times\{(\alpha+\beta)\log(\alpha+\beta)-\alpha\log\alpha\}+g(\beta) \quad ...(6)$$

Putting $\alpha = 0$ in (1) yields $G(0, \beta) = 0$. Next, putting $\alpha = 0$ and $G(0,\beta) = 0$ in (6) gives

$$0 = (\pi/2)\times\{\beta\log\beta - \lim_{\alpha\to 0}\alpha\log\alpha\}+g(\beta) \quad ...(7)$$

But, $$\lim_{\alpha\to 0}\alpha\log\alpha = \lim_{\alpha\to 0}\frac{\log\alpha}{(1/\alpha)} = \lim_{\alpha\to 0}\frac{(1/\alpha)}{(-1/\alpha)} = 0, \text{ using L' Hospital' s rule} \quad ...(8)$$

Using (8), (7) reduces to

$$0 = (\pi/2)\times\beta\log\beta + g(\beta) \quad \text{so that} \quad g(\beta) = -(\pi/2)\times\beta\log\beta \quad ...(9)$$

Using (9), (6) reduces to $$G(\alpha,\beta) = (\pi/2)\times\{(\alpha+\beta)\log(\alpha+\beta)-\alpha\log\alpha-\beta\log\beta\}$$

or $$G(\alpha,\beta) = (\pi/2)\times\{\log(\alpha\times\beta)^{\alpha+\beta} - \log\alpha^{\alpha} - \log\beta^{\beta}\}$$

or $$\int_0^\infty \frac{\tan^{-1}\alpha x\tan^{-1}\beta x}{x^2}dx = \frac{\pi}{2}\log\left\{\frac{(\alpha+\beta)^{\alpha+\beta}}{\alpha^\alpha\beta^\beta}\right\}, \quad \text{using (1)}$$

EXERCISE 21(*B*)

Assuming the validity of differentiation udner the integral sign, show that

1. $\displaystyle\int_0^\infty \frac{\log(1+a^2x^2)}{1+b^2x^2}dx = \frac{\pi}{b}\log\left(1+\frac{a}{b}\right)$

2. $\displaystyle\int_0^\infty e^{-a^2x^2-b^2/x^2}\,dx = (\sqrt{\pi}/2a)\times e^{-2ab}, a>0, b\geq 0$

Deduce that $\int_0^\infty e^{-(x^2+a^2/x^2)}\,dx = (\sqrt{\pi}/2)\times e^{-2a},\ a>0$

3. $\int_0^1 \frac{x^a - x^b}{\log x}\,dx = \log\frac{a+1}{b+1}$

4. $\int_0^\infty \frac{\cos\lambda x}{x}(e^{-ax}-e^{-bx})\,dx = \frac{1}{2}\log\frac{b^2+\lambda^2}{a^2+\lambda^2},\ a>0,\ b>0$

5. $\int_0^{\pi/2} \frac{d\theta}{(a^2\cos^2\theta + b^2\sin^2\theta)^2} = \frac{\pi(a^2+b^2)}{4a^3b^3}$

21.6A. Evaluation of integral $\int_{g(\alpha)}^{h(\alpha)} f(x,\alpha)\,dx$, where $g(\alpha)$ and $h(\alpha)$ are functions of parameter α. Working rule.

Step 1. Let $$F(\alpha) = \int_{g(\alpha)}^{h(\alpha)} f(x,\alpha)\,dx \qquad \ldots(1)$$

Step 2. Differentiating both sides of (1) w.r.t. 'α' and using general Leibnitz's rule of differntiation under integral sign (see Art 21.3), we get

$$\frac{dF}{d\alpha} = \int_{g(\alpha)}^{h(\alpha)} \frac{\partial F}{\partial\alpha}\,d\alpha + \frac{dh}{d\alpha} f(h(\alpha),\alpha) - \frac{dg}{d\alpha} f(g(\alpha),\alpha) \qquad \ldots(2)$$

While writing (2), write $dh/d\alpha = 0$ if h is independent of α (or write $dg/d\alpha = 0$ if g is independent of α)

Steps 3 to 5. Read Art 21.4 A for complete discussion.

21.6B. Solved examples of type 3 base on Art.21.6A

Ex.1. *Assuming the validity of differntiation under the integral sign, show that*

$$\int_{\pi/2-\alpha}^{\pi/2} \sin\theta\cos^{-1}(\cos\alpha\,\text{cosec}\,\theta)\,d\theta = \frac{\pi}{2}(1-\cos\theta)$$ **[Delhi Maths (H) 2001, 03, 04, 09]**

Sol. Let $$F(\alpha) = \int_{\pi/2-\alpha}^{\pi/2} \sin\theta\cos^{-1}(\cos\alpha\,\text{cosec}\,\theta)\,d\theta \qquad \ldots(1)$$

Here the lower limit of integral is function of the parameter α while the upper limit is independent of α Hence, differentiating both sides of (1) w.r.t. 'α' and using the general form of Leibnitz's rule of differentiation under the integral sign, we have

$$\frac{dF}{d\alpha} = \int_{\pi/2-\alpha}^{\pi/2} \frac{\partial}{\partial\alpha}\left\{\sin\theta\cos^{-1}(\cos\alpha\,\text{cosec}\,\theta)\right\}d\theta + \frac{d(\pi/2)}{d\alpha}\sin\frac{\pi}{2}\cos^{-1}\left(\cos\alpha\,\text{cosec}\,\frac{\pi}{2}\right)$$
$$-\frac{d(\pi/2-\alpha)}{d\alpha}\sin(\pi/2-\alpha)\cos^{-1}\{\cos\alpha\,\text{cosec}\,(\pi/2-\alpha)\}$$

$$= \int_{\pi/2-\alpha}^{\pi/2} \frac{\sin\theta\,\text{cosec}\,\theta\sin\alpha}{\sqrt{1-\cos^2\alpha\,\text{cosec}^2\theta}}\,d\theta - (-1)\times\cos\alpha\times\cos^{-1}(1)$$

$$= \sin\alpha\int_{\pi/2-\alpha}^{\pi/2} \frac{\sin\theta\,d\theta}{\sqrt{\sin^2\theta-\cos^2\alpha}} = \sin\alpha\int_{\pi/2-\alpha}^{\pi/2} \frac{\sin\theta\,d\theta}{\sqrt{(1-\cos^2\theta)-(1-\sin^2\alpha)}},\ \text{as}\ \cos^{-1}(1)=0$$

$$= -\sin\alpha\int_{\sin\alpha}^{0} \frac{dt}{\sqrt{\sin^2\alpha - t^2}},\ \text{putting}\ \cos\theta = t\ \text{and}\ -\sin\theta\,d\theta = dt$$

$$= -\sin\alpha\left[\sin^{-1}\frac{t}{\sin\alpha}\right]_{\sin\alpha}^{0} = -(\sin\alpha)\times\left(0-\frac{\pi}{2}\right) = \frac{\pi}{2}\sin\alpha$$

Thus, $dF = (\pi/2)\times\sin\alpha\, d\alpha$

Integrating, $F(\alpha) = -(\pi/2)\times\cos\alpha + C$, C being an arbitrary constant ...(2)

Putting $\alpha = 0$ in (1) yields $F(0) = 0$. Next, putting $\alpha = 0$ and $F(0) = 0$ in (2) yields $C = -\pi/2$. Hence, (2) reduces to

$$F(\alpha) = -(\pi/2)\times\cos\alpha + \pi/2 \quad \text{or} \quad \int_{\pi/2-\alpha}^{\alpha} \sin\theta\cos^{-1}(\cos\alpha\,\mathrm{cosec}\,\theta)\,d\theta = \frac{\pi}{2}(1-\cos\alpha), \text{ by (1)}$$

Ex. 2. What are the points of the extrema of the function $y = \int_0^x \frac{\sin t}{t}\,dt,\ x > 0$?

(*a*) 0, $\pm\ n\pi$ (*b*) $\pm n\pi$ only (*c*) $n\pi$ only (*d*) 0, $n\pi$ only, where $n = 1, 2, 3, \ldots$ **[I.A.S. (Prel) 2009]**

Sol. **Ans. (*c*).** Given $y(x) = \int_0^x \frac{\sin t}{t}\,dt,\ x > 0$ ...(1)

Differentiating both sides of (1) w.r.t. 'x' and using the general form of Leibnitz's rule of differentiation under the integral sign, we obtain

$$\frac{dy}{dx} = \int_0^x \frac{\partial}{\partial x}\left(\frac{\sin t}{t}\right)dt + \frac{dx}{dx}\times\frac{\sin x}{x} - \frac{d0}{dx}\times\lim_{t\to 0}\frac{\sin t}{t}$$

or $\frac{dy}{dx} = 0 + \frac{\sin x}{x} - (0\times 1)$ or $\frac{dy}{dx} = \frac{\sin x}{x},\ x > 0$ ...(2)

For extremen values of y, we have $dy/dx = 0$, *i.e.*, $\sin x/x = 0$ or $\sin x = 0$, since $x > 0$.

Thus, $x = n\pi$, where $n = 1, 2, 3, \ldots$ Hence the choice (*c*) is correct.

Ex. 3 The function $f(x) = \int_0^{x^2} \frac{t^2 - 5t + 4}{2 + e^t}$ has (*a*) two maxima and two minima points (*b*) two maxima and three minima points (*c*) Three maxima and two minima points (*d*) One maximum point and one minimum point **[I.A.S. (Prel.) 2009]**

Sol. Ans. (*b*). Given $f(x) = \int_0^{x^2} \frac{t^2 - 5t + 4}{2 + e^t}\,dt$ ...(1)

Differentiating both sides of (1) w.r.t. 'x' and using the general form of Leibnitz's rule of differentiation under the integral sign, we obtain

$$f'(x) = \int_0^{x^2} \frac{\partial}{\partial x}\left(\frac{t^2 - 5t + 4}{2 + e^t}\right)dt + \frac{dx^2}{dx}\times\frac{x^4 - 5x^2 + 4}{2 + e^{x^2}} - \frac{do}{dx}\times\frac{4}{3}$$

or $$f'(x) = \frac{2x(x^4 - 5x^2 + 4)}{2 + e^{x^2}} = \frac{2(x^5 - 5x^3 + 4x)}{2 + e^{x^2}} \quad \ldots(2)$$

For maximum and minimum values of $f(x)$, we have

$f'(x) = 0 \Rightarrow x(x^4 - 5x^2 + 4) = 0$ or $x(x^2 - 1)(x^2 - 4) = 0$, giving, $x = 0, 1, -1, 2, -2$.

Differentiating boths sides of (2) w.r.t. 'x', we get

$$f''(x) = 2\times\frac{(5x^4 - 15x^2 + 4)(2 + e^{x^2}) - (x^5 - 5x^3 + 4x)\times e^{x^2}\times 2x}{(2 + e^{x^2})^2}$$

or $$f''(x) = \frac{2(5x^4 - 15x^2 + 4)(2 + e^{x^2}) - 4x^2 e^{x^2}(x-1)(x+1)(x-2)(x+2)}{(2 + e^{x^2})^2} \quad \text{...(3)}$$

Now, (3) $\Rightarrow$ $f''(0) > 0,$ $f''(1) < 0,$ $f''(-1) < 0,$ $f''(2) > 0$ and $f''(-2) > 0,$ showing that the given function $f(x)$ has maxima at two points (namely, at $x = 1$ and $x = -1$) and it has minima at three points (namely, at $x = 0$, $x = 2$ and $x = -2$).

Ex. 4. If $F(x) = (1/x^2) \times \int_4^x \{4t^2 - 3F'(t)\}\, dt$, then what is $F'(4)$?

(*a*) 32/19 (*b*) 64/3 (*c*) 64/19 (*d*) 16/3. **[I.A.S. (Prel,) 2009]**

Sol. Ans. (*c*). Given $$F(x) = (1/x^2) \times \int_4^x \{4t^2 - 3F'(t)\}\, dt \quad \text{...(1)}$$

Differentiating both sides of (1) w.r.t. '*x*', we obtain

$$F'(x) = -\frac{2}{x^3}\int_4^x \{4t^2 - 3F'(t)\}\, dt + \frac{1}{x^2}\frac{d}{dx}\int_4^x \{4t^2 - 3F'(t)\}\, dt \quad \text{...(2)}$$

Using the general from of Leibinitz's urule of differentiating under the integral sign in the second term on R.H.S of (2), (2) reduces to

$$F'(x) = -\frac{2}{x^3}\int_4^x \{4t^2 - 3F'(t)\}\, dt + \frac{1}{x^2}\Bigg[\int_4^x \frac{\partial}{\partial x}\{4t^2 - 3F'(t)\}\, dt$$
$$+ (dx/dx) \times \{4x^2 - 3F'(x)\} - (d4/dx) \times \{64 - 3F'(4)\}\Bigg]$$

Hence, $$F'(4) = 0 + 0 + \{64 - 3F'(4)\}/16 - 0 \quad \Rightarrow \quad F'(4) = 64/19$$

Exercise 21(C)

1. Show that $\displaystyle\int_0^a \frac{\log(1+ax)}{1+x^2}\, dx = \frac{1}{2}\log(1+a^2)\tan^{-1} a$. Deduce that $\displaystyle\int_0^a \frac{\log(1+x)}{1+x^2}\, dx = \frac{\pi}{8}\log 8$

2. Show that $y = \dfrac{1}{k}\displaystyle\int_0^x \sin k(x-t)\, dt$ satisfies the equation $d^2y/dx^2 + k^2 y = f(x)$, where k is a constant.

3. Show that $\displaystyle\int_{\pi/6a}^{\pi/2a} \frac{\sin ax}{x}\, dx =$ constant.

21.7A Determination of the value of an integral when certain standard known integral is given with its value. Working rule.

Let $$\int_a^b f(x, \alpha)\, dx = F(\alpha) \quad \text{...(1)}$$

be given, a and b being independent of the parameter α. Then, differentiating both sides of (1) and using Leibnitz's rule of differentiating under the integral sign, we get a new integral on L.H.S of (1) and its value on the R.H.S. of (1)

Remark. Sometimes one or more than one successive differentiation w.r.t. 'α' may be required to get the required results.

21.7B. Solved examples of type 4 based on Art. 21.7 A

Ex.1. Given $\displaystyle\int_0^x \frac{dx}{x^2 + a^2} = \frac{1}{a}\tan^{-1}\frac{x}{a}$. *Using the rule of differentiation under the integral* *sign, show that* $$\int_0^x \frac{dx}{(x^2+a^2)^2} = \frac{1}{2a^3}\tan^{-1}\frac{x}{a} + \frac{x}{2a^2(x^2+a^2)}$$

Sol. Given $$\int_0^x \frac{dx}{x^2+a^2} = \frac{1}{a}\tan^{-1}\frac{x}{a} \quad \text{...(1)}$$

Differentiating both sides of (1) w.r.t. 'x' and using Leibnitz's rule of differentiation under the integral sign, we obtain

$$\int_0^x \left\{-\frac{1}{(x^2+a^2)^2}\times 2a\right\}dx = \frac{1}{a}\times\frac{1}{1+(x/a)^2}\times\left(-\frac{x}{a^2}\right) - \frac{1}{a^2}\tan^{-1}\frac{x}{a}$$

Hence, $$\int_0^x \frac{dx}{(x^2+a^2)^2} = \frac{1}{2a^3}\tan^{-1}\frac{x}{a} + \frac{x}{2a^2(x^2+a^2)}$$

Ex. 2. *From the value of* $\int_0^l x^m\,dx$, *deduce the value of* $\int_0^l x^m(\log x)^n\,dx$, $m \geq 0$ *and n is a positive integer*

Sol. We have, $$\int_0^1 x^m\,dx = \left[x^{m+1}/(m+1)\right]_0^1 = 1/(m+1) = (m+1)^{-1} \quad \text{...(1)}$$

Differentiating both sides of (1) w.r.t. 'm' and using the Leilnitz's rule of differentiation under the integral sign, we get

$$\int_0^1 x^m \log x\,dx = (-1)(m+1)^{-2} \quad \text{...(2)}$$

Again, differentiations boths sides of (2) w.r.t. 'm', as before, we get

$$\int_0^1 x^m(\log x)^2\,dx = (-1)(-2)(m+1)^{-3} \quad \text{...(3)}$$

Continuing likewise till (1) is differentiated n times w.r.t. 'm', we finally obtain

$$\int_0^1 x^m(\log x)^n\,dx = (-1)(-2)(-3)\ldots(-n)(m+1)^{-(n+1)} = \frac{(-1)^n\,n!}{(m+1)^{n+1}}$$

Ex.3. *From the value of* $\int_0^\infty e^{-ax^2}\,dx$ *deduce the value of* $\int_0^\infty e^{-ax^2}x^{2n}\,dx$.

Sol. Let $ax^2 = t$, i.e., $x = (t/a)^{1/2}$ so that $dx = (1/2\sqrt{a})\times t^{-1/2}dt$. Then, we have

$$\int_0^\infty e^{-ax^2}dx = \int_0^\infty e^{-t}\times\frac{1}{2\sqrt{a}}\times t^{-1/2}dt = \frac{1}{2\sqrt{a}}\int_0^\infty e^{-t}t^{(1/2)-1}dt$$

or $$\int_0^\infty e^{-ax^2}dx = \frac{1}{2\sqrt{a}}\Gamma\left(\frac{1}{2}\right) = \frac{\sqrt{\pi}}{2}\times a^{-1/2} \quad \text{...(1)}$$

[$\because$ By definition of Gamma function, (refer Art 20.2) $\int_0^\infty e^{-x}x^{n-1}dx = \Gamma(n)$]

Differentiating both sides of (1) w.r.t. 'a' and using Leibnitz's rule of differentiation under the integral sign, we obtain

$$\int_0^\infty e^{-ax^2}(-x^2)dx = \frac{\sqrt{\pi}}{2}\times\left(-\frac{1}{2}\right)\times a^{-3/2} \quad \text{or} \quad \int_0^\infty e^{-ax^2}x^2\,dx = \frac{\sqrt{\pi}}{2}\times\frac{1}{2}\times a^{-3/2} \quad \text{...(2)}$$

Differentiating both sides of (2) w.r.t. 'a' and proceeding as before, we get

$$\int_0^\infty e^{-ax^2} \times (-1) \times (x^2)^2\, dx = \frac{\sqrt{\pi}}{2} \times \frac{1}{2} \times \left(-\frac{3}{2}\right) a^{-5/2} \quad \text{or} \quad \int_0^\infty e^{-ax^2} (x^2)^2\, dx = \frac{\sqrt{\pi}}{2} \times \frac{1}{2} \times \frac{3}{2} \times a^{-5/2}$$

Continuing likewise till (1) is differentiated n times w.r.t. 'a', we get

$$\begin{aligned}\int_0^\infty e^{-ax^2} (x^2)^n\, dx &= \frac{\sqrt{\pi}}{2} \times \frac{1}{2} \times \frac{3}{2} \times \ldots \times \frac{2n-1}{2} \times a^{-(1/2)-n} \\ &= \frac{\sqrt{\pi}}{2} \times \frac{1}{a^{n+1/2}} \times \frac{2n-1}{2} \times \frac{2n-3}{2} \times \ldots \times \frac{3}{2} \times \frac{1}{2} \\ &= \frac{1}{2a^{n+1/2}} \left(n-\frac{1}{2}\right)\left(n-\frac{3}{2}\right) \ldots \frac{3}{2} \cdot \frac{1}{2} \Gamma\left(\frac{1}{2}\right), \text{ as } \sqrt{\pi} = \Gamma\left(\frac{1}{2}\right) \\ &= \left[\Gamma(n+1/2)\right] / 2a^{n+1/2}, \qquad \text{as} \qquad (n-1)\Gamma(n-1) = \Gamma(n)\end{aligned}$$

EXERCISE 21 (D)

1. Strarting with $\displaystyle\int_0^\pi \frac{dx}{a+b\cos x} = \frac{\pi}{\sqrt{a^2-b^2}}$, $a > 0$, $|b| < a$, deduce that

(*i*) $\displaystyle\int_0^\pi \frac{dx}{(a+b\cos x)^2} = \frac{\pi a}{(a^2-b^2)^2}$ (*ii*) $\displaystyle\int_0^\pi \frac{\cos x\, dx}{(a+b\cos x)^2} = -\frac{\pi b}{(a^2-b^2)^{3/2}}$

2. Starting from $\displaystyle\int_0^\infty e^{-ax}\, dx = \frac{1}{a}$ for $a > 0$, deduce that $\displaystyle\int_0^\infty x^m e^{-ax}\, dx = (m!) / a^{m+1}$

3. Using the value of the integral $\displaystyle\int_0^\infty \frac{dx}{x^2+ax}$, show that

$$\int_0^\infty \frac{dx}{(x^2+a^2)^{n+1}} = \frac{\pi}{2} \cdot \frac{1\cdot3\cdot5\cdots(2n-1)}{2\cdot4\cdot6\cdots(2n)} \cdot \frac{1}{a^{n+1/2}}$$

MISCELLANEOUS PROBLEMS ON CHAPTER 21

Ex.1, If $f(x) = \int_0^x \sqrt{1+t^6}\, dt, x > 0$, Then find $f'(2)$. **(Pune 2010)**

Sol. Differentiating both sides of the given equation and using Leibnitz's rule of differentiation under integral sign, we have

$$f'(x) = \int_0^x \frac{\partial}{\partial x} \sqrt{1+t^6}\, dt + \sqrt{1+x^6} \times \frac{dx}{dx} - \sqrt{1+0} \times \frac{d0}{dx}$$

Thus, $f'(x) = \sqrt{1+x^6}$ and *so* $f'(2) = \sqrt{1+2^6} = \sqrt{65}$

Ex.2. The value of $\displaystyle\lim_{x\to0} \frac{x e^{x^2}}{\int_0^x e^{t^2}\, dt}$ is

(*a*) 0 (b) 1 (*c*) does not exist (*d*) −1 **(I.A.S. 2004)**

Sol. Ans. (b). We have,

$$\lim_{x\to 0} \frac{x e^{x^2}}{\int_0^x e^{t^2}\,dt} \qquad \text{Form } \left[\frac{\infty}{\infty}\right]$$

$$= \lim_{x\to 0} \frac{d(x e^{x^2})/dx}{\frac{d}{dx}\int_0^x e^{t^2}\,dt}, \text{ by L' Hospital's rule}$$

$$= \lim_{x\to 0} \frac{(1\times e^{x^2}) + (x\times e^{x^3}\times 2x)}{\int_0^x \frac{\partial}{\partial x}(e^{t^2})dt + e^{x^2}\times\frac{dx}{dx} - e^0\times\frac{d0}{dx}}$$

[Using the Leibnitz' rule of differentiation under the integral sign]

$$= \lim_{x\to 0} \frac{(1+x^2)e^{x^2}}{0+e^{x^2}-0} = \lim_{x\to 0}(1+x^2) = 1$$

Ex. 3 Let $f : \mathbf{R} \to \mathbf{R}$ be a continuous function with f(1) = 5 and f (3) = 11. If $g(x) = \int_1^3 f(x+t)\,dt$, then $g'(0)$ is equal to ...... **(GATE 2013)**

Sol. Ans 6. Given $\qquad g(x) = \int_1^3 f(x+t)\,dt \qquad$ ... (1)

Differentiating both sides of (1) w.r.t '*x*' and using Leibnitz's rule of differentiation under integral sign, we have

$$g'(x) = \int_1^3 \frac{\partial f(x+t)}{\partial x}\,dt = \int_1^3 \frac{df(x+t)}{d(x+t)}\frac{\partial(x+t)}{\partial x}\,dt = \int_{x+1}^{x+3} \frac{df(z)}{dz}\,dz, \text{ putting } x+t=z$$

or $\quad g'(x) = [f(z)]_{x+1}^{x+3} = f(x+3) - f(x+1)$ so that $\quad g'(0) = f(3) - f(1) = 11 - 5 = 6$

Chapter

22

Fourier Series

22.1. INTRODUCTION. In many scientific problems, specially in the study of periodic phenomenae in conduction of heat, optics, acoustics and electrodynamics it is necessary to expand a function $f(x)$ in a trigonometric series of sines and cosiner in form a series, known as *Fourier series.*

SOME IMPORTANT RESULTS AND DEFINITIONS

(*i*) **Periodic function. Definition.** A function $f: \mathbf{R} \to \mathbf{R}$ is said to be periodic if there exists a positive number T such that $f(x + \mathrm{T}) = f(x)$ for all real numbers x and T is called a period of $f(x)$. If a periodic function has a smallest positive period T, then T is called the **primitive period** or simply **the period.**

Sin x and cos x are periodic functions, each having the period 2π. Tan x and cot x are periodic functions, each having the period π.

(*ii*) **Even and odd functions. Definitions.**

A function $f(x)$ is known as an even function if $f(-x) = f(x)$. For example, $x^4 + x^2 + 1$, $\cos x$, $|\sin x|$, $|x|$ and $\sin^2 x$ are all even functions. The graph of an even function is symmetrical about the y-axis.

A function $f(x)$ is known as an odd function if $f(-x) = -f(x)$. For example, $x^3 + x$, $\sin x$ and $\tan x$ are all odd functions. The graph of an odd function is symmetrical about the origin.

We shall often use the following results:

$$\int_{-l}^{l} f(x)\, dx = 0, \text{ when } f(x) \text{ is an odd function}$$

and

$$\int_{-l}^{l} f(x)\, dx = 2\int_{0}^{l} f(x)\, dx, \text{ when } f(x) \text{ is an even function.}$$

(*iii*) **Some useful results.** If n is any integer, $\cos n\pi = (-1)^n$, $\sin n\pi = 0$, $\cos\{(2n+1)\pi\}/2 = 0$ and $\sin\{(2n+1)\pi\}/2 = (-1)^n$.

22.2. Euler's formulae for Fourier series expansion of a function $f(x)$

(Osmania 2004, Delhi 2012)

Theorem. Let *$f(x)$ be a periodic function with period 2π. Suppose $f(x)$ can be represented as a trigonometric series in the interval $c < x < c + 2\pi$. as follows*

$$f(x) = \frac{a_0}{2} + \sum_{n=1}^{\infty} a_n \cos nx + \sum_{n=1}^{\infty} b_n \sin nx. \qquad ...(1)$$

$$\left.\begin{aligned} \text{Then, } a_0 &= \frac{1}{\pi}\int_c^{c+2\pi} f(x)\, dx,\quad a_n = \frac{1}{\pi}\int_c^{c+2\pi} f(x)\cos nx\, dx,\quad n = 1, 2, 3,... \\ \text{and}\quad b_n &= \frac{1}{\pi}\int_c^{c+2\pi} f(x)\sin nx\, dx, \qquad n = 1, 2, 3 ...... \end{aligned}\right\} \qquad ...(2)$$

Proof. To prove the theorem, we shall first prove the following results. Let m and n be any integers. Then, we have (*i*) $\int_c^{c+2\pi} \sin mx\, dx = -\left[\frac{\cos mx}{m}\right]_c^{c+2\pi} = -\frac{1}{m}[\cos(mc+2m\pi) - \cos mc]$

$$= -(1/m) \times [\cos mc - \cos mc] = 0, \text{ if } m \neq 0. \quad ...(3)$$

Similarly, $\int_c^{c+2\pi} \cos mx\, dx = 0$ if $m \neq 0$. ...(4)

(*ii*) $\int_c^{c+2\pi} \sin^2 mx\, dx = \frac{1}{2}\int_c^{c+2\pi} (1 - \cos 2mx)\, dx = \frac{1}{2}\left[x - \frac{\sin 2mx}{2m}\right]_c^{c+2\pi}$

$$= \frac{1}{2}\left[(c+2\pi) - c - \frac{1}{2m}\{\sin 2m(c+2\pi) - \sin 2mc\}\right], \text{ if } m \neq 0$$

$$= (1/2) \times [2\pi - (1/2m) \times \{\sin(2mc + 4m\pi) - \sin 2mc\}]$$

$$= (1/2) \times [2\pi - (1/2m) \times (\sin 2mc - \sin 2mc)] = \pi, \text{ if } m \neq 0 \quad ...(5)$$

Similarly, $\int_c^{c+2\pi} \cos^2 mx\, dx = \pi$, if $m \neq 0$. ...(6)

(*iii*) $\int_c^{c+2\pi} \sin mx \cos nx\, dx = \frac{1}{2}\int_c^{c+2\pi} [\sin(m+n)x + \sin(m-n)x]\, dx$

$$= -\frac{1}{2}\left[\frac{\cos(m+n)x}{m+n} + \frac{\cos(m-n)x}{m-n}\right]_c^{c+2\pi} = 0, \text{ if } m \neq n. \quad ...(7)$$

(*iv*) $\int_c^{c+2\pi} \sin mx \sin nx\, dx = \frac{1}{2}\int_c^{c+2\pi} [\cos(m-n)x + \cos(m+n)x]\, dx$

$$= \frac{1}{2}\left[\frac{\sin(m-n)x}{m-n} - \frac{\sin(m+n)x}{m+n}\right]_c^{c+2\pi} = 0, \text{ if } m \neq n. \quad ...(8)$$

(*v*) $\int_c^{c+2\pi} \cos mx \cos nx\, dx = \frac{1}{2}\int_c^{c+2\pi} [\cos(m+n)x + \cos(m-n)x]\, dx$

$$= \frac{1}{2}\left[\frac{\sin(m+n)x}{m+n} + \frac{\sin(m-n)x}{m-n}\right]_c^{c+2\pi} = 0, \text{ if } m \neq n. \quad ...(9)$$

(*vi*) $\int_c^{c+2\pi} \sin nx \cos nx\, dx = \frac{1}{2}\int_c^{c+2\pi} \sin 2nx\, dx = \left[\frac{\cos 2nx}{-2n}\right]_c^{c+2\pi} = 0$, if $n \neq 0$. ...(10)

While finding the coefficients a_0, a_n and b_n, we assume that the series on the R.H.S. of (1) can be integrated term by term in the given interval $c < x < c + 2\pi$.

To find a_0. Integrating both sides of (1) w.r.t. x from c to $c + 2\pi$, we get

$$\int_c^{c+2\pi} f(x)\, dx = \frac{a_0}{2}\int_c^{c+2\pi} dx + \sum_{n=1}^{\infty} a_n \int_c^{c+2\pi} \cos nx\, dx + \sum_{n=1}^{\infty} b_n \int_c^{c+2\pi} \sin nx\, dx$$

$$= (a_0/2) \times (c + 2\pi - c) + 0 + 0, \text{ using (3) and (4)}$$

Thus, $a_0 = \frac{1}{\pi}\int_c^{c+2\pi} f(x)\, dx.$

To find a_n. Multiplying both sides of (1) by cos mx and then integrating both sides of the resulting equation w.r.t. x from c to $c + 2\pi$, we get

$$\int_c^{c+2\pi} f(x)\cos mx\, dx = \frac{a_0}{2}\int_c^{c+2\pi} \cos mx\, dx + \sum_{n=1}^{\infty} a_n \int_c^{c+2\pi} \cos nx \cos mx\, dx$$

$$+ \sum_{n=1}^{\infty} b_n \int_c^{c+2\pi} \sin nx \cos mx\, dx$$

$$= a_m \int_c^{c+2\pi} \cos^2 mx\, dx, \text{ using (7) and (9)}$$

$$= \pi a_m, \text{ using (6)}$$

Thus, $\quad a_m = \dfrac{1}{\pi}\displaystyle\int_c^{c+2\pi} f(x)\cos mx\, dx \quad$ so that $\quad a_n = \dfrac{1}{\pi}\displaystyle\int_c^{c+2\pi} f(x)\sin nx\, dx.$

To find b_n. Multiplying both sides of (1) by sin mx and then integrating both sides of the resulting equation w.r.t. x from c to $c + 2\pi$, we get

$$\int_c^{c+2\pi} f(x)\sin mx\, dx = \frac{a_0}{2}\int_c^{c+2\pi} \sin mx\, dx + \sum_{n=1}^{\infty} a_n \int_c^{c+2\pi} \cos nx \sin mx\, dx$$

$$+ \sum_{n=1}^{\infty} b_n \int_c^{c+2\pi} \sin nx \sin mx\, dx$$

$$= b_m \int_c^{c+2\pi} \sin^2 mx\, dx, \text{ using (7) and (8)}$$

$$= \pi b_m, \text{ using (5)}$$

Thus, $\quad b_m = \dfrac{1}{\pi}\displaystyle\int_c^{c+2\pi} f(x)\sin mx\, dx \quad$ so that $\quad b_n = \dfrac{1}{\pi}\displaystyle\int_c^{c+2\pi} f(x)\sin nx\, dx.$

The formulae for the coefficients a_0, a_n and b_n given by (2) are known as **Euler's formulae.**

The series (1) in which a_0, a_n and b_n are given by the Euler's formulae (2) is called the **Fourier series** of $f(x)$. Also the coefficients a_0, a_n and b_n are known as **Fourier coefficients.**

Note. In what follows we state the Dirichlet's conditions for the existence of Fourier series expansion for a periodic function $f(x)$ in $c < x < c + 2\pi$.

22.3. Theorem. Dirichlet's conditions for Fourier series. (Meerut 2013

The function $f(x)$ of period 2π defined in $(c, 2\pi)$ can be expanded in Fourier series in the form

$$f(x) = \frac{a_0}{2} + \sum_{n=1}^{\infty} a_n \cos nx + \sum_{n=1}^{\infty} b_n \sin nx,$$

where a_0, a_n, b_n $(n = 1, 2, 3, ...)$ are constants given by the Euler's formulae, provided

(*i*) $f(x)$ is periodic with period 2π, single valued and bounded in $(c, c + 2\pi)$

(*ii*) $f(x)$ has at most a finite number of maxima and minima in $(c, c + 2\pi)$

(*iii*) $f(x)$ is piecewise continuous with finite number of discontinuities in $(c, c + 2\pi)$

Remark. The condition (*i*), (*ii*) and (*iii*) imposed on $f(x)$ are sufficient but not necessary, that is, if the conditions are satisfied the convergence of Fourier series is guaranteed. However, if they are not satisfied the Fourier series may or may not converge. The above three conditions are generally satisfied in problems which arise in science and engineering.

22.4. Convergence of Fourier series. Let $f(x)$ be a periodic function of period 2π defined in $(c, c+2\pi)$ and satisfy Dirichlet's conditions. Then, Fourier series is given by

$$f(x) = \frac{a_0}{2} + \sum_{n=1}^{\infty} a_n \cos nx + \sum_{n=1}^{\infty} b_n \sin nx,$$ where a_0, a_n, b_n $(n = 1, 2, 3, ...)$ are Fourier coefficients

converges to

(*i*) $f(x)$ if x is a point of continuity

(*ii*) $\{f(x+0) + f(x-0)\}/2$, if x is a point of discontinuity

(*iii*) $\{f(c+0) + f(c+2\pi-0)\}/2$ at $x = c$ or $x = c + 2\pi$ which are end points of interval $(c, c+2\pi)$

Note. In the above results (*ii*) and (*iii*) we have used the notion of left hand limit and right hand limit which are denoted and defined as follows:

$$\text{Left hand limit} = f(a-) = f(a-0) = \lim_{h \to 0} f(a-h)$$

and $$\text{right hand limit} = f(a+) = f(a+0) = \lim_{h \to 0} f(a+h)$$

22.5. Some particular cases of theorem of Art. 22.2 for direct applications

Particular case (*i*). If $c = -\pi$, the interval becomes $(-\pi, \pi)$ and the formulae (1) and (2) of Art. 22.2 reduce to

$$f(x) = \frac{a_0}{2} + \sum_{n=1}^{\infty} a_n \cos nx + \sum_{n=1}^{\infty} b_n \sin nx, \text{ where}$$

$$a_0 = \frac{1}{\pi}\int_{-\pi}^{\pi} f(x)\,dx,\ a_n = \frac{1}{\pi}\int_{-\pi}^{\pi} f(x)\cos nx\,dx,\ b_n = \frac{1}{\pi}\int_{-\pi}^{\pi} f(x)\sin nx\,dx, \text{ where } n = 1, 2, 3, 4, ...$$

The above series is further simplified if $f(x)$ is of following two special forms:

Special form I. *When $f(x)$ is an odd function.* Then, $a_0 = 0$.

Since $\cos nx$ is an even function, hence $f(x)\cos nx$ is an odd function and so $a_n = 0$. Also, since $\sin nx$ is an odd function, so $f(x)\sin nx$ is an even function. Then b_n can be re-written as

$$b_n = \frac{1}{\pi}\int_{-\pi}^{\pi} f(x)\sin nx\,dx = \frac{2}{\pi}\int_{0}^{\pi} f(x)\sin nx\,dx$$

Let $f(x)$, defined in $(-\pi, \pi)$, be a periodic function with period 2π. Now, if $f(x)$ is an odd function, then its Fourier series contains only sine terms. Thus, we have

$$f(x) = \sum_{n=1}^{\infty} b_n \sin nx, \qquad \text{where} \qquad b_n = \frac{2}{\pi}\int_{0}^{\pi} f(x)\sin nx\,dx, \qquad n = 1, 2, 3, ...$$

Special form II. *When $f(x)$ is an even function.* Then, we have

$$a_0 = \frac{1}{\pi}\int_{-\pi}^{\pi} f(x)\,dx = \frac{2}{\pi}\int_{0}^{\pi} f(x)\,dx.$$

Since $\cos nx$ is an even function, so $f(x)\cos nx$ is an even function.

$$\therefore \qquad a_n = \frac{1}{\pi}\int_{-\pi}^{\pi} f(x)\cos nx\,dx = \frac{2}{\pi}\int_{0}^{\pi} f(x)\cos nx\,dx.$$

Since $\sin nx$ is an odd function, $f(x)\sin nx$ is an odd function. So $b_n = 0$.

Let $f(x)$, defined in $(-\pi, \pi)$, be a periodic function with period 2π. Now, if $f(x)$ is an even function, then its Fourier series contains only cosine terms. Thus, we have

$$f(x) = \frac{a_0}{2} + \sum_{n=1}^{\infty} a_n \cos nx, \quad \text{where } a_0 = \frac{2}{\pi}\int_0^{\pi} f(x)\,dx, \quad a_n = \frac{2}{\pi}\int_0^{\pi} f(x)\cos nx\,dx.$$

Particular case (*ii*). If $c = 0$, the interval becomes $(0, 2\pi)$ and the formulae (1) and (2) of Art. 22.2 reduce to

$$f(x) = \frac{a_0}{2} + \sum_{n=1}^{\infty} a_n \cos nx + \sum_{n=1}^{\infty} b_n \sin nx, \qquad \text{where}$$

$$a_0 = \frac{1}{\pi}\int_0^{2\pi} f(x)\,dx,\ a_n = \frac{1}{\pi}\int_0^{2\pi} f(x)\cos nx\,dx,\ b_n = \frac{1}{\pi}\int_0^{2\pi} f(x)\sin nx\,dx, \text{ where } n = 1, 2, 3, \ldots$$

The above series is further simplified if $f(x)$ is of following two special forms:

Special form I. *When $f(x)$ satisfies the condition $f(2\pi - x) = f(x)$.*

Then, $$a_0 = \frac{1}{\pi}\int_0^{2\pi} f(x)\,dx = \frac{2}{\pi}\int_0^{\pi} f(x)\,dx.$$

Now, let $F(x) = f(x)\cos nx$. Then we have

$F(2\pi - x) = f(2\pi - x)\cos\{n(2\pi - x)\} = f(x)\cos nx = F(x)$ and so

$$a_n = \frac{1}{\pi}\int_0^{2\pi} f(x)\cos nx\,dx = \frac{1}{\pi}\int_0^{2\pi} F(x)\,dx = \frac{2}{\pi}\int_0^{\pi} F(x)\,dx = \frac{2}{\pi}\int_0^{\pi} f(x)\cos nx\,dx.$$

Again, let $G(x) = f(x)\sin nx$. Then, we have

$G(2\pi - x) = f(2\pi - x)\sin\{n(2\pi - x)\} = -f(x)\sin nx = -G(x)$ and so

$$b_n = \frac{1}{\pi}\int_0^{2\pi} f(x)\sin nx\,dx = \frac{1}{\pi}\int_0^{2\pi} G(x)\,dx = 0$$

Let $f(x)$, defined in $(0, 2\pi)$, be a periodic function with period 2π. Now, if $f(2\pi - x) = f(x)$, then its Fourier series contains only cosine terms. So

$$f(x) = \frac{a_0}{2} + \sum_{n=1}^{\infty} a_n \cos nx, \quad \text{where } a_0 = \frac{2}{\pi}\int_0^{\pi} f(x)\,dx, \quad a_n = \frac{2}{\pi}\int_0^{\pi} f(x)\cos nx\,dx.$$

Special form II. *When $f(x)$ satisfies the condition $f(2\pi - x) = -f(x)$.*

Then $a_0 = 0$. Now, let $F(x) = f(x)\cos nx$. Then, we have

$$F(2\pi - x)\cos\{n(2\pi - x)\} = -f(x)\cos nx = -F(x) \text{ and so}$$

$$a_n = \frac{1}{\pi}\int_0^{2\pi} f(x)\cos nx\,dx = \frac{1}{\pi}\int_0^{2\pi} F(x)\,dx = 0$$

Again, let $G(x) = f(x)\sin nx$. Then, we have

$$G(2\pi - x) = f(2\pi - x)\sin\{n(2\pi - x)\} = f(x)\sin nx = G(x) \text{ and so}$$

$$b_n = \frac{1}{\pi}\int_0^{2\pi} f(x)\sin nx\,dx = \frac{1}{\pi}\int_0^{2\pi} G(x)\,dx = \frac{2}{\pi}\int_0^{\pi} G(x)\,dx = \frac{2}{\pi}\int_0^{\pi} f(x)\sin nx\,dx.$$

Let $f(x)$, defined in $(0, 2\pi)$, be a periodic function with period 2π. Now, if $f(2\pi - x) = -f(x)$, then its Fourier series contains only sine terms. So

$$\therefore \quad f(x) = \sum_{n=1}^{\infty} b_n \sin nx, \qquad \text{where} \qquad b_n = \int_0^{\pi} f(x)\sin nx\,dx.$$

FOR READY REFERENCE, THE FOLLOWING USEFUL RESULTS ARE TO BE COMMITTED TO MEMORY :

Result 1: Fourier series of $f(x)$ in interval $-\pi < x < \pi$ is given by

$$f(x) = \frac{a_0}{2} + \sum_{n=1}^{\infty} a_n \cos nx + \sum_{n=1}^{\infty} b_n \sin nx, \qquad \text{where}$$

where $a_0 = \frac{1}{\pi}\int_0^{\pi} f(x)\,dx,\ a_n = \frac{1}{\pi}\int_0^{\pi} f(x)\cos nx\,dx,\ b_n = \frac{1}{\pi}\int_0^{\pi} f(x)\sin nx\,dx,\ n = 1, 2, 3, ...$

Result 2. Fourier series of an odd function $f(x)$ in $-\pi < x < \pi$ is

$$f(x) = \sum_{n=1}^{\infty} b_n \sin nx, \text{ where } b_n = \frac{2}{\pi}\int_0^{\pi} f(x)\sin nx\,dx \quad n = 1, 2, 3, ...$$

Result 3. Fourier series of an even function $f(x)$ in $-\pi < x < \pi$ is

$$f(x) = \frac{a_0}{2} + \sum_{n=1}^{\infty} a_n \cos nx, \text{ where } a_0 = \frac{2}{\pi}\int_0^{\pi} f(x)\,dx,\ a_n = \frac{2}{\pi}\int_0^{\pi} f(x)\cos nx\,dx,\ n = 1, 2, ...$$

Result 4. Fourier series of $f(x)$ in $(0, 2\pi)$ is given by

$$f(x) = \frac{a_0}{2} + \sum_{n=1}^{\infty} a_n \cos nx + \sum_{n=1}^{\infty} b_n \sin nx, \qquad \text{where}$$

$$a_0 = \frac{1}{\pi}\int_0^{2\pi} f(x)\,dx, \qquad a_n = \frac{1}{\pi}\int_0^{2\pi} f(x)\cos nx\,dx, \qquad b_n = \frac{1}{\pi}\int_0^{2\pi} f(x)\sin nx\,dx$$

Result 5. If $f(2\pi - x) = -f(x)$, then Fourier series of $f(x)$ in $(0, 2\pi)$ is

$$f(x) = \sum_{n=1}^{\infty} b_n \sin nx, \qquad \text{where} \qquad b_n = \frac{2}{\pi}\int_0^{\pi} f(x)\sin nx\,dx.$$

Result 6. If $f(2\pi - x) = f(x)$, then Fourier series of $f(x)$ in $(0, 2\pi)$ is

$$f(x) = \frac{a_0}{2} + \sum_{n=1}^{\infty} a_n \cos nx,\ a_0 = \frac{2}{\pi}\int_0^{\pi} f(x)\,dx,\ a_n = \frac{2}{\pi}\int_0^{\pi} f(x)\cos nx\,dx, = 1, 2, 3, ...$$

22.6. Solved examples based on six results of Art. 22.5

Ex. 1. (*a*) *Find the series of sines and cosines of multiples of x for the function $f(x)$ = $|\sin x|$ in the internal $[-\pi, \pi]$.* **[Delhi 2007, 10]**

(*b*) *Find the Fourier series which represents the function $f(x) = |\cos x|$ in the interval $[-\pi, \pi]$.* **[Delhi 2011]**

Sol. (*a*) Since $f(x) = |\sin x|$ is an even function, the required Fourier series is given by (refer result 3, Art. 22.5)

$$f(x) = |\sin x| = \frac{a_0}{2} + \sum_{n=1}^{\infty} a_n \cos nx, \qquad \text{where} \qquad ...(1)$$

$$a_0 = \frac{2}{\pi}\int_0^{\pi} f(x)\,dx = \frac{2}{\pi}\int_0^{\pi} \sin x\,dx = \frac{2}{\pi}[-\cos x]_0^{\pi} = \frac{2}{\pi}(1+1) = \frac{4}{\pi}$$

and
$$a_n = \frac{2}{\pi}\int_0^{\pi} f(x)\cos nx\,dx = \frac{1}{\pi}\int_0^{\pi} (2\sin x\cos nx)\,dx \qquad ...(2)$$

or $a_n = \dfrac{1}{\pi}\int_0^{\pi}\{\sin(n+1)x - \sin(n-1)x\}\,dx.$

$$= \frac{1}{\pi}\left[-\frac{\cos(n+1)x}{(n+1)} + \frac{\cos(n-1)x}{(n-1)}\right]_0^{\pi} = \frac{1}{\pi}\left[-\frac{\cos(n+1)\pi}{(n+1)} + \frac{\cos(n-1)\pi}{(n-1)} + \frac{1}{n+1} - \frac{1}{n-1}\right]$$

$$= \frac{1}{\pi}\left[-\frac{(-1)^{n+1}}{n+1} + \frac{(-1)^{n-1}}{n-1} + \frac{1}{n+1} - \frac{1}{n-1}\right] = \frac{1}{\pi}\left[\frac{(-1)^n}{n+1} - \frac{(-1)^n}{n-1} + \frac{n-1-(n+1)}{(n+1)(n-1)}\right] = \frac{2\{1+(-1)^n\}}{\pi(n^2-1)}$$

$\therefore$ For $n > 1$, $a_n = \begin{cases} 0, \text{if } n \text{ is odd} \\ -4/\pi(n^2-1), \text{ if } n \text{ is even} \end{cases}$

For $n = 1$, (2) reduces to $a_1 = (1/\pi) \times \int_0^{\pi} \sin 2x\,dx = (1/\pi)\times[(-1/2)\times\cos 2x]_0^{\pi} = 0$

$$\therefore\ (1) \Rightarrow |\sin x| = \frac{2}{\pi} - \frac{4}{\pi}\sum_{n=1}^{\infty}\frac{\cos 2nx}{(2n)^2-1} = \frac{2}{\pi} - \frac{4}{\pi}\sum_{n=1}^{\infty}\frac{\cos 2nx}{(2n-1)(2n+1)}.$$

(*ii*) Do yourself **Ans.** $|\cos x| = \dfrac{2}{\pi} + \dfrac{4}{\pi}\sum_{n=1}^{\infty}\dfrac{\cos 2nx}{(2n-1)(2n+1)}.$

Ex. 2. *Express $f(x) = (\pi - x)/2$ in a Fourier series in $0 < x < 2\pi$.*

(Meerut 2018; Madurai Kamraj 2008)

Sol. Here $f(2\pi - x) = \{\pi - (2\pi - x)\}/2 = -(\pi - x)/2 = -f(x)$. Hence the required Fourier series is given by (refer result 5, Art. 22.5)

$$f(x) = \frac{\pi - x}{2} = \sum_{n=1}^{\infty} b_n \sin nx, \text{ where} \qquad ...(1)$$

$$b_n = \frac{2}{\pi}\int_0^{\pi} f(x)\sin nx\,dx = \left(\frac{2}{\pi}\right)\left(\frac{1}{2}\right)\int_0^{\pi}(\pi - x)\sin nx\,dx$$

$$= \frac{1}{\pi}\left[(\pi - x)\left(-\frac{\cos nx}{n}\right) - (-1)\left(-\frac{\sin nx}{n^2}\right)\right]_0^{\pi} = \left(\frac{1}{\pi}\right)\left(\frac{\pi}{n}\right) = \frac{1}{n}.$$

$$\therefore \quad (1) \Rightarrow \quad \frac{\pi - x}{2} = \sum_{n=1}^{\infty}\frac{\sin nx}{n} = \frac{\sin x}{1} + \frac{\sin 2x}{2} + \frac{\sin 3x}{3} + ...$$

Ex. 3. *Express Fourier series expansion of $f(x)$ defined by $f(x) = -\pi$, when $-\pi < x < 0$ and $f(x) = x$, when $0 < x < \pi$. Deduce that $1/1^2 + 1/3^2 + 1/5^2 + ... = \pi^2/8$.*

(Meerut 2015; Osmania 2010; Himachal 2011)

Sol. The required Fourier series expansion of the given function $f(x)$ is given by

$$f(x) = \frac{a_0}{2} + \sum_{n=1}^{\infty}(a_n \cos nx + b_n \sin nx), \quad \text{where} \qquad ...(1)$$

$$a_0 = \frac{1}{\pi}\int_{-\pi}^{\pi} f(x)\,dx = \frac{1}{\pi}\left[\int_{-\pi}^{0} f(x)\,dx + \int_0^{\pi} f(x)\,dx\right] = \frac{1}{\pi}\left[-\pi\int_{-\pi}^{0} dx + \int_0^{\pi} x\,dx\right]$$

[Using the given value of $f(x)$]

or $a_0 = (1/\pi) \times \{(-\pi) \times \pi + \pi^2/2\} = -(\pi/2)$ ...(2)

$$a_n = \frac{1}{\pi}\int_{-\pi}^{\pi} f(x)\cos nx\,dx = \frac{1}{\pi}\left[\int_{-\pi}^{0}(-\pi)\cos nx\,dx + \int_0^{\pi} x\cos nx\,dx\right]$$

$$= \frac{1}{\pi}\left\{-\pi\left[\frac{\cos nx}{n}\right]_{-\pi}^{0} + \left[(x)\left(\frac{\sin nx}{n}\right) - (1)\left(-\frac{\cos nx}{n^2}\right)\right]_0^{\pi}\right\}$$

[Using the generdised integration by parts while evaluating the second integral]

$$= (1/\pi) \times \{(1/n^2) \times \cos n\pi - (1/n^2)\} = -(1/\pi n^2) \times (1 - \cos n\pi)$$

$$= -\frac{1-(-1)^n}{\pi n^2} = \begin{cases} 0, \text{if } n \text{ is even} \\ -2/\pi n^2, \text{if } n \text{ is odd} \end{cases}$$

and $b_n = \frac{1}{\pi}\int_{-\pi}^{\pi} f(x) \sin nx\, dx = \frac{1}{\pi}\left[\int_{-\pi}^{0}(-\pi)\sin nx\, dx + \int_0^{\pi} x \sin nx\, dx\right]$

$$= \frac{1}{\pi}\left\{-\pi\left[-\frac{\cos nx}{n}\right]_{-\pi}^{0} + \left[(x)\left(-\frac{\cos nx}{n}\right) - (1) \times \left(-\frac{\sin nx}{n^2}\right)\right]_0^{\pi}\right\}, \text{ as before}$$

$$= (1/\pi) \times \{(\pi/n) \times (1 - \cos n\pi) - (\pi/n) \times \cos n\pi\} = \{1 - 2(-1)^n\}/n \qquad ...(4)$$

In $[-\pi, \pi]$, the point $x = 0$ is the only point of discontinuity of the function f. Therefore, when x is different from 0, (1) reduces to

$$f(x) = -\frac{\pi}{4} + \sum_{n=1}^{\infty}\left\{\frac{-2}{\pi(2n-1)^2}\right\}\cos nx + \sum_{n=1}^{\infty}\frac{1-2(-1)^n}{n}\sin nx$$

or $f(x) = -\frac{\pi}{4} - \frac{2}{\pi}\left(\cos x + \frac{\cos 3x}{3^2} + \frac{\cos 5x}{5^2} + ...\right) + 3\sin x - \frac{\sin 2x}{2} - \frac{3\sin 3x}{3} - \frac{\sin 4x}{4} + ... \qquad ...(5)$

Since $x = 0$ is a point of discontinuity of f, therefore the sum of the series (5) for $x = 0$.

$$= (1/2) \times \{f(0+0) + f(0-0)\} = (1/2) \times \left\{\lim_{h \to 0} f(0+h) + \lim_{h \to 0} f(0-h)\right\}$$

$$= \frac{1}{2}\left\{\lim_{h \to 0} h + \lim_{h \to 0}(-\pi)\right\} = -\frac{\pi}{2}$$

Therefore, at $x = 0$, (5) takes the form (refer result (*i*) of Art. 22.4)

$$-\frac{\pi}{2} = -\frac{\pi}{4} - \frac{2}{\pi}\left(\frac{1}{1^2} + \frac{1}{3^2} + \frac{1}{5^2} + ...\right) \quad \text{so that} \quad \frac{1}{1^2} + \frac{1}{3^2} + \frac{1}{5^2} + ... = \frac{\pi^2}{8}$$

Ex. 4. *Find the Fourier series expansion for $f(x) = e^x$ in $0 \le x \le 2\pi$.*

(Madurai Kamraj 2008)

Sol. Let $f(x) = e^x = \frac{a_0}{2} + \sum_{n=1}^{\infty} a_n \cos nx + \sum_{n=1}^{\infty} b_n \sin nx$, where ...(1)

$$a_0 = \frac{1}{\pi}\int_0^{2\pi} f(x)\, dx = \frac{1}{\pi}\int_0^{2\pi} e^x\, dx = \frac{1}{\pi}\left[e^x\right]_0^{2\pi} = \frac{e^{2\pi}-1}{\pi} \qquad ...(2)$$

$$a_n = \frac{1}{\pi}\int_0^{2\pi} f(x)\cos nx\, dx = \frac{1}{\pi}\int_0^{2\pi} e^x \cos nx\, dx$$

$$= \frac{1}{\pi}\left[\frac{e^x(\cos nx + n\sin nx)}{1+n^2}\right]_0^{2\pi}, \quad \text{as} \int e^{ax}\cos bx\, dx = \frac{e^{ax}(a\cos bx + b\sin bx)}{a^2+b^2}$$

or $$a_n = (e^{2\pi} - 1)/\pi(1 + n^2) \qquad \text{...(3)}$$

and $$b_n = \frac{1}{\pi}\int_0^{2\pi} f(x)\sin nx\,dx = \frac{1}{\pi}\int_0^{2\pi} e^x \sin nx\,dx$$

$$= \frac{1}{\pi}\left[\frac{e^x(\sin nx - n\cos nx)}{1+n^2}\right]_0^{2\pi}, \quad \text{as} \quad \int e^{ax}\sin bx\,dx = \frac{e^{ax}(a\sin bx - b\cos bx)}{a^2+b^2}$$

or $$b_n = (-n\,e^{2\pi} + n)/\pi(1+n^2) = -n\,(e^{2\pi} - 1)/\pi(1+n^2) \qquad \text{...(4)}$$

Substituting the above values of a_0, a_n and b_n in (1), we have

$$e^x = \frac{e^{2\pi}-1}{2}\left\{\frac{1}{2} + \sum_{n=1}^{\infty}\frac{\cos nx}{n^2+1} - \sum_{n=1}^{\infty}\frac{n\sin nx}{n^2+1}\right\}$$

Ex. 5. *Find the Fourier series of the periodic function of with period 2π defined as follows: $f(x) = 0$, for $-\pi < x \leq 0$ and $f(x) = x$, for $0 \leq x \leq \pi$. What is the sum of the series at $x = 0, \pm\pi, 4\pi$ and -5π.* **(Delhi B.A. (Prog.) III, 2008, 2013)**

Sol. Let $$f(x) = \frac{a_0}{2} + \sum_{n=1}^{\infty} a_n \cos nx + \sum_{n=1}^{\infty} b_n \sin nx, \text{ where} \qquad \text{...(1)}$$

$$a_0 = \frac{1}{\pi}\int_{-\pi}^{\pi} f(x)\,dx = \frac{1}{\pi}\left[\int_{-\pi}^{0} f(x)\,dx + \int_0^{\pi} f(x)\,dx\right] = \frac{1}{\pi}\left[\int_{-\pi}^{0}(0)\,dx + \int_0^{\pi} x\,dx\right] = \frac{\pi}{2}$$

[Using the given value of $f(x)$]

$$a_n = \frac{1}{\pi}\int_{-\pi}^{\pi} f(x)\,dx = \frac{1}{\pi}\left[\int_{-\pi}^{0}(0)\,dx + \int_0^{\pi} x\cos nx\,dx\right] = \frac{1}{\pi}\left[(x)\left(\frac{\sin nx}{n}\right) - (1)\left(-\frac{\cos nx}{n^2}\right)\right]_0^{\pi}$$

$$= \frac{1}{\pi}\left[\frac{\cos n\pi}{n^2} - \frac{1}{n^2}\right] = -\frac{1-(-1)^n}{\pi n^2} = \begin{cases} 0, & \text{if } n \text{ is even} \\ -2/\pi n^2, & \text{if is odd} \end{cases}$$

$$b_n = \frac{1}{\pi}\int_{-\pi}^{\pi} f(x)\sin nx\,dx = \frac{1}{\pi}\left[\int_{-\pi}^{0}(0)\,dx + \int_0^{\pi} x\sin nx\,dx\right] = \frac{1}{\pi}\left[(x)\left(-\frac{\cos nx}{n}\right) - (1)\left(-\frac{\sin nx}{n^2}\right)\right]_0^{\pi}$$

$$= (1/\pi) \times \{-(\pi/n) \times \cos n\pi\} = (-1/n) \times (-1)^n = (-1)^{n+1} \times (1/n)$$

The graph of the given function f is shown in the following figure. In interval $[-\pi, \pi]$, the points $\pm\pi$ are the only points of discontinuity of given function f. Hence at all points of the interval $[-\pi, \pi]$ other than $\pm\pi$, (1) reduces to

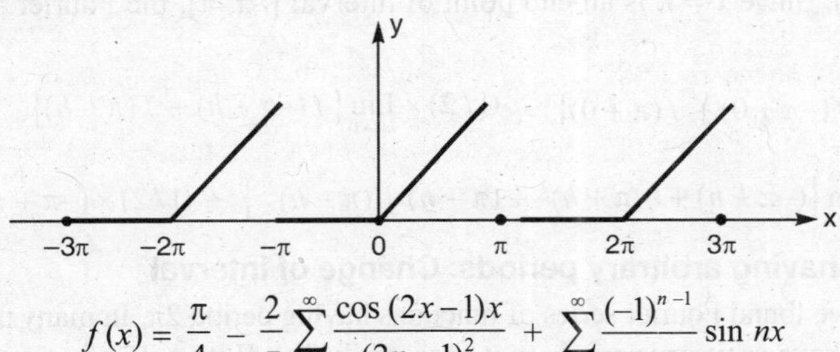

$$f(x) = \frac{\pi}{4} - \frac{2}{\pi}\sum_{n=1}^{\infty}\frac{\cos(2x-1)x}{(2n-1)^2} + \sum_{n=1}^{\infty}\frac{(-1)^{n-1}}{n}\sin nx$$

or $$f(x) = -\frac{\pi}{4} - \frac{2}{\pi}\left(\frac{\cos x}{1^2} + \frac{\cos 3x}{3^2} + \frac{\cos 5x}{5^2} + \ldots\right) + \left(\frac{\sin x}{1} - \frac{\sin 2x}{2} - \frac{\sin 3x}{3} + \ldots\right) \quad \ldots(2)$$

Since f is continuous at $x = 0$, the sum of the series $= f(0) = 0$. Because of periodicity, the value of the Fourier series (2) at $x = 4\pi$ is same as at $x = 0$.

At $x = \pm\pi$, points of discontinuity of f, the sum of series

$$= (1/2) \times [f(\pi - 0) + f(-\pi + 0)] = (1/2) \times \left[\lim_{h \to 0} f(\pi - h) + \lim_{h \to 0} f(-\pi + h)\right]$$

$$= (1/2) \times \left[\lim_{h \to 0} (\pi - h) + \lim_{h \to 0} 0\right] = (1/2) \times [\pi + 0] = \pi/2$$

Again because of periodicity, the sum of the series at $x = -5\pi$ is same as at $x = \pm\pi$.

Ex. 6. *Find the Fourier series for $f(x)$ in $[-\pi, \pi]$ where* $f(x) = \begin{cases} \pi + x, \text{ if } -\pi < x \le 0 \\ \pi - x, \text{ if } 0 \le x < \pi \end{cases}$

Sol. Re-writing, $f(x) = \pi - |x|$ and hence $f(x)$ is an even function. Hence, let (refer result 3, Art. 22.5)

$$f(x) = \frac{a_0}{2} + \sum_{n=1}^{\infty} a_n \cos nx, \quad \text{where} \quad \ldots(1)$$

$$a_0 = \frac{2}{\pi}\int_0^{\pi} f(x)\,dx = \frac{2}{\pi}\int_0^{\pi} (\pi - x)\,dx = \frac{2}{\pi}\left[\pi x - \frac{1}{2}x^2\right]_0^{\pi} = \frac{2}{\pi}\left(\pi^2 - \frac{1}{2}\pi^2\right) = \pi$$

and $$a_n = \frac{2}{\pi}\int_0^{\pi} f(x)\cos nx\,dx = \frac{2}{\pi}\int_0^{\pi} (\pi - x)\cos nx\,dx$$

$$= \frac{2}{\pi}\left[(\pi - x)\left(\frac{\sin nx}{n}\right) - (-1)\left(\frac{\cos nx}{n^2}\right)\right]_0^{\pi} = \frac{2}{\pi}\left[-\frac{\cos n\pi}{n^2} + \frac{1}{n^2}\right]$$

$$= \frac{2}{\pi n^2}\{1 - (-1)^n\} = \begin{cases} 0, \text{ if } n \text{ is even} \\ 4/\pi^2 n^2, \text{ if } n \text{ is odd} \end{cases}$$

$$\therefore \quad (1) \Rightarrow f(x) = \frac{\pi}{2} + \sum_{n=1}^{\infty} \frac{4\cos(2n-1)x}{\pi^2(2n-1)^2} = \frac{\pi}{2} + \frac{4}{\pi^2}\sum_{n=1}^{\infty}\frac{\cos(2n-1)x}{(2n-1)^2}$$

Ex. 7. Fourier series of the 2π – periodic function $f(x) = x + x^2$, $-\pi < x < \pi$ at $x = \pi$ converges to (*a*) π (*b*) 2π (*c*) π^2 (*d*) $\pi + \pi^2$. **(GATE 1997)**

Sol. Ans. (*c*). Since $x = \pi$ is an end point of interval $[-\pi, \pi]$, the Fourier Series of $f(x)$ at $x = \pi$ converges to

$$(1/2) \times \{f(-\pi + 0) + f(\pi + 0)\} = (1/2) \times \lim_{h \to 0}\{f(-\pi + h) + f(\pi - h)\}$$

$$= (1/2) \times \lim_{h \to 0}\{(-\pi + h) + (-\pi + h)^2 + (\pi - h) + (\pi - h)^2\} = (1/2) \times (-\pi + \pi^2 + \pi + \pi^2) = \pi^2$$

22.7. Functions having arbitrary periods. Change of interval

So far we have found Fourier series of functions having period 2π. In many practical problems the functions may have arbitrary periods (not necessarily 2π). With help of a transformation of the variable, we propose to get Euler's formulae for Fourier coefficients for functions having period $2l$, where l is any positive number.

Let $f(x)$ be given periodic function in interval $c < x < c + 2l$. To change the interval into one of length 2π, we put

$$x/l = z/\pi \quad \text{or} \quad x = \pi x/l \qquad ...(1)$$

$$\therefore \quad \text{when} \quad x = c, \qquad z = \pi c/l = d(\text{say}) \qquad ...(2)$$

$$\text{and} \quad \text{when} \quad x = c + 2l, \quad z = \pi(c + 2l)/l = (\pi c/l) + 2\pi = d + 2\pi \qquad ...(3)$$

Hence the function $f(x)$ of period $2l$ in $(c, c + 2l)$ is transformed to the function $f(lz/\pi) = \text{F}(z)$, say, of period 2π in $(d, d + 2\pi)$. Then F (z) can be expressed as Fourier series given by [Refer Theorem of Art. 22.2]

$$\text{F}(z) = \frac{a_0}{2} + \sum_{n=1}^{\infty} a_n \cos nz + \sum_{n=1}^{\infty} b_n \sin nz, \qquad \text{where} \qquad ...(4)$$

$$a_0 = \frac{1}{\pi}\int_d^{d+2\pi} \text{F}(z)\, dz, \qquad a_n = \frac{1}{\pi}\int_d^{d+2\pi} \cos nz\, dz, \quad b_n = \frac{1}{\pi}\int_d^{d+2\pi} \text{F}(z) \sin nz\, dz \qquad ...(5)$$

Now, putting $z = \pi x/l$ and $dz = (\pi/l)\, dx$ and noting that when $z = d$, $x = c$ and when $z = d + 2\pi$, $x = c + 2l$, (4) and (5) reduce to

$$\text{F}(\pi x/l) = f(x) = \frac{a_0}{2} + \sum_{n=1}^{\infty} a_n \cos\frac{n\pi x}{l} + \sum_{n=1}^{\infty} b_n \sin\frac{n\pi x}{l}, \qquad ...(6)$$

$$\text{where} \quad a_0 = \frac{1}{l}\int_c^{c+2l} f(x)\, dx, \; a_n = \frac{1}{l}\int_c^{c+2l} f(x)\cos\frac{n\pi x}{l}, \; b_n = \frac{1}{l}\int_c^{c+2l} f(x)\sin\frac{n\pi x}{l}\, dx \qquad ...(7)$$

Thus the Fourier series for $f(x)$ in $c < x < c + 2l$ is given by (6) and the corresponding Fourier coefficients are given by (7).

Particular Case I. Fourier series for $f(x)$ in $0 < x < 2l$ can by obtained by putting $c = 0$ in (6) and (7). Thus, we have

$$f(x) = \frac{a_0}{2} + \sum_{n=1}^{\infty} a_n \cos\frac{n\pi x}{l} + \sum_{n=1}^{\infty} b_n \sin\frac{n\pi x}{l}, \quad \text{where} \qquad ...(8)$$

$$a_0 = \frac{1}{l}\int_0^{2l} f(x)\, dx, \; a_n = \frac{1}{l}\int_0^{2l} f(x)\cos\frac{n\pi x}{l}\, dx, \; b_n = \frac{1}{l}\int_0^{2l} f(x)\sin\frac{n\pi x}{l}\, dx \qquad ...(9)$$

Particular Case II. Fourier series of $f(x)$ in $-l < x < l$ can be obtained by putting $c = -l$ in (6) and (7). Thus, we have

$$f(x) = \frac{a_0}{2} + \sum_{n=1}^{\infty} a_n \cos\frac{n\pi x}{l} + \sum_{n=1}^{\infty} b_n \sin\frac{n\pi x}{l}, \quad \text{where} \qquad ...(10)$$

$$a_0 = \frac{1}{l}\int_{-l}^{l} f(x)\, dx, \quad a_n = \frac{1}{l}\int_{-l}^{l} f(x)\cos\frac{n\pi x}{l}\, dx, \quad b_n = \frac{1}{l}\int_{-l}^{l} f(x)\sin\frac{n\pi x}{l}\, dx \qquad ...(11)$$

We now discuss two special forms of the above case II.

Special form I. Fourier series of $f(x)$ in $-l < x < l$ when $f(x)$ is an even function.

Note that if $f(x)$ is even function, then $f(x)\cos(n\pi x/l)$ is an even function whereas $f(x)\sin(n\pi x/l)$ is an odd function. Accordingly, (11) may be re-written as

$$a_0 = \frac{2}{l}\int_0^{l} f(x)\, dx, \qquad a_n = \frac{2}{l}\int_0^{l} f(x)\cos\frac{n\pi x}{l}\, dx \quad \text{and} \quad b_n = 0. \qquad ...(12)$$

Thus, Fourier series of an even function $f(x)$ in $(-l, l)$ is given by

$$f(x) = \frac{a_0}{2} + \sum_{n=1}^{\infty} a_n \cos\frac{n\pi x}{l}, \qquad \text{...(13)}$$

where a_0 and a_n are given by (12).

Special form II. Fourier series of $f(x)$ in $-l < x < l$ when $f(x)$ is an odd function.

Note that if $f(x)$ is odd function, then $f(x) \cos (n\pi x/l)$ is an odd function whereas $f(x) \sin (n\pi x/l)$ is an even function. Accordingly, (11) may be re-written as

$$a_0 = 0, \qquad a_n = 0 \qquad \text{and} \qquad b_n = \frac{2}{l}\int_0^l f(x) \sin\frac{n\pi x}{l}\,dx. \qquad \text{...(14)}$$

Thus, Fourier series of an odd function $f(x)$ in $(-l, l)$ is given by

$$f(x) = \sum_{n=1}^{\infty} b_n \sin\frac{n\pi x}{l} \qquad \text{...(15)}$$

where b_n is given by (14).

22.8. Solved examples based on results of Art. 22.7

Ex. 1. *Find the Fourier series of $f(x) = x^2$ in $(-1, 1)$*

Sol. Since $f(x) = x^2$ is an even function of x, the required Fourier series is given by [Refer results (12) and (13) of Art 22.7 and note that $l = 1$]

$$f(x) = x^2 = \frac{a_0}{2} + \sum_{n=1}^{\infty} a_n \cos n\pi x, \text{ where} \qquad \text{...(1)}$$

$$a_0 = 2\int_0^1 f(x)\,dx = 2\int_0^1 x^2\,dx = 2\left[x^3/3\right]_0^1 = 2/3$$

and $$a_n = 2\int_0^1 f(x)\cos n\pi x\,dx = 2\int_0^1 x^2 \cos n\pi x\,dx$$

$$= 2\left[(x^2)\left(\frac{\sin n\pi x}{n\pi}\right) - (2x)\left(-\frac{\cos n\pi x}{n^2\pi^2}\right) + (2)\left(-\frac{\sin n\pi x}{n^3\pi^3}\right)\right]_0^1$$

$$= 2\,(2/n^2\pi^2)\cos n\pi = 4\,(-1)^n/n^2\pi^2.$$

$\therefore$ From (1), $$x^2 = \frac{1}{3} + \frac{4}{\pi^2}\sum_{n=1}^{\infty}\frac{(-1)^n \cos n\pi x}{n^2}.$$

Ex. 2. *Find Fourier series for $f(x)$ define by $f(x) = x$ for $0 < x < 1$ and $f(x) = 1 - x$ for $1 < x < 2$. Deduce $1/1^2 + 1/3^2 + 1/5^2 + ... = \pi^2/8$.*

Sol. Let $$f(x) = \frac{a_0}{2} + \sum_{n=1}^{\infty} a_n \cos n\pi x + \sum_{n=1}^{\infty} b_n \sin n\pi x, \quad \text{where} \qquad \text{...(1)}$$

$$a_0 = \int_0^2 f(x)\,dx = \int_0^1 f(x)\,dx + \int_1^2 f(x)\,dx = \int_0^1 x\,dx + \int_1^2 (1-x)\,dx$$

$$= \left[x^2/2\right]_0^1 + \left[x - x^2/2\right]_1^2 = 1/2 + [(2-2) - (1-1/2)] = 0.$$

$$a_n = \int_0^2 f(x)\cos n\pi x\,dx = \int_0^1 x\cos n\pi x\,dx + \int_1^2 (1-x)\cos n\pi x\,dx$$

$$= \left[(x)\left(\frac{\sin n\pi x}{n\pi}\right) - (1)\left(-\frac{\cos n\pi x}{n^2\pi^2}\right)\right]_0^1 + \left[(1-x)\left(\frac{\sin n\pi x}{n\pi}\right) - (-1)\left(-\frac{\cos n\pi x}{n^2\pi^2}\right)\right]_1^2$$

$$= \left(\frac{(-1)^n}{n^2\pi^2} - \frac{1}{n^2\pi^2}\right) + \left(-\frac{1}{n^2\pi^2} + \frac{(-1)^n}{n^2\pi^2}\right) = \frac{2}{n^2\pi^2}[(-1)^n - 1]$$

$\therefore \quad a_n = 0$, if n is even; $\qquad a_n = -4/n^2\pi^2$, if n is odd.

Also, $b_n = \int_0^2 f(x) \sin n\pi x \, dx = \int_0^1 x \sin n\pi x \, dx + \int_1^2 (1-x) \sin n\pi x \, dx$

$$= \left[(x)\left(-\frac{\cos n\pi x}{n\pi}\right) - (1)\left(-\frac{\sin n\pi x}{n^2\pi^2}\right)\right]_0^1 + \left[(1-x)\left(-\frac{\cos n\pi x}{n\pi}\right) - (-1)\left(-\frac{\sin n\pi x}{n^2\pi^2}\right)\right]_1^2$$

$$= -\frac{(-1)^n}{n\pi} + \frac{1}{n\pi} = \frac{1-(-1)^n}{n\pi} = \begin{cases} 0, \text{ if } n \text{ is even} \\ 2/n\pi, \text{ if } n \text{ is odd} \end{cases}$$

$\therefore$ (1) $\Rightarrow$
$$f(x) = -\frac{4}{\pi^2}\sum_{n=1}^{\infty}\frac{\cos(2n-1)\pi x}{(2n-1)^2} + \frac{2}{\pi}\sum_{n=1}^{\infty}\frac{\sin(2n-1)\pi x}{2n-1} \quad ...(2)$$

Deduction. To get the required result, we now put $x = 1$ is (2). Since $x = 1$ is the point of discontinuity for $f(x)$, by Art. 22.4 the sum of the series (2) at $x = 1$ is

$$= (1/2) \times [f(1-0) + f(1+0)] \times \left[\lim_{h\to 0} f(1-h) + \lim_{h\to 0} f(1+h)\right]$$

$$= (1/2) \times \left[\lim_{h\to 0}(1-h) + \lim_{h\to 0}\{1-(1+h)\}\right] = (1/2) \times \{1+0\} = (1/2)$$

$\therefore$ (2) gives $\quad \dfrac{1}{2} = -\dfrac{4}{\pi^2}\displaystyle\sum_{n=1}^{\infty}\frac{(-1)^{2n-1}}{(2n-1)^2} \quad$ or $\quad \displaystyle\sum_{n=1}^{\infty}\frac{1}{(2n-1)^2} = \frac{\pi^2}{8}$

or $\qquad 1/1^2 + 1/3^2 + 1/5^2 + ... = \pi^2/8.$

22.9. Half range series

In several engineering and physical applications sometimes it is required to get the Fourier series expansion of a function in an interval $(0, l)$ where l is half the period. Clearly, when we require to expand a function in the interval $(0, l)$, then it is immaterial what the function may be outside the range $0 < x < l$. We are free to choose it arbitrarily in the interval $(-l, 0)$.

Half range sine series. Let $f(x)$ be defined in $(0,l)$. We now define a new function F(x) as follows:

$$F(x) = \begin{cases} f(x), \text{ if } 0 < x < l \\ -f(-x), \text{ if } -l < x < 0 \end{cases} \quad ...(1)$$

The function F (x) defined by (1) is an odd function in $(-l, l)$. Hence the Fourier series of $f(x)$ contains only sine terms. Also note that in the interval $(0, l)$, F $(x) = f(x)$ and hence the sine series of F (x) gives the required sine series of $f(x)$ in $(0, l)$. Thus, we have [Refer results (14) and (15) of Art. 22.7.]

$$f(x) = \sum_{n=1}^{\infty} b_n \sin\frac{n\pi x}{l} \quad \text{where} \quad b_n = \frac{2}{l}\int_0^l f(x)\sin\frac{n\pi x}{l}\,dx \quad ...(2)$$

Particular Case. If $f(x)$ is defined in $(0, \pi)$ so that $l = \pi$. Then the required half range sine series is given by

$$f(x) = \sum_{n=1}^{\infty} b_n \sin nx, \quad \text{where} \quad b_n = \frac{2}{\pi}\int_0^{\pi} f(x)\sin nx\,dx. \quad ...(3)$$

Half range cosine series. Let $f(x)$ be defined in $(0, l)$. We now define a new function G (x) as follow:

$$G(x) = \begin{cases} f(x), \text{ if } 0 < x < l \\ f(-x), \text{ if } -l < x < 0 \end{cases} \quad ...(4)$$

The function G (x) defined by (1) is an even function in $(-l, l)$. Hence the Fourier series of G (x) contains only cosine terms. Also note that in the interval $(0, l)$, G $(x) = f(x)$ and hence the cosine series of G (x) gives the required cosine series of $f(x)$ in $(0, l)$. Thus, we have [Refer results (12) and (13) of Art. 22.7]

$$f(x) = \frac{a_0}{2} + \sum_{n=1}^{\infty} a_n \cos \frac{n\pi x}{l}, \quad ...(5)$$

where $$a_0 = \frac{2}{l}\int_0^l f(x)\,dx \quad \text{and} \quad a_n = \frac{2}{l}\int_0^l f(x)\cos\frac{n\pi x}{l}\,dx. \quad ...(6)$$

Particular Case. If $f(x)$ is defined in $(0, \pi)$ so that $l = \pi$. Then the required half range cosine series is given by

$$f(x) = \frac{a_0}{2} + \sum_{n=1}^{\infty} a_n \cos n\pi x, \quad ...(7)$$

where $$a_0 = \frac{2}{\pi}\int_0^\pi f(x)\,dx \quad \text{and} \quad a_n = \frac{2}{\pi}\int_0^\pi f(x)\cos nx\,dx \quad ...(8)$$

22.10. Solve examples based on results of Art. 22.9

Ex. 1. *Find half-range sine series for e^x in $0 < x < 1$.* **(Osmania 2010)**

Sol. Let the required series be $$e^x = \sum_{n=1}^{\infty} b_n \sin n\pi x, \quad ...(1)$$

where $$b_n = 2\int_0^l e^x \sin n\pi x\,dx = 2\left[\frac{e^x(\sin n\pi x - n\pi\cos n\pi x)}{1+n^2\pi^2}\right]_0^1$$

$$= 2\left[\frac{e(-n\pi\cos n\pi)}{1+n^2\pi^2} - \frac{(-n\pi)}{1+n^2\pi^2}\right], \quad \text{as} \quad \int e^{ax}\sin bx\,dx = \frac{e^{ax}(a\sin bx - b\cos bx)}{a^2+b^2}$$

$$= \{2/(1+n^2\pi^2)\}\{-en\pi(-1)^n + n\pi\} = \{2n\pi/(1+n^2\pi^2)\}\{1 - e(-1)^n\}.$$

$\therefore$ From (1), $$e^x = 2\pi\sum_{n=1}^{\infty}\frac{n[1-e(-1)^n]}{1+n^2\pi^2}\sin n\pi x.$$

Ex. 2. *Find a series of cosines of multiples of x which will represent x sin x in $(0, \pi)$ and show that* $1 + 2/(1 \times 3) - 2/(3 \times 5) + 2/(5 \times 7) - ... = \pi/2$ **(Osmania 2004)**

Sol. Let $$f(x) = x\sin x = \frac{a_0}{2} + \sum_{n=1}^{\infty} a_n \cos nx, \quad \text{where} \quad ...(1)$$

$$a_0 = \frac{2}{\pi}\int_0^\pi f(x)\,dx = \frac{2}{\pi}\int_0^\pi x\sin x\,dx = \frac{2}{\pi}[(x)(-\cos x) - (1)(-\sin x)]_0^\pi = 2$$

$$a_n = \frac{2}{\pi}\int_0^\pi f(x)\cos nx\,dx = \frac{1}{\pi}\int_0^\pi x(2\sin x\cos nx)\,dx \quad ...(2)$$

or $a_n = \frac{1}{\pi}\int_0^{\pi} x\,[\sin(n+1)x - \sin(n-1)x]\,dx$

$$= \frac{1}{\pi}\left[x\left(-\frac{\cos(n+1)x}{n+1}\right) - (1)\left(-\frac{\sin(n+1)x}{n+1}\right)\right]_0^{\pi} - \frac{1}{\pi}\left[x\left(-\frac{\cos(n-1)x}{n-1}\right) - (1)\left(-\frac{\sin(n-1)x}{n-1}\right)\right]_0^{\pi}$$

$$= \frac{1}{\pi}\left[-\frac{\pi(-1)^{n+1}}{n+1}\right] - \frac{1}{\pi}\left[-\frac{\pi(-1)^{n-1}}{n-1}\right] = (-1)^n\left[\frac{1}{n+1} - \frac{1}{n-1}\right] = \frac{2(-1)^{n+1}}{(n-1)(n+1)},\ n \neq 1$$

When $n = 1$, by (2), $a_1 = \frac{2}{\pi}\int_0^{\pi} x\sin x\cos x\,dx = \frac{1}{\pi}\int_0^{\pi} x\sin 2x\,dx$

or $$a_1 = \frac{1}{\pi}\left[(x)\left(-\frac{1}{2}\cos 2x\right) - (1)\left(-\frac{1}{4}\sin 2x\right)\right]_0^{\pi} = \frac{1}{\pi}\left[-\frac{\pi\cos 2\pi}{2}\right] = \frac{1}{2}$$

$\therefore$ From (1), $$x\sin x = \frac{a_0}{2} + a_1\cos x + \sum_{n=2}^{\infty} a_n\cos nx$$

or $$x\sin x = 1 - \frac{1}{2}\cos x - 2\left(\frac{\cos 2x}{1\times 3} - \frac{\cos 3x}{2\times 4} + \frac{\cos 4x}{3\times 5} - \ldots\right) \quad \text{...(3)}$$

Putting $x = \pi/2$ in (3), we have

$$\frac{\pi}{2} = 1 - 2\left(-\frac{1}{1\times 3} + \frac{1}{3\times 5} - \frac{1}{5\times 7} + \ldots\right) \quad \text{or} \quad 1 + \frac{2}{1\times 3} - \frac{2}{3\times 5} + \frac{2}{5\times 7} - \ldots = \frac{\pi}{2}$$

22.11. Parseval's theorem. (Parseval's Identify)

Let $f(x)$ be a periodic function with period 2π defined in the interval $(c, c+2l)$. Then

$$\frac{1}{2\pi}\int_c^{c+2l}\{f(x)\}^2\,dx = \frac{a_0^2}{4} + \frac{1}{2}\sum_{n=1}^{\infty}(a_n^2 + b_n^2),$$

where a_0, a_n and b_n, are the Fourier coefficients of $f(x)$.

Proof. The Fourier serise of $f(x)$ defined in $(c, c+2l)$ is given by

$$f(x) = \frac{a_0}{2} + \sum_{n=1}^{\infty}\left(a_n\cos\frac{n\pi x}{l} + b_r\sin\frac{n\pi x}{l}\right), \text{ where} \quad \text{...(1)}$$

$$a_0 = \frac{1}{l}\int_c^{c+2l} f(x)\,dx, \quad a_n = \frac{1}{l}\int_c^{c+2l} f(x)\cos\frac{n\pi x}{l}\,dx, \quad b_n = \frac{1}{l}\int_c^{c+2l} f(x)\sin\frac{n\pi x}{l}\,dx \quad \text{...(2)}$$

Multiplying both sides of (1) by $f(x)$ and integrating term by term from c to $c+2l$, we have

$$\int_c^{c+2l}\{f(x)\}^2\,dx = \frac{a_0}{2}\int_c^{c+2l} f(x)\,dx + \sum_{n=2}^{\infty}\left[a_n\int_c^{c+2l} f(x)\cos\frac{n\pi x}{l}\,dx + b_n\int_c^{c+2l} f(x)\sin\frac{n\pi x}{l}\,dx\right]$$

or $$\int_c^{c+2l}\{f(x)\}^2\,dx = \frac{a_0}{2}(l\,a_0) + \sum_{n=1}^{\infty}[a_n(l\,a_n) + b_n(l\,b_n)], \text{ using (2)}$$

or $$\frac{1}{2l}\int_c^{c+2l}\{f(x)\}^2\,dx = \frac{a_0^2}{4} + \frac{1}{2}\sum_{n=1}^{\infty}(a_n^2 + b_n^2) \quad \text{...(3)}$$

Particular cases of Parseval's theorem

Case (*i*) when $l = \pi$ and $c = -\pi$, Parseval's theorem reduces

$$\frac{1}{2\pi}\int_{-\pi}^{\pi}\{f(x)\}^2\,dx = \frac{a_0^2}{4} + \frac{1}{2}\sum_{n=1}^{\infty}(a_n^2 + b_n^2), \text{ where} \qquad ...(7)$$

$$a_0 = \frac{1}{\pi}\int_{-\pi}^{\pi} f(x)\,dx,\ a_n = \frac{1}{\pi}\int_{-\pi}^{\pi} f(x)\cos nx\,dx,\ b_n = \frac{1}{\pi}\int_{-\pi}^{\pi} f(x)\sin nx\,dx\ ... \qquad ...(8)$$

Case (*ii*) when $l = \pi$ and $c = 0$, Parseval's theorem reduces to

$$\frac{1}{2\pi}\int_{0}^{2\pi}\{f(x)\}^2\,dx = \frac{a_0^2}{4} + \frac{1}{2}\sum_{n=1}^{\infty}(a_n^2 + b_n^2), \qquad \text{where} \qquad ...(9)$$

$$a_0 = \frac{1}{\pi}\int_{0}^{2\pi} f(x)\,dx,\ a_n = \frac{1}{\pi}\int_{0}^{2\pi} f(x)\cos nx\,dx,\ b_n = \frac{1}{\pi}\int_{0}^{2\pi} f(x)\sin nx\,dx\ ... \qquad ...(10)$$

Case (*iii*) Where $c = -l$, Parseval's theorem reduces to

$$\frac{1}{2l}\int_{-l}^{l}\{f(x)\}^2\,dx = \frac{a_0^2}{4} + \frac{1}{2}\sum_{n=1}^{\infty}(a_n^2 + b_n^2), \qquad \text{where} \qquad ...(11)$$

$$a_0 = \frac{1}{l}\int_{-l}^{l} f(x)\,dx, \quad a_n = \frac{1}{l}\int_{-l}^{l} f(x)\cos\frac{n\pi x}{l}\,dx, \quad b_n = \frac{1}{l}\int_{-l}^{l} f(x)\sin\frac{n\pi x}{l}\,dx\ ... \qquad ...(12)$$

Case (*iv*) When $c = 0$, Parseval, theorem reduces to

$$\frac{1}{2l}\int_{-l}^{l}\{f(x)\}^2\,dx = \frac{a_0^2}{4} + \frac{1}{2}\sum_{n=1}^{\infty}(a_n^2 + b_n^2), \qquad \text{where} \qquad ...(13)$$

$$a_0 = \frac{1}{l}\int_{0}^{2l} f(x)\,dx,\ a_n = \frac{1}{l}\int_{0}^{2l} f(x)\cos\frac{n\pi x}{l}\,dx,\ b_n = \frac{1}{l}\int_{0}^{2l} f(x)\sin\frac{n\pi x}{l}\,dx \qquad ...(14)$$

Case (*v*) If $f(x)$ is expressed in half range cosine series in $0 < x < l$ so that

$$f(x) = \frac{a_0}{2} + \sum_{n=1}^{\infty} a_n\cos\frac{n\pi x}{l},$$

where $$a_0 = \frac{2}{l}\int_{0}^{l} f(x)\,dx, \qquad a_n = \frac{2}{l}\int_{0}^{l} f(x)\cos\frac{n\pi x}{l}\,dx,$$

then Parseval's theorem becomes $$\frac{1}{l}\int_{0}^{l}\{f(x)\}^2\,dx = \frac{a_0^2}{4} + \frac{1}{2}\sum_{n=1}^{\infty} a_n^2 \qquad ...(15)$$

Case (*vi*) If $f(x)$ is expressed in half range cosine series in $0 < x < l$ so that

$$f(x) = \sum_{n=1}^{\infty} b_n\sin\frac{n\pi x}{l}, \qquad \text{where} \qquad b_n = \frac{2}{l}\int_{0}^{l} f(x)\sin\frac{n\pi x}{l}\,dx,$$

then Parseval's theorem becomes $$\frac{1}{l}\int_{0}^{l}\{f(x)\}^2\,dx = \frac{1}{2}\sum_{n=1}^{\infty} b_n^2 \qquad ...(16)$$

Note: The results (3), (7), (9), (11), (13), (15) and (16) are known as *Parseval's* identities.

SOME ILLUSTRATIVE SOLVED EXAMPLES BASED ON ART. 22.11

Ex.1. *Apply Parseval's identity to the function* $f(x) = x, -\pi \le x \le \pi$ and deduce that $\pi^2/6 = 1/1^2 + 1/2^2 + \ldots + 1/n^2 + \ldots$ **(Mumbai 2010)**

Sol. Let $f(x) = x, -\pi \le x \le \pi$. From case (*i*) of Art. 22.10, Parseval's identity is given by

$$\frac{1}{2\pi}\int_{-\pi}^{\pi}\{f(x)\}^2\,dx = \frac{a_0^2}{4} + \frac{1}{2}\sum_{n=1}^{\infty}(a_n^2 + b_n^2), \quad \text{where} \qquad \ldots(1)$$

$$a_0 = \frac{1}{\pi}\int_{-\pi}^{\pi} f(x)\,dx, \quad a_n = \frac{1}{\pi}\int_{-\pi}^{\pi} f(x)\cos nx\,dx, \quad b_n = \frac{1}{\pi}\int_{-\pi}^{\pi} f(x)\sin nx\,dx, \qquad \ldots(2)$$

Since $f(x) = x$, (2) yields $a_0 = 0$, $a_n = 0$ and

$$b_n = \frac{1}{\pi}\int_{-\pi}^{\pi} x\sin nx\,dx = \frac{2}{\pi}\int_0^{\pi} x\sin nx\,dx, \text{ as } x\sin nx \text{ is an even function}$$

$$= \frac{2}{\pi}\left[(x)\left(-\frac{\cos nx}{n}\right) - (1)\left(-\frac{\sin nx}{n^2}\right)\right]_0^{\pi}, \text{ using generalised rule of integration by parts}$$

$= (2/\pi) \times \{-(\pi/n) \times \cos n\pi\} = -(2/n) \times (-1)^n = (-1)^{n+1} \times (2/n)$

Again, $$\frac{1}{2\pi}\int_{-\pi}^{\pi}\{f(x)\}^2\,dx = \frac{1}{2\pi}\int_{-\pi}^{\pi} x^2\,dx = 2 \times \frac{1}{2\pi}\int_0^{\pi} x^2\,dx = \frac{1}{\pi} \times \frac{\pi^3}{3} = \frac{\pi^2}{3}$$

Substituting the above values in (1), we have

$$\frac{\pi^2}{3} = \frac{1}{2}\sum_{n=1}^{\infty}\frac{4}{n^2} \quad \text{or} \quad \frac{\pi^2}{6} = \sum_{n=1}^{\infty}\frac{1}{n^2} \quad \text{or} \quad \frac{\pi^2}{6} = \frac{1}{1^2} + \frac{1}{2^2} + \ldots + \frac{1}{n^2} + \ldots$$

Ex.2. *Using the Fourier series for* $f(x) = |x|, -\pi \le x \le \pi$, *find the sum* $1/1^2 + 1/3^2 + 1/5^2 + \ldots$, *assuming that Fourier series of function f at* $x = 0$ *converges to* $f(0)$. *Using Parseval's identity calculate the sum* $1/1^4 + 1/3^4 + 1/5^4 + \ldots$.

(Delhi 2008, Mumbai 2010, Nagpur 2005; Himanchal 2009; Meerut 2013)

Sol. Since $f(-x) = |-x| = |x| = f(x)$, hence f is an even function. For such a function the required Fourier series is given by (refer result 3 of Art. 22.5)

$$f(x) = |x| = \frac{a_0}{2} + \sum_{n=1}^{\infty} a_n\cos nx\,dx, \quad \text{where} \qquad \ldots(1)$$

where $$a_0 = \frac{2}{\pi}\int_0^{\pi} f(x)\,dx = \frac{2}{\pi}\int_0^{\pi} x\,dx = \frac{2}{\pi} \times \left[\frac{x^2}{2}\right]_0^{\pi} = \pi \qquad \ldots(2)$$

$$a_n = \frac{2}{\pi}\int_0^{\pi} f(x)\cos nx\,dx = \frac{2}{\pi}\int_0^{\pi} x\cos nx\,dx = \frac{2}{\pi}\left[(x)\left(\frac{\sin nx}{n}\right) - (1)\left(-\frac{\cos nx}{n^2}\right)\right]_0^{\pi}$$

$$= \frac{2(\cos n\pi - 1)}{\pi n^2} = -\frac{2\{1-(-1)^n\}}{\pi n^2} = \begin{cases} 0, & \text{if } n \text{ is even} \\ -4/\pi n^2, & \text{if } n \text{ is odd} \end{cases} \qquad \ldots(3)$$

Also, note that for even function $f(x)$, $\qquad b_n = 0 \qquad \ldots(4)$

Using (2) and (3), (1) reduces to

$$f(x) = \frac{\pi}{2} - \frac{4}{\pi}\sum_{n=1}^{\infty}\frac{\cos(2n-1)x}{(2n-1)^2} \quad \text{or} \quad |x| = \frac{\pi}{2} - \frac{4}{\pi}\left(\cos x + \frac{\cos 3x}{3} + \frac{\cos 5x}{5} + \ldots\right) \qquad \ldots(5)$$

Since f at $x = 0$ converges to $f(0)$ and $f(0) = 0$, hence putting $x = 0$ in (5), we have
$0 = (\pi/2) - (4/\pi) \times (1 + 1/3^2 + 1/5^2 + ...)$ or $1 + 1/3^2 + 1/5^2 + ... = \pi^2/8$.

To find the sum $1/1^4 + 1/3^4 + 1/5^4 + ...$ with help of Parseval's identity

Since $-\pi \le x \le \pi$, the Parseval's identity (refer case (*i*) of Art. 6.11) is given by

$$\frac{1}{2\pi}\int_{-\pi}^{\pi}\{f(x)\}^2\,dx = \frac{a_0^{\ 2}}{4} + \frac{1}{2}\sum_{n=1}^{\infty}(a_n^{\ 2} + b_n^{\ 2}), \quad ...(6)$$

(2), (3) and (4) give $b_n = 0\ (n = 1, 2, 3, ...)$, $a_0 = \pi$, $a_n = 0$, if n is even and $a_n = -(4/\pi n^2)$, if n is odd. Hence (6) reduces to

$$\frac{1}{2\pi}\int_{-\pi}^{\pi}x^2\,dx = \frac{\pi^2}{4} + \frac{1}{2}\sum_{n=1}^{\infty}\frac{16}{\pi^2(2n-1)^4} \quad \text{or} \quad \frac{\pi^2}{3} = \frac{\pi^2}{4} + \frac{8}{\pi^2}\sum_{n=1}^{\infty}\frac{1}{(2n-1)^4}$$

or $$\sum_{n=1}^{\infty}\frac{1}{(2n-1)^4} = \frac{\pi^2}{8} \times \left(\frac{\pi^2}{3} - \frac{\pi^2}{4}\right) \quad \text{or} \quad \frac{1}{1^4} + \frac{1}{3^4} + \frac{1}{5^4} + ... = \frac{\pi^4}{96}$$

22.12. Some Particular Fourier Series. In what follows, we give some functions defining some special waveforms occuring in communication engineering and science.

(*i*) Square wave form

$$f(x) = \begin{cases} k, \text{ for } 0 < x < \pi \\ 0, \text{ for } -\pi < x < 0, \end{cases}$$

satisfying $f(x + 2\pi) = f(x)$. $f(x)$ is defined at origin arbitrarily but finitely.

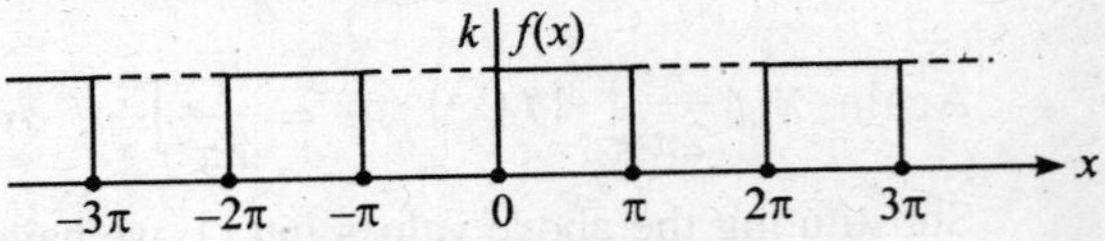

The graph of the function

Fourier series is $$f(x) = (k/2) + (2k/\pi) \times \sum_{n=1}^{\infty}(\sin nx)/n$$

(*ii*) Sawtooth wave form

$f(x) = x$ for $-\pi \le x \le \pi$

satisfying $f(x + 2\pi) = f(x)$

Forier series is $$f(x) = x = \sum_{n=1}^{\infty}\frac{(-1)^{n+1}\sin nx}{n}$$

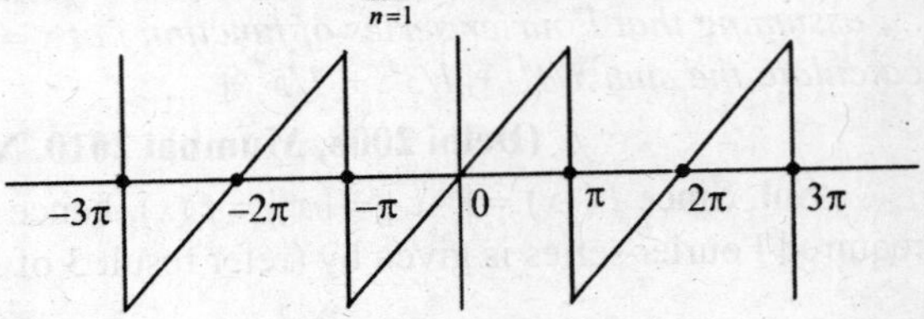

The graph of the function

(*iii*) Full wave rectifier

$$f(x) = \begin{cases} \sin x, \text{ for } 0 \le x \le \pi \\ -\sin x, \text{ for } -\pi \le x \le 0, \end{cases}$$

satisfying $f(x + \pi) = f(x)$

The graph of the function

Forier series is $$f(x) = (2/\pi) - \sum_{n=1}^{\infty}\{(\cos nx)/(n^2 - 1)\}$$

(*iv*) Half wave rectifier

$$f(x) = \begin{cases} \sin x, \text{ for } 0 \le x \le \pi \\ 0, \text{ for } -\pi \le x \le 0, \end{cases}$$

satisfying $f(x + \pi) = f(x)$

The graph of the function

The Fourier series is $$f(x) = (1/\pi) - (2/\pi)\sum_{n=1}^{\infty}\{(\cos 2nx)/(4n^2 - 1)\}$$

(v) Triangular Wave form

$$f(x) = \begin{cases} \pi + 2x, \text{ for } -\pi < x < 0 \\ \pi - 2x, \text{ for } 0 < x < \pi \end{cases}$$

satisfying $f(x + 2\pi) = f(x)$

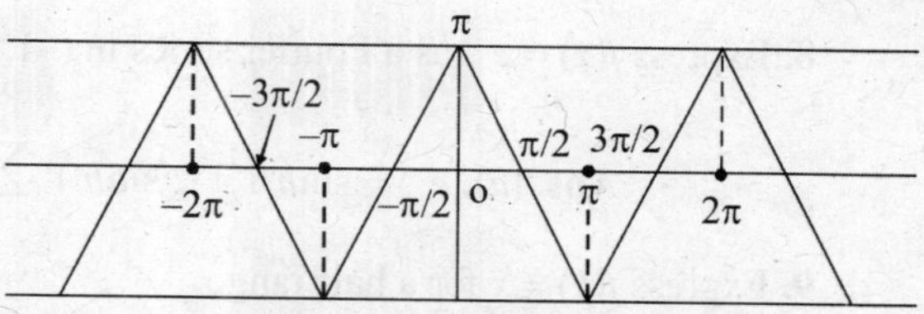

The graph of the function

The Fourier series is $f(x) = (4/\pi) \times \{(2/1^2) \times \cos x + (2/3^2) \times \cos 3x + (2/5^2) \times \cos 5x + ...\}$

EXERCISES

1. Obtain Fourier series for $f(x) = x$ in $(-\pi, \pi)$.

Ans. $x = 2\{\sin x - (1/2) \times \sin 2x + (1/3) \times \sin 3x + ...\}$

2. Find Fourier series for $f(x) = \sin mx$ in the interval $-\pi < x < \pi$, where m is neither zero nor an integer.

Ans. $\sin mx = (2/\pi) \times \sin m\pi\{(\sin x) / (1^2 - m^2) - (2 \sin 2x) / (2^2 - m^2) + (3 \sin 3x) / (3^2 - m^2) + ...\}$

3. If $f(x) = \begin{cases} -1, \text{ if } -\pi < x < 0 \\ 1, \text{ if } 0 \le x < \pi \end{cases}$ and $f(x + 2\pi) = f(x)$ for all x, obtain the Fourier series for $f(x)$.

Deduce that $1 - (1/3) + (1/5) ... = \pi/4$ **[Delhi B.A. (Prog.) III, 2014]**

Ans. $f(x) = (4/\pi) \times \{\sin x + (1/3) \times \sin 3x + (1/5) \times \sin 5x + ...\}$

4. (*a*) Obtain the Fourier series for $f(x)$ in $-\pi < x < \pi$ with period 2π where $f(x) = -1 + x$, for $-\pi < x < 0$ and $f(x) = 1 + x$, for $0 < x < \pi$.

(*b*) Obtain Fourier series for the function $f(x) = x^2$ in the interval $-\pi < x < \pi$ and deduce that $1/1^2 + 1/2^2 + 1/3^2 + ... = \pi^2/6$, $1 - 1/2^2 + 1/3^2 - ... = \pi^2/12$ and $1/1^2 + 1/3^2 + 1/5^2 + ... = \pi^2/8$.

(Kanpur 2012; Delhi B.A. (Prog.) III 2012)

Ans. (*a*). $f(x) = \frac{2}{n\pi}\sum_{n=1}^{\infty}\{1 - (1 + \pi)(-1)^n\} \sin nx$ **(*b*).** $x^2 = \frac{\pi^2}{3} + 4\sum_{n=1}^{\infty}\frac{(-1)^n \cos nx}{n^2}$

5. (*a*) Obtain the Fourier series to represent e^{ax} in $(-\pi, \pi)$

(*b*) Find the Fourier series for $f(x) = e^x$ in $(-\pi, \pi)$. **(Delhi 2008)**

Ans. (*a*) $e^{ax} = \frac{\sinh a\pi}{a\pi} + \frac{2a \sinh a\pi}{\pi}\sum_{n=1}^{\infty}\frac{(-1)^n \cos nx}{a^2 + n^2} - \frac{2 \sinh a\pi}{\pi}\sum_{n=1}^{\infty}\frac{n(-1)^n \sin nx}{a^2 + n^2}$

(*b*) $e^x = \frac{\sinh \pi}{\pi} + \frac{2 \sinh \pi}{\pi}\sum_{n=1}^{\infty}\frac{(-1)^n \cos nx}{1 + n^2} - \frac{2 \sinh \pi}{\pi}\sum_{n=1}^{\infty}\frac{n(-1)^n \sin nx}{1 + n^2}$

6. (*a*) Find the Fourier series to represent $f(x) = (\pi - x)^2/4$, $0 < x < 2\pi$. Deduce that $1/1^2 + 1/2^2 + 1/3^2 + ... = \pi^2/6$, $1/1^2 - 1/2^2 + 1/3^2 - ... = \pi^2/12$, and $1/1^2 + 1/3^2 + 1/5^2 + ... = \pi^2/8$.

(*b*) Find the Fourier series to represent the function $f(x)$ given by $f(x) = x$ for $0 \le x \le \pi$ and $f(x) = 2\pi - x$ for $\pi \le x \le 2\pi$. Deduce that $1/1^2 + 1/3^2 + 1/5^2 + ... = \pi^2/8$.

Ans. (*a*). $\frac{(\pi - x)^2}{4} = \frac{\pi^2}{12} + \sum_{n=1}^{\infty}\frac{\cos nx}{n^2}$ **(*b*).** $f(x) = \frac{\pi}{2} - \frac{4}{\pi}\sum_{n=1}^{\infty}\frac{\cos(2n-1)x}{(2n-1)^2}$

7. Find the Fourier series for the function $f(x) = 1$, if $0 < x < 1$ and $f(x) = 2$, if $1 < x < 2$.

Ans. $f(x) = \frac{3}{2} - \frac{2}{\pi}\sum_{n=1}^{\infty}\frac{\sin(2n-1)\pi x}{2n-1}$

8. Express $f(x) = e^{-x}$ as a Fourier series in (–1, 1).

$$\textbf{Ans. (a).}\ e^{-x} = \sinh 1 + 2 \sinh 1 \sum_{n=1}^{\infty} \frac{(-1)^n \cos n\pi x}{1+n^2\pi^2} + 2\pi \sinh 1 \sum_{n=1}^{\infty} \frac{n(-1)^n \sin n\pi x}{1+n^2\pi^2}$$

9. Express $f(x) = x$ for a half range
(*i*) sine series in $0 < x < 2$ **(Osmania 2004)** (*ii*) cosine series 1*n* $0 < x < 2$.

Ans. (*i*) $x = (4/\pi) \times \{\sin(\pi x/2) - (1/2) \times \sin(2\pi x/2) - (1/3) \times \sin(3\pi x/2) - ....\}$
(*ii*) $x = 1 - (8/\pi^2) \times \{\cos(\pi x/2) - (1/3^2) \times \cos(3\pi x/2) - (1/5^2) \times \sin(5\pi x/2) + ...\}$

10. Express $f(x) = \cos x$, $0 < x < \pi$ in half range sine series.

Ans. $\cos x = (8/\pi) \times \{(1/3) \times \cos 2x + (2/15) \times \cos 4x + (3/35) \times \cos 4x + ...\}$

11. Express the half range sine series for $f(x) = x(\pi - x)$ in $(0, \pi)$. Deduce that $1/1^3 - 1/3^3 + 1/5^3 - 1/7^3 + ... = \pi^3/32$.

12. Find the (*i*) half range sine and (*ii*) half range cosine series for $f(x) = x(l - x)$ in $0 < x < l$. **(Madurai Kamraj 2008)**

Ans. (*i*) $x(l - x) = (8l^2/\pi^3) \times \{\sin(\pi x/l) + (1/3^3) \times \sin(3\pi x/l) + (1/5^3) \times \sin(5\pi x/l) + ...\}$
(*ii*) $x(l - x) = l^2/6 + (4l^2/\pi^2) \times \{\cos(\pi x/l) + (1/3^2) \times \cos(3\pi x/l) + (1/5^2) \times \cos(5\pi x/l) + ...\}$

13. Find the series of sines multiples of x for the function $f(x)$ given by $f(x) = x$, if $0 < x \le 1$ and $f(x) = 2 - x$, if $1 < x < 2$. **(Delhi 2007)**

$$\textbf{Ans.}\ f(x) = \frac{8}{\pi^2} \sum_{n=1}^{\infty} \frac{(-1)^{n-1}}{(2n-1)^2} \sin \frac{(2n-1)\pi x}{2}$$

14. Find the Fourier series of the periodic function f with period 2π defined as $f(x) = x$, if $-\pi < x \le 0$ and $f(x) = 2x$, if $0 < x \le \pi$. Deduce that $\pi^2/8 = 1 + 1/3^2 + 1/5^2 + ...$ **(Delhi 2011)**

$$\textbf{Ans.}\ f(x) = \frac{\pi}{4} - \frac{2}{\pi}\left(\frac{\cos x}{1^2} + \frac{\cos 3x}{3^2} + \frac{\cos 5x}{5^2} + ...\right) + 3\left(\frac{\sin x}{1} - \frac{\sin 3x}{2} + \frac{\sin 3x}{3} - ...\right)$$

15. The function $f(x) = e^x$ has the Fourier series $e^x = \sum_{n=1}^{\infty} b_n \sin(nx)$ in the interval $(0, \pi)$.

Then $\sum_{n=1}^{\infty} (b_n)^2$ converges to

(*a*) $(e^{\pi} - 1)/\pi$ (*b*) $(e^{\pi} + 1)/\pi$ (*c*) $(e^{2\pi} - 1)/\pi$ (*d*) $(e^{2\pi} + 1)/\pi$ **(GATE 1999)**

16. For $x \in [-\pi, \pi]$, let $f(x) = (\pi + x)(\pi - x)$ and $g(x) = \begin{cases} \cos(1/x), & \text{if } x \ne 0 \\ 0, & \text{if } x = 0 \end{cases}$

Consider the statements P: The Fourier series of f converges uniformly to f on $[-\pi, \pi]$ Q: The Fourier series of g converges uniformly to g on $[-\pi, \pi]$. Then (*a*) P and Q are true (*b*) P is true but Q is false (*c*) P is false but Q is true (*d*) both P and Q are false. **(GATE 2008)**

17. The Fourier expansion in the interval [–4, 4] of the function $f(x) = -x$, if $-4 \le x \le 0$ and $f(x) = x$, if $0 < x < 4$ has

(*a*) no cosine term (*b*) no sine term
(*c*) both cosine and sine terms (*d*) none of these. **(GATE 2000)**

Sol. Ans. (*b*). Note that the given function is an even function.

18. Let $f(x) = \cos x$ for $0 < x < \pi$ and $f(x) = -\cos x$ for $-\pi < x < 0$. Show that the Fourier series which converges to $f(x)$ is

$$\frac{\pi}{4}\left(\frac{2}{1\times 3}\sin 2x+\frac{4}{3\times 5}\sin 4x+\frac{6}{5\times 7}\sin 6x+...\right)$$

Draw the graph of the sum function of the series for $-2\pi \le x \le 2\pi$

19. Show that when $-\pi \le x \le \pi$, $\cot kx = \frac{\sin kx}{\pi}\left(\frac{1}{k}-\frac{2k\cos x}{k^2-1^2}+\frac{2k\cos 2x}{k^2-2^2}+...\right)$,

k being non-integral. Reduce that

(*i*) $\pi\cot k\pi = \frac{1}{k}+2k\sum\frac{1}{k^2-n^2}$ (*ii*) $\frac{\pi}{\sin k\pi} = \sum(-1)^n\left(\frac{1}{n+k}+\frac{1}{n+k-k}\right)$

20. Obtain a sine series which will be equal to x^2 for $0 \le x < \pi$.
Draw a graph of the sum function of the series for $-2\pi \le x \le 2\pi$

21. Show that in $[0, \pi]$, $f(x) = \frac{\pi^2}{16}-2\sum_{n=1}^{\infty}\frac{\cos(4n-2)x}{(4n-2)^2}$,

where $f(x) = (\pi x)/4$ when $0 \le x \le \pi/2$ and $f(x) = \{\pi(\pi - x)\}/4$ when $\pi/2 < x \le \pi$

22. Show that $x+x^2 = \frac{1}{3}+\frac{2}{\pi}\sum_{n=1}^{\infty}(-1)^n\left(\frac{2\cos n\pi x}{n^2\pi^2}-\frac{Cos\, n\pi x}{n}\right)$, when $-1 < x < 1$

23. Show that for $0 \le x \le 1$, $x-\frac{1}{2}=-\frac{1}{\pi}\sum_{n=1}^{\infty}\frac{\sin 2n\pi x}{n}$, $x^2-x+\frac{1}{6}=\frac{1}{x^2}\sum_{n=1}^{\infty}\frac{\cos 2n\pi x}{n^2}$

24. If $f(x) = (a/4) - x$, when $0 \le x \le a/2$ and $f(x) = x - (3a/4)$, when $a/2 \le x \le a$, show that for all x in $0 \le x \le a$

$$f(x) = \frac{2a}{\pi^2}\sum_{n=1}^{a}\frac{1}{(2n-1)^2}\cos\frac{(4n-2)\pi x}{a}$$

25. Expand in Fourier series of the form $a_n \sin(n\pi x)$ a function f given by $f(x) = \sinh \pi x$. $0 \le x < 1/2$ and $f(x) = 0$ for $1/2 < x \le 1$. Deduce that

$$\sum_{n=1}^{\infty}\frac{1}{n^2+(n+1)^2}=\frac{\pi}{2}\tanh\frac{\pi}{2}$$

26. Show that the graph of the function

$$y^2 = \frac{a^2}{3}+\frac{16a^2}{\pi^2}\sum_{n=1}^{\infty}\frac{(-1)^{n-1}}{(2x-1)^3}\cos\frac{(2x-1)\pi x}{2a}+\frac{8a^2}{\pi^2}\sum_{n=1}^{\infty}\frac{(-1)^n}{4n^2}\cos\frac{n\pi x}{a}$$

in a series of circles of radii a connected by straight lines of length $2a$ the origin being the centre of one of teh cirlces.

Hint: Expand $f(x)$ in $[-2a, 2a]$ when $f(x) = a$ when $-2a \le x \le -a$, $f(x) = a^2 - x^2$ when $-a \le x \le a$ and $f(x) = 0$ when $a \le x \le 2a$.

22.13 SOME IMPORTANT THEOREMS

In Art 22.4, we have merely stated some results in connection with convergence of Fourier series. In what follows we propose to present proof of the results stated in Art. 22.4.

Theorem I. (*Bessel's theorem and Bessel's inequality*) **(Delhi 2011)**

If f is bounded and integrable on $[-\pi, \pi]$ *and* a_n, b_n *are its Fourier coefficients, then*

$$\sum_{n=1}^{\infty}(a_n^2+b_n^2) \text{ converges} \quad \text{and} \quad \sum_{n=1}^{\infty}(a_n^2+b_n^2) \le \frac{1}{\pi}\int_{-\pi}^{\pi}\{f(x)\}^2\,dx$$

Proof. The integral $$\int_{-\pi}^{\pi}\{f(x)-\sum_{n=1}^{m}(a_n\cos nx+b_n\sin nx)\}^2 \ge 0 \qquad ...(1)$$

because its integrand is never negative. Re-writing (1), we have

$$\int_{-\pi}^{\pi}\{f(x)\}^2\,dx+\int_{-\pi}^{\pi}\sum_{n=1}^{m}(a_n\cos nx+b_n\sin nx\}^2\,dx$$

$$-2\sum_{n=1}^{\infty}\{a_n\int_{-\pi}^{\pi}f(x)\cos nx\,dx+b_n\int_{-\pi}^{\pi}f(x)\sin nx\,dx\}\ge 0 \qquad ...(2)$$

The Fourier coefficients a_n, b_n are given by

$$a_n=\frac{1}{\pi}\int_{-\pi}^{\pi}f(x)\cos nx\,dx \quad \text{and} \quad b_n=\frac{1}{\pi}\int_{-\pi}^{\pi}f(x)\sin nx\,dx \qquad ...(3)$$

Again, $$\int_{-\pi}^{\pi}\sum_{n=1}^{m}(a_n\cos nx+b_n\sin nx)^2\,dx=\sum_{n=1}^{m}\int_{-\pi}^{\pi}(a_n\cos nx+b_n\sin nx)^2\,dx$$

$$=\sum_{n=1}^{m}\left\{a_n^2\int_{-\pi}^{\pi}\cos^2 nx\,dx+b_n^2\int_{-\pi}^{\pi}\sin^2 nx\,dx+2a_nb_n\int_{-\pi}^{\pi}\cos nx\sin nx\,dx\right\} \qquad ...(4)$$

Now, $$\int_{-\pi}^{\pi}\cos^2 nx\,dx=2\int_0^{\pi}\cos^2 nx\,dx=\int_0^{\pi}(1+\cos 2nx)\,dx=\left[x+\frac{1}{2}\sin 2nx\right]_0^{\pi}=\pi \qquad ...(5)$$

Similarly, $$\int_{-\pi}^{\pi}\sin^2 nx\,dx=\pi \quad \text{and} \quad \int_{-\pi}^{\pi}\cos nx\sin nx\,dx=0 \qquad ...(6)$$

Using (5) and (6), (4) yields $$\int_{-\pi}^{\pi}\sum_{n=1}^{m}(a_n\cos nx+b_n\sin nx)^2\,dx=\sum_{n=1}^{m}(\pi a_n^2+\pi b_n^2) \qquad ...(7)$$

Using (3) and (7), (2) may be re-written as

$$\int_{-\pi}^{\pi}\{f(x)\}^2\,dx+\pi\sum_{n=1}^{m}a_n^2+\pi\sum_{n=1}^{m}b_n^2-2\pi\sum_{n=1}^{m}a_n^2-2\pi\sum_{n=1}^{m}b_n^2\ge 0$$

Thus, $$\sum_{n=1}^{m}(a_n^2+b_n^2)\le\frac{1}{\pi}\int_{-\pi}^{\pi}\{f(x)\}^2\,dx \qquad ...(8)$$

Taking limit as $m\to\infty$, (8) yields $$\sum_{n=1}^{\infty}(a_n^2+b_n^2)\le\frac{1}{\pi}\int_{-\pi}^{\pi}\{f(x)\}^2\,dx, \qquad ...(9)$$

which is known as *Bessel's inequality*.

From (9), it follows that the series $\sum_{n=1}^{\infty}(a_n^2+b_n^2)$ is convergent.

Corollary. $\lim_{n \to \infty} \int_{-\pi}^{\pi} f(x) \cos nx \, dx = 0$ and $\lim_{n \to \infty} \int_{-\pi}^{\pi} f(x) \sin nx \, dx = 0$

Proof. By Bessel's theorem, $\sum_{n=1}^{\infty} (a_n^2 + b_n^2)$ is convergent. This implies that

$$\lim_{n \to \infty} (a_n^2 + b_n^2) = 0 \quad \Rightarrow \quad a_n \to 0 \quad \text{and} \quad b_n \to 0 \quad \text{as} \quad n \to \infty$$

Hence, $\frac{1}{\pi} \int_{-\pi}^{\pi} f(x) \cos nx \, dx \to 0$ and $\frac{1}{\pi} \int_{-\pi}^{\pi} f(x) \sin nx \, dx \to 0$ as $n \to \infty$

$$\Rightarrow \quad \lim_{n \to \infty} \int_{-\pi}^{\pi} f(x) \cos nx \, dx = 0 \quad \text{and} \quad \lim_{n \to \infty} \int_{-\pi}^{\pi} f(x) \sin nx \, dx = 0$$

Theorem II. (Riemann - Lebesgue theorem)

If function ϕ is bounded and integrable in $a \le x \le b$ and if

$$A_n = \int_a^b \phi(x) \cos nx \, dx \quad \textit{and} \quad B_n = \int_a^b \phi(x) \sin nx \, dx,$$

then A_n and B_n tend to zero as n tends to infinity.

Proof. To establish the theorem, we shall consider the following two cases:

Case I: Let a and b belong to the interval of the form $-k\pi \le x \le k\pi$ (k being an integer). Define a periodic function $f(x)$ with period 2π such that $f(x) = \phi(x)$ in (a, b) and $f(x) = 0$ outside (a, b). Hence, the function f is defined for all real values of x. Let a_n and b_n be the Fourier coefficients of $f(x)$ in $[-\pi, \pi]$. Then we have

$$A_n = \int_a^b \phi(x) \cos nx \, dx = \int_{-\pi}^{\pi} f(x) \cos nx \, dx = \pi \, a_n \quad \text{...(1)}$$

and $$B_n = \int_a^b \phi(x) \sin nx \, dx = \int_{-\pi}^{\pi} f(x) \sin nx \, dx = \pi \, b_n \quad \text{...(2)}$$

But by corollary of theorem I, $a_n \to 0$ and $b_n \to 0$ as $n \to \infty$ ...(3)

From (1), (2) and (3), it follows that $A_n \to 0$ and $B_n \to 0$ as $n \to \infty$

Case II: If a and b do not belong to some interval of the form $-k\pi \le x \le k\pi$ (k being an integer), then we can split up the interval $[a, b]$ into a finite number of sub-intervals, each of which lies in an interval of the form $-k\pi \le x \le k\pi$. Then, A_n and B_n can be expressed as the sum of a finite number of terms each tending to zero as $n \to \infty$. Hence $A_n \to 0$ and $B_n \to 0$ as $n \to \infty$.

Deduction. *Let ϕ be bounded and integrable function. Then, the limits of the so called Dirichlet integrals are equal, that is,*

$$\lim_{n \to \infty} \int_0^a \phi(x) \frac{\sin nx}{\sin x} dx = \lim_{n \to \infty} \int_0^a \phi(x) \frac{\sin nx}{x} dx$$

Proof. Assign the value zero to the function $\{(1/\sin x) - (1/x)\}$ at $x = 0$. Then, clearly this function is continuous and bounded in $[0, a]$, $a < \pi$. Hence, the function $\phi(x) \{(1/\sin x) - (1/x)\}$ is bounded and integrable on $[0, a]$. Hence, by the above theorem, it follows that

$$\lim_{n \to \infty} \int_0^a \phi(x) \left(\frac{1}{\sin x} - \frac{1}{x} \right) \sin nx \, dx \quad \Rightarrow \quad \lim_{n \to \infty} \int_0^a \phi(x) \frac{\sin nx}{\sin x} dx = \lim_{n \to \infty} \int_0^a \phi(x) \frac{\sin nx}{x} dx$$

Theorem III. ***Let function f be bounded and integrable in [0, a], a > 0 and let it be monotonic in some interval [0, α] where α, in any positive number less then a. Then***

$$\lim_{n \to \infty} \int_0^a f(x) \frac{\sin nx}{x} dx = f(0+) \int_0^{\infty} \frac{\sin x}{x} dx \quad \textbf{(Delhi 2007)}$$

Proof. Since f is bounded and monotonic in $[0, \alpha]$, therefore $f(0+)$ will exist. In order to prove the theorem, we shall consider the following two cases:

Case I: Let $f(0+) = 0$. Also, without affecting in any manner, whatsoever, we can suppose that $f(0) = 0$.

Let h be any positive number less than α, By the second mean value theorem, there exists $h_1 \in [0, h]$ such that

$$\int_0^h f(x)\frac{\sin nx}{x}\,dx = f(0)\int_0^{h_1}\frac{\sin nx}{x}\,dx + f(h)\int_{h_1}^{h}\frac{\sin nx}{x}\,dx$$

$$= f(h)\int_{h_1}^{h}\frac{\sin nx}{x}\,dx, \text{ since by hypothesis, } f(0) = 0$$

$$= f(h)\int_{nh_1}^{nh}\frac{\sin y}{y}\,dy, \text{ putting } nx = y \text{ and } n\,dx = dy$$

Thus, $$\int_0^h f(x)\frac{\sin nx}{x}\,dx = f(h)\int_{nh_1}^{nh}\frac{\sin x}{x}\,dx \qquad ...(1)$$

Since $\int_0^\infty \frac{\sin x}{x}\,dx$ is convergent, there exists $k > 0$ such that for all $t \geq 0$, we have

$$\left|\int_0^t \frac{\sin x}{x}\,dx\right| \leq k$$

which implies $$\left|\int_{nh_1}^{nh}\frac{\sin x}{x}\,dx\right| = \left|\int_0^{nh}\frac{\sin x}{x}\,dx - \int_0^{nh_1}\frac{\sin x}{x}\,dx\right| \leq 2k \qquad ...(2)$$

Also, $f(h) \to 0$ as $h \to 0 + 0$, therefore for any $\varepsilon > 0$, there exists $\delta > 0$ such that

$$|f(h)| < \varepsilon/4k, \qquad \text{where} \qquad 0 < h < \delta \qquad ...(3)$$

From (1), (2) and (3), it follows that $\varepsilon > 0$ being given, there exists $\delta > 0$ such that

$$\left|\int_0^h f(x)\frac{\sin nx}{x}\,dx\right| \leq |f(h)| \times 2k < \varepsilon/2, \qquad \text{where} \qquad 0 < h < \delta \qquad ...(4)$$

Again, $$\int_0^a f(x)\frac{\sin nx}{x}\,dx = \int_0^h f(x)\frac{\sin nx}{x}\,dx + \int_h^a f(x)\frac{\sin nx}{x}\,dx \qquad ...(5)$$

Since by theorem II, the second integral on right hand side of (5) tends to zero as $n \to \infty$, it follows that there exists a positive integer m such that for all $n \geq m$, we have

$$\left|\int_0^a f(x)\frac{\sin nx}{x}\,dx\right| < \varepsilon/2 + \varepsilon/2 = \varepsilon$$

which implies that $$\lim_{n\to\infty}\int_0^a f(x)\frac{\sin nx}{x}\,dx = 0$$

Thus, the theorem is proved when $f(0+) = 0$.

Case II: Let $f(0+) \neq 0$. Let us define a new function $F(x)$ given by

$$F(x) = f(x) - f(0+), \qquad ...(6)$$

showing that $F(0+) = 0$ and F is bounded and integrable function on $[0, a]$ and monotonic on $[0, h]$ where $0 < h < a$. Hence, by case I, we obtain

$$\lim_{n\to\infty}\int_0^a F(x)\frac{\sin nx}{x}\,dx = 0$$

$\Rightarrow$ $$\lim_{n\to\infty}\int_0^a \{f(x)-f(0+)\}\frac{\sin nx}{x}\,dx = 0, \text{ using (6)}$$

$\Rightarrow$ $$\lim_{n\to\infty}\int_0^a f(x)\frac{\sin nx}{x}\,dx = \lim_{n\to\infty}\int_0^a f(0+)\frac{\sin nx}{x}\,dx = f(0+)\lim_{n\to\infty}\int_0^a \frac{\sin nx}{x}\,dx$$

$$= f(0+)\lim_{n\to\infty}\int_0^{na}\frac{\sin y}{y}\,dy, \text{ on putting } nx = y$$

Thus, $$\lim_{n\to\infty}\int_0^a f(x)\frac{\sin nx}{x}\,dx = f(0+)\int_0^\infty \frac{\sin x}{x}\,dx$$

Corollary 1. *If f satisfies the conditions of the above theorem and* $0<\alpha<\pi$, *then*

$$\lim_{n\to\infty}\int_0^a f(x)\frac{\sin nx}{\sin x}\,dx = f(0+)\int_0^\infty \frac{\sin x}{x}\,dx$$

Proof. We write $$f(x)\frac{\sin nx}{\sin x} = \frac{x\,f(x)}{\sin x}\times\frac{\sin nx}{x} = g(x)\frac{\sin nx}{\sin x},$$

where $$g(x) = \{x\,f(x)\}/\sin x,$$

and let us assign the values 1 and n to $(x/\sin x)$ and $(\sin nx)/x$ respectively for $x = 0$. Now, we know that $(x/\sin x)$ is positive and monotonically increasing in $[0, \alpha]$, $(\alpha < 1/2\pi)$. Therefore, g is also monotonically increasing in $[0, \alpha]$. Hence, by the above theorem, we have

$$\lim_{n\to\infty}\int_0^a f(x)\frac{\sin nx}{x}\,dx = \lim_{n\to\infty}\int_0^a g(x)\frac{\sin nx}{x}\,dx = g(0+)\int_0^\infty \frac{\sin x}{x}\,dx = f(0+)\int_0^\infty\frac{\sin x}{x}\,dx.$$

If f is decreasing, so that $-f$ is increasing, then proceeding as above, we obtain

$$\lim_{n\to\infty}\int_0^a \{-f(x)\}\frac{\sin nx}{\sin x}\,dx = -f(0+)\int_0^\infty\frac{\sin x}{x}\,dx$$

or $$\lim_{n\to\infty}\int_0^a f(x)\frac{\sin nx}{\sin x}\,dx = f(0+)\int_0^\infty\frac{\sin x}{x}\,dx$$

Corollary 2. (*i*) $\displaystyle\lim_{n\to\infty}\int_0^a f(x)\frac{\sin nx}{x}\,dx = \frac{\pi}{2}f(0+)$ (*ii*) $\displaystyle\lim_{n\to\infty}\int_0^a f(x)\frac{\sin nx}{\sin x}\,dx = \frac{\pi}{2}f(0+)$

Proof. (*i*) By theorem 3, we have

$$\lim_{n\to\infty}\int_0^a f(x)\frac{\sin nx}{x}\,dx = f(0+)\int_0^\infty\frac{\sin x}{x}\,dx = \frac{\pi}{2}f(0+), \quad \text{as } \int_0^\infty\frac{\sin x}{x}\,dx = \frac{\pi}{2}$$

(*ii*) By corollary 1, we have

$$\lim_{n\to\infty}\int_0^a f(x)\frac{\sin nx}{\sin x}\,dx = f(0+)\int_0^\infty\frac{\sin x}{x}\,dx = \frac{\pi}{2}f(0+), \quad \text{as } \int_0^\infty\frac{\sin x}{x}\,dx = \frac{\pi}{2}$$

Theorem IV: *Let f be bounded and integrable in* $[-\pi, \pi]$ *and monotonic in* $[-\alpha, 0]$ *and* $[0, \alpha]$ (*not necessarily in the same sense*), *where* α *is some positive number less than* π, *then*

$$\frac{1}{2}a_0 + \sum_{n=1}^{\infty} a_n = \frac{f(0+)+f(0-)}{2},$$

where a_n, $n = 0, 1, 2, \ldots$ *denote the Fourier coefficients of f.* **(Delhi 2009)**

Proof. The Fourier coefficients a_0 and a_n ($n = 1, 2, 3, \ldots$) are given by

$$a_0 = \frac{1}{\pi}\int_{-\pi}^{\pi} f(x)\,dx \qquad \text{and} \qquad a_n = \frac{1}{\pi}\int_{-\pi}^{\pi} f(x)\cos nx\,dx \qquad \ldots(1)$$

Now, $\frac{1}{2}a_0 + \sum_{n=1}^{m} a_n$

$$= \frac{1}{2\pi}\int_{-\pi}^{\pi} f(x)\,dx + \frac{1}{\pi}\sum_{n=1}^{m}\int_{-\pi}^{\pi} f(x)\cos nx\,dx, \text{ using (1)}$$

$$= \frac{1}{\pi}\int_{-\pi}^{\pi} f(x)\left\{\frac{1}{2} + \sum_{n=1}^{m}\cos nx\right\}dx = \frac{1}{\pi}\int_{-\pi}^{\pi} f(x)\frac{\sin(m+1/2)x}{2\sin(x/2)}\,dx$$

$$= \frac{1}{2\pi}\int_{0}^{\pi} f(-x)\frac{\sin(m+1/2)x}{2\sin(x/2)}\,dx + \frac{1}{2\pi}\int_{0}^{2\pi} f(x)\frac{\sin(m+1/2)x}{\sin(x/2)}\,dx$$

$$\Rightarrow \frac{1}{2}a_0 + \sum_{n=1}^{m} a_n = \frac{1}{2\pi}\times 2\int_{0}^{\pi/2} f(-2x)\frac{\sin(2m+1)x}{\sin x}\,dx + \frac{1}{2\pi}\times 2\int_{0}^{\pi} f(2x)\frac{\sin(2m+1)x}{\sin x}\,dx \qquad \ldots(2)$$

Hence, letting $m \to \infty$, (2) reduces to

$$\frac{1}{2}a_0 + \sum_{n=1}^{\infty} a_n = \frac{1}{\pi}\{f(0+) + f(0-)\}\int_{0}^{\infty}\frac{\sin x}{x}\,dx \qquad \ldots(3)$$

We now proceed to evaluate the value of $\int_{0}^{\infty}\frac{\sin x}{x}\,dx$.

Taking $f(x) = 1$ for all $x \in \mathbf{R}$, (1) yields

$$a_0 = \frac{1}{2}\int_{-\pi}^{\pi} dx = 2 \qquad \text{and} \qquad a_n = \frac{1}{2}\int_{-\pi}^{\pi}\cos nx\,dx = \frac{1}{\pi}\left[\frac{\sin nx}{n}\right]_0^{\pi} = 0$$

Again, $f(x) = 1$ for all $x \in \mathbf{R}$ $\Rightarrow$ $f(0+) = f(0-) = 1$. Hence, (3) reduces to

$$\left(\frac{1}{2}\times 2\right) + 0 = \frac{1}{\pi}(1+1)\times\int_{0}^{\infty}\frac{\sin nx}{x}\,dx \qquad \text{giving} \qquad \int_{0}^{\infty}\frac{\sin nx}{x}\,dx = \frac{\pi}{2} \qquad \ldots(4)$$

Using (4), (3) reduces to

$$\frac{1}{2}a_0 + \sum_{n=1}^{\infty} a_n = \frac{1}{\pi}\{f(0+) + f(0-)\}\times\frac{\pi}{2} = \frac{f(0+) + f(0-)}{2}$$

Theorem V Fourier - Dirichlet theorem (or simply Fourier theorem). ***(Meerut 2012)***
Point-wise convergence of Fourier series.

Let $f: [-\pi, \pi] \to \mathbf{R}$ be bounded, integrable, periodic of period 2π and piecewise monotonic on $[-\pi, \pi]$, then*

$$\frac{1}{2}a_0 + \sum_{n=1}^{\infty}(a_n\cos n\xi + b_n\sin n\xi) = \begin{cases}\{f(\xi+) + f(\xi-)\}/2, & \text{for } -\pi < \xi < \pi\\ \{f(-\pi+) + f(\pi-)\}/2, & \text{for } \xi = -\pi \text{ and } \xi = \pi\end{cases}$$

where a_0, a_n and b_n ($n = 1, 2, 3, \ldots$) are Fourier coefficients.

---

* **Piece-wise monotone function. Definition.** A function $f: [a, b] \to \mathbf{R}$ is said to be piece-wise monotone if there exists a partition of $[a, b]$ into subintervals such that the function f is monotone on each of the subinterval.

Proof. In order to prove the theorem, we need the following lemma.

Lemma: *If f is bounded and integrable in every interval and is periodic with period* 2π *as its period, then* $\int_{-\pi}^{\pi} f(x)\,dx = \int_{-\pi}^{\pi} f(a+x)\,dx$, *a being any number whatsoever.*

Proof of Lemma: Putting $a + x = y$ and $dx = dy$, we have

$$\int_{-\pi}^{\pi} f(a+x)\,dx = \int_{a-\pi}^{a+\pi} f(y)\,dy = \int_{a-\pi}^{-\pi} f(y)\,dy + \int_{-\pi}^{\pi} f(y)\,dy + \int_{\pi}^{a+\pi} f(y)\,dy \qquad ...(1)$$

Putting $y = z - 2\pi$ and $dy = dz$ in the first integral on R.H.S. of (1), we get

$$\int_{a-\pi}^{-\pi} f(y)\,dy = \int_{a+\pi}^{\pi} f(z-2\pi)\,dz = \int_{a+\pi}^{\pi} f(z)\,dz, \text{ since } f \text{ is periodic with period } 2\pi$$

Thus, $$\int_{a-\pi}^{-\pi} f(y)\,dy = \int_{a+\pi}^{\pi} f(y)\,dy = -\int_{\pi}^{a+\pi} f(y)\,dy \qquad ...(2)$$

Using (2), (1) reduces to $\int_{-\pi}^{\pi} f(a+x)\,dx = \int_{-\pi}^{\pi} f(y)\,dy = \int_{-\pi}^{\pi} f(x)\,dx.$

Proof of the main theorem. Fourier coefficients a_0, a_n and b_n are given by

$$a_0 = \frac{1}{\pi}\int_{-\pi}^{\pi} f(x)\,dx, \quad a_n = \frac{1}{\pi}\int_{-\pi}^{\pi} f(x)\cos nx\,dx, \quad b_n = \frac{1}{\pi}\int_{-\pi}^{\pi} f(x)\sin nx\,dx, \; n=1,2,3... \qquad ...(3)$$

Considering the m **th partial sum at** ξ**, we have**

$$\frac{a_0}{2} + \sum_{n=1}^{m}(a_n \cos n\xi + b_n \sin n\xi)$$

$$= \frac{1}{2\pi}\int_{-\pi}^{\pi} f(x)\,dx + \sum_{n=1}^{m}\frac{1}{\pi}\int_{-\pi}^{\pi} f(x)\{\cos nx \cos n\xi + \sin nx \sin n\xi\}\,dx, \text{ using (3)}$$

$$= \frac{1}{2\pi}\int_{-\pi}^{\pi} f(x)\{1 + 2\sum_{n=1}^{m}\cos n(x-\xi)\}\,dx$$

$$= \frac{1}{2\pi}\int_{-\pi}^{\pi} f(x+\xi)\left\{1 + 2\sum_{n=1}^{m}\cos nx\right\}dx, \text{ using lemma}$$

$$= \frac{1}{2\pi}\int_{-\pi}^{\pi} f(x+\xi)\frac{\sin(m+1/2)x}{\sin(x/2)}\,dx$$

$$= \frac{1}{2\pi}\int_{-\pi}^{0} f(x+\xi)\frac{\sin(m+1/2)x}{\sin(x/2)}\,dx + \frac{1}{2\pi}\int_{0}^{\pi} f(x+\xi)\frac{\sin(m+1/2)x}{\sin(x/2)}\,dx \qquad ...(4)$$

Let us make the substitution $x = -2y$ in the first integral and $x = 2y$ in the second integral on R.H.S. of (4). Then (4) takes the form

$$\frac{a_0}{2} + \sum_{n=1}^{m}(a_n \cos n\xi + b_n \sin n\xi)$$

$$= \frac{1}{\pi}\int_{0}^{\pi/2} f(-2y+\xi)\frac{\sin(2m+1)y}{\sin y}\,dy + \frac{1}{\pi}\int_{0}^{\pi/2} f(2y+\xi)\frac{\sin(2m+1)y}{\sin y}\,dy \qquad ...(5)$$

Now, proceeding to limit as $m \to \infty$ and using theorem III, we have

$$\frac{1}{2}a_0 + \sum_{n=1}^{m}(a_n \cos n\xi + b_n \sin n\xi) = \frac{1}{\pi}\left\{\frac{\pi}{2}f(\xi-) + \frac{\pi}{2}f(\xi+)\right\} = \frac{f(\xi-) + f(\xi+)}{2}$$

Note. The result arrived at above may be re-stated as follows:

If f be bounded and integrable in [–π, π] and if it be possible to divide [–π, π] into a finite number of open subintervals in each of which f is monotonic, then the Fourier series, corresponding to f converges for all x ∈ ***R*** *and if σ denotes the sum function of series, then*

$$\sigma(x) = \{f(x+) + f(x-)\}/2, \quad \text{for all } x \in [-\pi, \pi]$$

$$\sigma(x) = \{f(\pi-) + f(-\pi+)\}/2, \quad \text{for } x = \pm\pi$$

and
$$\sigma(x + 2\pi) = \sigma(x)$$

The relation $\sigma(x + 2\pi) = \sigma(x)$ enables us to determine the value of the sum function at a point which does not belong to $[-\pi, \pi]$

Corollary. *At a point ξ of continuity of f, we have*

$$\frac{1}{2}a_0 + \sum_{n=1}^{m}(a_n \cos n\xi + b_n \sin n\xi) = \frac{f(\xi+) + f(\xi-)}{2} = \frac{f(\xi) + f(\xi)}{2} = f(\xi)$$

Thus, we see that if f satisfies the conditions of theorem V in $[-\pi, \pi]$, then the sum of the Fourier series corresponding to f is actually $f(x)$ at all such points x of $[-\pi, \pi]$, where f is continuous; at points x of discontinuity, the sum function of the series is $\{f(x+) + f(x-)\}/2$

Remark 1. Theorem V establishes the point-wise convergence of Fourier series of f to the function f and not the uniform convergence.

Remark 2. The Fourier series expression of a function is unique.

Remark 3. Two different functions can generate the same Fourier series.

22.14 SOME TYPICAL SOLVED EXAMPLES

Ex. 1. *Obtain an expression in a mixed series of sines and cosines of multiples of x which is zero between –π and 0 and is equal to e^x between o and π and give its values at three limits.*

Sol. Let the given function be denoted by $f(x)$ and hence

$$f(x) = \begin{cases} 0, \text{ for } -\pi < x < 0 \\ e^x, \text{ for } 0 < x < \pi \end{cases} \qquad ...(1)$$

Let
$$f(x) = \frac{a_0}{2} + \sum_{n=1}^{\infty}(a_n \cos nx + b_n \sin nx), \quad \text{where} \qquad ...(2)$$

$$a_0 = \frac{1}{\pi}\int_{-\pi}^{\pi} f(x)\,dx = \frac{1}{\pi}\left[\int_{-\pi}^{0} f(x)\,dx + \int_{0}^{\pi} f(x)\,dx\right] = \frac{1}{\pi}\int_{0}^{\pi} e^x\,dx = \frac{e^{\pi}-1}{\pi}, \text{ by (1)}$$

$$a_n = \frac{1}{\pi}\int_{-\pi}^{\pi} f(x)\cos nx\,dx = \frac{1}{\pi}\left[\int_{-\pi}^{0} f(x)\cos nx\,dx + \int_{0}^{\pi} f(x)\cos nx\,dx\right]$$

$$= \frac{1}{\pi}\int_{0}^{\pi} e^x \cos nx\,dx = \frac{1}{\pi}\left[\frac{e^x(\cos nx + n\sin nx)}{1+n^2}\right]_0^{\pi} = \frac{e^{\pi}\cos n\pi - 1}{\pi(1+n^2)}$$

$$b_n = \frac{1}{\pi}\int_{-\pi}^{\pi} f(x)\sin nx\,dx = \frac{1}{\pi}\left[\int_{-\pi}^{0} f(x)\sin nx\,dx + \int_{0}^{\pi} f(x)\sin nx\,dx\right]$$

$$= \frac{1}{\pi}\int_{0}^{\pi} e^x \sin nx\,dx = \frac{1}{\pi}\left[\frac{e^x(\sin nx - n\cos nx)}{1+n^2}\right]_0^{\pi} = -\frac{n(e^{\pi}\cos n\pi - 1)}{\pi(1+n^2)}$$

Substituting the above values of a_0, a_n and b_n in (1), we have

$$f(x) = \frac{e^{\pi}-1}{2\pi} + \frac{1}{n}\sum_{n=1}^{\infty}\frac{(-1)^n e^{\pi} - 1}{1+n^2}(\cos nx - n\sin nx) \qquad ...(3)$$

Since $x = 0$ is a point of discontinuity, hence at $x = 0$, the series (3) converges to

$$\frac{1}{2}\{f(0+)+f(0-)\}=\frac{1}{2}\left[\lim_{h\to 0} f(0+h)+\lim_{h\to 0}(0-h)\right]=\frac{1}{2}\left[\lim_{h\to 0} e^h+0)\right]=\frac{1}{2}$$

At the end points $x = \pi$ and $x = -\pi$, the Fourier series (3) converges to

$$\frac{1}{2}\{f(-\pi+)+f(\pi-)\}=\frac{1}{2}\left[\lim_{h\to 0} f(-\pi+h)+\lim_{h\to 0} f(\pi-h)\right]=\frac{1}{2}\left[0+\lim_{h\to 0} e^{\pi-h}\right]=\frac{1}{2}e^{\pi}$$

Ex. 2. *Find a series of sines and cosines of multiples of x, which will represent* $f(x) = (\pi/2 \sinh \pi) \times e^x$ *in the interval* $-\pi < x < \pi$. *Find also the sum of the series for* $x = \pm\pi$.

Sol. Let $$f(x)=\frac{1}{2}a_0+\sum_{n=1}^{\infty}(a_n \cos nx+b_n \sin nx), \quad \text{where} \quad \ldots(1)$$

$$a_0=\frac{1}{\pi}\int_{-\pi}^{\pi} f(x)\,dx=\frac{1}{2\sinh\pi}\int_{-\pi}^{\pi} e^x\,dx=\frac{e^{\pi}-e^{-\pi}}{e^{\pi}-e^{-\pi}}=1, \quad \text{since} \quad \sinh\pi=\frac{e^{\pi}-e^{-\pi}}{2}$$

$$a_n=\frac{1}{\pi}\int_{-\pi}^{\pi} f(x)\cos nx\,dx=\frac{1}{2\sinh\pi}\int_{-\pi}^{\pi} e^x\cos nx\,dx=\frac{1}{2\sinh\pi}\left[\frac{e^x(\cos nx+n\sin nx}{1+n^2}\right]_{-\pi}^{\pi}$$

$$=\frac{\cos n\pi(e^{\pi}-e^{-\pi})}{(1+n^2)(e^{\pi}-e^{-\pi})}=\frac{(-1)^n}{1+n^2}$$

$$b_n=\frac{1}{\pi}\int_{-\pi}^{\pi} f(x)\sin nx\,dx=\frac{1}{2\sinh\pi}\int_{-\pi}^{\pi} e^x\sin nx\,dx=\frac{1}{2\sinh\pi}\left[\frac{e^x(\sin nx-n\cos nx}{1+n^2}\right]_{-\pi}^{\pi}$$

$$=-\frac{n\cos n\pi(e^{\pi}-e^{-\pi})}{(1+n^2)(e^{\pi}-e^{-\pi})}=-\frac{n(-1)^n}{1+n^2}$$

Substituting the above values of a_0, a_n and b_n in (1), we get

$$f(x)=\frac{\pi e^x}{2\sinh\pi}=\frac{1}{2}+\sum_{n=1}^{\infty}\frac{(-1)^n}{1+n^2}(\cos nx-n\sin nx)$$

At end points $x = \pm\pi$, the series converges to

$$\frac{1}{2}\{f(-\pi+)+f(\pi-)\}=\frac{1}{2}\times\frac{\pi}{2\sinh\pi}(e^{\pi}+e^{-\pi})=\frac{2\pi\cosh\pi}{4\sinh\pi}=\frac{\pi}{2}\coth\pi$$

Ex. 3. *For all values of x between* $\pi/2$ *and* $-(\pi/2)$, *prove that*

$$x = (4/\pi) \times \{\sin x - (\sin 3x)/9 + (\sin 5x)/25 + \ldots\}$$

Sol. Here we shall change the interval $[-\pi/2, \pi/2]$ to $[0, \pi]$. To this end, let $x + \pi/2 = z$ so that z lies between 0 and π. Now, we shall express z as a cosine series. Suppose that

$$f(z)=z=\frac{1}{2}a_0+\sum_{n=1}^{\infty}a_n\cos nz, \quad \text{where} \quad \ldots(1)$$

$$a_0=\frac{2}{\pi}\int_0^{\pi} f(z)\,dz=\frac{2}{\pi}\int_0^{\pi} z\,dz=\frac{2}{\pi}\times\frac{\pi^2}{2}=\pi$$

and $$a_n=\frac{2}{\pi}\int_0^{\pi} f(z)\cos nz\,dz=\frac{2}{\pi}\int_0^{\pi} z\cos z\,dz=\frac{2}{\pi}\left[(z)\left(\frac{\sin nz}{n}\right)-(1)\left(-\frac{\cos nz}{n^2}\right)\right]_0^{\pi}$$

$$=\frac{2}{\pi}\left[\frac{\cos n\pi}{n^2}-\frac{1}{n^2}\right]=\frac{2\{(-1)^n-1\}}{\pi n^2}=-\frac{\{1-(-1)^n\}}{\pi n^2}$$

Hence, (1) reduces to $$z = \frac{\pi}{2} - \frac{2}{\pi}\sum_{n=1}^{\infty}\frac{1-(-1)^n}{n^2}\cos nz$$

or $$z = (\pi/2) - (4/\pi) \times \{\cos z + (1/3^2) \times \cos 3z + (1/5^2) \times \cos 5z + ...\} \qquad ...(2)$$

Since $z = \pi/2 + x$, (2) reduces to

$$x + \frac{\pi}{2} = \frac{\pi}{2} - \frac{4}{\pi}\left(-\sin x + \frac{1}{3^2}\sin 3x - \frac{1}{5^2}\sin 5x + ...\right) \quad \text{or} \quad x = \frac{4}{\pi}\left(\sin x + \frac{\sin 3x}{3^2} - \frac{\sin 5x}{5^2} + ...\right)$$

Ex. 4. *If the function defined by $y = x^2$ from 0 to $\pi/2$ and $y = 0$ from $\pi/2$ to π be represented by a series of sines of multiples of x, show that the coefficient of sin nx is given by*

$$(4/\pi n^3 - \pi/2n)\cot(n\pi/2) + (2/n^2)\sin(n\pi/2) - (4/\pi n^3)$$

Sol. Let the given function be $f(x)$. Then, we have

$$y = f(x) = \begin{cases} x^2, & \text{for } 0 < x < \pi/2 \\ 0, & \text{for } \pi/2 < x < \pi \end{cases} \qquad ...(1)$$

Let $$f(x) = \sum_{n=1}^{\infty} b_n \sin nx, \quad \text{where} \qquad ...(2)$$

$$b_n = \frac{2}{\pi}\int_0^{\pi} f(x)\sin nx\, dx = \frac{2}{\pi}\left[\int_0^{\pi/2} x^2 \sin nx\, dx + \int_{\pi/2}^{\pi}(0)\sin nx\, dx\right], \text{ using (1)}$$

$$= \frac{2}{\pi}\left[(x^2)\left(-\frac{\cos nx}{n}\right) - (2x)\left(-\frac{\sin nx}{n^2}\right) + (2)\left(\frac{\cos nx}{n^3}\right)\right]_0^{\pi/2}$$

$$= \frac{2}{\pi}\left[-\frac{\pi^2}{4n}\cos\frac{n\pi}{2} + \frac{\pi}{n^2}\sin\frac{n\pi}{2} + \frac{2}{n^3}\cos\frac{n\pi}{2} - \frac{2}{n^3}\right] = \left(\frac{4}{\pi n^3} - \frac{\pi}{2n}\right)\cos\frac{n\pi}{2} + \frac{2}{n^2}\sin\frac{n\pi}{2} - \frac{4}{\pi n^3}$$

and this is the coefficient of sin nx. The series at $x = \pi/2$ converges to

$(1/2) \times \{f(\pi/2 + 0) + f(\pi/2 - 0\}$

$$= (1/2) \times \left\{\lim_{h\to 0} f(\pi/2 + h) + \lim_{h\to 0} f(\pi/2 - h)\right\} = (1/2) \times \left\{0 + \lim_{h\to 0}(\pi/2 - h)^2\right\} = \pi^2/8$$

Ex. 5. *Find the function f(x) which shall be periodic with period 2l and shall be = l/4 from –l to –l/2; = x^2/l from –l/2 to l/2; = l/4 from l/2 to l. Draw the graph showing the function represented by the series.*

Sol. Here $f(x)$ is defined in $(-l, l)$ as follows:

$$f(x) = \begin{cases} l/4, & \text{for } -l < x < -l/2 \\ x^2/l, & \text{for } -l/2 < x < l/2 \\ l/4, & \text{for } \ l/2 < x < l \end{cases} \qquad ...(1)$$

Obviously, if x is changed to $-x$, $f(x)$ remains unaltered. Hence, $f(x)$ is an even function. Accordingly, the Fourier series for $f(x)$ will be of the form (refer special form I in Art. 22.7)

$$f(x) = \frac{1}{2}a_0 + \sum_{n=1}^{\infty} a_n \cos nx, \quad \text{where} \qquad ...(2)$$

$$a_0 = \frac{1}{l}\int_{-l}^{l} f(x)\, dx = \frac{1}{l}\left[\int_{-l}^{-l/2} f(x)\, dx + \int_{-l/2}^{l/2} f(x)\, dx + \int_{l/2}^{l} f(x)\, dx\right]$$

$$= \frac{1}{l}\left[\int_{-l}^{-l/2}\frac{l}{4}\,dx + \int_{-l/2}^{l/2}\frac{x^2}{l}\,dx + \int_{l/2}^{l}\frac{l}{4}\,dx\right] = \frac{l}{3}, \text{ using (1) and simplifying}$$

$$a_n = \frac{1}{l}\int_{-l}^{l} f(x)\cos\frac{n\pi x}{l}\,dx = \frac{1}{l}\left[\int_{-l}^{-l/2}\frac{l}{4}\cos\frac{n\pi x}{l}\,dx + \int_{-l/2}^{l/2}\frac{x^2}{l}\cos\frac{n\pi x}{l}\,dx + \int_{l/2}^{l}\frac{l}{4}\cos\frac{n\pi x}{4}\,dx\right]$$

$$= \frac{1}{4}\left[\frac{l}{n\pi}\sin\frac{n\pi x}{l}\right]_{-l}^{-l/2} + \frac{2}{l^2}\int_0^{l/2} x^2\cos\frac{n\pi x}{l}\,dx + \frac{1}{4}\left[\frac{l}{n\pi}\sin\frac{n\pi x}{l}\right]_{l/2}^{l}$$

$$= -2\times\frac{l}{4n\pi}\sin\frac{n\pi}{2} + \frac{2}{l^2}\left[(x^2)\left(\frac{l}{n\pi}\sin\frac{n\pi x}{l}\right) - (2x)\left(-\frac{l^2}{n^2\pi^2}\cos\frac{n\pi x}{l}\right) + (2)\left(-\frac{l^3}{n^3\pi^3}\sin\frac{n\pi x}{l}\right)\right]_0^{l/2}$$

$$= -\frac{l}{2n\pi}\sin\frac{n\pi}{2} + \frac{l}{2n\pi}\sin\frac{n\pi}{2} + \frac{2l}{n^2\pi^2}\cos\frac{n\pi}{2} - \frac{4l}{n^3\pi^3}\sin\frac{n\pi}{2}$$

$$= (2l/n^2\pi^2)\times\{\cos(n\pi/2) - (2/n\pi)\times\sin(n\pi/2)\}$$

Substituting the above values of a_0 and a_n in (2), we have

$$f(x) = \frac{l}{6} + \frac{2l}{\pi^2}\sum_{n=1}^{\infty}\left\{\frac{1}{n^2}\cos\frac{n\pi}{2} - \frac{2}{n^3\pi}\sin\frac{n\pi}{2}\right\}\cos\frac{n\pi x}{2}$$

The required graph consists of equal arcs of a parabola BOC ($y = x^2/l$) in ($-l/2$, $l/2$), straight line AB ($y = l/4$) in ($-l$, $-l/2$) and straight CD ($y = l/4$) in ($l/2$, l) as shown in the adjoining diagram.

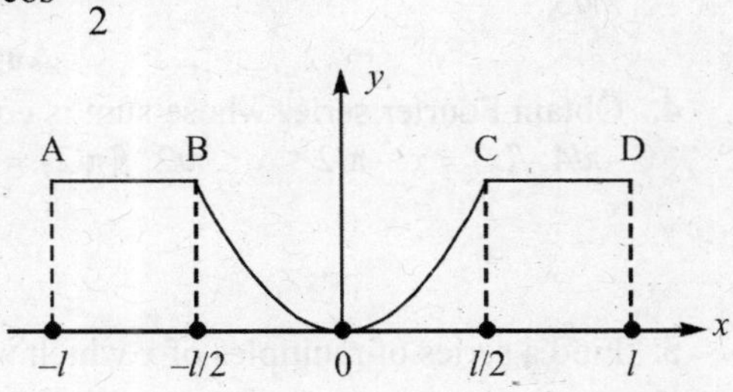

Ex. 6. *Show that the series*

$$\frac{v_1 + v_2 + v_3}{3} + \frac{1}{\pi}\sum_{n=1}^{\infty}\frac{1}{n}\sin\frac{n\pi}{3}\left\{2(v_3 - v_1)\sin\frac{2n\pi}{3}\sin\frac{n\pi x}{l} + (2v_3 - v_1 - v_2)\cos\frac{n\pi x}{l}\right\}$$

is equal to f(x), where $f(x) = \begin{cases} v_1, \text{ for } -l < x < -l/3 \\ v_2, \text{ for } l/3 < x < l/3 \\ v_3, \text{ for } l/3 < x < l \end{cases}$ *where v_1, v_2 and v_3 are constants.*

Sol. Let $$f(x) = \frac{a_0}{2} + \sum_{n=1}^{\infty}\left(a_n\cos\frac{n\pi x}{l} + b_n\sin\frac{n\pi x}{l}\right), \quad \text{where} \qquad ...(1)$$

$$a_0 = \frac{1}{l}\int_{-l}^{l} f(x)\,dx = \frac{1}{l}\left[\int_{-l}^{-l/3} f(x)\,dx + \int_{-l/3}^{l/3} f(x)\,dx + \int_{l/3}^{l} f(x)\,dx\right]$$

$$= \frac{1}{l}\left[\int_{-l}^{-l/3} v_1\,dx + \int_{-l/3}^{l/3} v_2\,dx + \int_{l/3}^{l} v_3\,dx\right] = \frac{2(v_1 + v_2 + v_3)}{3}$$

[Using the given value of $f(x)$ and simplifying]

$$a_n = \frac{1}{l}\int_{-l}^{l} f(x)\cos\frac{n\pi x}{l} = dx\,\frac{1}{l}\left[\int_{-l}^{-l/3} v_1\cos\frac{n\pi x}{l}\,dx + \int_{-l/3}^{l/3} v_2\cos\frac{n\pi x}{l}\,dx + \int_{l/3}^{l} v_3\cos\frac{n\pi x}{l}\,dx\right]$$

$$= \frac{v_1}{n\pi}\left(\sin n\pi - \sin\frac{n\pi}{3}\right) + \frac{2v_2}{n\pi}\sin\frac{n\pi}{3} + \frac{v_3}{n\pi}\left(\sin n\pi - \sin\frac{n\pi}{3}\right) = \frac{2v_3 - v_1 - v_2}{n\pi}\sin\frac{n\pi}{3}$$

$$b_n = \frac{1}{l}\int_{-l}^{l} f(x)\sin\frac{n\pi x}{l}\,dx = \frac{1}{l}\left[\int_{-l}^{-l/3} v_1 \sin\frac{n\pi x}{l}\,dx + \int_{-l/3}^{l/3} v_2 \sin\frac{n\pi x}{l}\,dx + \int_{l/3}^{l} v_3 \sin\frac{n\pi x}{l}\,dx\right]$$

$$= \frac{1}{n\pi}\left(\cos\frac{n\pi}{3} - \cos n\pi\right)(v_3 - v_1) = \frac{2}{n\pi}(v_3 - v_1)\sin\frac{2n\pi}{3}\sin\frac{n\pi}{3}$$

Substituting the above values of a_0, a_n and b_n in (1), we get the required series.

MISCELLANEOUS PROBLEMS BASED ON CHAPTER 22

1. Show that two different functions can generate the same Fourier series. **(Delhi 2010)**
2. Are the following trigonometric series, the Fourier series ?

(i) $\sum_{n=1}^{\infty} \frac{\sin nx}{n^2}$ (ii) $\sum_{n=1}^{\infty} \frac{\cos nx + \sin nx}{\sqrt{n}}$

(iii) $\sum_{n=1}^{\infty} \frac{\sin nx}{\sqrt{n}}$ (iv) $\frac{\cos x}{\sqrt{1}} + \frac{\sin 2x}{\sqrt{2}} + \frac{\cos 3x}{\sqrt{3}} + \frac{\sin 4x}{\sqrt{4}} + \ldots$

[**Ans.** (*i*) Yes (*ii*) No (*iii*) No (*iv*) No]

3. Find the Fourier series in $[0, \pi]$ for the function given by $f(x) = \pi/3$, for $0 < x < \pi/3$; $f(x) = 0$, for $\pi/3 < x < 2\pi/3$ and $f(x) = -\pi/3$, for $2\pi/3 < x < \pi$. Find the sum of the series when $x = \pi$, $2\pi/3$

Ans. $(\sin x)/1 + (\sin 4x)/2 + (\sin 8x)/4 + (\sin 10x)/5 + \ldots$

4. Obtain Fourier series whose sum is equal to $f(x)$, where $f(x) = 0$, $-\pi \le x < -\pi/2$, $f(-\pi/2) = -\pi/4$; $f(x) = x$, $-\pi/2 < x < \pi/2$, $f(\pi/2) = \pi/4$; $f(x) = 0$, $\pi/2 < x \le \pi$.

Ans. $f(x) = \frac{2}{\pi}\sum_{n=1}^{\infty}\left(\frac{1}{n^2}\sin\frac{n\pi}{2} - \frac{\pi}{2n}\cos\frac{n\pi}{2}\right)\sin nx$

5. Find a series of multiples of x which will represent $f(x)$ in the interval $(0, \pi)$, where $f(x) = \pi/3$, $0 < x < \pi/3$; $f(x) = 0$, $\pi/3 < x < 2\pi/3$; $f(x) = -\pi/3$, $2\pi/3 < x < \pi$. Can you represent this function by a series of cosines of multiples of x as well ? Explain your answer. Draw graphs of series and find the values of sine and cosine series where $x = -\pi/3, -2\pi/3, -\pi$.

Ans. $f(x) = 2\left(\frac{\sin 2x}{2} + \frac{\sin 4x}{4} + \frac{\sin 8x}{8} + \ldots\right)$; $f(x) = \frac{2}{\sqrt{3}}\left(\cos x - \frac{\cos 5x}{5} + \frac{\cos 7x}{7} + \ldots\right)$

6. If $f(x)$ $-\pi/4$ when $-\pi < x < 0$; $f(x) = \pi/4$ when $0 < x < \pi$. $f(-\pi) = f(0) = f(\pi) = 0$ and $f(x + 2\pi) = f(x)$ for all x, show that for all x, $f(x) = \sin x + (1/3) \times \sin 3x + (1/5) \times \sin 5x + \ldots$ Deduce that $\pi/4 = 1 - (1/3) + (1/5) - (1/7) + \ldots$
7. If $f(x) = cx$ when $-\pi < x < \pi$, $f(-\pi) = f(\pi) = 0$ and $f(x + 2\pi) = f(x)$ for all x, show that $f(x) = 2c\ \{\sin x - (1/2) \times \sin 2x + (1/3) \times \sin 3x - (1/4) \times \sin 4x + \ldots\}$ for all x. Draw the graph of $f(x)$.
8. Obtain the Fourier series, in the interval $[-1/2, 1/2]$ of the function f given by

$$f(x) = \begin{cases} x - [x] - (1/2), & \text{when } x \text{ is not an integer} \\ 0, & \text{otherwise} \end{cases}$$

where $[x]$ is the greatest integer $\le x$. **Ans.** $x - [x] - (1/2) = -\sum_{n=1}^{\infty}\{(1/n\pi)\times\sin(2n\pi x)\}$

9. State and prove Bessel's theorem. **(Delhi 2011)**
10. State and prove Riemann - Lebesgue theorem.
11. State and prove Fourier - Dirichlet theorem.

22.15. Additional solved examples

Ex. 1. *Let* $f(x) = \begin{cases} -3\pi, \text{ if } -\pi < x < 0 \\ 3\pi, \text{ if } \ 0 < x < \pi \end{cases}$ *be a periodic function of period* 2π. *The coefficient of* $\sin 3x$ *in the Fourier series expansion of* $f(x)$ *on the interval* $[-\pi, \pi]$ *is ....* **[GATE 2014]**

Sol. Ans. 4. From particular case (*i*) of Art. 22.5, page 22.4, we know that the Fourier series expansion of a function $f(x)$ on the interval $[-\pi, \pi]$ is given by

$$f(x) = \frac{a_0}{2} + \sum_{n=1}^{\infty} a_n \cos nx + \sum_{n=1}^{\infty} b_n \sin nx, \quad \text{where, } b_n = \frac{1}{\pi}\int_{-\pi}^{\pi} f(x) \sin nx \, dx, \ n = 1, 2, 3, ...$$

Hence, the required coefficient b_3 of $\sin 3x$ is given by

$$b_3 = \frac{1}{\pi}\int_{-\pi}^{\pi} f(x) \sin 3x \, dx = \frac{1}{\pi}\left[\int_{-\pi}^{o} f(x) \sin 3x \, dx + \int_{0}^{\pi} f(x) \sin 3x \, dx\right]$$

$$= \frac{1}{\pi}\int_{-\pi}^{0} (-3\pi) \sin 3x \, dx + \frac{1}{\pi}\int_{0}^{\pi} (3\pi) \sin 3x \, dx, \text{ using the given value of } f(x)$$

$$= -3\left[-\frac{\cos 3x}{3}\right]_{-\pi}^{0} + 3\left[-\frac{\cos 3x}{3}\right]_{0}^{\pi} = [\cos 3x]_{-\pi}^{0} - [\cos 3x]_{0}^{\pi} = (1+1) - (-1-1) = 4.$$

Ex. 2. *Find the Fourier series for the function* $f(x) = x$ *on* $-l \le x \le l$. **[Meerut 2011]**

Sol. Since $f(x) = x$ is an odd function of x, the required Fourier series is given by (Refer results (14) and (15) of Art. 22.7)

$$f(x) = x = \sum_{n=1}^{\infty} b_n \sin \frac{n\pi x}{l}, \quad (-l \le x \le l) \qquad ...(1)$$

where $$b_n = \frac{2}{l}\int_{0}^{l} f(x) \sin \frac{n\pi x}{l} dx = \frac{2}{l}\int_{0}^{l} x \sin \frac{n\pi x}{l} dx$$

$$= \frac{2}{l}\left[(x)\left\{\frac{\cos(n\pi x/l)}{-(n\pi/l)}\right\} - (1)\left\{\frac{\sin(n\pi x/l)}{-(n\pi/l)^2}\right\}\right]_0^l$$

[Using chain rule of integration by parts]

Thus, $$b_n = \frac{2}{l} \times \left\{-\frac{l^2}{n\pi}\cos n\pi\right\} = -\frac{2l}{n\pi} \times (-1)^n = -\frac{2l}{n\pi} \times (-1)^{n-1} \qquad ...(2)$$

Substituting the value of b_n given by (2) in (1), the required Fourier series is given by

$$x = \frac{2l}{\pi}\sum_{n=1}^{\infty} \frac{(-1)^{n-1}}{n} \sin \frac{n\pi x}{l}$$

Ex. 3(*a*). *Periodic function* $f(x)$ *with period* 2π *is defined as follows:* $f(x) = x^2$, $-\pi \le x < \pi$. *Determine its Fourier series.* **[Kanpur 2014]**

(*b*). *From part (a) deduce that*

(*i*) $1/1^2 + 1/2^2 + 1/3^2 + ... = \pi^2/6$ (*ii*) $1/1^2 - 1/2^2 + 1/3^2 - ... = \pi^2/12$

(*iii*) $1/1^2 + 1/3^2 + 1/5^2 + ... = \pi^2/8$

Sol. Part (*a*): Since $f(x) = x^2$ is an even function, the required Fourier series is given by (Refer special form II in Art 22.5)

$$f(x) = x^2 = \frac{a_0}{2} + \sum_{n=1}^{\infty} a_n \cos nx, \ (-\pi \le x \le \pi) \qquad ...(1)$$

where $$a_0 = \frac{2}{\pi} \int_0^{\pi} f(x)\, dx \qquad ...(2)$$

and $$a_n = \frac{2}{\pi} \int_0^{\pi} f(x) \sin nx\, dx, \quad n = 1, 2, 3, ... \qquad ...(3)$$

From (2), $$a_0 = \frac{2}{\pi} \int_0^{\pi} x^2\, dx = \frac{2}{\pi}\left[\frac{x^3}{3}\right]_0^{\pi} = \frac{2\pi^2}{3} \qquad ...(4)$$

and from (3), $$a_n = \frac{2}{\pi} \int_0^{\pi} x^2 \cos nx\, dx = \frac{2}{\pi}\left[(x^2)\left(\frac{\sin nx}{n}\right) - (2x)\left(-\frac{\cos nx}{n^2}\right) - (2)\left(-\frac{\sin nx}{n^3}\right)\right]_0^{\pi}$$

[Using chain rule of integration by parts]

or $$a_n = (2/\pi) \times (2\pi/n^2) \times \cos n\pi = (4/n^2) \times (-1)^n \qquad ...(5)$$

Substituting the values of a_0 and a_n given by (4) and (5) in (1), we get

$$x^2 = \frac{\pi^2}{3} + 4\sum_{n=1}^{\infty} \frac{(-1)^n}{n^2} \cos nx \qquad ...(6)$$

Part (*b*): (i) Putting $x = \pi$ in (6) and noting that $\cos n\pi = (-1)^n$, we obtain

$$\pi^2 = \frac{\pi^2}{3} + 4\sum_{n=1}^{\infty} \frac{(-1)^{2n}}{n^2} \text{ or } 4\left(\frac{1}{1^2} + \frac{1}{2^2} + \frac{1}{3^2} + ...\right) = \pi^2 - \frac{\pi^2}{3} \text{ or } \frac{1}{1^2} + \frac{1}{2^2} + \frac{1}{3^2} + ... = \frac{\pi^2}{6} \qquad ...(7)$$

(ii) Putting $x = 0$ in (6) yields $\quad 0 = \pi^2/3 + 4\sum_{n=1}^{\infty} \{(-1)^n / n^2\}$

or $$4\left(-\frac{1}{1^2} + \frac{1}{2^2} - \frac{1}{3^2} + ...\right) = -\frac{\pi^2}{3} \quad \text{or} \quad \frac{1}{1^2} - \frac{1}{2^2} + \frac{1}{3^2} - ... = \frac{\pi^2}{12} \qquad ...(8)$$

(iii) Adding (7) and (8), $2(1/1^2 + 1/3^2 + 1/5^2 + ...) = \pi^2/4$ or $1/1^2 + 1/3^2 + 1/5^2 + ... = \pi^2/8$

Ex. 4. *Express $f(x) = |x|$ by Fourier series in the interval $[-\pi, \pi]$, Deduce that*

$$\pi^2/8 = 1/1^2 + 1/3^2 + 1/5^2 + ...$$ **[Meerut 2013]**

Hint: Refer solved example 2, page 22.11.'

Ex. 5. *Find Fourier series for function in the interval $[-2, 2]$ if $f(x) = 0$ for $-2 < x < 0$ and $f(x) = 1$ for $0 < x < 2$.* **[Kanpur 2013]**

Sol. Refer result of particular case II given on page 22.11 and note that for the present problem $l = 2$. Thus, the required Fourier series for f is given by

$$f(x) = \frac{a_0}{2} + \sum_{n=1}^{\infty} a_n \cos\frac{n\pi x}{2} + \sum_{n=1}^{\infty} b_n \sin\frac{n\pi x}{2}, \qquad -2 \le x \le 2 \qquad ...(1)$$

where $$a_0 = \frac{1}{2}\int_{-2}^{2} f(x)\, dx,\ a_n = \frac{1}{2}\int_{-2}^{2} f(x) \cos\frac{n\pi x}{2}\, dx,\ b_n = \frac{1}{2}\int_{-2}^{2} f(x) \sin\frac{n\pi x}{2}\, dx,\ n = 1, 2, \ ...(2)$$

Given that $$f(x) = \begin{cases} 0, \text{ for } -2 < x < 0 \\ 1, \text{ for } \ 0 < x < 2 \end{cases} \qquad ...(3)$$

Using (3), (2) yields

$$a_0 = \frac{1}{2}\left[\int_{-2}^{0} f(x)\,dx + \int_{0}^{2} f(x)\,dx\right] = \frac{1}{2}\left[\int_{-2}^{0} (0)\,dx + \int_{0}^{2} dx\right] = \frac{1}{2}[x]_0^2 = 1$$

$$a_n = \frac{1}{2}\left[\int_{-2}^{0} f(x)\cos\frac{n\pi x}{2}\,dx + \int_{0}^{2} f(x)\cos\frac{n\pi x}{2}\,dx\right] = \frac{1}{2}\left[\int_{0}^{2}\cos\frac{n\pi x}{2}\,dx\right] = \frac{1}{2}\left[\frac{\sin(n\pi x/2)}{(n\pi/2)}\right]_0^2 = 0$$

$$b_n = \frac{1}{2}\left[\int_{-2}^{0} f(x)\sin\frac{n\pi x}{2}\,dx + \int_{0}^{2} f(x)\sin\frac{n\pi x}{2}\,dx\right] = \frac{1}{2}\left[\int_{0}^{2}\sin\frac{n\pi x}{2}\,dx\right] = \frac{1}{2}\left[-\frac{\cos(n\pi x/2)}{(n\pi/2)}\right]_0^2$$

$$= (1/n\pi) \times (-\cos n\pi + 1) = (1/n\pi) \times \{1 - (-1)^n\}$$

$$= \begin{cases} 0, & \text{if } n \text{ is even, say } n = 2m, \text{ where } m = 1, 2, 3, \ldots \\ 2/\{(2m-1)\pi\}, & \text{if } n \text{ is odd, say } n = 2m-1, \text{ where } m = 1, 2, 3, \ldots \end{cases}$$

Substituting the above values of a_0, a_n and b_n in (1), the required Fourier series is given by

$$f(x) = \frac{1}{2} + \sum_{m=1}^{\infty} \frac{2}{(2m-1)\pi}\sin\frac{(2m-1)\pi x}{2} \quad \text{or } f(x) = \frac{1}{2} + \frac{2}{\pi}\left(\sin\frac{\pi x}{2} + \frac{1}{3}\sin\frac{3\pi x}{2} + \frac{1}{5}\sin\frac{5\pi x}{2} + \ldots\right)$$

Ex. 6. *Obtain the Fourier series for the function f(x) given by f(x) = x, for 0 < x < 1 and f (x) = 2 – x, for 1 ≤ x < 2. Also find f(2), f(2.5) and f(–2.5).* **[Meerut 2011]**

Sol. Given
$$f(x) = \begin{cases} x, & \text{for } 0 < x < 1 \\ 2 - x, & \text{for } 1 \le x < 2 \end{cases} \qquad \ldots(1)$$

Refer particular case I on page 22.11 and note that here $2l = 2$ so that $l = 1$. The required Fourier series for $f(x)$ is given by

$$f(x) = \frac{a_0}{2} + \sum_{n=1}^{\infty} a_n \cos n\pi x + \sum_{n=1}^{\infty} b_n \sin n\pi x \quad (0 < x < 2) \qquad \ldots(2)$$

where $a_0 = \int_0^2 f(x)\,dx$, $a_n = \int_0^2 f(x)\cos n\pi x\,dx$, $b_n = \int_0^2 f(x)\sin n\pi x\,dx$, where $n = 1, 2, \ldots$ ...(3)

Using (1), (3) yields

$$a_0 = \int_0^1 f(x)\,dx + \int_1^2 f(x)\,dx = \int_0^1 x\,dx + \int_1^2 (2-x)\,dx = \left[x^2/2\right]_0^1 + \left[2x - x^2/2\right]_1^2$$

$$= (1/2) + (4 - 2) - (2 - 1/2) = 1$$

$$a_n = \int_0^1 f(x)\cos n\pi x\,dx + \int_1^2 f(x)\cos n\pi x\,dx = \int_0^1 x\cos n\pi x\,dx + \int_1^2 (2-x)\cos n\pi x\,dx$$

$$= \left[(x)\left(\frac{\sin n\pi x}{n\pi}\right) - (1)\left(-\frac{\cos n\pi x}{n^2\pi^2}\right)\right]_0^1 + \left[(2-x)\left(\frac{\sin n\pi x}{n\pi}\right) - (-1)\left(-\frac{\cos n\pi x}{n^2\pi^2}\right)\right]_1^2$$

$$= \frac{\cos n\pi - 1}{n^2\pi^2} + \frac{\cos n\pi - 1}{n^2\pi^2} = -\frac{2\{1-(-1)^n\}}{n^2\pi^2}$$

$$= \begin{cases} -4/(2m-1)^2\pi^2, & \text{if } n \text{ is odd, say } n = 2m-1, \text{ where } m = 1, 2, 3, \ldots \\ 0, & \text{if } n \text{ is even, say } n = 2m, \text{ where } m = 1, 2, 3, \ldots \end{cases}$$

and
$$b_n = \int_0^1 f(x)\sin n\pi x\,dx + \int_1^2 f(x)\sin n\pi x\,dx = \int_0^1 x\sin n\pi x\,dx + \int_1^2 (2-x)\sin n\pi x\,dx$$

$$= \left[(x)\left(-\frac{\cos n\pi x}{n\pi}\right)-(1)\left(-\frac{\sin n\pi x}{n^2\pi^2}\right)\right]_0^1 + \left[(2-x)\left(-\frac{\cos n\pi x}{n\pi}\right)-(-1)\left(-\frac{\sin n\pi x}{n^2\pi^2}\right)\right]_1^2$$

$$= -(1/n\pi)\times\cos n\pi + (1/n\pi)\times\cos n\pi = 0$$

Substituting the above values of a_0, a_n and b_n in (2), we have

$$f(x) = \frac{1}{2} - \frac{4}{\pi^2}\sum_{m=1}^{\infty}\frac{\cos(2m-1)\pi x}{(2m-1)^2} \quad \text{or} \quad f(x) = \frac{1}{2} - \frac{4}{\pi^2}\left(\frac{\cos \pi x}{1^2} + \frac{\cos 3\pi x}{3^2} + \frac{\cos 5\pi x}{5^2} + \ldots\right)$$

Determination of values of $f(2)$, $f(2.5)$ and $f(-2.5)$

Refer note and corollary on page 22.28.

At $x = 2$, which is a point of discontinuity of f, the value $f(2)$ is given by

$$f(2) = (1/2)\times\{f(0+0)+f(2-0)\} = (1/2)\times\left\{\lim_{h\to 0} f(0+h) + \lim_{h\to 0} f(2-h)\right\}$$

$$= (1/2)\times\left[\lim_{h\to 0}(0+h) + \lim_{h\to 0}\{2-(2-h)\}\right] = (1/2)\times(0+0) = 0, \text{ using (1)}$$

Since f is continuous at $x = 2.5$ and f is periodic function of period 2, hence

$$f(2.5) = f(0.5) = 0.5, \text{ using (1)}$$

Similarly, $f(-2.5) = f(1.5) = 2 - 1.5 = 0.5$, using (1).

Ex. 7. *State and prove Fourier theorem.* **[Meerut 2012]**

Hint. Refer theorem V on page 22.26.

Ex. 8. *Find the Fourier series of the function f defined by*

$$f(x) = \begin{cases} x, & -\pi \le x < 0 \\ -x, & 0 \le x \le \pi \end{cases}$$ **[Himachal 2011]**

Sol. Here, $f(x) = \begin{cases} x, & -\pi < x < 0 \\ -x, & 0 < x < \pi \end{cases}$ and $f(-x) = \begin{cases} -x, & \pi > x > 0 \\ x, & 0 > x > -\pi \end{cases}$ ...(1)

From (1), it follows that $f(x)$ is an even function in $(-\pi, \pi)$. Hence, by special form on page 22.4, the required Fourier series is given by

$$f(x) = \frac{a_0}{2} + \sum_{n=1}^{\infty} a_n \cos nx \quad \text{...(2)}$$

where $a_0 = \frac{2}{\pi}\int_0^{\pi} f(x)\,dx$ and $a_n = \frac{2}{\pi}\int_0^{\pi} f(x)\cos nx\,dx$, $(n = 1, 2, 3, \ldots)$ ...(3)

From (1) and (3), we have
$$a_0 = \frac{2}{\pi}\int_0^{\pi}(-x)\,dx = -\frac{2}{\pi}\left[\frac{x^2}{2}\right]_0^{\pi} = -\pi \quad \text{...(4)}$$

and
$$a_n = \frac{2}{\pi}\int_0^{\pi}(-x)\cos nx\,dx = \frac{2}{\pi}\left[(-x)\left(\frac{\sin nx}{n}\right)-(-1)\left(-\frac{\cos nx}{n^2}\right)\right]_0^{\pi}$$

or
$$a_n = \frac{2}{\pi}\left(-\frac{\cos n\pi}{n^2} + \frac{1}{n^2}\right) = \frac{2\{1-(-1)^n\}}{\pi n^2}$$

[Using the chain rule of integration by parts]

Thus,
$$a_n = \begin{cases} 0, & \text{if } n = 2m \text{ (even), where } m = 1, 2, 3, \ldots \\ 4/\{\pi(2m-1)^2\}, & \text{if } n = 2m-1 \text{ (odd), where } m = 1, 2, 3, \ldots \end{cases} \quad \text{...(5)}$$

Substituting the values of a_0 and a_n given by (4) and (5) in (2), the required Fourier series is given by

$$f(x) = -\frac{\pi}{2} + \frac{4}{\pi}\sum_{m=1}^{\infty}\frac{\cos(2m-1)x}{(2m-1)^2} \qquad \text{or} \qquad f(x) = -\frac{\pi}{2} + \frac{4}{\pi}\left(\frac{\cos x}{1^2} + \frac{\cos 3x}{3^2} + \frac{\cos 5x}{5^2} + \ldots\right)$$

22.16. Some important properties of Fourier coefficients

Property I. *The coefficients a_n and b_n in the Fourier series involving both cosines and sines for a piecewise continuous function f in $(-\pi, \pi)$ are the same as the coefficients in the Fourier cosine and sine series respectively for certain related functions on $0 < x < \pi$.*

Proof. Let $$f(x) = g(x) + h(x), \qquad -\pi < x < \pi \qquad \ldots(1)$$

where $$g(x) = \{f(x) + f(-x)\}/2, \qquad -\pi < x < \pi \qquad \ldots(2a)$$

and $$h(x) = \{f(x) - f(-x)\}/2, \qquad -\pi < x < \pi \qquad \ldots(2b)$$

From (2a), it follows the function $g(x)$ is even whereas the function $h(x)$ is odd, that is,

$$g(-x) = g(x) \qquad \text{and} \qquad h(-x) = -h(x) \qquad \ldots(3)$$

for each point x in the interval $-\pi < x < \pi$ at which these functions are defined.

Using special forms I and II of particular (i) of Ar. 22.5, we have

$$g(x) = \frac{a_0}{2} + \sum_{n=1}^{\infty} a_n \cos nx, \quad 0 < x < \pi \qquad \ldots(4)$$

where $$a_n = \frac{2}{\pi}\int_0^{\pi} g(x)\cos nx\, dx, \quad n = 0, 1, 2, 3, \ldots \qquad \ldots(5)$$

and $$h(x) = \sum_{n=1}^{\infty} b_n \sin nx, \qquad 0 < x < \pi \qquad \ldots(6)$$

where $$b_n = \frac{2}{\pi}\int_0^{\pi} h(x)\sin nx\, dx, \quad n = 0, 1, 2, 3, \ldots \qquad \ldots(7)$$

Observe that when (4) is valid for $0 < x < \pi$, the equation also holds on the interval $-\pi < x < 0$ since each side of the equation is an even function. Likewise, when (6) is valid for $0 < x < \pi$, the equation also holds on the interval $-\pi < x < 0$ since each side of the equation is an odd function. Since $f(x)$ is the sum of $g(x)$ and $h(x)$, it follows that

$$f(x) = \frac{a_0}{2} + \sum_{n=1}^{\infty}(a_n \cos nx + b_n \sin nx), \qquad -\pi < x < \pi \qquad \ldots(8)$$

may be valid under certain conditions.

Using (2a), (5) may be re-written as

$$a_n = \frac{2}{\pi}\int_0^{\pi}\frac{f(x) + f(-x)}{2}\cos nx\, dx = \frac{1}{\pi}\left\{\int_0^{\pi} f(x)\cos nx\, dx + \int_0^{\pi} f(-y)\cos ny\, dy\right\}$$

Putting $x = -y$ and $dx = -dy$ in the second integral on the R.H.S. of the above equation, we obtain

$$a_n = \frac{1}{\pi}\left\{\int_0^{\pi} f(x)\cos nx\, dx - \int_0^{-\pi} f(x)\cos nx\, dx\right\} = \frac{1}{\pi}\left\{\int_0^{\pi} f(x)\cos nx\, dx + \int_{-\pi}^{0} f(x)\cos nx\, dx\right\}$$

Thus, $$a_n = \frac{1}{\pi}\int_{-\pi}^{\pi} f(x)\cos nx\, dx, \qquad n = 0, 1, 2, \ldots \qquad \ldots(9)$$

Next, using (2b), (7) may be re-written as

$$b_n = \frac{2}{\pi}\int_0^{\pi} \frac{f(x)-f(-x)}{2}\sin nx\, dx = \frac{1}{\pi}\left\{\int_0^{\pi} f(x)\sin nx\, dx - \int_0^{\pi} f(-y)\sin ny\, dy\right\}$$

Putting $x = -y$ and $dx = -dy$ in the second integral on the R.H.S. of the above equation, we obtain

$$b_n = \frac{1}{\pi}\left\{\int_0^{\pi} f(x)\sin nx\, dx - \int_0^{-\pi} f(x)\sin nx\, dx\right\} = \frac{1}{\pi}\left\{\int_0^{\pi} f(x)\sin nx\, dx + \int_{-\pi}^{0} f(x)\sin nx\, dx\right\}$$

Thus, $$b_n = \frac{1}{\pi}\int_{-\pi}^{\pi} f(x)\sin nx\, dx,\ n = 1, 2, \ldots \quad \ldots(10)$$

From (5), (7), (9) and (10), the required result follows.

Property II. *For a piecewise continuous function f defined in the interval (0, π), the coefficients a_n and b_n in the cosine and sine series respectively always tend to zero as n tends to infinity.*

Proof. We know that (refer page 20.14) Fourier half range cosine series is given by

$$f(x) = \frac{a_0}{2} + \sum_{n=1}^{\infty} a_n \cos nx, \qquad 0 < x < \pi \quad \ldots(1)$$

where $$a_0 = \frac{2}{\pi}\int_0^{\pi} f(x)\, dx \quad \text{and} \quad a_n = \frac{2}{\pi}\int_0^{\pi} f(x)\cos nx\, dx, \quad n = 1, 2, \ldots \quad \ldots(2)$$

Let $S_p(x)$ denote the partial sum consisting of the first $p + 1$ ($p \geq 1$) terms in the series (1) so that

$$S_p(x) = \frac{a_0}{2} + \sum_{n=1}^{p} a_n \cos nx \quad \ldots(3)$$

Then, $$\int_0^{\pi}\{f(x) - S_p(x)\}^2 dx = \int_0^{\pi}\{f(x)\}^2 dx - 2\int_0^{\pi} f(x)\, S_p(x)\, dx + \int_0^{\pi}\{S_p(x)\}^2 dx \quad \ldots(4)$$

Multiplying both sides of (3) by $f(x)$ and then integrating both sides of the resulting equation from 0 to π, we have

$$\int_0^{\pi} f(x)\, S_p(x)\, dx = \frac{a_0}{2}\int_0^{\pi} f(x)\, dx + \int_0^{\pi}\left\{\sum_{n=1}^{p} a_n\, f(x)\cos nx\, dx\right\} \quad \ldots(5)$$

On interchanging the order of integration and summation in the second term on R.H.S of (5), we obtain

or $$\int_0^{\pi} f(x)\, S_p(x)\, dx = \frac{a_0}{2}\int_0^{\pi} f(x)\, dx + \sum_{n=1}^{p} a_n\left\{\int_0^{\pi} f(x)\cos nx\, dx\right\}$$

or $$\int_0^{\pi} f(x)\, S_p(x)\, dx = \frac{a_0}{2}\times\frac{\pi a_0}{2} + \sum_{n=1}^{p} a_n \times\left(\frac{\pi a_n}{2}\right), \text{ using (2)}$$

or $$\int_0^{\pi} f(x)\, S_p(x)\, dx = \frac{\pi}{2}\left(\frac{a_0^2}{2} + \sum_{n=1}^{p} a_n^2\right) \quad \ldots(6)$$

We now proceed to evaluate $\int_0^{\pi}\{S_p(x)\}^2 dx$.

Multiplying both sides of (3) by $S_p(x)$ and then integrating both sides of the resulting equation from 0 to π, we have

$$\int_0^{\pi} \{S_p(x)\}^2\, dx = \frac{a_0}{2}\int_0^{\pi} S_p(x)\, dx + \int_0^{\pi}\left\{\sum_{n=1}^{p} a_n\, S_p(x) \cos nx\right\} dx \qquad ...(7)$$

On interchanging the order of integration and summation in the second term on R.H.S of (7), we obtain

$$\int_0^{\pi} \{S_p(x)\}^2\, dx = \frac{a_0}{2}\int_0^{\pi} S_p(x)\, dx + \sum_{n=1}^{p} a_n \left\{\int_0^{\pi} S_p(x) \cos nx\, dx\right\} \qquad ...(8)$$

Integrating both sides of (3) from 0 to π, we obtain

$$\int_0^{\pi} S_p(x)\, dx = \int_0^{\pi} \frac{a_0}{2}\, dx + \int_0^{\pi}\left\{\sum_{n=1}^{p} a_n \cos nx\right\} dx \quad \text{or} \quad \int_0^{\pi} S_p(x)\, dx = \frac{a_0 \pi}{2} + \sum_{n=1}^{p} a_n \left\{\int_0^{\pi} \cos nx\, dx\right\}$$

On interchanging the order of integration and summation in the second term on R.H.S. of the above equation, we have

$$\int_0^{\pi} S_p(x)\, dx = \frac{a_0 \pi}{2} + \sum_{n=1}^{p} a_n \left[\frac{\sin nx}{n}\right]_0^{\pi} \quad \text{or} \quad \int_0^{\pi} S_p(x)\, dx = \frac{a_0 \pi}{2} \qquad ...(9)$$

Finally, we proceed to evaluate $\int_0^{\pi} S_p(x) \cos nx\, dx$. Using m as the index of summation in expression (3), we have

$$S_p(x) = \frac{a_0}{2} + \sum_{m=1}^{p} a_m \cos mx \qquad ...(10)$$

Multiplying both sides of (10) by cos nx, we obtain

$$S_p(x) \cos nx = \frac{a_0}{2} \cos nx + \sum_{m=1}^{p} a_m \cos mx \cos nx, \qquad ...(11)$$

where n has any one of the values $n = 1, 2, 3, , ..., p$.

Integrating both sides of (11) from 0 to π, we have

$$\int_0^{\pi} S_p(x) \cos nx\, dx = \frac{a_0}{2}\int_0^{\pi} \cos nx\, dx + \int_0^{\pi}\left\{\sum_{m=1}^{p} a_m \cos mx \cos nx\right\} dx \qquad ...(12)$$

On interchanging the order of integration and summation in the second term on R.H.S. of (12), we get

$$\int_0^{\pi} S_p(x) \cos nx\, dx = \frac{a_0}{2}\left[\frac{\sin nx}{n}\right]_0^{\pi} + \sum_{m=1}^{p} a_m \left\{\int_0^{\pi} \cos mx \cos nx\, dx\right\} \qquad ...(13)$$

If m and n are natural numbers, then we know that

$$\int_0^{\pi} \cos mx \cos nx\, dx = \begin{cases} 0, \text{ when } m \neq n \\ \pi/2, \text{ when } m = n \end{cases} \qquad ...(14)$$

Using (14), (13) reduces to $\int_0^{\pi} S_p(x) \cos nx\, dx = \frac{\pi}{2} a_n,\ n = 1, 2, ..., p$ ...(15)

Using (9) and (15) in (8), we have

$$\int_0^{\pi} \{S_p(x)\}^2\, dx = \frac{a_0}{2} \times \frac{a_0 \pi}{2} + \sum_{n=1}^{p}\left(a_n \times \frac{\pi}{2} a_n\right) = \frac{\pi}{2}\left(\frac{a_0^2}{2} + \sum_{n=1}^{p} a_n^2\right) \qquad ...(16)$$

Using (6) and (16), (4) reduces to

$$\int_0^{\pi}\{f(x)-S_p(x)\}^2\,dx=\int_0^{\pi}\{f(x)\}^2\,dx-2\times\frac{\pi}{2}\times\left(\frac{a_0^2}{2}+\sum_{n=1}^{p}a_n^2\right)+\frac{\pi}{2}\left(\frac{a_0^2}{2}+\sum_{n=1}^{p}a_n^2\right)$$

or $$\int_0^{\pi}\{f(x)-S_p(x)\}^2\,dx=\int_0^{\pi}\{f(x)\}^2\,dx-\frac{\pi}{2}\left(\frac{a_0^2}{2}+\sum_{n=1}^{p}a_n^2\right) \qquad ...(17)$$

Since the value of the integral on L.H.S. of (17) is non-negative, it follows that

$$\int_0^{\pi}\{f(x)\}^2\,dx-\frac{\pi}{2}\left(\frac{a_0^2}{2}+\sum_{n=1}^{p}a_n^2\right)\geq 0$$

so that $$\frac{a_0^2}{2}+\sum_{n=1}^{p}a_n^2\leq\frac{2}{\pi}\int_0^{\pi}\{f(x)\}^2\,dx,\qquad p=1,2,.... \qquad ...(18)$$

which is known as *Bessel's inequality* for the coefficients given by (2)

Now, we shall prove that $a_n\to 0$ as $n\to\infty$. Observe that the R.H.S. of the inequality (18) is independent of the natural number p, and as p increases on the L.H.S, the sums of the squares there form a sequence that is bounded and non decreasing. Such a sequence always converges and since this particular sequence is the sequence of the partial sums of the series whose terms are $a_0^2/2$ and a_n^2 $(n = 1, 2, ...)$, that series must converge. From Art. 6.2, we know that the nth term of a convergent series tends to zero as $n\to\infty$. Hence, it follows that $a_n\to 0$ as $n\to\infty$.

We now proceed to show that the coefficients b_n in sine series tend to zero as $n\to\infty$. We know that (refer page 20.13) Fourier half range sine series is given by

$$f(x)=\sum_{n=1}^{\infty}b_n\sin nx,\qquad 0<x<\pi \qquad ...(19)$$

where $$b_n=\frac{2}{\pi}\int_0^{\pi}f(x)\sin nx\,dx,\quad \text{where } n=1,2,3,... \qquad ...(20)$$

Let $s_p(x)$ denote the partial sum consisting of the first $p+1$ $(p\geq 1)$ terms in the series (19) so that

$$s_p(x)=\sum_{n=1}^{p}b_n\sin nx \qquad ...(21)$$

Then, $$\int_0^{\pi}\{f(x)-s_p(x)\}^2\,dx=\int_0^{\pi}\{f(x)\}^2\,dx-2\int_0^{\pi}f(x)\,s_p(x)\,dx+\int_0^{\pi}\{s_p(x)\}^2\,dx \qquad ...(22)$$

Multiplying both sides of (21) by $f(x)$ and then integrating both sides of the resulting equation from 0 to π, we have

$$\int_0^{\pi}f(x)\,s_p(x)\,dx=\int_0^{\pi}\left\{\sum_{n=1}^{p}b_n\,f(x)\sin nx\right\}dx=\sum_{n=1}^{p}b_n\left\{\int_0^{\pi}f(x)\sin nx\,dx\right\}$$

[On interchanging the order of integration and summation]

or $$\int_0^{\pi}f(x)\,s_p(x)\,dx=\sum_{n=1}^{p}\left(b_n\times\frac{\pi b_n}{2}\right)=\frac{\pi}{2}\sum_{n=1}^{p}b_n^2,\ \text{using (20)} \qquad ...(23)$$

Multiplying both sides of (21) by $s_p(x)$ and then integrating both sides of the resulting equation from 0 to π, we have

$$\int_0^{\pi} \{s_p(x)\}^2\, dx = \int_0^{\pi} \left\{ \sum_{n=1}^{p} b_n\, s_p(x) \sin nx \right\} dx = \sum_{n=1}^{p} b_n \int_0^{\pi} s_p(x) \sin nx\, dx \qquad ...(24)$$

Using m as the index of summation in expression (21), we have

$$s_p(x) = \sum_{m=1}^{p} b_m \sin mx \qquad ...(25)$$

Multiplying both sides of (25) by sin nx, we obtain

$$s_p(x) \sin nx = \sum_{m=1}^{p} b_m \sin mx \sin nx, \qquad ...(26)$$

where n has any of the values $n = 1, 2, 3, , ..., p$.

Integrating both sides of (26) from 0 to π, we have

$$\int_0^{\pi} s_p(x) \sin nx\, dx = \int_0^{\pi} \left\{ \sum_{m=1}^{p} b_m \sin mx \sin nx \right\} dx = \sum_{m=1}^{p} b_m \left\{ \int_0^{\pi} \sin mx \sin nx\, dx \right\} \qquad ...(27)$$

[On intercharging the order of integeration and summation]

If m and n are natural numbers, then we know that

$$\int_0^{\pi} \sin mx \sin nx\, dx = \begin{cases} 0, \text{ when } m \neq n \\ \pi/2, \text{ when } m = n \end{cases} \qquad ...(28)$$

Using (28), (27) reduces to $\qquad \int_0^{\pi} s_p(x) \sin nx\, dx = \dfrac{\pi}{2} b_n, \qquad n = 1, 2, ..., p \qquad ...(29)$

Using (29), (24) yields $\qquad \int_0^{\pi} \{s_p(x)\}^2\, dx = \sum_{n=1}^{p} \left(b_n \times \dfrac{\pi b_n}{2} \right) = \dfrac{\pi}{2} \sum_{n=1}^{p} b_n^2 \qquad ...(30)$

Using (23) and (30), (22) reduces to

$$\int_0^{\pi} \{f(x) - s_p(x)\}^2\, dx = \int_0^{\pi} \{f(x)\}^2\, dx - \pi \sum_{n=1}^{p} b_n^{\ 2} + \frac{\pi}{2} \sum_{n=1}^{p} b_n^{\ 2}$$

Thus, $\qquad \int_0^{\pi} \{f(x) - s_p(x)\}^2\, dx = \int_0^{\pi} \{f(x)\}^2\, dx - \dfrac{\pi}{2} \sum_{n=1}^{p} b_n^{\ 2} \qquad ...(31)$

Since the value of the integral on L.H.S. of (31) is non-negative, it follows that

$$\int_0^{\pi} \{f(x)\}^2\, dx - \frac{\pi}{2} \sum_{n=1}^{p} b_n^2 \geq 0 \quad \text{so that} \quad \sum_{n=1}^{p} b_n^2 \leq \frac{2}{\pi} \int_0^{\pi} \{f(x)\}^2\, dx, \; p = 1, 2, ... \qquad ...(32)$$

which is known as Bessel's inequality for the coefficients given by (20).

Now, we shall prove that $b_n \to 0$ as $n \to \infty$. Observe that the R.H.S. of the inequality (32) is independent of the natural number p, and as p increases on the L.H.S, the sums of the squares there form a sequence that is bounded and non decreasing. Such a sequence always converges and since this particular sequence is the sequence of the partial sums of the series whose terms are b_n^2 $(n = 1, 2, ...)$, that series must converge. Since the nth term of a convergent series tends to zero as $n \to \infty$, it follows that $b_n \to 0$ as $n \to \infty$.

Finally, from property I, we know that the coefficient a_n and b_n in the Fourier series involving both cosines and sines for a piecewise continuous f in the interval $(-\pi, \pi)$ are the same as the coefficients in the Fourier cosine series and Fourier sine series respectively, for certain related functions on $0 < x < \pi$. Hence, it follows that these coefficients must tend to zero as n tends to infinity.

Property III. *If a_n and b_n are the coefficients in the Fourier series*

$$f(x) = \frac{a_0}{2} + \sum_{n=1}^{\infty}(a_n \cos nx + b_n \sin nx)$$

for a piecewise continuous function f defined in the interval $(-\pi, \pi)$, then $\lim_{n\to\infty} a_n = 0$ *and* $\lim_{n\to\infty} b_n = 0$.

Proof. Let $$f(x) = g(x) + h(x), \quad -\pi < x < \pi \qquad ...(1)$$

where $$g(x) = \{f(x) + f(-x)\}/2, \quad -\pi < x < \pi \qquad ...(2a)$$

and $$h(x) = \{f(x) - f(-x)\}/2, \quad -\pi < x < \pi \qquad ...(2b)$$

Recall that in property I we have already proved that the coefficients a_n and b_n in the Fourier series

$$f(x) = \frac{a_0}{2} + \sum_{n=1}^{\infty}(a_n \cos nx + b_n \sin nx), \qquad -\pi < x < \pi \qquad ...(3)$$

are the same as the coefficients in Fourier cosine and Fourier sine series for $g(x)$ and $h(x)$ respectively on interval $0 < x < \pi$.

Proceeding as in property II, we arrive at the following two Bessel's inequalities:

$$\frac{a_0^2}{2} + \sum_{n=1}^{p} a_n^2 \le \frac{2}{\pi}\int_0^{\pi}\{g(x)\}^2\,dx, \qquad p = 1, 2, 3, \ldots \qquad ...(4)$$

and $$\sum_{n=1}^{p} b_n^2 \le \frac{2}{\pi}\int_0^{\pi}\{h(x)\}^2\,dx, \qquad p = 1, 2, 3, \ldots \qquad ...(5)$$

Adding (4) and (5), we obtain

$$\frac{a_0^2}{2} + \sum_{n=1}^{p}(a_n^2 + b_n^2) \le \frac{2}{\pi}\int_0^{\pi}\left[\{g(x)\}^2 + \{h(x)\}^2\right]dx, \; p = 1, 2, 3, \ldots$$

$$= \frac{2}{\pi}\int_0^{\pi}\left[\left\{\frac{f(x)+f(-x)}{2}\right\}^2 + \left\{\frac{f(x)-f(-x)}{2}\right\}^2\right]dx, \; p = 1, 2, 3, \ldots$$

$$= \frac{1}{\pi}\int_0^{\pi}\left[\{f(x)\}^2 + \{f(-x)^2\right]dx, \qquad p = 1, 2, 3, \ldots$$

Re-writing the above inequality, we have

$$\frac{a_0^2}{2} + \sum_{n=1}^{p}(a_n^2 + b_n^2) \le \frac{1}{\pi}\left[\int_0^{\pi}\{f(x)\}^2\,dx + \int_0^{\pi}\{f(-y)\}^2\,dy\right], \quad p = 1, 2, 3, \ldots \qquad ...(6)$$

Putting $x = -y$ and $dx = -dy$, we have

$$\int_0^{\pi}\{f(-y)\}^2\,dy = -\int_0^{-\pi}\{f(x)\}^2\,dx = \int_{-\pi}^{0}\{f(x)\}^2\,dx \qquad ...(7)$$

Using (7), (6) reduces to

$$\frac{a_0^2}{2} + \sum_{n=1}^{p}(a_n^2 + b_n^2) \le \frac{1}{\pi}\left[\int_0^{\pi}\{f(x)\}^2\,dx + \int_{-\pi}^{0}\{f(x)\}^2\,dx\right], \quad p = 1, 2, 3, \ldots$$

or $$\frac{a_0^2}{2} + \sum_{n=1}^{p}(a_n^2 + b_n^2) \le \frac{1}{\pi}\int_{-\pi}^{\pi}\{f(x)\}^2\,dx, \quad p = 1, 2, 3, \ldots \qquad ...(8)$$

which is known as *Bessel inequality*.

Observe that the R.H.S. of (8) is independent of the the natural number p, and as p increases on the L.H.S. of (8), the sum of the squares there form a sequence that is bounded and non-decreasing. Since such a sequence always converges and since this particular sequence is the sequence of the partial sums of the series whose terms are $a_0^2/2$, a_n^2 and b_n^2 $(n = 1, 2, ...)$, that series must converge. Since the nth term of a convergent series always tends to zero as $n \to \infty$ (refer Art 6.2), we have

$$\lim_{n\to\infty} (a_n^2 + b_n^2) = 0 \quad \text{giving} \quad \lim_{n\to\infty} a_n = 0 \quad \text{and} \quad \lim_{n\to\infty} b_n = 0$$

22.17. An important lemma

Let f denote a function such that
(i) f is continuous on the interval $-\pi \le x \le \pi$
(ii) $f(-\pi) = f(\pi)$
(iii) its derivative f' *is piecewise continuous on the interval* $-\pi < x < \pi$
If a_n *and* b_n *are the Fourier coefficients*

$$a_n = \frac{1}{\pi}\int_{-\pi}^{\pi} f(x)\cos nx\, dx,\ n = 0, 1, 2, \ldots \text{ and } b_n = \frac{1}{\pi}\int_{-\pi}^{\pi} f(x)\sin nx\, dx,\ n = 1, 2, \ldots$$

for f, then the series $\sum_{n=1}^{\infty} \sqrt{a_n^2 + b_n^2}$ *converges.*

Proof. Give that $$a_n = \frac{1}{\pi}\int_{-\pi}^{\pi} f(x)\cos nx\, dx,\ n = 0, 1, 2, 3, \ldots \quad \text{...(1)}$$

and $$b_n = \frac{1}{\pi}\int_{-\pi}^{\pi} f(x)\sin nx\, dx,\ n = 1, 2, 3, \ldots \quad \text{...(2)}$$

The class of functions satisfying conditions (i), (ii) and (iii) is a subspace of piecewise smooth functions on the interval $-\pi < x < \pi$.

Since f' is piecewise continuous, Fourier coefficients for f' exist and are given by

$$\alpha_0 = \frac{1}{\pi}\int_{-\pi}^{\pi} f'(x)\, dx, \qquad \alpha_n = \frac{1}{\pi}\int_{-\pi}^{\pi} f'(x)\cos nx\, dx,\ n = 1, 2, 3, \ldots \quad \text{...(3)}$$

and $$\beta_n = \frac{1}{\pi}\int_{-\pi}^{\pi} f'(x)\sin nx\, dx,\ n = 1, 2, 3, \ldots \quad \text{...(4)}$$

From (3), $\alpha_0 = \frac{1}{\pi}[f(x)]_{-\pi}^{\pi} = \frac{1}{\pi}[f(\pi) - f(-\pi)] = 0$, using given condition (ii)

and $$\alpha_n = \frac{1}{\pi}\left\{[\cos nx\ f(x)]_{-\pi}^{\pi} + n\int_{-\pi}^{\pi} \sin nx\ f(x)\, dx\right\}, \text{ on integrating by parts}$$

$$= \frac{1}{\pi}[\cos n\pi\ f(\pi) - \cos(-n\pi)\ f(-\pi)] + \frac{n}{\pi}\int_{-\pi}^{\pi} f(x)\sin nx\, dx$$

$$= \frac{\cos n\pi}{\pi}[f(\pi) - f(-\pi)] + nb_n, \text{ using (2)}$$

$= 0 + nb_n$, using the given condition (ii)

Thus, $$b_n = \alpha_n/n, \quad n = 1, 2, 3, \ldots \quad \text{...(5)}$$

From (4), $$\beta_n = \frac{1}{\pi}\left\{[\text{sins}\, nx\ f(x)]_{-\pi}^{\pi} - n\int_{-\pi}^{\pi} \cos nx\ f(x)\, dx\right\}, \text{ as before}$$

$= 0 - na_n$, using (1)

Thus, $$a_n = -(\beta_n/n), \quad n = 1, 2, 3, \ldots \quad \text{...(6)}$$

Let us consider the series $$\sum_{n=1}^{\infty}\sqrt{a_n^2+b_n^2} \quad ...(7)$$

Let S_m denote the sum of the first m terms of the infinite series (7). Then, we have

$$S_m=\sum_{n=1}^{m}\sqrt{a_n^2+b_n^2}=\sum_{n=1}^{m}(1/n)\times\left(\sqrt{\alpha_n^2+\beta_n^2}\right), \text{ using (5) and (6)} \quad ...(8)$$

In what follows, we shall use Cauchy's inequality whose statement is given below

If p_n (n = 1, 2, 3, ..., m) and q_n (n = 1, 2, 3, ..., m) be two sets of real numbers, then

$$\left(\sum_{n=1}^{m}p_n\,q_n\right)^2\le\left(\sum_{n=1}^{m}p_n^2\right)\left(\sum_{n=1}^{m}q_n^2\right) \quad ...(9)$$

Using Cauchy's inequality (9), we have

$$\left[\sum_{n=1}^{m}\left\{1/n)\times\sqrt{\alpha_n^2+\beta_n^2}\right\}\right]^2\le\left[\sum_{n=1}^{m}(1/n^2)\right]\times\left[\sum_{n=1}^{m}(\alpha_n^2+\beta_n^2)\right], \qquad m=1,2,3$$

or
$$S_m^2\le\left[\sum_{n=1}^{m}(1/n^2)\right]\times\left[\sum_{n=1}^{m}(\alpha_n^2+\beta_n^2)\right], \qquad m=1,2,... \quad ...(10)$$

Now, consider the series $$\sum_{n=1}^{\infty}(1/n^2) \quad ...(11)$$

Then, the sequence of sums $\sum_{n=1}^{m}(1/n^2)$ $(m = 1, 2, ...)$ occurring on R.H.S. of (10) is clearly bounded since each sum is a partial sum of convergent series (11)

Since α_n $(n = 0, 1, 2, ...)$ and β_n $(n = 1, 2, 3, ...)$ are the Fourier coefficients for f' on the interval $-\pi < x < \pi$, hence these coefficients must satisfy Bessel inequality (refer property III)

$$\sum_{n=1}^{m}(\alpha_n^2+\beta_m^2)\le\frac{1}{\pi}\int_{-\pi}^{\pi}\{f'(x)\}^2\,dx, \qquad m=1,2,3,... \quad ...(12)$$

From (12), it follows that the sequence $\sum_{n=1}^{m}(\alpha_n^2+\beta_n^2)$ $(m = 1, 2, ...)$ is bounded.

Since $\sum_{n=1}^{m}(1/n^2)$ and $\sum_{n=1}^{m}(\alpha_n^2+\beta_n^2)$ are both bounded for $m = 1, 2, 3, ...$, hence (10) shows that the sequence S_m^2 $(m = 1, 2, ...)$ is both bounded and non-decreasing. Hence it converges and this implies that the sequence S_m $(m = 1, 2, ...)$ converges. Thus the series (7) converges.

22.18. Absolute and uniform convergence of Fourier series

Theorem. *Let f denote a function such that*

(i) f is continuous on the interval $-\pi \le x \le \pi$

(ii) $f(-\pi) = f(\pi)$

(iii) its derivative f′ is piecewise continuous on the interval $-\pi < x < \pi$.

Then the Fourier series $\frac{a_0}{2}+\sum_{n=1}^{\infty}(a_n\cos nx+b_n\sin nx)$ *for f with coefficients*

$$a_n=\frac{1}{\pi}\int_{-\pi}^{\pi}f(x)\cos nx\,dx,\ n=0,1,2,..., \qquad b_n=\frac{1}{\pi}\int_{-\pi}^{\pi}f(x)\sin nx\,dx,\ n=1,2,...$$

converges absolutely and uniformly to f(x) on the interval $-\pi \le x \le \pi$.

Proof. For definition of absolute convergence of an infinite series, refer Art. 7.2. Again, for definition and test for uniform convergence of series of functions, refer Art. 15.7 and Art 15.8.

Consider the series $$\frac{a_0}{2}+\sum_{n=1}^{\infty}(a_n \cos nx + b_n \sin nx) \quad ...(1)$$

where $a_n = \frac{1}{\pi}\int_{-\pi}^{\pi} f(x) \cos nx\, dx,\ n = 0, 1, 2, ...$ and $b_n = \frac{1}{\pi}\int_{-\pi}^{\pi} f(x) \sin nx\, dx,\ n = 1, 2, ...$ ...(2)

Observe that the given conditions (i), (ii) and (iii) on f ensure the continuity of the periodic extension of f for all x. Therefore, the series (1) must converge to $f(x)$ everywhere in the interval $-\pi \le x \le \pi$.

Now,
$$|a_n \cos nx + b_n \sin nx| \le |a_n \cos nx| + |b_n \sin nx|$$
or
$$|a_n \cos nx + b_n \sin nx| \le |a_n|\,|\cos nx| + |b_n|\,|\sin nx|$$
or
$$|a_n \cos nx + b_n \sin nx| \le |a_n| + |b_n|, \quad \text{since} \quad |\cos nx| \le 1 \quad \text{and} \quad |\sin nx| < 1$$
or
$$|a_n \cos nx + b_n \sin nx| \le \sqrt{a_n^2 + b_n^2} + \sqrt{a_n^2 + b_n^2}$$

Then,
$$|a_n \cos nx + b_n \sin nx| \le 2\sqrt{a_n^2 + b_n^2}, \qquad n = 1, 2, ... \quad ...(3)$$

From lemma in Art. 22.17, it follows that the series $\sum_{n=1}^{\infty} 2\sqrt{a_n^2 + b_n^2}$ is convergent. Again, $(a_n \cos nx + b_n \sin nx)$ is nth term of (1). Hence, by Weierstrass M-test (refer Art. 15.8.2, page 15.14), it follows that the given series (1) is uniformly convergent on the interval $-\pi \le x \le \pi$. Again, by definition of absolute convergence, it follows that the given series (1) is also absolutely convergent on the interval $-\pi \le x \le \pi$.

22.19. Differentiation of Fourier Series

Theorem. *Let f denote a function such that*

(i) f is continuous on the interval $-\pi \le x \le \pi$

(ii) $f(-\pi) = f(\pi)$

(iii) its derivative f' is piecewise continuous on the interval $-\pi < x < \pi$.

Then the Fourier series representation

$$f(x) = \frac{a_0}{2}+\sum_{n=1}^{\infty}(a_n \cos nx + b_n \sin nx), \qquad -\pi \le x \le \pi$$

where $a_n = \frac{1}{\pi}\int_{-\pi}^{\pi} f(x) \cos nx\, dx,\ n = 0, 1, 2, ...,\quad b_n = \frac{1}{\pi}\int_{-\pi}^{\pi} f(x) \sin nx\, dx,\ n = 1, 2, 3, ...$

is differentiable at each point x in the interval $-\pi < x < \pi$ at which the second order derivative f'' exists:

$$f'(x) = \sum_{n=1}^{\infty} n(-a_n \sin nx + b_n \cos nx)$$

Proof. Consider a point x $(-\pi < x < \pi)$ at which f'' exists. Hence, it follows that f' is continuous at x. Therefore, on applying the Fourier theorem (refer page 22.26) to the function f', we obtain

$$f'(x) = \frac{\alpha_0}{2}+\sum_{n=1}^{\infty}(\alpha_n \cos nx + \beta_n \sin nx) \quad ...(1)$$

where $\alpha_n = \frac{1}{\pi}\int_{-\pi}^{\pi} f'(x) \cos nx\, dx,\ (n = 0, 1, 2, ...)$ and $\beta_n = \frac{1}{\pi}\int_{-\pi}^{\pi} f'(x) \sin nx\, dx\ \ n = 1, 2, ...$ ...(2)

Observe that f and f' satisfy all the conditions stated in lemma in Art 22.17. Proceed exactly as in Art. 22.17 and prove that

$$\alpha_0 = 0, \qquad \alpha_n = n\, b_n \qquad \text{and} \qquad \beta_n = -\, n\, a_n \qquad (n = 1, 2, ...) \qquad ...(3)$$

Substituting the values of α_0, α_n and β_n given by (2) in (1), we arrive at the following required result

$$f'(x) = \sum_{n=1}^{\infty} n(-a_n \sin nx + b_n \cos nx)$$

Remark. In the following example, we show that not all Fourier series are differentiable.

Illustrative example. The Fourier series for the function $f(x) = x\ (-\pi < x < \pi)$ is given by (prove your self)

$$x = 2\sum_{n=1}^{\infty} \frac{(-1)^{n+1}}{n} \sin nx \qquad (-\pi < x < \pi) \qquad ...(1)$$

However the differential series

$$2\sum_{n=1}^{\infty} (-1)^{n+1} \cos nx$$

does not converge since its nth term fails to approach to zero as n tends to infinity.

Remark. In the above theorem, we have given sufficient conditions for differentiability.

22.20. Integration of Fourier Series

Let f be a function that is piecewise continuous on the interval $-\pi < x < \pi$. Let Fourier series for be is given by

$$f(x) = \frac{a_0}{2} + \sum_{n=1}^{\infty} (a_n \cos nx + b_n \sin nx) \qquad (-\pi < x < \pi) \qquad ...(i)$$

$$\text{where } a_0 = \frac{1}{\pi}\int_{-\pi}^{\pi} f(x)\, dx,\ a_n = \frac{1}{\pi}\int_{-\pi}^{\pi} f(x) \cos nx\, dx,\ b_n = \frac{1}{\pi}\int_{-\pi}^{\pi} f(x) \sin nx\, dx,\ (n = 1, 2, ...) \qquad ...(ii)$$

Then, regardless whether series on R.H.S. of (i) converges or not, the following equation is valid when $-\pi < x < \pi$.

$$\int_{-\pi}^{\pi} f(s)\, ds = \frac{a_0}{2}(x + \pi) + \sum_{n=1}^{\infty} \frac{1}{n}[a_n \sin nx - b_n \{\cos nx + (-1)^{n+1}\}] \qquad \textbf{[Meerut 2012]}$$

Proof. Given

$$f(x) = \frac{a_0}{2} + \sum_{n=1}^{\infty} (a_n \cos nx + b_n \sin nx), \qquad -\pi < x < \pi \qquad ...(1)$$

$$\text{where } a_0 = \frac{1}{\pi}\int_{-\pi}^{\pi} f(x)\, dx,\ a_n = \frac{1}{\pi}\int_{-\pi}^{\pi} f(x) \cos nx\, dx,\ b_n = \frac{1}{\pi}\int_{-\pi}^{\pi} f(x) \sin nx\, dx,\ n = 1, 2, ... \qquad ...(2)$$

Consider the function F(x) given by

$$F(x) = \int_{-\pi}^{x} f(s)\, ds - \frac{a_0}{2} x, \qquad -\pi < x < \pi \qquad ...(3)$$

Since f is piecewise continuous, F(x) is continuous. Again,

$$F'(x) = f(x) - (a_0/2), \qquad -\pi < x < \pi \qquad ...(4)$$

except at points where f is discontinuous. Therefore, F′ is piecewise continuous on the interval $-\pi < x < \pi$. Here Fourier series for F(x) is given by

$$F(x) = \frac{A_0}{2} + \sum_{n=1}^{\infty} (A_n \cos nx + B_n \sin nx), \qquad -\pi < x < \pi \qquad \text{...(5)}$$

where $$A_n = \frac{1}{\pi}\int_{-\pi}^{\pi} F(x) \cos nx\, dx, \; n = 0, 1, 2, \ldots \qquad \text{...(6)}$$

and $$B_n = \frac{1}{\pi}\int_{-\pi}^{\pi} F(x) \sin nx\, dx, \; n = 1, 2, \ldots \qquad \text{...(7)}$$

From (3), $$F(-\pi) = \int_{-\pi}^{-\pi} f(s)\, ds + \frac{a_0 \pi}{2} = 0 + \frac{a_0\pi}{2} = \frac{a_0\pi}{2} \qquad \text{...(8)}$$

Again from (2) and (3), $$F(\pi) = \int_{-\pi}^{\pi} f(s)\, ds - \frac{a_0\pi}{2} = a_0\pi - \frac{a_0\pi}{2} = \frac{a_0\pi}{2} \qquad \text{...(9)}$$

From (8) and (9), we have $$F(-\pi) = F(\pi), \qquad \text{...(10)}$$
showing that the representation (5) is also valid at the end points of the open interval $-\pi < x < \pi$ and, therefore at each point of the closed interval $-\pi \le x \le \pi$.

When $n \ge 1$, we proceed to integrate integrals (6) and (7) by parts, using the fact that F is continuous and F′ is piecewise continuous. Thus, we obtain

$$A_n = \frac{1}{\pi}\left\{\left[F(x)\frac{\sin nx}{n}\right]_{-\pi}^{\pi} - \int_{-\pi}^{\pi} F'(x)\frac{\sin nx}{n}\, dx\right\} = -\frac{1}{n\pi}\int_{-\pi}^{\pi} F'(x) \sin nx\, dx$$

$$= -\frac{1}{n\pi}\int_{-\pi}^{\pi}\left\{f(x) - \frac{a_0}{2}\right\} \sin nx\, dx, \text{ using (4)}$$

$$= -\frac{1}{n\pi}\int_{-\pi}^{\pi} f(x) \sin nx\, dx + \frac{a_0}{2n\pi}\int_{-\pi}^{\pi} \sin nx\, dx = -\frac{b_n}{n} + \frac{a_0}{2n\pi}\left[-\frac{\cos nx}{n}\right]_{-\pi}^{\pi}, \text{ using (2)}$$

$$= -(b_n/n) - (a_0/2n^2\pi) \times \{\cos n\pi - \cos(-n\pi)\} = -(b_n/n)$$

and $$B_n = \frac{1}{\pi}\left\{\left[-F(x)\frac{\cos nx}{n}\right]_{-\pi}^{\pi} + \int_{-\pi}^{\pi} F'(x)\frac{\cos nx}{n}\, dx\right\}$$

$$= \frac{1}{n\pi}\{-F(\pi)\cos n\pi + F(-\pi)\cos(-n\pi)\} + \frac{1}{n\pi}\int_{-\pi}^{\pi} F'(x) \cos nx\, dx$$

$$= \frac{\cos n\pi}{n\pi}\{F(-\pi) - F(\pi)\} + \frac{1}{n\pi}\int_{-\pi}^{\pi}\left(f(x) - \frac{a_0}{2}\right)\cos nx\, dx, \text{ using (4)}$$

$$= 0 + \frac{1}{n\pi}\int_{-\pi}^{\pi} f(x) \cos nx\, dx - \frac{a_0}{2n\pi}\int_{-\pi}^{\pi} \cos nx\, dx, \text{ using (10)}$$

$$= \frac{a_n}{n} - \frac{a_0}{2n^2\pi}[\sin nx]_{-\pi}^{\pi}, \text{ using (2)}$$

$$= a_n/n$$

Thus, we obtain $$A_n = -(b_n/n) \quad \text{and} \quad B_n = a_n/n \qquad \text{...(11)}$$

We now proceed to determine value of A_0. Since (5) is valid when $x = \pi$, we have

$$F(\pi) = \frac{A_0}{2} + \sum_{n=1}^{\infty} A_n \cos n\pi \quad \text{or} \quad \frac{a_0 \pi}{2} = \frac{A_0}{2} - \sum_{n=1}^{\infty} \frac{(-1)^n b_n}{n}, \text{ by (9) and (11)}$$

Thus,
$$A_0 = a_0 \pi - 2\sum_{n=1}^{\infty} \frac{(-1)^{n+1}}{n} b_n \qquad \text{...(12)}$$

Substituting the values of A_0, A_n and B_n given by (11) and (12) in (5), we obtain

$$F(x) = \frac{a_0 \pi}{2} + \sum_{n=1}^{\infty} \frac{1}{n} [a_n \sin nx - b_n \{\cos nx + (-1)^{n+1}\}] \qquad \text{...(13)}$$

Substituting the value of $F(x)$ given by (3) in (13), we obtain

$$\int_{-\pi}^{x} f(s)\, ds - \frac{a_0 x}{2} = \frac{a_0 \pi}{2} + \sum_{n=1}^{\infty} \frac{1}{n} [a_n \sin nx - b_n \{\cos nx + (-1)^{n+1}\}]$$

or
$$\int_{-\pi}^{x} f(s)\, ds = \frac{a_0}{2}(x + \pi) + \sum_{n=1}^{\infty} \frac{1}{n} [a_n \sin nx - b_n \{\cos nx + (-1)^{n+1}\}]$$

APPENDIX

ADDITIONAL RESULTS AND PROBLEMS BASED ON CHAPTER 15

A.1. Some useful theorems

Theorem 1. *If a sequence* $< f_n >$ *converges uniformly to* f *in* $[a, b]$ *and* x_0 *is a point of* $[a, b]$ *such that* $\lim_{x \to x_0} f_n(x) = a_n \ \forall\, n \in \mathbf{N}$, *then*

(*i*) $< a_n >$ *converges* and (*ii*) $\lim_{x \to x_0} f(x) = \lim_{n \to \infty} a_n$. **[Delhi B.A/B.Sc. (Hons.) 2008]**

Proof. (*i*). Since the sequence $< f_n >$ converges uniformly to f in $[a, b]$, hence for $\varepsilon > 0$, there exists a positive integer m(independent of x) such that

$$\left| f_{n+p}(x) - f_n(x) \right| < \varepsilon/2, \quad \forall\, n \geq m, \quad \forall\, p \geq 1 \quad \text{and} \quad \forall\, x \in [a, b] \qquad \text{...(1)}$$

Given that

$$\lim_{x \to x_0} f_n(x) = a_n, \qquad \forall\, n \in \mathbf{N} \qquad \text{...(2)}$$

Keeping n, p fixed, letting $x \to x_0$ and using (2), (1) yields

$$\left| a_{n+p} - a_n \right| \leq \varepsilon/2 < \varepsilon, \qquad \forall\, n \geq m \quad \text{and} \quad \forall\, p \geq 1,$$

showing that the sequence $< a_n >$ converges. Suppose that

$$\lim_{n \to \infty} a_n = l \qquad \text{...(3)}$$

(*ii*). Since the sequence $< f_n >$ converges to f in $[a, b]$, hence for $\varepsilon > 0$, there exists a positive integer m_1 such that

$$\left| f_n(x) - f(x) \right| < \varepsilon/3, \qquad \forall n \geq m_1 \quad \text{and} \quad \forall x \in [a, b] \qquad \text{...(4)}$$

Since the sequence $< a_n >$ converges to l (refer part (*i*)), hence for $\varepsilon > 0$, there exists a positive integer m_2 such that

$$\left| a_n - l \right| < \varepsilon/3, \quad \forall n \geq m_2 \qquad \text{...(5)}$$

Let $m' = \max(m_1, m_2)$. From (2) it follows that $\lim_{x \to x_0} f_{m'}(x) = a_{m'}$. Therefore for $\varepsilon > 0$, there exists a $\delta > 0$ such that for $| x - x_0 | < \delta$, we have

$$| f_{m'}(x_0) - a_n | < \varepsilon/3 \qquad \text{...(6)}$$

Hence, for $| x - x_0 | < \delta$, we have

$$\begin{aligned} | f(x) - l | &= | \{f(x) - f_{m'}(x)\} + \{f_{m'}(x) - a_{m'}\} + (a_{m'} - l) | \\ &\leq | f(x) - f_{m'}(x) | + | f_{m'}(x) - a_{m'} | + | a_{m'} - l | \\ &< \varepsilon/3 + \varepsilon/3 + \varepsilon/3, \text{ using (4), (5) and (6)} \end{aligned}$$

Thus, $\qquad | f(x) - l | < \varepsilon \qquad$ whenever $\qquad | x - x_0 | < \delta$,

showing that $\lim_{x \to x_0} f(x)$ exists and equals l. Therefore, using (3), we obtain

$$\lim_{x \to x_0} f(x) = \lim_{n \to \infty} a_n \qquad \text{...(7)}$$

Remark. Result (7) may be re-written as: $\lim_{x \to x_0} \lim_{n \to \infty} f_n(x) = \lim_{n \to \infty} \lim_{x \to x_0} f_n(x)$

Theorem 2. *If a series* $\sum_{n=1}^{\infty} f_n$ *converges uniformly to* f *in* $[a, b]$, *and* x_0 *is a point in* $[a, b]$ *such that* $\lim_{x \to x_0} f_n(x) = a_n, \forall n \in \mathbf{N}$, *then* (*i*) $\sum_{n=1}^{\infty} a_n$ *converges and* (*ii*) $\lim_{x \to x_0} f(x) = \sum_{n=1}^{\infty} a_n$

Proof. (*i*). Since the series $\sum_{n=1}^{\infty} f_n$ converges to f in $[a, b]$, therefore for $\varepsilon > 0$, there exists a positive integer m such that

$$\left| \sum_{r=n+1}^{n+p} f_r(x) \right| < \varepsilon/2, \qquad \forall\, n \ge m, \qquad \forall\, p \ge 1 \qquad \text{and} \qquad \forall\, x \in [a, b] \qquad \text{...(1)}$$

Given that $$\lim_{x \to x_0} f_n(x) = a_n, \qquad \forall\, n \in \mathbf{N} \qquad \text{...(2)}$$

Keeping n, p fixed, letting $x \to x_0$ and using (2), (1) reduces to

$$\left| \sum_{r=n+1}^{n+p} a_r \right| \le \varepsilon/2 < \varepsilon, \qquad \forall\, n \ge m$$

Hence the series $\sum_{r=1}^{\infty} a_n$ converges to l, say. Then, we have

$$\sum_{n=1}^{\infty} a_n = l \qquad \text{...(3)}$$

(*ii*). Since the series $\sum_{n=1}^{\infty} f_n$ converges to f in $[a, b]$, hence for $\varepsilon > 0$, there exists a positive integer m_1 such that

$$\left| \sum_{r=1}^{n} f_r(x) - f(x) \right| < \varepsilon/3, \qquad \forall n \ge m_1 \qquad \text{and} \qquad \forall x \in [a, b] \qquad \text{...(4)}$$

Since the series $\sum_{n=1}^{\infty} a_n$ converges to l (refer result (3)), hence for $\varepsilon > 0$, there exists a positive integer m_2 such that

$$\left| \sum_{r=1}^{n} a_r - l \right| < \varepsilon/2, \qquad \forall n \ge m_2 \qquad \text{...(5)}$$

Let $m' = \max(m_1, m_2)$. Furthermore, since (2) holds, hence for the chosen $\varepsilon > 0$, it is possible to choose $\delta > 0$ such that for $n = 1, 2, 3, \ldots, m'$, we have (taking $\delta = \min(\delta_1, \delta_2, \ldots, \delta_{m'})$

$$| f_n(x) - a_n | < \varepsilon/3m' \qquad \text{for} \qquad | x - x_0 | < \delta.$$

Hence, $$\left| \sum_{r=1}^{m'} f_r(x) - \sum_{r=1}^{m'} a_r \right| \le \sum_{r=1}^{m'} |f_r(x) - a_r| < m' \times (\varepsilon/3m') = \varepsilon/3 \quad \text{for} \quad |x - x_0| < \delta \qquad \text{...(6)}$$

Thus, for $| x - x_0 | < \delta$, we have

$$|f(x) - l| = \left| \left\{ f(x) - \sum_{r=1}^{m'} f_r(x) \right\} + \left\{ \sum_{r=1}^{m'} f_r(x) - \sum_{r=1}^{m'} a_r \right\} + \left(\sum_{r=1}^{m'} a_r - l \right) \right|$$

$$\leq \left| f(x) - \sum_{r=1}^{m'} f_r(x) \right| + \left| \sum_{r=1}^{m'} f_r(x) - \sum_{r=1}^{m'} a_r \right| + \left| \sum_{r=1}^{m'} a_r - l \right|$$

$< \varepsilon/3 + \varepsilon/3 + \varepsilon/3$, using (4), (5) and (6)

Hence, $\quad |f(x) - l| = \varepsilon \quad$ whenever $\quad |x - x_0| < \delta$,

showing that $\lim_{x \to x_0} f(x)$ exists and equals l. Hence, using (3), we have

$$\lim_{x \to x_0} f(x) = \sum_{n=1}^{\infty} a_n \qquad ...(7)$$

Note: Re-writing (7), $\quad \lim_{x \to x_0} \sum_{n=1}^{\infty} f_n(x) = \sum_{n=1}^{\infty} \lim_{x \to x_0} f_n(x),$

i.e., the limit of the sum function of a series = the sum of the series of limits of functions.

AN ILLUSTRATIVE SOLVED EXAMPLE

Let $f(x) = \sum_{n=1}^{\infty} \frac{sin(nx)}{n^2}$. *Then* *(a)* $\lim_{x \to 0} f(x) = 0$ *(b)* $\lim_{x \to 0} f(x) = 1$

(c) $\lim_{x \to 0} f(x) = \pi^2/6$ *(d)* $\lim_{x \to 0} f(x)$ *does not exist* **(GATE 2013)**

Solution. Ans. (*a*). Let $f_n(x) = \{\sin nx)\}/n^2$. Then, we have

$$|f_n(x)| = \left| \frac{\sin(nx)}{n^2} \right| = \frac{|\sin(nx)|}{n^2} \leq \frac{1}{n^2} = M_n, \text{ say } \quad \forall x \in \mathbf{R}$$

Since $\sum_{n=1}^{\infty} M_n = \sum_{n=1}^{\infty} \frac{1}{n^2}$ is convergent, hence by Weierstrass's M-test, it follows that the series $\sum_{n=1}^{\infty} f_n(x)$, *i.e.*, $\sum_{n=1}^{\infty} \frac{\sin(nx)}{n^2}$ is uniformly convergent for all real values of x. Hence, from theorem 2 of A.1, we obtain

$$\lim_{x \to 0} f(x) = \lim_{x \to 0} \sum_{n=1}^{\infty} \frac{\sin(nx)}{n^2} = \sum_{n=1}^{\infty} \left\{ \lim_{x \to 0} \frac{\sin(nx)}{n^2} \right\} = 0.$$

A.2. Uniform convergence and continuity

Theorem 1. *If* $< f_n >$ *is a sequence of continuous functions on an interval* $[a, b]$ *and if* $f_n \to f$ *uniformly on* $[a, b]$, *then f is continuous on* $[a, b]$.

Proof. Refer Art. 15.4, page 15.7 of chapter 15.

Theorem 2. *If a series* $\sum_{n=1}^{\infty} f_n$ *converges uniformly to f in an interval* $[a, b]$, *and its terms* f_n *are continuous at a point* x_0 *of the interval* $[a, b]$, *then the sum function f is also continuous at* x_0.

Proof. (*i*) Let $S_n(x) = n$th partial sum $= f_1(x) + f_2(x) + ... + f_n(x)$ and let $\varepsilon > 0$ be given.

Since $\sum_{n=1}^{\infty} f_n$ converges uniformly to f on $[a, b]$, there exists a positive integer m such that

$$|S_n(x) - f(x)| < \varepsilon/3, \quad \forall\, n \geq m \text{ and } \forall\, x \in [a, b] \qquad ...(1)$$

Since x_0 is a point of $[a, b]$, (1) yields

$$|S_n(x_0) - f(x_0)| < \varepsilon/3, \quad \forall\, n \geq m \qquad ...(2)$$

Since f_n is continuous on $[a, b]$ $\forall\, n \in \mathbf{N}$, it follows that $S_n = f_1 + f_2 + ... + f_n$ is also continuous on $[a, b]$ $\forall\, n \in \mathbf{N}$,

Since $S_n(x)$ is continuous at a point x_0 of $[a, b]$, there exists $\delta > 0$ such that

$$|S_n(x) - S_n(x_0)| < \varepsilon/3, \qquad \text{whenever} \qquad |x - x_0| < \delta \qquad ...(3)$$

Now, $|f(x) - f(x_0)| = |\{f(x) - S_n(x)\} + \{S_n(x) - S_n(x_0)\} + \{S_n(x_0) - f(x_0)\}|$

$\leq |f(x) - S_n(x)| + |S_n(x) - S_n(x_0)| + |S_n(x_0) - f(x_0)|$

$< \varepsilon/3 + \varepsilon/3 + \varepsilon/3$, using (1), (2) and (3)

Thus, $|f(x) - f(x_0)| < \varepsilon$ whenever $|x - x_0| < \delta$

Hence f is continuous at $x = x_0$, *i.e.*, the sum function is continuous at $x = x_0$.

Corollary. *If a series* $\sum_{n=1}^{\infty} f_n$ *converges uniformly to f on an interval* $[a, b]$ *and if the functions* f_n *are all continuous throughout the interval, then so is the sum function f.*

[Delhi B.Sc. (Prog.) II, 2015; Himachal 2017]

Proof. First of all repeat the entire proof of the above theorem 2. Then proceed as follows:

Since x_0 is an arbitrary point of $[a, b]$, the above theorem 2 is true for all points of $[a, b]$. Therefore, the sum function is continuous on $[a, b]$.

Remarks. Uniform convergence of a sequence $< f_n >$ is only a sufficient but not a necessary condition for the continuity of the limit function f, that is, if the limit function f is continuous on $[a, b]$, then it is not necessary that the sequence $< f_n >$ is uniformly convergent on $[a, b]$. In fact, the above theorem 1 shows that if the limit function f is discontinuous then the sequence $< f_n >$ of continuous functions cannot be uniformly convergent on $[a, b]$. Thus, the above theorem 1 provides a useful negative test for uniform convergence of a sequence of functions.

Similarly, the above theorem 2 shows that if the sum function f is discontinuous, then the series $\sum_{n=1}^{\infty} f_n$ of continuous functions cannot be uniformly convergent on $[a, b]$. Thus the above theorem 2 provides a useful negative test for uniform convergence of a series of functions.

A.3. Dini's theorems on uniform convergence

In the following theorem 1 we propose to discuss a special class of sequences for which uniform convergence is equivalent to the continuity of the limit function of the sequence. Similarly, in the next theorem 2, we propose to discuss a special class of series for which uniform convergence is equivalent to the continuity of the sum function of the series.

Theorem 1. *If a sequence of continuous functions* $< f_n >$, *defined on* $[a, b]$ *is monotonic increasing and converges (point wise) to a continuous function f, then the convergence is uniform on* $[a, b]$.

Proof. Since the sequence $< f_n >$ is monotonic increasing and converges to f on $[a, b]$, hence for any $\varepsilon > 0$ and $\forall\, x \in [a, b]$, there exists a positive integer m such that

$$0 < f(x) - f_n(x) < \varepsilon \qquad ...(1)$$

Assume that $\quad R_n(x) = f(x) - f_n(x), \quad \forall\, n \in \mathbf{N} \qquad ...(2)$

From (2), it follows that the sequence $< R_n >$ is monotonic decreasing, *i.e.*,

$$R_1(x) \geq ...\, R_2(x) \geq ... \geq R_n(x) \geq ... \qquad ...(3)$$

and is bounded below by zero. Hence the sequence $< R_n >$ converges point-wise to zero on $[a, b]$. Furthermore, note that if (1) and (3) hold $\forall\, x \in [a, b]$, and m is independent of x, then the convergence is uniform.

Suppose, if possible, that for a certain $\varepsilon_0 > 0$, no such m independent of x exists. Then, in this situation for $\forall\, n \in \mathbf{N}$, there exists $x_n \in [a, b]$ such that

$$R_n(x_n) \geq \varepsilon_0, \quad ...(4)$$

showing that the sequence $< x_n >$ is bounded and hence it must possess at least one limit point ξ in $[a, b]$. It follows that there exists a subsequence $< x_{n_k} >$ of $< x_n >$, converging to ξ.

Since R_n is the difference of two continuous functions f and f_n, R_n is also continuous. Therefore, for each fixed m, we have a relation of the form

$$\lim_{k \to \infty} R_m(x_{n_k}) = R_m(\xi)$$

Now for any sufficiently large k and $\forall\, m \in \mathbf{N}$, we have $n_k \geq k > m$. Therefore, using (3) and (4), it follows that

$$R_m(x_{n_k}) \geq R_{n_k}(x_{n_k}) \geq \varepsilon_0 \quad ...(5)$$

Taking limits as $k \to \infty$, (5) yields

$$R_m(\xi) \geq \varepsilon_0 \text{ for any } m,$$

which contradicts the relation $\lim_{m \to \infty} R_m(\xi) = 0$, implied by the point-wise convergence of $< R_n >$ on $[a, b]$. Hence our hypothesis is wrong and therefore the sequence $< f_n >$ converges uniformly to f on $[a, b]$

Theorem 2. *If the sum function of a series* $\sum_{n=1}^{\infty} f_n$ *with non-negative continuous terms defined on an interval* $[a, b]$, *is continuous on* $[a, b]$, *then the series is uniformly convergent on the interval.*

Proof. Let $S_n(x) = \sum_{r=1}^{n} f_r(x)$. Clearly, the partial sums $S_n(x)$, with non-negative continuous terms f_r, form a non-decreasing sequence of continuous functions converging point-wise to a continuous function f. Hence, in view of the above theorem 1, the sequence $< S_n(x) >$ converges uniformly and therefore the series $\sum_{n=1}^{\infty} f_n$ is also uniformly convergent.

A.4. Uniform convergence and integration

Theorem 1. *If a sequence* $< f_n >$ *converges uniformly to* f *on* $[a, b]$ *and each* f_n *is integrable, then* f *is integrable on* $[a, b]$, *and the sequence* $< \int_a^b f_n(x)\, dx >$ *converges uniformly to* $\int_a^b f(x)\, dx$ *on* $[a, b]$, *that is,*

$$\lim_{n \to \infty} \int_a^b f_n(x)\, dx = \int_a^b f(x)\, dx$$

Proof. Let $\varepsilon > 0$ be given. Since $< f_n >$ converges uniformly to f on $[a, b]$, there exists a positive integer m such that

$$|f_n(x) - f(x)| < \varepsilon/2(b-a), \quad \forall\, n \in m \quad \text{and} \quad x \in [a, b], \quad a \neq b \quad ...(1)$$

Now, (1) $\Rightarrow f(x) - \{\varepsilon/2(b-a)\} < f_n(x) < f(x) + \{\varepsilon/2(b-a)\}$, $\forall\, n \geq m$ and $\forall\, x \in [a, b]$

Thus, $f(x) < f_n(x) + \{\varepsilon/2(b-a)\}$ and $f(x) > f_n(x) - \{\varepsilon/2(b-a)\}$, $\forall\, n \geq m, \forall\, x \in [a, b]$ ...(2)

Since f_n is integrable on $[a, b]$ for each $n \in \mathbf{N}$, we have

$$\int_a^b f_n(x)\,dx = \overline{\int_a^b} f_n(x)\,dx = \underline{\int_a^b} f_n(x)\,dx \qquad \text{...(3)}$$

From (2), $\overline{\int_a^b} f_n(x)\,dx < \overline{\int_a^b}\left\{f_n(x) + \frac{\varepsilon}{2(b-a)}\right\}dx = \overline{\int_a^b} f_n(x)\,dx + \frac{\varepsilon}{2(b-a)} \times (b-a)$

Thus, we obtain $\overline{\int_a^b} f(x)\,dx < \int_a^b f_n(x)\,dx + \frac{\varepsilon}{2}$, using (3) ...(4)

Again, (2) $\Rightarrow$ $\underline{\int_a^b} f(x)\,dx > \underline{\int_a^b}\left\{f_n(x) - \frac{\varepsilon}{2(b-a)}\right\}dx = \underline{\int_a^b} f_n(x)\,dx - \frac{\varepsilon}{2(b-a)} \times (b-a)$

Thus, we obtain $-\underline{\int_a^b} f(x)\,dx < -\int_a^b f_n(x)\,dx + \frac{\varepsilon}{2}$, using (3) ...(5)

Adding (4) and (5) yields $0 \le \overline{\int_a^b} f(x)\,dx - \underline{\int_a^b} f(x)\,dx < \varepsilon$ ...(6)

Since ε is arbitrary, (6) $\Rightarrow$ $\overline{\int_a^b} f(x)\,dx - \underline{\int_a^b} f(x)\,dx = 0$ $\Rightarrow$ $\overline{\int_a^b} f(x)\,dx = \underline{\int_a^b} f(x)\,dx,$

showing that f is integrable on $[a, b]$. Again, from (3), (4) and (5), we have

$$\int_a^b f_n(x)\,dx - \frac{\varepsilon}{2} < \int_a^b f(x)\,dx < \int_a^b f_n(x)\,dx + \frac{\varepsilon}{2}, \ \forall\, n \ge m$$

$$\Rightarrow \quad \int_a^b f(x)\,dx - \frac{\varepsilon}{2} < \int_a^b f_n(x)\,dx < \int_a^b f(x)\,dx + \frac{\varepsilon}{2}, \ \forall\, n \ge m$$

$$\Rightarrow \quad \left|\int_a^b f_n(x)\,dx - \int_a^b f(x)\,dx\right| < \frac{\varepsilon}{2} < \varepsilon, \ \forall\, n \ge m \quad \Rightarrow \quad \lim_{n\to\infty} \int_a^b f_n(x)\,dx = \int_a^b f(x)\,dx$$

Theorem 2. *If a series of functions* $\sum_{n=1}^{\infty} f_n$ *converges uniformly to f on [a, b] and each function* f_n *is integrable on [a, b], then f is integrable on [a, b] and* $\sum_{n=1}^{\infty} \int_a^b f_n(x)\,dx$ *converges uniformly to* $\int_a^b f(x)\,dx$, *that is,* $\sum_{n=1}^{\infty} \int_a^b f_n(x)\,dx = \int_a^b f(x)\,dx,$
that is, the series is term by term integrable.

Proof. Let $S_n = \sum_{r=1}^{n} f_r$. Then $< S_n >$ denotes the sequence of partial sums of the series $\sum_{n=1}^{\infty} f_n$. Since $\sum_{n=1}^{\infty} f_n$ converges uniformly to f on $[a, b]$, so by definitior the sequence $< S_n >$ converges uniformly to f on $[a, b]$. Now, S_n being the sum of n integrable functions is integrable $\forall\, n \in \mathbf{N}$. Therefore, by the theorem 1, it follows that f is integrable on $[a, b]$ and we have

$$\lim_{n \to \infty} \int_a^b S_n(x)\, dx = \int_a^b f(x)\, dx, \quad \text{i.e.,} \quad \sum_{n=1}^{\infty} \int_a^b f_n(x)\, dx = \int_a^b f(x)\, dx$$

Remark 1. The hypothesis of uniform convergence in the above theorems 1 and 2 is only sufficient condition but not a necessary condition. Hence, a series (sequence) may converge to an integrable limit without being uniformly convergent. On the other hand, if the limit is not integrable or if integrable is not equal to the limit of the series (sequence) of integrals, the convergence cannot be uniform.

Remark 2. Suppose $< f_n >$ is a sequence of integrable functions converging to f on $[a, b]$ and if $\lim_{n \to \infty} \int_a^b f_n(x)\, dx \neq \int_a^b f(x)\, dx$, then $< f_n >$ cannot converge uniformly to f.

A.5. Uniform convergence and integration. (Particular cases of theorems 1 and 2 discussed in Art. A.4.)

Theorem 1. *If a sequence* $< f_n >$ *converges uniformly to f on* $[a, b]$ *and each function* f_n *is continuous on* $[a, b]$, *then f is integrable and the sequence* $< \int_a^b f_n(x)\, dx >$ *converges to* $\int_a^b f(x)\, dx$,

that is,

$$\int_a^b f(x)\, dx = \lim_{n \to \infty} \int_a^b f_n(x)\, dx.$$

Proof. Since the sequence $< f_n >$ converges uniformly to f on $[a, b]$ and each f_n is continuous on $[a, b]$, hence f is continuous and therefore integrable on $[a, b]$.

Since the sequence $< f_n >$ converges uniformly to f, hence for a given $\varepsilon > 0$, there must exist a positive integer m such that

$$| f_n(x) - f(x) | < \varepsilon/(b - a), \quad \forall\, n \geq m \text{ and } \forall\, x \in [a, b] \tag{1}$$

Now, $\forall\, x \in [a, b]$ and for $n \geq m$, we have

$$\left| \int_a^b f(x)\, dx - \int_a^b f_n(x)\, dx \right| = \left| \int_a^b \{f(x) - f_n(x)\}\, dx \right| \leq \int_a^b | f(x) - f_n(x) |\, dx$$

$$< \{\varepsilon/(b - a)\} \times (b - a) = \varepsilon, \text{ using (1)}$$

showing that $< \int_a^b f_n(x)\, dx >$ converges uniformly to $\int_a^b f(x)\, dx$ on $[a, b]$, i.e.,

$$\int_a^b f(x)\, dx = \lim_{n \to \infty} \int_a^b f_n(x)\, dx$$

Theorem 2. *If a series* $\sum_{n=1}^{\infty} f_n$ *uniformly converges to f on* $[a, b]$ *and each* f_n *is continuous on* $[a, b]$, *then f is integrable on* $[a, b]$ *and the series* $\sum_{n=1}^{\infty} \left(\int_a^b f_n(x)\, dx \right)$ *converges uniformly to* $\int_a^b f(x)\, dx$ *on* $[a, b]$, *i.e.,*

$$\int_a^b f(x)\, dx = \sum_{n=1}^{\infty} \int_a^b f_n(x)\, dx$$ **(Agra 2012)**

Proof. Since $\sum_{n=1}^{\infty} f_n$ converges uniformly to f on $[a, b]$ and each f_n is continuous on $[a, b]$, it follows that the sum function f is continuous and therefore integrable on $[a, b]$.

Since each f_n is continuous, it follows that the sum of a finite number of functions, namely, $\sum_{r=1}^{n} f_r$ is also continuous and therefore integrable on $[a, b]$. Furthermore, we have

$$\sum_{r=1}^{n} \int_a^b f_r(x)\,dx = \int_a^b \sum_{r=1}^{n} f_r(x)\,dx$$

Now, since the series $\sum_{n=1}^{\infty} f_n$ converges uniformly to f, for a given $\varepsilon > 0$, there exists a positive integer m such that $\forall\, x \in [a, b]$

$$\left| f(x) - \sum_{r=1}^{n} f_r(x) \right| < \varepsilon/(b-a), \ \forall\, n \geq m \qquad \text{...(1)}$$

Again, $\forall\, x \in [a, b]$ and for $n \geq m$, we have

$$\left| \int_a^b f(x)\,dx - \sum_{r=1}^{n} \int_a^b f_r(x)\,dx \right| = \left| \int_a^b \left\{ f(x) - \sum_{r=1}^{n} f_r(x) \right\} dx \right| \leq \int_a^b \left| f(x) - \sum_{r=1}^{n} f_r(x) \right| dx$$

$$< \{\varepsilon/(b-a)\} \times (b-a) = \varepsilon, \text{ using (1)}$$

showing that $\sum_{n=1}^{\infty} \left(\int_a^b f_n(x)\,dx \right)$ converges uniformly to $\int_a^b f(x)\,dx$ on $[a, b]$, that is

$$\int_a^b f(x)\,dx = \sum_{n=1}^{\infty} \int_a^b f_n(x)\,dx$$

A.6. Uniform convergence and differentiation

Theorem 1. *Suppose that the sequence $< f_n >$ satisfies the following conditions:*

(i) Let each f_n have a finite derivative at each point of $[a, b]$

(ii) $< f_n(x_0) >$ converges for at least one point $x_0 \in [a, b]$

(iii) Let there exist a function g such that $f_n' \to g$ uniformly on $[a, b]$

Then, *(a) There exists a function f such that $f_n \to f$ uniformly on $[a, b]$*

(b) For each x in $[a, b]$, the derivative $f'(x)$ exists and equals $g(x)$, that is,

$$\lim_{n \to \infty} f_n'(x) = g(x) = f'(x).$$ ***(Meerut 2013***

Proof. From the given conditions (*ii*) and (*iii*), since $< f_n(x_0) >$ is convergent and $< f_n' >$ converges uniformly, it follows that for $\varepsilon > 0$, there exists a positive integer m such that $\forall x \in [a, b]$, we have

$$| f_{n+p}(x_0) - f_n(x_0) | < \varepsilon/2, \quad \forall\, n \geq m \quad \text{and} \quad \forall\, p \geq 1 \qquad \text{...(1)}$$

and
$$| f'_{n+p}(x) - f_n'(x) | < \varepsilon/2(b-a), \quad \forall\, n \geq m \quad \text{and} \quad \forall\, p \geq 1 \qquad \text{...(2)}$$

Applying Lagrange's mean value theorem to the function $(f_{n+p} - f_n)$ for any two points x and y of $[a, b]$, we have $x < c < y$, $\forall\, n \geq m$ and $\forall\, p \geq 1$

$$| f_{n+p}(x) - f_n(x) - \{f_{n+p}(y) - f_n(y)\} | = | x - y | \times | f'_{n+p}(c) - f'(c) |$$

$$< | x - y | \times \{\varepsilon/2(b-a)\}, \text{ using (2)}$$

Thus,
$$| f_{n+p}(x) - f_n(x) - f_{n+p}(y) + f_n(y) | < \varepsilon/2 \qquad \text{...(3)}$$

Since y is any point of $[a, b]$, setting $y = x_0$ in the above result yields

$$| f_{n+p}(x) - f_n(x) - f_{n+p}(x_0) + f_n(x_0) | < \varepsilon/2 \qquad \text{...(4)}$$

Now, $| f_{n+p}(x) - f_n(x) | = | \{f_{n+p}(x) - f_n(x) - f_{n+p}(x_0) + f_n(x_0)\} + \{f_{n+p}(x_0) - f_n(x_0)\} |$

$$\leq | f_{n+p}(x) - f_n(x) - f_{n+p}(x_0) + f_n(x_0) | + | f_{n+p}(x_0) - f_n(x_0) |$$

$$< \varepsilon/2 + \varepsilon/2, \text{ using (1) and (4)}$$

Thus, $\quad | f_{n+p}(x) - f_n(x) | = \varepsilon, \quad \forall\, p \geq m, \qquad \forall\, p \geq 1 \quad$ and $\quad \forall x \in [a, b]$,

showing that the sequence $< f_n >$ converges uniformly on $[a, b]$

Suppose that the sequence $< f_n >$ converges to f, say,

Now, for a fixed $x \in [a, b]$ and $y \in [a, b]$ $y \neq x$, define two functions $h_n(y)$ and $h(c)$ as follows:

$$h_n(y) = \{f_n(y) - f_n(x)\}/(y - x), \qquad \forall\, n \geq \mathbf{N} \qquad ...(5)$$

and
$$h(y) = \{f(y) - f(x)\}/(y - x) \qquad ...(6)$$

From condition (*i*) of the theorem, each f_n is differentiable. Hence, for $\forall\, n \geq \mathbf{N}$, from (5), we have

$$\lim_{y \to x} h_n(y) = \lim_{y \to x} \frac{f_n(y) - f_n(x)}{y - x} = f_n{}'(x) \qquad ...(7)$$

Again, using (5), we have

$$| h_{n+p}(y) - h_n(y) | = \frac{1}{|y - x|} \left| f_{n+p}(y) - f_{n+p}(x) - f_n(y) + f_n(x) \right|$$

$$= \frac{1}{|x - y|} \left| f_{n+p}(x) - f_n(x) - \{f_{n+p}(y) - f_n(y)\} \right|$$

$$< \varepsilon/2(b - a), \ \forall\, n \geq m \text{ and } \forall\, p \geq 1 \qquad \text{[using (3)]}$$

showing that the sequence $< h_n(y) >$ converges uniformly on $[a, b]$ for $y \neq x$

Now, $$\lim_{n \to \infty} h_n(y) = \lim_{n \to \infty} \frac{f_n(y) - f_n(x)}{y - x}, \text{ using (5)}$$

$$= \{f(y) - f(x)\}/(y - x), \text{ since } \lim_{n \to \infty} f_n = f$$

Hence, $$\lim_{n \to \infty} h_n(y) = h(y), \text{ using (6)}$$

Thus, $< h_n(y) >$ converges uniformly to $h(y)$ on $[a, b]$ for $y \in [a, b]$, $y \neq x$.

Applying theorem 1 of Art. A.1 to the uniformly convergent sequence $< h_n(y) >$ and using (7), we obtain

$$\lim_{y \to x} h(y) = \lim_{n \to \infty} f_n{}'(x) = g(x), \text{ by hypothesis,}$$

showing that $\lim_{y \to x} h(y)$ exists. Then, form (6), it follows that f is differentiable and $\lim_{y \to x} h(y) = f\,'(x)$. Thus, we have

$$f\,'(x) = g(x) = \lim_{n \to \infty} f_n'(x)$$

Note: If f_n' is continuous $\forall\, n \in \mathbf{N}$, there is an easy proof of the above theorem, as discussed in the next theorem 2.

Theorem 2. *Suppose that the sequence* $< f_n >$ *satisfies the following conditions:*

(*i*) *Let each* f_n *have a finite derivative at each point of* $[a, b]$

(*ii*) *Each* f_n' *is continuous on* $[a, b]$.

(*iii*) *The sequence* $< f_n >$ *converges pointwise on* $[a, b]$ *to* f.

(*iv*) *The sequence* $< f_n' >$ *converges uniformly on* $[a, b]$ *to* g.

Then, (*a*) $g(x) = f'(x) \ \forall x \in [a, b]$ (*b*) $\lim_{n \to \infty} f_n'(x) = f'(x) \ \forall x \in [a, b]$

Proof. By conditions (*ii*) and (*iv*), the sequence $< f_n' >$ of continuous functions converges uniformly on [*a*, *b*] to *g*. Hence, by theorem 1 of Art. A.2, *g* is also continuous on [*a*, *b*].

Let *y* be an arbitrary point of [*a*, *b*]. Since $< f_n' >$ converges uniformly to *g* on [*a*, *b*], hence by theorem 1 of Art. A.4, we have

$$\lim_{n \to \infty} \int_a^y f_n'(x)\, dx = \int_a^y g(x)\, dx \qquad ...(1)$$

Now, by the fundamental theorem of Integral calculus (refer Art. 13.15), we have

$$\int_a^y f_n'(x)\, dx = f_n(y) - f_n(a)$$

Hence, $$\lim_{n \to \infty} \int_a^y f_n'(x)\, dx = \lim_{n \to \infty} f_n(y) - \lim_{n \to \infty} f_n(a)$$

Thus, $$\lim_{n \to \infty} \int_a^y f_n'(x)\, dx = f(y) - f(a), \text{ using condition } (iii) \qquad ...(2)$$

Using (2) in (1), we have $$f(y) - f(a) = \int_a^y g(x)\, dx, \text{ where } a \le y \le b \qquad ...(3)$$

Now, using result of theorem II of Art. 13.14, (3) yields

$$f'(y) = g(y), \quad \text{where} \quad a \le y \le b \qquad ...(4)$$

Since (4) is true for each $y \in [a, b]$, we have $f'(x) = g(x) \ \forall x \in [a, b]$ ...(5)

By the given condition (*iv*), since $< f_n' >$ converges uniformly on [*a*, *b*] and $f'(x) = g(x) \ \forall x \in [a, b]$, it follows that $f_n'(x)$ converges to $f'(x) \ \forall x \in [a, b]$, that is,

$$\lim_{n \to \infty} f_n'(x) = f'(x), \quad \forall x \in [a, b]$$

Theorem 3. *Suppose that the series* $\sum_{n=1}^{\infty} f_n$ *satisfies the following conditions:*

(*i*) *The derivative* $f_n'(x)$ *exists for each* $x \in [a, b]$

(*ii*) *For at least one point* x_0 *in* [*a*, *b*], *the series* $\sum_{n=1}^{\infty} f_n(x_0)$ *converges*

(*iii*) *There exists a function g such that the series* $\sum_{n=1}^{\infty} f_n'$ *converges to g on* [*a*, *b*].

Then, (*a*) *There exists a function f such that* $\sum_{n=1}^{\infty} f_n$ *converges uniformly to f on* [*a*, *b*].

(*b*) *If* $x \in [a, b]$, *the derivative* $f'(x)$ *exists and equals* $\sum_{n=1}^{\infty} f_n'(x)$, *that is,*

$$\frac{d}{dx} \sum_{n=1}^{\infty} f_n(x) = \sum_{n=1}^{\infty} \frac{d}{dx} f_n(x),$$

showing that term-by-term differentiation of the series is valid.

Proof. Consider the sequence $< S_n >$ of partial sum of the given series where $S_n = f_1 + f_2 + ... + f_n$. We now proceed to prove that the sequence $< S_n >$ satisfies all the conditions of theorem 1 of Art. A.6 so that the proof follows by the same theorem.

Since each of f_1, f_2, ... f_n is differentiable in $[a, b]$, hence their finite sum S_n is also differentiable in $[a, b]$. Therefore, S_n' exists for $\forall x \in [a, b]$. Let $x_0 \in [a, b]$ so that

$$S_n(x_0) = f_1(x_0) + f_2(x_0) + ... + f_n(x_0).$$

By the given condition (*ii*), the series $\sum_{n=1}^{\infty} f_n(x)$ converges at $x = x_0$. Hence the sequence $< S_n(x_0) >$ converges.

Now, we have $$S_n'(x_0) = f_1'(x) + f_2'(x) + ... + f_n'(x).$$

By the given condition (*iii*), $\sum_{n=1}^{\infty} f_n'(x)$ converges to $g(x)$ uniformly on $[a, b]$. Hence, $S_n' \to g$ uniformly on $[a, b]$.

Thus, we have shown that the sequence $< S_n >$ satisfies all the three conditions of theorem 1. Therefore, by the same theorem, it follows that there exists a function f such that $S_n \to f$ uniformly on $[a, b]$. This implies that the series $\sum_{n=1}^{\infty} f_n$ converges uniformly to f on $[a, b]$. Thus, we have proved part (*a*) of the present theorem.

Again by theorem 1, for each x, $f'(x)$ exists and $f'(x) = \sum_{n=1}^{\infty} f_n'(x)$, which proves part (*b*) of the present theorem.

Note: If f_n' is continuous $\forall\, n \geq \mathbf{N}$, there is an easy proof of the above theorem 3 as discussed in the next theorem 4.

Theorem 4. *Suppose that the series* $\sum_{n=1}^{\infty} f_n$ *satisfies the following conditions:*

(*i*) *The series* $\sum_{n=1}^{\infty} f_n$ *converges pointwise to f on* $[a, b]$

(*ii*) $f_n'(x)$ *exists for* $\forall x \in [a, b]$

(*iii*) *For each* $n \in \mathbf{N}$, f_n' *is continuous on* $[a, b]$.

(*iv*) $\sum_{n=1}^{\infty} f_n'$ *converges uniformly to g on* $[a, b]$

Then, $$\frac{d}{dx}\sum_{n=1}^{\infty} f_n(x) = \sum_{n=1}^{\infty} \frac{d}{dx} f_n(x), \quad \text{where} \quad a \leq x \leq b$$

Proof. By the given conditions (*iii*) and (*iv*), the series $\sum_{n=1}^{\infty} f_n'$ of continuous functions converges uniformly to g on $[a, b]$. Therefore, the sum function g of $\sum_{n=1}^{\infty} f_n'$ is also continuous on $[a, b]$. Hence, the function $\int_a^x g(t)\,dt$ is differentiable and

$$\frac{d}{dx}\int_a^x g(t)\,dt = g(x), \ \forall x \in [a, b] \qquad ...(1)$$

Now, by the given condition (*i*), $$f(x) = \sum_{n=1}^{\infty} f_n(x), \ \forall x \in [a, b] \qquad ...(2)$$

By the given condition (*iii*), each function f_n' is continuous on $[a, b]$. Hence each function f_n' integrable on $[a, b]$. Therefore, using the fundamental theorem of Integral calculus, we obtain

$$\int_a^x f_n'(t)\,dt = f_n(x) - f_n(a), \qquad \forall\, n \ge 1 \quad \text{and} \quad \forall\, x \in [a, b]$$

Hence, $$\sum_{n=1}^{\infty} \int_a^x f_n'(t)\,dt = \sum_{n=1}^{\infty} f_n(x) - \sum_{n=1}^{\infty} f_n(a), \ \forall\, x \in [a, b]$$

or $$\sum_{n=1}^{\infty} \int_a^x f_n'(t)\,dt = f(x) - f(a), \ \forall\, x \in [a, b], \text{ using (2)} \qquad \text{...(3)}$$

Now, since the series $\sum_{n=1}^{\infty} f_n'$ of integrable functions, converges uniformly to g on $[a, b]$, therefore by theorem 2 of Art. A.4, term by term integration is valid, *i.e.*,

$$\int_a^x g(t)\,dt = \sum_{n=1}^{\infty} \int_a^x f_n'(t)\,dt, \ \forall\, x \in [a, b] \qquad \text{...(4)}$$

From (3) and (4), $$f(x) - f(a) = \int_a^x g(t)\,dt, \ \forall\, x \in [a, b] \qquad \text{...(5)}$$

From (5), $f'(x) = \dfrac{d}{dx}\displaystyle\int_a^x g(t)\,dt$ so that $f'(x) = g(x), \ \forall\, x \in [a, b]$,

Therefore, $$\frac{d}{dx}\sum_{n=1}^{\infty} f_n(x) = \sum_{n=1}^{\infty} \frac{d}{dx} f_n(x), \quad \text{where} \quad a \le x \le b.$$

A.7. Miscellaneous problems on chapter 15

Ex. 1. *Show that $x = 0$ is a point of non-uniform convergence of the series whose n^{th} term is $n^2 x\, e^{-n^2 x^2} - (n-1)^2\, x\, e^{-(n-1)^2 x^2}$* **(Agra 2012)**

Sol. Suppose that $$u_n(x) = n^2 x\, e^{-n^2 x^2} - (n-1)^2\, x\, e^{-(n-1)^2 x^2} \qquad \text{...(1)}$$

Also, let $S_n = u_1 + u_2 + \dots + u_n$ so that the sequence $< S_n >$ is the sequence of partial sums of the given series. From (1), we have

$$u_1(x) = x\, e^{-x^2} - 0$$

$$u_2(x) = 2^2 x\, e^{-2^2 x^2} - x\, e^{-x^2}$$

$$u_3(x) = 3^2 x\, e^{-3^2 x^2} - 2^2 x\, e^{-2^2 x^2}$$

....

$$u_n(x) = n^2 x\, e^{-n^2 x^2} - (n-1)^2\, x\, e^{-(n-1)^2 x^2}$$

Adding these, we have $$S_n(x) = \sum_{n=1}^{\infty} u_n(x) = n^2 x\, e^{-n^2 x^2}$$

Now, by definition $$S(x) = \lim_{n \to \infty} S_n(x) = 0 \ \text{ for } 0 \le x \le 1$$

Therefore, $$|S_n(x) - S(x)| = n^2 x\, e^{-n^2 x^2} = y, \text{ say}$$

where $$y = n^2 x\, e^{-n^2 x^2} \qquad \text{...(2)}$$

Form (2), $$dy/dx = n^2 \{e^{-n^2 x^2} + x(-2n^2 x)e^{-n^2 x^2}\} = n^2 e^{-n^2 x^2}(1 - 2n^2 x^2) \qquad \text{...(3)}$$

For maximum and minimum values of y, we have

$dy/dx = 0$ so that $\quad n^2 e^{-n^2x^2}(1-2n^2x^2)=0 \quad$ giving $\quad x = 1/n\sqrt{2}$

From (3), $\quad d^2y/dx^2 = n^2(-2n^2x)e^{-n^2x^2}(1-2n^2x^2) + n^2e^{-n^2x^2}(-4n^2x^2)$

so that $\quad d^2y/dx^2 = -2n^4 x\, e^{-n^2x^2}(3-2n^2x^2)$

When $x = 1/n\sqrt{2}$, $\quad d^2y/dx^2 = -4n^4(1/n\sqrt{2})\, e^{-(1/2)} < 0,$

showing that y is maximum when $x = 1/n\sqrt{2}$ and hence from (2), we have

$$\text{the maximum value of } y = n^2 \times (1/n\sqrt{2}) \times e^{-(1/2)} = n/\sqrt{2e} \qquad ...(4)$$

Now, $M_n = \sup\{|S_n(x) - S(x)| : x \in [0, 1]\} = \sup\{|y| : x \in [0, 1]\} = n/\sqrt{2e}$, using (4)

Hence, $M_n \to \infty$ as $n \to \infty$. Since M_n does not tend to 0 as $n \to \infty$, it follows that $< S_n(x) >$ is not uniformly convergent on [0, 1] and therefore the given series $\sum_{n=1}^{\infty} u_n$ is not uniformly convergent.

Here $x\, (= 1/n\sqrt{2}) \to 0$ as $n \to \infty$ and so $x = 0$ is a point of non-uniform convergence of the given series.

Ex. 2. *Show that the sequence* $< f_n >$ where $f_n(x) = nx/(1 + n^2x^2)$ is point-wise convergent but it is not uniformly convergent on any interval containing zero.

(Agra 2008; Delhi 2009, 12; G.N.D.U. Amritsar 2002; Kanpur 2007; Meerut 2006)

Sol. Given $\quad f_n(x) = nx/(1 + n^2x^2). \qquad ...(1)$

For $x \neq 0$, $\quad f_n(x) = \dfrac{1/nx}{1+(1/nx)^2} \to 0 \quad$ as $\quad n \to \infty$

Again, from (1) $\quad f_n(0) = 0, \ \forall\, n \in \mathbf{N}$

Thus, $\quad \lim_{n\to\infty} f_n(x) = 0, \ \forall\, x \in \mathbf{R}.$

Therefore, the sequence $< f_n >$ converges point-wise to zero on $\mathbf{R}$ and $f(x) = 0$ is the limit function of $< f_n >$ on $\mathbf{R}$

We now proceed to show that $< f_n >$ is not uniformly convergent on any interval containing zero.

Here, $\quad f(x) = \lim_{n\to\infty} f_n(x) = \lim_{n\to\infty} \dfrac{nx}{1+n^2x^2} = \lim_{n\to\infty} \dfrac{(x/n)}{(1/n^2)+x^2} = 0, \ \forall\, x \in \mathbf{R}$

Hence $\quad |f_n(x) - f(x)| = \left|\dfrac{nx}{1+n^2x^2} - 0\right| = \left|\dfrac{nx}{1+n^2x^2}\right| \qquad ...(2)$

Let $\quad y = (nx\, /\, (1 + n^2x^2) \qquad ...(3)$

From (3), $\quad \dfrac{dy}{dx} = \dfrac{n \times (1+n^2x^2) - (nx \times 2n^2x)}{(1+n^2x^2)^2} = n \times \dfrac{1-n^2x^2}{(1+n^2x^2)^2} \qquad ...(4)$

For maximum and minimum value of y, $dy/dx = 0$ giving $x = 1/n$

From (3), $\quad \dfrac{d^2y}{dx^2} = \dfrac{-2n^3x(1+n^2x^2)^2 - 4n^3x(1+n^2x^2)(1-n^2x^2)}{(1+n^2x^2)^4} = -\dfrac{4n^3x(3n^2x^2-1)}{(1+n^2x^2)^3}$

When $x = 1/n$, $d^2y/dx^2 = -n^2 < 0$ so that y is maximum when $x = 1/n$

Hence, from (3), the maximum value of y is given by $y = 1/2$ ...(5)

Also, $x\ (= 1/n) \to 0$ as $n \to \infty$. Consider an interval $[a, b]$ containing 0.

Here, $M_n = \sup\{| f_n(x) - f(x) | : x \in [a, b]\} = \sup \{| y | : x \in [a, b]\} = 1/2$, using (5)

Therefore, $M_n \to 1/2$ as $n \to \infty$. Since M_n does not tend to 0 as $n \to \infty$, it follows that the sequence $< f_n >$ is not uniformly convergent in any interval containing zero.

Ex. 3. *Let $< f_n >$ be a sequence of continuous functions on $[a, b]$ and suppose $f_n \to f$ uniformly on $[a, b]$. If $\phi_n(x) = \int_a^x f_n(t)\, dt$ for $a \le x \le b$, then prove that the sequence $< \phi_n(x) >$ is uniformly continuous on $[a, b]$.*

Sol. Since $f_n \to f$ uniformly on $[a, b]$, given $\varepsilon > 0$, there exists an $m \in \mathbf{N}$ such that

$$| f_n(t) - f(t) | < \varepsilon/(b - a) \text{ for } \forall n \ge m \text{ and } \forall t \in [a, b] \quad \text{... (1)}$$

Now,
$$\left| \int_a^x f_n(t)\, dt - \int_a^x f(t)\, dt \right| = \left| \int_a^x \{ f_n(t) - f(t)\}\, dt \right|$$

Thus,
$$\left| \int_a^x f_n(t)\, dt - \int_a^x f(t)\, dt \right| \le \int_a^x |f_n(t) - f(t)|\, dt \quad \text{... (2)}$$

Using (1), (2) reduces to

$$\left| \int_a^x f_n(t)\, dt - \int_a^x f(t)\, dt \right| < \frac{\varepsilon}{(b-a)} \times (x - a) < \varepsilon \quad \forall n \ge m \text{ and } \forall x \in [a, b]$$

From this it follows that the sequence $< \phi_n(x) >$ converges uniformly on $[a, b]$.

MISCELLANEOUS RESULTS AND PROBLEMS BASED ON THE ENTIRE BOOK

Ex. 1. *Show that the infimum of the set* $S = \{1/(n+1) : n \in \mathbf{N}\}$ is zero. **(Delhi 2012)**

Sol. Clearly 0 is a lower bound for the given set because $x \geq 0$ for all $x \in S$. Any real number greater than 0 cannot be a lower bound for S as shown below:

Let $l > 0$. However small l may be, there exists $n \in \mathbf{N}$ such that $1/(n+1) < l$. Thus there exists $1/(n+1) \in S$ such that $1/(n+1) < l$. Therefore l is not a lower bound for S. Thus, 0 is a lower bound for S and no real number greater than 0 can be a lower bound for S. Hence, by definition, inf S = 0. Observe that here $0 \notin S$.

Ex. 2. *State and prove the density theorem.* **(Delhi B.Sc. (Hons.) I, 2011, 2016)**

Sol. Statement of the density theorem. If *x and y are any two real numbers with* $x < y$, *then there exists a rational number r such that* $x < r < y$.

For proof, refer theorem I, page 2.21.

Ex. 3. *If* $A = \{1, 1/2, 1/3, ..., 1/100\}$, *then*

(*a*) D(A) = A (*b*) D(A) = ϕ (*c*) D(A) $\neq \phi$ (*d*) *None of these.* **(Agra 2012)**

Sol. Ans. (*b*). Note that given set A is a finite set. Hence proceed as in illustration 6 on page 3.6 to show that D(A) = ϕ.

Ex. 4. *If A is the set of rational numbers, then find the set of boundary points of A.* **(Agra 2012)**

Sol. For definition of the set of boundary points, refer Art. 3.12. In what follows we shall use the following alternative definition of the set of boundary points.

Alternative definition of the set of boundary points of a set A The set of boundary points of a set A is a set of all those elements which belong neither to the interior of A nor to the interior of its complement X ~ A. Symbolically, we write

$$b(A) = \partial A = (X \sim A^\circ) \cup (X \sim A)^\circ$$

In the given problem, A = $\mathbf{Q}$ = set of all rational numbers.

Hence, by the above alternative definition, are have

$$\partial A = \partial \mathbf{Q} = (\mathbf{R} \sim \mathbf{Q}^\circ) \cup (\mathbf{R} - \mathbf{Q})^\circ \qquad ...(1)$$

Since $\mathbf{Q}$ is not nbd of any of its points and so $\mathbf{Q}^\circ = \phi$. Similarly $(\mathbf{R} - \mathbf{Q})^\circ = \phi$.

Hence, (1) $\Rightarrow$ $\partial A = (\mathbf{R} - \phi) \cup \phi = \mathbf{R}$ = set of all real numbers.

Ex. 5. *Let* $x_1 > 1$ *and* $x_{n+1} = 2 - (1/x_n)$ *for* $\forall\, n \in \mathbf{N}$. *Show that the sequence* $< x_n >$ *is bounded and monotone. Find limit* x_n. **(Delhi B.Sc. (Hons.) I, 2016)**

Sol. Given that $x_1 > 1$ and $x_{n+1} = 2 - (1/x_n), \quad \forall\, n \in \mathbf{N}$ ...(1)

Using mathematical induction, we shall prove that $x_n > 1, \ \forall\, n \in \mathbf{N}$ ...(2)

Since $x_1 > 1$, therefore (2) is true for $n = 1$

Let (2) be true for some positive integer m, that is, let $x_m > 1$ ...(3)

Now, (3) $\Rightarrow 1/x_m < 1 \Rightarrow -(1/x_m) > -1 \Rightarrow 2 - (1/x_m) > 1 \Rightarrow x_{m+1} > 1$, by (1)

showing that if (2) is true for $n = m$, then (2) is also true for $n = m + 1$. Also, (2) is true for $n = 1$. Hence, by mathematical induction, (2) is true $\forall\, n \in \mathbf{N}$.

Now, from (1), $x_1 > 1$ and $x_2 = 2 - (1/x_1)$. Therefore, we have

$x_1 - x_2 = x_1 - (2 - 1/x_1) = (x_1 - x_2)^2/x_1 > 0$ and so $x_2 < x_1$ ...(4)

Using mathematical induction, we now prove that $x_{n+1} < x_n, \forall n \in \mathbf{N}$ ...(5)

From (4), it follows that (5) is true for $n = 1$.

Let (5) be true for some positive integer m, *i.e.*, let $x_{m+1} < x_m$ ...(6)

Now, (6) $\Rightarrow 1/x_{m+1} > 1/x_m \Rightarrow -(1/x_{m+1}) < -(1/x_m) \Rightarrow 2 - (1/x_{m+1}) < 2 - (1/x_m)$

Thus, $x_{m+2} < x_{m+1}$, using (1)

showing that if (5) is true for $n = m$, then (5) is also true for $n = m + 1$. Also, (5) is true for $n = 1$. Hence, by mathematical induction, it follows that (5) is true $\forall n \in \mathbf{N}$

From (2) and (5), it follows that $< x_n >$ is monotonically decreasing sequence and bounded below and hence $< x_n >$ is convergent

Let $\lim x_n = l$ so that $\lim x_{n+1} = l$

Then, $x_{n+1} = 2 - (1/x_n) \Rightarrow \lim x_{n+1} = 2 - (1/\lim x_n) \Rightarrow l = 2 - (1/l)$

Thus, $l^2 - 2l + 1 = 0$ or $(l + 1)^2 = 0$ giving $l = 1$ so that $\lim x_n = 1$.

Ex. 6. *Let $< s_n >$ be a sequence that converges. Show that if $s_n \geq a$ for all but finitely many n, then show that $\lim (s_n) \geq a$.* **(Delhi 2011)**

Ex. 7. *Let $< s_n >$ be a sequence in* $\mathbf{R}$. *Prove that $\lim s_n = 0$ if and only if $\lim | s_n | = 0$.* **(Delhi 2011)**

Sol. $\lim s_n = 0 \Leftrightarrow$ for a given $\varepsilon > 0$, there exists a positive integer m such that

$| s_n - 0 | < \varepsilon, \forall n \in m$, *i.e.*, $| s_n | < \varepsilon, \forall n \in m$

$\Leftrightarrow$ for a given $\varepsilon > 0$, there exists a positive integer m such that $| |s_n| - 0 | < \varepsilon$

$\Leftrightarrow \lim | s_n | = 0$.

Ex. 8. *Using squeze theorem or otherwise, find the limit of the sequence $< n^{1/n^2} >$.* **(Delhi 2011)**

Hint: Use $\dfrac{1}{n} - \dfrac{1}{n+1} = \dfrac{1}{n(n+1)} < \dfrac{1}{n^2} < \dfrac{1}{n}$ and $\lim n^{1/n} = 1$. **Ans.** 1.

Ex. 9. $\lim_{n \to \infty} (1 + 1/n)^n$ *equals* (*a*) $1/e$ (*b*) 0 (*c*) 1 (*d*) *e*. **(Agra 2012)**

Sol. Ans. (*d*). Refer solved Ex. 3, page 5.41

Ex. 10. $\lim_{n \to \infty} \log \dfrac{2n}{2n+3}$ *equals* (*a*) $\log \dfrac{3}{2}$ (*b*) $\log 3$ (*c*) $\log 0$ (*d*) $\log \dfrac{2}{3}$. **(Agra 2012)**

Sol. Ans. (*c*). $\lim_{n \to \infty} \log \dfrac{2n}{2n+3} = \lim_{n \to \infty} \dfrac{2}{2 + (3/n)} = \log 1 = 0$, as $\lim_{n \to \infty} \dfrac{3}{n} = 0$

Ex. 11. *Let $X = < x_n >$ be a sequence of real numbers that converges to x and suppose that $x_n \geq 0$. Then prove that the sequence $< \sqrt{x_n} >$ of positive square roots converges and $\lim (\sqrt{x_n}) = \sqrt{x}$* **(Delhi B.Sc. (Hons.) I, 2011, 2017)**

Sol. From theorem II on page 5.20, it follows that $x = \lim x_n \geq 0$. Two cases arise.

Case (*i*). Let $x = 0$. Let $\varepsilon > 0$ be given. Since $x_n \to 0$, there exists a positive integer m such that if $n \geq m$, then

$0 \leq x_n = x_n - 0 < \varepsilon^2$ so that $0 \leq \sqrt{x_n} < \varepsilon, \forall n \in m$. ...(1)

Since $\varepsilon > 0$ is arbitrary, (1) $\Rightarrow \sqrt{x_n} \to 0$.

Case (*ii*). Let $x > 0$. Then $\sqrt{x} > 0$ and we obtain

$$\sqrt{x_n} - \sqrt{x} = \frac{(\sqrt{x_n} - \sqrt{x})(\sqrt{x_n} + \sqrt{x})}{\sqrt{x_n} + \sqrt{x}} = \frac{x_n - x}{\sqrt{x_n} + \sqrt{x}} \quad ...(2)$$

Since $\sqrt{x_n} + \sqrt{x} \geq \sqrt{x} > 0$, (2) yields

$$\left|x_n - \sqrt{x}\right| \leq (1/\sqrt{x}) \times \left|x_n - x\right| \quad ...(3)$$

But $x_n \to x$ and hence (3) $\Rightarrow$ $\sqrt{x_n} \to \sqrt{x}$.

Ex. 12. *Let $< s_n >$ and $< t_n >$ be the following sequences that repeat in cycles of four:* $< s_n > = < 0, 1, 2, 1, 0, 1, 2, 1, 0, 1, 2, 1, ... >$ and $< t_n > = < 2, 1, 1, 0, 2, 1, 1, 0, 2, 1, 1, 0, ... >$. Find lim inf$(s_n + t_n)$ and lim sup$(s_n + t_n)$. **(Delhi 2011)**

Sol. By definition, $< s_n + t_n > = < 2, 2, 3, 1, 2, 2, 3, 1, 2, 2, 3, 1, ... >$

Hence, the set of limit points of the sequence $< s_n + t_n >$ is $\{2, 3, 1\}$.

Therefore, by definition, $\liminf(s_n + t_n) = 1$ and $\limsup(s_n + t_n) = 3$.

Ex. 13. *Show that the sequence $< a_n > = < (-1)^n >$ does not converge.* **(Delhi 2011)**

Sol. Let, if possible, $< a_n >$ be convergent. Taking $\varepsilon = 1/2$, there must exist a positive integer m such that

$$| a_n - a_m | < 1/2, \qquad \forall\, n \geq m \quad ...(1)$$

Now, on taking $n = m + 1$, (1) reduces to

$| a_n - a_m | = | a_{m+1} - a_m | = | (-1)^{m+1} - (-1)^m | = | (-1)^m (-1-1) | = | (-1)^m | \times | -2 | = 2$, which contradicts (1). Hence, our hypothesis is wrong and therefore the sequence $< (-1)^n >$ cannot converge.

Ex. 14. *Show that* $\lim (\sqrt{n} + 7) = +\infty$ **(Delhi 2011)**

Sol. Let $\Delta > 0$ be given and let $f_n = \sqrt{n} + 7$. Then, we have

$$f_n > \Delta \quad \Rightarrow \quad \sqrt{n} + 7 > \Delta \quad \Rightarrow \quad \sqrt{n} > \Delta - 7 \quad \Rightarrow \quad n > (\Delta - 7)^2$$

Let m be a positive integer such that $m > (\Delta - 7)^2$, then

$$f_n > \Delta, \ \forall\, n \in m \quad \Rightarrow \quad \lim_{n \to \infty} f_n = +\infty \quad \Rightarrow \quad \lim_{n \to \infty} (\sqrt{n} + 7) = +\infty$$

Ex. 15. *Let $X = < x_n >$ be a bounded increasing sequence. Show that X is convergent and* lim $x_n = sup\{x_n : n \in \mathbf{N}\}$. **(Delhi 2011)**

Sol. Since $< x_n >$ is bounded, there exists a real number K such that $x_n \leq K$, $\forall\, n \in \mathbf{N}$. According to the completeness property, the supremum $x^* = \sup\{x_n : n \in \mathbf{N}\}$ exists in $\mathbf{R}$. We now proceed to prove that $x^* = \lim x_n$.

If $\varepsilon > 0$ is given, then $x^* - \varepsilon$ is not an upper bound of the set $\{x_n : n \in \mathbf{N}\}$, and therefore there exists a member x_m of the set such that $x^* - \varepsilon < x_m$. Since the given sequence $< x_n >$ is increasing, therefore $x_m \leq x_n$ whenever $n \geq m$ and hence we have

$$x^* - \varepsilon < x_m \leq x_n \leq x^* + \varepsilon, \ \forall\, n \in m$$

so that

$$| x_n - x^* | < \varepsilon, \ \forall\, n \in m \quad ...(1)$$

Since $\varepsilon > 0$ is arbitrary, (1) $\Rightarrow$ $< x_n >$ converges to x^*.

Ex. 16. (Monotone convergence theorem). A monotone sequence of real numbers is convergent if it is bounded. Further,

(*i*) If the sequence $< x_n >$ is a bounded increasing, then $\lim x_n = \sup\{x_n : n \in \mathbf{N}\}$

(*ii*) If the sequence $< y_n >$ is a bounded decreasing, then $\lim y_n = \inf\{y_n : n \in \mathbf{N}\}$

Sol. Left as an exercise for the reader.

Ex. 17. *Find the convergence of the series whose nth term u_n is given by*
$u_n = (2n^2 - 1)^{1/3}/(3n^3 + 2n + 5)^{1/4}$ **(Agra 2011)**

Sol. Here $u_n = \dfrac{n^{2/3}(2 - 1/n^2)^{1/3}}{n^{3/4}(3 + 2/n^2 + 5/n^3)^{1/4}} = \dfrac{1}{n^{1/12}} \times \dfrac{(2-1/n^2)^{1/3}}{(3+2/n^2+5/n^3)^{1/4}}$

Let $v_n = 1/n^{1/12}$. Then, we have $\lim_{n\to\infty} (u_n / v_n) = 2^{1/3}/3^{1/4}$,

which is finite and non-zero. Again, $\Sigma v_n = \Sigma(1/n^{1/12})$ is divergent (refer λ-series on page 6.6). Therefore, the given series Σu_n is also divergent.

Ex. 18. *Test for convergence:* $\displaystyle\sum_{n=4}^{\infty} \frac{1}{n(\log n)(\log\log n)}$ **(Delhi 2011)**

Sol. Let $u_n = 1/\{n(\log n)(\log\log n)\} = u(n), \ \forall\, n \geq 4$

Hence $u(x) = 1/\{x(\log x)(\log\log x)\}, \ \forall\, x \geq 4$

For $x \geq 4$, $f(x)$ is non-negative and decreasing function. Hence, by Cauchy's integral test,

$\displaystyle\sum_{n=4}^{\infty} u_n$ and $\displaystyle\int_4^{\infty} u(x)\,dx$ converge or diverge together. Now, we have

$$\int_4^t u(x)\,dx = \int_4^t \frac{dx}{x(\log x)(\log\log x)} = \int_4^t \frac{1/(x\log x)}{\log(\log x)}\,dx = [\log(\log x)]_4^t = \log(\log t) - \log(\log 4)$$

Hence, $\displaystyle\int_4^{\infty} u(x)\,dx = \lim_{t\to\infty}\int_4^t u(x)\,dx = \lim_{t\to\infty}[\log(\log t) - \log(\log 4)] = \infty,$

showing that $\displaystyle\int_4^{\infty} u(x)\,dx$ is divergent and hence the series $\displaystyle\sum_{n=4}^{\infty} u_n$ is also divergent. Thus, the given series diverges.

Ex. 19. *Shcw that* $\displaystyle\sum_{n=1}^{\infty} \frac{1}{2^n + n}$ *is not convergent.* **(Delhi 2011)**

Ex. 20. *Give an example of a convergent series Σa_n for which Σa_n^2 diverges. Also, give an example of a divergent series Σa_n for which Σa_n^2 converges. Justify your answer.* **(Delhi 2011)**

Sol. Consider the series Σa_n where $a_n = (-1)^{n+1}/\sqrt{n}$. Then by Leibnitz test, Σa_n is convergent. But $\Sigma a_n^2 = \Sigma(1/n)$ is divergent.

Second part: Consider the series Σa_n where $a_n = 1/n$. Then, Σa_n is divergent whereas $\Sigma a_n^2 = \Sigma 1/n^2$ is convergent.

Ex. 21. *Show that* $\Sigma\{(-1)^n n!\}/2^n$ is divergent. **(Delhi 2011)**

Ex. 22. *Discuss the type of discontinuity, if any, of the function on* $[0, 1]$ *as follows:* $f(0) = 0$, $f(x) = (1/2) - x$, $0 < x < 1/2$, $f(1/2) = 1/2$, $f(x) = (3/2) - x$ if $1/2 < x < 1$, $f(1) = 1$.

(Delhi B.A. (Prog.) III 2011)

Sol. From the given function, we have

$$f(0) = 0 \quad \text{and} \quad f(0+0) = \lim_{h\to 0} f(0+h) = \lim_{h\to 0}(½ - h) = ½ \quad \text{...(1)}$$

$$\left.\begin{aligned} f(½ - 0) &= \lim_{h\to 0} f(½ - h) = \lim_{h\to 0}\{½ - (½ - h)\} = 0, \quad f(½) = ½ \\ \text{and} \quad f(½ + 0) &= \lim_{h\to 0} f(½ + h) = \lim_{h\to 0}\{½ - (½ + h)\} = 0 \end{aligned}\right\} \quad \text{...(2)}$$

Also, $\quad f(1-0) = \lim_{h \to 0} f(1-h) = \lim_{h \to 0} \{3/2 - (1-h)\} = ½ \quad$ and $\quad f(1) = 1$ ...(3)

From (1), since $f(0+0) \neq f(0)$, it follows that f has discontinuity of the first kind from right at $x = 0$.

From (2), since $f(½ - 0) \neq f(½)$ and $f(½ + 0) \neq f(½)$, it follows that there is discontinuity of the first kind from both sides at $x = ½$.

From (3), since $f(1-0) \neq f(1)$, it follows that there is discontinuity of the first kind from the left at $x = 1$.

Ex. 23. *Which of the following functions are uniformly continuous on their domain: (i) $f(x) = x$, $0 \le x < \infty$ (ii) $f(x) = x^2$, $0 \le x \le 5$ (iii) $f(x) = 1/x$, $0 < x \le 1$.* **(Delhi B.A. (Prog.) III 2012)**

Sol. (*i*) Let $x_1, x_2 \in [0, \infty[$. Then, $\quad |f(x_2) - f(x_1)| = |x_2 - x_1|$, as $f(x) = x$.

Let $\varepsilon > 0$ be given. Then, we have

$$|f(x_2) - f(x_1)| = \varepsilon \qquad \text{whenever} \qquad |x_2 - x_1| < \varepsilon, \quad \forall\, x_1, x_2 \in [0, \infty[\qquad \text{...(1)}$$

Hence, if we choose $\delta = \varepsilon$, (1) reduces to

$$|f(x_2) - f(x_1)| < \varepsilon \qquad \text{whenever} \qquad |x_2 - x_1| < \delta, \quad \forall\, x_1, x_2 \in [0, \infty[\qquad \text{...(2)}$$

From (2), it follows that f is uniformly continuous on $[0, \infty[$.

(*ii*) Proceed exactly as in Ex. 2, page 8.45 by taking $a = 0$ and $b = 5$.

(*iii*) Refer page 8.44 for solution.

Ex. 24. *Prove that (a)* $\lim_{x \to \infty} (1 + 1/x)^x = e$ *(b)* $\lim_{x \to 0} (1 + x)^{1/x} = e$ **(G.N.D.U. Amritsar 2010)**

Sol. (*a*) Since $x \to \infty$, we can assume x to be positive. Then there exists a natural number n such that $n \le x < n + 1$. Now, we have

$$n \le x < n+1 \quad \Rightarrow \quad \frac{1}{n+1} < \frac{1}{x} \le \frac{1}{n} \quad \Rightarrow \quad 1 + \frac{1}{n+1} < 1 + \frac{1}{x} \le 1 + \frac{1}{n}$$

Thus, $$\left(1 + \frac{1}{n+1}\right)^n < \left(1 + \frac{1}{x}\right)^x \le \left(1 + \frac{1}{n}\right)^{n+1}, \quad \text{since } n \le x < n+1 \qquad \text{...(1)}$$

From Ex. 3, page 5.41, we know that

$$\lim_{n \to \infty} \left(1 + \frac{1}{n}\right)^n = \lim_{n \to \infty} \left(1 + \frac{1}{n+1}\right)^{n+1} = e \qquad \text{...(2)}$$

From our choice, when $x \to \infty$, n will also tend to ∞. Hence, letting $n \to \infty$, (1) yields

$$\frac{\lim_{n \to \infty} \left(1 + \dfrac{1}{n+1}\right)^{n+1}}{\lim_{n \to \infty} \left(1 + \dfrac{1}{n+1}\right)} \le \lim_{x \to \infty} \left(1 + \frac{1}{x}\right)^x \le \lim_{n \to \infty} \left(1 + \frac{1}{n}\right)^n \left(1 + \frac{1}{n}\right) \qquad \text{...(3)}$$

Using (2), (3) yields $$e \le \lim_{x \to \infty} (1 + 1/x)^x \le e \Rightarrow \lim_{x \to \infty} (1 + 1/x)^x = e \qquad \text{...(4)}$$

(*b*) Proceed as in part to get $$\lim_{y \to \infty} (1 + 1/y)^y = e \qquad \text{...(i)}$$

Putting $y = 1/x$ in (*i*) and noting that $x \to 0$ as $y \to \infty$, we obtain

$$\lim_{x \to 0} (1 + x)^{1/x} = e$$

Ex. 25. *Discuss the continuity of the function at* $x = 0$

$$f(x) = \begin{cases} (1+7x)^{1/x}, \text{ when } x \neq 0 \\ e^7, \text{ when } x = 0 \end{cases}$$ **(Agra 2012)**

Sol. From the given function, we have $f(0) = e^7$ ...(1)

Now, $\lim_{x \to 0} f(x) = \lim_{x \to 0} (1+7x)^{1/x} = \lim_{t \to 0} (1+t)^{7/t}$, on putting $7x = t$

$$= \left\{ \lim_{t \to 0} (1+t)^{1/t} \right\}^7 = e^7, \text{ since from Ex. 24 (b), } \lim_{x \to 0} (1+x)^{1/x} = e$$

Thus, $\lim_{x \to 0} f(x) = e^7 = f(0)$, using (1)

Hence, by definition, $f(x)$ is continuous at $x = 0$.

Ex. 26. *Evaluate* $\lim_{x \to 1} (x - [x])$, *where* $[x]$ *is the greatest integer function.* **(Agra 2011)**

Sol. Let $f(x) = x - [x]$. Then, we have

$$f(1-0) = \lim_{h \to 0} f(1-h) = \lim_{h \to 0} (1-h-[1-h]) = \lim_{h \to 0} (1-h-0) = 1$$

and $$f(1+0) = \lim_{h \to 0} f(1+h) = \lim_{h \to 0} (1+h-[1+h]) = \lim_{h \to 0} (1+h-1) = 0$$

Since $f(1-0) \neq f(1+0)$, hence $\lim_{x \to 1} f(x)$, *i.e.,* $\lim_{x \to 1} (x-[x])$ does not exist.

Ex. 27. (*a*) *Let* $f(x) = x$ *on* $[0, 1]$. *Calculate* $\underline{\int_0^1} x\, dx$ *and* $\overline{\int_0^1} x\, dx$ *by dissecting* $[0, 1]$ *into n equal parts and show that* $f \in R[0, 1]$. **(Meerut 2012)**

Sol. Let $P = \{0, 1/n, 2/n, ..., (r-1)/n, r/n, ..., n/n = 1\}$. Then, the rth subinterval of the partition P $= I_r = [(r-1)/n, r/n]$, where $r = 1, 2, 3, ..., n$. If δ_r denotes the length of I_r, then $\delta_r = r/n - (r-1)/n = 1/n$. Let $f(x) = x$ and let M_r and m_r be respectively the supremum and infimum of f in I_r. Since f in increasing on $[0, 1]$, we have

$$M_r = r/n \quad \text{and} \quad m_r = (r-1)/n, \quad \text{since} \quad f(x) = x, \ \forall\, x \in [0, 1]$$

By definition, $$U(P, f) = \sum_{r=1}^{n} M_r\, \delta_r = \sum_{r=1}^{n} (r/n) \times (1/n) = (1/n^2) \times \sum_{r=1}^{n} r$$

Thus, $$U(P, f) = \frac{1}{n^2}(1+2+3+...+n) = \frac{1}{n^2} \times \frac{n(n+1)}{2} = \frac{n+1}{2n}$$

Again, $$L(P, f) = \sum_{r=1}^{n} m_r\, \delta_r = \sum_{r=1}^{n} \left(\frac{r-1}{n} \times \frac{1}{n} \right) = \frac{1}{n^2}\{1+2+3+...+(n-1)\} = \frac{1}{n^2} \times \frac{(n-1)n}{2} = \frac{n-1}{2n}$$

Now, by definition, we have

$$\underline{\int_0^1} f(x)\, dx = \underline{\int_0^1} x\, dx = \lim_{n \to \infty} \text{U(P, f)} = \lim_{n \to \infty} (1/2 + 1/2n) = 1/2$$

and $$\overline{\int_0^1} f(x)\, dx = \overline{\int_0^1} x\, dx = \lim_{n \to \infty} \text{L(P, f)} = \lim_{n \to \infty} (1/2 - 1/2n) = 1/2$$

Since $\underline{\int_0^1} x\, dx = \overline{\int_0^1} x\, dx$, it follows that $f \in$ R[0, 1].

Ex. 27. *(b) Find L(P, f) and U(P, f) for a partition P of [0, 1] for the function defined by*

$$f(x) = \begin{cases} 1, \text{ if } x \in [0,1] \text{ and } x \text{ is rational} \\ -1, \text{ if } x \in [0,1] \text{ and } x \text{ is irrational} \end{cases}$$

and show that $f \notin R[0, 1]$. **(Meerut 2012)**

Sol. Since $-1 \le f(x) \le 1,\ \forall\ x \in [0, 1]$, it follows that f is bounded on [0, 1]. Let $P = \{0 = x_0, x_1, x_2, \ldots, x_n = 1\}$ be any partition of [0, 1] and let its subinterval be $I_r = [x_{r-1}, x_r]$ for $r = 1, 2, 3, \ldots, n$. Here δ_r = length of $I_r = x_r - x_{r-1}$.

Let M_r and m_r be respectively the supremum and infimum of the function f in I_r. Since rational and irrational points are everywhere dense, it follows that every subinterval I_r will contain rational and irrational numbers. Hence by the given definition of the function f, we have

$$m_r = -1 \quad \text{and} \quad M_r = 1, \quad \text{for} \quad r = 1, 2, 3, \ldots, n.$$

By definition, $$U(P, f) = \sum_{r=1}^{\infty} M_r\,\delta_r = \sum_{r=1}^{\infty} \delta_r = \sum_{r=1}^{n} (x_r - x_{r-1})$$

$$= (x_1 - x_0) + (x_2 - x_1) + \ldots + (x_n - x_{n-1}) = x_n - x_0 = 1 - 0 = 1$$

and $$L(P, f) = \sum_{r=1}^{\infty} m_r\,\delta_r = -\sum_{r=1}^{\infty} \delta_r = -(1-0) = -1 \text{ as before}$$

Now, $\overline{\int_0^1} f(x)\,dx = \lim_{n\to\infty} U(P, f) = 1$ and $\underline{\int_0^1} f(x)\,dx = \lim_{n\to\infty} L(P, f) = -1$

Since $\overline{\int_0^1} f(x)\,dx \ne \underline{\int_0^1} f(x)\,dx$ it follows that $f \notin R[0, 1]$

Ex. 27. (*c*) Show that the greatest integer function $[x]$ is integrable on [0, 4] and $\int_0^4 [x]\,dx = 6$.

Sol. By definition of the greatest integer function $[x]$, we have

$$f(x) = [x] = \begin{cases} 0, \text{ if } 0 \le x < 1 \\ 1, \text{ if } 1 \le x < 2 \\ 2, \text{ if } 2 \le x < 3 \\ 3, \text{ if } 3 \le x < 4 \end{cases} \qquad \ldots(1)$$

From (1), it follows that f is bounded and it has four points of finite discontinuity at $x = 1, 2, 3$ and 4. Since the points of discontinuity on [0, 4] are finite in number, therefore f is integrable on [0, 4]. Now, using (1), we have

$$\int_0^4 f(x)\,dx = \int_0^1 f(x)\,dx + \int_1^2 f(x)\,dx + \int_2^3 f(x)\,dx + \int_3^4 f(x)\,dx$$

$$= \int_0^1 (0)\,dx + \int_1^2 (1)\,dx + \int_2^3 (2)\,dx + \int_3^4 (3)\,dx$$

Thus, $$\int_0^4 [x]\,dx = 0 + (2-1) + 2(3-2) + 3(4-3) = 6.$$

Ex. 27. (*d*) *Let* $f(x) = 1$ *for* $x \in [0, 1]$ *and* $f(x) = 2$ *for* $x \in [1, 2]$. *Prove that f is integrable and evaluate* $\int_0^2 f(x)\,dx$. **(Agra 2011)**

Sol. Given that $$f(x) = \begin{cases} 1, \text{ if } 0 \le x < 1 \\ 2, \text{ if } 1 \le x < 2 \end{cases} \qquad \ldots(1)$$

From (1), it follows that f is bounded and it has only one point of discontinuity at $x = 1$ in the interval [0, 2]. Hence, f is integrable on [0, 2]. Using (1), we have

$$\int_0^2 f(x)\,dx = \int_0^1 f(x)\,dx + \int_1^2 f(x)\,dx = \int_0^1 (1)\,dx + \int_1^2 (2)\,dx = 3.$$

Ex. 28. *Let g be a function defined on [a, b] as follows:*

$$g(x) = \begin{cases} 0, & \text{if } x \in [a, b] \text{ and } x \text{ is rational} \\ 1, & \text{if } x \in [a, b] \text{ and } x \text{ is irrational} \end{cases}$$

Then $\int_a^b g(x)\,dx$ (*a*) *does not exists* (*b*) *exists*

(*c*) $\underline{\int_a^b} g(x)\,dx$ *exists* (*d*) $\overline{\int_a^b} g(x)\,dx$ *exists* **(Agra 2011)**

Sol. Ans. (*a*). Refer solved example 5, page 13.12

Ex. 29. *Let* $f(x) = (\sin x)/x$, *if* $x \neq 0$ and $f(x) = 1$, *if* $x = 0$. *Prove that* $\int_0^1 f(x)\,dx$ *exists.*

Sol. Hence $(\sin x)/x$ is continuous for $x \neq 0$. Again, we have

$$\lim_{x \to 0+} \frac{\sin x}{x} = \lim_{x \to 0-} \frac{\sin x}{x} = 1 = f(0),$$

showing that f is continuous in [0, 1]. Hence f is Riemann integrable in [0, 1]. Therefore $\int_0^1 f(x)\,dx$ exists.

Ex. 30. (*a*) *Define zero measure of a set of real numbers with some examples.*

Sol. Zero measure of a set of real numbers. Definition.

Let S be a set of real numbers. If S can be covered by a finite or countably infinitely number of open intervals, length of which is less than an arbitrary positive number ε, then S is said to have *measure zero*.

Some examples of sets having measure zero

Example 1. *Every finite set has measure zero*

Proof. Let S = $\{a_1, a_2, ..., a_n\}$ be a finite set and $\varepsilon > 0$ be given. Then, $a_1, a_2, ..., a_n$ respectively can be covered by the following open intervals

$$\left(a_1 - \frac{\varepsilon}{2(n+1)}, a_1 + \frac{\varepsilon}{2(n+1)}\right), \left(a_2 - \frac{\varepsilon}{2(n+1)}, a_2 + \frac{\varepsilon}{2(n+1)}\right), ..., \left(a_n - \frac{\varepsilon}{2(n+1)}, a_n + \frac{\varepsilon}{2(n+1)}\right)$$

Hence, the total length of the above n intervals = $\{n/(n+1)\} \times \varepsilon < \varepsilon$, showing that the finite set S has measure zero.

Example 2. *Every countable set has measure zero.*

Proof. Let S = $\{a_1, a_2, ..., a_n, ...\}$ be a countable set. Let $\varepsilon > 0$ be given. Then, a_n can be covered by open interval $(a_n - \varepsilon/2^{n+2}, a_n + \varepsilon/2^{n+2})$ for each $n \in \mathbf{N}$. If l be the total length of these open itervals, then

$$l = \sum_{n=1}^{\infty} \frac{\varepsilon}{2^{n-1}} = \varepsilon\left(\frac{1}{2^2} + \frac{1}{2^3} + \frac{1}{2^4} + ...\right) = \varepsilon \times \frac{1/2^2}{1-(1/2)} = \frac{\varepsilon}{2} < \varepsilon,$$

showing that the given countable set S has measure zero.

Example 3. *Every infinite set S having finitely many limit points has measure zero.*

Proof. Let $\varepsilon > 0$ be given. Now, the finite number of limit points of S can be covered by finite number of open intervals, length of which is less that $\varepsilon/2$. Then, the remaining finitely many points of the set S can be covered by open intervals of total length less that $\varepsilon/2$. Combining all the above two types of open intervals, it follows that the set S can be covered by open intervals, length of which is iess than ($\varepsilon/2 + \varepsilon/2$), *i.e.*, ε. Therefore, by definition, the given set S has measure zero.

Ex. 30. **(*b*).** *Define continuous almost everywhere.*

(*c*). *State Lebesgue - Vitali theorem.*

(*d*). *Examine whether the following functions are Riemann integrable.*

(*i*) $f(x) = x^2$ if x is rational and $f(x) = 0$ if x is irrational.

(*ii*) $f(x) = \sin(1/x)$ for $0 < x \le 1$ and $f(0) = 2$.

(*iii*) $f(x) = n$, if $x = 1/n$ when $n = 1, 2, 3, \ldots$ and $f(x) = 0$ otherwise for $x \in [0, 1]$

(*iv*) $f(x) = 0$ for x in $[0, 1]$ and $f(x) = 1$ for $x \in \{0, 1/10, 2/10, 3/10, \ldots\}$

Sol. (*b*). Continuous almost everywhere. Definition.

A function $f : [a, b] \to \mathbf{R}$ is said to be continuous almost everywhere if the set of all points of discontinuity has measure zero.

(*c*). Lebesgue - Vitali theorem. A bounded function f is Riemann integrable over $[a, b]$ if and only if f is continuous almost everywhere on $[a, b]$

(*d*). Part (*i*): For each $x \in [0, 1]$, f is discontinuous so that f is discontinuous throughout the interval. Therefore, by Reimann -Vitali theorem, it follows that f is not Riemann integrable on $[0, 1]$

Part (*ii*): Proceed like example on page 8.18 and show that $x = 0$ is the only point of discontinuity of f in $[0, 1]$. Hence by definition f is continuous almost everywhere in $[0, 1]$. Therefore, by Lebesgue - Vitali theorem, f is Riemann integrable in $[0, 1]$.

Part (*iii*): From the given function f, it follows that f is continuous in $[0, 1]$ except at the points $\{0, 1, 1/2, 1/3, \ldots\}$. Since these points of discontinuity form a countable set and each countable set is a set of measure zero, it follows that f is continuous almost everywhere on $[0, 1]$. Therefore, by Lebesgue - Vitali theorem, it follows that f is integrable.

Part (*iv*): From the given function f, it follows that f is continuous in $[0, 1]$ except at the points $\{0, 1/10, 2/10, 3/10, \ldots\}$. Since these points of discontinuity form a countable set and each countable set is a set of measure zero, it follows that f is continuous almost everywhere on $[0, 1]$. Therefore, by Lebesgue - Vitali theorem, it follows that f is integrable.

Ex. 30. (*e*). *Let* $f(x) = 1$, *when* $x = 1/5, 2/5, 3/5, 4/5$ *and* $f(x) = 0$, *otherwise. Show that f is integrable on* $[0, 1]$ *and* $\int_0^1 f = 0$ **(Delhi 2012)**

Ex. 31. *Write comparison test for the convergence of improper integral of first kind and prove that the integral* $\int_0^\infty e^{-x^2}\,dx$ *is convergent.* **(Meerut 2012)**

Sol. For the first part, refer Art. 16.12, page 16.19

Test for convergence of $\int_0^\infty e^{-x^2}\,dx$.

Let $$\int_0^\infty e^{-x^2}\,dx = \int_0^1 e^{-x^2}\,dx + \int_1^\infty e^{-x^2}\,dx = I_1 + I_2, \text{ say} \qquad \ldots(1)$$

Here I_1 is a proper integral because its range of integration is finite and the integrand e^{-x^2} is bounded throughout the given interval. Therefore, I_1 is convergent.

We now proceed to test I_2 for convergence. Let $f(x) = e^{-x^2}$ and $g(x) = xe^{-x^2}$ *so that* $g(x)$ is positive throughout the interval $[1, \infty[$. Here $|f(x)| = e^{-x^2} \le xe^{-x^2}$ since $x \ge 1$. Therefore, $|f(x)| \le g(x)$ throughout the interval $[1, \infty[$. Now, we have

$$\int_1^\infty xe^{-x^2}\,dx = \lim_{t\to\infty}\int_1^t xe^{-x^2}\,dx = \lim_{t\to\infty}\left[-\frac{1}{2}e^{-x^2}\right]_1^t = \lim_{t\to\infty}\left[-\frac{1}{2}e^{-t^2}+\frac{1}{2}e^{-1}\right] = \frac{1}{2e},$$

which is finite. Hence $\int_1^\infty xe^{-x^2}\,dx$, *i.e.*, $\int_1^\infty g(x)\,dx$ is convergent and hence by comparison test, $\int_1^\infty f(x)\,dx$, *i.e.*, I_2 is convergent. Therefore from (1), it follows that $\int_0^\infty e^{-x^2}\,dx$ is convergent because it is the sum of two convergent integrals I_1 and I_2.

Ex. 32. *Show that the improper integral* $\int_0^\infty \frac{\sin x}{x^p}\,dx$ *is convergent for* $-1 < p < 2$.

(Delhi B.A. (Prog.) III 2012)

Sol. Let
$$\int_0^\infty \frac{\sin x}{x^p}\,dx = \int_0^1 \frac{\sin x}{x^p}\,dx + \int_1^\infty \frac{\sin x}{x^p}\,dx = I_1 + I_2, \text{ say} \qquad \ldots(1)$$

Discussion of convergent of I_1. If p is negative or zero, the given integral is a proper integra and hence I_1 is convergent when $p \le 0$. Again, when $p > 0$, the only point of infinite discontinuity is at $x = 0$. Let $f(x) = (\sin x)/x^p$ and $g(x) = 1/x^\mu$. Then, we have

$$\lim_{x\to 0+}\frac{f(x)}{g(x)} = \lim_{x\to 0+} x^{\mu-p}\sin x = \lim_{x\to 0+} x^{\mu-p+1}\left(\frac{\sin x}{x}\right) = \begin{cases}1, & \text{if } \mu - p + 1 = 0\\ 0, & \text{if } \mu - p + 1 > 0\\ \infty, & \text{if } \mu - p + 1 < 0\end{cases}$$

By taking $0 < \mu < 1$ and $\mu = p - 1$, we have $0 < p - 1 < 1$, *i.e*, $1 < p < 2$.

Then $\int_0^1 g(x)\,dx$ is convergent and hence $\int_0^1 f(x)\,dx$, *i.e.*, I_1 is convergent if $1 < p < 2$.

Discussion of convergence of I_2. Let $F(x) = \sin x$ and $G(x) = 1/x^p$. Then, we have

$$\left|\int_1^t F(x)\,dx\right| = \left|\int_1^t \sin x\,dx\right| = |\cos 1 - \cos t| \le |\cos 1| + |\cos t| \le 2,$$

showing that $\int_1^t F(x)\,dx$ is bounded for all $t \ge 1$. Also, $G(x)$ is bounded and monotonically decreasing function tending to zero as $x \to \infty$ when $p > 0$. Hence, by Dirichlet's test $\int_1^\infty F(x)G(x)\,dx$, *i.e.*, I_2 is convergent when $p > 0$.

Thus, from (1) it follows that the given integral is convergent when $1 < p < 2$.

Ex. 33. $\int_a^b f(x)\,dx$ *is called improper if (a) is* $-\infty$ (*b*) b is $+\infty$ (*c*) $f(x)$ is unbounded (*d*) All the above cases. **(Agra 2012)**

Sol. Ans. (*d*). Refer Art. 16.1

Ex. 34. *Define proper and improper integrals.* **(Agra 2009)**

Sol. Proper integral. Definition. The definite integral $\int_a^b f(x)\,dx$ is said to be a *proper integral* if the range of integration is finite and the integrand $f(x)$ is bounded.

For example, $\int_0^1 \frac{\sin x}{x}\,dx$ is a proper integral since $\lim_{x\to 0+}\frac{\sin x}{x} = 1$

Improper integral. Definition. The integral $\int_a^b f(x)\,dx$ is called an improper integral if

(*i*) $a = -\infty$ or $b = \infty$ or both, *i.e.*, one or both integration limits is infinite.

(*ii*) The integrand $f(x)$ is unbounded at one or more points of $a \le x \le b$,

For example $\int_0^\infty \frac{dx}{1+x^2}$ and $\int_0^1 \frac{1}{x^2}\,dx$ are improper integrals.

Different kinds of improper integrals.

(I) Improper integrals of the first kind

The integral $\int_a^b f(x)\,dx$ in which the range of integration is finite and the integrand $f(x)$ is unbounded at one or more points of $a \le x \le b$. Such points are called singularities of $f(x)$.

For example $\int_0^1 \frac{1}{x^2}\,dx$ is an improper integral of the first kind since its range of integration is finite and the integrand $1/x^2$ is unbounded at $x = 0$.

(II) Improper integrals of the second kind or infinite integrals

The definite integral $\int_a^b f(x)\,dx$ is called an improper integral of the first kind if $a = -\infty$ or $b = \infty$, or both *i.e.*, one or both integration limits is infinite and the integrand $f(x)$ is bounded. Thus, $\int_0^\infty \frac{dx}{1+x^2}$ is an improper integral of the second kind since the upper limit of integration is infinite and the integral $1/(1+x^2)$ is bounded.

(III) Improper integrals of the third kind or mixed kind

The improper integral $\int_a^b f(x)\,dx$ is called an improper integral of the third kind if the following two conditions are satisfied.

(*a*) $a = -\infty$ or $b = \infty$ or both, *i.e.*, one or both integration limits is infinite.

(*b*) $f(x)$ is unbounded at one are more points of $a \le x \le b$.

For example, $\int_0^\infty \frac{e^{-x}}{\sqrt{x}}\,dx$ is an improper integral of the third kind since range of integration is infinite and the integrand $1/\sqrt{x}$ is unbounded at $x = 0$.

Improper integrals of the third kind can be expressed in terms of improper integrals of the first and second kinds. Hence convergence of improper integrals can be easily done. For example,

$$\mathrm{I} = \int_0^\infty \frac{e^{-x}}{\sqrt{x}}\,dx = \int_0^1 \frac{e^{-x}}{\sqrt{x}}\,dx + \int_1^\infty \frac{e^{-x}}{\sqrt{x}}\,dx = \mathrm{I}_1 + \mathrm{I}_2,\ \text{say}$$

where I_1 and I_2 are improper integrals of second and first kinds respectively. So we test I_1 and I_2 for convergence of I.

Ex. 35. *Define general and Cauchy principal values of improper integrals of first kind.* **(Agra 2009)**

Sol. Let $\int_a^b f(x)\,dx$ be an improper integral of the first kind.

General value of an improper integral. Definition. If $f(x)$ becomes unbounded only at an interior point $x = c$ of the interval $a \le x \le b$, then we define the general value of the integral as follows.

$$\int_a^b f(x)\,dx = \lim_{\varepsilon_1 \to 0+} \int_a^{c-\varepsilon_1} f(x)\,dx + \lim_{\varepsilon_2 \to 0+} \int_{c+\varepsilon_2}^b f(x)\,dx \qquad \text{...(1)}$$

The integral on the left side of (1) converges or diverges according as the limits on the right exist or do not exist. If the limits on the right of (1) does exist, we call this value the general value of the integral on the left.

Cauchy principal value of an improper integral. In case the limits on the right of (1) do not exist when ε_1 and ε_2 approach zero independently, then in such a case it is possible that by choosing $\varepsilon_1 = \varepsilon_2 = \varepsilon$ in (1) *i.e.,* writing

$$\int_a^b f(x)\,dx = \lim_{\varepsilon \to 0+} \left\{ \int_a^{c-\varepsilon} f(x)\,dx + \int_{c+\varepsilon}^b f(x)\,dx \right\} \qquad \text{...(2)}$$

the limit does exist. If the limit on the right of (2) does exist, we call this limit value the Cauchy principal value of the integral on the left.

Note. The mere existence of the Cauchy's principal value does not imply the convergence of the improper integral.

Ex. 36. *Find the general and principal values of the integral* $\int_0^3 \frac{dx}{(x-1)^3}$ **(Agra 2008)**

Sol. Here $f(x) = 1/(x-1)^3$ which is unbounded at $x = 1$. Then, the general value of the given improper integral is given by

$$\int_0^3 \frac{dx}{(x-1)^3} = \lim_{\varepsilon_1 \to 0+} \int_0^{1-\varepsilon_1} \frac{dx}{(x-1)^3} + \lim_{\varepsilon_2 \to 0+} \int_{1+\varepsilon_2}^3 \frac{dx}{(x-1)^3}$$

$$= \lim_{\varepsilon_1 \to 0+} \left[-\frac{1}{2(x-1)^2} \right]_0^{1-\varepsilon_1} + \lim_{\varepsilon_2 \to 0+} \left[-\frac{1}{2(x-1)^2} \right]_{1+\varepsilon_2}^3$$

$$= \lim_{\varepsilon_1 \to 0+} \left(\frac{1}{2} - \frac{1}{2\varepsilon_1^2} \right) + \lim_{\varepsilon_2 \to 0+} \left(\frac{1}{2\varepsilon_2^2} - \frac{1}{8} \right) \qquad \text{...(1)}$$

and since the limits do not exist, the integral does not converge in the usual sense. Hence, the general value of the given improper integral does not exist.

Determination of the principal value, Consider

$$\int_0^3 \frac{dx}{(x-1)^3} = \lim_{\varepsilon \to 0+} \left(\int_0^{1-\varepsilon} \frac{dx}{(x-1)^3} + \int_{1+\varepsilon}^3 \frac{dx}{(x-1)^3} \right)$$

$$= \lim_{\varepsilon \to 0+} \left(\left[-\frac{1}{2(x-1)^2} \right]_0^{1-\varepsilon} + \left[-\frac{1}{2(x-1)^2} \right]_{1+\varepsilon}^3 \right)$$

$$= \lim_{\varepsilon \to 0+} \left\{ \left(\frac{1}{2} - \frac{1}{2\varepsilon^2} \right) + \left(\frac{1}{2\varepsilon^2} - \frac{1}{8} \right) \right\} = \frac{3}{8} \qquad \text{...(2)}$$

Since the limit on the right of (2) does exists, principal value of the given improper integral exists and equals 3/8.

Ex. 37. *Find the general value of the integral* $\int_{-1}^1 \frac{dx}{x^{2/3}}$ *and hence show that the given integral is convergent.*

Sol. Here integrand $1/x^{2/3}$ becomes unbounded only at an interior point $x = 0$ of the interval $-1 \le x \le 1$. The general value of the given integral is given by

$$\int_{-1}^{1} \frac{dx}{x^{2/3}} = \lim_{\varepsilon_1 \to 0+} \int_{-1}^{0-\varepsilon_1} \frac{dx}{x^{2/3}} + \lim_{\varepsilon_2 \to 0+} \int_{0+\varepsilon_2}^{1} \frac{dx}{x^{2/3}} = \lim_{\varepsilon_1 \to 0+} \left[3x^{1/3}\right]_{-1}^{-\varepsilon_1} + \lim_{\varepsilon_2 \to 0+} \left[3x^{1/3}\right]_{\varepsilon_2}^{1}$$

$$= \lim_{\varepsilon_1 \to 0+} (-3\varepsilon_1^{1/3} + 3) + \lim_{\varepsilon_2 \to 0+} (3 - 3\varepsilon_2^{1/3}) = 3 + 3 = 6,$$

which is a finite number. Hence, the general value of the given improper integral is 6 and the given integral is convergent.

Ex. 38. *Prove or disprove: The map $(a, b) \to (a + ib)$ for real plane $\mathbf{R}^2$ onto the complex place $\mathbf{C}$ is an isometry.* **(Delhi B.Sc. (Hons) III 2012)**

Sol. Let $f : \mathbf{R}^2 \to \mathbf{C}$ such that $\qquad f(x, y) = x + iy, \ \forall (x, y) \in \mathbf{R}^2 \qquad$ ...(1)

Consider the metric spaces $(\mathbf{R}^2, d_1)$ and $(\mathbf{C}, d_2)$ defined as follows:

$d_1(a, b) = \{(a_1 - b_1)^2 + (a_2 - b_2)^2\}^{1/2}, \ \forall a, b \in \mathbf{R}^2$ were $a = (a_1, a_2)$ and $b = (b_1, b_2)$ ...(2)

and $\qquad d_2(z_1, z_2) = | z_1 - z_2 |, \ \forall z_1, z_2 \in \mathbf{C}$ ...(3)

We now proceed to prove that f is an isometry:

(*i*) By hypothesis f is onto

(*ii*) $f(a_1, b_1) = f(a_2, b_2) \ \Rightarrow \ a_1 + ib_1 = a_2 + ib_2$, using (1)

$\Rightarrow \ a_1 = a_2$ and $b_1 = b_2 \ \Rightarrow \ (a_1, b_1) = (a_2, b_2)$,

showing that f is one-one.

(*iii*) We now prove that $\qquad d_1(a, b) = d_2(f(a), f(b)), \ \forall a, b \in \mathbf{R}^2$ ...(3)

R.H.S of (1) $= d_2(a_1 + ia_2, b_1 + ib_2)$ since $a = (a_1, a_2)$ and $b = (b_1, b_2)$

$= | (a_1 + ia_2) - (b_1 + ib_2)|$, using (3)

$= | (a_1 - b_1) + i(a_2 - b_2) | = \{(a_1 - b_1)^2 + (a_2 - b_2)^2\}^{1/2}$

$= d_1(a, b)$, using (2)

$=$ L.H.S. of (4)

Thus, (4) is true. From (*i*), (*ii*) and (*iii*), it follows that f is an isometry.

Ex. 39. *Consider the set* F *of all bounded real - valued functions from* [0, 1] to [0, 1]. *For each $f, g \in$* F, *define $d(f, g) = \sup \{| f(x) - g(x) | : x \in [0, 1]\}$. Show that d is a metric on* F.

(Delhi B.Sc. (Hons) III 2012)

Sol. To show that $d(f, g)$ is well defined. Since f and g are bounded, there exists real numbers k_1 and k_2 such that $| f(x) | \le k_1$ and $| g(x) | \le k_2, \ \forall x \in [0, 1]$. Thus, we have

$$| f(x) - g(x) | \le | f(x) | + | g(x) | \le k_1 + k_2, \ \forall x \in [0, 1]$$

Since $\{| f(x) - g(x) | : x \in [0, 1]\}$ is a non-empty set of real numbers bounded above by $k_1 + k_2$, it follows that $\sup \{| f(x) - g(x) | : x \in [0, 1]\}$ exists and hence $d(f, g)$ is well defined $\forall f, g \in \mathrm{F}$.

To show that d is a metric on F

Given that $\qquad d(f, g) = \sup \{| f(x) - g(x) | : x \in [0, 1]\}$ ...(1)

(*i*) $| f(x) - g(x) | \ge 0, \ \forall x \in [0, 1] \ \Rightarrow \ d(f, g) \ge 0$, using (1)

(*ii*) $d(f, g) = 0 \ \Leftrightarrow \ \sup \{| f(x) - g(x) | : x \in [0, 1]\} = 0$, using (1)

$\Leftrightarrow \ | f(x) - g(x) | = 0, \ \forall x \in [0, 1]$

$\Leftrightarrow \ f(x) - g(x) = 0, \ \forall x \in [0, i] \Leftrightarrow f = g$

(*iii*) $\quad d(f, g) = \sup \{| f(x) - g(x) | : x \in [0, 1]\}$, using (1)

$= \sup \{| g(x) - f(x) | : x \in [0, 1]\} = d(g, f)$, using (1) (*iv*) Let $f, g, h \in$ F. Then, we have

$$\begin{aligned}|f(x) - h(x)| &= |\{f(x) - g(x)\} + \{g(x) - h(x)\}|\\ &\le |f(x) - g(x)| + |g(x) - h(x)|\\ &\le \sup\{|f(x) - g(x)| : x \in [0, 1]\} \cup \sup\{|g(x) - h(x)| : x \in [0, 1]\}\end{aligned}$$

Thus, $\quad |f(x) - h(x)| \le d(f, g) + d(g, h), \ \forall\, x \in [0, 1],$ [using (1)]

showing that $|f(x) - h(x)|$ is bounded above by $d(f, g) + d(g, h)$. Hence, we have

$\sup\{|f(x) - h(x)| : x \in [0, 1]\} \le d(f, g) + d(g, h)$ or $d(f, h) \le d(f, g) + d(g, h)$

From (*i*), (*ii*), (*iii*) and (*iv*), it follows that d is a metric on F.

Remark. The metric space (F, d) is known as the *space of bounded real functions* on [0, 1] and is denoted by $\mathbb{R}$ [0, 1].

Ex. 40. *Prove or disprove. On a discrete metric space X, for each $x \in X$, the open balls $b(x, r)$ for $r \in]0, 1]$ are all equal to X.* **(Delhi B.Sc. (Hons.) III 2012)**

Sol. The discrete metric space (X, d) is defined as follows

$$d(x, y) = \begin{cases} 0, \text{ when } x = y \\ 1, \text{ when } x \neq y \end{cases} \qquad \text{...(1)}$$

Let $x \in$ X. Then, for any $r > o$, an open sphere of radius r centred at x is denoted and defined by $\quad S(x, r) = b(x, r) = \{x' \in X : d(x', x) < r\}$ ...(2)

In the discrete metric space (X, d), (2) takes the form

$$b(x, r) = \begin{cases} \{x\}, \text{ if } 0 < r \le 1 \\ X, \text{ if } r > 1 \end{cases} \qquad \text{...(3)}$$

For the present problem $r \in]0, 1]$, *i.e.*, $0 < r \le 1$ and hence (3) reduces to $b(x, r) = \{x\}$, showing that the conclusion $b(x, r) = X$ is not correct.

Remark. In the discrete metric space (X, d), closed ball $b[x, r]$ for $x \in X$ is given by

$$S[x, r] = b[x, r] = \begin{cases} \{x\}, \text{ if } 0 < x < 1 \\ X, \text{ if } r \ge 1 \end{cases}$$

Ex. 41. *If d is a usual metric on **R** and* A = {1, 2, 3} *and* B = {5, 6, 7}, *then distance between A and B is* (*a*) 4 (*b*) 1 (*c*) 2 (*d*) 5 **(Agra 2012)**

Sol. Ans. (*c*). Let d(A, B) denote distance between the given sets A and B.

Then, $d(A, B) = \inf\{d(x, y) : x \in A \text{ and } y \in B\}$, refer Art 19.6, page 19.4

$= \inf\{d(1, 5), d(1, 6), d(1, 7), d(2, 5), d(2, 6), d(2, 7), d(3, 5), d(3, 6), d(3, 7)\}$

$= \inf\{4, 5, 6, 3, 4, 5, 2, 3, 4\} = 2$

Ex. 42. *Prove that* $\left(\int_0^\infty x\, e^{-x^8} dx\right) \times \left(\int_0^\infty x^2\, e^{-x^4}\, dx\right) = \pi/16\sqrt{2}$ **(Delhi B.Sc. (Hons.) III 2012)**

Sol. Let $I_1 = \int_0^\infty x\, e^{-x^8}\, dx = \int_0^\infty x\, e^{-y} \dfrac{1}{8x^7}\, dx$ [On putting $x^8 = y$ and $dx = (1/8x^7)dy$]

$$= \frac{1}{8}\int_0^\infty e^{-y} x^{-6}\, dy = \frac{1}{8}\int_0^\infty e^{-y} (y^{1/8})^{-6}\, dy, \text{ as } x^8 = y$$

$$= \frac{1}{8}\int_0^\infty e^{-y} y^{-(3/4)}\, dy = \frac{1}{8}\int_0^\infty e^{-y} y^{(1/4)-1}\, dy = \frac{1}{8}\Gamma(1/4) \qquad \text{...(1)}$$

$$\left[\because \Gamma(n) = \int_0^\infty e^{-x} x^{n-1}\, dx\right]$$

Again, let $I_2 = \int_0^\infty x^2 e^{-x^4}\, dx = \int_0^\infty \frac{x^2 e^{-z}}{4x^3}\, dz = \frac{1}{4}\int_0^\infty e^{-z} x^{-1}\, dz$

[On putting $x^4 = z$ and $dx = (1/4x^3)dz$]

$$= \frac{1}{4}\int_0^\infty e^{-z} (z^{1/4})^{-1}\, dz = \frac{1}{4}\int_0^\infty e^{-z} z^{(3/4)-1}\, dz = \frac{1}{4}\Gamma(3/4) \quad ...(2)$$

From (1) and (2), we have

$$I_1 I_2 = (1/8)\times\Gamma(1/4)\times(1/4)\times\Gamma(3/4) = (1/32)\times\Gamma(1/4)\,\Gamma(1-1/4)$$

or $$\left(\int_0^\infty x\, e^{-x^8}\, dx\right)\times\left(\int_0^\infty x^2\, e^{-x^4}\, dx\right) = I_1 I_2 = \frac{1}{32}\times\frac{\pi}{\sin(\pi/4)} = \frac{\pi\sqrt{2}}{32} = \frac{\pi}{16\sqrt{2}}$$

$[\because \Gamma(n)\,\Gamma(1-n) = \pi/\sin n\pi, \text{ if } 0 < n < 1]$

Ex. 43. *The value of the limit* $\lim_{n\to\infty} \dfrac{2^{-n^2}}{\sum_{k=n+1}^{\infty} 2^{-k^2}}$ *is*

(a) 0 *(b) some c ∈ (0, 1)* *(c) 1* *(d) ∞* **(GATE 2013)**

Sol. Ans. (*a*). Let us suppose that $\sum_{k=n+1}^{\infty} u_k = \sum_{k=n+1}^{\infty} 2^{-k^2}$ so that

$$u_k = 2^{-k^2} = 1/2^{k^2} \quad \text{and hence} \quad u_{k+1} = 1/2^{(k+1)^2}.$$

Now, $$\lim_{k\to\infty}\frac{u_k}{u_{k+1}} = \lim_{k\to\infty}\frac{2^{(k+1)^2}}{2^{k^2}} = \lim_{k\to\infty}\frac{2^{k^2+2k+1}}{2^{k^2}} = \lim_{k\to\infty} 2^{2k+1} = \infty,$$

showing that the series Σu_k is convergent. Let $\Sigma u_k = l$, where l is a finite positive real number. Therefore, we have

$$\lim_{n\to\infty}\frac{2^{-n^2}}{\sum_{k=n+1}^{\infty} 2^{-k^2}} = \lim_{n\to\infty}\frac{2^{-n^2}}{l} = \frac{1}{l}\lim_{n\to\infty}\frac{1}{2^{n^2}} = 0$$

Ex. 44. *Let* $x_0 = 0$. *Define* $x_{n+1} = \cos x_n$ *for every* $n \geq 0$. *Then*
(a) $\{x_n\}$ *is increasing and converget* *(b)* $\{x_n\}$ is decreasing and convergent *(c)* $\{x_n\}$ *is convergent and* $x_{2n} < \lim_{n\to\infty} x_n < x_{2n+1}$ *for every* $n \in \mathbf{N}$ *(d)* $\{x_n\}$ *is not convergent* **(GATE 2013)**

Sol. Ans. (*c*) Left as an exercise.

Ex. 45. *Let* $\{a_n\}$ *be the sequence of consecutive positive solutions of the equation* $\tan x = x$ *and let* $\{b_n\}$ *be the sequence of consecutive positive solutions of* $\tan\sqrt{x} = x$. *Then*

(a) $\sum_{n=1}^{\infty}(1/a_n)$ *converges but* $\sum_{n=1}^{\infty}(1/b_n)$ *diverges* *(b)* $\sum_{n=1}^{\infty}(1/a_n)$ *diverges but* $\sum_{n=1}^{\infty}(1/b_n)$ *converges*

(c) Both $\sum_{n=1}^{\infty}(1/a_n)$ *and* $\sum_{n=1}^{\infty}(1/b_n)$ *converge* *(d) Both* $\sum_{n=1}^{\infty}(1/a_n)$ *and* $\sum_{n=1}^{\infty}(1/b_n)$ *diverge*

(GATE 2013)

Sol. Ans. (*b*) Left as an exercise.

Ex. 46. *Multiple choice question with exactly one correct answer.*

Let $f_n(x) = x^{1/n}$ *for* $x \in [0, 1]$. *Then*

(*A*) $\lim_{n \to \infty} f_n(x)$ *exists for all* $x \in [0, 1]$

(*B*) $\lim_{n \to \infty} f_n(x)$ *defines a continuous function on* $[0, 1]$

(*C*) $\{f_n\}$ *converges uniformly on* $[0, 1]$

(*D*) $\lim_{n \to \infty} f_n(x) = 0$, *for all* $x \in [0, 1]$ **(CSIR–UGC (NET), June 2011)**

Hint: Ans. (A) Refer chapter 15. Left as an exercise.

Ex. 47. *Multiple choice question with exactly one correct answer.*

Let $A = \{x^2 : 0 < x < 1\}$ *and* $B = \{x^3 : 1 < x < 2\}$. *Which of the following statements is true ?*

(*A*) *There is a one to one, onto function from A to B*

(*B*) *There is no one to one, onto function from A to B taking rationals to rationals*

(*C*) *There is no one to one function from A to B which is onto*

(*D*) *There is no onto function from A to B which is one to one*

(CSIR–UGC (NET), June 2011)

Hint: Ans. (A) Refer Art. 1.5, chapter 1. Left as an exercise.

Ex. 48. *Multiple choice question with exactly one correct answer.*

The set $\{(1/n) \times \sin(1/n) : n \in \mathbf{N}\}$ *has*

(*A*) *One limit point and it is 0* (*B*) *One limit point and it is 1*

(*C*) *One limit point and it is –1* (*D*) *Three limit points and these are –1, 0, 1*

(CSIR–UGC (NET), June 2011)

Hint: Ans. (*A*). Refer chapter 5. Left as an exercise.

Ex. 49. *Multiple choice question with exactly one correct answer.*

The power series $\sum_{0}^{\infty} 2^{-n} z^{2n}$ *converges if*

(*A*) $|z| \le 2$ (*B*) $|z| < 2$ (*C*) $|z| \ge \sqrt{2}$ (*D*) $|z| < \sqrt{2}$ **(CSIR–UGC (NET), June 2011)**

Sol. Ans. (*D*). Let $u_n = 2^{-n} z^{2n}$ so that $u_{n+1} = 2^{-(n+1)} z^{2n+2}$

Hence, $$\lim_{n\to\infty}\left|\frac{u_{n+1}}{u_n}\right| = \lim_{n\to\infty}\left|\frac{z^{2n+2}}{2^{n+1}} \times \frac{2^n}{z^{2n}}\right| = \frac{1}{2} \times |z|^2$$

Therefore, by D' Alembert's ratio test, the given power series converges (absolutely), if

$$|z|^2/2 < 1 \quad \Rightarrow \quad |z|^2 < 2 \quad \Rightarrow \quad |z| < \sqrt{2}$$

Ex. 50. *Multiple choice question with multiple correct options.*

Consider the function $f(x) = |\cos x| + |\sin(2 - x)|$. *At which of the following points is f not differentiable ?*

(*A*) $\{(2n + 1) \times (\pi/2) : n \in \mathbf{Z}\}$ (*B*) $\{n\pi : n \in \mathbf{Z}\}$

(*C*) $\{n\pi + 2 : n \in \mathbf{Z}\}$ (*D*) $\{(n\pi/2) : n \in \mathbf{Z}\}$ **(CSIR–UGC (NET), June 2011)**

Hint: Ans. (A) and (C). Refer chapter 9. Left as an exercise.

Ex. 51. *Multiple choice question with multiple correct options.*
Which of the following is/are metrics on $\boldsymbol{R}$?

(*A*) $d(x, y) = \min (x, y)$ (*B*) $d(x, y) = |x - y|$

(*C*) $d(x, y) = |x^2 - y^2|$ (*D*) $d(x, y) = |x^3 - y^3|$ **(CSIR–UGC (NET), June 2011)**

Sol. Ans. (*B*) and (*D*). In what follows we shall use Art. 19.2.

Consider option (*A*) : Given $d(x, y) = \min (x, y)$, where $x, y \in \boldsymbol{R}$ ... (1)

Here, $d(1, -1) = \min (1, -1) = -1 < 0$, showing that the property $d(x, y) \geq 0$, $\forall x, y \in \boldsymbol{R}$ is not satisfied. Hence, the option (*A*) is not correct.

Next, consider option (*B*) : Given $d(x, y) = |x - y|$, where $x, y \in \boldsymbol{R}$ ... (2)

Then, by Ex. 1, page 19.2. d is a metric on $\boldsymbol{R}$. So option (*B*) is correct.

Consider option (*C*); Given $d(x, y) = |x^2 - y^2|$, where $x, y \in \boldsymbol{R}$ ... (3)

Here, $d(1, -1) = |1^2 - (-1)^2| = 0$ while $1 \equiv -1$, showing that the property $d(x, y) = 0 \Leftrightarrow x = y$, $\forall x, y \in \boldsymbol{R}$ is not satisfied. So option (*C*) is not correct.

Finally, consider option (*D*): Given $d(x, y) = |x^3 - y^3|$, where $\forall x, y \in \boldsymbol{R}$ ... (4)

Here, (*i*) $|x^3 - y^3| \geq 0 \ \forall x, y \in \boldsymbol{R} \ \Rightarrow \ d(x, y) \geq 0 \ \forall x, y \in \boldsymbol{R}$

(*ii*) $|x^3 - y^3| = 0 \Leftrightarrow x^3 - y^3 = 0 \Leftrightarrow x = y$ so that $d(x, y) = 0 \Leftrightarrow x = y$

(*iii*) $|x^3 - y^3| = |y^3 - x^3| \ \forall x, y \in \boldsymbol{R} \ \Rightarrow \ d(x, y) = d(y, x) \ \forall x, y \in \boldsymbol{R}$

(*iv*) $|x^3 - y^3| = |(x^3 - z^3) + (z^3 - y^3)| \leq |x^3 - z^3| + |z^3 - y^3| \ \forall x, y \in \boldsymbol{R}$

$\Rightarrow \quad d(x, y) \leq d(x, z) + d(z, y) \ \forall x, y \in \boldsymbol{R}$

From (*i*) – (*iv*), we see that d defined by (4) is a metric on $\boldsymbol{R}$. So option (*D*) is correct.

Ex. 52. *Multiple choice question with multiple correct options.*
Which of the following is/are correct ?

(*A*) $n \log\left(1 + \dfrac{1}{n+1}\right) \to 1$ as $n \to \infty$ (*B*) $(n+1)\log\left(1 + \dfrac{1}{n+1}\right) \to 1$ as $n \to \infty$

(*C*) $n^2 \log\left(1 + \dfrac{1}{n}\right) \to 1$ as $n \to \infty$ (*D*) $n \log\left(1 + \dfrac{1}{n^2}\right) \to 1$ as $n \to \infty$

(CSIR–UGC (NET), June 2011)

Sol. Ans. (*A*) and (*B*). We have

$$\lim_{n \to \infty} n \log\left(1 + \frac{1}{n+1}\right) = \lim_{n \to \infty} n\left\{\frac{1}{n+1} - \frac{1}{2(n+1)^2} + \frac{1}{3(n+1)^3} - \dots\right\}$$

$$= \lim_{n \to \infty} \frac{1}{1 + (1/n)}\left\{1 - \frac{1}{2(n+1)} + \frac{1}{3(n+1)^2} - \dots\right\} = 1$$

$$\lim_{n \to \infty} (n+1) \log\left(1 + \frac{1}{n+1}\right) = \lim_{n \to \infty} (n+1)\left\{\frac{1}{n+1} - \frac{1}{2(n+1)^2} + \frac{1}{3(n+1)^3} - \dots\right\}$$

$$= \lim_{n \to \infty} \left\{1 - \frac{1}{2(n+1)} + \frac{1}{3(n+1)^2} - \dots\right\} = 1$$

$$\lim_{n \to \infty} n^2 \log\left(1 + \frac{1}{n}\right) = \lim_{n \to \infty} n^2\left(\frac{1}{n} - \frac{1}{2n^2} + \frac{1}{3n^3} - \dots\right) = \lim_{n \to \infty} n\left(1 - \frac{1}{2n} + \frac{1}{3n^2} - \dots\right) = \infty$$

$$\lim_{n\to\infty} n\log\left(1+\frac{1}{n^2}\right) = \lim_{n\to\infty} n\left(\frac{1}{n^2}-\frac{1}{2n^4}+\frac{1}{3n^6}-\dots\right) = \lim_{n\to\infty}\left(\frac{1}{n}-\frac{1}{2n^3}+\frac{1}{3n^5}-\dots\right)=0$$

Therefore, options (*A*) and (*B*) are correct.

Ex. 53. *Multiple choice question with multiple correct options.*

If $\{x_n\}$ and $\{y_n\}$ are sequences of real numbers, which of the following is/are true ?

(*A*) $\lim\sup_n (x_n + y_n) \le \lim\sup_n x_n + \lim\sup_n y_n$

(*B*) $\lim\sup_n (x_n + y_n) \ge \lim\sup_n x_n + \lim\sup_n y_n$

(*C*) $\lim\inf_n (x_n + y_n) \le \lim\inf_n x_n + \lim\inf_n y_n$

(*D*) $\lim\inf_n (x_n + y_n) \ge \lim\inf_n x_n + \lim\inf_n y_n$ **(CSIR–UGC (NET), June 2011)**

Hint: Ans. (*A*) and (*D*). Refer theorem V, page 5.55, chapter 5.

Ex. 54. *Multiple choice question with exactly one correct answer.*

Which of the following statements is true ?

(*A*) $\lim_{x\to\infty}\frac{\log x}{x^{1/2}}=0$ and $\lim_{x\to\infty}\frac{\log x}{x}=\infty$ (*B*) $\lim_{x\to\infty}\frac{\log x}{x^{1/2}}=\infty$ and $\lim_{x\to\infty}\frac{\log x}{x}=0$

(*C*) $\lim_{x\to\infty}\frac{\log x}{x^{1/2}}=0$ and $\lim_{x\to\infty}\frac{\log x}{x}=0$ (*D*) $\lim_{x\to\infty}\frac{\log x}{x^{1/2}}=0$ but $\lim_{x\to\infty}\frac{\log x}{x}$ does not exist

(CSIR–UGC (NET), June 2011)

Sol. Ans. (*C*). Using L' Hospital's rule, we have

Here $$\lim_{x\to\infty}\frac{\log x}{x^{1/2}}\left[\frac{\infty}{\infty}\right]=\lim_{x\to\infty}\frac{1/x}{1/2\sqrt{x}}=\lim_{x\to\infty}\frac{2}{\sqrt{x}}=0$$

Again, $$\lim_{x\to\infty}\frac{\log x}{x}\left[\frac{\infty}{\infty}\right]=\lim_{x\to\infty}\frac{1/x}{1}=\lim_{x\to\infty}\frac{1}{x}=0$$

Ex. 55. *Multiple choice question with multiple correct options.*

*Which of the following real-valued functions f defined on **R** have the property that f is continuous and $f\,0\,f = f$?*

(*A*) $f(x)=\begin{cases}|x|, \text{ if } x\in[-1,1]\\ x^2, \text{ if } x\notin[-1,1]\end{cases}$ (*B*) $f(x)=\begin{cases}x, \text{ if } x\in[0,1]\\ 1, \text{ if } x\ge 1\\ 0, \text{ if } x\le 0\end{cases}$

(*C*) $f(x)=\begin{cases}x, \text{ if } x\in[-1,1]\\ 1, \text{ if } x\ge 1\\ -1, \text{ if } x\le -1\end{cases}$ (*D*) $f(x)=\begin{cases}1, \text{ if } x\in[-23,27]\\ 22+x, \text{ if } x\le -23\\ -26+x, \text{ if } x\ge 27\end{cases}$

Hint. Ans. (B) and (C). Refer chapter 8. **[CSIR-UGC (NET), December 2011]**

Ex. 56. *Multiple choice question with multiple correct options.*

Let $f : \mathbf{R}\to\mathbf{R}$ be a continuous function such that $\int_0^\infty f(x)\,dx$ exists. Which of the following statements are correct ?

(*A*) *if* $\lim_{x\to\infty} f(x)$ *exists, then* $\lim_{x\to\infty} f(x)=0$

(*B*) *the limit* $\lim_{x\to\infty} f(x)$ *must exist and is equal to zero*

(*C*) *in case f is a non-negative function,* $\lim_{x \to \infty} f(x)$ *must exist*

(*D*) *in case f is a differentiable function,* $\lim_{x \to \infty} f'(x)$ *must exist and is zero*

(CSIR–UGC (NET), December 2011)

Hint. Ans. (A) and (C). Refer chapter 13. Left as an exercise.

Ex. 57. *Multiple choice question with multiple correct options.*

It is given that the series $\sum_{n=1}^{\infty} a_n$ *is convergent but not absolutely convergent and* $\sum_{n=1}^{\infty} a_n = 0.$

Denote by s_k *the partial sum* $\sum_{n=1}^{k} a_n$, $k = 1, 2, ...$ *Then*

(*A*) $s_k = 0$ *for infinitely many k*

(*B*) $s_k > 0$ *for infinitely many k and* $s_k < 0$ *for infinitely many k*

(*C*) *it is possible that* $s_k > 0$ *for all k*

(*D*) *it is possible that* $s_k > 0$ *for all but a finite number of values of k.*

(CSIR–UGC (NET), December 2011)

Hint. Ans. (B). Refer chapters 6 and 7. Left as an exercise

Ex. 58. *Multiple choice question with multiple correct options.*

Define $f : \mathbf{R} \to \mathbf{R}$ *by* $$f(x) = \begin{cases} x^2, \text{ if } x < 0 \\ 2x + x^2, \text{ if } x \geq 0 \end{cases}$$

Then which of the following statements are correct ?

(*A*) $f''(x) = 2$ *for all* $x \in \mathbf{R}$ (*B*) $f''(0)$ *does not exist*

(*C*) $f'(x)$ *exists for each* $x \neq 0$ (*D*) $f'(0)$ *does not exist*

(CSIR–UGC (NET), December 2011)

Hint. Ans. (*B*), (*C*) and (*D*). Refer chapters 9.

Ex. 59. *Multiple choice question with multiple correct options.*

Let $f : \mathbf{R} \to \mathbf{R}$ *be a differentiable function such that* f' *is bounded. Given a closed and bounded interval* $[a, b]$, *a partition* $P = \{a_0 = a < a_1 < ... < a_n = b\}$ *of* $[a, b]$, *let* $M(f, P)$ *and* $m(f, P)$ *denote respectively, the upper Riemann sum and the lower Riemann sum of f with respect to P. Then*

(*A*) $\left| M(f, P) - \int_a^b f(x)\,dx \right| \leq (b - a) \sup \{|f(x)| : x \in [a, b]\}$

(*B*) $\left| m(f, P) - \int_a^b f(x)\,dx \right| \leq (b - a) \inf \{|f(x)| : x \in [a, b]\}$

(*C*) $\left| M(f, P) - \int_a^b f(x)\,dx \right| \leq (b - a)^2 \sup \{|f'(x)| : x \in [a, b]\}$

(*D*) $\left| m(f, P) - \int_a^b f(x)\,dx \right| \leq (b - a)^2 \inf \{|f'(x)| : x \in [a, b]\}$

(CSIR–UGC (NET), December 2011)

Hint. Ans. (A) and (C). Refer chapter 13. Left as an exercise.

Ex. 60. *Multiple choice question with exactly one correct answer.*

Let $f_n(x) = \begin{cases} 1 - nx, \text{ for } x \in [0, 1/n] \\ 0, \quad \text{for } x \in [1/n, 1] \end{cases}$. *Then*

(*A*) $\lim_{n \to x} f_n(x)$ defines a continuous function on [0, 1]

(*B*) $\{f_n\}$ converges uniformly on [0, 1] (*C*) $\lim_{n \to x} f_n(x) = 0$ for all $x \in [0, 1]$

(*D*) $\lim_{n \to x} f_n(x)$ exists for all $x \in [0, 1]$ **(CSIR–UGC (NET), June 2012)**

Hint. Ans. (D). Refer chapters 8 and 15. Left as an exercise.

Ex. 61. *Multiple choice question with exactly one correct answer.*

Let $a_n = \sin(\pi/n)$. For the sequence $a_1, a_2, \ldots$ the supremum is

(*A*) 0 and it is attained (*B*) 0 and it is not attained

(*C*) 1 and it is attained (*D*) 1 and it is not attained

(CSIR–UGC (NET), June 2012)

Hint. Ans. (*C*). Refer chapter 5. Left as an exercise.

Ex. 62. *Multiple choice question with exactly one correct answer.*

The power series $\sum_{n=0}^{\infty} 3^{-n}(z-1)^{2n}$ *converges if*

(*A*) $|z| \le 3$ (*B*) $|z| < \sqrt{3}$ (*C*) $|z-1| < \sqrt{3}$ (*D*) $|z-1| \le \sqrt{3}$

(CSIR–UGC (NET), June 2012)

Sol. Ans. (*C*). Shifting the origin to 1, i.e., changing the variable $z - 1$ to x, the given power series may be rewritten as $\overset{\infty}{\underset{n=1}{\Sigma}} 3^{-n} x^{2n}$ or $\overset{\infty}{\underset{n=1}{\Sigma}} u_n$ so that

$$u_n = 3^{-n} x^{2n} \quad \text{and so} \quad u_{n+1} = 3^{-(n+1)} x^{2x+2}$$

Hence, $$\lim_{n\to\infty}\left|\frac{u_{n+1}}{u_n}\right| = \lim_{n\to\infty}\left|\frac{x^{2n+2}}{3^{n+1}} \times \frac{3^n}{x^{2n}}\right| = \frac{1}{3}|x|^2$$

Therefore, by *D*' Alembert's ratio test, the given series converges (absolutely), if

$$|x|^2/3 < 1 \quad \text{or} \quad |x| < \sqrt{3} \quad \text{or} \quad |z-1| < \sqrt{3}, \quad \text{since } x = z - 1 \text{ by our hypothesis.}$$

Ex. 63. *Multiple choice question with multiple correct options.*

Consider the function: $f(x) = \cos(|x - 5|) + \sin(|x - 3|) + |x + 10|^3 - (|x| + 4))^2$. At which of the following points is f not differentiable ?

(*A*) $x = 5$ (*B*) $x = 3$ (*C*) $x = -10$ (*D*) $x = 0$ **(CSIR–UGC (NET), June 2012)**

Hint. Ans. (B) and (D). Refer chapter 9. Left as an exercise.

Ex. 64. *Multiple choice question with multiple correct options.*

Which of the following metrics on $C = \{f : [0, 1] \to \mathbf{R}\}$ is a continuous function ?

(*A*) $d(f, g) = \sup\{|f(x) - g(x)| : x \in [0, 1]\}$ (*B*) $d(f, g) = \inf\{|f(x) - g(x)| : x \in [0, 1]\}$

(*C*) $d(f, g) = \int_0^1 |f(x) - g(x)|\, dx$

(*D*) $d(f, g) = \sup\{|f(x) - g(x)| : x \in [0, 1]\} + \int_0^1 |f(x) - g(x)|\, dx$

(CSIR–UGC (NET), June 2012)

Sol. Ans. (*A*), (*C*) and (*D*). Refer chapter 19.

Ex. 65. *Multiple choice question with multiple correct options.*

Which of the following is/are correct ?

(A) $\left(1+\dfrac{1}{n}\right)^{n+1} \to e$ as $n \to \infty$ (B) $\left(1+\dfrac{1}{n+1}\right)^{n} \to e$ as $n \to \infty$

(C) $\left(1+\dfrac{1}{n}\right)^{n^2} \to e$ as $n \to \infty$ (D) $\left(1+\dfrac{1}{n^2}\right)^{n} \to e$ as $n \to \infty$

(CSIR–UGC (NET), June, 2012)

Sol. Ans. (A) and (B). We know that $\lim\limits_{n\to\infty} (1+1/n)^n = e$ ... (1)

Now, $\lim\limits_{n\to\infty} (1+1/n)^{n+1} = \lim\limits_{n\to\infty} \{(1+1/n)^n \times (1+1/n)\}$

$= \lim\limits_{n\to\infty} (1+1/n)^n \times \lim\limits_{n\to\infty} (1+1/n) = e \times 1 = e$, using result (1)

$$\lim_{n\to\infty}\left(1+\frac{1}{n+1}\right)^n = \lim_{n\to\infty}\frac{\left(1+\dfrac{1}{n+1}\right)^{n+1}}{\left(1+\dfrac{1}{n+1}\right)} = \frac{\lim\limits_{n\to\infty}\left(1+\dfrac{1}{n+1}\right)^{n+1}}{\lim\limits_{n\to\infty}\left(1+\dfrac{1}{n+1}\right)} = \frac{e}{1} = e, \text{ using (1)}$$

Let $l = \lim\limits_{n\to\infty} (1+1/n)^{n^2}$ $[1^\infty]$ ... (2)

Hence, $\log_e l = \lim\limits_{n\to\infty} n^2 \log_e\left(1+\dfrac{1}{n}\right) = \lim\limits_{n\to\infty} n^2\left(\dfrac{1}{n}-\dfrac{1}{2n^2}+\dfrac{1}{3n^3}-\cdots\right)$

$\Rightarrow \log_e l = \infty \Rightarrow l = e^\infty = \infty \Rightarrow \lim\limits_{n\to\infty}(1+1/n)^{n^2} = \infty$, using (2)

Let $l' = \lim\limits_{n\to\infty} (1+1/n^2)^n$ $[1^\infty]$ ... (3)

Hence, $\log_e l' = \lim\limits_{n\to\infty} n \log_e\left(1+\dfrac{1}{n^2}\right) = \lim\limits_{n\to\infty} n\left(\dfrac{1}{n^2}-\dfrac{1}{2n^4}+\dfrac{1}{3x^6}-\ldots\right)$,

$\Rightarrow \log_e l' = 0 \Rightarrow l' = e^\circ = 1 \Rightarrow \lim\limits_{n\to\infty}(1+1/n^2)^n = 1$, using (3)

Thus, options (A) and (B) are correct whereas options (C) and (D) are not correct.

Ex. 66. *Multiple choice question with exactly one correct answer.*

Let f be a twice differentiate function on $\mathbf{R}$. *Given that* $f''(x) > 0$ *for all* $x \in \mathbf{R}$. *Then*

(A) $f(x) = 0$ *has exactly two solutions on* $\mathbf{R}$

(B) $f(x) = 0$ *has a positive solution if* $f(0) = 0$ *and* $f'(0) = 0$

(C) $f(x) = 0$ *has no positive solution if* $f(x) = 0$ *and* $f'(0) > 0$

(D) $f(x) = 0$ *has no positive solution if* $f(0) = 0$ *and* $f'(0) < 0$

(CSIR–UGC (NET), December 2012)

Hint. Ans. (C). Refer chapters 9 and 10. Left as an exercise.

Ex. 67. *Multiple choice question with exactly one correct answer.*

Let f be a continuously differentiable real - valued function on $[a, b]$ *such that* $|f'(x)| < K$ *for all* $x \in [a, b]$. *For a partition* $P = \{a = a_0 < a_1 < \ldots < a_n = b\}$, *let* $U(f, P)$ *and* $L(f, P)$ *denote the upper and lower Riemann sums of f with respect to P. Then*

(A) $|L(f, P)| \le K(b-a) \le |U(f, P)|$ (B) $U(f, P) - L(f, P) \le K(b-a)$

(C) $U(f, P) - L(f, P) \le K\|P\|$, *where* $\|P\| = \max\limits_{0\le i\le n-1}(a_{i+1}-a_i)$ *is the norm of the partition*

(D) $U(f, P) - L(f, P) \le K\|P\|\,(b-a)$ **(CSIR–UGC (NET), December 2012)**

Hint. Ans. (D). Refer chapter 13. Left as an exercise.

Ex. 68. *Multiple choice question with exactly one correct answer.*

Let f be a monotone non-decreasing real-valued function on **R**. *Then*

(*A*) $\lim_{x \to a} f(x)$ *exists at each point a* (*B*) *If* $a < b$, *then* $\lim_{x \to a+} f(x) \le \lim_{x \to b-} f(x)$

(*C*) *f is an unbounded function* (*D*) *the function* $g(x) = e^{-f(x)}$ *is a bounded function*

(CSIR–UGC (NET), December 2012)

Hint. Ans. (*B*). Refer chapter 8. Left as an exercise.

Ex. 69. *Multiple choice question with exactly one correct answer.*

Which of the following real-valued function on (0, 1) is uniformly continuous ?

(*A*) $f(x) = 1/x$ (*B*) $f(x) = (\sin x)/x$ (*C*) $f(x) = \sin(1/x)$ (*D*) $f(x) = (\cos x)/x$

(CSIR–UGC (NET), December 2012)

Hint. Ans. (*B*) Refer Art. 8.19, page 8.43.

Ex. 70. *Multiple choice question with exactly one correct answer.*

Let X be a metric space and $A \subseteq X$ *be a connected set with at least two distinct points Then the number of distinct points in A is*

(*A*) *2* (*B*) *more than 2, but finite* (*C*) *countably infinite* (*D*) *uncountable*

(CSIR–UGC (NET), December 2012)

Hint. Ans. (*D*). Refer chapter 19. Left as an exercise.

Ex. 71. *Multiple choice question with exactly one correct answer.*

Consider the power series $\sum_{n=1}^{\infty} z^n$. *The radius of convergence of this series is*

(*A*) 0 (*B*) ∞ (*C*) 1 (*D*) a real number greater than 1 **(CSIR–UGC (NET), December 2012)**

Hint. Ans. (C). Refer Art. 17.5, chapter 17.

Ex. 72. *Multiple choice question with multiple correct options.*

Let $\{f_n\}$ *be a sequence of continuous real-valued functions defined on* $[0, \infty)$. *Suppose* $f_n(x) \to f(x)$ *for all* $x \in [0, \infty)$ *and that f is integrable. Then*

(*A*) $\int_0^{\infty} f_n(x)\,dx \to \int_0^{\infty} f(x)\,dx$ as $n \to \infty$

(*B*) If $f_n \to f$ uniformly on $[0, \infty)$, then $\int_0^1 f_n(x)\,dx \to \int_0^1 f(x)\,dx$

(*C*) if $f_n \to f$ uniformly on $[0, \infty)$, then $\int_0^{\infty} f_n(x)\,dx \to \int_0^{\infty} f(x)\,dx$

(*D*) if $\int_0^1 |f_n(x) - f(x)|\,dx \to 0$, $f_n \to f$ uniformly on [0, 1]

(CSIR–UGC (NET), December 2012)

Hint. Ans. (*B*). Refer chapter 15. Left as an exercise.

Ex. 73. *Multiple choice question with multiple correct options.*

Let $f(x) = \sin x - x + (x^3/3!)$ *and* $g(x) = \cos x - 1 + (x^2/2!)$ *for* $x \in$ **R**. *Which of the following statements are correct ?*

(*A*) $f(x) \ge 0$ for all $x > 0$ (*B*) *g* is an increasing function on $[0, \infty)$

(*C*) *g* is a decreasing function on $[0, \infty)$ (*D*) *f* is a decreasing function on $[0, \infty)$

(CSIR–UGC (NET), December 2012)

Sol. Ans. (*A*) and (*B*). Consider $f(x) = \sin x - x + (x^3/3!)$, for $0 \le x < 1$ ... (1)

Let $x \ge 0$, Taylor's theorem with remainder after three terms is given by

$$f(x) = f(0) + xf'(0) + (x^2/2!) \times f''(01 + (x^3/3!) \times f'''(\theta x),\ 0 < \theta < 1 \quad ... (2)$$

Let $f(x) = \sin x$ so that $f'(x) = \cos x$, $f''(x) = -\sin x$ and $f'''(x) = -\cos x$ ... (3)

From (3), $f(0) = 0$, $f'(0) = 1$, $f''(o) = 0$, and $f'''(\theta x) = -\cos \theta x$ ... (4)

Using (4), (2) yields $\sin x = x - (x^3/6) \times \cos(\theta x)$, where $0 < \theta < 1$ ... (5)

We know that $\cos(\theta x) \leq 1$ whatever θx may be. Therefore, for $x \geq 0$, we have

$$-(x^3/6) \times \cos(\theta x) \geq -(x^3/6) \Rightarrow x - (x^3/6) \times \cos(\theta x) \geq 1 - (x^3/6!)$$

$$\Rightarrow \sin x \geq x - (x^3/6!), \text{ using (5)}$$

$$\Rightarrow \sin x - x + (x^3/6!) \geq 0, \text{ for } 0 \leq x < \infty$$

$$\Rightarrow f(x) \geq 0 \text{ for } 0 \leq x < \infty, \text{ using (1)} \qquad \text{... (6)}$$

(6) shown that option (*A*) is correct. But, (6) implies that f in an increasing function. Hence, option (*D*) is not correct

Next, consider $g(x) = \cos x - 1 + (x^2/2!)$, for $0 \leq x < 1$ ... (7)

Let $x \geq o$. Taylor's theorem with remainder after two terms is given by

$$g(x) = g(0) + xg'(0) + (x^2/2!) \times g''(\theta x), \text{ for } 0 < \theta < 1 \qquad \text{... (8)}$$

Let $g(x) = \cos x$ so that $g'(x) = -\sin x$ and $g''(x) = -\cos x$ ... (9)

From (9) $g(0) = 1$ and $g''(\theta x) = -\cos(\theta x)$ ... (10)

Using (10), (8) reduces to $\cos x = 1 - (x^2/2!) \times \cos(\theta x)$, $0 < \theta < 1$ ... (11)

We know that $\cos(\theta x) \leq 1$ whatever θx may be. Therefore, for $x \geq 0$, we have

$$-(x^2/2!) \times \cos(\theta x) \geq -(x^2/2!) \Rightarrow 1 - (x^2/2!) \times \cos(\theta x) \geq 1 - (x^2/2!)$$

$$\Rightarrow \cos x \geq 1 - (x^2/2!), \text{ using (11)}$$

$$\Rightarrow \cos x - 1 + (x^2/2!) \geq 0, \text{ for } 0 \leq x < \infty$$

$$\Rightarrow g(x) \geq 0, \text{ for } 0 \leq x < \infty, \text{ using (7)} \qquad \text{... (12)}$$

$$\Rightarrow g \text{ is an increasing function on } [0, \infty)$$

Therefore, option (*B*) in correct whereas option (*C*) is not correct.

Ex. 74. *Multiple option question with exactly one correct answer.*
Which of the following series is convergent ?

(*A*) $\displaystyle\sum_{n=1}^{\infty} \frac{1}{\sqrt{n+1} - \sqrt{n}}$ (*B*) $\displaystyle\sum_{n=1}^{\infty} \frac{\sin n}{n^2}$ (*C*) $\displaystyle\sum_{n=1}^{\infty} (-1)^n \log n$ (*D*) $\displaystyle\sum_{n=1}^{\infty} \frac{\log n}{n}$

(CSIR–UGC (NET), June 2013)

Sol. Ans. (*B*). Here, we have $u_n = (\sin n)/n^2$

Hence, $|u_n| = |\sin n|/n^2 \leq 1/n^2$, since $|\sin n| \leq |$

Again, $\displaystyle\sum_{n=1}^{\infty} (1/n^2)$ is convergent. Hence, by the comparison test, the series $\displaystyle\sum_{n=1}^{\infty} |u_n|$ is cnvergent.

This implies that the given series is absolutely convergent. Since every absolutely convergent series is also covergent, it follows that the series $\displaystyle\sum_{n=1}^{\infty} u_n$ *i.e.*, $\displaystyle\sum_{n=1}^{\infty} (\sin n)/n^2$ is convergent.

Ex. 75. *Multiple option question with exactly one correct answer.*

The power series $\displaystyle\sum_{n=0}^{\infty} \frac{[2+(-1)^n]^n}{3^n} x^n$ *converges*

(*A*) only for $x = 0$ (*B*) for all $x \in \boldsymbol{R}$
(*C*) only for $-1 < x < 1$ (*D*) only for $-1 < x \leq 1$ **(CSIR–UGC (NET), June 2013)**

Hint. Ans. (A). Refer chapter 17.

Ex. 76. *Multiple option question with exactly one correct answer.*

In which of the following cases, there is no continuous function f from the set S onto the set T ?

(*A*) S = [0, 1], T = **R** (*B*) S = (0, 1), T = **R**

(*C*) S = (0, 1), T = (0, 1] (*D*) S = **R**, T = (0, 1) **(CSIR–UGC (NET), June 2013)**

Hint. Ans. (A). Refer Art. 1.5, page 1.6, chapter 1. Left as an exercise.

Ex. 77. *Multiple option question with exactly one correct answer.*

The number of limit points of the set $\{1/m + 1/n : m, n \in N\}$ *is*

(*A*) 1 (*B*) 2 (*C*) finitely many (*D*) infinitely many **(CSIR–UGC (NET), June 2013)**

Sol. Ans. (*D*). Let $S = \{1/m + 1/n : m, n \in N\}$

Keep *m* fixed and vary *n*. Then, as *n* gets bigger, $1/n$ gets nearer to *o* so that $1/m + 1/n$ gets nearer to $1/m$ and so by definition, $1/m$ is a limit point of *S*. Since *m* is an arbitrary natural number, it follows that all the points of the set $\{1/m: m \in N\}$ are limit points of the given set *S*.

Again, if we vary both *m* and *n* so that $1/m + 1/n$ gets nearer and nearer to *o*. Therefore, by definition, *o* is also a limit point of *S*. Thus, all limit points of the set *S* are given by $\{1/m: m \in N\}$ U {0}, which is an infinite set. Therefore, option (***D***) is correct.

Ex. 78. *Multiple option question with multiple correct options.*

Which of the following define a metric on ***R***

(*A*) $d(x, y) = |x - y|/\{1 + |x - y|\}$ (*B*) $d(x, y) = |x - 2y| + |2y - x|$

(*C*) $d(x, y) = |x^2 - y^2|$ (*D*) $d(x, y) = |x^3 - y^3|$

[CSIR-UGC (NET), June 2013]

Sol. Ans. (*A*) and (*D*). We shall use Art. 19.2 for testing given options.

Consider option (*A*): Given $d(x, y) = |x - y|/\{1 + |x - y|\},\ \forall x, y \in \boldsymbol{R}$ ... (1)

(*i*) from (1), it follows that $d(x, y) \geq 0$

(*ii*) $d(x, y) = 0 \Leftrightarrow |x - y| / \{1 + |x - y|\} = 0 \Leftrightarrow x = y$

(*iii*) $d(x, y) = |x - y| / \{1 + |x - y|\} = |y - x| / \{1 + |y - x|\} = d(y, x)$, using (1)

Thus, $d(x, y) = d(y, x),\ \forall x, y \in \boldsymbol{R}$

(*iv*) For all *x, y, z* in ***R***, we have

$$d(x, y) + d(y, z) = \frac{|x-y|}{1+|x-y|} + \frac{|y-z|}{1+|y-z|}$$

$$\geq \frac{|x-y|}{1+|x-y|+|y-z|} + \frac{|y-z|}{1+|x-y|+|y-z|}$$

$$= \frac{|x-y|+|y-z|}{1+|x-y|+|y-z|} = 1 - \frac{1}{1+|x-y|+|y-z|} \quad \text{... (2)}$$

Now, $|x - y| + |y - z| \geq |x - z| \Rightarrow 1 + |x - y| + |y - z| \geq 1 + |x - z|$

Hence, $\dfrac{1}{1+|x-y|+|y-z|} \leq \dfrac{1}{1+|x-z|}$

$$\Rightarrow \quad 1 - \frac{1}{1+|x-y|+|y-z|} \geq 1 - \frac{1}{1+|x-z|} = \frac{|x-z|}{1+|x-z|} = d(x,z), \text{ using (1)} \quad \text{... (3)}$$

Then, (2) and (3) $\Rightarrow$ $d(x, y) + d(y, z) \geq d(x, z)$

From (*i*) – (*iv*). it follows that *d* is a metric on ***R***. So option (***A***) is correct.

Next, consider option (***B***). Given $d(x, y) = |x - 2y| + |2y - x|,\ \forall x, y \in \boldsymbol{R}$ ... (4)

Using (4), we have $d(1, 2) = |1 - 4| + |4 - 1| = 6$ and $d(2, 1) = |2 - 2| + |2 - 2| = 0$

Hence, the property $d(x, y) = d(y, x)\ \forall x, y \in \boldsymbol{R}$ is not satisfied. Therefore, the option (*B*) is not correct.

Now, consider option (*C*). Given $$d(x, y) = |x^2 - y^2|,\ \forall x, y \in \boldsymbol{R} \qquad \text{... (5)}$$

Then, as explained in Ex. 51, $d(x, y)$ given by (5) is not a metric on $\boldsymbol{R}$. Hence, option (*C*) is not correct.

Finally, consider option (*D*): Given $$d(x, y) = |x^3 - y^3|,\ \forall x, y \in \boldsymbol{R} \qquad \text{... (6)}$$

Then, as explained in Ex. 51, $d(x, y)$ given by (6) is metric on $\boldsymbol{R}$. Hence, option (*D*) is correct.

Ex. 79. *Multiple option question with multiple correct options.*

Consider $\boldsymbol{Q}$, *the set of rational numbers, with the metric* $d(p, q) = |p - q|$. *Then which of the following are true ?*

(*A*) $\{q \in \mathbf{Q} \mid 2 < q^2 < 3\}$ is closed (*B*) $\{q \in \mathbf{Q} \mid 2 \le q^2 \le 4\}$ is compact

(*C*) $\{q \in \mathrm{Q} \mid 2 \le q^2 \le 4\}$ is closed (*D*) $\{q \in \mathbf{Q} \mid q^2 \ge 1\}$ is compact

(CSIR–UGC (NET), June 2013)

Hint. Ans. (*A*) and (*C*). Refer chapter 19. Left as an exercise.

Ex. 80. *Multiple option question with multiple correct options.*

Which of the following functions f is uniformly continuous on the interval (0, 1) ?

(*A*) $f(x) = 1/x$ (*B*) $f(x) = x \sin(1/x)$

(*C*) $f(x) = \sin(1/x)$ (*D*) $f(x) = (\sin x)/x$ **(CSIR–UGC (NET), June 2013)**

Hint. Ans. (*B*) and (*D*). Refer Art. 8.19, page 8.43.

Ex. 81. *Multiple option question with multiple correct options.*

Let $f : \boldsymbol{R} \to \boldsymbol{R}$ *be a continuous function with period* $p > 0$. *Then* $g(x) = \int_x^{x+p} f(t)\, dt$ *is*

(*A*) *constant function* (*B*) *continuous function*

(*C*) *continuous function but not differentiable* (*D*) *neither continuous nor differentiable*

(CSIR–UGC (NET), June 2013)

Sol. Ans. (*A*), (*B*) and (*C*). Given $$g(x) = \int_x^{x+p} f(t)\, dt, \qquad \text{...(1)}$$

where f is a periodic function with period $p > 0$, so that $f(x + p) = f(x),\ \forall\ x \in \boldsymbol{R}$ ...(2)

Differentiating both sides of (1) w.r.t. 'x' and using Leibnitz's rule of differentiation under the integral sign (refer Art. 21.3, page 21.2), we get

$$g'(x) = \int_x^{x+p} \frac{\partial f(t)}{\partial x}\, dt + f(x+p)\frac{d(x+p)}{dx} - f(x)\frac{dx}{dx} \text{ or } g'(x) = f(x+p) - f(x) = 0, \text{ using (2).}$$

Hence, g is a constant function and therefore g is continuous as well as differentiable. Thus, options (*A*), (*B*) and (*C*) are correct.

Ex. 82. *What is the difference between limit point of a sequence and limit of a sequence.*

Sol. In what follows we shall use the definition of limit point of a sequence (refer Art. 5.3) and the definition of limit of a sequence (refer Art. 5.4).

If $l \in \mathbf{R}$ is the limit of a sequence $\langle f_n \rangle$, then for a given $\varepsilon > 0$, there exists $m \in \mathbf{N}$, such that $|f_n - l| < \varepsilon\ \forall\ n \ge m$. Thus, every neighourhood of l contains all except a finite number of terms of the sequence On the other hand, if $l \in \mathbf{R}$ is a limit point of the sequence $\langle f_n \rangle$, then every neighourhood of l containing infinitely many terms of the sequence does not exclude the possibility of an infinite number of terms of the sequence lying outside that neighbourhood. Therefore, limit of a sequence is a limit point of the sequence but a limit point of a sequence need not always be the limit of the sequence.

Ex. 83. *Multiple choice question with exactly one correct answer*

$$\lim_{n\to\infty}\frac{1}{\sqrt{n}}\left(\frac{1}{\sqrt{1}+\sqrt{3}}+\frac{1}{\sqrt{3}+\sqrt{5}}+\dots+\frac{1}{\sqrt{2n-1}+\sqrt{2n+1}}\right) \text{ equals}$$

(*A*). $\sqrt{2}$ (*B*). $1/\sqrt{2}$ (*C*). $1+\sqrt{2}$ (*D*). $1/(1+\sqrt{2})$ **[CSIR – UGC (NET), June 2014]**

Sol. Ans. (*B*). We have,

$$\frac{1}{\sqrt{2n-1}+\sqrt{2n+1}}=\frac{(\sqrt{2n+1}-\sqrt{2n-1})}{(\sqrt{2n+1}+\sqrt{2n-1})(\sqrt{2n+1}-\sqrt{2n-1})}=\frac{1}{2}\times(\sqrt{2n+1}-\sqrt{2n-1})$$

Now, $\displaystyle\lim_{n\to\infty}\frac{1}{\sqrt{n}}\left(\frac{1}{\sqrt{1}+\sqrt{3}}+\frac{1}{\sqrt{3}+\sqrt{5}}+\dots\dots+\frac{1}{\sqrt{2n+1}+\sqrt{2n-1}}\right)$

$$=\lim_{n\to\infty}(1/2\sqrt{n})\times\{(\sqrt{3}-\sqrt{1})+(\sqrt{5}-\sqrt{3})+\dots\dots+(\sqrt{2n+1}-\sqrt{2n-1})\}$$

[Using result (1) for $n = 1, 2, 3, \dots, n$]

$$=\lim_{n\to\infty}(1/2\sqrt{n})\times(\sqrt{2n+1}-1)=\lim_{n\to\infty}(1/2)\times\{\sqrt{2+1/n}-1/\sqrt{n}\}=(1/2)\times\sqrt{2}=1/\sqrt{2}$$

Ex. 84. *Let d_1, d_2 and d_3 be metrics on a set X with at least two elements. Which of the following is not a metric on X?*

(*a*) *min* $(d_1, 2)$ (*b*) *max* $(d_2, 2)$ (*c*) $d_3/(1+d_3)$ (*d*) $(d_1+d_2+d_3)/3$ **[GATE 2014]**

Sol. Ans. (*b*). Let $d = \max(d_2, 2)$. Since d_2 is a metric on X, we have

$$d_2(x, y) = 0 \quad\Leftrightarrow\quad x = y \qquad \dots(1)$$

Hence, when $x = y$, $d(x, y) = \max(d_2(x, y), 2) = \max(0, 2) = 2 \neq 0$, using (1)

Thus, $x = y$ does not imply that $d(x, y) = 0$. Therefore, d cannot be a metric on X because d does not satisfy the following axiom for d to be a metric on X: $d(x, y) = 0 \Leftrightarrow x = y$.

Ex. 85. *Multiple choice question with exactly one correct option.*

Let p(x) be a polynomial in the real variable x of degree 5. Then, $\displaystyle\lim_{n\to\infty}\{p(n)/2^n\}$ *is*

(*A*). 5 (*B*). 1 (*C*). 0 (*D*). ∞ **[CSIR – UGC (NET), June 2014]**

Sol. Ans. (*C*). Let $p(x) = a_5x^5 + a_4x^4 + a_3x^3 + a_2x^2 + a_1x + a_0$, where $a_5, a_4, \dots, a_0$ are constants. In what follows, we shall use L' Hospital's rule. Thus, we have

$$\lim_{n\to\infty}\frac{p(n)}{2^n}=\lim_{n\to\infty}\frac{a_5n^5+a_4n^4+a_3n^3+a_2n^2+a_1n+a_0}{2^n} \qquad \left[\text{Form }\frac{\infty}{\infty}\right]$$

$$=\lim_{n\to\infty}\frac{5a_5n^4+4a_4n^3+3a_3n^2+2a_2n+a_1}{(\log_e 2)\times 2^n} \qquad \left[\text{Form }\frac{\infty}{\infty}\right]$$

$$=\lim_{n\to\infty}\frac{20a_5n^3+12a_4n^2+6a_3n+2a_2}{(\log_e 2)^2\times 2^n} \qquad \left[\text{Form }\frac{\infty}{\infty}\right]$$

$$=\lim_{n\to\infty}\frac{60a_5n^2+24a_4n+6a_3}{(\log_e 2)^3\times 2^n} \qquad \left[\text{Form }\frac{\infty}{\infty}\right]$$

$$=\lim_{n\to\infty}\frac{120a_5n+24a_4}{(\log_e 2)^4\times 2^n} \qquad \left[\text{Form }\frac{\infty}{\infty}\right]$$

$$=\lim_{n\to\infty}\frac{120a_5}{(\log_e 2)^5\times 2^n}=0 \qquad \left[\text{Form }\frac{\infty}{\infty}\right]$$

Ex. 86. *Multiple choice question with exactly one correct option.*

Let $f : X \to Y$ be a function from a metric space X to another metric Y. For any Cauchy sequence $\{x_n\}$ in X,

*(**A**). If f is continuous, then $\{f(x_n)\}$ is a Cauchy sequence in Y*

*(**B**). If $\{f(x_n)\}$ is Cauchy, then $\{f(x_n)\}$ is always convergent in Y*

*(**C**). If $\{f(x_n)\}$ is Cauchy in Y, then f is continuous*

*(**D**). $\{x_n\}$ is always convergent in X* **[CSIR – UGC (NET), June 2014]**

Sol. Ans. (*D*). Refer theorem II, page 5.30.

Ex. 87. *Multiple choice question with multiple correct options.*

*Let $\{f_n\}$ be a sequence of continuous functions on **R***

*(**A**). If $\{f_n\}$ converges to f point-wise on **R**, then* $\lim_{n\to\infty}\int_{-\infty}^{\infty} f_n(x)\,dx = \int_{-\infty}^{\infty} f(x)\,dx$

*(**B**). If $\{f_n\}$ converges to f uniformly on **R**, then* $\lim_{n\to\infty}\int_{-\infty}^{\infty} f_n(x)\,dx = \int_{-\infty}^{\infty} f(x)\,dx$

*(**C**). If $\{f_n\}$ converges to f uniformly on **R**, then f is continuous on **R***

*(**D**). There exists a sequence of continuous functions $\{f_n\}$ on **R**, such that $\{f_n\}$ converges to f uniformly on **R** but* $\lim_{n\to\infty}\int_{-\infty}^{\infty} f_n(x)\,dx \neq \int_{-\infty}^{\infty} f(x)\,dx$ **[CSIR – UGC (NET), June 2014]**

Sol. Ans. (*B*) and (*C*). From theorem on page 15.9, it follows that option (***B***) is correct while options (***A***) and (***D***) are not correct options. Again, from theorem on page 15.7, it follows that option (***C***) is correct.

Ex. 88. *Multiple choice question having exactly one correct option.*

Let $\{a_n\}$, $\{b_n\}$ be sequences of real numbers satisfying $|a_n| \le |b_n|$ for all $n \ge 1$. Then,

*(**A**).* $\sum_{n=1}^{\infty} a_n$ *converges whenever* $\sum_{n=1}^{\infty} b_n$ *converges*

*(**B**).* $\sum_{n=1}^{\infty} a_n$ *converges absolutely whenever* $\sum_{n=1}^{\infty} b_n$ *converges absolutely*

*(**C**).* $\sum_{n=1}^{\infty} b_n$ *converges whenever* $\sum_{n=1}^{\infty} a_n$ *converges*

*(**D**).* $\sum_{n=1}^{\infty} b_n$ *converges absolutely whenever* $\sum_{n=1}^{\infty} a_n$ *converges absolutely*

Sol. Ans. (*B*). Refer Ex. 8 on page 7.19. **[CSIR – UGC (NET), December 2013]**

Ex. 89. *Multiple choice question with exactly one correct option.*

Let $f(x)\begin{cases}(\sin x)/x, & \text{if } x \neq 0\\ 1 & ,\text{if } x = 0\end{cases}$. *Then, f is*

*(**A**). discontinuous* *(**B**). continuous but not differentiable*

*(**C**). differentiable only once* *(**D**). differentiable more than once*

[CSIR – UGC (NET), December 2013]

Sol. Ans. (*D*). Let $x = c$ be an arbitrary non-zero real number. Since $\sin x$ and x are both differentiable at $x = c$ and $f(c) \neq 0$, hence $(\sin x)/x$ is also differentiable at $x = c$ (refer theorem III, page 9.3). Thus, $f(x)$ is differentiable at $x = c$ where c is an arbitrary real number. Thus, option (***D***) is correct.

Ex. 90. *Multiple choice question with more than one correct answer*

The function $f : \mathbf{R} \to \mathbf{R}$ *is given by* $f(x) = e^{|x| + x^2} + |x^2 - 1|$. *Which of the following is true about the function f ?*

(*A*). *It is not differentiable exactly at three points of* $\mathbf{R}$

(*B*). *It is not differentiable at* $x = 0$ (*C*) *It is differentiable at* $x = 2$

(*D*). *It is not differentiable at* $x = 1$ *and* $x = -1$ **[CSIR – UGC (NET), December 2013]**

Sol. Ans. (*A*), (*B*), (*C*), (*D*). Given $$f(x) = e^{|x| + x^2} + |x^2 - 1| \qquad ...(1)$$

***Testing* (1) *for differentiability at x* = 0.** We have

$$Rf'(0) = \lim_{h \to 0} \frac{f(0+h) - f(0)}{h} = \lim_{h \to 0} \frac{e^{|h| + h^2} + |h^2 - 1| - 2}{h} = \lim_{h \to 0} \frac{e^{h + h^2} + (1 - h^2) - 2}{h} \left[\text{Form } \frac{0}{0}\right]$$

$$= \lim_{h \to 0} \frac{e^{h + h^2} + (1 + 2h) - 2h}{1}, \text{ using L' Hospotal's rule}$$

Thus, $$Rf'(0) = 1 \qquad ...(2)$$

$$Lf'(0) = \lim_{h \to 0} \frac{f(0-h) - f(0)}{-h} = \lim_{h \to 0} \frac{e^{|-h| + h^2} + |h^2 - 1| - 2}{-h} = \lim_{h \to 0} \frac{e^{h + h^2} + (1 - h^2) - 2}{-h} \left[\text{Form } \frac{0}{0}\right]$$

$$= \lim_{h \to 0} \frac{e^{h + h^2} + (1 + 2h) - 2h}{-1} = -1, \text{ using L' Hospotal's rule}$$

Since $Rf'(0) \neq Lf'(0)$, the given function is not differentiable at $x = 0$. Therefore, option (*B*) is correct.

***Testing* (1) *for differentiability at x* = 1.** We have

$$Rf'(1) = \lim_{h \to 0} \frac{f(1+h) - f(1)}{h} = \lim_{h \to 1} \frac{e^{|1+h| + (1+h)^2} + |(1+h)^2 - 1| - e^2}{h}$$

$$= \lim_{h \to 0} \frac{e^{1 + h + (1+h)^2} + (h^2 + 2h) - e^2}{h} \qquad \left[\text{Form } \frac{0}{0}\right]$$

$$= \lim_{h \to 0} \frac{e^{1 + h + (1+h)^2} \times \{1 + 2(1+h)\} + (2h + 2)}{1} = 3e^2 + 2$$

and $$Lf'(1) = \lim_{h \to 0} \frac{f(1-h) - f(1)}{-h} = \lim_{h \to 0} \frac{e^{|1-h| + (1-h)^2} + |(1-h)^2 - 1| - e^2}{-h}$$

$$= \lim_{h \to 0} \frac{e^{1 - h + (1-h)^2} + (h^2 - 2h) - e^2}{-h} \qquad \left[\text{Form } \frac{0}{0}\right]$$

$$= \lim_{h \to 0} \frac{e^{1 - h + (1-h)^2} \times \{1 + 2(1-h)\} + (2h - 2)}{-1} = 2 - 3e^2$$

Since $Rf'(1) \neq Lf'(1)$, the given function is not differentiable at $x = 1$.

***Testing for differentiability at x* = –1.** We have

$$Rf'(-1) = \lim_{h \to 0} \frac{f(-1+h) - f(-1)}{h} = \lim_{h \to 0} \frac{e^{|-1+h| + (-1+h)^2} + |(-1+h)^2 - 1| - e^2}{h}$$

$$= \lim_{h \to 0} \frac{e^{1 - h + (h-1)^2} + (2h^2 - h) - e^2}{h} \qquad \left[\text{Form } \frac{0}{0}\right]$$

$$= \lim_{h \to 0} \frac{e^{1-h+(h-1)^2} \times \{-1+2(h-1)\} + (2-2h)}{1} = 2 - 3e^2$$

and $$\text{L}f'(-1) = \lim_{h \to 0} \frac{f(-1-h) - f(-1)}{-h} = \lim_{h \to 0} \frac{e^{|-1-h|+(-1-h)^2} + |(-1-h)^2 - 1| - e^2}{-h}$$

$$= \lim_{h \to 0} \frac{e^{1+h+(1+h)^2} + (2h+h^2) - e^2}{-h} \qquad \left[\text{Form } \frac{0}{0}\right]$$

$$= \lim_{h \to 0} \frac{e^{1+h+(1+h)^2} \times \{1+2(1+h)\} + (2+2h)}{-1} = -(2+3e^2)$$

Since $\text{R}f'(-1) \neq \text{L}f'(-1)$, the given function is not differentiable at $x = -1$.

Thus, the given function is not differentiable at $x = 1$ and $x = -1$. Hence option (***D***) is correct.

Testing* (1) *for differentiability at x = c, where c ≠ 0, c ≠ 1 and c ≠ −1

Let $\phi(x) = e^{|x|+x^2}$ and $\Psi(x) = |x^2 - 1|$. Then, $\qquad f(x) = \phi(x) + \Psi(x) \qquad$ ...(1)

Since ϕ and Ψ are both differentiable at $x = c$, it follows that their sum $\phi + \Psi$, i.e., f is also differentiable at $x = c$. (Refer theorem I, page 9.3). Hence options (***A***) and (***C***) are correct.

Ex. 91. *If* $\sum_{n=1}^{\infty} a_n$ *converges, prove that the following series are convergent where* $a_n > 0$

(***a***) $\sum_{n=1}^{\infty} a_n^2$ (***b***) $\sum_{n=1}^{\infty} \{a_n/(1+a_n)\}$ (***c***) $\sum_{n=1}^{\infty} \{a_n/(1-a_n)\},\ a_n \neq 1$

Sol. (*a*). Since $\sum_{n=1}^{\infty} a_n$ is convergent, $a_n \to 0$ as $n \to \infty$ (refer theorem on page 6.21).

Therefore, by definition, taking $\varepsilon = 1$, there exists $m \in \mathbf{N}$ such that

$a_n < 1$ for all $n \geq m$. This implies that $\qquad a_n^{\ 2} < a_n \qquad$ for all $\quad n \geq m$.

Therefore, by comparison test $\sum_{n=1}^{\infty} a_n^{\ 2}$ is convergent.

(***b***). Let $b_n = a_n/(a_n + 1)$. Then, $\qquad \lim_{n \to \infty} (a_n/b_n) = \lim_{n \to \infty} (a_n + 1) = 0$

$$\left[\because \sum_{n=1}^{\infty} a_n \text{ is convergent} \Rightarrow \lim_{n \to \infty} a_n = 0\right]$$

Therefore, by comparison test (refer Art. 6.8, page 6.9), the two series $\sum_{n=1}^{\infty} a_n$ and $\sum_{n=1}^{\infty} b_n$ either converge or diverge together. Since $\sum_{n=1}^{\infty} a_n$ is convergent, it follows that $\sum_{n=1}^{\infty} b_n$, i.e., $\sum_{n=1}^{\infty} \{a_n/(1+a_n)\}$ is also convergent.

(***c***). Proceed exactly as in part (***b***).

Ex. 92. *Multiple choice question with exactly one correct answer.*

If $\sum_{n=1}^{\infty} a_n$ *is absolutely convergent, then which of the following is NOT true.*

(*A*). $\sum_{m=n}^{\infty} a_m \to 0$ *as* $n \to \infty$ (*B*). $\sum_{n=1}^{\infty} a_n \sin n$ *is convergent*

(*C*). $\sum_{n=1}^{\infty} e^{a_n}$ *is divergent* (*D*). $\sum_{n=1}^{\infty} a_n^{\ 2}$ *is divergent* **[CSIR – UGC (NET), December 2013]**

Sol. Ans. (*D*). Since $\sum_{n=1}^{\infty} a_n$ is absolutely convergent, $\sum_{n=1}^{\infty} |a_n|$ must be convergent.

Since $\sum_{n=1}^{\infty} |a_n|$ is convergent, $|a_n| \to 0$ as $n \to \infty$ (refer theorem on page 6.21). Therefore, by definition, taking $\varepsilon = 1$, there exists $m \in \mathbf{N}$ such that

$|a_n| < 1$ for all $n \geq m$. This implies that $|a_n|^2 < a_n$ for all $n \geq m$.

Therefore, by comparison test $\sum_{n=1}^{\infty} |a_n|^2$ i.e., $\sum_{n=1}^{\infty} a_n^{\ 2}$ is convergent.

Ex. 93. *Multiple choice question with exactly one correct answer.*
If $f : [0, 1] \to (0, 1)$ *is a continuous mapping then which of the following is NOT true?*
(*A*). $F \subseteq [0, 1]$ *is a closed set implies* $f(F)$ *is closed in* $\mathbf{R}$
(*B*). *If* $f(0) < f(1)$, *then* $f([0, 1])$ *must be equal to* $[f(0), f(1)]$
(*C*). *There must exist* $x \in (0, 1)$ *such that* $f(x) = x$ (*D*). $f([0, 1]) \neq (0, 1)$
Sol. Ans. (B). Left as an exercise **[CSIR – UGC (NET), December 2013]**

Ex. 94. *Multiple choice question having more than one correct answer.*
Let A be a subset of $\mathbf{R}$ *with more than one element. Let* $\alpha \in A$. *If* $A \backslash \{\alpha\}$ *is compact, then*
(*A*). *A is compact* (*B*). *every subsets of A must be compact*
(*C*). *A must be finite set* (*D*). *A is disconnected.*
Sol. Ans. (A), (D). Try yourself **[CSIR – UGC (NET), December 2013]**

Ex. 95. *Multiple choice question with exactly one correct answer.*
Let $A \subseteq \mathbf{R}$ *and* $f : A \to \mathbf{R}$ *be given by* $f(x) = x^2$. *Then f is uniformly continuous if*
(*A*). *A is a bounded subset of* $\mathbf{R}$ (*B*). *A is a dense subset of* $\mathbf{R}$
(*C*). *A is an unbounded and connected subset of* $\mathbf{R}$
(*D*). *A is an unbounded and open subset of* $\mathbf{R}$ **[CSIR – UGC (NET), June 2014]**
Sol. Ans. (A). Try yourself

Ex. 96. *Multiple choice question having exactly one correct answer.*
For a continuous function $f : \mathbf{R} \to \mathbf{R}$, *let* $Z(f) = \{x \in \mathbf{R} : f(x) = 0\}$. *Then* $Z(f)$ *is*
(*A*). *compact* (*B*). *open* (*C*). *connected* (*D*). *closed* **[CSIR – UGC (NET), June 2014]**
Sol. Ans. (D). Try yourself

Ex. 97. *Multiple choice question having multiple correct options.*
Let $\{a_n\}_{n \geq 1}$ *be a sequence of positive numbers such that* $a_1 > a_2 > a_3 > .....$ *Then, which of following is/are always true ?*

(*A*). $\lim_{n \to \infty} a_n = 0$ (*B*). $\lim_{n \to \infty} \frac{a_n}{n} = 0$ (*C*). $\sum_{n=1}^{\infty} \frac{a_n}{n}$ *converges* (*D*). $\sum_{n=1}^{\infty} \frac{a_n}{n^2}$ *converges*

[CSIR – UGC (NET), December 2013]

Sol. Ans. (*A*), (*B*) and (*D*). Here $\{a_n\}$ is a monotonically decreasing sequence which is bounded below, such that its greatest lower bound is 0. Hence, by part (*ii*) of theorem I on page 5.38, we have $\lim_{n \to \infty} a_n = 0$. Hence, option (*A*) is correct. Again, we have

$\lim_{n \to \infty} a_n = 0$ and $\lim_{n \to \infty} (1/n) = 0 \Rightarrow \lim_{n \to \infty} (a_n / n) = 0$, using result of theorem III, page 5.12

Hence, option (*B*) is correct.

Since $\sum_{n=1}^{\infty} a_n$ is a positive monotonically decreasing sequence and $\sum_{n=1}^{\infty}(1/n^2)$ is a convergent series, $\sum_{n=1}^{\infty}(a_n/n^2)$ converges (refer Ablet's theorem, page 7.13). Hence, option (***D***) is correct.

Ex. 98. *Show that the sequence $\{f_n\}$ where $f_n(x) = x^n$ is uniformly convergent in [0, k] where k is a number less than 1 and only point-wise convergent in [0, 1].* **[Agra 2003; Bhopal 2000; Delhi B.Sc. (Prog) II 2008, Purvanchal 2006; Ravishankar 2001; Rewa 2001]**

Sol. Here, $$f(x) = \lim_{n\to\infty} f_n(x) = \begin{cases} 0, \textit{ when } 0 \le x < 1 \\ 1, \textit{ when } x = 1 \end{cases}$$

Thus, the sequence $\{f_n\}$ converges point-wise to f on [0, 1].

To examine whether the sequence $\{f_n\}$ is uniformly convergent, let $\varepsilon > 0$ be given.

For $0 < x < 1$, we have

$$|f_n(x) - f(x)| = |x^n - 0| = x^n < \varepsilon$$

$$\Rightarrow (1/x)^n > 1/\varepsilon \Rightarrow n \log(1/x) > \log(1/\varepsilon) \Rightarrow n > \frac{\log(1/\varepsilon)}{\log(1/x)}$$

[Note that $0 \le x < 1 \Rightarrow 1/x > 1 \Rightarrow \log(1/x) > 0$]

Clearly, the number $\dfrac{\log(1/\varepsilon)}{\log(1/x)}$ increases with x having maximum value $\dfrac{\log(1/\varepsilon)}{\log(1/k)}$ on [0, k], where $k < 1$.

If we choose a positive integer m just greater or equal to $\{\log(1/\varepsilon)\}/\{\log(1/k)\}$, then

$$|f_n(x) - f(x)| < \varepsilon \qquad \forall\ n \ge m \qquad \text{and} \qquad 0 < x < 1$$

Again, at $x = 0$ $\qquad |f_n(x) - f(x)| = |0 - 0| = 0 < \varepsilon \qquad \forall\ n \ge 1$

Therefore, there exists a positive integer m such that

$$|f_n(x) - f(x)| < \varepsilon \qquad \forall\ n \ge m \qquad \text{and} \qquad \forall\ x \in [0, k], \qquad k < 1,$$

showing that $\{f_n\}$ is uniformly convergent on [0, k], $k < 1$.

When $x \to 1$, the number $\{\log(1/\varepsilon)\}/\{\log(1/x)\} \to \infty$. Therefore, it is not possible to determine a positive integer m such that

$$|f_n(x) - f(x)| < \varepsilon \qquad \forall \qquad n \ge m \qquad \text{and} \qquad \forall\ x \in [0, 1]$$

Hence, the sequence $\{f_n\}$ is not uniformly convergent on any interval containing 1 and particular on [0, 1].

Ex. 99. *Multiple choice question having exactly one correct answer.*

Let $f_n : [1, 2] \to [0, 1]$ be given by $f_n(x) = (2 - x)^n$ for all non-negative integer n. Let $f(x) = \lim_{n\to\infty} f_n(x)$ for $1 \le x \le 2$. Then, which of the following is true ?

(***A***). *f is continuous on [1, 2]* (***B***). *f_n converges uniformly to f on [1, 2] as $n \to \infty$*

(***C***). $\lim_{n\to\infty} \int_1^2 f_n(x)\,dx = \int_1^2 f(x)\,dx$ (***D***). *for any $a \in (1, 2)$, we have* $\lim_{n\to\infty} f_n'(a) = f'(a)$

[CSIR – UGC (NET), December 2013]

Sol. Ans. (*D*). Here, $$f(x) = \lim_{n\to\infty} f_n(x) = \begin{cases} 0, \textit{ when } 0 < x \le 2 \\ 1, \textit{ when } x = 1 \end{cases}$$

Thus, the sequence $\{f_n\}$ converges point-wise to f on [0, 1].

In order to examine whether $\{f_n\}$ is uniformly convergent, let $\varepsilon > 0$ be given

When $x = 2$, $\quad |f_n(x) - f(x)| = |0 - 0| = 0 < \varepsilon \quad \forall \quad n \geq 1$

For $1 < x < 2$, $\quad |f_n(x) - f(x)| = |(2-x)^n - 0| = (2-x)^n < \varepsilon$

$$\Rightarrow \left(\frac{1}{2-x}\right)^n > \frac{1}{\varepsilon} \quad \Rightarrow \quad n \log\left(\frac{1}{2-\varepsilon}\right) > \log\left(\frac{1}{\varepsilon}\right) \quad \Rightarrow \quad n > \frac{\log(1/\varepsilon)}{\log\{1/(2-x)\}}$$

[Not that $1 < x < 2 \Rightarrow 0 < 2 - x < 1 \Rightarrow 1/(2-x) > 1 \Rightarrow \log\{1/(2-x)\} > 0$]

When $x \to 1$, the number $\log(1/\varepsilon)/\log\{1/(2-x)\} \to \infty$. Hence, it is not possible to find a positive integer m such that

$$|f_n(x) - f(x)| < \varepsilon \quad \forall \; n \geq m \quad \text{and} \quad \forall \; x \in [1, 2]$$

Therefore, the sequence $\{f_n\}$ is not uniformly convergent on any interval containing 1 and in particular on [1, 2]

Since $\{f_n\}$ is not uniformly convergent, option (***B***) is not correct.

Now, the fact that $\{f_n\}$ is not uniformly continuous on [1, 2[and result of theorem on page 15.7 together show that option (***A***) is not correct.

Again the fact that $\{f_n\}$ is not uniformly continuous on [1, 2[and result of theorem on page 15.9 together show that option (***C***) is not correct. Hence, option (***D***) is correct.

Ex. 100. *Multiple choice question having more than one correct answers.*

Let $f : (0, 1) \to \boldsymbol{R}$ be continuous. Suppose that

$|f(x) - f(y)| \leq \{\sin x - \sin y\}$ fir all x, $y \in (0, 1)$ Then

(***A***). f is discontinuous at least at one point in (0, 1)

(***B***). *f is continuous everywhere on (0, 1), but not uniformly continuous on (0. 1)*

(***C***). *f is uniformly continuous on (0, 1)*

(***D***). $\lim_{x \to 0+} f(x)$ *exists* **[CSIR – UGC (NET), June 2014]**

Sol. Ans. (C) and (D). Try yourself

Ex. 101. *Multiple choice question having more than one correct answer.*

Let $p_n(x) = a_n x^2 + b_n x + c_n$ be a sequence of quadratic polynomials when a_n, b_n, $c_n \in \boldsymbol{R}$, for all $n \geq 1$. Let λ_0, λ_1, λ_2 be distinct real numbers such that $\lim_{n \to \infty} p_n(\lambda_0) = A_0$, $\lim_{n \to \infty} p_n(\lambda_1) = A_1$ and $\lim_{n \to \infty} p_n(\lambda_2) = A_2$. Then

(***A***). $\lim_{n \to \infty} p_n(x)$ *exists for all $x \in \boldsymbol{R}$* (***B***). $\lim_{n \to \infty} p_n'(x)$ *exists for all $x \in R$*

(***C***). $\lim_{n \to \infty} p_n\left(\frac{\lambda_0 + \lambda_1 + \lambda_2}{3}\right)$ *does not exist* (***D***). $\lim_{n \to \infty} p_n'\left(\frac{\lambda_0 + \lambda_1 + \lambda_2}{3}\right)$ *does not exist*

Sol. Ans. (A) and (B). Try yourself **[CSIR – UGC (NET), June 2014]**

Ex. 102. *Multiple choice question having more than one correct answer.*

Ley $\{a_n\}$, $\{b_n\}$ be given bounded sequences of positive real numbers. Then (Here $a_n \uparrow a$ means a_n increases to a as n goes to ∞, similarly, $b_n \downarrow b$ means b_n decreases to b as n goes to ∞)

(***A***). *if $a_n \uparrow a$, then $\sup_{n \geq 1} (a_n b_n) = a\,(\sup_{n \geq 1} b_n)$*

(***B***). *if $a_n \uparrow a$, then $\sup_{n \geq 1} (a_n b_n) < a\,(\sup_{n \geq 1} b_n)$*

(***C***). *if $b_n \downarrow b$, then $\inf_{n \geq 1} (a_n b_n) = (\inf_{n \geq 1} a_n)\, b$*

(***D***). *if $b_n \downarrow b$, then $\inf_{n \geq 1} (a_n b_n) > (\inf_{n \geq 1} a_n)\, b$*

Sol. Ans. (A) and (C). Try yourself **[CSIR – UGC (NET), June 2014]**

Ex. 103. *Multiple choice question having exactly one correct answer.*

Let $f : [0, 1] \to [0, 1]$ be any twice differentiable function satisfying $f(ax + (1 - a)y) \le a f(x) + (1 - a) f(y)$ for all $x, y \in [0, 1]$ and any $a \in [0, 1]$. Then for all $x \in (0, 1)$.

(A) $f'(x) \ge 0$ (B) $f''(x) \ge 0$ (C) $f'(x) \le 0$ (D) $f''(x) \le 0$

Sol. Ans. (B). Try yourself **[CSIR – UGC (NET), December 2013]**

Ex. 104. *Prove that Cauchy sequence in **R** is convergent.* **[Kanpur 2014]**

Hint: Refer converse of theorem II, page 5.30.

Ex. 105. *Prove that every monotonic sequence converges.* **[Kanpur 2013]**

Hint: Refer matter under the heading "the condition is sufficient" of theorem II, page 5.38.

Ex. 106. *Test for convergence:* $1 + (x/2^2) + (x^2/3^2) + (x^3/4^2) + \ldots$ **[Kanpur 2013]**

Hint: Do like Ex. 2(*i*), page 6.18. **Ans.** Convergent if $x \le 1$; divergent if $x > 1$.

Ex. 107. *Test for convergence:* $x + (3x^2/5) + (8x^3/10) + (15x^4/17) + \ldots + \{(n^2 - 1)/(n^2 + 1)\} x^n + \ldots$ **[Kanpur 2012]**

Hint: Do like Ex. 2(*i*), page 6.18. **Ans.** Convergent if $x < 1$; divergent if $x \ge 1$.

Ex. 108. *Discuss the nature of discontinuity of the following function at origin:*

$$f(x) = (e^{1/x} - e^{-1/x})/(e^{1/x} + e^{-1/x}),\ x \ne 0, \quad f(0) = 0.$$ **[Kanpur 2013]**

Sol. Refer solved Ex. 1, page 8.18 and get $f(0 - 0) \ne f(0 + 0)$. This shows that f has a discontinuity of the first kind at $x = 0$.

Ex. 109. *Prove that* $e^x \sin x = x + x^2 - (2x^3/3) + \ldots + 2^{n/2} \sin (n\pi/4) \times (x^n/n)$.

Hint: Proceed as in Art 10.15, page 10.31. **[Kanpur 2013]**

Ex. 110. *Expand $e^x \cos y$ near the point $(1, \pi/4)$ by Taylor's theorem.* **[Kanpur 2014]**

Ex. 111. *Find Taylor's expansion of $f(x) = \log (1 + x)$, $-1 < x \le 1$ with Lagrange's form of remainder.* **[Kanpur 2012]**

Ex. 112. *Let $f(x) = x$ on $[0, 1]$. Calculate $\underline{\int_0^1} x\, dx$ and $\overline{\int_0^1} x\, dx$ by dissecting $[0, 1]$ into n equal parts and hence show that $f \in R[0, 1]$.* **[Meerut 2012, Purvanchal 2011]**

Sol. Let $P = \left\{0, \dfrac{1}{n}, \dfrac{2}{n}, \ldots, \dfrac{r-1}{n}, \dfrac{r}{n}, \ldots, \dfrac{n}{n} = 1\right\}$ be any portion of [0, 1].

Then, its rth sub-interval $= I_r = [(r - 1)n, r/n]$, $r = 1, 2, 3, \ldots, n$.

If δr be the length of I_r, then $\delta r = r/n - (r - 1)n = 1/n$, $r = 1, 2, \ldots, n$

Let M_r and m_r be respectively the supremum and infimum of f in I_r. Since f is increasing on [0, 1], we have

$M_r = r/n$ and $m_r = (r - 1)/n$ as $f(x) = x \quad \forall\ x \in [0, 1]$

$$\text{Now, } U(P, f) = \sum_{r=1}^{n} M_r\ \delta r = \sum_{r=1}^{n}\left(\frac{r}{n} \times \frac{1}{n}\right) = \frac{1}{n^2}\sum_{r=1}^{n} r = \frac{1}{n^2} \times (1 + 2 + 3 + \ldots + n)$$

$$= (1/n^2) \times \{n(n + 1)/2\} = (n + 1)/2n = 1/2 + (1/2n)$$

$$\text{Hence,} \quad \overline{\int_0^1} f\, dx = \lim_{n\to\infty} U(P, f) = \lim_{n\to\infty} \{(1/2) + (1/2n)\} = 1/2 \qquad \ldots(1)$$

$$\text{Next, } L(P, f) = \sum_{r=1}^{n} m_r\ \delta r = \sum_{r=1}^{n}\left(\frac{r-1}{n} \times \frac{1}{n}\right) = \frac{1}{n^2}\sum_{r=1}^{n} (r - 1) = \frac{1}{n^2}\{1 + 2 + \ldots + (n - 1)\}$$

$$= (1/n^2) \times \{n(n - 1)/2\} = (n - 1)/2n = 1/2 - (1/2n)$$

$$\text{Hence,} \quad \underline{\int_0^1} f\, dx = \lim_{n\to\infty} L(P, f) = \lim_{n\to\infty} \{(1/2) - (1/2n)\} = 1/2 \qquad \ldots(2)$$

From (1) and (2), $\overline{\int_0^1} f\,dx = \underline{\int_0^1} f\,dx = \dfrac{1}{2}$. Hence, $f \in [0, 1]$ and $\int_0^1 f\,dx = \int_0^1 x\,dx = \dfrac{1}{2}$.

Ex. 113. *Let $f(x) = x^2$ on $[0, a]$ $a > 0$. Show that $f \in R[0, a]$ and find $\int_0^a f$.*

[Rohikhand 2010]

Sol. Let $P = \left\{0, \dfrac{a}{n}, \dfrac{2a}{n}, \ldots, \dfrac{(r-1)a}{n}, \dfrac{ra}{n}, \ldots, \dfrac{na}{n} = a\right\}$ be any partition of $[0, a]$.

Then, its rth sub-interval $= I_r = [(r-1)a/n,\ ra/n]$, $\quad r = 1, 2, \ldots, n$.
If δr be the length of I_r, then $\delta r = (ra/n) - \{(r-1)a/n\} = a/n$, $r = 1, 2, \ldots, n$
Let M_r and m_r be respectively supremum and infimum of f in I_r. Then,

$$M_r = (ra)/n \quad \text{and} \quad m_r = (r-1)a/n$$

Now,
$$U(P, f) = \sum_{r=1}^{n} M_r\, \delta r = \sum_{r=1}^{1}\left(\frac{r^2a^2}{n^2} \times \frac{a}{n}\right) = \frac{a^3}{n^3}\sum_{r=1}^{n} r^2 = \frac{a^3}{n^3} \times \frac{n(n+1)(2n+1)}{6}$$

Hence,
$$\overline{\int_0^a} f\,dx = \lim_{n\to\infty} U(P, f) = \lim_{n\to\infty} \frac{a^3}{6}\left(1 + \frac{1}{n}\right)\left(2 + \frac{1}{n}\right) = \frac{a^3}{3} \qquad \ldots(1)$$

Next,
$$L(P, f) = \sum_{r=1}^{n} m_r\, \delta r = \sum_{r=1}^{n}\left\{\frac{(r-1)^2 a^2}{n^2} \times \frac{a}{n}\right\} = \frac{a^3}{n^3}\sum_{r=1}^{n}(r-1)^2 = \frac{a^3}{n^3} \times \frac{(n-1)\,n(2n-1)}{6}$$

Hence,
$$\underline{\int_0^a} f\,dx = \lim_{n\to\infty} L(P, f) = \lim_{n\to\infty} \frac{a^3}{6}\left(1 - \frac{1}{n}\right)\left(2 - \frac{1}{n}\right) = \frac{a^3}{3} \qquad \ldots(2)$$

From (1) and (2), $\overline{\int_0^a} f = \underline{\int_0^a} f$. Hence, $f \in R[0, a]$ and $\int_0^a f = \dfrac{a^3}{3}$.

Ex. 114. *Show that the function represented by $\sum_{n=1}^{\infty} \dfrac{\sin nx}{n^3}$ is differentiable for every x and its derivative is $\sum_{n=1}^{\infty} \dfrac{\cos nx}{n^2}$.* **[Kanpur 2012]**

Sol. Let $f_n(x) = (\sin nx)/n^3$ so that $f_n'(x) = (\cos nx)/n^2$.

Here,
$$|f_n(x)| = \left|\frac{\sin nx}{n^3}\right| = \frac{|\sin nx|}{n^3} \le \frac{1}{n^3} = M_n, \text{ say } \forall\ x \in \mathbf{R}$$

Since $\sum_{n=1}^{\infty} M_n = \sum_{n=1}^{\infty} (1/n^3)$ is convergent, so by Weierstrass M-test, the given series $\sum_{n=1}^{\infty} f_n(x)$ converges uniformly $\forall\ x \in \mathbf{R}$.

Again,
$$|f_n'(x)| = |(\cos nx)/n^2| = |\cos nx|/n^2 \le 1/n^2 = M_n^*, \text{ say } \forall\ x \in \mathbf{R}$$

Since $\sum_{n=1}^{\infty} M_n^* = \sum_{n=1}^{\infty} (1/n^2)$ is convergent, hence as before $\sum_{n=1}^{\infty} f_n'(x)$ is also uniformly convergent. Therefore, by theorem 4, page A.11 (refer Appendix), term-by-term differentiation of the given series $\sum_{n=1}^{\infty} f_n(x)$ is valid. Accordingly, we have

$$\frac{d}{dx}\sum_{n=1}^{\infty} f_n(x) = \sum_{n=1}^{\infty}\frac{d}{dx} f_n(x) \qquad \text{so that} \qquad \frac{d}{dx}\left[\sum_{n=1}^{\infty}\frac{\sin nx}{n^3}\right] = \sum_{n=1}^{\infty}\frac{d}{dx}\left(\frac{\sin nx}{n^3}\right) = \sum_{n=1}^{\infty}\frac{\cos nx}{n^2}$$

Ex. 115. *Discuss the convergence of* $\int_0^{\pi/2}\frac{\log(\sin x)}{(\sin x)^n}\,dx,\ n < 1.$ **[Kanpur 2010, 12]**

Sol. Let $$\mathrm{I} = \int_0^{\pi/2}\frac{\log(\sin x)}{(\sin x)^n}\,dx, \text{ where } n < 1 \qquad \text{...(1)}$$

Putting $\sin x = t$ so that $\cos x\, dx = dt$, (1) reduces to

$$\mathrm{I} = \int_0^1 \frac{\log t}{t^n(1-t^2)^{1/2}}\,dt, \qquad \text{since} \qquad dx = \frac{dt}{\cos x} = \frac{dt}{(1-\sin^2 x)^{1/2}} = \frac{dt}{(1-t^2)^{1/2}} \qquad \text{...(2)}$$

Here $t = 0$ and $t = 1$ are points of infinite discontinuity of $f(t)$ where

$$f(t) = (\log t)/\{t^n(1-t^2)^{1/2}\}$$

Then, (2) becomes $\mathrm{I} = \int_0^1 f(t)\,dt = \int_0^a f(t)\,dt + \int_a^1 f(t)\,dt$, where $0 < a < 1$ ...(3)

Convergence at t = 0. Let $g(t) = (\log t)/t^n$

Then, $$\lim_{t\to 0}\frac{f(t)}{g(t)} = \lim_{t\to 0}\frac{1}{(1-t^2)^{1/2}} = 1 = \text{finite}$$

Hence, $\int_0^a f(t)\,dt$ and $\int_0^a g(t)\,dt$ both have the same nature. But $\int_0^a g(t)\,dt = \int_0^a \{(\log t)/t^n\}\,dt$, $n < 1$ is convergent. Therefore, $\int_0^a f(t)\,dt$ is also convergent.

Convergence at *t* = 1. Let $h(t) = 1/(1-t^2)^{1/2}$

Now, $(1-t^2)^{1/2} > (t^2)^{1/2} = t$ for $t < 1 \quad \Rightarrow \quad 1/(1-t^2)^{1/2} < 1/t$

Taking $\phi(t) = 1/t$, we have $$h(t) < \phi(t) \qquad \text{...(4)}$$

Again, $$\int_a^1 \phi(t)\,dt = \int_a^1 \frac{1}{t}\,dt = [\log t]_a^1 = -\log a = \text{finite}$$

Hence, $$\int_a^1 \phi(t)\,dt \text{ is convergent} \qquad \text{...(5)}$$

From (4) and (5), it follows that $\int_a^1 h(t)\,dt$ is convergent.

But, $$\lim_{t\to 1}\frac{f(t)}{h(t)} = \lim_{t\to 1}\frac{\log t}{t^n} = 0 = \text{finite}, \qquad \text{when} \qquad n = 1,$$

showing $\int_a^1 f(t)\,dt$ and $\int_a^1 h(t)\,dt$ both have the same nature.

Since $\int_a^1 h(t)\,dt$ is convergent, hence $\int_a^1 f(t)\,dt$ is also convergent.

Since $\int_0^a f(t)\,dt$ and $\int_a^1 f(t)\,dt$ are both convergent, (3) shows their sum, namely, $\int_0^1 f(t)\,dt$ is also convergent. Therefore, the given integral is convergent.

Ex. 116. *Prove that* $\int_{-\infty}^{\infty} \frac{dx}{1+x^2} = \pi$ **[Kanpur 2008]**

Hint: Proceed as in examples given on page 16.17.

Ex. 117. *Test the convergence of* $\int_0^{\infty} e^{-x} \frac{\sin x}{x}\, dx$ **[Kanpur 2011, 13]**

Hint: Proceed as in Ex. 3, page 16.30 by taking $f(x) = (\sin x)/x$ *and* $\phi(x) = e^{-x}$.

Ex. 118. *Show that* $\int_0^{\infty} \frac{\sin x}{x^2}\, dx$ *convergent.*

[Bundelkhand 2011, Purvanchal 2010. Rhikhand 2012]

Hint: Proceed as in part (*iv*) of solved Ex. 1, page 16.20.

Ex. 119. *Show that the power series* $1 + 2x + 3x^2 + 4x^3 + \ldots$ *has radius of convergence equal to 1.* **[Kanpur 2012]**

Sol. The given power series is $\sum_{n=1}^{\infty} (n+1)\, x^n = \sum_{n=1}^{\infty} a_n\, x^n$, say

Hence, here $a_n = n + 1$. If R be the required radius of convergence, then

$1/R = \lim_{n\to\infty} |a_n|^{1/n} = \lim_{n\to\infty} (n+1)^{1/n} = \lim_{n\to\infty} n^{1/n} (1+1/n)^{1/n} = 1\times 1 = 1,$

showing that the radius of convergence of the given series = R = 1.

Ex. 120. (*a*) *Define power series of complex variable and give formula for its radius of convergence.* **[Kanpur 2006]**

(*b*) Find the radius of convergence of the power series $\sum_{n=1}^{\infty} (z^n / \sqrt{n+1})$.

[Jabalpur 2000, Kanpur 2014, Sagar 199]

Sol. (*a*). A series of the form $\sum_{n=1}^{\infty} a_n\, z^n$ is known as a power series of complex variable, where z is a complex variable and a_n is complex constant (i.e. a_n is free from $z\ \forall\ n \in \mathbf{N}$). The radius of convergence R of the series $\sum_{n=1}^{\infty} a_n\, z^n$ is given by

$$\frac{1}{R} = \lim_{n\to\infty} \frac{a_{n+1}}{a_n} \qquad \text{or} \qquad \frac{1}{R} = \lim_{n\to\infty} |a_n|^{1/n}$$

(*b*). Let the given power series be denotes by $\sum_{n=1}^{\infty} a_n\, z^n$. Then, here

$$a_n = 1/\sqrt{n+1} \qquad \text{so that} \qquad a_{n+1} = 1/\sqrt{n+2}.$$

Hence, the required radius of convergence R of the given series is given by

$$R = \lim_{n\to\infty} \left|\frac{a_n}{a_{n+1}}\right| = \lim_{n\to\infty} \left|\left(\frac{n+2}{n+1}\right)^{1/2}\right| = \lim_{n\to\infty} \left|\left(\frac{1+2/n}{1+1/n}\right)^{1/2}\right| = 1.$$

Ex. 121. *Determine the radius of convergence of the following series*

(*i*) $\sum_{n=1}^{\infty} \frac{(2n)!}{(n!)^2} z^n$ (*ii*) $\sum_{n=1}^{\infty} \frac{z^{2n+1}}{2n+1}$ **[Meerut 2012]** **Ans.** (*i*) 1/4 (*ii*) 1.

Ex. 122. *Show that the power series $x - (x^2/2) + (x^3/3) - (x^4/4) + ...$ is convergent for $-1 < x \le 1$.* **[Meerut 2013]**

Sol. The given power series is $\sum_{n=1}^{\infty} \frac{(-1)^{n-1}}{n} x^n = \sum_{n=1}^{\infty} a_n$, say so that $a_n = \frac{(-1)^{n-1}}{n}$

Let R be the radius of convergence of the given series. Then,

$$R = \lim_{n\to\infty} \left|\frac{a_n}{a_{n+1}}\right| = \lim_{n\to\infty} \left|\frac{n+1}{n}\right| = \lim_{n\to\infty} \left|1+\frac{1}{n}\right| = 1$$

Therefore, the given series is convergent for $\{x\} < 1$, i.e., for $-1 < x < 1$. Again, for $x = 1$, the given series reduces to $\sum_{n=1}^{\infty} \{(-1)^{n-1}/n\} = \sum_{n=1}^{\infty} (-1)^{n-1} u_x$, say. Then, here $u_n = 1/n$

so that $\lim_{n\to\infty} u_n = 0$ and $1/n > 1/(n+1) \Rightarrow u_n > u_{n+1} \ \forall n \in N$

Hence, by Leibnitz's test, $\sum_{n=1}^{\infty} (-1)^n u_n$, i.e., $\sum_{n=1}^{\infty} \{(-1)^{n+1}/n\}$ is convergent.

For $x = -1$, the given series is $-(1 + 1/2 + 1/3 + ...)$, which is divergent.

Hence, the given series is convergent for $-1 < x \le 1$.

Ex. 123. *Give an example to show that in a metric space the union of an infinite collection of closed sets is not necessarily closed.* **[Meerut 2010, Kanpur 2014]**

Sol. Consider the usual metric space $(\mathbf{R}, d)$ where $d(x, y) = |x - y|$, $\forall\ x, y \in \mathbf{R}$. Then, consider an infinite collection of closed sets $A_n = [1/n, 1]$ where $n \in \mathbf{N}$. However, $\bigcup_{n=1}^{\infty} A_n =]0, 1]$, which is not a closed set.

Ex. 124. *Show that the closure $\overline{A}$ of a subset A of a metric space X is the set of points whose distance from A is zero, i.e., $\overline{A} = \{x : d(x, A) = 0\}$.* **[Meerut 2010, 12]**

Sol. If $d(x, A) = 0$, then x is an adherent point of A, i.e., $x \in \overline{A}$.

On the other hand, if $d(x, A) = r > 0$, then the open sphere $S(x, r/2)$ contains no point of A. Hence, x is not an adherent point of A, i.e., $x \notin \overline{A}$. Therefore, we have $\overline{A} = \{x : d(x, A) = 0\}$.

Ex. 125. *Let (X, d) and (Y, ρ) be metric spaces and $f : X \to Y$ be a function. Then, f is continuous if and only if $\overline{f^{-1}(B)} \subset f^{-1}(\overline{B})$ for every subset B of Y.*

Sol. Let f be continuous and let $A = f^{-1}(B)$. Then we have

$$f(A) \subset B \Rightarrow \overline{f(A)} \subset \overline{B} \Rightarrow f(\overline{A}) \subset \overline{B}, \text{ by theorem III, page 19.8.}$$

Hence, $\overline{A} \subset f^{-1}(\overline{B}) \Rightarrow \overline{f^{-1}(B)} \subset f^{-1}(\overline{B})$, since $A = f^{-1}(B)$

Conversely, let $\overline{f^{-1}(B)} \subset f^{-1}(\overline{B})$, for every subset B of Y. Let F be any closed set in Y so that $\overline{F} = F$. Now, by the hypothesis, we have

$$\overline{f^{-1}(F)} \subset f^{-1}(\overline{F}) = f^{-1}(F), \quad \text{since} \quad \overline{F} = F \qquad ...(1)$$

But, we have
$$f^{-1}(F) \subset \overline{f^{-1}(F)} \qquad ...(2)$$

From (1) and (2), $\overline{f^{-1}(F)} = f^{-1}(F)$. Therefore, $f^{-1}(F)$ is closed in X.

Hence, by theorem II, page 19.18, it follows that f is continuous.

Ex. 126. *Let s = <s_n> be a Cauchy sequence in a metric space (X, d) and let < s_{n_i} > be a subsequence of s, then show that* $\lim_{n \to \infty} (s_n, s_{n_i}) = 0$ **[Meerut 2010]**

Sol. Since <s_n> is a Cauchy sequence, for each ε > 0 there exits a positive integer *p* such that

$$m, n \geq p \quad \Rightarrow \quad d(s_{nr}, s_n) < \varepsilon$$

Again, since $s_{n_i} \geq n$, it follows that

$$n \geq p \quad \Rightarrow \quad d(s_{n_i}, s_n) < \varepsilon, \quad \text{because} \quad n_i \geq n \geq p$$

Hence, we have $\lim_{n \to \infty} d(s_{n_i}, s_n) = 0$ or $\lim_{n \to \infty} d(s_n, s_{n_i}) = 0$

Ex. 127. *Let {b_1, b_2, ...} be a Cauchy sequence in a metric space X and let {a_1, a_2, ...} be a sequence in X such that d(a_n, b_n) < 1/n for every n ∈ N, then show that*

(i) {a_n} is a Cauchy sequence in X

(ii) {a_n} converges to some point p ∈ X if and only if {b_n} converges to p. **[Meerut 2010]**

Sol. Part (*i*). Since {b_n} is a Cauchy sequence, for each ε > 0 there exists a positive integer q_1 such that

$$m, n \geq q_1 \quad \Rightarrow d(b_m, b_n) < \varepsilon/3$$

Again, for this ε we may find a positive integer q_2 such that

$$m, n \geq q_2 \quad \Rightarrow \quad 1/m < \varepsilon/3 \quad \text{and} \quad 1/n < \varepsilon/3$$

Thus, if we choose $q = \max \{q_1, q_2\}$, then

$$m, n \geq q \quad \Rightarrow \quad d(b_m, b_n) < \varepsilon/3,\ 1/m < \varepsilon/3 \text{ and } 1/n < \varepsilon/3 \qquad ...(1)$$

Now, by triangle inequality, for *m*, *n* ≥ *q*, we have

$$d(a_m, a_n) \leq d(a_m, b_m) + d(b_m, b_n) + d(b_n, a_n) < 1/m < d(b_m, b_n) + 1/n$$

$$[\because d(b_m, a_m) < 1/m \text{ and } d(b_n, a_n) < 1/n]$$

$$< \varepsilon/3 + \varepsilon/3 + \varepsilon/3, \text{ using (1)}$$

Thus, $m, n \geq q \Rightarrow d(a_m, a_n) < \varepsilon \Rightarrow \{a_n\}$ is a Cauchy sequence.

Part (*ii*). Let $b_n \to p$. Then for each ε > 0 there exists m_1 ∈ **N** such that

$$n \geq m_1 \quad \Rightarrow \quad d(b_n, p) < \varepsilon/2$$

Again, for this ε we may find m_2 ∈ **N** such that

$$n \geq m_2 \quad \Rightarrow \quad 1/n < \varepsilon/2$$

Thus, if we choose $m = \max \{m_1, m_2\}$, then

$$n \geq m \Rightarrow d(b_n, p) < \varepsilon/2 \quad \text{and} \quad 1/n < \varepsilon/2 \qquad ...(2)$$

Now, by triangle inequality, for *n* ≥ *m*, we have

$$d(a_n, p) \leq d(a_n, b_n) + d(b_n, p) = d(b_n, a_n) + d(b_n, p) < 1/n + d(b_n, p)$$

$$\Rightarrow d(a_n, p) < \varepsilon/2 + \varepsilon/2, \quad \text{since} \quad 1/n < \varepsilon/2 \quad \text{and} \quad d(b_n, p) < \varepsilon, \text{ by (2)}$$

Thus, for *n* ≥ *m*, we have $d(a_n, p) < \varepsilon \Rightarrow a_n \to p$ whenever $b_n \to p$

Similarly, we can show that $b_n \to p$ whenever $a_n \to p$.

Ex. 128. *Show that a metric space (X, d) is countably compact if and only if every countable family of non-empty closed sets with finite inter section property has non-empty intersection.*

Sol. Let (X, *d*) be a countably compact and let {F_n : *n* ∈ **N**} be any countable family of closed sets in (X, *d*) with finite intersection property. Let, if possible,

$$\cap \{F_n : n \in \mathbf{N}\} = \phi \qquad ...(1)$$

On taking complements in X and using De-Morgan's law, (1) yields

$$\cup \{F_n^c : n \in \mathbf{N}\} = X, \qquad ...(2)$$

showing that $C = \{F_n^c : n \in \mathbf{N}\}$ is an open cover of X. Since X is countably compact, C is reducible to a finite subcover $\{F_{n_i}^C : i = 1, 2, ..., n\}$ such that

$$\cup \{F_{n_i}^c : i = 1, 2, ..., n\} = X \qquad ...(3)$$

On taking complements in X and using De-Morgan's law, (3) yields

$$\cap \{F_{n_i}^c : i = 1, 2, ..., n\} = \phi.$$

This is a contradiction of the fact that $\{F_n : n \in \mathbf{N}\}$ has the finite intersection property. Since the contradiction arises by assuming that $\cap \{F_n : n \in \mathbf{N}\} = \phi$, it follows that

$$\cap \{F_n : n \in \mathbf{N}\} \neq \phi$$

Conversely, let every countable family of closed sets in (X, *d*) with finite intersection property has non-empty intersection. We are to show that (X, *d*) is countably compact. Let $\{G_n : n \in \mathbf{N}\}$ be any countable open cover of X so that

$$\cup \{G_n : n \in \mathbf{N}\} = X \qquad ...(4)$$

On taking complements in X and using De-Morgan's law, (4) yields

$$\cap \{G_n^c : n \in \mathbf{N}\} = \phi,$$

showing that $\{G_n^c : n \in \mathbf{N}\}$ is a countable family of closed sets in X whose intersection is empty. Hence, by hypothesis, this family cannot have finite intersection property. Therefore, there exists a sub family $\{G_{n_i}^c : i = 1, 2, ..., n\}$ such that

$$\cap \{G_{n_i}^c : i = 1, 2, ..., n\} = \phi, \qquad ...(5)$$

On taking complements in *X* and using De-Morgan's law, (5) yields

$$\cup \{G_{n_i} : i = 1, 2, 3, ..., n\} = X,$$

showing that there exists a finite subcover $\{G_{n_i} : i = 1, 2, ..., n\}$ of countable cover $\{G_n : n \in \mathbf{N}\}$ and hence X is countably compact.

Ex. 129. *Show that a countably compact metric space has BWP.* **[Meerut 2010]**

Sol. Let (X, *d*) be a countably compact and suppose, if possible that it does not have BWP. Then, there must exist an infinite set S having no limit point. We know that every infinite set contains a countably infinite subset (refer chapter 4). Let A be a countably infinite subset of S. Then, A has no limit point. It follows that A is a closed set. Again, for each $a_n \in A$, a_n is not a limit point of A. Hence, there exists an open set G_n such that $a_n \in G$ and $G_n \cap A = \{a_n\}$.

Therefore, the collection $\{G_n : n \in \mathbf{N}\} \cup (X - A)$ is a countable cover of X. This cover has no finite subcover because if we remove a single G_n, it will not be cover of X since then a_n will not be covered. Therefore, X is not countably compact. But this contradicts the hypothesis. Hence, (X, *d*) must have BWP.

Ex. 130. *Let (X, d) be a metric space. Show that the following statements are equivalent.*

(a) X is compact *(b) X is countably compact*

(c) X has BWP *(d) X is sequentially compact* **[Meerut 2012]**

Sol. *(a)* ⇒ *(b)*. It follows from theorem IV, page 19.34 and Ex. 128.

(b) ⇒ *(c)*. It follows from Ex. 129.

(c) ⇒ *(d)*. It follows from theorem V, page 19.35.

(d) ⇒ *(a)*. It follows from theorem XIV, page 19.37.

Thus, *(a)* ⇒ *(b)* ⇒ *(c)* ⇒ *(d)* ⇒ *(a)*, showing that the four statements *(a)*, *(b)*, *(c)* and *(d)* are equivalent.

Ex. 131. *Let (X, d) be a metric space. Show that the following statements are equivalent.*

(a) X is compact *(b) X has BWP* *(c) X is sequentially compact*

Sol. Left as an exercise.

Ex. 132. *Prove that the integral function of an even function is always an odd function.* **[Kanpur 2013]**

Sol. Let g be the integral function of an even function f. Then, by definition, we have

$$g(x) = \int f(x)\, dx, \qquad \ldots (1)$$

where $$f(-x) = f(x), \quad \text{since } f \text{ is an even function} \qquad \ldots (2)$$

Putting $x = -t$ and $dx = -dt$, (1) reduces to $$g(-t) = -\int f(-t)\, dt \qquad \ldots (3)$$

From (3), we have $$g(-x) = -\int f(-x)\, dx = -\int f(x)\, dx, \text{ using (2)}$$

Thus, $$g(-x) = -g(x), \text{ using (1)} \qquad \ldots (4)$$

From (4), it follows that the integral function g of an even function f is always an odd function.

Ex. 133. *Show that every bounded monotonic sequence converges.* **[Kanpur 2012]**

Sol. For proof, refer the second part of the theorem on page 5.38

Ex. 134. *If f is sectionally continuous on an interval $(0, c)$ and b_n denotes the coefficient in the Fourier sine series for f on that interval, then using Bessel's inequality, prove that the series of terms b_n converges and that* $$\sum_{n=1}^{\infty} b_n^2 \le \frac{2}{c}\int_0^c \{f(x)\}^2\, dx$$ **[Meerut 2012]**

Sol. Proceed as in property II on page 22.38

Ex. 135. *Show that* $\displaystyle\int_a^{\infty} \frac{\cos \alpha x - \cos \beta x}{x}\, dx$ *is convergent if $a > 0$.* **[Rohilkhand 2011]**

Sol. Here, $$\int_a^{\infty} \frac{\cos \alpha x - \cos \beta x}{x}\, dx = \int_a^{\infty} \frac{\cos \alpha x}{x}\, dx - \int_a^{\infty} \frac{\cos \beta x}{x}\, dx = I_1 - I_2, \text{ say} \qquad \ldots (1)$$

where $$I_1 = \int_a^{\infty} \frac{\cos \alpha x}{x}\, dx \quad \text{and} \quad I_2 = \int_a^{\infty} \frac{\cos \beta x}{x}\, dx \qquad \ldots (2)$$

Let us first examine I_1 for its convergence. Let us suppose that

$$f(x) = \cos \alpha x \quad \text{and} \quad \phi(x) = 1/x \qquad \ldots (3)$$

Then, $$I_1 = \int_a^{\infty} \frac{\cos \alpha x}{x}\, dx = \int_a^{\infty} f(x)\,\phi(x)\, dx \qquad \ldots (4)$$

Now, $\phi(x)$ is monotonic and it tends to zero as $x \to \infty$. Again, using the inequality $|x - y| \le |x| + |y|$, we have

$$\left|\int_a^t f(x)\, dx\right| = \left|\int_a^t \cos \alpha x\, dx\right| = |(1/\alpha) \times (\sin \alpha t - \sin \alpha a)| \le (1/|\alpha|)\,[|\sin \alpha t| + |\sin \alpha a|]$$

Thus, $$\left|\int_a^t f(x)\, dx\right| \le 2/|\alpha| \quad \Rightarrow \quad \int_a^t f(x)\, dx \text{ is bounded for all } x \ge a$$

Therefore, by Dirichlet test $\int_a^{\infty} f(x)\,\phi(x)\, dx$, *i.e.*, $\int_a^{\infty} [(\cos \alpha x)/x]\, dx$ is convergent. Thus, from (4), it follows that I_1 is convergent.

Proceeding likewise, we see that I_2 is also convergent. Since both integrals occuring on the right hand side of (1) are convergent, it follows that the given integral on the left hand side of (1) is also convergent.

Ex. 136. *The definite integral* $\int_a^b f(x)\,dx$ *is said to be a proper integral if*

(*a*). *the range of integration is finite* (*b*). *the integrand f(x) is bounded*

(*c*). *both (a) and (b)* (*d*). *neither (a) not (b)*

Sol. Ans. (*c*). Refer Art. 16.1, page 16.1

Ex. 137. *The integral* $\int_a^\infty \frac{1}{x^n}\,dx$, *where a > 0, is convergent when*

(*a*). $n = 1$ (*b*). $n < 1$ (*c*). $n \le 1$ (*d*). $n > 1$ **[Meerut 2013]**

Sol. Ans. (*d*). Refer Art. 16.13, page 16.19

Ex. 138. *What is the value of μ in the μ-test of convergence for the integral* $\int_0^\infty \frac{x}{(1+x)^3}\,dx.$

(*a*). *1* (*b*). *–1* (*c*). *2* (*d*). *3* **[Meerut 2013]**

Sol. Ans. (*c*). For μ-test, refer page 16.20. Here, take $f(x) = x/(1+x)^3$

Then, we have $$x^\mu f(x) = x^\mu \times \frac{x}{(1+x)^3} = \frac{x^{\mu-2}}{(1+1/x)^3},$$

which tends to 1 as $x \to \infty$ if we choose $\mu = 2$

Thus, $\lim_{x\to\infty} x^\mu f(x) = \lim_{x\to\infty} x^\mu \times \{x/(1+x)^3\}$ is finite if $\mu = 2 > 1$

Therefore, by μ-test the given integral is convergent if we choose $\mu = 2$.

Ex. 139. *The integral* $\int_0^\infty \frac{e^{-x}\sin x}{x}\,dx$ *is*

(*a*). *divergent* (*b*). *not divergent* (*c*). *convergent* (*d*). *not convergent* **[Meerut 2013]**

Sol. Ans. (*c*). Taking $a = 1$, proceed as in Ex. 3, page 16.30.

Ex. 140. *Evaluate* $\int_1^\infty \frac{dx}{\sqrt{x}}\,dx$

(*a*). 2 (*b*). –2 (*c*). *1* (*d*). ∞ **[Meerut 2013]**

Sol. Ans. (*d*). By definition, we have (refer Art. 16.10, page 16.16)

$$\int_1^\infty \frac{dx}{\sqrt{x}} = \lim_{t\to\infty}\int_1^t \frac{dx}{\sqrt{x}} = \lim_{t\to\infty}\left[2\sqrt{x}\right]_1^t = \lim_{t\to\infty}\left[2\sqrt{t}-2\right] = \infty.$$

Ex. 141. *If* $\int_a^b f(x)\,dx$ *is convergent and* $\phi(x)$ *is bounded and monotonic for* $x > a$, *then* $\int_a^\infty f(x)\,\phi(x)\,dx$ *is*

(*a*). *divergent* (*b*). *bounded* (*c*). *convergent* (*d*). *monotome* **[Meerut 2013]**

Sol. Ans. (*c*). Refer Dirichlet test on page 16.27.

Ex. 142. *Define dense and perfect sets with examples.* **[Kanpur 2012]**

Sol. Refer Art. 3.17 and illustration 1 on page 3.29 for a dense set. Refer Art. 3.17 and illustrations 5 and 8 on page 3.29 for a perfect sets.

Ex. 143. *Prove that Cauchy sequence in* ***R*** *is convergent.* **[Kanpur 2014]**

Sol. For proof, refer theorem II, page 5.30.

Ex. 144. *Prove that an open interval is incomplete.* **[Meerut 2010]**

Sol. Let $X = (0, 1)$ and let $d(x, y) = |x - y|$, $\forall\, x, y \in X$. Let $<x_n>$ be the sequence in X defined by setting $x_n = 1/n$, $\forall\, n \in \mathbf{N}$ and let ε be any arbitrary positive number. Then, we have

$$d(x_m, x_n) = |x_m - x_n| = |1/m - 1/n| \leq 1/m + 1/n \qquad \text{... (1)}$$

Let p be any positive integer greater than $2/\varepsilon$. Then, we have

$$1/m + 1/n < \varepsilon/2 + \varepsilon/2 \quad \text{or} \quad 1/m + 1/n < \varepsilon, \quad \forall \quad m, n \geq p, \qquad \text{... (2)}$$

From (1) and (2), it follows that $d(x_m, x_n) < \varepsilon$ whenever $m, n \geq p$, showing that $<x_n>$ is a Cauchy sequence in X.

Now, let $x \in X$. By the archimedian property of real numbers (see page 2.20), there exists a positive integer n such that

$$n \leq 1/x < n + 1 \qquad \text{so that} \qquad 1/n \geq x > 1/(n + 1)$$

Two cases arise:

Case I: If $x = 1/n$, then $(1/(n - 1), 1/(n + 1))$ is a neighbourhood of x in $(0, 1)$ which contains only one element of the sequence $<1/n>$

Case II: If $x \neq 1/n$, then $(1/(n + 1), 1/n)$ is a neighbourhood of x in $(0, 1)$ which does not contain any element of the sequence $<1/n>$

In either case the sequence $<x_n>$, *i.e.*, $<1/n>$ cannot coverage to x. Since x is any point in X, it follows that the sequence $<x_n>$ cannot coverage to any point of X.

Hence (X, d) is an incomplete metric space *i.e.*, an open interval $(0, 1)$ is incomplete.

Ex. 145. *Let f be an one-to-one continuous function from a compact subset X into a Hausdroff space Y. Show that X and f(X) are homeomorphic.* **[Meerut 2010]**

Ex. 146. *Test for uniform convergence of the series of continuous functions for which the nth partial sum $S_n(x) = 1/(nx + 1)$, $0 \leq x \leq 1$.* **[Kanpur 2013]**

Sol. In what follows, we shall proceed as explained in examples on page 15.22.

The sum function $S(x)$ of the given series is given by

$$S(x) = \lim_{n\to\infty} S_n(x) = \lim_{n\to\infty} \frac{1}{nx+1} = \begin{cases} 1, \text{ if } x = 0 \\ 0, \text{ if } 0 < x \leq 1 \end{cases} \qquad \text{... (1)}$$

From (1) it follows that $S(x)$ is discontinuous at $x = 0$ and hence the given series cannot be uniformly continuous on $[0, 1]$.

Ex. 147. *Let E be the set of all rationals ρ such that $2 < \rho^2 < 3$. Then, E is*

(*a*). *compact in Q* (*b*). *closed and bounded in Q*

(*c*). *not compact in Q* (*d*). *closed and unbounded in Q* **[GATE 2000]**

Ex. 148. *The Fourier expansion in the interval [–4, 4] of the function*

$$f(x) = \begin{cases} x, \text{ for } \; 4 \leq x \leq 0 \\ x, \text{ for } 0 \leq x \leq 4 \end{cases} \quad \textit{has}$$

(*a*). *no cosine term* (*b*). *no sine term*

(*c*). *both cosine and sine terms* (*d*). *none of these* **[GATE 2000]**

Sol. Ans. (*b*). We have $$f(x) = \begin{cases} x, & -4 \le x \le 0 \\ -x, & 0 \le x \le 4 \end{cases} \qquad \dots (1)$$

From (1), $$f(-x) = \begin{cases} -x, & 4 \ge x \ge 0 \\ x, & 0 \ge x \ge -4 \end{cases} \quad \text{or} \quad f(-x) = \begin{cases} x, & -4 \le x \le 0 \\ -x, & 0 \le x \le 4 \end{cases} \qquad \dots (2)$$

From (1) and (2), we have $f(x) = f(-x)$ for all $x \in [-4, 4]$. Hence, $f(x)$ is an even function on $[-4, 4]$. Therefore, from special case I on page 22.11, it follows that there will be no sine term in the Fourier series (expansion) of $f(x)$ given by (1).

Ex. 149. *The series* $\displaystyle\sum_{n=1}^{\infty} \frac{z^n}{n\sqrt{n+1}}$, $|z| \le 1$ *is*

(*a*). *uniformly but not absolutely convergent* (*b*). *uniformly and absolutely convergent*
(*c*). *absolutely convergent but not uniformly convergent*
(*d*). *convergent but not uniformly convergent* **[GATE 2001]**

Sol. Ans. (*b*). Let the given series be denoted by $\displaystyle\sum_{n=1}^{\infty} u_n$. Then, we have

$$\lim_{n\to\infty} \frac{|u_n|}{|u_{n+1}|} = \lim_{n\to\infty} \frac{|z|^n}{n\sqrt{n+1}} \frac{(n+1)\sqrt{n+2}}{|z|^{n+1}} = \lim_{n\to\infty} \frac{(1+1/n)(1+2/n)^{1/2}}{(1+1/n)^{1/2}} \frac{1}{|z|} = \frac{1}{|z|} > 1, \text{ if } |z| < 1.$$

By ratio test, $\displaystyle\sum_{n=1}^{\infty} |u_n|$ is convergent and hence the given series is absolutely convergent when $|z| < 1$.

Again, if $|z| = 1$, then $$|u_n| = \frac{|z|^n}{n\sqrt{1+n}} = \frac{1}{n^{3/2}(1+1/n)^{1/2}}$$

If we choose $|v_n| = (1/n^{3/2})$, then we have

$$\lim_{n\to\infty} \frac{|u_n|}{|v_n|} = \lim_{n\to\infty} \frac{1}{(1+|1/n)^{1/2}} = 1, \text{ which is finite and non-zero}$$

Now, $\displaystyle\sum_{n=1}^{\infty} |v_n| = \sum_{n=1}^{\infty} (1/n^{3/2})$ is convergent and hence (by comparison test, page 6.9) $\Sigma|u_n|$ is also convergent. Therefore, the given series $\Sigma|u_n|$ is absolutely convergent when $|z| = 1$.

Combining the above two cases, it follows that the given series is absolutely convergent when $|z| \le 1$.

We now apply Weierstrass's M-test (refer page 15.14) for testing uniform convergence of the given series.

Here, $$|u_n| = \left|\frac{z^n}{n\sqrt{n+1}}\right| \le \frac{1}{n\sqrt{n+1}} = M_n, \text{ (say)}, \qquad \text{since} \quad |z| \le 1$$

We now show that $\displaystyle\sum_{n=1}^{\infty} M_n = \sum_{n=1}^{\infty} \frac{1}{n\sqrt{n+1}} = \sum_{n=1}^{\infty} \frac{1}{n^{3/2}(1+1/n)^{1/2}}$ is convergent.

Let $V_n = 1/n^{3/2}$. Since $M_n = 1/\{n^{3/2}(1 + 1/n)^{1/2}\}$, we get

$$\lim_{n\to\infty} \frac{M_n}{V_n} = \lim_{n\to\infty} \frac{1}{(1+1/n)^{1/2}} = 1, \text{ which is finite and non-zero.}$$

Since $\sum_{n=1}^{\infty} V_n = \sum_{n=1}^{\infty} (1/n^{3/2})$ is convergent, it follows that ΣM_n is also convergent. Therefore, by Weierstrass's M-test, it follows that the given series is uniformly convergent.

Thus, the given series is uniformly convergent and absolutely convergent. Hence, the option (*b*) is correct.

Ex. 150. *Pick out the largest of the sets given below on which the sequence of functions* $\left\{e^{-n\cos^2 x}\right\}_{r=1}^{\infty}$ *converges uniformly*

(*a*). *[(0, 9π/20) ∪ (11π/20, π]* (*b*). *[0, π/2) ∪ (π/2, π]*
(*c*). *[0, π/2 –δ) ∪ (π/2 + δ, π], 0 < δ < π/100* (*d*). *[0, π]* **[GATE 2002]**

Sol. Ans. (*b*). Left as an exercise.

Ex. 151. *Let X = (0, 1) ∪ (2, 3) be an open set in* ***R****. Let f be a continuous function on X such that the derivative f′ (x) = 0 for all x. Then, the range of f has*

(*a*). *uncountable number of points* (*b*). *countably infinite number of points*
(*c*). *atmost 2 points* (*d*). *atmost 1 point* **[GATE 2003]**

Sol. Ans. (*d*). Try yourself.

Ex. 152. *The continuous function f :* ***R*** *→* ***R*** *defined by* $f(x) = (x^2 + 1)^{2003}$ *is*

(*a*). *onto but not one-one* (*b*). *one-one but not onto*
(*c*). *both one-one and onto* (*d*). *neither one-one nor onto* **[GATE 2003]**

Sol. Ans. (*d*). Given that $f(x) = (x^2 + 1)^{2003}$ ... (1)

From (1), $f(1) = 2^{2003} = f(-1)$. Hence, two different elements 1 and –1 of the domain *R* of *f* have the same image and therefore *f* cannot be one-one.

From (1), it is clear that the pre-image of the element –1 belonging to the co-domain **R** of *f* does not exist. Hence, *f* cannot be onto function. Thus, *f* is neither one-one nor onto and therefore the option (*d*) is correct.

Ex. 153. *Let X = {x in* ***Q*** *|0 < x < 1|} be the metric space with standard metric from* ***R****. The completion of X is*

(*a*). *{x in* ***Q*** *|0 < x < 1|}* (*b*). *{x in* ***R*** *|0 < x < 1|}*
(*c*). *{x in* ***Q*** *|0 ≤ x ≤ 1|}* (*d*). *{x in* ***R*** *|0 ≤ x ≤ 1|* **[GATE 2003]**

Sol. Ans. (*c*). Refer chapter 19.

Ex. 154. *In the interval [–1, 1] the series* $\sum_{n=1}^{\infty} (-1)^{n-1} \frac{x^2 + n^2}{n^3}$ *is*

(*a*). *uniformly and absolutely convergent*
(*b*). *absolutely convergent but not uniformly convergent*
(*c*). *neither uniformly nor absolutely convergent*
(*d*). *uniformly convergent but not absolutely convergent* **[GATE 2003]**

Sol. Ans. (*d*). Try yourself.

Ex. 155. *Let* $\sin z = \sum_{n=0}^{\infty} a_n (z - \pi/4)^n$, *then* a_6 *equals*

(*a*). *0* (*b*). *1/720* (*c*). $1/(720\sqrt{2})$ (*d*). $-\{1/(720\sqrt{2})\}$ **[GATE 2004]**

Sol. Ans. (*d*). Let $$f(z) = \sin z = \sum_{n=0}^{\infty} a_n (z - \pi/4)^n \qquad \text{... (1)}$$

Then, by Taylor's theorem (refer page 17.13), we have

$$a_n = \frac{1}{n!} f^{(n)}(\pi/4) \qquad \text{so that} \qquad a_6 = \frac{1}{6!} f^{(6)}(\pi/4) \qquad \text{... (2)}$$

Now, from (1), $f'(z) = \cos z \quad \Rightarrow f''(z) = -\sin z \quad \Rightarrow f'''(z) = -\cos z$

$\Rightarrow f^{(4)}(z) = \sin z \quad \Rightarrow \quad f^{(5)}(z) = \cos z \quad \Rightarrow \quad f^{(6)}(z) = -\sin z$ ... (3)

Then, from (3), $$f^{(6)}(\pi/4) = -\sin(\pi/4) = -(1/\sqrt{2}) \qquad \text{... (4)}$$

Now, from (2) and (4), $$a_6 = \frac{1}{6!} \times \left(-\frac{1}{\sqrt{2}}\right) = -\frac{1}{720\sqrt{2}}.$$

Ex. 156. *Let* $f : [0, \infty) \to \mathbf{R}$ *be defined by* $$f(x) = \begin{cases} 1/\sqrt{x}, & x \neq 0 \\ 0, & x = 0 \end{cases}$$

Consider two improper integrals $I_1 = \int_0^1 f(x)\, dx$ *and* $I_2 = \int_1^\infty f(x)\, dx$. *Then,*

(*a*). *both* I_1 *and* I_2 *exist* (*b*). I_1 *exists but* I_2 *does not*

(*c*). I_1 *does not exist but* I_2 *does* (*d*). *neither* I_1 *nor* I_2 *exist* **[GATE 2004]**

Sol. Ans. (*b*). Using result of Art. 16.2, page 16.1, we have

$$\lim_{\varepsilon \to 0} \int_\varepsilon^1 f(x)\, dx = \lim_{\varepsilon \to 0} \int_\varepsilon^1 \frac{1}{\sqrt{x}}\, dx, \text{ using the given value of } f(x)$$

$$= \lim_{\varepsilon \to 0} \left[2\sqrt{x}\right]_\varepsilon^1 = \lim_{\varepsilon \to 0} \left[2 - 2\sqrt{\varepsilon}\right] = 2 = \text{finite number,}$$

showing that $\lim_{\varepsilon \to 0} \int_\varepsilon^1 f(x)\, dx$ exists and hence I_1 exists.

Again, using result of Art. 16.10, page 16.16, we have

$$\lim_{t \to \infty} \int_1^t f(x)\, dx = \lim_{t \to \infty} \int_1^t \frac{dx}{\sqrt{x}}, \text{ using the given value of } f(x)$$

$$= \lim_{t \to \infty} \left[2\sqrt{x}\right]_1^t = \lim_{t \to \infty} (2\sqrt{t} - 2) = \infty,$$

showing that $\lim_{t \to \infty} \int_1^t \frac{dx}{\sqrt{x}}$ does not exists and hence I_2 does not exists.

Thus, I_1 exists but I_2 does not exist and hence the option (*b*) is correct.

Ex. 157. *Let* $\sigma > 1$ *and* $g(x) = \sum_{n=1}^{\infty} (1/n^x)$, $\sigma \le x < \infty$. *Then, g(x) is*

(*a*). *not continuous* (*b*). *continuous but not differentiable*
(*c*). *differentiable but not continuously differentiable* (*d*). *continuously differentiable*

Sol. Ans. (*d*). Left as an exercise. **[GATE 2004]**

Ex. 158. *Consider two sequences* $\{f_n\}$ *and* $\{g_n\}$ *of functions where* $f_n : [0, 1] \to \mathbf{R}$ *and* $g_n : \mathbf{R} \to \mathbf{R}$ *are defined by*

$$f_n(x) = x^n \qquad \text{and} \qquad g_n(x) = \begin{cases} \cos\{(x-n)\pi/2\}, \text{ if } x \in [n-1, n+1] \\ 0, \text{ otherwise} \end{cases}. \text{ Then}$$

(*a*). *neither* $\{f_n\}$ *nor* $\{g_n\}$ *is uniformly convergent*
(*b*). $\{f_n\}$ *is not uniformly convergent but* $\{g_n\}$ *is*
(*c*). $\{g_n\}$ *is not uniformly convergent but* $\{f_n\}$ *is*
(*d*). *both* $\{f_n\}$ *and* $\{g_n\}$ *are uniformly convergent* **[GATE 2004]**

Sol. Ans. (*a*). Refer chapter 15.

Ex. 159. *The set of all continuous functions* $f : [0, 1] \to \mathbf{R}$ *satisfying*

$$\int_0^1 t^n f(t)\, dt = 0, \ n = 0, 1, 2, \ldots$$

(*a*). *is empty* (*b*). *contains a single element*
(*c*). *is countably infinite* (*d*). *is uncountably infinite* **[GATE 2004]**

Sol. Ans. (*b*). Refer solved example 1, page 15.30.

Ex. 160. *Let* $f : [0, 1] \to \mathbf{R}$ *and* $g : [0, 1] \to \mathbf{R}$ *be two functions defined by*

$$f(x) = \begin{cases} 1/n, \text{ if } x = 1/n, n \in \mathbf{N} \\ 0, \text{ otherwise} \end{cases} \quad \text{and} \quad g(x) = \begin{cases} n, \text{ if } x = 1/n, n \in \mathbf{N} \\ 0, \text{ otherwise} \end{cases} \quad \text{Then,}$$

(*a*). *both f and g are Riemann–integrable* (*b*). *f is Riemann–integrable but g is not*
(*c*). *g is Riemann–integrable but f is not*
(*d*). *neither f nor g is Riemann–integrable* **[GATE 2004]**

Sol. Ans. (*a*). In what follows, we shall use the method adopted in solved Ex. 30(*iii*), page M.9. From the definition of the given function *f*, it follows that *f* is continuous in [0, 1] except at the points $x = 0, 1, 1/2, 1/3, \ldots\ldots$ Since these points of discontinuity form a countable set and each countable set is a set of measure zero, it follows that *f* is continuous almost everywhere on [0, 1]. Therefore, by Lebesgue–Vitali theorem, it follows that *f* is Riemann–integrable.

Similarly, from the definition of the given function *g*, it follows that *g* is continuous in [0, 1] except at the points $x = 0, 1, 1/2, 1/3, \ldots$ Hence, as discussed above for the function *f*, the given function *g* is also Riemann–integrable.

Thus, both *f* and *g* are Riemann–integrable and hence the option (*a*) is correct.

Ex. 161. *Let* $S = \{1/n : n \in N\} \cup \{0\}$ *and* $T = \{n + 1/n : n \in N\}$ *be the subsets of the metric space* $\mathbf{R}$ *with usual metric. Then*

(*a*). *S is complete but not T* (*b*). *T is complete but not S*
(*c*). *both T and S are complete* (*d*). *neither T nor S is complete* **[GATE 2005]**

Sol. Ans. (*c*). Refer chapter 19. Try yourself.

Ex. 162. *Let $f : \mathbf{R} \to \mathbf{R}$ be a non-zero function such that $|f(x)| \le 1/(1 + 2x^2)$ for all $x \in \mathbf{R}$. Define real valued functions f_n on $\mathbf{R}$ for all $n \in \mathbf{N}$ by $f_n(x) = f(x + n)$. Then, the series $\sum_{n=1}^{\infty} f_n(x)$ converges uniformly*

(*a*). *on [0, 1] but not on [–1, 0]* (*b*). *on [–1, 0] but not on [0, 1]*

(*c*). *on both [–1, 0] and [0, 1]* (*d*). *neither on [–1, 0] nor on [0, 1]* **[GATE 2005]**

Sol. Ans. (*d*). Refer chapter 15. Try yourself.

Ex. 163. *Consider the improper integrals*

$$I_1 = \int_1^{\infty} \frac{dx}{(1+x^2)^{1/2}} \quad \text{and} \quad I_2 = \int_1^{\infty} \frac{dx}{(1+x^2)^{3/2}}. \text{ Then}$$

(*a*). *I_1 converges but not I_2* (*b*). *I_2 converges but not I_1*

(*c*). *both I_1 and I_2 converge* (*d*). *neither I_1 nor I_2 converges* **[GATE 2005]**

Sol. Ans. (*b*). In what follows, we shall use results of Art. 16.12 and Art. 16.13 given on page 16.19.

Let $I_1 = \int_1^{\infty} \frac{dx}{(1+x^2)^{1/2}} = \int_1^{\infty} f(x)\,dx$, where $f(x) = \frac{1}{(1+x^2)^{1/2}} = \frac{1}{x(1+1/x^2)^{1/2}}$

Taking $\phi(x) = 1/x$, we have

$$\lim_{x\to\infty} \frac{f(x)}{\phi(x)} = \lim_{x\to\infty} \frac{1}{x(1+1/x^2)^{1/2}} \times x = 1, \text{which is finite and non-zero,}$$

showing that $\int_1^{\infty} f(x)\,dx$ and $\int_1^{\infty} \phi(x)\,dx$ converge or diverge together.

Since $\int_1^{\infty} \phi(x)\,dx = \int_1^{\infty} \frac{1}{x}\,dx$ diverges, it follows that $\int_1^{\infty} f(x)\,dx$ *i.e.*, I_1 also diverges.

Next, let $I_2 = \int_1^{\infty} \frac{dx}{(1+x^2)^{3/2}} = \int_1^{\infty} g(x)\,dx$, where $g(x) = \frac{1}{(1+x^2)^{3/2}} = \frac{1}{x^3(1+1/x^2)^{3/2}}$

Taking $\Psi(x) = 1/x^3$, we have

$$\lim_{x\to\infty} \frac{g(x)}{\Psi(x)} = \lim_{x\to\infty} \frac{1}{x^3(1+1/x^2)^{3/2}} \times x^3 = 1, \text{ which is finite and non-zero,}$$

showing that $\int_1^{\infty} g(x)\,dx$ and $\int_1^{\infty} \Psi(x)\,dx$ converge or diverge together.

Since $\int_1^{\infty} \Psi(x)\,dx = \int_1^{\infty} \frac{1}{x^3}\,dx$ converges, it follows that $\int_1^{\infty} g(x)\,dx$, *i.e.*, I_2 also converges.

Thus, I_1 is not convergent whereas I_2 is convergent. Hence, the option (*b*) is correct.

Ex. 164. *Let f be a function on **R** defined by*

$$f(x) = \begin{cases} 1, & \text{if } x \text{ is rational} \\ -1, & \text{if } x \text{ is irrational} \end{cases}, \quad \text{known as Dirichlet function.}$$

*Show that f is discontinuous at every point of **R**.*

[Delhi B.A. (Prog.) III, 2011; Kanpur 2011, 13]

Sol. Let $h > 0$ be arbitrary. Then, two cases arise:

Case I: Let $x = a$ be any rational number, so that $f(a) = 1$

Again, by the given definition of f, we have

$$f(a+h) = \begin{cases} 1, \text{ if } a+h \text{ is rational} \\ -1, \text{ if } a+h \text{ is irrational} \end{cases}$$

Hence, $$\lim_{h \to 0} f(a+h) = \begin{cases} 1, \text{ if } a+h \text{ is rational} \\ -1, \text{ if } a+h \text{ is irrational} \end{cases} \quad \text{... (1)}$$

From (1), it follows that $\lim_{h \to 0} f(a+h)$ does not exist and hence f is discontinuous at each rational number $x = a$.

Case II: Let $x = a$ be any irrational number, so that $f(a) = -1$

Now, by the given definition of f, we have

$$f(a+h) = \begin{cases} 1, \text{ if } a+h \text{ is rational} \\ -1, \text{ if } a+h \text{ is irrational} \end{cases}$$

Hence, $$\lim_{h \to 0} f(a+h) = \begin{cases} 1, \text{ if } a+h \text{ is rational} \\ -1, \text{ if } a+h \text{ is irrational} \end{cases} \quad \text{... (2)}$$

From (2), it follows that $\lim_{h \to 0} f(a+h)$ does not exist and hence f is discontinuous at each irrational number $x = a$.

Thus, f is discontinuous at each rational and irrational points of **R** and hence f is discontinuous at every point of **R**.

Ex. 165. *Show that the function f defined by*

$$f(x) = \begin{cases} x, \textit{ if } x \textit{ is rational} \\ 0, \textit{ if } x \textit{ is irrational} \end{cases}$$

is continuous at $x = 0$ *but discontinuous when* $x = \lambda \neq 0$.

Sol. Let us first examine f for its continuity at $x = 0$. Since each neighbourhood of the rational number $x = 0$ contains an infinite number of rationals and irrationals, it follows that $0 + h$ and $0 - h$ may be rational or irrational, where $h > 0$. Now, from the given definition of function f, we have $f(0) = 0$ and

$$f(0+h) = \begin{cases} h, \text{ if } (0+h) \text{ is rational} \\ 0, \text{ if } (0+h) \text{ is irrational} \end{cases}$$

In either case, we have $$\lim_{h \to 0} f(0+h) = 0 \quad \text{... (1)}$$

Again, from the given definition of f, we have

$$f(0-h) = \begin{cases} -h, \text{ if } (0-h) \text{ is rational} \\ 0, \quad \text{if } (0-h) \text{ is irrational} \end{cases}$$

In either case, we have $$\lim_{h \to 0} f(0-h) = 0 \quad \text{... (2)}$$

From (1) and (2), $\lim_{h \to 0} f(0+h) = \lim_{h \to 0} f(0-h) = f(0)$, since $f(0) = 0$

Hence, f is continuous at $x = 0$

Next, let us examine f for its continuity at $x = \lambda \neq 0$.

Since each neighbourhood of any real number λ contains an infinite number of rational and irrational numbers, it follows that $\lambda + h$ and $\lambda - h$ may be rational or irrational, where $h > 0$.

Now, from the given definition of f, we have

$$f(\lambda + h) = \begin{cases} \lambda + h, \text{ if } (\lambda + h) \text{ is rational} \\ 0, \text{ if } (\lambda + h) \text{ is irrational} \end{cases} \quad \text{... (3)}$$

From (3), we have
$$\lim_{h \to 0} f(\lambda + h) = \begin{cases} \lambda, \text{ if } (\lambda + h) \text{ is rational} \\ 0, \text{ if } (\lambda + h) \text{ is irrational} \end{cases} \quad \text{... (4)}$$

From (4), it follows that $\lim_{h \to 0} f(\lambda + h)$ does not exist and hence f is discontinuous at $x = \lambda \neq 0$. Thus, f is continuous at $x = 0$ but discontinuous when $x = \lambda \neq 0$.

Ex. 166. *Show that the function f defined by*

$$f(x) = \begin{cases} x, \textit{ if } x \textit{ is rational} \\ -x, \textit{ if } x \textit{ is irrational} \end{cases}$$

is continuous only at x = 0.

Sol. Let x be any real number. Then, for each $n \in \mathbf{N}$, there exist a rational number a_n and an irrational number b_n such that

$$x - 1/n < a_n < x + 1/n \quad \text{and} \quad x - 1/n < b_n < x + 1/n$$

giving
$$|a_n - x| < 1/n \quad \text{and} \quad |b_n - x| < 1/n \quad \text{... (1)}$$

Taking limit as $n \to \infty$, (1) yields
$$\lim_{n \to \infty} a_n = x = \lim_{n \to \infty} b_n \quad \text{... (2)}$$

If f is continuous at x, then by Heine's definition of continuity (refer Art. 8.15, page 8.29), we must have

$$\lim_{n \to \infty} f(a_n) = f(x) = \lim_{n \to \infty} f(b_n) \quad \text{... (3)}$$

Since a_n is rational and b_n is irrational, from the given definition of the given function f, it follows that

$$f(a_n) = a_n \quad \text{and} \quad f(b_n) = -b_n \quad \text{... (4)}$$

Using (4), (3) reduces to

$$\lim_{n \to \infty} a_n = f(x) = -\lim_{n \to \infty} (-b_n) \quad \text{or} \quad x = f(x) = -x, \text{ using (2)}$$

Therefore, $2x = 0$ so that $x = 0$. Thus, $x = 0$ is the only possible point of continuity of the given function f.

Ex. 167. *Consider the function* $f : \mathbf{R} \to \mathbf{R}$ *defined by*

$$f(x) = \begin{cases} x^4, \textit{ if } x \textit{ is rational} \\ 2x^2 - 1, \textit{ if } x \textit{ is irrational} \end{cases}$$

Let S be the set of points where f is continuous. Then,

(a). $S = \{1\}$ *(b).* $S = \{-1\}$ *(c).* $S = \{1, -1\}$ *(d).* $S = \phi$ **[GATE 2007]**

Sol. Ans. (*c*). Let x be any real number. Then, for each $n \in \mathbf{N}$, there exist a rational number a_n and an irrational number b_n such that

$$x - 1/n < a_n < x + 1/n \quad \text{and} \quad x - 1/n < b_n < x + 1/n$$

giving $$|a_n - x| < 1/n \quad \text{and} \quad |b_n - x| < 1/n \qquad \text{... (1)}$$

Taking limit as $n \to \infty$, (1) yields $$\lim_{n\to\infty} a_n = x = \lim_{n\to\infty} b_n \qquad \text{... (2)}$$

If f is continuous at x, then by Heine's definition of continuity (refer Art. 8.15, page 8.29), we must have

$$\lim_{n\to\infty} f(a_n) = f(x) = \lim_{n\to\infty} f(b_n) \qquad \text{... (3)}$$

Since a_n is rational and b_n is irrational, from the given definition of the given function f, it follows that

$$f(a_n) = a_n^4 \quad \text{and} \quad f(b_n) = 2b_n^2 - 1 \qquad \text{... (4)}$$

Using (4), (3) reduces to $$\lim_{n\to\infty} a_n^4 = f(x) = \lim_{n\to\infty} (2b_n^2 - 1)$$

so that $$\left\{\lim_{n\to\infty} a_n\right\}^4 = 2\lim_{n\to\infty} b_n^2 - 1 \quad \text{or} \quad \left\{\lim_{n\to\infty} a_n\right\}^4 = 2\left\{\lim_{n\to\infty} b_n\right\}^2 - 1 \qquad \text{... (5)}$$

Using (2), (5) yields $x^4 = 2x^2 - 1$ or $(x^2 - 1)^2 = 0$ giving $x = 1$ and $x = -1$.

Hence, the given function f is continuous only at two points $x = 1$ and $x = -1$ and therefore $S = \{1, -1\}$. Hence, the option (*a*) is correct.

Ex. 168. *Discuss the continuity and differentiability of the function f defined by*

$$f(x) = \begin{cases} 2x + 1, & \textit{if } x \textit{ is rational} \\ x^2 - 2x + 5, & \textit{if } x \textit{ is irrational} \end{cases}$$

Sol. Let us first discuss the continuity of the given function f. Let x be any real number. Then, for each $n \in \mathbf{N}$, there exist a rational number a_n and an irrational number b_n such that

$$x - 1/n < a_n < x + 1/n \quad \text{and} \quad x - 1/n < b_n < x + 1/n$$

giving $$|a_n - x| < 1/n \quad \text{and} \quad |b_n - x| < 1/n \qquad \text{... (1)}$$

Taking limit as $n \to \infty$, (1) yields $$\lim_{n\to\infty} a_n = x = \lim_{n\to\infty} b_n \qquad \text{... (2)}$$

If f is continuous at x, then by Heine's definition of continuity (refer Art. 8.5, page 8.29), we must have

$$\lim_{n\to\infty} f(a_n) = f(x) = \lim_{n\to\infty} f(b_n) \qquad \text{... (3)}$$

Since a_n is rational and b_n is irrational, from the given definition of the given function f, we must have

$$f(a_n) = 2a_n + 1 \quad \text{and} \quad f(b_n) = b_n^2 - 2b_n + 5 \qquad \text{... (4)}$$

Using (4), (3) yields $$\lim_{n\to\infty} (2a_n + 1) = f(x) = \lim_{n\to\infty} (b_n^2 - 2b_n + 5)$$

Hence, $$2\lim_{n\to\infty} a_n + 1 = \left\{\lim_{n\to\infty} b_n\right\}^2 - 2\lim_{n\to\infty} b_n + 5 \quad \text{or} \quad 2x + 1 = x^2 - 2x + 5, \text{ using (2)}$$

or $x^2 - 4x + 4 = 0$ or $(x - 2)^2 = 0$ giving $x = 2$. Thus, $x = 2$ is the only possible point of continuity of the given function f.

Next, let us discuss the differentiability of the given function f. In the above discussion, we have shown that f is continuous only at $x = 2$. But from Art. 9.2, we know that if f is differentiable at $x = c$, then f must be continuous at $x = c$. Therefore, f cannot be differentiable at any point except possibly at $x = 2$. Hence, in what follows, we shall test the given function f at $x = 2$ only.

For differentiability of f at $x = 2$, using the given values of f, we obtain

$$\frac{f(x)-f(2)}{x-2} = \frac{(2x+1)-5}{x-2} = 2\left(\frac{x-2}{x-2}\right), \text{ when } x \text{ is rational} \quad \text{... (5)}$$

and
$$\frac{f(x)-f(2)}{x-2} = \frac{(x^2-2x+5)-5}{x-2} = x\left(\frac{x-2}{x-2}\right), \text{ when } x \text{ is irrational} \quad \text{... (6)}$$

From (5) and (6), we find that

$$\lim_{x \to 2} \frac{f(x)-f(2)}{x-2} = 2, \text{ in either case.}$$

Hence, f is differentiable at $x = 2$.

Ex. 169. *Let $f : (0, 2) \to$ **R** be defined by*

$$f(x) = \begin{cases} x^2, \text{ if } x \text{ is rational} \\ 2x-1, \text{ if } x \text{ is irrational} \end{cases}. \text{ Then,}$$

(a). f is differentiable exactly at one point *(b). f is differentiable exactly at two points*
(c). f is not differentiable at any point in (0, 2)
(d). f is differentiable at every point in (0, 2) **[GATE 2005]**

Sol. Ans. (*a*). We first discuss the continuity of the given function f. Let x be any real number. Then, for each $n \in$ **N**, there exist a rational number a_n and an irrational number b_n such that

$$x - 1/n < a_n < x + 1/n \quad \text{and} \quad x - 1/n < b_n < x + 1/n$$

giving
$$|a_n - x| < 1/n \quad \text{and} \quad |b_n - x| < 1/n \quad \text{... (1)}$$

Taking limit as $n \to \infty$, (1) yields
$$\lim_{n\to\infty} a_n = x = \lim_{n\to\infty} b_n \quad \text{... (2)}$$

If f is continuous at x, then by Heine's definition of continuity (refer Art. 8.5, page 8.29), we must have

$$\lim_{n\to\infty} f(a_n) = f(x) = \lim_{n\to\infty} f(b_n) \quad \text{... (3)}$$

Since a_n is rational and b_n is irrational, from the given definition of the given function f, it follows that

$$f(a_n) = a_n^2 \quad \text{and} \quad f(b_n) = 2b_n - 1 \quad \text{... (4)}$$

Using (4), (3) yields
$$\lim_{n\to\infty} a_n^2 = f(x) = \lim_{n\to\infty} (2b_n - 1)$$

Hence,
$$\left\{\lim_{n\to\infty} a_n\right\}^2 = 2\lim_{n\to\infty} b_n - 1 \quad \text{or} \quad x^2 = 2x - 1, \text{ using (2)}$$

giving $x^2 - 2x + 1 = 0$ or $(x-1)^2 = 0$ so that $x = 1$. Thus, $x = 1$ is the only possible point of continuity of the given function.

Next, let us discuss the differentiability of the given function f. In the above discussion, we have shown that f is continuous only at $x = 1$. But from Art. 9.2, we know that if f is differentiable at $x = c$, then f must be continuous at $x = c$. Therefore, f cannot be differentiable at any point except possibly at $x = 1$. Hence, in what follows, we shall test the given function f at $x = 1$ only.

For differentiability of f at $x = 1$, using the given values of f, we get

$$\frac{f(x)-f(1)}{x-1} = \frac{x^2-1}{x-1} = (x+1)\times\left(\frac{x-1}{x-1}\right), \text{ if } x \text{ is rational} \quad \text{... (5)}$$

and
$$\frac{f(x)-f(1)}{x-1} = \frac{(2x-1)-1}{x-1} = 2\left(\frac{x-1}{x-1}\right), \text{ if } x \text{ is irrational} \quad \text{... (6)}$$

From (5) and (6), we find that

$$\lim_{x\to 1}\frac{f(x)-f(1)}{x-1} = 2, \text{ in either case,}$$

showing that f is differentiable exactly at one points namely, $x = 1$. Hence, the option (*a*) is correct.

Ex. 170. *Let S, T,* $\subseteq$ ***R*** *be given by* $S = \{x \in \mathbf{R} : 2x^2 \cos(1/x) = 1\}$ *and* $T = \{x \in \mathbf{R} : 2x^2 \cos(1/x) \leq 1\} \cup \{0\}$. *Then, under the usual metric on* ***R***,

(*a*). *S is complete but not T* (*b*). *T is complete but not T*
(*c*). *both S and T are complete* (*d*). *neither S nor T is complete* **[GATE 2006]**

Sol. Ans. (*a*). Refer chapter 19. Solution left as an exercise.

Ex. 171. *Let* $f : \mathbf{R} \to \mathbf{R}$ *be defined by*

$$f(x) = \begin{cases} n, \text{ if } x = n, n \in \mathbf{N} \\ 0, \text{ otherwise} \end{cases} \quad \text{and} \quad T = \mathbf{N} \cup \{n + 1/n : n \in \mathbf{N}\}$$

Then, under the usual metric on ***R***, *f is uniformly continuous on*

(*a*). ***N*** *but not T* (*b*). *T but not* ***N***
(*c*). *both* ***N*** *and T* (*d*). *neither* ***N*** *nor T* **[GATE 2006]**

Sol. Ans. (c). Using chapter 19, try yourself.

Ex. 172. *For each* $n \in \mathbf{N}$ *and* $n > 1$ *denote* $f_n : [0, 1] \to \mathbf{R}$

$$f_n(x) = \begin{cases} |nx-1|, & \text{for } 0 \leq x \leq 2/n \\ 1, & \text{for } 2/n \leq x \leq 1 \end{cases}$$

Let $g_1, g_2 : [0, 1] \to \mathbf{R}$ *be defined by*

$$g_1(x) = \begin{cases} 1, \text{ for } 0 < x < 1 \\ 0, \text{ for } x = 0 \end{cases} \quad \text{and} \quad g_2(x) = 1 \text{ for } 0 \leq x \leq 1. \quad \text{Then, on } [0, 1]$$

(*a*). $f_n \to g_1$ *pointwise but not uniformly* (*b*). $f_n \to g_2$ *pointwise but not uniformly*
(*c*). $f_n \to g_1$ *uniformly* (*d*). $f_n \to g_2$ *uniformly* **[GATE 2006]**

Sol. Ans. (*b*). Using chapter 15, try yourself.

Ex. 173. *Let E be a connected subset of* ***R*** *with at least two elements. Then, the number of elements in E is*

(*a*). *exactly two* (*b*). *more than two but finite*
(*c*). *countably infinite* (*d*). *uncountable* **[GATE 2008]**

Sol. Ans. (*d*). Using chapter 19, try yourself.

Ex. 174. *Which one of the following subsets of* ***R*** *with usual metric is not complete ?*

(*a*). $[1, 2] \cup [3, 4]$ (*b*). $[0, \infty)$ (*c*). $[0, 1]$ (*d*). $\{0\} \cup \{1/n : n \in \mathbf{N}\}$ **[GATE 2008]**

Sol. Ans. (*d*). Using chapter 19, try yourself.

Ex. 175. *Which one of the following statements holds ?*

(*a*). *The series* $\sum_{n=0}^{\infty} x^n$ *converges for each* $x \in [-1, 1]$

(*b*). *The series* $\sum_{n=0}^{\infty} x^n$ *converges uniformly in* $[-1, 1]$

(*c*). *The series* $\sum_{n=1}^{\infty} (x^n / n)$ *converges for each* $x \in [-1, 1]$

(*d*). *The series* $\sum_{n=1}^{\infty} (x^n / n)$ *converges uniformly in* $[-1, 1]$ **[GATE 2008]**

Hint: Using results of chapters 6, 7 and 15, try yourself.

Ex. 176. *If* $f : [1, 2] \to \mathbf{R}$ *is a non-negative Riemann–integrable function such that*

$$\int_1^2 \frac{f(x)}{\sqrt{x}}\, dx = k \int_1^2 f(x)\, dx \neq 0, \text{ then } k \text{ belongs to the interval}$$

(*a*). *[0, 1/3]* (*b*). *(1/3, 2/3]* (*c*). *(2/3, 1]* (*d*). *(1, 4/3]* **[GATE 2010]**

Sol. Ans. (*c*). Recall the second mean value theorem (refer page 13.41) which states that if $\int_1^2 \phi(x)\, dx$ and $\int_1^2 f(x)\, dx$ both exist and $\phi(x)$ keeps the same sign, positive or negative, throughout the interval of integration, then there exists a number k, lying between the bounds of ϕ such that

$$\int_1^2 \phi(x)\, f(x)\, dx = k \int_1^2 f(x)\, dx \qquad \text{... (1)}$$

Here $\phi(x) = 1/\sqrt{x}$. Hence, we have

$$1 \le x \le 2 \quad \Rightarrow \quad 1/2 \le 1/x \le 1 \quad \Rightarrow \quad 1/\sqrt{2} \le 1/\sqrt{x} \le 1 \quad \Rightarrow \quad 1/\sqrt{2} \le \phi(x) \le 1$$

Therefore, by second mean value theorem, we have

$$\int_1^2 \frac{f(x)}{\sqrt{x}}\, dx = k \int_1^2 f(x)\, dx, \text{ where } k \in [1/\sqrt{2}, 1] \quad \text{or} \quad k \in (2/3, 1], \quad \text{as } \frac{2}{3} < \frac{1}{\sqrt{2}}$$

Ex. 177. *Consider the power series* $\sum_{n=1}^{\infty} (x^n / \sqrt{n})$ *and* $\sum_{n=1}^{\infty} (x^n / n)$. *Then,*

(*a*). *both converge on (–1, 1]* (*b*). *both converge on [–1, 1)*

(*c*). *exactly one of them converge on [–1, 1)*

(*d*). *None of them converges on [–1, 1)* **[GATE 2010]**

Sol. Ans. (*b*). Let us test $\sum_{n=1}^{\infty} (x^n / \sqrt{n})$ for its convergence. Let $u_n = x^n / \sqrt{n}$.

Then,
$$\lim_{n \to \infty} \frac{|u_n|}{|u_{n+1}|} = \lim_{n \to \infty} \frac{|x|^n}{\sqrt{n}} \times \frac{\sqrt{n+1}}{|x|^{n+1}} = \lim_{n \to \infty} \left(1 + \frac{1}{n}\right)^{1/2} \frac{1}{|x|} = \frac{1}{|x|}$$

Now, by ratio test, $\lim_{n \to \infty} |u_n|$ is convergent if $1/|x| > 1$, *i.e.*, if $|x| < 1$ or $-1 < x < 1$

Therefore, $\sum_{n=1}^{\infty} |u_n|$ is absolutely convergent if $-1 < x < 1$ and hence $\sum_{n=1}^{\infty} u_n$, *i.e.*, $\sum_{n=1}^{\infty} (x^n/\sqrt{n})$ is convergent if $-1 < x < 1$

When $x = 1$, then $\sum_{n=1}^{\infty} (x^n/\sqrt{n}) = \sum_{n=1}^{\infty} (1/n^{1/2})$, which is divergent

Next, if $x = -1$, then $\sum_{n=1}^{\infty} (x^n/\sqrt{n}) = \sum_{n=1}^{\infty} \{(-1)^n/\sqrt{n}\} = \sum_{n=1}^{\infty} (-1)^n v_n$, (say),

which is an alternating series. For this series, we have

$v_n = 1/\sqrt{n}$ so that $\lim_{n\to\infty} v_n = 0$

Again, $1/\sqrt{n} > 1/\sqrt{n+1} \quad \Rightarrow \quad v_n > v_{n+1}$ for each $n \in \mathbf{N}$

Hence, by Leibnitz's test, $\sum_{n=1}^{\infty} (-1)^n v_n$, *i.e.*, $\sum_{n=1}^{\infty} \{(-1)^n/\sqrt{n}\}$ is convergent.

From the above facts, it follows that $\sum_{n=1}^{\infty} (x^n/\sqrt{n})$ converges on $[-1, 1)$.

Next, let us test $\sum_{n=1}^{\infty} (x^n/n)$ for its convergence. Let $U_n = x^n/n$.

Then, $$\lim_{n\to\infty} \frac{|U_n|}{|U_{n+1}|} = \lim_{n\to\infty} \frac{|x|^n}{|x|^{n+1}} \times \frac{n+1}{n} = \lim_{n\to\infty} \left[\left(1+\frac{1}{n}\right) \times \frac{1}{|x|}\right] = \frac{1}{|x|}$$

Now, by ratio test, $\sum_{n=1}^{\infty} |U_n|$ is convergent if $1/|x| > 1$ or $|x| < 1$ or $-1 < x < 1$

Therefore, $\sum_{n=1}^{\infty} |U_n|$ is absolutely convergent if $-1 < x < 1$ and hence $\sum_{n=1}^{\infty} U_n$, *i.e.*, $\sum_{n=1}^{\infty} (x^n/n)$ is convergent if $-1 < x < 1$

When $x = 1$, then $\sum_{n=1}^{\infty} (x^n/n) = \sum_{n=1}^{\infty} (1/n)$, which is divergent

Next, if $x = -1$, then $\sum_{n=1}^{\infty} (x^n/n) = \sum_{n=1}^{\infty} \{(-1)^n/n\} = \sum_{n=1}^{\infty} (-1)^n V_n$, say,

which is an alternating series. For this series, we have

$V_n = 1/n$ so that $\lim_{n\to\infty} V_n = 0$

Again, $1/n > 1/(n+1) \quad \Rightarrow \quad V_n > V_{n+1}$, for each $n \in \mathbf{N}$

Hence, by Leibnitz's test, $\sum_{n=1}^{\infty} (-1)^n V_n$, *i.e.*, $\sum_{n=1}^{\infty} \{(-1)^n/n\}$ is convergent. From the above facts, it follows that $\sum_{n=1}^{\infty} \{(-1)^n/n\}$ converges on $[-1, 1)$.

Thus both the given series converge on $[-1, 1)$ and hence option (*b*) is correct.

Ex. 178. *The series* $\sum_{m=1}^{\infty} x^{\log_e m}, x > 0$ *is convergent on the interval*

(*a*). *(0, 1/e)* (*b*). *(1/e, e)* (*c*). *(0, e)* (*d*). *(1, e)* **[GATE 2011]**

Sol. Ans. (*a*). Using logarithmic test (see Art. 6.14), try yourself.

Ex. 179. *Let* $f_n(x) = x/[\{(n-1)x+1\}\ (nx+1)]$ *and* $s_n = \sum_{i=1}^{n} f_i(x)$ *for* $x \in [0, 1]$. *Then sequence* $< s_n >$

(*a*). *converges uniformly on [0, 1]* (*b*). *converges pointwise on [0, 1] but not uniformly*

(*c*). *converges pointwise for x = 0 but not for x ∈ (0, 1)*

(*d*). *does not converge for x ∈ (0, 1)* **[GATE 2011]**

Sol. Ans. (*b*). Resolving into partial fractions, we have

$$f_n(x) = \frac{x}{\{(n-1)x+1\}\,(nx+1)} = \frac{1}{(n-1)x+1} - \frac{1}{nx+1}, \text{ for each } n \in \mathbf{N} \quad \text{... (1)}$$

Also, given that $$s_n = \sum_{i=1}^{n} f_i(x) = f_1(x) + f_2(x) + \ldots + f_n(x)$$

or $$s_n = \left(1 - \frac{1}{x+1}\right) + \left(\frac{1}{x+1} - \frac{1}{2x+1}\right) + \ldots + \left\{\frac{1}{(n-1)x+1} - \frac{1}{nx+1}\right\}, \text{ using (1)}$$

Thus, we have $$s_n = 1 - \{1/(nx+1)\} \quad \text{... (2)}$$

From (2), we have $$s(x) = \lim_{n \to \infty} s_n = \begin{cases} 1, & \text{if } x \neq 0 \\ 0, & \text{if } x = 0 \end{cases} \quad \text{... (3)}$$

Therefore, the sequence $<s_n>$ is pointwise convergent in [0, 1] and the point-wise limit $s(x)$ is given by (3).

Since the limit-function s given by (3) is discontinuous at $x = 0$, it follows that the sum function s is discontinuous in [0, 1]. Thus, $<s_n>$ is a sequence of continuous functions and its limit function s is discontinuous in [0, 1]. Hence, from note 1 on page 15.21, it follows that the sequence $<s_n>$ is not uniformly convergent on [0, 1].

From the above discussion it follows that the sequence $<s_n>$ converges point-wise on [0, 1] but not uniformly. Hence, the option (*b*) is correct.

Ex. 180. *Which of the following statements are true ?*

P. *The set* $\{x \in \boldsymbol{R} : |\cos x| \le 1/2\}$ *is compact*

Q. *The set {x ∈ **R** : tan x is not differential} is compact*

R. *The set* $\{x \in \boldsymbol{R} : \sum_{n=0}^{\infty} \frac{(-1)^n x^{2n+1}}{(2n+1)!}$ *is convergent} is bounded*

S. *The set {x ∈ **R** : f(x) = cos x has a local maxima} is closed*

(*a*). *P and Q* (*b*). *R and S* (*c*). *Q and S* (*d*). *P and S* **[GATE 2012]**

Sol. Ans. (*c*). Try yourself.

Ex. 181. *Let* $f(x) = \sum_{n=1}^{\infty}(1/n^2)\times \sin nx$. *Then,*

(*a*). $\lim_{x\to 0} f(x) = 0$ (*b*). $\lim_{x\to 0} f(x) = 1$

(*c*). $\lim_{x\to 0} f(x) = \pi^2/6$ (*d*). $\lim_{x\to 0} f(x)$ *does not exist* **[GATE 2013]**

Sol. Ans. (*a*). Try yourself.

Ex. 182. *Let* $S = \{x \in \mathbf{R} : x \geq 0, \sum_{n=1}^{\infty} x^{\sqrt{n}} < \infty\}$. *Then, the supremum of S is*

(*a*). *1* (*b*). *1/e* (*c*). *0* (*d*). ∞ **[GATE 2013]**

Sol. Ans. (*a*). Try yourself.

Ex. 183. *The radius of convergence of the power series* $\sum_{n=1}^{\infty} 4^{(-1)^n n} z^{2n}$ *is ...* **[GATE 2014]**

Sol. Ans. 1/2. Using chapter 17, try yourself.

Ex. 184. *Obtain the interval of convergence of the series* $\sum_{n=0}^{\infty} a_n x^n$, *where* $a_{2n-1} = 1/3^{n+1}$ *and* $a_{2n} = 1/3^n$.

Sol. Re-writing the values of a_{2n} and a_{2n-1}, we have

$$a_{2n} = \left(\frac{1}{3}\right)^n = \left(\frac{1}{\sqrt{3}}\right)^{2n} \quad \text{and} \quad a_{2n-1} = \frac{1}{3^{n+1}} = \left(\frac{1}{\sqrt{3}}\right)^{2(n+1)} = \left(\frac{1}{\sqrt{3}}\right)^{(2n-1)+3} = \left(\frac{1}{\sqrt{3}}\right)^{2n-1} \times \left(\frac{1}{\sqrt{3}}\right)^{3}$$

$$\text{Hence,} \quad \limsup_n |a_n|^{1/n} = \frac{1}{\sqrt{3}}, \quad \text{since } \limsup_n \left(\frac{1}{\sqrt{3}}\right)^{3/(2n-1)} = \lim_{n\to\infty}\left(\frac{1}{3\sqrt{3}}\right)^{1/(2n-1)} = 1,$$

showing that the radius of convergence R of the given series is given by

$$1/\text{R} = \limsup_n |a_n|^{1/n} = 1/\sqrt{3} \quad \text{so that} \quad \text{R} = \sqrt{3}.$$

Therefore, the required interval of convergence is $(-\sqrt{3}, \sqrt{3})$

Ex. 185. *Consider the power series* $\sum_{n=0}^{\infty} a_n z^n$, *where* $a_n = 1/3^n$ *if n is even and* $a_n = 1/5^n$ *if n is odd. The radius of convergence of the series is equal to .....* **[GATE 2015]**

Sol. Ans. 3. Proceed as in the above solved Ex. 184.

Ex. 186. *Let* d_1 *and* d_2 *denote the usual and the discrete metrics on* **R**, *respectively. Let* $f : (\mathbf{R}, d_1) \to (\mathbf{R}, d_2)$ *be defined by* $f(x) = x$, $x \in \mathbf{R}$. *Then*

(*a*). *f is continuous but* f^{-1} *is not continuous*

(*b*). f^{-1} *is continuous but f is not continuous*

(*c*). *both f and* f^{-1} *are continuous*

(*d*). *neither f nor* f^{-1} *is continuous* **[GATE 2015]**

Sol. Ans. (*b*). Observe that $f^{-1} : (\mathbf{R}, d_2) \to (\mathbf{R}, d_1)$ is defined by $f^{-1}(x) = x$, $x \in \mathbf{R}$.

Testing f for its continuity. Consider a d_2-open subset B of **R** which is not d_1-open. By definition of f, $f^{-1}(B) = B$, which is not d_1-open by our mode of selection of the subset B. Thus, we see that the inverse image under f of a d_2-open set need not be d_1-open. Hence, from theorem I, page 19.17, it follows that f is not continuous.

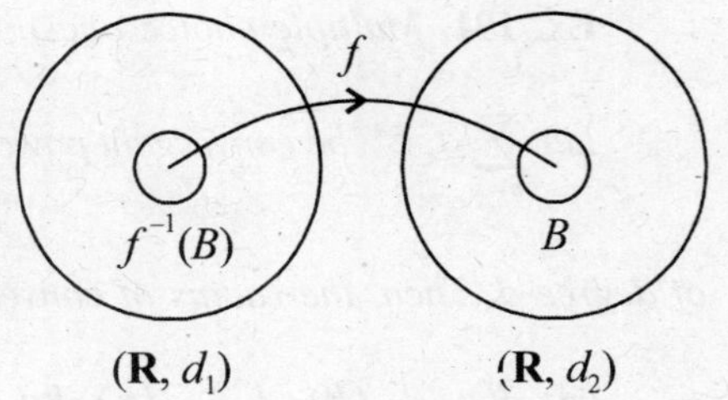

Testing f^{-1} for its continuity. Let A be a d_1-open subset of **R**. Now, by definition of f^{-1}, $(f^{-1})^{-1}(A) = A$, which is d_2-open subset of **R**, since every subset of discrete space (**R**, d_2) is open (refer Ex. 9, page 19.57). Thus, the inverse image under f^{-1} of every d_1-open subset of **R** is d_2-open. Therefore, from theorem I, page 19.17, it follows that f^{-1} is continuous.

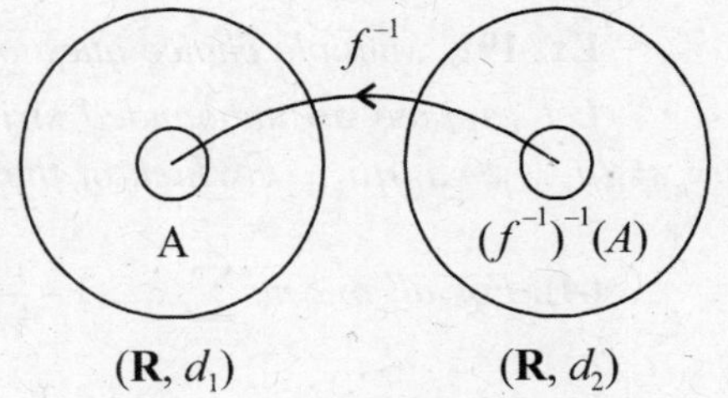

Ex. 187. *Define* $f_1, f_2 : [0, 1] \to \mathbf{R}$ *by* $\quad f_1(x) = \sum_{n=i}^{\infty} (x/n^2) \times \sin(n^2 x)$, *and*

$f_2(x) = \sum_{n=1}^{\infty} (x^2)(1 - x^2)^{n-1}$. *Then*

(*a*). *f_1 is continuous but f_2 is not continuous* (*b*). *f_2 is continuous but f_1 is not continuous*
(*c*). *both f_1 and f_2 are continuous* (*d*). *neither f_1 nor f_2 is continuous*

Sol. Ans. (*a*). Let as an exercise. **[GATE 2015]**

Ex. 188. *Let* $f : [0, \infty) \to \mathbf{R}$ *be defined by* $f(x) = \int_0^x \sin^2(t^2)\, dt$. *Then the function f is*

(*a*). *uniformly continuous on [0, 1) but not on (0, ∞)*
(*b*). *uniformly continuous on (0, ∞) but not on [0, 1)*
(*c*). *uniformly continuous on both [0, 1) and (0, ∞)*
(*d*). *neither uniformly continuous on [0, 1) nor uniformly continuous on (0, ∞)*

Sol. Ans. (c). Let as an exercise. **[GATE 2015]**

Ex. 189. *Let* $H = \{< x_n > \in l_2 : \sum_{n=1}^{\infty} (x_n / n) = 1\}$. *Then H*

(*a*). *is bounded* (*b*). *is closed* (*c*). *is a subspace* (*d*). *has an interior point*

Sol. Ans. (*b*). Let as an exercise. **[GATE 2015]**

Ex. 190. *Multiple choice question with exactly one correct option*

Let $\{b_n\}$ and $\{c_n\}$ be sequences of real numbers. Then a necessary and sufficient condition for the sequence of polynomials $f_n(x) = a_n x + b_n x^2$ to converge uniformly to 0 on the real line is

(*A*). $\lim_{n\to\infty} b_n = 0$ *and* $\lim_{n\to\infty} c_n = 0$ (*B*). $\sum_{n=1}^{\infty} |b_n| < \infty$ *and* $\sum_{n=1}^{\infty} |c_n| < \infty$

(**C**). *There exists a positive integer N such that $b_n = 0$ and $c_n = 0$ for all $n > N$*

(*D*). $\lim_{n\to\infty} c_n = 0$ **[CSIR – UGC (NET), December 2014]**

Sol. Ans. (c). Left as an exericse.

Ex. 191. *Multiple choice question with exactly one correct option*

Let $\sum_{n=0}^{\infty} a_n z^n$ *be convergent power series such that* $\lim_{n \to \infty} \dfrac{a_{n+1}}{a_n} = R > 0$. *Let p be a polynomial of degree d. Then, the radius of convergence of the power series* $\sum_{n=1}^{\infty} p(n) a_n z^n$ *is*

(*A*). *R* **(*B*).** *d* **(*C*).** *Rd* **(*D*).** *R + d* **[CSIR – UGC (NET), December 2014]**

Sol. Ans. (*A*). Left as an exercise.

Ex. 192. *Multiple choice question with multiple correct options*

Let $\{a_k\}$ *be an unbounded strictly increasing sequence of positive real numbers and* $x_k = (a_{k+1} - a_k)/a_{k+1}$. *Which of the following statements is/are correct ?*

(*A*). *For all* $n \geq m$, $\sum_{k=m}^{n} x_k > 1 - \dfrac{a_m}{a_n}$ **(*B*).** *There exists* $n \geq m$ *such that* $\sum_{k=m}^{n} x_k > \dfrac{1}{2}$

(*C*). $\sum_{k=1}^{n} x_k$ *converges to a finite limit* **(*D*).** $\sum_{k=1}^{n} x_k$ *diverges to* ∞

Sol. Ans. (*A*), (*B*) and **(*C*).** **[CSIR – UGC (NET), December 2014]**

Ex. 193. *Multiple choice question with multiple correct options*

For a non-empty subset S and a point x in a connected metric space (X, d), let $d(x, S) = \inf \{d(x, y) : y \in S\}$. *Which of the following statements is/are correct ?*

(*A*). *If S is closed and* $d(x, S) > 0$, *then x is not an accumulation point of S*

(*B*). *If S is open and* $d(x, S) > 0$, *then x is not an accumulation point of S*

(*C*). *If S is closed and* $d(x, S) > 0$, *then S does not contain x*

(*D*). *If S is open and* $d(x, S) > 0$, *then* $x \in S$ **[CSIR – UGC (NET), December 2014]**

Sol. Ans. (*A*), (*B*) and **(*C*).** Left as an exercise.

Ex. 194. *Multiple choice question with multiple correct options*

Let f be a continuously differentiable function on ***R***. *Suppose that* $L = \lim_{x \to \infty} \{f(x) + f'(x)\}$ *exists. If* $0 < L < \infty$, *then which of the following statements is/are correct ?*

(*A*). *If* $\lim_{x \to \infty} f'(x)$ *exists, then it is 0* **(*B*).** *If* $\lim_{x \to \infty} f(x)$ *exists, then it is L*

(*C*). *If* $\lim_{x \to \infty} f'(x)$ *exists, then* $\lim_{x \to \infty} f(x) = 0$ **(*D*).** *If* $\lim_{x \to \infty} f(x)$ *exists, then* $\lim_{x \to \infty} f'(x) = L$

[CSIR – UGC (NET), December 2014]

Sol. Ans. (*A*) and **(*B*).** Given $\lim_{x \to \infty} \{f(x) + f'(x)\} = L$, where $0 < L < \infty$ ...(1)

Re-writing f(x) as $\{e^x f(x)\} / e^x$, we have

$$\lim_{x \to \infty} f(x) = \lim_{x \to \infty} \frac{e^x f(x)}{e^x} \qquad \left[\text{Form } \frac{\infty}{\infty}\right]$$

$$= \lim_{x \to \infty} \frac{d\{e^x f(x)\}/dx}{d(e^x)/dx}, \text{ using } L' \text{ Hospital's rule}$$

$$= \lim_{x \to \infty} \frac{e^x \{f(x) + f'(x)\}}{e^x} = \lim_{x \to \infty} \{f(x) + f'(x)\} = L, \text{ using (1)}$$

Thus, $\lim_{x \to \infty} f(x) = L$ and hence (1) yields $\lim_{x \to \infty} f'(x) = 0,$

showing that the options (*A*) and (*B*) are correct while the options (*C*) and (*D*) are not true.

Ex. 195. *Multiple choice question having exactly one correct option*

The number of surjective maps from a set of 4 elements to a set of 3 elements is

(*a*). *36* (*b*). *64* (*c*). *69* (*d*). *81* **[CSIR – UGC (NET), December 2014]**

Sol. Ans. (*a*). We know that the number of surjective (onto) maps from a set *A* containing *m* elements to a set *B* having *n* elements ($n \le m$) is $\sum_{r=1}^{n} (-1)^{n-r}\, {}^nC_r\, r^m$.

Hence, the required number of surjective maps from a set containing 4 elements to a set containing 3 elements

$= \sum_{r=1}^{3} (-1)^{3-r}\, {}^3C_r\, r^4$, noting that here $m = 4$ and $n = 3$

$= (-1)^2 \times {}^3C_1 + (-1)^1 \times {}^3C_2 \times 2^4 + (-1)^0 \times {}^3C_3 \times 3^4 = 3 - 48 + 81 = 36$

Ex. 196. *Multiple choice question having exactly one correct option*

Let k be a positive integer. The radius of convergence of the series $\sum_{n=0}^{\infty} \frac{(n!)^k}{(kn)!} z^n$ *is*

(*A*). k **(*B*).** k^{-k} **(*C*).** k^k **(*D*).** ∞ **[CSIR – UGC (NET), December 2014]**

Sol. Ans. (*C*). Let the given series be denoted by $\sum_{n=0}^{\infty} a_n z^n$ so that

$$a_n = \frac{(n!)^k}{(kn)!} \text{ and } a_{n+1} = \frac{\{(n+1)!\}^k}{\{k(n+1)\}!} = \frac{(n+1)^k\,(n!)^k}{(kn+k)!}, \text{ as } (n+1)! = (n+1)\,n! \qquad \text{... (1)}$$

Let *R* be the radius of convergence of the given series. Then,

$$\frac{1}{R} = \lim_{n\to\infty} \frac{a_{n+1}}{a_n} = \lim_{n\to\infty} \left[\frac{(n+1)^k\,(n!)^k}{(kn+k)!} \times \frac{(kn)!}{(n!)^k}\right] = \lim_{n\to\infty} \frac{(n+1)^k\,(kn)!}{(kn+k)!}$$

or
$$\frac{1}{R} = \lim_{n\to\infty} \frac{n^k\,(1+1/n)^k\,(kn)!}{(kn+k)\,(kn+k-1)\,.....\,\{kn+k-(k-1)\}\,(kn)\,(kn-1)\,(kn-2)\,....3.2.1}$$

or
$$\frac{1}{R} = \lim_{n\to\infty} \frac{n^k\,(1+1/n)^k\,(kn)!}{(kn+k)\,(kn+k-1)\,(kn+k-2)\,....\,(kn+1)\,(kn)!}$$

or
$$\frac{1}{R} = \lim_{n\to\infty} \frac{n^k\,(1+1/n)^k}{(kn)\times(1+1/n)\times(kn)\times\{1+(k-1)/kn\}\times....\times(kn)\,(1+1/kn)}$$

or
$$\frac{1}{R} = \lim_{n\to\infty} \frac{n^k\,(1+1/n)^k}{k^k.n^k\,(1+1/n)\,\{1+(k-1)/kn\}\times....\times(1+1/kn)} = \frac{1}{k^k}$$

Hence, $R = k^k$ and therefore the option (*C*) is correct.

Ex. 197. *Multiple choice question having exactly one correct option*

Let $\{a_n : n \ge 1\}$ *be a sequence of real numbers such that* $\sum_{n=1}^{\infty} a_n$ *is convergent and* $\sum_{n=1}^{\infty} |\, a_n\,|$

is divergent. Let R be the radius of convergence of the power series $\sum_{n=1}^{\infty} a_n x^n$. *Then, we can conclude that*

(*A*). $0 < R < 1$ (*B*). $R = 1$ (*C*). $1 < R < \infty$ (*D*). $R = \infty$

[CSIR – UGC (NET), December 2014]

Sol. Ans. (*B*). We shall give an example to find the value of radius of convergence of $\sum_{n=1}^{\infty} a_n x^n$.

Let $a_n = (-1)^{n-1}/n$. Then, from example on page 7.4., we know that $\sum_{n=1}^{\infty} a_n$ is convergent. Again, $\sum_{n=1}^{\infty} |a_n| = \sum_{n=1}^{\infty} (1/n)$, which is divergent.

Thus, if $a_n = (-1)^{n-1}/n$, then both conditions of the given prublem are satisfied because $\sum_{n=1}^{\infty} a_n$ is convergent while $\sum_{n=1}^{\infty} |a_n|$ is divergent.

Then,
$$R = \lim_{n \to \infty} \left| \frac{a_n}{a_{n+1}} \right| = \lim_{n \to \infty} \left| \frac{(-1)^{n-1}/n}{(-1)^n/(n+1)} \right| = \lim_{n \to \infty} \left| \frac{n+1}{n} \right| = 1$$

Therefore, the option (*B*) is correct.

Ex. 198. *Multiple choice question with multiple correct options*

Let A be a subset of **R**. *Which of the following properties imply that A is compact ?*

(*A*). *Every continuous function from A to* **R** *is bounded*

(*B*). *Every sequence* $\{x_n\}$ *in A has a convergent subsequence converging to a point in A*

(*C*). *There exists a continuous function from A onto [0, 1]*

(*D*). *There is no one-one and continuous function from A onto (0, 1)*

[CSIR – UGC (NET), December 2014]

Sol. Ans. (*A*) and (*B*). From Art. 8.3, page 8.2, we know that a function whose range is bounded is said to be a bounded function. Again, from Ex. 21, page 19.50, we know that if f is continuous function from a compact metric space A into any metric space **R**, then $f(A)$ is bounded. Therefore, the option (A) is correct.

Now, using the definition of sequentially compact metric space (refer page 19.34) and theorem VI on page 19.35, we have

A is compact $\Rightarrow$ A is sequentially compact

$\Rightarrow$ every sequence $\{x_n\}$ in A has a convergent subsequence converging to a point in A

Hence, it follows that the option (B) is also correct.

Ex. 199. *Multiple choice question with exactly one correct option*

Let $f : X \to X$ *such that* $f(f(x)) = x$ *for all* $x \in X$. *Then*

(*A*). *f is one-to-one and onto* (*B*). *f is one-to-one but not onto*

(*C*). *f is onto but not one-to-one* (*D*). *f need not be either one-to-one or onto*

[CSIR – UGC (NET), June 2015]

Sol. Ans. (*A*) Given $f : X \to X$ such that $f(f(x)) = x \quad \forall x \in X$ ... (1)

(*1*) $\Rightarrow$ $fof(x) = x \quad \forall \in X$ $\Rightarrow$ fof is an identity function (see page 1.7)

Now, $fof = I_X \Rightarrow f = f^{-1}$, since $fof^{-1} = f^{-1}of = I_X$,

where I_X is an identity function. Thus, f^{-1} exists.

From Art. 1.7 (page 1.8), we know that a function is invertible if and only if it is one-one onto. Therefore, f must be one-one onto and hence the option (*A*) is correct.

Ex. 200. *Multiple choice question with multiple correct options*

Which of the following sets of functions are uncountable ? (*N stands for the set of natural numbers*)

(*A*). $\{f \mid f : N \to [1, 2]\}$ (*B*). $\{f \mid f : [1, 2] \to N\}$

(*C*). $\{f \mid f : [1, 2] \to N, f(1) \le f(2)\}$ (*D*). $\{f \mid f : N \to [1, 2], f(1) \le f(2)\}$

Sol. Ans. (*A*) and (*D*). Left as an exercise. **[CSIR – UGC (NET), June 2015]**

Ex. 201. *Multiple choice question with multiple correct options*

*Let f be a bounded function on **R** and $a \in$ **R**. For $\delta > 0$, let*

$$w(a, \delta) = \sup |f(x) - f(a)|, \; x \in [a - \delta, a + \delta]. \quad \text{Then}$$

(*A*). $w(a, \delta_1) \le w(a, \delta_2)$, if $\delta_1 \le \delta_2$ (*B*). $\lim_{\delta \to 0+} w(a, \delta) = 0$ *for all* $a \in$ **R**

(*C*). $\lim_{\delta \to 0+} w(a, \delta)$ *need not exist* (*D*). $\lim_{\delta \to 0+} w(a, \delta) = 0$ *if f is continuous at a*

Sol. Ans. (*A*) and (*D*). Left as an exercise. **[CSIR – UGC (NET), June 2015]**

Ex. 202. *Multiple choice question with multiple correct options*

For $n \ge 1$, let $g_n(x) = \sin^2(x + 1/n)$, $x \in [0, \infty)$ and $f_n(x) = \int_a^x g_n(t)\, dt$. Then,

(*A*). *$\{f_n\}$ converges pointwise to a function f on $[0, \infty)$, but does not converge uniformly on $[0, \infty)$*

(*B*). *$\{f_n\}$ does not converge pointwise to any function on $[0, \infty)$*

(*C*). *$\{f_n\}$ converges uniformly on $[0, 1]$* (*D*). *$\{f_n\}$ converges uniformly on $[0, \infty)$*

Sol. Ans. (*C*) and (*D*). Left as an exercise. **[CSIR – UGC (NET), June 2015]**

Ex. 203. *Multiple choice question with multiple correct options*

Let $\{a_0, a_1, a_2, \ldots\ldots\}$ be a sequence of real numbers. For any $k \ge 1$, let $S_n = \sum_{k=0}^{n} a_{2k}$. Which of the following statements are correct ?

(*A*). *If* $\lim_{n \to \infty} S_n$ *exists, then* $\sum_{m=0}^{\infty} a_m$ *exists* (*B*). *If* $\lim_{n \to \infty} S_n$ *exists, then* $\sum_{m=0}^{\infty} a_m$ *need not exist*

(*C*). *If* $\sum_{m=0}^{\infty} a_m$ *exists, then* $\lim_{n \to \infty} S_n$ *exists* (*D*). *If* $\sum_{m=0}^{\infty} a_m$ *exists, then* $\lim_{n \to \infty} S_n$ *need not exist*

Sol. Ans. (*B*) and (*D*). Left as an exercise. **[CSIR – UGC (NET), June 2015]**

Ex. 204. *Multiple choice question with exactly one correct option*

The limit $\lim_{x \to 0} \dfrac{1}{x} \int_x^{2x} e^{-t^2}\, dt$

(*A*). *does not exist* (*B*). *is infinite*

(*C*). *exists and equals 1* (*D*). *exists and equals 0* **[CSIR – UGC (NET), June 2015]**

Sol. Ans. (*C*). Re-writing the given expression, we have

$$\lim_{x \to 0} \frac{1}{x} \int_x^{2x} e^{-t^2}\, dt = \lim_{x \to 0} \frac{\int_x^{2x} e^{-t^2}\, dt}{x} \qquad \left[\text{Form } \frac{0}{0}\right]$$

or $$\lim_{x\to 0}\frac{1}{x}\int_x^{2x} e^{-t^2}\,dt = \lim_{x\to 0}\frac{\dfrac{d}{dx}\displaystyle\int_x^{2x} e^{-t^2}\,dt}{dx/dx}, \text{ using L' Hospital's rule}$$

$$= \lim_{x\to 0}\left[\int_x^{2x}\frac{\partial e^{-t^2}}{\partial x}\,dt + e^{-4x^2}\frac{d(2x)}{dx} - e^{-x^2}\frac{dx}{dx}\right], \text{ using Leibnitz's}$$

rule of differentiation under the integral sign (see page 21.2)

$$= \lim_{x\to 0}[0 + 2e^{-4x^2} - e^{-x^2}] = 2 - 1 = 1$$

Ex. 205. *Multiple choice question with multiple correct options*
Let a be a positive real number. Which of the following integrals are convergent ?

(***A***). $\displaystyle\int_0^a \frac{1}{x^4}\,dx$ (***B***). $\displaystyle\int_0^a \frac{1}{\sqrt{x}}\,dx$ (***C***). $\displaystyle\int_a^\infty \frac{1}{x\log_e x}\,dx$ (***D***). $\displaystyle\int_a^\infty \frac{1}{x(\log_e x)^2}\,dx$

[CSIR – UGC (NET), June 2015]

Sol. Ans. (*B*) and (*D*). From part (*i*) of Art 16.6, page 16.7, it follows that $\displaystyle\int_0^a \frac{dx}{(x-0)^n}$, *i.e.*, $\displaystyle\int_0^a \frac{dx}{x^n}$ converges only if $n < 1$. Hence, $\displaystyle\int_0^a \frac{dx}{x^4}$ is divergent while $\displaystyle\int_0^a \frac{dx}{x^{1/2}}$ is convergent. Therefore, the option (A) is not correct while the option (B) is correct.

Next, putting $\log_e x = t$ and $(1/x)\,dx = dt$, *we have*

$$\int_a^\infty \frac{1}{x\log_e x} = \int_{\log_e a}^\infty \frac{dt}{t} \quad\text{and}\quad \int_a^\infty \frac{dx}{x(\log_e x)^2} = \int_{\log_e a}^\infty \frac{dt}{t^2} \qquad \dots (1)$$

From Art. 16.13, page 16.19, we know that $\displaystyle\int_{\log_e a}^\infty \frac{dt}{t^n}$ is convergent only if $n > 1$. Therefore, from (1), it follows that $\displaystyle\int_a^\infty \frac{dx}{x\log_e x}$ is divergent while $\displaystyle\int_a^\infty \frac{dx}{x(\log_e x)^2}$ is convergent. Hence, the option (C) is not correct while the option (D) is correct.

Ex. 206. *Multiple choice question with multiple correct options*
For $n \geq 2$, let $a_n = 1/(n \log n)$. Then,

(***A***). *The sequence* $\{a_n\}_{n=2}^\infty$ *is convergent* (***B***). *The series* $\displaystyle\sum_{n=2}^\infty a_n$ *is convergent*

(***C***). *The series* $\displaystyle\sum_{n=2}^\infty a_n^2$ *is convergent* (***D***). *The series* $\displaystyle\sum_{n=2}^\infty (-1)^n a_n$ *is convergent*

[CSIR – UGC (NET), June 2015]

Sol. Ans. (*A*), (*C*) and (*D*). Let us examine the sequence $\{a_n\}_{n=2}^\infty$ for its convergence. Let $b_n = 1/n$ and $c_n = 1/\log n$ so that we have

$$\{a_n\}_{n=2}^\infty = \{1/(n\log n)\}_{n=2}^\infty = \{b_n\, c_n\}_{n=2}^\infty \qquad \dots (1)$$

We now proceed to show that $\lim_{n\to\infty} b_n = 0$ and $\lim_{n\to\infty} c_n = 0$

Let $\varepsilon > 0$ be given. Then, $|b_n - 0| = |1/n - 0| = 1/n < \varepsilon$ if $n > 1/\varepsilon$

If m is a positive integer greater than $1/\varepsilon$, then we have

$$|b_n - 0| < \varepsilon \quad \forall\, n \geq m \quad \Rightarrow \quad \lim_{n\to\infty} b_n = 0 \qquad \text{... (2)}$$

Next, let ε' be given. Then, $|c_n - 0| = |1/\log n - 0| = 1/\log n < \varepsilon'$ if $\log n > 1/\varepsilon'$ or $n > e^{1/\varepsilon'}$. If m' is a positive integer greater than $e^{1/\varepsilon'}$, then we have

$$|c_n - 0| < \varepsilon' \quad \forall\, n \geq m' \quad \Rightarrow \quad \lim_{n\to\infty} c_n = 0 \qquad \text{... (3)}$$

From theorem III, page 5.12, it follows that

$$\lim_{n\to\infty} a_n = \lim_{n\to\infty} (b_n\, c_n) = \left(\lim_{n\to\infty} b_n\right) \times \left(\lim_{n\to\infty} c_n\right) = 0, \text{ using (2) and (3)}$$

Hence the sequence $\{a_n\}_{n=2}^{\infty}$, *i.e.*, $\{1/(n\log n)\}_{n=2}^{\infty}$ is convergent.

Therefore, the option (A) is correct.

We now examine the series $\sum_{n=2}^{\infty} a_n$, *i.e.*, $\sum_{n=2}^{\infty} (1/n\log n)$ for its convergence.

From solved Ex. 3, page 6.46, we know that the series $\sum_{n=2}^{\infty} \{1/n(\log n)^p\}$ is convergent if $p > 1$ and divergent if $0 < p \leq 1$. On comparing the series under consideration with the above series, we find that $p = 1$. Hence, the series $\sum_{n=2}^{\infty} a_n$ is not convergent.

Therefore, the option (B) is not correct.

Next, let us examine $\sum_{n=2}^{\infty} a_n^2$, *i.e.*, $\sum_{n=2}^{\infty} \{1/n^2(\log n)^2\}$ for its convergence. In what follows, we shall proceed as in solved example on page 6.49.

Let a be a positive integer such that $a > 1$

Here, the nth term of the series $= \dfrac{1}{n^2(\log n)^2} = f(n)$, say ... (4)

From (4), $$a^n f(a^n) = a^n \times \frac{1}{(a^n)^2(\log a^n)^2} = \frac{1}{a^n\, n^2 (\log a)^2}$$

Let $$\sum_{n=1}^{\infty} u_n = \sum_{n=1}^{\infty} a^n f(a^n) = \sum_{n=1}^{\infty} \frac{1}{a^n\, n^2(\log a)^2}$$

Then, $$u_n = \frac{1}{a^n\, n^2(\log a)^2} \quad \text{and} \quad u_{n+1} = \frac{1}{a^{n+1}(n+1)^2(\log a)^2}$$

Hence, $$\lim_{n\to\infty} \frac{u_n}{u_{n+1}} = \lim_{n\to\infty} a\frac{(n+1)^2}{n^2} = \lim_{n\to\infty} a\left(1+\frac{1}{n}\right)^2 = a > 1$$

Therefore, by D'Alembert's ratio test, the series $\sum_{n=1}^{\infty} u_n$, *i.e.*, $\sum_{n=1}^{\infty} a^n f(a^n)$ is convergent. Hence, by Cauchy's condensation test (refer Art. 6.20, page 6.47), it follows that the given series $\sum_{n=1}^{\infty} f(n)$, *i.e.*, $\sum_{n=1}^{\infty} \{1/n^2(\log n)^2\}$ is also convergent. Thus, $\sum_{n=1}^{\infty} a_n^2$ is convergent and hence the option (C) is correct.

Finally, we now test $\sum_{n=2}^{\infty} (-1)^n a_n$, *i.e.*, $\sum_{n=2}^{\infty} \{(-1)^n/(n \log n)\}$ for its convergence with help of Leibnitz's test (refer Art. 7.3, page 7.2). Hence, we have

$a_n = 1/(n \log n) > 0 \ \forall n \geq \mathbf{N}$. Also, we have $\lim_{n \to \infty} a_n = 0$

We now proceed to show that $a_{n+1} < a_n$ for each $n \in \mathbf{N}$

Let $f(x) = \dfrac{1}{x \log x}, \ \forall x \geq 2$ so that $f'(x) = -\dfrac{1 + \log x}{(x \log x)^2} < 0 \quad \forall \quad x \geq 2$

Now, $f'(x) < 0 \ \forall x \leq 2 \quad \Rightarrow \quad f(x)$ is a decreasing function $\forall \ x \geq 2$

$\Rightarrow \quad a_n$ is a decreasing function $\forall \ n \geq 2$

$\Rightarrow \quad a_{n+1} < a_n \ \forall \ n \geq 2$

Thus, all the three conditions of Leibnitz's test are satisfied and hence the given series $\sum_{n=2}^{\infty} (-1)^n a_n$ is convergent. Therefore, the option (D) is correct.

Ex. 207. *Let $<a_n>$ be a sequence where $a_n = (1 + 1/n)^n$. Show that $<a_n>$ is bounded and monotonically increasing.* **[Delhi B.A. (Prog.) III, 2014]**

Sol. Refer solved Ex. 2, page 5.41.

Ex. 208. *Give an example to show that a conditionally convergent series need not be absolutely convergent.* **[Delhi B.A. (Prog.) III, 2014]**

Sol. Consider the series $\Sigma u_n = 1 - (1/2) + (1/3) - (1/4) + \dots$ ... (1)

From example on page 7.4, it follows that the series (1) is convergent

Again, the series $\Sigma |u_n| = 1 + (1/2) + (1/3) + (1/4) + \dots = \Sigma(1/n)$

is not a convergent series because $\sum_{n=1}^{\infty} (1/n^{\lambda})$ is convergent only when $\lambda > 1$ (refer Art. 6.5, page 6.6). Therefore, by definition, (1) is conditionally convergent (refer Art. 7.2, page 7.1) but (1) is not absolutely convergent (refer Art 7.2, page 7.1).

Ex. 209. *Show that the function $f(x) = x^2$ is uniformly continuous on $[-1, 1]$ but not on $[0, \infty)$* **[Delhi B.A. (Prog.) III, 2013]**

Sol. Taking $[-1, 1]$ in place of $[a, b]$ and $[0, \infty)$ in place of $\mathbf{R}$ in solved Ex. 2, page 8.45, proceed exactly in the same way to prove the required result.

Ex. 210. *Find the Fourier series of the function defined by*

$$f(x) = \begin{cases} 1, & for \ -\pi < x < 0 \\ -2, & for \quad 0 < x < \pi \end{cases}$$ **[Delhi B.A. (Prog.) III, 2014]**

Sol. In what follows, we shall proceed as in solved Ex. 3, page 22.7. The required Fourier series expansion of the given function $f(x)$ is given by

$$f(x) = \frac{a_0}{2} + \sum_{n=1}^{\infty} a_n \cos nx + \sum_{n=1}^{\infty} b_n \sin nx, \qquad \ldots (1)$$

where $a_0 = \frac{1}{\pi}\int_{-\pi}^{\pi} f(x)\,dx = \frac{1}{\pi}\left[\int_{-\pi}^{0} f(x)\,dx + \int_0^{\pi} f(x)\,dx\right] = \frac{1}{\pi}\left[\int_{-\pi}^{0} dx - 2\int_0^{\pi} dx\right]$

[Using the given values of $f(x)$]

Thus, $\quad a_0 = (1/\pi) \times \left\{[x]_{-\pi}^{0} - 2[x]_0^{\pi}\right\} = (1/\pi) \times (\pi - 2\pi) = -1 \qquad \ldots (2)$

Again, $\quad a_n = \frac{1}{\pi}\int_{-\pi}^{\pi} f(x)\cos nx\,dx = \frac{1}{\pi}\left[\int_{-\pi}^{0} f(x)\cos nx\,dx + \int_0^{\pi} f(x)\cos nx\,dx\right]$

or $\quad a_n = \frac{1}{\pi}\left[\int_{-\pi}^{0} \cos nx\,dx - 2\int_0^{\pi} \cos nx\,dx\right] = \frac{1}{\pi}\left\{\left[\frac{\sin nx}{n}\right]_{-\pi}^{0} - 2\left[\frac{\sin nx}{n}\right]_0^{\pi}\right\} = 0 \qquad \ldots (3)$

Also, $\quad b_n = \frac{1}{\pi}\int_{-\pi}^{\pi} f(x)\sin nx\,dx = \frac{1}{\pi}\left[\int_{-\pi}^{0} f(x)\sin nx\,dx + \int_0^{\pi} f(x)\sin nx\,dx\right]$

or $\quad b_n = \frac{1}{\pi}\left[\int_{-\pi}^{0} \sin nx\,dx - 2\int_0^{\pi} \sin nx\,dx\right] = \frac{1}{\pi}\left\{\left[-\frac{\cos nx}{n}\right]_{-\pi}^{0} - 2\left[-\frac{\cos nx}{n}\right]_0^{\pi}\right\}$

or $\quad b_n = (1/n\pi) \times \{-(1 - \cos n\pi) + 2(\cos n\pi - 1)\} = (3/n\pi) \times (\cos n\pi - 1)$

Thus, $\quad b_n = \frac{3}{n\pi}\{(-1)^n - 1\} = \begin{cases} 0, \text{ if } n \text{ is even} \\ -(6/n\pi), \text{ if } n \text{ is odd} \end{cases} \qquad \ldots (4)$

Using (2), (3) and (4), (1) yields the required Fourier series given by

$$f(x) = -\frac{1}{2} - \frac{6}{\pi}\sum_{n=1}^{\infty} \frac{\sin(2n-1)x}{(2n-1)}$$

Ex. 211. *Define divergent sequence and an oscillatory sequence.*

[Delhi B.A. (Prog.) III, 2014]

Sol. Divergent sequence. Definition.

A sequence which diverges to either $+\infty$ or $-\infty$ is known as a divergent sequence.

For examples, refer illustrations 1, 2, 3 and 4 on page 5.15.

Oscillatory sequence. Definition.

A sequence is said to oscillate if it neither converges nor diverges.

For examples, refer illustrations 5, 6, 7 and 8 on page 5.15.

Ex. 212. *If* $\lim_{n\to\infty} a_n = a$ *and* $a \neq 0$, *show that there exists a positive number k and a positive integer m such that* $|a_n| > k$ *whenever* $n \geq m$. **[Delhi B.A. (Prog.) III, 2014]**

Sol. Since $a \neq 0$, it follows that we may choose $\quad \varepsilon = |a|/2 \qquad \ldots (1)$

as a given positive number. Again $a_n \to a$ implies that there must exist a positive integer m such that

$$|a_n - a| < \varepsilon \quad \text{whenever} \quad n \geq m \qquad \dots (2)$$

Writing $a = (a - a_n) + a_n$ and using the inequality $|x + y| \leq |x| + |y|$, we get

$$|a| = |(a - a_n) + a_n| \leq |a - a_n| + |a_n| < \varepsilon + |a_n|, \quad \text{whenever } n \geq m$$

or $\quad |a| < |a|/2 + |a_n|, \quad$ whenever $\quad n \geq m, \quad$ using (1)

Thus, $\quad |a_n| > |a|/2, \quad$ whenever $\quad n \geq m, \qquad \dots (3)$

showing that there exists a positive number $k(= |a|/2)$ and a positive integer m, such that $|a_n| > k$, whenever $n \geq m$.

Ex. 213. *Define non-uniform continuity.* **[Kanpur 2013]**

Sol. A function f is said to be not uniformly continuous on I, if there exists some $\varepsilon > 0$ for which no $\delta > 0$ serves, *i.e.,* for any $\delta > 0$, there exist $x, y \in I$ such that

$$|f(x) - f(y)| \geq \varepsilon \quad \text{whenever} \quad |x - y| < \delta$$

Alternatively, a function f is said to be not uniformly continuous on I, if there exists some $\varepsilon > 0$ and two sequences $<x_n>$ and $<y_n>$ in I such that

$$\lim_{n \to \infty} (x_n - y_n) = 0 \quad \text{and} \quad |f(x_n) - f(y_n)| \geq \varepsilon \quad \text{for all} \quad n \in \mathbf{N}$$

Remark. We can apply the above alternative definition to show that $f(x) = 1/x$ is not uniformly continuous on $(0, \infty)$. For, if $x_n = 1/n$ and $y_n = 1/(n + 1)$, then we have

$$\lim_{n \to \infty} (x_n - y_n) = 0 \quad \text{but} \quad |f(x_n) - f(y_n)| = 1 \quad \text{for all} \quad n \in \mathbf{N}$$

Ex. 214. *A function f is said to be contraction map if there exists some $0 < \alpha < 1$ such that for any x, y in the domain $|f(x) - f(y)| < \alpha\, |x - y|$. Show that a contraction map is uniformly continuous.* **[Delhi B.A. (Prog.) III, 2014]**

Sol. Let f be a contaction map whose domain is I. Then, by the given definition of a contraction map there exists some $0 < \alpha < 1$ such

$$|f(x) - f(y)| < \alpha\, |x - y|, \qquad \forall \;\; x, y \in I \qquad \dots (1)$$

If condition (1) is satisfied, then given $\varepsilon > 0$, we can take $\delta = \varepsilon/\alpha$. If $x, y \in I$ satisfy $|x - y| < \delta$, then (1) yields

$$|f(x) - f(y)| < \alpha \times (\varepsilon/\alpha) \quad \text{or} \quad |f(x) - f(y)| < \varepsilon,$$

showing that f is uniformly continuous on I.

Ex. 215. *Discuss the continuity at $x = 2$ of the function f defined by $f(x) = x - [x]$ $\forall\, x \geq 0$, where $[x]$ is the greatest integer $\leq x$. Also state which kind of discontinuity does the function f have.* **[Delhi B.A. (Prog.) III, 2013]**

Sol. Given $\quad f(x) = x - [x], \;\; \forall \;\; x \geq 0 \qquad \dots (1)$

From (1), we have $\quad f(2) = 2 - [2] = 2 - 2 = 0 \qquad \dots (2)$

Now, $\quad f(2 - 0) = \lim_{h \to 0+} f(2 - h) = \lim_{h \to 0+} \{2 - h - [2 - h]\}$, using (1)

or $\quad f(2 - 0) = \lim_{h \to 0+} \{(2 - h) - 1\}$, by definition of $[x]$ $\quad$ or $\quad f(2 - 0) = 1 \qquad \dots (3)$

Next, $\quad f(2 + 0) = \lim_{h \to 0+} f(2 + h) = \lim_{h \to 0+} \{2 + h - [2 + h]\}$, using (1)

or $\quad f(2 + 0) = \lim_{h \to 0+} \{(2 + h) - 2\}$, by definition of $[x]$ $\quad$ or $\quad f(2 + 0) = 0 \qquad \dots (4)$

From (2), (3) and (4), we have $\quad f(2 - 0) \neq f(2 + 0) = f(0),$

showing that f is right continuous at $x = 2$ and has a discontinuity of the first kind from the left at $x = 2$.

Ex. 216. *Show that* $\displaystyle\int_0^{\pi/2} \sin^p x \cos^q x = \frac{1}{2} B\left(\frac{p+1}{2}, \frac{q+1}{2}\right)$, *p > –1, q > –1*

and deduce that $\displaystyle\int_0^{\pi/2} \sqrt{\sin\theta}\, d\theta \times \int_0^{\pi/2} \frac{d\theta}{\sqrt{\sin\theta}} = \pi.$ **[Delhi B.A. (Prog.) III, 2015]**

Sol. By definition of Beta function, we have

$$B(m, n) = \int_0^1 x^{m-1} (1-x)^{n-1}\, dx, \quad \text{where} \quad m > 0, n > 0 \quad \text{... (1)}$$

Replacing m and n by $(p+1)/2$ and $(q+1)/2$ respectively in (1) we have

$$B\left(\frac{p+1}{2}, \frac{q+1}{2}\right) = \int_0^1 x^{(p-1)/2} (1-x)^{(q-1)/2}\, dx, \quad \text{where} \quad p > -1, q > -1 \quad \text{... (2)}$$

Putting $x = \sin^2\theta$ and $dx = 2\sin\theta \cos\theta\, d\theta$ in (2), we have

$$B\left(\frac{p+1}{2}, \frac{q+1}{2}\right) = 2\int_0^{\pi/2} (\sin^2\theta)^{(p-1)/2} (\cos^2\theta)^{(q-1)/2} \sin\theta\cos\theta\, d\theta$$

Thus, $$\frac{1}{2} B\left(\frac{p+1}{2}, \frac{q+1}{2}\right) = \int_0^{\pi/2} \sin^p\theta \cos^q\theta\, d\theta \quad \text{... (3)}$$

Rewritting (3), $$\int_0^{\pi/2} \sin^p x \cos^q x\, dx = \frac{1}{2} B\left(\frac{p+1}{2}, \frac{q+1}{2}\right), \; p > -1, q > -1$$

Deduction. Replacing p by 1/2 and q by 0 in (3), we have

$$\int_0^{\pi/2} \sqrt{\sin\theta}\, d\theta = \frac{1}{2} B(3/4, 1/2) \quad \text{... (4)}$$

Again, replacing p by –(1/2) and q by 0 in (3), we have

$$\int_0^{\pi/2} \frac{d\theta}{\sqrt{\sin\theta}} = \frac{1}{2} B(1/4, 1/2) \quad \text{... (5)}$$

From (4) and (5), we have

$$\int_0^{\pi/2} \sqrt{\sin\theta}\, d\theta \times \int_0^{\pi/2} \frac{d\theta}{\sqrt{\sin\theta}} = \frac{1}{4} \times B(3/4, 1/2)\, B(1/4, 1/2)$$

$$= \frac{1}{4} \times \frac{\Gamma(3/4)\,\Gamma(1/2)}{\Gamma(3/4+1/2)} \times \frac{\Gamma(1/4)\,\Gamma(1/2)}{\Gamma(1/4+1/2)} = \frac{1}{4} \times \frac{\Gamma(3/4)\sqrt{\pi}}{\Gamma(5/4)} \times \frac{\Gamma(1/4)\sqrt{\pi}}{\Gamma(3/4)}$$

$$= \frac{\pi\,\Gamma(1/4)}{4\Gamma(1+1/4)} = \frac{\pi\,\Gamma(1/4)}{4 \times (1/4) \times \Gamma(1/4)} = \pi.$$

Ex. 217. *State and prove Weierstrass's M-test. Is the converse true ?*

[Delhi B.A. (Prog.) III, 2014]

Sol. For statement and proof of Weierstrass's *M*-test, refer page 15.14. Since from the uniform convergence or divergence of $\sum_{n=1}^{\infty} f_n(x)$, as $x \in [a, b]$, we cannot say anything about the convergence or divergence of ΣM_n, the converse of the theorem is not necessarily true.

Ex. 218. *For a sequence of functions show that uniform convergence implies pointwise convergence. Is the converse true ? Justify with an example.* **[Delhi B.A. (Prog.) III, 2013]**

Sol. Let us recall the definitions of pointwise convergence and uniform convergence of sequences of functions.

Let $\{f_n\}$ be a sequence of functions which converge pointwise on a set S to a limit function f. Then, by definition, for each point x in S and for each $\varepsilon > 0$, there exists a positive integer m depending on x and ε such that

$$n > m(x, \varepsilon) \quad \Rightarrow \quad |f_n(x) - f(x)| < \varepsilon \qquad \text{... (1)}$$

Suppose we are able to find a positive integer m which is the same for every x in S, then the convergence is said to be uniform on S. More precisely, we define it as follows.

A sequence of functions $\{f_n\}$ is said to converge uniformly to f on S, if for every $\varepsilon > 0$, there exists a positive integer m depending only on ε such that

$$n > m(\varepsilon) \quad \Rightarrow \quad |f_n(x) - f(x)| < \varepsilon \quad \text{for all } x \in S \qquad \text{... (2)}$$

Note carefully that in the case of pointwise convergence for every $\varepsilon > 0$, we get different m for different x in S, whereas in the case of uniform convergence, for every $\varepsilon > 0$, there is only one m independent of x in S satisfying (2)

Using the above definitions, we now prove the following result:

Every uniformly convergent sequence is also pointwise convergent

Proof: Let the sequence of functions $\{f_n\}$ be uniformly convergent. Hence, by the above definition (2), given $\varepsilon > 0$, there exists a positive integer m depending only on ε such that

$$|f_n(x) - f(x)| < \varepsilon \qquad \text{for all } n \geq m \qquad \text{... (3)}$$

Since (3) is true for all x in S, it follows that

$$|f_n(x) - f(x)| < \varepsilon \qquad \text{for all } n \geq m$$

is true at every point x in S. Hence f_n converges pointwise to f on S

The converse of the above result is not true as shown in the following example.

Example. *Show that the sequence $\{f_n\}$ of functions, where $f_n(x) = 1/(nx + 1)$ is pointwise convergent but not uniformly convergent in $(0, 1)$.*

Proof: Here, $|f_n(x)| = \dfrac{1}{|nx+1|} < \dfrac{1}{nx}$ so that $f_n(x) \to 0$ as $n \to \infty$ at every x in $(0, 1)$, showing that $\{f_n\}$ is point-wise convergent and that the point-wise limit is the function f such that $f(x) = 0$, $\forall\, x \in (0, 1)$.

We now show that the sequence $\{f_n\}$ is not uniformly convergent. To this end, let $\varepsilon > 0$ be given. We have

$$|f_n(x) - f(x)| = |f_n(x)| < 1/(nx) < \varepsilon \qquad \text{if} \qquad n > 1/(x\varepsilon)$$

Choose m to be a positive integer just greater than $1/x\varepsilon$. Then, we have

$$|f_n(x) - f(x)| < \varepsilon \qquad \text{for all} \qquad n \geq m$$

Since m depends both on ε and x in $(0, 1)$, f_n does not converge uniformly on $(0, 1)$.

Ex. 219. *Give an example of each of the following kind of functions:*

(*i*) bounded but not Riemann–integrable

(*ii*) Riemann integrable but not continuous **[Delhi B.A. (Prog.) III, 2014]**

Sol. (*i*) For an example of a bounded function which is not Riemann integrable, refer solved Ex. 5, page 13.12.

(*ii*) For an example of a Riemann integrable function which is not continuous, refer solved Ex. 2(*a*), page 13.27. Since the function f has an infinite number of points of discontinuity, it follows that the function f under consideration is not continuous in [0, 1] whereas f is Riemann integrable in [0, 1].

Ex. 220 (*a*). *Given that the power series* $\sum_{n=0}^{\infty} a_n x^n$ *has radius of convergence R, find the radius of convergence of each of the following series:*

(*i*). $\sum_{n=0}^{\infty} a_n^m x^n$ (*ii*). $\sum_{n=0}^{\infty} a_n x^{mn}$ (*iii*). $\sum_{n=0}^{\infty} a_n x^{n^2}$, *m being a fixed positive integer.*

Sol. Part (*i*): Since R is the radius of convergence of $\sum_{n=0}^{\infty} a_n x^n$, we have

$$R = \lim_{n\to\infty} \left|\frac{a_n}{a_{n+1}}\right| \qquad \text{... (1)}$$

Let R_1 be the radius of convergence of series (*i*). Hence, we have

$$R_1 = \lim_{n\to\infty} \left|\frac{a_n^m}{a_{n+1}^m}\right| = \lim_{n\to\infty} \left|\frac{a_n}{a_{n+1}}\right|^m = \left(\lim_{n\to\infty} \left|\frac{a_n}{a_{n+1}}\right|\right)^m = R^m, \text{ using (1)}$$

Part (*ii*): Since R is the radius of convergence of $\sum_{n=0}^{\infty} a_n x^n$, we get

$$1/R = \lim_{n\to\infty} |a_n|^{1/n} \qquad \text{... (2)}$$

Let R_2 be the radius of convergence of series (*ii*). Using Cauchy's root test, we see that the series (*ii*) is absolutely convergent if

$$\lim_{n\to\infty} \left|a_n x^{mn}\right|^{1/n} < 1 \quad \text{or} \quad \lim_{n\to\infty} \left(\left|a_n\right|^{1/n} \left|x\right|^m\right) < 1$$

$$\text{or} \quad \left[\lim_{n\to\infty} \left|a_n\right|^{1/n}\right] \times \left|x\right|^m < 1 \quad \text{or} \quad (1/R) \times |x|^m < 1, \text{ using (2)}$$

Thus, $|x|^m < R$ so that $|x| < R^{1/m}$,

showing that the series (*ii*) converges absolutely for $|x| < R^{1/m}$ and diverges for $|x| > R^{1/m}$. Hence, by the definition of the radius of convergence of a power series, it follows that the radius of convergence of the series (*ii*) is given by $R_2 = R^{1/m}$.

Part (*iii*): Since R is the radius of convergence of $\sum_{n=0}^{\infty} a_n x^n$, we have

$$1/R = \lim_{n\to\infty} |a_n|^{1/n}. \qquad \text{... (3)}$$

Let R_3 be the radius of convergence of series (*iii*). Then

$$1/R_3 = \lim_{n\to\infty} |a_n|^{1/n^2} \qquad \text{... (4)}$$

We now proceed to find the limit on the R.H.S. of (4) as follows:

Let $$x = |a_n|^{1/n^2} = \left[|a_n|^{1/n}\right]^{1/n} \qquad \text{... (5)}$$

From (5), $\log x = (1/n) \times \log (|a_n|^{1/n})$ giving $\lim_{n\to\infty} \log x = \left[\lim_{n\to\infty} (1/n)\right] \times \left[\lim_{n\to\infty} \log |a_n|^{1/n}\right]$

Thus, $$\lim_{n\to\infty} \log |a_n|^{1/n^2} = 0 \times (1/R), \text{ using (3) and (5)}$$

$$\Rightarrow \quad \log\left[\lim_{n\to\infty} |a_n|^{1/n^2}\right] = 0 \quad \text{giving} \quad \lim_{n\to\infty} |a_n|^{1/n^2} = e^0 = 1 \qquad \text{... (6)}$$

Using (6) in (4), we have $1/R_3 = 1$ so that $R_3 = 1$.

Ex. 220 (*b*). *Find the radius of convergence of the following series*

(*i*) $\displaystyle\sum_{n=0}^{\infty} \frac{(n!)^2}{(2n)!} x^{2n}$. **[Delhi B.A. (Prog.) III, 2014]**

(*ii*) $\displaystyle\sum_{n=0}^{\infty} 2^{-n} x^{3n}$ **[Delhi B.Sc. (Hons.) II 2019]**

Sol. (*i*) Re-writing, the given series may be written as $$\sum_{n=0}^{\infty} a_n x^{2n}, \qquad \text{... (1)}$$

where $$a_n = (n!)^2/(2n)! \qquad \text{... (2)}$$

Let R be the radius of convergence of the series $\displaystyle\sum_{n=0}^{\infty} a_n x^n$ ... (3)

Then, $$a_n = \frac{(n!)^2}{(2n)!} \text{ and } a_{n+1} = \frac{[(n+1)!]^2}{[2(n+1)]!} = \frac{[(n+1)\, n!]^2}{(2n+2)\,(2n+1)\,(2n)!}, \text{ using (2)} \qquad \text{... (4)}$$

Hence, $$R = \lim_{n\to\infty}\left|\frac{a_n}{a_{n+1}}\right| = \lim_{n\to\infty} \frac{(2n+2)\,(2n+1)}{(n+1)^2}, \text{ using (4)}$$

or $$R = \lim_{n\to\infty} \frac{4n^2(1+1/n)\,(1+1/2n)}{n^2(1+1/n)^2} = \lim_{n\to\infty} \frac{4(1+1/2n)}{(1+1/n)} = 4.$$

Comparing the given series (1) with $\displaystyle\sum_{n=0}^{\infty} a_n x^{mn}$, here $m = 2$. Hence, by the result of part (*ii*) of Ex. 220 (*a*), it follows that the radius of convergence of the given series $= 4^{1/m} = 4^{1/2} = 2$.

(*ii*). Proceed as in part (*i*). **Ans.** Radius of convergence $= 2^{1/3}$.

Ex. 221. *Find the radius of convergence of the following series:*

(***i***). $\displaystyle\sum_{n=0}^{\infty} k^n x^n$ (***ii***). $\displaystyle\sum_{n=1}^{\infty} \{(-1)^n / n\}\, x^{n(n+1)}$ (***iii***). $\displaystyle\sum_{n=0}^{\infty} x^{n!}$, *k being a constant.*

Sol. Part (*i*). Let the series (*i*) be denoted by $\displaystyle\sum_{n=0}^{\infty} a_n x^n$; where $a_n = k^n$. If R_1 be the radius of convergence of the given series, then, we have

$$R_1 = \lim_{n\to\infty}\left|\frac{a_n}{a_{n+1}}\right| = \lim_{n\to\infty}\left|\frac{k^n}{k^{n+1}}\right| = \lim_{n\to\infty} \frac{1}{k} = \frac{1}{k}.$$

Part (*ii*): Let R_2 be the radius of convergence of the series (*ii*). Since the coefficient of $x^{n(n+1)}$ of the series (*ii*) is $(-1)^n/n$, we have

$$\frac{1}{R_2} = \lim_{n\to\infty}\left|\frac{(-1)^n}{n}\right|^{1/\{n(n+1)\}} = \lim_{n\to\infty}\frac{1}{n^{1/\{n(n+1)\}}} \quad \text{so that} \qquad R_2 = \lim_{n\to\infty} n^{1/\{n(n+1)\}}$$

Hence, $$\log R_2 = \lim_{n\to\infty}\log n^{1/\{n(n+1)\}} = \lim_{n\to\infty}\frac{\log n}{n(n+1)} \qquad \left[\text{Form } \frac{\infty}{\infty}\right]$$

or $$\log R_2 = \lim_{n\to\infty}\frac{(1/n)}{2n+1}, \text{ using L' Hospital's rule}$$

Thus, $\log R_2 = 0$ so that $R_2 = e^0 = 1$.

Part (*iii*): Re-writing the given series (*iii*) in terms of coefficients, we get

$$\sum_{n=0}^{\infty} x^{n!} = 1 + x + x^2 + x^6 + x^{24} + \ldots\ldots \qquad \ldots (1)$$

Comparing the series (1) with $\sum_{n=0}^{\infty} a_n x^n$, we find that

$a_0 = a_1 = a_2 = 1,\ a_3 = a_4 = a_5 = 0,\ a_6 = 1,\ a_7 = a_8 = \ldots = a_{23} = 0,\ a_{24} = 1$ and so on.

Thus, we find that the infinite number of coefficients a_n's are zero. Hence, we proceed as follows to find the required radius of convergence R_3 of series (*iii*).

Clearly, $< |a_n|^{1/n} >$ is a bounded sequence with $|a_n|^{1/n} \le 1$.

Hence, $M_n = l.u.b.\ \{|a_n|^{1/n}, |a_{n+1}|^{1/(n+1)}, \ldots\} = 1$ for all n so that

$$1/R_3 = \limsup_{n\to\infty} |a_n|^{1/n} = \limsup_{n\to\infty} (1) = 1 \quad \Rightarrow \quad R_3 = 1.$$

Ex. 222. *Suppose that* $f : [0, 2] \to$ ***R*** *is continuous and differentiable on* (0, 2) *and that* $f(0) = 0, f(1) = 1, f(2) = 1$. **(*i*).** *Show that there exists* $c_1 \in (0, 1)$ *such that* $f'(c_1) = 1$ **(*ii*).** *Show that there exists* $c_2 \in (1, 2)$ *such that* $f'(c_2) = 0$ **(*iii*).** *Show that there exists* $c_3 \in (0, 2)$ *such that* $f'(c_3) = 1/3$. **[Delhi B.Sc. (Hons.) I 2011]**

Sol. (*i*): Since f is continuous on [0, 1] and differentiable on (0, 1), hence by Lagrange's mean value theorem (refer page 10.9), there exists $c_1 \in (0, 1)$ such that

$\{f'(1) - f(0)\}/(1-0) = f'(c_1)$ giving $f'(c_1) = 1$, as $f(1) = 1$ and $f(0) = 0$.

(*ii*): Since f is continuous on [1, 2] and differentiable on (1, 2), hence as in part (*i*),

$f(2) - f(1)\}/(2-1) = f'(c_2)$ where $c_2 \in (1, 2)$. Thus, $f'(c_2) = 0$, as $f(1) = f(2) = 1$.

(*iii*): Since f is differentiable in the closed interval [0, 2], $f'(c_1) \ne f'(c_2)$ and 1/3 is a number between $f'(c_1)$ and $f'(c_2)$, therefore by Darboux property (refer page 9.16), there exists at least one point $c_3 \in (0, 2)$ such that $f'(c_3) = 1/3$.

Ex. 223. *State and prove boundedness, theorem. Show that conclusion of the theorem fails if any one of the hypothesis is relaxed.*

Sol. Statement of boundedness theorem: *If a function f is continuous in a closed bounded interval [a, b] then it is bounded.*

For proof of this theorem, refer page 8.37.

In what follows, we now present examples that show that the conclusion of the above theorem fails if any of the hypothesis is relaxed.

(*i*). *The function f must be continuous.* Let the function f be defined on the closed and bounded interval [0, 1] as follows:

$$f(x) = \begin{cases} 1/x, \text{ if } 0 < x \le 1 \\ 0, \text{ if } x = 0 \end{cases} \qquad \text{... (1)}$$

Clearly, function f defined by (1) is discontinuous at $x = 0$ and unbounded on [0, 1].

(ii). *The interval must be closed.* Let the function f be defined by $f(x) = 1/x$, for $0 < x \le 1$. Then, f is continuous on the half-open interval (0, 1] but not bounded on (0, 1].

(iii). *The interval must be bounded.* Let the function f be defined on unbounded closed interval $[0, \infty)$ as follows. $f(x) = x$, for all $x \in [0, \infty)$. Then, f is continuous on $[0, \infty)$ but it is not bounded on $[0, \infty)$.

Ex. 224. *If function f is continuous and bounded on **R**, is f uniformly continuous on **R** ?*

Sol. With help of the following counter example, we shall prove that f cannot be uniformly continuous.

Let $$f(x) = \sin x^2, \text{ for } x \in \mathbf{R} \qquad \text{... (1)}$$

Clearly, the function f defined by (1) is continuous and bounded on **R**. Let $\varepsilon = 2$. Since f satisfies all the conditions of Lagrange's mean value theorem, we obtain

$$f(x_1) - f(x_2) = (x_1 - x_2) f'(c), \qquad \text{where} \qquad x_1 < c < x_2$$

Using (1), this $\Rightarrow$ $f(x_1) - f(x_2) = (x_1 - x_2)\,[2x \cos x^2]_{x=c}$ for $x_1 < c < x_2$.

$\Rightarrow$ $|f(x_1) - f(x_2)| = |x_1 - x_2| \times 2c\,|\cos c^2|$, where $x_1 < c < x_2$

$\Rightarrow$ $|f(x_1) - f(x_2)| \le |x_1 - x_2| \times 2c$, (since $|\cos c^2| \le 1$), for $x_1 < c < x_2$ ... (2)

If there exists a $\delta > 0$ such that $|x_1 - x_2| < \delta$, then (2) yields

$$|f(x_1) - f(x_2)| < 2c\delta < 2,$$

showing that $c\delta < 1$ for $x_1 < c < x_2$. Since x_1 and x_2 are arbitrary points of **R** and $x_1 < c < x_2$, therefore $c\,\delta < 1$ is not always true. Hence, for the function defined by (1), corresponding to $\varepsilon = 2$, there is no δ which will work for all $\forall x \in \mathbf{R}$. Therefore, function f given by (1) is not uniformly continuous on **R**.

Ex. 225. *Let f be a continuous function on [a, b] and let ϕ be a differentiable function such that $\phi'(x) = f(x)$ for all $x \in [a, b]$. Prove that* $\int_a^b f(x)\,dx = \phi(b) - \phi(a)$. **[Agra 2006, Meerut 2010]**

Sol. Since f is continuous on $[a, b]$, hence by theorem I, page 13.25, it follows that f is integrable on $[a, b]$. Now, proceed as in Art. 13.15, page 13.40, and show that

$$\int_a^b f(x)\,dx = \phi(b) - \phi(a).$$

Ex. 226. *By an example, prove that result* $\int_a^b f'(x)\,dx = f(b) - f(a)$ *may fail to hold.*

Sol. Consider a function f defined as follows:

$$f(x) = \begin{cases} x^2 \cos(\pi/x^2), \text{ if } 0 < x \le 1 \\ 0, \text{ if } x = 0 \end{cases} \qquad \text{... (1)}$$

Then, clearly f is differentiable on [0, 1] and

$$f'(x) = \begin{cases} 2x \cos(\pi/x^2) + (2\pi/x) \times \sin(\pi/x^2), \text{ if } 0 < x \le 1 \\ 0, \text{ if } x = 0 \end{cases} \qquad \text{... (2)}$$

From (2), it follows that f' is not bounded on [0, 1]. Hence f' is not Riemann integrable, i.e., $\int_0^1 f'(x)\,dx$ does not exist.

Thus, $\int_0^1 f'(x)\,dx = f(1) - f(0)$ fails to hold.

Ex. 227. *If P^* is a refinement of P, then prove that*
$U(P^*, f, \alpha) - L(P^*, f, \alpha) \le U(P, f, \alpha) - L(P, f, \alpha)$. **[Himachal 2015]**

Sol. Proceed as in theorem I, page 14.2 and obtain

$$L(P, f, \alpha) \le L(P^*, f, \alpha) \quad \text{... (1)}$$

and
$$U(P^*, f, \alpha) \le U(P, f, \alpha) \quad \text{... (2)}$$

Multiplying both sides of (1) by –1, we have

$-L(P, f, \alpha) \ge -L(P^*, f, \alpha)$ giving $-L(P^*, f, \alpha) \le -L(P, f, \alpha)$ ... (3)

Then, (2) and (3) $\Rightarrow$ $U(P^*, f, \alpha) - L(P^*, f, \alpha) \le U(P, f, \alpha) - L(P, f, \alpha)$.

Ex. 228. *Show that series of functions Σf_n defined on [a, b] converges uniformly on [a, b] if and only if for each $\varepsilon > 0$ and for all $x \in [a, b]$, there exists a positive integer m such that*

$$|f_{n+1}(x) + f_{n+2}(x) + \dots + f_{n+p}(x)| < \varepsilon, \ \forall\ n \ge m \quad \text{and} \quad \forall p \ge 1.$$ **[Himachal 2014]**

Sol. Σf_n is uniformly convergent on [a, b]

$\Leftrightarrow$ The sequence $< S_n >$ of the partial sums is uniformly convergent an [a, b]

$\Leftrightarrow$ By Cauchy's general principle of convergence of a sequence (refer Art. 15.2, page 15.2) for each $\varepsilon > 0$ and for all $x \in [a, b]$, there exists a positive integer m(depending on ε and not on x) such that

$$|S_{n+p}(x) + S_n(x)| < \varepsilon, \quad \forall n \ge m, \quad \forall p \ge 1 \quad \text{and} \quad \forall x \in [a, b]$$

$$\Leftrightarrow |f_{n+1}(x) + f_{n+2}(x) + \dots + f_{n+p}(x)| < \varepsilon, \quad \forall n \ge m, \quad \forall p \ge 1 \quad \text{and} \quad \forall x \in [a, b]$$

Ex. 229. *Suppose f is bounded on [a, b], f has only finitely many points of discontinuity on [a, b] and α is continuous at every point at which f is discontinuous. Show that $f \in R(\alpha)$ on [a, b].* **[Himachal 2014]**

Sol. Let $\varepsilon > 0$ be given. Write $M = \sup |f(x)|$ ... (1)

Let A denote the set of all points at which f is discontinuous. Since A is finite and α is continuous at every point of A, it is possible to cover A by finitely many disjoint intervals $[u_j, v_j] \subset [a, b]$ such that

$$\sum_{j=1}^{n} [\alpha(v_j) - \alpha(u_j)] < \varepsilon \quad \text{... (2)}$$

Again, we can arrange these intervals in such a way that every point of $A \cap (a, b)$ lies in the interior of some $[u_j, v_j]$.

Write
$$B = [a, b] - \bigcup_{i=1}^{n} (u_j, v_j) \quad \text{... (3)}$$

Then, B is a compact set and hence f is uniformly continuous on B. Consequently, there exists $\delta > 0$ such that

$$|x - y| < \delta \text{ and } x, y \in B \quad \Rightarrow \quad |f(x) - f(y)| < \varepsilon \quad \text{... (4)}$$

We now construct a partition $P = \{a = x_0, x_1, ..., x_n = b\}$ of [a, b] satisfying the following three properties:

(*i*) Each u_j occurs in P and each v_j occurs in P

(*ii*) No point of any segment (u_j, v_j) occurs in P

(*iii*) If x_{i-1} is not one of the u_j, then $\delta_i < \delta$. Here $< \delta$. Here, $\delta_i = x_i - x_{i-1}$

Let $M_i = \sup f(x)$, $x_{i-1} \le x \le x_i$ and $m_i = \inf f(x)$, $x_{i-1} \le x \le x_i$

Now, for every i, we have

$$M_i - m_i = |M_i - m_i| \le |M_i| + |m_i| \le 2|M_i| = 2 \sup_{x \in [x_{i-1}, x_i]} |f(x_i)| \le 2 \sup_{x \in [a, b]} |f(x)|$$

Hence, for every i, $\qquad M_i - m_i = 2M$, using (1) ... (5)

Again, we have $\qquad M_i - m_i \le \varepsilon$ unless x_{i-1} is one of the u_j ... (6)

Now, $\qquad U(P, f, \alpha) - L(P, f, \alpha) = \sum_{i=1}^{n} (M_i - m_i)\,\delta\alpha_i$, where $\delta\alpha_i = \alpha(x_i) - \alpha(x_{i-1})$ ... (7)

Keeping in view results (5) and (6), (7) yields

$$U(P, f, \alpha) - L(P, f, \alpha) \le 2M\,\varepsilon + \varepsilon \sum_{i=1}^{n} \delta\,\alpha_i \qquad \text{... (8)}$$

But $\qquad \sum_{i=1}^{n} \delta\,\alpha_i = \sum_{i=1}^{n} [\alpha(x_i) - \alpha(x_{i-1})] = [\alpha(x_1) - \alpha(x_0)] + \ldots + [\alpha(x_n) - \alpha(x_{n-1})]$

or $\qquad \sum_{i=1}^{n} \delta\alpha_i = \alpha(x_n) - \alpha(x_0) = \alpha(b) - \alpha(a)$, as $x_0 = a$ and $x_n = b$

Hence, (8) reduces to $\qquad U(P, f, \alpha) - L(P, f, \alpha) \le 2M\,\varepsilon + \varepsilon\,[\alpha(b) - \alpha(a)]$ ... (9)

Since ε is arbitrary, by theorem on page 14.5, it follows that f is integrable with respect to α, i.e., $f \in R(\alpha)$.

Ex. 230. *Multiple choice question having exactly one correct option*

$$\lim_{n \to \infty} \frac{1}{\sqrt{n}} \left(\frac{1}{\sqrt{2} + \sqrt{4}} + \frac{1}{\sqrt{4} + \sqrt{6}} + \ldots + \frac{1}{\sqrt{2n} + \sqrt{2n+2}} \right) \textit{ is}$$

A. $\sqrt{2}$ **B.** $1/\sqrt{2}$ **C.** $1 + \sqrt{2}$ **D.** $1/(1 + \sqrt{2})$ **[CSIR–UGC (NET), December 2015]**

Sol. Ans. (B). We are to evaluate $\qquad \lim_{n \to \infty} (1/\sqrt{n}) \times \sum_{n=1}^{n} u_n,$

$$\text{where } u_n = \frac{1}{\sqrt{2n+2} + \sqrt{2n}} = \frac{\sqrt{2n+2} - \sqrt{2n}}{(\sqrt{2n+2} + \sqrt{2n})\,(\sqrt{2n+2} - \sqrt{2n})} = (1/2) \times (\sqrt{2n+2} - \sqrt{2n}) \quad \text{... (1)}$$

Hence, the required limit

$$= \lim_{n \to \infty} (1/\sqrt{n}) \times \sum_{n=1}^{n} u_n = \lim_{n \to \infty} (1/\sqrt{n}) \times \sum_{n=1}^{n} (1/2) \quad (\sqrt{2n+2} - \sqrt{2n}), \text{ using (1)}$$

$$= \lim_{n \to \infty} (1/2\sqrt{n}) \times \{(\sqrt{4} - \sqrt{2}) + (\sqrt{6} - \sqrt{4}) + \ldots + (\sqrt{2n+2} - \sqrt{2n})\}$$

$$= \lim_{n \to \infty} \frac{\sqrt{2n+2} - \sqrt{2}}{2\sqrt{n}} = \lim_{n \to \infty} \frac{\sqrt{2n+2}}{2\sqrt{n}} - \lim_{n \to \infty} \frac{\sqrt{2}}{2\sqrt{n}} = \lim_{n \to \infty} \frac{\sqrt{2 + 2/n}}{2} - 0 = \frac{\sqrt{2}}{2} = \frac{1}{\sqrt{2}}$$

Thus, the required limit = $1/\sqrt{2}$ and hence the option (B) is correct.

Ex. 231. *Multiple choice question with multiple correct options*

Consider all sequences $\{f_n\}$ of real valued continuous functions on $[0, \infty)$. Identify which of the following statements are correct.

A. *If $\{f_n\}$ converges to f pointwise on [0, ∞), then* $\lim_{n\to\infty} \int_0^\infty f_n(x)\,dx = \int_0^\infty f(x)\,dx$

B. *If $\{f_n\}$ converges to f uniformly on [0, ∞), then* $\lim_{n\to\infty} \int_0^\infty f_n(x)\,dx = \int_0^\infty f(x)\,dx$

C. *If $\{f_n\}$ converges to f uniformly on [0, ∞), then f is continuous on [0, ∞)*

D. *There exists a sequence of continuous functions $\{f_n\}$ on [0, ∞) such that $\{f_n\}$ converges to f uniformly on [0, ∞) but*

$$\lim_{n\to\infty} \int_0^\infty f_n(x)\,dx \neq \int_0^\infty f(x)\,dx.$$ **[CSIR–UGC (NET), December 2015]**

Sol. Ans. (B) and **(C).** From theorem of Art. 15.4, page 15.7, it follows that the option (C) is true. Again, from theorem of Art. 15.5, page 15.9, it follows that the option (B) is correct whereas options (A) and (D) are not correct.

Ex. 232. *Multiple choice question having exactly one correct option*

Let $f : \mathbf{R} \to \mathbf{R}$ be a twice continuous differentiable function with $f(0) = f(1) = f'(0) = 0$. Then,

A. *f'' is the zero function* **B.** *$f''(0)$ is zero*

C. *$f''(x) = 0$ for some $x \in (0, 1)$* **D.** *f'' never vanishes*

[CSIR–UGC (NET), December 2015]

Sol. Ans. (C). Since $f(1) = 0$, it follows that $(x - 1)$ is a factor of $f(x)$.

Again, since $f(0) = f'(0) = 0$, it follows that $(x - 0)^2$, *i.e.*, x^2 is a factor of $f(x)$.

In view of the above facts, we take $f(x) = kx^2(x - 1)$ or $f(x) = k(x^3 - x^2)$ ... (1)

where k is a constant. Differentiating (1) twice with respect to x yields

$$f''(x) = k(6x - 2) = 2k(3x - 1) \quad \text{so that} \quad f''(x) = 0 \quad \Rightarrow \quad x = 1/3 \in (0, 1)$$

Thus, the option (C) is correct while options (A), (B) and (D) are not correct.

Ex. 233. *Multiple choice question having multiple correct options*

Let $f : (0, 1) \to \mathbf{R}$ be continuous. Suppose that

$|f(x) - f(y)| \le |\cos x - \cos y|$ for all $x, y \in (0, 1)$. Then,

A. *f is discontinuous at least at one point in (0, 1)*

B. *f is continuous everywhere on (0, 1) but not uniformly continuous on (0, 1)*

C. *f is uniformly continuous on (0, 1)*

D. $\lim_{x\to 0} f(x)$ *exists.* **[CSIR–UGC (NET), December 2015]**

Sol. Ans. (C) and **(D)** Given that for all $x, y \in (0, 1)$, we have

$$|f(x) - f(y)| \le |\cos x - \cos y|$$

$$\Rightarrow \quad |f(x) - f(y)| \le \left|2\sin\frac{y+x}{2}\sin\frac{y-x}{2}\right| = 2\times\left|\sin\frac{x+y}{2}\right|\times\left|\sin\frac{y-x}{2}\right|$$

$$\Rightarrow \quad |f(x) - f(y)| \le 2\times\frac{|x-y|}{2}, \text{ as } \left|\sin\frac{x+y}{2}\right| \le 1 \text{ and } \left|\sin\frac{y-x}{2}\right| \le \frac{|y-x|}{2} = \frac{|x-y|}{2}$$

Thus, $|f(x) - f(y)| \le |x - y|$, for all $x, y \in (0, 1)$ ... (1)

Now, for a given $\varepsilon > 0$, let us choose $\delta = \varepsilon$. Then, (1) yields

$$|f(x) - f(y)| < \varepsilon \quad \text{whenever} \quad |x - y| < \varepsilon \quad \text{for all } x, y \in (0, 1)$$

This shows that f is uniformly continuous on (0, 1) and hence the option (C) is correct. Again,

f is uniformly continuous on (0, 1)

$\Rightarrow f$ is continuous on (0, 1) (by theorem I, page 8.43) $\Rightarrow$ $\lim_{x \to 0} f(x)$ exists.

Therefore, the option (D) is also correct.

Ex. 234. *Multiple choice question having exactly one correct option*

$\lim_{n \to 0} (1 - 1/n^2)^n$ *equals* **A.** *1* **B.** $e^{-1/2}$ **C.** e^{-2} **D.** e^{-1}

[CSIR–UGC (NET), December 2015]

Sol. Ans. (A). In what follows, we shall use the following standard result:

$$\lim_{n \to \infty} (1 + x/n)^n = e^x \qquad \text{... (1)}$$

Now, $\lim_{n \to \infty} (1 - 1/n^2)^n = \lim_{n \to \infty} \{(1 + 1/n)(1 - 1/n)\}^n = \lim_{n \to \infty} (1 + 1/n)^n \times \lim_{n \to \infty} (1 - 1/n)^n$

$= e \times e^{-1} = e^0 = 1$, using result (1) for $x = 1$ and $x = -1$

Hence, the option (A) is correct.

Ex. 235. *Multiple choice question having multiple correct options*

Let $x_1 = 0$, $x_2 = 1$ *and for* $n \geq 3$, *define* $x_n = (x_{n-1} + x_{n-2})/2$.

Which of the following is/are true ?

A. *$\{x_n\}$ is a monotone sequence* **B.** $\lim_{n \to \infty} x_n = 1/2$

C. *$\{x_n\}$ is a Cauchy sequence* **D.** $\lim_{n \to \infty} x_n = 2/3$ **[CSIR–UGC (NET), June 2016]**

Sol. Ans. (C) and (D). Given that $x_1 = 0, \quad x_2 = 1$ ... (1)

and $x_n = (x_{n-1} + x_{n-2})/2$, for all $n \geq 3$ ... (2)

Using (1) and (2), we have

$$x_3 = \frac{x_2 + x_1}{2} = \frac{1}{2} = \frac{2 - 2^0}{2}, \qquad x_4 = \frac{x_3 + x_4}{2} = \frac{(1/2) + 1}{2} = \frac{3}{4} = \frac{2^2 - 2^1 + 2^0}{2^2},$$

$$x_5 = \frac{x_4 + x_3}{2} = \frac{(3/4) + (1/2)}{2} = \frac{5}{8} = \frac{2^3 - 2^2 + 2 - 2^0}{2^3}$$ and so on. Thus, in general, we have

$$x_n = \frac{2^{n-2} - 2^{n-3} + 2^{n-4} + \dots + (-1)^{n-2} \times 2^0}{2^{n-2}} = 1 + \left(-\frac{1}{2}\right) + \left(-\frac{1}{2}\right)^2 + \dots + + \left(-\frac{1}{2}\right)^{n-2}, \qquad \text{... (3)}$$

which is a geometric series whose first term $= a = 1$, common ratio $= r = -(1/2)$ and number of terms $= n - 1$.

Hence, using the usual formula for sum of n terms of a geometric series namely, $S_n = \{a(1 - r^n)/n\}$, we have

$$x_n = \frac{1 - (-1/2)^{n-1}}{1 - (-1/2)} = \frac{2}{3} \times \left[1 - \frac{(-1)^{n-1}}{2^{n-1}}\right] \quad \Rightarrow \quad \lim_{n \to \infty} x_n = \frac{2}{3} \qquad \text{... (4)}$$

From (4), it follows that the option (D) is correct. Again, (4) also shows that the sequence $\{x_n\}$ is a convergent sequence.

Since every convergent sequence is also a Cauchy sequence (refer theorem II, page 5.30), it follows that $\{x_n\}$ is a Cauchy sequence. Thus, the option (C) is also true.

Ex. 236. *Multiple choice question having multiple correct options*

Let A be any set. Let P(A) be the power set of A, that is, the set of all subsets of A; P(A) = {B : B ⊆ A}. Then, which of the following is/are true about the set P(A).

A. *P(A) = φ for any set A* **B.** *P(A) is finite set for some A*

C. *P(A) is a countable set for some A* **D.** *P(A) is a uncountable set for some A*

[CSIR–UGC (NET), June 2016]

Sol. Ans. (B) and **(D).** If $A = \phi$, then $P(A) = P(\phi) = \{\phi\}$, which contains one set, namely, φ. Here we have used the fact that φ is subset of every set. Thus, the option *A* is not true.

Next, let *A* contain finite number of *n* elements. Then, *P*(*A*) will contain 2^n subsets (refer solved Ex. 1, page 1.10). Thus, *P*(*A*) is finite when *A* is itself a finite set. Hence, the option (B) is correct.

Finally, let *A* be countable infinite set. Then, by Contor's theorem, the power set *P*(*A*) of *A* is uncountable infinite set. For example, the set of all natural numbers **N** is countable whereas the power set *P*(**N**) is uncountable. Hence, the option (D) is correct whereas the option (C) is not correct.

Ex. 237. *Multiple choice question having exactly one correct option*

Consider the improper Riemann integral $\int_0^x y^{-1/2}\, dy$. *The integral is*

A. *continuous in [0, ∞)* **B.** *continuous only in (0, ∞)*

C. *discontinuous in (0, ∞)* **D.** *discontinuous only in (1/2, ∞)*

[CSIR–UGC (NET), June 2016]

Sol. Ans. (A). Let $$f(x) = \int_0^x y^{-1/2}\, dy,$$

which is improper integral of the second kind. Using definition (refer Art. 16.2, page 16.1), we have

$$f(x) = \lim_{\varepsilon \to 0+} \int_\varepsilon^x y^{-1/2}\, dy = \lim_{\varepsilon \to 0+} \left[2\sqrt{y}\right]_\varepsilon^x = \lim_{\varepsilon \to 0+} [2\sqrt{x} - 2\sqrt{\varepsilon}] = 2\sqrt{x} \quad \text{... (1)}$$

From (1), we have $f(0) = 0$ and $f(0+0) = \lim_{h \to 0} f(0+h) = \lim_{h \to 0} 2\sqrt{h} = 0$

Since $f(0) = f(0+0)$, it follows that *f* is continuous at the left end $x = 0$ of the interval [0, ∞). Clearly, *f* is also continuous in (0, ∞). Thus, *f* is continuous in [0, ∞). Hence the option (A) is correct.

Ex. 238. *Multiple choice question with multiple correct options*

Which of the following functions is/are uniformly continuous on the interval (0, 1) ?

A. *1/x* **B.** *sin (1/x)* **C.** *x sin (1/x)* **D.** *(sin x)/x* **[CSIR–UGC (NET), June 2016]**

Sol. Ans. (C) and **(D).** Let us examine the option (A). Write, $f(x) = 1/x$.

Let **N** denote the set of all natural numbers. Suppose that $x_1 = 1/n$ and $x_2 = 1/2n$, where $n \in \mathbf{N}$. Then, $x_1, x_2 \in (0, 1)$. Let $\delta > 1/n$. Then, we have

$$|x_1 - x_2| = |(1/n) - (1/2n)| = 1/2n < 1/n < \delta, \quad \text{as} \quad \delta > 1/n$$

and $$|f(x_1) - f(x_2)| = |1/x_1 - 1/x_2| = |n - 2n| = n > 1/2, \quad \text{as } n \in \mathbf{N} \Rightarrow n \geq 1$$

Hence, if we choose $\varepsilon = 1/2 > 0$, then whenever $\delta > 0$, there exists $x_1, x_2 \in (0, 1)$ such that

$$|f(x_1) - f(x_2)| > \varepsilon \quad \text{whenever} \quad |x_1 - x_2| < \delta,$$

showing that *f* is not uniformly continuous on (0, 1) (refer note 2 on page 8.43). Hence, the option (A) is not correct.

We now examine option (B). Let $$g(x) = \sin(1/x).$$

Let $\varepsilon > 0$ be given. Then, in order to prove that g is not uniformly continuous, we shall show that for each $\delta > 0$, there exist. $x_1, x_2 \in (0, 1)$ such that

$$|x_1 - x_2| < \delta \quad \Rightarrow \quad |g(x_1) - g(x_2)| \geq \varepsilon \qquad \text{... (1)}$$

Now, for each $\delta > 0$, there exists $m \in \mathbf{N}$ such that

$$1/\{2n(n\pi + \pi/2\} < \delta, \quad \text{for all} \quad n \geq m \qquad \text{... (2)}$$

Let $x_1 = 1/(m\pi + \pi/2)$ and $x_2 = 1/(m\pi)$. Then, $x_1, x_2 \in (0, 1)$ such that

$$|x_1 - x_2| = \left|\frac{1}{m\pi + \pi/2} - \frac{1}{m\pi}\right| = \left|\frac{(-\pi/2)}{m\pi(m\pi + \pi/2)}\right| = \frac{1}{2m(m\pi + \pi/2)} < \delta, \text{ using (2)}$$

But $\quad |g(x_1) - g(x_2)| = |\sin(m\pi + \pi/2) - \sin m\pi| = |\cos m\pi| = |(-1)^m| = 1 = \varepsilon,$

where we chosen $\varepsilon = 1$. Thus, we have shown that

$$|x_1 - x_2| < \delta \quad \Rightarrow \quad |g(x_1) - g(x_2)| = \varepsilon,$$

showing that g is not uniformly continuous on (0, 1). Hence, the option (B) is also not correct.

Reader can verify that functions given in the options (C) and (D) are uniformly continuous. Therefore, the options (C) and (D) are true.

Ex. 239. *Multiple choice question having multiple correct options.*

Define f an [0, 1] by $f(x) = \begin{cases} x^2, \text{ if } x \text{ is rational} \\ x^3, \text{ if } x \text{ is irrational} \end{cases}$. *Then*

A. *f is not Riemann integrable on [0, 1]* **B.** *f is Riemann integrable and* $\int_0^1 f(x)\,dx = 1/4$

C. *f is Riemann integrable and* $\underline{\int_0^1} f(x)\,dx = 1/3$

D. $1/4 = \underline{\int_0^1} f(x)\,dx < \overline{\int_0^1} f(x)\,dx = 1/3$, *where* $\underline{\int_0^1} f(x)\,dx$ *and* $\overline{\int_0^1} f(x)\,dx$ *are lower and upper integrals of f.* **[CSIR–UGC (NET), June 2016]**

Sol. Ans. (A) and **(D).** In what follows we shall proceed as in Ex. 7, page 13.13.

We have, $\quad x^2 > x^3 \quad$ for all $\quad x \in (0, 1)$

Let M_r and m_r be the supremum and infimum of the given function f in the rth subinterval I_r of any partition P of [0, 1]. Then, for all values of r, $M_r = x^2$ and $m_r = x^3$. Now, by definition, we have

$$\text{the upper Riemann integral} = \overline{\int_0^1} f(x)\,dx = \int_0^1 x^2\,dx = \frac{1}{3} \qquad \text{... (1)}$$

and

$$\text{the lower Riemann integral} = \underline{\int_0^1} f(x)\,dx = \int_0^1 x^3\,dx = \frac{1}{4} \qquad \text{... (2)}$$

From (1) and (2), we have $\quad \overline{\int_0^1} f(x)\,dx \neq \underline{\int_0^1} f(x)\,dx,$

showing that f is not Riemann integrable on [0, 1]. Hence, the option (A) is correct whereas the option (B) and (C) are not correct.

From (1) and (2) we have $\qquad 1/4 = \int_0^1 f(x)\,dx < \int_0^1 f(x)\,dx = 1/3,$

showing that the option (D) is correct.

Ex. 240. *A function f:* $\mathbf{R} \to \mathbf{R}$ *is said to be periodic on* $\mathbf{R}$ *it there exists a number* $p > 0$ *such that* $f(x + p) = f(x)$ *for all* $x \in \mathbf{R}$. *Prove that a continuous periodic function on* $\mathbf{R}$ *is bounded and uniformly continuous.*

Sol. Since f is bounded on $[0, p]$, it follow, that f is a bounded function on **R**.

Since f is continuous on the closed interval $[-1, p + 1]$, it must be uniformly continuous on $[-1, p + 1]$ (refer theorem II, page 8.44). Again, this implies that f is uniformly continuous on **R**.

Ex. 241. *Multiple choice question having multiple correct options.*

Let f: $\mathbf{R} \to \mathbf{R}$ *be a continuous function on* $\mathbf{R}$ *and* $f(x + 1) = f(x)$ *for all* $x \in R$. *Then*

A. *f is bounded above, but not bounded below*

B. *f is bounded above and below, but may not attain its bounds*

C. *f is bounded above and below, and f attains its bounds*

D. *f is uniformly continuous.* **[CSIR–UGC (NET), June 2016]**

Sol. Ans. (C) and **(D).** This is a particular case of the above Ex. 240. Note that here $p = 1$. Proceed as in Ex. 240 to show that f is a bounded function on **R**. Hence, by definition, f is bounded above and below and f attains its bounds. Therefore, the option (C) is correct whereas options (A) and (B) are not correct.

Again, proceed as in Ex. 240 to show that f is uniformly continuous on **R**. Hence, the option (D) is also correct.

Ex. 242. (*a*). *Let A be a closed and bounded set of real numbers. Show that every sequence in A has a limit point in A.*

(*b*). *Let I be a closed interval. Then, show that every sequence in I has a limit point in I.*

Sol. (*a*). Let $\{x_n\}$ be a sequence in A. Since A is bounded, therefore, $\{x_n\}$ is bounded and hence it has a limit point, say l (refer theorem I, page 5.3).

Now l cannot belong to $R - A$, for if $l \in R - A$, then $R - A$ is a neighbourhood of l that contains no x_n and this contradicts the fact that l is a limit point of $\{x_n\}$. Hence, $l \in A$.

Part (*b*). Since I is a closed interval, therefore I is a closed and bounded set. Hence, writing I in place of set A in part (*a*), we see that every sequence in I has a limit point in I.

Ex. 243. *Multiple choice question having exactly one correct option*

Consider the interval $[-1, 1]$ *and a sequence* $\{\alpha_n\}_{n=1}^{\infty}$ *of elements in it. Then*

A. *Every limit point of* $\{\alpha_n\}$ *is in* $(-1, 1)$

B. *Every limit point of* $\{\alpha_n\}$ *is in* $[-1, 1]$

C. *The limit point of* $\{\alpha_n\}$ *can only be in* $\{-1, 0, 1\}$

D. *The limit point of* $\{\alpha_n\}$ *cannot be in* $\{-1, 0, 1\}$ **[CSIR–UGC (NET), June 2016]**

Sol. Ans. (B). Since $[-1, 1]$ is a closed interval, hence it is a bounded set. Therefore, taking $A = [-1, 1]$ in part (*a*) of Ex. 242, it follows that every sequence in $[-1, 1]$ has a limit point in $[-1, 1]$. Thus, the option (B) is correct whereas the options (A), (C) and (D) are not correct.

Ex. 244. *Let f be monotonic on (a, b). Then show that the set of points of (a, b) at which f is discontinuous is at most countable.*

Sol. Suppose, for the sake of definiteness, that f is a monotonic increasing function. Let A denote the set of points at which f in discontinuous.

Let us associate a rational number $r(x)$ corresponding to each point x to A such that

$$r(x - 0) < r(x) < r(x + 0).$$

Now, $x_1 < x \Rightarrow r(x_1 + 0) \le r(x_2 - 0)$ and hence it follows that

$$r(x_1) \ne r(x_2) \text{ if } x_1 \ne x_2.$$

We have thus arrived at a one-one correspondence between the set A and a subset (say, B) of **Q**, where **Q** in the set of all rational numbers. Now, by theorem VII, page 4.4, **Q** is a countable set and hence B is also countable (since by theorem II, page 4.2, every subset of a countable set is countable). Thus, we have obtained a one-one correspondence between the sets A and B. Since B is a countable set, it follows that A is also a countable set. Hence, A is at most countable.

Ex. 245. *Multiple choice question having exactly one correct option*

*Let F: **R** $\to$ **R** be a monotone function. Then*

A. *F has no discontinuities* **B.** *F has only finitely many discontinuities*

C. *F can have at most countably many discountinuities*

D. *F can have uncountably many discontinuities* **[CSIR–UGC (NET), June 2016]**

Sol. Ans. (C). Proceeding as in Ex. 244, it follows that the function F can have at most countably many discountinuities. Hence, the option (C) is correct.

Ex. 246. *Multiple choice question having exactly one correct option*

Which one of the following statements is true for the sequence of functions

$f_n(x) = 1/(n^2 + x^2)$, $n = 1, 2, \ldots,$ $x \in [1/2, 1]$?

A. *The sequence is monotonic and has 0 as the limit for all $x \in [1/2, 1]$ as $n \to \infty$*

B. *The sequence is not monotonic but has $f(x) = 1/x^2$ as the limit as $n \to \infty$*

C. *The sequence is monotonic and has $f(x) = 1/x^2$ as the limit as $n \to \infty$*

D. *The sequence is not monotonic but has 0 as the limit.* **[CSIR–UGC (NET), June 2016]**

Sol. Ans. (A). Here $f_n(x) = 1/(n^2 + x^2)$ and $f_{n+1}(x) = 1/\{(n+1)^2 + x^2\}$. Hence, we have

$$f_n(x) - f_{n+1}(x) = \frac{1}{n^2 + x^2} - \frac{1}{(n+1)^2 + x^2} = \frac{(2n+1)}{(n^2 + x^2)\{(n+1)^2 + x^2\}} > 0 \text{ for all } n \in \mathbf{N},$$

where **N** is the set of all natural numbers. From this, it follows that

$$f_n(x) > f_{n+1}(x) \ \forall n \in \mathbf{N} \text{ and hence } \{f_n\} \text{ is a monotonic sequence.}$$

Again, $$\lim_{n \to \infty} f_n(x) = \lim_{n \to \infty} \frac{1}{n^2 + x^2} = 0 \text{ for all } x \in [1/2, 1]$$

Thus, the sequence $\{f_n\}$ is monotonic and has 0 as the limit for all $x \in [1/2, 1]$ as $n \to \infty$. Hence, the option (A) is correct.

Ex. 247. *Let $p(n)$ be a polynomial of degree $d \ge 2$. The radius of convergence of the power series $\sum_{n=0}^{\infty} p(n)\, z^n$ is*

A. *0* **B.** *1* **C.** ∞ **D.** *depends on d.* **[CSIR–UGC (NET), June 2016]**

Sol. Ans. (B). Since $p(n)$ is a polynomial of degree $d \ge 2$, we write

$$p(n) = a_0 + a_1 n + a_2 n^2 + \ldots + a_{d-2} n^{d-2} + a_{d-1} n^{d-1} + a_d n^d \quad \ldots (1)$$

where $d \ge 2$ and $a_0, a_1, a_2, \ldots, a_{d-1}, a_d$ are constants free from n.

$$(1) \Rightarrow p(n+1) = a_0 + a_1 (n+1) + a_2 (n+1)^2 + \ldots + a_{d-1} (n+1)^{d-1} + a_d (n+1)^d \quad \ldots (2)$$

Comparing the given power series with $\sum_{n=0}^{\infty} c_n z^n$, here $c_n = p(n)$.

Let R be the radius of convergence of the given power series. Then, we have

$$R = \lim_{n\to\infty}\left|\frac{c_n}{c_{n+1}}\right| = \lim_{n\to\infty}\left|\frac{p(n)}{p(n+1)}\right| = \lim_{n\to\infty}\left|\frac{a_0 + a_1 n + ... + a_{d-1}n^{d-1} + a_d n^d}{a_0 + a_1(n+1) + ... + a_{d-1}(n+1)^{d-1} + a_d(n+1)^d}\right|$$

or $$R = \lim_{n\to\infty}\left|\frac{a_0 \times (1/n)^d + a_1(1/n)^{d-1} + ... + a_{d-1}\times(1/n) + a_d}{a_0\times(1/n)^d + a_1(1+1/n)\times(1/n)^{d-1} + ... + a_{d-1}(1+1/n)^{d-1}\times(1/n) + a_d(1+1/n)^d}\right|$$

Thus, $R = (a_d/a_d) = 1.$ $\Rightarrow$ The option (B) is correct.

Ex. 248. *Multiple choice question having exactly one correct option*

Let $\sum_{n=0}^{\infty} a_n z^n$ be a convergent power series such that $\lim_{n\to\infty}(a_n/a_{n+1}) = R > 0$. Let p be a polynomial of degree d. Then, the radius of convergence of the power series $\sum_{n=0}^{\infty} p(n)\, a_n z^n$ is

(A). R **(B).** d **(C).** Rd **(D).** $R + d$ **[CSIR–UGC (NET), December 2014]**

Sol. Ans. (A). Since $p(n)$ is a polynomial of degree d, we write

$$p(n) = \alpha_0 + \alpha_1 n + \alpha_2 n^2 + ... + \alpha_{d-2} n^{d-2} + \alpha_{d-1} n^{d-1} + \alpha_d n^d, \quad ... (1)$$

where $\alpha_0, \alpha_1, \alpha_2, ..., \alpha_{d-1}, \alpha_d$ are constants free from n.

$$(1) \Rightarrow p(n+1) = \alpha_0 + \alpha_1 (n+1) + \alpha_2 (n+1)^2 + ... + \alpha_{d-1}(n+1)^{d-1} + \alpha_d (n+1)^d \quad ... (2)$$

Let R' be the radius of convergence of the power series $\sum_{n=0}^{\infty} p(n)\, a_n z^n$. Comparing this power series with $\sum_{n=0}^{\infty} c_n z^n$, we have

$$c_n = p(n)\, a_n \quad \text{and} \quad c_{n+1} = p(n+1)\, a_{n+1} \quad ... (3)$$

Then, $$R' = \lim_{n\to\infty}\left|\frac{c_n}{c_{n+1}}\right| = \lim_{n\to\infty}\left|\frac{p(n)\, a_n}{p(n+1)\, a_{n+1}}\right| = \lim_{n\to\infty}\left|\frac{a_n}{a_{n+1}}\right| \times \lim_{n\to\infty}\left|\frac{p(n)}{p(n+1)}\right|, \text{ using (3)}$$

Given that $\lim_{n\to\infty}(a_n/a_{n+1}) = R > 0$. Hence, we obtain

$$R' = R\lim_{n\to\infty}\left|\frac{p(n)}{p(n+1)}\right| = R\lim_{n\to\infty}\left|\frac{\alpha_0 + \alpha_1 n + ... + \alpha_{d-1}n^{d-1} + \alpha_d n^d}{\alpha_0 + \alpha_1(n+1) + ... + \alpha_{d-1}(n+1)^{d-1} + \alpha_d(n+1)^d}\right|$$

$$\Rightarrow R' = R\lim_{n\to\infty}\left|\frac{\alpha_0\times(1/n)^d + \alpha_1(1/n)^{d-1} + ... + \alpha_{d-1}\times(1/n) + \alpha_d}{\alpha_0\times(1/n)^d + \alpha_1(1+1/n)\times(1/n)^{d-1} + ... + \alpha_{d-1}(1+1/n)^{d-1}\times(1/n) + \alpha_d(1+1/n)^d}\right|$$

Thus, $R' = R \times (\alpha_d/\alpha_d) = R$ $\Rightarrow$ The option (A) is correct.

Ex. 249. *Multiple choice question having multiple correct options*

Consider the integral $A = \int_0^1 x^n (1-x)^n\, dx$. Pick each correct statement from below

A. *A is not a rational number* **B.** $0 < A < 4^{-n}$

C. *A is a natural number* **D.** *A^{-1} is a natural number* **[CSIR–UGC (NET), June 2016]**

Sol. Ans. (B) and **(D).** From Art. 20.2 and Art. 20.11, it follows that

$$\int_0^1 x^{m-1}(1-x)^{n-1}\,dx = B(m, n) = \frac{\Gamma(m)\,\Gamma(n)}{\Gamma(m+n)},\ m>0, n>0 \qquad \text{... (1)}$$

Using (1), we have
$$A = \int_0^1 x^{(n+1)-1}(1-x)^{(n+1)-1}\,dx = \frac{\Gamma(n+1)\,\Gamma(n+1)}{\Gamma(n+1+n+1)}$$

Thus, we have
$$A = \frac{n!\,n!}{(2n+1)!} \qquad \text{... (2)}$$

From (2), it follows that A is a rational number and so the option (A) is not correct. Again, clearly A is not a natural number whereas A^{-1} is a natural number. Hence the option (C) is not correct whereas the option (D) is correct.

We now proceed to use mathematical induction for proving that the option (B) is correct. Therefore, we shall prove that

$$A < 4^{-n} \text{ for all } n \in \mathbf{N}, \qquad \text{... (3)}$$

where **N** is the set of all natural number. For $n = 1$, (2) reduces to $A = 1/(3!) = 1/6 < 1/4$ so that $A < 4^{-1}$. Thus, we have shown that the statement (3) is true for $n = 1$.

Now, suppose that the statement (3) is true for $n = m$ where $m \in \mathbf{N}$. Then, from (2) and (3),

$$A < 4^{-m}, \qquad i.e., \qquad \frac{m!\,m!}{(2m+1)!} < 4^{-m} \qquad \text{... (4)}$$

We now proceed to verify that the statement (3) is also true for $n = m + 1$, that is, for $n = m + 1$,

$$A < 4^{-(m+1)}$$

Now, for $n = m + 1$, (2) reduces to

$$A = \frac{(m+1)!\,(m+1)!}{(2m+3)!} = \frac{(m+1)m!\,(m+1)m!}{(2m+3)\,(2m+3)\,(2m+1)!}$$

or
$$A = \frac{(m+1)\,(m+1)}{2(2m+3)\,(m+1)} \times \frac{m!\,m!}{(2m+1)!} = \frac{(m+1)}{4(m+3/2)} \times \frac{m!\,m!}{(2m+1)!} \qquad \text{... (5)}$$

Note that $m + 1 < m + 3/2$ so that
$$\frac{m+1}{m+3/2} < 1 \qquad \text{... (6)}$$

Using (4) and (6), (5) reduces to

$$A < 4^{-1} \times 4^{-m} \qquad \text{or} \qquad A < 4^{-(m+1)} \qquad \text{... (7)}$$

Thus, the statement (3) is true for $n = 1$. Again, from (4) and (7) it follows that the statement (3) is true for $n = m + 1$ whenever the statement (3) is **true for *n* =** m. Hence, by mathematical induction, the statement (3) is true for all $n \in \mathbf{N}$. Clearly, $A > 0$. Thus, $0 < A < 4^{-n}$ is true. Hence, the option (B) is correct.

Ex. 250. *Let $a_1 = 1$ and $a_n = a_{n-1} + 4,\ n \geq 2$. Then*

$$\lim_{n\to\infty}\left(\frac{1}{a_1a_2} + \frac{1}{a_2a_3} + \dots + \frac{1}{a_{n-1}a_n}\right) \text{ is equal to ....}$$ **[Gate 2016]**

Sol. Ans. 0.25. Given $a_1 = 1$ and $a_n - a_{n-1} = 4$, for $n \geq 2$ ... (1)

From (1), $a_2 - a_1 = a_3 - a_2 = ... = a_n - a_{n-1} = 4$, ... (2)

showing that $a_1, a_2, a_3, ..., a_n$ are in arithmetic progression whose first term $= a_1 = 1$ and whose common difference $d = 4$. Hence, the nth term a_n of the above arithmetic progression is given by the usual formula

$$a_n = a_1 + (n-1)d = 1 + (n-1) \times 4 \quad \text{so that} \quad a_n = 4n - 3 \qquad ... (3)$$

Now, $$\lim_{n \to \infty} \left(\frac{1}{a_1 a_2} + \frac{1}{a_2 a_3} + ... + \frac{1}{a_{n-1} a_n} \right) = \lim_{n \to \infty} \frac{1}{4} \left(\frac{4}{a_1 a_2} + \frac{4}{a_2 a_3} + ... + \frac{4}{a_{n-1} a_n} \right)$$

$$= \lim_{n \to \infty} \frac{1}{4} \left(\frac{a_2 - a_1}{a_1 a_2} + \frac{a_3 - a_2}{a_2 a_3} + ... + \frac{a_n - a_{n-1}}{a_{n-1} a_n} \right), \text{ using (2)}$$

$$= \lim_{n \to \infty} \frac{1}{4} \left[\left(\frac{1}{a_1} - \frac{1}{a_2} \right) + \left(\frac{1}{a_2} - \frac{1}{a_3} \right) + ... + \left(\frac{1}{a_{n-1}} - \frac{1}{a_n} \right) \right] = \lim_{n \to \infty} \frac{1}{4} \left(\frac{1}{a_1} - \frac{1}{a_n} \right)$$

$$= \lim_{n \to \infty} \frac{1}{4} \left(1 - \frac{1}{4n-3} \right), \text{ using (1) and (3)}$$

$$= (1/4) \times (1 - 0) = 1/4 = 0.25.$$

Ex. 251. *Let* $f(x) = a_0 + \sum_{n=1}^{\infty} (a_n \cos nx + b_n \sin nx)$ *be the Fourier series of the* 2π *periodic function defined by* $f(x) = x^2 + 4\sin x \cos x$, $-\pi \leq x \leq \pi$. *Then* $\left| \sum_{n=0}^{\infty} a_n - \sum_{n=1}^{\infty} b_n \right|$ *is equal to ......*

[GATE 2016]

Sol. Ans. 2. Given $$f(x) = a_0 + \sum_{n=1}^{\infty} (a_n \cos nx + b_n \sin nx), \quad -\pi \leq x \leq \pi \qquad ... (1)$$

Comparing (1) with the particular case (*i*) on page 22.4, we note that here we have a_0 in place of $a_0/2$. Accordingly, the coefficients a_0, a_n and b_n are given by (observe that here we are given that $f(x) = x^2 + 2\sin 2x$)

$$a_0 = \frac{1}{2\pi} \int_{-\pi}^{\pi} f(x)\, dx = \frac{1}{2\pi} \int_{-\pi}^{\pi} (x^2 + 2\sin 2x)\, dx = \frac{1}{2\pi} \left[\frac{x^3}{3} - \cos 2x \right]_{-\pi}^{\pi} = \frac{\pi^2}{3} \qquad ... (2)$$

$$a_n = \frac{1}{\pi} \int_{-\pi}^{\pi} f(x) \cos nx\, dx = \frac{1}{\pi} \int_{-\pi}^{\pi} (x^2 + 2\sin 2x) \cos nx\, dx = \frac{2}{\pi} \int_0^{\pi} x^2 \cos nx\, dx,$$

where we have used the fact that $x^2 \cos nx$ and $2\sin 2x \cos nx$ are respectively even and odd functions of x. Using chain rule of integration by parts, we get

$$a_n = \frac{2}{\pi} \left[(x^2) \left(\frac{\sin nx}{n} \right) - (2x) \left(-\frac{\cos nx}{n^2} \right) + (2) \left(-\frac{\sin nx}{n^3} \right) \right]_0^{\pi}$$

or $$a_n = \frac{2}{\pi} \times \frac{2\pi \cos n\pi}{n^2} = \frac{4(-1)^n}{n^2} = -\frac{4(-1)^{n-1}}{n^2}, \quad n = 1, 2, 3, ... \qquad ... (3)$$

Finally, $b_n = \frac{1}{\pi}\int_{-\pi}^{\pi} f(x)\sin nx\,dx = \frac{1}{\pi}\int_{-\pi}^{\pi}(x^2 + 2\sin 2x)\sin nx\,dx = \frac{2}{\pi}\int_0^{\pi} 2\sin 2x \sin nx\,dx,$ ... (4)

where we used the fact that $x^2 \sin nx$ and $2\sin 2x \sin nx$ are respectively odd and even functions of x. When $n \neq 2$, (4) may be re-written as

$$b_n = \frac{2}{\pi}\int_0^{\pi}\{\cos(n-2)x - \cos(n+2)x\,dx = \frac{2}{\pi}\left[\frac{\sin(n-2)x}{n-2} - \frac{\sin(n+2)x}{n+2}\right]_0^{\pi} = 0, \text{ if } n \neq 2 \quad \text{... (5)}$$

When $n = 2$, (4) reduces to

$$b_n = \frac{2}{\pi}\int_0^{\pi} 2\sin^2 2x\,dx = \frac{2}{\pi}\int_0^{\pi}(1-\cos 4x)\,dx = \frac{2}{\pi}\left[x - \frac{\sin 4x}{4}\right]_0^{\pi} = 2 \quad \text{... (6)}$$

Using (2), (3), (5) and (6), we have

$$\left|\sum_{n=0}^{\infty} a_n - \sum_{n=1}^{\infty} b_n\right| = \left|a_0 + \sum_{n=1}^{\infty} a_n - \sum_{n=1}^{\infty} b_n\right| = \left|\frac{\pi^2}{3} - 4\sum_{n=1}^{\infty}\frac{(-1)^{n-1}}{n^2} - 2\right| \quad \text{... (7)}$$

From standard results, we have $$\sum_{n=1}^{\infty}\frac{(-1)^{n-1}}{n^2} = 1 - \frac{1}{2^2} + \frac{1}{3^2} - \frac{1}{4^2} + ... = \frac{\pi^2}{12} \quad \text{... (8)}$$

Using (8), (7) yields $$\left|\sum_{n=0}^{\infty} a_n - \sum_{n=1}^{\infty} b_n\right| = \left|\frac{\pi^2}{3} - \left(4\times\frac{\pi^2}{12}\right) - 2\right| = |-2| = 2.$$

Ex. 252. *The number of roots of the equation $x^2 - \cos x = 0$ in the interval $[-\pi/2, \pi/2]$ is equal to ....* **[GATE 2016]**

Sol. Ans. 2. Let $$y = x^2 - \cos x \quad \text{... (1)}$$
Let us draw the graph of the function y given by (1).

x	y
$-\pi/2$	$\pi^2/4$
0	-1
$\pi/2$	$\pi^2/4$

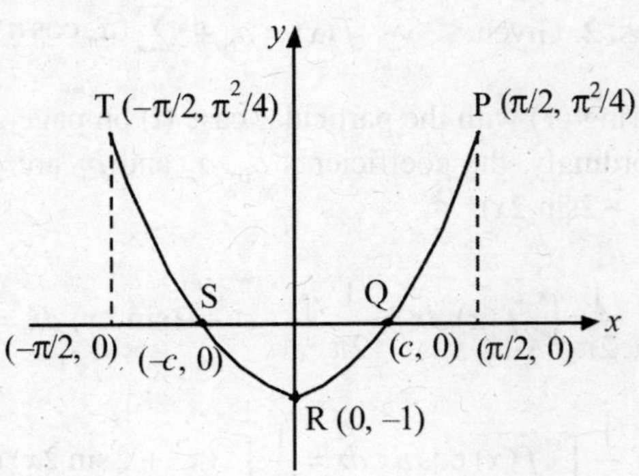

PQRST is the graph of curve (1) for $-\pi/2 \leq x \leq \pi/2$. From (1), we have
$$dy/dx = 2x + \sin x, \quad \text{where} \quad -\pi/2 \leq x \leq \pi/2$$
showing that y is a monotonic increasing function in the interval $[0, \pi/2]$

From (1), we have $y(0) = -1 < 0$ and $y(\pi/2) = \pi^2/4 > 0$. Since y is a continuous function in the closed interval $[0, \pi/2]$ and $y(0)$, $y(\pi/2)$ are of opposite signs, it follows by the location of roots theorem (see page 8.38), that there exists $c \in [0, \pi/2]$ such that $y(c) = 0$. In other words, there must be a root of the equation $y = 0$, i.e., $x^2 - \cos x = 0$ in $[0, \pi/2]$. As already proved, y is an increasing function in $[0, \pi/2]$ and hence there can be only one root of $x^2 - \cos x = 0$ in $[0, \pi/2]$.

Again from (1), it is clear that y is an even function of x in $[-\pi/2, \pi/2]$. Therefore, there must exists one more root of $x^2 - \cos x = 0$ in the closed interval $[-\pi/2, 0]$. Thus, there can be only two roots of the given equation $x^2 - \cos x = 0$ in the interval $[-\pi/2, \pi/2]$.

Ex. 253. *If $<a_n>$ and $<b_n>$ are two sequences such that $\lim_{n\to\infty} a_n = a$ and $\lim_{n\to\infty} b_n = b$, then show that $\lim_{n\to\infty} (a_n b_n) = ab$. What about the converse ? Justify.* **[Delhi B.Sc. (Prog) II, 2015]**

Sol. For first part, refer theorem IV, page 5.13

Second Part: The converse of the above theorem is not necessarily true, i.e., the existence of $\lim_{n\to\infty} (a_n b_n)$ does not necessarily imply that the two limits $\lim_{n\to\infty} a_n$ and $\lim_{n\to\infty} b_n$ also exist.

For example, consider $a_n = b_n = (-1)^n$, then both limits $\lim_{n\to\infty} a_n$ and $\lim_{n\to\infty} b_n$ do not exist. But

$$a_n b_n = (-1)^n (-1)^n = (-1)^{2n} = 1 \;\forall n \in \mathbf{N} \quad \Rightarrow \quad \lim_{n\to\infty} (a_n b_n) = 1 \text{ exists.}$$

Ex. 254. *Find the limit points of the sequence $< \cos (n\pi/4) >$.* **[Delhi B.Sc. (Prog) II, 2015]**

Sol. Let $a_n = \cos (n\pi/4)$. Then, we have

$$a_n = \cos\left(\frac{n\pi}{4}\right) = \begin{cases} 1/\sqrt{2}, \text{ if } n = 4m-3 \\ 0, \text{ if } n = 4m-2 \\ -(1/\sqrt{2}), \text{ if } n = 4m-1 \\ -1, \text{ if } n = 4m \end{cases}, \quad \text{where } m \in \mathbf{N} \qquad \text{... (1)}$$

Thus, $$< a_n > = < 1/\sqrt{2}, 0, -1/\sqrt{2}, -1, 1/\sqrt{2}, 0, -1/\sqrt{2}, -1, ... > \qquad \text{... (2)}$$

From (2), it follows that terms $1/\sqrt{2}, 0, -(1/\sqrt{2}), -1$ are repeated infinitely many times in the given sequence $< a_n >$. Hence, by definition of limit point of a sequence, it follows that $1/\sqrt{2}, 0, -(1/\sqrt{2})$ and -1 are four limit points of the given sequence $< \cos (n\pi/4) >$.

Ex. 255. *Find the radius of convergence and exact interval of convergence of the power series $\sum_{n=0}^{\infty} \{(x-1)^n / 2^n\}$.* **[Delhi B.Sc. (Prog) II, 2016]**

Sol. Let the given power series be denoted by $\sum_{n=0}^{\infty} a_n (x-1)^n$. Then, we have

$$a_n = 1/2^n \qquad \text{and} \qquad a_{n+1} = 1/2^{n+1}.$$

Let R be the radius of convergence of the given power series. Then, we have

$$R = \lim_{n\to\infty} \left|\frac{a_n}{a_{n+1}}\right| = \lim_{n\to\infty} \left|\frac{1/2^n}{1/2^{n+1}}\right| = \lim_{n\to\infty} 2 = 2.$$

Since the given power series is about the point $x = x_0 = 1$, the interval of convergence of the power series is given by

$$x_0 - R < x < x_0 + R \qquad \text{or} \qquad 1 - 2 < x < 1 + 2 \qquad \text{or} \qquad -1 < x < 3.$$

At $x = 3$, the given series reduces to $\sum_{n=0}^{\infty} (2^n / 2^n)$, *i.e.*, $\sum_{n=0}^{\infty} 1$, which is clearly a divergent series. Hence, the given power series is not convergent at $x = 3$.

Next, at $x = -1$, the given series reduces to $\sum_{n=0}^{\infty} (-2)^n / 2^n$, *i.e.*, $\sum_{n=0}^{\infty} (-1)^n$ which is an oscillatory series. Hence, the given power series is not convergent at $x = -1$.

Thus, the exact interval of convergence of the given series is open interval $(-1, 3)$.

Ex. 256. *Test the convergence of the series* $\sum_{n=1}^{\infty} e^{-n^2}$. **[Delhi B.Sc. (Prog) II, 2015]**

Sol. Let the given series be denoted by $\sum_{n=1}^{\infty} u_n$ so that $u_n = e^{-n^2}$.

Now, $e^x > x, \forall x > 0 \Rightarrow e^{n^2} > n^2 \ \forall n \in \mathbf{N} \Rightarrow e^{-n^2} < 1/n^2 \ \forall n \in \mathbf{N}$ ... (1)

Let $v_n = 1/n^2$. Then, from (1), $u_n < v_n \ \forall n \in \mathbf{N}$

Since $\sum_{n=1}^{\infty} v_n = \sum_{n=1}^{\infty} (1/n^2)$, it follows that Σv_n is convergent. Hence, by comparison test, the given series $\sum_{n=1}^{\infty} u_n$, *i.e.*, $\sum_{n=1}^{\infty} e^{-n^2}$ is also convergent.

Ex. 257. *Let* $< x_n >$ *be a sequence defined by* $x_1 = 1$, $x_{n+1} = (3 + 2x_n)/(2 + x_n)$, $n \geq 1$. *Show that the sequence* $< x_n >$ *is convergent. Also, find its limit.* **[Delhi B.Sc. (Prog) II, 2016]**

Sol. Given $x_1 = 1$ and $x_{n+1} = (3 + 2x_n)/(2 + x_n), n \geq 1$ ... (1)

We now proceed to use Mathematical Induction to prove that the $< x_n >$ is a monotonic sequence, i.e.,

$$x_{n+1} > x_n, \qquad \forall n \in \mathbf{N} \qquad \text{... (2)}$$

From (1), $x_1 = 1$ and $x_2 = (3 + 2x_1)/(2 + x_1) = 5/3$. Thus, $x_2 > x_1$, showing that the statement (2) is true for $n = 1$.

Let the statement (2) be true for $n = m$ so that

$$x_{m+1} > x_m \qquad \text{or} \qquad x_{m+1} - x_m > 0 \qquad \text{... (3)}$$

We now proceed to show that the statement (2) is true for $n = m + 1$, i.e., $x_{m+2} > x_{m+1}$. Now, using (1), we have

$$x_{m+2} - x_{m+1} = \frac{3 + 2x_{m+1}}{2 + x_{m+1}} - \frac{3 + 2x_m}{2 + x_m} = \frac{x_{m+1} - x_m}{(2 + x_{m+1})(2 + x_m)} > 0, \text{ using (3)}$$

Thus, $x_{m+2} - x_{m+1} > 0 \Rightarrow x_{m+2} > x_{m+1}$

Thus, have shown that the statement (2) is true for $n = 1$ and it is also true for $n = m + 1$ whenever it is true for $n = m$. Therefore, by Mathematical Induction, (2) is true for all $x \in \mathbf{N}$. Thus, the given sequence is monotonic increasing.

Now, $$x_{n+1} = \frac{3 + 2x_n}{2 + x_n} = \frac{2(2 + x_n) - 1}{(2 + x_n)} = 2 - \frac{1}{2 + x_n} < 2, \ \forall \ n \geq 1$$

Also, $x_1 = 1 < 2$. Thus, $x_n < 2 \ \forall n \in \mathbf{N}$ and so $< x_n >$ is bounded above. Since $< x_n >$ is monotonic increasing and bounded, so by theorem, I, page 5.38 it is a convergent sequence. Let $\lim_{n \to \infty} x_n = l$. Then, $\lim_{n \to \infty} x_{n+1} = l$. Now we have

$$x_{n+1} = \frac{3 + 2x_n}{2 + x_n} \Rightarrow \lim_{n \to \infty} x_{n+1} = \lim_{n \to \infty} \frac{3 + 2x_n}{2 + x_n} = \frac{3 + 2 \lim_{n \to \infty} x_n}{2 + \lim_{n \to \infty} x_n}$$

or $\quad l = (3 + 2l)/(2 + l) \quad \Rightarrow \quad 2l + l^2 = 3 + 2l \quad \Rightarrow \quad l^2 = 3 \quad \Rightarrow \quad l = \pm\sqrt{3}.$

Since all the terms of the given sequence are positive, it follows that l cannot be negative.

Hence, $\lim_{n \to \infty} x_n = \sqrt{3}.$

Ex. 258. *Prove that every convergent sequence of real numbers is a Cauchy sequence. Is the sequence $< x_n >$ a Cauchy sequence, where $x_n = (n + 1)/n$. Justify your answer.*

[Delhi B.Sc. (Hons) I, 2016]

Sol. For first part, refer theorem II, page 5.30.

Second Part: Given $\quad x_n = (n + 1)/n = 1 + (1/n) \quad$ ... (1)

Let $\varepsilon > 0$ be given. Then for $n \geq m$, using (1), we have

$$|x_n - x_m| = \left|\left(1+\frac{1}{n}\right) - \left(1+\frac{1}{m}\right)\right| = \left|\frac{1}{n} - \frac{1}{m}\right| = \frac{n-m}{n} \times \frac{1}{m} < \frac{1}{m}, \text{ as } \quad 0 < \frac{n-m}{n} < 1 \quad \text{... (2)}$$

If we choose $m \in \mathbf{N}$ such that $m > 1/\varepsilon$, i.e., $1/m < \varepsilon$, then (2) reduces to $|x_n - x_m| < \varepsilon, \forall n \geq m$.

Thus, for any given $\varepsilon > 0$, there exists a positive integer m such that

$$|x_n - x_m| < \varepsilon, \ \forall n \geq m.$$

Hence, by the definition of a Cauchy sequence, $< x_n >$ is a Cauchy sequence.

Ex. 259. *Discuss the absolute and convergence for the series* $\sum_{n=1}^{\infty} (-1)^n \{n/(n^3 + 1)\}$.

[Delhi B.Sc. (Hons) I, 2016]

Sol. Let the given series be denoted by $\sum_{n=1}^{\infty} u_n$. Then, $\qquad u_n = (-1)^n \{n/n^3 + 1)\}$.

Now, $\qquad |u_n| = \left|(-1)\frac{n}{n^3+1}\right| = \frac{n}{n^3+1} = \frac{1}{1+(1/n^3)} \times \frac{1}{n^2}$. Take $v_n = \dfrac{1}{n^2}$.

Then, $\qquad \lim_{n \to \infty} \frac{|u_n|}{v_n} = \lim_{n \to \infty} \frac{1}{1+(1/n^3)} = 1$, which is finite and non-zero.

Since $\sum_{n=1}^{\infty} (1/n^2)$ converges, hence by comparison test $\sum_{n=1}^{\infty} |u_n|$ is also convergent. Since $\sum_{n=1}^{\infty} |u_n|$ is convergent, by definition of absolutely convergent series the given series $\sum_{n=1}^{\infty} u_n$ is absolutely convergent.

But every absolutely convergent series is convergent (refer page 7.2). Hence $\sum_{n=1}^{\infty} u_n$ being an absolutely convergent series implies that $\sum_{n=1}^{\infty} u_n$ is convergent.

Thus, the given series is absolutely convergent and convergent.

Ex. 260. *Let $< x_n >$ be a sequence of real numbers and let $x \in \boldsymbol{R}$. If $<a_n>$ is a sequence of positive real numbers with $\lim a_n = 0$ and if for some constant $C > 0$ and some $m \in \boldsymbol{N}$, we have $|x_n - x| \leq C\, a_n \ \forall n \geq m$, then prove that $\lim x_n = x$.* **[Delhi B.Sc. (Hons.) I, 2016]**

Sol. If $\varepsilon > 0$ is given, then since $\lim a_n = 0$, we know that there exists $k = k(\varepsilon/C)$ such that $n \geq k$ implies

$$a_n = |a_n - 0| < \varepsilon/C, \text{ where } C > 0$$

Thus, $$a_n < \varepsilon/C, \quad \forall n \geq k \qquad \text{... (1)}$$

Also, given $$|x_n - x| \leq C\, a_n \;\forall n \geq m \qquad \text{... (2)}$$

Let $p = \max\{k, m\}$. Then, (1) and (2) may be re-written as

$$a_n < \varepsilon/C, \quad \forall n \geq p \qquad \text{... (3)}$$

and $$|x_n - x| \leq C\, a_n \;\forall n \geq p \qquad \text{... (4)}$$

Then, from (3) and (4), $$|x_n - x| < C \times (\varepsilon/C), \;\forall n \geq p$$

or $$|x_n - x| < \varepsilon, \;\forall n \geq p$$

Hence, by definition of limit of a sequence, $\lim x_n = x$.

Ex. 261. *Prove that if $c > 0$, then $\lim_{n\to\infty} c^{1/n} = 1$.* **[Delhi B.Sc. (Hons.) I, 2016]**

Sol. Three cases arise:

Case I: Let $c = 1$. Then $< c^{1/n} >$ is the constant sequence $<1, 1, 1, ... >$, which converges to 1. Hence, $\lim c^{1/n} = 1$ when $c = 1$.

Case II: Let $c > 1$. Then, we may write $c^{1/n} = 1 + h$, where $h > 0$.

Hence by Bernoulli's inequality (refer Art. 2.15), we have

$$c = (1 + h)^n \geq 1 + nh \text{ for some } n \in \mathbf{N} \qquad \text{... (1)}$$

(1) $\Rightarrow c - 1 \geq nh$ so that $h < (c - 1)/n$. Hence, we have

$$|c^{1/n} - 1| = h \leq (c - 1) \times (1/n) \text{ for some } n \in \mathbf{N} \qquad \text{... (2)}$$

Using the result of Ex. 260, (2) implies that $\lim c^{1/n} = 1$ when $c > 1$.

Case III: Let $0 < c < 1$. Then $c^{1/n} = 1/(1 + h)$ for some $h > 0$. Hence, using Bernoulli's inequality, we have

$$c = \frac{1}{(1+h)^n} \leq \frac{1}{1+nh} < \frac{1}{nh} \qquad \text{... (3)}$$

From (3), we have $0 < h < 1/nc$ for some $n \in \mathbf{N}$ ... (4)

Now, $0 < 1 - c^{1/n} = 1 - \dfrac{1}{1+h} = \dfrac{h}{1+h} < h < \dfrac{1}{nc}$, using (4)

Thus, $$|c^{1/n} - 1| = (1/c) \times (1/n) \text{ for some } n \in \mathbf{N} \qquad \text{... (5)}$$

As before, using the result of Ex. 2.60, (5) implies that $\lim c^{1/n} = 1$ where $0 < c < 1$.

Ex. 262. *If $\lim a_n = a$ and $a_n \geq 0$ for all n, then prove that $a \geq 0$. Deduce that if $< a_n >$ and $< b_n >$ are two sequences such that $a_n \geq b_n$ for all n, then $\lim a_n \geq \lim b_n$.*

[Delhi B.Sc. (Prog.) II, 2016]

Sol. For the first part, refer theorem II, page 5.20

Deduction. Given that $a_n \geq b_n$ for all n.

Let $c_n = a_n - b_n$. Then $c_n \geq 0$ and so by first part, we have

$$\lim c_n \geq 0 \;\Rightarrow\; \lim (a_n - b_n) \geq 0 \;\Rightarrow\; \lim a_n - \lim b_n \geq 0 \;\Rightarrow\; \lim a_n \geq \lim b_n$$

Ex. 263. *Test the convergence of the following series:*

(*i*) $1 - \frac{1}{3} + \frac{1}{9} - \frac{1}{27} + \frac{1}{81} - \frac{1}{243} + ...$ (*ii*) $\sum \frac{1}{(2n+1)^2}$ (*iii*) $\sum \frac{n+1}{2^n}$

[Delhi B.Sc. (Hons.) I, 2016]

Sol. Part (*i*): Re-writing, the given series is $\sum_{n=1}^{\infty} (-1)^{n-1} \times (1/3^{n-1})$. ... (1)

Let the given series (1) be denoted by $\Sigma(-1)^{n-1} u_n$.

Then, $u_n = 1/3^{n-1}$ and $u_{n+1} = 1/3^n$. Hence, $u_{n+1} < u_n \ \forall\, n \in \mathbf{N}$

Also, $\lim_{n\to\infty} u_n = \lim_{n\to\infty} (1/3)^{n-1} = 0$. Hence, by Leibnitz test, the given series is convergent.

Part (*ii*): Ans. Convergent. Hint: Use comparison test (see Art. 6.8)

Part (*iii*): Use D' Alembert's ratio test. (see Art. 6.10). **Ans.** Convergent.

Ex. 264. *Use the definition of limit of a sequence to find the following:*

(*i*) $\lim_{n\to\infty} \frac{n^2-1}{2n^2+3}$ (*ii*) $\lim_{n\to\infty} \frac{1}{\sqrt{n+7}}$ **[Delhi B.Sc. (Hons.) I, 2016]**

Sol. Part (*i*): Using results of Art. 5.5, page 5.11, we have

$$\lim_{n\to\infty} \frac{n^2-1}{2n^2+3} = \lim_{n\to\infty} \frac{1-(1/n^2)}{2+(3/n^2)} = \frac{1-\lim(1/n^2)}{2+\lim(3/n^2)} = \frac{1}{2}$$

We now proceed to use the definition of limit of sequence to prove that

$$\lim_{n\to\infty} \frac{n^2-1}{2n^2+3} = \frac{1}{2}.$$

Let $f_n = (n^2-1)/(2n^2+3)$ and let $\varepsilon > 0$ be given. Then, we have

$$\left| f_n - \frac{1}{2} \right| = \left| \frac{n^2-1}{2n^2+3} - \frac{1}{2} \right| = \frac{5}{2(2n^2+3)} < \frac{1}{2n^2+3} < \frac{1}{n^2} < \varepsilon \quad \text{if } n > \frac{1}{\sqrt{\varepsilon}}.$$

Let m be any positive integer greater than $1/\sqrt{\varepsilon}$.

Then for $\varepsilon > 0$, there exists a positive integer m such that

$$|f_n - 1/2| < \varepsilon, \quad \forall\, n \geq m \quad \Rightarrow \quad \left| \frac{n^2-1}{2n^2+3} - \frac{1}{2} \right| < \varepsilon, \quad \forall n \geq m,$$

showing that $$\lim_{n\to\infty} \frac{n^2-1}{2n^2+3} = \frac{1}{2}.$$

Part (*ii*): Using results of Art. 5.5, page 5.11, we have

$$\lim_{n\to\infty} \frac{1}{\sqrt{n+7}} = \lim_{n\to\infty} \frac{1}{(1+7/n)^{1/2}} \times \frac{1}{\sqrt{n}} = 1 \times 0 = 0$$

We now proceed to show that $\lim_{n\to\infty} 1/\sqrt{n+7} = 0.$

Let $g_n = 1/\sqrt{n+7}$ and let $\varepsilon > 0$ be given. Then, we have

$$|g_n - 0| = \left|\frac{1}{\sqrt{n+7}} - 0\right| = \frac{1}{\sqrt{n+7}} < \frac{1}{\sqrt{n}} < \varepsilon \qquad \text{if } n > \frac{1}{\varepsilon^2}$$

Let m be any positive integer greater than $1/\varepsilon^2$.

Then, for $\varepsilon > 0$ there exists a positive integer m such that

$$|g_n - 0| < \varepsilon, \quad \forall n \geq m \quad \Rightarrow \quad \left|1/\sqrt{n+7} - 0\right| < \varepsilon, \quad \forall n \geq m,$$

showing that $\lim_{n\to\infty} (1/\sqrt{n+7}) = 0$.

Ex. 265. *Define limit point of a set. Find the limit points of [0, 1].*

[Delhi B.Sc. (Hons.) I, 2016]

Sol. For definition of limit point, refer Art. 3.4, page 3.4.

Second Part: Let p be any real number Three cases arise:

Case I: Let $p < 0$. Then for $0 < \varepsilon < 0 - p$, there exists a nbd $]p - \varepsilon, p + e[$ of p containing no point of [0, 1]. Hence, any real number $p(< 0)$ cannot be a limit point of [0, 1].

Case II: Let $p > 1$. Then for $0 < \varepsilon < p - 1$, there exists a nbd $]p - \varepsilon, p + \varepsilon[$ of p containing no point of [0, 1]. Hence, any real number $p(> 1)$ cannot be a limit point of [0, 1].

Case III: Let $p \in [0, 1]$. Then, for any $\varepsilon > 0$, $]p - \varepsilon, p + \varepsilon[$ is a nbd of p such that

$]p - \varepsilon, p + \varepsilon[\cap [0, 1] =]c, d[$, where $c = \max\{p - \varepsilon, 0\}$ and $d = \min\{p + \varepsilon, 1\}$

Then since $]c, d[$ contains infinitely many points of [0, 1], it follows that p is a limit point of [0, 1]. Since p is any point of [0, 1], hence each point of [0, 1] is a limit point of [0, 1].

Thus, the set of all limit points of [0, 1] is [0, 1].

Ex. 266. *Define limit point of a set. Prove that the only limit point of the set $\{1 + 1/n : n \in \mathbf{N}\}$ is 1.* **[Delhi B.Sc. (Prog) II, 2015]**

Sol. First part: For definition of limit point, refer Art. 3.4, page 3.4

Second part: Let $S = \{1 + 1/n : n \in \mathbf{N}\} = \{1 + 1, 1 + 1/2, 1 + 1/3, ...\}$

Let p be any real number. Five cases arise:

Case I: Let $p < 1$. Then, there exists a nbd $(-\infty, 1)$ of p containing no point of S. Hence, any real number $p(< 1)$ cannot be a limit point of S.

Case II: Let $p > 1 + 1$ i.e., $p > 2$. Then, there exists a nbd $(2, \infty)$ of p containing no point of S. Hence, any real number $p(> 2)$ cannot be a limit point of S.

Case III: Let $p = 1 + 1 = 2$. Then $(1 + 1/2, \infty)$ is a nbd of p which contains only one element of S, namely $p = 1 + 1 = 2$. Hence 2 cannot be a limit point of S.

Case IV: Let $1 < p < 1 + 1$. Then, $0 < p - 1 < 1 \Rightarrow 1/(p - 1) > 0$

Hence, there exists unique $m \in \mathbf{N}$ such that

$$m < \frac{1}{p-1} < m + 1 \quad \Rightarrow \quad \frac{1}{m} > p - 1 > \frac{1}{m+1} \quad \Rightarrow \quad 1 + \frac{1}{m+1} < p < 1 + \frac{1}{m},$$

showing that $\left(1 + \frac{1}{m+1}, 1 + \frac{1}{m}\right)$ is a nbd of p containing no point of S.

Hence, if $1 < p < 2$, then p cannot be a limit point of S.

Case V: Let $p = 1$. Then, for every $\varepsilon > 0$, there exists $m \in \mathbf{N}$ such that $m\varepsilon > 1$, i.e., $\varepsilon > 1/m$.

Hence, $1 + \varepsilon > 1 + 1/m \Rightarrow 1 - \varepsilon < 1 < 1 + (1/m) < 1 + \varepsilon$,

showing that every nbd $(1 - \varepsilon, 1 + \varepsilon)$ of 1 contains a point $1 + (1/m)$ of S which is different from 1. Hence, 1 is a limit point of S.

Thus, 1 is the only limit point of the given set S.

Ex. 267. *Show that if* $\sum_{k=1}^{\infty} a_k$ *satisfies the Cauchy criterion for convergence of series, then for any* $\varepsilon > 0$*, there is an* n_0 *such that*

$$\left|\sum_{k=m}^{n} a_k\right| < \varepsilon \quad \text{for any} \quad n \geq m \geq n_0$$

[Delhi B.Sc. (Hons.) I, 2016]

Sol. We say that a series $\sum_{k=1}^{\infty} a_k$ satisfies the Cauchy criterion if its sequence $< S_n >$ of partial sums is a Cauchy sequence so that for each $\varepsilon > 0$ there exists a number n_0 such that

$$m, n \geq n_0 \quad \Rightarrow \quad |S_n - S_m| < \varepsilon \qquad \text{... (1)}$$

Nothing is lost in this definition if we impose the restriction $n > m$. Moreover, it is only a notational matter to work with $m - 1$ where $m \leq n$ instead of m where $m \leq n$. Therefore, (1) is equivalent to

for each $\varepsilon > 0$ there exists a number n_0 such that

$$n \geq m > n_0 \quad \Rightarrow \quad |S_n - S_{m-1}| < \varepsilon \qquad \text{... (2)}$$

Since $S_n - S_{m-1} = \sum_{k=m}^{n} a_k$, condition (2) can be re-written as

for each $\varepsilon > 0$ there exists a number n_0 such that

$$n \geq m > n_0 \quad \Rightarrow \quad \left|\sum_{k=m}^{n} a_k\right| < \varepsilon$$

Ex. 268. *Let* Σa_n *and* Σb_n *be two series of positive terms. If the sequence* $< a_n/b_n >$ converges with positive limit, then prove that Σa_n *converges if and only if* Σb_n *converges.*

[Delhi B.Sc. (Hons.) I, 2016]

Sol. For proof, refer Art. 6.8, page 6.9

Ex. 269. *Give an example of an unbounded sequence that has a convergent subsequence.*

[Delhi B.Sc. (Hons.) I, 2016]

Sol. Consider the sequence $< x_n >$, where

$$x_n = \begin{cases} n, \text{ if } n \text{ is an odd natural number} \\ 1/n, \text{ if } n \text{ is an even natural number} \end{cases}$$

Then, we have $< x_n > = <1, 1/2, 3, 1/4, 5, 1/6, ...>$ ... (1)

The sequence given by (1) is bounded below but is not bounded above. Hence, (1) is an unbounded sequence.

Clearly, $<1/2, 1/4, 1/6, ...>$ is a subsequence of $< x_n >$ and this subsequence converges to 0. Thus, $< x_n >$ is an example of an unbounded sequence that has a convergent subsequence.

Ex. 270. *Suppose that* $x_n \geq 0 \ \forall n \in \mathbf{N}$ *and that* $\lim (-1)^n x_n$ *exists. Show that* $< x_n >$ *converges.* **[Delhi B.Sc. (Hons.) I, 2016]**

Sol. Let $\lim (-1)^n x_n = l$. Hence, by definition, for a given $\varepsilon > 0$, there exists a positive integer m such that

$$|(-1)^n x_n - l| < \varepsilon, \qquad \forall n \geq m \qquad \text{... (1)}$$

Now, $|(-1)^n x_n - l| \geq |\,|(-1)^n x_n| - |l|\,|$, as $|x - y| \geq |\,|x| - |y|\,|$

Thus, $|(-1)^n x_n - l| \geq |\,|x_n| - |l|\,|$ ... (2)

Since by hypothesis $x_n \geq 0\ \forall n \in \mathbf{N}$, we have $l \geq 0$. Hence, (2) yields

$|(-1)^n x_n - l| \geq |x_n - l|$ or $|x_n - l| < |(-1)^n x_n - l|$ ... (3)

From (1) and (3), we have $|x_n - l| < |(-1)^n x_n - l| < \varepsilon,\ \forall n \geq m$

Thus, $|x_n - l| < \varepsilon,\ \forall n \geq m \Rightarrow \lim x_n = l \Rightarrow \langle x_n \rangle$ is convergent.

Ex. 271. *Write a note on introduction of the trigonometric functions in an analytical manner.*

Sol. In elementary courses, the trigonometric functions are usually introduced on a geometrical basis in terms of either triangles or the unit circle. We now propose to introduce the trigonometric functions in an analytical manner.

Definition of sine function $S(x)$ and cosine function $C(x)$ in terms of power series.

[Delhi B.Sc. (Prog.) II, 2015, 2016]

Theorem. *The two series* $\sum_{n=0}^{\infty} \frac{(-1)^n x^{2n+1}}{(2n+1)!}$ *and* $\sum_{n=0}^{\infty} \frac{(-1)^n x^{2n}}{(2n)!}$ *are uniformly convergent in every interval* $[a, b]$.

Proof: Let M be any positive number greater than $|a|$ as well as $|b|$. Then, $\forall x \in [a, b]$,

$$\left|\frac{(-1)^n x^{2n+1}}{(2n+1)!}\right| \leq \frac{M^{2n+1}}{(2n+1)!} = M_n, \text{ say} \qquad \forall x \in [a, b]$$

We now test the convergence of ΣM_n, where $M_n = M^{2n+1}/(2n+1)!$

Now, $\lim_{n\to\infty} \frac{M_{n+1}}{M_n} = \lim_{n\to\infty} \frac{M^{2n+3}}{(2n+3)!} \times \frac{(2n+1)!}{M^{2n+1}} = \lim_{n\to\infty} \frac{M^2}{(2n+3)(2n+2)} = 0$. which is < 1

Hence, the series ΣM_n is convergent and therefore, by Weierstran's M-test, it follows that the power series $\sum_{n=0}^{\infty} \frac{(-1)^n x^{2n+1}}{(2n+1)!}$ is uniformly convergent for all $x \in [a, b]$

Proceeding likewise, it easily follows that the power series

$\sum_{n=0}^{\infty} \frac{(-1)^n x^{2n}}{(2n)!}$ is also uniformly convergent for all $x \in [a, b]$.

Definitions. The above theorem justifies the following two definitions of sine function $S(x)$ and cosine function $C(x)$.

$$S(x) = x - \frac{x^3}{3!} + \frac{x^5}{5!} - \frac{x^7}{7!} + \ldots = \sum_{n=0}^{\infty} \frac{(-1)^n x^{2n+1}}{(2n+1)!} \qquad \text{... (1)}$$

$$C(x) = 1 - \frac{x^2}{2!} + \frac{x^4}{4!} - \frac{x^6}{6!} + \ldots = \sum_{n=0}^{\infty} \frac{(-1)^n x^{2n}}{(2n)!} \qquad \text{... (2)}$$

In what follows we shall show that functions $S(x)$ and $C(x)$ coincide with the functions $\sin x$ and $\cos x$ whose definitions are usually based on geometrical considerations.

Some basic properties of $C(x)$ and $S(x)$

Property I: $S(0) = 0$ and $C(0) = 1$

These results follow from (1) and (2) on putting $x = 0$

Property II: $S(-x) = -S(x)$ and $C(-x) = C(x)$

From (1), $S(-x) = -x - \dfrac{(-x)^3}{3!} + \dfrac{(-x)^5}{5!} - \cdots = -\left[x - \dfrac{x^3}{3!} + \dfrac{x^5}{5!} - \cdots\right] = -S(x)$

and $C(-x) = 1 - \dfrac{(-x)^2}{2!} + \dfrac{(-x)^4}{4!} - \cdots = 1 - \dfrac{x^2}{2!} + \dfrac{x^4}{4!} - \cdots = C(x)$

Property III: The function $S(x)$ and $C(x)$ are continuous for all x

Proof. Since $S(x)$ and $C(x)$ are defined by (1) and (2) and the power series (1) and (2) are uniformly convergent, it follows that their sum functions $S(x)$ and $C(x)$ must be continuous for all x (refer theorem 2 of Art. A.2 on page A.3)

Property IV: The functions $S(x)$ and $C(x)$ are derivable for all x and

$$S'(x) = \cos x \qquad \text{and} \qquad C'(x) = -S(x)$$

Proof. Since the power series (1) and (2) are uniformly convergent, hence term by term differentiation is valid for them (refer result III of Art 15.10, page 15.21)

If we differentiate term by term the uniformly convergent series (1) and (2), we have

$$S'(x) = 1 - \frac{3x^2}{3!} + \frac{5x^4}{5!} - \cdots = 1 - \frac{x^2}{2!} + \frac{x^4}{4!} - \cdots = C(x), \text{ using (2)}$$

and $$C'(x) = -\frac{2x}{2!} + \frac{4x^3}{4!} - \frac{6x^5}{6!} - \cdots = -\left[x - \frac{x^3}{3!} + \frac{x^5}{5!} - \cdots\right] = -S(x), \text{ using (1)}$$

Property V: The addition theorem. To show that

$$S(x + y) = S(x)\, C(y) + C(x)\, S(y),\ \forall x, y \in \mathbf{R} \qquad \text{... (3)}$$

$$C(x + y) = C(x)\, C(y) - S(x)\, S(y),\ \forall x, y \in \mathbf{R} \qquad \text{... (4)}$$

Proof. Giving any value to y and then keeping it fixed, we write

$$f(x) = S(x + y) - S(x)\, C(y) - C(x)\, S(y) \qquad \text{... (5)}$$

and $$g(x) = C(x + y) - C(x)\, C(y) + S(x)\, S(y) \qquad \text{... (6)}$$

Differentiating (5) and (6) with respect to 'x', we obtain

$$f'(x) = C(x + y) - C(x)\, C(y) + S(x)\, S(y) = g(x), \text{ using (6)} \qquad \text{... (7)}$$

and $$g'(x) = -S(x + y) + S(x)\, C(y) + C(x)\, S(y) = -f(x), \text{ using (5)} \qquad \text{... (8)}$$

Now, $$\frac{d}{dx}\{f^2(x) + g^2(x)\} = 2f(x)\, f'(x) + 2g(x)\, g'(x)$$

$$= 2f(x)\, g(x) - 2g(x)\, f(x), \text{ using (7) and (8)}$$

Thus, $\dfrac{d}{dx}\{f^2(x) + g^2(x)\} = 0,\ \forall x \in \mathbf{R} \quad \Rightarrow \quad f^2(x) + g^2(x)$ is a constant.

Hence, $\forall x \in \mathbf{R}$, $\quad f^2(x) + g^2(x) = f^2(0) + g^2(0)$ ... (9)

Putting $x = 0$ in (5) and (6) and using the results $S(0) = 0$ and $C(0) = 1$, we have $f(0) = 0$ and $g(0) = 0$. Hence, (9) yields

$$f^2(x) + g^2(x) = 0, \qquad \forall x \in \mathbf{R}$$

which implies that $f(x) = 0$ and $g(x) = 0$. Setting $f(x) = 0$ and $g(x) = 0$ in (5) and (6) we get the required results (3) and (4) respectively.

Property VI: $C^2(x) + S^2(x) = 1 \ \forall x \in \mathbf{R}$ **[Delhi B.Sc. (Prog.) II, 2015]**

Proof. Replacing y by $-x$ in (4), we have

$$C(0) = C(x)\, C(-x) - S(x)\, S(-x) \quad \Rightarrow \quad 1 = C^2(x) + S^2(x),$$

where used the results $C(0) = 1$, $C(-x) = C(x)$ and $S(-x) = -S(x)$

Property VII: $|C(x)| \le 1$ **and** $|S(x)| \le 1$

From property VI, $C^2(x) + S^2(x) = 1 \Rightarrow C^2(x) \le 1$ and $S^2(x) \le 1$

Hence, we have $|C(x)| \le 1$ and $|S(x)| \le 1$

Property VIII: $S(2x) = 2S(x)\, C(x)$

and $C(2x) = C^2(x) - S^2(x)$. **[Delhi B.Sc. (Prog.) II, 2015]**

Proof. Changing y to x in (3) and (4), we obtain

$$S(x + x) = S(x)\, C(x) + C(x)\, S(x) \quad \text{giving} \quad S(2x) = 2S(x)\, C(x)$$

and $$C(x + x) = C(x)\, C(x) - S(x)\, S(x) \quad \text{giving} \quad C(2x) = C^2(x) - S^2(x)$$

Property IX: *To show that*

$$S(x - y) = S(x)\, C(y) - C(x)\, S(y), \ \forall x, y \in \mathbf{R} \quad \text{... (10)}$$

$$C(x - y) = C(x)\, C(y) + S(x)\, S(y), \ \forall x, y \in \mathbf{R} \quad \text{... (11)}$$

Proof. Changing y to $-y$ in (3) and (4), we have

$$S(x - y) = S(x)\, C(-y) + C(x)\, S(-y) = S(x)\, C(y) - C(x)\, S(y)$$

and $$C(x - y) = C(x)\, C(-y) - S(x)\, S(-y) = C(x)\, C(y) + S(x)\, S(y),$$

where we used the results of property II.

Ex. 272. *Define sine function S(x) and cosine function C(x) in terms of power series. Specify the domains of convergence of the respective power series.* **[Delhi B.Sc. (Prog.) II, 2016]**

Sol. By definition, we have

$$S(x) = \sum_{n=0}^{\infty} \frac{(-1)^n}{(2n+1)!} x^{2n+1} \quad \text{and} \quad C(x) = \sum_{n=0}^{\infty} \frac{(-1)^n}{(2n)!} x^{2n} \quad \text{... (1)}$$

In what follows, we shall proceed as explained in note 2, page 17.3. Let Σu_n denote the power series $\sum_{n=0}^{\infty} \{(-1)^n x^{2n+1}/(2n+1)!\}$. Then, we have

$$u_n = \frac{(-1)^n x^{2n+1}}{(2n+1)!} \quad \text{and} \quad u_{n+1} = \frac{(-1)^{n+1} x^{2n+3}}{(2n+3)!} = \frac{(-1)^{n+1} x^2 \times x^{2n+1}}{(2n+3)\,(2n+2)\,(2n+1)!}$$

Hence, $$\lim_{n \to \infty} \left| \frac{u_{n+1}}{u_n} \right| = \lim_{n \to \infty} \left| -\frac{x^2}{(2n+3)\,(2n+2)} \right| = |x|^2 \times 0 = 0 \ \forall x \in \mathbf{R}$$

Hence, by D' Alembert's ratio test, it follows that $\Sigma |u_n|$ is convergent $\forall x \in \mathbf{R}$. Hence, the series Σu_n is absolutely convergent $\forall x \in \mathbf{R}$. But every absolutely convergent series is also convergent (refer page 7.2). Hence, Σu_n is convergent for all $\forall x \in \mathbf{R}$. Therefore, the domain of convergence of the power series Σu_n defining $S(x)$ is $\mathbf{R}$.

Proceeding likewise, prove yourself that the domain of convergence of the power series defining $C(x)$ is also $\mathbf{R}$.

Ex. 273. *Let* $u_n = (n!)/\{1 \times 3 \times 5 \times \ldots \times (2n-1)\}, n \in \mathbf{N}$. *Then* $\lim_{n \to \infty} u_n$ *is equal to* ... **[GATE 2019]**

Sol. 0. From the given value of u_n, we have

$$u_{n+1} = \frac{(n+1)!}{1 \times 3 \times 5 \times \quad \times (2n-1)\,(2n+1)} = \frac{(n+1) \times n!}{1 \times 3 \times 5 \quad \times (2n-1)\,(2n+1)}$$

Hence, $$\lim_{n \to \infty} \frac{u_{n+1}}{u_n} = \lim_{n \to \infty} \frac{n+1}{2n+1} = \lim_{n \to \infty} \frac{1 + 1/n}{2 + 1/n} = \frac{1}{2} < 1$$

Therefore, by theorem VIII, Art. 5.8, it follows that $\lim_{n \to \infty} u_n = 0$.

INDEX

Abel's
  lemma, 15.18
  tes+, 15.18, 16.26
  theorem, 7.13
absolute
  convergence, 7.1, 16.15, 16.25
  value, 2.26
accumulation point, 3.5, 19.10
adherent point, 3.25, 19.12
algebra of,
  continuous functions, 8.26
  convergent sequences, 5.5
  derivatives, 9.3
  limits, 8.4
  operations on functions, 8.2
  R-S integrable functions, 14.5
  sets, 14.18
alternating series, 6.6, 7.2
Archimedian,
  ordered field, 2.21
  property of real numbers, 2.20
arithmetic continua, 2.29
associative law for sets, 1.5
Baire's category theorem, 19.31
Bernoulli's inequality, 2.29
Bessel'
  inequatity, 22.22
  theorem, 22.22
Beta function, 20.1
Bolzano-Weierstrass
  property, 19.34
  theorem for sequences, 5.3
  theorem for sets, 3.8
Banach's fixed point theorem, 19.30
bijection, 1.7
binary,
  operation, 1.8
  representation of real number, 4.7
Borei's theorem, 8.40
boundary
  point, 3.22, 19.16
  of a set, 3.22, 19.16
bounded
  above sequence, 5.2
  below sequence, 5.2
  function, 8.2
  metric space, 19.4
  sequence, 5.2
  set, 2.10
Cantor,
  intersection theorem, 19.27
  nested interval theorem, 3.30
  set, 4.8
cardinal number, 4.1
cartesian product of sets, 1.6
Caearo's theorem, 5.26
Cauchy
  condensation test, 6.47
  criterion for convergence
  of sequence, 5.30
  criterion for finite limits, 8.15
  first theorem on limits, 5.21
  general principal of convergence
  for series, 6.3, 7.10
  general principal of convergence for
  infinite series of functions, 15.14
  general principle of convergence for
  sequence, 5.31
  general principle of uniform
  convergence of sequence of
  functions, 15.2
  Hadamard theorem, 17.2, 17.18
  integral test, 6.41
  mean value theorem, 10.24
  n th root test, 6.19
  Schwarz inequality, 19.1
  second theorem on limits, 5.23
  sequence, 5.29, 19.23
  test, 16.14
change of variable,
  in an integral, 13.44
  for R-S integrals 14.29
chain rule, 9.4
closed,
  ball, 19.5
  interval, 2.7
  sphere, 19.5
  set, 19.10
closure of a set, 3.25, 19.12
cluster point of a,
  sequence, 5.2
  set, 3.5, 19.10
commutative law for sets, 1.5
compact metric spaces, 19.32
compact set, 3.31
comparison test, 16.6, 16.19
complement of a set, 1.5
complete metric space, 19.24
complete ordered field, 2.20
components of a metric space, 19.46
composite of functions, 1.7, 8.28
condensation point, 3.5, 5.2, 19.10
conditionally convergence, 16.15, 16.25

conditionally convergent series, 7.1
connected,
 metric space, 19.40
 sets, 19.40
connectives, 1.2
contact point, 3.25, 19.12
continuous,
 almost everywhere, M.9
 at an end point, 8.17
 at an interior point, 8.16
 function, 19.17
 in an interval, 8.17
contraction mapping, 19.30
constant function, 1.7
convergence of improper integral,
 of first kind, 16.1
 of second kind, 16.16
convergent
 sequence, 5.5, 19.20
 series, 6.1
countable set, 4.1
countably compact metric space, 19.34
cover of a set, 3.31
covering of a set, 3.31
critical,
 point, 11.2
 value, 11.2
D'Alembert's ratio test, 6.14
Darboux
 sums, 13.2
 theorem, 9.16, 13.7
decreasing function, 10.10
Dedekind property, 2.19
deleted neighbourhood of a point, 3.1, 8.3
De-Morgan and Bertrand test, 6.25
De-Morgan's laws for sets, 1.5
dense in itself, 3.29, 19.15
dense set, 3.29, 19.15
denseness property of a set, 2.21
density theorem, M.1
denumerable set, 4.1
derivability, 9.1
 at the end point, 9.1
 in a closed interval, 9.1
derived set, 3.5, 19.10
diameter of set, 19.4
difference of sets, 1.5
Dini's theorem, A.4
Dirichlet's
 conditions, 22.3
 test, 15.19, 16.27
 theorem, 7.12
disconnected sets, 19.40
discontinuity of
 function, 8.17
 the first kind, 8.17
 the second kind, 8.18
discrete
 metric space, 19.2
 set, 3.29
dissection of a closed interval, 13.1
distance
 between two sets, 19.4
 of a point from a set, 19.4
distributive law for sets, 1.5
divergent series, 6.1
Double sequence, 18.1
 series, 182
Empty set, 1.4
enumerable set, 4.1
equality of sets, 1.4
equivalent
 metrics, 19.7
 sets, 4.1
Euclidean plane, 19.3
Euler's formulae, 22.3
Even function, 22.1
everywhere dense, 3.29
existential quantifiers, 1.3
exterior
 of a set, 3.22, 19.16
 point, 3.22, 19.16
extreme value, 11.1
extremum, 11.1
Field of numbers, 2.5
finite intersection property, 19.34
finite set, 4.1
finite subcover, 3.31
first category, 19.31
first mean value theorem for
 R-integrable functions 13.37
 R-S-integrable functions, 14.16
fixed point, 19.30
fixed point theorem, 8.46, 19.30
Fourier
 coefficients, 22.3
 convergence of, 22.4
 Dirichlet theorem, 22.26
 series, 22.3
 series expansion, 22.1
frontier
 of a set, 19.16
 point, 19.16
full wave form, 22.18
function of function, 1.6, 8.28
functional limits, 8.16
fundamental
 sequence, 5.29
 theorem of Integral Calculus, 13.41
Gamma function, 20.1
Gauss' test, 6.37

geometric
  Continua, 2.29
  series, 6.6
geometric (interpretation/meaning) of
  derivative, 9.6
  mean value theorem, 10.11
  Rolle's theorem, 10.2
generalized mean value theorem, 10.25, 13.41
greatest lower bound (g.l.b) of a set, 2.7
greatest member of a set, 2.7
Half-range,
  cosine series, 22.13
  sine series, 22.13
Half-wave form, 22.18
harmonic series, 6.6
Heine-Borel theorem, 3.33
Heine's definition of continuity, 8.29
Hereditary property, 19.76
homeomorphic spaces, 19.19
homeomorpnism, 19.19
Housdroff
  property, 3.4
  theorem, 19.57
Indepotent laws, 1.5
identity function, 1.7
identity laws. 1.5
improper integral, 16.1, M.11
  general value of, M.11
  principal value of, M.12
  of first kind, 16.1, M.11
  of mixed kind, M.11
  of second kind, 16.1, M.11
  of third kind, M.11
incomplete metric space, 19.24
increasing function, 10.10
indeterminate forms, 12.1
infimum, 2.7
infinite,
  discontinuity, 8.19
  integrals, M.11
  limits, 8.10
  series, 6.1, 15.14
  set, 4.1
injection, 1.6
integer, 2.3
integral function, 13.39
integration by parts, 13.45
interior
  of a set, 3.22, 19.16
  point, 3.22, 19.16
intermediate value theorem, 8.39, 19.45
intersection of sets, 1.5
interval of convergence, 17.2
into mapping, 1.6
irrational number, 2.6
isolated point, 3.29
isometry, 19.19
inverse
  function, 1.8
  function theorem, 9.5
invertible functions, 8.48
Jump discontinuity, 8.17
Kummer's test, 6.39
Lagrange's mean value theorem, 10.9
least upper bound (l.u.b), 2.7
Lebesgue
  Covering lemma, 19.36
  number, 19.36
  Vitali theorem, M.9
left hand
  derivative, 9.1
  limit, 8.8
Legendre-duplication theorem, 20.21
Leibnitz's theorem, 7.2, 21.1
L'Hopital's theorem, 12.2
limit
  at infinity, 8.10
  inferior of a sequence, 5.52
  of a function, 8.3
  of a sequence, 5.5
  point of a sequence, 5.5
  point of a set, 3.4, 19.10
  superior of a set, 5.52
locally compact space, 19.38
logarithmic test, 6.25
lower Darboux sum, 13.2
lower Riemann
  sum, 13.2
  integral, 13.6
  Stieltjes sum, 14.1
  Stieltjes integral, 14.4
Maclaurin's
  infinite series, 10.30
  theorem, 10.29
many-one mapping, 1.6
mapping, 1.6
Mathematical Induction, 2.29
maximum value, 11.1
Merten's theorem, 7.26
metric space0
  bounded, 19.4
  compact. 19.32
  complete, 19.24
  connected, 19.40
  countably compact, 19.34
  disconnected, 19.40
  discrete, 19.2
  incomplete, 19.24
  locally compact, 19.38
  n-dimensional Euclidean, 19.3
  separable, 19.15

sequentially compact, 19.34
totally bounded, 19.35
unbounded, 19.5
usual, 19.2
minimum value, 11.1
Minkowski's inequality, 19.2
mixed discontinuity, 8.19
Mn-test, 15.5
modulus of real number, 2.26
monotone
convergence theorem, M.3
function, 8.2, 10.10
sequence, 5.37
μ-test, 16.7, 16.20
Natural number, 2.1
negation, 1.2
neighbourhood of a point, 19.5
nested sequence, 19.28
net of a closed interval, 13.1
non-dense set, 3.29, 19.15
norm of a partition, 13.1
nowhere dense set, 3.29, 19.15
null set, 1.4
Odd function, 22.1
one-one function, 1.6
one sided limits, 8.8
onto function, 1.6
open,
ball 19.5
interval, 2.7
sphere, 19.5
set, 19.5
real numbers, 2.16
ordered field of numbers, 2.5
ordered pair, 1.6
oscillatory
series, 6.1
sum, 13.2
Parseval's
identity, 22.15
theorem, 22.15
Partial integration formula, 14.28
partition of a closed interval, 13.1
perfect set, 3.29, 19.15
Period, 22.1
Periodic function, 22.1
Piecewise monotonic, 22.7
pointwise
convergence, 15.1
limit, 15.1
power series representation, 10.29
primitive, 13.40
principle of finite induction, 2.2
Pringhein theorem 18.1
product of
functions, 1.7
metric spaces, 19.16
proper
integral, M.10
set, 1.4
proper set, 1.4
pseudo-metric space, 19.1
Raabe's test, 6.23
radius of convergence, 17.2
range of a sequence, 5.1
rational number, 2.4
re-arrangement of series, 7.14
reduction of R-S integral into
R-integral, 14.26
refinement of a partition, 13.2
removable discountinuity, 8.17
repeated double limits 18.1
Riemann
integral, 13.6
Lebseque theorem, 22.23
sum, 13.2
Stieltjes integral, 14.4
Stieltjes sum, 14.1
theorem, 7.20
right hand
derivative, 9.1
limit, 8.8
Rolle's theorem, 10.1
root function, 8.50
Sandwitch theorem, 5.21
sawtooth wave form, 22.18
second category, 19.31
second logarithm ratio test, 6.27
second mean value theorem, 13.43, 14.29
semi-closed interval, 2.7
semi-open interval, 2.7
separable space, 19.15
separated sets, 19.40
sequence, 19.20
of partial sums of a series, 6.1
sequentially compact, 19.34
set, 1.4
bounded below, 2.7
bounded above, 2.7
of first species, 3.5
of second species, 3.5
sign of derivative, 9.15
singleton, 1.5
smallest member of a set, 2.7
square wave form, 22.18
squeeze principle, 5.21, 8.6
statement, 1.1
stationary,
point, 11.2
value, 11.2
strictly
decreasing function, 8.3, 10.10
increasing function, 8.3, 10.10

monotone function, 10.10
subsequence, 5.48
subset, 1.4
subspace of metric space, 19.11
sum of an infinite series, 6.1
summation of an infinite series, 13.19
super set, 1.4
supremum, 2.7
surjection, 1.6
Taylor's infinite series, 10.29
Taylor's theorem for power series, 18.11
Taylor's theorem with
Cauchy form of remainder, 10.28
Lagrange's form of remainder, 10.28
Schlimlch and Roche form of
remainder, 10.27
totally bounded, 19.35
triangle inequality, 19.1
triangular wave form, 22.18
trivial metric space, 19.2
turning value, 11.1
Upper Daraboux sum, 13.2
upper Riemann
integral, 13.6
sum, 13.2
upper Riemann-Stieltjes
integral, 14.4
sum, 14.1
unbounded
above, 2.8, 5.2
below, 2.8, 5.2
function, 8.2
metric space, 19.5
sequence, 5.1
uniform
continuity theorem, 8.44
continuous, 8.43, 19.18
convergence of sequence, 15.1
union of sets, 1.5
universal set, 1.4
Urysohn lemma, 19.61
usual metric, space, 19.2
Void set, 1.4
Vortex function, 10.42
Vortex function 10.42
Weierstrass
approximation theorem, 15.29
M-test, 15.14
Zero measure of set, M.17